Foreword

Please join us as we celebrate fifty successful years in Asia! The University of Maryland University College is an award-winning academic leader with a worldwide commitment to educating military and civilian students. We're pioneers in the field of distance education, providing students globally with the opportunity to take their courses in-class and on-line. Our dedication and support for military students and their dependents is a core part of our University culture, our past and our future.

As we pause to reflect on our achievements, we're most proud of the thousands of students whose lives we've touched over the past fifty years. At nearly every speaking engagement, I meet professionals who proudly describe that their education started with UMUC overseas, and provided them with a chance to succeed. The fifty year celebration provides me with an opportunity to thank all of our UMUC students and alumni. Of course, the faculty and administrative teams that support the students in Asia and in Maryland deserve our grateful praise for support often beyond the call of duty.

With great pride the University of Maryland University College celebrates this fifty year anniversary, with a commitment to continue the legacy and partnership with the US Military for many years in the future.

Susan C. Aldridge, Ph.D.
President
University of Maryland University College

University of Maryland University College (UMUC) first offered courses in Asia during Academic Year 1956-1957. During the five decades since, hundreds of thousands of students—with the significant majority being active-duty American service members—have studied with the Asian Division.

Thousands have earned certificates and degrees.

During these 50 years, Maryland classes have been offered in traditional classroom format in Japan, Okinawa, Korea, Guam, Hong Kong, Singapore, Thailand, Australia, New Zealand, Vietnam, Laos, Taiwan, Guam, China, the Marshall Islands, the Philippines, Malaysia, Diego Garcia and on Midway Island. During the last 10 years, students all over the world have enrolled in the Asian Division's online distance education program.

UMUC has taken great pride in its 50-year partnership with the United States Military in Asia and the Pacific and is most pleased to have provided educational opportunities wherever and whenever requested by the Military Services. In cooperation with educational partners such as Thomson Learning, UMUC also greatly looks forward in the future to continuing to meet the educational needs of the American Military Communities in Asia and the Pacific.

Joe Arden
Director
Asian Division
University of Maryland University College

BASIC CONCEPTS IN BIOLOGY

SIXTH EDITION

CECIE STARR

Christine A. Evers

Lisa Starr

FROM **BIOLOGY: CONCEPTS AND APPLICATIONS**

THOMSON

BROOKS/COLE

Australia • Canada • Mexico • Singapore
Spain • United Kingdom • United States

PUBLISHER Jack C. Carey

VICE-PRESIDENT, EDITOR-IN-CHIEF Michelle Julet

SENIOR DEVELOPMENT EDITOR Peggy Williams

ASSOCIATE DEVELOPMENT EDITOR Suzannah Alexander

EDITORIAL ASSISTANT Chris Ziemba

COPY EDITOR Diana Starr

TECHNOLOGY PROJECT MANAGER Keli Amann

MARKETING MANAGER Ann Caven

MARKETING ASSISTANT Leyla Jowza

ADVERTISING PROJECT MANAGER Nathaniel Michelson

PROJECT MANAGER, EDITORIAL PRODUCTION Andy Marinkovich

PRINT/MEDIA BUYER Karen Hunt

PERMISSIONS EDITOR Joohee Lee

PRODUCTION SERVICE Grace Davidson & Associates

TEXT AND COVER DESIGNER Gary Head

PHOTO RESEARCHERS Myrna Engler, Terri Wright

ILLUSTRATORS Leif Buckley, Seth Gold, Gary Head,
Chris Keeney, and Lisa Starr

COVER PRINTER Phoenix Color Corp

COMPOSITOR Lachina Publishing Services

PRINTER R.R. Donnelley/Willard

COVER IMAGE One of the Hawaiian honeycreepers, living
representatives of a spectacular adaptive radiation in the
Hawaiian Islands. Frans Lanting/Minden Pictures

Printed in the United States of America
1 2 3 4 5 6 7 09 08 07 06 05

Library of Congress Control Number: 2004114895

ISBN 0-534-42029-X

For more information about our products, contact us at:
Thomson Learning Academic Resource Center

1-800-423-0563

For permission to use material from this text or product, submit
a request online at http://www.thomsonrights.com.
Any additional questions about permissions can be submitted
by email to thomsonrights@thomson.com.

BOOKS IN THE BROOKS/COLE BIOLOGY SERIES

Biology: The Unity and Diversity of Life, Tenth, Starr/Taggart
Engage Online for Biology: The Unity and Diversity of Life
Biology: Concepts and Applications, Sixth, Starr
Basic Concepts in Biology, Sixth, Starr
Biology Today and Tomorrow, Starr
Biology, Seventh, Solomon/Berg/Martin
Human Biology, Sixth, Starr/McMillan
Biology: A Human Emphasis, Sixth, Starr
Human Physiology, Fifth, Sherwood
Fundamentals of Physiology, Second, Sherwood
Human Physiology, Fourth, Rhoades/Pflanzer

Laboratory Manual for Biology, Fourth, Perry/Morton/Perry
Laboratory Manual for Human Biology, Morton/Perry/Perry
Photo Atlas for Biology, Perry/Morton
Photo Atlas for Anatomy and Physiology, Morton/Perry
Photo Atlas for Botany, Perry/Morton
Virtual Biology Laboratory, Beneski/Waber
Introduction to Cell and Molecular Biology, Wolfe
Molecular and Cellular Biology, Wolfe
Biotechnology: An Introduction, Second, Barnum

Introduction to Microbiology, Third, Ingraham/Ingraham
Microbiology: An Introduction, Batzing
Genetics: The Continuity of Life, Fairbanks/Anderson
Human Heredity, Seventh, Cummings
Current Perspectives in Genetics, Second, Cummings
Gene Discovery Lab, Benfey

Animal Physiology, Sherwood, Kleindorf, Yarcey
Invertebrate Zoology, Seventh, Ruppert/Fox/Barnes
Mammalogy, Fourth, Vaughan/Ryan/Czaplewski
Biology of Fishes, Second, Bond
Vertebrate Dissection, Ninth, Homberger/Walker

Plant Biology, Second, Rost/Barbour/Stocking/Murphy
Plant Physiology, Fourth, Salisbury/Ross
Introductory Botany, Berg

General Ecology, Second, Krohne
Essentials of Ecology, Third, Miller
Terrestrial Ecosystems, Second, Aber/Melillo
Living in the Environment, Fourteenth, Miller
Environmental Science, Tenth, Miller
Sustaining the Earth, Seventh, Miller
Case Studies in Environmental Science, Second, Underwood
Environmental Ethics, Third, Des Jardins
Watersheds 3—Ten Cases in Environmental Ethics, Third,
Newton/Dillingham

Problem-Based Learning Activities for General Biology, Allen/Duch
The Pocket Guide to Critical Thinking, Second, Epstein

Thomson Higher Education
10 Davis Drive
Belmont, CA 94002-3098
USA

Asia (including India)
Thomson Learning
5 Shenton Way
#01-01 UIC Building
Singapore 068808

Australia/New Zealand
Thomson Learning Australia
102 Dodds Street
Southbank, Victoria 3006
Australia

Canada
Thomson Nelson
1120 Birchmount Road
Toronto, Ontario M1K 5G4

UK/Europe/Middle East/Africa
Thomson Learning
High Holborn House
50/51 Bedford Row
London WC1R 4LR
United Kingdom

CONTENTS IN BRIEF

INTRODUCTION

1   Invitation to Biology

UNIT I   PRINCIPLES OF CELLULAR LIFE

2   Life's Chemical Basis

3   Molecules of Life

4   How Cells Are Put Together

5   How Cells Work

6   Where It Starts—Photosynthesis

7   How Cells Release Chemical Energy

UNIT II   PRINCIPLES OF INHERITANCE

8   How Cells Reproduce

9   Meiosis and Sexual Reproduction

10   Observing Patterns in Inherited Traits

11   Chromosomes and Human Genetics

12   DNA Structure and Function

13   From DNA to Proteins

14   Controls Over Genes

15   Studying and Manipulating Genomes

UNIT III   PRINCIPLES OF EVOLUTION

16   Processes of Evolution

17   Evolutionary Patterns, Rates, and Trends

18   The Origin and Early Evolution of Life

UNIT IV   EVOLUTION AND BIODIVERSITY

19   Prokaryotes and Viruses

20   The Simplest Eukaryotes—Protists and Fungi

21   Plant Evolution

22   Animal Evolution—The Invertebrates

23   Animal Evolution—The Vertebrates

24   Plants and Animals—Common Challenges

UNIT V   HOW ANIMALS WORK

25   Animal Reproduction and Development

UNIT VI   PRINCIPLES OF ECOLOGY

26   Population Ecology

27   Community Structure and Biodiversity

28   Ecosystems

29   The Biosphere

30   Behavioral Ecology

DETAILED CONTENTS

INTRODUCTION

1 Invitation to Biology

IMPACTS, ISSUES *What Am I Doing Here?* 2

1.1 LIFE'S LEVELS OF ORGANIZATION 4
From Small to Smaller 4
From Smaller to Vast 4

1.2 OVERVIEW OF LIFE'S UNITY 6
DNA, The Basis of Inheritance 6
Energy, The Basis of Metabolism 6
Life's Responsiveness to Change 7

1.3 IF SO MUCH UNITY, WHY SO MANY SPECIES? 8

1.4 AN EVOLUTIONARY VIEW OF DIVERSITY 10

1.5 THE NATURE OF BIOLOGICAL INQUIRY 11
Observations, Hypotheses, and Tests 11
About the Word "Theory" 11

1.6 THE POWER OF EXPERIMENTAL TESTS 12
An Assumption of Cause and Effect 12
Example of an Experimental Design 12
Example of a Field Experiment 12
Bias in Reporting Results 13

1.7 THE LIMITS OF SCIENCE 14

UNIT I   PRINCIPLES OF CELLULAR LIFE

2 Life's Chemical Basis

IMPACTS, ISSUES *What Are You Worth?* 18

2.1 START WITH ATOMS 20

2.2 FOCUS ON SCIENCE *Radioisotopes* 21

2.3 WHAT HAPPENS WHEN ATOM BONDS WITH ATOM? 22
Electrons and Energy Levels 22
From Atoms to Molecules 22

2.4 BONDS IN BIOLOGICAL MOLECULES 24
Ion Formation and Ionic Bonding 24
Covalent Bonding 24
Hydrogen Bonding 25

2.5 WATER'S LIFE-GIVING PROPERTIES 26
Polarity of the Water Molecule 26
Water's Temperature-Stabilizing Effects 26
Water's Solvent Properties 27
Water's Cohesion 27

2.6 ACIDS AND BASES 28
The pH Scale 28
How Do Acids and Bases Differ? 28
Salts and Water 29
Buffers Against Shifts in pH 29

3 Molecules of Life

IMPACTS, ISSUES *Science or the Supernatural?* 32

3.1 MOLECULES OF LIFE—FROM STRUCTURE TO FUNCTION 34
Carbon's Bonding Behavior 34
Functional Groups 34
How Do Cells Actually Build Organic Compounds? 35

3.2 FOCUS ON THE ENVIRONMENT *Bubble, Bubble, Toil and Trouble* 36

3.3 THE TRULY ABUNDANT CARBOHYDRATES 38
The Simple Sugars 38
Short-Chain Carbohydrates 38
Complex Carbohydrates 38

3.4 GREASY, OILY—MUST BE LIPIDS 40
Fats and Fatty Acids 40
Phospholipids 41
Waxes 41
Cholesterol and Other Sterols 41

3.5 PROTEINS—DIVERSITY IN STRUCTURE AND FUNCTION 42

3.6 WHY IS PROTEIN STRUCTURE SO IMPORTANT? 44
Just One Wrong Amino Acid . . . 44
. . . And You Get Sickle-Shaped Cells! 44
Denaturation 44

3.7 NUCLEOTIDES AND THE NUCLEIC ACIDS 46

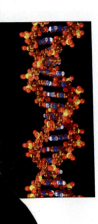

4   How Cells Are Put Together

IMPACTS, ISSUES   *Where Did Cells Come From?*   50

4.1   SO WHAT IS "A CELL"?   52
Components of All Cells   52
Why Aren't Cells Bigger?   52

4.2   FOCUS ON SCIENCE   *How Do We "See" Cells?*   54
The Cell Theory   54
Some Modern Microscopes   54

4.3   ALL LIVING CELLS HAVE MEMBRANES   56

4.4   INTRODUCING PROKARYOTIC CELLS   58

4.5   INTRODUCING EUKARYOTIC CELLS   60

4.6   THE NUCLEUS   61

4.7   THE ENDOMEMBRANE SYSTEM   62
Endoplasmic Reticulum   62
Golgi Bodies   62
Membranous Sacs With Diverse Functions   63

4.8   MITOCHONDRIA AND CHLOROPLASTS   64
Mitochondria   64
Chloroplasts   64

4.9   VISUAL SUMMARY OF EUKARYOTIC CELL COMPONENTS   65

4.10   THE CYTOSKELETON   66
Moving Along With Motor Proteins   66
Cilia, Flagella, and False Feet   67

4.11   CELL SURFACE SPECIALIZATIONS   68
Eukaryotic Cell Walls   68
Matrixes Between Animal Cells   69
Cell Junctions   69

5   How Cells Work

IMPACTS, ISSUES   *Alcohol, Enzymes, and Your Liver*   72

5.1   INPUTS AND OUTPUTS OF ENERGY   74
The One-Way Flow of Energy   74
Up and Down the Energy Hills   74
ATP—The Cell's Energy Currency   75

5.2   INPUTS AND OUTPUTS OF SUBSTANCES   76
The Nature of Metabolic Reactions   76
Redox Reactions   76
Types of Metabolic Pathways   77

5.3   HOW ENZYMES MAKE SUBSTANCES REACT   78

5.4   ENZYMES DON'T WORK IN A VACUUM   80
Help From Cofactors   80
Controls Over Enzymes   80
Effects of Temperature, pH, and Salinity   81

5.5   DIFFUSION, MEMBRANES, AND METABOLISM   82
What Is a Concentration Gradient?   82
What Determines Diffusion Rates?   83
Membrane Crossing Mechanisms   83

5.6   HOW THE MEMBRANE TRANSPORTERS WORK   84
Passive Transport   84
Active Transport   85

5.7   WHICH WAY WILL WATER MOVE?   86
Movement of Water   86
Effects of Tonicity   86
Effects of Fluid Pressure   86

5.8   MEMBRANE TRAFFIC TO AND FROM THE CELL SURFACE   88
Endocytosis and Exocytosis   88
Membrane Cycling   89

6   Where It Starts—Photosynthesis

IMPACTS, ISSUES   *Pastures of the Seas*   92

6.1   SUNLIGHT AS AN ENERGY SOURCE   94
Properties of Light   94
Pigments—The Rainbow Catchers   94

6.2   WHAT IS PHOTOSYNTHESIS, AND WHERE DOES IT HAPPEN?   96

Two Stages of Reactions 96
A Look Inside the Chloroplast 96
Photosynthesis Changed the Biosphere 96

6.3 LIGHT-DEPENDENT REACTIONS 98
Transducing the Absorbed Energy 98
Making ATP and NADPH 98

6.4 FOCUS ON SCIENCE A Case of Controlled
Energy Release 100

6.5 LIGHT-INDEPENDENT REACTIONS:
THE SUGAR FACTORY 101

6.6 FOCUS ON THE ENVIRONMENT Different Plants,
Different Carbon-Fixing Pathways 102

7 How Cells Release Chemical Energy

IMPACTS, ISSUES When Mitochondria Spin Their
Wheels 106

7.1 OVERVIEW OF ENERGY-RELEASING
PATHWAYS 108
Comparison of the Main Types of
Energy-Releasing Pathways 108
Overview of Aerobic Respiration 108

7.2 FIRST STAGE: GLYCOLYSIS 110

7.3 SECOND STAGE OF AEROBIC
RESPIRATION 112
Acetyl–CoA Formation 112
The Krebs Cycle 112

7.4 THIRD STAGE OF AEROBIC RESPIRATION—
THE BIG ENERGY PAYOFF 114
Electron Transfer Phosphorylation 114
Summing Up: The Energy Harvest 115

7.5 FERMENTATION PATHWAYS 116
Alcoholic Fermentation 116
Lactate Fermentation 117

7.6 ALTERNATIVE ENERGY SOURCES IN THE
BODY 118
The Fate of Glucose at Mealtime and
In Between Meals 118
Energy From Fats 118
Energy From Proteins 118

7.7 FOCUS ON EVOLUTION Perspective
on Life 120

8 How Cells Reproduce

IMPACTS, ISSUES Henrietta's Immortal Cells 124

8.1 OVERVIEW OF CELL DIVISION MECHANISMS 126
Mitosis, Meiosis, and the Prokaryotes 126
Key Points About Chromosome
Structure 126

8.2 INTRODUCING THE CELL CYCLE 128
The Wonder of Interphase 128
Mitosis and the Chromosome Number 129

8.3 A CLOSER LOOK AT MITOSIS 130

8.4 DIVISION OF THE CYTOPLASM 132
Cleavage in Animals 132
Cell Plate Formation in Plants 133
Appreciate the Process! 133

8.5 FOCUS ON HEALTH When Control
Is Lost 134
The Cell Cycle Revisited 134
Checkpoint Failure and Tumors 134
Characteristics of Cancer 135

9 Meiosis and Sexual Reproduction

IMPACTS, ISSUES Why Sex? 138

9.1 AN EVOLUTIONARY VIEW 140

9.2 OVERVIEW OF MEIOSIS 140
Think "Homologues" 140
Two Divisions, Not One 141

9.3 VISUAL TOUR OF MEIOSIS 142

9.4 HOW MEIOSIS PUTS VARIATION
IN TRAITS 144
Crossing Over in Prophase I 144
Metaphase I Alignments 145

9.5 FROM GAMETES TO OFFSPRING 146
Gamete Formation in Plants 146
Gamete Formation in Animals 146
More Shufflings at Fertilization 146

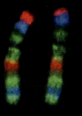

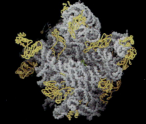

10 Observing Patterns in Inherited Traits

IMPACTS, ISSUES *Menacing Mucus* 150

10.1 MENDEL'S INSIGHT INTO INHERITANCE PATTERNS *152*
Mendel's Experimental Approach *152*
Terms Used in Modern Genetics *153*

10.2 MENDEL'S THEORY OF SEGREGATION *154*
Monohybrid Cross Predictions *154*
Testcrosses *155*

10.3 MENDEL'S THEORY OF INDEPENDENT ASSORTMENT *156*

10.4 MORE PATTERNS THAN MENDEL THOUGHT *158*
ABO Blood Types—A Case of Codominance *158*
Incomplete Dominance *158*
Single Genes With a Wide Reach *159*
When Products of Gene Pairs Interact *159*

10.5 COMPLEX VARIATIONS IN TRAITS *160*
Regarding the Unexpected Phenotype *160*
Continuous Variation in Populations *160*

10.6 GENES AND THE ENVIRONMENT *162*

11 Chromosomes and Human Genetics

IMPACTS, ISSUES *Strange Genes, Tortured Minds* 166

11.1 THE CHROMOSOMAL BASIS OF INHERITANCE *168*
A Rest Stop on Our Conceptual Road *168*
Autosomes and Sex Chromosomes *168*
Sex Determination in Humans *169*

11.2 FOCUS ON SCIENCE *Karyotyping Made Easy* 170

11.3 IMPACT OF CROSSING OVER ON INHERITANCE *171*

11.4 HUMAN GENETIC ANALYSIS *172*

11.5 EXAMPLES OF HUMAN INHERITANCE PATTERNS *174*
Autosomal Dominant Inheritance *174*
Autosomal Recessive Inheritance *174*
X-Linked Inheritance *175*

11.6 FOCUS ON HEALTH *Too Young, Too Old* 176

11.7 ALTERED CHROMOSOMES *176*
The Main Categories of Structural Change *176*
Duplication *176*
Inversion *176*
Deletion *176*
Translocation *177*
Does Chromosome Structure Evolve? *177*

11.8 CHANGES IN THE CHROMOSOME NUMBER *178*
Autosomal Change and Down Syndrome *178*
Changes in the Sex Chromosome Number *179*
Female Sex Chromosome Abnormalities *179*
Male Sex Chromosome Abnormalities *179*

11.9 FOCUS ON HEALTH *Prospects in Human Genetics* 180
Bioethical Questions *180*
Some of the Options *180*
Genetic Screening *180*
Phenotypic Treatments *180*
Prenatal Diagnosis *181*
Genetic Counseling *181*
Regarding Abortion *181*
Preimplantation Diagnosis *181*

12 DNA Structure and Function

IMPACTS, ISSUES *Goodbye, Dolly* 184

12.1 THE HUNT FOR FAME, FORTUNE, AND DNA *186*
Early and Puzzling Clues *186*
Confirmation of DNA Function *186*
Enter Watson and Crick *187*

12.2 THE DISCOVERY OF DNA'S STRUCTURE *188*
DNA's Building Blocks *188*
Patterns of Base Pairing *189*

12.3 FOCUS ON BIOETHICS *Rosalind's Story* *190*

12.4 DNA REPLICATION AND REPAIR *190*

12.5 FOCUS ON SCIENCE *Reprogramming DNA
To Clone Mammals* *192*

---

13 From DNA to Proteins

IMPACTS, ISSUES *Ricin and Your Ribosomes* *194*

13.1 HOW IS RNA TRANSCRIBED
FROM DNA? *196*
The Nature of Transcription *196*
Finishing Touches on mRNA Transcripts *197*

13.2 DECIPHERING mRNA TRANSCRIPTS *198*
The Genetic Code *198*
The Other RNAs *199*

13.3 TRANSLATING mRNA INTO PROTEIN *200*

13.4 MUTATED GENES AND THEIR PROTEIN
PRODUCTS *202*
Common Mutations *202*
How Do Mutations Arise? *203*
The Proof Is in the Protein *203*

---

14 Controls Over Genes

IMPACTS, ISSUES *Between You and Eternity* *206*

14.1 SOME CONTROL MECHANISMS *208*

14.2 PROKARYOTIC GENE CONTROL *208*
Negative Control of the Lactose Operon *208*
Positive Control of the Lactose Operon *208*

14.3 EUKARYOTIC GENE CONTROLS *210*
Same Genes, Different Cell Lineages *210*
When Controls Come Into Play *210*

14.4 EXAMPLES OF GENE CONTROL *212*
Homeotic Genes and Body Plans *212*
X Chromosome Inactivation *212*

14.5 FOCUS ON SCIENCE *There's a Fly in My
Research* *214*
Drosophila! *214*

Clues to Gene Control *214*
Genes and Patterns in Development *215*

---

15 Studying and Manipulating Genomes

IMPACTS, ISSUES *Golden Rice or Frankenfood?* *218*

15.1 FOCUS ON SCIENCE *Tinkering With the
Molecules of Life* *220*
Emergence of Molecular Biology *220*
The Human Genome Project *221*

15.2 A MOLECULAR TOOLKIT *222*
The Scissors: Restriction Enzymes *222*
Cloning Vectors *222*
cDNA Cloning *223*

15.3 HAYSTACKS TO NEEDLES *224*
Isolating Genes *224*
PCR *225*

15.4 FOCUS ON SCIENCE *First Just Fingerprints,
Now DNA Fingerprints* *226*

15.5 AUTOMATED DNA SEQUENCING *227*

15.6 PRACTICAL GENETICS *228*
Designer Plants *228*
Barnyard Biotech *229*

15.7 FOCUS ON BIOETHICS *Weighing Benefits and
Risks* *230*
Who Gets Well? *230*
Who Gets Enhanced? *230*
Knockout Cells and Organ Factories *231*
Regarding "Frankenfood" *231*

15.8 BRAVE NEW WORLD *232*
Genomics *232*
DNA Chips *232*

UNIT III PRINCIPLES OF EVOLUTION

---

16 Processes of Evolution

IMPACTS, ISSUES *Rise of the Super Rats* *236*

16.1 EARLY BELIEFS, CONFOUNDING
DISCOVERIES *238*
Questions From Biogeography *238*

Questions From Comparative Morphology *239*

Questions About Fossils *239*

16.2 A FLURRY OF NEW THEORIES *240*

Squeezing New Evidence Into Old Beliefs *240*

Voyage of the *Beagle* *240*

16.3 DARWIN'S THEORY TAKES FORM *242*

Old Bones and Armadillos *242*

A Key Insight—Variation in Traits *242*

Natural Selection Defined *243*

16.4 THE NATURE OF ADAPTATION *244*

Salt-Tolerant Tomatoes *244*

No Polar Bears in the Desert *244*

Adaptation to What? *245*

16.5 INDIVIDUALS DON'T EVOLVE, POPULATIONS DO *246*

Variation in Populations *246*

The "Gene Pool" *246*

Stability and Change in Allele Frequencies *246*

16.6 MUTATIONS REVISITED *248*

16.7 DIRECTIONAL SELECTION *248*

Pesticide Resistance *248*

Antibiotic Resistance *248*

Coat Color in Desert Mice *249*

16.8 SELECTION AGAINST OR IN FAVOR OF EXTREME PHENOTYPES *250*

Stabilizing Selection *250*

Disruptive Selection *251*

16.9 MAINTAINING VARIATION IN A POPULATION *252*

Sexual Selection *252*

Sickle-Cell Anemia—Lesser of Two Evils? *252*

16.10 GENETIC DRIFT—THE CHANCE CHANGES *254*

Bottlenecks and the Founder Effect *254*

Genetic Drift and Inbred Populations *255*

16.11 GENE FLOW *255*

17 Evolutionary Patterns, Rates, and Trends

IMPACTS, ISSUES *Measuring Time* *258*

17.1 FOSSILS—EVIDENCE OF ANCIENT LIFE *260*

How Do Fossils Form? *260*

Fossils in Sedimentary Rock Layers *261*

Interpreting the Fossil Record *261*

17.2 FOCUS ON SCIENCE *Dating Pieces of the Puzzle* *262*

Radiometric Dating *262*

Placing Fossils in Geologic Time *262*

17.3 EVIDENCE FROM BIOGEOGRAPHY *264*

An Outrageous Hypothesis *264*

Drifting Continents, Changing Seas *265*

17.4 EVIDENCE FROM COMPARATIVE MORPHOLOGY *266*

Morphological Divergence *266*

Morphological Convergence *267*

17.5 EVIDENCE FROM PATTERNS OF DEVELOPMENT *268*

17.6 EVIDENCE FROM BIOCHEMISTRY *270*

Protein Comparisons *270*

Nucleic Acid Comparisons *270*

Molecular Clocks *271*

17.7 REPRODUCTIVE ISOLATION, MAYBE NEW SPECIES *272*

17.8 THE MAIN MODEL FOR SPECIATION *274*

The Nature of Allopatric Speciation *274*

Allopatric Speciation on Archipelagos *274*

17.9 OTHER SPECIATION MODELS *276*

Sympatric Speciation *276*

Evidence From Cichlids in Africa *276*

Polyploidy's Impact *276*

Parapatric Speciation *277*

17.10 PATTERNS OF SPECIATION AND EXTINCTION *278*

Branching and Unbranched Evolution 278
Evolutionary Trees and Rates of Change 278
Adaptive Radiations 278
Extinctions—The End of the Line 279

17.11 HOW CAN WE ORGANIZE THE EVIDENCE? 280
Naming, Identifying, and Classifying Species 280
What's in a Name? A Cladistics View 281

17.12 AN EVOLUTIONARY TREE OF LIFE 282

18 The Origin and Early Evolution of Life

IMPACTS, ISSUES *Looking for Life in All the Odd Places* 286

18.1 IN THE BEGINNING . . . 288
Conditions on the Early Earth 288
Abiotic Synthesis of Organic Compounds 289

18.2 HOW DID CELLS ORIGINATE? 290
Origin of Agents of Metabolism 290
Origin of the First Plasma Membranes 290
Origin of Self-Replicating Systems 291

18.3 THE FIRST CELLS 292

18.4 FOCUS ON SCIENCE *Where Did Organelles Come From?* 294
Origin of the Nucleus and ER 294
Origin of Mitochondria and Chloroplasts 294
Evidence of Endosymbiosis 295

18.5 TIMELINE FOR LIFE'S ORIGIN AND EVOLUTION 296

UNIT IV EVOLUTION AND BIODIVERSITY

19 Prokaryotes and Viruses

IMPACTS, ISSUES *West Nile Virus Takes Off* 300

19.1 CHARACTERISTICS OF PROKARYOTIC CELLS 302
Body Plans, Shapes, and Sizes 302
Metabolic Diversity 302
Growth and Reproduction 303
Classification 303

19.2 THE BACTERIA 304
Representative Diversity 304
Regarding the "Simple" Bacteria 305

19.3 THE ARCHAEA 306
The Third Domain 306
Here, There, Everywhere 306

19.4 VIRUSES AND VIROIDS 308

19.5 FOCUS ON HEALTH *Evolution and Infectious Diseases* 310
The Nature of Disease 310
Drug-Resistant Strains 310
Foodborne Diseases and Mad Cows 310

20 The Simplest Eukaryotes— Protists and Fungi

IMPACTS, ISSUES *Tiny Critters, Big Impacts* 314

20.1 CHARACTERISTICS OF PROTISTS 316

20.2 THE MOST ANCIENT GROUPS 316
Flagellated Protozoans 316
Euglenoids 316
Amoeboid Protozoans 317

20.3 THE ALVEOLATES 318
Ciliated Protozoans 318
Apicomplexans 318
Dinoflagellates 319

20.4 FOCUS ON THE ENVIRONMENT *Algal Blooms* 320

20.5 THE STRAMENOPILES 320
Oomycotes 320
Chrysophytes 321
Brown Algae 321

20.6 RED ALGAE 322

20.7 GREEN ALGAE 323

20.8 FOCUS ON THE ENVIRONMENT *Environmental Escape Artists* 324
Consider a Green Alga 324
Consider a Slime Mold 324

20.9 CHARACTERISTICS OF FUNGI 326

20.10 FUNGAL DIVERSITY 328

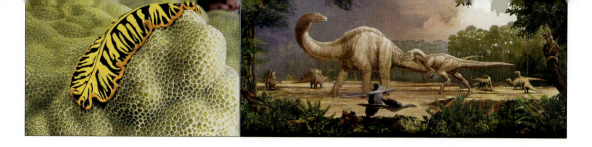

20.11 FOCUS ON HEALTH *The Unloved Few* 329

20.12 FUNGAL SYMBIONTS 330
   Fungal Endophytes 330
   Lichens 330
   Mycorrhizae 330
   As Fungi Go, So Go the Forests 331

21   Plant Evolution

IMPACTS, ISSUES *Beginnings, and Endings* 334

21.1 EVOLUTIONARY TRENDS AMONG PLANTS 336
   Roots, Stems, and Leaves 336
   From Haploid to Diploid Dominance 336
   Evolution of Pollen and Seeds 337

21.2 THE BRYOPHYTES—NO VASCULAR TISSUE 338

21.3 SEEDLESS VASCULAR PLANTS 340
   Lycophytes 340
   Horsetails 341
   Ferns 341

21.4 FOCUS ON THE ENVIRONMENT *Ancient Carbon Treasures* 342

21.5 THE RISE OF SEED-BEARING PLANTS 343

21.6 GYMNOSPERMS—PLANTS WITH NAKED SEEDS 344

21.7 ANGIOSPERMS—THE FLOWERING PLANTS 346

21.8 A GLIMPSE INTO FLOWERING PLANT DIVERSITY 348

21.9 FOCUS ON THE ENVIRONMENT *Deforestation Revisited* 349

22   Animal Evolution—The Invertebrates

IMPACTS, ISSUES *Old Genes, New Drugs* 352

22.1 OVERVIEW OF THE ANIMAL KINGDOM 354

General Characteristics 354
Clues in Body Plans 354

22.2 ANIMAL ORIGINS 356

22.3 SPONGES—SUCCESS IN SIMPLICITY 357

22.4 CNIDARIANS—SIMPLE TISSUES, NO ORGANS 358

22.5 FLATWORMS—THE SIMPLEST ORGAN SYSTEMS 360

22.6 ANNELIDS—SEGMENTS GALORE 362
   Advantages of Segmentation 362
   Annelid Adaptations—A Case Study 362

22.7 THE EVOLUTIONARILY PLIABLE MOLLUSKS 364
   Hiding Out, Or Not 364
   A Cephalopod Need for Speed 365

22.8 ROUNDWORMS 366

22.9 ARTHROPODS—THE MOST SUCCESSFUL ANIMALS 367

22.10 A LOOK AT THE CRUSTACEANS 368

22.11 SPIDERS AND THEIR RELATIVES 369

22.12 A LOOK AT INSECT DIVERSITY 370

22.13 THE PUZZLING ECHINODERMS 372

23   Animal Evolution—The Vertebrates

IMPACTS, ISSUES *Interpreting and Misinterpreting the Past* 376

23.1 THE CHORDATE HERITAGE 378
   Chordate Characteristics 378
   Invertebrate Chordates 378
   A Word of Caution 379

23.2 EVOLUTIONARY TRENDS AMONG THE VERTEBRATES 380
   Early Craniates 380
   Key Innovations 381
   Major Vertebrate Groups 381

23.3 JAWED FISHES AND THE RISE OF TETRAPODS 382

Cartilaginous Fishes 382

"Bony Fishes" 383

23.4 AMPHIBIANS—THE FIRST TETRAPODS ON LAND 384

23.5 FOCUS ON THE ENVIRONMENT Vanishing Acts 385

23.6 THE RISE OF AMNIOTES 386

The "Reptiles" 386

The Age of Dinosaurs 386

23.7 EXISTING REPTILIAN GROUPS 388

23.8 BIRDS—THE FEATHERED ONES 390

23.9 MAMMALS 392

23.10 FROM EARLY PRIMATES TO HOMINIDS 394

Trends in Primate Evolution 394

Origins and Early Divergences 395

23.11 EMERGENCE OF EARLY HUMANS 396

23.12 EMERGENCE OF MODERN HUMANS 398

Early Big-Time Walkers 398

Where Did Modern Humans Originate? 398

24 Plants and Animals—Common Challenges

IMPACTS, ISSUES Too Hot To Handle 402

24.1 LEVELS OF STRUCTURAL ORGANIZATION 404

From Cells to Multicelled Organisms 404

Growth Versus Development 404

Structural Organization Has a History 404

The Body's Internal Environment 405

Start Thinking "Homeostasis" 405

24.2 RECURRING CHALLENGES TO SURVIVAL 406

Gas Exchange in Large Bodies 406

Internal Transport in Large Bodies 406

Maintaining the Water–Solute Balance 406

Cell-to-Cell Communication 407

On Variations in Resources and Threats 407

24.3 HOMEOSTASIS IN ANIMALS 408

Negative Feedback 408

Positive Feedback 409

24.4 DOES HOMEOSTASIS OCCUR IN PLANTS? 410

Walling Off Threats 410

Sand, Wind, and the Yellow Bush Lupine 410

About Rhythmic Leaf Folding 411

24.5 HOW CELLS RECEIVE AND RESPOND TO SIGNALS 412

UNIT V   HOW ANIMALS WORK

25 Animal Reproduction and Development

IMPACTS, ISSUES Mind-Boggling Births 416

25.1 REFLECTIONS ON SEXUAL REPRODUCTION 418

Sexual Versus Asexual Reproduction 418

Costs of Sexual Reproduction 418

25.2 STAGES OF REPRODUCTION AND DEVELOPMENT 420

25.3 EARLY MARCHING ORDERS 422

Information in the Egg 422

Cleavage—The Start of Multicellularity 422

Cleavage Patterns 423

25.4 HOW DO SPECIALIZED TISSUES AND ORGANS FORM? 424

Cell Differentiation 424

Morphogenesis 425

25.5 PATTERN FORMATION 426

Embryonic Induction 426

A Theory of Pattern Formation 426

Evolutionary Constraints on Development 427

25.6 REPRODUCTIVE SYSTEM OF HUMAN MALES 428

When Gonads Form and Become Active 428

Structure and Function of the Reproductive System 428

Cancers of the Prostate and Testes 429

25.7 SPERM FORMATION 430

25.8 REPRODUCTIVE SYSTEM OF HUMAN FEMALES 432

Components of the System 432

Overview of the Menstrual Cycle 432

25.9 PREPARATIONS FOR PREGNANCY 434

Cyclic Changes in the Ovary 434

Cyclic Changes in the Uterus 435

25.10 VISUAL SUMMARY OF THE MENSTRUAL CYCLE 436

25.11 PREGNANCY HAPPENS 437

Sexual Intercourse 437

Fertilization 437

25.12 FOCUS ON BIOETHICS Control of Human Fertility 438

The Issue 438

Some Options 438

Seeking or Ending Pregnancy 439

25.13 FOCUS ON HEALTH Sexually Transmitted Diseases 440

25.14 FORMATION OF THE EARLY EMBRYO 442

Cleavage and Implantation 442

Extraembryonic Membranes 442

25.15 EMERGENCE OF THE VERTEBRATE BODY PLAN 444

25.16 WHY IS THE PLACENTA SO IMPORTANT? 445

25.17 EMERGENCE OF DISTINCTLY HUMAN FEATURES 446

25.18 FOCUS ON HEALTH Mother as Provider, Protector, Potential Threat 448

Nutritional Considerations 448

Infectious Diseases 448

Alcohol, Tobacco, and Other Drugs 449

25.19 FROM BIRTH ONWARD 450

Giving Birth 450

Nourishing the Newborn 451

Postnatal Development 451

25.20 FOCUS ON SCIENCE Why Do We Age and Die? 452

Programmed Life Span Hypothesis 452

Cumulative Assaults Hypothesis 452

UNIT VI PRINCIPLES OF ECOLOGY

26 Population Ecology

IMPACTS, ISSUES The Human Touch 456

26.1 CHARACTERISTICS OF POPULATIONS 458

26.2 FOCUS ON SCIENCE Elusive Heads to Count 459

26.3 POPULATION SIZE AND EXPONENTIAL GROWTH 460

Gains and Losses in Population Size 460

From Zero to Exponential Growth 460

What Is the Biotic Potential? 461

26.4 LIMITS ON THE GROWTH OF POPULATIONS 462

What Are the Limiting Factors? 462

Carrying Capacity and Logistic Growth 462

Density-Independent Factors 463

26.5 LIFE HISTORY PATTERNS 464

Life Tables 464

Patterns of Survival and Reproduction 464

26.6 FOCUS ON SCIENCE Natural Selection and Life Histories 466

26.7 HUMAN POPULATION GROWTH 468

26.8 FERTILITY RATES AND AGE STRUCTURE 470

26.9 POPULATION GROWTH AND ECONOMIC EFFECTS 472

Demographic Transitions 472

A Question of Immigration Policies 472

A Question of Resource Consumption 472

Impacts of No Growth 473

27 Community Structure and Biodiversity

IMPACTS, ISSUES Fire Ants in the Pants 476

27.1 WHICH FACTORS SHAPE COMMUNITY STRUCTURE? 478

The Niche 478

Categories of Species Interactions 478

27.2 MUTUALISM 479

27.3 COMPETITIVE INTERACTIONS 480

Competitive Exclusion 480

Resource Partitioning 481

27.4 PREDATOR–PREY INTERACTIONS 482

Coevolution of Predators and Prey 482

Models for Predator–Prey Interactions 482

The Canadian Lynx and Snowshoe Hare 482

27.5 FOCUS ON EVOLUTION An Evolutionary Arms Race 484

Adaptations of Prey 484

Adaptive Responses of Predators 485

27.6 PARASITE–HOST INTERACTIONS 486

Parasites and Parasitoids 486

Uses as Biological Controls 487

27.7 FOCUS ON EVOLUTION Cowbird Chutzpah 487

27.8 FORCES CONTRIBUTING TO COMMUNITY STABILITY 488

A Succession Model 488

The Climax Pattern Model 489

Cyclic, Nondirectional Changes 489

27.9 FORCES CONTRIBUTING TO COMMUNITY INSTABILITY 490

The Role of Keystone Species 490

Species Introductions Tip the Balance 491

27.10 FOCUS ON THE ENVIRONMENT Exotic Invaders 492

The Plants That Ate Georgia 492

The Alga Triumphant 492

The Rabbits That Ate Australia 493

27.11 BIOGEOGRAPHIC PATTERNS 494

Mainland and Marine Patterns 494

Island Patterns 495

27.12 THREATS TO BIODIVERSITY 496

On the Newly Endangered Species 496

Habitat Losses and Fragmentation 496

Conservation Biology 497

27.13 SUSTAINING BIODIVERSITY 498

Identifying Areas At Risk 498

Economic Factors and Sustainable Development 498

28 Ecosystems

IMPACTS, ISSUES Bye-Bye, Blue Bayou 502

28.1 THE NATURE OF ECOSYSTEMS 504

Overview of the Participants 504

Structure of Ecosystems 505

28.2 THE NATURE OF FOOD WEBS 506

28.3 FOCUS ON THE ENVIRONMENT DDT in Food Webs 508

28.4 STUDYING ENERGY FLOW THROUGH ECOSYSTEMS 509

What Is Primary Productivity? 509

Ecological Pyramids 509

28.5 FOCUS ON SCIENCE Energy Flow Through Silver Springs 510

28.6 OVERVIEW OF BIOGEOCHEMICAL CYCLES 511

28.7 GLOBAL CYCLING OF WATER 512

The Hydrologic Cycle 512

The Water Crisis 512

28.8 CARBON CYCLE 514

28.9 FOCUS ON THE ENVIRONMENT Greenhouse Gases, Global Warming 516

28.10 NITROGEN CYCLE 518

The Cycling Processes 518

Human Impact on the Nitrogen Cycle 519

28.11 PHOSPHORUS CYCLE 520

29 The Biosphere

IMPACTS, ISSUES Surfers, Seals, and the Sea 524

29.1 GLOBAL AIR CIRCULATION PATTERNS 526

Climate and Temperature Zones 526

Harnessing the Sun and Wind 527

29.2 FOCUS ON THE ENVIRONMENT Air Circulation Patterns and Human Affairs 528

A Fence of Wind and Ozone Thinning 528
No Wind, Lots of Pollutants, and Smog 528
Winds and Acid Rain 529

29.3 THE OCEAN, LANDFORMS, AND CLIMATES 530
Ocean Currents and Their Effects 530
Rain Shadows and Monsoons 531

29.4 REALMS OF BIODIVERSITY 532

29.5 MOISTURE-CHALLENGED BIOMES 534
Deserts, Natural and Man-Made 534
Dry Shrublands, Dry Woodlands,
and Grasslands 534

29.6 MORE RAIN, MORE TREES 536
Broadleaf Forests 536
Coniferous Forests 536

29.7 BRIEF SUMMERS AND LONG, ICY WINTERS 538

29.8 DON'T FORGET THE SOILS 539

29.9 FRESHWATER PROVINCES 540
Lake Ecosystems 540
Stream Ecosystems 541

29.10 LIFE AT LAND'S END 542
Wetlands and the Intertidal Zone 542
Rocky and Sandy Coastlines 543
Coral Reefs 543

29.11 FOCUS ON THE ENVIRONMENT Coral Bleaching 544

29.12 THE OPEN OCEAN 544
Surprising Diversity 544
Upwelling and Downwelling 544

29.13 FOCUS ON SCIENCE Applying Knowledge
of the Biosphere 546

30 Behavioral Ecology

IMPACTS, ISSUES My Pheromones Made Me Do It 550

30.1 BEHAVIOR'S HERITABLE BASIS 552
Genes and Behavior 552

Hormones and Behavior 552
Instinctive Behavior 553

30.2 LEARNED BEHAVIOR 554

30.3 THE ADAPTIVE VALUE OF BEHAVIOR 555

30.4 COMMUNICATION SIGNALS 556
The Nature of Communication Signals 556
Communication Displays 556
Illegitimate Signalers and Receivers 557

30.5 MATES, OFFSPRING, AND REPRODUCTIVE
SUCCESS 558
Sexual Selection and Mating Behavior 558
Parental Care 559

30.6 COSTS AND BENEFITS OF SOCIAL GROUPS 560
Cooperative Predator Avoidance 560
The Selfish Herd 560
Cooperative Hunting 560
Dominance Hierarchies 561
Regarding the Costs 561

30.7 WHY SACRIFICE YOURSELF? 562
Social Insects 562
Social Mole Rats 562
Indirect Selection for Altruism 563

30.8 AN EVOLUTIONARY VIEW OF HUMAN
SOCIAL BEHAVIOR 564
Human Pheromones 564
Hormones and Bonding Behavior 564
Evolutionary Questions 564

EPILOGUE Biological Principles and the Human
Imperative 567

Appendix I   Classification System
Appendix II   Units of Measure
Appendix III Answers to Self-Quizzes
Appendix IV Answers to Genetics Problems
Appendix V   Closer Look at Some Major
             Metabolic Pathways
Appendix VI The Amino Acids
Appendix VII Periodic Table of the Elements

PREFACE

What a triumph! Today, biologists offer us a sweeping story of life's unity and diversity—how living things are built, how they work, how they got that way, and where they came from. In that story are fabulous clues to human health, reproduction, our connections with everything else on Earth, even our collective survival.

With this book, we offer a coherent introduction to that story. We weave examples of problem solving and experiments through its pages to show the power of thinking critically about the natural world. We highlight core concepts, current understandings, and research trends for major fields of biological inquiry. We explain the structure and function of a broad sampling of organisms in enough detail so students can develop a working vocabulary about life's parts and processes. We selectively introduce applications that may help students sense the value of learning the core concepts.

Teachers of many millions of students already know about the effectiveness of an approach that integrates current topics with take-home lessons. They recognize, as we do, that nearly all students will stick with lively, relevant, easy-to-follow writing. Students can "get" the big picture of life—and become confident enough to think critically about the past, present, and future on their own. This is biology's greatest gift.

Make It Relevant

Biological inquiry now reaches into our lives in many direct and indirect ways. *What students learn today will have impact on how they make decisions tomorrow*—in the voting booth as well as in their personal lives.

HOW WOULD YOU VOTE?   Each chapter opens with a current issue that relates to its content. For instance, protein synthesis chapter starts with how a bioweapon in the news—ricin—kills by unraveling ribosomes. The microevolution chapter starts with how a selective agent —a rodenticide—favors "super rats." Recognizing that the current generation is at ease with music videos, we created a custom *videoclip* about each introductory issue. These unique videos, in the student CD *and* instructor's Media Manager, are dynamic lecture launchers.

We ask students, *How would you vote on research or on an application related to such issues?* We return to the question in more depth at the end of the chapter. All over the country, students will vote *on-line* and access campuswide, statewide, and nationwide tallies. This interactive approach to issues reinforces the premise that individual actions can make a difference.

Make It Easy To Follow

THE BIG PICTURE   As students start each chapter, they get "*the big picture*"—a mini-overview on-page and in video format. Each big-picture concept has a boxed green title. We repeat these titles at the top of appropriate text pages as ongoing reminders of how core concepts fit together.

HIGHLIGHTED KEY POINTS   Students often complain that their textbooks are dry and boring. We opt for a conversational style that holds their attention, without sacrificing clarity. We encourage them to focus on the flow of content rather than hunting for bits they may be tested on. How? By highlighting the key points for them, in blue boxes at the end of each section. We keep them on target with a *running in-text summary*, and with a *section-by-section summary* at the chapter's end.

READ-ME-FIRST ART   *Read-me-first diagrams* help students who are comfortable with visual learning as well as with reading. For a preview of where their reading will take them, students can first walk step by step through the art in each text section. We offer the same art in narrated, animated form on the Media Manager lecture tool and the student CD. Repeated exposure to visual material reinforces understanding, engages multiple learning styles, and allows self-paced review of challenging biological concepts.

Make It Brief, With Clear Explanations

To keep book length manageable, we were selective about which topics to include while allocating enough space to explain those topics clearly. If something is worth reading about, why reduce it to a factoid? Factoids invite mind-numbing memorization, a study habit that will not help students develop their capacity to think critically about the world and their place in it.

For instance, you can safely bet that most nonmajors simply do not want to memorize each catalytic step of crassulacean acid metabolism. They *do* want to learn about the biological basis of sex, and many female students want to know what will be going on inside them if and when they get pregnant. Good explanations can help them make their own informed decisions on many biology-related issues, including STDs, fertility drugs, prenatal diagnoses of genetic disorders, gene therapies, and abortions. Over the years, a number of student readers have written to tell us they devoured such material even when it was not assigned.

Our choices for which topics to condense, expand, or delete were not arbitrary. They reflect three decades of feedback from teachers throughout the world.

Offer Easy-to-Use Media Tools

Each chapter ends with a *Media Menu*. The menu directs students to the free CD, which starts with an issues-oriented *video* and an animation of *the big picture*. The menu also lists all *read-me-first animations*, and most of the additional animations and interactions.

Our readers have free access to an exclusive online database—*InfoTrac College Edition*, a full-text library of more than 4,000 periodicals. The Media Menu also lists sample articles, examples of *web sites* relevant to that chapter, and an expanded *How-Would-You-Vote* question.

A free *BiologyNow* CD-ROM in each book is a portal to all CD- and web-based learning tools. It is an easy, integrated way for students to do homework and assess how well they understand it. After reading a chapter, students go to http://biology.brookscole.com/starr6, enter the access code packaged with their book, and take a *diagnostic pre-test*. Based on their individual answers, a personalized learning plan directs them to text sections, art, and animations that explain questions they answer incorrectly. Students can e-mail pre-test and *post-test* results to instructors. Ideal for homework assignments, the self-grading pre-tests also can flow directly into WebCT or Blackboard gradebooks when the instructor uses the WebTutor™ ToolBox course management tool that is offered free with this book.

vMentor is a free online biology tutorial service available from 6 AM to 12 PM Monday through Saturday. For review, *interactive flashcards* define all of the book's boldface terms and have audio pronunciation guides.

InfoTrac and an annotated list of web sites make research on the pros and cons of an issue a snap. After students research the *How-Would-You-Vote* question, they can cast their vote and also see at once how others have voted. They can email information about their research into the issue, and their vote, to instructors.

ACKNOWLEDGMENTS

Thanks to the advisors listed below for their ongoing impact on the book's content. John Jackson and Walt Judd both deserve special recognition for their deep commitment to excellence in education. This latest edition still reflects the influential contributions of the instructors listed on the following page, who helped shape our thinking. *Impacts/Issues* sections, custom videos, *the big picture* overviews—such features are responses to their insights from the classroom. Lisa Starr and Christine Evers are invaluable partners in research, writing, creating art, and page make-up.

Starting with Susan Badger, Thomson Learning proved why it's one of the world's foremost publishers; Sean Wakely, Michelle Julet, Kathie Head, thank you again. Keli Amann created a fine web site and managed the CD production. Peggy Williams brought tenacity, intelligence, and humor to the project. Both Andy Marinkovich and Grace Davidson took my world-class compulsivity in stride. Gary Head created functional designs and great graphics for the book; Steve Bolinger did so for the media tools. Star MacKenzie, Ann Caven, Karen Hunt, Suzannah Alexander, Diana Starr, Chris Ziemba, Myrna Engler—the list goes on. Yet no listing conveys how this team interacted to create something extraordinary. And thank you, Jack Carey, for being the first to identify the need for features, including student voting, that can further biology education.

CECIE STARR, *November 2004*

MEDIA TOOLS

Media Menu End-of-chapter menu listing art on the CD-Rom, InfoTrac links, web site links, how-would-you vote question instructor, by section.

Resources Integrator All of the media for the book integrated for instructors, by section.

BiologyNow™ Diagnoses which topics students have not yet mastered and creates customized learning plans to focus their study and review.

How Would You Vote? Invitation to research and vote on a controversial question.

vMentor Free on-line help with homework, through two-way voice communication and through a computer whiteboard. Restrictions apply.*

InfoTrac College Edition Exclusive online, searchable database of periodical, plus exercises that can help guide student research.

Interactive Flashcards Definitions and audio guides to pronouncing all boldface terms in the text.

Web Sites Section-by-section annotated lists to the best biology on the Web. Internet exercises to guide research.

MAJOR ADVISORS AND REVIEWERS

JOHN D. JACKSON *North Hennepin Community College*
WALTER JUDD *University of Florida*

JOHN ALCOCK *Arizona State University*
CHARLOTTE BORGESON *University of Nevada*
DEBORAH C. CLARK *Middle Tennessee University*
MELANIE DeVORE *Georgia College and State University*
DANIEL FAIRBANKS *Brigham Young University*
TOM GARRISON *Orange Coast College*
DAVID GOODIN *The Scripps Research Institute*
PAUL E. HERTZ *Barnard College*
TIMOTHY JOHNSTON *Murray State University*
EUGENE KOZLOFF *University of Washington*
ELIZABETH LANDECKER–MOORE *Rowan University*
KAREN MESSLEY *Rock Valley College*
THOMAS L. ROST *University of California, Davis*
LAURALEE SHERWOOD *West Virginia University*
E. WILLIAM WISCHUSEN *Louisiana State University*
STEPHEN L. WOLFE *University of California, Davis*

CONTRIBUTORS OF INFLUENTIAL REVIEWS AND CLASS TESTS

ADAMS, DARYL   *Minnesota State University, Mankato*
ANDERSON, DENNIS   *Oklahoma City Community College*
BENDER, KRISTEN   *California State University, Long Beach*
BOGGS, LISA   *Southwestern Oklahoma State University*
BORGESON, CHARLOTTE   *University of Nevada*
BOWER, SUSAN   *Pasadena City College*
BOYD, KIMBERLY   *Cabrini College*
BRICKMAN, PEGGY   *University of Georgia*
BROWN, EVERT   *Casper College*
BRYAN, DAVID W.   *Cincinnati State College*
BURNETT, STEPHEN   *Clayton College*
BUSS, WARREN   *University of Northern Colorado*
CARTWRIGHT, PAULYN   *University of Kansas*
CASE, TED   *University of California, San Diego*
COLAVITO, MARY   *Santa Monica College*
COOK, JERRY L.   *Sam Houston State University*
DAVIS, JERRY   *University of Wisconsin, LaCrosse*
DENGLER, NANCY   *University of California, Davis*
DESAIX, JEAN   *University of North Carolina*
DIBARTOLOMEIS, SUSAN   *Millersville University of Pennsylvania*
DIEHL, FRED   *University of Virginia*
DONALD-WHITNEY, CATHY   *Collin County Community College*
DUWEL, PHILIP   *University of South Carolina, Columbia*
EAKIN, DAVID   *Eastern Kentucky University*
EBBS, STEPHEN   *Southern Illinois University*
EDLIN, GORDON   *University of Hawaii, Manoa*
ENDLER, JOHN   *University of California, Santa Barbara*
ERWIN, CINDY   *City College of San Francisco*
FOX, P. MICHAEL   *SUNY College at Brockport*
FOREMAN, KATHERINE   *Moraine Valley Community College*
GIBLIN, TARA   *Stephens College*
GILLS, RICK   *University of Wisconsin, La Crosse*
GREENE, CURTIS   *Wayne State University*
GREGG, KATHERINE   *West Virginia Wesleyan College*
HARLEY, JOHN   *Eastern Kentucky University*
HARRIS, JAMES   *Utah Valley Community College*
HELGESON, JEAN   *Collin County Community College*
HESS, WILFORD M.   *Brigham Young University*
HOUTMAN, ANNE   *Cal State, Fullerton*
HUFFMAN, DAVID   *Southwestern Texas University*
HUFFMAN, DONNA   *Calhoun Community College*
INEICHER, GEORGIA   *Hinds Community College*
JOHNSTON, TAYLOR   *Michigan State University*
JUILLERAT, FLORENCE   *Indiana University, Purdue University*
KENDRICK, BRYCE   *University of Waterloo*
HOUTMAN, ANNE   *Cal State, Fullerton*
KETELES, KRISTEN   *University of Central Arkansas*
KIRKPATRICK, LEE A.   *Glendale Community College*
KREBS, CHARLES   *University of British Columbia*
LANZA, JANET   *University of Arkansas, Little Rock*
LEICHT, BRENDA   *University of Iowa*
LOHMEIER, LYNNE   *Mississippi Gulf Coast Community College*
LORING, DAVID   *Johnson County Community College*
MACKLIN, MONICA   *Northeastern State University*
MANN, ALAN   *University of Pennsylvania*
MARTIN, KATHY   *Central Connecticut State University*
MARTIN, TERRY   *Kishwaukee College*
MASON, ROY B.   *Mount San Jacinto College*
MATTHEWS, ROBERT   *University of Georgia*
MAXWELL, JOYCE   *California State University, Northridge*
MCNABB, ANN   *Virginia Polytechnic Institute and State University*
MEIERS, SUSAN   *Western Illinois University*

MEYER, DWIGHT H.   *Queensborough Community College*
MICKLE, JAMES   *North Carolina State University*
MINOR, CHRISTINE V.   *Clemson University*
MCCLURE, JERRY   *Miami University*
MILLER, G. TYLER   *Wilmington, North Carolina*
MITCHELL, DENNIS M.   *Troy University*
MONCAYO, ABELARDO C.   *Ohio Northern University*
MOORE, IGNACIO   *Virginia Tech*
MORRISON-SHETTLER, ALLISON   *Georgia State University*
MORTON, DAVID   *Frostburg State University*
NELSON, RILEY   *Brigham Young University*
NICKLES, JON R.   *University of Alaska Anchorage*
NOLD, STEPHEN   *University of Wisconsin- Stout*
PADGETT, DONALD   *Bridgewater State College*
PENCOE, NANCY   *State University of West Georgia*
PERRY, JAMES   *University of Wisconsin, Center Fox Valley*
PITOCCHELLI, DR. JAY   *Saint Anselm College*
PLETT, HAROLD   *Fullerton College*
POLCYN, DAVID M.   *California State University, San Bernardino*
PURCELL, JERRY   *San Antonio College*
REID, BRUCE   *Kean College of New Jersey*
RENFROE, MICHAEL   *James Madison University*
REZNICK, DAVID   *California State University, Fullerton*
RICKETT, JOHN   *University of Arkansas, Little Rock*
ROHN, TROY   *Boise State University*
ROIG, MATTIE   *Broward Community College*
ROSE, GRIEG   *West Valley College*
SANDIFORD, SHAMILI A.   *College of Du Page*
SCHREIBER, FRED   *California State University, Fresno*
SELLERS, LARRY   *Louisiana Tech University*
SHONTZ, NANCY   *Grand Valley State University*
SHAPIRO, HARRIET   *San Diego State University*
SHOPPER, MARILYN   *Johnson County Community College*
SIEMENS, DAVID   *Black Hills State University*
SMITH, BRIAN   *Black Hills State University*
SMITH, JERRY   *St. Petersburg Junior College, Clearwater Campus*
STEINERT, KATHLEEN   *Bellevue Community College*
SUNDBERG, MARSHALL D.   *Emporia State University*
SUMMERS, GERALD   *University of Missouri*
SVENSSON, PETER   *West Valley College*
SWANSON, ROBERT   *North Hennepin Community College*
SWEET, SAMUEL   *University of California, Santa Barbara*
SZYMCZAK, LARRY J.   *Chicago State University*
TERHUNE, JERRY   *Jefferson Community College, University of Kentucky*
TAYLOR, JANE   *Northern Virginia Community College*
TIZARD, IAN   *Texas A&M University*
TRAYLER, BILL   *California State University at Fresno*
TROUT, RICHARD E.   *Oklahoma City Community College*
TURELL, MARSHA   *Houston Community College*
TYSER, ROBIN   *University of Wisconsin, LaCrosse*
VAJRAVELU, RANI   *University of Central Florida*
VANDERGAST, AMY   *San Diego State University*
VERHEY, STEVEN   *Central Washington University*
VICKERS, TANYA   *University of Utah*
VOGEL, THOMAS   *Western Illinois University*
WARNER, MARGARET   *Purdue University*
WEBB, JACQUELINE F.   *Villanova University*
WELCH, NICOLE TURRILL   *Middle Tennessee State University*
WELKIE, GEORGE W.   *Utah State University*
WENDEROTH, MARY PAT   *University of Washington*
WINICUR, SANDRA   *Indiana University, South Bend*
WOLFE, LORNE   *Georgia Southern University*
YONENAKA, SHANNA   *San Francisco State University*
ZAYAITZ, ANNE   *Kutztown University of Pennsylvania*

Introduction

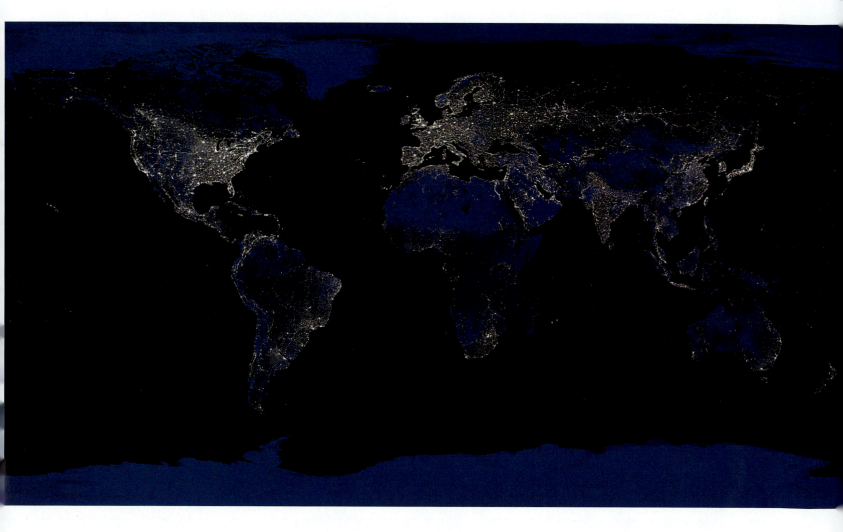

Current configurations of the Earth's oceans and land masses—the geologic stage upon which life's drama continues to unfold. This composite satellite image reveals global energy use at night by the human population. Just as biological science does, it invites you to think more deeply about the world of life—and about our impact upon it.

IMPACTS, ISSUES *What Am I Doing Here?*

Leaf through a newspaper on any given Sunday and you may get an uneasy feeling that the world is spinning out of control. There's a lot about the Middle East, where

World Trade Center

great civilizations have come and gone. You won't find much on the amazing coral reefs in the surrounding seas, especially at the northern end of the Red Sea. Now the news is about oil and politics, terrorists, and war.

Think back on the 1991 Persian Gulf conflict, when thick smoke from oil fires blocked out sunlight, and black rain fell. Iraqis deliberately released about 460 million gallons of crude oil into the Gulf. Uncounted numbers of reef organisms died. So did thousands of birds.

Today Kuwaitis wonder if the oil fires caused their higher cancer rates. They join New Yorkers who are worried about developing lung problems from breathing dense, noxious dust that filled the air after the horrific terrorist attack on the World Trade Center.

Nature, too, seems to have it in for us. Cholera, the flu, and SARS pose global threats. An AIDS pandemic is unraveling the very fabric of African societies. Forests burn fiercely. Storms, droughts, and heat waves are often monstrous. Polar ice caps and once-vast glaciers are melting too rapidly, and the whole atmosphere is warming up.

It's enough to make you throw down the paper and long for the good old days, when things were so much simpler.

Of course, read up on the good old days and you'll find they weren't so good. Bioterrorists were around in 1346, when soldiers catapulted the corpses of bubonic

the big picture

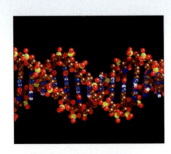

Life's Underlying Unity Life shows a hierarchy of organization, extending from the molecular level through the biosphere. Shared features at the molecular level are the basis of life's unity.

Life's Diversity Life also shows spectacular diversity. Several million kinds of organisms already have been named, past and present, each with some traits that make it unique from all the others.

plague victims into a walled city under siege. Infected people and rats fled the city and helped fuel the Black Death, a plague that left 25 million dead in Europe. In 1918, the Spanish flu raced around the world and left somewhere between 30 and 40 million people dead. Between 1945 and 1949, about 100,000 people in the United States contracted polio, a disease that left many permanently paralyzed. In those times, too, many felt helpless in a world that seemed out of control.

What it boils down to is this: For a couple of million years, we humans and our immediate ancestors have been trying to make sense of the natural world and what we're doing in it. We observe it, come up with ideas, then test the ideas. But the more pieces of the puzzle we fit together, the bigger the puzzle gets. We now know that it is almost overwhelmingly big.

You could walk away from the challenge and simply not think. You could let others tell you what to think. Or you could choose to develop your own understanding of the puzzle.

Maybe you're interested in the pieces that affect your health, the food you eat, or your children, should you choose to reproduce. Maybe you just find life fascinating. No matter what your focus might be, you can deepen your perspective. You can learn ways to sharpen how you interpret the natural world, including human nature. This is the gift of biology, the scientific study of life.

 **How Would You Vote?**

The warm seas of the Middle East support some of the world's most spectacular coral reef ecosystems. Should the United States provide funding to help preserve the reefs? See the Media Menu for details, then vote online.

Explaining Unity in Diversity Evolutionary theories, especially the theory of evolution by natural selection, help us see a profound connection between life's underlying unity and its diversity.

How We Know Biologists find out about life by observing, asking questions, and formulating and testing hypotheses in nature or the laboratory. They report results in ways that others can test.

1.1 Life's Levels of Organization

The world of life shows levels of organization, from the simple to the complex. Take time to see how these levels connect to get a sense of how the topics of this book are organized and where they will take you.

FROM SMALL TO SMALLER

Imagine yourself on the deck of a sailing ship, about to journey around the world. The distant horizon of a vast ocean beckons, and suddenly you sense that you are just one tiny part of the great scheme of things.

Now imagine one of your red blood cells can think. It realizes it's only a tiny part of the great scheme of your body. A string of 375 cells like itself would fit across a straightpin's head—and you have trillions of cells. One of the fat molecules at the red blood cell's surface is thinking about how small it is. A string of 1,200,000 molecules like itself would stretch across that pinhead. Now a hydrogen atom in the molecule is pondering the great scheme of a fat molecule. Figure 1.1*a* depicts one. It would take 53,908,355 side-by-side hydrogen atoms to stretch across the head of a pin!

FROM SMALLER TO VAST

With that single atom, you have reached the entry level of nature's great pattern of organization. Like nonliving things, all organisms are made of building blocks called atoms. At the next level are molecules. Life's unique properties emerge when certain kinds of molecules are organized into cells. These "molecules of life" are complex carbohydrates, complex fats and other lipids, proteins, DNA, and RNA (Figure 1.1*b*). The **cell** is the smallest unit of organization with the

b molecule ⟶ **c** cell ⟶ **d** tissue ⟶ **e** organ ⟶ **f** organ system

b molecule
Two or more joined atoms of the same or different elements. "Molecules of life" are complex carbohydrates, lipids, proteins, DNA, and RNA. Only living cells now make them.

c cell
Smallest unit that can live and reproduce on its own or as part of a multicelled organism. It has an outer membrane, DNA, and other components.

d tissue
Organized aggregation of cells and substances interacting in a specialized activity. Many cells (*white*) made this bone tissue from their own secretions.

e organ
Structural unit made of two or more tissues interacting in some task. A parrotfish eye is a sensory organ used in vision.

f organ system
Organs interacting physically, chemically, or both in some task. Parrotfish skin is an integumentary system with tissue layers, organs such as glands, and other parts.

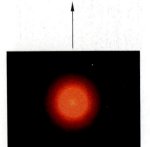

a atom
Smallest unit of an element that still retains the element's properties. Electrons, protons, and neutrons are its building blocks. This hydrogen atom's electron zips around a proton in a spherical volume of space.

Figure 1.1 Increasingly complex levels of organization in nature, extending from subatomic particles to the biosphere.

capacity to survive and reproduce on its own, given raw materials, energy inputs, information encoded in its DNA, and suitable conditions in its environment.

At the next level of organization are multicelled organisms made of specialized, interdependent cells, often organized as tissues, organs, and organ systems. A higher level of organization is the **population**, a group of single-celled or multicelled individuals of the same species occupying a specified area. A school of fish is a population (Figure 1.1*h*), as are all of the single-celled amoebas in an isolated lake.

Next comes the **community**, all populations of all species occupying one area. Its extent depends on the area specified. It might be the Red Sea, an underwater cave, or a forest in South America. It might even be a community of tiny organisms that live, reproduce, and die quickly inside the cupped petals of a flower.

The next level of organization is the **ecosystem**, or a community together with its physical and chemical environment. Finally, the **biosphere** is the highest level of life. It encompasses all regions of the Earth's crust, waters, and atmosphere in which organisms live.

This book is a journey through the globe-spanning organization of life. So take a moment to study Figure 1.1. You can use it as a road map of where each part fits in the great scheme of things.

Nature shows levels of organization, from the simple to the increasingly complex.

Life's unique characteristics originate at the atomic and molecular level. They extend through cells, populations, communities, ecosystems, and the biosphere.

Read Me First!
and watch the narrated animation on life's levels of organization

GULF OF AQABA

RED SEA

g multicelled organism
Individual made of different types of cells. Cells of most multicelled organisms, including this Red Sea parrotfish, are organized as tissues, organs, and organ systems.

h population
Group of single-celled or multicelled individuals of the same species occupying a specified area. This is a fish population in the Red Sea.

i community
All populations of all species occupying a specified area. This is part of a coral reef in the Gulf of Aqaba at the northern end of the Red Sea.

j ecosystem
A community that is interacting with its physical environment. It has inputs and outputs of energy and materials. Reef ecosystems flourish in warm, clear seawater throughout the Middle East.

k the biosphere
All regions of the Earth's waters, crust, and atmosphere that hold organisms. In the vast universe, Earth is a rare planet. Without its abundance of free-flowing water, there would be no life.

1.2 Overview of Life's Unity

"Life" isn't easy to define. It's just too big, and it's been changing for 3.9 billion years! Even so, you can frame a definition in terms of its unity and diversity. Here's the unity part: All living things grow and reproduce with the help of DNA, energy, and raw materials. They sense and respond to what is going on. But details of their traits differ among many millions of kinds of organisms. That's the diversity part—variation in traits.

DNA, THE BASIS OF INHERITANCE

You will never, ever find a rock made of nucleic acids, proteins, and complex carbohydrates and lipids. In the natural world, only living cells make these molecules. And the signature molecule of life is the nucleic acid called DNA. No chunk of granite or quartz has it.

DNA holds information for building proteins from smaller molecules, the amino acids. By analogy, if you follow suitable instructions and invest enough energy in the task, you might organize a pile of a few kinds of ceramic tiles (representing amino acids) into diverse patterns (representing proteins), as in Figure 1.2.

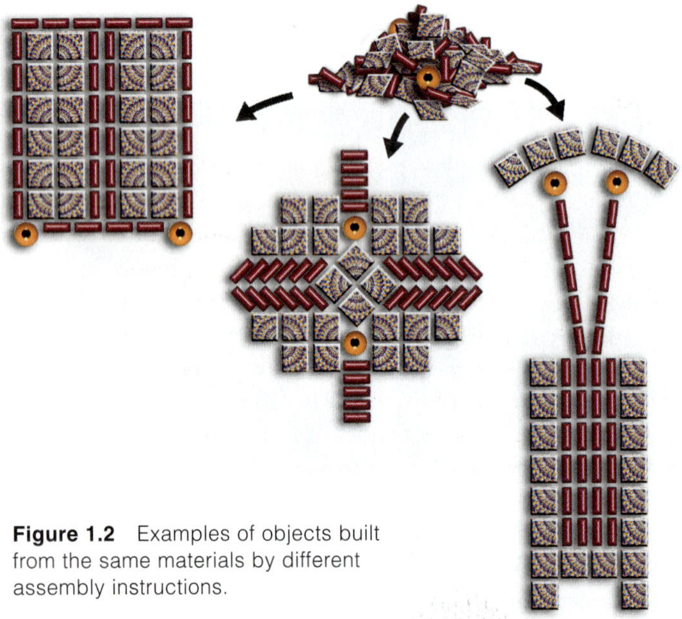

Why are proteins so important? Many are structural materials, regulators of cell activities, and enzymes. Enzymes are the cell's main worker molecules. They build, split, and rearrange the molecules of life in ways that keep cells alive. Without enzymes, nothing much could be done with DNA's information. There would be no new organisms.

In nature, each organism inherits its DNA—and its traits—from parents. *Inheritance* means an acquisition of traits after parents transmit their DNA to offspring. Think about it. Baby storks look like storks and not like pelicans because they inherited stork DNA, which isn't exactly the same as pelican DNA.

Reproduction refers to actual mechanisms by which parents transmit DNA to offspring. For frogs, humans, trees, and other organisms, the information in DNA guides *development*—the transformation of the first cell of a new individual into a multicelled adult, typically with many different tissues and organs (Figure 1.3).

ENERGY, THE BASIS OF METABOLISM

Becoming alive and maintaining life processes requires energy—the capacity to do work. Each normal living cell has ways to obtain and convert energy from its surroundings. By a process called **metabolism**, every cell acquires and uses energy to maintain itself, grow, and make more cells.

Where does the energy come from? Nearly all of it flows from the sun into the world of life, starting with **producers**. Producers are plants and other organisms that make their own food molecules from simple raw materials. Animals and decomposers are **consumers**. They cannot make their own food; they survive by feeding on tissues of producers and other organisms.

Figure 1.2 Examples of objects built from the same materials by different assembly instructions.

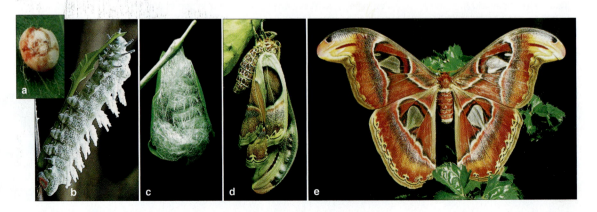

Figure 1.3 "The insect"—actually a series of stages of development guided largely by instructions in DNA. Here, a silkworm moth, from a fertilized egg (**a**), to a larval stage called a caterpillar (**b**), to a pupal stage (**c**), to the winged form of the adult (**d**,**e**).

When, say, zebras browse on plants, some energy stored in plant tissues is transferred to them. Later on, energy is transferred to a lion as it devours the zebra. And it gets transferred again as decomposers go to work, acquiring energy from the remains of zebras, lions, and other organisms.

Decomposers are mostly the kinds of bacteria and fungi that break down sugars and other molecules to simpler materials. Some of the breakdown products are cycled back to producers as raw materials. Over time, energy that plants originally captured from the sun returns to the environment.

Energy happens to flow in one direction, from the environment, through producers, then consumers, and then back to the environment (Figure 1.4). These are the energy exchanges that maintain life's organization. Later on, you will see how life's interconnectedness relates to modern-day problems, including major food shortages, AIDS, cholera, acid rain, global warming, and rapid losses in biodiversity.

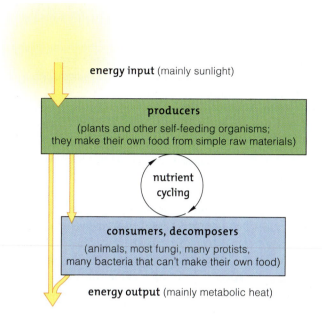

energy input (mainly sunlight)

producers
(plants and other self-feeding organisms;
they make their own food from simple raw materials)

nutrient
cycling

consumers, decomposers
(animals, most fungi, many protists,
many bacteria that can't make their own food)

energy output (mainly metabolic heat)

Figure 1.4 The one-way flow of energy and cycling of materials in the world of life.

LIFE'S RESPONSIVENESS TO CHANGE

It's often said that only living things respond to the environment. Yet even a rock shows responsiveness, as when it yields to gravity's force and tumbles down a hill or changes its shape slowly under the repeated batterings of wind, rain, or tides.

The difference is this: Living things sense changes in their surroundings, and they make compensatory, controlled responses to them. How? With receptors. Receptors are molecules and structures that detect stimuli, which are specific kinds of energy. Different receptors respond to different stimuli. A stimulus may be sunlight energy, chemical potential energy (as when a substance is more concentrated outside a cell than inside), or the mechanical energy of a bite (Figure 1.5).

Switched-on receptors can trigger changes in cell activities. As a simple example, after you finish eating a piece of fruit, sugars leave your gut and enter your bloodstream. Think of blood and the fluid around cells as an *internal* environment, which must be kept within tolerable limits. Too much or too little sugar in blood changes that internal environment. This can cause diabetes and other medical problems. Normally, when there is too much sugar in blood, your pancreas starts secreting more insulin. Most living cells in your body have receptors for this hormone, which stimulates them to take up more sugar. When enough cells do so, the blood sugar level returns to normal.

In such ways, organisms keep the internal environment within a range that cells can tolerate. This state is called **homeostasis**, and it is one of the key defining characteristics of life.

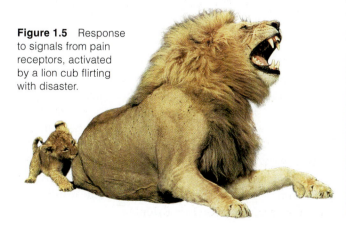

Figure 1.5 Response to signals from pain receptors, activated by a lion cub flirting with disaster.

Organisms build proteins based on instructions in DNA, which they inherit from their parents. Organisms reproduce, grow, and stay alive by way of metabolism—ongoing energy conversions and energy transfers at the cellular level.

Organisms interact through a one-way flow of energy and a cycling of materials. Collectively, their interdependencies have global impact.

Organisms sense and respond to changing conditions in controlled ways. The responses help them maintain tolerable conditions in their internal environment.

1.3 If So Much Unity, Why So Many Species?

Although unity pervades the world of life, so does diversity. Organisms differ enormously in body form, in the functions of their body parts, and in behavior.

Superimposed on life's unity is tremendous diversity. Millions of kinds of organisms, or **species**, live on Earth. Many more lived during the past 3.9 billion years, but their lineages vanished; about 99.9 percent of all species have become extinct.

For centuries, scholars have tried to make sense of diversity. In 1735, a physician named Carolus Linnaeus devised a scheme for classifying organisms by assigning a two-part name to each species. The first part designates the *genus* (plural, genera). Each genus is one or more species grouped together on the basis of a number of traits that are unique to that group alone. The second part of the name refers to a particular species within the genus. Today, biologists attempt to sort out the relationships among species not only on the basis of observable traits, but also using evidence of descent from a common ancestor.

For instance, *Scarus gibbus* is the scientific name for the humphead parrotfish (Figure 1.1g). Another species in the same genus is *S. coelestinus*, the midnight parrotfish. We abbreviate a genus name once it's been spelled out in a document.

Biologists are still working out how to group the organisms. Most now favor a classification system with three domains: Bacteria, Archaea, and Eukarya (Figure 1.6). As shown in Figure 1.7, the third domain includes protists, plants, fungi, and animals.

The **archaea** and **bacteria** are single-celled. They are *prokaryotic,* meaning they do not have a nucleus (a membrane-bound sac that keeps DNA separated from the rest of the cell's interior). Prokaryotes include diverse producers or consumers. Of all groups, theirs shows the greatest metabolic diversity.

Archaea live in boiling ocean water, freezing desert rocks, sulfur-rich lakes, and other habitats as harsh as those thought to have prevailed when life originated.

protists Diverse single-celled and multicelled eukaryotic species that range from microscopic single cells to giant seaweeds. Even this tiny sampling conveys why many biologists now believe the "protists" are many separate lineages.

archaea These prokaryotes are evolutionarily closer to the eukaryotes than to bacteria. This is a colony of methane-producing cells.

Figure 1.7 A few representatives of life's diversity.

Bacteria are sometimes called eubacteria, which means "true bacteria," to distinguish them from archaea. They are far more common than archaeans, and they live throughout the world in diverse habitats.

Plants, fungi, animals, and protists are members of the group **eukarya**, which means they have nuclei. Eukaryotes are generally larger and far more complex than the prokaryotes. The differences among protistan lineages are so great that they could be divided into several separate groups, which would result in a major reorganization of the domain.

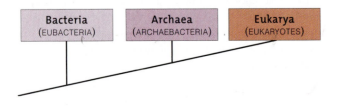

| Bacteria (EUBACTERIA) | Archaea (ARCHAEBACTERIA) | Eukarya (EUKARYOTES) |

Figure 1.6 Three domains of life.

Read Me First!
and watch the narrated animation on life's diversity

plants Generally, photosynthetic, multicelled eukaryotes, many with roots, stems, and leaves. Plants are the primary producers for ecosystems on land. Redwoods and flowering plants are examples.

fungi Single-celled and multicelled eukaryotes; mostly decomposers, also many parasites and pathogens. Without the fungal and bacterial decomposers, communities would become buried in their own wastes.

animals Multicelled eukaryotes that ingest tissues or juices of other organisms. Like this basilisk lizard, most actively move about during at least part of their life.

c	d	e	f
PROTISTS	PLANTS	FUNGI	ANIMALS

EUKARYA

a — ARCHAEA

b — BACTERIA

origin of life

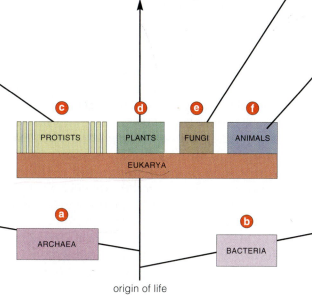

bacteria By far the most common prokaryotes; collectively, these single-celled species are the most metabolically diverse organisms on Earth.

Plants are multicelled, photosynthetic producers. They can make their own food by using simple raw materials and sunlight as an energy source.

Most **fungi**, such as the mushrooms sold in grocery stores, are multicelled decomposers and consumers with a distinct way of feeding. They secrete enzymes that digest food outside the fungal body, then their individual cells absorb the digested nutrients. **Animals** are multicelled consumers that ingest tissues of other organisms. Different kinds are herbivores (grazers), carnivores (meat eaters), scavengers, and parasites.

All develop by a series of embryonic stages, and they actively move about during their life.

Pulling this information together, are you getting a sense of what it means when someone says that life shows unity *and* diversity?

To make the study of life's diversity more manageable, we group organisms related by descent from a shared ancestor. We recognize three domains—archaea, bacteria, and eukarya (protists, fungi, plants, and animals).

1.4 An Evolutionary View of Diversity

How can organisms be so much alike and still show staggering diversity? A theory of evolution by natural selection explains this. For now, simply think about how it starts with a simple observation: Individuals of a population show variation in the details of their shared traits.

Your traits make you and 6.3 billion other individuals members of the human population. Traits are different aspects of an organism's form, function, or behavior. For example, humans show a range of height and hair color. All natural populations have differences among their individuals.

What causes variation in traits? Mutations. These are heritable changes in DNA. Some mutations lead to novel traits that make an individual better able to secure food, a mate, hiding places, and so on. We call these *adaptive* traits.

An adaptive form of a trait tends to become more common over generations, because it gives individuals a better chance to live and bear more offspring than

WILD ROCK DOVE

individuals who don't have it. When different forms of a trait are becoming more or less common, evolution is under way. To biologists, **evolution** simply means heritable change in a line of descent. Mutations, the source of new traits, provide the variation that serves as the raw material for evolution.

"Diversity" refers to variations in traits that have accumulated in lines of descent. Later chapters show the actual mechanisms that bring it about. For now, start thinking about what a great naturalist, Charles Darwin, discovered about evolution:

First, populations tend to increase in size, past the capacity of their environment to sustain them, so their members must compete for resources (food, shelter).

Second, individuals of natural populations differ from one another in the details of their shared traits. Most variation has a heritable basis.

Third, when individuals differ in their ability to survive and reproduce, the traits that help them do so tend to become more common in the population over time. This outcome is called **natural selection**.

Take a look at the pigeons in Figure 1.8. They differ in feather color, size, and other traits. Suppose pigeon breeders are looking for, say, pigeons with black, curly-tipped feathers. They allow only the pigeons with the darkest and curliest-tipped feathers to mate. In time, only pigeons with black, curly-tipped feathers make up the breeders' captive population. Lighter, less curly feathers will become less common.

Pigeon breeding is a case of *artificial* selection. One form of a trait is favored over others in an artificial environment under contrived, manipulated conditions. Darwin saw that breeding practices could be an easily understood model for *natural* selection, a favoring of some forms of a given trait over others in nature.

Just as breeders are "selective agents" promoting reproduction of particular captive pigeons, different agents operate across the range of variation in the wild. Pigeon-eating peregrine falcons are among them (Figure 1.8). Swifter or better camouflaged pigeons are more likely to avoid peregrine falcons and live long enough to reproduce, compared with not-so-swift or too-conspicuous pigeons.

Figure 1.8 Outcome of artificial selection. Just a few of the more than 300 varieties of domesticated pigeons, all descended from captive populations of wild rock doves. By contrast, peregrine falcons are one of the agents of natural selection in the wild.

Traits are variations in form, function, or behavior that arise as a result of mutations in DNA. Some traits are more adaptive than others to prevailing conditions.

Natural selection is an outcome of differences in survival and reproduction among individuals of a population that vary in one or more heritable traits. The process of evolution, or change in lines of descent, gives rise to life's diversity.

1.5 The Nature of Biological Inquiry

The preceding sections introduced some big concepts. Consider approaching this or any other collection of "facts" with a critical attitude. "Why should I accept that they have merit?" The answer requires a look at how biologists make inferences about observations, then test their inferences against actual experience.

OBSERVATIONS, HYPOTHESES, AND TESTS

To get a sense of "how to do science," you might start with practices that are common in scientific research:

1. Observe some aspect of nature, carefully check what others have found out about it, then frame a question or identify a problem related to your observation.

2. Formulate **hypotheses**, or educated guesses, about possible answers to questions or solutions to problems.

3. Using hypotheses as your guide, make a **prediction** —a statement of what you should find in the natural world if you were to go looking for it. This is often called the "if–then" process. *If* gravity does not pull objects toward the Earth, *then* it should be possible to observe apples falling up, not down, from a tree.

4. Devise ways to **test** the accuracy of predictions, as by making systematic observations, building models, and conducting experiments. **Models** are theoretical, detailed descriptions or analogies that might help us visualize something that hasn't been directly observed.

5. If your tests do not confirm a prediction, check to see what might have gone wrong. It may be that you overlooked a factor that had an impact on the results. Or maybe a hypothesis is not a good one.

6. Repeat the tests or devise new ones—the more the better, because hypotheses that withstand many tests are likely to have a higher probability of being useful.

7. Objectively analyze and report the test results as well as the conclusions you drew from them.

You might hear someone refer to these practices as "the scientific method," as if all scientists march to the drumbeat of an absolute, fixed procedure. They do not. Many observe, describe, and report on some aspect of nature, then leave the hypothesizing to others. Some scientists are lucky; they stumble onto information that they are not even looking for. Of course, it isn't always a matter of luck. Chance seems to favor a mind that has already been prepared, by education, experience, or both, to recognize what the information might mean. So it is not a single method that scientists have in common. It is a critical attitude about being shown rather than being told—that is, by accepting ideas supported by tests, and by taking a logical approach to problem solving.

ABOUT THE WORD "THEORY"

Suppose no one has disproved a hypothesis after years of rigorous tests. Suppose scientists use it to interpret more data or observations, which could involve more hypotheses. When a hypothesis meets these criteria, it may become accepted as a **scientific theory**.

You may hear people apply the word "theory" to a speculative idea, as in the expression "It's just a theory." But a scientific theory differs from speculation for this reason: *After testing the predictive power of a scientific theory many times and in many ways in the natural world, researchers have yet to find evidence that disproves it.* This is why the theory of natural selection is respected. It successfully explains diverse issues, such as how life originated, how river dams can alter ecosystems, and why antibiotics aren't working.

Maybe a well-tested theory is as close to the truth as scientists can get with known evidence. For instance, after more than a century of many thousands of tests, Darwin's theory holds, with only minor modification. We can't prove it holds under all possible conditions; that would take an infinite number of tests. As for any theory, we can only say *it has a high probability of being a good one.* Biologists do keep looking for information and devising tests that might disprove its premises.

A scientific approach to studying nature is based on asking questions, formulating hypotheses, making predictions, testing the predictions, and objectively reporting the results.

A scientific theory is a long-standing hypothesis, supported by tests, that explains the cause or causes of a broad range of related phenomena. All scientific theories remain open to tests, revision, and tentative acceptance or rejection.

1.6 The Power of Experimental Tests

Experiments are tests that simplify observation in nature, because conditions under which observations are made can be controlled. Well-designed experiments help you predict what you'll find in nature when a hypothesis is a good one—or won't find if it is wrong.

AN ASSUMPTION OF CAUSE AND EFFECT

A scientific experiment starts with a key premise: *Any aspect of nature has an underlying cause that can be tested by observation.* This premise is what sets science apart from faith in the supernatural ("beyond nature"). It means a scientific hypothesis must be testable in the natural world in ways that might well disprove it.

Most aspects of nature are complex, an outcome of many interacting variables. A **variable** is a specific aspect of an object or event that can differ among individuals or changes over time. Scientists simplify their observation of complex phenomena by designing experiments to test one variable at a time. They define a **control group**—a standard for comparison with one or more **experimental groups**. There are two kinds of control groups. A control group can be identical to an experimental group; except for *one variable event*, it is tested the same way as the experimental group. A control group may also differ from an experimental group in *one variable aspect*; in this case, it is tested exactly the same way as the experimental group.

EXAMPLE OF AN EXPERIMENTAL DESIGN

In 1996, the FDA approved Olestra®, a type of synthetic fat replacement made from sugar and vegetable oil, for use as a food additive. The first Olestra-containing product to reach consumers in the United States was a potato chip. Soon controversy raged. Some people complained of severe gastrointestinal distress after eating the chips. In 1998, medical researchers at Johns Hopkins University performed an experiment to test whether the new chips were indeed causing problems. Their prediction was this: *If Olestra causes intestinal problems, then people who eat products that contain Olestra will end up with gastrointestinal cramps.*

A suburban Chicago multiplex theater was chosen as the "laboratory" for this experiment. More than 1,100 people were invited to watch a movie and eat their fill of potato chips while they were there. They ranged between 13 and 88 years old. Unmarked bags each contained a family-size portion of potato chips. Some of the bags held Olestra potato chips, and the others held regular potato chips.

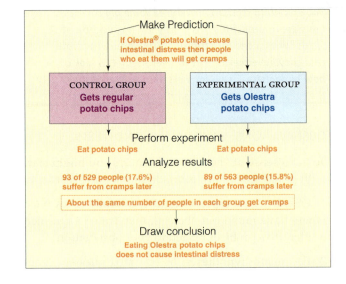

Figure at right shows steps:

Make Prediction — If Olestra® potato chips cause intestinal distress then people who eat them will get cramps

CONTROL GROUP — Gets regular potato chips
EXPERIMENTAL GROUP — Gets Olestra potato chips

Perform experiment — Eat potato chips / Eat potato chips

Analyze results

93 of 529 people (17.6%) suffer from cramps later
89 of 563 people (15.8%) suffer from cramps later

About the same number of people in each group get cramps

Draw conclusion — Eating Olestra potato chips does not cause intestinal distress

Figure 1.9 Example of a typical sequence of steps taken in a scientific experiment.

In this experiment, both control and experimental groups consisted of a random sample of moviegoers; each group got different chips (the variable event).

Later, the researchers telephoned the moviegoers at home and tabulated reports of gastrointestinal distress. They found that 89 of 563 people (15.8 percent) who ate Olestra chips complained of stomach cramps. Of 529 people, 93 (17.6 percent) who ate the regular chips did as well. They concluded that eating Olestra potato chips—at least during one sitting—does not cause gastrointestinal distress (Figure 1.9).

EXAMPLE OF A FIELD EXPERIMENT

Consider that many toxic or unpalatable species are vividly colored, often with distinctive patterning. Predators learn to avoid individuals that display particular visual cues after eating a few of them and suffering ill consequences.

In 1879, a naturalist named Fritz Müller formulated a hypothesis about unrelated species of distasteful butterflies that show striking resemblance to one another. A visual similarity between different species that may confuse potential predators (or prey) is called **mimicry**. Müller thought such a resemblance benefits individuals of both butterfly species because they share the burden of educating predatory birds.

Durrell Kapan, an evolutionary biologist, tested the hypothesis in 2001 with a field experiment in the rain forests of Ecuador. There are two forms of *Heliconius cydno*, an unpalatable species of butterfly. One has yellow markings on its wings; the other does not.

Figure 1.10 *Heliconius* butterflies. (**a**) Two forms of *H. cydno* and (**b**) *H. eleuchia*.

(**c**) Kapan's experiment with *Heliconius* butterflies in an Ecuadoran rain forest. *H. cydno* butterflies with or without yellow markings on their wings were captured and transferred to a habitat of *H. eleuchia*, a species that also has yellow wing markings. Local predatory birds, familiar with untasty yellow *H. eleuchia*, avoided the *H. cydno* butterflies with yellow markings but ate the white ones.

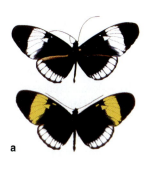
a

b

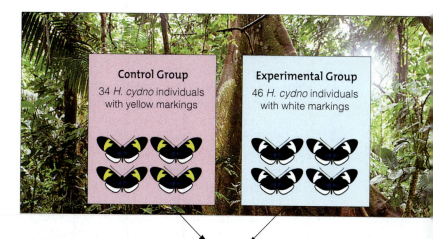

Control Group
34 *H. cydno* individuals with yellow markings

Experimental Group
46 *H. cydno* individuals with white markings

Experiment
Both yellow and white forms of *H. cydno* butterflies are introduced into isolated rain forest habitat of yellow *H. eleuchia* butterflies. Numbers of individuals resighted recorded on a daily basis for two weeks.

one of the agents of selection

Results
Experimental group (*H. cydno* individuals without yellow wing markings) is selected against. 37 of the original group of 46 white butterflies disappear (80%), compared with 20 of the 34 yellow controls (58%).

c

Both resemble another unpalatable species that lives nearby, *H. eleuchia*, which also has yellow in its wings (Figure 1.10). Kapan made a prediction: Birds that had already learned not to prey on *H. eleuchia* would also avoid *H. cydno* butterflies with yellow markings.

He captured both forms of *H. cydno*. The form with no yellow markings was the experimental group, and the form with the yellow markings was the control. He released both groups into parts of the forest that held isolated populations of *H. eleuchia* butterflies. He made daily counts of how many of the transplanted butterflies survived during the next two weeks, the approximate life span of the butterflies.

Kapan found that individuals of the experimental group were less likely to survive in the new habitat (Figure 1.10c). Resident birds familiar with *H. eleuchia* butterflies most likely ate them because they did not bear the familiar visual cue—yellow markings—that signaled bad taste. The control group did better, as you can see from the test results listed in Figure 1.10c. Local birds probably had an idea of how the new butterflies would taste, and avoided them.

Kapan's test results confirmed his prediction, and it also turned out to be evidence of natural selection.

BIAS IN REPORTING RESULTS

Experimenters run a risk of interpreting data in terms of what they wish to prove or dismiss. That is why scientists prefer *quantitative* reports of experiments, with numbers or some other precise measurement. Such data give other experimenters an opportunity to confirm tests, and, perhaps more importantly, allow others to check their conclusions.

This last point gets us back to the value of thinking critically. Scientists must keep asking themselves: *Will observations or experiments show that a hypothesis is false?* They expect one another to put aside pride or bias by testing ideas in ways that may prove them wrong. Even if someone won't, others will—because science is a cooperative yet competitive community. Ideally, individuals share ideas, knowing it's as useful to expose errors as to applaud insights. They can and often do change their mind when evidence contradicts their ideas. And therein lies the strength of science.

Experiments simplify observations in nature by restricting a researcher's focus to one variable at a time.

Tests are based on the premise that any aspect of nature has one or more underlying causes. Scientific hypotheses can be tested in ways that might disprove them.

1.7 The Limits of Science

Beyond the realm of scientific inquiry, some events are unexplained. Why do we exist, for what purpose? Why do we have to die at a particular moment? Such questions lead to *subjective* answers, which come from within as an integrated outcome of all the experiences and mental connections that shape our consciousness. People differ enormously in this regard. That is why subjective answers do not readily lend themselves to scientific analysis and experiments.

This is not to say subjective answers are without value. No human society can function for long unless its individuals share a commitment to standards for making judgments, even if they are subjective. Moral, aesthetic, philosophical, and economic standards vary from one society to the next. But they all guide people in deciding what is important and good, and what is not. All attempt to give meaning to what we do.

Every so often, scientists stir up controversy when they explain something that was thought to be beyond natural explanation, or belonging to the supernatural. This is often the case when a society's moral codes are interwoven with religious interpretations of the past. Exploring a long-standing view of the natural world from a scientific perspective might be misinterpreted as questioning morality, even though the two are not the same thing.

As one example, centuries ago in Europe, Nikolaus Copernicus studied the planets and concluded that the Earth circles the sun. Today this seems obvious. Back then, it was heresy. The prevailing belief was that the Creator made the Earth—and, by extension, humans—the immovable center of the universe. One respected scholar, Galileo Galilei, studied the Copernican model of the solar system, thought it was a good one, and said so. He was forced to retract his statement, on his knees, and put the Earth back as the fixed center of things. (Word has it that he also muttered, "Even so, it *does* move.") Later, Darwin's theory of evolution also ran up against prevailing belief.

Today, as then, society has sets of standards. Those standards might be questioned when a new, natural explanation runs counter to supernatural beliefs. This doesn't mean scientists who raise questions are less moral, less lawful, less sensitive, or less caring than anyone else. It only means one more standard guides their work. Their ideas about nature must be tested in the external world, in ways that can be repeated.

The external world, not internal conviction, is the testing ground for the theories generated in science.

Summary

Section 1.1 Life shows many levels of organization. All things, living and nonliving, are made of atoms. The properties of life emerge in cells. An organism may be a single cell or multicelled. In most multicelled species, cells are organized as tissues, organs, and organ systems.

A population consists of individuals of the same species in a specified area. A community consists of all populations occupying the same area. An ecosystem is a community and its environment. The biosphere includes all regions of Earth's atmosphere, waters, and land where we find living organisms.

Section 1.2 Life shows unity. All organisms have DNA, which holds instructions for building proteins. They inherit the instructions from their parents and pass them on to offspring. All require energy and raw materials from the environment to grow and reproduce. All sense changes in the surroundings and respond to them in controlled ways (Table 1.1).

Section 1.3 Life shows tremendous diversity. Many millions of species exist; many more lived in the past. Each is unique in some aspects of its body plan, function, and behavior. We group species that are related by descent from a common ancestor. A current classification system puts species in three domains: archaea, bacteria, and eukarya. Protists, plants, fungi, and animals are eukaryotes.

Table 1.1 Summary of Life's Characteristics
Shared characteristics that reflect life's unity
1. All life forms contain "molecules of life" (complex carbohydrates, lipids, proteins, and nucleic acids).
2. Organisms consist of one or more cells.
3. Cells are constructed of the same kinds of atoms and molecules according to the same laws of energy.
4. Organisms acquire and use energy and materials to survive and reproduce.
5. Organisms sense and make controlled responses to conditions in their internal and external environments.
6. Heritable information is encoded in DNA.
7. Characteristics of individuals in a population can change over generations; the population can evolve.
Foundations for life's diversity
1. Mutations in DNA give rise to variations in traits, or details of body form, function, and behavior.
2. Traits enhancing survival and reproduction become more common in a population over generations. This process is called natural selection.
3. Diversity is the sum total of variations that accumulated in different lines of descent over the past 3.9 billion years.

Section 1.4 Mutations change DNA and give rise to new variations of heritable traits. Natural selection occurs if a variation affects survival and reproduction. A population is evolving by natural selection when an adaptive form of a trait is becoming more common.

Section 1.5 Scientific methods are varied, but all are based on a logical approach to explaining nature. Scientists observe some aspect of nature, then develop a hypothesis about what might have caused it. They use the hypothesis to make predictions that can be tested by making more observations, building models, or doing experiments.

Scientists analyze test results, draw conclusions from them, and share this information with other scientists. A hypothesis that does not hold up under repeated testing is modified or discarded. A scientific theory is a long-standing hypothesis that explains a broad range of related phenomena and has been supported by many different tests.

Section 1.6 Science cannot answer all questions. It deals only with aspects of nature that lend themselves to systematic observation, hypotheses, predictions, and experiments.

Most aspects of nature are complex, an outcome of many interacting variables. A scientific experiment allows a scientist to change one variable at a time and observe what happens. Experiments are designed so experimental groups can be compared with a control group. Scientists share their results so others can check their conclusions.

Self-Quiz

Answers in Appendix III

1. The smallest unit of life is the __cell__ .

2. __Metab__ is the capacity of cells to extract energy from sources in the environment, and use it to live, grow, and reproduce.

3. __Homeostasis__ is a state in which the internal environment is being maintained within a tolerable range.

4. A trait is __adaptiv__ if it improves an organism's ability to survive and reproduce in a given environment.

5. Differences in heritable traits arise through __mutation__

6. Researchers assign all species to one of three __Domain__

7. __A__ secure energy from their surroundings.
   a. Producers   c. Decomposers
   b. Consumers   d. All of the above

8. DNA __D__ .
   a. contains instructions for building proteins
   b. undergoes mutation
   c. is transmitted from parents to offspring
   d. all of the above

9. __D__ is the acquisition of traits after parents transmit their DNA to offspring.
   a. Metabolism   c. Homeostasis
   b. Reproduction   d. Inheritance

10. A control group is __A__ .
    a. a standard against which experimental groups can be compared
    b. an experiment that gives conclusive results

11. Match the terms with the most suitable descriptions.
    __C__ adaptive trait
    __e__ natural selection
    __D__ scientific theory
    __E__ hypothesis
    __A__ prediction

    a. statement of what you can expect to observe in nature
    b. proposed explanation; an educated guess
    c. improves chances of surviving and reproducing
    d. related set of hypotheses that form a broadly useful, testable explanation
    e. outcome of differences in survival, reproduction among individuals of a population that differ in the details of one or more traits

Critical Thinking

1. A scientific theory about some aspect of nature rests upon inductive logic—inference of a generalized conclusion from particular instances. The assumption is that, because an outcome of some event has been observed to happen with great regularity, it will happen again. However, we can't know this for certain, because there is no way to account for all possible variables that may affect the outcome. To illustrate this point, Garvin McCain and Erwin Segal offer a parable:

Once there was a highly intelligent turkey. The turkey lived in a pen, attended by a kind, thoughtful master, and it had nothing to do but reflect upon the world's wonders and regularities. Morning always began with the sky getting light, followed by the clop, clop, clop of its master's friendly footsteps, which was followed by the appearance of delicious food. Other things varied—sometimes the morning was warm and sometimes cold—but food always followed footsteps. The sequence of events was so predictable that it eventually became the basis of the turkey's theory about the goodness of the world.

One morning, after more than a hundred confirmations of the goodness theory, the turkey listened for the clop, clop, clop, heard it, and had its head chopped off.

The turkey learned the hard way that explanations about the world only have a high or low probability of being correct. Today, some people take this uncertainty to mean that "facts are irrelevant—facts change." If that is so, should we just stop doing scientific research? Why or why not?

2. Witnesses in a court of law are asked to "swear to tell the truth, the whole truth, and nothing but the truth." What are some of the problems inherent in the question? Can you think of a better alternative?

3. Many popular magazines publish an astounding array of articles on diet, exercise, and other health-related topics. Some authors recommend a diet or dietary supplement. What kinds of evidence do you think the articles should include so that you can decide whether to accept their recommendations?

a Natalie, blindfolded, randomly plucks a jelly bean from a jar of 120 green and 280 black jelly beans. That's a ratio of 30 to 70 percent.

b The jar is hidden before she removes her blindfold. She observes a single green jelly bean in her hand and assumes the jar holds only green jelly beans.

c Still blindfolded, Natalie randomly picks 50 jelly beans from the jar and ends up with 10 green and 40 black ones.

d The larger sample leads her to assume one-fifth of the jar's jelly beans are green and four-fifths are black (a ratio of 20 to 80). Her larger sample more closely approximates the jar's green-to-black ratio. The more times Natalie repeats the sampling, the greater the chance she will come close to knowing the actual ratio.

Figure 1.11 A simple demonstration of sampling error.

4. Rarely can experimenters observe all individuals of a group. They select subsets or samples of populations, events, and other aspects of nature. They must avoid *sampling error*, which means obtaining misleading results by using subsets that aren't really representative of the whole (Figure 1.11). Test results are less likely to be distorted when a sampling is large and the test is repeated. Explain how sampling error might have affected the results of the butterfly experiment described in Section 1.6.

5. The Olestra® potato chip experiment in Section 1.6 was a *double-blind* study: Neither the subjects of the experiment nor the researchers knew which potato chips were in which bag until after all the subjects had reported. What do you think are some of the challenges for researchers performing a double-blind study?

6. In 1988 Dr. Randolph Byrd and his colleagues undertook a study of 393 patients admitted to the San Francisco General Hospital Coronary Care Unit. In the experiment, "born-again" Christian volunteers were asked to pray daily for a patient's rapid recovery and for prevention of complications and death. None of the patients knew if he or she was being prayed for or not, and none of the volunteers or patients knew each other. How each patient fared in the hospital was classified by Byrd as "good," "intermediate," or "bad." Byrd determined that the patients who had been prayed for fared a little better than those who had not. His was the first experiment to document, in a scientific fashion, statistically significant results in support of the prediction that prayer has beneficial effects on the outcome of seriously ill patients. Publication of these results engendered a storm of criticism, mostly from scientists who cited bias in Byrd's experimental design. For instance, Byrd classified the patients after the experiment had been finished. Think about how bias might play a role in interpreting medical data. Why do you think this experiment generated a dramatic response from the rest of the scientific community?

I Principles of Cellular Life

Staying alive means securing energy and raw materials from the environment. Shown here, a large living cell called Stentor. A protist, Stentor has hairlike projections around a cavity in its body, which is about two millimeters long. Its "hairs" of fused-together cilia beat the surrounding water. They create a current that wafts food into its cavity, which is filled with symbiotic algae called chlorella (bright green).

IMPACTS, ISSUES   *What Are You Worth?*

Hollywood thinks Leonardo DiCaprio is worth $20 million a picture, the Yankees think shortstop Alex Rodriguez is worth $217 million per decade, and the United States thinks the average teacher is worth $44,367 per year. Chemically, though, how much is a human body really worth (Figure 2.1a)?

Think about it. The human body is a collection of **elements**, or types of atoms. **Atoms** are fundamental substances that have mass and take up space, and cannot be broken apart by everyday means. Keep grinding up a chunk of copper, and the smallest bit you will end up with will be a lone atom of copper. Atoms are the smallest units of an element that still retain the element's properties.

Oxygen, hydrogen, carbon, and nitrogen are the most abundant elements in organisms. Next are phosphorus, potassium, sulfur, calcium, and sodium. *Trace* elements make up less than 0.01 percent of body weight. Selenium is an example.

Wait a minute! Selenium, mercury, arsenic, lead, and many other elements in the body are toxic, right? Maybe, or maybe not. As researchers

Element	Amount	Value
Oxygen (O)	43.00 kg	$0.021739
Carbon (C)	16.00 kg	6.400000
Hydrogen (H)	7.00 kg	0.028315
Nitrogen (N)	1.80 kg	9.706929
Calcium (Ca)	1.00 kg	15.500000
Phosphorus (P)	780.00 g	68.198594
Potassium (K)	140.00 g	4.098737
Sulfur (S)	140.00 g	0.011623
Sodium (Na)	100.00 g	2.287748
Chlorine (Cl)	95.00 g	1.409496
Magnesium (Mg)	19.00 g	0.444909
Iron (Fe)	4.20 g	0.054600
Fluorine (F)	2.60 g	7.917263
Zinc (Zn)	2.30 g	0.088090
Silicon (Si)	1.00 g	0.370000
Rubidium (Rb)	0.68 g	1.087153
Strontium (Sr)	0.32 g	0.177237
Bromine (Br)	0.26 g	0.012858
Lead (Pb)	0.12 g	0.003960
Copper (Cu)	72.00 mg	0.012961
Aluminum (Al)	60.00 mg	0.246804
Cadmium (Cd)	50.00 mg	0.010136
Cerium (Ce)	40.00 mg	0.043120
Barium (Ba)	22.00 mg	0.028776
Iodine (I)	20.00 mg	0.094184
Tin (Sn)	20.00 mg	0.005387
Titanium (Ti)	20.00 mg	0.010920
Boron (B)	18.00 mg	0.002172
Nickel (Ni)	15.00 mg	0.031320
Selenium (Se)	15.00 mg	0.037949

Element	Amount	Value
Chromium (Cr)	14.00 mg	0.003402
Manganese (Mg)	12.00 mg	0.001526
Arsenic (As)	7.00 mg	0.023576
Lithium (Li)	7.00 mg	0.024233
Cesium (Cs)	6.00 mg	0.000016
Mercury (Hg)	6.00 mg	0.004718
Germanium (Ge)	5.00 mg	0.130435
Molybdenum (Mo)	5.00 mg	0.001260
Cobalt (Co)	3.00 mg	0.001509
Antimony (Sb)	2.00 mg	0.000243
Silver (Ag)	2.00 mg	0.013600
Niobium (Nb)	1.50 mg	0.000624
Zirconium (Zr)	1.00 mg	0.000830
Lanthanium (La)	0.80 mg	0.000566
Gallium (Ga)	0.70 mg	0.003367
Tellurium (Te)	0.70 mg	0.000722
Yttrium (Y)	0.60 mg	0.005232
Bismuth (Bi)	0.50 mg	0.000119
Thallium (Tl)	0.50 mg	0.000894
Indium (In)	0.40 mg	0.000600
Gold (Au)	0.20 mg	0.001975
Scandium (Sc)	0.20 mg	0.058160
Tantalum (Ta)	0.20 mg	0.001631
Vanadium (V)	0.11 mg	0.000322
Thorium (Th)	0.10 mg	0.004948
Uranium (U)	0.10 mg	0.000103
Samarium (Sm)	50.00 µg	0.000118
Beryllium (Be)	36.00 µg	0.000218
Tungsten (W)	20.00 µg	0.000007

Grand Total: $118.63

Figure 2.1 (**a**) What are you worth, chemically speaking? (**b**) Proportions of the most common elements in a human body, Earth's crust, and seawater. How are they similar? How do they differ?   **a**

the big picture

Atoms and Elements

All substances are made of one or more elements. Atoms are the smallest units of matter that still retain the element's properties. They are composed of protons, neutrons, and electrons.

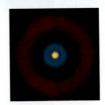

Why Electrons Matter

Whether an atom will interact with other atoms depends on how many electrons it has and how they are arranged. Chemical bonds unite two or more atoms.

Human		Earth's Crust		Ocean	
Oxygen	61.0%	Oxygen	46.0%	Oxygen	85.7%
Carbon	23.0	Silicon	27.0	Hydrogen	10.8
Hydrogen	10.0	Aluminum	8.2	Chlorine	2.0
Nitrogen	2.6	Iron	6.3	Sodium	1.1
Calcium	1.4	Calcium	5.0	Magnesium	0.1
Phosphorus	1.1	Magnesium	2.9	Sulfur	0.1
Potassium	0.2	Sodium	2.3	Calcium	0.04
Sulfur	0.2	Potassium	1.5	Potassium	0.03

b

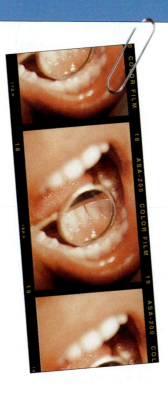

decipher chemical processes peculiar to life, they are finding that many trace elements considered to be poisons actually perform essential biological functions. For instance, large doses of chromium damage nerves and cause cancer, but one form works with insulin, a hormone that helps control the glucose level in blood. A little selenium is toxic, but too little causes heart and thyroid problems. An intricate balance of the right kinds of elements keeps the body functioning properly.

One more point: Earth's crust contains the same elements as the human body, but we're not just dirt. Like all living things, the proportions and organization of our elements are unique (Figure 2.1b). And building and maintaining that organization takes tremendous input of energy (just ask any pregnant woman).

You could buy all of the elements in a 150-pound human body for about $118.63. But constructing any living thing requires a remarkably complex interplay of energy and biological molecules that is far beyond the scope of any laboratory to duplicate, at least for now.

✓ How Would You Vote?

Fluoride has been proven to help prevent tooth decay. But too much wrecks bones and teeth, and causes birth defects. A lot can kill you. Many communities in the United States add fluoride to their drinking water. Do you want it in yours? See the Media Menu for details, then vote online.

Atoms Bond

The molecular organization and the activities of every living thing arise from ionic, covalent, and hydrogen bonds between atoms.

No Water, No Life

Water's unique characteristics, including temperature-stabilizing effects, cohesion, and solvent properties, make life possible on Earth.

Hydrogen Ions Rule

Life is adapted to the properties of water and to changing concentrations of hydrogen ions and other substances dissolved in water.

2.1 Start With Atoms

Life's chemical properties start with protons, neutrons, and electrons. The unique character of each element actually begins with the number of protons, which is the same in all of its atoms.

Atoms, again, are the smallest units that retain the properties of an element. All atoms are made of three kinds of subatomic particles: protons, neutrons, and electrons (Figure 2.2). Each **proton** carries a positive *charge*, or a defined amount of electricity. Protons are symbolized as p^+. An atom's nucleus (core) holds one or more protons. It also holds **neutrons**, which have no charge. Zipping about the nucleus are one or more **electrons**, which carry a negative charge (e^-).

The positive charge of a proton and the negative charge of an electron balance each other. So an atom that has the same number of electrons and protons has no net electrical charge.

Each element has a unique **atomic number**, or the number of protons in the nucleus of its atoms. For example, the atomic number for hydrogen, which has one proton, is 1. For carbon, with six protons, it is 6.

Each element also has a **mass number**, equal to the total number of protons and neutrons in the atomic nucleus. For example, carbon, with six protons and six neutrons, has a mass number of 12.

Why bother with atomic and mass numbers? If you know how many electrons, protons, and neutrons the atoms of an element contain, you can predict what the

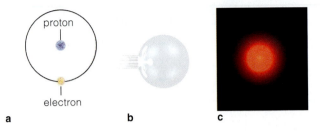

a   b   c

Figure 2.2 Different ways of representing atoms, using hydrogen (H) as the example. (**a**) A shell model shows the number of electrons and their relative distances from the nucleus. (**b**) Balls show relative sizes of atoms. (**c**) Electron density clouds show electron distribution around the nucleus.

chemical behavior of that element will probably be under different conditions.

Elements were being classified in terms of chemical similarities long before their subatomic particles were discovered. In 1869, Dmitry Mendeleev, known more for his extravagant hair than his discoveries (he cut it only once per year), arranged the known elements into a repeating pattern based on their chemical properties. Using gaps in this **periodic table**, Mendeleev was able to predict correctly the existence of other elements that had yet to be discovered.

The elements fall into order in the periodic table according to their atomic number (Figure 2.3). Those in the same column of the table have the same number of electrons available for interaction with other atoms. As a result, they behave in a remarkably similar way. For instance, helium, neon, radon, and other gases in the vertical column farthest to the right are called *inert* elements because none of their electrons is available for chemical interaction. Consequently, they rarely do much; they exist mostly as solitary atoms.

Not all of the elements in the periodic table occur in nature. The elements after atomic number 92 are so highly unstable that they have been produced only in very small quantities in the laboratory—sometimes no more than a single atom. They wink out of existence that fast. Some elements still haven't been made.

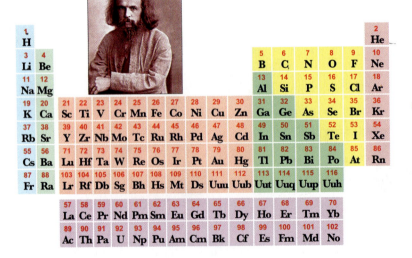

Figure 2.3 Periodic table of the elements and Dmitry Mendeleev, who created it. Some of the symbols for elements are abbreviations for their Latin names. For instance, Pb (lead) is short for *plumbum;* the word "plumbing" is related, because ancient Romans used lead to make their water pipes.

Atoms are the smallest units of an element, or fundamental substance, that still retain the properties of that element. Ninety-two elements occur naturally on Earth.

One or more positively charged protons, negatively charged electrons, and (except for hydrogen) neutrons make up atoms.

An element's chemical properties are a direct consequence of the number of electrons it has available for interacting with other atoms.

2.2 Radioisotopes

All elements are defined by the number of protons in their atoms—but an element's atoms can differ in their number of neutrons. We call such atoms isotopes of the same element. And some are radioactive.

Henri Becquerel discovered radioactivity by accident in 1896. He put some crystals of phosphorescent uranium salts on top of an unexposed photographic plate inside a desk drawer. Between the uranium and the plate were several sheets of opaque black paper, a coin, and a metal screen. A day later, he used the film and developed it. Surprisingly, a negative image of the coin and screen appeared on it. Energy emitted by the uranium had exposed the film all around the metal. Becquerel concluded that uranium salts emit some form of "radiation" capable of going through things that light cannot penetrate. What was it?

As we now know, most elements in nature have two or more kinds of isotopes. Carbon has three, nitrogen has two, and so on. A superscript number to the left of an element's symbol is the isotope's mass number (combined number of protons and neutrons). For instance, carbon's three natural isotopes are ^{12}C (or carbon 12, the most common form, with six protons, six neutrons), ^{13}C (six protons, seven neutrons), and ^{14}C (six protons, eight neutrons).

Too many or too few neutrons in the nucleus of an atom can cause it to be unstable, or radioactive. A radioactive atom spontaneously emits energy as subatomic particles and x-rays when its nucleus disintegrates. This process, called **radioactive decay**, transforms one element into another. ^{13}C and ^{14}C are radioactive isotopes, or **radioisotopes**, of carbon. Each radioisotope decays with a particular amount

of energy into a predictable, more stable product. For example, after 5,700 years, about half of the atoms in a sample of ^{14}C will have turned into ^{14}N (nitrogen) atoms. As you'll see in Chapter 17, researchers use radioactive decay to estimate the age of fossils.

Different isotopes of an element are still the same element. For the most part, carbon is carbon, regardless of how many neutrons it has. Living systems use ^{12}C the same way as ^{14}C. Knowing this, researchers or clinicians studying a certain type of molecule make **tracers** in which a radioisotope gets substituted for a stable element in that molecule. They deliver tracers into a cell, a multicelled body, or an ecosystem. Energy from radioactive decay is like a shipping label; it helps us track the molecule of interest with instruments that detect radioactivity.

Melvin Calvin and his colleagues used a tracer, carbon dioxide gas made with ^{14}C, to discover the specific steps of photosynthesis. By steeping plants in the radioactive gas, they were able to follow the path of the radioactive carbon atoms through each reaction step in the formation of sugars and starches.

Radioisotopes also are used in medicine. *PET* (short for *Positron-Emission Tomography*) uses radioisotopes to form images of body tissues. Clinicians attach a radioisotope to glucose or another sugar. They inject this tracer into a patient, who is moved into a PET scanner (Figure 2.4a). Cells throughout the body absorb the tracer at different rates. The scanner then detects radiation caused by energy from the decay of the radioisotope, and that radiation is used to form an image. Such images can reveal variations and abnormalities in metabolic activity (Figure 2.4d).

detector ring inside PET scanner   body section inside ring

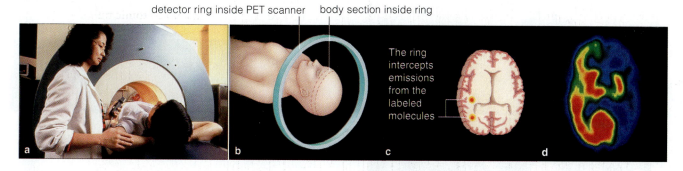

The ring intercepts emissions from the labeled molecules

Figure 2.4 (**a**) Patient moving into a PET scanner. (**b**,**c**) Inside, a ring of detectors intercepts radioactive emissions from labeled molecules that were injected into the patient. Computers analyze and color-code the number of emissions from each location in the scanned body region.

(**d**) Different colors in a brain scan signify differences in metabolic activity. Cells of this brain's left half absorbed and used the labeled molecules at expected rates. However, cells in the right half showed little activity. The patient was diagnosed as having a neurological disorder.

2.3 What Happens When Atom Bonds With Atom?

Atoms acquire, share, and donate electrons. Atoms of some elements do this easily; others do not. Why is this so? To come up with an answer, look to the number and arrangement of electrons in atoms.

ELECTRONS AND ENERGY LEVELS

In our world, simple physics explains the motion of an apple falling from a tree. Tiny electrons belong to a strange world where everyday physics doesn't apply. (If electrons were as big as apples, you'd be about 3.5 times taller than our solar system is wide.) Different forces bring about the motion of electrons, which can get from here to there without going in between!

We can calculate where an electron is, although not exactly. The best we can do is say that it's somewhere in a fuzzy cloud of probability density. Where it can go depends on how many other electrons are buzzing about an atom's nucleus. As it turns out, electrons can occupy orbitals, which are volumes of space around the nucleus. There are many orbitals, with different three-dimensional shapes.

An atom has about same number of electrons as protons. For most atoms, that's a lot of electrons. How are these electrons arranged, given that they repel each other? Think of an atom as a multilevel apartment building with lots of vacant rooms to rent to electrons, and a nucleus in the basement. Each "room" is one orbital, and it rents out to two electrons at most. An orbital holding one electron only has a vacancy; another electron can move in.

Each floor in that atomic apartment building corresponds to an energy level. There is only one room on the first floor (one orbital at the lowest energy level, closest to the nucleus), and it fills first. For hydrogen, the simplest atom, that room has a single electron (Figure 2.5). For helium, it has two. In other words, helium has no vacancies at the first (lowest) energy level. In larger atoms, more electrons rent second-floor rooms. If the second floor is filled, additional electrons rent third-floor rooms, and so on. *They fill orbitals at successively higher energy levels.*

The farther an electron is from the basement (the nucleus), the greater its energy. An electron in a first-floor room can't move to the second or third floor, let alone the penthouse, unless a boost of energy gets it there. Suppose it absorbs the right amount of energy from, say, sunlight, to get excited about moving up. Move it does. If nothing fills that lower room, though, the electron will quickly go back to it, emitting extra energy as it does. Later, you'll see how cells in plants and in your eyes can harness and use that energy.

FROM ATOMS TO MOLECULES

In shell models, nested "shells" correspond to energy levels. They offer us an easy way to check for electron vacancies in various atoms (Figure 2.6). Bear in mind, atoms do not look like these flat diagrams. The shells are not three-dimensional volumes of space, and they certainly don't show the electron orbitals.

Atoms that have vacancies in the outermost "shell" tend to give up, acquire, or share electrons with other atoms. This kind of electron-swapping between atoms is known as **chemical bonding** (Section 2.4). Atoms with zero vacancies rarely bond with other atoms. By contrast, the most common atoms in organisms—such as oxygen, carbon, hydrogen, nitrogen, and calcium—have vacancies in orbitals at their outermost energy level. And they do bond with other atoms.

third energy level (second floor)

second energy level (first floor)

first energy level (closest to the basement)

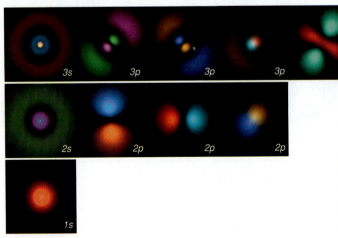

Figure 2.5 First, second, and third levels of the atomic apartment building. Each picture is a three-dimensional approximation of an electron orbital. Colors are most intense in locations where electrons are most likely to be. Orbitals farthest from the nucleus have greater energy and are more complex.

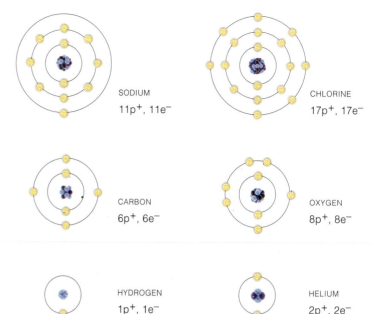

c **Third shell** shows the third set of orbitals: one *s* orbital, three *p* orbitals, and five *d* orbitals, a total of nine orbitals with room for 18 electrons. Sodium has one electron in the third shell of orbitals, and chlorine has seven. Both have vacancies, so they form chemical bonds.

SODIUM
$11p^+$, $11e^-$

CHLORINE
$17p^+$, $17e^-$

b **Second shell** shows the second energy level, which combines a set of one *s* orbital plus three *p* orbitals. The second shell of orbitals has room for a total of eight electrons. Carbon has six electrons, two in the first shell and four in the second shell. It has four vacancies. Oxygen has two vacancies. Both carbon and oxygen form chemical bonds.

CARBON
$6p^+$, $6e^-$

OXYGEN
$8p^+$, $8e^-$

a **First shell** shows the first energy level, containing a single orbital (*1s*). Hydrogen has only one electron in this orbital that can hold two. Hydrogen gives up its electron easily, becoming a chemically reactive free proton. A helium atom has two electrons in the *1s* orbital. Having no vacancies, helium does not usually form chemical bonds.

HYDROGEN
$1p^+$, $1e^-$

HELIUM
$2p^+$, $2e^-$

Figure 2.6 Shell model. Using this model, it is easy to see the vacancies in each atom's outer orbitals. Each circle represents all of the orbitals on one energy level. Larger circles correspond to higher energy levels. This model is highly simplified; a more realistic rendering would show the electrons as fuzzy clouds of probability density about ten thousand times bigger than the nucleus.

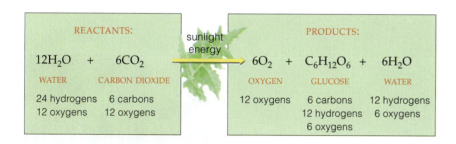

REACTANTS:

sunlight energy

PRODUCTS:

$$12H_2O \ + \ 6CO_2 \ \longrightarrow \ 6O_2 \ + \ C_6H_{12}O_6 \ + \ 6H_2O$$

WATER   CARBON DIOXIDE   OXYGEN   GLUCOSE   WATER

24 hydrogens   6 carbons
12 oxygens   12 oxygens

12 oxygens   6 carbons   12 hydrogens
            12 hydrogens   6 oxygens
            6 oxygens

Figure 2.7 Chemical bookkeeping. Chemical equations are representations of reactions, or interactions between atoms and molecules. Substances entering a reaction are to the left of a reaction arrow (reactants), and products are to the right, as shown by this chemical equation for photosynthesis.

A **molecule** is simply two or more atoms of the same or different elements joined in a chemical bond. You can write a molecule's chemical composition as a formula that uses symbols for elements. A formula shows the number of each kind of atom in a molecule (Figure 2.7). Water has the chemical formula H_2O. The subscript number tells you that two hydrogen (H) atoms are present for each oxygen (O) atom.

Compounds are molecules that consist of two or more different elements in proportions that never do vary. Water is an example. All water molecules have one oxygen atom bonded to two hydrogen atoms. The ones in rain clouds, the seas, a Siberian lake, a flower's petals, your bathtub, or anywhere else always have twice as many hydrogen as oxygen atoms.

In a **mixture**, two or more molecules intermingle without chemically bonding. For instance, you can make a mixture by swirling water and sugar together. The proportions of elements in a mixture can vary.

Electrons occupy orbitals, or defined volumes of space around an atom's nucleus. Successive orbitals correspond to levels of energy, which become higher with distance from the nucleus.

One or at most two electrons can occupy any orbital. Atoms with vacancies in their highest level orbitals can interact with other atoms.

A molecule is two or more atoms joined in a chemical bond. Atoms of two or more elements are bonded together in compounds. A mixture consists of intermingled molecules.

2.4 Bonds in Biological Molecules

The distinctive properties of biological molecules start with atoms interacting at the level of electrons.

ION FORMATION AND IONIC BONDING

An electron, recall, has a negative charge equal to a proton's positive charge. When an atom has as many electrons as protons, these charges balance each other, so the atom will have a net charge of zero.

Atoms with more electrons than protons carry a net negative charge, and those with more protons than electrons carry a net positive charge. An atom that has either a positive or negative charge is known as an **ion**. Ions form when atoms gain or lose electrons.

Example: An uncharged chlorine atom has seven electrons, hence one vacancy, in the third orbital level. Chlorine tends to grab an electron from other places. That extra electron will make it a chloride ion (Cl^-), with a net negative charge. A sodium atom has a lone electron in the same orbital level, but it is easier to give that one up than to acquire seven more. If it does, it will only have second-level orbitals, and they will be full of electrons—so no vacancy. It becomes a sodium ion with a net positive charge (Na^+).

What happens when one atom gives up an electron that another accepts? The two resulting ions may stay close together, because they have opposite charges that attract each other. A close association of ions is an **ionic bond**. Figure 2.8*a* shows a crystal of table salt, or NaCl. In such crystals, ionic bonds hold the ions in an orderly arrangement.

COVALENT BONDING

In an ionic bond, one atom donates an extra electron that the other accepts. What if both atoms want an extra electron? They can *share* one of their electrons in a hybrid orbital that spans both nuclei. Each atom's vacancy becomes partly filled with a shared electron. When atoms share one or more electrons, they are joined in a **covalent bond** (Figure 2.8*b*). Such bonds are stable and are much stronger than ionic bonds.

Unlike chemical formulas, structural formulas show how atoms are physically arranged in a molecule— they reveal the bonding pattern. A single line that connects two atoms in a structural formula represents two shared electrons in one covalent bond. Molecular hydrogen, with one covalent bond, is written H—H.

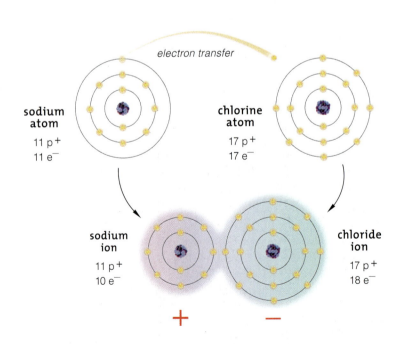

a Ionic bonding. A sodium atom donates its extra electron to a chlorine atom.

In each crystal of table salt, or NaCl, many sodium and chloride ions stay close together because of the mutual attraction of opposite charges. Their ongoing interaction is a case of ionic bonding.

Figure 2.8 Important bonds in biological molecules.

Two atoms can share two electron pairs in a *double* covalent bond. Molecular oxygen (O=O) is like this. In a *triple* covalent bond, two atoms share three pairs, as they do in molecular nitrogen (N≡N). Each time you breathe in, a stupendous number of gaseous O_2 and N_2 molecules flows into your lungs.

In a *nonpolar* covalent bond, two identical atoms share electrons equally, and the molecule shows no difference in charge between its two ends. Molecular hydrogen (H_2) has such symmetry, as do O_2 and N_2.

A *polar* covalent bond forms between atoms of different elements. One of the atoms pulls the shared electrons a little toward one end of the bond. Because the electrons spend extra time there, that part of the molecule bears a slight negative charge. The opposite end bears a slight positive charge. A water molecule (H—O—H) has two polar covalent bonds; the oxygen is negatively charged, and the hydrogens are positive.

HYDROGEN BONDING

A hydrogen atom taking part in a polar covalent bond bears a slight positive charge, so it attracts negatively charged atoms. When the negatively charged atom is bound to a different molecule or to a different part of the same molecule, the interaction between it and the hydrogen atom is called a **hydrogen bond**.

Because they are weak, hydrogen bonds form and break easily. They play crucial roles in the structure and function of biological molecules, especially with water (Section 2.5). They often form between different parts of very large molecules that have folded over on themselves, and hold them in a particular shape. They are also what holds the two nucleotide strands of large DNA molecules together. You can get a sense of these interactions from Figure 2.8c.

Ions form when atoms acquire a net charge by gaining or losing electrons. Two ions of opposite charge attract each other. They can associate in an ionic bond.

In a covalent bond, atoms share a pair of electrons. When atoms share the electrons equally, the bond is nonpolar. When the sharing is not equal, the bond is polar—slightly positive at one end, slightly negative at the other.

In a hydrogen bond, a covalently bound hydrogen atom attracts a negatively charged atom taking part in a different covalent bond.

Read Me First!
and watch the narrated animation on how atoms bond

Two hydrogen atoms, each with one proton, share two electrons in a single nonpolar covalent bond.

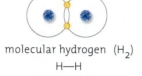

molecular hydrogen (H_2)
H—H

Two oxygen atoms, each with eight protons, share four electrons in a nonpolar double covalent bond.

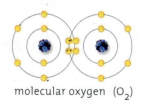

molecular oxygen (O_2)
O=O

Oxygen has vacancies for two electrons in its highest energy level orbitals. Two hydrogen atoms can each share an electron with oxygen. The resulting two polar covalent bonds form a water molecule.

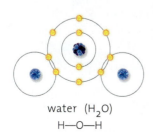

water (H_2O)
H—O—H

b Covalent bonding. Each atom becomes more stable by sharing electron pairs in hybrid orbitals.

Two molecules interacting weakly in one H bond, which can form and break easily.

hydrogen bond

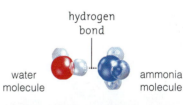

water molecule   ammonia molecule

H bonds helping to hold part of two large molecules together.

Many H bonds hold DNA's two strands together along their length. Individually they are weak, but collectively stabilize DNA's large structure.

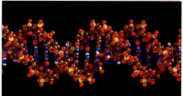

c Hydrogen bonds. Such bonds can form at a hydrogen atom that is already covalently bonded in a molecule. The atom's slight positive charge weakly attracts an atom with a slight negative charge that is already covalently bonded to something else. As shown, this can happen between one of the hydrogen atoms of a water molecule and the nitrogen atom of an ammonia molecule.

2.5 Water's Life-Giving Properties

No sprint through basic chemistry is complete unless it leads to the collection of molecules called water. Life originated in water. Organisms still live in it or they cart water around with them inside cells and tissue spaces. Many metabolic reactions use water. Cell shape and cell structure absolutely depend on it.

POLARITY OF THE WATER MOLECULE

Figure 2.9a shows the structure of a water molecule. Two hydrogen atoms have formed polar covalent bonds with an oxygen atom. The molecule has no net charge. Even so, the oxygen pulls the shared electrons more than the hydrogen atoms do. Thus, the molecule of water has a slightly negative "end" that's balanced out by its slightly positive "end."

A water molecule's polarity attracts other water molecules. Also, it is so attractive to sugars and other polar molecules that hydrogen bonds readily form between them. That is why polar molecules are known as **hydrophilic** (water-loving) substances.

That same polarity repels oils and other nonpolar molecules, which are **hydrophobic** (water-dreading) substances. Shake a bottle filled with water and salad oil, then set it on a table. Soon, new hydrogen bonds replace the ones that the shaking broke. The reunited water molecules push out oil molecules, which cluster as oil droplets or as an oily film at the water's surface.

The same kinds of interactions occur at the thin, oily membrane between the water inside and outside cells. Membrane organization, and life itself, starts with hydrophilic and hydrophobic interactions. You'll be reading about membrane structure in Chapter 4.

WATER'S TEMPERATURE-STABILIZING EFFECTS

Cells are mostly water, and they also release a lot of metabolic heat. Without water's hydrogen bonds, cells would cook in their own juices. How? All molecules vibrate nonstop, and they move more as they absorb heat. **Temperature** is a measure of molecular motion. Compared to most other fluids, water absorbs more heat energy before it gets measurably hotter. So water acts as a heat reservoir, and its temperature remains relatively stable. In time, increases in heat step up the motion within water molecules. Before that happens, however, much of the heat will go into disrupting hydrogen bonds between molecules.

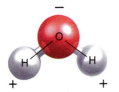

slight negative charge on the oxygen atom

The + and – ends balance each other; the whole molecule carries no net charge, overall.

slight positive charge on the hydrogen atoms

a

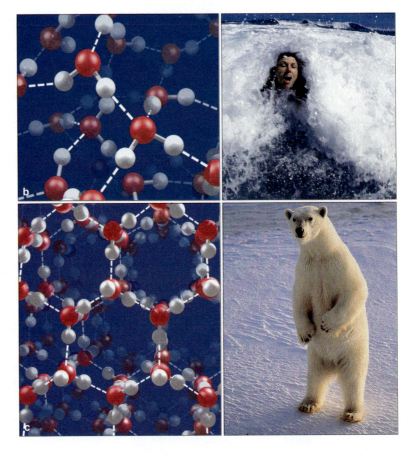

Figure 2.9 Water, a substance essential for life.

(**a**) Polarity of an individual water molecule.

(**b**) Hydrogen bonding pattern among water molecules in liquid water. Dashed lines signify hydrogen bonds, which break and reform rapidly.

(**c**) Hydrogen bonding in ice. Below 0°C, every water molecule hydrogen-bonds to four others, in a rigid three-dimensional lattice. The molecules are farther apart, or less dense, than they are in liquid water. As a result, ice floats on water.

Thanks partly to rising levels of methane and other greenhouse gases that are contributing to global warming, the Arctic ice cap is melting. At current rates, it will be gone in fifty years. So will the polar bears. Already their season for hunting seals is shorter, bears are thinner, and they are giving birth to fewer cubs.

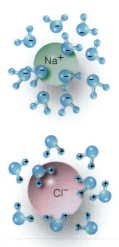

Figure 2.10 Two spheres of hydration.

Figure 2.11 Examples of water's cohesion. (**a**) When a pebble hits liquid water and forces molecules away from the surface, the individual water molecules don't fly off every which way. They stay together in droplets. Why? Countless hydrogen bonds exert a continuous inward pull on individual molecules at the surface.

(**b**) And just how does water rise to the very top of trees? Cohesion, and evaporation from leaves, pulls it upward.

With a fairly stable water temperature, hydrogen bonds form as fast as they break. Energy inputs can increase the molecular motion so much that the bonds stay broken, and individual molecules at the water's surface escape into air. By this process, **evaporation**, heat energy converts liquid water to a gas. An energy input has overcome the attraction between molecules of water, which break free. The surface temperature of water decreases during evaporation.

Evaporative water loss helps you and some other mammals cool off when you sweat on hot, dry days. Sweat, about 99 percent water, evaporates from skin.

Below 0°C, water molecules don't move enough to break their hydrogen bonds, and they become locked in the latticelike bonding pattern of ice (Figure 2.9c). Ice is less dense than water. During winter freezes, ice sheets may form near the surface of ponds, lakes, and streams. The ice blanket "insulates" the liquid water beneath it and helps protect many fishes, frogs, and other aquatic organisms against freezing.

WATER'S SOLVENT PROPERTIES

Water is an excellent *solvent*, meaning ions and polar molecules easily dissolve in it. A dissolved substance is known as a **solute**. In general, a substance is said to be *dissolved* after water molecules cluster around ions or molecules of it and keep them dispersed in fluid.

Water molecules cluster around a solute, thereby forming a *sphere of hydration*. Spheres form around any solute in cellular fluids, tree sap, blood, the fluid in your gut, and every other fluid associated with life. Watch it happen after you pour table salt (NaCl) into a cup of water. In time, the crystals of salt separate into ions of sodium (Na$^+$) and chloride (Cl$^-$). Each Na$^+$ attracts the negative end of some water molecules even as Cl$^-$ attracts the positive end of others (Figure 2.10). Spheres of hydration formed this way keep the ions dispersed in fluid.

WATER'S COHESION

Still another life-sustaining property of water is its cohesion. **Cohesion** means something is showing a capacity to resist rupturing when it is stretched, or placed under tension. You see its effect when a tossed pebble breaks the surface of a lake, a pond, or some other body of liquid water (Figure 2.11a). At or near the surface, uncountable numbers of hydrogen bonds are exerting a continuous, inward pull on individual molecules. Bonding creates a high surface tension.

Cohesion is in play inside organisms, too. Plants, for example, absorb nutrient-laden water while they grow. Very narrow columns of liquid water rise inside pipelines of vascular tissues, which extend from roots to leaves and other plant parts. On sunny days, water evaporates from leaves as molecules break free and diffuse into the air (Figure 2.11b). The cohesive force of hydrogen bonds pulls replacements into the leaf cells, in ways you'll read about in Section 26.3.

Being slightly polar, water molecules hydrogen bond to one another and to other polar (hydrophilic) substances. They tend to repel nonpolar (hydrophobic) substances.

The unique properties of liquid water make life possible. Water has cohesion, temperature-stabilizing effects, and a capacity to dissolve many substances.

2.6 Acids and Bases

Ions are dissolved in fluids inside and outside a cell, and they affect its structure and function. Among the most influential are hydrogen ions. They have far-reaching effects largely because they are chemically active and there are so many of them.

THE PH SCALE

At any instant in liquid water, some water molecules split into ions of hydrogen (H^+) and hydroxide (OH^-). These ions are the basis of the **pH scale**. The scale is a way to measure the relative amount of hydrogen ions in solutions such as seawater, blood, or sap. The greater the H^+ concentration, the lower the pH. Pure water (not rainwater or tap water) always has as many H^+ as OH^- ions. This state is neutrality, or pH 7.0 (Figure 2.12).

A one unit decrease from neutrality corresponds to a tenfold increase in H^+ concentration, and an increase by one unit corresponds to a tenfold decrease in H^+ concentration. One way to get a sense of the range is to taste baking soda (pH 9), water (pH 7), and lemon juice (pH 2).

HOW DO ACIDS AND BASES DIFFER?

Substances called **acids** *donate* hydrogen ions and **bases** *accept* hydrogen ions when dissolved in water. *Acidic* solutions, such as lemon juice, gastric fluid, and coffee, release H^+; their pH is below 7. *Basic* solutions, such as seawater, baking soda, and egg white, combine with H^+. Basic solutions (also known as alkaline solutions) have a pH above 7.

Nearly all of life's chemistry occurs near pH 7. Most of your body's internal environment (tissue fluids and blood) is between pH 7.3 and 7.5. Seawater is more basic than body fluids of the organisms living in it.

Acids and bases can be weak or strong. The weak acids, such as carbonic acid (H_2CO_3), are stingy H^+ donors. Strong acids readily give up H^+ in water. An example is the hydrochloric acid that dissociates into H^+ and Cl^- inside your stomach. The H^+ makes your gastric fluid far more acidic, which in turn activates protein-digesting enzymes.

Too much HCl can cause an *acid stomach*. Antacids taken for this condition, including milk of magnesia, release OH^- ions that combine with H^+ to reduce the pH of stomach contents.

High concentrations of strong acids or bases can disrupt ecosystems and make it impossible for cells

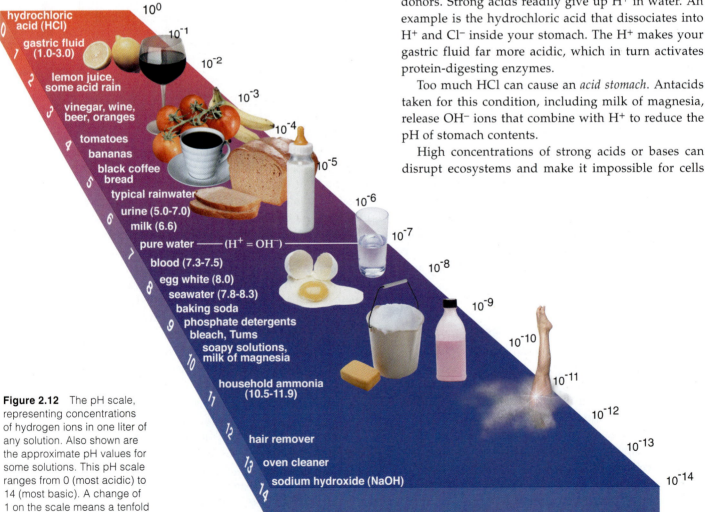

Figure 2.12 The pH scale, representing concentrations of hydrogen ions in one liter of any solution. Also shown are the approximate pH values for some solutions. This pH scale ranges from 0 (most acidic) to 14 (most basic). A change of 1 on the scale means a tenfold change in H^+ concentration.

Scale labels:

hydrochloric acid (HCl) — 0
gastric fluid (1.0–3.0) — 1
lemon juice, some acid rain — 2
vinegar, wine, beer, oranges — 3
tomatoes, bananas — 4
black coffee, bread — 5
typical rainwater, urine (5.0–7.0), milk (6.6) — 6
pure water ——— ($H^+ = OH^-$) — 7
blood (7.3–7.5) — 7
egg white (8.0) — 8
seawater (7.8–8.3), baking soda — 8
phosphate detergents, bleach, Tums — 9
soapy solutions, milk of magnesia — 10
household ammonia (10.5–11.9) — 11
hair remover — 12
oven cleaner — 13
sodium hydroxide (NaOH) — 14

Concentration markers: 10^0, 10^{-1}, 10^{-2}, 10^{-3}, 10^{-4}, 10^{-5}, 10^{-6}, 10^{-7}, 10^{-8}, 10^{-9}, 10^{-10}, 10^{-11}, 10^{-12}, 10^{-13}, 10^{-14}

Figure 2.13 Emissions of sulfur dioxide from a coal-burning power plant. Airborne pollutants such as sulfur dioxide dissolve in water vapor to form acidic solutions. They are a component of acid rain.

to survive. Read the labels on containers of ammonia, drain cleaner, and other products commonly stored in households. Many cause severe *chemical burns*. So does sulfuric acid in car batteries. Fossil fuel burning and nitrogen fertilizers release strong acids that lower the pH of rainwater (Figure 2.13). Some regions are quite sensitive to this *acid rain*. Alterations in the chemical composition of soil and water can harm organisms. We return to this topic in Section 42.2.

SALTS AND WATER

A **salt** is any compound that dissolves easily in water and releases ions *other than* H^+ and OH^-. It commonly forms when an acid interacts with a base. For example:

$$HCl \text{ (acid)} + NaOH \text{ (base)} \rightleftharpoons NaCl \text{ (salt)} + H_2O$$

HYDROCHLORIC ACID   SODIUM HYDROXIDE   SODIUM CHLORIDE

$$Na^+ \quad Cl^- \text{ (ionization)}$$

Bidirectional arrows indicate that the reaction goes in both directions. Many of the ions released when salts dissolve in fluid are important components of cellular processes. For example, ions of sodium, potassium, and calcium help nerve and muscle cells function and help plant cells take up water from soil.

BUFFERS AGAINST SHIFTS IN PH

Cells must respond fast to even slight shifts in pH, because excess H^+ or OH^- can alter the functions of biological molecules. Responses are rapid with **buffer systems**. Think of such a system as a dynamic chemical partnership between a weak acid or base and its salt. These two related chemicals work in equilibrium to counter slight shifts in pH. For example, if a small amount of a strong base enters a buffered fluid, the weak acid partner can neutralize the excess OH^- ions by donating some H^+ ions to the solution.

Most body fluids are buffered. Why? Enzymes, receptors, and all other essential biological molecules function properly only within a narrow range of pH. Deviation from the range halts cellular processes.

Carbon dioxide, a by-product of many reactions, combines with water in the blood to compose a buffer system of carbonic acid and bicarbonate ions. When blood pH rises a bit, the carbonic acid neutralizes the excess OH^- by releasing some hydrogen ions, which combine with the OH^- to form water:

$$OH^- + H_2CO_3 \longrightarrow HCO_3^- + H_2O$$

CARBONIC ACID   BICARBONATE (SALT)   WATER

When blood becomes more acidic, this salt mops up the excess H^+ and so shifts the balance of the buffer system toward the acid:

$$HCO_3^- + H^+ \longrightarrow H_2CO_3$$

BICARBONATE   CARBONIC ACID

Buffer systems can neutralize only so many excess ions. With even a slight excess above that point, the pH swings widely. When the blood pH (7.3–7.5) falls even to 7, the individual may fall into a *coma*, an often irreversible state of unconsciousness. This happens in *respiratory acidosis*. Carbon dioxide accumulates, too much carbonic acid forms, and blood pH plummets. By contrast, when the blood pH increases even to 7.8, *tetany* may occur; skeletal muscles cannot be released from contraction. In *alkalosis*, a rise in blood pH can't be reversed. Such conditions can be lethal.

Ions dissolved in fluids on the inside and outside of cells have key roles in cell function. Acidic substances release hydrogen ions, and basic substances accept them. Salts are compounds that release ions other than H^+ and OH^-.

Acid–base interactions help maintain pH, which is the H^+ concentration in a fluid. Buffer systems help control the body's acid–base balance at levels suitable for life.

Summary

Introduction Chemistry helps us understand the nature of all substances that make up cells, organisms, and the Earth, its waters, and the atmosphere. Table 2.1 summarizes some key chemical terms that you will encounter throughout this book.

Table 2.1	Summary of Important Players in the Chemical Basis of Life
ATOM	Fundamental form of matter that occupies space, has mass, and cannot be broken apart by ordinary physical or chemical means.
Proton (p^+)	Positively charged particle of the atomic nucleus.
Electron (e^-)	Negatively charged particle that can occupy a volume of space (orbital) around the nucleus.
Neutron	Uncharged particle of the atomic nucleus. For a given element, the mass number is the sum of the number of protons and neutrons in the nucleus.
ELEMENT	Type of atom defined by the number of protons, which is its atomic number. Each element has unique chemical properties.
MOLECULE	Unit of matter in which two or more atoms of the same element, or different ones, are bonded together by shared electrons.
Compound	Molecule composed of two or more different elements in unvarying proportions. Water is an example.
Mixture	Intermingling of two or more elements or compounds in proportions that vary.
ISOTOPE	One of two or more forms of an element that differ in the number of neutrons in their nuclei.
Radioisotope	Unstable isotope, having an unbalanced number of protons and neutrons, that emits particles and energy.
Tracer	Molecule of a substance to which a radioisotope is attached. Together with tracking devices, it is used to identify movement or destination of the substance in a metabolic pathway, the body, or some other system.
ION	Atom in which the number of electrons differs from the number of protons; negatively or positively charged. A proton without an electron zipping around it is a hydrogen ion (H^+).
SOLUTE	Any molecule or ion dissolved in some solvent.
Hydrophilic substance	Polar molecule or molecular region that can readily dissolve in water.
Hydrophobic substance	Nonpolar molecule or molecular region that strongly resists dissolving in water.
ACID	Substance that donates H^+ when dissolved in water.
BASE	Substance that accepts H^+ when dissolved in water.
SALT	Compound that releases ions other than H^+ or OH^- when dissolved in water.

Section 2.1 All substances consist of one or more elements. Ninety-two elements are naturally occurring. An atom consists of one or more positively charged protons, negatively charged electrons, and (except for hydrogen atoms) one or more uncharged neutrons. Protons and neutrons occupy the core region, or nucleus. In elements, all of the atoms have the same number of protons.

Section 2.2 Most elements have isotopes, which are two or more forms of atoms that have the same number of protons but different numbers of neutrons. An atom is radioactive when its nucleus is unstable. All elements have one or more radioactive isotopes.

Section 2.3 Whether an atom interacts with others depends on the number and arrangement of its electrons, which occupy orbitals (volumes of space) around the atomic nucleus. When an atom has one or more vacancies in orbitals at its highest energy level, it can interact with other atoms by donating, accepting, or sharing electrons.

Section 2.4 An atom may lose or gain one or more electrons and thus become an ion, which has a positive or negative charge.

Generally, a chemical bond is a union between the electron structures of atoms.

a. In an ionic bond, a positive ion and negative ion stay together by mutual attraction of opposite charges.

b. Atoms often share one or more pairs of electrons in covalent bonds. Electron sharing is equal in nonpolar covalent bonds, and it is unequal in polar covalent bonds. Interacting atoms have no net charge overall, even though the bond can be slightly negative at one end and slightly positive at the other.

c. In a hydrogen bond, one covalently bonded atom (e.g., oxygen) that has a slight negative charge is weakly attracted to the slight positive charge of a hydrogen atom taking part in a different polar covalent bond.

Section 2.5 Polar covalent bonds join together three atoms in a water molecule (two hydrogens and one oxygen). The water molecule's polarity invites extensive hydrogen bonding between molecules in bodies of water. Such bonding is the basis of liquid water's ability to resist temperature changes (more than other fluids do), display internal cohesion, and easily dissolve polar or ionic substances. These properties make life possible.

Section 2.6 The pH of a solution indicates its hydrogen ion concentration. A typical pH range is from 0 (highest H^+ concentration, most acidic) to 14 (lowest H^+ concentration, most basic). At pH 7, or neutrality, H^+ and OH^- concentrations are equal. Acids release H^+ ions in water; bases combine with them. Buffer systems help maintain a favorable pH in internal environments. This is important because most biological processes operate only within a narrow range of pH.

Self-Quiz

Answers in Appendix III

1. Is this statement true or false: Every type of atom consists of protons, neutrons, and electrons. **T.**

2. Electrons carry a __b__ charge.
   a. positive   b. negative   c. zero

3. A(n) __d__ is any molecule to which a radioisotope has been attached for research or diagnostic purposes.
   a. ion   c. element
   b. isotope   d. tracer

4. Atoms share electrons unequally in a(n) __c__ bond.
   a. ionic   c. polar covalent
   b. hydrogen   d. nonpolar covalent

5. In a hydrogen bond, a hydrogen atom covalently bonded to one molecule weakly interacts with a __a__ part of a neighboring molecule.
   a. polar   b. nonpolar   c. hydrophobic

6. Liquid water shows __f__ .
   a. polarity   d. cohesion
   b. hydrogen-bonding capacity   e. b through d
   c. notable heat resistance   f. all of the above

7. Hydrogen ions (H$^+$) are __f__ .
   a. the basis of pH values   d. dissolved in blood
   b. unbound protons   e. both a and b
   c. targets of certain buffers   f. a through d

8. When dissolved in water, a(n) __ACID__ donates H$^+$; however, a(n) __BASE__ accepts H$^+$.

9. A(n) __c__ is a dynamic chemical partnership between a weak acid and a weak base.
   a. ionic bond   c. buffer system
   b. solute   d. solvent

10. Match the terms with their most suitable description.
    __e__ trace element   a. atomic nucleus components
    __d__ salt   b. two atoms sharing electrons
    __b__ covalent   c. any polar molecule that readily
    bond   dissolves in water
    __c__ hydrophilic   d. releases ions other than H$^+$ and
    substance   OH$^-$ when dissolved in water
    __a__ protons,   e. makes up less than 0.001
    neutrons   percent of body weight

Critical Thinking

1. By weight, oxygen is the most abundant element in organisms, ocean water, and Earth's crust. Predict which element is the most abundant in the whole universe.

2. *Ozone* is a chemically active form of oxygen gas. High in Earth's atmosphere, a vast layer of it absorbs about 98 percent of the sun's harmful rays. Normal oxygen gas consists of two oxygen atoms joined in a double nonpolar covalent bond: O=O. Ozone has three covalent bonds in this arrangement: O=O—O. It is highly reactive with a variety of substances, and it gives up an oxygen atom and releases gaseous oxygen (O=O). Using what you know about chemistry, explain why you think it is so reactive.

3. Some undiluted acids are less corrosive than when diluted with a little water. In fact, lab workers are told to wipe off splashes with a towel before washing. Explain.

4. Medieval scientists and philosophers called alchemists were the predecessors of modern-day chemists. Many of them tried to transform lead (atomic number 82) into gold (atomic number 79). Explain why they never succeeded.

5. David, an inquisitive three-year-old, poked his fingers into warm water in a metal pan on the stove and didn't sense anything hot. Then he touched the pan itself and got a nasty burn. Explain why water in a metal pan heats up far more slowly than the pan itself.

6. How do many insects, and the basilisk lizard shown in Figure 1.7, walk on water?

7. Why do you think H$^+$ is often written as H$_3$O$^+$?

Media Menu

Student CD-ROM

Impacts, Issues Video
   What Are You Worth?
Big Picture Animation
   Elements, bonding patterns, and pH
Read-Me-First Animation
   How atoms bond
Other Animations and Interactions
   The shell model of electron distribution
   Structure of water
   How salt dissolves
   The pH scale

InfoTrac

- One-Molecule Chemistry Gets Big Reaction. *Science News*, September 2000.
- What's Water Got to Do with It? *Astronomy*, August 2001.
- Walking on Water. *Natural History*, April 2000.

Web Sites

- Web Elements: www.webelements.com/
- Chemistry Review: web.mit.edu/esgbio/www/chem/review.html
- Water Science for Schools: ga.water.usgs.gov/edu/

How Would You Vote?

Fluoride has been proven to prevent tooth decay. However, a high intake of fluoride can discolor teeth, weaken bones, and cause birth defects. Really large amounts can kill you. Many communities in the United States add fluoride to their water supply. Do you want it added to yours?

3 MOLECULES OF LIFE

IMPACTS, ISSUES *Science or the Supernatural?*

About 2,000 years ago, in the mountains of Greece, the oracle of Delphi delivered prophecies from Apollo after inhaling sweet-smelling fumes that had collected in the sunken floor of her temple. Her prophecies tended to be rambling and cryptic. Why? She was babbling in a hydrocarbon-induced trance. Geologists recently found intersecting, earthquake-prone faults under the temple. When the faults slipped, methane, ethane, and ethylene

methane

ethane

ethylene

Figure 3.1 *Left:* In Greece, ruins of the Temple of Apollo, where hydrocarbon gases seep from the earth. *Above:* The oracle at Delphi, believed to dispense cryptic advice direct from Apollo to people who, like her, had no knowledge of chemistry. The invisible, hallucinogenic fumes that induced her babblings are fancifully depicted in this painting.

the big picture

No Carbon, No Life Living cells build carbohydrates, lipids, proteins, and nucleic acids from simpler organic compounds. These large molecules of life have a backbone of carbon atoms with attached functional groups that help dictate their structure and function.

Carbohydrates Carbohydrates are the most abundant biological molecules. Simple types function as quick energy sources or transportable forms of energy. Complex types function as structural materials or energy reservoirs.

seeped out from the depths. All three gases are mild narcotics. The sweet-smelling ethylene can bring on hallucinations (Figure 3.1).

Ancient Greeks thought that Apollo spoke to them through the oracle; they believed in the supernatural. Scientists looked for a natural explanation, and they found carbon compounds behind her words. Why is their explanation more compelling? It started with tested information about the structure and effects of the world's substances, and it was based on analysis of three gaseous substances drawn from the site.

All three gases are nothing more than a few carbon and hydrogen atoms; hence the name, hydrocarbons. Thanks to scientific inquiry, we now know a lot about them. For example, we know that methane was present when Earth first formed. We know it is released when volcanoes erupt, when we burn wood or peat or fossil fuels, and when termites and cattle pass gas. Methane collects in the atmosphere and in ocean depths along the continental shelves. We also know methane is one of the greenhouse gases, which you will read about in Chapter 41, and that it is a contributing factor in global warming.

In short, knowledge about lifeless substances can tell you a lot about life. It will serve you well when you turn your mind to almost any topic concerning the past, present, and future—from Greek myths, to health and disease, to forests, and to physical and chemical conditions that span the globe and affect life everywhere.

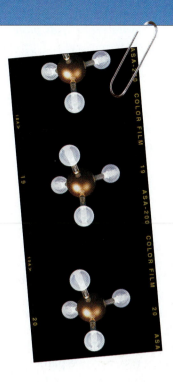

 How Would You Vote?

Undersea methane deposits might be developed as a vast supply of energy, but the environmental costs are unknown. Should we continue to move toward exploiting this resource? See the Media Menu for details, then vote online.

Lipids

Certain lipids function as energy reservoirs, others as structural components of cell membranes, as waterproofing or lubricating substances, and as signaling molecules.

Proteins

Structurally and functionally, proteins are the most diverse molecules of life. They include enzymes, structural materials, signaling molecules, and transporters.

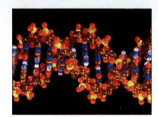

Nucleotides and Nucleic Acids

The nucleic acids DNA and RNA are made of a few kinds of nucleotide subunits. They store, retrieve, and translate genetic information that provides instructions for building proteins.

3.1 Molecules of Life—From Structure to Function

Under present-day conditions on Earth, only living cells can make complex carbohydrates and lipids, proteins, and nucleic acids. These are the molecules of life, and their structure holds clues to how each kind functions.

The molecules of life are **organic compounds**, which contain carbon and at least one hydrogen atom. They have a precise number of atoms arranged in specific ways. **Functional groups** are lone atoms or clusters of atoms that are covalently bonded to carbon atoms of organic compounds.

CARBON'S BONDING BEHAVIOR

Living things consist mainly of oxygen, hydrogen, and carbon. Most of the oxygen and hydrogen are in the form of water. Put water aside, and carbon makes up more than half of what's left.

Carbon's importance in life arises from its versatile bonding behavior. *Each carbon atom can covalently bond with as many as four other atoms.* Such bonds, in which two atoms share one, two, or three pairs of electrons, are relatively stable. They often join carbon atoms into "backbones" to which hydrogen, oxygen, and other elements are attached. The three-dimensional shapes of large organic compounds start with these bonds.

Methane, mentioned in the chapter introduction, is the simplest organic compound. It has four hydrogen atoms bonded covalently to a carbon atom (CH_4).

Figure 3.2*a* is a ball-and-stick model for glucose, an organic compound with hydrogen and oxygen bonded covalently to a backbone of six carbon atoms. Usually this backbone coils back on itself, with two carbons joined to form a ring structure (Figure 3.2*b*).

You can represent a carbon ring in different ways. A flat structural model may show the carbons but not the atoms bonded to it (Figure 3.2*c*).

We use insights into the structure of molecules to explore how cells and multicelled organisms function. For instance, virus particles can infect a cell if they dock at specific protein molecules of a cell membrane. Like Lego® blocks, membrane proteins have ridges, clefts, and charged regions that can match up with ridges, clefts, and charged regions of a protein at the surface of a virus particle.

a

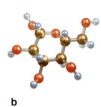

b

c

Figure 3.2 Some ways of representing organic compounds. (**a**) Ball-and-stick model for glucose, linear structure. (**b**) Glucose ring structure. (**c**) Two kinds of simplified six-carbon rings.

FUNCTIONAL GROUPS

Functional groups, again, are either atoms or clusters of atoms that impart distinct properties to a molecule. The number, kind, and arrangement of functional groups influences structural and chemical properties of carbohydrates, lipids, proteins, and nucleic acids (Figure 3.3).

For example, hydrocarbons (organic molecules of only carbon and hydrogen atoms) are hydrophobic, or nonpolar. Fatty acids have chains of them, which is why lipids with fatty acid tails resist dissolving. Sugars, a class of compounds called alcohols, contain one or more polar *hydroxyl* (—OH) groups. Molecules of water quickly form hydrogen bonds with these groups; that is why sugars dissolve fast in water.

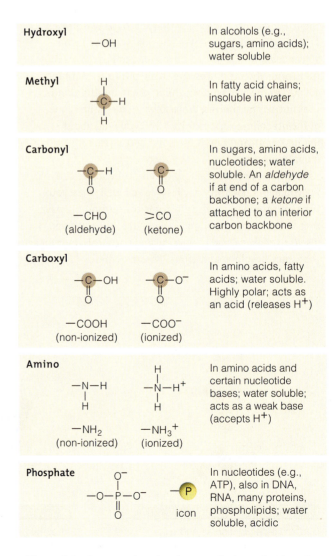

Figure 3.3 Common functional groups in the molecules of life, with examples of their chemical characteristics.

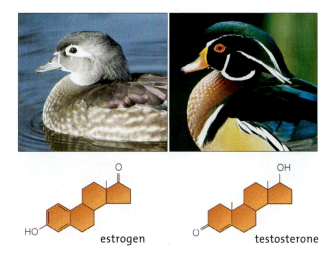

Figure 3.4 Observable differences in traits between female and male wood ducks (*Aix sponsa*), influenced by estrogen and testosterone. These two sex hormones have the same carbon ring structure. They differ only in the position of functional groups attached to the rings.

As another example, testosterone and estrogen are sex hormones. Both are remodeled versions of a type of lipid called cholesterol, and they differ slightly in their functional groups (Figure 3.4). That seemingly tiny difference has big consequences. Consider: Early on, a vertebrate embryo is neither male nor female; it just has a set of tubes that will slowly develop into a reproductive system. In the presence of testosterone, the tubes will become male sex organs, and male traits will develop. In the absence of testosterone, however, the tubes will become female sex organs that secrete estrogens. Those estrogens will guide the formation of distinctly female traits.

HOW DO CELLS ACTUALLY BUILD ORGANIC COMPOUNDS?

Cells build big molecules mainly from four families of small organic compounds: simple sugars, fatty acids, amino acids, and nucleotides. These compounds have two to thirty-six carbon atoms, at most. We refer to them as *monomers* when they are structural units of larger molecules. Molecules that contain repeating monomers are also known as polymers (*mono*–, one; *poly*–, many). As you will see shortly, starch can be considered a polymer of many glucose units.

How do cells actually do the construction work? At this point, just be aware that reactions by which a cell builds, rearranges, and splits apart all organic compounds require more than inputs of energy. They

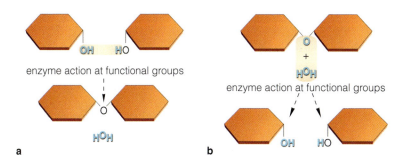

Figure 3.5 Two metabolic reactions with frequent roles in building organic compounds. (**a**) A condensation reaction, with two molecules being joined into a larger one. (**b**) Hydrolysis, a water-requiring cleavage reaction.

also require **enzymes**: proteins that make substances react faster than they would on their own. Different enzymes mediate the following classes of reactions:

1. *Functional-group transfer.* A functional group split away from one molecule is transferred to another.

2. *Electron transfer.* An electron split away from one molecule is donated to another.

3. *Rearrangement.* One type of organic compound is converted to another by a juggling of internal bonds.

4. *Condensation.* Covalent bonding between two small molecules results in a larger molecule.

5. *Cleavage.* A molecule splits into two smaller ones.

To get a sense of these cell activities, think of what happens in a **condensation reaction**. Enzymes split an —OH group from one molecule and an H atom from another. A covalent bond between the discarded parts forms H_2O (Figure 3.5*a*). Polymers such as starch form by repeated condensation reactions.

Another example: **Hydrolysis**, a type of cleavage reaction, is like condensation in reverse (Figure 3.5*b*). Enzymes split molecules at specific groups, then attach one —OH group and a hydrogen atom derived from water to the exposed sites.

Complex carbohydrates, complex lipids, proteins, and nucleic acids are the molecules of life.

Organic compounds have diverse, three-dimensional shapes and functions that arise from their carbon backbone and with functional groups covalently bonded to it.

Enzyme-mediated reactions build the molecules of life mainly from smaller organic compounds—simple sugars, fatty acids, amino acids, and nucleotides.

3.2 Bubble, Bubble, Toil and Trouble

Why include methane, a "lifeless" hydrocarbon, in a chapter on the molecules of life? Consider: Vast methane hydrate deposits beneath the ocean floor could explode at any time, as a colossal methane belch that could actually end life as we know it.

This story started long ago, when organic remains and wastes of countless marine organisms sank to the bottom of the ocean. Over time, more and more sediments accumulated above them. Today, a few kilometers under the seafloor, this organic collection nourishes methane-producing archaea. A tremendous amount of methane, the product of metabolic activity, bubbles upward and emerges into ocean water in places called *methane seeps* (Figures 3.6 and 3.7).

In the mud near seeps, methane pressure is high. There, unrelated microorganisms associate in tight clusters. Archaea inside the clusters use methane as

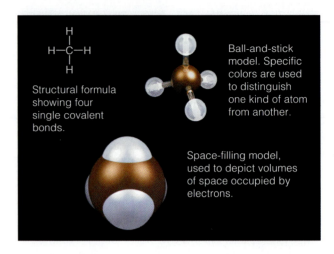

Structural formula showing four single covalent bonds.

Ball-and-stick model. Specific colors are used to distinguish one kind of atom from another.

Space-filling model, used to depict volumes of space occupied by electrons.

Figure 3.6 Molecular models for methane (CH_4). The ball-and-stick model is good for conveying bond angles and the distribution of a molecule's mass. The space-filling model is better for showing a molecule's surfaces.

Figure 3.7 (**a**) "Chimneys" of microorganisms and bubbles of methane gas almost 230 meters (750 feet) below sea level in the Black Sea.

(**b**) The seafloor methane cycle. Methane, formed by archaea far beneath the seafloor, seeps out through vents. Some is captured by clusters of microorganisms that release carbon dioxide and hydrogen sulfide as metabolic products. Other microbes use those products and become the basis of deep-sea food webs.

methane-eating archaea (*red*) and sulfate-eating bacteria (*green*) near seeps

methane-producing archaea

b

methane hydrates

Figure 3.8 A blob of methane hydrate on the seafloor. Notice the methane bubbles above it.

Figure 3.9 *Lystrosaurus*, about a meter long. This animal is now extinct but made it through the Permian mass extinction.

an energy source, and carbon dioxide and hydrogen are released as wastes. In a remarkable metabolic handoff, bacteria that surround them immediately use the hydrogen. During the process they convert sulfate dissolved in seawater to hydrogen sulfide.

The allied organisms accomplish a chemical feat that neither would be capable of on its own. But this doesn't account for all of the methane produced by the underground archaea. What happens to the rest? At methane seeps, high water pressure and low temperature "freezes" the bubbling methane into an icy material called *methane hydrate* (Figure 3.8).

Recently, scientists discovered vast deposits of methane hydrate around the world. They estimate that a thousand billion tons of methane are frozen on the seafloor. The deposits actually are the world's largest reserve of natural gas, but we don't have a safe, efficient way to retrieve it.

Here's the problem. The icy crystals are unstable. Methane hydrate instantly falls apart into methane gas and liquid water as soon as the temperature goes up or the pressure goes down. It doesn't take much, only a few degrees. Methane hydrate disintegration can be explosive. It can cause an irreversible chain reaction that can vaporize neighboring deposits. We have physical evidence of ancient methane hydrate explosions. Small ones pockmarked the ocean floor; immense ones caused underwater landslides that stretched from one continent to another.

The greatest of all mass extinctions occurred 250 million years ago; it marked the end of the Permian period. All but about five percent of marine species abruptly vanished. So did about 70 percent of the known plants, insects, and other species on land.

Chemical clues locked in fossils point to a huge spike in the atmospheric concentration of carbon dioxide—not just any carbon dioxide, but rather carbon dioxide that had been assembled by living things. Something caused lots of methane hydrate to disintegrate at once. In an abrupt, gargantuan burp, millions of tons of methane exploded from the seafloor. Methane-eating bacteria quickly converted most of it to carbon dioxide, which displaced most of the oxygen in the atmosphere and ocean.

Too much carbon dioxide, too little oxygen. Just imagine being instantaneously transported to the top of Mount Everest and trying to jog in the "thin air," with its lower oxygen concentration. You would pass out and die. Before the Permian's Great Dying, oxygen made up about 35 percent of the atmosphere. After the burp, its concentration plummeted to 12 percent. Nearly all of the animals on land and in the seas probably suffocated.

For a long time, scientists couldn't figure out why *Lystrosaurus* didn't become extinct along with nearly everything else. Lucky *Lystrosaurus*. Someone finally figured out that this mammal-like reptile did not suffocate because it was already adapted to the stale, oxygen-poor air of its underground burrows. As Figure 3.9 indicates, *Lystrosaurus* had a big chest cavity, big lungs, stout ribs, and a short route for gas flow inside its stubby nostrils.

The methane problem is closer than you might think. Not long ago, huge methane hydrate deposits were discovered about 96 kilometers (60 miles) off the coast of Newport, Oregon, and off the Atlantic seaboard. What is to become of small-lunged, thin-ribbed people after another methane burp?

3.3 The Truly Abundant Carbohydrates

*Which biological molecules are most plentiful in nature?
Carbohydrates. Most consist of carbon, hydrogen, and
oxygen in a 1:2:1 ratio. Cells use different carbohydrates
as structural materials and transportable or storable
forms of energy. Monosaccharides, oligosaccharides,
and polysaccharides are the main classes.*

THE SIMPLE SUGARS

"Saccharide" is from a Greek word that means sugar.
The *mono*saccharides (one sugar unit) are the simplest
carbohydrates. They have at least two —OH groups
bonded to their carbon backbone and one aldehyde or
ketone group. Most dissolve easily in water. Common
types have a backbone of five or six carbon atoms that
tends to form a ring structure when dissolved.

Ribose and deoxyribose are the sugar monomers
of RNA and DNA, respectively; each has five carbon
atoms. Glucose has six (Figure 3.10*a*). Cells use glucose
as an energy source, a structural unit, or a precursor

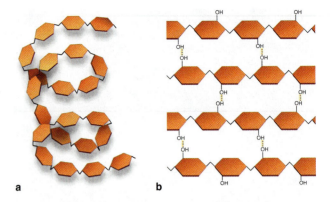

Figure 3.11 Bonding patterns for glucose units in (**a**) starch,
and (**b**) cellulose. In amylose, a form of starch, a series of
covalently bonded glucose units form a chain that coils. In
cellulose, bonds form between glucose chains. The pattern
stabilizes the chains, which can become tightly bundled.

(parent molecule) for other organic compounds, such
as vitamin C (a sugar acid) and glycerol, an alcohol
with three —OH groups.

SHORT-CHAIN CARBOHYDRATES

Unlike the simple sugars, an *oligo*saccharide is a short
chain of covalently bonded sugar monomers. (*Oligo*–
means a few.) The *di*saccharides consist of two sugar
monomers. Sucrose, the most plentiful sugar in nature,
contains a glucose and a fructose unit (Figure 3.10*b*).
Lactose, a disaccharide in milk, has one glucose and
one galactose unit. Table sugar is sucrose extracted
from sugarcane and sugar beets. Many proteins and
lipids have oligosaccharide side chains. Later in the
book, you will come across chains with essential roles
in self-recognition and immunity. You also will see
how such chains are part of receptors that function in
cell-to-cell communication.

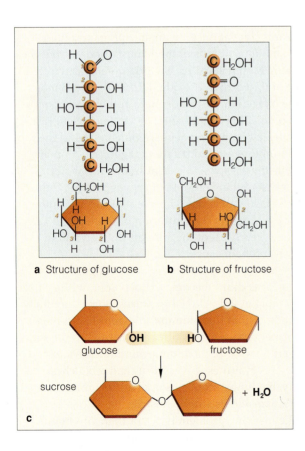

a Structure of glucose

b Structure of fructose

glucose

fructose

sucrose

$+ H_2O$

c

Figure 3.10 (**a**,**b**) Straight-chain and ring forms of glucose and fructose.
For reference purposes, carbon atoms of these simple sugars are numbered
in sequence, starting at the end closest to the molecule's aldehyde or ketone
group. (**c**) Condensation of two monosaccharides into a disaccharide.

COMPLEX CARBOHYDRATES

The "complex" carbohydrates, or *poly*saccharides, are
straight or branched chains of many sugar monomers
—often hundreds or thousands. Different kinds have
one or more types of monomers. The most common
polysaccharides are cellulose, starch, and glycogen.
Even though all three are made of glucose, they differ
a lot in their properties. Why? The answer starts with
differences in covalent bonding patterns between their
glucose units, which are joined together in chains.

In starch, the pattern of covalent bonding puts each
glucose unit at an angle relative to the next unit in
line. The chain ends up coiling like a spiral staircase

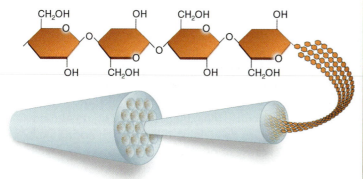

a Structure of amylose, a soluble form of starch. Cells inside tree leaves briefly store excess glucose monomers as starch grains in their chloroplasts, which are tiny, membrane-bound sacs that specialize in photosynthesis.

b Structure of cellulose. In cellulose fibers, chains of glucose units stretch side by side and hydrogen-bond at –OH groups. The many hydrogen bonds stabilize the chains in tight bundles that form long fibers. Few organisms produce enzymes that can digest this insoluble material. Cellulose is a structural component of plants and plant products, such as wood and cotton dresses.

c Glycogen. Animal cells build this polysaccharide as a storage form when the body has excess glucose. It is especially abundant in the liver and muscles of highly active animals, including fishes and people.

Figure 3.12 Molecular structure of starch (**a**), cellulose (**b**), and glycogen (**c**), and their typical locations in a few organisms. All three carbohydrates consist only of glucose units.

(Figure 3.11*a*). Many —OH groups project outward from the coiled chains and make the chains accessible to certain enzymes. This is important. For example, plants briefly store their photosynthetically produced glucose in the form of starch. When free glucose is scarce, enzymes quickly hydrolyze the starch.

In cellulose, glucose chains stretch side by side and hydrogen-bond to one another, as in Figure 3.11*b*. This bonding arrangement stabilizes the chains in a tightly bundled pattern that can resist hydrolysis by most enzymes. Long fibers of cellulose are a structural part of plant cell walls (Figure 3.12*b*). Like steel rods in reinforced concrete, these fibers are tough, insoluble, and resistant to weight loads and mechanical stress, such as strong winds against stems.

In animals, glycogen is the sugar-storage equivalent of starch in plants (Figure 3.12*c*). Muscle and liver cells store a lot of it. When the sugar level in blood falls, liver cells degrade glycogen, and the released glucose enters blood. Exercise strenuously but briefly, and muscle cells tap glycogen for a burst of energy.

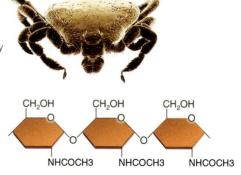

Figure 3.13 (*Right*) A tick's body covering is a protective cuticle reinforced with a polysaccharide called chitin (*below*). You may "hear" chitin when big spider legs clack across an aluminum oil pan on a garage floor.

Chitin has nitrogen-containing groups attached to its glucose units. This polysaccharide derivative strengthens the external skeletons and other hard body parts of many animals, including crabs, earthworms, insects, spiders, and ticks of the sort shown in Figure 3.13. It also strengthens the cell walls of fungi.

The simple sugars (such as glucose), oligosaccharides, and polysaccharides (such as starch) are carbohydrates. Each cell requires carbohydrates as structural materials, and as storable or transportable packets of energy.

3.4 Greasy, Oily—Must Be Lipids

If something is greasy or oily to the touch, you can bet it's a lipid or has lipid parts. Cells use different lipids as energy packages, structural materials, and signaling molecules. Fats, phospholipids, and waxes have fatty acid tails. Sterols have a backbone of four carbon rings.

FATS AND FATTY ACIDS

Lipids are nonpolar hydrocarbons. Although they do not dissolve in water, they mix with other nonpolar substances, as butter does in warm cream sauce.

The lipids called **fats** have one, two, or three fatty acids attached to a glycerol molecule. A **fatty acid** has a backbone of as many as thirty-six carbon atoms, a carboxyl group at one end, and hydrogen atoms at most or all of the remaining carbons (Figure 3.14). They stretch out like flexible tails from the glycerol. *Unsaturated* fatty acids have one or more double bonds. *Saturated* types have single bonds only.

Weak interactions keep many saturated fatty acids tightly packed in animal fats. These fats are solid at room temperature. Most plant fats stay liquid at room temperature, as "vegetable oils." Their packing isn't as stable because of rigid kinks in their fatty acid tails. That's why vegetable oils flow freely.

Neutral fats such as butter, lard, and vegetable oils, are mostly **triglycerides**. Each has three fatty acid tails linked to a glycerol (Figure 3.15). Triglycerides are the most abundant lipids inside your body and its richest reservoir of energy. Gram for gram, they yield more than twice as much energy as complex carbohydrates such as starches. All vertebrates store triglycerides as droplets in fat cells that make up adipose tissue.

Layers or patches of adipose tissue insulate the body and cushion some of its parts. Like many other kinds of

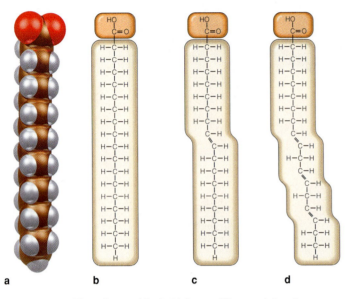

a   **b**   **c**   **d**

Figure 3.14 Three fatty acids. (**a,b**) Space-filling model and structural formula for stearic acid. The carbon backbone is fully saturated with hydrogen atoms. (**c**) Oleic acid, with a double bond in its backbone, is an unsaturated fatty acid. (**d**) Linolenic acid, also unsaturated, has three double bonds.

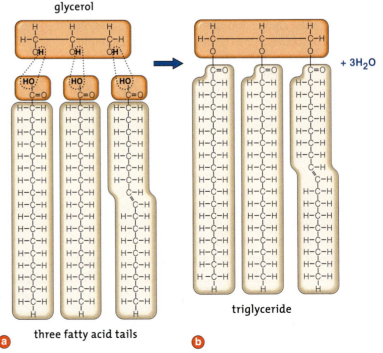

glycerol

three fatty acid tails   triglyceride   + 3H₂O

Figure 3.15 Condensation of (**a**) three fatty acids and one glycerol molecule into (**b**) a triglyceride. The photograph shows triglyceride-protected emperor penguins during an Antarctic blizzard.

a

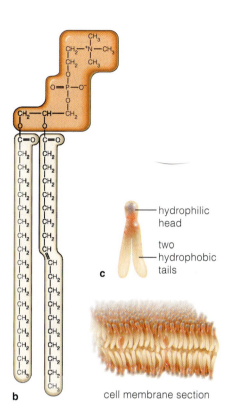

hydrophilic
head

two
hydrophobic
tails

c

cell membrane section

Figure 3.16 (a) Space-filling model, (b) structural formula, and (c) an icon for a phospholipid. This is the most common type in animal and plant cell membranes. Are its two tails saturated or unsaturated?

b

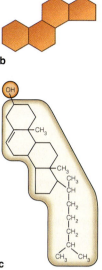

b

a

c

Figure 3.17 (a) Honeycomb: food warehouses and bee nurseries. Bees construct the compartments from their own water-repellent, waxy secretions. (b) Sterol backbone. (c) Structural formula for cholesterol, the main sterol of animal tissues. Your liver makes enough cholesterol for your body. A fat-rich diet may lead to clogged arteries.

animals, penguins of the Antarctic can keep warm in extremely cold winter months thanks to a thick layer of triglycerides beneath their skin (Figure 3.15).

PHOSPHOLIPIDS

Phospholipids have a glycerol backbone, two nonpolar fatty acid tails, and a polar head (Figure 3.16). They are a main component of cell membranes, which have two layers of lipids. The heads of one layer are dissolved in the cell's fluid interior; heads of the other layer are dissolved in the surroundings. Sandwiched between the two are the tails. You will read about membranes in Chapter 4.

WAXES

Waxes have long-chain fatty acids tightly packed and linked to long-chain alcohols or carbon rings. All have a firm consistency; all repel water. Surfaces of plants have a cuticle that contains waxes and another lipid, cutin. A plant cuticle restricts water loss and thwarts some parasites. Waxes also protect, lubricate, and lend pliability to skin and to hair. Birds secrete waxes, fats, and fatty acids that waterproof feathers. Bees use beeswax for honeycomb, which houses each new bee generation as well as honey (Figure 3.17a).

CHOLESTEROL AND OTHER STEROLS

Sterols are among the many lipids with no fatty acids. The sterols differ in the number, position, and type of their functional groups, but all have a rigid backbone of four fused-together carbon rings (Figure 3.17b).

Sterols are components of every eukaryotic cell membrane. The most common type in animal tissues is cholesterol (Figure 3.17c). Cholesterol also becomes remodeled into compounds as diverse as bile salts, steroids, and the vitamin D required for good bones and teeth. Bile salts assist in fat digestion in the small intestine. Sex hormones are vital for the formation of gametes and the development of secondary sexual traits. Such traits include the amount and distribution of hair in mammals, and feather color in birds.

Being largely hydrocarbon, lipids can intermingle with other nonpolar substances, but they resist dissolving in water.

Triglycerides, or neutral fats, have a glycerol head and three fatty acid tails. They are the major energy reservoirs. Phospholipids are a main component of cell membranes.

Sterols such as cholesterol serve as membrane components and precursors of steroid hormones and other compounds. Waxes are firm, yet pliable, components of water-repellent and lubricating substances.

3.5 Proteins—Diversity in Structure and Function

Of all large biological molecules, proteins are the most diverse. Some kinds speed reactions; others are the stuff of spider webs or feathers, bones, hair, and other body parts. Nutritious types abound in seeds and eggs. Many proteins move substances, help cells communicate, or defend against pathogens. Amazingly, cells assemble thousands of different proteins from only twenty kinds of amino acids!

An **amino acid** is a small organic compound with an amino group ($-NH_3^+$), a carboxyl group ($-COO^-$, the acid part), a hydrogen atom, and one or more atoms called its R group. In most cases, these components are attached to the same carbon atom (Figure 3.18a). Biological amino acids are shown in Appendix VI.

When a cell constructs a protein, it strings amino acids together, one after the other. Instructions coded

Read Me First!
and watch the narrated animation on peptide bond formation

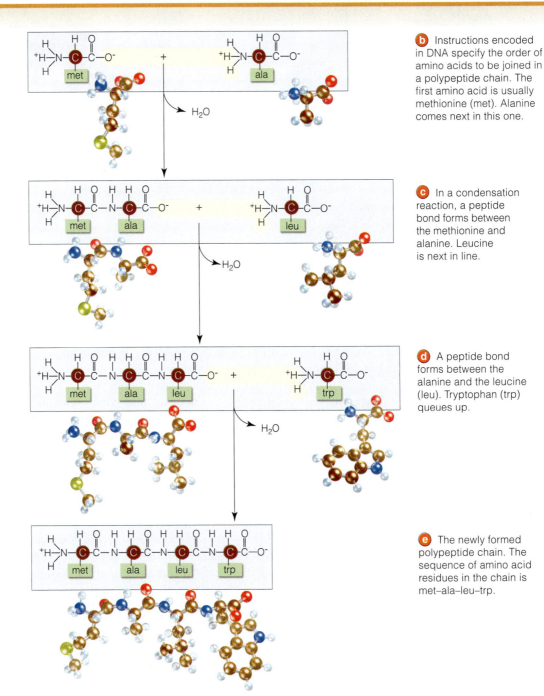

b Instructions encoded in DNA specify the order of amino acids to be joined in a polypeptide chain. The first amino acid is usually methionine (met). Alanine comes next in this one.

c In a condensation reaction, a peptide bond forms between the methionine and alanine. Leucine is next in line.

d A peptide bond forms between the alanine and the leucine (leu). Tryptophan (trp) queues up.

e The newly formed polypeptide chain. The sequence of amino acid residues in the chain is met–ala–leu–trp.

amino group   carboxyl group

R group (20 kinds, each with distinct properties)

a

Figure 3.18 (a) Generalized formula for amino acids. The green box highlights the R group, one of the side chains that include functional groups. (b–e) Peptide bond formation during protein synthesis. Chapter 13 provides a closer look at protein synthesis.

Figure 3.19 Three levels of protein structure. (**a**) Primary structure is a linear sequence of amino acids. (**b**) Many hydrogen bonds (dotted lines) along a polypeptide chain result in a helically coiled or sheetlike secondary structure. (**c**) Coils and sheets packed into stable domains represent a third structural level.

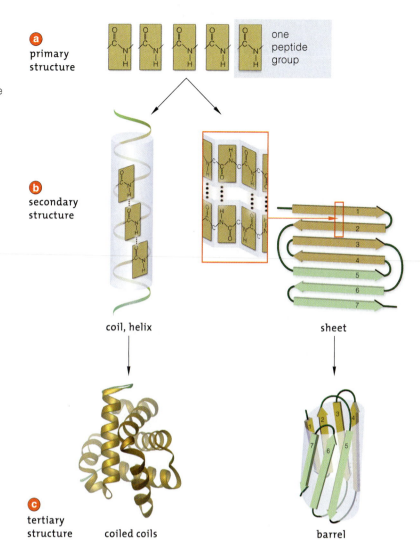

in DNA specify the order in which any of the twenty kinds of amino acids will occur in a given protein. A *peptide* bond forms as a condensation reaction joins the amino group of one amino acid and the carboxyl group of the next in line. Each **polypeptide chain** consists of three or more amino acids. The carbon backbone of this chain incorporates nitrogen atoms in this regular pattern: —N—C—C—N—C—C—.

The sequence of amino acids in a polypeptide chain is its *primary* structure. A new polypeptide chain twists, bends, loops, and folds, which is its *secondary* structure. Hydrogen bonds between R groups make some stretches of amino acids coil into a helical shape, a bit like a spiral staircase; they might make other regions form sheets or loops (Figure 3.19*b*). Bear in mind, the primary structure for each type of protein is unique in some respects, but similar patterns of coils, sheets, and loops do recur among them.

Much as an overly twisted rubber band coils back on itself, the coils, sheets, and loops of a protein fold up even more, into compact domains. A "domain" is a polypeptide chain or part of it that has become organized as a structurally stable unit. This third level of organization, a protein's *tertiary* structure, is what makes the protein a functional molecule. For instance, barrel-shaped domains of some proteins function as subway tunnels through membranes (Figure 3.19*c*).

Many proteins are two or more polypeptide chains that are bonded together or closely associated with one another. This is the fourth level of organization, or *quaternary* protein structure. Many enzymes and other proteins are globular, with multiple polypeptide chains folded into rounded shapes. Hemoglobin, described shortly, is a classic example of such a protein.

Protein structure doesn't stop here. Enzymes often attach short, linear, or branched oligosaccharides to a new polypeptide chain, making a *glyco*protein. Many glycoproteins occur at the cell surface or are secreted from cells. Lipids also get attached to many proteins. The cholesterol, triglycerides, and phospholipids that your body absorbs after a meal are transported about as components of *lipo*proteins.

Many proteins are fibrous, with polypeptide chains organized as strands or sheets. They contribute to cell shape and organization, and help cells and cell parts move about. Other proteins make up cartilage, hair, skin, and parts of muscles and brain cells.

A protein has primary structure, a sequence of amino acids covalently bonded as a polypeptide chain.

Local regions of a polypeptide chain become twisted and folded into helical coils, sheetlike arrays, and loops. These arrangements are the protein's secondary structure.

A polypeptide chain or parts of it become organized as structurally stable, compact, functional domains. Such domains are a protein's tertiary structure.

Many proteins show quaternary structure; they consist of two or more polypeptide chains.

3.6 Why Is Protein Structure So Important?

Cells are good at making proteins that are just what their DNA specifies. But sometimes a protein just turns out wrong. A different amino acid may lead to a misfolded shape that has far-reaching consequences.

JUST ONE WRONG AMINO ACID . . .

Four tightly packed polypeptides called globins make up each hemoglobin molecule. Each globin chain is folded into a pocket that cradles a **heme** group, a large organic molecule with an iron atom at its center (Figure 3.20a). Heme is an oxygen transporter. During its life span, each of the red blood cells in your body transports billions of oxygen molecules, all bound to the heme in globin molecules.

Globin comes in two slightly different forms, alpha and beta. Two of each form make up one hemoglobin molecule in adult humans. Glutamate is normally the sixth amino acid in the beta globin chain, but a DNA mutation sometimes puts a different amino acid—valine—in the sixth position instead (Figure 3.21b). Unlike glutamate, which carries an overall negative charge, valine has no net charge. As a result of that one substitution, a tiny patch of the protein changes from polar to nonpolar, which in turn causes the globin's behavior to change slightly. Hemoglobin with this mutation in its beta chain is designated HbS.

. . . AND YOU GET SICKLE-SHAPED CELLS!

Every human inherits two genes for beta globin, one from each of two parents. (Genes are units of DNA that encode heritable traits.) Cells use both genes when they make beta globin. If one is normal and the other has the valine mutation, a person makes enough normal hemoglobin and can lead a relatively normal life. But someone who inherits two mutant genes can only make the mutant hemoglobin HbS. The outcome is sickle-cell anemia, a severe genetic disorder.

As blood moves through lungs, the hemoglobin in red blood cells binds oxygen, then gives it up in body regions where oxygen levels are low. After oxygen is released, red blood cells quickly return to the lungs and pick up more. In the few moments when they have no bound oxygen, the hemoglobin molecules clump together just a bit. But HbS molecules do not form such clusters in places where oxygen levels are low. They form large, stable, rod-shaped aggregates.

Red blood cells containing these aggregates become distorted into a sickle shape (Figure 3.21c). The sickle cells clog tiny blood vessels called capillaries, which disrupts blood circulation. Tissues become oxygen-starved. Figure 3.21d lists the far-reaching effects of sickle-cell anemia on tissues and organs.

DENATURATION

The shape of a protein defines its biological activity. A globin molecule cradles heme, an enzyme speeds some reaction, a receptor transduces an energy signal. These proteins and others cannot function unless they stay coiled, folded, and packed in a precise way. Their shape depends on many hydrogen bonds and other interactions—which heat, shifts in pH, or detergents can disrupt. At such times, polypeptide chains unwind and change shape in an event called **denaturation**.

Consider albumin, a protein in the white of an egg. When you cook eggs, the heat does not disrupt the covalent bonds of albumin's primary structure. But it destroys albumin's weaker hydrogen bonds, and so the protein unfolds. When the translucent egg white turns opaque, we know albumin has been altered. For a few proteins, denaturation might be reversed if and when normal conditions return, but albumin isn't one of them. There is no way to uncook an egg.

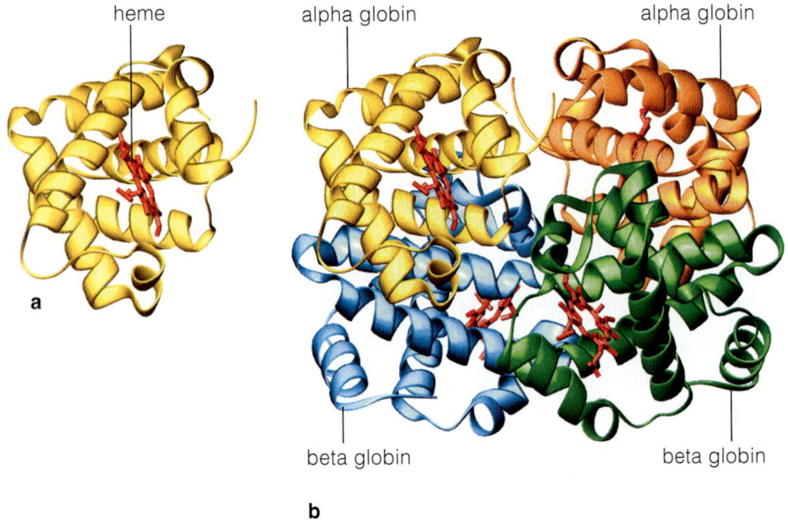

heme | alpha globin | alpha globin

a

beta globin | beta globin

b

Figure 3.20 (a) Globin. This coiled polypeptide chain cradles heme, a functional group that contains an iron atom. (b) Hemoglobin, an oxygen-transport protein in red blood cells. This is one of the proteins with quaternary structure. It consists of four globin molecules (two alphas and two betas) held together by hydrogen bonds.

Read Me First!

and watch the narrated animation on sickle-cell anemia

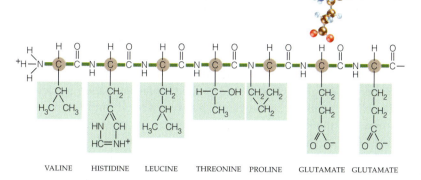

VALINE   HISTIDINE   LEUCINE   THREONINE   PROLINE   GLUTAMATE   GLUTAMATE

a Normal amino acid sequence at the start of a beta chain for hemoglobin.

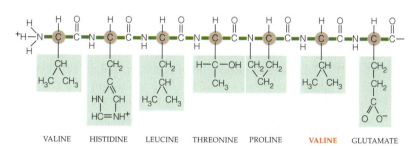

VALINE   HISTIDINE   LEUCINE   THREONINE   PROLINE   **VALINE**   GLUTAMATE

b One amino acid substitution results in the abnormal beta chain in HbS molecules. During protein synthesis, valine was added instead of glutamate at the sixth position of the growing polypeptide chain.

c Glutamate has an overall negative charge; valine has no net charge. This difference gives rise to a water-repellent, sticky patch on HbS molecules. They stick together because of that patch, forming rod-shaped clumps that distort normally rounded red blood cells into sickle shapes. (A sickle is a farm tool that has a crescent-shaped blade.)

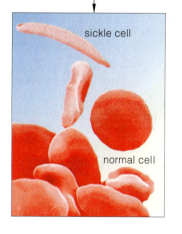

sickle cell

normal cell

Clumping of cells in bloodstream

Circulatory problems, damage to brain, lungs, heart, skeletal muscles, gut, and kidneys

Heart failure, paralysis, pneumonia, rheumatism, gut pain, kidney failure

Spleen concentrates sickle cells

Spleen enlargement

Immune system compromised

Rapid destruction of sickle cells

Anemia, causing weakness, fatigue, impaired development, heart chamber dilation

Impaired brain function, heart failure

d *Above left:* Melba Moore, celebrity spokesperson for sickle-cell anemia organizations. *Above right:* Range of symptoms for a person with two mutated genes (Hb^S) for hemoglobin's beta chain.

Figure 3.21 Sickle-cell anemia's molecular basis and its main symptoms. Sections 16.9 and 33.3 touch upon the evolutionary and ecological aspects of this genetic disorder.

So what is the big take-home lesson? Hemoglobin, hormones, enzymes, transporters—these are the kinds of proteins that help us survive. Twists and folds in their polypeptide chains form anchors, or membrane-spanning barrels, or jaws that can grip enemy agents in the body. Mutation can alter the chains enough to block or enhance an anchoring, transport, or defensive function. Sometimes the consequences are awful. Yet changes in the sequences and functional domains also give rise to variation in traits—the raw material of evolution. *Learn about the structure and function of proteins, and you are on your way to comprehending life in its richly normal and abnormal expressions.*

The structure of proteins dictates function. Mutations that alter a protein's structure can have drastic consequences on its function, and the health of organisms harboring them.

3.7 Nucleotides and the Nucleic Acids

Certain small organic compounds called nucleotides are energy carriers, enzyme helpers, and messengers. Some are the building blocks for DNA and RNA. They are, in short, central to metabolism, survival, and reproduction.

Nucleotides have one sugar, at least one phosphate group, and one nitrogen-containing base. Deoxyribose or ribose is the sugar. Both sugars have a five-carbon ring structure; ribose has an oxygen atom attached to carbon 2 of the ring and deoxyribose does not. The bases have a single or double carbon ring structure.

The nucleotide **ATP** (adenosine triphosphate) has a row of three phosphate groups attached to its sugar (Figure 3.22). ATP can readily transfer the outermost phosphate group to many other molecules and make them reactive. Such transfers are vital for metabolism.

Other nucleotides have different metabolic roles. Some are **coenzymes**, necessary for enzyme function. They move electrons and hydrogen from one reaction site to another. NAD$^+$ and FAD are major kinds.

Still other nucleotides act as chemical messengers within and between cells. Later in the book, you will read about one of these messengers, which is known as cAMP (cyclic adenosine monophosphate).

Certain nucleotides also function as monomers for single- and double-stranded molecules called **nucleic acids**. In such strands, a covalent bond forms between the sugar of one nucleotide and the phosphate group of the next (Figure 3.23). The nucleic acids DNA and RNA store and retrieve heritable information.

All cells start out life and maintain themselves with instructions in their double-stranded molecules of deoxyribonucleic acid, or **DNA**. This nucleic acid is made of four kinds of deoxyribonucleotides. Figure 3.23a shows their structural formulas. As you can see, the four differ only in their component base, which is adenine, guanine, thymine, or cytosine.

Figure 3.24 shows how hydrogen bonds between bases join the two strands along the length of a DNA molecule. Think of every "base pairing" as one rung of a ladder, and the two sugar–phosphate backbones as the ladder's two posts. The ladder twists and turns in a regular pattern, forming a double helical coil.

The sequence of bases in DNA encodes heritable information about all the proteins that give each new cell the potential to grow, maintain itself, and even to reproduce. Part of that sequence is unique for each species. Some parts are identical, or nearly so, among many species. We return to DNA's structure and its function in Chapter 12.

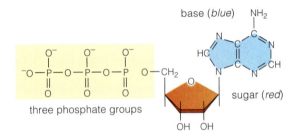

base (*blue*)   NH$_2$

three phosphate groups

sugar (*red*)

Figure 3.22
Structural formula for an ATP molecule.

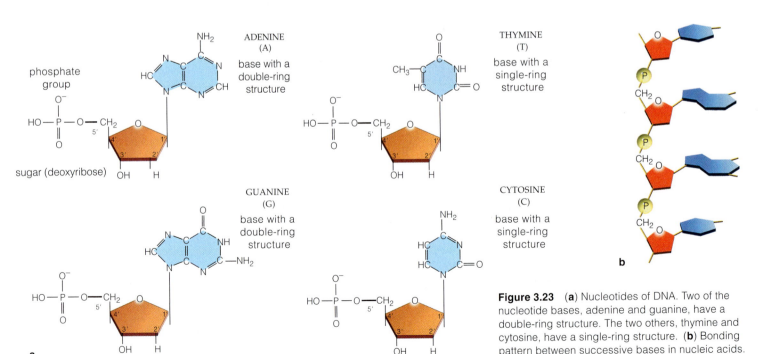

phosphate group

sugar (deoxyribose)

ADENINE (A) base with a double-ring structure

GUANINE (G) base with a double-ring structure

THYMINE (T) base with a single-ring structure

CYTOSINE (C) base with a single-ring structure

a

b

Figure 3.23 (**a**) Nucleotides of DNA. Two of the nucleotide bases, adenine and guanine, have a double-ring structure. The two others, thymine and cytosine, have a single-ring structure. (**b**) Bonding pattern between successive bases in nucleic acids.

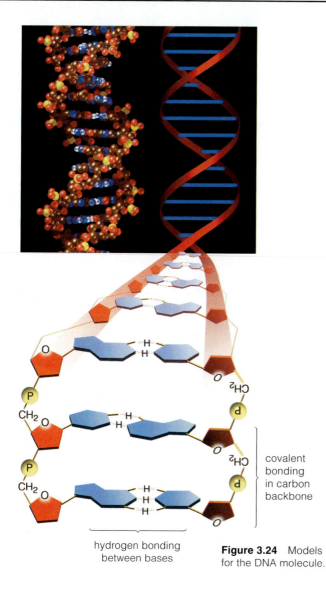

covalent bonding in carbon backbone

hydrogen bonding between bases

Figure 3.24 Models for the DNA molecule.

The **RNAs** (ribonucleic acids) have four kinds of ribonucleotide monomers. Unlike DNA, most RNAs are single strands, and one base is uracil instead of thymine. One type of RNA is a messenger that carries eukaryotic DNA's protein-building instructions out of the nucleus and into the cytoplasm, where they are translated by other RNAs. Chapter 13 returns to RNA and its role in protein synthesis.

Different nucleotides serve as coenzymes, subunits of nucleic acids, energy carriers, and chemical messengers.

The nucleic acid DNA consists of two nucleotide strands joined by hydrogen bonds and twisted as a double helix. Its nucleotide sequence contains heritable instructions about how to build all of a cell's proteins.

RNA is a single-stranded nucleic acid. Different RNAs have roles in the processes by which a cell retrieves and uses genetic information in DNA to build proteins.

Summary

Section 3.1 Organic compounds consist of carbon and at least one hydrogen atom. Carbon atoms bond covalently with up to four other atoms, often forming long chains or rings. Functional groups attached to a carbon backbone contribute to an organic compound's properties. Enzyme-driven reactions synthesize carbohydrates, proteins, lipids, and nucleic acids from smaller organic subunits. Table 3.1 on the next page summarizes these compounds.

Section 3.2 Methane gas produced by archaea far below the seafloor is partially metabolized by diverse microorganisms. Millions of tons of the remaining methane has become frozen into unstable, potentially explosive methane hydrate deposits on the seafloor.

Section 3.3 Carbohydrates include simple sugars, oligosaccharides, and polysaccharides. Living cells use carbohydrates as energy sources, transportable or storage forms of energy, and structural materials.

Section 3.4 Lipids are greasy or oily compounds that tend not to dissolve in water but mix easily with nonpolar compounds, such as other lipids. Neutral fats (triglycerides), phospholipids, waxes, and sterols are lipids. Cells use lipids as major sources of energy and as structural materials.

Section 3.5 Structurally and functionally, proteins are the most diverse molecules of life. Their primary structure is a sequence of amino acids—a polypeptide chain. Such chains twist, coil, and bend into functional domains. Many proteins, including hemoglobin and most enzymes, consist of two or more chains. Certain protein aggregates form hair, muscle, connective tissue, cytoskeleton, and other materials.

Section 3.6 A protein's overall structure determines its function. Sometimes a mutation in DNA results in an amino acid substitution that can drastically alter a protein. Such changes can cause genetic diseases, including sickle-cell anemia. Weak bonds that hold a protein's shape are disrupted by temperature, pH shifts, or exposure to detergent, and usually result in the protein unfolding permanently.

Section 3.7 Nucleotides consist of sugar, phosphate, and a nitrogen-containing base. They have essential roles in metabolism, survival, and reproduction. ATP energizes many kinds of molecules by phosphate-group transfers. Other nucleotides are coenzymes or chemical messengers. DNA and RNA are nucleic acids, each composed of four kinds of nucleotide subunits.

The sequence of nucleotide bases in DNA encodes instructions for how to construct all of a cell's proteins. Different kinds of RNA molecules interact in the translation of DNA's genetic information.

Table 3.1 Summary of the Main Organic Compounds in Living Things

Category	Main Subcategories	Some Examples and Their Functions	
CARBOHYDRATES . . . contain an aldehyde or a ketone group, and one or more hydroxyl groups	**Monosaccharides** (simple sugars)	Glucose	Energy source
	Oligosaccharides (short-chain carbohydrates)	Sucrose (a disaccharide)	Most common form of sugar; the form transported through plants
	Polysaccharides (complex carbohydrates)	Starch, glycogen Cellulose	Energy storage Structural roles
LIPIDS . . . are mainly hydrocarbon; generally do not dissolve in water but do dissolve in nonpolar substances, such as other lipids	**Lipids with fatty acids**		
	Glycerides: Glycerol backbone with one, two, or three fatty acid tails	Fats (e.g., butter), oils (e.g., corn oil)	Energy storage
	Phospholipids: Glycerol backbone, phosphate group, one other polar group, and (often) two fatty acids	Phosphatidylcholine	Key component of cell membranes
	Waxes: Alcohol with long-chain fatty acid tails	Waxes in cutin	Conservation of water in plants
	Lipids with no fatty acids		
	Sterols: Four carbon rings; the number, position, and type of functional groups differ among sterols	Cholesterol	Component of animal cell membranes; precursor of many steroids and vitamin D
PROTEINS . . . are one or more polypeptide chains, each with as many as several thousand covalently linked amino acids	**Fibrous proteins**		
	Long strands or sheets of polypeptide chains; often tough, water-insoluble	Keratin Collagen Enzymes	Structural component of hair, nails Structural component of bone Great increase in rates of reactions
	Globular proteins		
	One or more polypeptide chains folded into globular shapes; many roles in cell activities	Hemoglobin Insulin Antibodies	Oxygen transport Control of glucose metabolism Tissue defense
NUCLEIC ACIDS (AND NUCLEOTIDES) . . . are chains of units (or individual units) that each consist of a five-carbon sugar, phosphate, and a nitrogen-containing base	**Adenosine phosphates**	ATP cAMP (Section 20.8)	Energy carrier Messenger in hormone regulation
	Nucleotide coenzymes	NAD^+, $NADP^+$, FAD	Transfer of electrons, protons (H^+) from one reaction site to another
	Nucleic acids		
	Chains of thousands to millions of nucleotides	DNA, RNAs	Storage, transmission, translation of genetic information

---

Self-Quiz   *Answers in Appendix III*

1. Name the molecules of life and the families of small organic compounds from which they are built.

2. Each carbon atom can share pairs of electrons with as many as _____ other atoms.
   a. one   b. two   c. three   d. four

3. Sugars are a class of _____ , which have one or more _____ groups.
   a. proteins; amino   c. alcohols; hydroxyl
   b. acids; phosphate   d. carbohydrates; carboxyl

4. _____ is a simple sugar (a monosaccharide).
   a. Glucose   c. Ribose   e. both a and b
   b. Sucrose   d. Chitin   f. both a and c

5. The fatty acid tails of unsaturated fats incorporate one or more _____ .
   a. single covalent bonds   b. double covalent bonds

6. Sterols are among the many lipids with no _____ .
   a. saturation   c. phosphates
   b. fatty acids   d. carbons

7. Which of the following is a class of molecules that encompasses all of the other molecules listed?
   a. triglycerides   c. waxes   e. lipids
   b. fatty acids   d. sterols f. phospholipids

8. _____ are to proteins as _____ are to nucleic acids.
   a. Sugars; lipids   c. Amino acids; hydrogen bonds
   b. Sugars; proteins   d. Amino acids; nucleotides

9. A denatured protein has lost its _____ .
   a. hydrogen bonds   c. function
   b. shape   d. all of the above

10. Nucleotides occur in _____ .
    a. ATP   b. DNA   c. RNA   d. all are correct

11. Which of the following nucleotides is *not* found in DNA?
    a. adenine   b. uracil   c. thymine   d. guanine

12. Match the molecule with the most suitable description.
    ____ long sequence of amino acids   a. carbohydrate
    ____ a rechargeable battery in cells   b. phospholipid
    ____ glycerol, fatty acids, phosphate   c. polypeptide
    ____ two strands of nucleotides   d. DNA
    ____ one or more sugar monomers   e. ATP

Critical Thinking

1. In the following list, identify which is the carbohydrate, the fatty acid, the amino acid, and the polypeptide:

   a. $^+NH_3$—CHR—COO$^-$

   c. (glycine)$_{20}$

   b. $C_6H_{12}O_6$

   d. $CH_3(CH_2)_{16}COOH$

2. A clerk in a health-food store tells you that "natural" vitamin C extracts from rose hips are better than synthetic tablets of this vitamin. Given what you know about the structure of organic compounds, what would be your response? How would you design an experiment to test whether a natural and synthetic version of a vitamin differ?

3. It seems there are "good" and "bad" unsaturated fats. The double bonds of both put a bend in their fatty acid tails. But the bend in *trans* fatty acid tails keeps them aligned in the same direction along their whole length. The bend in *cis* fatty acid tails makes them zigzag (Figure 3.25).

   Some *trans* fatty acids occur naturally in beef. But most form by industrial processes that solidify vegetable oils for margarine, shortening, and the like. These substances are widely used in prepared foods (such as cookies) and in french fries and other fast-food products. *Trans* fatty acids are linked to heart attacks. Speculate on why your body might have an easier time dealing with *cis* fatty acids than *trans* fatty acids.

4. The shapes of protein domains often are clues to functions. For example, shown at left is an HLA, a type of recognition protein at the surface of vertebrate body cells. Certain cells of the immune system use HLAs to distinguish self (the body's own cells) from nonself. Each HLA has a jawlike region (*arrow*) that can bind and display nonself fragments, thus alerting the immune system to the presence of an invader or some other threat. Speculate on what may happen if a mutation caused the jawlike region to misfold.

5. Cholesterol from food or synthesized in the liver is too hydrophobic to circulate in blood; complexes of protein and lipids ferry it around. Low density lipoprotein, or *LDL*, transports cholesterol out of the liver and into cells. High density lipoprotein, or *HDL*, ferries the cholesterol that is released from dead cells back to the liver.

   High LDL levels are implicated in atherosclerosis, heart disease, and stroke. The main protein in LDL is called ApoA1, and a mutant form of it has the wrong amino acid (cysteine instead of arginine) at one location in its primary sequence. Carriers of this LDL mutation have very low levels of HDL, which is typically predictive of heart disease, but paradoxically they have no heart problems.

   Some heart patients received injections of the mutant LDL, which acted like a drain cleaner; it quickly reduced the size of cholesterol deposits in the patients' arteries.

   A few years from now, it might be possible to reverse years of damage with such treatment. However, many caution that a low-fat, low-cholesterol diet is still the best preventive measure for long-term health. Would you opt for artery-rooting treatments over a healthy diet?

cis fatty acid   *trans* fatty acid

Figure 3.25 Maybe rethink the french fries?

Media Menu

Student CD-ROM

Impacts, Issues Video
   Science or the Supernatural?
Big Picture Animation
   The chemistry of organic compounds
Read-Me-First Animation
   Peptide bond formation
   Sickle-cell anemia
Other Animations and Interactions
   Condensation and hydrolysis
   Triglyceride formation
   Structure of hemoglobin
   DNA structure

InfoTrac

- Sweet Medicines. *Scientific American*, July 2002.
- The Form Counts: Proteins, Fats, and Carbohydrates. *Consumers' Research Magazine*, August 2001.
- Sorting Fat from Fiction. *Prepared Foods*, October 2002.
- Proteins Rule. *Scientific American*, April 2002.

Web Sites

- General Chemistry Online: antoine.frostburg.edu/chem/senese/101/index.shtml
- The Molecules of Life: biop.ox.ac.uk/www/mol_of_life/index.html
- The Protein Data Bank: www.rcsb.org/pdb

How Would You Vote?

Huge methane reservoirs lie off the east coast of the United States and in arctic regions of North America. The environmental impact of disturbing the reserves is not known but might be significant. Yet using them could lessen our dependence on foreign oil. Should the government encourage use of these reserves for research and development?

IMPACTS, ISSUES *Where Did Cells Come From?*

Do you ever think of yourself as being close to 1/1,000 of a kilometer tall? Probably not. Yet that's how we think of cells. We measure them in micrometers—in millionths of a millimeter, which is a thousandth of a meter, which is a thousandth of a kilometer. The bacteria in Figure 4.1 are a few micrometers "tall."

Somewhere in the distant past, between 3.9 billion and 2.5 billion years ago, cells no bigger than this first appeared on Earth. They were prokaryotic, meaning their DNA was exposed to the rest of their insides. They had no nucleus. At the time, the atmosphere had little free oxygen. The earliest cells probably extracted energy from their food by way of anaerobic reactions, which don't use free oxygen. Plenty of food—simple organic compounds—had already accumulated in the environment through natural geologic processes.

By 2.1 billion years ago, in tidal flats and freshwater habitats, tiny cells were slowly changing the world. Vast populations were making food by a photosynthetic pathway that released oxygen as a by-product. At first, all of that oxygen combined with iron in rocks, forming rust. When oxygen saturated all of the exposed iron deposits on Earth, free oxygen became more and more concentrated in water, then in air.

Oxygen, a reactive gas, attacks organic compounds, including the ones making up cells. The oxygen-enriched

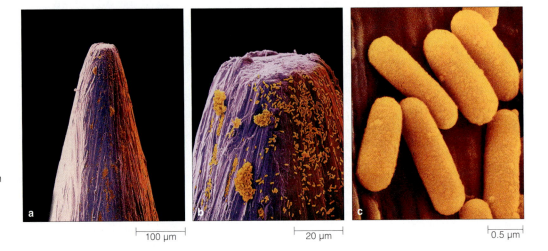

Figure 4.1 How small are cells? Shown here, bacterial cells peppering the tip of a household pin.

a   ├─ 100 µm ─┤

b   ├─ 20 µm ─┤

c   ├ 0.5 µm ┤

the big picture

Basic Cell Features   Nearly all cells are microscopic in size. They all start out life with a plasma membrane, a semifluid interior called cytoplasm, and an inner region of DNA. The plasma membrane helps control the flow of specific substances into and out of cells.

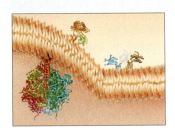

Cell Membrane Features   Two layers of phospholipid molecules are the structural basis of cell membranes. Proteins in this bilayer and those attached to its surfaces carry out many different membrane functions.

atmosphere became a selection pressure of global dimensions. Cell lineages that couldn't neutralize oxygen never left mud and other anaerobic habitats. In other lineages, though, mutations changed metabolic steps in ways that could neutralize oxygen, then *use* it. Aerobic respiration, an oxygen-requiring pathway, had emerged in most groups, and it proved handy in the growing competition for resources.

With so much oxygen around, the free organic compounds that were food for microorganisms became scarce. Compounds made by living cells became the sought-after sources of carbon and energy. Coinciding with this major change in available food sources, novel predators, parasites, and partners evolved. The first eukaryotic cells were among them.

The new cells ran reactions and stored things inside tiny sacs and other compartments made of membranes. One sac, the nucleus, controlled access to their DNA. Others, called mitochondria, yielded far more energy from metabolism than anaerobic reactions, enough to power more active life-styles and build larger, more complex bodies. Without such innovations in small cells, big plants and animals never would have evolved.

This chapter introduces the key defining features of prokaryotic and eukaryotic cells. It invites you to reflect on the earliest, simplest ancestors of you and all other eukaryotic forms of life. Why bother? Science is close to *creating* simple forms of life in test tubes. A bioethical line is about to be crossed.

How Would You Vote?

Researchers are modifying prokaryotes in efforts to make the simplest form of life possible. They are creating "new" organisms by removing genes one at a time. Should this research continue? See the Media Menu for details, then vote online.

Prokaryotic Cells
Compared to eukaryotic cells, prokaryotes have little internal complexity, and no nucleus. However, when taken as a group, they are the most metabolically diverse organisms. All species have prokaryotic ancestors.

Eukaryotic Cells
Eukaryotic cells contain organelles. These membrane-bound compartments divide the cell interior into functional regions for specialized tasks. A major organelle, the nucleus, keeps the DNA away from cytoplasmic machinery.

4.1 So What Is "A Cell"?

Inside your body and at all of its moist surfaces, many trillions of cells live in interdependency. In northern forests, four-celled structures called pollen grains drift down from pine trees. In scummy pond water, free-living single cells called bacteria and amoebas move about. How are these cells alike, and how do they differ?

COMPONENTS OF ALL CELLS

The **cell** is the smallest unit with the properties of life: a capacity for metabolism, controlled responses to the environment, growth, and reproduction. Cells differ in size, shape, and activities, yet are alike in three respects. They start out life with a plasma membrane, a region of DNA, and cytoplasm (Figure 4.2):

1. **Plasma membrane**. A thin, outermost membrane maintains a cell as a distinct entity. It lets metabolic events proceed in controllable ways, separated from the outside environment. Yet this plasma membrane does not isolate the cell interior. It's a bit like a house with many windows and doors that don't open for just anyone. Water, oxygen, and carbon dioxide cross

in and out freely. Other substances, such as nutrients and ions, get escorted across.

2. **Nucleus** or **nucleoid**. DNA occupies a membrane-bound sac (nucleus) inside the cell or, in the simplest kinds of cells, a nucleoid (part of the cytoplasm).

3. **Cytoplasm**. Cytoplasm is everything between the plasma membrane and the region of DNA. It consists of a semifluid matrix and other components, such as **ribosomes**, the structures on which proteins are built.

In structural terms, prokaryotes are the simplest cells; nothing separates their DNA from the cytoplasm. Bacteria and archaea are the only prokaryotes. All other organisms—from amoebas and trees to puffball mushrooms and elephants—are eukaryotic. Internal membranes divide the cytoplasm of eukaryotic cells into functional compartments. One compartment, the nucleus, is the key defining feature of these cells.

WHY AREN'T CELLS BIGGER?

Can any cells be observed with the unaided eye? Just a few types can. They include "yolks" of bird eggs, cells in watermelon tissues, and fish eggs. These get big because they aren't doing much, metabolically speaking, at maturity. Most metabolically active cells are too tiny to be seen by the unaided eye (Figure 4.3).

So why aren't all cells big? A physical relationship called the **surface-to-volume ratio** constrains increases in cell size. By this relationship, an object's volume increases with the cube of its diameter, but the surface area increases only with the square.

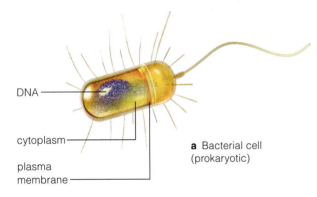

DNA
cytoplasm
plasma membrane

a Bacterial cell (prokaryotic)

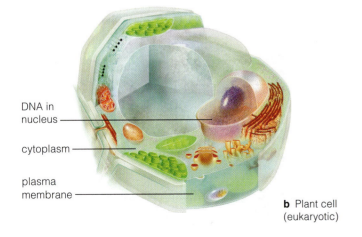

DNA in nucleus
cytoplasm
plasma membrane

b Plant cell (eukaryotic)

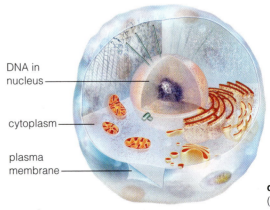

DNA in nucleus
cytoplasm
plasma membrane

c Animal cell (eukaryotic)

Figure 4.2 Overview of the general organization of prokaryotic cells and eukaryotic cells. The three cells are not drawn to the same scale.

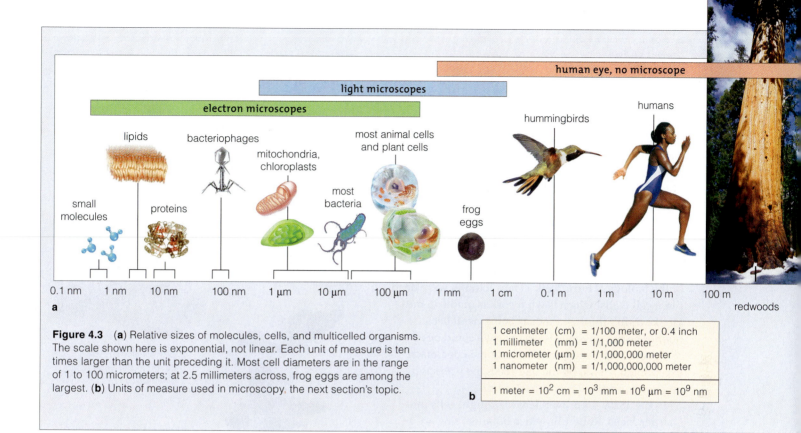

Figure 4.3 (**a**) Relative sizes of molecules, cells, and multicelled organisms. The scale shown here is exponential, not linear. Each unit of measure is ten times larger than the unit preceding it. Most cell diameters are in the range of 1 to 100 micrometers; at 2.5 millimeters across, frog eggs are among the largest. (**b**) Units of measure used in microscopy, the next section's topic.

1 centimeter (cm)	= 1/100 meter, or 0.4 inch
1 millimeter (mm)	= 1/1,000 meter
1 micrometer (µm)	= 1/1,000,000 meter
1 nanometer (nm)	= 1/1,000,000,000 meter

b   1 meter = 10^2 cm = 10^3 mm = 10^6 µm = 10^9 nm

diameter (cm):	0.5	1.0	1.5
surface area (cm²):	0.79	3.14	7.07
volume (cm³):	0.06	0.52	1.77
surface-to-volume ratio:	13.17:1	6.04:1	3.99:1

Figure 4.4 One example of the surface-to-volume ratio. This physical relationship between increases in volume and surface area puts constraints on cell sizes and shapes.

Apply this constraint to a round cell. As Figure 4.4 shows, *if a cell expands in diameter during growth, then its volume will increase faster than its surface area does.* Suppose you induce a round cell to grow four times wider. Its volume increases 64 times (4^3). However, its surface area increases only 16 times (4^2). This means each unit of plasma membrane must service four times as much cytoplasm as before. A lot more substances have to get in and out!

If a cell's diameter is too great, the inward flow of nutrients and the outward flow of wastes just won't be fast enough to keep up with metabolic activity, and you'll end up with a dead cell.

A big, round cell also would have trouble moving materials throughout its cytoplasm. Random motions of molecules distribute materials through tiny cells. If a cell isn't tiny, you can expect it to be long or thin, or have outfoldings or infoldings that increase its surface relative to its volume. *The smaller or narrower or more frilly-surfaced the cell, the more efficiently materials cross its surface and become distributed through the interior.*

Surface-to-volume constraints also shape the body plans of multicelled species. For example, small cells attach end to end in strandlike algae, so each interacts directly with its surroundings. Cells in your muscles are as long as the muscle itself, but each one is thin enough to facilitate diffusion.

All living cells have an outermost plasma membrane, an internal region called cytoplasm, and an internal region where DNA is concentrated.

Bacteria and archaea are prokaryotic cells. Unlike eukaryotic cells, they do not have a nucleus.

As cells grow, their volume increases faster than their surface area. A surface-to-volume ratio is a physical relationship that affects metabolic activity, and thus constrains cell size. It also constrains cell shape and body plans of multicelled organisms.

4.2 How Do We "See" Cells?

Like their centuries-old forerunners, modern microscopes are our best windows on the cellular world.

THE CELL THEORY

Early in the seventeenth century, Galileo Galilei put two glass lenses inside a cylinder and peered at the patterns of an insect's eyes. He was one of the first to record a biological observation with a microscope. The study of the cellular basis of life was about to begin, first in Italy, then in France and England.

At midcentury, Robert Hooke focused a microscope on thinly sliced cork from a mature tree and saw tiny compartments (Figure 4.5). He gave them the Latin name *cellulae*, meaning small rooms—hence the origin of the biological term "cell." They actually were dead plant cell walls, which is what cork is made of, but Hooke didn't think of them as being dead because neither he nor anyone else knew cells could be alive.

Antony van Leeuwenhoek, a shopkeeper, made exceptional lenses. By the late 1600s, he was spying on sperm, protists, even a bacterium.

In the 1820s, improved lenses brought cells into sharper focus. Robert Brown, a botanist, saw an opaque spot in cells and called it a nucleus. Later, the botanist Matthias Schleiden wondered if a plant cell develops as an independent unit even though it's part of the plant. By 1839, after years of studying animal tissues, the zoologist Theodor Schwann reported that cells and cell products make up animals as well as plants—and that cells have an individual life of their own even when they are part of a multicelled species.

Rudolf Virchow, a physiologist, completed his own studies of a cell's growth and reproduction—that is, its division into daughter cells. Every cell, he decided, comes from a cell that already exists.

And so, microscopic analysis yielded three generalizations, which constitute the **cell theory**. *First*, every organism consists of one or more cells. *Second*, the cell is the smallest unit that still displays the properties of life. *Third*, the continuity of life arises directly

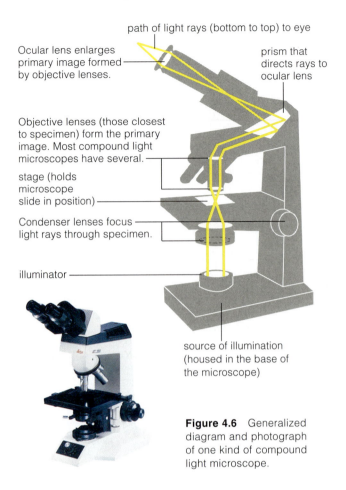

Figure 4.6 Generalized diagram and photograph of one kind of compound light microscope.

path of light rays (bottom to top) to eye

Ocular lens enlarges primary image formed by objective lenses.

prism that directs rays to ocular lens

Objective lenses (those closest to specimen) form the primary image. Most compound light microscopes have several.

stage (holds microscope slide in position)

Condenser lenses focus light rays through specimen.

illuminator

source of illumination (housed in the base of the microscope)

from the growth and division of single cells. Today, microscopy still supports all three insights.

SOME MODERN MICROSCOPES

Like the earlier instruments, many microscopes still use light rays to make images. Picture a series of waves moving across an ocean. Each **wavelength** is the distance from one wave's peak to the peak of the wave behind it. Light also travels in waves as it moves. In a *compound light microscope* (Figure 4.6), two or more sets of glass lenses bend light waves passing through a cell or some other specimen in ways that form an enlarged view of it.

Cells are visible under the microscope when they are thin enough for light to pass through them, but most are nearly colorless and look uniformly dense. Some colored dyes stain cells nonuniformly, and are used to make their components visible.

The best light microscopes can enlarge cells up to 2,000 times. Beyond that, cell parts appear larger but not clearer. Why? Parts that are smaller than one-half

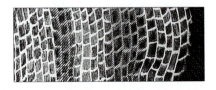

Figure 4.5 Robert Hooke's microscope and his drawing of cell walls from cork tissue.

Figure 4.7   Generalized diagram of an electron microscope.

You can get an idea of the diameter of the lenses from this photograph of a transmission electron microscope (TEM). A beam of electrons from an electron gun moves down the microscope column and is focused by magnets.

In transmission electron microscopy, electrons pass through a thin slice of specimen, then illuminate a fluorescent screen on the monitor. The shadows cast by the specimen's internal details appear, as in Figure 4.8c.

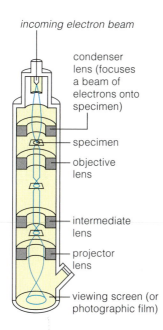

incoming electron beam

condenser lens (focuses a beam of electrons onto specimen)

specimen

objective lens

intermediate lens

projector lens

viewing screen (or photographic film)

of a wavelength of light are too small to make light bend, so they don't show up.

Electron microscopes use magnetic lenses to bend and focus beams of electrons, which can't be focused through a glass lens (Figure 4.7). Electrons travel in wavelengths that are about 100,000 times shorter than those of visible light. That is why an electron microscope can bring into focus objects 100,000 times smaller than you can see with a light microscope.

In a *transmission* electron microscope, electrons pass through a sample and are focused into an image of the specimen's internal details (Figure 4.8c). In *scanning* electron microscopes, a beam of electrons moves back and forth across a specimen that has a thin coating of metal. The metal responds by emitting its own electrons and x-rays, which can be converted into an image of the surface. The images can have fantastic detail. Figure 4.8d shows an example.

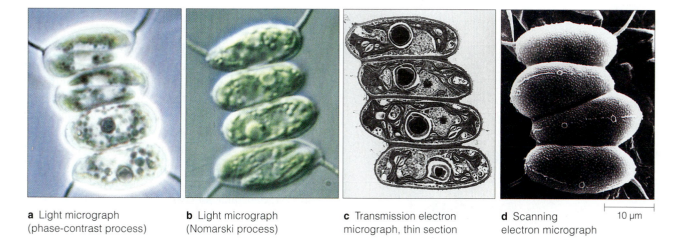

a Light micrograph (phase-contrast process)

b Light micrograph (Nomarski process)

c Transmission electron micrograph, thin section

d Scanning electron micrograph

10 µm

Figure 4.8   How different microscopes reveal different aspects of the same organism—a green alga (*Scenedesmus*). All four images are at the same magnification. (**a**,**b**) Light micrographs. (**c**) Transmission electron micrograph. (**d**) Scanning electron micrograph. A horizontal bar below a micrograph, as in (**d**), provides a visual reference for size. One micrometer (µm) is 1/1,000,000 of 1 meter. Using the scale bar, can you estimate the length and width of a *Scenedesmus* cell?

4.3 All Living Cells Have Membranes

Cell membranes consist of a lipid bilayer in which many different kinds of proteins are embedded. The membrane is a continuous boundary layer across which the flow of substances is selectively controlled.

Think back on the phospholipids, the most abundant components of cell membranes (Section 3.4). Each has one phosphate-containing head and two fatty acid tails attached to a glycerol backbone (Figure 4.9*a*). The head is hydrophilic; it dissolves fast in water. The two tails are hydrophobic; water repels them.

When you immerse many phospholipid molecules in water, they interact with water molecules and with one another until they spontaneously cluster into a sheet or film at the water's surface. Some even line up as two layers, with their fatty acid tails sandwiched between their outward-facing hydrophilic heads. This arrangement, called a **lipid bilayer**, is the structural basis of every cell membrane (Figure 4.9*b,c*).

Figure 4.10 shows the **fluid mosaic model**. By this model, a cell membrane has a mixed composition—a

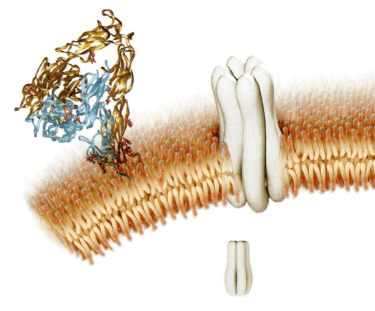

adhesion proteins

Adhesion proteins project outward from plasma membranes of multicelled species especially. They help cells of the same type stick together in the proper tissues.

communication proteins

Communication proteins of two adjoining cells match up, forming a direct channel between their cytoplasms. Signals flow through the channel. This channel is part of a gap junction between two heart muscle cells.

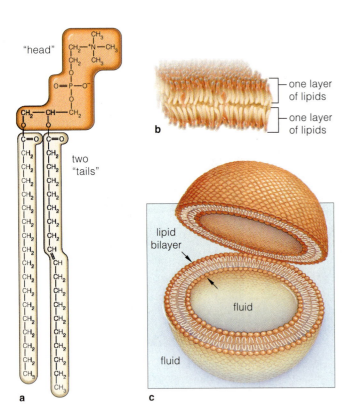

"head"

two "tails"

b one layer of lipids / one layer of lipids

lipid bilayer

fluid

fluid

a **c**

Figure 4.9 Lipid bilayer organization of cell membranes. (**a**) One of the phospholipids, the most abundant membrane components. (**b,c**) These lipids and others are arranged as two layers. Their hydrophobic tails are sandwiched between their hydrophilic heads in the bilayer.

mosaic—of phospholipids, glycolipids, sterols, and proteins. Its phospholipids are diverse, with different kinds of heads and with tails that vary in length and saturation. Unsaturated fatty acids have one or more double covalent bonds in their carbon backbone, and fully saturated fatty acids have none (Section 3.4).

Also by this model, a membrane is *fluid* because of the motions and interactions of its components. Most phospholipids drift sideways, spin on their long axes, and flex their tails, so they don't bunch up as a solid layer. Most membrane phospholipids have at least one kinked (unsaturated) fatty acid tail.

Hydrogen bonds and other weak interactions help proteins associate with the phospholipids. Many of the membrane proteins span the bilayer, with hydrophilic parts extending past both of its surfaces. Others are anchored to underlying cell structures.

Read Me First!
and watch the
narrated animation
on cell membranes

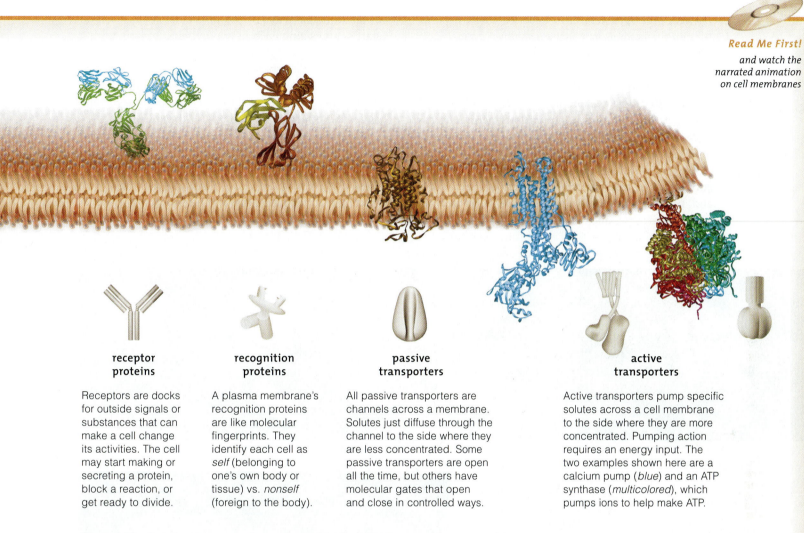

receptor proteins

Receptors are docks for outside signals or substances that can make a cell change its activities. The cell may start making or secreting a protein, block a reaction, or get ready to divide.

recognition proteins

A plasma membrane's recognition proteins are like molecular fingerprints. They identify each cell as *self* (belonging to one's own body or tissue) vs. *nonself* (foreign to the body).

passive transporters

All passive transporters are channels across a membrane. Solutes just diffuse through the channel to the side where they are less concentrated. Some passive transporters are open all the time, but others have molecular gates that open and close in controlled ways.

active transporters

Active transporters pump specific solutes across a cell membrane to the side where they are more concentrated. Pumping action requires an energy input. The two examples shown here are a calcium pump (*blue*) and an ATP synthase (*multicolored*), which pumps ions to help make ATP.

Figure 4.10 Part of a plasma membrane. It consists of a lipid bilayer and far more proteins than we can show here. Later chapters use the icons below these protein ribbon models.

The lipid bilayer functions mainly as a barrier to water-soluble substances. Proteins carry out nearly all other membrane tasks. Many proteins are receptors for signals. Others transport specific solutes across the bilayer. Some transport the solutes passively; others require an energy input. Still other proteins function as enzymes that mediate events at the membrane.

Especially among multicelled species, the plasma membrane bristles with diverse proteins. Recognition proteins identify the cell as belonging to the body, and other kinds help defend it against attacks. Special proteins even help different cells communicate with one another or stick together in tissues. Figure 4.10 introduces important categories of membrane proteins and gives a brief description of their functions.

The fluid mosaic model is a good starting point for thinking about cell membranes. But keep in mind that membranes have different types and arrangements of molecules. Even the two surfaces of the same bilayer are not exactly alike. For example, carbohydrate side chains that are attached to many proteins and lipids project from the cell, not into it. All such differences among plasma membranes and internal membranes correlate with their functions.

All cell membranes consist of two layers of lipids—mainly phospholipids—and diverse proteins. Hydrophobic parts of the lipids are sandwiched between hydrophilic parts, which are dissolved in cytoplasmic fluid or in extracellular fluid.

All cell membranes have protein receptors, transporters, and enzymes. The plasma membrane also incorporates adhesion, communication, and recognition proteins.

4.4 Introducing Prokaryotic Cells

The word prokaryote is taken to mean "before the nucleus." The name reminds us that bacteria and then archaea originated before cells with a nucleus evolved.

Prokaryotes are the smallest known cells. As a group they are the most metabolically diverse forms of life on Earth. Different kinds can exploit energy and raw materials in nearly all environments, from dry deserts to hot springs to mountain ice.

We recognize two domains of prokaryotic cells: the eubacteria, or true bacteria, and archaea (Sections 1.3 and 19.1). Cells of both groups are alike in outward appearance, size, and where they live. Even so, they

differ in major ways at the molecular level. Bacteria start synthesizing each new polypeptide chain with a modified amino acid, formylmethionine. Archaea, like eukaryotes, start chains with methionine. They also have a few proteins called histones that interact with their DNA. Eukaryotic DNA has a great many histone molecules attached; bacterial DNA has none.

Most prokaryotic cells are not much wider than one micrometer; rod-shaped species are at most a few micrometers long (Figures 4.11 and 4.12). Structurally, these are the simplest cells. A semirigid or rigid wall around the plasma membrane helps impart shape to most species. Also, just under the plasma membrane,

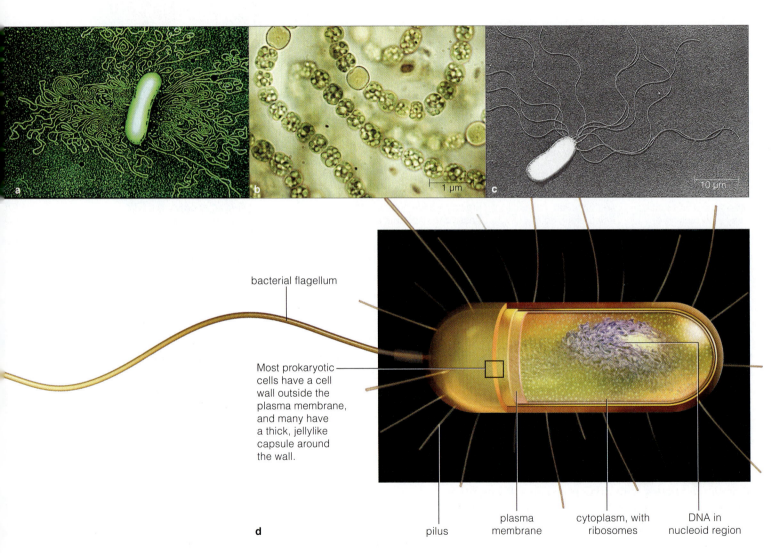

bacterial flagellum

Most prokaryotic cells have a cell wall outside the plasma membrane, and many have a thick, jellylike capsule around the wall.

pilus | plasma membrane | cytoplasm, with ribosomes | DNA in nucleoid region

d

Figure 4.11 (**a**) Micrograph of *Escherichia coli*. Researchers manipulated this bacterial cell to release its single, circular molecule of DNA. (**b**) Cells of various bacterial species are shaped like balls, rods, or corkscrews. Ball-shaped cells of *Nostoc*, a photosynthetic bacteria, stick together in a thick, jellylike sheath of their own secretions. Chapter 19 gives other examples. (**c**) Like this *Pseudomonas marginalis* cell, many species have one or more bacterial flagella that propel the cell body in fluid environments. (**d**) Generalized sketch of a typical prokaryotic body plan.

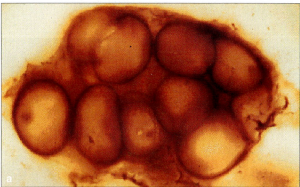

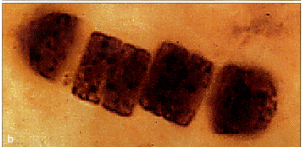

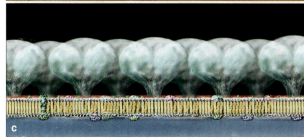

Figure 4.12 From Bitter Springs, Australia, fossilized bacteria dating back to about 850 million years ago, in Precambrian times: (**a**) a colonial form, most likely *Myxococcoides minor*, and (**b**) cells of a filamentous species (*Palaeolyngbya*).

(**c**) One of the structural adaptations seen among archaea. Many of these prokaryotic species live in extremely hostile habitats, such as the ones thought to have prevailed when life originated. Most archaea and some bacteria have a dense lattice of proteins anchored to the outer surface of their plasma membranes. The unique composition of some of these lattices may help cells withstand the insults of extreme environments, such as the near-boiling, mineral-rich water spewing from hydrothermal vents on the ocean floor.

arrays of protein filaments in the cytoplasm compose a simple internal "skeleton," a bit like the cytoskeleton of eukaryotic cells.

Sticky polysaccharides that often envelop bacterial cell walls help them attach to interesting surfaces such as river rocks, teeth, and the vagina. Many disease-causing (pathogenic) bacteria have a thick protective capsule of jellylike polysaccharides around their wall.

All cell walls are permeable to dissolved substances, which are free to move to and away from the plasma membrane. However, eukaryotic cell walls differ in their structure, as you'll see in Section 4.11.

One or more bacterial flagella often project above the cell wall. Bacterial flagella are motile but differ in structure from eukaryotic flagella (Section 4.10); they do not have an orderly, inner array of microtubules. They help cells move about in fluid habitats, including animal body fluids. Other surface projections include pili (singular, pilus). These protein filaments help many kinds of bacterial cells attach to surfaces and to one another, sometimes for transfer of genetic material.

Like eukaryotic cells, bacteria and archaea depend on their plasma membrane to selectively control the flow of substances into and out of the cytoplasm. The lipid bilayer bristles with diverse protein channels, transporters, and receptors. It incorporates built-in machinery for reactions. For example, photosynthesis proceeds at the plasma membrane in many bacterial species. Organized arrays of proteins harness light energy and convert it to chemical energy in the form of ATP, which is used to build sugars.

The cytoplasm holds many ribosomes on which polypeptide chains are built. DNA is concentrated in an irregularly shaped region of cytoplasm called the nucleoid. Prokaryotic cells inherit one molecule of DNA, in the form of a circle. We call it a bacterial chromosome. The cytoplasm of some species also holds

plasmids: far smaller circles of DNA that carry just a few genes. Typically, plasmid genes confer selective advantages, such as antibiotic resistance.

One more intriguing point: In cyanobacteria, part of the plasma membrane projects into the cytoplasm, where it repeatedly folds back on itself. As it happens, pigments and other molecules of photosynthesis are embedded in the membrane, as they are in the inner membrane of chloroplasts. Were ancient cyanobacteria the forerunners of chloroplasts? Section 18.4 looks at this possibility. It is one aspect of a remarkable story about how prokaryotes gave rise to all protists, plants, fungi, and animals.

Bacteria and archaea are different groups of prokaryotic cells; their DNA is not housed inside a nucleus. Most have a permeable cell wall around their plasma membrane that structurally supports and imparts shape to the cell.

These are the simplest cells, but as a group they show the most metabolic diversity. Their metabolic activities proceed at the plasma membrane and within the cytoplasm.

4.5 Introducing Eukaryotic Cells

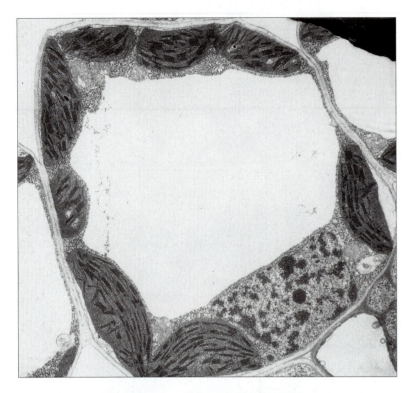

Figure 4.13 Transmission electron micrograph of a plant cell, cross-section. This is a photosynthetic cell from a blade of timothy grass.

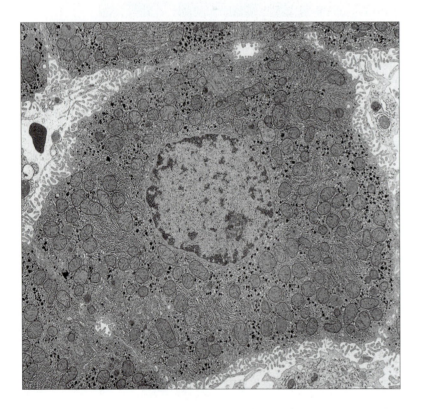

Figure 4.14 Transmission electron micrograph of an animal cell, cross-section. This is a cell from a rat liver.

All cells engage in biosynthesis, dismantling tasks, and energy production, but eukaryotic cells compartmentalize these operations. Their interior is subdivided into a nucleus and other organelles having specialized functions.

Like the prokaryotes, eukaryotic cells have ribosomes in the cytoplasm. Unlike them, eukaryotic cells have an intricate internal skeleton of proteins; we call it a cytoskeleton. They also start out life with **organelles**: internal compartments such as the nucleus. *Eu–* means true; and *karyon*, meaning kernel, refers to a nucleus. Figures 4.13 and 4.14 show two eukaryotic cells.

What advantages do organelles offer? Their outer membrane encloses and sustains a microenvironment. Membrane components selectively control the types and amounts of substances entering or leaving. Their action concentrates substances for metabolic reactions, isolates toxic or disruptive ones, and exports others.

For instance, organelles called mitochondria and chloroplasts concentrate hydrogen ions in ways that lead to the formation of ATP molecules. Enzymes in lysosomes can digest large organic compounds, and they would digest the whole cell if they escaped.

In addition, just as your organ systems interact in controlled ways to keep your whole body running, specialized organelles interact in ways that keep the whole cell running. Ions and molecules move out of one organelle and into another. They move to and from the plasma membrane.

Some substances move through the cytoplasm by a series of organelles. One series functions as a *secretory* pathway. It moves new polypeptide chains from ribosomes through organelles known as ER, then through Golgi bodies, then on to the plasma membrane for release from the cell. Another series is an *endocytic* pathway; it moves substances into the cell. The substances don't travel unescorted; they are enclosed in sacs (vesicles) that have pinched off from organelle membranes or the plasma membrane. Section 4.9 is a visual summary of eukaryotic cell components.

All eukaryotic cells start out life with a nucleus and other organelles, as well as ribosomes and a cytoskeleton. Specialized cells typically incorporate additional kinds of organelles and structures.

Organelles physically separate chemical reactions, many of which are incompatible.

Organelles organize events, as when they assemble, store, or move substances along pathways to and from the plasma membrane or to specific destinations in the cytoplasm.

4.6 The Nucleus

Constructing, operating, and reproducing cells can't be done without carbohydrates, lipids, proteins, and nucleic acids. It takes a class of proteins—enzymes—to build and use these molecules. Said another way, a cell's structure and function start with proteins. And instructions for building proteins are located in DNA.

Unlike prokaryotes, eukaryotic cells have their genetic material distributed among a number of linear DNA molecules of different lengths. The term **chromosome** refers to one double-stranded DNA molecule together with the many histones and other protein molecules attached to it. Each human body cell, for instance, has forty-six chromosomes; frog cells have twenty-six. **Chromatin** is the name for the collection of DNA and proteins in any nucleus of a eukaryotic cell.

The nucleus has two functions. First, it isolates the cell's DNA from potentially damaging reactions in the cytoplasm. Second, it allows or restricts access to DNA's hereditary information through controls over receptors, transport proteins, and pores at its surface. This structural and functional separation makes it far easier to keep DNA molecules organized and also to copy them before a cell divides.

When eukaryotic cells are not dividing, you can't see individual DNA molecules. The nucleus just looks grainy in micrographs, as in Figures 4.14 and 4.15. When cells divide, they duplicate their DNA. During actual division stages, the duplicated DNA molecules become more condensed and compact, like tiny rods. At such times, the DNA no longer looks grainy; each molecule becomes visible in micrographs.

A **nuclear envelope** encloses the semifluid interior of the nucleus (nucleoplasm). It consists of two lipid bilayers studded with proteins. Many of the proteins are organized in complexes that form pores across the envelope (Figure 4.15b). A nucleus also contains at least one nucleolus (plural, nucleoli), a construction site where large and small subunits of ribosomes are assembled from RNA and proteins. The subunits pass through pores and enter the cytoplasm. There, large and small subunits join briefly as intact ribosomes.

The outer envelope of the nucleus keeps DNA molecules separated from the cytoplasmic machinery and thus controls access to a cell's hereditary information.

With this separation, DNA is easier to keep organized and to copy before a parent cell divides into daughter cells.

Pores across the nuclear envelope help control the passage of many substances between the nucleus and cytoplasm.

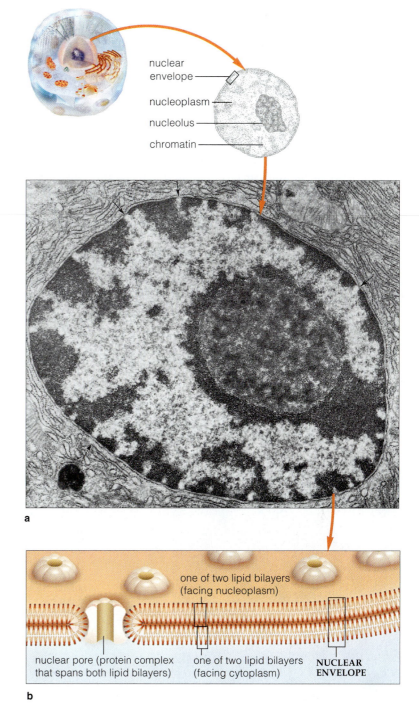

Figure 4.15 (**a**) Pancreatic cell nucleus. Small arrows on the transmission electron micrograph point to pores where control systems selectively restrict or allow passage of specific substances across the nuclear envelope. (**b**) Sketch of part of the nuclear envelope. Each pore is an organized cluster of membrane proteins.

Labels in figure: nuclear envelope, nucleoplasm, nucleolus, chromatin, one of two lipid bilayers (facing nucleoplasm), nuclear pore (protein complex that spans both lipid bilayers), one of two lipid bilayers (facing cytoplasm), NUCLEAR ENVELOPE

4.7 The Endomembrane System

New polypeptide chains become folded into proteins.
Some proteins are stockpiled in the cytoplasm or used
at once. Others enter flattened sacs and tubes of the
endomembrane system: *ER, Golgi bodies, and vesicles.*
All proteins that are destined for export or for insertion
into cell membranes pass through this system.

ENDOPLASMIC RETICULUM

Endoplasmic reticulum, or **ER**, is a channel that starts
at the nuclear envelope and extends through part of
the cytoplasm. Here, polypeptide chains are processed
into final proteins, and lipids are assembled. Vesicles
deliver many proteins and lipids to Golgi bodies.

Rough ER consists of flattened sacs and tubes with
ribosomes attached to their outer surface, as in Figure
4.16c. Newly forming polypeptide chains enter it or
become inserted into its membrane. They can do so
only if they contain a built-in signal (a special sequence

of fifteen to twenty amino acids). Enzymes in the
channel often modify polypeptide chains into final
form. You'll see a lot of rough ER in cells that make,
store, and secrete proteins. Example: ER-rich gland
cells in your pancreas make and secrete enzymes that
end up in your small intestine and help digest meals.

Smooth ER is ribosome-free (Figure 4.16d). It makes
lipid molecules that become part of cell membranes.
The ER also takes part in fatty acid breakdown and
degrades some toxins. Sarcoplasmic reticulum, a type
of smooth ER, functions in muscle contraction.

GOLGI BODIES

Patches of ER membrane bulge and break away as
vesicles, each with proteins inside or incorporated in
its membrane. Many vesicles fuse with **Golgi bodies**.
These organelles are folded into flattened, membrane-
bound sacs (Figure 4.16e). Golgi bodies attach sugar

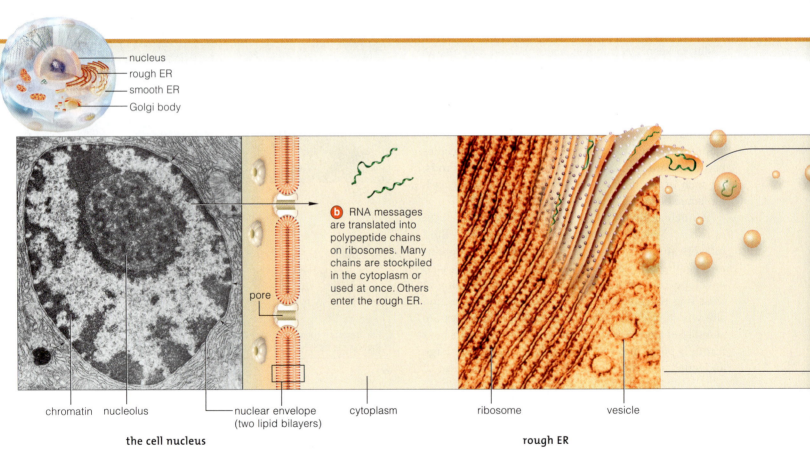

nucleus
rough ER
smooth ER
Golgi body

b RNA messages are translated into polypeptide chains on ribosomes. Many chains are stockpiled in the cytoplasm or used at once. Others enter the rough ER.

pore

chromatin   nucleolus   nuclear envelope (two lipid bilayers)   cytoplasm   ribosome   vesicle

the cell nucleus   **rough ER**

a DNA instructions for making proteins are transcribed in the nucleus and moved to the cytoplasm. RNAs are the messengers and protein builders.

c Flattened sacs of rough ER form one continuous channel between the nucleus and smooth ER. Polypeptide chains that enter the channel undergo modification. They will be inserted into organelle membranes or will be secreted from the cell.

Figure 4.16 Endomembrane system. Here, many proteins are processed, lipids are assembled, and both products are sorted and shipped to cellular destinations or to the plasma membrane for export.

side chains to ER proteins and lipids. They also cleave some proteins. The finished products are packaged in vesicles and then shipped to lysosomes, to the plasma membrane, or to the outside of the cell.

MEMBRANOUS SACS WITH DIVERSE FUNCTIONS

Vesicles help organize metabolic activities. Different kinds bud from ER, Golgi, or plasma membranes. For instance, the **lysosomes** that bud from Golgi bodies contain enzymes that digest carbohydrates, proteins, nucleic acids, and lipids. Different vesicles transport proteins and other substances between organelles or to the outer membrane where they are expelled.

The vesicles called **peroxisomes** hold enzymes that digest fatty acids and amino acids. An important function is the breakdown of hydrogen peroxide, a toxic product of metabolism. Enzyme action converts

hydrogen peroxide to water and oxygen or uses it in reactions that break down alcohol and other toxins. Drink alcohol, and peroxisomes of liver and kidney cells normally will degrade nearly half of it.

Some vesicles fuse and form large vacuoles, such as the **central vacuole** of mature plant cells. Ions, amino acids, sugars, and toxic substances accumulate in the fluid-filled interior of a central vacuole, which expands and forces the pliable cell wall to enlarge. One benefit is an increase in cell surface area.

> *Endoplasmic reticulum is a membrane-bound channel where polypeptide chains are processed and lipids are assembled. Golgi bodies further modify many of the proteins and lipids.*
>
> *Vesicles help integrate cell activities. Different kinds transport substances around the cell, and break down nutrients and toxins. In plant cells, the central vacuole functions in storage and in increasing the cell surface area.*

Read Me First!
and watch the narrated animation on the endomembrane system

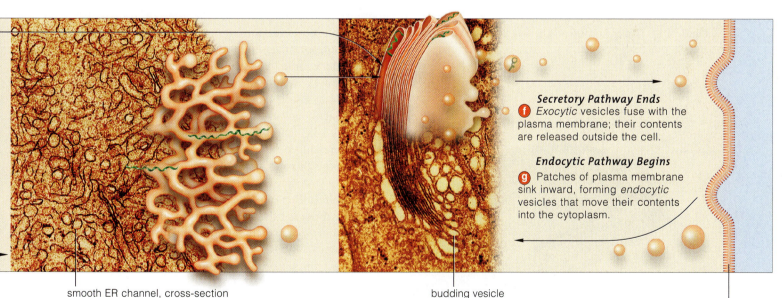

Secretory Pathway Ends
f *Exocytic* vesicles fuse with the plasma membrane; their contents are released outside the cell.

Endocytic Pathway Begins
g Patches of plasma membrane sink inward, forming *endocytic* vesicles that move their contents into the cytoplasm.

smooth ER channel, cross-section

budding vesicle

smooth ER

d Some proteins in the channel continue on to smooth ER, becoming membrane proteins or smooth ER enzymes. Some of these enzymes make lipids and inactive toxins.

Golgi body

e A Golgi body receives, processes, and repackages substances that arrive in vesicles from the ER. Different vesicles transport the substances to other parts of the cell.

plasma membrane

h Exocytic vesicles release cell products and wastes to the outside. Endocytic vesicles move nutrients, water, and other substances into the cytoplasm from outside (Section 5.5).

4.8 Mitochondria and Chloroplasts

ATP, recall, is the energy carrier that jump-starts most of the reactions in cells. So how do cells get ATP in the first place? All of them make ATP by aerobic respiration, which is completed in mitochondria. Algae and photosynthetic plant cells also contain ATP-making organelles called chloroplasts.

MITOCHONDRIA

ATP-forming reactions of aerobic respiration end in the **mitochondrion** (plural, mitochondria). Compared to prokaryotic cells, these organelles make far more ATP from the same compounds. A much-folded inner membrane divides the interior of the mitochondrion into two compartments (Figure 4.17). Hydrogen ions released from the breakdown of organic compounds accumulate in the inner compartment by operation of transport systems, of the sort described in Sections 5.6 and 6.3. As they flow back to the outer compartment, they drive the formation of ATP. Oxygen accepts the spent electrons and keeps the reactions going. Each time you breathe in, you are securing oxygen mainly for the mitochondria in your many trillions of cells.

All eukaryotic cells have one or more mitochondria. The cells that require a great deal of energy, such as those of your liver, heart, and skeletal muscles, may each contain a thousand or more.

CHLOROPLASTS

Many plant cells have plastids, which are organelles of photosynthesis, storage, or both. **Chloroplasts** are important plastids, and only photosynthetic eukaryotic cells have them. In these organelles, energy from the sun drives the formation of ATP and NADPH, which are then used in the formation of organic compounds.

A chloroplast has two outer membranes around a semifluid interior, the stroma, which bathes an inner membrane. Often this single membrane is folded back on itself as a series of stacked, flattened disks (Figure 4.18). Each stack is called a thylakoid. Embedded in the thylakoid membrane are light-trapping pigments, including chlorophylls, as well as enzymes and other proteins with roles in photosynthesis. Glucose, then sucrose, starch, and other organic compounds are built from carbon dioxide and water in the stroma.

Both mitochondria and chloroplasts are a lot like bacteria in size, structure, and biochemistry. Both have their own DNA, RNA, and ribosomes. Coincidence? Probably not, according to the theories discussed in Section 18.4.

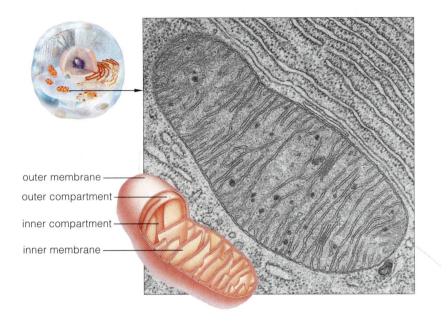

outer membrane
outer compartment
inner compartment
inner membrane

Figure 4.17 Typical mitochondrion. This organelle specializes in producing ATP.

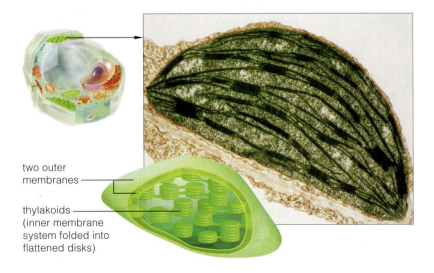

two outer membranes

thylakoids (inner membrane system folded into flattened disks)

Figure 4.18 Typical chloroplast, the key defining feature of photosynthetic eukaryotic cells.

Reactions that release energy from organic compounds occur at the compartmented, internal membrane of mitochondria. The reactions, which require oxygen, produce far more ATP than can be produced by any other cellular reaction.

Photosynthetic eukaryotic cells contain chloroplasts, which specialize in making sugars and other carbohydrates.

4.9 Visual Summary of Eukaryotic Cell Components

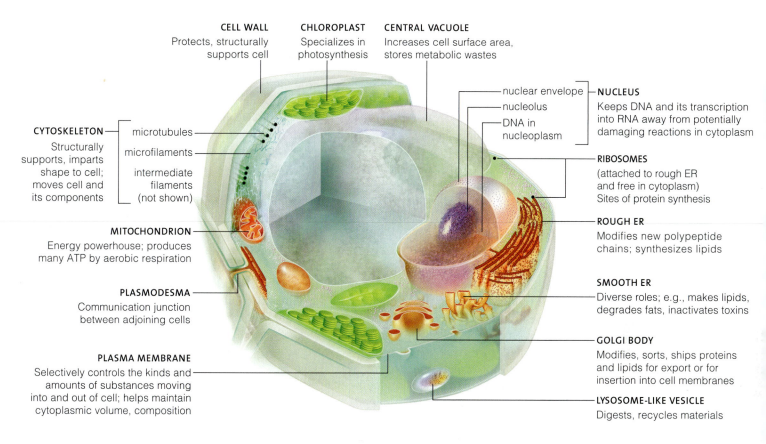

CELL WALL
Protects, structurally supports cell

CHLOROPLAST
Specializes in photosynthesis

CENTRAL VACUOLE
Increases cell surface area, stores metabolic wastes

nuclear envelope
nucleolus
DNA in nucleoplasm

NUCLEUS
Keeps DNA and its transcription into RNA away from potentially damaging reactions in cytoplasm

CYTOSKELETON
Structurally supports, imparts shape to cell; moves cell and its components
microtubules
microfilaments
intermediate filaments (not shown)

RIBOSOMES
(attached to rough ER and free in cytoplasm) Sites of protein synthesis

ROUGH ER
Modifies new polypeptide chains; synthesizes lipids

MITOCHONDRION
Energy powerhouse; produces many ATP by aerobic respiration

PLASMODESMA
Communication junction between adjoining cells

SMOOTH ER
Diverse roles; e.g., makes lipids, degrades fats, inactivates toxins

GOLGI BODY
Modifies, sorts, ships proteins and lipids for export or for insertion into cell membranes

PLASMA MEMBRANE
Selectively controls the kinds and amounts of substances moving into and out of cell; helps maintain cytoplasmic volume, composition

LYSOSOME-LIKE VESICLE
Digests, recycles materials

Figure 4.19 Typical organelles and structures of plant cells.

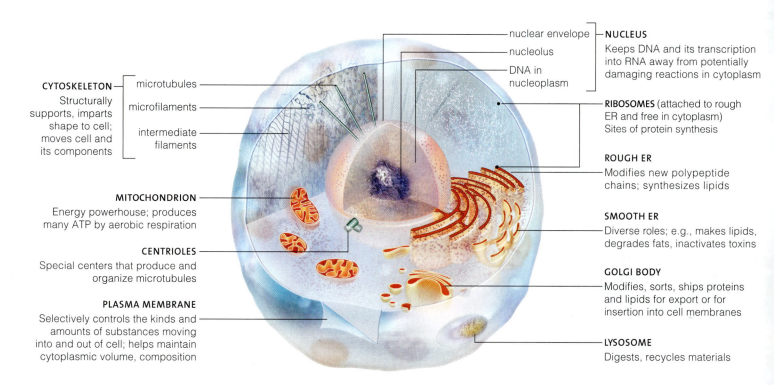

CYTOSKELETON
Structurally supports, imparts shape to cell; moves cell and its components
microtubules
microfilaments
intermediate filaments

nuclear envelope
nucleolus
DNA in nucleoplasm

NUCLEUS
Keeps DNA and its transcription into RNA away from potentially damaging reactions in cytoplasm

RIBOSOMES (attached to rough ER and free in cytoplasm) Sites of protein synthesis

ROUGH ER
Modifies new polypeptide chains; synthesizes lipids

MITOCHONDRION
Energy powerhouse; produces many ATP by aerobic respiration

CENTRIOLES
Special centers that produce and organize microtubules

SMOOTH ER
Diverse roles; e.g., makes lipids, degrades fats, inactivates toxins

GOLGI BODY
Modifies, sorts, ships proteins and lipids for export or for insertion into cell membranes

PLASMA MEMBRANE
Selectively controls the kinds and amounts of substances moving into and out of cell; helps maintain cytoplasmic volume, composition

LYSOSOME
Digests, recycles materials

Figure 4.20 Typical organelles and structures of animal cells.

4.10 The Cytoskeleton

Like you, all eukaryotic cells have an internal structural framework—a skeleton. Unlike your skeleton, theirs has elements that are not permanently rigid; they assemble and disassemble at different times.

In between the nucleus and plasma membrane of all eukaryotic cells is a **cytoskeleton**—an interconnected system of many protein filaments. Different parts of the system reinforce, organize, and move structures, and often the whole cell. Many parts are permanent; others form at certain times in a cell's life. Figure 4.21 shows an example from one kind of animal cell. The two major cytoskeletal elements—microtubules and microfilaments—have diverse functions. Another type, intermediate filaments, strengthens some animal cells.

Microtubules are long, hollow cylinders of tubulin subunits (Figure 4.21a). They organize the cell interior and form a dynamic framework that moves structures such as chromosomes to specific locations. Controls govern which microtubules grow or fall apart at any given time. Those growing in a specific direction— say, the forward end of a prowling amoeba—might get a protein cap that keeps them intact. Those at the trailing end aren't used, aren't capped, and fall apart.

Colchicine, made by the autumn crocus (*Colchicum autumnale*), is a poison. It blocks microtubule assembly, so the cells of animals that eat the plant can't divide. Western yews (*Taxus brevifolia*) make taxol, another microtubule poison. Taxol can stop the uncontrolled cell divisions that give rise to some kinds of cancers.

Microfilaments consist of two coiled-up polypeptide chains of actin monomers, as in Figure 4.21b. They often reinforce cell shape or cause it to change. For example, crosslinked, bundled, and gel-like arrays of microfilaments make up a reinforcing **cell cortex** that underlies the plasma membrane. Also, microfilaments anchor proteins and assist in muscle contraction. An animal cell divides as microfilaments around its midsection contract, pinching the cell in two. Although prokaryotic cells lack a cytoskeleton, some types do have microfilament-like proteins that reinforce the cell body.

The microtubules and microfilaments found in all eukaryotic cells are similar. How can they do so many different things if they are so uniform? Other proteins assist them. Among these *accessory* proteins are **motor proteins**, which move cell parts along microtubules when repeatedly energized by ATP.

Intermediate filaments are the most stable parts of some cytoskeletons (Figure 4.21c). They strengthen and help maintain cell structures. One type, the lamins, anchor actin and myosin of contractile units found in muscle cells. Other types anchor cells in tissues.

One or at most two kinds of intermediate filaments occur in certain animal cells. Researchers use them to identify the type of cell. *Cell typing* is a useful tool in diagnosing the tissue origin of diverse cancers.

MOVING ALONG WITH MOTOR PROTEINS

Think about the bustle at a train station during the busiest holiday season, and you get an idea of what goes on in cells. Microtubules and microfilaments are the cell's tracks. Kinesins, dyneins, myosins, and other motor proteins are the freight engines (Figure 4.22). ATP is the fuel for movement.

Some motor proteins move chromosomes. Others slide one microtubule over another, or chug along tracks inside nerve cells that extend from your spine to your toes. Many engines are organized in series, each moving some vesicle partway along the track before giving it up to the next in line. Kinesins in

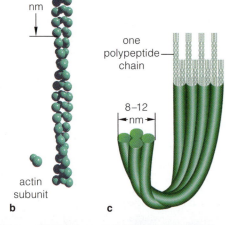

one polypeptide chain

8–12 nm

5–7 nm

actin subunit

a

b

c

Figure 4.21 Subunits and structure of (**a**) microtubules, (**b**) microfilaments, and (**c**) one kind of intermediate filament. The micrograph at left shows intermediate filaments (*red*) of cultured kangaroo rat cells. The *blue*-stained organelle inside each cell is the nucleus.

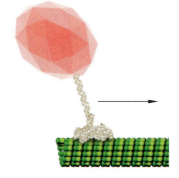

Figure 4.22 A motor protein, kinesin, on a microtubule. Kinesin scoots along the length of the microtubule in a hand-over-hand motion. If the microtubule is anchored near the cell's center, the kinesin moves its freight away from the center.

Figure 4.23 (**a**) Internal organization of flagella and cilia. Inside both motile structures is a 9+2 array: a ring of nine pairs of microtubules around one pair at the core. All are connected by spokes and linking elements that restrict the range of sliding. (**b**) Cilia (*gold*) on cells lining an airway that leads to human lungs.

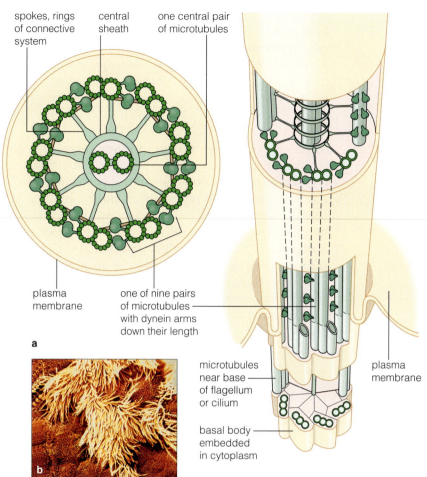

spokes, rings of connective system

central sheath

one central pair of microtubules

plasma membrane

one of nine pairs of microtubules with dynein arms down their length

microtubules near base of flagellum or cilium

plasma membrane

basal body embedded in cytoplasm

a

b

plant cells drag chloroplasts to new positions that are more efficient for light interception as the angle of the sun changes overhead.

Different myosins can move structures along microfilaments or slide one microfilament over another. For example, muscle cells contain long fibers divided into contractile units. Each unit has side-by-side arrays of microfilaments and myosin filaments. When ATP activates it, myosin slides all microfilaments in directions that shorten each unit. When all of the units shorten, the cell itself shortens; it contracts.

CILIA, FLAGELLA, AND FALSE FEET

Besides moving internal parts, many cells move their body or extend parts of it. First, consider **flagella** (singular, flagellum) and **cilia** (singular, cilium). Both are motile structures that project from the cell surface.

Eukaryotic flagella usually are longer and not as profuse as cilia. Many eukaryotic cells swim with the help of whiplike flagella. Sperm do this. The ciliated protozoans swim by beating many cilia in synchrony. In the airways to your lungs, cilia beat nonstop; their coordinated movement sweeps out airborne bacteria and particles that otherwise might reach the lungs (Figure 4.23*b*).

Inside these motile structures is a ring of nine pairs of microtubules around a central pair. Protein spokes and links stabilize the *9+2 array*, which starts at a **centriole** (Figure 4.23*a*). This barrel-shaped structure produces and organizes microtubules, then it remains positioned below the finished array as a **basal body**.

Flagella and cilia move by a sliding mechanism. All pairs of microtubules extend the same distance into the motile structure's tip. Stubby dynein arms project from each pair in the outer ring. When ATP energizes them, the arms grab the microtubule pair in front of them, tilt in a short, downward stroke, then let go. As the bound pair slides down, its arms bind the pair in front of it, forcing it to slide down also—and so on around the ring. The microtubules can't slide too far, but each *bends* a bit. Their sliding motion is converted to a bending motion.

As a final example, some free-living cells, such as macrophages and amoebas, form **pseudopods** ("false feet"). These temporary, irregular lobes project from the cell and function in locomotion and prey capture. Pseudopods move as microfilaments elongate inside them. Motor proteins attached to the microfilaments drag the plasma membrane with them.

A cytoskeleton of protein filaments is the basis of eukaryotic cell shape, internal structure, and movement. Accessory proteins extend the range of functions for those filaments.

Microtubules move cell components. Microfilaments form flexible, linear bundles and networks that reinforce and restructure the cell surface. Intermediate filaments strengthen and maintain shapes of some animal cells.

Cell contractions and migrations, chromosome movements, and other forms of cell movements arise at organized arrays of microtubules, microfilaments, and accessory proteins.

When energized by ATP, motor proteins move in specific directions, along tracks of microtubules and microfilaments. They deliver cell components to new locations.

4.11 Cell Surface Specializations

Our survey of eukaryotic cells concludes with a look at cell walls and other specialized surface structures. Many of these architectural marvels are made primarily of cell secretions. Others are clusters of membrane proteins that connect neighboring cells, structurally and functionally.

EUKARYOTIC CELL WALLS

Single-celled eukaryotic species are directly exposed to the environment. Many have a **cell wall**, a structural component that encloses the plasma membrane. A cell wall protects and physically supports a cell. It's porous, so water and solutes easily move to and from the plasma membrane. A cell would die without these exchanges. Many single-celled protists have a wall around their plasma membranes. So do plant cells and many types of fungal cells.

For example, in the growing parts of multicelled plants, young cells secrete molecules of pectin and other glue-like polysaccharides, as well as cellulose. The cellulose molecules are laid down in the gluey matrix as ropelike strands. All of these materials are components of the plant cell's **primary wall** (Figure 4.24). The sticky primary wall cements abutting cells together. Being thin and pliable, it permits the cell to enlarge under the pressure of incoming water.

Cells that have only a thin primary wall retain the capacity to divide or change shape as they grow and develop. Many types stop enlarging when they are mature. Such cells secrete material on the primary wall's inner surface. These deposits form a lignified, rigid **secondary wall** that reinforces cell shape (Figure 4.24*d*). Secondary wall deposits are extensive and contribute more to structural support.

In woody plants, up to 25 percent of the secondary wall is made of lignin. This organic compound makes plant parts more waterproof, less susceptible to plant-attacking organisms, and stronger.

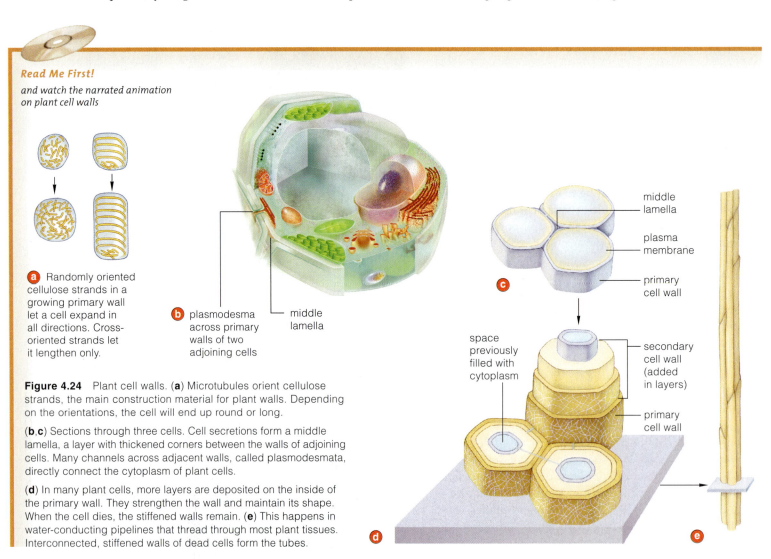

Read Me First!

and watch the narrated animation on plant cell walls

a Randomly oriented cellulose strands in a growing primary wall let a cell expand in all directions. Cross-oriented strands let it lengthen only.

b plasmodesma across primary walls of two adjoining cells

middle lamella

middle lamella

plasma membrane

primary cell wall

c

space previously filled with cytoplasm

secondary cell wall (added in layers)

primary cell wall

d

e

Figure 4.24 Plant cell walls. (**a**) Microtubules orient cellulose strands, the main construction material for plant walls. Depending on the orientations, the cell will end up round or long.

(**b,c**) Sections through three cells. Cell secretions form a middle lamella, a layer with thickened corners between the walls of adjoining cells. Many channels across adjacent walls, called plasmodesmata, directly connect the cytoplasm of plant cells.

(**d**) In many plant cells, more layers are deposited on the inside of the primary wall. They strengthen the wall and maintain its shape. When the cell dies, the stiffened walls remain. (**e**) This happens in water-conducting pipelines that thread through most plant tissues. Interconnected, stiffened walls of dead cells form the tubes.

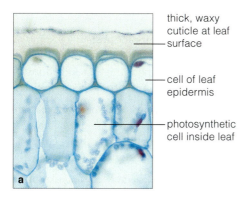

thick, waxy
cuticle at leaf
surface

cell of leaf
epidermis

photosynthetic
cell inside leaf

Figure 4.25 (**a**) Section through a plant cuticle, an outer surface layer of cell secretions. (**b**) A living cell inside bone tissue, the stuff of vertebrate skeletons.

At plant surfaces exposed to air, waxes and other cell secretions build up, forming a protective cuticle. This type of semitransparent surface covering limits water losses from aboveground parts during hot, dry days (Figure 4.25*a*).

MATRIXES BETWEEN ANIMAL CELLS

Animal cells have no cell walls. Intervening between many of them are matrixes made of cell secretions and of materials absorbed from the surroundings. For example, cartilage at the knobby ends of leg bones consists of scattered cells and protein fibers embedded in a ground substance of firm polysaccharides. Living cells also secrete the extensive, hardened matrix that we call bone tissue (Figure 4.25*b*).

CELL JUNCTIONS

Even when a wall or some other structure imprisons a cell in its own secretions, the cell still has contact with the outside world at its plasma membrane. Also, in multicelled species, membrane components extend into adjoining cells or the surrounding matrix. Among the components are **cell junctions**: molecular structures where a cell sends or receives signals or materials, or recognizes and cements itself to cells of the same type.

In plants, for instance, channels extend across the primary wall of adjacent living cells and interconnect the cytoplasm of both (Figure 4.24*b*). Each channel is a plasmodesma (plural, plasmodesmata). Substances flow quickly from cell to cell across these junctions.

In most animal tissues, three types of cell-to-cell junctions are common (Figure 4.26). *Tight* junctions link cells of most body tissues, including epithelia that line outer surfaces, internal cavities, and organs. The junctions seal abutting cells together so water-soluble substances cannot pass between them. That is why gastric fluid does not leak across the stomach lining

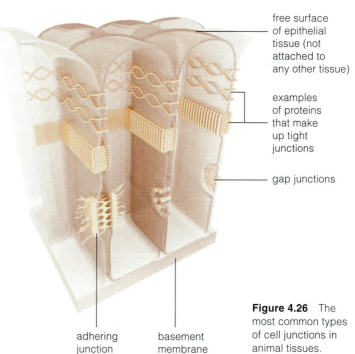

free surface
of epithelial
tissue (not
attached to
any other tissue)

examples
of proteins
that make
up tight
junctions

gap junctions

adhering
junction

basement
membrane

Figure 4.26 The most common types of cell junctions in animal tissues.

and damage internal tissues. *Adhering* junctions occur in skin, the heart, and in other organs subjected to stretching. *Gap* junctions link the cytoplasm of certain adjoining cells. They are open channels for a rapid flow of substances, most notably in heart muscle.

A variety of protistan, plant, and fungal cells have a porous wall that surrounds the plasma membrane.

Young plant cells have a thin primary wall pliable enough to permit expansion. Some mature cells also deposit a lignin-reinforced secondary wall that affords structural support.

Animal cells have no walls, but they and many other cells often secrete substances that help form matrixes of tissues. Junctions often occur between cells of multicelled organisms.

Summary

Section 4.1 The cell is the smallest unit that still displays the properties of life. The surface-to-volume ratio constrains size increases. All cells start out life with an outer plasma membrane, cytoplasm, and a nucleus or nucleoid area that contains DNA.

Section 4.2 Most cells are microscopically small. Different microscopes reveal cell shapes and structures.

Section 4.3 Cell membranes consist mainly of lipids and proteins. A lipid bilayer gives a membrane its fluid properties and prevents water-soluble substances from freely crossing it. Proteins embedded in the bilayer or positioned at its surfaces carry out many functions.

Section 4.4 Bacteria and archaea are prokaryotic. They have no nucleus and are the smallest, structurally simplest cells known (Table 4.1).

Sections 4.5, 4.6 Eukaryotic cells generally have diverse organelles: membranous sacs that divide the cell's interior into functional compartments (Table 4.1). Organelles physically separate chemical reactions from the rest of the cell. A nucleus isolates and protects the cell's genetic material.

Section 4.7 In the endomembrane system's ER and Golgi bodies, new polypeptide chains take on final form and lipids are assembled; both get packaged into vesicles for transport, storage, and other cell activities.

Section 4.8 The final reactions of aerobic respiration occur in mitochondria, where many ATP form. The chloroplasts of photosynthetic plant and algal cells use energy from the sun to make sugars.

Section 4.10 Eukaryotic cells have a cytoskeleton of microtubules, microfilaments, and intermediate filaments. It imparts shape and supports and moves cell parts, motile structures, and often the whole cell.

Section 4.11 Many bacterial, protistan, fungal, and plant cells have a wall around the plasma membrane. In multicelled organisms, adjoining cells form diverse structural and functional connections.

Table 4.1 Summary of Typical Components of Prokaryotic and Eukaryotic Cells

Cell Component	Function	Prokaryotic Bacteria, Archaea	Eukaryotic Protists	Eukaryotic Fungi	Eukaryotic Plants	Eukaryotic Animals
Cell wall	Protection, structural support	✔*	✔*	✔	✔	None
Plasma membrane	Control of substances moving into and out of cell	✔	✔	✔	✔	✔
Nucleus	Physical separation and organization of DNA	None	✔	✔	✔	✔
DNA	Encoding of hereditary information	✔	✔	✔	✔	✔
RNA	Transcription, translation of DNA messages into polypeptide chains of specific proteins	✔	✔	✔	✔	✔
Nucleolus	Assembly of subunits of ribosomes	None	✔	✔	✔	✔
Ribosome	Protein synthesis	✔	✔	✔	✔	✔
Endoplasmic reticulum (ER)	Initial modification of many of the newly forming polypeptide chains of proteins; lipid synthesis	None	✔	✔	✔	✔
Golgi body	Final modification of proteins, lipids; sorting and packaging them for use inside cell or for export	None	✔	✔	✔	✔
Lysosome	Intracellular digestion	None	✔	✔*	✔*	✔
Mitochondrion	ATP formation	**	✔	✔	✔	✔
Photosynthetic pigments	Light–energy conversion	✔*	✔*	None	✔	None
Chloroplast	Photosynthesis; some starch storage	None	✔*	None	✔	None
Central vacuole	Increasing cell surface area; storage	None	None	✔*	✔	None
Bacterial flagellum	Locomotion through fluid surroundings	✔*	None	None	None	None
Flagellum or cilium with 9+2 microtubular array	Locomotion through or motion within fluid surroundings	None	✔*	✔*	✔*	✔
Complex cytoskeleton	Cell shape; internal organization; basis of cell movement and, in many cells, locomotion	Rudimentary***	✔*	✔*	✔*	✔

* Known to be present in cells of at least some groups.
** Many groups use oxygen-requiring (aerobic) pathways of ATP formation, but mitochondria are not involved.
*** Protein filaments form a simple scaffold that helps support the cell wall in at least some species.

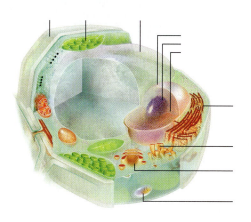
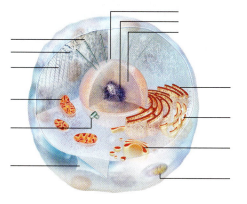

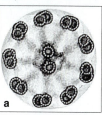
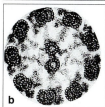

Figure 4.27 Cross-sections through the flagellum of a sperm cell from (**a**) a male with Kartagener syndrome and (**b**) an unaffected male. Notice the dynein arms that extend from the paired microtubules.

Self-Quiz
Answers in Appendix III

1. Cell membranes consist mainly of a _____ .
   a. carbohydrate bilayer and proteins
   b. protein bilayer and phospholipids
   c. lipid bilayer and proteins

2. Identify the components of the cells shown above.

3. Organelles _____ .
   a. are membrane-bound compartments
   b. are typical of eukaryotic cells, not prokaryotic cells
   c. separate chemical reactions in time and space
   d. All of the above are features of organelles.

4. Cells of many protists, plants, and fungi, but not animals, commonly have _____ .
   a. mitochondria   c. ribosomes
   b. a plasma membrane d. a cell wall

5. Is this statement true or false: The plasma membrane is the outermost component of all cells. Explain your answer.

6. Unlike eukaryotic cells, prokaryotic cells _____ .
   a. lack a plasma membrane c. have no nucleus
   b. have RNA, not DNA   d. all of the above

7. Match each cell component with its function.
   ____ mitochondrion   a. protein synthesis
   ____ chloroplast   b. initial modification of new polypeptide chains
   ____ ribosome
   ____ rough ER   c. modification of new proteins; sorting, shipping tasks
   ____ Golgi body
                         d. photosynthesis
                         e. formation of many ATP

Critical Thinking

1. Why is it likely that you will never meet a two-ton amoeba on a sidewalk?

2. Your professor shows you an electron micrograph of a cell with many mitochondria, Golgi bodies, and a lot of rough ER. What kinds of cellular activities would require such an abundance of the three kinds of organelles?

3. *Kartagener syndrome* is a genetic disorder caused by a mutated form of the protein dynein. Affected people have chronically irritated sinuses, and mucus builds up in the airways to their lungs. Bacteria form huge populations in the thick mucus. Their metabolic by-products and the inflammation they trigger combine to damage tissues. Males affected by the syndrome make sperm but are infertile (Figure 4.27). Some have become fathers with the help of a procedure that injects sperm cells directly into eggs. Explain how an abnormal dynein molecule could cause the observed effects.

IMPACTS, ISSUES *Alcohol, Enzymes, and Your Liver*

Consider the cells that are supposed to keep a heavy drinker alive. It makes little difference whether a drinker gulps down 12 ounces of beer, 5 ounces of wine, or 1–1/2 ounces of eighty-proof vodka. Each drink has the same amount of "alcohol" or, more precisely, ethanol.

Ethanol molecules—CH_3CH_2OH—have water-soluble and fat-soluble components, which the stomach and small intestine quickly absorb. The bloodstream moves more than 90 percent of these components to the liver, where enzymes speed their breakdown to a nontoxic form called acetate (acetic acid). However, the liver's alcohol-metabolizing enzymes can detoxify only so much in a given hour.

One of the enzymes you'll read about in this chapter is catalase, a foot soldier against toxin attacks on the body (Figure 5.1). Catalase is thought to assist alcohol dehydrogenase. When alcohol circulates in blood, these enzymes convert it to acetaldehyde. Reactions can't end there, however, because acetaldehyde is toxic at high concentrations. In healthy people at least, another kind of enzyme speeds its breakdown to nontoxic forms.

Given the liver's central role in alcohol metabolism, habitually heavy drinkers gamble with alcohol-induced liver diseases. Over time, the capacity to tolerate alcohol diminishes because there are fewer and fewer liver cells —hence fewer enzymes—for detoxification.

In *alcoholic hepatitis*, inflammation and destruction of liver tissue is widespread. Another disease, *alcoholic cirrhosis*, permanently scars the liver. In time, the liver just stops working, with devastating effects.

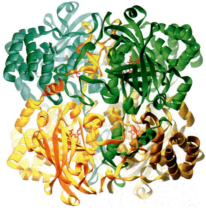

Figure 5.1 Ribbon model of catalase, an enzyme that helps detoxify many substances that can damage the body, such as the alcohol in beer, martinis, and other drinks.

the big picture

The One-Way Flow of Energy Energy, the capacity to do work, can be converted from one form to another but can't be created from scratch. It flows in one direction, from usable to less usable forms. Some is lost as heat with each conversion.

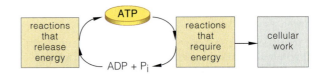

How Cells Use Energy Energy flows into the web of life, mainly from the sun, and flows out of it. Cells tap into the one-way flow by energy-acquiring processes, starting with photosynthesis. They convert inputs of energy to forms that keep them alive and working properly.

The liver is the largest gland in the human body, and its activity impacts everything else. You'd have a hard time digesting and absorbing food without it. Your cells would have a hard time synthesizing and taking up carbohydrates, lipids, and proteins, and staying alive.

There's more. The liver makes plasma proteins. These proteins circulate in blood and are vital for blood clotting, immunity, maintaining the fluid volume of the internal environment, and other tasks. Also, liver enzymes get rid of a lot more toxic compounds than acetaldehyde.

Binge drinking—consuming large amounts of alcohol in a brief period—is now the most serious drug problem on campuses in the United States. Consider: 44 percent of nearly 17,600 students surveyed at 140 colleges and universities are caught up in the culture of drinking. They report having five alcoholic drinks a day, on average.

Binge drinking does more than damage the liver. Put aside the related 500,000 injuries from accidents, the 70,000 cases of date rape, and the 400,000 cases of (whoops) unprotected sex among students in an average year. Binge drinking can kill before you know what hit you. Drink too much, too fast, and you can abruptly stop the beating of your heart. Think about it.

With this example, we turn to **metabolism**, the cell's capacity to acquire energy and use it to build, degrade, store, and release substances in controlled ways. At times, the activities of your cells may seem remote from your interests. But they help define who you are and what you will become, liver and all.

 How Would You Vote?

Some people have damaged their liver because they drank too much alcohol. Others have a diseased liver. There aren't enough liver donors for all the people waiting for liver transplants. Should life-style be a factor in deciding who gets a transplant? See the Media Menu for details, then vote online.

How Enzymes Work   Without enzymes, substances would not react fast enough to maintain living cells, hence life itself. Controls over enzyme action also maintain life through adjustments in the concentration of substances moving across cell membranes.

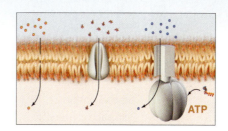

Membranes and Metabolism   Cells have built-in mechanisms that increase and decrease concentrations of substances across their membranes. The adjustments are essential for metabolic reactions and metabolic pathways.

5.1 Inputs and Outputs of Energy

Cells secure energy from their surroundings and use it for thousands of tasks. Energy drives metabolism—chemical work that stockpiles, builds, rearranges, and breaks down substances. It drives the mechanical work of moving cell parts, body parts, or the whole organism. It drives the electrochemical work of moving charged substances across membranes, as happens when cells make ATP.

THE ONE-WAY FLOW OF ENERGY

Energy is a capacity to do work, and you can't create it out of nothing. By the **first law of thermodynamics**, any isolated system has a finite amount of energy that cannot be added to or lost. Energy *can* be converted from one form to another. However, the total amount in the system stays the same. Motion, chemical bonds, heat, electricity, sound, nuclear forces, and gravity are examples of different forms of energy.

"Entropy" is a measure of the degree of a system's disorder. By the **second law of thermodynamics**, the entropy, or disorder, of the universe always increases. Think of Egyptian pyramids—once highly organized, now crumbling, and thousands of years from now, dust. According to the second law, pyramids and all other things are on their way toward maximum entropy.

Energy is part of this big picture. It spontaneously flows toward its most disorganized form—heat. Why? Converting energy from one form to another is never 100 percent efficient. Although energy is conserved in any exchange, at least some of it dissipates as heat. It is not easy to convert heat to a different form of energy.

Can life be one glorious pocket of resistance to this depressing flow toward maximum entropy? After all, new bonds hold atoms together in orderly patterns in each new organism. Molecules get more organized and have richer stores of energy, not poorer.

Even so, the second law does apply to life on Earth. Life's main energy source is the sun, which has been losing energy ever since it formed about 5 billion years ago. Photosynthetic cells intercept light energy from the sun and convert it to chemical bond energy in sugars, starches, and other compounds. Organisms that eat plants get at the stored chemical energy by breaking and rearranging chemical bonds. With each conversion, however, a bit of energy escapes as heat. Cells don't convert that heat to other forms of energy. They simply can't use it to do work.

Overall, then, energy flows in one direction. Life can maintain its astounding organization only because it is being continually resupplied with energy that is being lost from someplace else (Figure 5.2).

UP AND DOWN THE ENERGY HILLS

Cells store and retrieve energy when they convert one molecule to another. In photosynthetic cells, sunlight energy drives ATP formation, then energy from ATP drives glucose formation (Figure 5.3a). Six molecules of carbon dioxide (CO_2) and six of water (H_2O) are converted to one molecule of glucose ($C_6H_{12}O_6$) and six of oxygen (O_2). Photosynthetic reactions require energy input; they are *endergonic* (meaning energy in).

ENERGY LOST
Energy continually flows from the sun.

ENERGY GAINED
Sunlight energy reaches environments on Earth. Producers of nearly all ecosystems secure some and convert it to stored forms of energy. They and all other organisms convert stored energy to forms that can drive cellular work.

producers

NUTRIENT CYCLING

ENERGY LOST
With each conversion, there is a one-way flow of a bit of energy back to the environment. Nutrients cycle between producers and consumers.

consumers

Figure 5.2 A one-way flow of energy into ecosystems compensates for the one-way flow of energy out of it.

We think of glucose as a high-energy compound because it can be converted to more stable molecules for a net gain of energy. It does take an investment of energy to get the conversion reactions started, but the formation of more stable end products releases more energy than the amount invested. For example, CO_2 and H_2O are all that's left of glucose at the end of aerobic respiration. Both still have energy stored in covalent bonds, but the two products are so stable that cells cannot gain energy by converting them to something else. It's as if carbon dioxide and water are at the base of an "energy hill."

Aerobic respiration releases energy bit by bit, with many conversion steps, so cells can capture some of it efficiently. This metabolic process is like a downhill run, from high-energy glucose to low-energy carbon dioxide and water (Figure 5.3b). Such reactions, which show a net energy release, are said to be *exergonic*.

ATP—THE CELL'S ENERGY CURRENCY

All cells stay alive by *coupling* energy inputs to energy outputs, mainly with adenosine triphosphate, or **ATP**. This nucleotide consists of a five-carbon sugar (ribose), a base (adenine), and three phosphate groups (Figure 5.4a). ATP readily gives up a phosphate group to other molecules and primes them to react. Such phosphate-group transfers are known as **phosphorylations**.

ATP is the currency in a cell's economy. Cells earn it by investing in energy-releasing reactions. They spend it in energy-requiring reactions that keep them alive. We use a cartoon coin to symbolize ATP.

Because ATP is the main energy carrier for so many reactions, you might infer—correctly—that cells have ways to renew it. When ATP gives up a phosphate group, ADP (adenosine diphosphate) forms. ATP can re-form when ADP binds to inorganic phosphate (P_i) or to a phosphate group that was split from a different molecule. Regenerating ATP by this **ATP/ADP cycle** helps drive most metabolic reactions (Figure 5.4b).

> *Energy is the capacity to do work. It flows in one direction, from more usable to less usable forms. Heat is the least usable form of energy. Organisms maintain complex organization by being resupplied with energy from someplace else.*
>
> *All organisms secure energy from outside sources. The sun is the primary source of energy for the web of life. All organisms use and store energy in chemical bonds.*
>
> *ATP is the main energy carrier in all living cells. It couples energy-releasing and energy-requiring reactions. ATP primes molecules to react by transferring a phosphate group to them.*

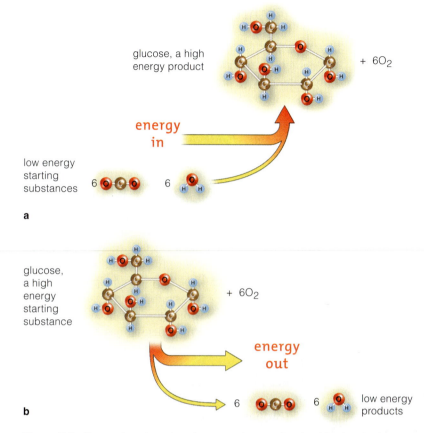

Figure 5.3 Two main categories of energy changes involved in chemical work. (**a**) Endergonic reactions, which won't run without an energy input. (**b**) Exergonic reactions, which end with a net release of usable energy.

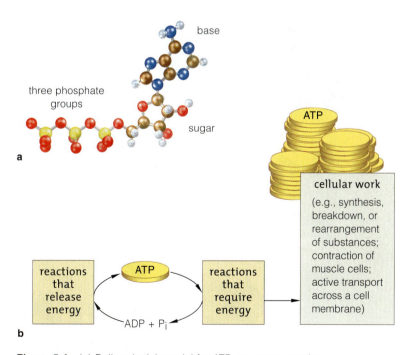

Figure 5.4 (**a**) Ball-and-stick model for ATP, an energy carrier. (**b**) ATP couples energy-releasing reactions with energy-requiring ones. In the ATP/ADP cycle, recurring phosphate-group transfers turn ATP into ADP, and back again to ATP.

5.2 Inputs and Outputs of Substances

How cells get energy is only one aspect of metabolism. Another is the accumulation, conversion, and disposal of materials by energy-driven reactions. Most reactions are part of stepwise metabolic pathways.

THE NATURE OF METABOLIC REACTIONS

For any metabolic reaction, the starting substances are called *reactants*. Substances formed during a reaction sequence are *intermediates*, and those left at the end are the *products*. ATP and other *energy carriers* activate enzymes and other molecules by making phosphate-group transfers. *Enzymes* are catalysts: They can speed specific reactions enormously. *Cofactors* are metal ions and coenzymes such as NAD^+. They help enzymes by moving functional groups, atoms, and electrons from one reaction site in an enzyme to another. *Transport proteins* help solutes across membranes. Controls over transport proteins adjust concentrations of substances required for reactions, and so influence the timing and direction of metabolism.

Bear in mind, metabolic reactions don't always run from reactants to products. They might start out in this "forward" direction. But most also run in reverse, with products being converted back to reactants. Such reversible reactions tend to run spontaneously toward **chemical equilibrium**, when the reaction rate is about the same in both directions. For most reactions, the amounts of reactant and product molecules differ at that time (Figure 5.5). It is like a party where people drift between two rooms. The number in each room stays the same—say, thirty in one and ten in the other—even as individuals move back and forth.

Why bother to think about this? *Each cell can bring about big changes in its activities by controlling a few steps of reversible metabolic pathways.*

For instance, when your cells need a quick bit of energy, they rapidly split glucose into two pyruvate molecules. They do so by a sequence of nine enzyme-mediated steps of a pathway called glycolysis. When glucose supplies are too low, cells quickly reverse this pathway and build glucose from pyruvate and other substances. How? Six steps of the pathway happen to be reversible, and the other three are bypassed. An input of energy from ATP drives the bypass reactions in the uphill (energetically unfavorable) direction.

What if cells did not have this reverse pathway? They wouldn't be able to build glucose fast enough to compensate for episodes of starvation, when glucose supplies in blood become dangerously low.

REDOX REACTIONS

Energy flows from the environment through all living things by way of photosynthesis and other metabolic pathways. Individual cells capture free energy, store it, then release it in manageable bits. They *control* the energy they require to grow and reproduce.

Cells release energy efficiently by electron transfers, or **oxidation–reduction reactions**. In these "redox" reactions, one molecule gives up electrons (is oxidized) and another gains them (is reduced). Commonly, hydrogen atoms are released at the same time, thus becoming H^+. Being attracted to the opposite charge of the electrons, H^+ tags along with them.

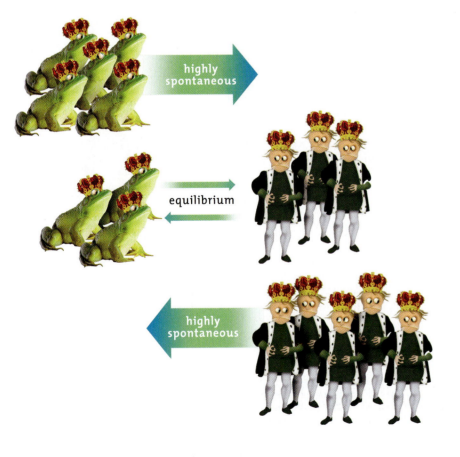

Figure 5.5 Chemical equilibrium. With a high concentration of reactant molecules (represented as wishful frogs), a reaction runs most strongly in the forward direction, to products (the princes). When the concentration of product molecules is high, it runs most strongly in reverse. At equilibrium, the rates of the forward and reverse reactions are the same.

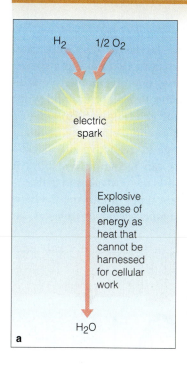

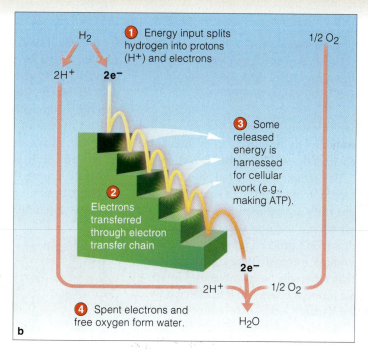

a

electric spark

Explosive release of energy as heat that cannot be harnessed for cellular work

H_2O

b

1 Energy input splits hydrogen into protons (H^+) and electrons

$2H^+$ $2e^-$

2 Electrons transferred through electron transfer chain

3 Some released energy is harnessed for cellular work (e.g., making ATP).

$2e^-$

$2H^+$ $1/2 O_2$

H_2O

4 Spent electrons and free oxygen form water.

Read Me First!
and watch the narrated animation on controlling energy release

Figure 5.6 Uncontrolled versus controlled energy release. (**a**) Free hydrogen and oxygen exposed to an electric spark react and release energy all at once. (**b**) Electron transfer chains let the same reaction proceed in small, more manageable steps that can access the released energy.

Start thinking about redox reactions, because they are central to photosynthesis and aerobic respiration. In the next two chapters, you'll see how coenzymes pick up electrons and H^+ stripped from substrates, then deliver them to **electron transfer chains**. Such chains are membrane-bound arrays of enzymes and other molecules that accept and give up electrons in sequence. Electrons are at a higher energy level when they enter a chain than when they leave. Think of the electrons as descending a staircase and stingily losing a bit of energy at each step (Figure 5.6). For these two pathways, stepwise electron transfers concentrate H^+ in ways that contribute to ATP formation.

TYPES OF METABOLIC PATHWAYS

We've mentioned metabolic pathways in passing, but let's now formally define them. **Metabolic pathways** are enzyme-mediated sequences of reactions in cells. Many are *biosynthetic* (or anabolic), and they require energy inputs. Examples are the assembly of glucose, starch, proteins, and other high-energy molecules from small molecules. The main biosynthetic pathway in the biosphere is photosynthesis (Figure 5.7).

Degradative (or catabolic) pathways are exergonic, overall. These reactions can break down molecules to smaller, lower energy products. Aerobic respiration releases a lot of usable energy (ATP) in the step-by-step enzymatic breakdown of glucose. It is the main degradative pathway in the biosphere (Figure 5.7).

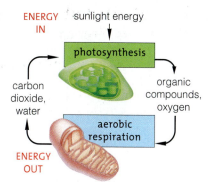

ENERGY IN   sunlight energy

photosynthesis

carbon dioxide, water

organic compounds, oxygen

aerobic respiration

ENERGY OUT

Figure 5.7 The main metabolic pathways in ecosystems. Energy input from the sun drives photosynthesis, and aerobic respiration yields a lot of usable energy. ATP forms in both pathways by way of redox reactions.

Not all metabolic pathways are linear, a straight line from the reactants to products. In cyclic pathways, the final step regenerates a reactant that is the point of entry for the reaction sequence. In branched pathways, reactants or intermediates are channeled into two or more different reaction sequences.

Metabolic pathways are orderly, enzyme-mediated reaction sequences, some biosynthetic, others degradative.

Control over a key step of a metabolic pathway can bring about rapid shifts in cell activities.

Many aspects of metabolism involve electron transfers, or oxidation–reduction reactions. Electron transfer chains are important sites of energy exchange in both photosynthesis and aerobic respiration.

5.3 How Enzymes Make Substances React

What would happen if you left a cupful of glucose out in the open? Not much. Years would pass before you would see evidence of its conversion to carbon dioxide and water. Yet that same conversion takes only a few seconds in your body. Enzymes make the difference.

Enzymes, again, are catalytic molecules; they speed rates of specific reactions by hundreds to millions of times. Enzymes chemically recognize, bind, and alter specific reactants. They remain unchanged, and so can mediate the same reaction over and over again. Except for a few RNAs, enzymes are proteins.

Regardless of whether a reaction is spontaneous or enzyme-mediated, it won't proceed unless the starting substances have enough internal energy to overcome repulsive forces that otherwise keep molecules apart. All molecules have internal energy that is affected by temperature and pressure. **Activation energy** refers to the minimum amount of internal energy that molecules must have before a reaction gets going.

Activation energy is an energy barrier—something like a hill or a brick wall (Figures 5.8 and 5.9). One way or another, that barrier must be surmounted before the reaction will proceed. Enzymes lower the barrier. How? *Compared to the surrounding environment,*

they offer a stable microenvironment that is more favorable for reaction. Enzymes are far larger than **substrates**, another name for the reactants that bind to a specific enzyme. Each enzyme has one or more **active sites**: pockets or crevices where substrates bind and where specific reactions are catalyzed (Figure 5.10).

Part of the substrate is complementary in shape, size, solubility, and charge to the active site. Because of this fit, each enzyme can recognize and bind its substrate among thousands of substances in cells.

Think back on the main types of enzyme-mediated reactions (Section 3.1). In *functional group transfers,* one

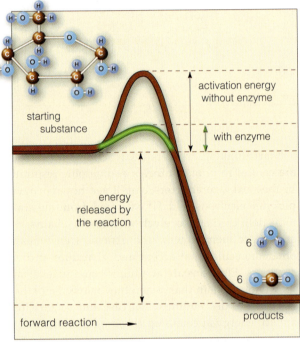

Figure 5.8 Activation energy. Reactants must have a minimum amount of internal energy before a given reaction will run to products. Sometimes they need an input of energy to get there. An enzyme enhances the reaction rate by lowering the amount of activation energy required. It makes the energy hill smaller.

Figure 5.9 A simple way to think about the energy required to get a reaction going without an enzyme (**a**) and with the help of an enzyme (**b**).

Read Me First!

*and watch the narrated
animation on catalase action*

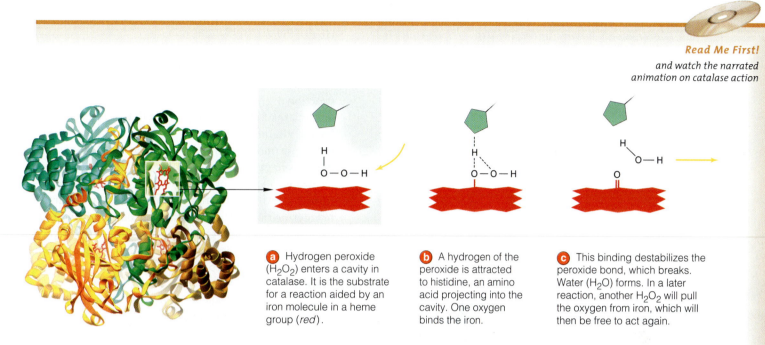

a Hydrogen peroxide (H_2O_2) enters a cavity in catalase. It is the substrate for a reaction aided by an iron molecule in a heme group (*red*).

b A hydrogen of the peroxide is attracted to histidine, an amino acid projecting into the cavity. One oxygen binds the iron.

c This binding destabilizes the peroxide bond, which breaks. Water (H_2O) forms. In a later reaction, another H_2O_2 will pull the oxygen from iron, which will then be free to act again.

Figure 5.10 How catalase works. This enzyme has four polypeptide chains and four heme groups.

molecule gives up a functional group to another. In *electron transfers*, one or more electrons stripped from one molecule are donated to another. In *rearrangements*, a juggling of internal bonds converts one molecule to another. In *condensation*, two molecules are covalently bound together as a larger molecule. Finally, in *cleavage* reactions, a larger molecule splits into smaller ones.

When we talk about activation energy, *we really are talking about the energy it takes to align reactive chemical groups, briefly destabilize electric charges, and break bonds.*

Enzymes lower activation energy by restraining a reactant molecule. Binding to the active site stretches and squeezes the reactant into a certain shape, maybe next to another molecule or reactive group. This puts a substrate at its **transition state**, meaning its bonds are at the breaking point and the reaction can run easily to product.

The binding between an enzyme and its substrate is weak, and temporary (that's why the reaction does not change the enzyme). But energy is released when weak bonds form. This "binding energy" stabilizes the transition state long enough to keep the enzyme and its substrate together for the reaction.

Four mechanisms work alone or in combination to get substrates to the transition state:

Helping substrates get together. Substrate molecules rarely react at low concentrations. Binding to an active site boosts local substrate concentration by as much as ten millionfold.

Orienting substrates in positions favoring reaction. On their own, substrates collide from random directions. By contrast, weak but extensive bonds at an active site put reactive groups close together.

Shutting out water. Because of its ability to form hydrogen bonds so easily, water can interfere with the breaking and formation of bonds during reactions. Some active sites contain mostly nonpolar amino acids. The hydrophobic groups keep water away from the active site and reactions.

Inducing changes in enzyme shape. Weak interactions between the enzyme and its substrate may induce the enzyme to change its shape. By the **induced-fit model**, a substrate is not quite complementary to an active site. The enzyme bends and optimizes the fit; in doing so, it pulls the substrate to the transition state.

On their own, chemical reactions occur too slowly to sustain life. Enzymes greatly increase reaction rates by lowering the activation energy—the minimum amount of energy required to align reactive groups, destabilize electric charges, and break bonds so that products can form from reactants.

Enzymes drive their substrates to a transition state, when the reaction can most easily run to completion. This happens in the enzyme's active site.

In the active site, substrates move to the transition state by mechanisms that concentrate and orient them, that exclude water, and that induce an optimal fit with the active site.

5.4 Enzymes Don't Work In a Vacuum

Controls over enzyme function help cells respond quickly to changing conditions by triggering adjustments in metabolic reactions. Feedback mechanisms that can activate or inhibit enzymes conserve resources. Cells synthesize what conditions require—no more, no less.

HELP FROM COFACTORS

Cofactors (specific metal ions or coenzymes) help out at the active site of enzymes or taxi electrons, H+, or functional groups to a different location. **Coenzymes** are a class of organic compounds that may or may not have a vitamin component.

One or more metal ions assist nearly a third of all known enzymes. Metal ions easily give up and accept electrons. As part of coenzymes, they help products form by shifting electron arrangements in substrates or intermediates. That is what goes on at the hemes in catalase. Heme, an organic ring structure, incorporates iron at its center. Figure 5.10 shows how iron atoms in heme coenzymes help catalase break down hydrogen peroxide to water.

Like vitamin E, catalase is one of the **antioxidants**: It helps neutralize free radicals. *Free radicals* are atoms with unpaired electrons—reactive, unbound fragments left over from reactions. As we age, we make less and less catalase, so free radicals accumulate. They attack the structure of DNA and other biological molecules.

Some coenzymes are tightly bound to an enzyme. Others, such as NAD+ and NADP+, can diffuse freely through a cell membrane or cytoplasm. Either way, coenzymes participate intimately in a reaction. Unlike enzymes, many become modified during the reaction, but they are regenerated elsewhere.

CONTROLS OVER ENZYMES

Many controls over enzymes maintain, lower, and raise the concentrations of substances. Others adjust how fast enzyme molecules are synthesized, and they activate or inhibit the ones already built.

In some cases, a molecule that acts as an activator or inhibitor reversibly binds to its own *allosteric* site, not the active site, on the enzyme (*allo–*, other; *steric*, structure). Binding alters the enzyme's shape in a way that hides or exposes the active site (Figure 5.11).

Picture a bacterial cell making tryptophan and other amino acids—the building blocks for proteins. Even when the cell has made enough proteins, tryptophan synthesis continues until its increasing concentration causes **feedback inhibition**. This means a change that results from a specific activity *shuts down the activity*.

A feedback loop starts and ends at many allosteric enzymes. In this case, unused tryptophan binds to an allosteric site on one of the enzymes in a tryptophan biosynthesis pathway. It blocks the active site, so less tryptophan is made (Figure 5.12). At times when not many tryptophan molecules are around, more active

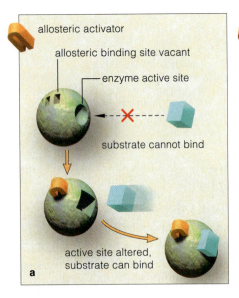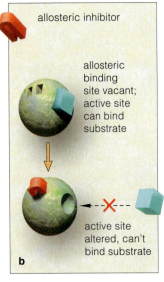

Figure 5.11 Allosteric control over enzyme activity. (**a**) An active site is unblocked when an activator binds to a vacant allosteric site. (**b**) An active site is blocked when an inhibitor binds to a vacant allosteric site.

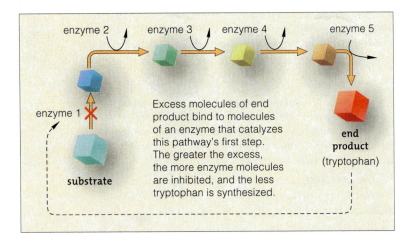

Figure 5.12 Feedback inhibition of a metabolic pathway. Five kinds of enzymes act in sequence to convert a substrate to tryptophan.

sites remain exposed, and so the synthesis rate picks up. In such ways, feedback loops quickly adjust the concentrations of substances.

In humans and other multicelled species, enzyme controls are just amazing. They keep individual cells functioning in ways that benefit the whole body!

EFFECTS OF TEMPERATURE, PH, AND SALINITY

Temperature is a measure of molecular motion. As it rises, it boosts reaction rates both by increasing the likelihood that a substrate will bump into an enzyme and by raising a substrate molecule's internal energy. Remember, the more energy a reactant molecule has, the closer it gets to jumping that activation energy barrier and taking part in a reaction.

Above or below the range of temperature that an enzyme can tolerate, weak bonds break, and enzyme shape changes. Substrates no longer can bind to the active site, and the reaction rate falls sharply (Figure 5.13). Such declines typically occur with fevers above 44°C (112°F), which people usually cannot survive.

Enzyme action is also affected by pH (Figure 5.14). In the human body, most enzymes work best when the pH is between 6 and 8. For instance, trypsin is active in the small intestine (pH of 8 or so).

One of the notable exceptions is pepsin, a protein-digesting enzyme. Pepsin is a nonspecific protease; it chews up any proteins. It is produced in inactive form and normally becomes activated only in gastric fluid, in the stomach. Gastric fluid happens to be a highly acidic environment (pH 1–2). It's a good thing that activated pepsin is confined to the stomach. If it were to leak out (as happens with peptic ulcers), it could digest a lot of you instead of proteins in your food.

Most enzymes don't work well when the fluids in which they are dissolved are saltier or less salty than their range of tolerance. Too much or too little salt interferes with the hydrogen bonds that help hold an enzyme in its three-dimensional shape. By doing so, it inactivates the enzyme.

Many enzymes are assisted by cofactors, which are specific metal ions or coenzymes.

Enzyme action adjusts the concentrations and kinds of substances available in cells. Controls over enzymes enhance or inhibit their activity.

Enzymes work best when the cellular environment stays within limited ranges of temperature, pH, and salinity. The actual ranges differ from one type of enzyme to the next.

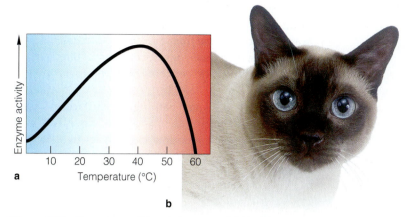

Figure 5.13 Enzymes and the environment. (**a**) How increases in temperature affect one enzyme's activity. (**b**) Temperature outside the body affects the fur color of Siamese cats. Epidermal cells that give rise to the cat's fur produce a brownish-black pigment, melanin. Tyrosinase, an enzyme in the melanin production pathway, is heat-sensitive in the Siamese. It becomes inactive in warmer parts of the cat's body, which end up with less melanin, and lighter fur. Put this cat in booties for a few weeks and its warm feet will turn white.

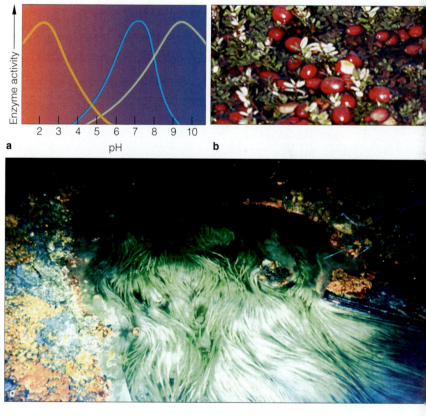

Figure 5.14 Enzymes and the environment. (**a**) How pH values affect three enzymes. (**b**) Cranberry plants grow best in acidic bogs. Unlike most plants, they have no nitrate reductase. This enzyme converts nitrate (NO_3) found in typical soils to metabolically useful ammonia (NH_3). Nitrogen in highly acidic soils is already in the form of ammonia (NH_4^+). (**c**) Life in wastewater from a copper mine in California. The slime streamers are microbial communities dominated by an archaean, which makes unique enzymes that help it live in this toxic, highly acidic environment.

5.5 Diffusion, Membranes, and Metabolism

What determines whether a substance will move one way or another to and from a cell, across that cell's membranes, or through the cell itself? Part of the answer has to do with something called diffusion.

Think about the water bathing the surfaces of a cell membrane. Plenty of substances are dissolved in it, but the kinds and amounts close to its two surfaces differ. The membrane itself set up the difference and is busy maintaining it. How? Each cell membrane has

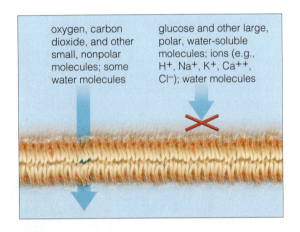

Figure 5.15 shows the arrangement:

- oxygen, carbon dioxide, and other small, nonpolar molecules; some water molecules
- glucose and other large, polar, water-soluble molecules; ions (e.g., H^+, Na^+, K^+, Ca^{++}, Cl^-); water molecules

Figure 5.15 Selective permeability of cell membranes. Small, nonpolar molecules and some water molecules can cross the lipid bilayer. Ions and large, polar, water-soluble molecules and the water dissolving them cross with the help of transport proteins.

selective permeability. Its molecular structure allows some substances but not others to cross it in certain ways, at certain times.

Lipids of a membrane's bilayer are mostly nonpolar, so they let small, nonpolar molecules such as O_2 and CO_2 slip across. Water molecules are polar, but some slip through gaps that form as the hydrophobic tails of many lipids flex and bend. The bilayer itself is not permeable to ions or large, polar molecules such as glucose; these cross with the help of proteins. Water often crosses with them (Figure 5.15).

Membrane barriers and crossings are vital, because metabolism depends on the cell's capacity to increase, decrease, and maintain concentrations of molecules and ions required for reactions. They also supply cells or organelles with raw materials, get rid of wastes, and collectively maintain the cell's volume and pH.

WHAT IS A CONCENTRATION GRADIENT?

Now picture molecules or ions of some substance near a membrane. They move constantly, collide at random, and bounce off one another. When the concentration in one region is not the same as in an adjoining region, this condition is a *gradient*. A **concentration gradient** is a difference in the number per unit volume of ions or molecules of a substance between adjoining regions.

In the absence of other forces, a substance tends to move from a region where it is more concentrated to a region where it is less concentrated. At temperatures characteristic of life, *thermal energy that is inherent in molecules drives this movement*. Although the molecules are colliding and careening back and forth millions of times per second, their *net* movement is away from the place where they are most concentrated.

Diffusion is the name for the net movement of like molecules or ions down a concentration gradient. It is a factor in how substances move into, through, and out of cells. In multicelled species, it moves substances between body regions and between the body and its environment. For instance, when oxygen builds up in leaf cells, it may diffuse into air inside the leaf, then into air outside, where its concentration is lower.

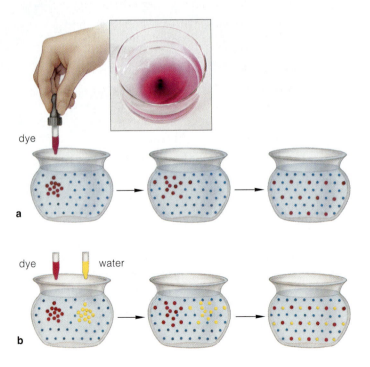

dye

dye   water

Figure 5.16 Two cases of diffusion. (**a**) A drop of dye enters a bowl filled with water. Gradually, the dye molecules become evenly dispersed through the molecules of water. (**b**) The same thing happens with water molecules. Here, dye (*red*) and water (*yellow*) are added to the same bowl. Each substance will show a net movement down its own concentration gradient.

Like other substances, oxygen tends to diffuse in a direction set by its *own* concentration gradient, not by gradients of other solutes. You can see the outcome of this tendency by squeezing a drop of dye into water. The dye molecules diffuse into the region where they are not as concentrated, and water molecules move into the region where *they* are not as concentrated. Figure 5.16 shows simple examples of diffusion.

WHAT DETERMINES DIFFUSION RATES?

How fast a particular solute diffuses depends on the steepness of its concentration gradient, its size, the temperature, and electric or pressure gradients that may be present.

First, rates are high with steep gradients, because more molecules are moving out of a region of greater concentration compared to the number moving into it. Second, more heat energy in warmer regions makes molecules move faster and collide more often. Third, smaller molecules diffuse faster than large ones do.

Fourth, an electric gradient may alter the rate and direction of diffusion. An **electric gradient** is simply a difference in electric charge between adjoining regions. For example, each ion dissolved in fluids bathing a cell membrane contributes to an electric charge at one side or the other. Opposite charges attract. Therefore, the fluid with more negative charge overall exerts the greatest pull on positively charged substances, such as sodium ions. Later chapters explain how many cell activities, including ATP formation and the sending and receiving of signals in nervous systems, are based on the force of electric and concentration gradients.

Fifth, as you will see shortly, diffusion also may be affected by a **pressure gradient**. This is a difference in the exerted force per unit area in two adjoining regions.

MEMBRANE CROSSING MECHANISMS

Before getting into the actual mechanisms that move substances across membranes, study the overview in Figure 5.17. These mechanisms help supply cells and organelles with raw materials and get rid of wastes. Collectively, they help maintain the volume and pH of cells or organelles within functional ranges.

Small, nonpolar molecules such as oxygen diffuse across the membrane's lipid bilayer. Polar molecules and ions diffuse through the interior of transport proteins that span the bilayer. Passive transporters simply allow a substance to follow its concentration gradient across a membrane. The mechanism is called *passive transport*, or "facilitated" diffusion.

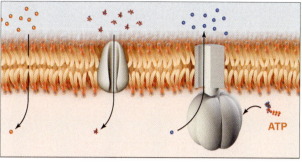
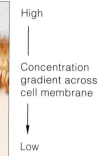

High

Concentration gradient across cell membrane

Low

| Diffusion of lipid-soluble substances across bilayer | Passive transport of water-soluble substances through channel protein; no energy input needed | Active transport through ATPase; requires energy input from ATP |

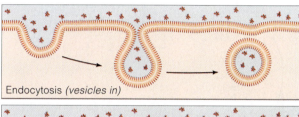

Endocytosis (vesicles in)

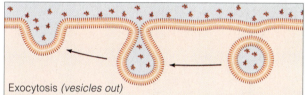

Exocytosis (vesicles out)

Figure 5.17 Overview of membrane crossing mechanisms.

Polar molecules cross the membrane through the interior of active transporters. The net direction of movement is against the concentration gradient, and it requires an input of energy. We call this mechanism *active transport*. Energy-activated transporters move a substance against its concentration gradient.

Other mechanisms move substances in bulk into or out of cells. *Exocytosis* involves fusion of the plasma membrane and a membrane-bound vesicle that formed inside the cytoplasm. *Endocytosis* involves an inward sinking of a patch of plasma membrane, which seals back on itself to form a vesicle inside the cytoplasm.

Diffusion is the net movement of molecules or ions of a substance into an adjoining region where they are not as concentrated.

The force of a concentration gradient can drive the directional movement of a substance across membranes. The gradient's steepness, temperature, molecular size, and electric and pressure gradients affect diffusion rates.

Cellular mechanisms increase and decrease concentration gradients across cell membranes.

5.6 How the Membrane Transporters Work

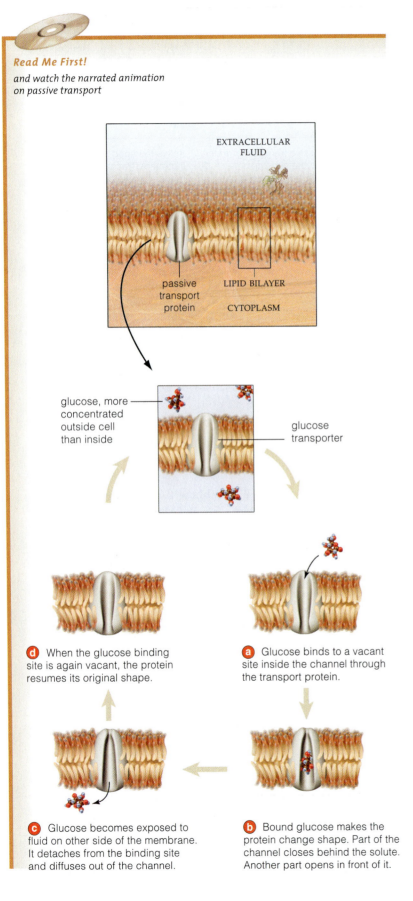

*Large, polar molecules and ions can't cross a lipid
bilayer; they require the help of transport proteins.*

Many kinds of solutes cross a membrane by diffusing
through a channel or tunnel inside transport proteins.
When one solute molecule or ion enters the channel
and weakly binds to the protein, the protein's shape
changes. The channel closes behind the bound solute
and opens in front of it, which exposes the solute to
the fluid environment on the opposite side of the
membrane. There, the binding site reverts to what it
was before, so the solute is released.

PASSIVE TRANSPORT

In **passive transport**, a concentration gradient, electric
gradient, or both drive the diffusion of a substance
across a membrane, through the interior of a transport
protein. The protein does not require an energy input
to assist the directional movement. That is why this
mechanism is also known as facilitated diffusion.

Some passive transporters are open channels, and
some are channels with gates that can be opened or
closed when conditions change. Others, including the
glucose transporter illustrated in Figure 5.18, assist
solutes across by undergoing reversible changes in
their shape.

The *net* direction of movement depends on how
many molecules or ions of the solute are randomly
colliding with the transporters. Encounters are more
frequent on the side of the membrane where the solute
concentration is greatest. The solute's *net* movement
tends to be toward the side of the membrane where it is
less concentrated.

If nothing else were going on, passive transport
would continue until concentrations on both sides of
a cell membrane became equal. But other processes
affect the outcome. For instance, glucose transporters
help glucose from blood move into cells, which use it
for biosynthesis and for quick energy. How? As fast
as glucose molecules are diffusing into cells, others
are being used up. By using up glucose, then, these
cells maintain the gradient that favors the uptake of
more glucose.

d When the glucose binding site is again vacant, the protein resumes its original shape.

a Glucose binds to a vacant site inside the channel through the transport protein.

c Glucose becomes exposed to fluid on other side of the membrane. It detaches from the binding site and diffuses out of the channel.

b Bound glucose makes the protein change shape. Part of the channel closes behind the solute. Another part opens in front of it.

Figure 5.18 Passive transport. This model shows one of the glucose transporters that span the plasma membrane. Glucose crosses in both directions. The *net* movement is down its concentration gradient until its concentrations are equal on both sides of the membrane.

ACTIVE TRANSPORT

Only in a dead cell have solute concentrations become equal on both sides of membranes. Living cells never stop expending energy to pump solutes into and out of their interior. With **active transport**, energy-driven protein motors help move a specific solute across the cell membrane *against* its concentration gradient.

Only specific solutes can bind to functional groups that line the interior channel of an active transporter. When the solute enters the channel and binds to one of those groups, the transporter accepts a phosphate group from an ATP molecule. The phosphate-group transfer changes the transporter's shape in a way that releases the solute on the other side of the membrane.

Figure 5.19 focuses on a **calcium pump**. This active transporter helps keep the concentration of calcium in a cell at least a thousand times lower than outside. A different active transporter, the **sodium–potassium pump**, mediates the movement of two kinds of ions, in opposite directions. Sodium ions (Na^+) from the cytoplasm diffuse into the open channel of the pump, where they bind to functional groups. A phosphate-group transfer by ATP prompts the pump to change shape and release the sodium ions outside the cell.

The channel through the activated pump is now open to the outside of the cell. Potassium ions (K^+) diffuse into the pump and bind to functional groups inside. The phosphate group is released from the pump, which reverts to its original shape. When it does, the potassium ions are released to the cytoplasm.

Active transport systems help maintain membrane gradients that are essential to many processes, such as muscle contraction and nerve cell (neuron) function.

Some transport proteins are open or gated channels across cell membranes. Others change shape when solutes bind to them.

In passive transport, a solute simply diffuses through the interior of a transporter; an energy input is not necessary.

In active transport, the net diffusion of a specific solute is against its concentration gradient. The transporter must be activated by an energy input from ATP to counter the force inherent in the gradient.

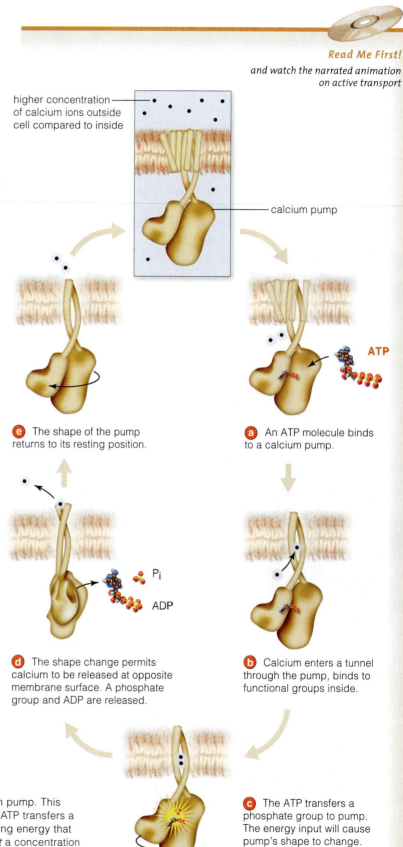

Read Me First!
and watch the narrated animation on active transport

higher concentration of calcium ions outside cell compared to inside

calcium pump

ATP

e The shape of the pump returns to its resting position.

a An ATP molecule binds to a calcium pump.

P_i

ADP

d The shape change permits calcium to be released at opposite membrane surface. A phosphate group and ADP are released.

b Calcium enters a tunnel through the pump, binds to functional groups inside.

c The ATP transfers a phosphate group to pump. The energy input will cause pump's shape to change.

Figure 5.19 Active transport by a calcium pump. This sketch shows its channel for calcium ions. ATP transfers a phosphate group to the pump, thus providing energy that can drive the movement of calcium *against* a concentration gradient across the cell membrane.

5.7 Which Way Will Water Move?

By far, more water diffuses across cell membranes than any other substance, so the main factors that influence its directional movement deserve special attention.

MOVEMENT OF WATER

Something as trickly as a running faucet or as mighty as Niagara Falls demonstrates **bulk flow**, or the mass movement of one or more substances in response to pressure, gravity, or another external force. Bulk flow accounts for some water movement in big multicelled organisms. A beating heart generates fluid pressure that pumps blood, which is mostly water. Sap flows inside tubes in trees, and this, too, is bulk flow.

What about the movement of water into and out of cells and organelles? If the concentration of water is not equal across a cell membrane, osmosis tends to occur. **Osmosis** is the diffusion of water across a selectively permeable membrane, to a region where the water concentration is lower.

You may be asking: How can water, a liquid, be more or less "concentrated"? Its concentration actually is influenced by the concentration of *solutes* on both sides of the membrane. If you pour glucose or some other solute into a glass of water, you will increase the volume of liquid in the glass. Now the same number of water molecules will become less concentrated than they were before; they will diffuse through the larger volume of space.

Now suppose you divide the interior of a glass container with a selectively permeable membrane, one that permits the diffusion of water but not glucose (a large, polar molecule) across it. You have created a water concentration gradient. More water molecules will diffuse across the membrane, into the solution, than will diffuse back (Figure 5.20).

In cases of osmosis, "solute concentration" refers to the *total number* of molecules or ions in a specified volume of a solution. It doesn't matter whether the dissolved substance is glucose, urea, or anything else; the *type* of solute doesn't dictate water concentration.

EFFECTS OF TONICITY

Suppose you decide to test the statement that water tends to move to a region where solutes are more concentrated. You make three sacs from a membrane that water but not sucrose can cross. You fill each sac with a solution that's 2 percent sucrose, then immerse one in a liter of water. You immerse another sac in a solution that is 10 percent sucrose. And you immerse the third sac in a solution that is 2 percent sucrose.

In each experiment, tonicity dictates the extent and direction of water movement across the membrane, as Figure 5.21 shows. *Tonicity* refers to the relative solute concentrations of two fluids. When two fluids that are on opposing sides of a membrane differ in their solute concentrations, the **hypotonic solution** is the one with fewer solutes. The one having more solutes is a **hypertonic solution**. And water tends to diffuse from a hypotonic fluid to a hypertonic one. **Isotonic solutions** show no net osmotic movement.

Normally, the fluid inside your cells is isotonic with tissue fluid outside. If the fluid outside becomes far too hypotonic, too much water will diffuse into those cells and make them burst. If it gets too hypertonic, water will diffuse out, and the cells will shrivel.

Most cells have built-in mechanisms that adjust to changes in tonicity. Red blood cells don't. Figure 5.21 shows what happens to them when tonicity changes.

EFFECTS OF FLUID PRESSURE

Selective transport of solutes across the plasma membrane keeps animal cells from bursting. Cells of plants and many protists, fungi, and bacteria avoid bursting with the help of pressure on their cell walls.

water molecules   protein molecules

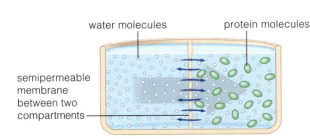

semipermeable
membrane
between two
compartments

Figure 5.20 Solute concentration gradients and osmosis. A membrane divides this container. Water, but not proteins, can cross it. Pour 1 liter of water in the left compartment and 1 liter of a protein-rich solution in the right one. The proteins occupy some of the space in the right one. The net diffusion of water in this case is from left to right (large *gray* arrow).

Pressure differences as well as solute concentrations influence osmosis. Take a look at Figure 5.22. It shows how water will diffuse across a membrane between a hypotonic and a hypertonic solution until the solute concentration is the same on both sides. As you can see, the *volume* of the formerly hypertonic solution has increased (because its solutes cannot diffuse out).

Hydrostatic pressure is the force that any volume of fluid exerts against a wall, a membrane, or some other structure enclosing it. (In plants, this pressure is called *turgor*.) Hydrostatic pressure that has built up in a cell can counter the further inward diffusion of water. This **osmotic pressure** is the amount of force preventing any further increase in volume.

Think of the pliable primary wall of a young plant cell. As it matures, many vesicles start to coalesce into a large central vacuole. During cell growth, water diffuses into the vacuole and puts more fluid pressure on the cell wall. The wall expands, so the cell volume increases. Continued expansion of the wall (and of the cell) ends when enough internal fluid pressure develops to counter the water uptake.

Plant cells are vulnerable to water losses, which can occur when soil dries out or becomes too salty. Water stops diffusing in and starts diffusing out, so internal fluid pressure falls and the cytoplasm shrinks.

In later chapters, you'll see how fluid pressure has a role in the distribution of water and solutes inside the body of plants and animals.

Osmosis is a net diffusion of water between two solutions that differ in solute concentration and are separated by a selectively permeable membrane. The greater the number of molecules and ions dissolved in a given amount of water, the lower the water concentration will be.

Water tends to move osmotically to regions of greater solute concentration (from hypotonic to hypertonic solutions). There is no net diffusion between isotonic solutions.

Fluid pressure that a solution exerts against a membrane or wall influences the osmotic movement of water.

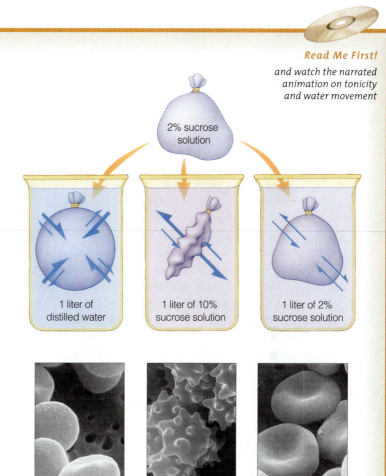

Read Me First! and watch the narrated animation on tonicity and water movement

2% sucrose solution

1 liter of distilled water   1 liter of 10% sucrose solution   1 liter of 2% sucrose solution

Hypotonic Conditions
Water diffuses into red blood cells, which swell up

Hypertonic Conditions
Water diffuses out of the cells, which shrink

Isotonic Conditions
No net movement of water, no change in cell size or shape

Figure 5.21 Tonicity and the direction of water movement. In each of three containers, arrow widths signify the direction and the relative amounts of flow. The micrographs below each sketch show the shape of a human red blood cell that is immersed in fluids of higher, lower, or equal concentrations of solutes. The solutions inside and outside red blood cells are normally balanced. This type of cell has no way to adjust to drastic change in solute levels in its fluid surroundings.

Figure 5.22 Experiment showing an increase in fluid volume as an outcome of osmosis. A semipermeable membrane separates two compartments. Over time, the net diffusion will be the same in both directions across the membrane, but the fluid volume in the second compartment will be greater because there are more solute molecules in it.

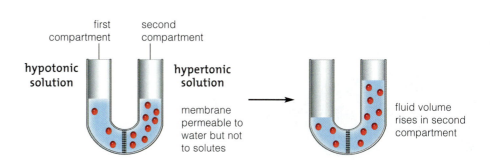

first compartment   second compartment

hypotonic solution   **hypertonic solution**

membrane permeable to water but not to solutes

fluid volume rises in second compartment

5.8 Membrane Traffic To and From the Cell Surface

We leave this chapter with another look at exocytosis and endocytosis. By these mechanisms, vesicles move substances to and from the plasma membrane. Vesicles help the cell take in and expel materials in amounts that are more than transport proteins can handle.

ENDOCYTOSIS AND EXOCYTOSIS

Think back on the membrane traffic to and from a cell surface (Figure 5.17). By **exocytosis**, a vesicle moves to the surface, and the protein-studded lipid bilayer of its membrane fuses with the plasma membrane. As this exocytic vesicle is losing its identity, its contents are released to the outside (Figures 5.23 and 5.24).

There are three pathways of **endocytosis**, but all take up substances near the cell surface. A small patch of plasma membrane balloons inward and pinches off inside the cytoplasm, forming an endocytic vesicle that moves its contents to some organelle or stores them in a cytoplasmic region (Figure 5.23).

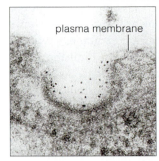 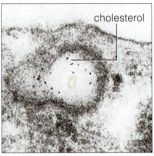

Figure 5.23 Endocytosis of cholesterol molecules.

With *receptor-mediated* endocytosis, receptors at the membrane bind to molecules of a hormone, vitamin, mineral, or another substance. A tiny pit forms in the plasma membrane beneath the receptors. The pit sinks into the cytoplasm and closes back on itself, and in this way it becomes a vesicle (Figure 5.24).

Phagocytosis ("cell eating") is a common endocytic pathway. Phagocytes such as amoebas engulf microbes, food particles, or cellular debris. In multicelled species, macrophages and some other white blood cells do this to pathogenic viruses or bacteria, cancerous body cells, and other threats.

Phagocytosis also involves receptors. Receptors that bind a target cause microfilaments to form a mesh just beneath the phagocyte's plasma membrane. When the microfilaments contract, they squeeze some cytoplasm toward the margins of the cell, forming a bulging lobe called a pseudopod (Figure 5.25). Pseudopods flow all around the target and form a vesicle. This sinks into the cell and fuses with lysosomes, the organelles of intracellular digestion. Lysosomes digest the trapped items into fragments and smaller, reusable molecules.

Bulk-phase endocytosis is not as selective. A vesicle forms around a small volume of the extracellular fluid regardless of the kinds of substances dissolved in it. This pathway continually removes patches of plasma membrane, balancing the steady additions that arrive in the form of exocytic vesicles from the cytoplasm.

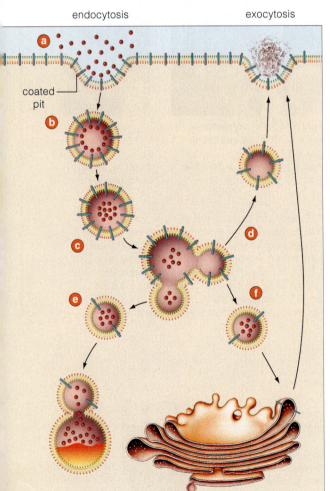

endocytosis   exocytosis

a Molecules get concentrated inside coated pits of plasma membrane.

coated pit

b Endocytic vesicles form from the pits.

c Enclosed molecules are sorted and often released from receptors.

d Many sorted molecules are cycled back to the plasma membrane.

e,f Many other sorted molecules are delivered to lysosomes and stay there or are degraded. Still others are routed to spaces in the nuclear envelope and inside ER membranes, and others to Golgi bodies.

Figure 5.24 Cycling of membrane lipids and proteins. This sketch starts with receptor-mediated endocytosis. Patches of the plasma membrane form endocytic vesicles. New membrane arrives as exocytic vesicles that budded from ER membranes and Golgi bodies. The membrane initially used for endocytic vesicles will cycle receptor proteins and lipids back to the plasma membrane.

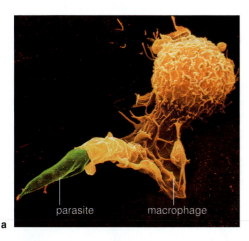

b   bacterium   phagocytic vesicle

Figure 5.25 (**a**) A macrophage engulfing *Leishmania mexicana*. This parasitic protozoan causes leishmaniasis, a disease that can be fatal. Bites from infected sandflies transmit the parasite to humans.

(**b**) Phagocytosis. Lobes of an amoeba's cytoplasm surround a target. The plasma membrane of the extensions fuses to form a phagocytic vesicle. In the cytoplasm, this endocytic vesicle fuses with lysosomes, which digest its contents.

MEMBRANE CYCLING

For as long as a cell remains alive, exocytosis and endocytosis continually replace and withdraw patches of its plasma membrane, as in Figure 5.24. And they apparently do so at rates that can maintain the plasma membrane's total surface area.

As one example, neurons release neurotransmitters in bursts of exocytosis. Each neurotransmitter is a type of signaling molecule that acts on neighboring cells. An intense burst of endocytosis counterbalances each major burst of exocytosis.

Whereas transport proteins in a cell membrane deal only with ions and small molecules, exocytosis and endocytosis move large packets of materials across a plasma membrane.

By exocytosis, a cytoplasmic vesicle fuses with the plasma membrane, and its contents are released outside the cell.

By endocytosis, a small patch of the plasma membrane sinks inward and seals back on itself, forming a vesicle inside the cytoplasm. Membrane receptors often mediate this process.

Summary

Section 5.1 Cells engage in metabolism, or chemical work. They obtain and use energy to stockpile, build, rearrange, and break apart substances.

Energy in biological systems flows in one direction, from usable to less usable forms. Life maintains its complex organization by being resupplied with energy lost from someplace else. Sunlight is the ultimate energy source for the web of life.

ATP, the main energy carrier, couples reactions that release energy with reactions that require it. It primes molecules to react through phosphate-group transfers.

Section 5.2 Metabolic pathways are orderly, enzyme-mediated reaction sequences. Photosynthesis and other energy-requiring, *biosynthetic* pathways build large molecules with high energy from smaller ones. Energy-releasing, *degradative* pathways such as aerobic respiration break down large molecules to small products with lower bond energies. Table 5.1 lists the participants.

Cells increase, maintain, and lower concentrations of substances by coordinating thousands of reactions. They rapidly shift rates of metabolism by controlling a few steps of reversible pathways.

Electron transfers, or oxidation–reduction reactions, often proceed in series at cell membranes.

Section 5.3 Enzymes are catalysts; they enormously enhance rates of specific reactions and are not altered by their function. Pockets or cavities in these big molecules create favorable microenvironments for the reaction; these are the active sites.

Table 5.1	Summary of the Main Participants in Metabolic Reactions
Reactant	Substance that enters a metabolic reaction or pathway; also called the substrate of a specific enzyme
Intermediate	Substance formed between the reactants and end products of a reaction or pathway
Product	Substance at the end of a reaction or pathway
Enzyme	A protein that greatly enhances reaction rates; a few RNAs also do this
Cofactor	Coenzyme (such as NAD^+) or metal ion; assists enzymes or taxis electrons, hydrogen, or functional groups between reaction sites
Energy carrier	Mainly ATP; couples energy-releasing reactions with energy-requiring ones
Transport protein	Protein that passively assists or actively pumps specific solutes across a cell membrane

Activation energy is the minimum internal energy that reactant molecules must have for a reaction to occur. Enzymes lower it by boosting substrate concentrations in the active site, by orienting substrates, by shutting out most or all water, and by inducing a precise fit with them.

Section 5.4   Cofactors (metal ions, coenzymes, or both) help an enzyme catalyze a reaction. Controls over enzyme action influence the kinds and amounts of substances available. Enzymes function best within a limited range of temperature, pH, and salinity.

Section 5.5   Diffusion is the movement of molecules or ions toward an adjoining region where they are less concentrated. The steepness of the concentration gradient, temperature, molecular size, and gradients in electrical charge and pressure influence diffusion rates.

Built-in cellular mechanisms work with and against gradients to move solutes across membranes.

Molecular oxygen, carbon dioxide, and other small nonpolar molecules diffuse across a membrane's lipid bilayer. Ions and large, polar molecules such as glucose cross it with the help of transport proteins. Some water moves through proteins and some through the bilayer.

Section 5.6   Many solutes cross membranes through transport proteins that act as open or gated channels or that reversibly change shape. Passive transport does not require energy input; a solute is free to follow its own concentration gradient across the membrane. Active transport requires an energy input from ATP to move a specific solute against its concentration gradient.

Section 5.7   Osmosis is the diffusion of water across a selectively permeable membrane, down the water concentration gradient. Pressure gradients can affect it.

Section 5.8   By exocytosis, a cytoplasmic vesicle fuses with the plasma membrane, and its contents are released outside. By endocytosis, a patch of plasma membrane forms a vesicle that sinks into the cytoplasm.

Self-Quiz

Answers in Appendix III

1. _____ is life's primary source of energy.
   a. Food   b. Water   c. Sunlight   d. ATP

2. Which of the following statements is *not* correct? A metabolic pathway _____ .
   a. has an orderly sequence of reaction steps
   b. is mediated by only one enzyme that starts it
   c. may be biosynthetic or degradative, overall
   d. all of the above

3. An enzyme _____ .
   a. is a protein
   b. lowers the activation energy of a reaction
   c. is destroyed by the reaction it catalyzes
   d. a and b

4. Immerse a living cell in a hypotonic solution, and water will tend to _____ .
   a. diffuse into the cell   c. show no net movement
   b. diffuse out of the cell   d. move in by endocytosis

5. _____ can readily diffuse across a lipid bilayer.
   a. Glucose   c. Carbon dioxide
   b. Oxygen   d. b and c

6. Sodium ions cross a membrane at transport proteins that receive an energy boost. This is a case of _____ .
   a. passive transport   c. facilitated diffusion
   b. active transport   d. a and c

7. Vesicle formation occurs in _____ .
   a. membrane cycling   c. endocytosis, exocytosis
   b. phagocytosis   d. all of the above

8. The rate of diffusion is affected by _____ .
   a. temperature   c. molecular size
   b. electrical gradients   d. all of the above

9. Match the substance with its suitable description.
   ____ coenzyme or metal ion   a. reactant
   ____ adjusts gradients at membrane   b. enzyme
   ____ substance entering a reaction   c. cofactor
   ____ substance formed during   d. intermediate
        a reaction   e. product
   ____ substance at end of reaction   f. energy carrier
   ____ enhances reaction rate   g. transporter
   ____ mainly ATP   proteins

Critical Thinking

1. Cyanide, a toxic compound, binds irreversibly to an enzyme that is a component of electron transfer chains. The outcome is *cyanide poisoning*. Binding prevents the enzyme from donating electrons to a nearby acceptor molecule in the system. What effect will this have on ATP formation? From what you know of ATP's function, what effect will this have on a person's health?

2. In cells, superoxide dismutase (*below*) has a quaternary structure—it consists of two polypeptide chains. In each chain, a strandlike domain is arrayed as a barrel around a copper ion and a zinc ion (coded *red* and *blue*).

   Which part of the barrel is probably hydrophobic? Which part is hydrophilic? Do you suppose substrates bind inside or outside the barrels? Do the metal ions have a role in catalysis?

3. Catalase breaks down hydrogen peroxide, a reactive by-product of aerobic metabolism, to water and oxygen (Figure 5.10). It is a very efficient enzyme: One molecule of catalase can break down 6 million hydrogen peroxide molecules every minute. It is found in most organisms that live under aerobic conditions because hydrogen

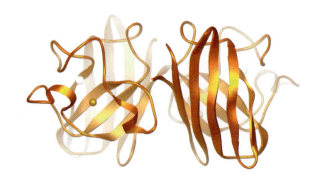

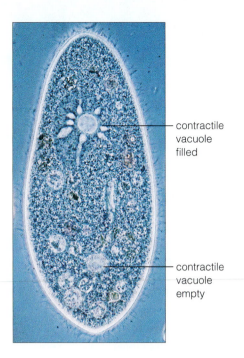

contractile
vacuole
filled

contractile
vacuole
empty

Figure 5.26 *Paramecium* contractile vacuoles.

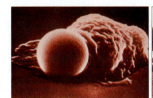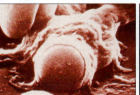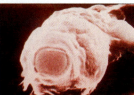

Figure 5.27 Go ahead, name the mystery membrane mechanism.

peroxide is toxic—cells must dispose of it quickly or they risk being damaged. Peroxide is catalase's substrate; but by a neat trick, it also can inactivate other toxins, including alcohol. Can you guess what the trick is?

4. Nutritional supplements often include plant enzymes. Explain why it is not likely that plant enzymes will aid your digestion.

5. Why does applying lemon juice to sliced apples keep them from turning brown?

6. Explain why hydrogen peroxide bubbles when you dribble it on an open cut but does not bubble on skin that is unbroken.

7. Most of the cultivated fields in California are heavily irrigated. Over the years, most of the imported water has evaporated from the soil, leaving behind solutes. What problems will the altered soil cause plants?

8. Water moves osmotically into *Paramecium*, a single-celled aquatic protist. If unchecked, the influx would bloat the cell and rupture its plasma membrane, killing the cell. An energy-requiring mechanism that involves contractile vacuoles expels excess water (Figure 5.26). Water enters the vacuole's tubelike extensions and collects inside. A full vacuole contracts and squirts water out of the cell through a pore. Are *Paramecium*'s surroundings hypotonic, hypertonic, or isotonic?

9. Imagine you're a juvenile shrimp living in an *estuary*, where freshwater draining from the land mixes with saltwater from the sea. Many people own homes near a lake and want boat access to the sea. They ask their city for permission to build a canal to your estuary. If they succeed, what may happen to you?

10. Is the white blood cell shown in Figure 5.27 disposing of a worn-out red blood cell by endocytosis, phagocytosis, or both?

Media Menu

 Student CD-ROM

Impacts, Issues Video
    Alcohol, Enzymes, and Your Liver
Big Picture Animation
    Energy, enzymes, and movement across membranes
Read-Me-First Animation
    Controlling energy release
    Catalase action
    Passive transport
    Active transport
    Tonicity and water movement
Other Animations and Interactions
    Activation energy interaction
    Allosteric activation
    Feedback inhibition

 InfoTrac

- Harnessing the Energy. *World and I*, October 2001.
- Ion Channel Protein Contraceptive Target. *Drug Discovery & Technology News*, October 2001.
- Drug Abuse During 1970s and 1980s May Explain Doubling of Deaths from Alcoholic Liver Disease. *Hepatitis Weekly*, August 2002.

 Web Sites

- National Institute on Alcohol Abuse and Alcoholism: www.niaaa.nih.gov
- Adenosine Triphosphate—ATP: www.bris.ac.uk/Depts/Chemistry/MOTM/atp/atp1.htm
- Introduction to Enzymes: www.worthington-biochem.com/introBiochem/introEnzymes.html
- Pumping Ions: www.mbl.edu/publications/LABNOTES/10.1/pumping_ions.html

How Would You Vote?

The only cure for liver failure, regardless of its cause, is a liver transplant. A shortage of livers means many potential transplant recipients die waiting. How should these organs be allocated? Should people who invited liver failure by their own abusive life-style be a lower priority for transplants than those with failure brought on by a transfusion or a genetic disorder?

IMPACTS, ISSUES *Pastures of the Seas*

Think about the last bit of apple, celery, chicken, pizza, or any other food you ate. Where did it come from? Look past the refrigerator, the market or restaurant, or even the farm. Look to individual plants, the starting point for nearly all of your food.

Plants, and many bacteria and protists, are "self-nourishing" organisms, or **autotrophs**. They tap into an environmental energy source and use it to *make* food from simple materials. By contrast, most bacteria and protists, all fungi, and all animals are **heterotrophs**. They are not self-nourishing; they cannot make their own food. They must eat autotrophs, one another, and organic wastes. (*Hetero–* means other; in this case, "being nourished by others.")

Plants do something you'll never do. By the process of **photosynthesis**, they make food by using no more than sunlight energy, water, and carbon dioxide (CO_2). Each year, they produce 220 billion tons of sugar, enough to make 300 quadrillion sugar cubes. These photosynthetic autotrophs have been doing so for more than a billion years. That is a LOT of sugar.

Uncountable numbers of photosynthesizers also abound in the seas. A cupful of seawater may hold more than 24 million microscopically small cells of different species! Most are bacteria and protists that form "pastures of the seas"—the producers that feed most other marine organisms. Like plants, they too "bloom" in spring, when nutrients churned up from the deep support rapid population growth. Figure 6.1 is a record of an algal bloom that stretched from North Carolina to Spain.

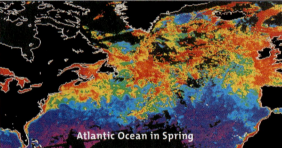

Figure 6.1 Satellite images that convey the magnitude of photosynthetic activity during springtime in the North Atlantic Ocean. Sensors responded to concentrations of chlorophyll, which were greatest in regions coded *red*.

the big picture

Catching the Rainbow Energy enters the world of life when chlorophyll and other photosynthetic pigments absorb energy in the sun's rays.

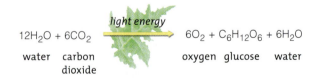

$$12H_2O + 6CO_2 \xrightarrow{\text{light energy}} 6O_2 + C_6H_{12}O_6 + 6H_2O$$

water carbon dioxide   oxygen glucose water

Overview of Photosynthesis Photosynthesis occurs in two stages in chloroplasts. Energy from the sun is converted to chemical energy and stored in ATP and NADPH. These molecules are later used to assemble sugars from carbon dioxide and water.

Imagine zooming in on just one small patch of "pasture" in an Antarctic sea. There, tiny shrimplike crustaceans are rapidly eating tinier photosynthesizers, including algal cells of the sort shown in the filmstrip. Dense concentrations of such crustaceans, known as krill, are feeding other animals, including fishes, penguins, seabirds, and the immense blue whale. A single, mature whale is straining four tons of krill from the water today, as it has been doing for months. And before they themselves were eaten, the four tons of krill had munched through *1,200 tons* of the pasture!

Another point: Collectively, photosynthetic cells on land and in the seas handle staggering numbers of reactant and product molecules. By doing so, they even help shape the global climate. They also sponge up nearly half of the CO_2 we humans release each year, as by burning fossil fuels. Without them, CO_2 would accumulate faster and warm the atmosphere, which already is warming too fast.

In short, *photosynthesis is the main pathway by which energy and carbon enter the web of life.* Photosynthetic autotrophs make, use, and store organic compounds, the food for most heterotrophs. And *all* organisms release that stored energy for cellular work, mainly by aerobic respiration.

There are different types of photoautotrophs, and they perform photosynthesis in different ways. In this chapter we focus on oxygenic (oxygen-producing) photosynthesis in plants and algae.

✓ How Would You Vote?

Crop plants feed most of the human population. Limits on the activity of some enzymes can limit crop production. Should we genetically engineer plants to boost photosynthesis and get higher crop yields? See the Media Menu for details, then vote online.

Making ATP and NADPH   In the first stage of photosynthesis, sunlight energy becomes converted to chemical bond energy of ATP. Water molecules are broken apart, NADPH forms, and oxygen escapes into the air.

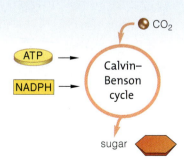

Making Sugars   The second stage is the "synthesis" part of photosynthesis. ATP delivers energy to reaction sites where sugars are built with atoms of hydrogen (delivered by NADPH), carbon, and oxygen (from carbon dioxide in the air).

6.1 Sunlight As an Energy Source

*Photosynthesis runs on a fraction of the electromagnetic spectrum, or the full range of energy radiating from the sun. Radiant energy undulates across space, something like waves crossing a sea. The horizontal distance between two successive waves is a **wavelength**.*

PROPERTIES OF LIGHT

Our story starts with energy in the sun's rays—not all of it, just light of wavelengths between 380 and 750 nanometers. These are wavelengths of visible light, the ones that drive photosynthesis.

Light is made of *photons*, which are individual packets of electromagnetic energy traveling in waves. The shorter a photon's wavelength, the higher its energy (Figure 6.2). For example, blue light has a shorter wavelength and more energy than red light:

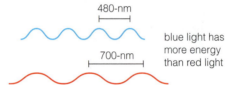

480-nm

700-nm

blue light has more energy than red light

Photons with wavelengths shorter than violet light are energetic enough to disrupt DNA of living cells.

PIGMENTS—THE RAINBOW CATCHERS

Pigments are a class of molecules that absorb photons with particular wavelengths. Photons that a pigment cannot absorb bounce off or continue on through it; they are reflected or transmitted.

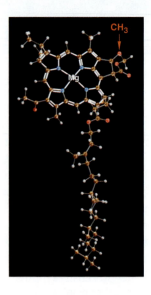

Figure 6.3 Ball-and-stick model for chlorophylls *a* and *b*, which differ by only a single functional group. In chlorophyll *b*, the group is —COO⁻, not the —CH₃ shown. The light-catching portion is the flattened ring structure—which is similar to a heme except it holds a magnesium atom instead of iron. The hydrocarbon backbone readily dissolves in the lipid bilayers of cell membranes.

Certain pigments are the molecular bridges from sunlight to photosynthesis. **Chlorophyll *a*** is the most abundant type in plants, green algae, and a number of photoautotrophic bacteria (Figure 6.3). It is the best at absorbing red and violet wavelengths. Chlorophyll *b* absorbs light at slightly different wavelengths. It is an *accessory* pigment, meaning that it enhances efficiency of photosynthesis reactions by capturing additional wavelengths. All chlorophylls reflect or transmit green wavelengths, which is why plant parts that are rich in chlorophylls appear green to us.

Accessory pigments include the **carotenoids**, which absorb blue-violet and blue-green wavelengths, and reflect red, orange, and yellow ones. Beta-carotene is a carotenoid that colors carrots and other plant parts

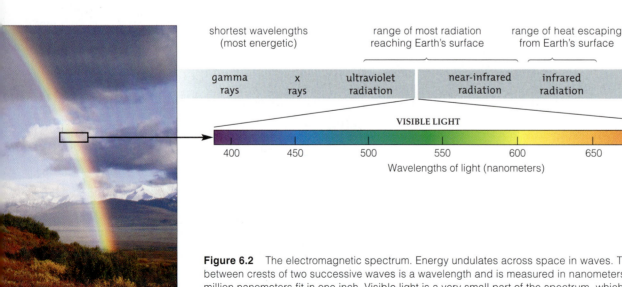

Figure 6.2 The electromagnetic spectrum. Energy undulates across space in waves. The distance between crests of two successive waves is a wavelength and is measured in nanometers. About 2.5 million nanometers fit in one inch. Visible light is a very small part of the spectrum, which includes all electromagnetic waves. Like many other organisms, we perceive visible light wavelengths as colors.

Figure 6.4 (**a**) Absorption spectra reveal the efficiency with which chlorophylls *a* and *b* absorb wavelengths of visible light. Peaks in the graphs reveal the wavelengths that each type of pigment absorbs best.

(**b**) Absorption spectra for beta-carotene (a carotenoid) and phycobilin. As Figure 6.6 describes, before scientists devised ways to measure absorption efficiencies, the botanist T. Engelmann figured out which colors of light were best at driving photosynthesis in a green alga.

Collectively, these and other photosynthetic pigments can capture almost the entire spectrum of visible light.

Figure 6.5 Leaf color. In spring and summer, intensely green leaves have an abundance of chlorophylls, which mask the presence of carotenoids, xanthophylls, and other accessory pigments.

In many kinds of plants, chlorophyll synthesis lags behind its breakdown in autumn, so more stable pigments show through. Cold, sunny days trigger the production of water-soluble anthocyanins in leaf cells. The anthocyanins act like a sunscreen; they protect leaves from ultraviolet radiation.

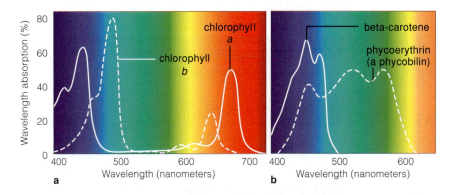

Figure 6.6 T. Engelmann's study of photosynthesis in *Spirogyra*, a strandlike green alga. A long time ago, most people assumed plants withdrew raw materials for photosynthesis from soil. By 1882 a few chemists suspected that plants use light, water, and something in the air. Engelmann wondered: What parts of sunlight do plants favor?

As he knew, free oxygen is released during photosynthesis. He also knew some bacteria use oxygen during aerobic respiration, as most organisms do. He hypothesized: If bacteria require oxygen, then we can expect them to gather in places where the most photosynthesis is going on. He put a water droplet containing bacterial cells on a microscope slide with the green alga *Spirogyra*. He used a crystal prism to break up a beam of sunlight and cast a spectrum of colors across the slide.

Bacteria gathered mostly where violet and red light fell on the green alga. Algal cells released more oxygen in the part illuminated by light of those colors—the very best light for photosynthesis. Compare Figure 6.4.

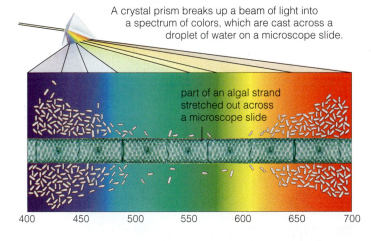

A crystal prism breaks up a beam of light into a spectrum of colors, which are cast across a droplet of water on a microscope slide.

part of an algal strand stretched out across a microscope slide

orange. **Xanthophylls** are yellow, brown, purple, or blue accessory pigments; **phycobilins** are red or blue-green. Absorption spectra (singular, spectrum) give us a picture of how such photosynthetic pigments absorb different wavelengths of visible light (Figure 6.4).

The chlorophyll content in the leaves of deciduous species declines in autumn and lets the carotenoids, xanthophylls, and **anthocyanin**, a red-purple pigment, show through (Figure 6.5). Each year in New England, tourists spend a billion dollars to watch a three-week display of red, orange, and gold leaves of maples and other trees. We also can thank the deep red to purple anthocyanins for the visual appeal of many flowers and food, including blueberries, red grapes, cherries, red cabbage, and rhubarb.

Such photosynthetic pigments do not work alone. Organized arrays of them work together and harvest energy from the sun. For now, start thinking about the structure of that chlorophyll molecule in Figure 6.3—particularly the flattened ring. Here, alternating single and double covalent bonds share electrons. And these are the electrons which, when excited by inputs of energy, get photosynthesis going.

Light from the sun travels through space in waves, and wavelengths of visible light correspond to specific colors.

Chlorophyll a and diverse accessory pigments absorb specific wavelengths of visible light. They are the molecular bridge between the sun and photosynthesis.

6.2 What Is Photosynthesis and Where Does It Happen?

Sit outdoors on a warm, sunny day and you will never build your own food but you will get hot. Plants plug into the sun's energy usefully, without getting cooked.

TWO STAGES OF REACTIONS

Photosynthesis proceeds through two reaction stages. The first stage, the **light-dependent reactions**, converts light energy to chemical bond energy of ATP. Also, water molecules are split, and the coenzyme NADP+ picks up the released electrons and hydrogen. We call its reduced form NADPH. The oxygen atoms released from water molecules escape into the surroundings.

In the second stage, called the **light-independent reactions**, energy from ATP jump-starts reactions that form glucose and other carbohydrates. At these same sites, NADPH gives up electrons and hydrogen ions, which bond with carbon and oxygen to form glucose.

Here is a simple way to summarize the reactions of photosynthesis:

$$12H_2O + 6CO_2 \xrightarrow[\text{enzymes}]{\text{light energy}} 6O_2 + C_6H_{12}O_6 + 6H_2O$$

water carbon dioxide   oxygen glucose water

A LOOK INSIDE THE CHLOROPLAST

Photosynthetic reactions differ among certain bacteria, protists, and plants. For now, focus on what goes on in **chloroplasts**, the organelles of photosynthesis in plants and algae. Each chloroplast has two outer membranes, which enclose a semifluid interior, the stroma (Figure 6.7c). Inside the stroma is the **thylakoid membrane**, a third membrane folded in ways that form a single compartment. Often the folds look like flattened channels between stacks of flattened sacs (thylakoids). The space inside the sacs is part of one continuous compartment. Sugars are built outside this compartment, in the stroma (Figure 6.8).

Embedded in all the thylakoids are **photosystems**: clusters of 200 to 300 pigments and other molecules that trap energy from the sun. Chloroplasts have two types of photosystems, called I and II (Figure 6.7d).

PHOTOSYNTHESIS CHANGED THE BIOSPHERE

Before zooming down further, to the mechanisms of photosynthesis, zoom out in your mind to the global impact of one of the steps involved. About 3.2 billion

leaf's upper surface   photosynthetic cells   central vacuole

vein   stoma (gap) in lower epidermis   chloroplast

a Section from the leaf, showing its internal organization

b One photosynthetic cell inside the leaf

Figure 6.7 Zooming in on sites of photosynthesis in a typical plant leaf. Two thousand chloroplasts, lined up single file, would be no wider than a dime. Think of all the chloroplasts in a corn or rice plant—each a tiny sugar-making factory—to get a sense of the magnitude of metabolic events required to feed you and every other living thing.

years ago, oxygen released by photosynthetic bacteria started accumulating in the atmosphere, which before then held little of it. As the atmosphere changed, so did the world of life. All that free oxygen favored the evolution of a novel pathway—aerobic respiration—that efficiently releases a great deal of energy from organic compounds. An oxygen-rich atmosphere was a key environmental factor in the evolution of large, active animals, which the aerobic pathway sustains. Breathing in oxygen helps keep them alive.

In the first stage of photosynthesis, sunlight energy drives ATP and NADPH formation, and oxygen is released. In chloroplasts, this stage occurs at the thylakoid membrane.

Embedded in the membrane are photosystems—clusters of pigments and other molecules—where light energy is captured.

The second stage occurs in the stroma. Energy from ATP drives the synthesis of sugars. Carbon dioxide provides carbon and oxygen atoms for the reactions. NADPH delivers the required electrons and hydrogen atoms.

The atmosphere was free of oxygen before photosynthesis evolved. Oxygen released by emerging photosynthesizers slowly accumulated. It changed the atmosphere and became a selective force in the evolution of aerobic respiration.

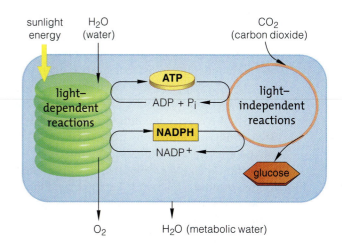

Figure 6.8 Overview of the two stages of photosynthesis in a chloroplast. The first stage, the light-dependent reactions, occurs at the thylakoid membrane. The second stage (light-independent reactions that produce sugars) occurs in the stroma.

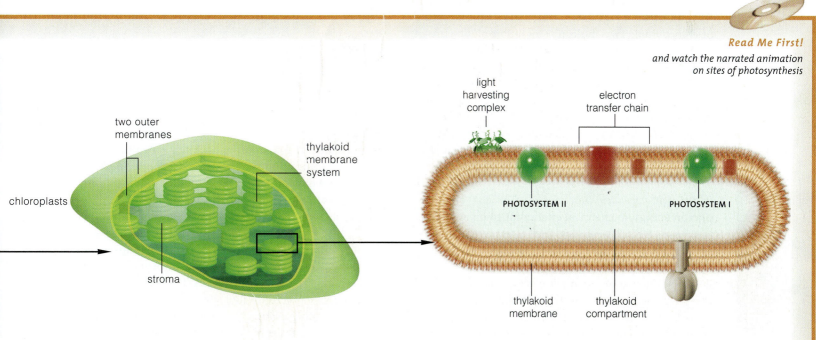

Read Me First!
and watch the narrated animation on sites of photosynthesis

c Closer look at one chloroplast. It has two outer membranes and an inner *thylakoid* membrane in its semifluid interior (the stroma). In many cells, the inner membrane resembles stacks of flattened sacs connected by channels. The interiors of all sacs and channels interconnect, forming a single compartment.

d Components of the thylakoid membrane system that carry out the first stage of photosynthesis—the light-dependent reactions. Light-harvesting complexes capture photon energy and pass it to two types of photosystems. Electron transfer chains embedded in the membrane have roles in ATP and NADPH formation.

6.3 Light-Dependent Reactions

In the first stage of photosynthesis, the sunlight energy harvested at photosystems drives ATP formation. Water molecules are split, and their oxygen diffuses away. NADP+, a coenzyme, picks up the electrons and hydrogen, which will be used in the second stage to form sugars.

TRANSDUCING THE ABSORBED ENERGY

Suppose a photon collides with a pigment molecule that absorbs it. The photon's energy will boost one of the pigment's electrons to a higher energy level. If nothing else happens, the electron quickly will drop back to its unexcited state, losing the extra energy as it does. The energy is emitted as heat, or as another photon. Photon emissions from electrons losing extra energy are visible as fluorescent light.

In the thylakoid membrane, however, energy that excited electrons give up is kept in play. Embedded in the membrane are many photosystems. Surrounding them are hundreds of light-harvesting complexes, or circular clusterings of pigments and other proteins (Figure 6.9a). Pigments in light-harvesting complexes also absorb photon energy, but they don't waste it. Electrons of these pigments can hold on to energy by passing it back and forth, like a volleyball. The energy released from one cluster is passed to another, which passes it on to another, and so on until it reaches a photosystem—a reaction center.

Look back on Figure 6.3, which shows the structure of chlorophyll. Two molecules of chlorophyll *a* are at the center of every photosystem. Their flat rings face each other so closely that electrons in both rings are destabilized. When light-harvesting neighbors pass on photon energy to a photosystem, electrons come right off of that special pair of chlorophylls.

The freed electrons immediately enter an electron transfer chain positioned next to the photosystem in the thylakoid membrane. *The entry of electrons from a photosystem into an electron transfer chain is the first step in the light-dependent reactions*—in the conversion of photon energy to chemical energy for photosynthesis.

MAKING ATP AND NADPH

Let's use Figure 6.9 to track electrons that a type II photosystem gives up. **Electron transfer chains**, recall, are cell membrane components. Each is an organized array of enzymes, coenzymes, and other proteins through which electrons are transferred step-by-step. In the process of moving electrons, molecules of the chain pick up hydrogen ions (H^+) from the stroma, cart them across the thylakoid membrane, and release

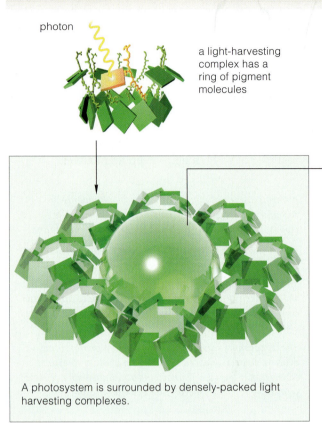

photon

a light-harvesting complex has a ring of pigment molecules

A photosystem is surrounded by densely-packed light harvesting complexes.

a In the thylakoid membrane of chloroplasts, rings of pigments can intercept photons coming from any direction. They pass captured photon energy to nearby photosystems. Each photosystem collects energy from hundreds of light-harvesting complexes surrounding it; a few are shown here.

Figure 6.9 How ATP and NADPH form during photosynthesis in chloroplasts.

them to the inner compartment. Their repeated action causes concentration and electric gradients to build up across the membrane. The combined force of those gradients attracts H^+ back toward the stroma.

The H^+ ions can only cross the membrane with the help of **ATP synthases**, as explained in Section 4.3. Ion flow through these membrane proteins causes the attachment of inorganic phosphate to a molecule of ADP in the stroma. In this way, ATP forms.

As long as electrons flow through transfer chains, the cell can keep on producing ATP. But where do the electrons come from in the first place? By the process of *photolysis*, photosystem II replaces its lost electrons by pulling them from water molecules—which then dissociate into hydrogen ions and molecular oxygen. The free oxygen diffuses out of the chloroplast, then

you are here

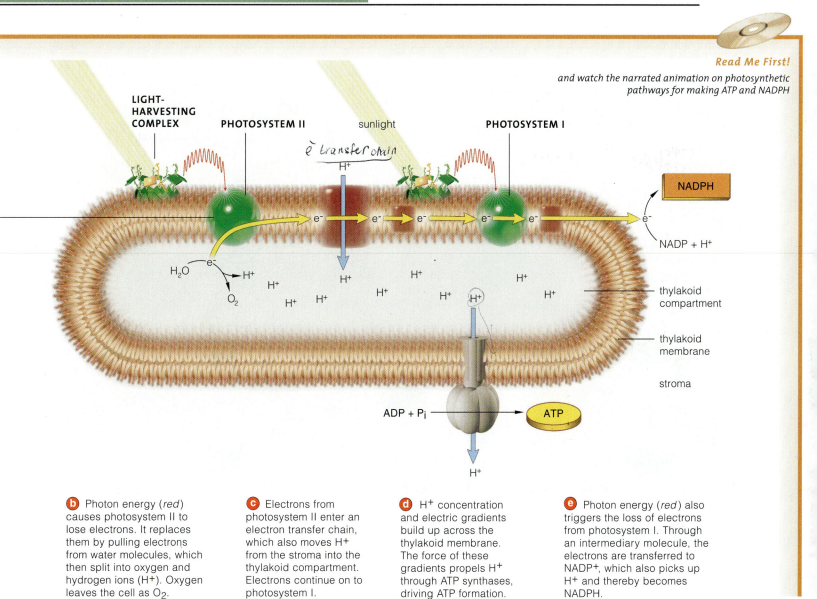

Read Me First!
and watch the narrated animation on photosynthetic pathways for making ATP and NADPH

b Photon energy (*red*) causes photosystem II to lose electrons. It replaces them by pulling electrons from water molecules, which then split into oxygen and hydrogen ions (H+). Oxygen leaves the cell as O_2.

c Electrons from photosystem II enter an electron transfer chain, which also moves H+ from the stroma into the thylakoid compartment. Electrons continue on to photosystem I.

d H+ concentration and electric gradients build up across the thylakoid membrane. The force of these gradients propels H+ through ATP synthases, driving ATP formation.

e Photon energy (*red*) also triggers the loss of electrons from photosystem I. Through an intermediary molecule, the electrons are transferred to NADP+, which also picks up H+ and thereby becomes NADPH.

out of the cell and into the air. Hydrogen ions remain in the thylakoid compartment, and they contribute to the gradients that drive ATP formation.

So where do the electrons end up? After passing through the electron transfer chain, they continue on to photosystem I. There, light-harvesting complexes volley energy to a special pair of chlorophylls at the photosystem's reaction center, causing them to release electrons. An intermediary molecule transfers them to NADP+, which attracts hydrogen ions at the same time. In this way, NADPH forms.

Photosystem I also runs independently in a more ancient cyclic pathway. Electrons freed from it enter an adjoining transfer chain, which moves hydrogen ions into the thylakoid compartment. As before, the resulting gradient drives ATP formation. At the end

of this chain, however, electrons are cycled back to photosystem I, and no NADPH forms.

These "noncyclic" and "cyclic" pathways operate at the same time in many photosynthetic organisms. Which one dominates at a particular time depends on metabolic demands for ATP and NADPH.

In the light-dependent reactions, sunlight energy drives the formation of ATP, NADPH, or both.

Both ATP and NADPH form by a noncyclic pathway in which electrons are pulled from water molecules, then flow through two types of photosystems, and finally to NADP+. This is the photosynthetic pathway that releases free oxygen.

ATP alone forms in a cyclic pathway that starts and ends at photosystem I.

6.4 A Case of Controlled Energy Release

One of the themes threading through this book is that organisms convert one form of energy to another in highly controlled ways. The light-dependent reactions are a classic example of such conversions. Figure 6.10 walks you step-by-step through these conversions.

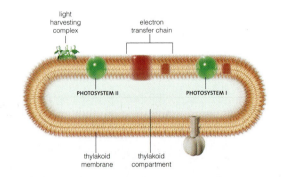

Read Me First!

and watch the narrated animation on energy release in photosynthesis

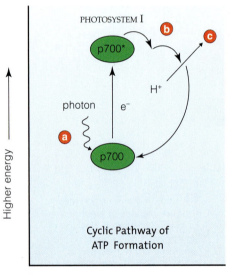

Cyclic Pathway of ATP Formation

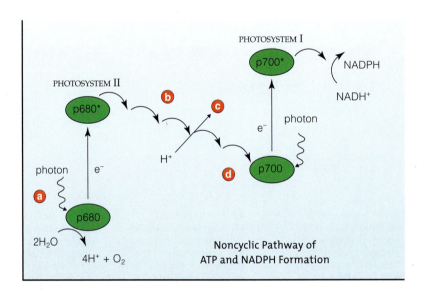

Noncyclic Pathway of ATP and NADPH Formation

a Photosystem I gets a boost of photon energy from a light-harvesting cluster. It loses an electron.

b The electron passes from one molecule to another in an electron transfer chain that is embedded in the thylakoid membrane. It loses a little energy with each transfer, and ends up being reused by photosystem I.

c Molecules in the transfer chain ferry H^+ across the thylakoid membrane into the inner compartment. Hydrogen ions accumulating in the compartment create an electrochemical gradient across the membrane that drives ATP synthesis, as shown in Figure 6.9.

a Photosystem II gets a boost of photon energy from a light-harvesting cluster, then loses an electron. Here, too, the electron moves through a different electron transfer chain and loses a little energy with each transfer. It ends up at photosystem I.

b Photosystem I gets a boost of photon energy from a light-harvesting complex, then loses electrons. The freed electrons, along with hydrogen ions, are used in the formation of NADPH from $NADH^+$.

c As in the cyclic pathway, operation of the electron transfer chain puts hydrogen ions into the thylakoid compartment. In this case, hydrogens released from dissociated water molecules also enter the compartment. The H^+ concentration and electric gradient across the membrane are tapped for ATP formation (Figure 6.9).

d Electrons lost from photosystem I are replaced by the electrons lost from photosystem II. Electrons lost from photosystem II are replaced by electrons from water. (Photolysis pulls water molecules apart into electrons, H^+ and O_2.)

Figure 6.10 Energy transfers in the light-dependent reactions. The pair of chlorophyll *a* molecules at the center of photosystem I is designated p700. The pair in photosystem II is designated p680. The pairs respond most efficiently to wavelengths of 700 and 680 nanometers, respectively.

6.5 Light-Independent Reactions: The Sugar Factory

The chloroplast is a sugar factory, and the Calvin–Benson cycle is its machinery. These cyclic, light-independent reactions are the "synthesis" part of photosynthesis.

Sugars form in the **Calvin–Benson cycle**, which runs inside the stroma of chloroplasts (Figure 6.11). This cyclic pathway uses ATP and NADPH from the light-dependent reactions. We call them light-*independent* because they also can run in the dark, as long as ATP and NADPH are available.

ATP energy drives these sugar-building reactions, and NADPH donates hydrogen and electrons. Plants get carbon and oxygen building blocks from carbon dioxide (CO_2) in the air. Algae of aquatic habitats get them from CO_2 dissolved in water.

Rubisco, an enzyme, transfers a carbon from CO_2 to five-carbon ribulose biphosphate, or RuBP. The resulting unstable compound is the entry point for the Calvin–Benson cycle. It splits at once into two stable molecules of phosphoglycerate (PGA), each having a backbone of three carbons. The process of securing carbon from the environment by incorporating it in a stable organic compound is called **carbon fixation**.

Each PGA gets a phosphate group from ATP, and hydrogen and electrons from NADPH. For every six CO_2 fixed, twelve phosphoglyceraldehydes (PGAL) form. Ten PGAL become rearranged in a way that regenerates RuBP. The other two combine to make a six-carbon glucose with a phosphate group attached.

Most of the glucose is converted at once to sucrose or starch by other pathways that conclude the light-independent reactions. Sucrose is the main form in which carbohydrate is transported in plants; starch is the main storage form. Cells convert excess PGAL to starch, which they briefly store as starch grains in the stroma. After the sun goes down, starch is converted to sucrose for export to other cells in leaves, stems, and roots. Photosynthetic products and intermediates end up as energy sources or building blocks for all the lipids, amino acids, and other organic compounds that plants require for growth, survival, and reproduction.

Driven by ATP energy, the light-independent reactions make sugars with hydrogen and electrons from NADPH, and with carbon and oxygen from carbon dioxide.

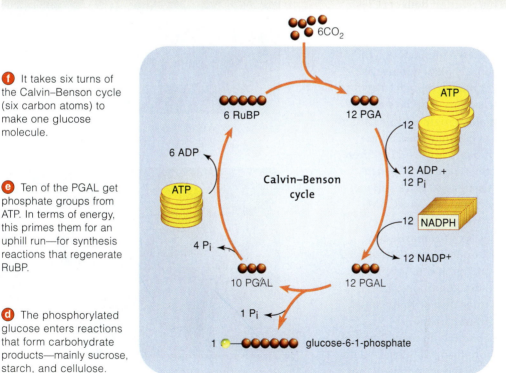

Read Me First! and watch the narrated animation on the Calvin–Benson cycle

f It takes six turns of the Calvin–Benson cycle (six carbon atoms) to make one glucose molecule.

e Ten of the PGAL get phosphate groups from ATP. In terms of energy, this primes them for an uphill run—for synthesis reactions that regenerate RuBP.

d The phosphorylated glucose enters reactions that form carbohydrate products—mainly sucrose, starch, and cellulose.

a CO_2 in air spaces inside a leaf diffuses into a photosynthetic cell. Rubisco attaches the carbon atom of CO_2 to RuBP, which starts the Calvin–Benson cycle. Each resulting intermediate splits at once into two PGAs.

b Each PGA molecule gets a phosphate group from ATP, plus hydrogen and electrons from NADPH. The resulting intermediate, PGAL, is thus primed for reaction.

c Two of the twelve PGAL molecules combine to form one molecule of glucose with an attached phosphate group.

Figure 6.11 Light-independent reactions of photosynthesis. *Brown* circles signify carbon atoms. Appendix V details the reaction steps.

6.6 Different Plants, Different Carbon-Fixing Pathways

If sunlight intensity, air temperature, rainfall, and soil composition never varied, photosynthesis might be the same in all plants. But environments differ, and so do details of photosynthesis. For example, you see such differences on hot days when water is scarce.

Take a look at the leaves in Figure 6.12. They all have a waxy cuticle that restricts water loss. Water and gases move into and out of leaves across tiny openings called **stomata** (singular, stoma). Stomata close on hot, dry days. Water and O_2 can't get out, and CO_2 can't get in. A plant's capacity to make sugars declines when its photosynthetic cells are exposed to too much O_2 and not enough CO_2.

That's why beans, sunflowers, and many other plants don't grow well in hot, dry climates unless they are steadily irrigated. We call them **C3 plants**, because the *three*-carbon PGA is the first stable intermediate of the Calvin–Benson cycle. Remember the enzyme that fixes carbon for this cycle? When oxygen builds up in leaves of C3 plants, rubisco uses oxygen—not CO_2—in an alternate reaction that yields only one molecule of PGA (Figure 6.12*a*).

In **C4 plants**, *four*-carbon oxaloacetate forms first in reactions that fix carbon twice (Figure 6.12*b*). In mesophyll cells, the C4 cycle fixes carbon no matter how much O_2 there is. This reaction delivers CO_2 directly to bundle-sheath cells, where it enters the Calvin–Benson cycle (Figure 6.12*b*). The C4 cycle keeps the CO_2 level near rubisco high enough to stop the competing reaction. C4 plants do use an extra ATP. Compared to C3 plants, though, they lose less water and make more sugar when days are dry.

CAM plants open stomata at night and fix carbon by repeated turns of a C4 cycle, then the Calvin–Benson cycle runs the next day (Figure 6.12*c*). These plants include cacti and other succulents, which have juicy, water-storing tissues and thick surface layers adapted to hot, dry climates. Some CAM plants survive prolonged droughts by closing stomata even at night. They fix CO_2 released by aerobic respiration, which supports slow growth.

In short, C3 plants, C4 plants, and CAM plants respond differently to hot, dry conditions, when their photosynthetic cells must deal with too much oxygen and not enough carbon dioxide.

The C4 cycle evolved separately in many lineages, over millions of years. Before then, CO_2 levels in air were higher, so C3 plants had the advantage in hot climates. Which cycle will be best in the future? CO_2 levels have been rising for decades and may double in fifty years. C3 plants may again have the edge.

Leaves of basswood (*Tilia americana*), a typical C3 plant.

Leaves of corn (*Zea mays*), a typical C4 plant

Beavertail cactus (*Opuntia basilaris*), one of the CAM plants

Figure 6.12 Comparison of carbon-fixing adaptations in three kinds of plants.

Read Me First!
and watch the narrated animation
on carbon-fixing adaptations

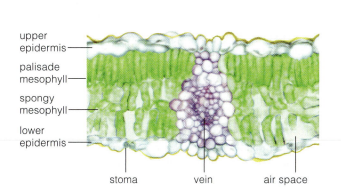

upper epidermis
palisade mesophyll
spongy mesophyll
lower epidermis

stoma   vein   air space

Basswood leaf, cross-section.

stomata closed, no CO$_2$ uptake

RuBP   PGA
Calvin–Benson cycle

sugar

a Many C3 plants evolved in moist temperate zones. The basswood tree is one of them. On hot, dry days, it can't grow as well as C4 plants because its rubisco uses O$_2$ in an inefficient reaction that competes with the Calvin–Benson cycle. Not as many sugars are produced.

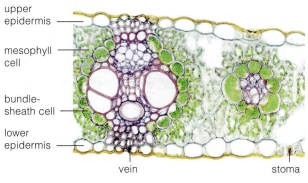

upper epidermis
mesophyll cell
bundle-sheath cell
lower epidermis

vein   stoma

Corn leaf, cross-section.

stomata closed, no CO$_2$ uptake

C4 cycle   oxaloacetate   mesophyll cell

CO$_2$

RuBP   PGA
Calvin–Benson cycle   bundle-sheath cell

sugar

b How C4 plants fix carbon in hot, dry weather, when there is too little CO$_2$ and too much O$_2$ inside leaves. A C4 cycle is common in grasses, corn, and other plants that evolved in the tropics. In their mesophyll cells, CO$_2$ gets fixed by an enzyme that ignores O$_2$. That reaction releases carbon dioxide in adjoining bundle sheath cells, where the Calvin–Benson cycle runs.

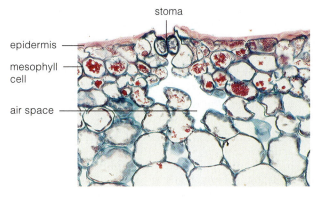

stoma
epidermis
mesophyll cell
air space

Cacti have photosynthetic, fleshy stems, not leaves. This cross-section shows a stoma and mesophyll cells inside.

CO$_2$ uptake at night only

C4 cycle   runs at night

Calvin–Benson cycle   runs during day   mesophyll cell

sugar

c How CAM plants fix carbon in hot, dry climates. Their stomata limit water loss by opening only at night. That is when CO$_2$ enters and O$_2$ departs. The CO$_2$ is fixed by a C4 cycle that runs at night. The fixed carbon enters the Calvin–Benson cycle in the same cell during daylight hours.

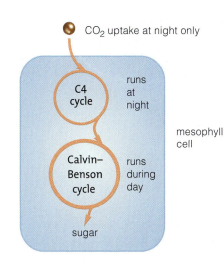

Summary

Section 6.1   Photosynthesis starts with the absorption of light energy by pigment molecules. Chlorophyll *a*, the main photosynthetic pigment, is best at absorbing violet and red wavelengths. Carotenoids and other pigments absorb characteristic wavelengths.

Section 6.2   Photosynthesis has two stages: the light-dependent and the light-independent reactions. This equation and Figure 6.13 summarize the process:

$$12H_2O + 6CO_2 \xrightarrow{\text{light energy}} 6O_2 + C_6H_{12}O_6 + 6H_2O$$

water carbon   oxygen glucose water
      dioxide

In plants and algae, the light-dependent reactions occur at the thylakoid membrane, which forms a continuous compartment in the semifluid interior (stroma) of the chloroplast. Starting long ago, oxygen released by these reactions has accumulated in the atmosphere. Without it, aerobic respiration would not have evolved.

Sections 6.3, 6.4   Many clusters of pigments in the thylakoid membrane absorb photons and pass the energy to many photosystems. The light-dependent reactions start when electrons released from photosystems enter electron transfer chains. Operation of the transfer chains results in the formation of the energy carrier ATP and the reduced coenzyme NADPH.

The released electrons can move through a noncyclic or a cyclic pathway. In the noncyclic reactions, electrons lost from photosystem II enter an electron transfer chain. Electron flow through the chain causes hydrogen ions to accumulate in the thylakoid compartment. Light energy prompts photosystem I to lose electrons which, along with hydrogen ions, convert NADP$^+$ to NADPH.

Electrons lost from photosystem I are replaced by electrons from photosystem II. Photosystem II replaces its lost electrons by pulling them away from water molecules. This dissociates them into H$^+$ and O$_2$, a process called photolysis.

In the cyclic reactions, electrons from photosystem I enter a different electron transfer chain, then are recycled back to the same photosystem.

In both pathways, H$^+$ accumulation in the thylakoid compartment forms concentration and electric gradients across the thylakoid membrane. H$^+$ flows in response to the gradients, through ATP synthases. The flow causes P$_i$ to be attached to ADP in the stroma, forming ATP.

Section 6.5   Light-independent reactions, the synthesis part of photosynthesis, occur in the stroma. In C3 plants, the enzyme rubisco attaches carbon to RuBP to start the Calvin–Benson cycle. In this cyclic pathway, energy from ATP, carbon and oxygen from CO$_2$, and hydrogen and electrons from NADPH are used to make glucose, which immediately enters reactions that form the end products of photosynthesis (e.g., sucrose, cellulose, and starch).

Section 6.6   On hot, dry days, plants close stomata and conserve water, so oxygen from photosynthesis builds up in leaves. When that happens, rubisco uses oxygen instead of CO$_2$, which slows sugar production. C4 plants fix carbon twice, in two cell types. CAM plants close stomata during the day and fix carbon at night.

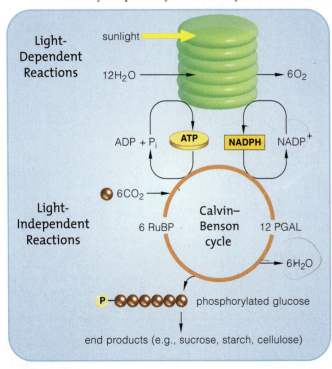

Figure 6.13   Visual summary of photosynthesis.

Self-Quiz

Answers in Appendix III

1. Photosynthetic autotrophs use _____ from the air as a carbon source and _____ as their energy source.

2. Light-*dependent* reactions in plants occur at the _____ .
   a. thylakoid membrane   c. stroma
   b. plasma membrane   d. cytoplasm

3. In the light-*dependent* reactions, _____ .
   a. carbon dioxide is fixed   c. CO$_2$ accepts electrons
   b. ATP and NADPH form   d. sugars form

4. What accumulates inside the thylakoid compartment during the light-*dependent* reactions?
   a. glucose   c. hydrogen ions
   b. RuBP   d. carbon dioxide

5. Light-*independent* reactions proceed in the _____ .
   a. cytoplasm   b. plasma membrane   c. stroma

6. The Calvin–Benson cycle starts when _____ .
   a. light is available
   b. carbon dioxide is attached to RuBP
   c. electrons leave photosystem II

7. Match each event with its most suitable description.
   _____ ATP formation only   a. rubisco required
   _____ CO$_2$ fixation   b. ATP, NADPH required
   _____ PGAL formation   c. electrons cycled back
                               to photosystem I

Critical Thinking

1. Imagine walking through a garden of red, white, and blue petunias. Explain each of the colors in terms of which wavelengths of light the flower is absorbing.

2. While gazing into an aquarium, you observe bubbling from an aquatic plant (Figure 6.14). What's going on?

3. In the laboratory, Krishna invites plants to take up a carbon radioisotope ($^{14}CO_2$). In which compound will the labeled carbon appear first?

4. About 200 years ago, Jan Baptista van Helmont did experiments on the nature of photosynthesis. He wanted to know where growing plants get the materials necessary for increases in size. He planted a tree seedling weighing 5 pounds in a barrel filled with 200 pounds of soil. He watered the tree regularly. Five years passed. Then van Helmont weighed the tree and the soil. The tree weighed 169 pounds, 3 ounces. The soil weighed 199 pounds, 14 ounces. Because the tree gained so much weight and the soil lost so little, he concluded the tree had gained weight by absorbing the water he had added to the barrel. Given what you know about the composition of biological molecules, why was he misguided? Knowing what you do about photosynthesis, what really happened?

5. The green alga in Figure 6.15a lives in seawater. Its main pigments absorb red light. Its accessory pigments help harvest energy in sunlit waters, and some shield it against ultraviolet radiation. Other green algae live in ponds, lakes, on rocks, even in snow. The red alga in Figure 6.15b grows on tropical reefs in clear, warm water. Its phycobilins absorb green and blue-green wavelengths that penetrate deep water. Some of its relatives live in deep, dimly lit waters; their pigments are nearly black.

   If wavelengths are such a vital source of energy, why aren't all pigments black?

   Hint: If photoautotrophs first evolved in the seas, then their pigment molecules must also have evolved in the seas. We have evidence that life arose near hydrothermal vents on the seafloor. Survival may have depended on being able to move away from weak infrared radiation (heat energy), which has been measured at vents, to keep from being boiled alive. Millions of years later, bacterial descendants were evolving near the sea surface. By one hypothesis, light-sensing machinery in deep-sea bacteria became modified for shallow-water photosynthesis.

6. Only about eight classes of pigment molecules are known, but this limited group gets around in the world. For example, animals synthesize the brownish-black melanin and some other pigments, but not carotenoids. Photoautotrophs make carotenoids, which move up through food webs, as when tiny aquatic snails graze on green algae and then flamingos eat the snails.

   Flamingos modify ingested carotenoids in plenty of ways. For instance, their cells split beta-carotene to form two molecules of vitamin A. This vitamin is the precursor of retinol, a visual pigment that transduces light into electric signals in the flamingo's eyes. Beta-carotene also gets dissolved in fat reservoirs under the skin. From there they are taken up by cells that give rise to bright pink feathers. Choose an animal and do some research into its life cycle and diet. Use your research to identify possible sources for the pigments that color its surfaces.

Figure 6.14 Leaves of *Elodea*, an aquatic plant.

Figure 6.15 (**a**) A green alga (*Codium*) from shallow coastal waters. (**b**) Red alga from a tropical reef.

Media Menu

Student CD-ROM

Impacts, Issues Video
Pastures of the Seas
Big Picture Animation
Harnessing light energy to build sugars
Read-Me-First Animation
Sites of photosynthesis
Photosynthetic pathways for making ATP and NADPH
Energy release in photosynthesis
The Calvin–Benson cycle
Carbon-fixing adaptations
Other Animations and Interactions
T. Englemann's study
Overview of the stages of photosynthesis

InfoTrac

- Light of Our Lives. *Geographical*, January 2001.
- Sunlight at Southall Green. *Perspectives in Biology and Medicine*, Summer 2001.
- Scripps Research Gives Tiny Phytoplankton Large Role in Earth's Climate System; Study Shows Microscopic Plants Keep Planet Warm, Offers New Considerations for Iron Fertilization Efforts in Oceans. *Ascribe Higher Education News Service*, November 6, 2002.

Web Sites

- ASU Photosynthesis Center: photoscience.la.asu.edu/photosyn
- NASA Earth Observatory: earthobservatory.nasa.gov
- Calvin Nobel Prize: www.nobel.se/chemistry/laureates/1961/index.html

How Would You Vote? The carbon-fixing enzyme rubisco is the world's most abundant protein. Plants require a lot of this enzyme because it's not that efficient. Some scientists think that using genetic engineering to modify the rubisco of plants could help increase world crop production. Would you support the use of tax dollars to fund this research?

IMPACTS, ISSUES *When Mitochondria Spin Their Wheels*

In the early 1960s, Swedish physician Rolf Luft reflected on some odd symptoms of a young patient. The woman felt weak and too hot all the time. Even on the coldest winter days, she couldn't stop sweating and her skin was flushed. She was thin in spite of a huge appetite.

Luft inferred that his patient's symptoms pointed to a metabolic disorder. Her cells seemed to be spinning their wheels. They were active, but a lot of activity was being dissipated as metabolic heat. He ordered tests designed to detect her metabolic rates. The patient's oxygen consumption was the highest ever recorded!

Microscopic examination of a tissue sample from the patient's skeletal muscles revealed mitochondria, the

cell's ATP-producing powerhouses. But there were far too many of them, and they were abnormally shaped. Other studies showed that the mitochondria were engaged in aerobic respiration—yet very little ATP was forming.

The disorder, now called *Luft's syndrome*, was the first to be linked directly to a defective organelle. By analogy, someone with this mitochondrial disorder functions like a city with half of its power plants shut down. Skeletal and heart muscles, the brain, and other hard-working body parts with the highest energy demands are hurt the most.

More than a hundred other mitochondrial disorders are now known. One, a heritable genetic disease called *Friedreich's ataxia*, runs in families (Figure 7.1). Affected people develop weak muscles, loss of coordination (ataxia), and visual problems. Many die when they are young adults because of heart muscle irregularities.

Figure 7.1 Photogenic siblings with Friedreich's ataxia. Leah (*left*) began to lose balance and coordination when she was 5. She was in a wheelchair by the time she was 11. She is now diabetic, and has lost part of her hearing. Joshua (*right*) was 3 when his symptoms began. By the time he was 11, he was unable to walk. He is now legally blind.

Both young people have a heart condition called hypertrophic cardiomyopathy. Both had spinal fusion surgery. Although they have lost a large part of their fine motor skills, with the aid of adaptive equipment they continue to go to school and work in productive jobs. Leah has a part-time modeling career.

the big picture

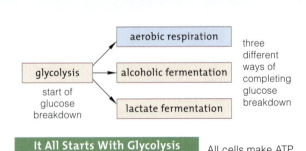

aerobic respiration

glycolysis → alcoholic fermentation

start of glucose breakdown

lactate fermentation

three different ways of completing glucose breakdown

It All Starts With Glycolysis All cells make ATP by breaking down organic compounds, which releases energy stored in chemical bonds. The main pathways all start in the cytoplasm, with glycolysis.

How the Aerobic Route Ends In aerobic respiration alone, glucose breakdown is completed in mitochondria. This pathway has the greatest net energy yield from each glucose molecule.

A mutant gene and its abnormal protein product give rise to Friedreich's ataxia. The abnormal protein causes iron to accumulate inside mitochondria. Iron is required for electron transfers that drive ATP formation. But too much invites an accumulation of **free radicals**—unbound molecular fragments with the wrong number of electrons. These highly reactive fragments can attack all of the molecules of life.

Type 1 diabetes, atherosclerosis, amyotrophic lateral sclerosis (Lou Gehrig's disease), Parkinson's, Alzheimer's, and Huntington's diseases—defective mitochondria contribute to every one of these age-related problems. So when you consider mitochondria in this chapter, don't assume they are too remote from your interests. Without them, you wouldn't make enough ATP even to read about how they do it.

The preceding chapter described how plants and all other photosynthetic organisms get energy from the sun. You and all the other heterotrophs around you get some of the energy that they captured secondhand, thirdhand, and so on. Regardless of its source, energy must first be put into a form that can drive thousands of different life-sustaining reactions. That form is ATP's chemical bond energy.

You already read about the way ATP molecules form during photosynthesis. Turn now to how all organisms make ATP by tapping into the chemical bond energy of organic compounds—especially glucose.

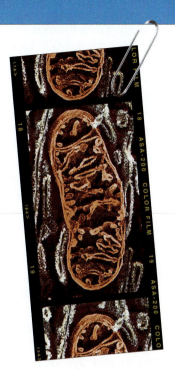

 How Would You Vote?

Developing new drugs is costly. There's little incentive for pharmaceutical companies to target ailments, such as Friedreich's ataxia, that affect relatively few individuals. Should the government provide some funding to private companies that search for cures for diseases affecting only a small number of people? See the Media Menu for details, then vote online.

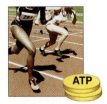

How Fermentation Routes End

In lactate and alcoholic fermentation, glucose breakdown starts *and* ends in the cytoplasm. The net energy yield is small.

What Cells Do With Food

Big meals, small meals, no meals—cells shunt carbohydrates, lipids, and proteins into breakdown pathways.

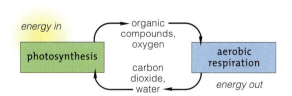

Evolutionary Connections

Aerobic respiration and photosynthesis are connected on a global scale, and that connection can be traced to the evolution of novel metabolic pathways.

7.1 Overview of Energy-Releasing Pathways

Plants make ATP during photosynthesis and use it to synthesize glucose and other carbohydrates. But all organisms, plants included, can make ATP by breaking down carbohydrates, lipids, and proteins.

Organisms stay alive by taking in energy. Plants and all other photosynthetic autotrophs get energy from the sun. Heterotrophs get energy by eating plants and one another. Regardless of its source, the energy must be in a form that can drive thousands of diverse life-sustaining reactions. Energy that becomes converted into chemical bond energy of adenosine triphosphate —ATP—serves that function.

COMPARISON OF THE MAIN TYPES OF ENERGY-RELEASING PATHWAYS

The first energy-releasing metabolic pathways were operating billions of years before Earth's oxygen-rich atmosphere evolved, so we can expect that they were *anaerobic*; the reactions did not use free oxygen. Many prokaryotes and protists still live in places where oxygen is absent or not always available. They make ATP by fermentation and other anaerobic pathways. Many eukaryotic cells still use fermentation, including skeletal muscle cells. However, the cells of nearly all species extract energy efficiently from glucose by way of **aerobic respiration**, an oxygen-dependent pathway. Each breath you take provides your actively respiring cells with a fresh supply of oxygen.

Make note of this point: *In all cells, all of the main energy-releasing pathways start with the same reactions in the cytoplasm.* During the initial reactions, **glycolysis**, enzymes cleave and rearrange a glucose molecule into two molecules of **pyruvate**, an organic compound that has a three-carbon backbone.

Once glycolysis is over, energy-releasing pathways differ. Only the aerobic pathway continues and ends in mitochondria (Figure 7.2). There, oxygen accepts and removes electrons that drove the reactions.

Only aerobic respiration delivers enough ATP to build and maintain big multicelled organisms, including redwoods and highly active animals, such as people and Canada geese.

As you examine the energy-releasing pathways in sections to follow, keep in mind that enzymes catalyze each step, and intermediates formed at one step serve as substrates for the next enzyme in the pathway.

OVERVIEW OF AEROBIC RESPIRATION

Of all energy-releasing pathways, aerobic respiration gets the most ATP for each glucose molecule. Whereas anaerobic routes have a net yield of two ATP, aerobic respiration typically yields thirty-six or more. If you were a bacterium, you would not require much ATP. Being far larger, more complex, and highly active, you depend on the aerobic pathway's high yield. When a molecule of glucose is used as the starting material, aerobic respiration can be summarized this way:

$$C_6H_{12}O_6 + 6O_2 \longrightarrow 6CO_2 + 6H_2O$$

glucose   oxygen   carbon dioxide   water

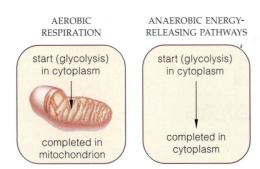

Figure 7.2 Where the main energy-releasing pathways of ATP formation start and finish.

AEROBIC RESPIRATION

start (glycolysis) in cytoplasm

completed in mitochondrion

ANAEROBIC ENERGY-RELEASING PATHWAYS

start (glycolysis) in cytoplasm

completed in cytoplasm

Read Me First!
and watch the narrated animation
on aerobic respiration

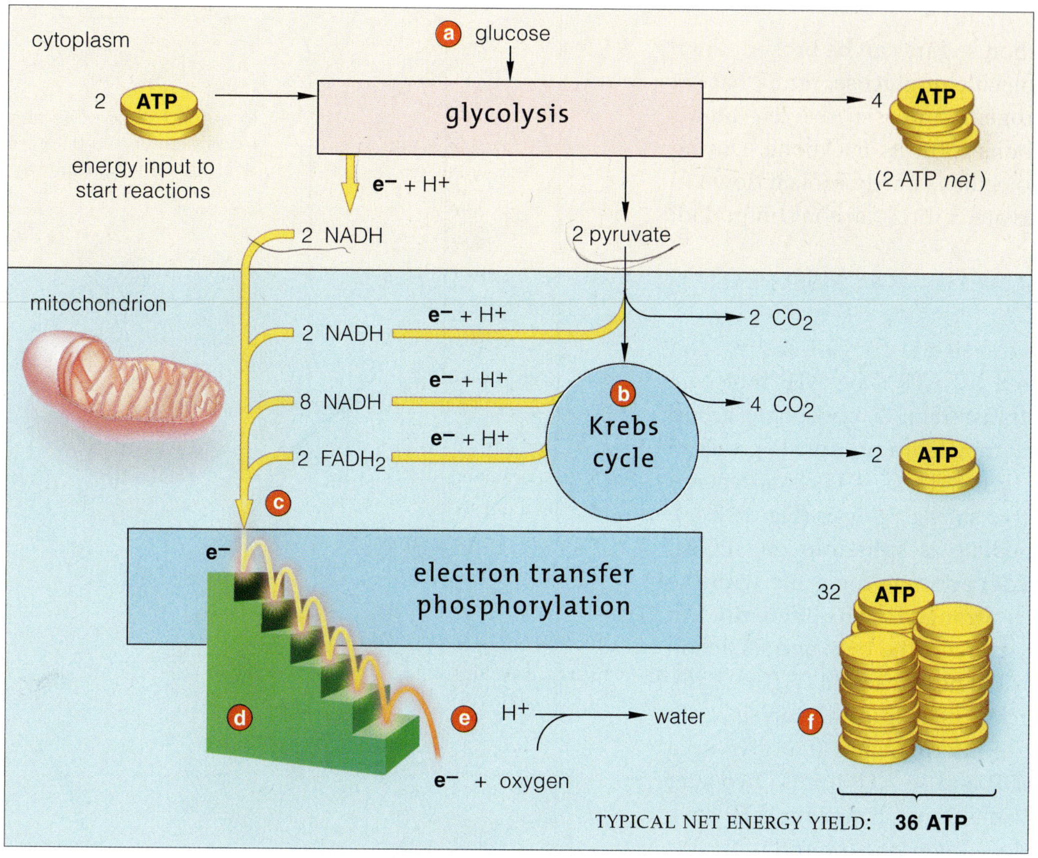

Figure 7.3 Overview of aerobic respiration. Reactions start in the cytoplasm and end in mitochondria.

(**a**) In the first stage, glycolysis, enzymes partly break down glucose to pyruvate.

(**b**) In the second stage, enzymes break down pyruvate to carbon dioxide.

(**c**) NAD^+ and FAD pick up the electrons and hydrogen stripped from intermediates in both stages.

(**d**) The final stage is electron transfer phosphorylation. The reduced coenzymes NADH and $FADH_2$ give up electrons to electron transfer chains. H^+ tags along with electrons. Electron flow through the chains sets up H^+ gradients, which are tapped to make ATP.

(**e**) Oxygen accepts electrons at the end of the third stage, forming water.

(**f**) From start to finish, a typical net energy yield from a glucose molecule is thirty-six ATP.

Figure 7.3 diagram:

- cytoplasm
- (a) glucose
- 2 ATP — energy input to start reactions
- glycolysis
- 4 ATP (2 ATP *net*)
- $e^- + H^+$
- 2 NADH
- 2 pyruvate
- mitochondrion
- $e^- + H^+$ → 2 CO_2
- 2 NADH
- $e^- + H^+$
- 8 NADH
- (b) Krebs cycle → 4 CO_2
- $e^- + H^+$
- 2 $FADH_2$
- (c)
- 2 ATP
- (d) e^-
- electron transfer phosphorylation
- 32 ATP
- (e) H^+ → water
- e^- + oxygen
- (f)
- TYPICAL NET ENERGY YIELD: **36 ATP**

However, as you can see, the summary equation only tells us what the substances are at the start and finish of the pathway. In between are three reaction stages.

Figure 7.3 is your overview of aerobic respiration. Glycolysis, again, is the first stage. The second stage is a cyclic pathway, the **Krebs cycle**. Enzymes break down pyruvate to carbon dioxide and water, which releases many electrons and hydrogen atoms.

As you track the reactions, you'll come across two enzyme helpers, the coenzymes **NAD⁺** (nicotinamide adenine dinucleotide) and **FAD** (flavin adenine dinucleotide). Both accept electrons and hydrogen derived from intermediates of glucose breakdown. Unbound hydrogen atoms are simply hydrogen ions (H^+), or naked protons. When the two coenzymes are carrying electrons and hydrogen, they are in a reduced form and may be abbreviated NADH and $FADH_2$.

Few ATP form during glycolysis or the Krebs cycle. The big energy harvest comes in the third stage, after the coenzymes give up the electrons and hydrogen to

electron transfer chains. The chains are the machinery of **electron transfer phosphorylation**. They set up H^+ concentration and electric gradients, which drive ATP formation at nearby membrane proteins. It is in this final stage that so many ATP molecules are produced. As it ends, oxygen inside the mitochondrion accepts the "spent" electrons from the last component of each transport system. Oxygen picks up H^+ at the same time and thereby forms water.

Nearly all metabolic reactions run on energy released from glucose and other organic compounds. The main energy-releasing pathways start in the cytoplasm with glycolysis, a stage of reactions that break down glucose to pyruvate.

Anaerobic pathways have a small net energy yield, typically two ATP per glucose.

Aerobic respiration, an oxygen-dependent pathway, runs to completion in mitochondria. From start (glycolysis) to finish, it typically has a net energy yield of thirty-six ATP.

7.2 The First Stage: Glycolysis

Let's track what happens to a glucose molecule in the first stage of aerobic respiration. Remember, the same steps happen in anaerobic energy-releasing pathways.

Any of several six-carbon sugars can be broken down in glycolysis. Each molecule of glucose, recall, has six carbon, twelve hydrogen, and six oxygen atoms (Section 3.3). The carbons form its backbone. During glycolysis, this one molecule is partly broken down to two molecules of pyruvate, a three-carbon compound:

The initial steps of gycolysis are *energy-requiring*, and that energy is delivered by ATP. One ATP molecule activates glucose by transferring a phosphate group to it. Then another ATP transfers a phosphate group to the intermediate that forms. Thus, it takes an energy investment of two ATP to start glycolysis (Figure 7.4a).

The second intermediate is split into one PGAL (phosphoglyceraldehyde) and one molecule with the same number of atoms arranged a bit differently. An enzyme can reversibly convert the two, and its action delivers two PGAL for the next reaction (Figure 7.4b).

In the first *energy-releasing* step of glycolysis, both PGALs are converted to intermediates that give up a phosphate group to ADP, and so ATP forms. Two later intermediates do the same thing. Thus four ATP have been formed by **substrate-level phosphorylation**. We define this metabolic event as the direct transfer of a phosphate group from the substrate of a reaction to some other molecule—in this case, to ADP.

Meanwhile, the coenzyme NAD$^+$ accepts electrons and hydrogens from each PGAL, becoming NADH.

By this time, a total of four molecules of ATP have formed, but remember that two ATP were invested to get the reactions going. The *net* yield of glycolysis is two ATP and two NADH.

To summarize, glycolysis converts the bond energy of glucose to bond energy of ATP—a transportable form of energy. The electrons and hydrogen stripped from glucose and picked up by NAD$^+$ will enter the next stage of reactions. And so will the end products of glycolysis—two pyruvate molecules.

Glycolysis is a series of reactions that partially break down glucose or other six-carbon sugars to two molecules of pyruvate. It takes two ATP to jump-start the reactions.

Two NADH and four ATP form. However, when we subtract the two ATP required to start the reactions, the net energy yield of glycolysis is two ATP from one glucose molecule.

GLUCOSE

Figure 7.4 Glycolysis. This first stage of the main energy-releasing pathways occurs in the cytoplasm of all prokaryotic and eukaryotic cells. Glucose is the reactant in this example. Appendix V gives the structural formulas of intermediates that form during its breakdown. Two pyruvate, two NADH, and four ATP form in glycolysis. Cells invest two ATP to start the reactions however, so the *net* energy yield is two ATP.

Depending on the type of cell and environmental conditions, the pyruvate may enter the second set of reactions of the aerobic pathway, including the Krebs cycle. Or it may be used in other reactions, such as those of fermentation.

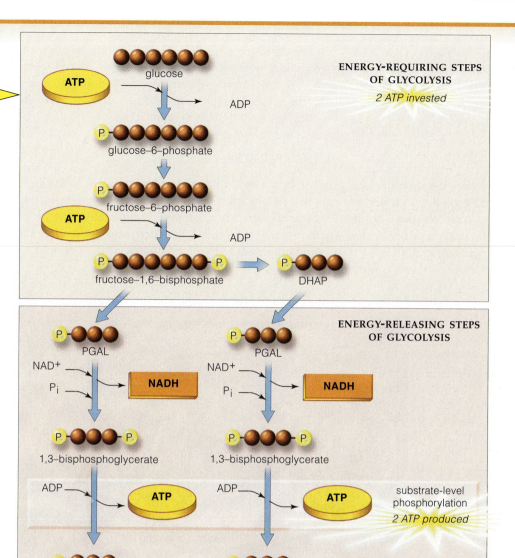

ENERGY-REQUIRING STEPS OF GLYCOLYSIS

2 ATP invested

ATP

glucose

ADP

P glucose–6–phosphate

P fructose–6–phosphate

ATP

ADP

P fructose–1,6–bisphosphate P

P DHAP

ENERGY-RELEASING STEPS OF GLYCOLYSIS

P PGAL

NAD+
Pi

NADH

P PGAL

NAD+
Pi

NADH

P 1,3–bisphosphoglycerate P

P 1,3–bisphosphoglycerate P

ADP

ATP

ADP

ATP

substrate-level phosphorylation

2 ATP produced

P 3–phosphoglycerate

P 3–phosphoglycerate

P 2–phosphoglycerate

P 2–phosphoglycerate

H₂O

H₂O

P PEP

P PEP

ADP

ATP

ADP

ATP

substrate-level phosphorylation

2 ATP produced

pyruvate

pyruvate

to second set of reactions

Read Me First!

and watch the narrated animation on glycolysis

Track the six carbon atoms (*brown circles*) of glucose. Glycolysis requires an energy investment of two ATP:

a One ATP transfers a phosphate group to glucose, jump-starting the reactions.

b Another ATP transfers a phosphate group to an intermediate, causing it to split into two three-carbon compounds: PGAL and DHAP (dihydroxyacetone phosphate). Both have the same atoms, arranged differently. They are interconvertible, but only PGAL can continue on in glycolysis. DHAP gets converted, so two PGAL are available for the next reaction.

c Two NADH form when each PGAL gives up two electrons and a hydrogen atom to NAD⁺.

d Two intermediates each transfer a phosphate group to ADP. *Thus, two ATP have formed by direct phosphate group transfers.* The original energy investment of two ATP is now paid off.

e Two more intermediates form. Each gives up one hydrogen atom and an —OH group. These combine as water. Two molecules called PEP form by these reactions.

f Each PEP transfers a phosphate group to ADP. *Once again, two ATP have formed by substrate-level phosphorylation.*

In sum, glycolysis has a net energy yield of two ATP for each glucose molecule. Two NADH also form during the reactions, and two molecules of pyruvate are the end products.

7.3 Second Stage of Aerobic Respiration

The two pyruvate molecules formed by glycolysis can leave the cytoplasm and enter a mitochondrion. There they enter reactions that get the Krebs cycle going. Many coenzymes pick up the electrons and hydrogens released when the two pyruvates are dismantled.

ACETYL–CoA FORMATION

Start with Figure 7.5, which shows the structure of a typical mitochondrion. Figure 7.6 zooms in on part of the interior where the second-stage reactions occur. At the start of these reactions, enzyme action strips one carbon atom from each pyruvate and attaches it

to oxygen, forming CO_2. Each two-carbon fragment combines with a coenzyme (designated A) and forms acetyl–CoA, a type of cofactor that can get the Krebs cycle going. The initial breakdown of each pyruvate also yields one NADH (Figure 7.7).

THE KREBS CYCLE

The two acetyl–CoA molecules enter the Krebs cycle separately. Each transfers its two-carbon acetyl group to four-carbon oxaloacetate. Incidentally, this cyclic pathway is also called the citric acid cycle, after the first intermediate that forms (citric acid, or citrate).

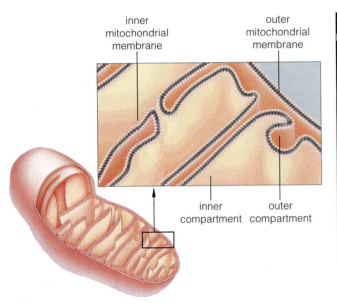

Figure 7.5 *Above*, functional zones inside the mitochondrion. An inner membrane system divides this organelle's interior into an inner and an outer compartment. The second and third stages of aerobic respiration play out at this membrane.

Figure 7.6 *Right:* Overview of the number of ATP molecules and coenzymes that form in the second stage of aerobic respiration. The reactions start with two molecules of pyruvate from glycolysis. Pyruvate moves from the cytoplasm, then across the outer and inner mitochondrial membranes, into the inner compartment where the reactions take place.

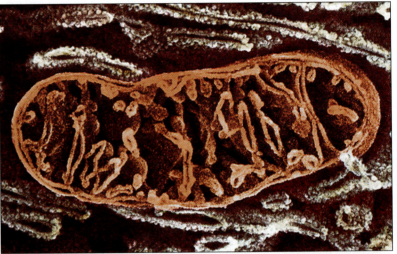

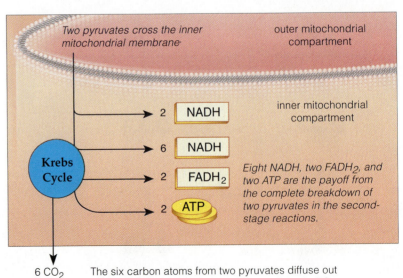

Read Me First!
and watch the
narrated animation
on the Krebs cycle

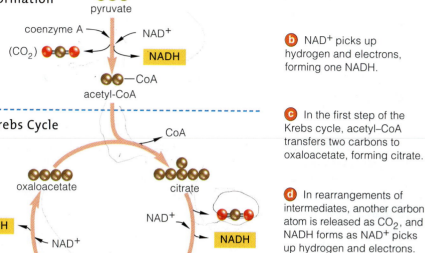

a One carbon atom is stripped from each pyruvate and is released as CO_2. The remaining fragment binds with coenzyme A, forming acetyl–CoA.

b NAD^+ picks up hydrogen and electrons, forming one NADH.

c In the first step of the Krebs cycle, acetyl–CoA transfers two carbons to oxaloacetate, forming citrate.

h The final steps regenerate oxaloacetate. NAD^+ picks up hydrogen and electrons, forming NADH. *At this point in the cycle, three NADH and one $FADH_2$ have formed.*

d In rearrangements of intermediates, another carbon atom is released as CO_2, and NADH forms as NAD^+ picks up hydrogen and electrons.

g $FADH_2$ forms as the coenzyme FAD picks up electrons and hydrogen.

e Another carbon atom is released as CO_2. Another NADH forms. *Three carbon atoms now have been released.* This balances out the three carbons that entered (in one pyruvate).

f A phosphate group is attached to ADP. At this point, one ATP has formed by substrate-level phosphorylation.

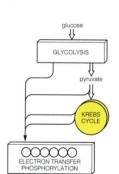

Figure 7.7 Aerobic respiration's second stage: formation of acetyl–CoA and the Krebs cycle. The reactions occur in a mitochondrion's inner compartment. It takes two turns of the cycle to break down the two pyruvates from glucose. A total of two ATP, eight NADH, two $FADH_2$, and six CO_2 molecules form. Organisms release the CO_2 from the reactions into their surroundings.

It takes two turns of the Krebs cycle to completely break down two molecules of pyruvate to CO_2 and water. Only two ATP form, which doesn't add much to the small net yield from glycolysis. However, in addition to those two NADH produced during the formation of acetyl–CoA, six more NADH and two $FADH_2$ are produced in the cycle. With their cargo of electrons and hydrogen atoms, these ten coenzymes constitute a big potential payoff for the cell.

Four more CO_2 molecules form as the Krebs cycle turns. In total, *six* carbon atoms (from two pyruvates) depart during the second stage of aerobic respiration, in six molecules of CO_2 (Figures 7.6 and 7.7). And so glucose from glycolysis has lost all of its carbons; it has become fully oxidized.

For interested students, Appendix V has a closer look at the steps of these remarkable reactions.

Aerobic respiration's second stage starts after two pyruvate molecules from glycolysis move from the cytoplasm, across the outer and inner mitochondrial membranes, and into the inner mitochondrial compartment.

Here, pyruvate is converted to acetyl–CoA, which starts the Krebs cycle. A total of two ATP and ten coenzymes (eight NADH, two $FADH_2$) form. All of pyruvate's carbons depart, in the form of carbon dioxide.

Together with two coenzymes (NADH) that formed during glycolysis, the ten from the second stage will deliver electrons and hydrogen to the third and final stage.

7.4 Third Stage of Aerobic Respiration—The Big Energy Payoff

In the aerobic pathway's third stage, energy release goes into high gear. Coenzymes from the first two stages provide the hydrogen and electrons that drive the formation of many ATP. Electron transfer chains and ATP synthases function as the machinery.

ELECTRON TRANSFER PHOSPHORYLATION

The third stage starts as coenzymes donate electrons to electron transfer chains that are located in the inner mitochondrial membrane (Figure 7.8). The flow of electrons through the chains drives the attachment of phosphate to ADP molecules. Hence the name *electron transfer phosphorylation.*

Incremental energy release, recall, is more efficient than one big burst of energy that would result in little more than a lot of unusable heat (Section 5.2). When electrons flow through transfer chains, they give up energy bit by bit, in usable parcels, to substances that can briefly store it.

The two NADH that formed in the cytoplasm (by glycolysis) can't reach the ATP-producing machinery directly. They give up their electrons and hydrogen to transport proteins, which shuttle them into the inner compartment. There, NAD^+ or FAD inside pick them up. Eight NADH and two $FADH_2$ from the second stage are already inside.

When all of these coenzymes turn over electrons to transfer chains, they release hydrogen ions (H^+) at the same time. Electrons passing through the chains lose a bit of energy at each step. In three parts of the chain, that energy drives the pumping of H^+ into the outer compartment. There, accumulation of H^+ sets up an electrochemical gradient across the inner membrane.

H^+ can't diffuse across membranes. The only way it can follow the gradients, which lead back to the inner compartment, is by flowing through the interior of ATP synthases (Figures 7.8 and 7.9). H^+ flow through these transport proteins drives the formation of ATP from ADP and unbound phosphate.

The last molecules in the electron transfer chains pass electrons to gaseous oxygen, which forms water after combining with H^+. *Oxygen is the final acceptor of electrons stripped from glucose.* In oxygen-starved cells, electrons in the transfer chain have nowhere to go. The whole chain backs up with electrons all the way to NADPH, so no H^+ gradient forms, and no ATP is made. Cells of complex organisms don't survive long without oxygen, because they can't produce enough ATP to sustain life processes.

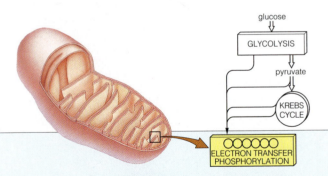

Figure 7.8 Electron transfer phosphorylation, the third and final stage of aerobic respiration.

At the inner mitochondrial membrane, NADH and $FADH_2$ give up electrons to the transfer chains. When electrons are transferred through the chains, unbound hydrogen (H^+) is shuttled across the membrane to the outer compartment:

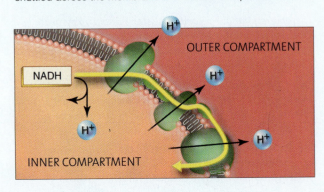

H^+ concentration is now greater in the outer compartment. Concentration and electric gradients across the membrane have been set up. H^+ follows these gradients through the interior of ATP synthases. The flow drives the formation of ATP from ADP and unbound phosphate (P_i).

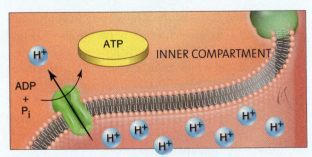

Do these events sound familiar? They should. ATP forms in much the same way inside chloroplasts. H^+ concentration and electric gradients across the inner thylakoid membrane drive ATP formation. In thylakoids, H^+ flows in the opposite direction compared to the flow in chloroplasts.

Read Me First!
and watch the
narrated
animation on
third-stage
reactions

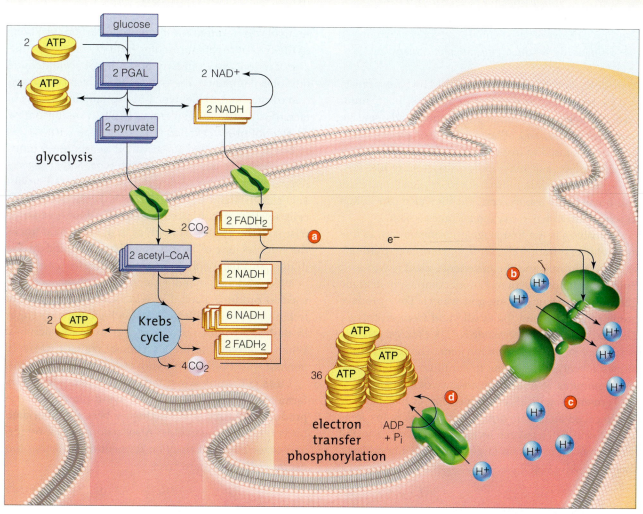

a Electrons and hydrogen from NADH and FADH₂ that formed during the first and second stages enter electron transfer chains.

b As electrons are transferred through the chains, H⁺ ions are shuttled across the inner membrane, into the outer compartment.

c H⁺ concentration becomes greater in the outer compartment than the inner one. Chemical and electrical gradients have been established.

d Hydrogen ions follow the gradients through the interior of ATP synthases, driving ATP synthesis from ADP and phosphate (P$_i$).

Figure 7.9 Summary of the transfers of electrons and hydrogen from coenzymes involved in ATP formation in mitochondria.

SUMMING UP: THE ENERGY HARVEST

Thirty-two ATP typically form in the third stage. Add in the four produced in the first and second stages, and aerobic respiration has netted thirty-six ATP from one glucose molecule. That's a lot of ATP! Anaerobic pathways may use up eighteen glucose molecules to get the same net yield.

The actual yield varies. Shifting concentrations of reactants, intermediates, and products affect it. So does the shuttling of electrons and hydrogen from NADH that forms in the cytoplasm. Shuttling mechanisms are not the same in all cell types.

In aerobic respiration's third stage, electrons and hydrogen from coenzymes (NADH and FADH₂) interact with electron transfer chains in the mitochondrion's inner membrane.

Electron flow through transfer chains makes H⁺ accumulate in the outer mitochondrial compartment. The resulting chemical and electrical gradients across the inner membrane drive the synthesis of thirty-two ATP.

Aerobic respiration typically nets thirty-six ATP molecules from each glucose molecule metabolized.

7.5 Fermentation Pathways

We turn now to the use of glucose as a substrate for fermentation pathways. These are anaerobic pathways. They don't use oxygen as the final acceptor of electrons that ultimately drive the ATP-forming machinery.

Diverse organisms are fermenters. Many are protists and bacterial species that live in marshes, bogs, mud, ocean sediments, the animal gut, canned foods, sewage treatment ponds, and other oxygen-free places. Some die when exposed to free oxygen. Bacterial species that cause botulism and many other diseases are like this. Other fermenters are indifferent to oxygen's presence. Still other kinds use oxygen, but they also can switch to fermentation when oxygen becomes scarce.

Glycolysis is the first stage of fermentation, as it is in aerobic respiration (Figure 7.4). Here, too, pyruvate and NADH form, and the net energy yield is two ATP. But fermentation reactions cannot completely degrade glucose (to carbon dioxide and water). They produce no more ATP beyond the small yield from glycolysis. *The final steps simply regenerate NAD+, the coenzyme that is essential for the breakdown reactions.* The regeneration allows glycolysis reactions to continue production of small amounts of ATP in the absence of oxygen.

Fermentation yields enough energy to sustain many single-celled anaerobic organisms. It even helps some aerobic cells when oxygen levels are stressfully low. But it isn't enough to sustain large, multicelled organisms, this being why you'll never see anaerobic elephants.

ALCOHOLIC FERMENTATION

In **alcoholic fermentation**, the three-carbon backbone of the two pyruvate molecules from glycolysis is split. The reactions result in two molecules of acetaldehyde (an intermediate having a two-carbon backbone), and two of carbon dioxide. Next, the acetaldehydes accept electrons and hydrogen from NADH, thus becoming an alcohol product called ethanol (Figure 7.10).

Some single-celled fungi called yeasts are famous for their use of this pathway. One type, *Saccharomyces cerevisiae*, makes bread dough rise. Bakers mix it with sugar, then blend both into dough. Fermenting yeast cells release carbon dioxide, and the dough expands (rises) as the gas forms bubbles in it. Oven heat forces the bubbles out of spaces they had occupied in the dough, and the alcohol product evaporates away.

Wild and cultivated strains of *Saccharomyces* are used to produce alcohol in wine. Crushed grapes are left in vats along with the yeast, which converts sugar in the juice to ethanol. Ethanol is toxic to microbes. When a fermenting brew's ethanol content nears 10 percent, yeast cells start dying and fermentation ends.

Birds get drunk when they eat too many naturally fermented berries. Landscapers avoid planting berry-producing shrubs along highways because inebriated birds fly into windshields. Also, wild turkeys, robins, and other birds get tipsy on fermenting fruit that has dropped from orchard trees.

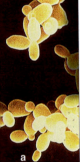

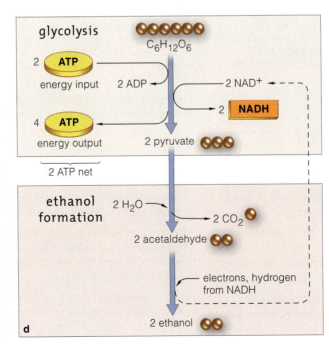

Figure 7.10 A look at alcoholic fermentation. (**a**) Yeasts, single-celled organisms, make ATP by this anaerobic pathway.

(**b**) A vintner examining the color and clarity of one fermentation product of *Saccharomyces*. Strains of this yeast live on the sugar-rich tissues of ripe grapes.

(**c**) Carbon dioxide released from cells of *S. cerevisiae* makes bread dough rise in this bakery.

(**d**) Alcoholic fermentation. The intermediate acetaldehyde functions as the final electron acceptor. The end product of the reactions is ethanol.

Figure diagram (d) labels:
glycolysis — $C_6H_{12}O_6$ — 2 ATP energy input — 2 ADP — 2 NAD+ — 2 NADH — 4 ATP energy output — 2 pyruvate — 2 ATP net — ethanol formation — 2 H_2O — 2 CO_2 — 2 acetaldehyde — electrons, hydrogen from NADH — 2 ethanol

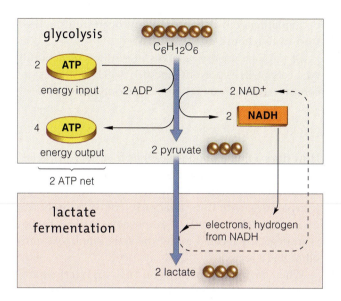

Figure 7.11 Lactate fermentation. In this anaerobic pathway, the product (lactate) is the final acceptor of electrons originally stripped from glucose. The reactions have a net energy yield of two ATP (from glycolysis).

Figure 7.12 Sprinters, calling upon lactate fermentation in their muscles. The micrograph is a cross-section through part of a muscle showing three types of fibers. The lighter fibers contribute to muscle speed by producing ATP with lactate fermentation when demands for energy are high. Darker fibers use aerobic respiration and support greater levels of endurance.

LACTATE FERMENTATION

In **lactate fermentation**, NADH gives up electrons and hydrogen to two pyruvate molecules from glycolysis. The transfer converts each pyruvate to lactate, a three-carbon compound (Figure 7.11). You've probably heard of lactic acid, the non-ionized form of this compound, but lactate is by far the most common form inside living cells, which is our focus here.

Lactobacillus and some other bacteria use lactate fermentation. Their fermenting action can spoil food, yet some species have commercial uses. For instance, huge populations that break down glucose in milk give us cheeses, yogurt, buttermilk, and other dairy products. Fermenters also help in curing meats and in pickling some fruits and vegetables, such as sauerkraut. Lactate is an acid; it gives these foods a sour taste.

Lactate fermentation as well as aerobic respiration yields ATP for muscles that are partnered with bones. These skeletal muscles contain a mixture of cell types. Cells composing *slow-twitch* muscle fibers support light, steady, prolonged activity, as during marathon runs or bird migrations. Slow-twitch muscle cells make ATP only by the aerobic respiration pathway, and so they have many mitochondria. They are dark red because they contain large amounts of myoglobin, a pigment related to hemoglobin that is used to store oxygen for aerobic respiration.

By contrast, cells of pale *fast-twitch* muscle fibers have few mitochondria and no myoglobin, and use lactate fermentation to produce ATP. They are useful when demands for energy are immediate and intense, such as in weight lifting or sprints (Figure 7.12). Lactate fermentation works quickly but not for long—it does not produce enough ATP to sustain activity. That is one reason you don't see migrating chickens. Flight muscles in a chicken are the white breast meat, containing mostly fast-twitch fibers.

Short bursts of flight evolved in the ancestors of chickens, perhaps as a way of escaping predators or improving agility during territorial battles. Chickens do walk or sprint; hence the "dark meat" (slow-twitch muscle) in their thighs and legs. So what sort of breast meat can you expect to find in a migrating duck?

Section 32.5 is an overview of alternative energy pathways for muscle cells.

In fermentation pathways, an organic substance that forms during the reactions is the final acceptor of electrons originally derived from glucose.

Alcoholic fermentation and lactate fermentation both have a net energy yield of two ATP for each glucose molecule metabolized. That ATP forms during glycolysis. The remaining reactions regenerate NAD$^+$, the coenzyme that keeps these pathways operating.

7.6 Alternative Energy Sources in the Body

So far, you've looked at what happens after glucose molecules enter an energy-releasing pathway. Now start thinking about what cells do when they have too much or too little glucose.

THE FATE OF GLUCOSE AT MEALTIME AND IN BETWEEN MEALS

What happens to glucose at mealtime? While you and all other mammals are eating, glucose and other small organic molecules are being absorbed across the gut lining, and your blood is transporting them through the body. The rising glucose concentration in blood prompts an organ, the pancreas, to secrete insulin. This hormone makes cells take up glucose faster.

Cells trap the incoming glucose by converting it to glucose–6–phosphate. This is the first intermediate of glycolysis, formed by a phosphate group transfer from ATP (Figures 7.4 and 7.13). Phosphorylated glucose can't be transported out of the cell.

When glucose intake exceeds cellular demands for energy, the body's ATP-producing machinery goes into high gear. Unless a cell is using ATP rapidly, the ATP concentration in cytoplasm rises, and glucose–6–phosphate is diverted into a biosynthesis pathway. Glucose gets stored as glycogen, one of the storage polysaccharides found in animals (Section 3.3). Liver cells and muscle cells especially favor this alternative pathway. Together, these two types of cells maintain the largest stores of glycogen in the body.

Between meals, the blood level of glucose declines. If the decline were not countered, that would be bad news for the brain, your body's glucose hog. At any time, your brain is taking up more than two-thirds of the freely circulating glucose. Why? The brain's many hundreds of millions of nerve cells (neurons) use this sugar as their preferred energy source.

The pancreas responds to glucose decline by secreting glucagon. This hormone causes liver cells to convert stored glycogen to glucose and send it back to the blood. Only liver cells do this; muscle cells won't give it up. The blood glucose level rises, and brain cells keep on functioning. Thus, *hormones control whether your cells use free glucose as an energy source or tuck it away.*

Don't let this explanation lead you to believe that your cells squirrel away huge amounts of glycogen. Glycogen makes up only 1 percent or so of the adult body's total energy reserves, the energy equivalent of two cups of cooked pasta. Unless you eat on a regular basis, you'll end up depleting your liver's small glycogen stores in less than twelve hours.

Of the total energy reserves in, say, a typical adult who eats well, 78 percent (about 10,000 kilocalories) is concentrated in body fat and 21 percent in proteins.

ENERGY FROM FATS

How does a human body access its fat reservoir? A fat molecule, recall, has a glycerol head and one, two, or three fatty acid tails. The body stores most fats as triglycerides, with three tails each. Triglycerides build up inside of the fat cells of adipose tissue. This tissue is strategically located under the skin of buttocks and other body regions.

When the blood glucose level falls, triglycerides are tapped as an energy alternative. Enzymes in fat cells cleave bonds between glycerol and fatty acids, which both enter the blood. Enzymes in the liver convert the glycerol to PGAL. And PGAL, recall, is one of the key intermediates for glycolysis (Figure 7.4). Nearly all cells of your body take up circulating fatty acids, and enzymes inside them cleave the fatty acid backbones. The fragments are converted to acetyl–CoA, which can enter the Krebs cycle.

Compared to glucose, a fatty acid tail has far more carbon-bound hydrogen atoms, so it yields more ATP. Between meals or during steady, prolonged exercise, fatty acid conversions supply about half of the ATP that muscle, liver, and kidney cells require.

What happens if you eat too many carbohydrates? Aerobic respiration, remember, converts glucose to pyruvate, then to acetyl–CoA, which enters the Krebs cycle. When too much glucose is circulating through the body, acetyl–CoA is diverted to a pathway that synthesizes fatty acids. *Too much glucose ends up as fat.*

ENERGY FROM PROTEINS

Some enzymes in your digestive system split dietary proteins into their amino acid subunits, which are then absorbed into the bloodstream. Cells use amino acids to build other proteins or nitrogen-containing compounds. Even so, if you eat more protein than your body needs, amino acids will be broken down further. Their $-NH_3^+$ group is pulled off, forming ammonia (NH_3). Depending on the types of amino acids, the leftover carbon backbones are broken down to either acetyl–CoA, pyruvate, or one of the intermediates of the Krebs cycle. Your cells can funnel any of these compounds into the Krebs cycle.

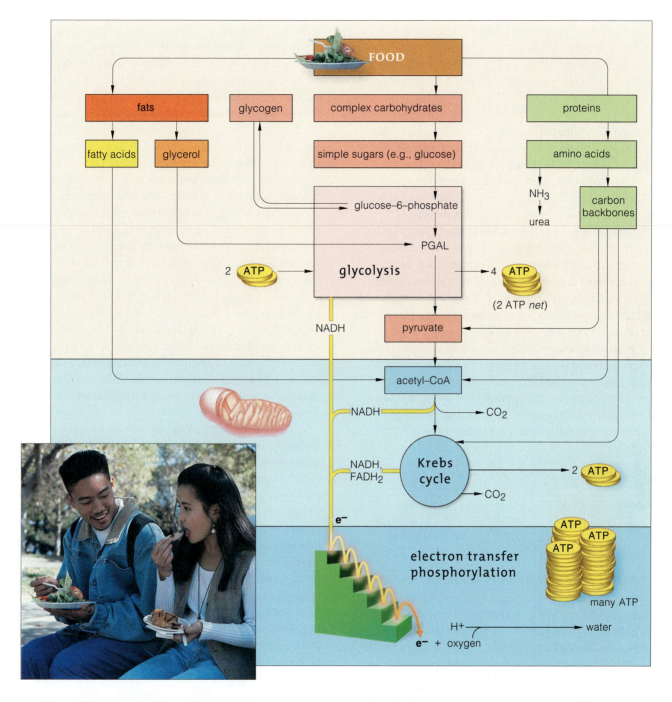

Figure 7.13 Reaction sites where a variety of organic compounds can enter the different stages of aerobic respiration. The compounds shown are alternative energy sources in humans and other mammals.

Notice how complex carbohydrates, fats, and proteins cannot enter the aerobic pathway directly. The digestive system, then individual cells, must first break apart these molecules to simpler compounds that the pathway can dismantle further.

As you can see, maintaining and accessing energy reserves is complicated business. Controlling the use of glucose is special because it is the fuel of choice for the brain. However, providing all of your cells with energy starts with the kinds of food you eat.

In humans and other mammals, the entrance of glucose or other organic compounds into an energy-releasing pathway depends on the kinds and proportions of carbohydrates, fats, and proteins in the diet.

7.7 Perspective on Life

In this unit you read about photosynthesis and aerobic respiration—the main pathways by which cells trap, store, and release energy. What you may not know is that the two pathways became linked, on a grand scale, over evolutionary time.

When life originated long ago, the atmosphere held little free oxygen. We can expect that those first cells had to make ATP by reactions similar to glycolysis, and fermentation pathways probably dominated. More than a billion years passed before the oxygen-evolving pathway of photosynthesis emerged.

Oxygen slowly accumulated in the atmosphere. Some cells now used it to accept electrons, perhaps as a chance outcome of mutated proteins in electron transfer chains. In time, some of their descendants abandoned anaerobic habitats. Among them were the forerunners of all bacteria, protists, plants, fungi, and animals that now engage in aerobic respiration.

With aerobic respiration, a great flow of carbon, hydrogen, and oxygen through metabolic pathways of living organisms came full circle. For the final products of this aerobic pathway—carbon dioxide and water—are the same materials necessary to build organic compounds in photosynthesis:

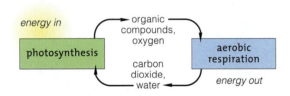

Perhaps you have difficulty seeing the connection between yourself—a highly intelligent being—and such remote-sounding events as energy flow and the cycling of carbon, hydrogen, and oxygen. Is this really the stuff of humanity?

Think back on the structure of a water molecule. Two hydrogen atoms sharing electrons with oxygen may not seem close to your daily life. Yet, through that sharing, water molecules show polarity and hydrogen-bond with one another. Their chemical behavior is a beginning for the organization of lifeless matter that leads in turn to the organization of all living things.

For now you can visualize other diverse molecules interspersed through water. Nonpolar kinds resist interaction with water; polar kinds dissolve in it. On their own, the phospholipids among them assemble into a two-layered film. Such lipid bilayers, recall, are the framework of cell membranes, hence all cells.

From the beginning, the cell has been the basic *living* unit. The essence of life is not some mysterious force; it is molecular organization and metabolic control. With a membrane to contain them, reactions *can* be controlled. With molecular mechanisms built into membranes, cells respond to energy changes and shifting concentrations of solutes in the environment. Response mechanisms operate by "telling" proteins—enzymes—when and what to build or tear down.

And it is not some mysterious force that creates proteins. DNA, the double-stranded treasurehouse of inheritance, has the chemical structure—*the chemical message*—that allows molecule to reproduce molecule, one generation after the next. In your body, DNA strands tell trillions of cells how countless molecules must be built or torn apart for their stored energy.

So yes, carbon, hydrogen, oxygen, and other atoms of organic molecules are the stuff of you, and us, and all of life. Yet it takes more than molecules to complete the picture. Life continues as long as a continuous flow of energy sustains its organization. Molecules become assembled into cells, cells into organisms, organisms into communities, and so on through the biosphere.

It takes energy inputs from the sun to maintain the levels of biological organization. And energy flows through time in one direction—from organized to less organized forms. Only as long as energy continues to flow into the great web of life can life continue in all its rich expressions.

So life is no more *and no less* than a marvelously complex system for prolonging order. Sustained with energy transfusions from the sun, life continues by its capacity for self-reproduction. With the hereditary instructions of DNA, energy and materials become organized, generation after generation. Even with the death of individuals, life elsewhere is prolonged. With each death, molecules are released and may be recycled as raw materials for new generations.

With this flow of energy and cycling of material through time, each birth is affirmation of our ongoing capacity for organization, each death a renewal.

Summary

Section 7.1 All organisms, including photosynthetic types, make ATP by the breakdown of glucose and other organic compounds. Glycolysis, the initial breakdown of one glucose to two pyruvate molecules, takes place in the cytoplasm. It is the first stage of all the main energy-releasing pathways, and it doesn't require free oxygen.

Anaerobic pathways end in the cytoplasm, and the net yield of ATP is small. An oxygen-requiring pathway called aerobic respiration continues in mitochondria, and it releases far more ATP energy from glucose.

Section 7.2 The first steps of glycolysis require an energy input of 2 ATP. Phosphate-group transfers from ATP drive the breakdown of a molecule of glucose (or another sugar) to two molecules of pyruvate, each with a three-carbon backbone. Two molecules of the coenzyme NAD^+ pick up electrons and hydrogen stripped from reaction intermediates, forming two NADH. Four ATP form during glycolysis, but the net energy yield is two ATP (because two ATP had to be invested up front).

Section 7.3 If the two pyruvates from glycolysis enter a mitochondrion, they will be fully degraded, as part of the second and third stages of aerobic respiration.

The second stage consists of acetyl–CoA formation and the Krebs cycle. Two pyruvates are converted to acetyl–CoA, and two carbon atoms depart in the form of CO_2. In two turns of the Krebs cycle (one for each pyruvate), intermediates are degraded; four more carbons escape (as CO_2). Coenzymes NAD^+ and FAD pick up electrons and hydrogen from intermediates. Two ATP form.

In total, eight NADH, two $FADH_2$, two ATP, and six CO_2 form during the aerobic pathway's second stage.

Section 7.4 The third stage of aerobic respiration proceeds at electron transfer chains and ATP synthases in the inner mitochondrial membrane. Electron transfer chains accept electrons and hydrogen from the NADH and $FADH_2$ that formed in the first two stages. Electron flow through the chains causes H^+ to accumulate in the inner mitochondrial compartment, so H^+ concentration and electric gradients build up across the membrane. H^+ flows down the gradients, through the interior of ATP synthases. This flow drives the attachment of unbound phosphate to ADP, forming many ATP.

Free oxygen picks up the electrons at the end of the transfer chains and combines with H^+, forming water.

Aerobic respiration has a typical net energy yield of thirty-six ATP for each glucose molecule metabolized.

Section 7.5 Fermentation pathways do not require oxygen, and they take place only in the cytoplasm. They use the pyruvate and ATP that formed during the first stage of reactions (glycolysis). The remaining reactions regenerate NAD^+. No more ATP forms. The net energy yield is only the two ATP that formed in glycolysis.

In alcoholic fermentation, the two pyruvates from glycolysis are converted to two acetaldehyde and two CO_2 molecules. When NADH transfers electrons and hydrogen to acetaldehyde, two ethanol molecules form and NAD^+ is regenerated.

In lactate fermentation, NAD^+ is regenerated when electrons and hydrogen are transferred from NADH to the two pyruvate molecules from glycolysis, which forms two lactate molecules as end products.

Slow-twitch and fast-twitch skeletal muscle cells support different levels of activity. Aerobic respiration and lactate fermentation occur in different cells that make up these muscles.

Section 7.6 In the human body, simple sugars from carbohydrates, glycerol and fatty acids from fats, and carbon backbones of amino acids from proteins can enter the aerobic pathway as alternative energy sources.

Section 7.7 Life's diversity, interconnections, and continuity arise from its unity at the molecular level.

Self-Quiz

Answers in Appendix III

1. Glycolysis starts and ends in the _____ .
   a. nucleus   c. plasma membrane
   b. mitochondrion   d. cytoplasm

2. Which of the following molecules does not form during glycolysis?
   a. NADH   b. pyruvate   c. $FADH_2$ d. ATP

3. Aerobic respiration is completed in the _____ .
   a. nucleus   c. plasma membrane
   b. mitochondrion   d. cytoplasm

4. In the third stage of aerobic respiration, _____ is the final acceptor of electrons from glucose.
   a. water   b. hydrogen   c. oxygen   d. NADH

5. Fill in the blanks in the diagram below.

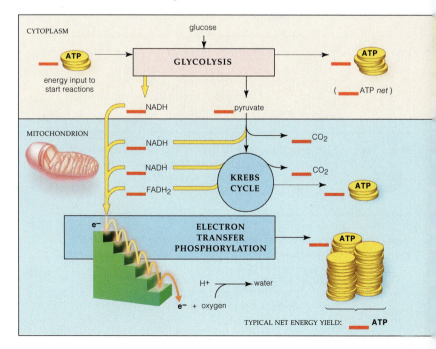

6. In alcoholic fermentation, _____ is the final acceptor of electrons stripped from glucose.
   a. oxygen
   b. pyruvate
   c. acetaldehyde
   d. sulfate

7. Fermentation pathways produce no more ATP beyond the small yield from glycolysis. The remaining reactions _____ .
   a. regenerate FAD
   b. regenerate NAD^+
   c. regenerate NAD
   d. regenerate $FADH_2$

8. In certain organisms and under certain conditions, _____ can be used as an energy alternative to glucose.
   a. fatty acids
   b. glycerol
   c. amino acids
   d. all of the above

9. Match the event with its most suitable description.
   _____ glycolysis
   _____ fermentation
   _____ Krebs cycle
   _____ electron transfer phosphorylation

   a. ATP, NADH, $FADH_2$, CO_2, and water form
   b. glucose to two pyruvates
   c. NAD^+ regenerated, two ATP net
   d. H^+ flows through ATP synthases

Media Menu

Student CD-ROM

Impacts, Issues Video
   When Mitochondria Spin Their Wheels
Big Picture Animation
   Energy-releasing pathways and links to photosynthesis
Read-Me-First Animation
   Aerobic respiration
   Glycolysis
   The Krebs cycle
   Third-stage reactions
Other Animations and Interactions
   Comparison of energy-releasing pathways
   Structure and function of a mitochondrion
   Fermentation pathways
   Alternative energy sources

InfoTrac

- My Personal Challenge. *The Exceptional Parent*, August 1998.
- Mitochondria: Cellular Energy Co.—Researchers Strive to Keep the Energy Pipeline Open in the Face of Damaging Cellular Insults. *The Scientist*, June 2002.

Web Sites

- United Mitochondrial Disease Foundation: www.umdf.org
- Friedreich's Ataxia Research Alliance: www.frda.org
- National Organization for Rare Disorders: www.raredisorders.org

How Would You Vote?

Friedreich's ataxia is devastating but relatively rare. In the United States, it affects 1 individual in 50,000. This is good news for most of us, but means that there is relatively little incentive for companies to develop treatments. Who should fund this research? Should we provide tax incentives to companies that work to find cures for rare diseases?

Critical Thinking

1. Living cells of your body absolutely do not use their nucleic acids as alternative energy sources. Suggest why.

2. Suppose you start a body-building program. You are already eating plenty of carbohydrates. Now a qualified nutritionist recommends that you start a protein-rich diet that includes protein supplements. Speculate on how extra dietary proteins will be put to use, and in which tissues.

3. Each year, Canada geese lift off in precise formation from their northern breeding grounds. They head south to spend the winter months in warmer climates. Then they make the return trip in spring. As is the case for other migratory birds, their flight muscle cells are efficient at using fatty acids as an energy source. Remember, the carbon backbone of fatty acids can be cleaved into small fragments that can be converted to acetyl–CoA for entry into the Krebs cycle.

   Suppose a lesser Canada goose from Alaska's Point Barrow has been steadily flapping along for about three thousand kilometers and is approaching Klamath Falls, Oregon. It looks down and notices a rabbit sprinting from a coyote with a taste for rabbit.

   With a stunning burst of speed, the rabbit reaches the safety of its burrow.

   Which energy-releasing pathway predominated in muscle cells in the rabbit's legs? Why was the Canada goose relying on a different pathway for most of its journey? And why wouldn't the pathway of choice in goose flight muscle cells be much good for a rabbit making a mad dash from its enemy?

4. At high altitudes, oxygen levels are low. Mountain climbers risk altitude sickness, which is characterized by shortness of breath, weakness, dizziness, and confusion.

   Curiously, early symptoms of *cyanide poisoning* are similar to altitude sickness. This highly toxic poison binds tightly to a cytochrome, the last molecule in mitochondrial electron transfer chains. When cyanide becomes bound to it, cytochrome can't transfer electrons to the next component of the chain. Explain why cytochrome shutdown might cause the same symptoms as altitude sickness.

5. ATP form in mitochondria. In warm-blooded animals, so does a lot of heat, which can be circulated in ways that help regulate body temperature. Cells of brown adipose tissue (fat) make a protein that disrupts the formation of electron transfer chains in mitochondrial membranes. H^+ gradients are affected, so fewer ATP form; electrons in the transfer chains give up more of their energy as heat. Because of this, some researchers are hypothesizing that brown adipose tissue may not function like white adipose tissue, which is an energy reservoir. Brown adipose tissue may function in thermogenesis, or heat production.

   Mitochondria, recall, contain their own DNA, which may have mutated independently in human populations that evolved in the Arctic and in the hot tropics. If that is so, then mitochondrial function may be adapted to climate.

   How do you suppose such a mitochondrial adaptation might affect people living where the temperature range no longer correlates with their ancestral heritage? Would you expect people whose ancestors evolved in the Arctic to be more or less likely to put on a lot of weight than those whose ancestors lived in the tropics? See *Science*, January 9, 2004: 223–226 for more information.

II Principles of Inheritance

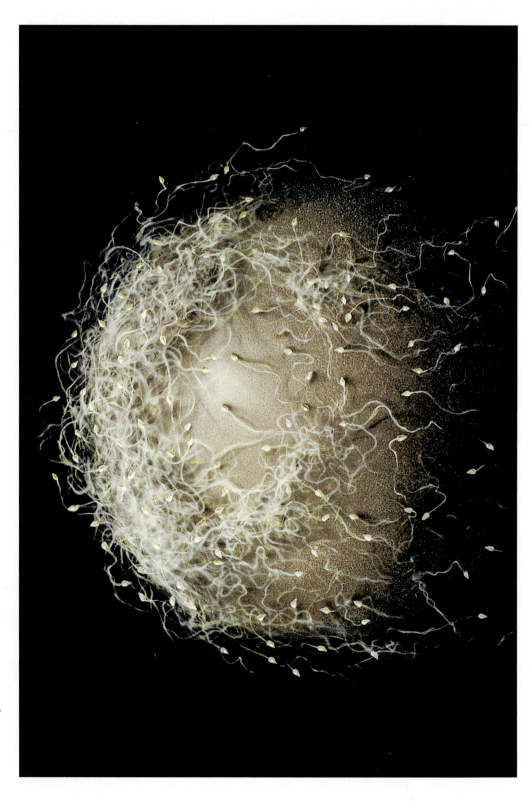

Human sperm, one of which will penetrate this mature egg and so set the stage for the development of a new individual in the image of its parents. This exquisite art is based on a scanning electron micrograph.

IMPACTS, ISSUES *Henrietta's Immortal Cells*

Each human starts out as a fertilized egg. By the time of birth, cell divisions and other processes have given rise to a body of about a trillion cells. Even in the adult, billions of cells are still dividing and replacing their damaged or worn-out predecessors.

In 1951, George and Margaret Gey of Johns Hopkins University were trying to develop a way to keep human cells dividing outside the body. An "immortal" cell lineage could help researchers study basic life processes as well as cancer and other diseases. Using cells to study cancer would be a far better alternative than experimenting directly on patients and risking their lives.

For almost thirty years, the Geys tried to grow normal and diseased human cells. But they could not stop the cellular descendants from dying within a few weeks.

Mary Kubicek, a lab assistant, tried again and again to establish a self-perpetuating lineage of cultured human cancer cells. She was about to give up, but she prepared one last sample and named them HeLa cells. The code name signified the first two letters of the patient's first and last names.

Those HeLa cells began to divide. Four days later, there were so many cells that the researchers subdivided them into more culture tubes. The cells grew at a phenomenal rate; they divided every twenty-four hours and coated the surface of the tubes within days.

Sadly, cancer cells in the patient were dividing just as often. Six months after she had been diagnosed with cancer, malignant cells had infiltrated tissues all through her body. Two months after that, Henrietta Lacks, a young woman from Baltimore, was dead.

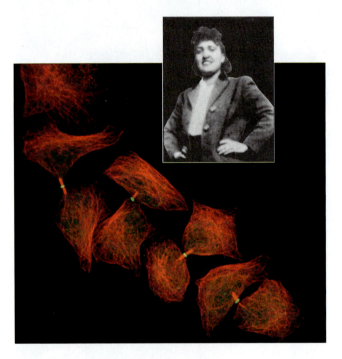

Figure 8.1 Dividing HeLa cells—a legacy of Henrietta Lacks, who was a casualty of cancer. Her cellular contribution to science is still helping others every day.

the big picture

What Divides, and When Eukaryotic cells reproduce by duplicating their chromosomes, getting them into genetically identical parcels by mitosis or meiosis, and dividing the parcels as well as cytoplasm among daughter cells. Prokaryotic cells divide by a different mechanism.

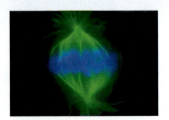

Mitosis Mitosis, a nuclear division mechanism, has four continuous stages: prophase, metaphase, anaphase, and telophase. During these stages, a microtubular spindle moves duplicated chromosomes so that they end up in two genetically identical nuclei.

Although Henrietta passed away, her cells lived on in the Geys' laboratory (Figure 8.1). In time, HeLa cells were shipped to research laboratories all over the world. The Geys used HeLa cells to identify precisely the viral strains that cause polio, which was rampant at the time. Tissue culture techniques developed in their laboratory were used to grow a vaccine. Other scientists used the cells to study mechanisms of cancer, viral growth, the effects of radiation, protein synthesis, and more. Some HeLa cells even traveled into space for experiments on the *Discoverer XVII* satellite. Each year, hundreds of important research projects move forward, thanks to Henrietta's immortal cells.

Henrietta was only thirty-one when runaway cell divisions killed her. Decades later, her legacy continues to help humans everywhere, through her cellular descendants that are still dividing day after day.

Understanding cell division—and, ultimately, how new individuals are put together in the image of their parents—starts with answers to three questions. *First*, what kind of information guides inheritance? *Second*, how is the information copied in a parent cell before being distributed into daughter cells? *Third*, what kinds of mechanisms actually parcel out the information to daughter cells?

We will need more than one chapter to survey the nature of cell reproduction and other mechanisms of inheritance. This chapter introduces the structures and mechanisms that cells use to reproduce.

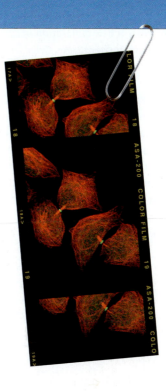

✔ How Would You Vote?

It is illegal to sell your organs, but you can sell your cells, including eggs, sperm, and blood cells. HeLa cells continue to be sold all over the world by cell culture firms. Should the family of Henrietta Lacks share in the profits? See the Media Menu for details, then vote online.

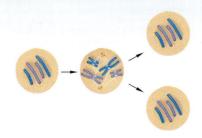

Cytoplasmic Division
After nuclear division, the cytoplasm divides in a way that typically puts a nucleus in each daughter cell. The cytoplasm of an animal cell is simply pinched in two. In a plant cell, a cross-wall forms in the cytoplasm and divides it.

The Cell Cycle and Cancer
The cell cycle has built-in checkpoints, or mechanisms that monitor and control the timing and rate of cell division. On rare occasions, these surveillance mechanisms fail, and cell division becomes uncontrollable. Tumor formation is the outcome.

8.1 Overview of Cell Division Mechanisms

*The continuity of life depends on **reproduction**. By this process, parents produce a new generation of cells or multicelled individuals like themselves. Cell division is the bridge between generations.*

A dividing cell faces a challenge. Each of its daughter cells must get information encoded in the parental DNA and enough cytoplasm to start up its own operation. DNA "tells" it which proteins to build. Some of the proteins are structural materials; others are enzymes that speed construction of organic compounds. If the cell does not inherit all of the required information, it will not be able to grow or function properly.

In addition, the parent cell's cytoplasm already has enzymes, organelles, and other metabolic machinery. When a daughter cell inherits what looks like a blob of cytoplasm, it really is getting start-up machinery that will keep it running until it can use information in DNA for growing on its own.

MITOSIS, MEIOSIS, AND THE PROKARYOTES

Eukaryotic cells can't just split in two, because a single nucleus holds the DNA. They do split their cytoplasm into daughter cells. But they don't do this until *after* their DNA has been copied, sorted out, and packaged by way of mitosis or meiosis.

Mitosis is a nuclear division mechanism that occurs in *somatic* cells (body cells) of multicelled eukaryotes. It is the basis of increases in body size during growth, replacements of worn-out or dead cells, and tissue repair. Many plants, animals, fungi, and single-celled protists also reproduce asexually, or make copies of themselves, by way of mitosis (Table 8.1).

Meiosis is a different nuclear division mechanism. It functions only in sexual reproduction, and it precedes the formation of gametes (such as sperm and eggs) or spores. In complex animals, gametes form from *germ* cells. As you will see in this chapter and the one that follows, meiosis and mitosis have a lot in common, but the outcomes differ.

What about prokaryotic cells—the archaea and the eubacteria? They reproduce asexually by an entirely different mechanism called prokaryotic fission. We will consider prokaryotic fission later, in Section 19.1.

KEY POINTS ABOUT CHROMOSOME STRUCTURE

Every eukaryotic cell has a characteristic number of DNA molecules, each with many attached proteins. Together, a molecule of DNA and its proteins are one **chromosome**. Chromosomes are duplicated before the cell enters nuclear division. Each chromosome and its copy stay attached to each other as **sister chromatids** until late in the nuclear division process. Think of each chromatid as one arm and leg of a sunbather stretched out on the sand (Figure 8.2).

Early in mitosis or meiosis, a chromosome coils back on itself repeatedly, to a highly condensed form, by interactions between its proteins and DNA. At high magnification, you can see the histone proteins, which look like beads on a string (Figure 8.3d). DNA winds

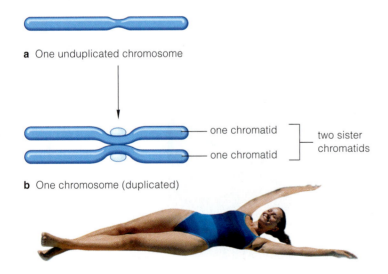

a One unduplicated chromosome

one chromatid ⎫
one chromatid ⎬ two sister chromatids

b One chromosome (duplicated)

Figure 8.2 A simple way to visualize a eukaryotic chromosome in the unduplicated state and duplicated state. Eukaryotic cells are duplicated before mitosis or meiosis.

Table 8.1	Cell Division Mechanisms
Mechanisms	**Functions**
Mitosis, cytoplasmic division	In *all* multicelled eukaryotes, the basis of the following: 1. Increases in body size during growth. 2. Replacement of dead or worn-out cells. 3. Repair of damaged tissues. In single-celled and many multicelled eukaryotes, *also* the basis of asexual reproduction.
Meiosis, cytoplasmic division	In single-celled and multicelled eukaryotes, the basis of sexual reproduction; precedes gamete or spore formation (Chapter 9).
Prokaryotic fission	In bacteria and archaea only, the basis of asexual reproduction (Section 19.1).

Read Me First!
and watch the narrated animation on chromosome structural organization

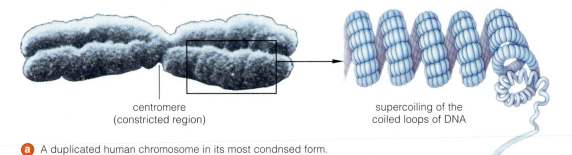

centromere
(constricted region)

supercoiling of the
coiled loops of DNA

b At times when a chromosome is most condensed, the proteins associated with it interact in ways that package loops of already coiled DNA into "supercoils."

a A duplicated human chromosome in its most condnsed form.

Figure 8.3 (**a**) Scanning electron micrograph of a duplicated human chromosome in its most condensed form. (**b**,**c**) Interacting proteins hold loops of coiled DNA in the supercoiled array of a cylindrical fiber. (**d**,**e**) The most basic unit of organization is the nucleosome: part of a DNA molecule looped twice around a core of histones. The transmission electron micrographs correspond to organizational levels (**c**) and (**d**).

twice around histone "spools." A histone–DNA spool is a **nucleosome**, a unit of structural organization.

While each duplicated chromosome is condensing, a pronounced constriction appears in a predictable location along its length. At this constriction, the **centromere**, the chromosome's sister chromatids are attached to each other (Figure 8.3). On its surface, we find kinetochores: docking sites for microtubules that will move the chromosome during nuclear division. The centromere's location is different for each type of chromosome and is one of its defining characteristics.

So what is the point of the structural organization? Tight packaging might help keep chromosomes from getting tangled up while they are moved and sorted out into parcels *during* nuclear division. Also, *between* divisions, nucleosome packaging can be loosened up in selected regions, giving enzymes access to required bits of information in the DNA.

When a cell divides, each daughter cell receives a required number of chromosomes and some cytoplasm. In eukaryotic cells, this involves nuclear and cytoplasmic division.

One nuclear division mechanism, mitosis, is the basis of bodily growth, cell replacements, tissue repair, and often asexual reproduction in eukaryotes.

Meiosis, another nuclear division mechanism, is the basis of sexual reproduction. It precedes gamete or spore formation.

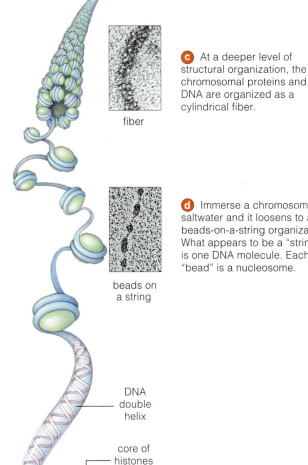

c At a deeper level of structural organization, the chromosomal proteins and DNA are organized as a cylindrical fiber.

fiber

d Immerse a chromosome in saltwater and it loosens to a beads-on-a-string organization What appears to be a "string" is one DNA molecule. Each "bead" is a nucleosome.

beads on a string

DNA double helix

core of histones

nucleosome

e A nucleosome consists of part of a DNA molecule looped twice around a core of histone proteins.

8.2 Introducing the Cell Cycle

Let's start thinking about cell reproduction as a recurring series of events, a cycle. This isn't the same as a life cycle, which is a sequence of stages through which individuals of a species pass during their lifetime.

A **cell cycle** is a series of events from one cell division to the next (Figure 8.4). It starts when a new daughter cell forms by mitosis and cytoplasmic division; it ends when the cell divides. Mitosis, cytoplasmic division, and then interphase constitute one turn of the cycle.

THE WONDER OF INTERPHASE

During **interphase**, a cell increases in mass, roughly doubles the number of its cytoplasmic components, and duplicates its DNA. For most cells, interphase is the longest portion of the cell cycle. Biologists divide it into three stages:

G1 Interval ("*Gap*") of cell growth and functioning before the onset of DNA replication

S Time of "*Synthesis*" (DNA replication)

G2 Second interval (*Gap*), after DNA replication when the cell prepares for division

G1, S, and G2 are code names for some events that are just amazing, considering how much DNA is stuffed in a nucleus. For example, if you could stretch out all the DNA molecules from one of your somatic cells in a single line, they would extend past the fingertips of an outstretched arm. A line of all the DNA from one salamander cell would stretch about 540 feet!

The wonder is, enzymes and other proteins in cells *selectively* access, activate, and silence information in all that DNA. They also make base-by-base copies of every DNA molecule before they divide. Most of this cellular work is completed in interphase.

G1, S, and G2 of interphase have distinct patterns of biosynthesis. Most of your cells remain in G1 while they are building proteins, carbohydrates, and lipids. Cells destined to divide enter S, when they copy their DNA and the proteins attached to it. During G2, they make the proteins that will drive mitosis.

Once S begins, DNA replication usually proceeds at about the same rate and continues during mitosis. The rate holds for all cells of a species, so you might well wonder if the cell cycle has built-in molecular brakes. It does. Apply the brakes that are supposed to work in G1, and the cycle stalls in G1. Lift the brakes, and the cell cycle runs to completion. Said another way, *control mechanisms govern the rate of cell division.*

Imagine a car losing its brakes just as it starts down a steep mountain road. As you will read later in the chapter, that's how cancer starts. Crucial controls over division are lost, and the cell cycle can't stop turning.

The cell cycle lasts about the same amount of time for cells of the same type but varies among different types. For example, all neurons (nerve cells) in your brain remain in G1 of interphase, and usually they will not divide again. By contrast, every second, 2 to 3 million precursors of red blood cells form to replace worn-out ones circulating in your body. Early in the development of a sea urchin embryo, the number of cells doubles every two hours.

Adverse conditions often disrupt the cell cycle. When deprived of a vital nutrient, for example, the free-living cells called amoebas do not leave interphase. Even so, when any cell moves past a certain point in interphase, the cycle normally will continue regardless of the conditions outside because of built-in controls over its duration.

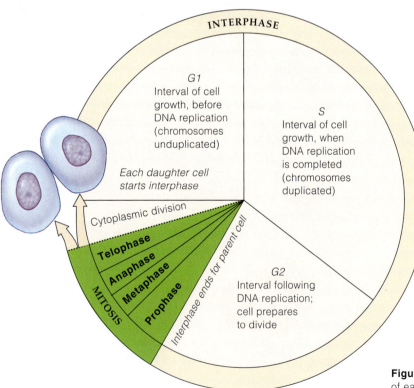

INTERPHASE

G1
Interval of cell growth, before DNA replication (chromosomes unduplicated)

Each daughter cell starts interphase

S
Interval of cell growth, when DNA replication is completed (chromosomes duplicated)

Cytoplasmic division

Telophase

Anaphase

Metaphase

Prophase

MITOSIS

Interphase ends for parent cell

G2
Interval following DNA replication; cell prepares to divide

Figure 8.4 Eukaryotic cell cycle, generalized. The length of each interval differs among different cell types.

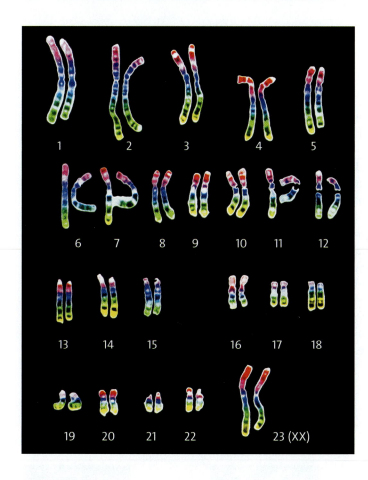

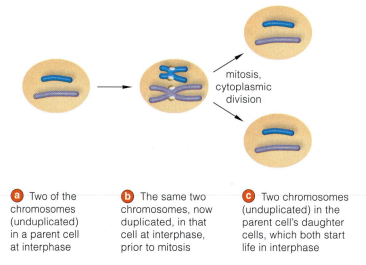

a Two of the chromosomes (unduplicated) in a parent cell at interphase

b The same two chromosomes, now duplicated, in that cell at interphase, prior to mitosis

c Two chromosomes (unduplicated) in the parent cell's daughter cells, which both start life in interphase

Figure 8.5 *Above:* One way to think about how mitosis maintains a parental chromosome number, one generation to the next.

Left: Here's an example. These are twenty-three pairs of metaphase chromosomes from a diploid cell of a human female. The last two are a pair of sex chromosomes. (In human females, such cells have an XX pairing; in males, they have an XY pairing.) When all goes well, each time a somatic cell in this female undergoes mitosis and cytoplasmic division, the daughter cells will always end up with an unduplicated set of these twenty-three pairs of chromosomes.

MITOSIS AND THE CHROMOSOME NUMBER

To know what mitosis does, you have to know that each species has a characteristic **chromosome number**, or the sum of all chromosomes in cells of a given type. Body cells of gorillas and chimpanzees have 48, pea plants have 14, and humans have 46 (Figure 8.5).

Actually, your cells have a **diploid number** ($2n$) of chromosomes; there are two of each type. Those 46 are like volumes of two sets of books numbered from 1 to 23. You have two volumes of, say, chromosome 22—*a pair of them.* Except for one sex chromosome pairing (XY), both have the same length and shape, and carry the same hereditary information about the same traits.

Think of them as two sets of books on how to build a house. Your father gave you one set. Your mother had her own ideas about wiring, plumbing, and so on. She gave you an alternate edition on the same topics, but it says slightly different things about many of them.

With mitosis, a diploid parent cell can produce two diploid daughter cells. This doesn't mean each merely gets forty-six or forty-eight or fourteen chromosomes. If only the total mattered, then one cell might get, say, two pairs of chromosome 22 and no pairs whatsoever of chromosome 9. But neither cell could function like its parent *without two of each type of chromosome.*

Mitosis has four stages: *prophase, metaphase, anaphase,* and *telophase.* All use a **bipolar mitotic spindle**. This dynamic structure is made of microtubules that grow or shrink as tubulin subunits are added or lost from their ends. The spindle forms as microtubules grow toward each other from two poles until they overlap. Some tether the duplicated chromosomes.

One chromatid of each chromosome gets attached to microtubules extending from one spindle pole, its sister gets attached to microtubules from the other pole, then they are dragged apart. A complete set of (now-unduplicated) chromosomes ends up in each half of the cell before the cytoplasm divides. That is how mitosis can maintain a parental chromosome number through turn after turn of the cell cycle (Figure 8.5).

microtubule of bipolar spindle

Interphase, mitosis, and cytoplasmic division constitute one turn of the cell cycle.

During interphase, a new cell increases its mass, roughly doubles the number of its cytoplasmic components, and duplicates its chromosomes. The cycle ends after the cell undergoes mitosis and then divides its cytoplasm.

8.3 A Closer Look at Mitosis

Let's focus on a "typical" animal cell to see how mitosis can keep the chromosome number constant, division after division, from one cell generation to the next.

By the time a cell enters **prophase**—the first stage of mitosis—its chromosomes are already duplicated, with sister chromatids joined at the centromere. They are in threadlike form, but now they start to twist and fold. By the end of prophase, they will be condensed into thick, compact, rod-shaped forms (Figure 8.6a–c).

Also before prophase, two barrel-shaped centrioles and two centrosomes started duplicating themselves next to the nucleus. A centriole, recall, gives rise to a cilium or flagellum. In animal cells, it is embedded in a centrosome, which it helps organize. A **centrosome** is a site where microtubules originate. In prophase, the duplicated centrioles move apart—as do the two centrosomes—until they are on opposite sides of the nucleus. Microtubules grow out of each centrosome. *These are the microtubules that form the bipolar spindle.*

As prophase ends, the nuclear envelope starts to break up into tiny flattened vesicles. The microtubules now interact with the chromosomes and one another. Some tether chromosomes at the docking sites called kinetochores. Others tether the chromosome arms. And still others keep on growing from centrosomes until they overlap midway between the two spindle poles. Driven by ATP energy, motor proteins (dyneins and kinesins) produce the force necessary to assemble the spindle, and to bind and move the chromosomes.

Microtubules from one pole tether one chromatid of each chromosome; microtubules from the opposite pole tether the other. They engage in a tug-of-war, growing and shrinking until they are the same length. At that point, **metaphase**, all duplicated chromosomes are aligned midway between the spindle poles. The alignment is crucial for the next stage of mitosis.

At **anaphase**, the kinetochores of sister chromatids detach from each other and take off toward opposite spindle poles. Driven by motor proteins, they move

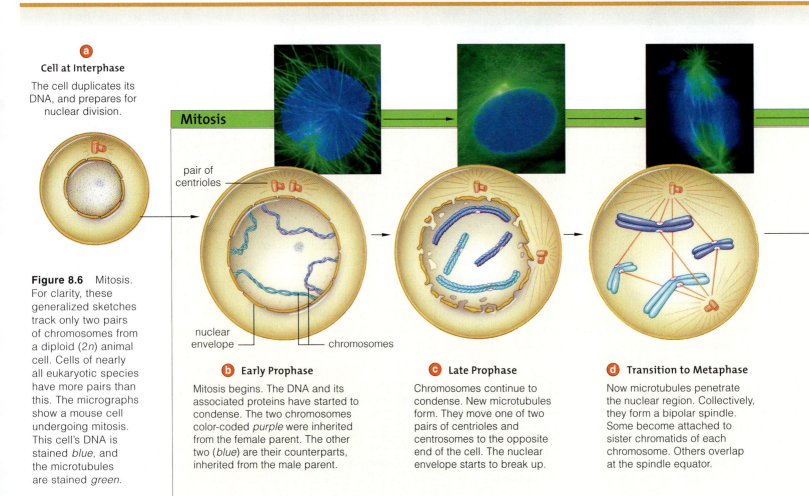

a Cell at Interphase

The cell duplicates its DNA, and prepares for nuclear division.

Mitosis

pair of centrioles

nuclear envelope

chromosomes

Figure 8.6 Mitosis. For clarity, these generalized sketches track only two pairs of chromosomes from a diploid (2n) animal cell. Cells of nearly all eukaryotic species have more pairs than this. The micrographs show a mouse cell undergoing mitosis. This cell's DNA is stained *blue*, and the microtubules are stained *green*.

b Early Prophase

Mitosis begins. The DNA and its associated proteins have started to condense. The two chromosomes color-coded *purple* were inherited from the female parent. The other two (*blue*) are their counterparts, inherited from the male parent.

c Late Prophase

Chromosomes continue to condense. New microtubules form. They move one of two pairs of centrioles and centrosomes to the opposite end of the cell. The nuclear envelope starts to break up.

d Transition to Metaphase

Now microtubules penetrate the nuclear region. Collectively, they form a bipolar spindle. Some become attached to sister chromatids of each chromosome. Others overlap at the spindle equator.

along microtubules toward the opposite spindle poles, dragging the chromatids with them. At the same time, the microtubules are shortening at both ends even as chromatids remain attached to them. The net effect is that sister chromatids are reeled in to opposite poles.

Also at the same time, microtubules that overlap midway between the spindle poles are ratcheting past one another. Motor proteins drive their interactions, which push the two spindle poles farther apart.

Sister chromatids, recall, are genetically identical. Once they detach from each other at anaphase, each is a separate chromosome in its own right.

Telophase gets under way when one of each type of chromosome reaches a spindle pole. Each half of the cell now contains two genetically identical clusters of chromosomes. Now all the chromosomes decondense. Vesicles derived from the old nuclear envelope fuse and form patches of membrane around each cluster. Patch joins with patch until a new nuclear envelope encloses each cluster. And so two nuclei form (Figure 8.6g). In our example, the parent cell had a diploid number of chromosomes. So does each nucleus.

Once two nuclei have formed, telophase is over—and so is mitosis.

Prior to mitosis, each chromosome in a cell's nucleus is duplicated, so it consists of two sister chromatids.

In prophase, chromosomes condense to rodlike forms, and microtubules form a bipolar spindle. The nuclear envelope breaks up. Some microtubules harness the chromosomes.

At metaphase, all chromosomes are aligned midway between the spindle's poles, at its equator.

At anaphase, microtubules move sister chromatids of each chromosome apart, to opposite spindle poles.

At telophase, a new nuclear envelope forms around each of two clusters of decondensing chromosomes.

Thus mitosis forms two daughter nuclei. Each has the same chromosome number as the parent cell's nucleus.

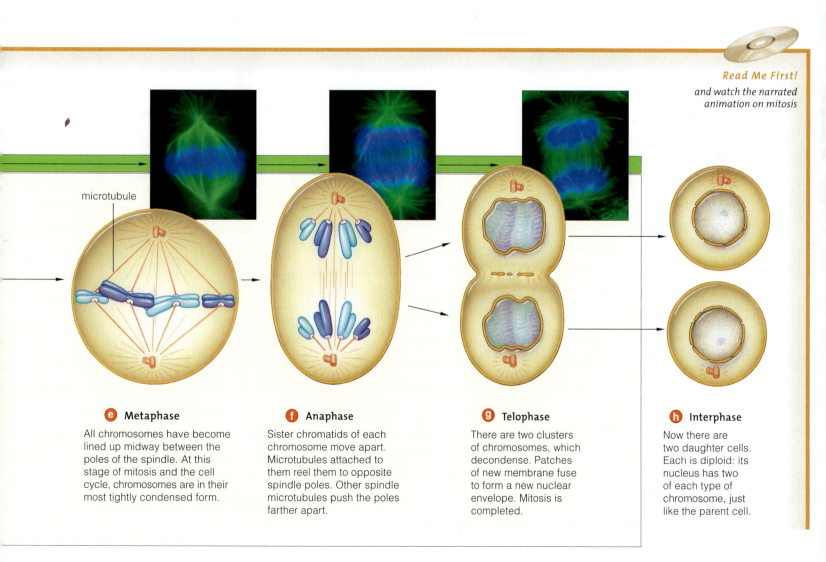

Read Me First!
and watch the narrated animation on mitosis

microtubule

e Metaphase

All chromosomes have become lined up midway between the poles of the spindle. At this stage of mitosis and the cell cycle, chromosomes are in their most tightly condensed form.

f Anaphase

Sister chromatids of each chromosome move apart. Microtubules attached to them reel them to opposite spindle poles. Other spindle microtubules push the poles farther apart.

g Telophase

There are two clusters of chromosomes, which decondense. Patches of new membrane fuse to form a new nuclear envelope. Mitosis is completed.

h Interphase

Now there are two daughter cells. Each is diploid: its nucleus has two of each type of chromosome, just like the parent cell.

8.4 Division of the Cytoplasm

The cytoplasm usually divides at some time between late anaphase and the end of telophase. The actual mechanism of cytoplasmic division—or, as it is often called, cytokinesis—differs among species.

CLEAVAGE IN ANIMALS

An animal cell divides by **cleavage**, a mechanism that pinches its cytoplasm in two. Typically, the plasma membrane starts to sink inward as a thin indentation about halfway between the cell's two poles (Figure 8.7a). This cleavage furrow is the first visible sign that

the cytoplasm in an animal cell is dividing. The furrow advances until it extends all the way around the cell. As it does, it deepens along a plane that corresponds to the equator of the former microtubular spindle.

How does this happen? In the cytoplasm just under the plasma membrane, microfilaments organized in a thin, ringlike band generate the contractile force for the cut (Figure 8.8). These cytoskeletal elements are attached to the plasma membrane. They slide past one another, as outlined in Section 4.10. As they do, they drag the plasma membrane deeper and deeper inward until the cytoplasm is partitioned. Each of the two

Read Me First!
and watch the narrated animation on cytoplasmic division

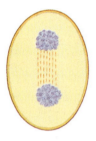 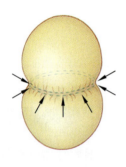 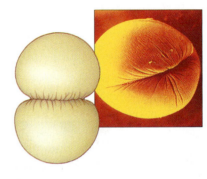 

1 Mitosis is over, and the spindle is disassembling.

2 At the former spindle equator, a ring of microfilaments attached to the plasma membrane contracts.

3 As the microfilament ring shrinks in diameter, it pulls the cell surface inward.

4 Contractions continue; the cell is pinched in two.

a Animal cell division

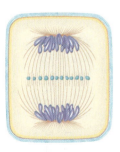 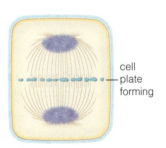 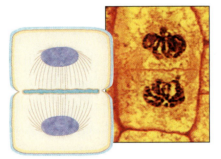  

cell plate forming

1 As mitosis ends, vesicles cluster at the spindle equator. They contain materials for a new primary cell wall.

2 Vesicle membranes fuse. The wall material is sandwiched between two new membranes that lengthen along the plane of a newly forming cell plate.

3 Cellulose is deposited inside the sandwich. In time, these deposits will form two cell walls. Others will form the middle lamella between the walls and cement them together.

4 A cell plate grows at its margins until it fuses with the parent cell plasma membrane. The primary wall of growing plant cells is still thin. New material is deposited on it.

b Plant cell division

Figure 8.7 Cytoplasmic division of an animal cell (**a**) and a plant cell (**b**).

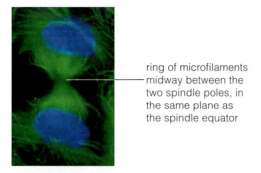

ring of microfilaments midway between the two spindle poles, in the same plane as the spindle equator

Figure 8.8 Cleavage. Inside this animal cell, a ring of microfilaments is pinching the cytoplasm in two.

daughter cells that forms this way ends up with a nucleus, some cytoplasm, and plasma membrane.

CELL PLATE FORMATION IN PLANTS

Plant cells cannot divide the same way your cells do, because most of them have a cell wall. That wall prevents their cytoplasm from simply pinching in two.

Instead, cytoplasmic division of plant cells involves **cell plate formation**, as shown in Figure 8.7b. By this mechanism, tiny vesicles packed with wall-building materials fuse with one another and with remnants of the microtubular spindle. Together, deposits of the materials form a disklike structure called a cell plate. Deposits of cellulose accumulate at the plate. In time, they thicken enough to form a cross-wall through the cell. New plasma membrane extends across both sides of it. This wall grows until it bridges the cytoplasm and divides the parent cell in two.

APPRECIATE THE PROCESS!

Take a moment to look closely at your hands. Visualize the cells making up your palms, thumbs, and fingers. Now imagine the mitotic divisions that produced all of the cell generations that preceded them while you were developing, early on, inside your mother (Figure 8.9). And be grateful for the astonishing precision of mechanisms that led to their formation at prescribed times, in prescribed numbers, for the alternatives can be terrible indeed.

Why? Good health and survival itself depend on the proper timing and completion of cell cycle events. Some genetic disorders arise as a result of mistakes that happened during the duplication or distribution of even one chromosome. Unchecked cell divisions often destroy surrounding tissues and, ultimately, the

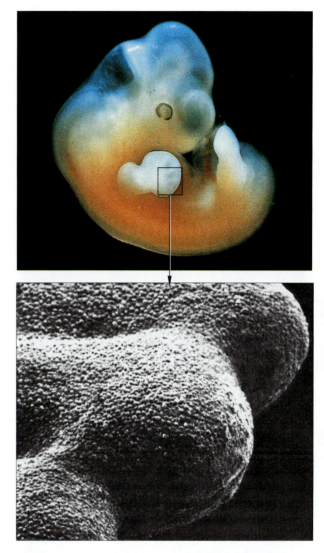

Figure 8.9 Transformation of a paddlelike structure into a human hand through mitosis, cytoplasmic divisions, and other processes of embryonic development. The scanning electron micrograph shows individual cells.

individual. Such losses can start in body cells. They can start in the germ cells that give rise to sperm and eggs, although rarely. The last section of this chapter can give you a sense of the consequences.

After mitosis, a separate mechanism cuts the cytoplasm into two daughter cells, each with a daughter nucleus.

Cleavage is a form of cytoplasmic division in animal cells. Microfilaments banded around a cell's midsection slide past one another in a way that pinches the cytoplasm in two.

Cytoplasmic division in plants often involves the formation of a cross-wall between the new plasma membranes of adjoining daughter cells.

8.5 When Control Is Lost

Growth and reproduction depend on controls over cell division. On rare occasions, something goes wrong in a somatic cell or germ cell. Cancer may be the outcome.

THE CELL CYCLE REVISITED

Millions of cells in your skin, bone marrow, gut lining, liver, and elsewhere divide and replace their worn-out, dead, and dying predecessors every second. They don't divide willy-nilly; many cellular mechanisms control cell growth, DNA replication, and division. They also control when the division machinery is put to rest.

What happens when something goes wrong? For example, if sister chromatids do not separate as they should during mitosis, one daughter cell may end up with too many chromosomes, the other with too few. Chromosomal DNA can be attacked by free radicals or peroxides, two metabolic by-products. It can be damaged by cosmic radiation, which bombards us all the time. Problems are frequent, inevitable, and must be corrected quickly.

The cell cycle has built-in checkpoints that keep errors from getting out of hand. Certain proteins— products of checkpoint genes—monitor whether the DNA gets fully replicated, whether it gets damaged, even whether nutrient concentrations are sufficient to support cell growth. The surveillance helps cells identify and correct problems.

Some checkpoint proteins make the cell cycle advance; their absence arrests it. The ones called *growth factors* invite transcription of genes that help the body grow. For instance, epidermal growth factor activates a kinase when it binds to cells in epithelial tissues; it is a signal to start mitotic cell divisions.

Other proteins inhibit cell cycle changes. Several checkpoint gene products put the brakes on mitosis when chromosomal DNA gets damaged (Figure 8.10). Some of the *kinases*, enzymes that phosphorylate other molecules, act as checkpoint gene products. When DNA is broken or incomplete, they activate other proteins in a cascade of signaling events that ultimately stop the cell cycle or induce cell death.

CHECKPOINT FAILURE AND TUMORS

Sometimes a checkpoint gene mutates so that its protein product no longer functions properly. When all checkpoint mechanisms for a particular process fail, the cell loses control over its replication cycle. Figures 8.11 through 8.13 show a few of the outcomes.

In some cases it gets stuck in mitosis and divides over and over again, with no interphase. In other cases, damaged chromosomes are replicated or cells don't die as they are supposed to, because signals calling for cell death are disabled. A growing mass of a cell's defective descendants may invade other tissues in the body, as a tumor.

In most tumor cells, at least one checkpoint protein is missing. That is why checkpoint gene products that *inhibit* mitosis are called tumor suppressors. Checkpoint genes encoding proteins that *stimulate* mitosis are called oncogenes. Mutations that affect oncogene products or the rate at which they form help transform a normal cell into a tumor cell. Mutant checkpoint genes are linked with increased risk of tumors, and sometimes they run in families.

Moles and other tumors are **neoplasms**—abnormal masses of cells that lost controls over how they grow

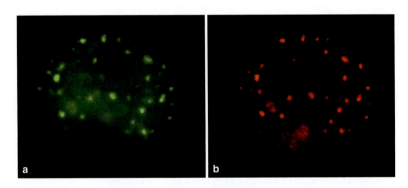

Figure 8.10 Protein products of checkpoint genes in action. DNA in the nucleus of this cell has been damaged by ionizing radiation. (**a**) *Green* spots pinpoint the location of *53BP1*, and (**b**) *red* spots pinpoint the location of *BRCA1*. Both proteins have clustered around the same chromosome breaks in a single cell nucleus. The integrated action of these proteins and others can arrest mitosis until the DNA breaks are fixed.

Figure 8.11 Scanning electron micrograph of a cervical cancer cell, the kind that killed Henrietta Lacks.

and divide. Ordinary skin moles and other *benign*, or noncancerous, neoplasms grow very slowly, and their cells retain the surface recognition proteins that are supposed to keep them in a home tissue (Figure 8.12). Unless a benign neoplasm grows too large or becomes irritating, it poses no threat to the body.

CHARACTERISTICS OF CANCER

Cancers are abnormally growing and dividing cells of a *malignant* neoplasm. They disrupt surrounding tissues, both physically and metabolically. Cancer cells are grossly disfigured. They can break loose from their home tissues. They can slip into and out of blood vessels and lymph vessels, and invade other tissues where they do not belong (Figure 8.12).

All cancer cells display four characteristics. *First*, they grow and divide abnormally. The controls on overcrowding in tissues are lost and cell populations reach abnormally high densities. The number of tiny blood vessels that service the growing cell mass also increases abnormally.

Second, the cytoplasm and plasma membrane of cancer cells become grossly altered. The membrane becomes leaky and has abnormal or lost proteins. The cytoskeleton shrinks, becomes disorganized, or both. Enzyme action shifts, as in an amplified reliance on ATP formation by glycolysis.

Third, cancer cells have a weakened capacity for adhesion. Recognition proteins are lost or altered, so they can't stay anchored in proper tissues. They break away and may establish growing colonies in distant tissues. *Metastasis* is the name for this process of abnormal cell migration and tissue invasion.

Fourth, cancer cells usually have lethal effects. Unless they are eradicated by surgery, chemotherapy, or other procedures, their uncontrollable divisions put an individual on a painful road to death.

Each year in the developed countries alone, cancers cause 15 to 20 percent of all deaths. And cancer is not just a human problem. Cancers are known to occur in most of the animal species studied to date.

Cancer is a multistep process. Researchers have already identified many of the mutant genes that contribute to it. They also are working to identify drugs that specifically target and destroy cancer cells or stop them from dividing.

HeLa cells, for instance, were used in early tests of taxol, an anticancer drug that stops spindles from disassembling. With this kind of research, we may one day have drugs that can put the brakes on cancer cells. We return to this topic in later chapters.

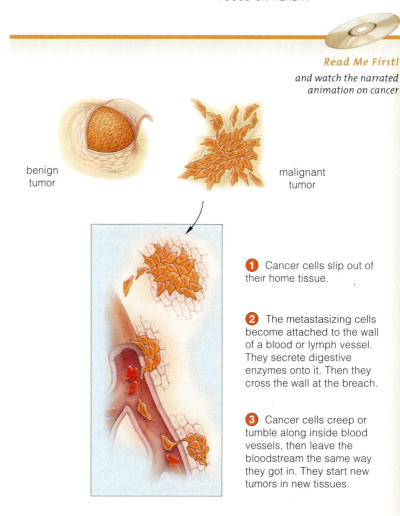

Read Me First!
and watch the narrated animation on cancer

1 Cancer cells slip out of their home tissue.

2 The metastasizing cells become attached to the wall of a blood or lymph vessel. They secrete digestive enzymes onto it. Then they cross the wall at the breach.

3 Cancer cells creep or tumble along inside blood vessels, then leave the bloodstream the same way they got in. They start new tumors in new tissues.

Figure 8.12 Comparison of benign and malignant tumors. Benign tumors typically are slow-growing and stay put in their home tissue. Cells of a malignant tumor can migrate abnormally through the body and establish colonies even in distant tissues.

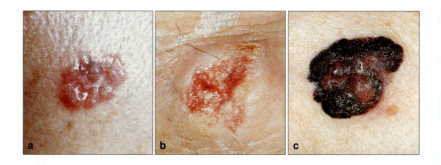

Figure 8.13 Skin cancers. (**a**) A *basal cell carcinoma* is the most common type. This slow-growing, raised lump is typically uncolored, reddish-brown, or black.

(**b**) The second most common form is *squamous cell carcinoma*. This pink growth, firm to the touch, grows fast under the surface of skin exposed to the sun.

(**c**) *Malignant melanoma* spreads fastest. Cells form dark, encrusted lumps. They may itch like an insect bite or bleed easily.

Summary

Section 8.1 The continuity of life depends on reproduction. By this process, parents produce a new generation of individuals like themselves. Cell division is the bridge between generations. When a cell divides, its daughters each receive a required number of DNA molecules and some cytoplasm.

Mitosis and meiosis occur only in eukaryotic cells. These nuclear division mechanisms sort out a parent cell's chromosomes into daughter nuclei. A separate mechanism divides the cytoplasm. Prokaryotic cells divide by a different mechanism.

Mitosis is the basis of multicellular growth, cell replacements, and tissue repair. Many eukaryotic organisms also reproduce *asexually* by mitosis.

Meiosis, the basis of sexual reproduction, precedes the formation of gametes or spores.

A chromosome is a molecule of DNA and associated proteins. When duplicated, it consists of two sister chromatids, both attached to its centromere region by kinetochores. These are docking sites for microtubules.

Section 8.2 Each cell cycle starts when a new cell forms. It proceeds through interphase and ends when the cell reproduces by nuclear and cytoplasmic division. A cell carries out its functions in interphase. Before it divides, it increases in mass, roughly doubles the number of its cytoplasmic components, then duplicates each of its chromosomes.

Section 8.3 The sum of all chromosomes in cells of a given type is the chromosome number. Human body cells have a diploid chromosome number of 46, or two copies of 23 types of chromosome. Mitosis maintains the chromosome number, one generation to the next.

Mitosis has four continuous stages:

a. Prophase. The duplicated, threadlike chromosomes start to condense. With the help of motor proteins, new microtubules start forming a bipolar mitotic spindle. The nuclear envelope starts to break apart into tiny vesicles.

Some microtubules growing from one spindle pole tether one chromatid of each chromosome; others that are growing from the opposite pole tether its sister chromatid. Still other microtubules extending from both poles grow until they overlap at the midpoint of the newly forming spindle.

b. Metaphase. At metaphase, all chromosomes have become aligned at the spindle's midpoint.

c. Anaphase. Kinetochores detach from chromosomes, dragging the chromatids with them along microtubules, which are shortening at both ends. The microtubules that overlap ratchet past each other, pushing the spindle poles farther apart. Different motor proteins drive the movements. One of each type of parental chromosome ends up clustered together at each spindle pole.

d. Telophase. Chromosomes decondense to threadlike form. A new nuclear envelope forms around each cluster. Both nuclei have the parental chromosome number.

Fill in the blanks of the diagram below to check your understanding of the four stages of mitosis.

Section 8.4 Cytoplasmic division mechanisms differ. Animal cells undergo cleavage. A microfilament ring under the plasma membrane contracts, pinching the cytoplasm in two. In plant cells, a cross-wall forms in the cytoplasm and divides it.

Section 8.5 Checkpoint gene products are part of mechanisms that control the cell cycle. The mechanisms can stimulate or arrest the cell cycle, or even prompt cell death. Mutant checkpoint genes cause tumors to form by disrupting normal cell cycle controls.

Cancer is a multistep process involving altered cells that grow and divide abnormally. Such cells have a weakened ability to stick to one another in tissues, and sometimes they migrate to new tissues.

Self-Quiz   Answers in Appendix III

1. Mitosis and cytoplasmic division function in _____ .
   a. asexual reproduction of single-celled eukaryotes
   b. growth, tissue repair, often asexual reproduction
   c. gamete formation in prokaryotes
   d. both a and b

2. A duplicated chromosome has _____ chromatid(s).
   a. one   b. two   c. three   d. four

3. The basic unit that structurally organizes a eukaryotic chromosome is the _____ .
   a. supercoil   c. nucleosome
   b. bipolar mitotic spindle   d. microfilament

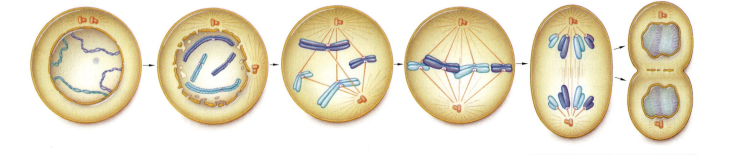

4. The chromosome number is _____ .
    a. the sum of all chromosomes in cells of a given type
    b. an identifiable feature of each species
    c. maintained by mitosis
    d. all of the above

5. A somatic cell having two of each type of chromosome has a(n) _____ chromosome number.
    a. diploid   b. haploid   c. tetraploid   d. abnormal

6. Interphase is the part of the cell cycle when _____ .
    a. a cell ceases to function
    b. a germ cell forms its spindle apparatus
    c. a cell grows and duplicates its DNA
    d. mitosis proceeds

7. After mitosis, the chromosome number of a daughter cell is _____ the parent cell's.
    a. the same as   c. rearranged compared to
    b. one-half   d. doubled compared to

8. Only _____ is not a stage of mitosis.
    a. prophase b. interphase c. metaphase d. anaphase

9. Match each stage with the events listed.
    ____ metaphase   a. sister chromatids move apart
    ____ prophase   b. chromosomes start to condense
    ____ telophase   c. daughter nuclei form
    ____ anaphase   d. all duplicated chromosomes are
                         aligned at the spindle equator

Critical Thinking

1. Figure 8.14 shows metaphase chromosomes. Name their levels of structural organization, starting with DNA molecules and histones.

2. Pacific yews (*Taxus brevifolius*) are among the slowest growing trees, which makes them vulnerable to extinction. People started stripping their bark and killing them when they heard that *taxol,* a chemical extracted from the bark, may work against breast and ovarian cancer. It takes bark from about six trees to treat one patient. Do some research and find out why taxol has potential as an anticancer drug and what has been done to protect the trees.

3. X-rays emitted from some radioisotopes damage DNA, especially in cells undergoing DNA replication. Humans exposed to high levels of x-rays face *radiation poisoning*. Hair loss and a damaged gut lining are early symptoms. Speculate why. Also speculate on why radiation exposure is used as a therapy to treat some cancers.

4. Suppose you have a way to measure the amount of DNA in a single cell during the cell cycle. You first measure the amount at the G1 phase. At what points during the remainder of the cycle would you predict changes in the amount of DNA per cell?

5. The cervix is part of the uterus, a chamber in which embryos develop. The *Pap smear* is a screening procedure that can detect *cervical cancer* in its earliest stages.

Treatments range from freezing precancerous cells or killing them with a laser beam to removal of the uterus (a hysterectomy). The treatments are more than 90 percent effective when this cancer is detected early. Survival chances plummet to less than 9 percent after it spreads.

Most cervical cancers develop slowly. Unsafe sex increases the risk. A key risk factor is infection by human

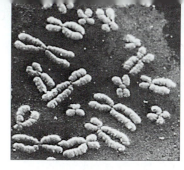

Figure 8.14 Human chromosomes at metaphase, each in the duplicated state.

papillomaviruses (HPV), which cause genital warts. Viral genes coding for the tumor-inducing proteins get inserted into the DNA of cervical cells. In one study, 91 percent of patients with cervical cancer had been infected with HPV.

Not all women request Pap smears. Many wrongly believe the procedure is costly. Many don't recognize the importance of abstinence or "safe" sex. Others simply don't want to think about whether they have cancer. Knowing what you've learned so far about the cell cycle and cancer, what would you say to a woman who falls into one or more of these groups?

Media Menu

Student CD-ROM

Impacts, Issues Video
    Henrietta's Immortal Cells
Big Picture Animation
    Normal cell division and cancer
Read-Me-First Animation
    Chromosome structural organization
    Mitosis
    Cytoplasmic division
    Cancer
Other Animations and Interactions
    The cell cycle

InfoTrac

• Cell Cycle Circuits Mapped. *Applied Genetics News,* October 2001.
• HIV Protein Stops Cell Division. *Virus Weekly,* April 2002.
• Familiar Proteins Play Unfamiliar Role In Cell Division. *Stem Cell Week,* January 2002.
• How You Can Lower Your Cancer Risk. *Harvard Health Letter,* August 2002.

Web Sites

• Talking Genetic Glossary: www.genome.gov/glossary.cfm
• Mitosis World: www.bio.unc.edu/faculty/salmon/lab/mitosis
• Animated Cell Cycle: www.cellsalive.com/cell_cycle.htm
• The Cytokinetic Mafia Home Page: www.bio.unc.edu/faculty/salmon/lab/mafia/

How Would You Vote? When she died, Henrietta Lacks left behind a husband and five children. The scientists who propagated the HeLa cell line never told her or her family how they were using her tissues. Today, HeLa cells are sold by cell culture firms around the world. Should her survivors get a share of the profits?

IMPACTS, ISSUES *Why Sex?*

Women and men, does and bucks, geese and ganders. Most of us take it for granted that it takes two to make offspring, and among eukaryotes it usually does—at least some of the time. Sexual reproduction combines DNA from individuals of two mating types. It started many hundreds of millions of years ago among tiny, single-celled eukaryotes, although no one knows how. An unsolved puzzle is why it happened at all. Asexual reproduction, whereby an individual makes offspring that are copies of itself, is far more efficient.

Protists and fungi routinely reproduce by mitosis. Plants and many invertebrates, including corals, sea stars, and flatworms, can form new individuals from parts that bud or break off. But almost all of these species also reproduce sexually.

For instance, some single-celled algae reproduce asexually again and again, by way of mitosis, and form huge populations exactly like themselves. Only when nitrogen is scarce do sexual cells of two types form. The fusion of two cells, one of each type, produces new individuals. The result is offspring that are not exact copies of one another or their parents.

Consider also the plant-sucking insects called aphids. In the summer nearly all are females. Each matures in less than a week and can produce as many as five new females a day from her unfertilized eggs (Figure 9.1). This process, called parthenogenesis, allows population sizes to soar rapidly. Only as autumn approaches do male aphids develop. Even then, females that manage to survive over the winter can do without the opposite

Figure 9.1 Sexual reproduction moments. (**a**) Mealybugs mating. (**b**) Poppy plant being helped by a beetle, which makes pollen deliveries for it. (**c**) Aphid giving birth. Like females of many other sexually reproducing species, this one also reproduces asexually, all by itself.

the big picture

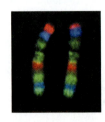

Sexual Reproduction Sexual reproduction requires meiosis, gamete formation, and fertilization. Meiosis is a nuclear division mechanism that halves the parental chromosome number for forthcoming gametes.

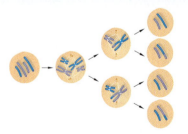

Meiosis Meiosis consists of two nuclear divisions, not one. There is no interphase between divisions. It results in four haploid nuclei. When cytoplasmic division follows, each of the resulting haploid cells may function as a gamete or spore.

sex. The next spring they begin another round of female production all by themselves.

There are even a few all-female species of fishes, reptiles, and birds. No mammal is parthenogenic in nature. In 2004, however, researchers at the University of Tokyo in Japan fused two mouse eggs in a test tube to produce an embryo with all-female DNA. The embryo developed into Kaguya—the world's first fatherless mammal. She grew to adulthood, mated with a male mouse, and produced offspring of her own, as shown in the filmstrip at right.

Does this mean males could soon be unneccessary? Hardly. It wasn't easy to produce a mouse that has all-female DNA. The researchers had to turn off genes in one egg. Even then, it took more than 600 attempts before they succeeded in producing two viable embryos. Besides, the prevalance of sexually reproducing species suggests that a division into two sexes must offer selective advantages.

With this chapter, we turn to the kinds of cells that serve as the bridge between generations of organisms. Three interconnected events—meiosis, the formation of gametes, and fertilization—are the hallmarks of sexual reproduction. As you will see in many chapters throughout the book, these events occur in the life cycle of almost all eukaryotic species. Through the production of offspring with new and unique traits, they have contributed immensely to the diversity of life.

✓ How Would You Vote?

Japanese researchers have successfully created a "fatherless" mouse that contains the genetic material from the eggs of two females. The mouse is healthy and fully fertile. Do you think researchers should be allowed to try the same process with human eggs? See the Media Menu for details, then vote online.

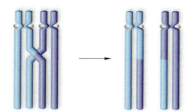

Gene Shufflings With Sex   During meiosis, crossing over and the random alignment of chromosomes at metaphase puts different mixes of maternal and paternal genes in gametes. More mixing occurs at fertilization. Such events introduce variation in traits among offspring.

Meiosis in Life Cycles   Plant and animal life cycles typically include fertilization, meiosis, and gamete formation. In plant life cycles, meiosis is followed by the formation of spores which, under favorable conditions, germinate and give rise to gamete-producing bodies.

9.1 An Evolutionary View

Asexual reproduction produces genetically identical copies of a parent. Sexual reproduction introduces variation in the details of traits among offspring.

When an orchid or aphid reproduces all by itself, what sort of offspring does it get? By the process of **asexual reproduction**, one alone produces offspring, and each offspring inherits the same number and kinds of genes as its parent. **Genes** are stretches of chromosomes—that is, of DNA molecules. The genes for each species contain all the heritable information necessary to make new individuals. Rare mutations aside, then, asexually produced individuals can only be *clones*, or genetically identical copies of the parent.

Inheritance gets far more interesting with **sexual reproduction**. The process involves meiosis, formation of gametes, and fertilization (union of the nuclei of two gametes). In most sexual reproducers, such as humans, the first cell of a new individual contains *pairs of genes* on pairs of chromosomes. Usually, one of each pair is maternal and the other paternal in origin (Figure 9.2).

If information in every pair of genes were identical down to the last detail, sexual reproduction would also produce clones. Just imagine—you, every person you know, the entire human population might be a clone, with everybody looking alike. But the two genes of a pair might *not* be identical. Why not? The molecular structure of a gene can change; it can mutate. So two genes that happen to be paired in a person's cells may "say" slightly different things about a trait. Each unique molecular form of the same gene is called an **allele**.

Such tiny differences affect thousands of traits. For example, whether your chin has a dimple depends on which pair of alleles you inherited at one chromosome location. One kind of allele at that location says "put a dimple in the chin." Another kind says "no dimple." This leads to one reason why the individuals of sexually reproducing species don't all look alike. *By sexual reproduction, offspring inherit new combinations of alleles, which lead to variations in the details of their traits.*

This chapter gets into the cellular basis of sexual reproduction. More importantly, it starts you thinking about far-reaching effects of gene shufflings at certain stages of the process. The process introduces variations in traits among offspring that are typically acted upon by agents of natural selection. Thus, *variation in traits is a foundation for evolutionary change.*

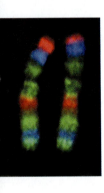

Figure 9.2 A maternal and a paternal chromosome. Any gene on one might be slightly different structurally than the same gene on the other.

> Sexual reproduction introduces variation in traits by bestowing novel combinations of alleles on offspring.

9.2 Overview of Meiosis

Meiosis is a nuclear division process that divides the parental chromosome number by half in specialized reproductive cells. It is central to sexual reproduction.

THINK "HOMOLOGUES"

Think back to the preceding chapter and its focus on mitotic cell division. Unlike mitosis, **meiosis** partitions chromosomes into parcels not once but *twice* prior to cytoplasmic division. Unlike mitosis, it is the first step leading to the formation of gametes. Male and female gametes, such as sperm and eggs, fuse to form a new individual. In most multicelled eukaryotes, cells that form in specialized reproductive structures or organs give rise to gametes. Figure 9.3 shows examples of where the cellular antecedents of gametes originate.

As you know, the **chromosome number** is the sum total of chromosomes in cells of a given type. If a cell has a **diploid number** ($2n$), it has a *pair* of each type of chromosome, often from two parents. Except for a pairing of nonidentical sex chromosomes, each pair has the same length, shape, and assortment of genes, and they line up with each other at meiosis. We call them **homologous chromosomes** (*hom–* means alike).

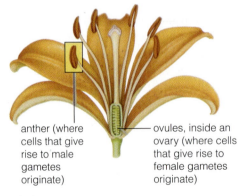

anther (where cells that give rise to male gametes originate)

ovules, inside an ovary (where cells that give rise to female gametes originate)

a Flowering plant

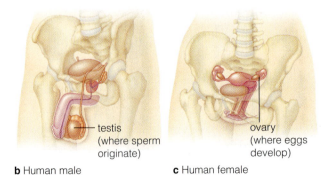

testis (where sperm originate)

ovary (where eggs develop)

b Human male   **c** Human female

Figure 9.3 Examples of reproductive organs, where cells that give rise to gametes originate.

Body cells of humans have 23 + 23 homologous chromosomes (Figure 9.4). So do the germ cells that give rise to human gametes. After a germ cell finishes meiosis, 23 chromosomes—one of each type—will end up in those gametes. Meiosis halves the chromosome number, so the gametes have a **haploid number** (*n*).

TWO DIVISIONS, NOT ONE

Meiosis is like mitosis in some ways, but the result is different. As in mitosis, a germ cell duplicates its DNA in interphase. The two DNA molecules and associated proteins stay attached at the centromere, the notably constricted region along their length. For as long as they remain attached, we call them **sister chromatids**:

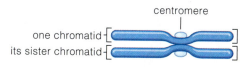

one chromosome in the duplicated state

As in mitosis, the microtubules of a spindle apparatus move the chromosomes in prescribed directions.

With meiosis, however, *chromosomes go through two consecutive divisions that end with the formation of four haploid nuclei*. There is no interphase between divisions, which we call meiosis I and meiosis II:

interphase (DNA replication before meiosis I)	MEIOSIS I	no interphase (no DNA replication before meiosis II)	MEIOSIS II
	PROPHASE I		PROPHASE II
	METAPHASE I		METAPHASE II
	ANAPHASE I		ANAPHASE II
	TELOPHASE I		TELOPHASE II

In meoisis I, each duplicated chromosome aligns with its partner, *homologue to homologue*. After the two chromosomes of every pair have lined up with each other, they are moved apart:

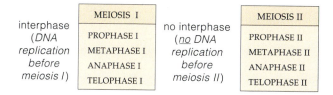

each homologue in the cell pairs with its partner

then the partners separate

The cytoplasm typically starts to divide at some point after each homologue detaches from its partner. The two daughter cells formed this way are haploid,

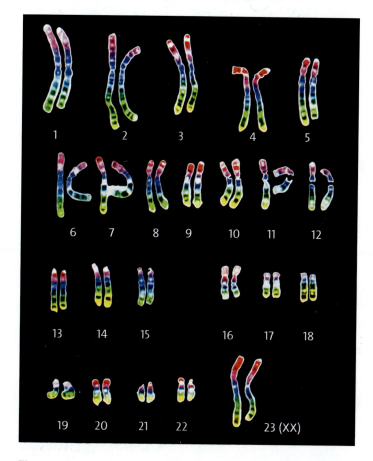

Figure 9.4 Another look at the twenty-three pairs of homologous human chromosomes. This example is from a human female, with two X chromosomes. Human males have a different pairing of sex chromosomes (XY). As in Figure 9.2, these chromosomees have been labeled with several fluorescent markers.

with *one* of each type of chromosome. Don't forget, these chromosomes are still in the duplicated state.

Next, during meiosis II, *the two sister chromatids of each chromosome are separated from each other*:

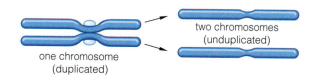

one chromosome (duplicated)

two chromosomes (unduplicated)

Each chromatid is now a separate chromosome. Next, four nuclei form, and the cytoplasm typically divides once more. The final outcome is four haploid cells. Figure 9.5, on the next two pages, offers a closer look at key events of meiosis and their consequences.

Meiosis, a nuclear division mechanism, reduces a parental cell's chromosome number by half—to a haploid number (n).

9.3 Visual Tour of Meiosis

Meiosis I

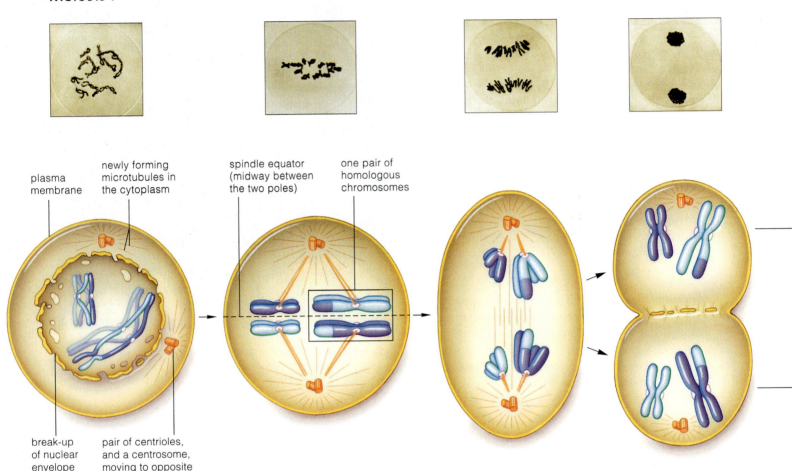

plasma membrane

newly forming microtubules in the cytoplasm

spindle equator (midway between the two poles)

one pair of homologous chromosomes

break-up of nuclear envelope

pair of centrioles, and a centrosome, moving to opposite sides of nucleus

ⓐ Prophase I

As interphase ends, the chromosomes are threadlike and duplicated. Now they start to condense. Each pairs with its homologue and the two usually swap segments, as in the larger chromosomes. Centrioles help organize centrosomes on both sides of the nuclear envelope, which is now starting to break apart. Microtubules of a bipolar spindle originate from the two centrosomes (Section 8.3).

ⓑ Metaphase I

Microtubules from one spindle pole tether one of each type of chromosome; microtubules from the other pole tether the homologue of each pair. A tug-of-war between microtubules aligns all of the chromosomes at the spindle equator. Motor proteins, activated by ATP, drive the movements.

ⓒ Anaphase I

Microtubules attached to each chromosome shorten at both ends in a way that reels it in toward a spindle pole. Other microtubules, which extend from the poles and overlap at the spindle equator, ratchet past each other and push the poles farther apart. Motor proteins drive the movements.

ⓓ Telophase I

One of each type of chromosome has now arrived at the spindle poles. The cytoplasm divides at some point, forming two haploid (*n*) cells. All of the chromosomes are still in the duplicated state.

Figure 9.5 Sketches of meiosis in a generalized animal cell. This is a nuclear division mechanism. It reduces the parental chromosome number in immature reproductive cells by half, to the haploid number, for forthcoming gametes. To keep things simple, we track only two pairs of homologous chromosomes. Maternal chromosomes are shaded *purple* and paternal chromosomes *blue*.

Of the four haploid cells that form by way of meiosis and cytoplasmic divisions, one or all may develop into gametes and function in sexual reproduction. In plants, cells that form by way of meiosis may develop into spores, which take part in a stage of the life cycle that precedes gamete formation.

Read Me First!
and watch the narrated animation on meiosis

Meiosis II

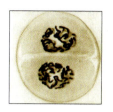

There is no DNA replication between the two nuclear divisions.

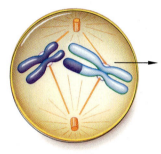
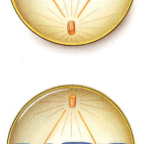
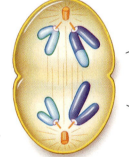
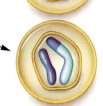

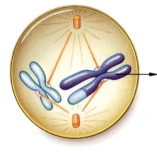

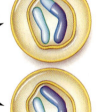

e Prophase II

Once again, a centriole pair and a centrosome have become positioned opposite from each other, and new microtubules form a bipolar spindle. Microtubules from one spindle pole tether one chromatid of each duplicated chromosome. Microtubules from the opposite spindle pole tether the sister chromatid.

f Metaphase II

Following a tug-of-war between microtubules from both spindle poles, all chromosomes have become positioned midway between the poles.

g Anaphase II

The attachment between sister chromatids of each chromosome breaks. Each is now a separate chromosome. It is still tethered to microtubules, which reel it toward a spindle pole. Other microtubules push the poles apart. A cluster of unduplicated chromosomes, one of each type found in the parent cell, are now clustered near each pole.

h Telophase II

In telophase II, a new nuclear envelope forms around the chromosome cluster, forming four daughter nuclei. After cytoplasmic division, each of the resulting daughter cells has a haploid (*n*) chromosome number.

9.4 How Meiosis Puts Variation in Traits

As Sections 9.2 and 9.3 make clear, the overriding function of meiosis is the reduction of a parental chromosome number by half. However, two other events that occur during meiosis have evolutionary consequences.

The preceding section mentioned in passing that pairs of homologous chromosomes swap parts of themselves during prophase I. It also showed how homologous chromosomes become aligned with their partner at metaphase I. Let's take a look at these two events, because they contribute enormously to variation in the traits of sexually reproducing species. They introduce novel combinations of alleles into the gametes that form after meiosis. Those combinations—and the way they are further mixed together at fertilization—are the start of a new generation of individuals that differ in the details of their shared traits.

CROSSING OVER IN PROPHASE I

Prophase I of meiosis is a time of much gene shuffling. Reflect on Figure 9.6a, which shows two chromosomes condensed to threadlike form. All chromosomes in a germ cell condense this way. When they do, each is drawn close to its homologue. Molecular interactions stitch homologues together point by point along their length with little space between them. The intimate, parallel orientation favors **crossing over**, a molecular interaction between a chromatid of one chromosome and a chromatid of its homologous partner. The two *nonsister* chromatids break at the same places along their length, then the two exchange corresponding segments; they swap genes.

Gene swapping would be pointless if each type of gene never varied. But remember, a gene can come in

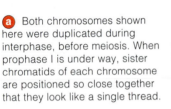

a Both chromosomes shown here were duplicated during interphase, before meiosis. When prophase I is under way, sister chromatids of each chromosome are positioned so close together that they look like a single thread.

b Each chromosome becomes zippered to its homologue, so all four chromatids are tightly aligned. If the two sex chromosomes have different forms, such as X paired with Y, they still get zippered together, but only in a tiny region at their ends.

c We show the pair of chromosomes as if they already condensed only to give you an idea of what goes on. They really are in a tightly aligned, threadlike form during prophase I.

d The intimate contact encourages one crossover (and usually more) to happen at various intervals along the length of nonsister chromatids.

e Nonsister chromatids exchange segments at the crossover sites. They continue to condense into thicker, rodlike forms. By the start of metaphase I, they will be unzipped from each other.

f Crossing over breaks up old combinations of alleles and puts new ones together in the cell's pairs of homologous chromosomes.

Figure 9.6 Key events of prophase I, the first stage of meiosis. For clarity, we show only one pair of homologous chromosomes and one crossover event. Typically, more than one crossover occurs. *Blue* signifies the paternal chromosome; *purple* signifies its maternal homologue.

slightly different forms: alleles. You can bet that some number of the alleles on one chromosome will *not* be identical to their partner alleles on the homologue. Every crossover is a chance to swap slightly different versions of hereditary instructions for gene products.

We will look at the mechanism of crossing over in later chapters. For now, just remember this: *Crossing over leads to recombinations among genes of homologous chromosomes, and to variation in traits among offspring.*

METAPHASE I ALIGNMENTS

Major shufflings of intact chromosomes start during the transition from prophase I to metaphase I. Suppose this is happening right now in one of your germ cells. Crossovers have already made genetic mosaics of the chromosomes, but put this aside to simplify tracking. Just call the twenty-three chromosomes you inherited from your mother the *maternal* chromosomes, and the twenty-three inherited from your father the *paternal* chromosomes.

At metaphase I, microtubules have tethered one chromosome of each pair to one spindle pole and its homologue to the other, and all are lined up at the spindle equator (Figure 9.5*b*). Have they tethered all maternal chromosomes to one pole and all paternal chromosomes to the other? Maybe, but probably not. As microtubules grow outward from the poles, they latch on to the first chromosome they contact. Because the tethering is random, there is no particular pattern to the metaphase I positions of maternal and paternal chromosomes. Carry this thought one step further. After a pair of homologous chromosomes are moved apart during anaphase I, *either one* of them can end up at either spindle pole.

Think of the possibilities while tracking just three pairs of homologues. By metaphase I, these three pairs may be arranged in any one of four possible positions (Figure 9.7). This means that eight combinations (2^3) are possible for forthcoming gametes.

Cells that give rise to human gametes have twenty-three pairs of homologous chromosomes, not three. So every time a human sperm or egg forms, you can expect a total of *8,388,608* (or 2^{23}) possible combinations of maternal and paternal chromosomes!

Moreover, in each sperm or egg, many hundreds of alleles inherited from the mother might not "say" the exact same thing about hundreds of different traits as alleles inherited from the father. Are you beginning to get an idea of why such fascinating combinations of traits show up the way they do among the generations of your own family tree?

Read Me First!
and watch the narrated animation on random alignment

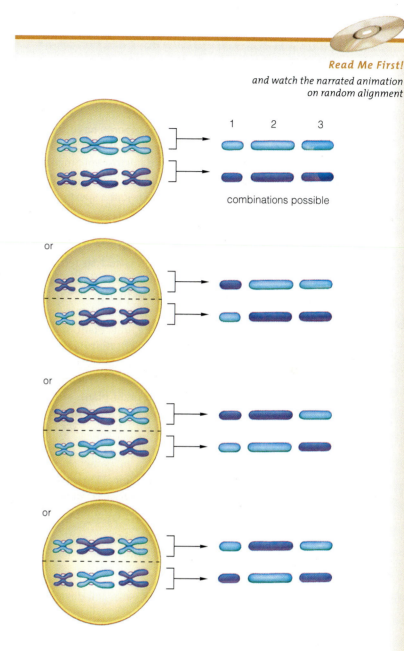

Figure 9.7 Possible outcomes for the random alignment of merely three pairs of homologous chromosomes at metaphase I. The three types of chromosomes are labeled 1, 2, and 3. With four alignments, eight combinations of maternal chromosomes (*purple*) and paternal chromosomes (*blue*) are possible in gametes.

Crossing over, an interaction between a pair of homologous chromosomes, breaks up old combinations of alleles and puts new ones together during prophase I of meiosis.

The random tethering and subsequent positioning of each pair of maternal and paternal chromosomes at metaphase I lead to different combinations of maternal and paternal traits in each new generation.

9.5 From Gametes to Offspring

What happens to the gametes that form after meiosis?
Later chapters have specific examples. Here, just focus on
where they fit in the life cycles of plants and animals.

Gametes are not all the same in their details. Human sperm have one tail, opossum sperm have two, and roundworm sperm have none. Crayfish sperm look like pinwheels. Most eggs are microscopic in size, yet an ostrich egg inside its shell is as big as a football. A flowering plant's male gamete is just a sperm nucleus.

GAMETE FORMATION IN PLANTS

Seasons vary for plants on land, and so fertilization must coincide with spring rains and other conditions that favor growth of the new individual. That is why life cycles of plants generally alternate between spore production and gamete production. Plant **spores** are haploid resting cells, often walled, that develop after meiosis (Figure 9.8*a*). They originate in reproductive structures of *sporophytes*, or spore-producing bodies. Pine trees, corn plants, and all other plants with roots, stems, and leaves are examples of sporophytes.

Plant spores stay dormant in dry or cold seasons. When they resume growth (germinate), they undergo mitosis and form *gametophytes*, or gamete-producing haploid bodies. For example, female gametophytes form on pine cone scales. In their tissues, gametes form by way of meiosis, as Chapter 27 explains.

GAMETE FORMATION IN ANIMALS

In animals, germ cells give rise to gametes. In a male reproductive system, a diploid germ cell develops into a large, immature cell: a primary spermatocyte. This cell enters meiosis and cytoplasmic divisions. Four haploid cells result, and they develop into spermatids (Figure 9.9). These cells undergo changes that include the formation of a tail. Each becomes a **sperm**, which is a common type of mature male gamete.

In female animals, a germ cell becomes an **oocyte**, or immature egg. Unlike sperm, an oocyte stockpiles many cytoplasmic components, and its four daughter cells differ in size and function (Figure 9.10).

As an oocyte divides after meiosis I, one daughter cell—the secondary oocyte—gets nearly all of the cytoplasm. The other cell, a first polar body, is small. Later, both of these haploid cells enter meiosis II, then cytoplasmic division. One of the secondary oocyte's daughter cells develops into a second polar body. The other receives most of the cytoplasm and develops into a gamete. A mature female gamete is called an ovum (plural, ova) or, more often, an **egg**.

And so we have one egg. The three polar bodies that formed don't function as gametes and aren't rich in nutrients or plump with cytoplasm. In time they degenerate. But the fact that they formed means the egg now has a haploid chromosome number. Also, by getting most of the cytoplasm, the egg has enough metabolic machinery to support the early divisions of the new individual.

MORE SHUFFLINGS AT FERTILIZATION

The chromosome number characteristic of the parents is restored at **fertilization**, a time when a female and male gametes unite and their haploid nuclei fuse. If meiosis did not precede it, fertilization would double

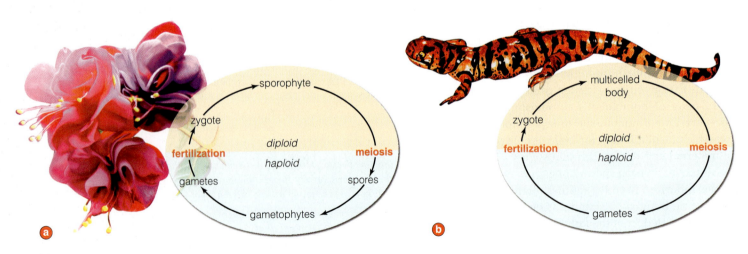

Figure 9.8 (**a**) Generalized life cycle for most plants. (**b**) Generalized life cycle for animals. The zygote is the first cell to form when the nuclei of two gametes fuse at fertilization.

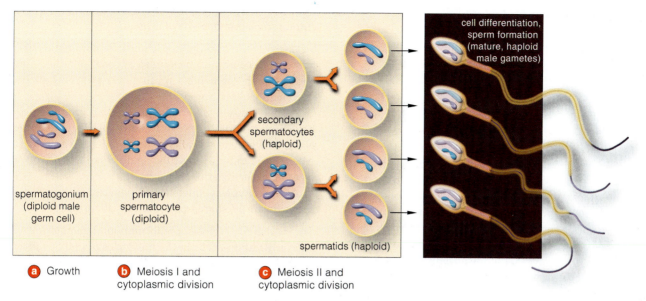

a Growth   **b** Meiosis I and cytoplasmic division   **c** Meiosis II and cytoplasmic division

Figure 9.9 Generalized sketch of sperm formation in animals. Figure 38.16 shows a specific example (how sperm form in human males).

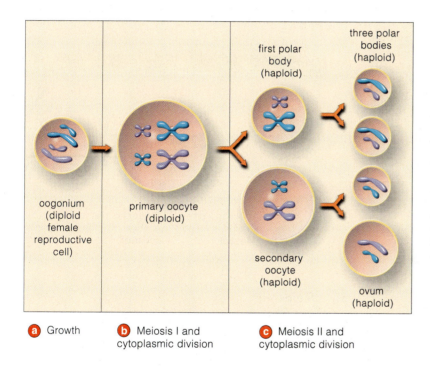

a Growth   **b** Meiosis I and cytoplasmic division   **c** Meiosis II and cytoplasmic division

Figure 9.10 Animal egg formation. Eggs are far larger than sperm and larger than the three polar bodies. The painting above, based on a scanning electron micrograph, depicts human sperm surrounding an ovum.

the chromosome number each generation. Doublings would disrupt hereditary information, usually for the worse. Why? That information is like a fine-tuned set of blueprints that must be followed exactly, page after page, to build a normal individual.

Fertilization also adds to variation among offspring. Reflect on the possibilities for humans alone. During prophase I, each human chromosome undergoes an average of two or three crossovers. Even without these crossovers, random positioning of pairs of paternal and maternal chromosomes at metaphase I results in one of millions of possible chromosome combinations in each gamete. And of all male and female gametes that are produced, *which* two actually get together is a matter of chance. The sheer number of combinations that can exist at fertilization is staggering!

The distribution of random mixes of chromosomes into gametes, random metaphase chromosome alignments, and fertilization contribute to variation in traits of offspring.

Summary

Section 9.1 Alleles are slightly different molecular forms of the same gene, and they specify different versions of the same gene product.

Sections 9.2, 9.3 Meiosis, which consists of two nuclear divisions, is central to sexual reproduction. It halves the parental chromosome number (Figure 9.11).

Meiosis, the first nuclear division, partitions homologous chromosomes into two clusters, both with one of each type of chromosome. All of the chromosomes were duplicated earlier, in interphase.

Prophase I: Chromosomes start condensing into rodlike forms. The nuclear envelope starts to break up. If duplicated pairs of centrioles are present, one pair moves to the opposite side of the nucleus along with a centrosome, from which new microtubules of a spindle originate. Crossing over occurs between homologues.

Metaphase I: All pairs of homologous chromosomes are positioned at the spindle equator. Microtubules have tethered the maternal or paternal chromosome of each pair to either pole, at random.

Anaphase I: Microtubules pull each chromosome away from its homologue, to opposite spindle poles.

Telophase I: Two haploid nuclei form. Cytoplasmic division typically follows.

In meiosis II, the second nuclear division, the sister chromatids of all the chromosomes are pulled away from each other and partitioned into two clusters. By the end of telophase II, four nuclei—each with a haploid chromosome number—have formed.

Section 9.4 In prophase I, *non*sister chromatids of homologous chromosomes break at corresponding sites and exchange segments. Crossing over puts new allelic combinations in chromosomes.

At metaphase I, maternal and paternal chromosomes have been randomly tethered to one spindle pole or the other, which mixes up allelic combinations even more. Alleles are randomly shuffled again when two gametes meet up at fertilization.

All three types of allele shufflings lead to variation in the details of shared traits among offspring.

Section 9.5 Meiosis, the formation of gametes, and fertilization occur in the life cycles of plants and animals. In plant sporophytes, meiosis is followed by haploid spore formation. Germinating spores give rise to gametophytes, where cells that give rise to gametes originate. In most animals, germ cells in reproductive organs give rise to sperm or eggs. Fusion of a sperm and egg nucleus at fertilization results in a zygote.

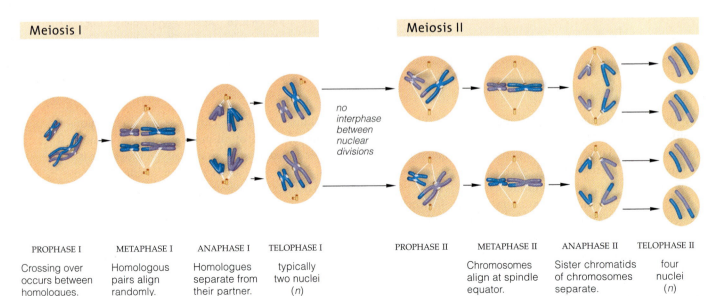

Figure 9.11 Comparison of the key features of mitosis and meiosis. We use a diploid cell with only two paternal and two maternal chromosomes. All of the chromosomes were duplicated during interphase, prior to nuclear division. Mitosis maintains the chromosome number. Meiosis halves it, to the haploid number.

Mitosis

PROPHASE	METAPHASE	ANAPHASE	TELOPHASE
	Chromosomes align at spindle equator.	Sister chromatids of chromosomes separate.	two nuclei (*2n*)

Meiosis I

PROPHASE I	METAPHASE I	ANAPHASE I	TELOPHASE I
Crossing over occurs between homologues.	Homologous pairs align randomly.	Homologues separate from their partner.	typically two nuclei (*n*)

no interphase between nuclear divisions

Meiosis II

PROPHASE II	METAPHASE II	ANAPHASE II	TELOPHASE II
	Chromosomes align at spindle equator.	Sister chromatids of chromosomes separate.	four nuclei (*n*)

Self-Quiz

Answers in Appendix III

1. Meiosis and cytoplasmic division function in __d,c__ .
   a. asexual reproduction of single-celled eukaryotes
   b. growth, tissue repair, often asexual reproduction
   c. sexual reproduction
   d. both b and c

2. A duplicated chromosome has __b__ chromatid(s).
   a. one   b. two   c. three   d. four

3. A somatic cell having two of each type of chromosome has a(n) __a__ chromosome number.
   a. diploid   b. haploid   c. tetraploid   d. abnormal

4. Sexual reproduction requires __a,d__ .
   a. meiosis   c. gamete formation
   b. fertilization   d. all of the above

5. Generally, a pair of homologous chromosomes __a,d__ .
   a. carry the same genes   c. interact at meiosis
   b. are the same length, shape d. all of the above

6. Meiosis __b__ the parental chromosome number.
   a. doubles   b. halves   c. maintains   d. corrupts

7. Meiosis is a division mechanism that produces __d__ .
   a. two cells   c. eight cells
   b. two nuclei   d. four nuclei

8. Pairs of duplicated, homologous chromosomes end up at opposite spindle poles during __a or d__ __c__ .
   a. prophase I   c. anaphase I
   b. prophase II   d. anaphase II

9. Sister chromatids of each duplicated chromosome end up at opposite spindle poles during __d__ .
   a. prophase I   c. anaphase I
   b. prophase II   d. anaphase II

10. Match each term with its description.
    __d__ chromosome   a. different molecular forms
        number   of the same gene
    __a__ alleles   b. none between meiosis I, II
    __c__ metaphase I   c. all chromosomes aligned
    __b__ interphase   at spindle equator
                        d. all chromosomes of a given type

Critical Thinking

1. Why can we expect meiosis to give rise to genetic differences between parent cells and their daughter cells in fewer generations than mitosis?

2. As mentioned in the chapter introduction, aphids can reproduce asexually and sexually at different times of year. How might their reproductive flexibility be an adaptation that allows them to avoid predators?

3. The bdelloid rotifer lineage started at least 40 million years ago (Figure 9.12). About 360 known species of these tiny animals are found in many aquatic habitats worldwide. They show tremendous genetic diversity. Speculate on why scientists were surprised to discover that all bdelloid rotifers are female.

4. Actor Viggo Mortensen inherited a gene that makes his chin dimple. Figure 9.13b shows what he might have looked like with an ordinary form of that gene. What is the name for alternative forms of the same gene?

Figure 9.12 Bdelloid rotifer.

Figure 9.13 Viggo Mortensen (**a**) with and (**b**) without a chin dimple.

Media Menu

Student CD-ROM

Impacts, Issues Video
    Why Sex?
Big Picture Animation
    Meiosis and sexual reproduction
Read-Me-First Animation
    Meiosis
    Crossing over
    Random alignment
Other Animations and Interactions
    Variation in life cycles
    Sperm formation
    Egg formation

InfoTrac
- Bdelloids: No Sex for over 40 Million Years. *Science News*, May 2000.
- Crossover Interference in Humans. *American Journal of Human Genetics*, July 2003.
- Tracking Down a Cheating Gene. *American Scientist*, March 2000.

Web Sites
- Meselson Lab: golgi.harvard.edu/meselson/research.html
- Mitosis vs. Meiosis: www.pbs.org/wgbh/nova/miracle/divide.html

How Would You Vote?

Japanese researchers have created a "fatherless" mouse from two eggs. Other scientists have coaxed unfertilized human eggs to develop into embryos. Some people object to the use of any human embryo for research purposes, and some worry about the potential to produce "fatherless" humans. Would you support a ban on this technique?

IMPACTS, ISSUES *Menacing Mucus*

Cystic fibrosis (CF) is a debilitating and ultimately fatal genetic disorder. In 1989, researchers identified the mutated gene that causes it. In 2001, the American College of Obstetricians and Gynecologists suggested that all prospective parents be screened for mutated versions of the gene. The suggestion led to the first mass screening for carriers of a genetic disorder.

CFTR, the gene's product, is a membrane transport protein. It helps chloride and water move into and out of cells that secrete mucus or sweat. More than 10 million people in the United States inherited one normal and one abnormal copy of the CFTR gene. They do have sinus problems, but no other symptoms develop. Most do not even know they carry the gene.

CF develops in anyone who inherits a mutant form of the gene from both parents. Thick, dry mucus clogs bronchial airways to their lungs and makes it hard to breathe (Figure 10.1). The mucus coating the airways is supposed to be thin enough to trap airborne particles and pathogens, so that ciliated cells lining the airways can sweep them out. Bacterial populations thrive in the thick mucus of CF patients, shown in the filmstrip.

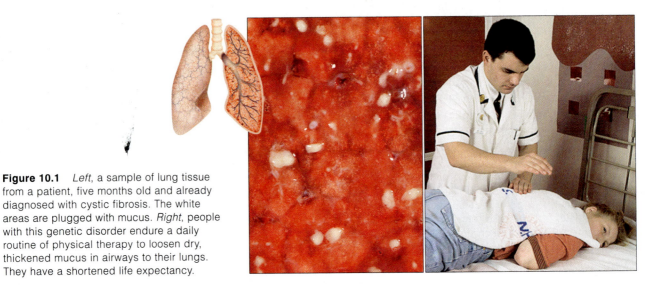

Figure 10.1 *Left*, a sample of lung tissue from a patient, five months old and already diagnosed with cystic fibrosis. The white areas are plugged with mucus. *Right*, people with this genetic disorder endure a daily routine of physical therapy to loosen dry, thickened mucus in airways to their lungs. They have a shortened life expectancy.

the big picture

An Experimental Approach   Experiments with pea plants yielded the first observable evidence that parents transmit genes—units of information about heritable traits—to offspring. The experiments also revealed some underlying patterns of inheritance.

Two Theories Emerge   As Mendel sensed, diploid organisms have pairs of genes and each gamete gets only one of the pairs. Also, genes on pairs of homologous chromosomes tend to be sorted out for distribution into gametes independently of gene pairs of other chromosomes.

Antibiotics help keep the pathogens under control but cannot get rid of them entirely. Also, to loosen the mucus, patients must go through daily routines of posture changes and thumps on the chest and back. Even with physiotherapy, most can expect lung failure. A double lung transplant can extend their life, but donor organs are scarce. Even if they do receive a transplant, few will live past their thirtieth birthday.

The severity of CF and the prevalence of carriers in the general population persuaded doctors to screen prospective parents—hundreds of thousands of them. By 2003, however, the law of unintended consequences took effect. Some people misunderstood the screening results. Some took unnecessary diagnostic tests to find out if their child would be normal. Confused by test results, a few may have aborted normal fetuses.

So here we are today, working our way through the genetic basis of our very lives. And where did it all start? It started in a small garden, with a monk named Gregor Mendel. By analyzing generation after generation of pea plants in experimental plots, he uncovered indirect but observable evidence of how parents bestow units of hereditary information—genes—on offspring.

This chapter starts out with the methods and some representative results of Mendel's experiments. His pioneering work remains a classic example of how a scientific approach can pry open important secrets about the natural world. To this day, it serves as the foundation for modern genetics.

☑ *How Would You Vote?*

Many advances in genetics, including the ability to detect mutant genes that cause severe disorders, raise bioethical questions. Should we encourage the mass screening of prospective parents for the alleles that cause cystic fibrosis? And should we as a society encourage women to give birth only if their child will not develop severe medical problems? See the Media Menu for details, then vote online.

Beyond Mendel The traits that Mendel studied happened to follow simple dominant-to-recessive patterns of gene expression. The expression of genes for most traits is not as straightforward. Incomplete dominance and codominance are cases in point.

Less Predictable Variation Although many genes have predictable, observable effects on traits, the expression of most genes is variable. Most traits are outcomes of interactions among the products of two or more genes. Environmental factors also influence gene expression.

10.1 Mendel's Insight Into Inheritance Patterns

We turn now to recurring inheritance patterns among humans and other sexually reproducing species. You already know meiosis halves the parental chromosome number, which is restored at fertilization. Here the story picks up with some observable outcomes of these events.

More than a century ago, people wondered about the basis of inheritance. As many knew, both sperm and eggs transmit information about traits to offspring, but few suspected that the information is organized in units (genes). By the prevailing view, the father's

Figure 10.2 Gregor Mendel, the founder of modern genetics.

carpel   stamen

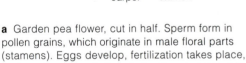

a Garden pea flower, cut in half. Sperm form in pollen grains, which originate in male floral parts (stamens). Eggs develop, fertilization takes place, and seeds mature in female floral parts (carpels).

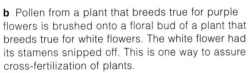

b Pollen from a plant that breeds true for purple flowers is brushed onto a floral bud of a plant that breeds true for white flowers. The white flower had its stamens snipped off. This is one way to assure cross-fertilization of plants.

c Later, seeds develop inside pods of the cross-fertilized plant. An embryo within each seed develops into a mature pea plant.

d Each new plant's flower color is indirect but observable evidence that hereditary material has been transmitted from the parent plants.

Figure 10.3 Garden pea plant (*Pisum sativum*), which can self-fertilize or cross-fertilize. Experimenters can control the transfer of its hereditary material from one flower to another.

blob of information "blended" with the mother's blob at fertilization, like milk into coffee.

Carried to its logical conclusion, blending would slowly dilute a population's shared pool of hereditary information until there was only a single version of each trait. Freckled children would never pop up in a family of nonfreckled people. In time, all of the colts and fillies that are descended from a herd of white stallions and black mares would be gray. But freckles do show up, and not all horses are gray. The blending theory could scarcely explain the obvious variation in traits that people could observe with their own eyes. Even so, few disputed the theory.

"Blending" proponents dismissed Charles Darwin's theory of natural selection. According to the theory's key premise, individuals of a population vary in the details of the traits they have in common. Over the generations, variations that help an individual survive and reproduce show up among more offspring than variations that do not. Less helpful variations might persist, but among fewer individuals. They may even disappear. It is not that some versions of a trait are "blended out" of the population. Rather, *the frequency of each version of a trait among all individuals of the population may persist or change over time.*

Even before Darwin presented his theory, someone was gathering evidence that eventually would help support it. A monk, Gregor Mendel (Figure 10.2), had already guessed that sperm and eggs carry distinct "units" of information about heritable traits. After analyzing certain traits of pea plants generation after generation, he found indirect but *observable* evidence of how parents transmit genes to offspring.

MENDEL'S EXPERIMENTAL APPROACH

Mendel spent most of his adult life in Brno, a city near Vienna that is now part of the Czech Republic. Yet he was not a man of narrow interests who accidentally stumbled onto dazzling principles.

Mendel's monastery was close to European capitals that were centers of scientific inquiry. Having been raised on a farm, he was keenly aware of agricultural principles and their applications. He kept abreast of literature on breeding experiments. He also belonged to a regional agricultural society and even won awards for developing improved varieties of vegetables and fruits. Shortly after he entered the monastery, Mendel took a number of courses in mathematics, physics, and botany at the University of Vienna. Few scholars of his time showed interest in both plant breeding *and* mathematics.

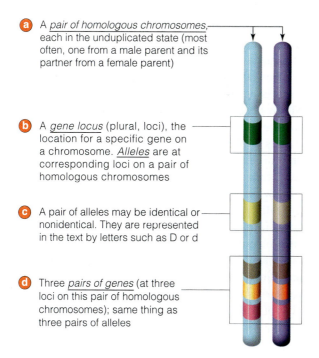

a A *pair of homologous chromosomes,* each in the unduplicated state (most often, one from a male parent and its partner from a female parent)

b A *gene locus* (plural, loci), the location for a specific gene on a chromosome. *Alleles* are at corresponding loci on a pair of homologous chromosomes

c A pair of alleles may be identical or nonidentical. They are represented in the text by letters such as D or d

d Three *pairs of genes* (at three loci on this pair of homologous chromosomes); same thing as three pairs of alleles

Figure 10.4 A few genetic terms. Garden peas and other species with a diploid chromosome number have pairs of genes, on pairs of homologous chromosomes. Most genes come in slightly different molecular forms called alleles. Different alleles specify different versions of the same trait. An allele at any given location on a chromosome may or may not be identical to its partner on the homologous chromosome.

Shortly after his university training, Mendel began studying *Pisum sativum,* the garden pea plant (Figure 10.3). This plant is self-fertilizing. Its male and female gametes—call them sperm and eggs—originate in the same flower, and fertilization can occur in the same flower. A lineage of pea plants can "breed true" for certain traits. This means successive generations will be just like parents in one or more traits, as when all offspring grown from seeds of self-fertilized, white-flowered parent plants also have white flowers.

Pea plants also cross-fertilize when pollen from one plant's flower reaches another plant's flower. Mendel knew he could open the flower buds of a plant that bred true for a trait, such as white flowers, and snip out its stamens. Pollen grains, in which sperm develop, originate in stamens. Then he could brush the buds with pollen from a plant that bred true for a *different* version of the same trait—say, purple flowers.

As Mendel hypothesized, such clearly observable differences might help him track a given trait through many generations. If there were patterns to the trait's inheritance, *then those patterns might tell him something about heredity itself.*

TERMS USED IN MODERN GENETICS

In Mendel's time, no one knew about genes, meiosis, or chromosomes. As we follow his thinking, we can clarify the picture by substituting some modern terms used in inheritance studies, as stated here and in Figure 10.4:

1. **Genes** are units of information about heritable traits, transmitted from parents to offspring. Each gene has a specific location (locus) on a chromosome.

2. Cells with a **diploid** chromosome number ($2n$) have pairs of genes, on pairs of homologous chromosomes.

3. **Mutation** alters a gene's molecular structure and its message about a trait. It may cause a trait to change, as when one gene for flower color specifies white and a mutant form specifies yellow. All molecular forms of the same gene are known as **alleles**.

4. When offspring inherit a pair of *identical* alleles for a trait generation after generation, we expect them to be a true-breeding lineage. Offspring of a cross between two individuals that breed true for different forms of a trait are **hybrids**; each one has inherited *nonidentical* alleles for the trait.

5. When a pair of alleles on homologous chromosomes are identical, this is a *homozygous* condition. When the two are not identical, this is a *heterozygous* condition.

6. An allele is *dominant* when its effect on a trait masks that of any *recessive* allele paired with it. We use capital letters to signify dominant alleles and lowercase letters for recessive ones. *A* and *a* are examples.

7. Pulling this all together, a **homozygous dominant** individual has a pair of dominant alleles (*AA*) for the trait being studied. A **homozygous recessive** individual has a pair of recessive alleles (*aa*). And a **heterozygous** individual has a pair of nonidentical alleles (*Aa*).

8. Two terms help keep the distinction clear between genes and the traits they specify. *Genotype* refers to the particular alleles that an individual carries. *Phenotype* refers to an individual's observable traits.

9. P stands for the parents, F_1 for their first-generation offspring, and F_2 for the second-generation offspring.

Mendel hypothesized that tracking clearly observable differences in forms of a given trait might reveal patterns of inheritance.

He predicted that hereditary information is transmitted from one generation to the next as separate units (genes) and is not "diluted" at fertilization.

10.2 Mendel's Theory of Segregation

Mendel used monohybrid crosses to test his hypothesis that pea plants inherit two "units" of information for a trait, one from each parent.

Monohybrid crosses use F_1 offspring of parents that breed true for different forms of a trait ($AA \times aa = Aa$). The experiment itself is a cross between two identical F_1 heterozygotes, which are the "monohybrids" ($Aa \times Aa$).

MONOHYBRID CROSS PREDICTIONS

Mendel tracked many traits over two generations. In one set of experiments, he crossed plants that bred true for purple *or* white flowers. All F_1 offspring had purple flowers. They self-fertilized, and some of the F_2 offspring had white flowers! What was going on?

Pea plants have pairs of homologous chromosomes. Assume one plant is homozygous dominant (AA) and another is homozygous recessive (aa) for flower color. Following meiosis in both plants, each sperm or egg that forms has only one allele for flower color (Figure 10.5). Thus, when a sperm fertilizes an egg, only one outcome is possible: $A + a = Aa$.

With his background in mathematics, Mendel knew about sampling error (Chapter 1). He crossed many thousands of plants. He also counted and recorded the number of dominant and recessive forms of traits. On average, three of every four F_2 plants were dominant, and one was recessive (Figure 10.6).

The ratio hinted that fertilization is a chance event having a number of possible outcomes. Mendel knew about probability, which applies to chance events *and so could help him predict possible outcomes of his genetic crosses*. **Probability** means this: The chance that each outcome of an event will occur is proportional to the number of ways in which the event can be reached.

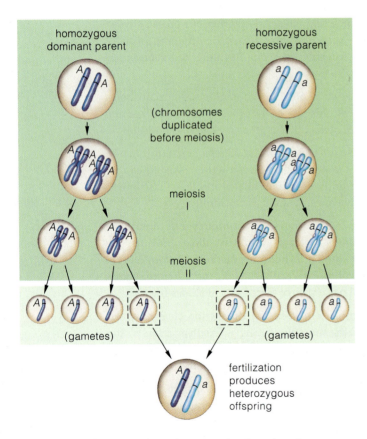

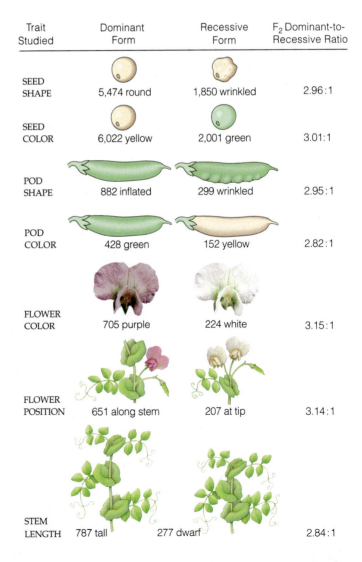

Trait Studied	Dominant Form	Recessive Form	F_2 Dominant-to-Recessive Ratio
SEED SHAPE	5,474 round	1,850 wrinkled	2.96:1
SEED COLOR	6,022 yellow	2,001 green	3.01:1
POD SHAPE	882 inflated	299 wrinkled	2.95:1
POD COLOR	428 green	152 yellow	2.82:1
FLOWER COLOR	705 purple	224 white	3.15:1
FLOWER POSITION	651 along stem	207 at tip	3.14:1
STEM LENGTH	787 tall	277 dwarf	2.84:1

Figure 10.5 One gene of a pair segregating from the other gene in a monohybrid cross. Two parents that breed true for two versions of a trait produce only heterozygous offspring.

Figure 10.6 *Right*: Some monohybrid cross experiments with pea plants. Mendel's counts of F_2 offspring having dominant or recessive hereditary "units" (alleles). On average, the 3:1 phenotypic ratio held for traits.

A **Punnett-square method**, explained and applied in Figure 10.7, shows the possibilities. If half of a plant's sperm or eggs are *a* and half are *A*, then we can expect four outcomes with each fertilization:

POSSIBLE EVENT	PROBABLE OUTCOME
sperm *A* meets egg *A*	1/4 *AA* offspring
sperm *A* meets egg *a*	1/4 *Aa*
sperm *a* meets egg *A*	1/4 *Aa*
sperm *a* meets egg *a*	1/4 *aa*

Each F_2 plant has 3 chances in 4 of inheriting at least one dominant allele (purple flowers). It has 1 chance in 4 of inheriting two recessive alleles (white flowers). That is a probable phenotypic ratio of 3:1.

Mendel's observed ratios were not *exactly* 3:1. Yet he put aside the deviations. To understand why, flip a coin several times. As we all know, a coin is as likely to end up heads as tails. But often it ends up heads, or tails, several times in a row. If you flip the coin only a few times, the observed ratio might differ a lot from the predicted ratio of 1:1. Flip it many times, and you are more likely to approach the predicted ratio.

That is why Mendel used rules of probability and counted so many offspring. He minimized sampling error deviations in the predicted results.

TESTCROSSES

Testcrosses supported Mendel's prediction. In such experimental tests, an organism shows dominance for a specified trait but its genotype is unknown, so it is crossed to a known homozygous recessive individual in a number of matings. Results may reveal whether it is homozygous dominant or heterozygous.

For example, Mendel crossed F_1 purple-flowered plants with true-breeding white-flowered plants. If all were homozygous dominant, then all the F_2 offspring would be purple flowered. If heterozygous, then only about half would. That is what happened. Half of the F_2 offspring had purple flowers (*Aa*) and half had white (*aa*). Go ahead and construct Punnett squares as a way to predict possible outcomes of this testcross.

Results from Mendel's monohybrid crosses became the basis of a theory of **segregation**, as stated here:

MENDEL'S THEORY OF SEGREGATION *Diploid cells have pairs of genes, on pairs of homologous chromosomes. The two genes of each pair are separated from each other during meiosis, so they end up in different gametes.*

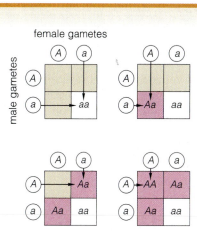

female gametes

male gametes

a Punnett-square method

Read Me First!
and watch the narrated animation on monohybrid crosses

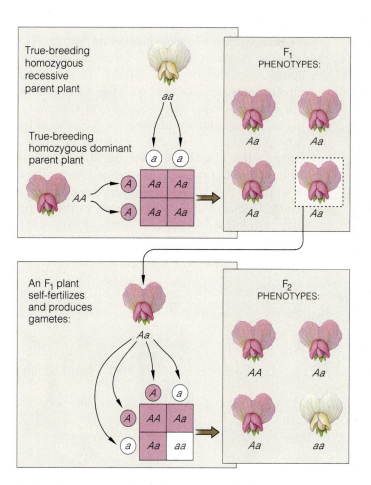

True-breeding homozygous recessive parent plant

True-breeding homozygous dominant parent plant

F_1 PHENOTYPES:

An F_1 plant self-fertilizes and produces gametes:

F_2 PHENOTYPES:

b Cross between two plants that breed true for different forms of a trait, followed by a monohybrid cross between their F_1 offspring

Figure 10.7 (**a**) Punnett-square method of predicting probable outcomes of genetic crosses. Circles signify gametes. *Italics* indicate dominant or recessive alleles. Possible genotypes among offspring are written in the squares. (**b**) Results from one of Mendel's monohybrid crosses. On average, the ratio of dominant-to-recessive that showed up among second-generation (F_2) plants was 3:1.

10.3 Mendel's Theory of Independent Assortment

In another set of experiments, Mendel used dihybrid crosses to explain how two pairs of genes assort into gametes.

*Di*hybrids are the offspring of parents that breed true for different versions of two traits. A **dihybrid cross** is an experimental intercross between F_1 dihybrids that are identically heterozygous for two pairs of genes.

Let's duplicate one of Mendel's dihybrid crosses for flower color (alleles *A* or *a*) and for height (*B* or *b*):

True-breeding parents: *AABB* X *aabb*

Gametes: *AB   AB   ab   ab*

F_1 hybrid offspring: *AaBb*

As Mendel would have predicted, F_1 offspring from this cross are all purple-flowered and tall (*AaBb*).

How will the two gene pairs assort into gametes in these F_1 plants? It depends partly on the chromosome locations of the two pairs. Assume that one pair of homologous chromosomes have the *Aa* alleles and a different pair have the *Bb* alleles. All chromosomes, recall, align midway between the spindle poles at metaphase I of meiosis (Figures 9.5 and 10.8). The one bearing the *A* or the *a* allele might be tethered to either pole. The same can happen to the chromosome bearing the *B* or *b* allele. Following meiosis, only four combinations of alleles are possible in the sperm or eggs that form: 1/4 *AB*, 1/4 *Ab*, 1/4 *aB*, and 1/4 *ab*.

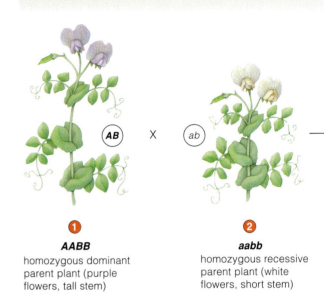

1
AABB
homozygous dominant parent plant (purple flowers, tall stem)

2
aabb
homozygous recessive parent plant (white flowers, short stem)

Figure 10.9 Results from Mendel's dihybrid cross starting with parent plants that bred true for different versions of two traits: flower color and plant height. *A* and *a* signify dominant and recessive alleles for flower color. *B* and *b* signify dominant and recessive alleles for height. The Punnett square shows the F_2 combinations possible:

- 9/16 or 9 purple flowered, tall
- 3/16 or 3 purple-flowered, dwarf
- 3/16 or 3 white-flowered, tall
- 1/16 or 1 white-flowered, dwarf

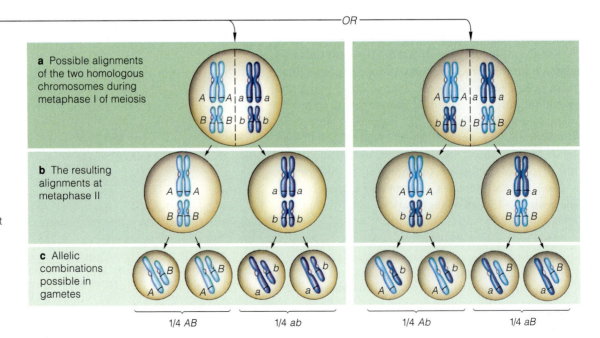

Nucleus of a diploid (2*n*) reproductive cell with two pairs of homologous chromosomes

Figure 10.8 An example of independent assortment at meiosis. Either chromosome of a pair may get tethered to either spindle pole. When just two pairs are tracked, two different metaphase I alignments are possible.

—OR—

a Possible alignments of the two homologous chromosomes during metaphase I of meiosis

b The resulting alignments at metaphase II

c Allelic combinations possible in gametes

1/4 AB   1/4 ab   1/4 Ab   1/4 aB

Read Me First!
*and watch the narrated
animation on dihybrid crosses*

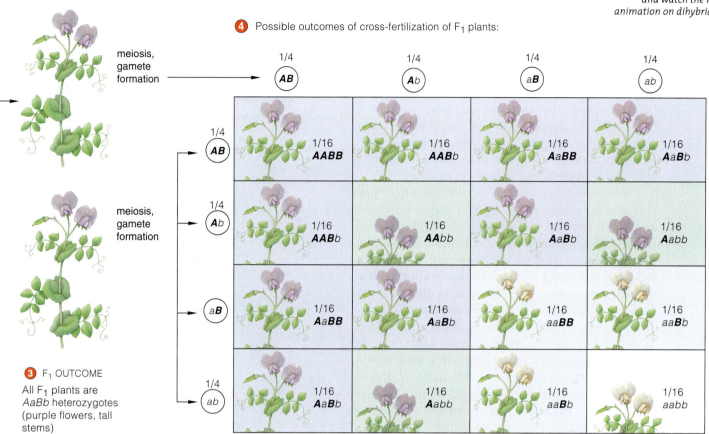

4 Possible outcomes of cross-fertilization of F_1 plants:

meiosis, gamete formation →

	1/4 **AB**	1/4 **Ab**	1/4 **aB**	1/4 **ab**
1/4 **AB**	1/16 **AABB**	1/16 **AAB**b	1/16 **A**a**BB**	1/16 **A**a**B**b
1/4 **Ab**	1/16 **AAB**b	1/16 **AA**bb	1/16 **A**a**B**b	1/16 **A**abb
1/4 a**B**	1/16 **A**a**BB**	1/16 **A**a**B**b	1/16 aa**BB**	1/16 aa**B**b
1/4 ab	1/16 **A**a**B**b	1/16 **A**abb	1/16 aa**B**b	1/16 aabb

meiosis, gamete formation

3 F_1 OUTCOME

All F_1 plants are *AaBb* heterozygotes (purple flowers, tall stems)

Given the alternative metaphase I alignments, many allelic combinations can result at fertilization. Simple multiplication (four sperm types × four egg types) tells us that sixteen combinations of genotypes are possible among F_2 offspring of a dihybrid cross (Figure 10.9).

Adding all possible phenotypes gives us a ratio of 9:3:3:1. We can expect to see 9/16 tall purple-flowered, 3/16 dwarf purple-flowered, 3/16 tall white-flowered, and 1/16 dwarf white-flowered F_2 plants. Results from one dihybrid cross were close to this ratio.

Mendel could only analyze numerical results from such crosses because he did not know seven pairs of homologous chromosomes carry a pea plant's "units" of inheritance. He could do no more than hypothesize that the two units for flower color were sorted out into gametes independently of the two units for height.

In time, his hypothesis became known as the theory of **independent assortment**. In modern terms, after meiosis ends, the genes on each pair of homologous chromosomes are assorted into gametes independently

of how all the other pairs of homologues are sorted out. Independent assortment and hybrid intercrosses give rise to genetic variation. In a monohybrid cross for one gene pair, three genotypes are possible: *AA*, *Aa*, and *aa*. We represent this as 3^n, where *n* is the number of gene pairs. The more pairs, the more combinations are possible. If parents differ in twenty gene pairs, for instance, the number approaches 3.5 billion!

In 1866, Mendel published his idea, but apparently he was read by few and understood by no one. Today his theory of segregation still stands. However, his theory of independent assortment does not apply to *all* gene combinations, as you will see in Chapter 11.

MENDEL'S THEORY OF INDEPENDENT ASSORTMENT *As meiosis ends, genes on pairs of homologous chromosomes have been sorted out for distribution into one gamete or another, independently of gene pairs of other chromosomes.*

10.4 More Patterns Than Mendel Thought

Mendel happened to focus on traits that have clearly dominant and recessive forms. However, expression of genes for most traits is not as straightforward.

ABO BLOOD TYPES—A CASE OF CODOMINANCE

In *codominance*, a pair of nonidentical alleles affecting two phenotypes are both expressed at the same time in heterozygotes. For example, red blood cells have a type of glycolipid at the plasma membrane that helps give them their unique identity. The glycolipid comes in slightly different forms. An analytical method, *ABO blood typing*, reveals which form a person has.

An enzyme dictates the glycolipid's final structure. Humans have three alleles for this enzyme. Two, I^A and I^B, are codominant when paired. The third allele, *i*, is recessive; a pairing with I^A or I^B masks its effect. (If the letter for an allele is superscript, it signifies a lack of dominance.) Here we have a **multiple allele system**, the occurrence of three or more alleles for one gene locus among individuals of a population.

Each of these glycolipid molecules was assembled in the endomembrane system (Figure 4.16). First, an oligosaccharide chain was attached to a lipid, then a sugar was attached to the chain. But alleles I^A and I^B specify different versions of the enzyme that attaches the sugar. The two attach *different* sugars, which gives a glycolipid molecule a different identity: A or B.

Which alleles do you have? If you have I^AI^A or I^Ai, your blood is type A. With I^BI^B or I^Bi, it is type B. With codominant alleles I^AI^B, it is AB—you have both versions of the sugar-attaching enzyme. If you are homozygous recessive (*ii*), the glycolipid molecules never did get a final sugar side chain, so your blood type is not A or B. It is O (Figure 10.10).

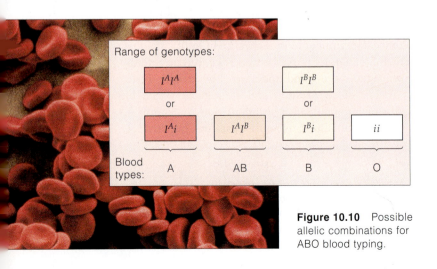

Figure 10.10 Possible allelic combinations for ABO blood typing.

Range of genotypes:

I^AI^A or I^Ai		I^BI^B or I^Bi	
I^Ai	I^AI^B	I^Bi	*ii*

Blood types: A · AB · B · O

homozygous parent x homozygous parent

All F$_1$ offspring heterozygous for flower color:

Cross two of the F$_1$ plants, and the F$_2$ offspring will show three phenotypes in a 1:2:1 ratio:

Figure 10.11 Incomplete dominance in heterozygous (*pink*) snapdragons, in which an allele that affects red pigment is paired with a "white" allele.

INCOMPLETE DOMINANCE

In *incomplete* dominance, one allele of a pair is not fully dominant over its partner, so the heterozygote's phenotype is *somewhere between* the two homozygotes. Cross true-breeding red and white snapdragons and their F$_1$ offspring will be pink-flowered. Cross two F$_1$ plants and you can expect to see red, white, and *pink* flowers in a certain ratio (Figure 10.11). Why the "odd" pattern? Red snapdragons have two alleles that let them make a lot of molecules of a red pigment. White snapdragons have two mutant alleles, and they are pigment-free. Pink snapdragons have one "red" allele and one "white" one. These heterozygotes make just enough pigment to color flowers pink, not red.

Two interacting gene pairs also can give rise to a phenotype that neither produces by itself. In chickens, interactions among *R* and *P* alleles specify the walnut, rose, pea, and single combs shown in Figure 10.12.

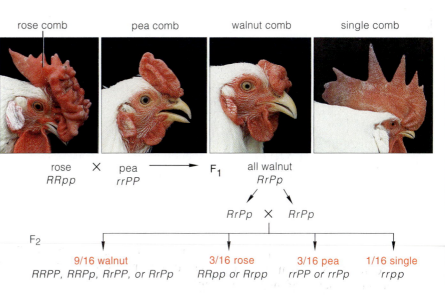

Figure 10.12 Interaction between two genes with variable effects on the comb on a chicken's head. The first cross is between a Wyandotte (rose comb) and a Brahma (pea comb). Check the outcomes by making a Punnett-square diagram.

rose comb   pea comb   walnut comb   single comb

rose × pea → F₁ all walnut
RRpp   rrPP   RrPp

RrPp × RrPp

F₂

9/16 walnut   3/16 rose   3/16 pea   1/16 single
RRPP, RRPp, RrPP, or RrPp   RRpp or Rrpp   rrPP or rrPp   rrpp

SINGLE GENES WITH A WIDE REACH

The alleles at one locus on a chromosome may affect two or more traits in good or bad ways. This outcome of the activity of one gene's product is **pleiotropy**. We see its effects in many genetic disorders, such as cystic fibrosis, sickle-cell anemia, and Marfan syndrome.

An autosomal dominant mutation in the gene for fibrillin causes *Marfan syndrome*. Fibrillin is a protein of connective tissues, the most abundant, widespread of all vertebrate tissues. We find many thin fibrillin strands, loose or cross-linked with the protein elastin, in the heart, blood vessels, and skin. They passively recoil after being stretched, as by the beating heart.

Altered fibrillin weakens the connective tissues in 1 of 10,000 men and women of any ethnicity. The heart, blood vessels, skin, lungs, and eyes are at risk. One of the mutations disrupts the synthesis of fibrillin 1, its secretion from cells, and its deposition. It skews the structure and function of smooth muscle cells inside the wall of the aorta, a big vessel carrying blood out of the heart. Cells infiltrate and multiply inside the wall's epithelial lining. Calcium deposits accumulate and the wall becomes inflamed. Elastic fibers split into fragments. The aorta wall, thinned and weakened, can rupture abruptly during strenuous exercise.

Until recent medical advances, Marfan syndrome killed most affected people before they were fifty years old. Flo Hyman was one of them (Figure 10.13).

WHEN PRODUCTS OF GENE PAIRS INTERACT

Traits also arise from interactions among products of two or more gene pairs. In some cases, two alleles can mask expression of another gene's alleles, and some expected phenotypes may not appear at all.

For example, several gene pairs govern fur color in Labrador retrievers. The fur appears black, yellow, or brown depending on how enzymes and other products of gene pairs synthesize melanin, a dark pigment, and deposit it in different body regions. Allele *B* (black) has a stronger effect and is dominant to *b* (brown). Alleles at another gene locus control how much melanin gets

Figure 10.13 Flo Hyman, at left, captain of the United States volleyball team that won an Olympic silver medal in 1984. Two years later, at a game in Japan, she slid to the floor and died. A dime-sized weak spot in the wall of her aorta had burst. We know at least two affected college basketball stars also died abruptly as a result of Marfan syndrome.

deposited in hair. Allele *E* permits full deposition. Two recessive alleles (*ee*) reduce it, so fur appears yellow.

Alleles at another locus (*C*) may override those two. They encode the first enzyme in a melanin-producing pathway. *CC* or *Cc* individuals do make the functional enzyme. An individual with two recessive alleles (*cc*) cannot. *Albinism*, the absence of melanin, is the result.

> Some alleles are fully dominant, incompletely dominant, or codominant with a partner on the homologous chromosome.
>
> With pleiotropy, alleles at a single locus have positive or negative impact on two or more traits.
>
> Gene effects do not always appear together but rather appear over time. A gene's product may alter one trait, which may cause alteration in another trait, and so on.

10.5 Complex Variations in Traits

For most populations or species, individuals show rich variation for many of the same traits. Variation arises from gene mutations, cumulative gene interactions, and variations in environmental conditions.

REGARDING THE UNEXPECTED PHENOTYPE

As Mendel found out, phenotypic effects of one or two pairs of certain genes show up in predictable ratios. Two or more gene pairs also can produce phenotypes in predictable ratios. However, track some genes over the generations, and you might find that the resulting phenotypes were not what you expected.

As one example, *camptodactyly*, a rare abnormality, affects the shape and movement of fingers. Some of the people who carry a mutant allele for this heritable trait have immobile, bent fingers on both hands. Others have immobile, bent fingers on the left or right hand only. Fingers of still other people who have the mutant allele are not affected in any obvious way at all.

What causes such odd variation? Remember, most organic compounds are synthesized by a sequence of metabolic steps. *Different enzymes, each a gene product, control different steps.* Maybe one gene has mutated in a number of ways. Maybe a gene product blocks some pathway or makes it run nonstop or not long enough. Maybe poor nutrition or another variable factor in the individual's environment influences a crucial enzyme in the pathway. Such variable factors often introduce less predictable variations in the phenotypes that we otherwise associate with certain genes.

CONTINUOUS VARIATION IN POPULATIONS

Generally, individuals of a population display a range of small differences in most traits. This characteristic of natural populations, called **continuous variation**, depends mainly on how many gene products affect a given trait and on how many environmental factors impact them. The greater the number of genes

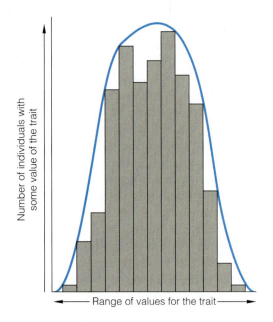

a Idealized bell-shaped curve for a population that displays continuous variation in a trait

The line of a bell-shaped curve reveals continuous variation in the population

b A bell-shaped curve that corresponds to the height distribution among individual females in the far-right photograph in (**c**)

and environmental factors, the more continuous is the distribution of all versions of the trait.

Look in a mirror at your eye color. The colored part is the iris, a doughnut-shaped, pigmented structure just under the cornea. The color results from several gene products. Some products help make and distribute the light-absorbing pigment melanin, the same pigment that affects coat color in mammals. Almost black irises have dense melanin deposits, and melanin molecules absorb most of the incoming light. Deposits are not as great in brown eyes, so some unabsorbed light is reflected out. Light brown or hazel eyes have even less melanin (Figure 10.14).

Green, gray, or blue eyes do not have green, gray, or blue pigments. The iris has some melanin, but not

Figure 10.14 Examples from a range of continuous variation in human eye color. Products of different gene pairs interact in making and distributing the melanin that helps color the iris. Different combinations of alleles result in small color differences. The frequency distribution for the eye-color trait is continuous over a range from black to light blue.

Read Me First!

*and watch the narrated animation
on continuous variation in traits*

5'3" 5'4" 5'5" 5'6" 5'7" 5'8" 5'9" 5'10" 5'11" 6'0" 6'1" 6'2" 6'3" 6'4" 6'5"
Height (feet/inches)

Figure 10.15 Continuous variation in body height, one of the traits that help characterize the human population.

(**a**) A bar graph can depict continuous variation in a population. The proportion of individuals in each category is plotted against the range of measured phenotypes. The curved line above this particular set of bars is an idealized example of the kind of bell-shaped curve that emerges for populations showing continuous variation in a trait.

(**b,c**) Jon Reiskind and Greg Pryor wanted to show the frequency distribution for height among biology students at the University of Florida. They divided students into two groups: male and female. For each group, they divided the range of possible heights, measured the students, and assigned each to the appropriate category.

4'11" 5'0" 5'1" 5'2" 5'3" 5'4" 5'5" 5'6" 5'7" 5'8" 5'9" 5'10" 5'11"
Height (feet/inches)

c Two examples of continuous variation: Biology students (males, *left*; females, *right*) organized by height.

much. Many or most of the blue wavelengths of light that do enter the eyeball are simply reflected out.

How can you describe the continuous variation of some trait in a group? Consider the students in Figure 10.15. They range from short to tall, with average heights more common than the extremes. Start out by dividing the full range of phenotypes into measurable categories—for instance, number of inches. Next, count how many students are in each category to get the relative frequencies of all phenotypes across the range of measurable values.

The chart in Figure 10.15*b* is a plot of the number of students in each height category. The shortest bars represent categories having the fewest individuals. The tallest bar signifies the category with the most. In this case, a graph line skirting the top of all the bars will be a bell-shaped curve. Such "bell curves" are typical of any trait that shows continuous variation.

> Enzymes and other gene products control each step of most metabolic pathways. Mutations, interactions among genes, and environmental conditions may affect one or more steps. The outcome is variation in phenotypes.
>
> For most traits, individuals of a population or species show continuous variation—a range of small differences.
>
> Usually, the greater the number of genes and environmental factors that influence a trait, the more continuous the distribution of versions of that trait.

10.6 Genes and the Environment

We have mentioned, in passing, that the environment often contributes to variable gene expression among a population's individuals. Now consider a few cases.

Possibly you have noticed a Himalayan rabbit's coat color. Like a Siamese cat, this mammal has dark hair in some parts of its body and lighter hair in others. The Himalayan rabbit is homozygous for the c^h allele of the gene specifying tyrosinase. Tyrosinase is one of the enzymes involved in melanin production. The c^h allele specifies a heat-sensitive form of this enzyme. And this form is active only when the air temperature around the body is below 33°C, or 91°F.

When cells that give rise to this rabbit's hair grow under warmer conditions, they cannot make melanin, so hairs appear light. This happens in body regions that are massive enough to conserve a fair amount of metabolic heat. The ears and other slender extremities tend to lose metabolic heat faster, so they are cooler. Figure 10.16 shows one experiment that demonstrated how environmental temperatures affect this allele.

One classic experiment identified environmental effects on yarrow plants. These plants can grow from cuttings, so they are a useful experimental organism. Why? Cuttings from the same plant all have the same genotype, so experimenters can discount genes as a basis for differences that show up among them.

In this case, three yarrow cuttings were planted at three elevations. Two plants that grew at the lowest elevation and highest elevation fared best; the one at the mid-elevation grew poorly (Figure 10.17).

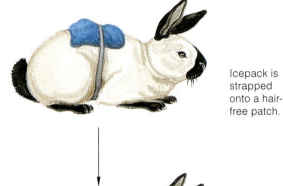

Icepack is strapped onto a hair-free patch.

New hair growing in patch exposed to cold is black.

Figure 10.16 Observable effect of an environmental factor that influences gene expression. A Himalayan rabbit normally has black hair only on its long ears, nose, tail, and leg regions farthest from the body mass. In one experiment, a patch of a rabbit's white coat was removed, then an icepack was secured over the hairless patch. Where the colder temperature had been maintained, the hairs that grew back were black.

Himalayan rabbits are homozygous for an allele of the gene for tyrosinase, an enzyme required to make melanin. As described in the text, this allele encodes a heat-sensitive form of the enzyme, which functions only when air temperature is below about 33°C.

a Mature cutting at high elevation (3,050 meters above sea level)

b Mature cutting at mid-elevation (1,400 meters above sea level)

c Mature cutting at low elevation (30 meters above sea level)

Figure 10.17 One experiment demonstrating the impact of environmental conditions of three different habitats on gene expression in yarrow (*Achillea millefolium*). Cuttings from the same parent plant were grown in the same soil batch but at three different elevations.

However, recall from Chapter 1 that sampling error can skew experimental results. The experimenters did the same growth experiments for *many* yarrow plants and found no consistent pattern; phenotypic variation was too great. For instance, a cutting from one plant did *best* at mid-elevation. The conclusion? For yarrow plants, at least, individuals with different genotypes react differently across a range of environments.

Similarly, plant a hydrangea in a garden and it may have pink flowers instead of the expected blue ones. Soil acidity affects the function of gene products that color hydrangea flowers.

What about humans? One of our genes codes for a transporter protein that moves serotonin across the plasma membrane of brain cells. This gene product has several effects, one of which is to counter anxiety and depression when traumatic events challenge us. For a long time, researchers have known that some people handle stress without getting too upset, while others spiral into a deep and lasting depression.

Mutation of the gene for the serotonin transporter compromises responses to stress. It is as if some of us are bicycling through life without an emotional helmet. Only when we take a fall does the phenotypic effect—depression—appear. Other genes also affect emotional states, but mutation of this one reduces our capacity to snap out of it when bad things happen.

And so we conclude this chapter, which introduces heritable and environmental factors that give rise to great variation in traits. What is the take-home lesson? Simply this: An individual's phenotype is an outcome of complex interactions among its genes, enzymes and other gene products, and the environment.

> Variation in traits arises not only from gene mutations and interactions, but also in response to variations in environmental conditions that each individual faces.

Summary

Section 10.1   Genes are heritable units of information about traits. Each gene has its own locus on a particular chromosome. Different molecular forms of the same gene are known as alleles. Diploid cells have two copies of each gene, usually one inherited from each of two parents, on homologous chromosomes.

Offspring of a cross between two individuals that breed true for different forms of a trait are hybrids; each has inherited nonidentical alleles for the trait.

An individual with two dominant alleles for a trait (AA) is homozygous dominant. A homozygous recessive has two recessive alleles (aa). A heterozygote has two nonidentical alleles (Aa) for a trait. A dominant allele may mask the effect of a recessive allele partnered with it on the homologous chromosome.

Section 10.2   Results from Mendel's monohybrid crosses between F_1 offspring of true-breeding pea plants in time led to a theory of segregation: Paired genes on homologous chromosomes separate from each other at meiosis and end up in different gametes. The theory is based on a pattern of dominance and recessiveness that showed up among F_2 offspring of monohybrid crosses:

	A	a
A	AA	Aa
a	Aa	aa

AA (dominant)
Aa (dominant)
Aa (dominant)
aa (recessive) } the expected phenotypic ratio of 3:1

Section 10.3   Mendel did dihybrid crosses between F_2 offspring of parents that bred true for two different traits. The dihybrid cross results were close to a 9:3:3:1 ratio:

  9 dominant for both traits
  3 dominant for A, recessive for b
  3 dominant for B, recessive for a
  1 recessive for both traits

His results support a theory of independent assortment: Meiosis assorts gene pairs of homologous chromosomes for forthcoming gametes independently of how gene pairs of the other chromosomes are sorted out. This is an outcome of the random alignment of all pairs of homologous chromosomes at metaphase I.

Section 10.4   Inheritance patterns are not always straightforward. Some alleles are codominant or not fully dominant. Products of gene pairs often interact in ways that influence the same trait. A single gene may have effects on two or more traits. Products of pairs of genes often interact in ways that influence the same trait. One gene may have positive or negative effects on two or more traits, a condition called pleiotropy.

Section 10.5   Mutations and interactions among gene products contribute to variation in traits among the individuals of a population. Some traits show a range of small, incremental differences—continuous variation.

Section 10.6 Environmental conditions can alter gene expression. The individuals of most populations show complex variation in traits, a combination of gene expression and exposure to environmental factors.

Self-Quiz

Answers in Appendix III

1. Alleles are _____ .
   a. different molecular forms of a gene
   b. different phenotypes
   c. self-fertilizing, true-breeding homozygotes

2. A heterozygote has a _____ for a trait being studied.
   a. pair of identical alleles
   b. pair of nonidentical alleles
   c. haploid condition, in genetic terms
   d. a and c

3. The observable traits of an organism are its _____ .
   a. phenotype   c. genotype
   b. sociobiology   d. pedigree

4. Second-generation offspring from a cross are the _____ .
   a. F_1 generation   c. hybrid generation
   b. F_2 generation   d. none of the above

5. F_1 offspring of the monohybrid cross $AA \times aa$ are _____ .
   a. all AA c. all Aa
   b. all aa d. 1/2 AA and 1/2 aa

6. Refer to Question 5. Assuming complete dominance, the F_2 generation will show a phenotypic ratio of _____ .
   a. 3:1   b. 9:1   c. 1:2:1   d. 9:3:3:1

7. Crosses between F_1 pea plants resulting from the cross $AABB \times aabb$ lead to F_2 phenotypic ratios close to _____ .
   a. 1:2:1   b. 3:1   c. 1:1:1:1   d. 9:3:3:1

8. Match each example with the most suitable description.
   ____ dihybrid cross   a. *bb*
   ____ monohybrid cross   b. *AABB* × *aabb*
   ____ homozygous condition   c. *Aa*
   ____ heterozygous condition d. *Aa* × *Aa*

Figure 10.18 Two albino organisms. By not posing his subjects as objects of ridicule, the photographer of human albinos is attempting to counter the notion that there is something inherently unbeautiful about them.

Genetics Problems

Answers in Appendix IV

1. A certain recessive allele *c* is responsible for *albinism*, an inability to produce or deposit melanin, a brownish-black pigment, in body tissues. Humans and a number of other organisms can have this phenotype. Figure 10.18 shows two stunning examples. In cases of albinism, what are the possible genotypes of the father, the mother, and their children?
   a. Both parents have normal phenotypes; some of their children are albino and others are unaffected.
   b. Both parents are albino and have albino children.
   c. The woman is unaffected, the man is albino, and they have one albino child and three unaffected children.

2. One gene has alleles *A* and *a*. Another has alleles *B* and *b*. For each genotype, what type(s) of gametes will form? Assume that independent assortment occurs.

   a. *AABB*   c. *Aabb*
   b. *AaBB*   d. *AaBb*

3. Refer to Problem 2. What will be the genotypes of offspring from the following matings? Indicate the frequencies of each genotype among them.
   a. *AABB* × *aaBB*   c. *AaBb* × *aabb*
   b. *AaBB* × *AABb*   d. *AaBb* × *AaBb*

4. Certain dominant alleles are so essential for normal development that an individual who is homozygous recessive for a mutant recessive form can't survive. Such recessive, *lethal alleles* can be perpetuated in the population by heterozygotes.
   Consider the Manx allele (M^L) in cats. Homozygous cats ($M^L M^L$) die when they are still embryos inside the mother cat. In heterozygotes ($M^L M$), the spine develops abnormally. The cats end up with no tail (Figure 10.19).
   Two $M^L M$ cats mate. What is the probability that any one of their *surviving* kittens will be heterozygous?

5. In one experiment, Mendel crossed a pea plant that bred true for green pods with one that bred true for yellow pods. All the F_1 plants had green pods. Which form of the trait (green or yellow pods) is recessive? Explain how you arrived at your conclusion.

6. Return to Problem 2. Assume you now study a third gene having alleles *C* and *c*. For each genotype listed, what type(s) of gametes will be produced?
   a. *AABBCC*   c. *AaBBCc*
   b. *AaBBcc*   d. *AaBbCc*

7. Mendel crossed a true-breeding tall, purple-flowered pea plant with a true-breeding dwarf, white-flowered plant. All F_1 plants were tall and had purple flowers. If an F_1 plant self-fertilizes, then what is the probability that a randomly selected F_2 offspring will be heterozygous for the genes specifying height and flower color?

8. *DNA fingerprinting* is a method of identifying individuals by locating unique base sequences in their DNA molecules (Section 15.4). Before researchers refined the method, attorneys often relied on the ABO blood-typing system to settle disputes over paternity. Suppose that you, as a geneticist, are asked to testify during a paternity case in which the mother has type A blood, the child has type O blood, and the alleged father has type B blood. How would you respond to the following statements?

a. Attorney of the alleged father: "The mother's blood is type A, so the child's type O blood must have come from the father. My client has type B blood; he could not be the father."

b. Mother's attorney: "Because further tests prove this man is heterozygous, he must be the father."

9. Suppose you identify a new gene in mice. One of its alleles specifies white fur. A second allele specifies brown fur. You want to determine whether the relationship between the two alleles is one of simple dominance or incomplete dominance. What sorts of genetic crosses would give you the answer? On what types of observations would you base your conclusions?

10. Your sister gives you a purebred Labrador retriever, a female named Dandelion. Suppose you decide to breed Dandelion and sell puppies to help pay for your college tuition. Then you discover that two of her four brothers and sisters show *hip dysplasia*, a heritable disorder arising from a number of gene interactions. If Dandelion mates with a male Labrador known to be free of the harmful alleles, can you guarantee to a buyer that her puppies will not develop the disorder? Explain your answer.

11. A dominant allele *W* confers black fur on guinea pigs. A guinea pig that is homozygous recessive (*ww*) has white fur. Fred would like to know whether his pet black-furred guinea pig is homozygous dominant (*WW*) or heterozygous (*Ww*). How might he determine his pet's genotype?

12. Red-flowering snapdragons are homozygous for allele R^1. White-flowering snapdragons are homozygous for a different allele (R^2). Heterozygous plants (R^1R^2) bear pink flowers. What phenotypes should appear among first-generation offspring of the crosses listed? What are the expected proportions for each phenotype?

a. $R^1R^1 \times R^1R^2$ c. $R^1R^2 \times R^1R^2$

b. $R^1R^1 \times R^2R^2$ d. $R^1R^2 \times R^2R^2$

(In cases of incomplete dominance, alleles are usually designated by superscript numerals, as shown here, not by the uppercase letters for dominance and lowercase letters for recessiveness.)

13. Two pairs of genes affect comb type in chickens (Figure 10.12). When both genes are recessive, a chicken has a single comb. A dominant allele of one gene, *P*, gives rise to a pea comb. Yet a dominant allele of the other (*R*) gives rise to a rose comb. An *epistatic* interaction occurs when a chicken has at least one of both dominants, *P__ R __*, which gives rise to a walnut comb.

Predict the ratios resulting from a cross between two walnut-combed chickens that are heterozygous for both genes (*PpRr*).

14. As Section 3.6 explains, a single mutant allele gives rise to an abnormal form of hemoglobin (Hb^S instead of Hb^A). Homozygotes (Hb^SHb^S) develop sickle-cell anemia. Heterozygotes (Hb^AHb^S) show few obvious symptoms.

Suppose a woman's mother is homozygous for the Hb^A allele. She marries a male who is heterozygous for the allele, and they plan to have children. For *each* of her pregnancies, state the probability that this couple will have a child who is:

a. homozygous for the Hb^S allele

b. homozygous for the Hb^A allele

c. heterozygous Hb^AHb^S

Figure 10.19 The Manx, a breed of cat that has no tail.

IMPACTS, ISSUES · *Strange Genes, Tortured Minds*

"This man is brilliant," was the entire text of a letter of recommendation from Richard Duffin, a mathematics professor at Carnegie Mellon University. Duffin wrote it in 1948 on behalf of John Forbes Nash, Jr. (Figure 11.1), who was twenty years old at the time and applying to Princeton University's graduate school.

In the next decade, Nash made brilliant contributions to the field of mathematics and was considered to be

Figure 11.1 John Forbes Nash, Jr., a prodigy who solved problems that had long baffled some of the greatest minds in mathematics. His early work in economic game theory won him a Nobel Prize. He is shown here at a premier of "A Beautiful Mind," an award-winning film based on his long, tormented battle with schizophrenia.

one of the nation's top scientists. Apart from his social awkwardness, which is common among highly gifted people, there was no warning that paranoid schizophrenia would debilitate him in his thirtieth year. Nash had to abandon his position at the Massachusetts Institute of Technology. Two decades would pass before he would return to his work in mathematics.

Of every hundred people worldwide, one is affected by *schizophrenia*, which is characterized by delusions, hallucinations, disorganized speech and behavior, and social dysfunction. Many researchers have speculated that extraordinary creativity is linked to schizophrenia and other neurobiological disorders (NBDs) including depression, bipolar disorder (manic depression), and autism.

Certainly not every individual with high IQ shows such a link, but a higher percentage of geniuses have NBDs compared to the general population. Creative writers alone are eighteen times more suicidal, ten times more likely to be depressed, and twenty times more likely to have bipolar disorder.

We now have evidence that highly creative, healthy people have more personality traits in common with the mentally ill than with normal, less creative people, particularly in their sensitivity to environmental stimuli. People with NBDs belong to an illustrious crowd that includes Socrates, Newton, Beethoven, Darwin, Lincoln, Poe, Dickens, Tolstoy, Van Gogh, Freud, Churchill, Einstein, Picasso, Woolf, Hemingway, and Nash.

the big picture

Focus on Chromosomes Males and females differ in their sex chromosomes, but all other homologous chromosomes are the same in both. All of their chromosomes are subject to crossing over and other changes, which diagnostic tools can detect.

Human Inheritance Patterns Family pedigrees often reveal patterns of autosomal dominant, autosomal recessive, and sex-linked inheritance. Such patterns underlie many genetic abnormalities and disorders.

Abnormal brain biochemistry underlies NBDs. For instance, people with *bipolar disorder* show extreme swings in mood, thoughts, energy, and behavior. In their brain cells, expression of some mitochondrial genes that control aerobic respiration and protein breakdown is markedly low.

Change in any step of a crucial biochemical pathway could impair the brain's wiring. Therefore, we can expect that alterations in genes contribute to the abnormal neurochemistry in NBDs. Indeed, NBDs tend to run in families. Geniuses and individuals with one or more types of NBD often appear in the same family.

We already know about several mutant genes that predispose individuals to neural disorders. We also know that their bearers do not always show severe symptoms. Individuals who push the envelope of human creativity walk a razor's edge of mental stability, and it may take interplays of gene products and environmental factors to knock them off.

This brief account of neurobiological disorders is a glimpse into the world of modern genetics research. It invites you to think about how far you have come in this unit of the book. You first looked at cell division, the starting point of inheritance. You looked at how chromosomes and the genes they carry are shuffled during meiosis, then at fertilization. You also mulled over Mendel's insights into patterns of inheritance and some exceptions to his conclusions. Turn now to the chromosomal basis of inheritance.

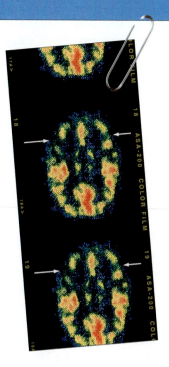

✓ How Would You Vote?

Diagnostic tests for predisposition to neurobiological disorders will soon be available. Individuals might use knowledge of their susceptibility to modify life-style choices. Insurance companies and employers might also use that information to exclude predisposed but otherwise healthy individuals. Would you support legislation governing these tests? See the Media Menu for details, then vote online.

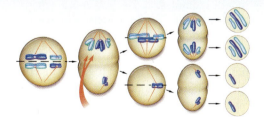

Chromosome Abnormalities   Certain genetic disorders arise from structural alterations of chromosomes or from abnormal changes in the chromosome number. Some changes occur spontaneously, and others result from exposure to harmful agents in the environment.

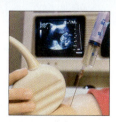

Prospects in Human Genetics   Some genetic disorders are treatable. Prospective parents who are at risk of transmitting a gene for a severe disorder often request genetic counseling or screening options, including prenatal and preimplantation diagnosis.

11.1 The Chromosomal Basis of Inheritance

You already know about chromosomes and what happens to them during meiosis. Now we'll start correlating chromosome structure to human inheritance patterns.

A REST STOP ON OUR CONCEPTUAL ROAD

Before driving on to the land of human inheritance, take a few minutes to check the following road map. It offers perspective on six important concepts:

1. A *gene*, again, is a unit of information about a heritable trait. The genes of eukaryotic cells are distributed among a number of chromosomes. Each gene has its own location, or locus, in one type of chromosome.

2. A cell with a diploid chromosome number, or 2*n*, has *pairs of homologous chromosomes*. All but one pair are normally the same in length, shape, and order of genes. The exception is a pairing of nonidentical sex chromosomes, such as X with Y in humans. Each chromosome becomes aligned with its homologous partner at metaphase I of meiosis.

3. Genes mutate, so a pair of genes on homologous chromosomes may or may not be the same. All of the slightly different molecular forms of a gene that occur among individuals of a population are called *alleles*.

4. A *wild-type* allele is a gene's most common form, in either a natural population or in a standardized, laboratory-bred strain of the species. A less common form of the gene is a *mutant* allele.

5. All genes on the same chromosome are physically connected. The farther apart any two genes are along the length of a chromosome, the more vulnerable they are to *crossing over*. By this event, a chromatid of one chromosome and a chromatid of its homologue swap corresponding segments (Figure 11.2). Crossing over between nonsister chromatids is a form of *genetic recombination* that introduces novel combinations of alleles in chromosomes.

6. On rare occasions, the *structure* of a chromosome or the parental *chromosome number* changes in mitosis or meiosis. Such chromosomal abnormalities can have severe phenotypic consequences.

AUTOSOMES AND SEX CHROMOSOMES

Some species show separation into sexes, and it all starts with genes on chromosomes. Say the species has a diploid chromosome number, so that body cells have pairs of homologous chromosomes. All but one pair are alike in length, shape, and gene sequence. A unique chromosome occurs in either females *or* males of many species, but not both.

For instance, a diploid cell in a human female has two X chromosomes (XX). A diploid cell in a human male has one X and one Y chromosome (XY). This is a common inheritance pattern among mammals, fruit flies, and many other animals. It is not the only one. In butterflies, moths, birds, and certain fishes, the two sex chromosomes are identical in males, and they are not identical in females.

Human X and Y chromosomes differ physically. The Y is a lot shorter, almost a remnant of the other in appearance. The two also differ in which genes they carry. They still synapse (zipper together briefly) in a small region. That bit of zippering allows them to interact as homologues during meiosis.

Human X and Y chromosomes fall into the more general category of **sex chromosomes**. When inherited in certain combinations, sex chromosomes determine a new individual's gender—whether a male or female will develop. All other chromosomes in a cell are the same in both sexes. We categorize them as **autosomes**.

Read Me First!

and watch the narrated animation on crossing over and genetic recombination

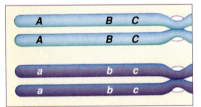

(a) A pair of duplicated homologous chromosomes (two sister chromatids each). In this example, nonidentical alleles occur at three gene loci (*A* with *a*, *B* with *b*, and *C* with *c*).

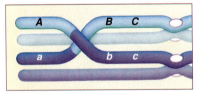

(b) In prophase I of meiosis, a crossover event occurs: Two nonsister chromatids exchange corresponding segments.

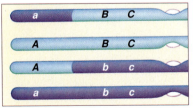

(c) What is the outcome of the crossover? Genetic recombination between nonsister chromatids (which are shown here, after meiosis, as two unduplicated, separate chromosomes).

Figure 11.2 Review of crossing over. As shown in Figure 9.6, this event occurs in prophase I of meiosis.

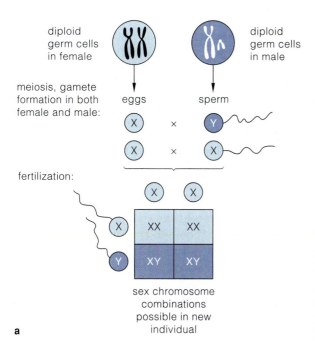

diploid germ cells in female

diploid germ cells in male

meiosis, gamete formation in both female and male:

eggs

sperm

X × Y

X × X

fertilization:

X X

sex chromosome combinations possible in new individual

a

Figure 11.3 (**a**) Punnett-square diagram showing the sex determination pattern in humans.

(**b**) Early on, a human embryo is neither male nor female. Then tiny ducts and other structures that can develop into male *or* female reproductive organs start forming. In an XX embryo, ovaries form *in the absence of the SRY gene on the Y chromosome*. In an XY embryo, the gene product triggers the formation of testes. A hormone secreted from testes calls for development of male traits. (**c**) External reproductive organs in human embryos.

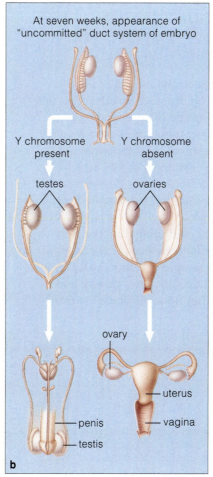

At seven weeks, appearance of "uncommitted" duct system of embryo

Y chromosome present

Y chromosome absent

testes

ovaries

ovary

uterus

penis

vagina

testis

b

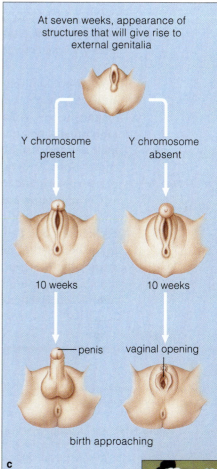

At seven weeks, appearance of structures that will give rise to external genitalia

Y chromosome present

Y chromosome absent

10 weeks

10 weeks

penis

vaginal opening

birth approaching

c

SEX DETERMINATION IN HUMANS

Each normal egg produced by a human female has one X chromosome. Half the sperm cells formed in a male carry an X chromosome, and half carry a Y. Say an X-bearing sperm fertilizes an X-bearing egg. The individual will develop into a female. If the sperm carries a Y chromosome, the individual will develop into a male (Figure 11.3*a*).

SRY, one of 330 genes in a human Y chromosome, is the master gene for male sex determination. Its expression in XY embryos triggers the formation of testes, the primary male reproductive organs (Figure 11.3*b*). Testes make testosterone, and this sex hormone governs the emergence of male sexual traits.

An XX embryo has no *SRY* gene, so primary female reproductive organs—ovaries—form instead. Ovaries make estrogens and other sex hormones that govern the development of female sexual traits.

The human X chromosome carries 2,062 genes. Like other chromosomes, it carries some genes associated with sexual traits, such as the distribution of body fat

and hair. But most of its genes deal with *nonsexual* traits, such as blood-clotting functions. Such genes can be expressed in males as well as in females. Males, remember, also inherit one X chromosome.

Diploid cells have pairs of genes, on pairs of homologous chromosomes. The alleles (alternative forms of a gene) at a given locus may be identical or nonidentical.

As a result of crossing over and other events, offspring inherit combinations of alleles not found on parental chromosomes.

Abnormal events at meiosis or mitosis can change the structure and number of chromosomes.

Autosomes are pairs of chromosomes that are the same in males and females of a species. One other pairing, the sex chromosomes, differs between males and females.

The SRY gene on the human Y chromosome dictates that a new individual will develop into a male. In the absence of the Y chromosome (and the gene), a female develops.

11.2 Karyotyping Made Easy

Karyotyping is a diagnostic tool that allows us to check images of the structure and number of chromosomes in an individual's somatic cells.

How do we know so much about an individual's autosomes and sex chromosomes? *Karyotyping* is one diagnostic tool. A **karyotype** is a preparation of an individual's metaphase chromosomes, sorted out by their defining visual features. Any abnormalities in chromosome structure or number can be detected by comparing a standard karyotype for the species.

Chromosomes are in their most condensed form and easiest to identify when a cell enters metaphase. Technicians don't count on finding dividing cells in the body—they culture cells and induce mitosis artificially. They put a sample of cells, usually from blood, into a solution that stimulates growth and mitotic cell division. They add colchicine to arrest the cell cycle at metaphase. Colchicine, recall, is a microtubule poison that blocks spindle formation.

The cell culture is centrifuged to isolate all the metaphase cells (Figure 11.4). A hypotonic solution makes the cells swell, by osmosis. The cells, along with their chromosomes, move apart. Then they are mounted on slides, fixed, and stained.

The chromosomes are viewed and photographed through a microscope. The photograph is cut, either with scissors or on a computer, and the individual chromosomes are lined up by their size and shape.

Spectral karyotyping uses a range of colored fluorescent dyes that bind to specific regions of chromosomes. Analysis of the resulting rainbow-hued karyotype often reveals crossovers and abnormalities that would not be otherwise visible. You will see an example of a multicolor spectral karyotype in Section 11.7.

Read Me First!

and watch the narrated animation on karyotype preparation

Figure 11.4 Karyotyping. With this type of diagnostic tool, an image of metaphase chromosomes is cut apart. Individual chromosomes are aligned by their centromeres and arranged according to size, shape, and length.

(**a**) A sample of cells from an individual is added to a medium that stimulates cell growth and mitotic cell division. The cell cycle is arrested at metaphase, with colchicine. (**b**) The culture is subjected to *centrifugation*, which works because cells have greater mass and density than the solution bathing them. A centrifuge's spinning force moves the cells farthest from the center of rotation, so they collect at the base of the centrifuge tubes.

(**c**) The culture medium is removed; a hypotonic solution is added. As the cells swell, the chromosomes move apart. (**d**) The cells are mounted on a microscope slide, fixed by air-drying, and stained. Chromosomes show up.

(**e**) A photograph of one cell's chromosomes is cut up and organized, as in the human karyotype in (**f**), which shows 22 pairs of autosomes and 1 pair of sex chromosomes—XX *or* XY. Scissors or computers do the cuts.

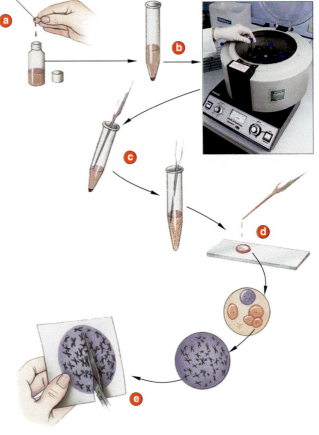

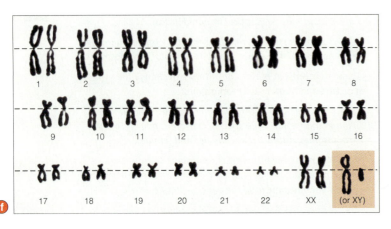

11.3 Impact of Crossing Over on Inheritance

*Crossing over between homologous chromosomes is
one of the main pattern-busting events in inheritance.*

We now know there are many genes on each type of
autosome and sex chromosome. All the genes on one
chromosome are called a **linkage group**. For instance,
the fruit fly (*Drosophila melanogaster*) has four linkage
groups, corresponding to its four pairs of homologous
chromosomes. Indian corn (*Zea mays*) has ten linkage
groups, corresponding to its ten pairs, and so on.

If linked genes stayed connected through meiosis,
then there would be no surprising mixes of parental
traits. You could expect parental phenotypes among,
say, F$_2$ offspring of dihybrid crosses to show up in a
predictable ratio. As early experiments with fruit flies
made clear, however, plenty of genes on the same
chromosomes do *not* stay together through meiosis.

In one experiment, mutant female flies that bred
true for white eyes and a yellow body were crossed
with wild-type males (red eyes and gray body). As
expected, 50 percent of the F$_1$ offspring had one or the
other parental phenotype. However, 129 of the 2,205
F$_2$ offspring were recombinants! They had white eyes
and a gray body, or red eyes and a yellow body.

Why? Some alleles tend to stay together more often
than others through meiosis. They are closer together
along the length of a chromosome and therefore less
vulnerable to a crossover. *The probability that crossing
over will disrupt the linkage between any two gene loci is
proportional to the distance between them.*

If, say, genes *A* and *B* are twice as far apart as
genes *C* and *D*, we can expect crossing over to disrupt
the linkage between *A* and *B* far more frequently:

Two genes are very closely linked when the distance
between them is small. Their combinations of alleles
nearly always end up in the same gamete. Linkage is
more vulnerable to crossing over when the distance
between two gene loci is greater (Figure 11.5). When
two loci are far apart, crossing over is so frequent that
the genes assort independently into gametes.

Human gene linkages were identified by tracking
phenotypes in families over generations. One thing is
clear from such studies: Crossovers are not rare. For
most eukaryotes, meiosis cannot even be completed
properly until at least one crossover occurs between
each pair of homologous chromosomes.

---

*All of the genes at different locations along the length of a
chromosome belong to the same linkage group. They do not
all assort independently at meiosis.*

*Crossing over between homologous chromosomes disrupts
gene linkages and results in nonparental combinations
of alleles in chromosomes.*

*The farther apart two genes are on a chromosome, the
greater will be the frequency of crossing over and genetic
recombination between them.*

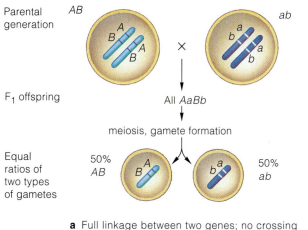

a Full linkage between two genes; no crossing
over. Half of the gametes have one parental
genotype, and half have the other. Genes that
are very close together along the length of a
chromosome typically stay together in gametes.

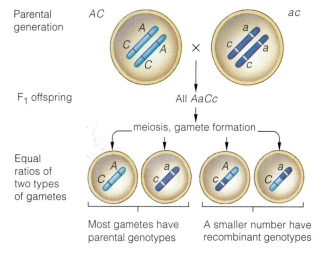

Most gametes have parental genotypes

A smaller number have recombinant genotypes

b Incomplete linkage; crossing over affected
the outcome. Genes that are far apart along
the length of a chromosome are more
vulnerable to crossing over.

Figure 11.5 Examples of outcomes of crossing over between two gene loci.

11.4 Human Genetic Analysis

Some organisms, including pea plants and fruit flies, are ideal for genetic analysis. They do not have a lot of chromosomes. They grow and reproduce fast in small spaces, under controlled conditions. It doesn't take long to track a trait through many generations. Humans, however, are another story.

Unlike fruit flies in the laboratory, we humans live under variable conditions in diverse environments, and we live as long as the geneticists who study our traits. Most of us select our own mates and reproduce if and when we want to. Most families are not large, which means there are not enough offspring available for researchers to make easy inferences.

Geneticists often gather information from several generations to increase the numbers for analysis. If a trait follows a simple Mendelian inheritance pattern, they can be confident about predicting the probability of its showing up again. The pattern also can be a clue to the past (Figure 11.6).

Such information is often displayed in **pedigrees**, or charts of genetic connections among individuals. Standardized methods, definitions, and symbols that represent different kinds of individuals are used to

Figure 11.6 An intriguing pattern of inheritance. Eight percent of the men in Central Asia carry nearly identical Y chromosomes, which implies descent from a shared ancestor. If so, then 16 million males living between northeastern China and Afghanistan—close to 1 of every 200 men alive today—belong to a lineage that started with the warrior and notorious womanizer Genghis Khan. In time, his offspring ruled an empire that stretched from China all the way to Vienna.

construct these charts. Figure 11.7 gives an example. Those who analyze pedigrees rely on their knowledge of probability and Mendelian inheritance patterns that may yield clues to a trait. As you will see, they have traced many genetic abnormalities and disorders to a dominant or recessive allele, even to its location on an autosome or a sex chromosome.

Bear in mind, a genetic *abnormality* is simply a rare or uncommon version of a trait, as when a person is born with six digits on each hand or foot instead of the usual five. Whether we view such a condition as disfiguring or merely interesting is subjective; there is nothing inherently life-threatening about it. A **genetic disorder**, however, is a heritable condition that sooner or later gives rise to mild to severe medical problems. A set of symptoms, or **syndrome**, characterizes each abnormality or disorder. Table 11.1 gives examples.

You might be thinking that a disease, too, has a set of symptoms that arises from an abnormal change in how the body functions. However, each **disease** is an illness that results from an infection, dietary problems, or environmental factors—*not* from a heritable mutation. It might be appropriate to call it a *genetic* disease only when such factors alter previously workable genes in a way that disrupts body functions.

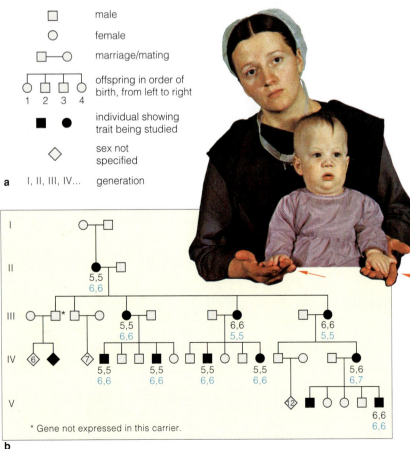

☐	male
○	female
☐—○	marriage/mating
○☐☐☐ (1 2 3 4)	offspring in order of birth, from left to right
■ ●	individual showing trait being studied
◇	sex not specified
I, II, III, IV...	generation

a

* Gene not expressed in this carrier.

b

Figure 11.7 (**a**) Some standardized symbols used in pedigrees. (**b**) A pedigree for *polydactyly*, characterized by extra fingers, toes, or both. *Black* numerals signify the number of fingers on each hand; *blue* numerals signify the number of toes on each foot. This condition recurs as one symptom of Ellis–van Creveld syndrome.

Table 11.1 Examples of Human Genetic Disorders and Genetic Abnormalities

Disorder or Abnormality	Main Symptoms	Disorder or Abnormality	Main Symptoms
Autosomal recessive inheritance		**X-linked recessive inheritance**	
Albinism	Absence of pigmentation	Androgen insensitivity syndrome	XY individual but having some female traits; sterility
Blue offspring	Bright blue skin coloration	Color blindness	Inability to distinguish among some or all colors
Cystic fibrosis	Excessive glandular secretions leading to tissue, organ damage	Fragile X syndrome	Mental impairment
Ellis–van Creveld syndrome	Extra fingers, toes, short limbs	Hemophilia	Impaired blood-clotting ability
Fanconi anemia	Physical abnormalities, bone marrow failure	Muscular dystrophies	Progressive loss of muscle function
Galactosemia	Brain, liver, eye damage	X-linked anhidrotic dysplasia	Mosaic skin (patches with or without sweat glands); other effects
Phenylketonuria (PKU)	Mental impairment		
Sickle-cell anemia	Adverse pleiotropic effects on organs throughout body	**Changes in chromosome number**	
		Down syndrome	Mental impairment; heart defects
Autosomal dominant inheritance		Turner syndrome	Sterility; abnormal ovaries, abnormal sexual traits
Achondroplasia	One form of dwarfism	Klinefelter syndrome	Sterility; mild mental impairment
Camptodactyly	Rigid, bent fingers	XXX syndrome	Minimal abnormalities
Familial hypercholesterolemia	High cholesterol levels in blood; eventually clogged arteries	XYY condition	Mild mental impairment or no effect
Huntington disease	Nervous system degenerates progressively, irreversibly	**Changes in chromosome structure**	
Marfan syndrome	Abnormal or no connective tissue	Chronic myelogenous leukemia (CML)	Overproduction of white blood cells in bone marrow; organ malfunctions
Polydactyly	Extra fingers, toes, or both	Cri-du-chat syndrome	Mental impairment; abnormally shaped larynx
Progeria	Drastic premature aging		
Neurofibromatosis	Tumors of nervous system, skin		

Alleles that give rise to severe genetic disorders are rare in populations, because they put their bearers at risk. Why don't they disappear? Rare mutations introduce new copies of the alleles into populations. Also, in heterozygotes, a normal allele is paired with a harmful one and may cover its functions, in which case the harmful allele can be transmitted to offspring.

With these qualifications in mind, we turn next to examples of chromosomal inheritance patterns in the human population. Figure 11.8 is an early introduction to one of these examples—an autosomal dominant disorder called Huntington disease.

Pedigree analysis often reveals simple Mendelian inheritance patterns. From such patterns, specialists infer the probability that offspring will inherit certain alleles.

A genetic abnormality is a rare or less common version of a heritable trait. A genetic disorder is a heritable condition that results in mild to severe medical problems.

Figure 11.8 Pedigree for *Huntington disease*, a progressive degeneration of the nervous system. Researcher Nancy Wexler and her team constructed this extended family tree for nearly 10,000 Venezuelans. Their analysis of unaffected and affected individuals revealed that a dominant allele on human chromosome 4 is the culprit. Wexler has a special interest in the disease; it runs in her family.

11.5 Examples of Human Inheritance Patterns

Some human phenotypes arise from a dominant or recessive allele on an autosome or X chromosome that is inherited in simple Mendelian patterns.

AUTOSOMAL DOMINANT INHERITANCE

Figure 11.9*a* shows a typical inheritance pattern for an autosomal dominant allele. If one of the parents is heterozygous and the other homozygous, any child of theirs has a 50 percent chance of being heterozygous. The trait usually appears in every generation because the allele is expressed even in heterozygotes.

Achondroplasia is a classic example. This autosomal dominant disorder affects approximately 1 in 10,000 people. The homozygous dominants often die before birth, but heterozygotes can still reproduce. Skeletal cartilage does not form properly in achondroplasiacs. Adults have abnormally short arms and legs relative to other body parts. They are about 4 feet, 4 inches tall, as in Figure 11.9*a*. The dominant allele often has no other phenotypic effects.

In *Huntington disease*, the nervous system slowly deteriorates, and involuntary muscle action becomes more frequent. Symptoms may not start until past age thirty; those affected die in their forties or fifties. Many unknowingly transmit the gene to children, before the onset of symptoms. The mutation causing the disorder changes a protein necessary for normal development of brain cells. It is an *expansion* mutation, which ends up as multiple repeats in the same DNA segment. The repeats disrupt gene function.

A few dominant alleles that cause severe problems persist in populations because expression of the allele may not interfere with reproduction, or affected people reproduce before the symptoms become severe. Rarely, spontaneous mutations reintroduce some of them.

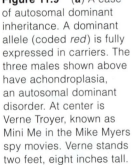

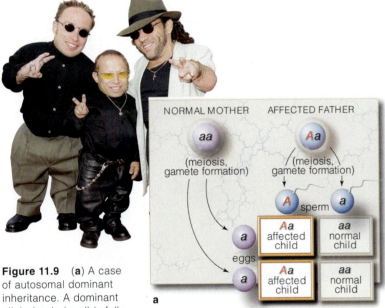

Figure 11.9 (**a**) A case of autosomal dominant inheritance. A dominant allele (coded *red*) is fully expressed in carriers. The three males shown above have achondroplasia, an autosomal dominant disorder. At center is Verne Troyer, known as Mini Me in the Mike Myers spy movies. Verne stands two feet, eight inches tall.

(**b**) An autosomal recessive pattern. In this case, both parents are heterozygous carriers of the recessive allele (coded *red*).

AUTOSOMAL RECESSIVE INHERITANCE

For some traits, inheritance patterns reveal clues that point to a recessive allele on an autosome. First, if both parents are heterozygous, any child of theirs will have a 50 percent chance of being heterozygous and a 25 percent chance of being homozygous recessive, as in Figure 11.9*b*. Second, if they are both homozygous recessive, any child of theirs will be, also.

About 1 in 100,000 newborns is homozygous for a recessive allele that causes *galactosemia*. They do not have working copies of one of the enzymes that digest lactose, so a reaction intermediate builds up to toxic levels. Normally, lactose is converted to glucose and galactose, then glucose–1–phosphate (which is broken down by glycolysis or converted to glycogen). The conversion is blocked in galactosemics (Figure 11.10).

High galactose levels can be detected in urine. The excess causes malnutrition, diarrhea, vomiting, and damage to the eyes, liver, and brain. When untreated, galactosemics typically die early. If they are quickly placed on a restricted diet excluding dairy products, they grow up symptom-free.

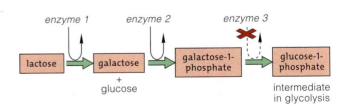

Figure 11.10 Blocked metabolic pathway in galactosemics.

Figure 11.11 One pattern for X-linked recessive inheritance. In this case, the mother carries a recessive allele on one of her X chromosomes (*red*).

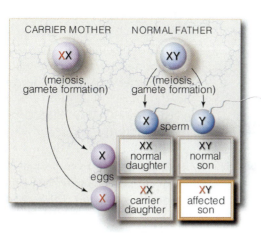

Figure 11.12 One of many standardized tests that can reveal color blindness. If you cannot see the red "29" inside this circle, then you may have some form of red–green color blindness.

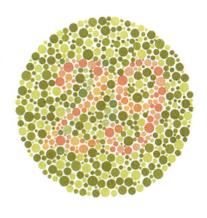

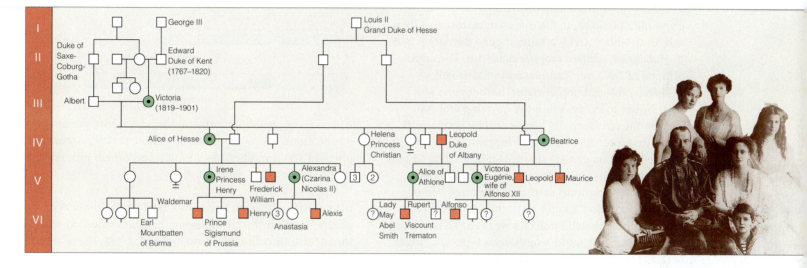

Figure 11.13 Partial pedigree for Queen Victoria's descendants, including carriers and affected males who inherited the X-linked allele for hemophilia A. At one time, the recessive allele was present in eighteen of Victoria's sixty-nine descendants, who sometimes intermarried. Of the Russian royal family members shown, the mother was a carrier; Crown Prince Alexis was hemophilic.

X-LINKED INHERITANCE

An X-linked gene is found only on the X chromosome. In X-linked genetic disorders, females are not affected as often as males, because a dominant allele on their other X chromosome can mask a recessive one (Figure 11.11). A son cannot inherit a X-linked allele from his father, but a daughter can. When she does, each of her sons has a 50 percent chance of inheriting it.

Color blindness is an inability to distinguish among some or all colors. It results from several common recessive disorders associated with X-linked genes. Mutant forms of the genes change the light-absorbing capacity of sensory receptors inside the eyes.

Normally, humans can detect differences among 150 colors. A person who is red–green color blind sees fewer than 25 colors; some or all of the receptors that respond to visible light of red and green wavelengths are weakened or absent. Others confuse red and green colors or see shades of gray instead of green. Tests can identify affected people (Figure 11.12).

The trait is more common in men, but heterozygous women also show symptoms. Can you explain why?

Hemophilia A, a blood-clotting disorder, is one case of X-linked recessive inheritance. Normally, a clotting mechanism quickly stops bleeding from minor injuries. Some clotting proteins are products of genes on the X chromosome. Bleeding is prolonged in males with one of these mutant X-linked genes. About 1 in 7,000 males is affected. In heterozygous females, clotting time is close to normal.

The frequency of hemophilia A was high in royal families of nineteenth-century Europe, in which close relatives often married (Figure 11.13).

Genetic analyses of family pedigrees have revealed simple Mendelian inheritance patterns for certain traits, as well as for many genetic disorders that arise from expression of alleles on an autosome or X chromosome.

11.6 Too Young, Too Old

FOCUS ON HEALTH

Sometimes textbook examples of the human condition seem a bit abstract, so take a moment to think about two boys who were too young to be old.

Imagine being ten years old with a mind trapped in a body that is getting a bit more shriveled, more frail—*old*—every day. You are barely tall enough to peer over the top of the kitchen counter; you weigh less than thirty-five pounds. Already you are bald and have a crinkled nose. Maybe you have a few more years to live. Would you, like Mickey Hayes and Fransie Geringer, still be able to laugh?

Of every 8 million newborn humans, one will grow old far too soon. On one of its autosomes, that rare individual carries a mutant gene that gives rise to *Hutchinson–Gilford progeria syndrome*. Through billions of DNA replications and mitotic cell divisions, information encoded in that gene was distributed to every cell in the growing embryo, then in the newborn. Its legacy will be accelerated aging and a terribly reduced life span.

The mutation grossly disrupts interactions among genes that bring about growth and development. Observable symptoms start to materialize before age two. Skin that should be plump and resilient starts to thin. Skeletal muscles weaken. Tissues in limb bones that should lengthen and grow stronger soften. Hair loss is pronounced; premature baldness is inevitable (Figure 11.14). There are no documented cases of progeria running in families, so we suspect it arises from spontaneous mutations. Probably the mutated gene is dominant over a normal allele on the homologous chromosome.

Most progeriacs expect to die in their early teens, from strokes or heart attacks. These final insults are brought on by a hardening of the wall of arteries, a condition typical of advanced age. When Mickey turned eighteen, he was the oldest living progeriac. Fransie was seventeen when he died.

Figure 11.14
Two boys who met at a gathering of progeriacs at Disneyland, California, when they were not yet ten years old.

11.7 Altered Chromosomes

Rarely, chromosome structure changes spontaneously or by exposure to chemicals or irradiation. Some changes can be detected. Many have severe or lethal outcomes.

THE MAIN CATEGORIES OF STRUCTURAL CHANGE

DUPLICATION Even normal chromosomes have gene sequences that have been repeated several to many thousands of times. These are **duplications**:

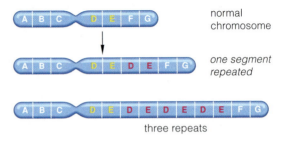

normal chromosome

one segment repeated

three repeats

Although no genetic information has been lost, certain duplications cause a variety of neural problems and physical abnormalities. As you will see, others proved useful over evolutionary time.

INVERSION With an **inversion**, part of the sequence of DNA within the chromosome becomes oriented in the reverse direction, with no molecular loss:

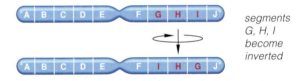

segments G, H, I become inverted

An inversion is not a problem if it does not disrupt a crucial gene region. But it mispairs during meiosis, so it can lead to chromosome deletions in gametes. Some people don't even know they have an inverted chromosome region until they have kids.

DELETION Whether it happens as a consequence of inversion or of an attack by an environmental agent, a **deletion** is the loss of a portion of a chromosome:

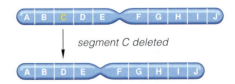

segment C deleted

In mammals, most deletions cause serious disorders, or are lethal. Why? Missing or incomplete genes disrupt the body's program of growth, development, and maintenance activities. For example, one deletion from human chromosome 5 results in an abnormally shaped larynx and mental impairment. When affected infants cry, they produce sounds like a cat's meow. Hence the name of the disorder, *cri-du-chat*, which means cat-cry in French. Sounds become normal later on. Figure 11.15 shows an affected boy.

TRANSLOCATION   In **translocation**, a broken part of a chromosome is attached to a different chromosome. Most translocations are reciprocal; both chromosomes exchange broken parts:

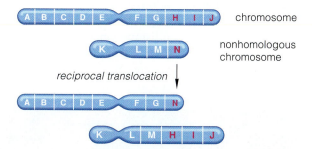

If the chromosome's genetic information does not get garbled, translocations may not pose a threat to the individual or its offspring. However, translocations can cause severe problems, including some sarcomas, lymphomas, myelomas, and leukemias. One notorious reciprocal translocation results in the Philadelphia chromosome, which is named after the city in which it was discovered. It is a killer (Figure 11.16).

DOES CHROMOSOME STRUCTURE EVOLVE?

Alterations in the structure of chromosomes generally are not good and tend to be selected against. Even so, over evolutionary time, many alterations with neutral effects became built into the DNA of all species.

We can expect that some of the duplications turned out to be adaptive. Perhaps some copies continued to specify an unaltered gene product even as others underwent modification. Think back on hemoglobin's polypeptide chains (Section 3.6). In humans and other primates, several globin genes are strikingly similar. They may have evolved as an outcome of duplications, mutations, and transpositions of the same gene. With small structural differences, the different globins have slightly different capacities to bind and then transport oxygen under a range of cellular conditions.

In addition, alterations in chromosome structure may have contributed to differences among closely related organisms, such as apes and humans. Consider this: Eighteen of the twenty-three pairs of human chromosomes are almost identical with chimpanzee

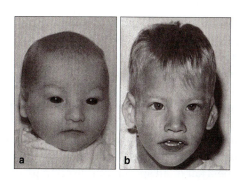

Figure 11.15 (a) Male infant who developed cri-du-chat syndrome. His ears are low on the side of the head relative to the eyes. (b) Same boy, four years later. By this age, affected humans stop making mewing sounds typical of the syndrome.

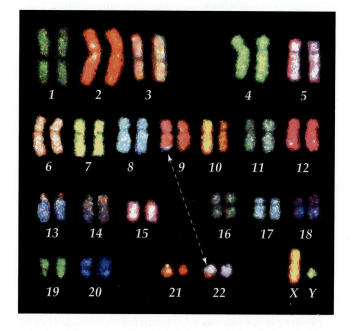

Figure 11.16 A reciprocal translocation, as revealed by spectral karyotyping. The Philadelphia chromosome is longer than its normal counterpart, human chromosome 9.

By chance, chromosomes 9 and 22 broke in a stem cell in bone marrow. Each broken part was reattached on the wrong one. At the broken end of chromosome 9, a gene with a role in cell division fused with the control region of a gene at chromosome 22's broken end. Overexpression of the mutant gene leads to uncontrolled divisions of white blood cells. A type of cancer, chronic myelogenous leukemia (CML), is the outcome.

and gorilla chromosomes. The other five chromosomes differ only at inverted and translocated regions.

On rare occasions, a segment of a chromosome may become duplicated, inverted, moved to a new location, or deleted.

Most chromosome changes are harmful or lethal. Others have been conserved over evolutionary time; they confer adaptive advantages or have had neutral effects.

11.8 Changes in the Chromosome Number

Occasionally, abnormal events occur before or during cell division, and gametes and new individuals end up with the wrong chromosome number. Consequences range from minor to lethal physical changes.

In **aneuploidy**, cells usually have one extra or one less chromosome. Autosomal aneuploidy is usually fatal for humans and is linked to most miscarriages. In **polyploidy**, cells have three or more of each type of chromosome. Half of all species of flowering plants, some insects, fishes, and other animals are polyploid.

Nearly all changes in chromosome number arise through **nondisjunction**, whereby one or more pairs of chromosomes don't separate as they should during mitosis or meiosis. Figure 11.17 shows an example.

The chromosome number also may change during fertilization. Suppose a normal gamete fuses by chance with an $n+1$ gamete, with one extra chromosome. The new individual will be trisomic ($2n+1$), with three of one type of chromosome and two of every other type. If an $n-1$ gamete fuses with a normal n gamete, the new individual will be monosomic ($2n-1$). Mitotic divisions perpetuate the mistake when the embryo is growing in size and developing.

AUTOSOMAL CHANGE AND DOWN SYNDROME

A few trisomics are born alive, but only trisomy 21 individuals reach adulthood. A newborn with three chromosomes 21 will develop *Down syndrome*. This autosomal disorder is the most frequent type of altered chromosome number in humans; it occurs once in every 800 to 1,000 births. It affects more than 350,000 people in the United States. Figure 11.17*b* shows a karyotype for a trisomic 21 female. About 95 percent of all cases arise through nondisjunction at meiosis.

Affected individuals have upward-slanting eyes, a fold of skin that starts at the inner corner of each eye, a deep crease across each palm and foot sole, one (not two) horizontal furrows on their fifth fingers, and somewhat flattened facial features. Not all of these symptoms develop in every individual.

That said, trisomic 21 individuals have moderate to severe mental impairment and heart defects. Their skeleton develops abnormally, so older children have shorter body parts, loose joints, and misaligned hip, finger, and toe bones. Muscles and reflexes are weak. Speech and other motor skills develop slowly. With medical care, they live fifty-five years, on average.

The incidence of nondisjunction rises as mothers become older (Figure 11.18). It may originate with the father, but less often. Trisomy 21 is one of hundreds

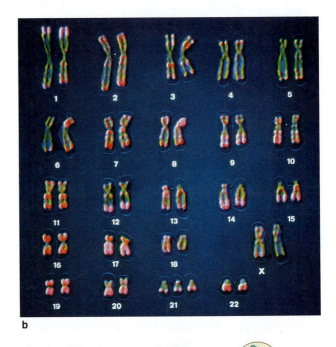

b

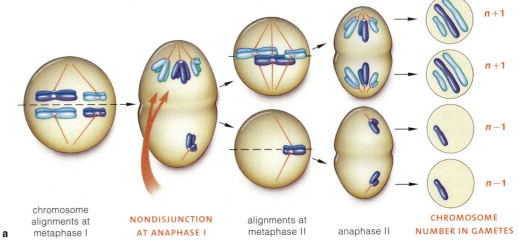

Figure 11.17 (**a**) One example of how nondisjunction arises. Of the two pairs of homologous chromosomes shown here, one fails to separate during anaphase I of meiosis. The chromosome number is altered in the gametes that form after meiosis.

(**b**) An actual case of nondisjunction. This karyotype reveals the trisomic 21 condition of a human female.

a chromosome alignments at metaphase I

NONDISJUNCTION AT ANAPHASE I

alignments at metaphase II   anaphase II

CHROMOSOME NUMBER IN GAMETES

$n+1$
$n+1$
$n-1$
$n-1$

of conditions that can be detected through prenatal diagnosis (Section 11.9). With early special training and medical intervention, individuals still can take part in normal activities. As a group, they tend to be cheerful and sociable.

CHANGES IN THE SEX CHROMOSOME NUMBER

Nondisjunction also causes most of the alterations in the number of X and Y chromosomes. The frequency of such changes is 1 in 400 live births. Most often they lead to difficulties in learning and motor skills, such as speech, although problems can be so subtle that the underlying cause is not even diagnosed.

FEMALE SEX CHROMOSOME ABNORMALITIES *Turner syndrome* individuals have an X chromosome and no corresponding X or Y chromosome (XO). About 1 in 2,500 to 10,000 newborn girls are XO. Nondisjunction originating with the father accounts for 75 percent of the cases. Yet cases are few, compared to other sex chromosome abnormalities. At least 98 percent of XO embryos may spontaneously abort early in pregnancy.

Despite the near lethality, XO survivors are not as disadvantaged as other aneuploids. On average, they are only four feet, eight inches high, but they are well proportioned (Figure 11.19). Most can't make enough sex hormones; they don't have functional ovaries. This affects development of secondary sexual traits, such as breast enlargement. A few eggs form in ovaries but are destroyed by the time these girls are two years old.

Another example: A few females inherit three, four, or five X chromosomes. The *XXX syndrome* occurs at a frequency of about 1 in 1,000 live births. Adults are fertile. Except for slight learning difficulties, most fall within the normal range of social behavior.

MALE SEX CHROMOSOME ABNORMALITIES About one of every 500 to 2,000 males inherits one Y and two or more X chromosomes, mainly through nondisjunction. Most have an XXY mosaic genotype. About 67 percent of those affected inherited the extra chromosome from their mother.

The resulting *Klinefelter syndrome* develops during puberty. XXY males tend to be overweight and tall. The testes and the prostate gland usually are smaller than average. Many XXY males are within the normal range of intelligence, although some have short-term memory loss and learning disabilities. They make less testosterone and more estrogen than normal males, with feminizing effects. Sperm counts are low. Hair is sparse, the voice is pitched high, and the breasts are a

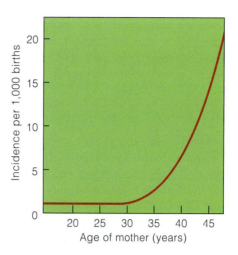

Figure 11.18 Relationship between the frequency of Down syndrome and mother's age at childbirth. The data are from a study of 1,119 affected children. The risk of having a trisomic 21 baby rises with the mother's age. This may seem odd, because about 80 percent of trisomic 21 individuals are born to women not yet 35 years old. However, these women are in the age categories with the highest fertility rates, and they simply have more babies.

bit enlarged. Testosterone injections starting at puberty can reverse the feminized traits.

About 1 in 500 to 1,000 males has one X and two Y chromosomes, an *XYY condition.* They tend to be taller than average, with mild mental impairment, but most are phenotypically normal. XYY males were once thought to be genetically predisposed to a life of crime. This misguided view was based on a sampling error—too few cases among narrowly selected groups, such as prison inmates. The same researchers gathered the karyotypes *and* personal histories. Fanning the stereotype was a report that a mass murderer of young nurses was XYY. He wasn't.

In 1976 a Danish geneticist reported on a study of 4,139 tall males, all twenty-six years old, who had reported to their draft board. Besides giving results of physical examinations and intelligence tests, those draft records held clues to their social and economic status, education, and criminal convictions, if any. Twelve of the males were XYY, which meant there were more than 4,000 males in the control group. The only finding was that mentally impaired, tall males who engage in criminal activity are just more likely to get caught—irrespective of karyotype.

The majority of XXY, XXX, and XYY children may not even be properly diagnosed. Some are dismissed unfairly as being underachievers.

Figure 11.19 One young girl with Turner syndrome.

Nondisjunction in germ cells, gametes, or early embryonic cells changes the number of autosomes or the number of sex chromosomes. The change affects development and the resulting phenotypes.

Nondisjunction at meiosis causes most sex chromosome abnormalities, which typically lead to subtle difficulties with learning, and speech and other motor skills.

11.9 Prospects in Human Genetics

With the arrival of their newborn, parents typically ask, "Is our baby normal?" Quite naturally, they want their baby to be free of genetic disorders, and most babies are. What are the options when they are not?

BIOETHICAL QUESTIONS

Humans do not view diseases and genetic disorders the same way. We attack diseases with antibiotics, surgery, and other tactics. But how do we attack a heritable "enemy" that can be transmitted to our own offspring?

Should we institute regional, national, or global programs to identify people who may carry harmful alleles? Should they be told that they are "defective" and might bestow some disorder on their children? Who decides which alleles are bad? Should society bear the cost of treating all genetic disorders of all individuals, before and after birth? If so, should society also have a say in whether an embryo that bears harmful alleles will be born at all, or aborted? An **abortion** is the expulsion of a pre-term embryo or fetus from the uterus.

As you most likely have learned by now, such questions are the tip of an ethical iceberg.

SOME OF THE OPTIONS

GENETIC SCREENING Through large-scale screening programs, affected individuals or carriers of some harmful allele can be detected early enough to start preventive measures before symptoms develop. For instance, most hospitals in the United States routinely screen newborns for PKU, described next, so we now see fewer individuals with symptoms of the disorder.

PHENOTYPIC TREATMENTS The symptoms of a number of genetic disorders can be minimized or alleviated by surgery, drugs, hormone replacement therapy, or in some cases by controlling diet.

For instance, dietary control works for individuals affected by *phenylketonuria*, or PKU. In this case, a homozygous recessive mutation impairs an enzyme that converts the amino acid phenylalanine to tyrosine. Phenylalanine builds up and is diverted into other pathways. Compounds that impair brain function form as a result. Affected people can lead relatively normal lives by restricting phenylalanine intake. For example, they can avoid soft drinks and other food products sweetened with aspartame, a compound that contains phenylalanine.

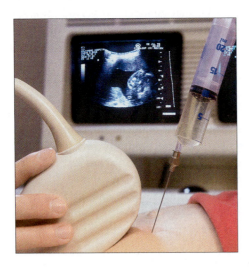

Figure 11.20 Amniocentesis, a prenatal diagnostic tool.

A pregnant woman's doctor holds an ultrasound emitter against her abdomen while drawing a sample of amniotic fluid into a syringe. He monitors the path of the needle with an ultrasound screen, in the background. Then he directs the needle into the amniotic sac that holds the developing fetus and withdraws twenty milliliters or so of amniotic fluid. The fluid contains fetal cells and wastes that can be analyzed for genetic disorders.

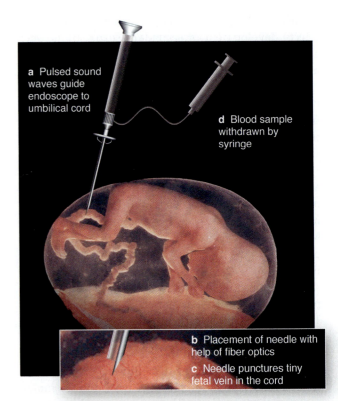

a Pulsed sound waves guide endoscope to umbilical cord

d Blood sample withdrawn by syringe

b Placement of needle with help of fiber optics

c Needle punctures tiny fetal vein in the cord

Figure 11.21 Fetoscopy for prenatal diagnosis.

PRENATAL DIAGNOSIS   Methods of *prenatal diagnosis* are used to determine the sex of embryos or fetuses and to screen for more than 100 genetic abnormalities. *Prenatal* means before birth. *Embryo* is a term that applies until eight weeks after fertilization, after which the term *fetus* is appropriate.

Suppose a forty-five-year-old woman is pregnant and worries about Down syndrome. Between 8 and 12 weeks after conception, such women often request *amniocentesis* (Figure 11.20). A tiny sample of fluid inside the amnion, a membranous sac enclosing the fetus, is withdrawn. Some cells shed by the fetus are suspended in the sample. The cells are analyzed for many genetic disorders, such as Down syndrome, sickle-cell anemia, and cystic fibrosis.

Chorionic villi sampling (CVS) is another procedure. A clinician withdraws a few cells from the chorion, a membranous sac that encloses the amnion and gives rise to the placenta. Unlike amniocentesis, CVS can yield results as early as eight weeks into pregnancy.

A developing fetus can be seen with an endoscope, a fiber-optic device. During *fetoscopy*, pulsed sound waves are used to scan the uterus and locate parts of the fetus, umbilical cord, or placenta (Figure 11.21). A sample of fetal blood can be drawn, so fetoscopy also is useful to diagnose blood cell disorders such as sickle-cell anemia and hemophilia.

There are risks to a fetus associated with all three procedures, including punctures or infection. Also, if the amnion does not reseal itself fast, too much fluid can leak out and endanger the fetus. Amniocentesis raises the risk of miscarriage by 1 to 2 percent. With CVS, placental development may be compromised, and 0.3 percent of newborns will have missing or underdeveloped fingers and toes. Fetoscopy raises the risk of a miscarriage by 2 to 10 percent.

GENETIC COUNSELING   Parents-to-be can seek *genetic counseling* to compare risks of diagnostic procedures against the risk that their child will be affected by a severe genetic disorder. But they also should be told about the small overall risk of 3 percent that *any* child might have some kind of birth disorder. And they should also consider whether the risk becomes greater with increased age of the potential mother or father.

Suppose a first child or close relative has a severe disorder. Genetic counseling may involve diagnosis of parental genotypes, pedigrees, and genetic testing for known disorders. Using this information, geneticists can predict the risk for disorders in future children. Counselors should remind prospective parents that the same risk usually applies to each pregnancy.

Figure 11.22   Eight-cell and multicelled stages of human development.

REGARDING ABORTION   What happens after prenatal diagnosis reveals a serious problem? Do prospective parents opt for an induced abortion? We can only say here that they must weigh awareness of the severity of the disorder against their ethical and religious beliefs. Worse, they must play out their personal tragedy on a larger stage dominated by a nationwide battle between highly vocal "pro-life" and "pro-choice" factions. We return to this topic in Section 38.12.

PREIMPLANTATION DIAGNOSIS   This procedure relies on **in vitro fertilization**. Sperm and eggs taken from prospective parents are mixed in a sterile culture medium. One or more eggs may be fertilized. If so, within forty-eight hours, mitotic cell divisions may convert it into a ball of eight cells (Figure 11.22 and Section 38.12). According to one view, the tiny, free-floating ball is a *pre*-pregnancy stage. Like unfertilized eggs discarded monthly from a woman, it has not attached to the uterus. All of its cells have the same genes; all are not yet committed to being specialized for any organ. Doctors take one of the undifferentiated cells and analyze its genes. If it has no detectable genetic defects, the ball is inserted into the uterus.

Some couples at risk of passing on cystic fibrosis, muscular dystrophy, or other genetic disorders have opted for the procedure. Many of the resulting *"test-tube" babies* have been born in good health.

Summary

Section 11.1 Human somatic cells are diploid ($2n$), with twenty-three pairs of homologous chromosomes. One is a pairing of sex chromosomes—XX in females, XY in males. All other chromosomes are autosomes, which are the same in both sexes. Crossing over in mitosis produces new combinations of alleles.

Section 11.2 Karyotyping is a diagnostic tool that makes a preparation of an individual's metaphase chromosomes and arranges them by their defining features, such as centromere location, length, and shape.

Section 11.3 The theory of independent assortment does not explain all gene combinations, because pairs of homologous chromosomes swap segments by crossing over during prophase I of meiosis. This frequent and expected event invites genetic recombination, or new combinations of alleles in chromosomes. The farther apart two genes are on the same chromosome, the more likely they will undergo crossing over.

Section 11.4 Geneticists often use pedigrees, or charts of genetic connections in a lineage over time, to estimate probabilities that offspring will inherit a given trait.

Sections 11.5, 11.6 Dominant or recessive alleles on either an autosome or X chromosome can be tracked when they are inherited in simple Mendelian patterns.

Section 11.7 On rare occasions, a chromosome's structure changes. A segment is deleted, inverted, moved to a new location (translocated), or duplicated. Most alterations are harmful or lethal. However, many have accumulated in the chromosomes of all species over evolutionary time. Either they had neutral effects or they later proved to be useful.

Section 11.8 The parental chromosome number can change, as by nondisjunction during meiosis. Aneuploids have one extra or one less chromosome; most autosomal aneuploids die before birth. About half of all flowering plants and some insects, fishes, and other animals are polyploid (three or more of each type of chromosome). More often, changes in number cause genetic disorders.

Section 11.9 Phenotypic treatments, genetic screening, genetic counseling, prenatal diagnosis, and preimplantation diagnosis are some options available for potential parents at risk of having children who will develop a genetic disorder.

Self-Quiz   *Answers in Appendix III*

1. The probability of a crossover occurring between two genes on the same chromosome is _____ .
   a. unrelated to the distance between them
   b. increased if they are close together
   c. increased if they are far apart

2. Genes on the same chromosome are _____ .
   a. linked   c. homologous   e. all of the
   b. identical alleles   d. autosomes   above

3. Chromosome structure can be altered by a _____ .
   a. deletion   c. inversion   e. all of the
   b. duplication   d. translocation   above

4. A recognized set of symptoms that characterize a specific disorder is a _____ .
   a. syndrome   b. disease   c. pedigree

5. Most genes for human traits are located on _____ .
   a. the X chromosome   c. autosomes
   b. the Y chromosome   d. dominant chromosomes

6. Nondisjunction at meiosis can result in _____ .
   a. karyotyping   c. duplications
   b. crossing over   d. aneuploidy

7. Turner syndrome (XO) is an example of _____ .
   a. dominance   c. aneuploidy
   b. polyploidy   d. gene linkage

8. Match the chromosome terms appropriately.
   ____ crossing over   a. number and defining
   ____ deletion   features of an individual's
   ____ nondisjunction   metaphase chromosomes
   ____ translocation   b. segment of a chromosome
   ____ karyotype   moves to a nonhomologous
   ____ linkage group   chromosome
                             c. disrupts gene linkages
                             d. one outcome: gametes with
                                wrong chromosome number
                             e. a chromosome segment lost
                             f. all genes on a chromosome

Genetics Problems   *Answers in Appendix IV*

1. Human females are XX and males are XY.
   a. Does a male inherit the X from his mother or father?
   b. With respect to X-linked alleles, how many different types of gametes can a male produce?
   c. If a female is homozygous for an X-linked allele, how many types of gametes can she produce with respect to that allele?
   d. If a female is heterozygous for an X-linked allele, how many types of gametes might she produce with respect to that allele?

2. Marfan syndrome follows a pattern of autosomal dominant inheritance. What is the chance that any child will inherit the dominant allele if one parent does not carry the allele and the other is heterozygous for it?

3. Somatic cells of individuals with Down syndrome usually have an extra chromosome 21; they contain forty-seven chromosomes.
   a. At which stages of meiosis I and II could a mistake alter the chromosome number?
   b. A few individuals have forty-six chromosomes, including two normal-appearing chromosomes 21 and a longer-than-normal chromosome 14. Speculate on how this chromosome abnormality may have arisen.

4. Much of what we know about human genetics comes from studies of experimental organisms. For example, the embryologist Thomas Morgan discovered a genetic basis for a relationship between sex determination and some nonsexual traits in *Drosophila melanogaster*. This fruit fly

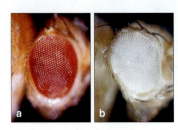

Figure 11.23 The common fruit fly *Drosophila melanogaster*, with (**a**) wild-type red eyes and (**b**) white eyes. (**c**) One experimental search for a *Drosophila* sex-linked gene.

recessive male

homozygous dominant female

all red-eyed F_1 offspring

gametes

gametes

White-eyed males show up in F_2 generation.

can live in small bottles on agar, cornmeal, molasses, and yeast. A female lays hundreds of eggs in a few days and offspring reproduce in less than two weeks. In a single year, Morgan was able to follow traits through nearly thirty generations of thousands of flies.

At first all the flies were wild-type for eye color; they had red eyes. Then mutation gave rise to a recessive allele for eye color, and a white-eyed male turned up. Morgan established true-breeding strains of white-eyed males and females for *reciprocal crosses*. In the first of such paired crosses, one parent displays the trait of interest. In the second cross, the other parent displays it.

Morgan let white-eyed males mate with homozygous red-eyed females. All F_1 offspring had red eyes, and some F_2 males had white eyes (Figure 11.23). Then Morgan mated true-breeding red-eyed males with white-eyed females. Half the F_1 offspring were red-eyed females, and half were white-eyed males. Also, of the F_2 offspring, 1/4 were red-eyed females, 1/4 white-eyed females, 1/4 red-eyed males, and 1/4 white-eyed males.

Test results pointed to a relationship between an eye-color gene and sex determination. Was the locus on a sex chromosome? Which one? Before answering, think about male and female sex chromosome pairings, and how a dominant allele can mask a recessive one.

5. Does the phenotype indicated by red circles and squares in this pedigree show a Mendelian inheritance pattern that's autosomal dominant, autosomal recessive, or X-linked?

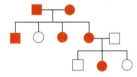

6. One of the *muscular dystrophies*, a category of genetic disorders, is due to a recessive X-linked allele. Usually, symptoms start in childhood. Gradual, progressive loss of muscle function leads to death, usually by age twenty or so. Unlike color blindness, the disorder is nearly always restricted to males. Suggest why.

7. In the human population, mutation of two genes on the X chromosome causes two types of X-linked hemophilia (A and B). In a few cases, a woman is heterozygous for both mutant alleles (one on each of the X chromosomes). All of her sons should have either hemophilia A or B.

However, on very rare occasions, one of these women gives birth to a son who does not have hemophilia, and his one X chromosome does not have either mutant allele. Explain how such an X chromosome could arise.

IMPACTS, ISSUES *Goodbye, Dolly*

In 1997, geneticist Ian Wilmut made headlines when he coaxed part of a specialized cell from an adult sheep into becoming part of the first cell of an embryo. His team had been removing the nucleus from unfertilized eggs and slipping a nucleus from a specialized cell of an adult animal into them. Of hundreds of modified eggs, one developed into a whole animal.

The cloned lamb, named Dolly, grew up (Figure 12.1). In time she gave birth to six lambs of her own. Since then, researchers all over the world have cloned other kinds of mammals, including mice, cows, pigs, rabbits, a mule, and cats.

Wilmut and Dolly were back in the limelight in early 2002 with bad news. By age five, Dolly had become fat and arthritic. Sheep usually don't get old-age symptoms until they are about ten years old. In 2003, Dolly developed a progressive lung infection and was put to sleep.

Did Dolly develop health problems simply because she was a clone? Earlier studies of her telomeres had raised suspicions. Telomeres are short segments that cap the ends of chromosomes and stabilize them. They get shorter and shorter as an animal ages. When Dolly was only two years old, her telomeres were as short as those of a six-year-old sheep—the exact age of the animal from which she had been cloned.

Cloning mammals is difficult. Not many nuclear transfers are successful. Most individuals that develop from modified eggs die before birth or shortly afterward.

Figure 12.1 (a) Where the molecular revolution started—James Watson and Francis Crick posing in 1953 by their newly unveiled structural model of DNA, the molecule of inheritance in all living cells. (b) The now-deceased Dolly. She helped awaken society to the implications of where the molecular revolution is taking us.

the big picture

DNA's Function

Early experiments with viruses and bacteria revealed that DNA is the hereditary material in living things. Biochemical analysis helped show that DNA has two strands, helically coiled in a precise pattern.

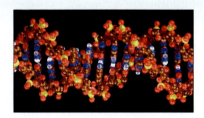

DNA's Structure

In all living things, the structure of DNA arises from base pairing between adenosine and thymine, guanine and cytosine. The sequence of bases along its two strands is unique for each species. It is the foundation for life's diversity.

It took almost seven hundred attempts to get one live clone of a guar, a wild ox on the endangered species list. Less than two days after his foster mother gave birth, he died from complications following an infection.

Surviving clones typically have health problems. Like Dolly, many become unusually overweight as they age. Other clones are exceptionally large from birth or have some enlarged organs. Cloned mice develop lung and liver problems, and almost all die prematurely. Cloned pigs have heart problems, they limp, and one never did develop a tail or, worse still, an anus.

Physically moving a nucleus from one cell to another is just part of the challenge. Most genes in an adult cell are inactive. They have to be reprogrammed or switched on in controlled ways in an unfertilized egg. So far, not all genes in clones are being properly activated.

Some people want to put a stop to cloning complex animals, saying the risk of bringing defective ones into the world troubles them deeply. Others want research to continue because the potential benefits are enormous. However, nearly all agree that cloning humans would be an outrage.

Think about these issues as you read through the rest of the chapters in this unit of the book. They deal with how cells replicate and repair their DNA, how genes are expressed, and what happens when things go wrong. They also invite you to reflect on how researchers are programming these molecular events in previously unimaginable ways.

 How Would You Vote?

Animal cloning experiments often produce abnormal animals, but cloning research may also result in new drugs and organ replacements for human patients. Should animal cloning be banned? See the Media Menu for details, then vote online.

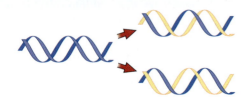

How DNA Is Replicated   During the DNA replication process, the two strands of a double-stranded DNA molecule unwind from each other. Each serves as the template for assembly of a new strand with a complementary base sequence.

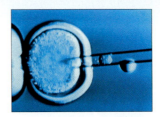

Reprogramming DNA   With recent advances in nuclear transfer techniques, researchers have been able to reprogram the DNA of specialized cells to make exact genetic replicas, or clones, of adult mammals.

12.1 The Hunt for Fame, Fortune, and DNA

With this chapter, we turn to investigations that led to our understanding of DNA. The chapter is more than a march through details of DNA's structure and function. It also reveals how ideas are generated in science.

EARLY AND PUZZLING CLUES

In the 1800s, Johann Miescher was collecting cells from the pus of open wounds and sperm cells from a fish. Such cells have little cytoplasm, which makes it easier to isolate their nuclear material. Miescher, a physician, wanted to identify the composition of the nucleus. In time, he isolated an acidic compound that had a bit of phosphorus. He had discovered what became known as **deoxyribonucleic acid**, or **DNA**.

Now fast-forward to 1928. An army medical officer, Frederick Griffith, wanted to develop a vaccine against the bacterium *Streptococcus pneumoniae*, a major cause of pneumonia. He did not succeed, but he isolated and cultured two strains that unexpectedly shed light on mechanisms of heredity. Colonies of one strain had a rough surface appearance; colonies of the other strain appeared smooth. Griffith designated the strains *R* and *S*. He then used them in a series of four experiments, as shown in Figure 12.2.

First, he injected mice with live *R* cells. The mice did not develop pneumonia. *The R strain was harmless.*

Second, he injected other mice with live *S* cells. The mice died. Blood samples from them teemed with live *S* cells. *The S strain was pathogenic; it caused the disease.*

Third, he killed *S* cells by exposing them to high temperature. *Mice injected with dead S cells did not die.*

Fourth, he mixed live *R* cells with heat-killed *S* cells. He injected them into mice. The mice died—*and blood samples drawn from them teemed with live S cells!*

What went on in the fourth experiment? Maybe heat-killed *S* cells in the mix weren't really dead. But if that were so, then the mice injected with just the heat-killed *S* cells in experiment 3 would have died. Or maybe the harmless *R* cells had mutated into a killer strain. But if that were so, then mice injected with just *R* cells in experiment 1 would have died.

The simplest explanation was this: *Heat had killed the S cells but did not destroy their hereditary material— including the part that specified "how to cause infection."* Somehow, that material had been transferred from the dead *S* cells into living *R* cells, which put it to use.

After later tests, it was clear that the transformation was permanently heritable. Even after a few hundred generations, *S* cell descendants were still infectious.

What was the hereditary material that caused the transformation? Scientists started looking in earnest, but most were thinking "proteins." Because heritable traits are diverse, they assumed the molecules of inheritance had to be structurally diverse, too. Proteins, they said, could be built from unlimited mixes of twenty kinds of amino acids. Other molecules just seemed too simple.

But Griffith's results intrigued Oswald Avery, who went on to transform the harmless bacterial cells with extracts of killed pathogens. When he added protein-digesting enzymes to the extracts, bacterial cells were still transformed. However, when he added an enzyme that digests DNA but not protein to the extracts, the cells were *not* transformed. DNA was looking good.

CONFIRMATION OF DNA FUNCTION

By the 1950s, molecular detectives were using viruses for experiments. **Bacteriophages**, which infect certain bacteria, were the viruses of choice. These infectious particles contain information on how to build new virus particles. At some point after they infect a host cell, viral enzymes take over its metabolic machinery. And then the machinery starts synthesizing substances necessary to make more virus particles.

As researchers knew, some bacteriophages consist only of DNA and a protein coat. Also, micrographs revealed that the coat remains on the *outer surface* of infected cells. Were the viruses injecting only hereditary material into cells? If so, was the material protein, DNA, or both? Figure 12.3 outlines just two of many experiments that pointed to DNA.

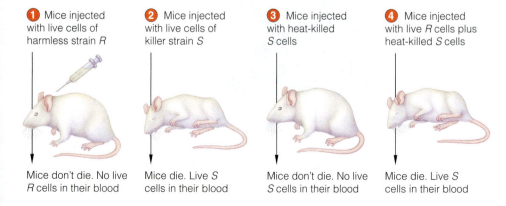

1 Mice injected with live cells of harmless strain *R*

Mice don't die. No live *R* cells in their blood

2 Mice injected with live cells of killer strain *S*

Mice die. Live *S* cells in their blood

3 Mice injected with heat-killed *S* cells

Mice don't die. No live *S* cells in their blood

4 Mice injected with live *R* cells plus heat-killed *S* cells

Mice die. Live *S* cells in their blood

Figure 12.2 Summary of results of Fred Griffith's experiments with *Streptococcus pneumoniae* and laboratory mice.

Read Me First!
and watch the narrated animation on
the Hershey–Chase experiments

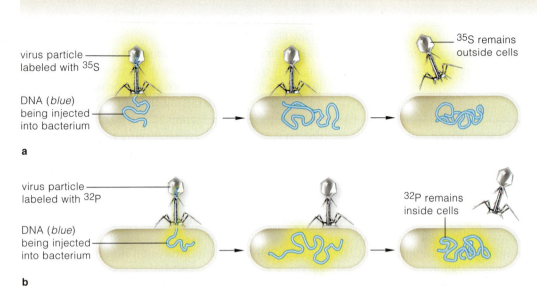

virus particle labeled with ^{35}S

DNA (*blue*) being injected into bacterium

^{35}S remains outside cells

a

virus particle labeled with ^{32}P

DNA (*blue*) being injected into bacterium

^{32}P remains inside cells

b

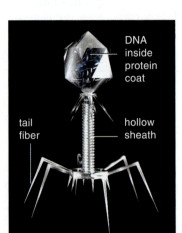

DNA inside protein coat

tail fiber

hollow sheath

micrograph of virus particles injecting DNA into an *E. coli* cell

Figure 12.3 Example of the landmark experiments that tested whether genetic material resides in bacteriophage DNA, proteins, or both. As Alfred Hershey and Martha Chase knew, sulfur (S) but not phosphorus (P) is present in proteins, and phosphorus but not sulfur is present in DNA.

(**a**) In one experiment, bacterial cells were grown on a culture medium with a tracer, the radioisotope ^{35}S. The cells used the ^{35}S when they built proteins. Bacteriophages infected the labeled cells, which started making viral proteins. So the proteins, and new virus particles, became labeled with the ^{35}S. The labeled virus particles infected a new batch of unlabeled cells. The mixture was whirred in a kitchen blender. Whirring dislodged the viral protein coats from infected cells. Chemical analysis revealed the presence of labeled protein in the solution but only traces of it inside the cells.

(**b**) In another experiment, bacteriophages infected cells that had taken up the radioisotope ^{32}P. The infected cells used the ^{32}P when they built viral DNA. The DNA became labeled, as did new virus particles. Later, the labeled viruses infected bacteria in solution, then were dislodged from them. Most of the labeled viral DNA stayed in the cells—evidence that DNA is the genetic material of this virus.

Then Linus Pauling did something no one had done before. With his training in biochemistry, a talent for model building, and a dose of intuition, he deduced the structure of a protein—collagen. His discovery was electrifying. If someone could pry open the secrets of proteins, then why not DNA? And if DNA's structural details were deduced, wouldn't they hold clues to how it functions? *Someone could go down in history as having discovered the secret of life!*

ENTER WATSON AND CRICK

Having a shot at fame and fortune quickens the pulse of men and women in any profession, and scientists are no exception. However, science is a community effort. Individuals share not only what they find but also what they do not understand. Even if an experiment does not yield an expected result, it may turn up something that others can use or raise questions others can answer.

And so scientists all over the world started sifting through all the clues. Among them were James Watson,

a postdoctoral student from Indiana University, and Francis Crick, a researcher at Cambridge University. They spent hours arguing over everything they read about DNA's size, shape, and bonding requirements. They fiddled with cardboard cutouts and badgered chemists to help them identify possible bonds they may have overlooked. They built models from thin bits of metal connected with suitably angled "bonds" of wire.

In 1953, they built a model that fit all the pertinent biochemical rules and insights they had gleaned from other sources. They had discovered DNA's structure. As you will see, the structure's breathtaking simplicity helped Crick answer another enormous riddle—*how life can show unity at the molecular level and still give rise to spectacular diversity at the level of whole organisms.*

DNA functions as the cell's treasurehouse of inheritance. Its molecular structure encodes the information required to reproduce parental traits in offspring.

12.2 The Discovery of DNA's Structure

Long before the bacteriophage studies were under way, biochemists knew that DNA contains only four kinds of nucleotides that are the building blocks of nucleic acids. But how were the nucleotides arranged in DNA?

DNA'S BUILDING BLOCKS

Each **nucleotide** consists of a five-carbon sugar (which, in DNA, is deoxyribose), a phosphate group, and one of the following nitrogen-containing bases:

adenine	guanine	thymine	cytosine
A	G	T	C

In all four types of nucleotides, the component parts are organized the same way (Figure 12.4). But T and C are pyrimidines, with a carbon backbone arranged as a single ring. A and G are purines, which are larger, bulkier molecules; they have two carbon rings.

By 1949, the biochemist Erwin Chargaff had shared with the scientific community two crucial insights into the composition of DNA. First, the amount of adenine relative to guanine differs from one species to the next. Second, the amounts of thymine and adenine in a DNA molecule are exactly the same, and so are the amounts of cytosine and guanine. We may show this as A = T and G = C. The symmetrical proportions of the four kinds of nucleotides had to mean something. Was this a tantalizing clue to how the nucleotides are arranged in the DNA molecule?

The first convincing evidence of that arrangement emerged from Maurice Wilkins's research laboratory in England. Rosalind Franklin, one of his colleagues, made good **x-ray diffraction images** of DNA. (Maybe for reasons given in Section 12.3, her contribution has only recently been acknowledged.) X-ray diffraction images are made after directing a beam of x-rays at a molecule, which scatters the x-rays in a pattern that can be captured on film. The pattern consists only of dots and streaks; in itself, it is not the structure of the molecule. Researchers use it to calculate the positions of the molecule's atoms.

Also, DNA must be processed first. A suspension of DNA molecules has to be spun rapidly, spooled onto a rod, and gently pulled into gossamer fibers, like cotton candy. When dry, the fibers twist and turn into two forms, which makes x-ray diffraction images too complicated to decipher. Wet fibers have only one.

Franklin was the first to make a spectacularly clear x-ray diffraction image of wet DNA fibers, as Figure 12.5 shows. She used it to calculate that a molecule of DNA is long and thin, with a 2-nanometer diameter. And she calculated that some molecular configuration is repeated every 0.34 nanometer along its length, and another every 3.4 nanometers.

Figure 12.5 Rosalind Franklin's superb x-ray diffraction image of DNA fibers.

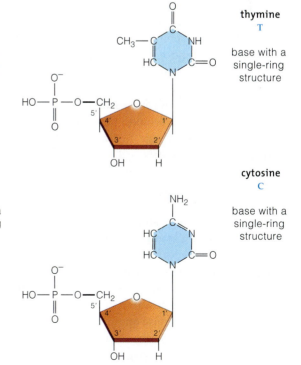

Figure 12.4 All of the chromosomes in a cell contain DNA. What does DNA contain? Four kinds of nucleotides: A, G, T, and C.

Could the sequence of DNA's nucleotide bases be twisting up in a repeating pattern, a bit like a circular stairway? Certainly Pauling thought so. After all, he discovered a helical shape in collagen. Like everyone else—including Wilkins, Watson, and Crick—he was thinking "helix." Watson later wrote, "We thought, why not try it on DNA? We were worried that *Pauling* would say, why not try it on DNA? Certainly he was a very clever man. He was a hero of mine. But we beat him at his own game. I still can't figure out why."

Pauling, it turned out, made a big chemical mistake. His model had all the negatively charged phosphate groups inside the DNA helix instead of outside. If they were that close together, they would repel each other too much to be stable.

PATTERNS OF BASE PAIRING

Franklin filed away the image of wet fibers, and then Watson and Crick took the lead. They perceived that DNA must consist of two strands of nucleotides, held together at their bases by hydrogen bonds (Figure 12.6). Such bonds form when the two strands run in opposing directions and twist to form a double helix. Two kinds of base pairings form along the molecule's length: A—T and G—C.

The bonding pattern accommodates variation in the order of bases. For instance, a stretch of DNA from a rose, a human, or any other organism might be:

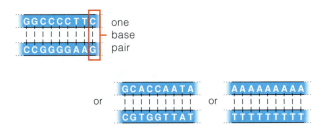

All DNA molecules show the same bonding pattern, but each species has a number of unique DNA base sequences. *This molecular constancy and variation among species is the foundation for the unity and diversity of life.*

Intriguingly, computer simulations show that if you want to pack a string into the least space, coil it into a helix. Was a space-saving advantage a factor in the molecular evolution of the DNA double helix? Maybe.

The pattern of base pairing between the two strands in DNA is constant for all species—A with T, and G with C. However, each species has a number of unique sequences of base pairs along the length of their DNA molecules.

Figure 12.6 Composite of different models for a DNA double helix. The two sugar–phosphate backbones run in *opposing* directions. Think of the sugar units (deoxyribose) of one strand as being upside down.

By comparing the numerals used to identify each carbon atom of the deoxyribose molecule (1′, 2′, 3′, and so on), you see that one strand runs in the 5′➛3′ direction and the other runs in the 3′➛5′ direction.

2-nanometer diameter overall

0.34-nanometer distance between each pair of bases

3.4-nanometer length of each full twist of the double helix

In all respects shown here, the Watson–Crick model for DNA structure is consistent with the known biochemical and x-ray diffraction data.

The pattern of base pairing (A only with T, and G only with C) is consistent with the known composition of DNA (A = T, and G = C).

12.3 Rosalind's Story

Watson and Crick got the attention of the world, in part by using Rosalind Franklin's data. She got cancer, most likely because of her intensive work with x-rays.

When Rosalind Franklin started at King's Laboratory of Cambridge University, she already had impressive credentials. She developed a refined x-ray diffraction method while studying the structure of coal. She took a new mathematical approach to interpreting x-ray diffraction images and, like Pauling, had built three-dimensional molecular models. At Cambridge, she was asked to create and run a state-of-the-art x-ray crystallography laboratory. Her assignment was to investigate the structure of DNA.

No one bothered to tell Franklin that, down the hall, Maurice Wilkins was working on the puzzle. Even the graduate student assigned to assist her didn't mention it. No one bothered to tell Wilkins about Franklin's assignment; he assumed she was a technician hired to do his x-ray crystallography work because he didn't know how to do it himself. And so the clash began. To Franklin, Wilkins seemed inexplicably prickly. To Wilkins, Franklin displayed an appalling lack of the deference that technicians usually show researchers.

Wilkins had a prized cache of crystalline DNA fibers —each with parallel arrays of hundreds of millions of DNA molecules—which he gave to his "technician." Five months later, Franklin gave a talk on what she had learned. DNA, she said, may have two, three, or four parallel chains twisted in a helix, with phosphate groups projecting outward.

With his crystallography background, Crick would have recognized the significance of her report—*if* he had been there. (A *pair* of chains oriented in opposing directions would be the same even if flipped 180 degrees. Two pairs of chains? No. DNA's density ruled that out. But one pair of chains? Yes!) Watson was in the audience but didn't know what Franklin was talking about.

Later, Franklin produced her outstanding x-ray diffraction image of wet DNA fibers. It fairly screamed *Helix!* She also worked out DNA's length and diameter. But she had been working with dry fibers for a long time and didn't dwell on her new data. Wilkins did.

In 1953, he let Watson see that image and reminded him of what Franklin had reported more than a year before. When Watson and Crick did focus on her data, they had the final bit of information they needed to build a DNA model—one with two helically twisted chains running in opposing directions.

Figure 12.7 Portrait of Rosalind Franklin arriving at Cambridge in style, from Paris.

12.4 DNA Replication and Repair

The discovery of DNA structure was a turning point in studies of inheritance. Crick understood at once how cells duplicate their DNA before they divide.

Until Watson and Crick presented their model, no one could explain **DNA replication**, or how the molecule of inheritance is duplicated before a cell divides.

Enzymes easily break the hydrogen bonds between the two nucleotide strands of a DNA molecule. When enzymes and other proteins act on the molecule, one strand unwinds from the other and exposes stretches of its nucleotide bases. Cells contain stockpiles of free nucleotides that can pair with the exposed bases.

Each parent strand stays intact, and a companion strand gets assembled on each one according to this base-pairing rule: **A** to **T**, and **G** to **C**. As soon as a stretch of a new, partner strand forms on a stretch of the parent strand, the two twist together into a double helix. Because the parent DNA strand is conserved during the replication process, half of every double-stranded DNA molecule is "old" and half is "new." Figures 12.8 and 12.9 describe this process, which we call *semiconservative* replication.

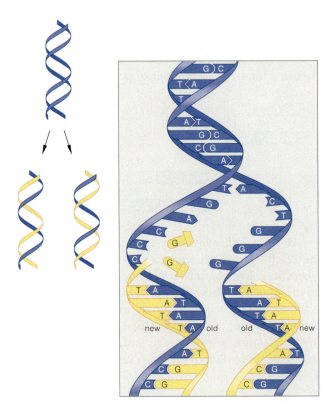

Figure 12.8 A simple picture of the semiconservative nature of DNA replication. The original two-stranded DNA molecule is coded *blue*. Each parent strand remains intact. One new strand (*gold*) is assembled on each of the parent strands.

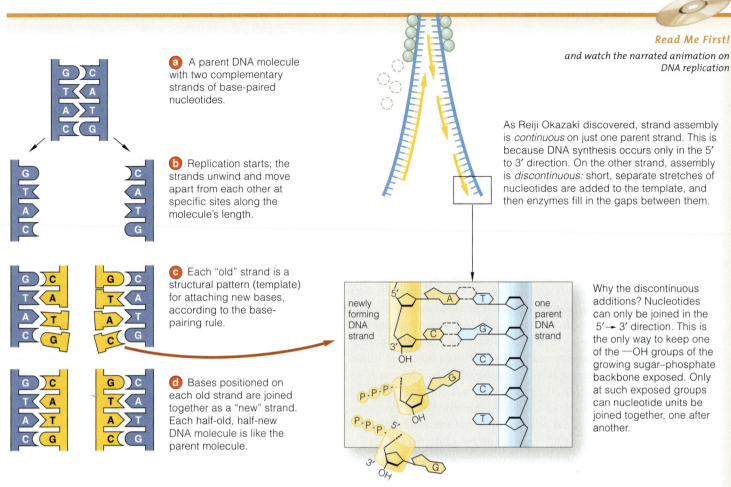

a A parent DNA molecule with two complementary strands of base-paired nucleotides.

b Replication starts; the strands unwind and move apart from each other at specific sites along the molecule's length.

c Each "old" strand is a structural pattern (template) for attaching new bases, according to the base-pairing rule.

d Bases positioned on each old strand are joined together as a "new" strand. Each half-old, half-new DNA molecule is like the parent molecule.

Figure 12.9 A closer look at DNA replication.

Read Me First!
and watch the narrated animation on DNA replication

As Reiji Okazaki discovered, strand assembly is *continuous* on just one parent strand. This is because DNA synthesis occurs only in the 5' to 3' direction. On the other strand, assembly is *discontinuous*: short, separate stretches of nucleotides are added to the template, and then enzymes fill in the gaps between them.

newly forming DNA strand

one parent DNA strand

Why the discontinuous additions? Nucleotides can only be joined in the 5'→ 3' direction. This is the only way to keep one of the —OH groups of the growing sugar–phosphate backbone exposed. Only at such exposed groups can nucleotide units be joined together, one after another.

DNA replication uses a team of molecular workers. In response to cellular signals, replication enzymes become active along the length of the DNA molecule. Along with other proteins, some enzymes unwind the strands in both directions and prevent them from rewinding. Enzyme action jump-starts the unwinding but is not required to unzip hydrogen bonds between the strands; hydrogen bonds are individually weak.

Now **DNA polymerases**, a class of enzymes, attach short stretches of free nucleotides to unwound parts of the parent template. Free nucleotides themselves drive the strand assembly. Each has three phosphate groups. A DNA polymerase splits off two, releasing energy that drives the attachments.

DNA ligases fill in the tiny gaps between the new short stretches and form one continuous strand. Then enzymes wind up the template and complementary strands together, forming a DNA double helix.

Sometimes a molecule of DNA breaks. Sometimes it is replicated incorrectly so that it does not exactly match the parent molecule; it acquired mismatched bases or has missing or extra segments. DNA repair processes minimize such damage. Ligases can fix the breaks, and specialized DNA polymerases can correct the mismatched base pairs or replace mutated ones.

The repair processes confer survival advantage on cells. Why? Mistakes in DNA can result in the altered or diminished function of encoded proteins and thus disrupt how cells operate. This is what happens with genetic disorders. Also, a broken strand of DNA may block replication and cause a cell to commit suicide by issuing signals for its own death.

DNA is replicated prior to cell division. Enzymes unwind its two strands. Each strand remains intact throughout the process—it is conserved—and enzymes assemble a new, complementary strand on each one.

Mistakes happen. Repair systems fix mismatched base pairs and help maintain the integrity of genetic information. They also bypass breaks, which helps keep replication from shutting down.

12.5 Reprogramming DNA To Clone Mammals

Knowledge of DNA structure and function opened up exciting, and troubling, research avenues—including cloning adult mammals from little more than DNA and a cell stripped of its nucleus.

Geneticists started out cloning some embryos from *in vitro* fertilization. Briefly, a sperm fertilizes an egg in a petri dish. Mitotic divisions produce two cells, then four, then eight. The eight-cell stage is gently split into two- or four-cell clusters that are implanted in surrogate mothers. There they grow and develop into genetically identical animals; they are clones.

Splitting such early stages, or *artificial twinning*, gives us clones that are identical to one another but not to sexually reproducing parents; they have genes from both parents. Such clones must grow up before researchers can find out whether genes for a desired maternal or paternal trait were inherited.

Using a differentiated cell is faster, because the desired genotype is already known. You may wonder what "differentiated" means. All cells descended from a fertilized egg have the same DNA, but different lineages start making unique selections from it during development. The selections commit them to becoming liver cells, blood cells, or other specialists in structure, composition, and function in the adult (Section 14.3).

A differentiated cell must be tricked into rewinding the clock. *Its DNA must be reprogrammed into starting over again and directing the development of a whole individual.* Nuclear transfer is one way to trick it. The nucleus of a differentiated cell from an animal to be cloned replaces an unfertilized egg's nucleus (Figure 12.10). Chemicals or electric shocks may induce the cell to divide. If all goes well, a cluster of embryonic cells forms and can be implanted in a surrogate mother.

In Dolly's case, the nucleus came from a cell in a sheep's udder. Nuclei from cumulus cells were used to clone mice and CC, the first cloned cat. Cumulus cells surround immature eggs in mammalian ovaries. Genetic tests confirmed that CC (short for Carbon Copy) is a clone, even though her coat patterning differs from that of the genetic donor (Figure 12.10). Here is visible evidence that environmental factors (in this case, in the uterus) can alter gene expression.

Variations in gene expression are less obvious but more of a problem in other clones. About 4 percent of all the genes tested in cloned mice were expressed at abnormal levels. Gene expression also was disturbed in cloned mice that had received genetic material from cells of entirely different tissues. Does this mean that nuclear transfer procedures invite defects? Probably.

Abnormal gene expression in clones isn't surprising. Genes switch on and off during normal development. Researchers are not yet rewinding the clock to cover all of the ticks—minutes or hours—between nuclear transfer and the first cell divisions.

  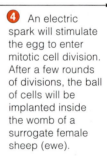 

① A microneedle is about to remove the nucleus from an unfertilized sheep egg (center).

② The microneedle has now emptied the sheep egg of its own nucleus, which held the DNA.

③ DNA from a donor cell is about to be deposited in the enucleated egg.

④ An electric spark will stimulate the egg to enter mitotic cell division. After a few rounds of divisions, the ball of cells will be implanted inside the womb of a surrogate female sheep (ewe).

the first cloned sheep

Figure 12.10 Steps in the nuclear transfer process that led to Dolly, and a gallery of famous firsts in the brave new world of mammalian cloning.

the first cloned mice   the first cloned pigs   CC (*left*), and her genetic donor, Rainbow.

Summary

Section 12.1 Experimental tests that used bacteria and bacteriophages offered the first solid evidence that DNA is the hereditary material in living organisms.

Sections 12.2, 12.3 DNA consists only of nucleotides, each with a five-carbon sugar (deoxyribose), a phosphate group, and one of four kinds of nitrogen-containing bases: adenine, thymine, guanine, or cytosine.

A DNA molecule consists of two nucleotide strands twisted together into a double helix. Bases of one strand hydrogen-bond with bases of the other.

Bases of the two DNA strands pair in a constant way. Adenine pairs with thymine (A–T), and guanine with cytosine (G–C). Which base follows another along a strand varies among species. The DNA of each species incorporates some number of unique stretches of base pairs that set it apart from the DNA of all other species.

Section 12.4 In DNA replication, enzymes unwind the two strands of a double helix and assemble a new strand of complementary sequence on each parent strand. Two double-stranded DNA molecules result. One strand of each molecule is old (is conserved); the other is new.

During replication, repair systems fix base-pairing mistakes and mutated bases. They also repair breaks in DNA strands that could shut down replication. DNA ligase and special DNA polymerases are involved.

Section 12.5 Embryo splitting and nuclear transfers are two methods that produce clones, or individuals that have identical DNA. Clones can show phenotypic differences if they are exposed to different factors that affect gene expression during development.

Self-Quiz

Answers in Appendix III

1. Which is *not* a nucleotide base in DNA?
   a. adenine   c. uracil   e. cytosine
   b. guanine   d. thymine   f. All are in DNA.

2. What are the base-pairing rules for DNA?
   a. A–G, T–C   c. A–U, C–G
   b. A–C, T–G   d. A–T, G–C

3. One species' DNA differs from others in its _____ .
   a. sugars   c. base sequence
   b. phosphates   d. all of the above

4. When DNA replication begins, _____ .
   a. the two DNA strands unwind from each other
   b. the two DNA strands condense for base transfers
   c. two DNA molecules bond
   d. old strands move to find new strands

5. DNA replication requires _____ .
   a. free nucleotides   c. many enzymes
   b. new hydrogen bonds   d. all of the above

6. Cell differentiation involves _____ .
   a. cloning   c. selective gene expression
   b. nuclear transfers   d. both b and c

Critical Thinking

1. Matthew Meselson and Frank Stahl's experiments supported a semiconservative model of DNA replication. These researchers obtained "heavy" DNA by growing *Escherichia coli* in a medium enriched with ^{15}N, a heavy isotope of nitrogen. They also prepared "light" DNA by growing *E. coli* in the presence of ^{14}N, the more common isotope. An available technique helped them identify which replicated molecules were heavy, light, or hybrid (one heavy strand and one light). Use pencils of two colors, one for heavy strands and one for light. Starting with a DNA molecule having two heavy strands, arrange them to show how daughter molecules would form after replication in a ^{14}N-containing medium. Show the four DNA molecules that would form if daughter molecules are replicated a second time in the ^{14}N medium.

2. Mutations, permanent changes in DNA base sequences, are the original source of genetic variation and the raw material of evolution. Yet how can mutations accumulate, given that cells have repair systems that fix structurally altered or discontinuous DNA strands during replication?

Media Menu

Student CD-ROM

Impacts, Issues Video
    Goodbye, Dolly
Big Picture Animation
    DNA structure and function
Read-Me-First Animation
    Hershey–Chase experiments
    DNA replication
Other Animations and Interactions
    Griffith's experimental transformation of bacteria
    DNA double helix
    Cloning by nuclear transfer

InfoTrac

- Beyond the Double Helix: Francis Crick. *Time*, February 2003.
- Combing Chromosomes. *American Scientist*, May 2002.
- Jumpstarting DNA Repair. *Environmental Health Perspectives*, December 2002.
- Ma's Eyes, Not Her Ways: Clones Can Vary in Behavioral—and Physical—Traits. *Scientific American*, April 2003.

Web Sites

- Dolan DNA Learning Center: www.dnalc.org
- DNA Structure: molvis.sdsc.edu/dna
- Nobel e-Museum: www.nobel.se/medicine/educational
- Roslin Institute: www.roslin.ac.uk

How Would You Vote?

Mammalian cloning using adult cells is a difficult process and often produces abnormal individuals. Many researchers argue that continued experimentation will allow them to refine methods and develop new drugs and organs for transplants. Some activists argue that cloning animals from adult cells should be banned. Should this cloning research continue?

IMPACTS, ISSUES *Ricin and Your Ribosomes*

In 2003, police acting on an intelligence tip stormed a London apartment, where they collected castor oil beans (Figure 13.1) and laboratory glassware. They arrested several young men and reminded the world of ricin's potential as a biochemical weapon.

Ricin is a protein product of the castor oil plant (*Ricinus communis*). A dose the size of a grain of salt can kill you; only plutonium and botulism toxin are more deadly. Researchers knew about it as long ago as 1888. Later, when Germany unleashed mustard gas against allied troops during World War II, the United States and England feverishly investigated whether ricin, too, could be used as a weapon. Both countries shelved the research when the war ended.

Fast-forward to 1969. Georgi Markov, a Bulgarian writer, defected to the West at the height of the Cold War.

Figure 13.1 Castor oil plant seeds, source of the ribosome-busting ricin.

As he strolled down a London street, an assassin used a modified umbrella to poke a tiny ball laced with ricin into one of his legs. Markov died in agony three days later.

Ricin is on stage once again. Traces of ricin showed up in hastily abandoned Afghanistan caves in 2001 and on a Chechen fighter killed in Moscow in 2003. In early 2004, traces turned up in a United States Senate mailroom, in a State Department building, and in an envelope addressed to the White House.

Each year, a lot of ricin-rich wastes form during castor oil production. The entire castor oil plant is poisonous, but ricin is most concentrated in seeds. How does ricin exert its deadly effects? It inactivates ribosomes, the cell's protein-building machinery.

Ricin is a protein. One of its two polypeptide chains, shown in the filmstrip, helps ricin insert itself into cells. The other chain, an enzyme, wrecks part of the ribosome where amino acids are joined together. It yanks adenine subunits from an RNA molecule that is a crucial part of the ribosome. The ribosome's three-dimensional shape unravels, protein synthesis stops, and cells spiral toward death. So does the individual; there is no antidote.

It's possible to get on with your life without knowing what a ribosome is or what it does. It also is possible to recognize that protein synthesis is not a topic invented to torture biology students. It is something worth knowing about and appreciating for how it keeps us alive—and for appreciating anti-terrorism researchers who are working to keep us that way.

the big picture

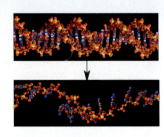

Making the Transcripts It takes two steps to get from DNA's protein-building information to a new protein molecule. In the first step, a strand of mRNA is transcribed from a gene region, a sequence of nucleotide bases in an unwound part of a DNA molecule.

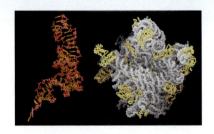

Readers of the Genetic Code Every three ribonucleotide bases along the length of an mRNA transcript is a "word" corresponding to a particular amino acid. Two other classes of RNAs recognize a range of these base triplets, which represent a genetic code.

So start with what you know about DNA, the book of protein-building information in cells. The alphabet used to write the book is simple enough—just A, T, G, and C, for the nucleotide bases adenine, thymine, guanine, and cytosine. But how do you get from an alphabet to a protein? The answer starts with the order, or sequence, of those four nucleotide bases in a DNA molecule.

As you already know, the two strands unwind from each other entirely when a cell is replicating its DNA. At other times, however, cells selectively unwind the two strands in certain regions and thereby expose the base sequences we call genes. Most of the genes encode information on building particular proteins.

You'll see from this chapter that it takes two steps, transcription and translation, to do something with the information in a gene. In all eukaryotic cells, the first step proceeds in the nucleus. A newly exposed DNA base sequence functions as a structural pattern, or a template, for making a strand of ribonucleic acid (RNA) from the cell's pool of free nucleotides.

The RNA then moves into the cytoplasm, where it is translated. In this second step of protein synthesis, the RNA guides the assembly of amino acids into a new polypeptide chain. The new chains become folded into the three-dimensional shapes of specific proteins.

In short, DNA guides the synthesis of RNA, then RNA guides the synthesis of proteins:

$$DNA \xrightarrow{transcription} RNA \xrightarrow{translation} PROTEIN$$

How Would You Vote?

A large-scale biochemical terrorist attack using ricin is unlikely, because it is very difficult to disperse in air. Scientists are developing a vaccine to protect against ricin exposure. Should we use the new vaccine to carry out mass immunizations? See the Media Menu for details, then vote online.

Translating the Transcripts   In the second step of protein synthesis, tRNAs and rRNAs interact to translate the mRNA transcript of a gene region into a polypeptide chain. The chain grows as one of the rRNA components of ribosomes catalyzes the bonding between amino acids.

Mutations and Proteins   Mutations change the genetic code words in the messages that specify particular proteins. When the protein is an essential part of cell architecture or metabolism, we can expect the outcome to be an abnormal cell.

13.1 How Is RNA Transcribed From DNA?

In transcription, the first step in protein synthesis, the base sequence in an unwound DNA region becomes a template for assembling a strand of ribonucleic acid, or RNA.

It takes three classes of RNA molecules to synthesize proteins. Most genes are transcribed into **messenger RNA**, or **mRNA**—the only kind that carries *protein-building* instructions. Other genes are transcribed into **ribosomal RNA**, or **rRNA**, a component of ribosomes.

A ribosome is a large molecular structure upon which polypeptide chains are assembled. Transcription of still other genes yields **transfer RNA**, or **tRNA**, which can deliver amino acids one at a time to a ribosome.

THE NATURE OF TRANSCRIPTION

An RNA molecule is almost but not quite like a single strand of DNA. It has four kinds of ribonucleotides, each with the five-carbon sugar ribose, a phosphate group, and a base. Three bases—adenine, cytosine, and guanine—are the same as those in DNA. In RNA, however, the fourth base is **uracil**, not thymine. Like thymine, uracil can pair with adenine. This means a new RNA strand can be built according to the same base-pairing rules as DNA (Figure 13.2).

Transcription *differs* from DNA replication in three respects. Part of a DNA strand, not the whole molecule, is used as the template. The enzyme **RNA polymerase**, not DNA polymerase, adds ribonucleotides one at a time to the end of a growing strand of RNA. And transcription results in one free strand of RNA, not a hydrogen-bonded double helix.

The many coding regions in DNA are transcribed separately, and each has its own START and STOP signal. A **promoter** is one sequence of bases in DNA that signals the start of a gene. RNA synthesis gets going as soon as RNA polymerases and other proteins attach to it. Each polymerase moves along the DNA strand, joining one ribonucleotide after another on

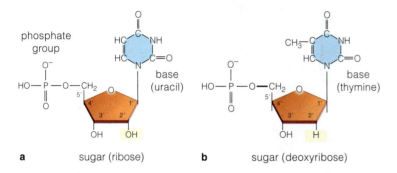

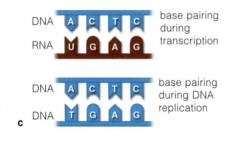

Figure 13.2 (**a**) Uracil, one of four ribonucleotides in RNA. The other three—adenine, guanine, and cytosine—differ only in their bases. Compare uracil to (**b**) thymine, a DNA nucleotide. (**c**) Base pairing of DNA with RNA at transcription, compared to base pairing during DNA replication.

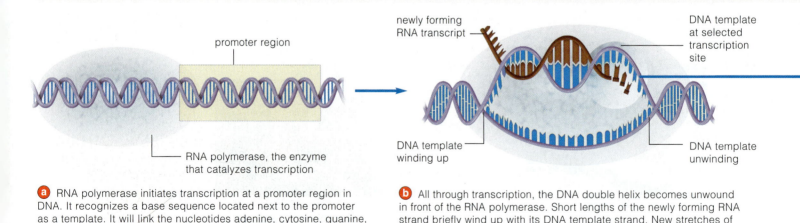

a RNA polymerase initiates transcription at a promoter region in DNA. It recognizes a base sequence located next to the promoter as a template. It will link the nucleotides adenine, cytosine, guanine, and uracil into a strand of RNA, in the order specified by DNA.

b All through transcription, the DNA double helix becomes unwound in front of the RNA polymerase. Short lengths of the newly forming RNA strand briefly wind up with its DNA template strand. New stretches of RNA unwind from the template (and the two DNA strands wind up again).

Figure 13.3 Gene transcription. By this process, an RNA molecule is assembled on a DNA template. (**a**) Gene region of DNA. The base sequence along one of DNA's two strands (not both) is used as the template. (**b–d**) Transcribing that region results in a molecule of RNA.

the DNA as a template (Figure 13.3). It soon reaches another base sequence in the DNA that signals "the end," and the new RNA is released as a free transcript.

FINISHING TOUCHES ON mRNA TRANSCRIPTS

In eukaryotic cells, each new mRNA molecule is not in final form. Just as a dressmaker may snip off some threads or put bows on a dress before it leaves the shop, so do these cells tailor their "pre-mRNA." For instance, some enzymes attach a modified guanine "cap" to the start of each pre-mRNA transcript. Others attach about 100 to 300 adenine ribonucleotides as a tail to the other end. Hence its name, poly-A tail.

Later, in the cytoplasm, the cap will help bind the mRNA to a ribosome. Enzymes will begin to nibble away at the tail. Each tail's length determines how long a particular mRNA molecule will last. It helps keep protein-building messages intact for as long as the cell requires them.

The protein-coding parts of eukaryotic genes are **exons**. In between them are one or more **introns**, or sequences that are removed before a mature mRNA transcript is translated. The introns are sites where the protein-building information can be snipped apart and spliced back together in more than one way. This *alternative splicing* lets cells use the same gene to make proteins that differ slightly in form and function. It may be a way to increase DNA's information-storing capacity, hence the capacity to make diverse proteins.

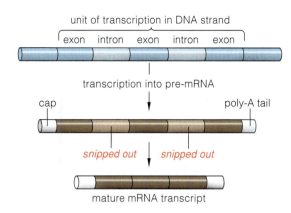

Figure 13.4 Pre-mRNA transcript processings. Some or all introns are removed before the transcript leaves the nucleus.

As Figure 13.4 shows, introns are transcribed along with exons. They are snipped out before the mature mRNA transcript leaves the nucleus. Think of it this way: The *ex*ons are *ex*ported from the nucleus, and *in*trons stay *in* the nucleus, where they are recycled.

> In gene transcription, a sequence of exposed bases on one of the two strands of a DNA molecule serves as a template.
>
> Using that template, RNA polymerase assembles a single strand of RNA from the four ribonucleotides, A, U, C, and G.
>
> Before leaving the nucleus, each new mRNA transcript, or pre-mRNA, undergoes modification into final form.

Read Me First!
and watch the narrated animation on transcription

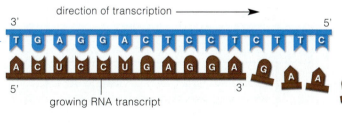

c What happened at the assembly site? RNA polymerase catalyzed the assembly of ribonucleotides, one after another, into an RNA strand, using exposed bases on the DNA as a template. Many other proteins assist this process.

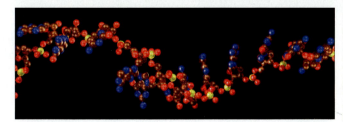

d At the end of the gene region, the last stretch of the new transcript is unwound and released from the DNA template. Shown below is a model for a transcribed strand of RNA.

13.2 Deciphering mRNA Transcripts

Like a strand of DNA, an mRNA transcript is a linear sequence of nucleotides. What are the protein-building "words" in its sequence? Each is a three-letter word, composed of three nucleotide bases.

THE GENETIC CODE

Marshall Nirenberg, Philip Leder, Severo Ochoa, and Gobind Korana deduced the correspondence between genes and proteins. Picture an mRNA transcript that was built on a DNA template, as in Figure 13.5*a*. To translate it into the amino acid sequence of a protein, you have to know just how many letters—nucleotide bases—correspond to a word (an amino acid). That is what the researchers figured out.

As you will see, after mRNA docks at a ribosome, its ribonucleotide bases are "read" *three at a time*, as triplets. Base triplets in mRNA transcripts were given the name **codons**. Figure 13.5*b* shows how the order of different codons in mRNA determines the order in which different kinds of amino acids will follow one another in a growing polypeptide chain.

Count the different codons in Figure 13.6, and you see there are sixty-four possible choices. That's a lot more than the twenty kinds of amino acids. But some amino acids are encoded by more than one codon. For instance, both GAA and GAG call for glutamate.

In most species, the first AUG in a transcript is a START signal for translating "three-bases-at-a-time." This means methionine is the first amino acid in all new polypeptide chains. UAA, UAG, and UGA do not call for any amino acid. They are STOP signals that block further additions of amino acids to a new chain.

The set of sixty-four different codons is a **genetic code**. Protein synthesis adheres to this code in nearly all cases. Mitochondria are one of the exceptions. They have their own "mitochondrial code," which includes a few unique codons. This is one clue that supports a theory of endosymbiotic origins for eukaryotic cells. By this theory, ancient aerobic bacteria were ingested by other cells. They resisted digestion, then evolved into mitochondria inside host cells (Section 18.4).

DNA

mRNA

a

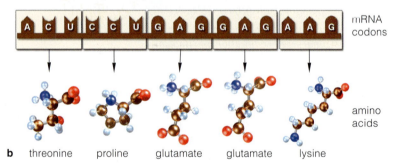

mRNA codons

amino acids

b   threonine   proline   glutamate   glutamate   lysine

Figure 13.5 Example of the correspondence between genes and proteins. (**a**) An mRNA transcript (*brown*) of a gene region of DNA (*blue*). Three nucleotide bases, equaling one codon, specify one amino acid. This series of codons (base triplets) specifies the sequence of amino acids shown in (**b**).

Figure 13.6 The near-universal genetic code. Each codon in mRNA is a set of three ribonucleotide bases. Sixty-one of these base triplets encode specific amino acids. Three are signals that stop translation.

The *left* vertical column (*dark brown*) lists choices for the first base of a codon. The *top* horizontal row (*light tan*) lists the second choices. The *right* vertical column (*dark tan*) lists the third.

To give three examples, reading from left to right, the triplet U G G corresponds to tryptophan. Both U U U and U U C correspond to phenylalanine.

first base	second base				third base
	U	C	A	G	
U	phenylalanine	serine	tyrosine	cysteine	U
	phenylalanine	serine	tyrosine	cysteine	C
	leucine	serine	STOP	STOP	A
	leucine	serine	STOP	tryptophan	G
C	leucine	proline	histidine	arginine	U
	leucine	proline	histidine	arginine	C
	leucine	proline	glutamine	arginine	A
	leucine	proline	glutamine	arginine	G
A	isoleucine	threonine	asparagine	serine	U
	isoleucine	threonine	asparagine	serine	C
	isoleucine	threonine	lysine	arginine	A
	methionine (or START)	threonine	lysine	arginine	G
G	valine	alanine	aspartate	glycine	U
	valine	alanine	aspartate	glycine	C
	valine	alanine	glutamate	glycine	A
	valine	alanine	glutamate	glycine	G

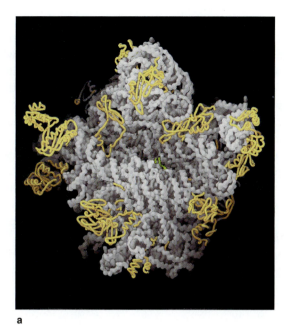

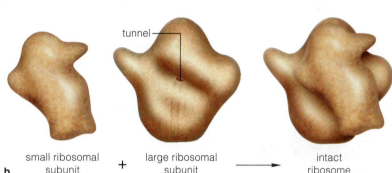

Figure 13.8 (**a**) Ribbon model for the large subunit of a bacterial ribosome. It has two rRNA molecules (*gray*) and thirty-one structural proteins (*gold*), which stabilize the structure. At one end of a tunnel through the large subunit, rRNA catalyzes polypeptide chain assembly. This is an ancient, highly conserved structure. Its role is so vital that the corresponding subunit of the eukaryotic ribosome, which is larger, may be similar in structure and function. (**b**) Model for the small and large subunits of a eukaryotic ribosome.

codon in mRNA transcript

anticodon in tRNA

amino acid

Figure 13.7 Model for a tRNA. The icon shown to the right is used in following illustrations. The "hook" at the lower end of this icon represents the binding site for a specific amino acid.

tunnel

small ribosomal subunit + large ribosomal subunit → intact ribosome

THE OTHER RNAs

In the cytoplasm of all cells are pools of free amino acids and tRNA molecules. Each tRNA has a molecular "hook," an attachment site for an amino acid. It has an **anticodon**, a ribonucleotide base triplet that can pair with an mRNA codon (Figure 13.7). When tRNAs bind to mRNA on a ribosome, their amino acid cargo will become automatically positioned in the order that the codons specify.

There are sixty-four codons but not as many kinds of tRNAs. How do tRNAs match up with more than one type of codon? According to base-pairing rules, adenine pairs with uracil, and cytosine with guanine. However, in codon–anticodon interactions, these rules can loosen for the third base in a codon. This freedom in codon–anticodon pairing at a base is known as the "wobble effect."

To give one example, AUU, AUC, and AUA specify isoleucine. All three codons can base-pair with one type of tRNA that hooks on to isoleucine.

Again, interactions between the tRNAs and mRNA take place at ribosomes. A ribosome has two subunits (Figure 13.8). They are built from rRNA and structural proteins in the nucleus, then shipped separately to the cytoplasm. There, a large and small subunit converge into an intact, functional ribosome only when mRNA is to be translated.

The genetic code is a set of sixty-four codons, which are ribonucleotide bases in mRNA that are read in sets of three. Different amino acids are specified by different codons.

Only mRNA carries DNA's protein-building instructions from the nucleus into the cytoplasm.

tRNAs deliver amino acids to ribosomes. Their anticodons base-pair with codons in the order specified by mRNA.

Polypeptide chains are built on ribosomes, each consisting of a large and small subunit made of rRNA and proteins.

13.3 Translating mRNA Into Protein

DNA's hereditary information must be stored intact, safely, in one place. Think of mRNA transcripts as intermediaries that deliver messages from DNA to ribosomes, which translate them into the polypeptide chains of proteins.

Translation has three stages: initiation, elongation, and termination. *Initiation* requires an initiator tRNA, the only one that can start transcription. It binds to a small ribosomal subunit. Then mRNA's START codon, AUG, joins with that tRNA's anticodon. A large ribosomal subunit now joins with the small one (Figure 13.9a–c). Together, the ribosome, mRNA, and initiator tRNA are an initiation complex. The next stage can begin.

In *elongation*, a polypeptide chain is synthesized as the mRNA passes between the two ribosomal subunits,

like a thread passing through the eye of a needle. tRNA molecules move amino acids to the ribosome, where they bind to the mRNA in the order specified by its codons. Part of an rRNA molecule at the center of the large ribosomal subunit functions as an enzyme. It catalyzes peptide bond formation between the amino acids (Figure 13.9d–f).

Figure 13.9g shows how one peptide bond forms between the most recently attached amino acid and the next one brought to the ribosome. Here, you might look once more at Section 3.5, which includes a step-by-step description of peptide bond formation during protein synthesis.

During the last stage of translation, *termination*, the ribosome reaches the mRNA's STOP codon. No tRNA

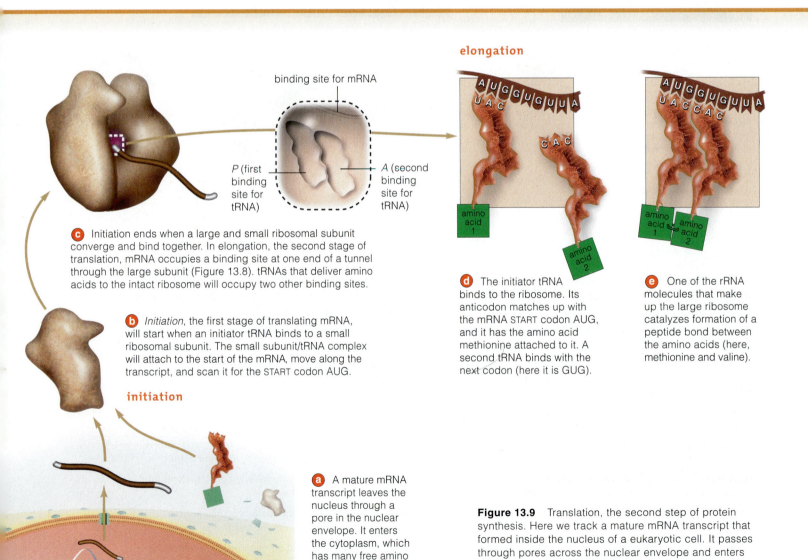

elongation

binding site for mRNA

P (first binding site for tRNA)   *A* (second binding site for tRNA)

c Initiation ends when a large and small ribosomal subunit converge and bind together. In elongation, the second stage of translation, mRNA occupies a binding site at one end of a tunnel through the large subunit (Figure 13.8). tRNAs that deliver amino acids to the intact ribosome will occupy two other binding sites.

b *Initiation*, the first stage of translating mRNA, will start when an initiator tRNA binds to a small ribosomal subunit. The small subunit/tRNA complex will attach to the start of the mRNA, move along the transcript, and scan it for the START codon AUG.

initiation

a A mature mRNA transcript leaves the nucleus through a pore in the nuclear envelope. It enters the cytoplasm, which has many free amino acids, tRNAs, and ribosomal subunits.

amino acid 1
amino acid 2

d The initiator tRNA binds to the ribosome. Its anticodon matches up with the mRNA START codon AUG, and it has the amino acid methionine attached to it. A second tRNA binds with the next codon (here it is GUG).

amino acid 1
amino acid 2

e One of the rRNA molecules that make up the large ribosome catalyzes formation of a peptide bond between the amino acids (here, methionine and valine).

Figure 13.9 Translation, the second step of protein synthesis. Here we track a mature mRNA transcript that formed inside the nucleus of a eukaryotic cell. It passes through pores across the nuclear envelope and enters the cytoplasm, which contains pools of many free amino acids, tRNAs, and ribosomal subunits.

has a corresponding anticodon. Proteins called release factors bind to the ribosome. Binding triggers enzyme activity that detaches the mRNA *and* the polypeptide chain from the ribosome (Figure 13.9*i–k*).

Unfertilized eggs and other cells that rapidly make many copies of different proteins usually stockpile mRNA transcripts in their cytoplasm. In cells that are quickly using or secreting proteins, you often observe many clusters of ribosomes (polysomes) on an mRNA transcript, all translating it at the same time.

Many newly formed polypeptide chains carry out their functions in the cytoplasm. Many others have a shipping label, a special sequence of amino acids. The label lets them enter the ribosome-studded, flattened sacs of rough ER (Section 4.7). In the organelles of the endomembrane system, the chains will take on final form before shipment to their ultimate destinations as structural or functional proteins.

Translation is initiated when a small ribosomal subunit and an initiator tRNA arrive at an mRNA transcript's START codon, and then a large ribosomal subunit binds to them.

tRNAs deliver amino acids to a ribosome in the order dictated by the linear sequence of mRNA codons. A polypeptide chain lengthens as peptide bonds form between the amino acids.

Translation ends when a STOP codon triggers events that cause the polypeptide chain and the mRNA to detach from the ribosome.

Read Me First!
and watch the narrated animation on translation

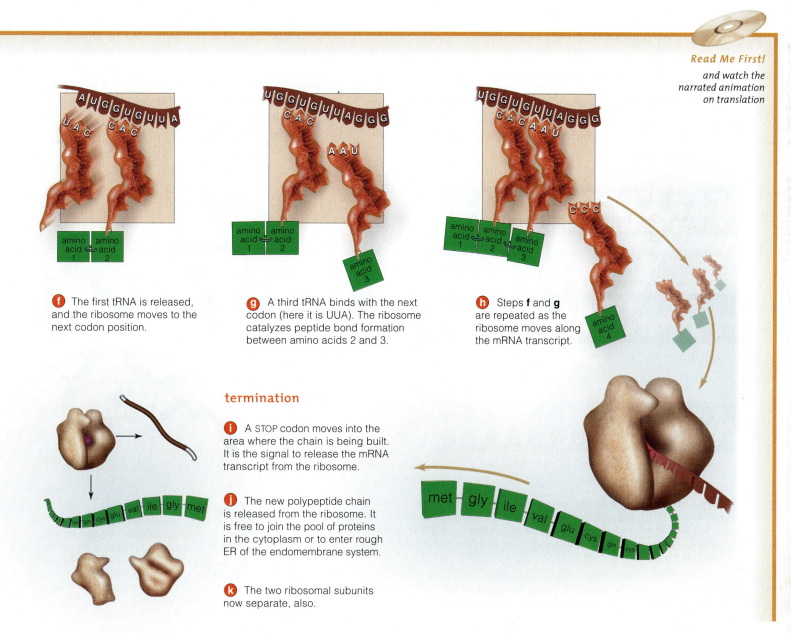

f The first tRNA is released, and the ribosome moves to the next codon position.

g A third tRNA binds with the next codon (here it is UUA). The ribosome catalyzes peptide bond formation between amino acids 2 and 3.

h Steps **f** and **g** are repeated as the ribosome moves along the mRNA transcript.

termination

i A STOP codon moves into the area where the chain is being built. It is the signal to release the mRNA transcript from the ribosome.

j The new polypeptide chain is released from the ribosome. It is free to join the pool of proteins in the cytoplasm or to enter rough ER of the endomembrane system.

k The two ribosomal subunits now separate, also.

13.4 Mutated Genes and Their Protein Products

When a cell taps its genetic code, it is making proteins with precise structural and functional roles that keep it alive. If something changes a gene, the protein that it encodes may change. If the protein has an essential role, we can expect the outcome to be an abnormal cell.

Gene sequences can change. Sometimes one base gets substituted for another in the nucleotide sequence. At other times, an extra base is inserted or one is lost. Such small-scale changes in the nucleotide sequence of a DNA molecule are **gene mutations**.

There is some leeway here, because more than one codon specifies the same amino acid. If UCU were changed to UCC, for example, it probably would not have dire effects, because both codons specify serine. However, many mutations give rise to proteins that function in altered ways or not at all. Repercussions are sometimes harmful or lethal.

COMMON MUTATIONS

In gene mutations called **base-pair substitutions**, one base is copied incorrectly during DNA replication. Its outcome? A protein may incorporate the wrong amino acid, or its synthesis may have been cut off too soon. In the example in Figure 13.10b, adenine *replaced* one thymine in a gene for beta hemoglobin. The mutant gene's product has a single amino acid substitution that causes sickle-cell disease (Section 3.6).

Figure 13.10c shows a different gene mutation, one in which a single base—thymine—was *deleted*. Again, DNA polymerases read base sequences in blocks of three. A deletion is one of the *frameshift* mutations; it shifts the "three-bases-at-a-time" reading frame. An altered mRNA is transcribed from the mutant gene, so an altered protein is the result.

Frameshift mutations fall in the broader categories of **insertions** and **deletions**. One or more base pairs become inserted into DNA or are deleted from it.

Other mutations arise from transposable elements, or **transposons**, that can jump around in the genome. Geneticist Barbara McClintock found that these DNA segments or copies of them move spontaneously to a new location in a chromosome or even to a different chromosome. When transposons land in a gene, they alter the timing or duration of its activity, or block it entirely. Their unpredictability can give rise to odd variations in traits. Figure 13.11 gives an example.

a

| THREONINE | PROLINE | GLUTAMATE | GLUTAMATE | LYSINE |

part of DNA template

mRNA transcribed from DNA

resulting amino acid sequence

b

| THREONINE | PROLINE | VALINE | GLUTAMATE | LYSINE |

base substitution in DNA

altered mRNA

altered amino acid sequence

c

| THREONINE | PROLINE | GLYCINE | ARGININE |

deletion in DNA

altered mRNA

altered amino acid sequence

Figure 13.10 Gene mutation. (**a**) Part of the gene, the mRNA, and the resulting amino acid sequence of the hemoglobin beta chain. (**b**) A base substitution in DNA replaces a thymine with an adenine. When the altered mRNA transcript is translated, valine replaces glutamate as the sixth amino acid of the new polypeptide chain. Sickle-cell anemia is the eventual outcome.

(**c**) Deletion of the same thymine would be a frameshift mutation. The reading frame for the rest of the mRNA shifts, a different protein product forms, and it causes thalassemia, a different type of red blood cell disorder.

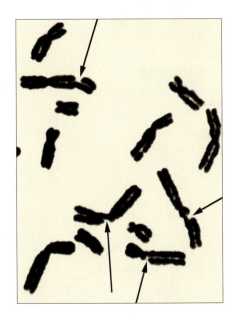

Figure 13.12 Chromosomes from a human cell exposed to gamma rays, a form of ionizing radiation. We can expect such broken pieces (*arrows*) to be lost in interphase, when DNA is being replicated. The extent of chromosome damage in an exposed cell typically depends on how much radiation it absorbed.

Figure 13.11 Barbara McClintock, who won a Nobel Prize for her research. She proved that transposons slip into and out of different locations in DNA. The curiously nonuniform coloration of kernels in strains of Indian corn (*Zea mays*) sent her on the road to discovery.

Several genes govern pigment formation and deposition in corn kernels, which are a type of seed. Mutations in one or more of these genes produce yellow, white, red, orange, blue, and purple kernels. However, as McClintock realized, *unstable* mutations can cause streaks or spots in *individual* kernels.

All of a corn plant's cells have the same pigment-encoding genes. But a transposon invaded a pigment-encoding gene before the plant started growing from a fertilized egg. While a kernel's tissues were forming, its cells couldn't make pigment. But the same transposon jumped back out of the pigment-encoding gene in some of its cells. Descendants of *those* cells could make pigment. The streaks and spots in individual kernels are evidence of those cell lineages.

HOW DO MUTATIONS ARISE?

Many mutations happen spontaneously while DNA is being replicated. This is not surprising, given the swift pace of replication (about twenty bases per second in humans and a thousand bases per second in certain bacteria). DNA polymerases can repair most of the mistakes (Section 12.4). Sometimes, however, they go on assembling a new strand right over an error. The bypass can result in a mutated DNA molecule.

Not all mutations are spontaneous. A number arise after DNA is exposed to mutation-causing agents. For instance, x-rays and similar high-energy wavelengths of **ionizing radiation** break chromosomes into pieces (Figure 13.12). Such radiation also indirectly damages DNA because it penetrates living tissue, leaving in its wake a potentially destructive trail of free radicals. Because of this, doctors and dentists use the lowest possible doses of x-rays in order to minimize damage to their patients' DNA.

Nonionizing radiation boosts electrons to a higher energy level. DNA absorbs one form, ultraviolet (UV) radiation. Two types of nucleotides in DNA, cytosine and thymine, are most vulnerable to excitation that can change base-pairing properties. For example, UV light can induce two thymine bases to pair (T–T, not A–T). At least seven gene products interact as a DNA repair mechanism to remove this error, which is a thymine dimer. Thymine dimers form in skin cells after exposure of unprotected skin to sunlight.

When they are not repaired, thymine dimers cause DNA polymerase to make additional errors during the next cycle of replication. They are the source of mutations that lead to certain cancers.

Natural and synthetic chemicals accelerate the rates of spontaneous mutations. **Alkylating agents** are one example. They transfer charged methyl or ethyl groups to reactive sites in DNA. At these sites, DNA is more vulnerable to base-pair changes that invite mutation. Cancer-causing agents in cigarette smoke and many substances exert their effects by alkylating DNA.

THE PROOF IS IN THE PROTEIN

When a mutation arises in a somatic cell, its good or bad effects do not endure, because it cannot be passed on to offspring. When a mutation arises in a germ cell or a gamete, however, it may enter the evolutionary arena. It also may do so if it arises during asexual reproduction. Either way, *a protein product of a heritable mutation will have harmful, neutral, or beneficial effects on the individual's ability to function in the prevailing environment.* The effects of uncountable mutations in millions of species have had spectacular evolutionary consequences. And that is a topic of later chapters.

A gene mutation is a change involving one or more bases in the nucleotide sequence of DNA. The most common types are base-pair substitutions, insertions, and deletions.

Exposure to harmful radiation and chemicals in the environment can cause mutations in DNA.

A protein specified by a mutated gene may have harmful, neutral, or beneficial effects on the ability of an individual to function in the environment.

Summary

Section 13.1 Transcription and translation are two steps of a process leading from genes to proteins.

$$\text{DNA} \xrightarrow{\text{transcription}} \text{RNA} \xrightarrow{\text{translation}} \text{PROTEIN}$$

In eukaryotic cells, transcription occurs in the nucleus (Figure 13.13). DNA's two strands unwind from each other at a selected region. RNA polymerases use the exposed DNA bases as a template on which an RNA molecule is built from free ribonucleotides (adenine, guanine, cytosine, and uracil). The mRNA transcript becomes modified before it leaves the nucleus.

Section 13.2 In translation, three types of RNAs interact to build polypeptide chains. Messenger RNA (mRNA) carries DNA's translated protein-building information from the nucleus to the cytoplasm. Its genetic message is written in codons, or sets of three nucleotides along an mRNA strand that specify an amino acid. There are sixty-four codons, a few of which act as START or STOP signals for translation. Ribosomes consist of ribosomal RNA (rRNA) and proteins that stabilize it. Transfer RNA (tRNA) molecules bind free amino acids in the cytoplasm and deliver them to ribosomes during protein synthesis. Different tRNAs bind different amino acids.

Section 13.3 During translation, amino acids are joined in the order specified by codons in mRNA.

Translation proceeds through three stages. In initiation, ribosomal subunits, an initiator tRNA, and an mRNA join to form an initiation complex. In elongation, tRNAs deliver amino acids to ribosomes, which synthesize a polypeptide chain from them. Part of the rRNA in the ribosomes catalyzes peptide bond formation between the amino acids. At termination, a STOP codon and other factors trigger the release of the mRNA and the new polypeptide chain, and they make the ribosome's subunits separate from each other.

Section 13.4 Gene mutations are heritable, small-scale changes in the base sequence of DNA. Major types are base-pair substitutions, insertions, and deletions. Many arise spontaneously as DNA is being replicated. Some occur when transposons jump around in a gene or after DNA is exposed to ionizing radiation or to chemicals in the environment. Mutations may cause changes in protein structure, protein function, or both.

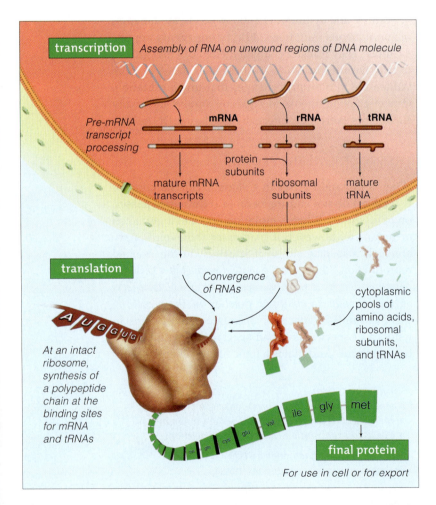

Figure 13.13 Summary of protein synthesis in eukaryotes. DNA is transcribed into RNA in the nucleus. RNA is translated in the cytoplasm. Prokaryotic cells don't have a nucleus; transcription and translation proceed in their cytoplasm.

Self-Quiz
Answers in Appendix III

1. DNA contains many different genes that are transcribed into different _____ .
   - a. proteins
   - b. mRNAs only
   - c. mRNAs, tRNAs, rRNAs
   - d. all of the above

2. An RNA molecule is typically _____ .
   - a. a double helix
   - b. single-stranded
   - c. double-stranded
   - d. triple-stranded

3. An mRNA molecule is synthesized by _____ .
   - a. replication
   - b. duplication
   - c. transcription
   - d. translation

4. Each codon specifies a (an) _____ .
   - a. protein
   - b. polypeptide
   - c. amino acid
   - d. mRNA

5. Anticodons pair with _____ .
   - a. mRNA codons
   - b. DNA codons
   - c. RNA anticodons
   - d. amino acids

6. Match the terms with the most suitable description.
   - ____ alkylating agent
   - ____ chain elongation
   - ____ exons
   - ____ genetic code
   - ____ anticodon
   - ____ introns
   - ____ codon

   - a. parts of mature mRNA transcript
   - b. base triplet for amino acid
   - c. second stage of translation
   - d. base triplet; pairs with codon
   - e. one environmental agent that induces mutation in DNA
   - f. set of 64 codons in mRNA
   - g. the parts removed from a pre-mRNA transcript before translation

Critical Thinking

1. *Antisense drugs* may help us fight cancer and viral diseases, including SARS. They are short mRNA strands that are complementary to the mRNAs associated with these illnesses. Speculate on how these drugs work.

2. A DNA polymerase made an error while a crucial gene region of DNA was being replicated. DNA repair enzymes didn't detect or repair the damage. Here is the part of the DNA strand that contains the error:

After the DNA molecule is replicated, two daughter cells form. One daughter cell is carrying the mutation and the other cell is normal. Develop a hypothesis to explain this observation.

3. *Neurofibromatosis* is a human autosomal dominant disorder caused by mutations in the *NF1* gene. It is characterized by the formation of soft, fibrous tumors in the peripheral nervous system and skin as well as abnormalities in muscles, bones, and internal organs (Figure 13.14).

Because the gene is dominant, an affected child usually has an affected parent. Yet in 1991, scientists reported on a boy who had neurofibromatosis whose parents did not. When they examined both copies of his *NF1* gene, they found the copy he had inherited from his father contained a transposon. Neither the father nor the mother had a transposon in any of the copies of their own *NF1* genes. Explain the cause of neurofibromatosis in the boy and how it arose.

4. Cigarette smoke is mostly carbon dioxide, nitrogen, and oxygen. The rest contains at least fifty-five different chemicals that have been identified as carcinogenic, or cancer-causing, by the International Agency for Research on Cancer (IARC). When these carcinogens enter the bloodstream, enzymes convert them to a series of chemical intermediates in an attempt to make them easier to excrete. Some of the intermediates bind irreversibly to DNA. Speculate on one mechanism by which smoking causes cancer.

5. Using the data in Figure 13.6, translate the following mRNA segment into an amino acid sequence:

5'-GGTTTCTTCAAGAGA-3'

6. The termination of DNA transcription by prokaryotic RNA polymerases depends in some cases on the structure of the newly forming RNA transcript. The terminal end of an mRNA chain often folds back on itself and makes a hairpin-looped structure like the one shown to the right.

Why do you suppose a "stem-loop" structure such as this one causes RNA polymerases to stop transcription when they reach it?

```
              C
          U—C
          G—C
          A—U
          C—G
          C—G
          G—C
          C—G
          C—G
...CCCACAG—CAUUUUU...
```

Figure 13.14 Soft skin tumors on a person with neurofibromatosis, an autosomal dominant disorder.

IMPACTS, ISSUES *Between You and Eternity*

You are in college, your whole life ahead of you. Your risk of developing cancer is as remote as old age, an abstract statistic that is easy to forget.

"There is a moment when everything changes—when the width of two fingers can suddenly be the total distance between you and eternity." Robin Shoulla wrote those words after being diagnosed with breast cancer. She was only seventeen. At an age when most young women are thinking about school, parties, and potential careers, Robin was dealing with a radical mastectomy, pleading with her oncologist not to use her jugular vein for chemotherapy, wondering if she would survive through the next year (Figure 14.1).

Robin became an annual statistic—one of 10,000 or so females and males under age forty who develop breast cancer. About 180,000 new cases are diagnosed each year in the United States population at large.

Cancers are as diverse as their underlying causes, but several gene mutations predispose individuals to developing certain kinds. Either the mutant genes are inherited or they mutate spontaneously in individuals after being assaulted by environmental agents, such as toxic chemicals and ultraviolet radiation.

One gene on chromosome 17 encodes *Her2*, a type of membrane receptor. *Her2* is part of a control pathway that governs the cell cycle—that is, when and how often cells divide. It also is one of the **proto-oncogenes**. When mutated or overexpressed, such genes help bring about cancerous transformations. Cells of about

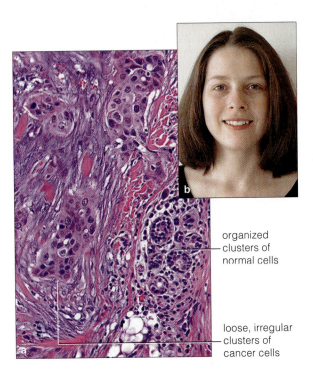

organized clusters of normal cells

loose, irregular clusters of cancer cells

Figure 14.1 Breast cancer. (**a**) Infiltrating ductal carcinoma cells form irregular clusters in breast tissue. (**b**) Robin Shoulla. Diagnostic tests revealed the presence of such cells in her body.

the big picture

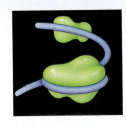

Types of Control Mechanisms Whether and how a gene is expressed depends on regulatory proteins that interact with DNA, RNA, proteins, and one another. It also depends on the attachment and detachment of certain functional groups to the DNA.

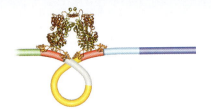

Gene Control in Prokaryotes Being tiny single cells and fast reproducers, prokaryotic cells make rapid responses to short-term changes in nutrient availability and other aspects of the environment. They compensate for the changes by quickly adjusting gene transcription rates.

25 percent of breast cancer patients have too many *Her2* receptors or extra copies of the gene itself. They divide too fast, and abnormal masses of cells result.

Proteins encoded by two different genes, *BRCA1* and *BRCA2*, are among the tumor suppressors that help keep benign or cancerous cell masses from forming. The filmstrip shows part of one of the proteins, which are crucial for DNA repair processes. When *BRCA1* or *BRCA2* is mutated, a cell's capacity to fix breaks in DNA or correct replication errors is compromised. Diverse mutations are free to accumulate throughout the DNA, and such an accumulation leads to cancer.

BRCA1 and *BRCA2* are known as *breast cancer genes*, because cancerous breast cells often hold mutated versions of them. A female in which a *BRCA* gene has mutated in one of three especially dangerous ways has about an 80 percent chance of developing breast cancer before reaching seventy.

Robin Shoulla survived. She may never know which mutation caused her cancer. Thirteen years later, she has what she calls a normal life—a career, husband, children. Her goal is to grow very old with grey hair and spreading hips, smiling.

Robin's story invites you to enter the world of **gene controls**, the molecular mechanisms that govern when and how fast specific genes will be transcribed and translated, and whether gene products will be switched on or silenced. You will be returning to the impact of such controls in many chapters throughout the book.

☑ *How Would You Vote?*

Some females at high risk of developing breast cancer opt for prophylactic mastectomy, the surgical removal of one or more breasts even before cancer develops. Many of them would never have developed cancer. Should the surgery be restricted to cancer treatment? See the Media Menu for details, then vote online.

Gene Control in Eukaryotes   Like prokaryotes, eukaryotic cells control short-term shifts in diet and activity. Unlike prokaryotes, they also control a long-term program of development, which is based largely on selective gene expression and cell differentiation.

Researching Gene Controls   In complex, multicelled eukaryotes, cascades of gene controls guide development of a single fertilized egg into a complete individual with a predictable body plan. A century of research with the common fruit fly has yielded clues about how these controls work.

14.1 Some Control Mechanisms

When, how, and to what extent any gene is expressed depends on the type of cell, its functions, its chemical environment, and signals from the outside.

Diverse mechanisms control gene expression through interactions with DNA, RNA, and new polypeptide chains or the final proteins. Some respond to rising or falling concentrations of specific substances in a cell. Others respond to external signaling molecules.

Control agents include **regulatory proteins** that intervene before, during, or after gene transcription or translation. They include signaling molecules such as hormones, which initiate changes in cell activities when they dock at suitable receptors.

With **negative control**, regulatory proteins slow or stop gene action; with **positive control**, they promote or enhance it. Some DNA base sequences that don't encode proteins are sites of transcriptional control. A **promoter** is a common type of noncoding sequence that marks where to start transcription. **Enhancers** are binding sites for some activator proteins.

Chemical modification offers more control. With methylation, for example, methyl groups ($-CH_3$) get attached to specific regions of newly replicated DNA and prevent access to them. Many heavily methylated genes are activated when the groups are stripped off. Attachment of acetyl groups to histones that organize DNA also exerts control (Section 8.1 and Figure 14.2).

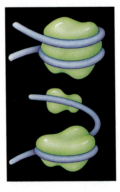

Figure 14.2 How loosening of the DNA–histone packaging in chromosomes may expose genes for transcription. Attachment of an acetyl group to a histone makes it loosen its grip on the DNA that is wound around it. Enzymes that are associated with transcription attach or detach acetyl groups.

You'll read about major signaling mechanisms later. Here, we will sample the events they set in motion.

Gene expression is controlled by regulatory proteins that interact with one another, with control elements built into the DNA, with RNA, and with newly synthesized proteins.

Control also is exerted through chemical modifications that inactivate or activate specific gene regions or the histones that organize the DNA.

14.2 Prokaryotic Gene Control

Think about the dot of the letter "i." About a thousand bacterial cells would stretch side by side across the dot— and each depends as much on gene controls as you do!

Prokaryotic cells grow and divide fast when nutrients are plentiful and other conditions also favor growth. At such times, controls promote the rapid synthesis of enzymes for nutrient absorption and other growth-related metabolic events. Genes that specify enzymes for a metabolic pathway often occur as a linear set in the DNA. And they all may be transcribed together, in a single RNA strand.

NEGATIVE CONTROL OF THE LACTOSE OPERON

With this bit of background, consider an example of how one kind of prokaryote responds to the presence or absence of lactose. *Escherichia coli* lives in the gut of mammals, where it dines on nutrients traveling past. Milk typically nourishes mammalian infants. It does not contain glucose, the sugar of choice for *E. coli*. It does contain lactose, a different sugar.

After being weaned, infants of most species drink little (if any) milk. Even so, *E. coli* cells can still use lactose if and when it shows up in the gut. They can activate a set of three genes for lactose-metabolizing enzymes. A promoter precedes all three genes in *E. coli* DNA, and operators flank it. An **operator** is a binding site for a repressor, a regulatory protein that stops transcription. Such an arrangement, in which a promoter and a set of operators control more than one bacterial gene, is called an **operon** (Figure 14.3).

In the absence of lactose, a repressor molecule binds to a set of operators. Binding causes the DNA region that contains the promoter to twist into a loop, as in Figure 14.3*b*. RNA polymerase, the workhorse that transcribes genes, can't bind to a looped-up promoter. So operon genes aren't used when they aren't required.

When lactose *is* in the gut, *E. coli* converts some of it to allolactose. This sugar binds to the repressor and changes its molecular shape. The altered repressor can't bind to operators. The looped DNA unwinds and RNA polymerase transcribes the genes, so lactose-degrading enzymes are produced when required.

POSITIVE CONTROL OF THE LACTOSE OPERON

E. coli cells pay far more attention to glucose than to lactose. They transcribe genes for its breakdown faster, and continuously. Even when lactose is in the gut, the lactose operon is not used much—unless there is no glucose. Such conditions call for an **activator** protein

Read Me First!
and watch the narrated animation
on the lactose operon

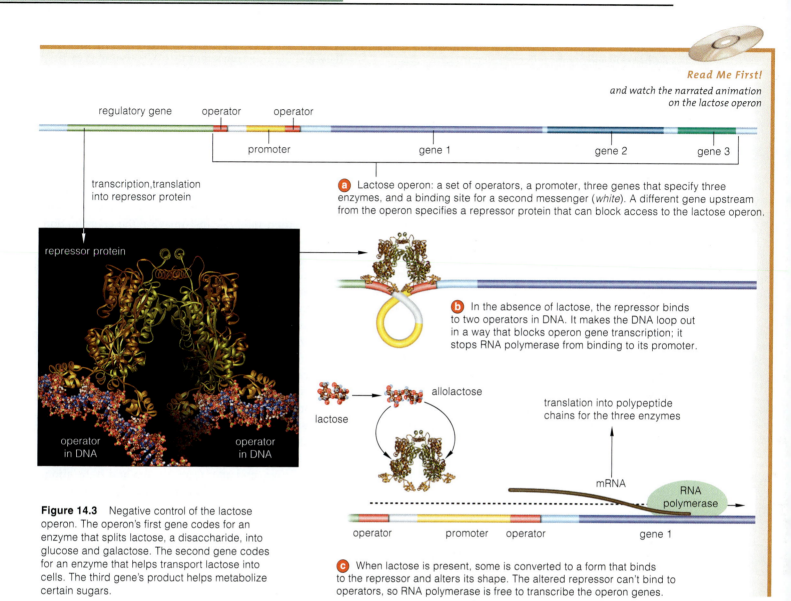

regulatory gene   operator   operator

promoter   gene 1   gene 2   gene 3

transcription,translation
into repressor protein

a Lactose operon: a set of operators, a promoter, three genes that specify three enzymes, and a binding site for a second messenger (*white*). A different gene upstream from the operon specifies a repressor protein that can block access to the lactose operon.

repressor protein

operator
in DNA

operator
in DNA

b In the absence of lactose, the repressor binds to two operators in DNA. It makes the DNA loop out in a way that blocks operon gene transcription; it stops RNA polymerase from binding to its promoter.

allolactose

lactose

translation into polypeptide
chains for the three enzymes

mRNA

RNA
polymerase

operator   promoter   operator   gene 1

Figure 14.3 Negative control of the lactose operon. The operon's first gene codes for an enzyme that splits lactose, a disaccharide, into glucose and galactose. The second gene codes for an enzyme that helps transport lactose into cells. The third gene's product helps metabolize certain sugars.

c When lactose is present, some is converted to a form that binds to the repressor and alters its shape. The altered repressor can't bind to operators, so RNA polymerase is free to transcribe the operon genes.

known as CAP (short for catabolite activator protein). This activator exerts positive control over the lactose operon by making a promoter far more inviting to RNA polymerase. But CAP can't issue the invitation until it is bound to a chemical messenger called cAMP (cyclic adenosine monophosphate). When cAMP and this activator join together and bind to the promoter, they make it far easier for RNA polymerase to start transcribing genes.

When glucose is plentiful, ATP forms by glycolysis, but synthesis of an enzyme necessary to make cAMP is blocked. Blocking is lifted when glucose is scarce and lactose becomes available. cAMP accumulates, CAP–cAMP complexes form, and the lactose operon genes are transcribed. The gene products allow lactose to be used as an alternative energy source.

Unlike cells of *E. coli*, many of us develop *lactose intolerance*. Cells making up the lining of our small intestine make and then secrete lactase into the gut. As many people age, however, concentrations of this lactose-digesting enzyme decline. Lactose accumulates and ends up in the large intestine (colon), where it promotes population explosions of resident bacteria. As the bacteria digest the lactose, a gaseous metabolic product accumulates, distends the colon, and causes pain. Short fatty acid chains released by the bacteria also lead to diarrhea, which can be severe.

Transcription rates of bacterial genes for nutrient-digesting enzymes are quickly adjusted downward and upward by control systems that respond to nutrient availability.

14.3 Eukaryotic Gene Control

Like bacteria, eukaryotic cells control short-term shifts in diet and in levels of activity. If those cells happen to be among hundreds or trillions of cells in a multicelled organism, long-term controls also enter the picture. They orchestrate gene interactions during development.

SAME GENES, DIFFERENT CELL LINEAGES

Later in the book, you will be reading about how you and other complex organisms developed from a single cell. For now, tentatively accept this premise: All cells of your body started out life with the same genes, because every one arose by mitotic cell divisions from the same fertilized egg. And they all transcribe many of the same genes, because they are alike in most aspects of structure and basic housekeeping activities.

In other ways, however, *nearly all of your body cells became specialized in composition, structure, and function.* This process of **cell differentiation** occurs during the development of all multicelled organisms. Differences arise among cells that use different subsets of genes. Specialized tissues and organs are the result.

For example, nearly all of your cells continually transcribe genes for the enzymes of glycolysis. Only immature red blood cells can transcribe the genes for hemoglobin. Your liver cells transcribe genes required to make enzymes that neutralize certain toxins, but they are the only ones that do. When your eyes first formed, certain cells accessed the genes necessary for synthesizing crystallin. No other cells can activate the genes for this protein, which helps make transparent fibers of the lens in each eye.

WHEN CONTROLS COME INTO PLAY

Ultimately, gene expression is all about controlling the amounts and kinds of proteins present in a cell in any specified interval. Just imagine the coordination that goes into making, stockpiling, using, exporting, and degrading thousands of types of proteins in the same moment of cellular time.

Most genes in complex, multicelled organisms are switched off, either permanently or part of the time. Expression of the rest is adjusted up and down. Why? Cells continually deliver and secrete substances into tissue fluids—the body's internal environment—and withdraw substances from it. Inputs and outputs cause slight shifts in the concentrations of nutrients, signaling molecules, metabolic products, and other solutes. Homeostasis is maintained as cells respond to these changes by adjusting gene expression.

Controls over gene expression work at certain stages before, during, and after transcription and translation. Figure 14.4 introduces the main control points.

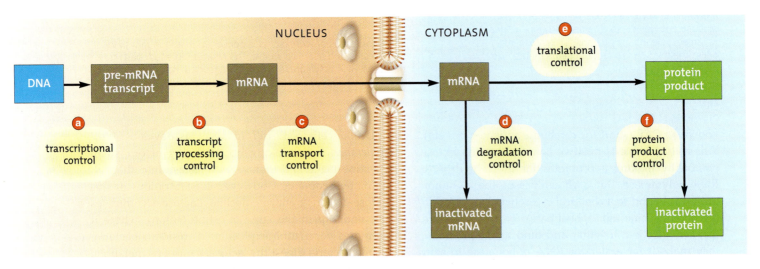

a Chemical modification of DNA restricts access to genes. Genes can be duplicated or rearranged.

b Pre-mRNA spliced in alternative ways can lead to different forms of a protein. Other modifications affect whether a transcript reaches the cytoplasm.

c Transport protein binding determines whether an mRNA becomes delivered to the correct region of cytoplasm for local translation.

d How long an mRNA lasts depends on the proteins that are attached to it and the length of its poly-A tail.

e Translation can be blocked. mRNA cannot attach to a ribosome when proteins bind to it. Initiation factors can be inactivated.

f Processing of new polypeptides may activate or disable them. Control here indirectly affects other activities.

Figure 14.4 Controls that influence whether, when, and how a eukaryotic gene will be expressed.

Figure 14.5 Polytene chromosomes. To sustain their rapid growth rate, *Drosophila* (fruit fly) larvae eat continuously and use a lot of saliva. Giant chromosomes in their salivary gland cells are produced by repeated mitotic DNA replication without cell division. Each strand contains many copies of the same chromosome, aligned side-by-side.

An insect hormone, ecdysone, serves as a regulatory protein; it promotes gene transcription. In response to the hormonal signal, these chromosomes loosen, and they puff out in the regions where genes are being transcribed. Puffs are largest and most diffuse where transcription is most intense.

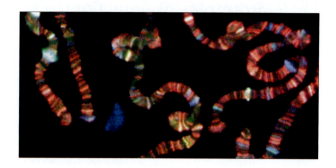

CONTROLS BEFORE TRANSCRIPTION   Remember how many histones and other proteins organize the DNA in a eukaryotic chromosome (Section 8.1)? They affect whether RNA polymerase can access genes and start transcription. Methyl, acetyl, and other functional groups attached to the DNA also can block access to genes. In diploid cells, either the maternal or paternal allele at a gene locus may get methylated, which can block the maternal or paternal influence on a trait.

Also before transcription, some controls trigger the duplication or rearrangement of gene sequences. In immature amphibian eggs and gland cells of certain insect larvae, chromosomes are copied repeatedly in an undividing cell. These multiple DNA replications produce *polytene* chromosomes that have hundreds or thousands of side-by-side gene copies (Figure 14.5).

CONTROL OF TRANSCRIPT PROCESSING   After genes are transcribed, several mechanisms control what the cell does with the RNA. Transcript processing steps dictate whether, when, and how pre-mRNA becomes translated (Section 13.1). For instance, in different kinds of muscle cells, enzymes remove different parts of the pre-RNA transcript for troponin, a contractile protein. After the remaining exons are spliced together, their protein-building message is unique in one tiny region. In each cell type, the resulting protein works in a distinctive way, which helps account for subtle variations in how different kinds of muscles function.

The nuclear membrane is a barrier between a new mRNA and the cellular machinery that can translate it. Only after the mRNA binds to certain proteins will nuclear pore complexes let it cross to the cytoplasm.

Once in the cytoplasm, an mRNA is guided about according to base sequences in its untranslated ends, which are like zip codes. A transport protein bound to a zip code region delivers the mRNA to a particular area of the cell, where it will be translated or stored. In immature eggs, uneven distribution of "maternal messages" and their protein products determines the head-to-tail polarity of the future developing embryo. Control over mRNA localization occurs in the form of binding proteins that attach to the zip code region. These delay or block delivery of an mRNA.

Some transcripts are shelved when the cytoplasm has too many Y-box proteins. When phosphorylated, these proteins bind and help stabilize an mRNA, but if too many of them become attached they block its translation. Thus, phosphorylation of Y-box proteins is a control point for mRNA inactivation. The mRNA stored in unfertilized eggs is bound to Y-box proteins.

CONTROLS AT TRANSLATION   The greatest range of controls over eukaryotic gene expression operates at translation. This process depends on the coordinated participation of many kinds of molecules, including ribosomal subunits and a host of initiation factors. Each kind of molecule is regulated independently of the other kinds.

The stability of mRNA transcripts is also a control point. The more stable a given transcript is, the more proteins can be produced from it. Enzymes typically start digesting mRNA within minutes, nibbling away at its poly-A tail (Section 13.1). How fast they do the deed depends on the tail's length, on base sequences in untranslated regions and other sequences in the coding region, and on attached proteins.

CONTROLS AFTER TRANSLATION   Lastly, control over gene expression is exerted when the protein products are modified, as when phosphate groups are attached to Y-box proteins. Diverse controls activate, inhibit, and stabilize enzymes and other molecules used in protein synthesis. A case in point is allosteric control of tryptophan synthesis, as Section 5.4 describes.

Cell differentiation arises when diverse populations of cells activate or suppress genes in selective, unique ways.

In the cells of complex, multicelled species, gene expression is controlled by mechanisms that govern events before, during, and after transcription and translation.

Most controls over gene expression occur at translation.

14.4 Examples of Gene Controls

Cells rarely use more than 5 to 10 percent of their genes at a given time; controls silence most of them. Which genes are active depends on the type of organism, the stage of growth and development it's passing through, and the controls operating at that stage.

The preceding section introduced you to an important idea. All differentiated cells in a complex, multicelled body use most of their genes in much the same way, but each type also uses a fraction of those genes in a unique, selective way. **Selective gene expression** has made each kind distinctive in one or more aspects of their structure, composition, and function. Here we consider two examples of the controls that guide their selections during embryonic development.

HOMEOTIC GENES AND BODY PLANS

Whether a particular gene gets transcribed depends in part on the action of regulatory proteins, which can bind with promoters, enhancers, or one another. For example, **homeotic genes** are a class of master genes in most eukaryotic organisms. They are transcribed in specific locations in the developing embryo, so their products form in local tissue regions. By interacting with one another and with other control elements, homeotic genes guide formation of organs and limbs by turning on other genes in precise areas, according to a basic body plan.

Homeotic genes were discovered through mutations that cause cells in a *Drosophila* embryo to develop into a body part that belongs somewhere else. For instance, the *antennapedia* gene is supposed to be transcribed where a thorax, complete with legs, should form. In all other regions, cells normally don't transcribe this gene. But Figure 14.6a shows what happens when a mutation allows the gene to be wrongly transcribed in the body region destined to become a head.

Do animals alone have homeotic genes? No. In corn plants, for instance, a different homeotic gene guides the formation of all leaf veins in straight, parallel lines. If the gene mutates, the veins will twist.

Homeotic genes code for regulatory proteins that include a "homeodomain," a sequence of about sixty amino acids. This sequence binds to control elements in promoters and enhancers (Figure 14.6b). More than 100 homeotic genes have been identified in diverse eukaryotes—and the same mechanisms control their transcription. Many of the genes are interchangeable among species as evolutionarily distant as yeasts and humans, so we can expect that they evolved in the most ancient eukaryotic cells. Their protein products often differ only in *conservative* substitutions. In other words, one amino acid has replaced another, but it has similar chemical properties.

X CHROMOSOME INACTIVATION

Diploid cells of female humans and female calico cats have two X chromosomes. One is in threadlike form. The other stays scrunched up, even during interphase. This scrunching is a programmed shutdown of all but about three dozen genes on *one* of two homologous X chromosomes. The shutdown is called **X chromosome inactivation**, and it happens in female embryos of all placental mammals and their marsupial relatives.

Figure 14.7a shows one condensed X chromosome in the nucleus of a cell at interphase. We also call this condensed structural form a Barr body (after Murray Barr, who first identified it).

One X chromosome gets inactivated when embryos are still a tiny ball of cells. In placental mammals, the shutdown is random, in that *either* chromosome could become condensed. The maternal X chromosome may be inactivated in one cell; the paternal or the maternal X chromosome may be inactivated in a cell next to it.

Figure 14.6 (a) Experimental evidence of controls over where body parts develop. In *Drosophila* larvae, activation of genes in one group of cells normally results in antennae on the head. A mutation that affects *antennapedia* gene transcription puts legs on the head. This is one of the genes controlled by regulatory proteins that have homeodomains. (b) Stick model for the binding of a homeodomain sequence to a transcriptional control site in DNA.

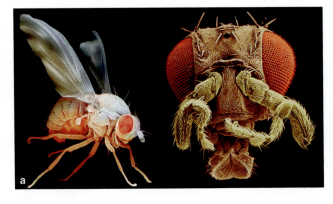
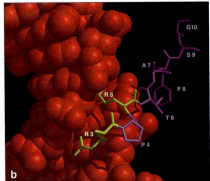

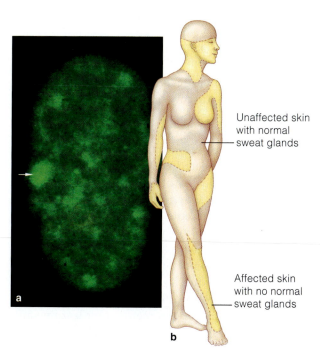

Figure 14.7 (**a**) In the somatic cell nucleus of a human female, a condensed X chromosome, also called a Barr body (*arrow*). The X chromosome in cells of human males is not condensed this way. (**b**) A mosaic tissue effect that shows up in anhidrotic ectodermal dysplasia.

Unaffected skin with normal sweat glands

Affected skin with no normal sweat glands

Figure 14.8 Why is this female cat "calico"? In her body cells, one of her two X chromosomes has a dominant allele for the brownish-black pigment melanin. Expression of the allele on her other X chromosome results in orange fur. When this cat was an embryo, one X chromosome was inactivated at random in each cell that had formed by then.

Patches of different colors reflect which allele was inactivated in cells that formed a given tissue region. White patches are an outcome of an interaction that involves a different gene, the product of which blocks melanin synthesis.

Once that random molecular decision is made in a cell, all of that cell's descendants make the exact same decision as they go on dividing to form tissues. What is the outcome? *A fully developed female has patches of tissue where genes of the maternal X chromosome are being expressed, and patches of tissue where genes of the paternal X chromosome are being expressed.* She has become a "mosaic" for X chromosome expression!

When alleles on two homologous X chromosomes are not identical, differences may occur among patches of tissues throughout the body. Mosaic tissues can be observed in human females who are heterozygous for a rare recessive allele that causes an absence of sweat glands. Sweat glands formed in only parts of their skin. Where the glands are absent, the recessive allele is on the X chromosome that was not shut down. This mosaic effect is one symptom of *anhidrotic ectodermal dysplasia* (Figure 14.7b), a heritable disorder that is characterized by abnormalities in the skin and in the structures derived from it, including teeth, hair, nails, and sweat glands.

A different mosaic tissue effect shows up in female calico cats, of the sort shown in Figure 14.8. These cats are heterozygous for a certain coat color allele on their X chromosomes.

The shutdown isn't an accident of evolution; it has an important function. In mammals, recall, males have one X and one Y chromosome. This means the females have twice as many X chromosome genes. Inactivating one of their X chromosomes balances gene expression between the sexes. The normal development of female embryos depends on this type of control mechanism, which is called **dosage compensation**.

How, in a single nucleus, does one X chromosome get shut down while the other does not? Several molecules participate, including histone methylases and an X chromosome gene called *XIST*. The *XIST* product, a large RNA molecule, binds chromosomal DNA like a gene-masking paint. Although the *XIST* gene is found on both X chromosomes, only one of them expresses it. As it does, it gets fully painted with RNA, and so its genes become inactivated. The other X chromosome does not express *XIST*, and does not become painted. Only the genes on this chromosome remain active and may be transcribed. Why only one of the two X chromosomes expresses *XIST* is not yet fully understood.

Controls over when, how, and to what extent a gene is expressed depend on the type of cell and its functions, on the cell's chemical environment, and on signals for change. Homeotic gene expression and dosage compensation are examples of control mechanisms in eukaryotic cells.

14.5 There's a Fly in My Research

Structural patterns emerge as the embryos of animals and plants develop, and they are both beautiful and fascinating. Researchers have correlated those patterns with the expression of specific genes at particular times, in particular tissues, for diverse organisms.

DROSOPHILA!

For about a hundred years, *Drosophila melanogaster* has been the fly of choice for laboratory experiments. It costs almost nothing to feed and house this tiny fruit fly. *D. melanogaster* reproduces fast in bottles, it has a short life cycle, and disposing of spent bodies after an experiment is a snap. Thanks to automated gene sequencing, we now know how its 13,601 genes are distributed among its four pairs of chromosomes.

Studies of *Drosophila* at the anatomical, cytological, biochemical, and genetic levels continue to reveal much about gene controls over how animal embryos develop. They also yield insights into the evolutionary connections among groups of animals.

CLUES TO GENE CONTROLS

Over the past ten years, *Drosophila* researchers have made remarkable discoveries about how embryos develop, especially through **knockout experiments**. In such experiments, individual genes are deleted from wild-type experimental organisms. Differences between the engineered and wild-type organisms, either morphological or behavioral, are clues to the function of the missing gene.

Knockout experiments have identified many hundreds of *Drosophila* genes, which tend to be named after what happens when they are missing. Many turned out to be homeotic genes. For instance, *eyeless* is a control gene expressed in fruit fly embryos. In its absence, no eyes form. Other named genes include *dunce* (a regulatory protein required for learning and memory), *wingless*, *wrinkled*, *tinman* (necessary for heart development), *minibrain*, and *groucho* (which, among other things, prevents overproduction of whisker bristles). Figures 14.9 and 14.10 show a small sampling of mutant flies.

More ambitious *Drosophila* experiments with deleted genes yield intriguing information about the controls over development. By adding special promoters to a gene, researchers can control its expression with external cues, such as temperature. They also can delete genes from one part of the *Drosophila* genome and insert them into another. This molecular sleight-of-hand with the *eyeless* gene demonstrated that its expression can induce an eye to form not only on the fly's head, but also on the legs, wings, and antennae (Figure 14.10).

Astonishingly, the *eyeless* gene has counterparts in humans (a gene named *Aniridia*), mice (*Pax-6*), and squids (also *Pax-6*). Humans who have no

Figure 14.9 A few *Drosophila* mutants. (**a**) Wild-type (normal) fruit fly. The photograph above it can give you an idea of a fruit fly's size relative to the surface of a peach. (**b**) Yellow miniature. (**c**) Curly wings. (**d**) Vestigial wings.

Figure 14.10 Two more *Drosophila* mutations. *Left*, an eye that formed on a fruit fly leg. *Right*, a fruit fly with a double thorax, the outcome of a homeobox gene mutation.

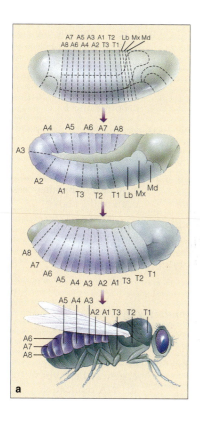

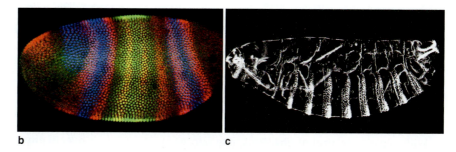

b

c

Figure 14.11 Genes and *Drosophila*'s segmented body plan. (**a**) Fate map for the surface of a *Drosophila* zygote. Such maps indicate where each differentiated cell type in the adult originated. The pattern starts with the polar distribution of maternal mRNA and proteins in the unfertilized egg. This polarity dictates the future body axis. A series of segments will develop along this axis. Genes specify whether legs, wings, eyes, or some other body parts will develop on a particular segment.

Briefly, here's how it happens: Maternal gene products prompt expression of gap genes. Different gap genes become activated in regions of the embryo with higher or lower concentrations of different maternal gene products. Gap gene products influence each other's expression as well. They form a primitive spatial map.

Depending where they occur relative to the concentrations of gap gene products, embryonic cells express different pair-rule genes. Products of pair-rule genes accumulate in seven transverse stripes that mark the onset of segmentation (**b**). They activate other genes, the products of which divide the body into units (**c**). These interactions influence the expression of homeotic genes, which collectively govern the identity of each segment.

functional *Aniridia* genes have eyes without irises. *Aniridia* or *Pax-6* inserted into an *eyeless* mutant fly has the same effect as the *eyeless* gene—it induces eye formation wherever it is expressed. Here is evidence that animals as evolutionarily distant as insects, cephalopods, and mammals are connected by a shared ancestor.

GENES AND PATTERNS IN DEVELOPMENT

Different cells become organized in different ways in a new embryo. They divide, differentiate, and live or die; they migrate or stick to cells of the same type in tissues. Descendant cells fill in the details in orderly patterns, in keeping with a master body plan.

Such master plans consist of genes expressed in certain places at certain times during development. The regional and temporal gene expression generates a three-dimensional map of many overlapping proteins, most of which are transcription factors.

As an embryo develops, certain proteins induce undifferentiated cells to develop into different body tissues, depending on where the cells start out on the map. One example is the development of segments in *Drosophila* embryos (Figure 14.11).

Figure 14.12 *Left*, seven spots in the embryonic wing of a moth larva identify the presence of a gene product that will induce the formation of seven "eyespots" in the wing of the adult (*right*).

Pattern formation is the name for the emergence of embryonic tissues and organs in predictable patterns, in places where we expect them to be. Figures 14.11 and 14.12 are graphic examples. In Section 38.5, you will be taking a closer look at the controlled gene interactions that fill in details of the body plan.

Summary

Section 14.1 Whether, when, and how a gene gets expressed depends on controls over transcription and translation, and on modifications to protein products. Regulatory proteins (e.g., activators and inhibitors of transcription) and hormones are examples of control agents. These controls interact with one another, with control elements built into DNA molecules, with RNA, and with gene products.

With negative control mechanisms, regulatory proteins slow or curtail gene activity. With positive control mechanisms, they promote gene activity.

Section 14.2 Like all cells, prokaryotes respond quickly to short-term shifts in nutrients and other environmental conditions. Most of their gene control mechanisms adjust transcription rates in response to nutrient availability. Bacterial operon systems are examples of prokaryotic gene regulation.

Section 14.3 All cells of complex multicelled eukaryotes inherit the same genes, but each cell type selectively activates or suppresses a fraction of the genes in ways that lead to one or more unique aspects of structure, composition, and function.

At any time, most genes in a eukaryotic cell are shut off, unused. Those that the cell uses for housekeeping purposes, ongoing metabolic functions, are switched on all the time, at low levels. Expression of the other genes is adjustable. When control mechanisms come into play depends on cell type, prevailing chemical conditions, and signals from other cell types that can change a target cell's activities.

Gene expression within a cell changes in response to external conditions and is subject to long-term controls over growth and development. Eukaryotic cells control gene expression at key points, including transcription, RNA processing, RNA transport, mRNA degradation, translation, and protein activity. Translation is the major control point for most eukaryotic genes because so many participating molecules are regulated.

Section 14.4 Selective gene expression is the basis of cell differentiation during growth and development. It gives rise to cells that differ from one another in structure and function.

Complex eukaryotic body plans are influenced by homeotic genes, the master genes that control the emergence of the basic body plan during development.

X-chromosome inactivation is an example of dosage compensation, a control mechanism that maintains a crucial balance of gene expression between the sexes.

Section 14.5 Experiments with *Drosophila* identified a host of control genes. In embryo development, spatial maps of regulatory proteins guide the formation of tissues and organs in expected patterns.

Self-Quiz   *Answers in Appendix III*

1. The expression of a given gene depends on the _____ .
   a. type of cell and its functions   c. environmental signals
   b. chemical conditions   d. all of the above

2. Hormones may _____ gene transcription in target cells.
   a. promote   c. participate in
   b. inhibit   d. both a and b

3. Eukaryotic genes guide _____ .
   a. fast short-term activities   c. development
   b. overall growth   d. all of the above

4. Gene expression adjusts in response to changing _____ .
   a. nutrient availability   c. signals from other cells
   b. solute concentrations   d. all of the above

5. Cell differentiation _____ .
   a. occurs in all complex multicelled organisms
   b. requires unique genes in different cells
   c. involves selective gene expression
   d. both a and c
   e. all of the above

6. Regulatory proteins interact with _____ .
   a. DNA   c. gene products
   b. RNA   d. all of the above

7. An operon typically governs _____ .
   a. bacterial genes   c. genes of all types
   b. eukaryotic genes   d. DNA replication

8. In prokaryotic cells but not eukaryotic cells, a(n) _____ is a type of base sequence that precedes genes of an operon.
   a. lactose molecule   c. operator
   b. promoter   d. both b and c

9. A nucleotide sequence that signals the start of a gene is a(n) _____ .
   a. promoter   b. operator   c. enhancer   d. activator

10. Eukaryotic cells in complex organisms regulate gene expression by controlling different processes in _____ .
    a. transcription   e. mRNA degradation
    b. RNA processing   f. protein activity
    c. translation   g. a and d
    d. RNA transport   h. all of the above

11. X chromosome inactivation _____ .
    a. is dosage compensation   c. makes calico cats
    b. balances gene expression   d. both a and b

12. Homeotic genes _____ .
    a. are part of a bacterial operon
    b. control eukaryotic body plans
    c. control X chromosome inactivation
    d. both a and c

13. A cell with a Barr body is _____ .
    a. prokaryotic   c. from a female mammal
    b. from a male mammal   d. infected by the Barr virus

14. Match the terms with the most suitable description.
    ____ homeotic gene   a. binding site for repressor
    ____ operator   b. specialization during
    ____ proto-oncogene   development
    ____ differentiation   c. inactivated X chromosome
    ____ Barr body   d. can cause cancer
                              e. body plan development

Critical Thinking

1. Distinguish between:
   a. repressor protein and activator protein
   b. promoter and operator

2. Define three types of gene controls. Do they work for both prokaryotic and eukaryotic cells?

3. Unlike most rodents, guinea pigs are well developed at the time of birth. Within a few days, they can eat grass, vegetables, and other plant material. Suppose a breeder decides to separate baby guinea pigs from their mothers after three weeks. He wants to keep the males and females in different cages. However, he has trouble identifying the sex of young guinea pigs. Suggest how a microscope can help him identify their sex.

4. A plant, a fungus, and an animal consist of diverse cell types. How might this diversity arise, given that body cells in each of these organisms inherit the same set of genetic instructions? As part of your answer, define cell differentiation and the general way that selective gene expression brings it about.

5. In what fundamental way do negative and positive controls of transcription differ? Is the effect of one or the other form of control (or both) reversible?

6. If all cells in your body start out life with the same inherited information on how to build proteins, then what caused the differences between a red blood cell and a white one? Between a white blood cell and a nerve cell?

7. *Duchenne muscular dystrophy*, a genetic disorder, affects boys almost exclusively. Early in childhood, muscles begin to atrophy (waste away) in affected individuals, who typically die in their teens or early twenties.

   Muscle biopsies of a few women who carry an allele that is associated with the disorder identified some body regions of atrophied muscle tissue. They also showed that muscles adjacent to a region of atrophy were normal or even larger and more chemically active, as if to compensate for the weakness of the adjoining region.

   Form a hypothesis about the genetic basis of Duchenne muscular dystrophy that includes an explanation of why it might appear in some body regions but not others.

8. The closer a mammalian species is to humans in its genetic makeup, the more useful information it yields in laboratory studies of the mechanisms of cancer. Do you support the use of any mammal for cancer research? Why or why not?

9. Geraldo isolated an *E. coli* strain in which a mutation has hampered the capacity of CAP to bind to a region of the lactose operon, as it would do normally. How will this mutation affect transcription of the lactose operon when the *E. coli* cells are exposed to the following conditions? Briefly state your answers:
   a. Lactose and glucose are both available.
   b. Lactose is available but glucose is not.
   c. Both lactose and glucose are absent.

10. Calico cats are almost always female. A male calico cat is usually sterile. Briefly explain why.

11. The *Drosophila* embryo in Figure 14.13 displays a repeating pattern of gene expression. Reflect on Figure 14.11, then think about the gene product that made the red rings. What type of gene specified this product?

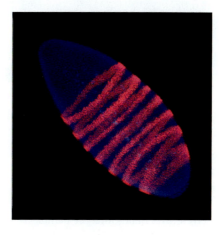

Figure 14.13 *Drosophila* zygote. The fluorescent rings are evidence that a gene is being expressed in certain regions.

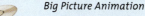

Media Menu

Student CD-ROM

Impacts, Issues Video
   Between You and Eternity
Big Picture Animation
   Regulating and researching gene expression
Read-Me-First Animation
   The lactose operon
Other Animations and Interactions
   X chromosome inactivation in a calico cat

InfoTrac

- How Hibernators Might One Day Solve Medical Problems. *The Lancet*, October 2001.
- Face Shift: How Sleeping Sickness Parasites Evade Human Defenses. *Scientific American*, May 2002.
- Researchers Discover DNA Packaging in Living Cells Is Dynamic. *Ascribe Higher Education News Service*, January 2003.
- Silence of the Xs. *Science News*, August 2001.

Web Sites

- The lac Repressor: www.rcsb.org/pdb/molecules/pdb39_1.html
- Genomes in Flux: opbs.okstate.edu/~melcher/MG/MGW3/MG3.html
- Mutant Fruit Flies: www.exploratorium.edu/exhibits/mutant_flies

How Would You Vote?

Women and men with particular gene mutations are far more susceptible to developing breast cancer than people who do not carry those mutations. Prophylactic mastectomy reduces their risk by the surgical removal of one or both breasts before cancer can develop. Statistically, most of the people who opt for prophylactic mastectomy would never have developed cancer in the first place. Should mastectomy be restricted to people who have already developed cancer?

IMPACTS, ISSUES   *Golden Rice or Frankenfood?*

Not too long ago, the World Health Organization made a conservative estimate that 124 million children around the world show vitamin A deficiencies. These children may become permanently blind and develop other disorders. Researchers began working on a solution. They transferred three genes into rice plants. The genes directed the plants to make beta-carotene, a yellow pigment that is a precursor for vitamin A. Eating just 300 grams per day of the new "golden rice" might be enough to prevent vitamin A deficiency.

No one wants children to suffer or die. But many people oppose the idea of genetically modified foods, including golden rice. Possibly they are unaware of the history of agrarian societies. It isn't as if our ancestors were twiddling their green thumbs. For thousands of years, their artificial selection practices coaxed new plants and new breeds of cattle, cats, dogs, and birds from wild ancestral stocks. Meatier turkeys, seedless watermelons, big juicy corn kernels from puny hard ones—the list goes on (Figure 15.1).

And we're newcomers at this! During the 3.8 billion years before we even made our entrance, nature busily conducted uncountable numbers of genetic experiments by way of mutation, crossing over, and gene transfers

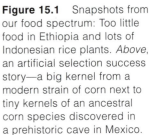

Figure 15.1 Snapshots from our food spectrum: Too little food in Ethiopia and lots of Indonesian rice plants. *Above*, an artificial selection success story—a big kernel from a modern strain of corn next to tiny kernels of an ancestral corn species discovered in a prehistoric cave in Mexico.

the big picture

The Genome Project   The discovery of the structure of DNA in 1953 sparked intense interest in creating technologies to manipulate that structure. Fifty years later, the entire sequence of bases in the human genome was completed.

Tools of the Trade   Scientists use DNA technologies to cut, identify, isolate, clone, copy, sequence, compare, and manipulate the DNA of any organism they wish to study. They put these technologies to practical use in genetic engineering.

between species. These processes introduced changes in the molecular messages of inheritance, and today we see their outcomes in the sweep of life's diversity.

Maybe the unsettling thing about the more recent human-directed changes is that the pace has picked up, hugely. We're getting pretty good at tinkering with the genetics of many organisms. We do this for pure research and for useful, practical applications.

Some say we must never alter the DNA of anything. The concern is that we as a species simply do not have the wisdom to bring about genetic changes without causing irreparable harm. One is reminded of our very human tendency to leap before we look.

And yet, we dream of the impossible. Something about the human experience gave us a capacity to imagine wings of our own making, and that capacity carried us to the frontiers of space. Someone else dreamed of turning plain rice into more nutritious food that might keep some children from going blind.

Many economic questions also remain unanswered. For example, will the patents on golden rice translate into higher production costs for the rural farmers of developing countries that urgently need the rice? Will transfer of beta-carotene genes disrupt a rice plant's messages of inheritance in some unexpected way?

The questions confronting you are these: Should we be more cautious, believing the risk takers may go too far? What do we stand to lose if risks are not taken?

☑ *How Would You Vote?*

Nutritional labeling is required on all packaged food in the United States, but genetically modified food products may be sold without labeling. Should food distributors be required to label all products made from genetically modified plants or livestock? See the Media Menu for details, then vote online.

Genetic Engineering Normal or modified genes are being inserted into individual organisms both for research and practical applications. Genetically modified organisms help farmers produce food more efficiently. They are also a source of biomaterials and pharmacologic products.

Bioethics The human genome has been sequenced, and the findings are being used for gene therapy and other applications. Many ethical and social issues remain unresolved as objections to the use of genetic engineering continue.

15.1 Tinkering With the Molecules of Life

In this unit, you started with cell division mechanisms that allow parents to pass on DNA to new generations. You moved to the chromosomal and molecular basis of inheritance, then on to gene controls that guide life's continuity. The sequence parallels the history of genetics. And now, you have arrived at the point where geneticists hold molecular keys to the kingdom of inheritance.

EMERGENCE OF MOLECULAR BIOLOGY

In 1953, James Watson and Francis Crick unveiled their model of the DNA double helix and ignited a global blaze of optimism about genetic research. The very book of life seemed to open up for scrutiny. In reality, it dangled just beyond reach. Major scientific breakthroughs are seldom accompanied by the simultaneous discovery of the tools necessary to study them. New methods of DNA research had to be invented before that book could become readable.

In 1972, Paul Berg and his associates were the first to make **recombinant DNA**. They fused fragments of DNA from one species into the genetic material from a different species, which they had grown in the laboratory. Their new recombinant DNA technique allowed them to isolate and replicate manageable subsets of DNA from any organism they wanted to study. The science of molecular biology was born, and suddenly everybody was worried about it.

Although researchers knew that DNA was not toxic, they could not predict with absolute certainty what would happen every time they fused genetic material from different organisms. Would they create new super-pathogens by accident? Could DNA from normally harmless organisms be fused to create a new form of life? What if their creation escaped into the environment and transformed other organisms?

In a remarkably quick and responsible display of self-regulation, scientists reached a consensus on safety guidelines for DNA research. Adopted at once by the National Institutes of Health (NIH), the guidelines listed laboratory procedural precautions. They covered the design and use of host organisms that could survive only under the narrow range of conditions that occur in the laboratory. Researchers stopped using the DNA from pathogenic or toxic organisms for recombination experiments until proper containment facilities were developed.

A golden age of recombinant DNA research soon followed. The emphasis had shifted from DNA's chemical and physical properties to its specific molecular structure. In 1977, Allan Maxam, Walter Gilbert, and Fred Sanger developed a method for determining the nucleotide sequence of cloned DNA fragments. The tools for reading the book of life, opened more than twenty years before, were now available for everyone to use.

DNA sequencing was cool, a visually rewarding, data-rich technique that entranced more than a few scientists. Unbelievable amounts of sequence data accumulated, from unbelievably diverse organisms. Computer technology at the time was advancing simultaneously, but it was barely keeping pace with

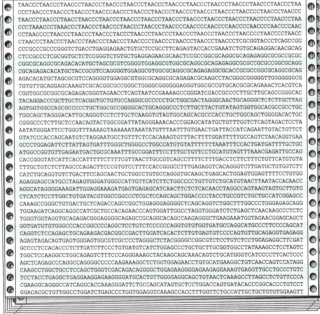

Figure 15.2 A few bases of the human genome—and a few of the supercomputers used to sequence it—at Celera Genomics in Maryland.

the tremendous demand for sequence data analysis and storage. In 1982, the NIH provided three million dollars to fund the first large-scale DNA database in the United States, one accessible to the public.

THE HUMAN GENOME PROJECT

Around 1986, everyone seemed to be arguing about sequencing the human genome. A **genome** is all the DNA in a haploid number of chromosomes. Many scientists insisted that the benefits for medicine and pure research would be incalculable. Others said the mapping would divert funding from other studies that had greater urgency as well as more likelihood of success.

At the time, sequencing three billion bases was a daunting and seemingly impossible task. With the techniques available at the time, it would have taken a worldwide consortium at least fifty years just to identify the sequence, even before deciphering what it meant. But techniques were getting better every year, and more bases were being sequenced and analyzed in less time. Automated (robotic) sequencing had just been invented, as had PCR, the polymerase chain reaction. Although both techniques were still cumbersome, expensive, and far from standardized, many sensed their potential for molecular biology.

Sequencing was still laborious. Waiting for faster methods seemed to be the most efficient means of sequencing the human genome, but exactly when was the technology going to be fast *enough*? Who would decide?

It was during this heated debate in 1987 that several independent organizations launched their own versions of the Human Genome Project. Among them was a company started by Walter Gilbert. He declared that his company would not only sequence the human genome, it would also patent the genome.

In early 1988, the NIH effectively annexed the entire Human Genome Project by hiring Watson as its head and providing researchers with 200 million dollars per year. A consortium formed between the NIH and other institutions working on different versions of the project.

Watson set aside 3 percent of the funding to study ethical and social issues arising from the research. He resigned in 1992 because of a disagreement with the NIH about patenting partial gene sequences. Francis Collins replaced him in 1993.

Amid ongoing squabbles over patent issues, the bulk of genome project sequencing in the United States continued at the NIH until 1998, when the

Figure 15.3 Everyone pitches in. The Dalai Lama prepares mouse DNA for sequencing during his 2003 visit to Whitehead Institute/MIT Center for Genome Research in Cambridge, Massachusetts.

scientist Craig Venter started Celera Genomics (Figure 15.2). Venter declared that *his* new company would be the first to complete the genome sequence. His challenge prompted the United States government to move its sequencing efforts into high gear.

Sequencing of the human genome was officially completed in 2003—fifty years after the discovery of DNA structure. About 99 percent of the coding regions in human DNA have been deciphered with a high degree of accuracy. A number of other genomes also have been fully sequenced (Figure 15.3).

What do we do with this vast amount of data? The next step is to investigate questions about precisely what that sequence means—where the genes are, where they are not, what the control mechanisms are and how they operate.

Recently, more than 21,000 human genes were identified. This doesn't mean we know what all those genes encode; it only means we know they are definitely genes. One of the many interesting discoveries is that the first intron and the last exon of most gene sequences are longer than the others. They may actually be part of an as yet undiscovered transcriptional control mechanism.

15.2 A Molecular Toolkit

Analysis of genes starts with manipulation of DNA. With molecular tools, researchers can cut DNA from different sources, then splice the fragments together.

THE SCISSORS: RESTRICTION ENZYMES

In 1970, Hamilton Smith and his colleagues were studying viral infection of *Haemophilus influenzae*, a species of bacteria. The bacteria protected themselves from infection by cutting up the invading viral DNA before it inserted itself into the bacterial chromosome.

Smith isolated one of the bacterial enzymes that was cutting up the viral DNA, the first known **restriction enzyme**. In time, several hundred strains of bacteria and a few eukaryotic cells yielded thousands more. Each restriction enzyme cuts double-stranded DNA at a specific base sequence between four and eight base pairs in length. Most of these recognition sites contain the same nucleotide sequence on both strands of the DNA. For instance, GAATTC is recognized on both strands and cut by the enzyme EcoRI.

Many restriction enzymes make staggered cuts that put a single-stranded "tail" on DNA fragments. Such cuts have a "sticky" end—a single-stranded tail. That tail can base-pair with a tail of another DNA molecule cut by the same enzyme, because the sticky ends of both fragments will match up (Figure 15.4a).

Tiny nicks remain when the fragments base-pair. A different enzyme, **DNA ligase**, seals the nicks, which results in a recombinant DNA molecule (Figure 15.4b). Recombinant DNA can consist of base sequences from different organisms of the same or different species.

CLONING VECTORS

Bacterial cells, recall, have only one chromosome—a circular DNA molecule. But many also have plasmids. A **plasmid** is a small circle of extra DNA with just a few genes (inset, *left*). It gets replicated right along with the bacterial chromosome. Bacteria normally can live without plasmids. Even so, some plasmid genes are useful, as when they confer resistance to antibiotics.

Under favorable conditions, bacteria divide often, so huge populations of genetically identical cells form swiftly. Before each division, replication enzymes copy both the chromosomal DNA *and* the plasmid DNA, in some cases repeatedly. This gave researchers an idea. Why not try to insert a fragment of foreign DNA into a plasmid and see if a bacterial cell replicates it?

A modified plasmid that accepts foreign DNA and slips into a host bacteria, yeast, or some other cell is a **cloning vector**. Cloning vectors usually have multiple cloning sites, which are several unique restriction enzyme sequences clustered in one part of the vector. As you'll see later, the vector also has genes that help researchers identify which cells it slips into, such as genes for antibiotic resistance (Figure 15.5).

A cell that takes up a cloning vector may found a huge population of descendant cells, each containing an identical copy of the vector and the foreign DNA inserted into it. Collectively, all of the identical cells hold many "cloned" copies of the foreign DNA.

Such DNA cloning is a tool that helps researchers amplify and harvest unlimited amounts of particular DNA fragments for their studies (Figure 15.6).

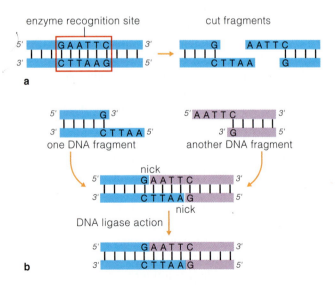

Figure 15.4 (**a**) Formation of restriction fragments and (**b**) splicing fragments into a recombinant DNA molecule.

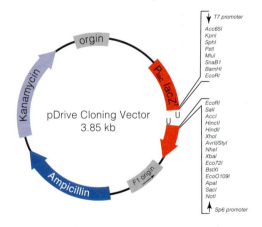

Figure 15.5 A commercially available cloning vector, with its useful restriction enzyme sites listed at right. This vector includes antibiotic resistance genes (*blue*) and the bacterial *lacZ* gene (*red*). These genes help researchers identify cells that take up recombinant molecules.

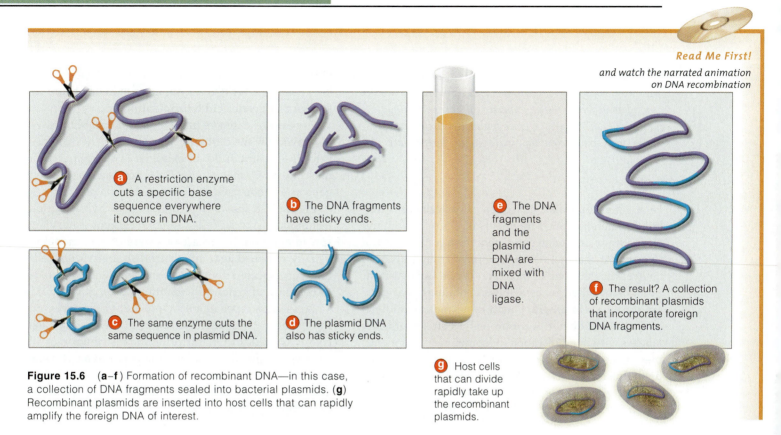

a A restriction enzyme cuts a specific base sequence everywhere it occurs in DNA.

b The DNA fragments have sticky ends.

c The same enzyme cuts the same sequence in plasmid DNA.

d The plasmid DNA also has sticky ends.

e The DNA fragments and the plasmid DNA are mixed with DNA ligase.

f The result? A collection of recombinant plasmids that incorporate foreign DNA fragments.

g Host cells that can divide rapidly take up the recombinant plasmids.

Figure 15.6 (**a–f**) Formation of recombinant DNA—in this case, a collection of DNA fragments sealed into bacterial plasmids. (**g**) Recombinant plasmids are inserted into host cells that can rapidly amplify the foreign DNA of interest.

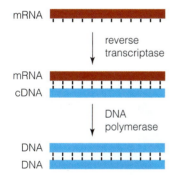

Figure 15.7 How to make cDNA. Reverse transcriptase catalyzes the assembly of a single DNA strand on an mRNA template, forming an mRNA–cDNA hybrid molecule. Next, DNA polymerase replaces the mRNA with another DNA strand. The result is double-stranded DNA.

cDNA CLONING

Chromosomal DNA usually contains introns (Section 13.1). Sometimes it's impossible to tell whether a gene sequence is part of an intron or exon or to pinpoint where it starts and ends. A researcher investigating gene products or gene expression focuses on mRNA, because the introns have already been snipped out of it. All that's left is coding sequence and some small signal sequences. Any time a gene is being expressed,

mRNA is being transcribed, so cells that are actually using a gene will also contain the mRNA it encodes.

Restriction enzymes do not cut RNA, so mRNA cannot be cloned until it has been translated first into DNA. Replication enzymes isolated from viruses or bacteria can be used to translate the mRNA *in vitro*, or inside a test tube. **Reverse transcriptase** is a viral enzyme that uses the mRNA as a template. Using free nucleotides, it assembles a single strand of **cDNA**, or *complementary* DNA, on the template (Figure 15.7). A hybrid molecule is the outcome; one strand of mRNA and one strand of cDNA are base-paired together.

DNA polymerase added to the mix strips the RNA from the hybrid molecule as it copies the first strand of cDNA into a second strand. The result is a double-stranded DNA copy of the original mRNA. And that copy may be used for cloning.

Molecular biologists manipulate DNA and RNA. Restriction enzymes cut DNA from organisms of the same or different species, and ligases glue the fragments into plasmids.

A recombinant plasmid is a cloning vector; it can slip into bacteria, yeast, or other cells that divide rapidly. Host cells make multiple, identical copies of the foreign DNA.

Reverse transcriptase uses mRNA as a template to make cDNA.

15.3 Haystacks to Needles

Any genome consists of thousands of genes. E. coli has 4,279; humans have about 30,000. To study or modify any one of those genes, researchers must first find it among all others in the genome, and it's like searching for a needle in a haystack. Once found, it must be copied many times to make enough material for experiments.

ISOLATING GENES

Each **gene library** is a collection of bacterial cells that house different cloned fragments of DNA. We call the cloned fragments of an entire genome a *genomic* library. By contrast, a *cDNA library* is derived from mRNA.

A particular gene of interest must be isolated from millions of other genes. Clones containing that gene are mixed up in a library with thousands or millions of others that do not. A **probe**, a short stretch of DNA labeled with a radioisotope (or sometimes a pigment), may be used to find a one-in-a-million clone. Probes distinguish one DNA sequence from all of the others in a library of clones or any other collection of mixed DNA. A radiolabeled probe base-pairs with DNA in the gene region of interest, then researchers can find it with devices that detect radiation. Such base-pairing between DNA (or RNA) from more than one source is known as **nucleic acid hybridization**.

How do researchers make a suitable probe? If they already know the desired gene sequence, they can use it to design and build an oligomer (a short stretch of nucleotides). Or they can use DNA from a closely related gene as a probe even if it isn't an exact match.

Figure 15.8 shows steps of one probe hybridization technique. Bacterial cells containing a gene library are spread apart on the surface of a solid growth medium, usually enriched agar, in a petri dish. Individual cells undergo repeated divisions, which form large clusters, or colonies, of genetically identical bacterial cells.

Press a piece of nylon or nitrocellulose paper on top of the petri dish and some of the cells from each colony will stick to it, mirroring the distribution of all colonies on the dish. Soaking the paper in an alkaline solution ruptures the cells, which releases their DNA. The solution also denatures the DNA, separating it into single strands that stick to the paper in the spots

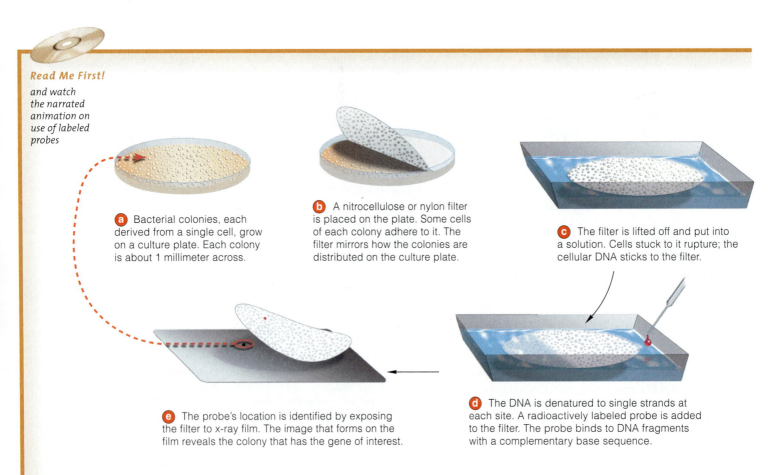

Read Me First!
and watch the narrated animation on use of labeled probes

a Bacterial colonies, each derived from a single cell, grow on a culture plate. Each colony is about 1 millimeter across.

b A nitrocellulose or nylon filter is placed on the plate. Some cells of each colony adhere to it. The filter mirrors how the colonies are distributed on the culture plate.

c The filter is lifted off and put into a solution. Cells stuck to it rupture; the cellular DNA sticks to the filter.

e The probe's location is identified by exposing the filter to x-ray film. The image that forms on the film reveals the colony that has the gene of interest.

d The DNA is denatured to single strands at each site. A radioactively labeled probe is added to the filter. The probe binds to DNA fragments with a complementary base sequence.

Figure 15.8 Use of a radioactive probe to identify a bacterial colony that contains a targeted gene.

Read Me First!
and watch the narrated animation on PCR

where the colonies were. When the probe is washed over the paper, it hybridizes with, or sticks to, only the DNA that has the target sequence. The hybridized probe makes a radioactive spot that can be detected with x-ray film. The position of the spot on the film reflects the position of the original colony on the petri dish. Cells from that colony alone are cultured to isolate the cloned gene of interest.

PCR

Researchers may replicate a gene, or part of it, with **PCR** (*Polymerase Chain Reaction*). PCR uses primers and a heat-tolerant polymerase for a hot–cold cycled reaction that replicates targeted DNA fragments. And it can replicate them by a billionfold. This technique can transform one needle in a haystack, that one-in-a-million DNA fragment, into a huge stack of needles with a little hay in it.

Figure 15.9 shows the reaction steps. **Primers** are synthetic nucleotide oligomers, usually between ten and thirty bases long. They are designed to base-pair with specific nucleotide sequences on either end of the fragment of interest. In a PCR reaction, researchers mix primers, DNA polymerase, nucleotides, and the DNA, which will act as a template for replication. Then the researchers expose the mixture to repeated cycles of high and low temperatures. The two strands of a DNA double helix separate into single strands at high temperature. When the mixture is cooled, some of the primers will hybridize with the DNA template.

The elevated temperatures required to separate DNA strands destroy typical DNA polymerases. But the heat-tolerant DNA polymerase employed for PCR reactions is from *Thermus aquaticus*, a bacterium that lives in superheated water springs (Chapter 19). Like all DNA polymerases, it recognizes primers bound to DNA as places to initiate synthesis.

Synthesis proceeds along the DNA template until the temperature cycles up and the DNA strands are separated again. When the temperature cycles down, primers rehybridize, and the reactions start all over. With each round of temperature cycling, the number of copies of targeted DNA can double. PCR quickly and exponentially amplifies even a tiny bit of DNA.

Probes may be used to help identify one particular gene among many in gene libraries.

The polymerase chain reaction (PCR) is a method of rapidly and exponentially amplifying DNA samples of interest.

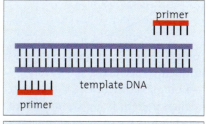
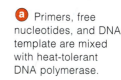

a Primers, free nucleotides, and DNA template are mixed with heat-tolerant DNA polymerase.

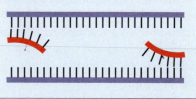

b When the mixture is heated, the DNA denatures. When it is cooled, some primers hydrogen-bond to the DNA template.

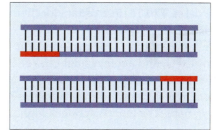

c *Taq* polymerase uses the primers to initiate synthesis, copying the DNA template. The first round of PCR is completed.

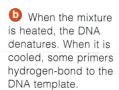

d The mixture is heated again. This denatures all the DNA into single strands. When the mixture is cooled, some of the primers hydrogen-bond to the DNA.

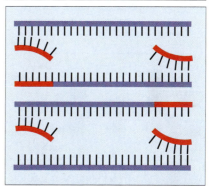
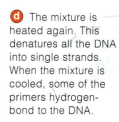

e *Taq* polymerase uses the primers to initiate synthesis, copying the DNA. The second round of PCR is complete. Each successive round of synthesis can double the number of DNA molecules.

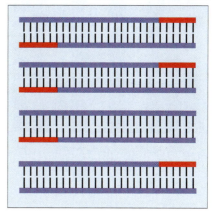

Figure 15.9 Two rounds of the polymerase chain reaction, or PCR. A bacterium, *Thermus aquaticus*, is the source of the *Taq* polymerase. Thirty or more cycles of PCR may yield a billionfold increase in the number of starting DNA template molecules.

15.4 First Just Fingerprints, Now DNA Fingerprints

Except for identical twins, no two people have exactly the same base sequence in their DNA. Scientists can distinguish one person from another on the basis of differences in those sequences.

Each human has a unique set of fingerprints. Like all other sexually reproducing species, each also has a **DNA fingerprint**—a unique array of DNA sequences inherited in a Mendelian pattern from parents. More than 99 percent of the DNA is the same in all humans, but the other 1 percent is unique to each individual. These unique stretches of DNA are sprinkled through the human genome as **tandem repeats**—many copies of the same short DNA sequences, positioned one after the other along the length of a chromosome.

For example, one person's DNA might contain four repeats of the bases TTTTC in a certain location. Another person's DNA might have them repeated fifteen times in the same location. One person might have five repeats of CGG, and another might have fifty. Such repetitive sequences slip spontaneously into the DNA during replication, and their numbers grow or shrink over time. The mutation rate is high in these regions.

DNA fingerprinting reveals differences in the tandem repeats among individuals. A restriction enzyme cuts their genomic DNA into an assortment of fragments. The sizes of those fragments are unique to the individual. They reveal genetic differences between individuals, and they can be detected as RFLPs (restriction fragment length polymorphisms).

The differences show up with **gel electrophoresis**. In this technique, an electric field pulls a sample of DNA fragments through a slab of a semisolid matrix, such as an agar gel. Fragments of different sizes migrate through the matrix at different rates. Larger molecules are hindered by the matrix more than smaller ones, much as elephants are slower than tigers at slipping between trees in a dense forest. In short, gel electrophoresis separates the fragments of DNA according to their length. After a time, the different fragments separate into distinct bands.

A banding pattern of genomic DNA fragments is the DNA fingerprint unique to the individual. It is identical only between identical twins. Otherwise, the odds of two people sharing an identical DNA fingerprint are one in three trillion.

PCR can also be used to amplify tandem-repeat regions. Differences in the size of the resulting amplified DNA fragments are again detected with gel electrophoresis. A few drops of blood, semen, or cells from a hair follicle at a crime scene or on a suspect's clothing yield enough DNA to amplify with PCR, and then generate a fingerprint.

DNA fingerprints help forensic scientists identify criminals, victims, and innocent suspects. Figure 15.10 shows some tandem repeat RFLPs that were separated by gel electrophoresis. Those samples of DNA had been taken from seven people and from a bloodstain left at a crime scene. One of the DNA fingerprints matched.

Defense attorneys initially challenged the use of DNA fingerprinting as evidence in court. Today, however, the procedure has been firmly established as accurate and unambiguous. DNA fingerprinting is routinely submitted as evidence in disputes over paternity, and it is being widely used to convict the guilty and exonerate the innocent. To date, such evidence has already helped release 143 innocent people from prison.

DNA fingerprint analysis also has confirmed that human bones exhumed from a shallow pit in Siberia belonged to five individuals of the Russian imperial family, all shot to death in secrecy in 1918. It also was used to identify the remains of those who died in the World Trade Center on September 11, 2001.

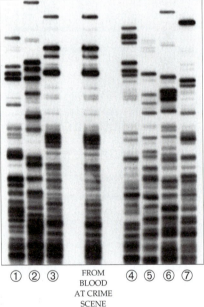

① ② ③   FROM BLOOD AT CRIME SCENE   ④ ⑤ ⑥ ⑦

Figure 15.10 Damning comparison of the DNA fingerprints from a bloodstain left behind at a crime scene and from blood samples of seven suspects (the circled numbers). Can you point out which of the seven is a match?

15.5 Automated DNA Sequencing

Sequencing reveals the order of nucleotides in DNA. This technique uses DNA polymerase to partially replicate a DNA template. In current research labs, manual methods have been replaced largely by automated techniques.

Automated DNA sequencing can reveal the sequence of a stretch of cloned or PCR-amplified DNA in just a few hours. Researchers use four standard nucleotides (T, C, A, and G). They also use four modified versions, which we represent as T*, C*, A*, and G*. Each of the four types of modified nucleotide has become labeled with a pigment that will fluoresce in a particular color when it passes through a laser beam.

Researchers mix all eight kinds of nucleotides with a single-stranded DNA template, a primer, and DNA polymerase. The polymerase uses the primer to copy the template DNA into new strands of DNA. One by one, it adds nucleotides in the order dictated by the sequence of the DNA template.

Each time, the polymerase randomly attaches one of the standard *or* one of the modified nucleotides to the DNA template. When one of the modified nucleotides becomes covalently bonded to the newly forming DNA strand, it stops further synthesis of that strand. After enough time passes, there will be some new strands that stop at each base in the DNA template sequence.

Eventually the mixture holds millions of copies of DNA fragments, all fluorescent-tagged. The fragments are now separated by gel electrophoresis, which is part of an automated sequencer. Shortest fragments migrate fastest and reach the end of the block of gel first; the longest fragment is last. Fragments of the same length migrate through the gel at the same speed, and they form observable bands (Figure 15.11a).

Each fragment passes through a laser beam, and the modified nucleotide attached to its tail end makes it fluoresce a certain color. The sequencer detects and records the fluorescent colors as the fragments pass through the end of the gel. Because each color codes for a particular nucleotide, the order of colored bands is the DNA sequence. The machine itself assembles the sequence data.

Figure 15.11b shows partial results from a run through an automated DNA sequencer. Each peak in the tracing represents the detection of one fluorescent color as the fragments reached the end of the gel. The sequence is shown beneath the graph line.

With automated DNA sequencing, the order of nucleotides in a DNA fragment that has been cloned or amplified can be determined rapidly.

Read Me First!

and watch the narrated animation on automated DNA sequencing

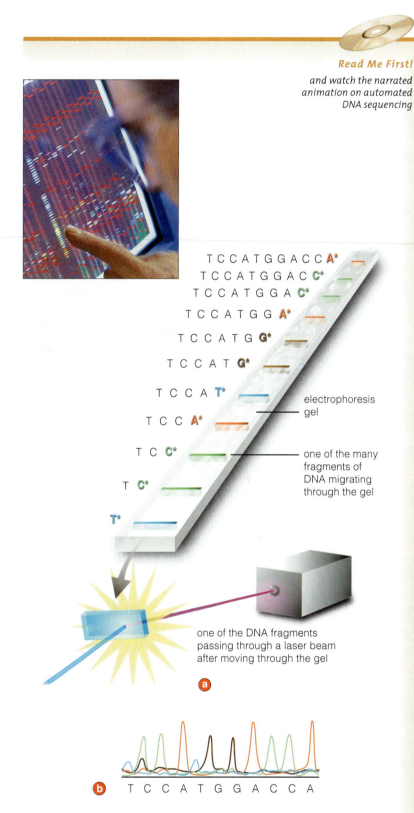

electrophoresis gel

one of the many fragments of DNA migrating through the gel

one of the DNA fragments passing through a laser beam after moving through the gel

a

b T C C A T G G A C C A

Figure 15.11 Automated DNA sequencing. (**a**) DNA fragments are synthesized using a template and fluorescent nucleotides. The bands are separated by gel electrophoresis. (**b**) The order of the fluorescent bands that appear in the gel is detected by the automated sequencer, and indicates the template DNA sequence. Today, researchers use sequence databases that are accessible globally via the Internet.

15.6 Practical Genetics

Even the tiniest living organisms are able to make complex organic compounds. Researchers harness this ability for practical purposes in genetic engineering.

DESIGNER PLANTS

As crop production expands to keep pace with human population growth, it puts unavoidable pressure on ecosystems everywhere. Irrigation leaves mineral and salt residues in soils. Tilled soil erodes, taking topsoil with it. Runoff clogs rivers, and fertilizer in it causes algae to grow so much that fish suffocate. Pesticides harm humans, other animals, and beneficial insects.

Pressured to produce more food at lower cost and with less damage to the environment, some farmers are turning to genetically engineered crop plants.

Genetic engineering is the process of changing the genetic makeup of an organism, often with intent to alter one or more aspects of phenotype. Researchers may accomplish this by transferring a gene from one species into another species, or by modifying a gene and inserting it into an organism of the same species.

Cotton plants with a built-in insecticide gene kill only the insects that eat it, so farmers that grow them don't have to use as many pesticides. Genetically modified wheat has double yields per acre. Certain transgenic tomato plants survive in salty soils that wither other plants; they also absorb and store excess salt in their leaves, thus purifying saline soil for future crops. *Transgenic* simply refers to an organism into which DNA from another species has been inserted, as in Figures 15.12 and 15.13.

a   b

Figure 15.12 Transgenic plants. (**a**) Cotton plant (*left*), and cotton plant with a gene for herbicide resistance (*right*). Both were sprayed with weed killer. (**b**) Genetically engineered aspen seedlings in which lignin biosynthesis has been partially blocked. Unmodified seedling is on the left.

The cotton plants in Figure 15.12*a* were genetically engineered for resistance to a relatively short-lived herbicide. Spraying fields with this herbicide will kill all weeds but not the engineered cotton plants. The practice means farmers can use fewer and less toxic chemicals. They also don't have to till the soil as much to control weeds, so there is less river-clogging runoff.

Aspen tree seedlings in which a lignin biosynthesis pathway has been modified still make lignin, but not as much—and root, stem, and leaf growth are greatly enhanced (Figure 15.12*b*). Wood from lignin-deficient trees makes it easier to manufacture paper and clean-burning fuels such as ethanol.

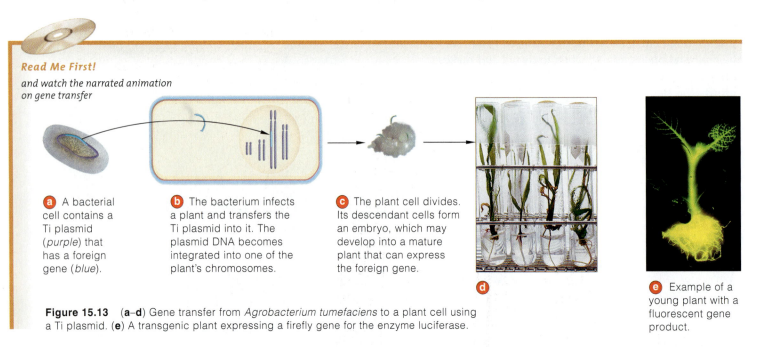

Read Me First!

and watch the narrated animation on gene transfer

a A bacterial cell contains a Ti plasmid (*purple*) that has a foreign gene (*blue*).

b The bacterium infects a plant and transfers the Ti plasmid into it. The plasmid DNA becomes integrated into one of the plant's chromosomes.

c The plant cell divides. Its descendant cells form an embryo, which may develop into a mature plant that can express the foreign gene.

d

e Example of a young plant with a fluorescent gene product.

Figure 15.13 (**a–d**) Gene transfer from *Agrobacterium tumefaciens* to a plant cell using a Ti plasmid. (**e**) A transgenic plant expressing a firefly gene for the enzyme luciferase.

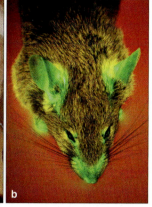
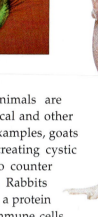

Figure 15.14 Two genetically modified animals: (**a**) Mira, a goat transgenic for human antithrombin III, an anticlotting factor. (**b**) This transgenic mouse has been engineered to produce green fluorescent protein (GFP).

(**c**) A featherless chicken breed developed by traditional cross-breeding methods in Israel. They thrive in desert environments where cooling systems are not an option. Chicken farmers in the United States have lost millions of feathered chickens at a time when temperatures skyrocketed.

Engineering plant cells starts with vectors that can carry genes into plant cells. *Agrobacterium tumefaciens* is a bacterial species that infects eudicots, including beans, peas, potatoes, and other vital crops. Its plasmid genes cause tumor formation on these plants; hence the name Ti plasmid (*Tumor-i*nducing). The Ti plasmid is used as a vector for transferring new or modified genes into plants.

Researchers excise the tumor-inducing genes, then insert a desired gene into the plasmid (Figure 15.13). Some plant cells cultured with the modified plasmid take it up. Whole plants may be regenerated.

Modified *A. tumefaciens* bacteria deliver genes into monocots that also are food sources, including wheat, corn, and rice. Researchers can also transfer genes into plants by way of electric shocks, chemicals, or blasts of microscopic particles coated with DNA.

BARNYARD BIOTECH

The first mammals enlisted for experiments in genetic engineering were laboratory mice. Transgenic mice appeared on the research scene in 1982 when scientists built a plasmid containing a gene for rat somatotropin (also known as growth hormone). They injected the recombinant DNA into fertilized mouse eggs, which were subsequently implanted into female mice. One-third of the resulting offspring grew much larger than their littermates—the rat gene had become integrated into their DNA and was being expressed.

Transgenic animals are now used routinely for medical research. The function and regulation of many gene products have been discovered using "knockout mice," in which targeted genes are inactivated. Defects in the resulting mice give clues about the gene. Strains of mice engineered to be susceptible to human diseases allow researchers to study both the diseases and their cures without experimenting on humans.

Genetically engineered animals are also sources of pharmacological and other valuable proteins. As a few examples, goats produce CFTR protein (for treating cystic fibrosis) and TPA protein (to counter the effects of a heart attack). Rabbits produce human interleukin-2, a protein that stimulates divisions of immune cells (T-lymphocytes). Cattle, too, may soon be producing human collagen that can be used to repair cartilage, bone, and skin. Goats make spider silk protein that might be used to make bullet-proof vests, medical supplies, and space equipment. Other goats make human antithrombin, used to treat people with blood clotting disorders (Figure 15.14*a*).

Genetic engineering has also given us dairy goats with healthier milk, pigs whose manure is easier on the environment, freeze-resistant salmon, extra-hefty sheep, low-fat pigs, mad cow disease-resistant cows, and even allergen-free cats.

Tinkering with the genetics of animals for the sake of human convenience does raise some serious ethical issues, particularly because failed experiments can have gruesome results. However, is transgenic animal research simply an extension of thousands of years of acceptable barnyard breeding practices (Figure 15.14*c*)? The techniques have changed, but not the intent. Like our ancestors, we continue to have a vested interest in improving our livestock.

Transgenic plants help farmers grow crops more efficiently and with less impact on the environment.

Transgenic animals are widely used in medical research. Some are sources of medically valued proteins and other biomaterials. Food animals are being altered to be more nutritious, disease resistant, or easier to raise.

15.7 Weighing the Benefits and Risks

We as a society continue to work our way through the ethical implications of DNA research even while we are applying the new techniques to medicine, industry, agriculture, and environmental remediation.

WHO GETS WELL?

More than 15,500 genetic disorders affect between 3 and 5 percent of all newborns, and they cause 20 to 30 percent of all infant deaths per year. They account for about 50 percent of the mentally impaired and nearly 25 percent of all hospital admissions.

Rhys Evans (Figure 15.15*a*) was born with a severe immune deficiency known as SCID-X1, which stems from mutations in gene *IL2RG*. Children affected by this disorder can live only in germ-free isolation tents, because they cannot fight infections.

In 1998, a virus was used to insert nonmutated copies of *IL2RG* into stem cells taken from the bone marrow of eleven boys with SCID-X1. *Stem* cells are still "uncommitted" and have the potential to differentiate into other types, including white blood cells of the immune system. Each child's modified stem cells were infused back into his bone marrow. Months afterward, ten of the children left their isolation tents for good. Their immune systems had been repaired by the gene therapy. Since then, many other SCID-X1 patients, including Rhys Evans, have been cured in other gene therapy trials.

In 2002, two children from the initial experiment in 1998 developed leukemia. Their illness surprised researchers, who had anticipated that any cancer related to the therapy would be extremely rare. An overproduction of white blood cells (T-lymphocytes) caused the leukemia in both children. The very gene targeted for repair work— *IL2RG*—may be the problem, particularly when combined with the viral vector used in the gene therapy. No other children in any gene therapy experiments for SCID-X1 have developed leukemia. Even so, our understanding of the human genome clearly lags behind our ability to modify it.

WHO GETS ENHANCED?

Modifying the human genome has profound ethical implications even beyond the unexpected risks. To many of us, human gene therapy to correct genetic disorders seems like a socially acceptable goal. Now take this idea one step further. Is it acceptable to change some genes of a normal human in order to alter or enhance traits?

Through gene transfers, researchers have already engineered strains of mice with enhanced memory and learning abilities. Maybe their work heralds help for Alzheimer's disease patients, perhaps even for those who just want more brain power.

The idea of being able to select desirable human traits is referred to as *eugenic engineering*. Yet who decides which forms of a trait are the most desirable? Realistically, cures for many severe but rare genetic disorders will not be pursued because the payback for research is not financially attractive. Eugenics, however, might turn a profit. Just how much would potential parents pay to engineer tall or blue-eyed or fair-skinned children? Would it be okay to engineer

Figure 15.15 Experimental gene therapy patients. (**a**) Rhys Evans was born with a gene that causes SCID-X1. His immune system never developed in a way that could fight infections. A gene transfer freed him from life in a germ-free isolation tent. (**b**) Max Randell smiles at his mother after receiving gene therapy in 2001 for Canavan's disease. This is a degenerative and fatal disease of the central nervous system. At the time of this writing, Max is alive and doing well.

"superhumans" who have breathtaking strength or intelligence? How about an injection that would help you lose extra weight, and keep it off permanently? The borderline between interesting and abhorrent is not the same for everyone.

In a survey conducted not long ago in the United States, more than 40 percent of those interviewed said it would be fine to use gene therapy to make smarter and better looking babies. In one poll of British parents, 18 percent were willing to use genetic enhancement to prevent their children from being aggressive, and 10 percent were willing to use it to keep them from growing up to be homosexual.

KNOCKOUT CELLS AND ORGAN FACTORIES

Each year, about 75,000 people are on waiting lists for an organ transplant, but human donors are in short supply. There is talk of harvesting organs from pigs (Figure 15.16), because pig organs function very much like ours do. Transferring an organ from one species into another is called **xenotransplantation**.

The human immune system battles anything that it recognizes as "nonself." It rejects a pig organ at once. A certain sugar molecule occurs on the plasma membrane of cells that make up a pig organ's blood vessels. Antibodies circulating in human blood latch on to that sugar quickly and doom the transplant. Within a few hours, cascading reactions lead to massive coagulation inside the organ's blood vessels, and failure is swift. Drugs can suppress this immune response, but there's a serious side effect: the drugs make organ recipients vulnerable to infections.

Pig DNA contains two copies of *Ggta1*, the gene for alpha-1,3-galactosyltransferase. This enzyme catalyzes a key step in biosynthesis of alpha-1,3-galactose, the pig sugar that human antibodies recognize. Researchers succeeded in knocking out both copies of the *Ggta1* gene in transgenic piglets. Without the gene, the pigs lack alpha-1,3-galactose. If one of their organs or tissues is transplanted, the human immune system might not recognize it. The tissues and organs from such animals could benefit millions of people, including those who suffer from diabetes or Parkinson's disease.

Critics of xenotransplantation are concerned that, among other things, pig–human transplants would invite pig viruses to cross species and infect humans, perhaps catastrophically. Their concerns are not unfounded. In 1918, an influenza pandemic killed twenty million people worldwide. It originated with a swine flu virus—in pigs.

Figure 15.16 Inquisitive transgenic pig at the Virginia Tech Swine Research facility.

REGARDING "FRANKENFOOD"

Genetically engineered food crops are widespread in the United States. At least 45 percent of cotton crops, 38 percent of soybean crops, and 25 percent of corn crops are now engineered to withstand weedkillers or to make their own pesticides. For years, modified corn and soybeans have been used in tofu, breakfast cereals, soy sauce, vegetable oils, beer, and soft drinks. They are fed to farm animals. Engineered crop plants hold down food production costs, reduce dependence on pesticides and herbicides, and enhance crop yields. Food plants are being designed for flavor, nutritional value, and extended shelf life.

In Europe especially, public resistance to modified food runs high. Besides arguing that modified foods might be toxic and have lower nutritional value, many people worry that designer plants might cross-pollinate wild plants and produce "superweeds."

The chorus of critics in Europe may provoke a trade war with the United States. The outcome is not small potatoes, so to speak. In 1998, the value of American agricultural exports was about 50 billion dollars. Restrictions will profoundly impact agriculture in the United States, and inevitably the impact will trickle down to what you eat and how much you pay for it.

All of which invites you to read scientific research and form your own opinions. The alternative is to be swayed by media hype (the term "Frankenfood," for instance), or by potentially biased reports from other groups that might have a different agenda (such as chemical pesticide manufacturers).

15.8 Brave New World

*The structural and comparative analysis of genomes is
yielding information about evolutionary trends as well
as potential therapies for genetic diseases.*

GENOMICS

Research into genomes of humans and other species
has converged into a new research field—**genomics**.
The *structural* genomics branch deals with the actual
mapping and sequencing of genomes of individuals.
The *comparative* genomics branch is concerned with
finding evolutionary relationships among groups of
organisms. Comparative genomics researchers analyze
similarities and differences among genomes.

Comparative genomics has practical applications
as well as potential for research. The basic premise is
that the genomes of all existing organisms are derived
from common ancestors. For instance, pathogens share
some conserved genes with human hosts even though
the lineages diverged long ago. Shared gene sequences,
how they are organized, and where they differ may
hold clues to where our immune defenses against
pathogens are strongest or the most vulnerable.

Genomics has potential for **human gene therapy**—
the transfer of one or more normal or modified genes
into a person's body cells to correct a genetic defect or
boost resistance to disease. However, even though the
human genome is fully sequenced, it is not easy to
manipulate within the context of a living individual.

Experimenters employ stripped-down viruses as
vectors that inject genes into human cells. Some gene
therapies deliver modified cells into a patient's tissue.
In many cases, therapies make a patient's symptoms
subside even when the modified cells are producing
just a small amount of a required protein.

However, no one can yet predict where virus-
injected genes will end up in a person's chromosomes.
The danger is that the insertion will disrupt other
genes, particularly those controlling cell division and
growth. One-for-one gene swaps with recombination
methods are possible but still experimental.

DNA CHIPS

Analysis of genomes is now advancing at a stunning
pace. Researchers pinpoint which genes are silent and
which are being expressed with the use of **DNA chips**.
These are microarrays of thousands of gene sequences
representing an entire genome—all stamped onto a
glass plate about the size of a smallish business card.

A fluorescent labeled cDNA probe is made using
mRNA, say, from cells of a cancer patient. Only the

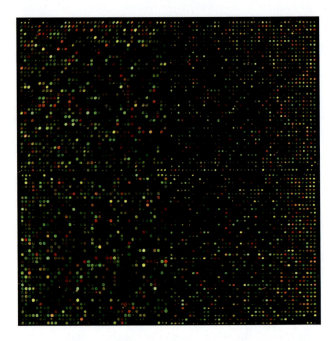

Figure 15.17 Complete yeast genome array on a DNA chip
that is about 19 millimeters (3/4 inch) across. Green spots
pinpoint the genes that are active during fermentation. Red
spots pinpoint the ones used during aerobic respiration.
Yellow spots indicate genes active during both pathways.

genes expressed at the time the cells are harvested
will be making mRNA, so they alone will make up
the resulting probe population. The labeled probe is
then incubated with a chip made from genomic DNA.
Wherever the probe binds with complementary base
sequences on the chip, there will be a spot that glows
under fluorescent light. Analysis of which spots on
the chip are glowing reveals which of the thousands
of genes inside the cells are active and which are not.

DNA chips are being used to compare different
gene expression patterns between cells. Examples are
yeasts grown in the presence and absence of oxygen,
and different types of cells from the same multicelled
individual. RNA from one set of cells is transformed
into green fluorescent cDNA, and RNA from the other
set into red fluorescent cDNA. The cDNAs are mixed
and incubated with a genomic DNA chip. Green or
red fluorescence indicates expression of genes in the
different cell types. Yellow is a mixture of both red
and green, and it indicates that both genes were being
expressed at the same time in a cell (Figure 15.17).

*In genomics, new techniques such as DNA chips allow
researchers to rapidly evaluate and compare genome-
spanning expression patterns.*

Summary

Section 15.1 Discovery of DNA's double helical structure sparked interest in deciphering its genetic messages. A global race to complete the sequence of the human genome spurred rapid development of new techniques to study and manipulate DNA. The entire human genome has been sequenced and is now being analyzed. Genomes of other organisms have been sequenced as well.

Section 15.2 Recombinant DNA technology uses restriction enzymes that cut DNA into fragments. The fragments may be spliced into cloning vectors by using DNA ligase. Recombinant plasmids are taken up by rapidly dividing cells, such as bacteria, to make multiple, identical copies of the foreign DNA. Reverse transcriptase copies mRNA into cDNA for cloning.

Section 15.3 A gene library is a mixed collection of cells that have taken up cloned DNA. A particular gene can be isolated from a library by using a probe, a short stretch of DNA that can base-pair with the gene and that is traceable with a radioactive or pigment label. Probes help researchers identify one particular clone among millions of others. Base-pairing between nucleotide sequences from different sources is called nucleic acid hybridization.

The polymerase chain reaction (PCR) is a way to rapidly copy particular pieces of DNA. A sample of a DNA template is mixed with nucleotides, primers, and a heat-resistant DNA polymerase. Each round of PCR proceeds through a series of temperature changes that amplifies the number of DNA molecules exponentially.

Section 15.4 Tandem repeats are multiple copies of a short DNA sequence that follow one another along a chromosome. The number and distribution of tandem repeats, unique in each person, can be revealed by gel electrophoresis; they form a DNA fingerprint.

Section 15.5 Automated DNA sequencing rapidly reveals the order of nucleotides in DNA fragments. As DNA polymerase is copying a template DNA, progressively longer fragments stop growing when one of four different fluorescent nucleotides becomes attached. Electrophoresis separates the resulting labeled fragments of DNA into bands according to length. The order of the colored bands as they migrate through the gel reflects which fluorescent base was added to the end of each fragment, and so indicates the template DNA base sequence.

Section 15.6 Genetic engineering is the directed modification of the genetic makeup of an organism, often with intent to modify its phenotype. Researchers insert normal or modified genes from one organism into another of the same or different species. Gene therapies also reinsert altered genes into individuals.

Genetically engineered bacteria produce medically valued proteins. Transgenic crop plants help farmers produce food more efficiently. Genetic engineering of animals allows commercial production of human proteins, as well as research into genetic disorders.

Section 15.7 Human gene therapy and modification of animals for xenotransplantation are examples of developing technologies. As with any new technology, potential benefits must be weighed against potential risks, including ecological and social repercussions.

Section 15.8 Genomics, the study of human and other genomes, is shedding light on evolutionary relationships and has practical uses. Human gene therapy transfers normal or modified genes into body cells to correct genetic defects. Gene chips are used to compare patterns of gene expression.

Self-Quiz

Answers in Appendix III

1. _____ is the transfer of normal genes into body cells to correct a genetic defect.
   - a. Reverse transcription
   - b. Nucleic acid hybridization
   - c. PCR
   - d. Gene therapy

2. DNA is cut at specific sites by _____ .
   - a. DNA polymerase
   - b. DNA probes
   - c. restriction enzymes
   - d. reverse transcriptase

3. Fill in the blank: A _____ is a small circle of bacterial DNA that is not part of the bacterial chromosome.

4. By reverse transcription, _____ is assembled on a(n) _____ template.
   - a. mRNA; DNA
   - b. cDNA; mRNA
   - c. DNA; ribosomes
   - d. protein; mRNA

5. PCR stands for _____ .
   - a. polymerase chain reaction
   - b. polyploid chromosome restrictions
   - c. polygraphed criminal rating
   - d. politically correct research

6. By gel electrophoresis, fragments of DNA can be separated according to _____ .
   - a. sequence
   - b. length
   - c. species

7. Automated DNA sequencing relies on _____ .
   - a. supplies of standard and labeled nucleotides
   - b. primers and DNA polymerases
   - c. gel electrophoresis and a laser beam
   - d. all of the above

8. _____ can be used to insert genes into human cells.
   - a. PCR
   - b. Modified viruses
   - c. Xenotransplantation
   - d. DNA microarrays

9. Match the terms with the most suitable description.
   - ____ DNA fingerprint
   - ____ Ti plasmid
   - ____ nature's genetic experiments
   - ____ nucleic acid hybridization
   - ____ eugenic engineering

   - a. selecting "desirable" traits
   - b. mutations, crossovers
   - c. used in some gene transfers
   - d. a person's unique collection of tandem repeats
   - e. base pairing of nucleotide sequences from different DNA or RNA sources

Figure 15.18 (**a**) ANDi, the first transgenic primate; his cells incorporate a jellyfish gene for bioluminescence. (**b**) The same gene was transferred into these zebrafish.

Media Menu

Student CD-ROM

Impacts, Issues Video
    Golden Rice or Frankenfood?
Big Picture Animation
    DNA technology, genetic engineering, and human
    applications
Read-Me-First Animation
    DNA recombination
    Use of labeled probes
    Polymerase chain reaction (PCR)
    Automated DNA sequencing
    Gene transfer
Other Animations and Interactions
    Action of restriction enzymes
    Making cDNA

InfoTrac
- Speed Reader. *Forbes*, November 2002.
- New Tools, Moon Tigers, and the Extinction Crisis. *BioScience*, September 2001.
- The Terminator's Back: Controversial Scheme Might Prevent Transgenic Spread. *Scientific American*, September 2002.
- Human Gene Therapy: Harsh Lessons, High Hopes. *FDA Consumer*, September 2000.

Web Sites
- National Center for Biotechnology Information: www.ncbi.nlm.nih.gov
- National Human Genome Research Institute: www.genome.gov
- Ag BioTech InfoNet: www.biotech-info.net
- Issues in Biotechnology: www.actionbioscience.org/biotech

How Would You Vote?

The United States is the world's leading producer and consumer of genetically modified organisms. Some people are uneasy about genetic engineering and would like to avoid products based on this technology. Should food that contains genetically modified plants or livestock be clearly identified on product labels?

Critical Thinking

1. What if it were possible to create life in test tubes? This is the question behind recent attempts to model and eventually create *minimal organisms*, which we define as living cells having the smallest set of genes required to survive and reproduce.

Craig Venter and Claire Fraser recently found that *Mycoplasma genitalium*, a bacterium that has 517 genes (and 2,209 transposons), is a good candidate for genetic research. By disabling its genes one at a time in the laboratory, they discovered that it may have only 265–350 essential protein-coding genes.

What if those genes were to be synthesized one at a time and inserted into an engineered cell consisting only of a plasma membrane and cytoplasm? Would the cell come to life? The possibility that it might prompted Venter and Fraser to seek advice from a panel of bioethicists and theologians. No one on the panel objected to synthetic life research. They felt that much good might come of it, provided scientists didn't claim to have found "the secret of life." The 10 December 1999 issue of *Science* includes an essay from the panel and an article on *M. genitalium* research. Read both, then write down your thoughts about creating life in a test tube.

2. Lunardi's Market put out a bin of tomatoes having vine-ripened redness, flavor, and texture. A sign identified them as genetically engineered produce. Most shoppers selected unmodified tomatoes in the adjacent bin even though those tomatoes were pale pink, mealy-textured, and tasteless. Which tomatoes would you pick? Why?

3. The sequence of the human genome has been completed, and knowledge about a number of the newly discovered genes is already being used to detect genetic disorders. Many women have refused to take advantage of genetic screening for a gene that is associated with the development of breast cancer.

Should medical records about people participating in genetic research and in genetic clinical services be made readily available to insurance companies, potential employers, and others? If not, how can such information be protected?

4. Scientists at Oregon Health Sciences University produced Tetra, the first primate clone. They also made the first transgenic primate by inserting a jellyfish gene into a fertilized egg of a rhesus monkey. (The gene encodes a bioluminescent protein that fluoresces green.) The egg was implanted in a surrogate monkey's uterus, where it developed into a male that was named ANDi (Figure 15.18).

The long-term goal of this gene transfer project is not to make glowing-green monkeys. It is the transfer of human genes into the primates whose genomes are most like ours. Transgenic primates could be studied to gain insight into genetic disorders, which might lead to the development of cures for those who are affected and vaccines for those at risk.

However, something more controversial is at stake. Will the time come when foreign genes can be inserted into human embryos? Would it be ethical to transfer a chimpanzee or monkey gene into a human embryo to cure a genetic defect? Or to bestow immunity against a potentially fatal disease such as AIDS?

III Principles of Evolution

Two male frigate birds (Fregata minor) in the Galápagos Islands, hundreds of miles off the coast of Ecuador. Each male inflates a gular sac, a balloon of red skin at his throat, in a display that may catch the eye of a female. The males lurk in the bushes together, gular sacs inflated, until a female flies overhead. Then they wag their head back and forth and call seductively to her. We can expect that, like other male structures used exclusively in courtship displays, the gular sac is an outcome of sexual selection—with the females being the selective agents.

IMPACTS, ISSUES *Rise of the Super Rats*

Slipping in and out of the pages of human history are rats—*Rattus*—the most notorious of mammalian pests. One kind or another has distributed pathogens and parasites that cause bubonic plague, typhus, and other deadly infectious diseases (Figure 16.1). The death toll from fleas that bit infected rats, then people, has exceeded the dying in all wars combined.

In contrast, rats thrive around us. By one estimate, there is one rat for every person in urban and suburban centers of the United States. Besides spreading diseases, they chew their way through the walls and wires of our homes and cities. In any given year, the economic losses they inflict typically approach nineteen billion dollars.

People have been fighting back for years with traps, ratproof storage facilities, and toxic poisons, including arsenic and cyanide. During the 1950s, scientists devised baits laced with warfarin. This chemical interferes with blood clotting mechanisms. Rats that ate the baits died within days from internal bleeding or minor scratches that would not heal.

Warfarin was extremely effective. Compared to other rat poisons, it had a lot less impact on harmless species. It quickly became the rodenticide of choice.

In 1958, a Scottish researcher reported that some rats were indifferent to the poison. Similar reports followed from other European countries. About twenty years

Figure 16.1 Medieval attempts to deal with a bubonic plague pandemic—the Black Death that may have killed half the people in Europe alone. Not knowing that rats spread the disease agent, Europeans tried to protect themselves by praying and dancing 'til they dropped; physicians wore bird masks. For the next 300 years, anyone accused of causing the sporadic outbreaks of the plague—no matter how absurd the evidence—was burned alive.

In this century, by using ever more potent poisons on rats, we unwittingly promoted the rise of super rats. Three centuries from now, how will people be viewing us?

the big picture

Evolutionary Views Emerge The world distribution of species, similarities and differences in body form, and the fossil record gave early evidence of evolution—of changes in lines of descent. Darwin and Wallace had an idea of how those changes occur.

Variation and Adaptation An adaptation is any heritable aspect of form, function, behavior, or development that promotes survival and reproduction. An outcome of microevolution, it enhances the fit between the individual and prevailing conditions.

later, 10 percent of the urban rats caught in the United States were warfarin resistant. What happened?

To find out, researchers compared warfarin-resistant rat populations with still-vulnerable rats. They traced the difference to a gene on one of the rat chromosomes. A dominant allele at that locus was common among warfarin-resistant rat populations but rare among the vulnerable ones. And the product of the dominant allele neutralizes warfarin's effect on blood clotting.

"What happened" was evolution. As warfarin started to exert pressure on rat populations, the previously rare dominant allele abruptly proved adaptive. The lucky rats that inherited the allele survived and produced more rats. The unlucky ones that inherited the recessive allele had no defense; they died. Over time, the dominant allele's frequency increased in all populations of rats exposed to the poison.

Of course, selection pressures can and often do change. In response to increasing warfarin resistance, people stopped using warfarin. And guess what: The dominant allele's frequency declined. Now the latest worry is the evolution of "super rats," which the newer and even more potent rodenticides can't seem to kill.

The point is, if you hear someone question whether life evolves, remember the word simply means heritable change in lines of descent. The actual mechanisms that bring about changes in organisms are the focus of this chapter. Later chapters highlight how these mechanisms contribute to the evolution of new species.

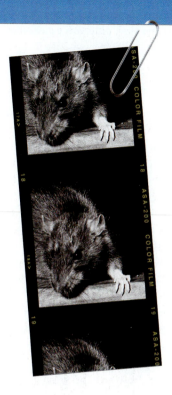

 How Would You Vote?

Antibiotic-resistant strains of bacteria are becoming dangerously pervasive. Standard animal husbandry practice includes continually dosing healthy animals with antibiotics—the same antibiotics prescribed for people. Should this practice stop? See the Media Menu for details, then vote online.

Think Gene Pools Individuals don't evolve; populations do. All individuals of a population represent a pool of genes, the frequencies of which can shift over the generations. Such shifts can change the characteristics that define the population, and the species.

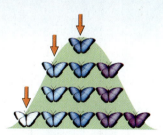

Microevolutionary Processes Mutation, genetic drift, natural selection, and gene flow are microevolutionary processes. By changing allele frequencies in a population, they change the observable characteristics that define the population and, more broadly, the species.

16.1 Early Beliefs, Confounding Discoveries

Prevailing beliefs can influence how we interpret clues to natural processes and their observable outcomes.

QUESTIONS FROM BIOGEOGRAPHY

At one time Europeans viewed nature as a great Chain of Being extending from the "lowest" forms of life to humans, and on to spiritual beings. Each kind of being, or **species** as it was called, was one separate link in the chain. All the links had been designed and forged at the same time at one center of creation. They had not changed since. Once all the links were discovered and described, the meaning of life would be revealed.

Then Europeans embarked on their globe-spanning explorations and discovered the world is a lot bigger than Europe. Tens of thousands of unique plants and animals turned up in Asia, Africa, the New World, and the Pacific islands.

In 1590, the naturalist Thomas Moufet attempted to sort through the bewildering arrays. He simply gave up and wrote such gems as this description of locusts and grasshoppers: "Some are green, some black, some blue. Some fly with one pair of wings; others with more; those that have no wings they leap; those that cannot fly they walk. Some there are that sing, others are silent." It was not a work of breathtaking insight.

Later on, Alfred Wallace and a few other naturalists saw *patterns* in the distribution of species. Not content with cataloging species, they looked for forces that shaped similarities and differences among them. They were pioneers in **biogeography**—the study of patterns in the geographic distribution of species. They found clues to the ecological and evolutionary forces that influence individual species and entire communities.

Some patterns were intriguing. For instance, many plants and animals are unique to islands in the middle of the ocean and other remote places. Similar-looking species often live in the same kinds of habitats, even when vast expanses of open ocean or high, impassable mountain ranges separate them.

Consider: Flightless, long-necked, long-legged birds are native to three continents (Figure 16.2a–c). Why are they so much alike? The plants in Figure 16.2d,e live on separate continents. Both have spines, tiny leaves, and short fleshy stems. Why are *they* so much alike?

Compare the global locations of the flightless birds and you'll find they all live in flat, open grasslands in dry climates. American and African desert plants are about the same distance from the equator, in regions where water is scarce. Their fleshy stems store water and have a thick cuticle, which keeps water in. Their rows of sharp spines deter thirsty, hungry animals.

Figure 16.2 Species that resemble one another, strikingly so, even though they are native to distant geographic realms.

(a) South American rhea, (b) Australian emu, and (c) African ostrich. All three types of birds live in similar habitats. They are unlike most birds in several traits, most notably in their inability to get airborne.

(d) A spiny cactus native to the hot deserts of the American Southwest. (e) A spiny spurge native to southwestern Africa.

coccyx

ankle bone

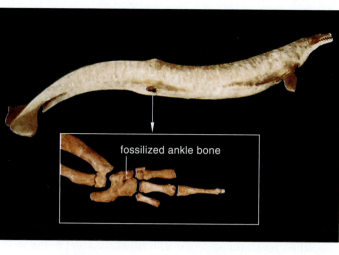

fossilized ankle bone

Figure 16.3 Body parts with no apparent function. *Above*, reconstruction of an ancient whale (*Basilosaurus*). This giant predator's head was as long as a sofa. The whale was fully aquatic, so it did not use its hindlimbs to support body weight as you do—yet it had ankle bones. We use our ankles, but not our coccyx bones.

Figure 16.4 (**a**) Fossilized ammonites and (**b**) a cutaway view of its shell, compared to that of a chambered nautilus (**c**). Ammonites, now extinct, lived hundreds of millions of years ago. Like the chambered nautilus, which exists today, ammonites were marine predators.

If all birds and plants were created in one place, then how did such similar kinds end up in the same kind of habitat in such remote and distant places?

QUESTIONS FROM COMPARATIVE MORPHOLOGY

The similarities and differences in body plans among groups of organisms raised questions that started yet another line of inquiry—**comparative morphology**. For example, bones of a human arm, whale flipper, and bat wing differ in size, shape, and function. As you'll see in Chapter 17, these bones are similarly located in the body and are constructed of the same kinds of tissues arranged in much the same patterns. And they develop in much the same way in embryos. Naturalists who deduced all of this wondered: How can there be animals that differ hugely in certain features yet are so much alike in other features?

By one hypothesis, body plans are so perfect there was no need to make a new design for each organism at the time of its creation. Yet if that were so, then why were useless body parts created? For instance, an ancient, ocean-dwelling whale had ankle bones but didn't walk (Figure 16.3). Why the bones? Our coccyx is like some tailbones in many other mammals. We do not have a tail. Why do we have parts of one?

QUESTIONS ABOUT FOSSILS

Geologists had been mapping layers of rock exposed by erosion or quarrying, and their discoveries added to the confusion. They found the same kinds of layers in different parts of the world. They thought the layers formed when sediments slowly collected, year after year, on the floor of ancient rivers and seas. If that were so, then the deepest layers were the oldest. And if that were so, then fossils embedded in successive layers could be clues to the past. **Fossils** came to be recognized as stone-hard evidence of earlier forms of life. Figures 16.3 and 16.4 show examples.

A puzzle: Many deep layers held fossils of simple marine life. Some layers above them contained fossils that were structurally similar but more intricate. In higher layers, fossils were like living species. What did the sequences in complexity among fossils of a given type mean? Were they evidence of lines of descent?

Taken as a whole, the findings from biogeography, comparative morphology, and geology did not fit with prevailing beliefs. Scholars floated novel hypotheses. If a simultaneous dispersal of all species from a center of creation was unlikely, *then perhaps species originated in more than one place*. If species had not been created in a perfect state—and fossil sequences and "useless" body parts among species implied they had not—*then perhaps species had become modified over time*. Awareness of evolution was in the wind.

> *Awareness of biological evolution emerged over centuries through the cumulative observations of many naturalists, biogeographers, comparative anatomists, and geologists.*

16.2 A Flurry of New Theories

Nineteenth-century naturalists found themselves trying to reconcile the evidence of change with a traditional conceptual framework that simply did not allow for it.

SQUEEZING NEW EVIDENCE INTO OLD BELIEFS

A respected anatomist, Georges Cuvier, was among those trying to make sense of the growing evidence for change. For years Cuvier had compared body plans of fossils and living organisms. He knew about fossils in sedimentary layers. He was the first to recognize that abrupt shifts in the fossil record mark mass extinctions.

By Cuvier's hypothesis, the changes resulted from periodic natural disasters, which he called revolutions. Except for those episodes, Earth was an unchanging stage for the human drama. A single time of creation had populated the world with all species. Many died in periodic global natural disasters, such as floods and earthquakes, and then the survivors repopulated the world. There never were any *new* species; naturalists simply hadn't yet found all of the fossils that would date back to the time of creation.

The hypothesis enjoyed support for some time. It even became elevated to the rank of theory, one that later became known as **catastrophism**.

Even so, many scholars kept at the puzzle. Jean Lamarck proposed a different hypothesis, known as inheritance of *acquired* characteristics. Environmental pressures and internal "needs" induced permanent changes in an individual's body form and functioning, then the individual passed on the changes, acquired during its lifetime, to offspring. Life, created long ago in a simple state, slowly improved as a result. The force for change was a drive toward perfection, up the Chain of Being. It was centered in nerves that directed an unknown "fluida" to body parts in need of change.

Apply his hypothesis to modern giraffes. Say they had a short-necked, hungry ancestor that had to keep stretching its neck to browse upon leaves beyond the reach of other animals. Big stretches directed fluida to its neck and made it lengthen permanently. Offspring inherited the longer neck, then stretched their necks, too. Generation after generation of straining to reach ever loftier leaves led to the modern giraffe.

As Lamarck correctly inferred, the environment *is* a factor in evolution. However, his hypothesis as well as others made at the time have not been supported by experimental tests. Environmental factors can alter an individual's traits, as when serious strength training builds huge muscles. But offspring of a muscle-bound parent won't be born muscle-bound. They can inherit genes, but not increased muscle mass.

VOYAGE OF THE *BEAGLE*

In 1831, in the midst of the confusion, Charles Darwin was twenty-two years old and wondering what to do with his life. Ever since he was eight, he had wanted to hunt, fish, collect shells, or just watch insects and birds—anything but sit in school. Later, at his father's

Figure 16.5 (**a**) Charles Darwin. (**b**) Replica of the *Beagle* sailing off a rugged coastline of South America. During one of his trips, Darwin ventured into the Andes. He discovered fossils of marine organisms in rock layers 3.6 kilometers above sea level.

(**c–e**) The Galápagos Islands are isolated in the ocean, far to the west of Ecuador. They arose by volcanic action on the seafloor about 5 million years ago. Winds and currents carried organisms to the once lifeless island. All the island's native species are descended from those travelers. At far right, a blue-footed booby, one of many species Darwin observed during his voyage.

insistence, he did attempt to study medicine in college. The crude, painful procedures being used on patients at that time sickened him. His exasperated father urged him to become a clergyman, so he packed for Cambridge. His grades were good enough to earn a degree in theology. Yet he spent most of his time with faculty members who embraced natural history.

John Henslow, a botanist, perceived Darwin's real interests. He hastily arranged for Darwin to become ship's naturalist aboard the *Beagle*, which was about to embark on a five-year voyage around the world. The young man who hated schoolwork, and who had no formal training as a naturalist, suddenly started to work with enthusiasm.

The *Beagle* sailed first to South America to finish work on mapping the coastline (Figure 16.5). During the Atlantic crossing, Darwin collected and studied marine life. He read Henslow's parting gift, the first volume of Charles Lyell's *Principles of Geology*. He saw diverse species in environments ranging from sandy shores of remote islands to high mountains. And he started circling the question of evolving life, which was now on the minds of many respected individuals.

Darwin started by mulling over a radical theory. As Lyell and other geologists were arguing, erosion and other gradual, natural processes of change had more impact on Earth history than rare catastrophes. Geologists for years had chipped away at sandstones, limestones, and other rocks that form after sediments accumulate in the beds of rivers and seas. They saw

how those beds often consist of a number of stacked layers. If, they hypothesized, deposition took place as slowly in the past as it did in their own era, then it must have taken many millions of years for thick stacks to form. They even incorporated earthquakes and other infrequent events into their theory. After all, immense floods, over a hundred big earthquakes, and twenty or so volcanic eruptions occur in a typical year, so catastrophes aren't that unusual.

The idea that gradual, repetitive change shaped the Earth became known as the **theory of uniformity**. It challenged the prevailing views of the Earth's age.

The theory bothered scholars who firmly believed that Earth could not be more than 6,000 years old. They thought people had recorded everything that happened during those 6,000 years—and in all that time, no one had ever mentioned seeing a species evolve. Yet, by Lyell's calculations, it must have taken millions of years to sculpt out the present landscape. *Wasn't that enough time for species to evolve in diverse ways?* Later, Darwin thought so. But exactly *how* did they evolve? He would end up devoting the rest of his life to that burning question.

Prevailing beliefs can influence how we interpret clues to natural processes and their observable outcomes.

Darwin's observations during a global voyage helped him think about species in a novel way.

16.3 Darwin's Theory Takes Form

Darwin's observations of thousands of species in different parts of the world helped him see how species may evolve.

OLD BONES AND ARMADILLOS

When Darwin returned to England, he had volumes of notes and thousands of specimens, although he had not done a great job of correlating them with habitats. Other naturalists helped him fill in some important blanks—and in time he arrived at an explanation of how all of that diversity evolved.

In Argentina, for example, Darwin found fossils of glyptodonts, now extinct. Of all animals on Earth, only living armadillos are like them (Figure 16.6). Of all places on Earth, armadillos live only in the same places where glyptodonts had lived. If the two kinds of animals had been created at the same time, lived in the same place, and were so much alike, why is only one still alive? Would it be reasonable to assume that glyptodonts were the early ancestors of armadillos? Many of their shared traits might have been retained over many thousands of generations. Other traits may have been modified in the armadillo branch of a family tree. *Descent with modification*—it did seem possible. What, then, could be the driving force for evolution?

A KEY INSIGHT—VARIATION IN TRAITS

While Darwin assessed his notes, an influential essay by Thomas Malthus, a clergyman and economist, made him consider a topic of social interest. Malthus had correlated population size with famine, disease, and war. Humans, said Malthus, run out of food, living space, and other resources because they reproduce too much. The larger a population gets, the more there are to reproduce. Resources dwindle, and struggles to live intensify. Many starve, get sick, or engage in war and other forms of competition for dwindling resources.

Darwin thought that *any* population has a capacity to produce more individuals than the environment can support. Case in point: A sea star can release 2,500,000 eggs a year, but the seas don't fill up with sea stars. Predators eat many of the eggs and larvae. Most of the survivors die of disease and other assaults.

Darwin also reflected on populations he had seen during his voyage. Their individuals were not alike down to the last detail. They varied in size, color, and other traits. *It dawned on him that variations in traits could affect an individual's ability to secure resources—and to survive and reproduce in particular environments.*

Did the Galápagos Islands reveal evidence of this? Between the islands and the South American coast are 900 kilometers of open ocean. The islands have diverse shoreline, desert, and mountain habitats. Nearly all of the species live nowhere else, although they resemble mainland species. Did wind and water put colonizing species on the Galápagos? Were species on different islands island-hopping descendants of the colonizers?

Try correlating variations in the Galápagos finches with different environmental challenges. Imagine yourself watching a large-billed, seed-cracking type (Figure 16.7). A few birds with a stronger bill crack seeds too tough for the others. If most of the seeds that form during a given season have hard coats, then the strong-billed birds will have a competitive edge and a better chance of surviving and producing offspring. If the trait is heritable, then the advantage will help their descendants, also. If environmental factors continue to "select" the most adaptive version of this trait, then the population may end up being mostly strong-billed birds. *And a population is evolving when forms of heritable traits change over the generations.*

Figure 16.6 (**a**) A modern armadillo. (**b**) A Pleistocene glyptodont, about as big as a Volkswagen Beetle, and now extinct. Glyptodonts shared unusual traits and a restricted distribution with the existing armadillos. Yet the two kinds of animals are widely separated in time. Their similarities were a clue that helped Darwin develop a theory of evolution by natural selection.

a

b

Figure 16.7 Four of thirteen finch species on the Galápagos Islands. (**a**) *Geospiza conirostris* and (**b**) *G. scandens*. Both eat cactus flowers and fruits; other finches eat its seeds. (**c**) *Certhidea olivacea*, a tree-dweller, uses its slender bill to probe for food. (**d**) *Camarhynchus pallidus* eats wood-boring insects. It learns to break and use cactus spines and twigs as probes. We now know that bill shape depends on when and where the protein BMP4, a signaling molecule, is switched on in bird embryos. Mutations in the gene for this protein, are an evolutionary source for the differences.

NATURAL SELECTION DEFINED

Here are Darwin's key observations and conclusions about evolution, expressed in modern terms:

1. *Observation:* Natural populations have an inherent reproductive capacity to increase in numbers through successive generations.

2. *Observation:* No population can indefinitely grow in size, because its individuals will run out of food, living space, and other resources.

3. *Inference:* Sooner or later, individuals will end up competing for dwindling resources.

4. *Observation:* Those individuals generally have the same genes encoding the same shared traits. Genes are the population's pool of heritable information.

5. *Observation:* Mutations have given rise to alleles, or slightly different molecular forms of genes, which are a source of differences in phenotypic details.

6. *Inferences:* Some phenotypes are better than others at helping an individual compete for resources, and to survive and reproduce. Alleles for those phenotypes increase in the population, and other alleles decrease. In time the genetic changes lead to increased **fitness**—an increase in adaptation to the environment.

7. *Conclusions:* **Natural selection** among individuals of a population is an outcome of variation in traits that affect which individuals survive and reproduce in each generation. This microevolutionary process results in adaptation, or increased fitness to the environment.

Darwin kept on looking for patterns in his data and filling in gaps in his reasoning. He waited too long. More than ten years after he wrote but did not publish his theory, Alfred Wallace sent him a letter (Figure 16.8). Wallace had been doing impressive work in the Amazon Basin and Malay Archipelago, Madagascar, and elsewhere. He had written earlier to Lyell and Darwin about the causes of species distributions. And he had arrived at the same theory!

Though daunted, Darwin encouraged Wallace to publish. Wallace and other colleagues thought Darwin should get credit. At a scientific meeting in 1858,

Figure 16.8 Alfred Wallace, one of the first to study island biogeography. For a stimulating view of the Darwin–Wallace story, read David Quamman's *The Song of the Dodo*.

with neither Darwin nor Wallace in attendance, the theory was presented and attributed to both. The next year, Darwin published *On the Origin of Species*, with detailed evidence in support of the theory.

You may have heard that Darwin's book fanned an intellectual firestorm, but most scholars were quick to accept the idea that diversity is the result of evolution. The theory of natural selection *was* fiercely debated. Decades passed before experimental evidence from a new field, genetics, led to its widespread acceptance.

As Darwin and Wallace perceived, natural selection is the outcome of variations in shared traits that influence which individuals of a population survive and reproduce in each generation. It can lead to increased fitness—that is, to an increase in adaptation to the environment.

16.4 The Nature of Adaptation

A word of caution: Observable traits are not always easy to correlate with conditions in an organism's environment.

"Adaptation" is one of those words that have different meanings in different contexts. An individual plant or animal often can quickly adjust its form, function, and behavior. Junipers in inhospitably windy places grow less tall than junipers of the same species in more sheltered places. A clap of thunder may make you lurch the first time you hear it, but then you may get used to the sound over time and ignore it. These are examples of *short-term* adaptations, because they last only as long as the individual does.

Over the long term, an **adaptation** is some heritable aspect of form, function, behavior, or development that improves the odds for surviving and reproducing in a given environment. It is an *outcome* of microevolution —natural selection especially—an enhancement of the fit between the individual and prevailing conditions.

Figure 16.9 (**a**) Severe, rapid wilting of one commercial tomato plant (*Solanum lycopersicum*) that absorbed salty water. (**b**) Galápagos tomato plant, *S. cheesmanii*, which stores most absorbed salts in its leaves, not in its fruits.

SALT-TOLERANT TOMATOES

As an example of long-term adaptation, compare how tomato species handle salty water. Tomatoes evolved in Ecuador, Peru, and the Galápagos Islands. The type sold most often in markets, *Solanum lycopersicum*, has eight close relatives in the wild. If you mix ten grams of table salt with sixty milliliters of water, then pour it into the soil around *S. lycopersicum*'s roots, the plant will wilt severely in less than thirty minutes (Figure 16.9*a*). Even if the soil has only 2,500 parts per million of salt, this species will grow poorly.

Yet the Galápagos tomato (*S. cheesmanii*) survives and reproduces in seawater-washed soils. We know that its salt tolerance is a heritable adaptation. How? F₁ crosses of a wild species with the commercial one

result in a small, edible hybrid. The hybrid tolerates irrigation water that is two parts fresh and one part salty. It's gaining interest in places where fresh water is scarce and where salts have built up in croplands.

It may take modification of only a few traits to get new salt-tolerant plants. Revving up just one gene for a sodium–hydrogen ion transporter helps the tomato plants use salty water and still bear edible fruits.

NO POLAR BEARS IN THE DESERT

You can safely bet that a polar bear (*Ursus maritimis*) is finely adapted to the icy Arctic, and that its form and function would be a flop in a desert (Figure 16.10). You

Figure 16.10 Which adaptations of a polar bear (*Ursus maritimus*) won't help in a desert? Which ones help an oryx (*Oryx beisa*)? For each animal, make a tentative list of possible structural and functional adaptations to the environment. Later, after you finish reading Unit VI, see how you can expand the list.

might be able to make some educated guesses about why that is so. However, detailed knowledge of its anatomy and physiology might make you view it—or any other animal or plant—with respect. How does a polar bear maintain its internal temperature when it sleeps on ice? How can its muscles function in frigid water? How often must it eat? How does it find food? Conversely, how can an oryx walk about all day in the blistering heat of an African desert? How does it get enough water when there is no water to drink? You will find some answers, or at least ideas about how to look for them, in the next three units of this book.

ADAPTATION TO WHAT?

Bear in mind, it isn't always easy to identify a direct relationship between adaptation and the environment. For instance, the prevailing environment may be very different from the one in which a trait evolved.

Consider the llama. It is native to the cloud-piercing peaks of the Andes in western South America (Figure 16.11). The llama lives 4,800 meters (16,000 feet) above sea level. Compared to humans at lower elevations, its lungs have more air sacs and blood vessels. The llama heart has larger chambers, so it pumps larger volumes of blood. Llamas don't have to make extra blood cells, as people do when they move permanently from the lowlands to high elevations. (Extra cells make blood "stickier," so the heart has to pump harder.) But the most publicized adaptation is this: Llama hemoglobin is better than ours at latching on to oxygen. It picks up oxygen in the lungs far more efficiently.

Superficially, at least, the oxygen-binding affinity of llama hemoglobin appears to be an adaptation to thin air at high altitudes. Is it? Apparently not.

Llamas belong to the same family as dromedary camels. Both share camelid ancestors that evolved in the Eocene grasslands and deserts of North America. Later, the ancestors of camels and llamas went their separate ways. Camel forerunners moved into Asia's low-elevation grasslands and deserts by a land bridge, which later submerged when the sea level rose. Llama forerunners moved in a different direction—down the Isthmus of Panama, and on into South America.

Intriguingly, a dromedary camel's hemoglobin also shows a high oxygen-binding capacity. So if the trait arose in a shared ancestor, then how was it adaptive at *low* elevations? We know camels and llamas didn't just *happen* to evolve in the same way. They are close kin, and their most recent ancestors lived in very different environments with different oxygen concentrations.

Who knows why the trait was originally favored? Eocene climates were alternately warm and cool, and hemoglobin's oxygen-binding capacity does go down as temperatures go up. Did it prove adaptive during a long-term shift in climate? Or were its effects neutral at first? What if the mutant gene for the trait became fixed in some ancestral population simply by chance?

Use all of these "what-ifs" as a reminder to think carefully about connections between form and function. Identifying such connections takes a lot of intuition, research, and experimental tests.

A long-term adaptation is any heritable aspect of form, function, behavior, or development that contributes to the fit between an individual and its environment.

An adaptive trait improves the odds of surviving and reproducing, or at least it did so under conditions that prevailed when genes for the trait first evolved.

16.5 Individuals Don't Evolve, Populations Do

Evolution starts with changes in the gene pool of a population, as brought about by mutation, natural selection, gene flow, and genetic drift.

VARIATION IN POPULATIONS

As Darwin and Wallace perceived, *individuals do not evolve; populations do.* By definition, a **population** is a group of individuals of the same species occupying a given area. To understand how it evolves, start with variation in the features that characterize it.

Individuals of a population share basic features. Jays have two feathered wings, three toes forward, one toe back, and so on. These are *morphological* traits (*morpho–*, form). The body parts of all individuals work much the same way in metabolism, growth, and reproduction. Such *physiological* traits help the body function in its environment. Individuals respond the same way to basic stimuli, as when babies imitate adult facial expressions. These are *behavioral* traits.

Especially for sexually reproducing species, details of most traits vary among individuals of populations. Pigeon feathers and butterfly wings differ in colors or patterns. Figure 16.12 hints at human variations in skin color and distribution, color, texture, and amount of hair. Almost every trait of every species is variable.

Many traits show *qualitative* differences; they have two or more distinct forms, or morphs. (Remember the purple or white pea plant flowers?) Two forms is a dimorphism; three or more is polymorphism. Other traits, such as height and eye color, show *quantitative* differences, which are small and incremental.

THE "GENE POOL"

Genes hold information on heritable traits. In general, a population's individuals all have the same number and kinds of genes. We say "in general" because in sexually reproducing species, the sex chromosomes of males and females are not alike.

Think of all the genes in a population as the **gene pool**—a pool of genetic resources that the individuals of the population, and their offspring, share. Often, each kind of gene in the pool is present in two or more slightly different molecular forms, or **alleles**.

Individuals have different combinations of alleles. This leads to variations in *phenotype*, or differences in details of traits. Whether you have black, brown, red, or blond hair depends upon the certain alleles that you inherited from your two parents.

Also, don't forget that offspring inherit genes, not phenotypes. Environmental conditions often alter gene expression (Section 10.6). Variation resulting from their effects lasts no longer than the individual.

Which alleles end up in a gamete and then the new individual? Five events, described in earlier chapters, shape the outcome. We summarize them here:

1. Gene mutation (produces new alleles)

2. Crossing over during meiosis I (introduces novel combinations of alleles in chromosomes)

3. Independent assortment at meiosis I (puts mixes of maternal and paternal chromosomes in gametes)

4. Fertilization (combines alleles from two parents)

5. Change in chromosome number or structure (leads to the loss, duplication, or repositioning of genes)

Only mutation *creates* new alleles. Other events shuffle *existing* alleles into different combinations—but what a shuffle! For example, a human gamete gets one of 10^{600} possible combinations. Not even 10^{10} humans are alive today. Unless you are an identical twin, it is extremely unlikely that any other person with your precise genetic makeup has ever lived, or ever will.

STABILITY AND CHANGE IN ALLELE FREQUENCIES

Researchers typically track **allele frequencies**, or the abundance of certain alleles in a population. They start at **genetic equilibrium**, a theoretical reference

Figure 16.12 A sampling of variation in human and snail populations.

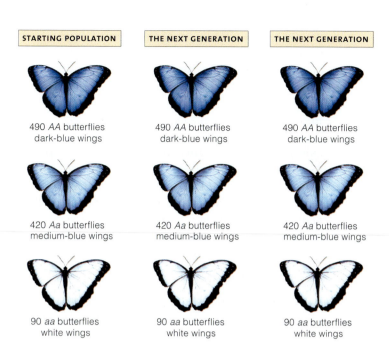

STARTING POPULATION	THE NEXT GENERATION	THE NEXT GENERATION
490 *AA* butterflies dark-blue wings	490 *AA* butterflies dark-blue wings	490 *AA* butterflies dark-blue wings
420 *Aa* butterflies medium-blue wings	420 *Aa* butterflies medium-blue wings	420 *Aa* butterflies medium-blue wings
90 *aa* butterflies white wings	90 *aa* butterflies white wings	90 *aa* butterflies white wings

Figure 16.13 How to find out if a population is evolving. Start with five assumptions of the *Hardy–Weinberg rule*: A population's allele frequencies don't change when there is no mutation, the population is infinitely large and isolated from others of the species, mating is random with respect to the alleles of interest, and all individuals survive and reproduce sexually (no selection).

Track a hypothetical pair of alleles that influence butterfly wing color. Butterflies are sexually reproducing organisms, with pairs of genes on pairs of homologous chromosomes. Allele *A* is associated with dark-blue, *a* with white, and *Aa* with medium-blue wings. At genetic equilibrium, the proportions of the wing-color genotypes are

$$p^2\ AA\ +\ 2pqAa\ +\ q^2\ aa\ =\ 1.0$$

where *p* and *q* are frequencies of alleles *A* and *a*. The frequencies of *A* and *a* must add up to 1.0. Example: If *A* occupies half of all loci for this gene in the population, then *a* must occupy the other half (0.5 + 0.5 = 1.0). If *A* occupies 90 percent of all the loci, then *a* must occupy 10 percent (0.9 + 0.1 = 1.0). No matter what the proportions,

$$p\ +\ q\ =\ 1.0$$

At meiosis, the alleles of each pair segregate and end up in different gametes. So the proportion of gametes having the *A* allele is *p*, and the proportion having the *a* allele is *q*. This Punnett square shows the genotypes possible in the next generation (*AA*, *Aa*, and *aa*):

	p Ⓐ	q ⓐ
p Ⓐ	*AA* (p^2)	*Aa* (pq)
q ⓐ	*Aa* (pq)	*aa* (q^2)

The frequencies add up to 1.0: $p^2\ +\ 2pq\ +\ q^2\ =\ 1.0$.

Let's say the population has 1,000 individuals, and that each one produces two gametes:

  490 *AA* individuals produce 980 *A* gametes
  420 *Aa* individuals produce 420 *A* and 420 *a* gametes
   90 *aa* individuals produce 180 *a* gametes

The frequency of alleles *A* and *a* among the 2,000 gametes is

$$A\ =\ \frac{2 \times 490\ +\ 420}{2,000\ alleles}\ =\ \frac{1,400}{2,000}\ =\ 0.7\ =\ p$$

$$a\ =\ \frac{2 \times 90\ +\ 420}{2,000\ alleles}\ =\ \frac{600}{2,000}\ =\ 0.3\ =\ q$$

At fertilization, gametes combine at random, forming a new generation. If the population stays at 1,000, you have 490 *AA*, 420 *Aa*, and 90 *aa* individuals. Because the allele frequencies for dark-blue, medium-blue, and white wings are the same as in the original gametes, they will give rise to the same phenotypic frequencies as the second generation.

As long as the five conditions of the Hardy–Weinberg rule hold, this pattern will persist. If traits show up in different proportions from one generation to the next, then one or more of the five assumptions is not being met. *The hunt can begin for one or more specific evolutionary forces driving the change.*

point. At this point, a population is *not* evolving with respect to the allele frequencies being studied. Why? Five conditions are prevailing: No mutations have occurred; the population is infinitely large; it is fully isolated from all other populations of the species; the product of the allele being studied has no effect on survival or reproduction; and all mating is random.

Figure 16.13 is an example. If you choose to pass on reading it, just know that the five conditions hardly ever prevail at the same time in the natural world. Why? Mutations are rare, but inevitable. Also, three processes—*natural selection, gene flow,* and *genetic drift*—drive such populations out of genetic equilibrium.

Microevolution refers to small-scale changes in population allele frequencies that arise from mutation, natural selection, gene flow, and genetic drift.

A population or species is characterized by morphological, physiological, and behavioral traits, most of which are heritable. Details of these traits vary among its individuals.

Different combinations of alleles among individuals of a population give rise to variations in phenotype—that is, to differences in the details of their shared structural, functional, and behavioral traits.

In sexually reproducing species, a population's individuals share a pool of genetic resources—that is, a gene pool.

Few populations ever achieve genetic equilibrium. Natural selection, gene flow, and genetic drift change a gene pool's allele frequencies.

16.6 Mutations Revisited

Reflect, for a moment, on the statement that mutations are the original source of alleles. These heritable changes in DNA usually give rise to altered gene products, and evolution starts with them.

We can't predict when or in which individual a given mutation will occur. We do know that each gene has a **mutation rate**, the probability of its mutating during or between DNA replications. For each gamete the rate is between 10^{-5} and 10^{-6} mutations per gene.

Mutations may give rise to structural, functional, or behavioral alterations that reduce an individual's chances of surviving and reproducing. Even a single biochemical change can be devastating. For instance, skin, bones, tendons, lungs, blood vessels, and many other vertebrate organs incorporate collagen. When the collagen gene mutates, drastic changes all through the body follow. Any mutation that severely changes phenotype usually causes death; it is a **lethal mutation**.

By comparison, a **neutral mutation** doesn't help or hurt. Natural selection neither increases nor decreases its frequency in a population, because it won't have a discernible effect on whether an individual survives and reproduces. If you carry a mutant gene that keeps earlobes attached to your head instead of swinging freely, this in itself shouldn't stop you from surviving and perhaps reproducing just as well as anybody else.

Every so often, a mutation proves useful. A mutant gene product that influences growth may make a corn plant grow larger or faster and thus give it the best access to sunlight and to nutrients. A neutral mutation may prove helpful after conditions in the environment change. Even when it bestows only a small advantage, chance events or natural selection might preserve the mutant gene in the DNA and favor its representation in the next generation.

Mutations are so rare they usually have little or no immediate effect on a population's allele frequencies. But both beneficial and neutral mutations have been accumulating in diverse lineages for billions of years. All that time, they have functioned as raw material for evolutionary change, for the staggering range of biodiversity, past and present. From an evolutionary view, the reason you don't look like a bacterium or an avocado or earthworm or even your neighbors down the street began with mutations that arose at different times, in different lines of descent.

Go ahead. Hit me with a pesticide.

Mutations occur at a predictable rate, but are themselves unpredictable. They may alter an individual's phenotype.

16.7 Directional Selection

We turn now to mechanisms of natural selection. In some cases, the range of values for a trait shifts in one direction. At other times, an existing range of values may be stabilized or disrupted, depending on conditions.

In cases of **directional selection**, allele frequencies that give rise to a range of variation in phenotype tend to shift in a consistent direction. The shift is a response to directional change in the environment or to one or more new conditions. A new mutation that benefits individuals also may cause a directional shift. Either way, forms at one end of a phenotypic range become more common than midrange forms (Figure 16.14).

PESTICIDE RESISTANCE

As the chapter introduction made clear, pesticides can bring about directional selection. A few individuals usually survive the initial applications. Some heritable aspect of body structure, physiology, or behavior helps them resist the chemicals. The most resistant ones are favored, so resistance becomes more common. Today, 450 species resist one or more pesticides.

Pesticides also kill natural predators of pests. Freed from natural constraints, resistant populations cause more damage than before. This outcome of directional selection is called *pest resurgence*. Crops of genetically engineered plants can resist pests. But these plants, too, will start exerting selection pressure on populations of pests. Individuals that can overcome the engineered defenses tend to be favored.

ANTIBIOTIC RESISTANCE

Natural and synthetic antibiotics can fight bacterial diseases. Natural **antibiotics** are toxins that some microorganisms release to kill bacterial competitors for nutrients. Streptomycins, for instance, inhibit bacterial protein synthesis. The penicillins disrupt covalent bonds that hold a bacterial cell wall together.

Antibiotics should be prescribed with restraint and care. Besides performing their intended function, they often disrupt resident populations in the intestines and vagina. Imbalances lead to secondary infections.

Antibiotics also have been overprescribed, often for simple infections that most people can fight on their own. Genetic variation in bacterial gene pools allows individuals with certain genotypes to survive while others do not. Overuse of antibiotics has thus favored resistant populations, which does not bode well for the millions of people who each year contract cholera, tuberculosis, and other bacterial diseases.

COAT COLOR IN DESERT MICE

Researchers Michael Nachman, Hopi Hoekstra, and Susan D'Agostino reported directional selection among rock pocket mice (*Chaetodipus intermedius*) living in the same part of Arizona's Sonoran Desert (Figure 16.15). Of more than eighty genes known to affect coat color in mice, they pinpointed one that governs a difference between two populations of this mouse species.

Rock pocket mice are small mammals that spend the day in underground burrows. At night they forage for seeds, scampering over the tawny-colored granite outcroppings. Here, individuals with tawny fur are camouflaged from predators (Figure 16.15b).

A smaller population of pocket mice lives in the same region, but these mice scamper over dark basalt of ancient lava flows. They have dark coats, so they, too, are camouflaged from predators (Figure 16.15c).

We can expect that night-flying predatory birds are selective agents that affect fur color. Earlier studies, for example, demonstrated that owls have an easier time seeing mice with fur that does not match the rocks. Nachman drew on genetic data about laboratory mice to formulate a hypothesis on coat color differences in the two populations. He predicted that a mutation of either the *Mclr* or *agouti* gene causes the difference.

He collected DNA from dark-colored pocket mice at a lava flow and from light-colored mice at adjacent granite outcroppings. DNA analysis showed that the *Mclr* gene sequence differs between the groups. The gene sequence for all dark-fur mice differed by four nucleotides from that of their light-furred neighbors.

With directional selection, allele frequencies underlying a range of variation tend to shift in a consistent direction in response to some change in the environment.

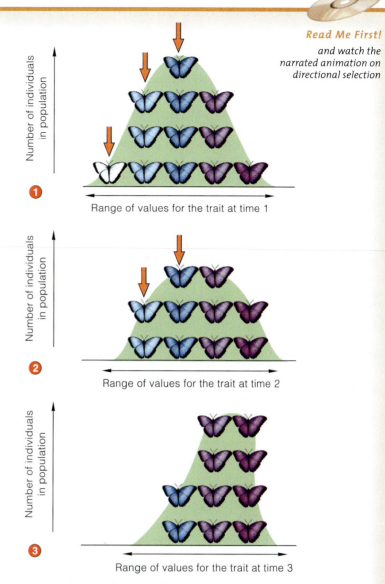

Read Me First!
and watch the narrated animation on directional selection

Figure 16.14 Directional selection in a butterfly population. This bell-shaped curve signifies a range of continuous variation in a wing color trait. *Medium-blue*, the most common form, is between two extremes—*white* and *dark purple*. The *orange* arrows signify which forms are being selected against over time.

Figure 16.15 Visible evidence of directional selection between two neighbor populations of rock pocket mice. (**a**) Lava basalt flow at the study site. The two color morphs of rock pocket mice, each posed on two different backgrounds: (**b**) tawny fur and (**c**) dark fur.

16.8 Selection Against Or in Favor of Extreme Phenotypes

Consider now two additional modes of natural selection. One works against phenotypes at the fringes of a range of variation; the other favors them (Figure 16.16).

STABILIZING SELECTION

With **stabilizing selection**, intermediate forms of a trait in a population are favored and alleles for the extreme forms are not. This mode of selection can counter mutation, gene flow, and genetic drift. It preserves the most common phenotypes.

As an example, prospects are not good for human babies who weigh far more or far less than average at birth. Also, pre-term instead of full-term pregnancies increase the danger, as reflected in Figure 16.17. Newborns weighing less than 5.51 pounds or born before thirty-eight weeks of pregnancy are completed tend to develop high blood pressure, diabetes, and heart disease when they are adults. The mother's blood concentration of cortisol, a stress hormone, may be linked to low birth weight and the illnesses that develop later in life.

Read Me First!

and watch the narrated animation on stabilizing and disruptive selection

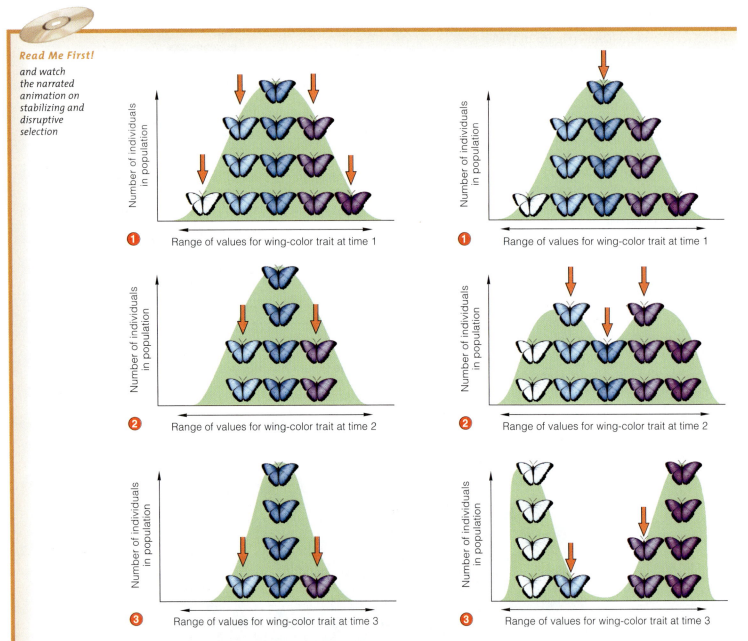

Figure 16.16 Selection against or in favor of extreme phenotypes, with a population of butterflies as the example. *Left*, stabilizing selection and *right*, disruptive selection. The *orange* arrows show forms of the trait being selected against.

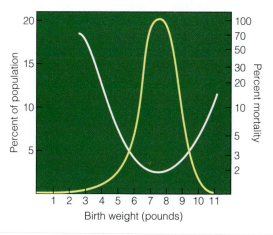

Figure 16.17 Weight distribution for 13,730 human newborns (*yellow* curve) correlated with death rate (*white* curve).

Figure 16.18 Adult sociable weaver (*Philetairus socius*), a native of the African savanna. These birds cooperate in constructing and using large communal nests in a region where trees and other good nesting sites are scarce.

Rita Covas and her colleagues gathered evidence of stabilizing selection on the body mass of juvenile and adult sociable weavers (*Philetairus socius*), as in Figure 16.18. Between 1993 and 2000, they captured, measured, tagged, released, and recaptured 70 to 100 percent of birds present in communal nests during the breeding season. Their field studies supported a prediction that body mass is a trade-off between risks of starvation and predation, giving intermediate-mass birds the selective advantage. Foraging is not easy in this habitat, and lean birds don't store enough fat to avoid starvation. We can expect that fat ones are more attractive to predators and not as good at escaping.

DISRUPTIVE SELECTION

With **disruptive selection**, forms at both ends of the range of variation are favored and intermediate forms are selected against.

Consider the black-bellied seedcracker (*Pyrenestes ostrinus*) of Cameroon. Females and males of these African finches have large or small bills—but no sizes in between (Figure 16.19). It's like everyone in Texas being four feet *or* six feet tall, with no one in between.

The pattern holds all through the geographic range. If unrelated to gender or geography, what causes it? If only two bill sizes persist, then disruptive selection may be eliminating birds with intermediate-size bills. Which factors affect feeding performance? Cameroon's swamp forests flood during the wet season, and fires sparked by lightning burn during the dry season. Two kinds of sedges dominate these forests. Sedges are fire-resistant, grasslike plants. One species produces hard seeds and the other, soft.

Obviously, a seedcracker's ability to crack seeds directly affects survival. It turns out that birds with

lower bill 12 mm wide   lower bill 15 mm wide

Figure 16.19 Disruptive selection in African finch populations. Selection pressures favor birds with bills that are about 12 or 15 millimeters wide. The difference is correlated with competition for scarce food resources during the dry season.

small bills prefer to eat soft seeds in the habitat, and birds with large bills are better at cracking the hard ones. In the dry season, all seeds are scarce and birds compete fiercely for them. Limited availability of two types of seeds during recurring periods of famine has had a disruptive effect on bill size in the seedcracker population; birds with intermediate sizes are selected against and all bills are 12 or 15 millimeters wide.

In the seedcracker, bills of a particular size have a genetic basis. In experimental crosses between two birds with the two optimal bill sizes, all offspring had a bill of one size or the other, nothing in between.

> With stabilizing selection, intermediate phenotypes are favored and extreme phenotypes at both ends of the range of variation are eliminated.
>
> With disruptive selection, intermediate forms of traits are selected against; extreme forms in the range of variation are favored.

16.9 Maintaining Variation in a Population

Natural selection theory helps explain diverse aspects of nature, including male–female differences, and the relationship between sickle-cell anemia and malaria.

SEXUAL SELECTION

The individuals of many sexually reproducing species show a distinct male or female phenotype, or **sexual dimorphism** (*dimorphos,* having two forms). Often the males are larger and flashier than females. Courtship rituals and male aggression are common.

These adaptations and behaviors seem puzzling. All take energy and time away from an individual's survival activities. Why do they persist if they do not contribute directly to survival? The answer is **sexual selection**: a form of natural selection in which the genetic winners are the ones that outreproduce others of the population. The most adaptive traits help individuals defeat same-sex rivals for mates or are the ones most attractive to the opposite sex.

Figure 16.20 One male bird of paradise in a flashy courtship display. He caught the eye (and, perhaps, the sexual interest) of the smaller, less colorful female. The males of this species compete fiercely for females, which are the selective agents. (Why do you suppose drab-colored females have been favored?) outcome of sexual selection. This (*Paradisaea raggiana*) is engaged

By choosing mates, one gender acts as an agent of selection on its own species. For example, the females of some species shop among a clustering of males, which differ in appearance and courtship behavior. The selected males, as well as the females making the selection, pass on their alleles to the next generation.

Flashy structures and behaviors are correlated with species in which males have little or nothing to do with raising offspring. The female chooses a male by observable signs of his health and vigor, which might improve the odds of producing healthy, vigorous offspring (Figure 16.20).

You might be wondering whether we can correlate genes with specific forms of sexual behavior. One of the most amazing demonstrations of this comes from sexual deception as practiced by an Australian orchid. The flowers of *Chiloglottis trapeziformis* attract male wasps by making a compound—a sex pheromone—that is identical to one released by the female wasps, the point being to get pollinated as the male is busy doing what otherwise would perpetuate its genes.

This orchid is stingy. It gives a male wasp nothing in return, not a single drop of nectar, even though it is the orchid's sole pollinator. The wingless female wasps hatch in soil. When males don't lift and carry them to a food source, they starve to death.

When *C. trapeziformis* puts out blooms, the male wasps waste precious time and metabolic energy trying to find females. Evolutionary biologist Florian Schiestl suggests that selection pressure is afoot for wasps that can make a new sex pheromone, one that the orchid can't duplicate.

And while the interaction exploits males, Wittko Francke thinks it might put pressure on their brains to evolve. In an orchid patch, the average tiny-brained male wasp copulates blindly with whatever smells right. It will try to copulate even with the head of a pin that has a few micrograms of pheromone sprayed on it. However, a few wasps with a slightly less robotic brain might be able to identify the females by other cues, such as visual ones. Alternatively, both species could face extinction, another pattern in nature.

SICKLE-CELL ANEMIA—LESSER OF TWO EVILS?

With *balancing* selection, two or more alleles for a trait are being maintained at frequencies above 1 percent in the population. Their persistence is called **balanced polymorphism** (*polymorphos,* having many forms). The allele frequencies might shift slightly, but they often return to the same values over the long term. We often see this balance when conditions favor heterozygotes.

In some way, their nonidentical alleles for a given trait give them higher fitness than homozygotes, which, recall, have identical alleles for the trait.

Consider the environmental pressures that favor an Hb^A/Hb^S pairing in humans. The Hb^S allele codes for a mutant form of hemoglobin, an oxygen-transporting protein in blood. Homozygotes (Hb^S/Hb^S) develop *sickle-cell anemia*, a genetic disorder (Section 3.6).

The Hb^S frequency is highest in subtropical and tropical regions of Asia and Africa. Often, Hb^S/Hb^S homozygotes die in their early teens or early twenties. Yet, in these same regions, heterozygotes (Hb^A/Hb^S) make up nearly a third of the population! Why is this combination maintained at such high frequency?

The balancing act is most pronounced in areas that have the highest incidence of *malaria* (Figure 16.21). Mosquitoes transmit the parasitic agent of malaria, *Plasmodium*, to human hosts. The parasite multiplies in the liver and, later, in red blood cells. The target cells rupture and release new parasites during severe, recurring bouts of infection (Section 20.3).

It turns out that the mutant hemoglobin interferes with the life cycle of the parasitic agent, so Hb^A/Hb^S heterozygotes are more likely to survive malaria than people who produce normal (Hb^A/Hb^A) hemoglobin. There are several potential survival mechanisms. In heterozygotes, infected cells have a sickle shape under normal conditions. The abnormal shape marks them as targets for the immune system, which proceeds to destroy them. Also, heterozygotes have one normal hemoglobin allele. Although they are not completely healthy, they produce enough normal hemoglobin to support body functions. As a result, they are more likely than the Hb^S/Hb^S homozygotes to survive and reach reproductive age.

So the persistence of the "harmful" Hb^S allele is a matter of relative evils. Natural selection has favored the Hb^A/Hb^S combination in malaria-ridden areas because heterozygotes show more resistance to the disease. In such environments, the combination has more survival value than either Hb^S/Hb^S or Hb^A/Hb^A. And malaria has been a selective force for thousands of years in tropical and subtropical areas of Asia, the Middle East, and Africa.

With sexual selection, some version of a gender-related trait gives the individual an advantage in reproductive success. Sexual dimorphism is one outcome of sexual selection.

In a population showing balanced polymorphism, natural selection is maintaining two or more alleles at frequencies greater than 1 percent over the generations.

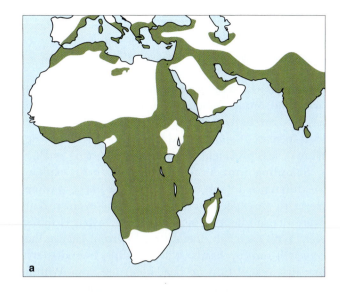

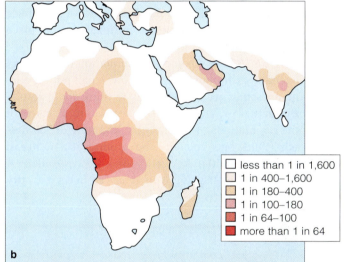

less than 1 in 1,600
1 in 400–1,600
1 in 180–400
1 in 100–180
1 in 64–100
more than 1 in 64

Figure 16.21 (**a**) Distribution of malaria cases in Africa, Asia, and the Middle East in the 1920s, before the start of programs to control mosquitoes, the vector for *Plasmodium*. (**b**) Distribution and frequency of people with the sickle-cell trait. Notice the close correlation between the maps. (**c**) Physician searching for *Plasmodium* larvae in Southeast Asia.

16.10 Genetic Drift—The Chance Changes

Random changes in allele frequencies can lead to a loss of genetic diversity in a population. The drift in those frequencies is greatest for small populations.

Genetic drift is a random change in allele frequencies over time, brought about by chance alone. It tends to have minor impact in very large populations. Even so, it increases the likelihood that an allele will become more or less prevalent when the population is small.

Sampling error, a rule of probability, helps explain the difference. By this rule, you are less likely to come closer to an expected outcome of some event if that event doesn't happen very often. For instance, flip a coin. With each flip, there is a 50 percent chance the coin will turn up heads. With ten flips, the odds are low that it will turn up heads half the time. With a thousand flips, you are more likely to come close to 500 heads and 500 tails. Sampling error also applies each time random mating and fertilization take place in a population.

Figure 16.22 is a computer simulation of the effect of genetic drift in one large and one small population. The outcomes are a simple way to think about results from actual experiments. The simulation starts with nine populations of 25 flies each and nine of 500 flies each. Which ones entered either group was a matter of chance. The initial frequency of wild-type allele *A* was 0.5. Some offspring were removed in each of fifty generations to maintain population size. In the end, *A* wasn't the only allele left in large groups. But it was fixed in five of the small groups. **Fixation** means only one kind of allele remains at a locus in a population. All individuals have become homozygous for it.

Thus, *in the absence of other forces, random change in allele frequencies leads to the homozygous condition and a loss of genetic diversity over the generations.* It happens in all populations. It just happens faster in small ones. Once alleles from a parent population are fixed, their frequencies will not change again unless mutation or gene flow introduces new alleles.

BOTTLENECKS AND THE FOUNDER EFFECT

Genetic drift is pronounced when a few individuals rebuild a population or start a new one. This happens after a **bottleneck**, a drastic reduction in population size brought about by severe pressure or a calamity. Suppose contagious disease, habitat loss, or hunting nearly wipes out a population. Even if a moderate number of individuals survive the bottleneck, allele frequencies will have been altered at random.

In the 1890s, hunters killed all but twenty of a large population of northern elephant seals. Government restrictions since then have allowed them to recover to a current population of about 130,000. Each of them is homozygous at every gene locus examined so far.

Read Me First!
and watch the narrated animation on genetic drift

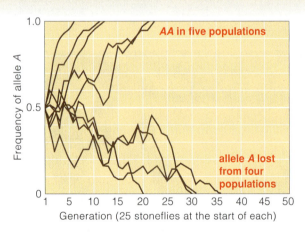

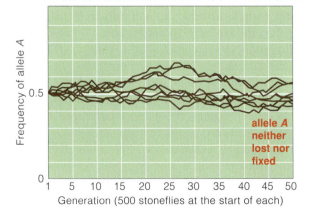

a The size of nine populations of flies was held constant at 25 breeding individuals in each generation, through fifty generations. The five graph lines reaching the top of this diagram tell you that allele *A* became fixed in five of these small populations. The four lines plummeting off the bottom of the diagram tell you that it was lost from four of them. As you can see, *alleles can be fixed or lost even in the absence of selection.*

b The size of nine different populations was kept at 500 individuals in each generation, through fifty generations. In these larger populations, allele *A* did not become fixed. The magnitude of genetic drift was much less in each generation than in the small populations tracked in (**a**).

Figure 16.22 Computer simulation of genetic drift's effect on allele frequencies in small and large populations of flies. Equal fitness is assumed for three simulations (*AA* = 1, *Aa* = 1, and *aa* = 1).

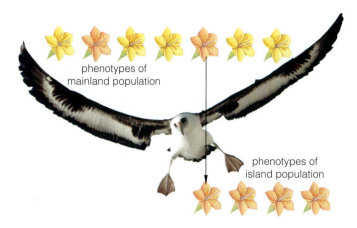

phenotypes of mainland population

phenotypes of island population

Figure 16.23 Founder effect. This wandering albatross carries seeds, stuck to its feathers, from the mainland to a remote island. By chance, most of the seeds carry an allele for orange flowers that are rare in the original population. Without further gene flow or selection for color, genetic drift will fix the allele on the island.

Genetic outcomes also can be unpredictable after a few individuals establish a new population. This form of bottlenecking is a **founder effect**. By chance, allele frequencies of the founders may differ in the original population. If there is no further gene flow, genetic drift will work on the small population. The effect can be pronounced on isolated islands (Figure 16.23).

GENETIC DRIFT AND INBRED POPULATIONS

Genetic drift is pronounced in an inbred population. **Inbreeding** is nonrandom mating among very close relatives, which share many identical alleles. It leads to the homozygous condition. It also lowers fitness if harmful recessive alleles are increasing in frequency.

Most human societies forbid or discourage incest (inbreeding between parents and children or siblings). But inbreeding among other close relatives is common in geographically or culturally isolated small groups. The Old Order Amish of Pennsylvania, for instance, are a highly inbred group having distinct genotypes. One outcome of inbreeding is a high frequency of the recessive allele that causes *Ellis–van Creveld syndrome*. Affected individuals have extra fingers, toes, or both and short limbs (Section 11.4). The allele might have been rare when a few founders entered Pennsylvania. Currently, about 1 in 8 individuals of the community are heterozygous and 1 in 200 are homozygous for it.

Genetic drift is the random change in allele frequencies over the generations, brought about by chance alone. The magnitude of its effect is greatest in small populations, such as one that endures a bottleneck.

16.11 Gene Flow

Individuals, and their alleles, move into and out of populations, and this physical flow counters changes introduced by other microevolutionary processes.

Individuals of the same species don't always stay put. A population will lose alleles whenever an individual permanently leaves it, an event called *emigration*. The population gains alleles whenever new individuals permanently move in, an event called *immigration*. In both cases, there is a **gene flow**—a physical flow of alleles between two or more populations. Gene flow tends to counter genetic differences that we expect to see developing by way of mutation, natural selection, and genetic drift. It helps keep separated populations genetically similar.

Think of the acorns that blue jays disperse when they gather nuts for the winter. Each fall the jays visit acorn-bearing oak trees repeatedly, then bury acorns in the soil of home territories that may be as much as a mile away (Figure 16.24). Alleles flowing in with the "immigrant acorns" help decrease genetic differences between stands of oak trees.

Figure 16.24 Blue jay, a mover of acorns that helps keep genes flowing between separate oak populations.

Or think of the millions of people from politically explosive, economically bankrupt countries who seek a more stable home. The scale of their emigrations is unprecedented, but the flow of genes is not. Human history is rich with cases of gene flow that minimized many of the genetic differences among geographically separate groups. Remember Genghis Khan? His genes flowed from China to Vienna (Figure 11.6). Similarly, the armies of Alexander the Great brought the genes for green eyes from Greece all the way to India.

Gene flow is the physical movement of alleles into and out of a population, through immigration and emigration. It tends to counter the effects of mutation, natural selection, and genetic drift.

Summary

Section 16.1 Awareness of evolution, or changes in lines of descent over time, emerged long ago from biogeography, comparative morphology, and geology.

Section 16.2 Prevailing cultural belief systems influence interpretation of natural events. In the nineteenth century, naturalists worked to reconcile traditional belief systems with a growing body of physical evidence in support of evolution.

Section 16.3 Charles Darwin and Alfred Wallace proposed a novel theory that natural selection in populations results in evolution. The theory of natural selection is this: Populations increase in size until resources dwindle and individuals compete for them. When individuals have forms of traits that make them more competitive, they tend to produce more offspring. Over generations, those forms increase in frequency. Nature "selects" variations in traits that are more effective at helping individuals survive and reproduce in particular environments.

Section 16.4 Long-term adaptations are heritable aspects of form, function, behavior, or development that improve the chance of surviving and reproducing.

Section 16.5 Individuals of a population generally have the same number and kinds of genes for the same traits. Mutations are the source of *new* alleles (different molecular forms of genes). Individuals who inherit different allele combinations vary in details of one or more traits. An allele at any locus may become more or less common relative to other kinds or may be lost.

Microevolution refers to changes in allele frequencies of a population brought about by mutation, natural selection, gene flow, and genetic drift (Table 16.1).

At genetic equilibrium, a population is not evolving. By the Hardy–Weinberg rule, this occurs only if there is no mutation, the population is infinitely large and isolated from other populations of the species, there is no selection, mating is random, and members survive and reproduce equally. Deviations from this theoretical baseline indicate microevolution is at work.

Section 16.6 Mutations are rare for any given individual, but they happen at a predictable rate. Where a mutation does occur is unpredictable.

Section 16.7 Natural selection acts on phenotypic variation within a population. Directional selection favors the forms at one end of the phenotypic range.

Section 16.8 Intermediate forms of a trait are favored by stabilizing selection. In disruptive selection, forms at both ends of a range of variation are favored over the intermediate forms.

Section 16.9 Sexual selection, by females or males, leads to forms of traits that favor reproductive success. Persistence in phenotypic differences between males and females (sexual dimorphism) is one outcome.

Selection may result in balanced polymorphism, with nonidentical alleles for a trait being maintained over time at frequencies greater than 1 percent.

Section 16.10 Genetic drift is a random change in allele frequencies over time due to chance alone. The random changes tend to lead to the homozygous condition and loss of genetic diversity through the generations. The effect of genetic drift is greatest in small populations, such as ones that pass through a bottleneck or arise from a small group of founders.

Section 16.11 Gene flow shifts allele frequencies by physically moving alleles into a population (by way of immigration) and out of it (by emigration). It tends to keep different populations of the same species alike by countering mutation, natural selection, and genetic drift.

Table 16.1	Microevolutionary Processes
Mutation	A heritable change in DNA
Natural selection	Change or stabilization of allele frequencies; an outcome of differences in survival and reproduction among variant individuals of a population
Genetic drift	Random fluctuation in allele frequencies over time due to chance occurrences alone
Gene flow	Individuals, and their alleles, move into and out of populations; the physical flow counters the effects of the other microevolutionary processes

Self-Quiz
Answers in Appendix III

1. Individuals don't evolve, _____ do.

2. Biologists define evolution as _____ .
   a. the origin of a species
   b. heritable change in a line of descent
   c. acquiring traits during the individual's lifetime
   d. all of the above

3. _____ is the original source of new alleles.
   a. Mutation   c. Genetic drift   e. All give rise
   b. Natural selection d. Gene flow   to new alleles

4. Natural selection may occur when there are _____ .
   a. differences in forms of traits
   b. differences in survival and reproduction among individuals that differ in one or more traits
   c. both a and b

5. Directional selection _____ .
   a. eliminates uncommon forms of alleles
   b. shifts allele frequencies in a consistent direction
   c. favors intermediate forms of a trait
   d. works against adaptive traits

6. Disruptive selection _____ .
   a. eliminates uncommon forms of alleles
   b. shifts allele frequencies in a consistent direction
   c. doesn't favor intermediate forms of a trait
   d. both b and c

Figure 16.25 Reconstruction, based on fossils discovered in Pakistan, of *Rodhocetus*. This cetacean lived 47 million years ago, along the shores of the Tethys Sea. Its ankle bones indicate a close evolutionary link between early whales and hoofed land mammals.

7. _____ tends to reduce allelic differences among populations of a species.

   a. Genetic drift   c. Mutation
   b. Gene flow   d. Natural selection

8. Match the evolution concepts.

   ____ gene flow   a. source of new alleles
   ____ natural   b. changes in a population's allele
         selection   frequencies due to chance alone
   ____ mutation   c. allele frequencies change owing to
   ____ genetic   immigration, emigration, or both
         drift   d. outcome of differences in survival,
                       reproduction among individuals
                       that vary in forms of shared traits

Critical Thinking

1. Martha is studying a population of tropical birds. Male birds have brightly colored tail feathers and the females don't. She suspects this difference is maintained by sexual selection. Design an experiment to test her hypothesis.

2. A few families in a remote region of Kentucky show a high frequency of *blue offspring*, an autosomal recessive disorder. Skin of affected individuals appears bright blue. Homozygous recessives lack an enzyme that maintains hemoglobin in its normal molecular form. Without it, a blue form of hemoglobin accumulates in blood and shows through the skin. Formulate a hypothesis to explain why the blue offspring trait recurs among a cluster of families but is rare in the human population at large.

3. For some time, evolutionists accepted that the ancestors of whales were four-legged animals that walked on land, then took up life in water about 55 million years ago. Fossils show gradual changes in skeletal features that made an aquatic life possible. But which four-legged mammals were its ancestors? The answer came from Philip Gingerich and Iyad Zalmout. While digging in Pakistan, they found fossils of early aquatic whales. Intact, sheep-like ankle bones *and* archaic whale skull bones were in the same fossilized skeletons (Figures 16.3 and 16.25).

Ankle bones of fossilized, early whales from Pakistan have the same form as the unique ankle bones of extinct and modern artiodactyls. Modern cetaceans no longer have even a remnant of an ankle bone. Here is evidence of an evolutionary link between certain aquatic mammals and a major group of mammals on land.

No one was around to witness the transition. Yet the fossils are real, just as the morphology and molecular makeup of living organisms are real. As you'll see in the next chapter, radiometric dating assigns fossils to places in time.

Because there were no witnesses, do you think there can be absolute proof of evolution? Is the circumstantial evidence of fossil morphology enough to convince you that the theory is valid?

Media Menu

Student CD-ROM

Impacts, Issues Video
   Rise of the Super Rats
Big Picture Animation
   Evolutionary views and processes
Read-Me-First Animation
   Directional selection
   Stabilizing and disruptive selection
   Genetic drift
Other Animations and Interactions
   Adaptation questions

InfoTrac

- Rats, "Super-Rats" and the Environment. *Biological Sciences Review*, November 2001.
- Portraits of Evolution: Studies of Coloration in Hawaiian Spiders. *BioScience*, July 2001.
- AIDS in Africa Has Potential to Affect Human Evolution. *AIDS Weekly*, June 2001.

Web Sites

- PBS Evolution: www.pbs.org/wgbh/evolution
- Talk.Origins: www.talkorigins.org
- Issues in Evolution: www.actionbioscience.org/evolution
- BBC Evolution: www.bbc.co.uk/education/darwin

How Would You Vote?

Deadly "super bacteria" are the outcome of decades of antibiotic overuse. The same antibiotics used to treat bacterial infection in people also help sick animals. On many farms, these antibiotics are used on a daily basis to prevent infection of healthy animals. This practice may have contributed to evolution of the new antibiotic-resistant human pathogens. One suggestion is to ban the preventive use of antibiotics of value to humans in farm animals. Would you support such a ban?

IMPACTS, ISSUES *Measuring Time*

How do you measure time? Is your comfort level with the past limited to your own generation? Probably you can relate to a few centuries of human events. But geologic time? Comprehending the distant past requires a huge intellectual leap from the familiar to the unknown.

Consider this: Asteroids are rocky, metallic bodies hurtling through space. They are a few meters to 1,000 kilometers across. When our solar system's planets were forming, their gravitational force swept most asteroids from the sky. At least 6,000 asteroids, including the one shown in Figure 17.1a, still orbit the sun in a belt between Mars and Jupiter. Millions more frequently zip past Earth. They are hard to spot because they don't emit light. We don't discover most of them unless they pass close by. Some have passed too close for comfort.

Big asteroid impacts altered the course of evolution. For instance, researchers found a thin layer of iridium around the world, and it dates to a mass extinction that wiped out the last of the dinosaurs (Figure 17.1b). Iridium is rare on Earth but not in asteroids. Above the layer, there are no more fossils of dinosaurs, anywhere.

It has only been about 100,000 years since the first modern humans (*Homo sapiens*) evolved. We know that dozens of humanlike species evolved in Africa during the 5 million years before our species even showed up. So why are we the only ones left?

Unlike today's large, globally dispersed populations of humans, the early species lived in small bands. What

Figure 17.1 (**a**) An asteroid 19 kilometers (about 12 miles) long, still hurtling through space. (**b**) Two views of the last few minutes of the Cretaceous. The filmstrip at far right shows a sample of the worldwide, iridium-rich layer of sediment (*black*) that dates precisely to the K–T boundary. It's evidence of an asteroid impact.

the big picture

Evidence of Evolution   Fossils are direct evidence of ancient life. Biogeography and comparisons of body form, developmental patterns, and biochemistry are helping us piece together and interpret the fossil record.

How Species Originate   Microevolution and plain luck have both contributed to the origin of species. Reproductively isolated subpopulations of a species diverge genetically. A new species is recognized when divergences are great enough to prevent successful interbreeding.

if most were casualties of the twenty asteroids that struck when they were alive? What if *our* ancestors were just plain lucky? About 2.3 million years ago, one huge object from space hit the ocean, west of what is now Chile. If it had collided with the rotating Earth just a few hours earlier, our ancestors in southern Africa might have been incinerated.

Now that we know what to look for, we are seeing more and more craters in satellite images of the Earth. Less than 4,000 years ago, in what is now Iraq, an impact released energy that was equivalent to the detonation of hundreds of nuclear weapons.

If we can figure out what an asteroid impact will do to us, we can figure out how impacts affected life in the past. We *can* comprehend life long before our own. This chapter introduces some tools and evidence used to interpret patterns, trends, and rates of change among life's major lineages. Along the way you will read about their causes, including good and bad cosmic luck.

This leap through time starts with the premise that any aspect of the natural world, past as well as present, has one or more underlying causes. We look for clues by studying physical and chemical aspects of the Earth, analyzing fossils, and comparing the morphology and biochemistry of species. We test our hypotheses with experiments, models, and advancing technologies. This shift from experience to inference—from the known to what can only be surmised—has given us astonishing glimpses into the past.

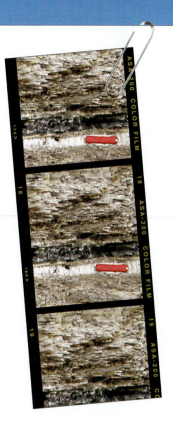

 *How Would You Vote?*

A major asteroid impact could obliterate civilization and much of Earth's biodiversity. Should nations around the world contribute resources to searching for and tracking asteroids? See the Media Menu for details, then vote online.

Big Evolutionary Events All species share genetic connections through ancient lineages that have changed over evolutionary time. New species emerged in response to opportunities that opened up following often-catastrophic challenges.

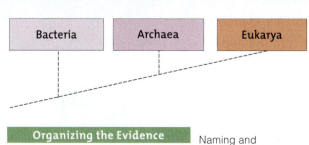

Organizing the Evidence Naming and classifying species helps us manage data on biodiversity. Evolutionary classification systems group species with respect to derived traits, which evolved only once in the most recent shared ancestor of two or more groups.

17.1 Fossils—Evidence of Ancient Life

The fossil record helps explain the connections between Earth's evolution and life's evolution.

About 500 years ago, Leonardo da Vinci was puzzled by seashells entombed in rocks of northern Italy's high mountains, hundreds of kilometers from the sea. How did they get there? By the prevailing belief, water from a stupendous, divinely invoked flood had surged up against the mountains, where it deposited the shells. But many of the shells were thin, fragile, and intact. If they had been swept across such great distances, then wouldn't they be battered to bits?

Leonardo also brooded about the rocks. They were stacked like cake layers. Some layers had shells, others had none. Then he remembered how large rivers swell with spring floodwaters and deposit silt in the sea. Did such depositions happen in ancient seasons? If so, then shells in the mountains could be evidence of layered communities of organisms that once lived in the seas!

By the 1700s, **fossils** were accepted as the remains and impressions of organisms that lived in the past. (*Fossil* comes from a Latin word for "something that was dug up.") People were still interpreting fossils through the prism of cultural beliefs, as when a Swiss naturalist unveiled the remains of a giant salamander and excitedly announced that they were the skeleton of a man who drowned in the great flood.

By midcentury, though, scholars were questioning these interpretations. Why? Mining, quarrying, and excavations for canals were under way. Diggers were finding similar rock layers and similar sequences of fossils in distant places, such as the nearshore cliffs on both sides of the English Channel. If those layers had been deposited with the passing of time, then the vertical sequence of fossils in them might be a record of past life—*a fossil record.*

HOW DO FOSSILS FORM?

Most fossils discovered so far are bones, teeth, shells, seeds, spores, and other hard parts (Figure 17.2). Fossilized feces (coprolites) hold residues of species that were eaten in ancient times. Imprints of leaves, stems, tracks, burrows, and other *trace* fossils provide further indirect evidence of past life.

Fossilization is a slow process that starts when an organism or traces of it become covered by volcanic ash or sediments. Water slowly infiltrates the remains, and metal ions and other inorganic compounds that are dissolved in it replace the minerals in bones and other hardened tissues. As sediments accumulate, they exert increasing pressure on the burial site. In time, the pressure and mineralization processes transform those remains into stony hardness.

Remains that become buried quickly are less likely to be obliterated by scavengers. Preservation is also favored when a burial site stays undisturbed. Usually, however, erosion and other geologic assaults deform, crush, break, or scatter the fossils. This is one reason fossils are relatively rare.

Figure 17.2 Representatives of more than 250,000 ancient species known from the fossil record. *Left*, fossilized parts of the oldest known land plant (*Cooksonia*). Its stems were less than seven centimeters tall. *Right*, fossilized skeleton of an ichthyosaur. This marine reptile lived 200 million years ago.

Figure 17.3 A slice through time—Butterloch Canyon, Italy, once at the bottom of a sea. Its sedimentary rock layers slowly formed over hundreds of millions of years. Later, geologic forces lifted the stacked layers above sea level. Later still, the erosive force of river water carved the canyon walls and exposed the layers. Scientists Cindy Looy and Mark Sephton are climbing to reach the Permian–Triassic boundary layer, where they will look for fossilized fungal spores.

Other factors affect preservation. Organic materials cannot decompose in the absence of free oxygen, for instance. They might endure if sap, tar, ice, mud, or another air-excluding substance protects them. Insects in amber and frozen woolly mammoths are examples.

FOSSILS IN SEDIMENTARY ROCK LAYERS

Stratified (stacked) layers of sedimentary rock formed long ago from deposits of volcanic ash, silt, sand, and other materials. Sand and silt piled up after rivers transported them from land to the sea, as Leonardo suspected. Sandstones formed from sand, and shales from silt. Depositions were sometimes interrupted, in part because the sea level changed as ice ages began. Tremendous volumes of water froze in glaciers, rivers dried up, and the depositions ended in some regions. Later in time, when the climate warmed and glaciers melted, the depositions resumed.

The formation of sedimentary rock layers is called **stratification**. The first to form are now the deepest layers, and those closest to the surface were the last. Most formed horizontally, as in Figure 17.3, because particles tend to settle in response to gravity. You may see tilted or ruptured layers, as along a road cut into a mountainside. Major crustal movements or upheavals caused them, much later in time.

We find most fossils in sedimentary rock. When you understand how rock layers form, it is obvious that the fossils in a particular layer formed at a given time in Earth history. Specifically, *the older the layer, the older the fossils*. Given that rock layers formed in sequence, their fossils are unique to sequential ages.

INTERPRETING THE FOSSIL RECORD

We have fossils for more than 250,000 known species. Judging from the current range of biodiversity, there must have been many, many millions more. Yet the fossil record will never be complete. Why is this so?

The odds are against finding signs of an extinct, ancient species. At least one individual had to be gently buried before it decomposed or something ate it. The burial site had to escape erosion, lava flows, and other geologic forces. The fossil had to end up in a place where someone could actually find it. Fossils often are found on the side of a canyon carved out by a river that exposed the layers of sedimentary rock.

Fossils did not form in many habitats, and most species didn't lend themselves to preservation. Unlike bony fishes and hard-shelled mollusks, jellyfishes and soft worms don't show up as much in the fossil record. Yet they probably were just as common, or more so.

Also think about population density and body size. One plant population might release millions of spores in a single season. The earliest humans lived in small bands and raised few offspring. What are the odds of finding even one fossilized human bone compared to spores of plants that lived at the same time?

Finally, imagine one line of descent, a **lineage**, that vanished when its habitat on a remote volcanic island sank into the sea. Or imagine two lineages, one lasting only briefly and the other for billions of years. Which is more likely to be represented in the fossil record?

Fossils are physical evidence of life in the remote past. Those embedded in sedimentary rock layers are a historical record of life. The deepest layers generally contain the oldest fossils.

The fossil record is incomplete. Geologic events obliterated much of it. The record is slanted toward species that had large bodies, hard parts, dense populations, and wide distribution, and that persisted for a long time.

Even so, the fossil record is now substantial enough for us to reconstruct patterns and trends in the history of life.

17.2 Dating Pieces of the Puzzle

How do we assign fossils to a place in time? In other words, how do we know how old they really are?

RADIOMETRIC DATING

For as long as they've been digging up rocks, people have been coming across fossils. At one time, they could assign only *relative* ages to their treasures, not absolute ones. For instance, a fossilized mollusk in a rock layer was said to be younger than a fossil below it and older than a fossil above it, and so on.

Things changed with **radiometric dating**. This is a way to measure the proportions of a daughter isotope and the parent radioisotope of some element trapped inside a rock since the time the rock formed. Again, a radioisotope is a form of an element with an unstable nucleus (Section 2.2). Radioactive atoms decay, or lose energy and subatomic particles until they reach a more stable form.

It is not possible to predict the exact instant of one atom's decay, but a predictable number of an isotope's atoms will decay over a period of time. Like the ticking of a perfect clock, the characteristic rate of decay for each isotope is constant. In other words, changes in pressure, temperature, or chemical state don't alter it. The time it takes for half of a quantity of a radioisotope's atoms to decay is its **half-life** (Figure 17.4a). For instance, uranium 238 has a half-life of 4.5 billion years. It decays into thorium 234, which in turn decays into something else, and so on through a series of intermediate daughter isotopes. The final, stable daughter element is lead. By measuring the ratio of uranium 238 to lead in the oldest rocks, geologists estimated that Earth formed more than 4.6 billion years ago.

Radiometric dating doesn't work for sedimentary rock. It works for volcanic rock or ashes, which hold the most fossils. The ratio of carbon 14 to carbon 12 is used to date recent fossils that still contain some carbon (Figure 17.4b–d). The only way to date older fossils is to determine their position relative to any volcanic rocks in the same area. This dating method has an error factor of less than 10 percent.

PLACING FOSSILS IN GEOLOGIC TIME

Early geologists carefully counted backward through layers of sedimentary rock, then used their counts to construct a chronology of Earth history as a **geologic time scale** (Figure 17.5). By comparing evidence from

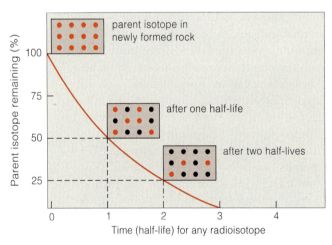

a A simple way to think about the decay of a radioisotope to a more stable form, as plotted against time.

Figure 17.4 (**a**) The decay of radioisotopes at a fixed rate to more stable forms. The half-life of each kind of radioisotope is the time it takes for 50 percent of a sample to decay. After two half-lives, 75 percent of the sample has decayed, and so on.

(**b–d**) Radiometric dating of a fossil. Carbon 14 (^{14}C) forms in the atmosphere. There, it combines with free oxygen, the result being carbon dioxide. Along with far greater quantities of its more stable isotopes, trace amounts of carbon 14 enter food webs by way of photosynthesis. All organisms incorporate carbon into body tissues.

b Long ago, trace amounts of ^{14}C and a lot more ^{12}C were incorporated into tissues of a living mollusk. The carbon was part of the organic compounds making up the tissues of its prey. As long as it lived, the proportion of ^{14}C to ^{12}C in its tissues remained the same.

c When the mollusk died, it stopped gaining carbon. Over time, proportion of ^{14}C to ^{12}C in its remains declined because of the radioactive decay of ^{14}C. Half of the ^{14}C had decayed in 5,370 years, half of what remained was gone in another 5,370 years, and so on.

d Fossil hunters find the fossil. They measure its $^{14}C/^{12}C$ ratio to determine the half-life reductions since death. The ratio turns out to be one-eighth of the $^{14}C/^{12}C$ ratio in living organisms. Thus the mollusk lived about 16,000 years ago.

Eon	Era	Period	Epoch	Millions of Years Ago	Major Geologic and Biological Events That Occurred Millions of Years Ago (mya)
PHANEROZOIC	CENOZOIC	QUATERNARY	Recent	0.01	1.8 mya to present. Major glaciations. Modern humans evolve. The most recent *extinction crisis* is under way.
			Pleistocene	1.8	
		TERTIARY	Pliocene	5.3	65–1.8 mya. Major crustal movements, collisions, mountain building. Tropics, subtropics extend poleward. When climate cools, dry woodlands, grasslands emerge. *Adaptive radiations* of flowering plants, insects, birds, mammals.
			Miocene	22.8	
			Oligocene	33.7	
			Eocene	55.5	
			Paleocene	65	65 mya. Asteroid impact; *mass extinction* of all dinosaurs and many marine organisms.
	MESOZOIC	CRETACEOUS	Late		99–65 mya. Pangea breakup continues, inland seas form. Adaptive radiations of marine invertebrates, fishes, insects, and dinosaurs. Origin of angiosperms (flowering plants).
				99	
			Early		145–99 mya. Pangea starts to break up. Marine communities flourish. *Adaptive radiations* of dinosaurs.
				145	145 mya. Asteroid impact? Mass extinction of many species in seas, some on land. Mammals, some dinosaurs survive.
		JURASSIC		213	248–213 mya. *Adaptive radiations* of marine invertebrates, fishes, dinosaurs. Gymnosperms dominate land plants. Origin of mammals.
		TRIASSIC		248	248 mya. *Mass extinction*. Ninety percent of all known families lost.
	PALEOZOIC	PERMIAN		286	286–248 mya. Supercontinent Pangea and world ocean form. On land, *adaptive radiations* of reptiles and gymnosperms.
		CARBONIFEROUS		360	360–286 mya. Recurring ice ages. On land, *adaptive radiations* of insects, amphibians. Spore-bearing plants dominate; cone-bearing gymnosperms present. Origin of reptiles.
		DEVONIAN		410	360 mya. *Mass extinction* of many marine invertebrates, most fishes.
					410–360 mya. Major crustal movements. Ice ages. *Mass extinction* of many marine species. Vast swamps form. Origin of vascular plants. *Adaptive radiation* of fishes continues. Origin of amphibians.
		SILURIAN		440	440–410 mya. Major crustal movements. *Adaptive radiations* of marine invertebrates, early fishes.
		ORDOVICIAN		505	505–440 mya. All land masses near equator. Simple marine communities flourish until origin of animals with hard parts.
		CAMBRIAN		544	544–505 mya. Supercontinent breaks up. Ice age. *Mass extinction*.
PROTEROZOIC					2,500–544 mya. Oxygen accumulates in atmosphere. Origin of aerobic metabolism. Origin of eukaryotic cells. Divergences lead to eukaryotic cells, then protists, fungi, plants, animals.
				2,500	
ARCHEAN AND EARLIER					3,800–2,500 mya. Origin of photosynthetic prokaryotic cells. 4,600–3,800 mya. Origin of Earth's crust, first atmosphere, first seas. Chemical, molecular evolution leads to origin of life (from proto-cells to anaerobic prokaryotic cells).

Figure 17.5 Geologic time scale. Major boundaries mark the times of the greatest mass extinctions. If the time spans listed were to the same scale, the Archean and Proterozoic portions would run off this page. Compare Figure 17.6.

around the world, they found four abrupt transitions in fossil sequences and used them as boundaries for four great intervals. They named the first interval the Proterozoic, after finds that predate fossils of animals. They named other intervals the Paleozoic, Mesozoic, and the "modern" era, the Cenozoic.

The current geologic time scale now correlates with **macroevolution**, or major patterns, trends, and rates of change among lineages. Also, being more immense than early researchers suspected, the Proterozoic is now subdivided into more intervals. Life originated in one of those intervals, the Archean eon.

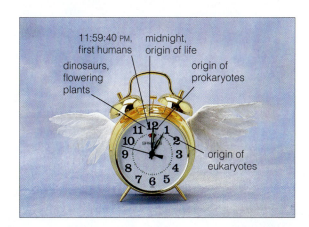

Figure 17.6 A geologic time clock. Think of the spans as minutes on a clock that runs from midnight to noon. If we say that life started at midnight, then the Paleozoic began at 10:04 A.M., the Mesozoic at 11:09 A.M., and the Cenozoic at 11:47 A.M. The recent epoch of the Cenozoic era started after the last 0.1 second before noon. And where does that put you?

17.3 Evidence From Biogeography

By clinking their hammers against the rocks, geologists discovered that the "solid" Earth hasn't stayed put.

As the early geologists were discovering fossils and mapping the record of Earth history, Charles Darwin was so taken with one of their theories that it helped shape his view of life. The **theory of uniformity** held that mountain building and erosion had repeatedly worked over the Earth's surface in precisely the same ways through time (Section 16.2). However, as more fossils were found, geologists realized that repetitive change was only part of the picture. Like life, Earth itself had changed irreversibly.

AN OUTRAGEOUS HYPOTHESIS

For instance, the Atlantic coasts of South America and Africa seemed to "fit" like jigsaw puzzle pieces. Were all continents once part of a bigger one that had split into fragments and drifted apart? One model for the proposed supercontinent—**Pangea**—took into account the world distribution of fossils and existing species. It also took into account glacial deposits, which held clues to ancient climate zones.

Most scientists did not accept the continental drift theory. Continents drifting about on their own across the Earth's mantle seemed to be an outrageous idea, and they preferred the theory of uniformity.

But more evidence kept piling up. Iron-rich rocks are molten when they form. Iron particles in them orient north–south in response to Earth's magnetic poles, and stay that way after the rocks harden. Yet in North and South America, the tiny iron compasses in rocks that had formed 200 million years ago did *not* point pole to pole. So scientists came up with a map. They made a north–south alignment work by joining North America and western Europe—and orienting them in a way that sure didn't look like modern maps.

More puzzles! Deep-sea probes showed that the seafloor is spreading away from mid-oceanic ridges (Figure 17.7). Molten rock spewing from a ridge flows sideways in both directions, then it hardens into new crust. The spreading new crust forces older crust into deep trenches elsewhere in the seafloor. All the ridges and trenches are actually edges of thin but enormous plates, like pieces of a cracked eggshell. They move with almost imperceptible slowness. However, over time, land masses take up new positions.

These findings put continental drift into a broader explanation of crustal movements, now known as the **plate tectonic theory**. Researchers soon found ways to use the new theory's predictive power.

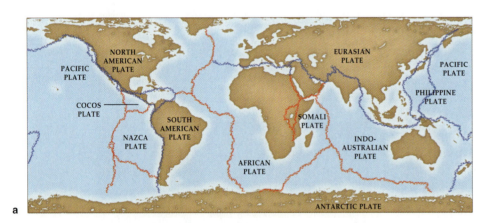

a

b island arc — oceanic crust — oceanic ridge — trench — continental crust — lithosphere (solid layer of mantle) — hot spot — athenosphere (plastic layer of mantle) — subducting plate

Figure 17.7 Some forces of geologic change.

(**a**) Present configuration of Earth's crustal plates. These immense, rigid parts of the crust split, drift apart, and collide at almost imperceptible rates. On the seafloor, *red* signifies the newest crust (less than 10 million years old), and *blue* is the oldest; it formed 180 million years ago.

(**b**) Huge plumes of molten material drive the movement. They well up from the interior, spread laterally under the crust, and rupture it at mid-oceanic ridges. At these deep ridges, molten material seeps out, cools, and slowly forces the seafloor away from the rupture, which displaces the plates from ridges.

The leading edge of one plate commonly plows under an adjoining plate and uplifts it. *Blue* lines in (**a**) show where this is now happening. The Cascades, Andes, and other great mountain ranges paralleling the coasts of continents formed this way.

Long ago, superplumes violently ruptured the crust at what are now called "hot spots" in the mantle. The Hawaiian Archipelago has been forming this way. Continents also rupture. Deep rifting and splitting are happening now in Missouri, at Lake Baikal in Russia, and in eastern Africa.

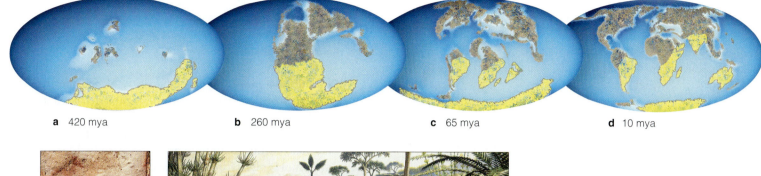

a 420 mya   **b** 260 mya   **c** 65 mya   **d** 10 mya

Figure 17.8 Reconstructions of drifting continents. (**a**) Gondwana (*yellow*) 420 million years ago. (**b**) Later in time, all major land masses had collided to form a supercontinent, Pangea. (**c**) Positions of the fragments after Pangea split apart 65 million years ago, and (**d**) their positions 10 million years ago.

About 260 million years ago, seed ferns and other plants lived nowhere except on the part of Pangea that had once been Gondwana. So did the therapsids, or mammal-like reptiles. (**e**) Fossilized leaf of one of the seed ferns, *Glossopteris*. (**f**) *Lystrosaurus*, a therapsid about 1 meter (3 feet) long. This tusked herbivore fed on the fibrous plants of dry floodplains.

For instance, the same series of glacial deposits, coal seams, and basalt are found in Africa, India, Australia, and South America. All four southern land masses hold fossils of *Glossopteris*, a seed fern. They also hold fossils of *Lystrosaurus*, a mammal-like reptile (Figure 17.8). Neither the plant's heavy seeds nor the reptile could have floated across a vast, open ocean. Researchers suspected they had evolved together on **Gondwana**, a supercontinent that preceded Pangea.

Antarctica, too, formed after Gondwana broke up. A geologist predicted that fossils of *Glossopteris* and *Lystrosaurus* would be discovered there, in a series of glacial deposits, coal seams, and basalt just like that on the other southern continents. Sure enough, Antarctic explorers did find the same series and fossils, which supported the prediction and plate tectonics theory.

DRIFTING CONTINENTS, CHANGING SEAS

Let's take stock. In the remote past, crustal movement put huge land masses on collision courses. In time, the masses converged to form supercontinents, which later split at deep rifts and formed new ocean basins. Gondwana drifted south from the tropics, across the south pole, then north until it slowly piled into other land masses. The outcome? A supercontinent, Pangea, that extended from pole to pole, with a single world ocean lapping against its coasts. All the while, erosive forces of water and wind resculpted the land surface. Asteroids and meteorites smacked into the crust. The impacts and their aftermath had long-term effects on global temperature and climate.

Such changes on land and in the ocean and atmosphere influenced life's evolution. Imagine early life in shallow, warm waters along continents. Shorelines vanished as continents collided and wiped out many lineages. Yet, even as habitats vanished, new ones opened up for survivors—and evolution took off in new directions.

Over the past 3.8 billion years, gradual as well as catastrophic events changed the Earth's crust, the atmosphere, and the ocean—and influenced the evolution of life.

17.4 Evidence From Comparative Morphology

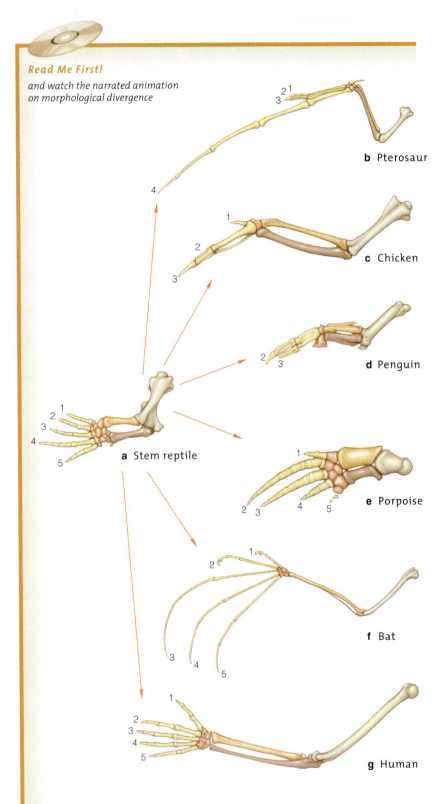

b Pterosaur

c Chicken

d Penguin

a Stem reptile

e Porpoise

f Bat

g Human

Figure 17.9 Morphological divergence among vertebrate forelimbs, starting with bones of a stem reptile. Similarities in the number and position of skeletal elements were preserved when diverse forms evolved. Some bones were lost over time (compare the numbers 1 through 5). The drawings are not to the same scale.

Evolution, remember, simply means heritable changes in lines of descent. Comparisons of the body form and structures of major groups of organisms yield clues to evolutionary trends.

Comparative morphology is the study of body forms and structures of major groups of organisms, such as vertebrates and flowering plants. Often it reveals similarities in one or more body parts that suggest inheritance from a common ancestor. Such body parts are **homologous structures** (*homo–* means the same). In such cases, genetically based similarities are there, even when different kinds of organisms are using the structures for different functions.

MORPHOLOGICAL DIVERGENCE

Populations of a species genetically diverge when gene flow ends between them (Chapter 16). In time, some morphological traits that help to define their species commonly diverge, also. Change from the body form of a common ancestor is a major macroevolutionary pattern called **morphological divergence**. The Greek *morpho–* means body form.

Even if the same body part of two related species became dramatically different, some other aspects of the species may remain alike. A careful look beyond unique modifications may reveal the shared heritage. For example, all vertebrates on land are descendants of the first amphibians. Divergences led to what we generally call reptiles, then to birds and mammals. We know about cotylosaurs, the "stem reptiles" that probably were ancestral to all of those groups. Their fossilized, five-toed limb bones tell us the cotylosaurs crouched low to the ground (Figure 17.9*a*). Their descendants diversified into many new land habitats. We now know a few kinds adapted to land returned to the seas when environmental conditions changed.

A five-toed limb was evolutionary clay. It became molded into different kinds of limbs with different functions. In lineages that eventually led to penguins and porpoises, it became modified into flippers used in swimming. In the lineage leading to modern horses, it became modified into long, one-toed limbs suitable for running fast. Among moles, it became stubby and useful for burrowing into dirt. Among elephants, it became strong and pillarlike, suitable for supporting a great deal of weight.

The five-toed limb also became modified into the human arm and hand. Later, a thumb evolved in opposition to the four fingers of the human hand; it was the basis of stronger and more precise motions.

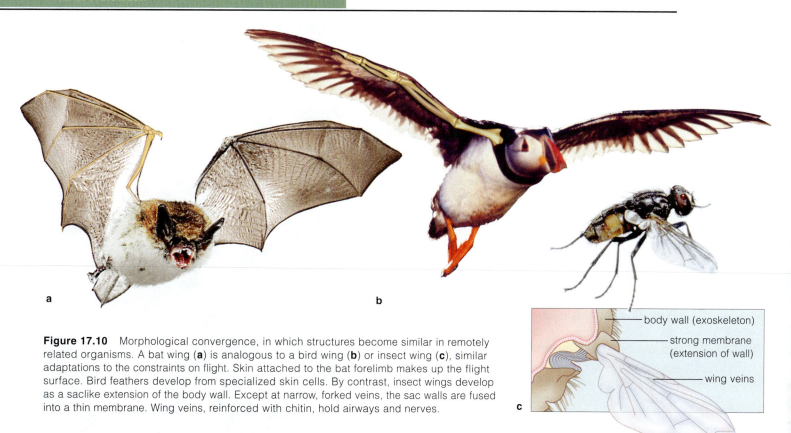

Figure 17.10 Morphological convergence, in which structures become similar in remotely related organisms. A bat wing (**a**) is analogous to a bird wing (**b**) or insect wing (**c**), similar adaptations to the constraints on flight. Skin attached to the bat forelimb makes up the flight surface. Bird feathers develop from specialized skin cells. By contrast, insect wings develop as a saclike extension of the body wall. Except at narrow, forked veins, the sac walls are fused into a thin membrane. Wing veins, reinforced with chitin, hold airways and nerves.

Even though vertebrate forelimbs are not the same in size, shape, or function from one group to the next, they clearly are alike in the structure and positioning of their bony elements. Also, the forelimbs are alike in the internal arrangement of nerves, blood vessels, and muscles that develop inside them. In addition, other comparisons of early vertebrate embryos reveal strong resemblances in patterns of bone development. Such similarities point to common ancestry.

MORPHOLOGICAL CONVERGENCE

Body parts with similar form or function in different lineages aren't *always* homologous. Sometimes they evolved independently in remote lineages. Parts that differed at first might have become similar because organisms were subjected to similar environmental pressures. **Morphological convergence** refers to cases where dissimilar body parts evolved in similar ways in evolutionarily distant lineages.

For instance, you just read about the homologous forelimbs of birds and bats. Bones aside, are bird and bat wings homologous, too? No. The flight surface of birds evolved as a sweep of feathers, all derived from skin. The forelimb structurally supports it. The flight surface of bats is a thin membrane, an extension of the skin itself. The bat wing is attached to reinforcing bony elements inside the forelimb (Figure 17.10a,b).

The insect wing, too, resembles bird and bat wings in its function—flight. Is it homologous with them? No. This wing develops as an extension of an outer body wall reinforced with chitin. It has no underlying bony elements to support it (Figure 17.10c).

The differences between bat, bird, and insect wings are evidence that each of these animal groups adapted independently to the same physical constraints that govern how a wing can function in the environment. The wings of all three are **analogous structures**. They are not modifications of comparable body parts in different lineages. They are three different responses of dissimilar parts to the same challenge. The Greek *analogos* means similar to one another.

With morphological divergence, comparable body parts became modified in different ways in different lines of descent from a common ancestor.

Such divergences resulted in homologous structures. Even if these body parts differ in size, shape, or function, they have an underlying similarity because of shared ancestry.

In morphological convergence, dissimilar body parts became similar in independent lineages that are not closely related.

Such body parts are analogous structures. They became similar only as a result of similar pressures; ancestry had nothing to do with it.

17.5 Evidence From Patterns of Development

Comparing the patterns of embryonic development often yields evidence of evolutionary relationships.

Over evolutionary time, mutations in lines of descent have resulted in built-in constraints on how a plant or an animal embryo can grow and develop. That is why most mutations and changes in chromosomes tend to be selected against. During embryonic development, certain steps can't occur unless earlier steps precede them in expected ways. Some mutations disrupt key steps. Every so often, though, a neutral or beneficial change moves a lineage past a constraint.

Let's look at how a developmental step might shift. Homeotic genes, remember, orchestrate how the body plan of multicelled organisms develops (Chapter 14). Mutations in homeotic genes can cause an organism to develop differently, sometimes drastically so. Most of the time the alteration causes real problems, but occasionally an altered form offers a survival benefit.

For example, homeotic genes guide the timing and pattern of flower formation. A single mutation in one of these genes, *Apetala1*, causes field mustard flowers to develop male floral structures, or anthers, where there are supposed to be petals (Figure 17.11a). In the laboratory at least, the abundantly anthered mutant is notably fertile. This master gene has been found in many other flowering plants (Figure 17.11d).

As another example, embryos of some vertebrate lineages are alike in early stages. Tissues form in similar ways as cells divide, differentiate, and interact. The gut and heart, bones, skeletal muscles, and other parts

Figure 17.11 How a single mutation in a plant homeotic gene affects flower form and function.

(**a**) Mutation in the *Apetala1* gene in field mustard (*Brassica oleracea*) results in the formation of a truly distorted flower. (**b**) Normal *Brassica* flower. (**c**) The *Apetala1* mutation in mouse-ear cress (*Arabidopsis thaliana*) results in flowers that have no petals. (**d**) Normal *Arabidopsis* flower.

Read Me First!

and watch the narrated animation on mutation and proportional changes

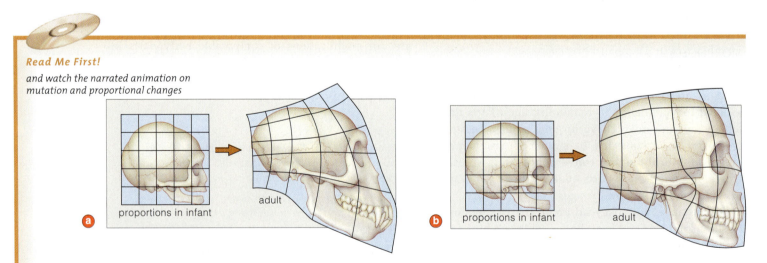

Figure 17.12 Differences between two primates, a possible outcome of mutations that changed the timing of steps in the body's development. The skulls are depicted as paintings on a rubber sheet divided into a grid. Stretching both sheets deforms the grid in a way that corresponds to differences in growth patterns between these primates. (**a**) Proportional changes in chimpanzee skull, and (**b**) human skull.

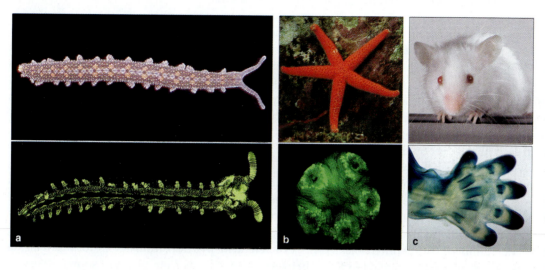

Figure 17.13 How many legs? Mutations in ancestral genes may help explain why animals differ in the number of legs and other appendages. *Dll* is a homeotic gene that initiates limb development, and other genes control its expression. Fluorescent *green* reveals *Dll* expression in (**a**) a velvet walking worm, and (**b**) a sea star; and (**c**) *blue* in a mouse embryo's foot. (**d**) Cambrian legs.

develop in orderly spatial patterns that are strikingly similar among these groups.

How, then, did adults of different groups get to be so different? We can expect that heritable changes in the onset, rate, or completion of developmental steps led to many of the differences. Some changes could have increased or decreased relative sizes of tissues and organs. Some changes could have put a stop to growth during a juvenile stage. In some cases, adults with some juvenile traits are still functional.

Altered growth rates might have caused the major proportional differences between the chimpanzee and human skull bones (Figure 17.12). For humans, facial bones and skull bones around the brain grow at fairly consistent rates, from infant to adult. The growth rate is faster for chimp facial bones, so the proportions of an infant skull and an adult skull differ significantly.

Did transposons cause some of the variation among lineages? As you read in Section 13.4, these short DNA segments can spontaneously and repeatedly slip into new places in genomes. Depending on where they end up, they can have powerful effects on gene expression.

Only primates carry the 300 base-pair transposons called *Alu* elements, and they have done so for at least 30 million years. *Alu* elements are noncoding, yet they have sequences that resemble intron–exon splice signals. When inserted into coding regions of DNA, they cause alternative splicing of genes. *Alu* elements have had profound effects on the expression of genes for estrogen, thyroid hormones, and other essential proteins that control growth and development, so we can expect that they were pivotal in primate evolution.

About one million *Alu* elements make up more than 10 percent of the human genome. Ancestors of humans and chimpanzees diverged between 6 and 4 million years ago, but more than 98 percent of human and chimpanzee DNA remains identical. Something about the remaining 2 percent accounts for the differences. Uniquely positioned *Alu* elements may be factors.

As a final example, appendages as diverse as crab legs, beetle legs, butterfly wings, sea star arms, fish fins, and mouse feet start out as buds from the body surface. The buds form wherever the *Dll* gene product is expressed. This product is a signal for clusters of dividing embryonic cells to "stick out from the body" in an expected pattern, as in Figure 17.13. Normally, *Hox* genes help sculpt the body by suppressing *Dll* expression where appendages aren't supposed to form.

The *Dll* gene is expressed in similar ways across many phyla, which is a strong case for its ancient origin. Indeed, some Cambrian fossils suggest that its early expression was unrestricted (Figure 17.13d). Over time, layers of gene controls evolved, resulting in the variable numbers and locations of appendages we see today. Control extends to all complex animals, including humans and other vertebrates.

Similarities in patterns of development are often clues to an evolutionary relationship among plant and animal lineages.

Heritable changes that alter key steps in a developmental program may be enough to bring about major differences between adult forms of related lineages. Transposons and single gene mutations can bring about such changes.

17.6 Evidence From Biochemistry

All species are a mix of ancestral and novel traits, which include biochemical traits. The kinds and numbers of traits they do or don't share are clues to relationships.

Each species, recall, has its own DNA base sequence, which encodes instructions for making RNAs and then proteins. We can expect that a number of genes have mutated over time in each line of descent. In addition, we can expect that fewer mutations have accumulated in lineages that originated recently compared to those that evolved much earlier in time. Because of this, the RNA and proteins of closely related species will be more similar than those of distantly related ones.

Identifying biochemical similarities and differences among species is now rapid, thanks to methods of automated gene sequencing (Section 15.5). Extensive sequence data of many genomes and proteins are compiled in internationally accessible databases. With such data, we know (for example) that 31 percent of the 6,000 yeast cell genes have counterparts in our genome. So do 40 percent of the 19,023 roundworm genes and 50 percent of the fruit fly genes.

PROTEIN COMPARISONS

When two species have many proteins with similar or identical amino acid sequences, we can expect them to be close relatives. When most of the sequences differ a lot, many mutations have accumulated, so a long time must have passed since the two shared a common ancestor.

A few essential genes have evolved very little; they are highly *conserved* across diverse species. One of them encodes cytochrome *c*. This protein component of electron transfer chains occurs in species that range from aerobic bacterial species to humans. In humans, its primary structure consists of only 104 amino acids. Figure 17.14 shows the striking similarity between the entire amino acid sequences for cytochrome *c* from a yeast, a plant, and an animal. And think about this: The *entire* amino acid sequence of human cytochrome

c is identical to that of chimpanzee cytochrome *c*. It differs by only 1 amino acid in rhesus monkeys, 18 in chickens, 19 in turtles, and 56 in yeasts. With this biochemical information in hand, would you predict that humans are more closely related to chimpanzees or to rhesus monkeys? Chickens or yeast?

NUCLEIC ACID COMPARISONS

Mutations that cause structural differences between species are often dispersed through the nucleotide sequences of their DNA. Some unique alterations have accumulated in each lineage.

Nucleic acid hybridization refers to base-pairing between DNA strands from different sources (Section 15.3). In a hybrid molecule, more hydrogen bonds form between matched bases than mismatched bases; the more matched bases, the stronger the association between the strands. The amount of heat required to separate two strands of a hybrid can be used as a comparative measure of their similarity. It takes more heat to disrupt hybrid DNA of closely related species.

Evolutionary distances are still being measured by DNA–DNA hybridizations, although automated DNA sequencing now gives faster, more quantifiable results. In a version of DNA fingerprinting (Section 15.4), restriction fragments of DNA from different species can be compared after they are separated by gel electrophoresis. All of these techniques are used to compare DNA isolated from nuclei, mitochondria, and chloroplasts (Figure 17.15).

Mitochondrial DNA (mtDNA) can be used to estimate diversity in eukaryotic populations because it mutates quickly. It is inherited in entirety from one parent in sexually reproducing species (typically, the mother), so any changes between maternally related individuals are due to mutations, not recombination. Genes encoding ribosomal RNA (rDNA) are used to compare species.

Computer programs quickly compare collections of DNA sequencing data. Often, evolutionary trees

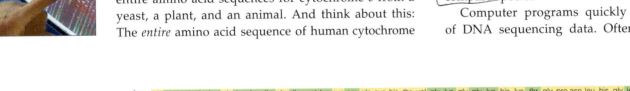

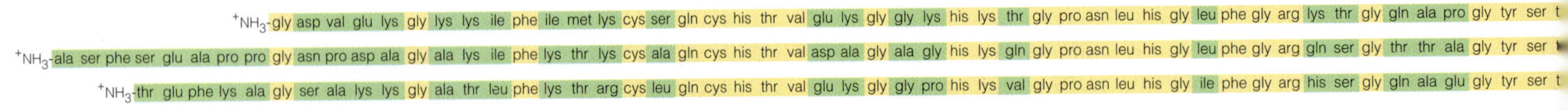

⁺NH₃-gly asp val glu lys gly lys lys ile phe ile met lys cys ser gln cys his thr val glu lys gly gly lys his lys thr gly pro asn leu his gly leu phe gly arg lys thr gly gln ala pro gly tyr ser t

⁺NH₃-ala ser phe ser glu ala pro pro gly asn pro asp ala gly ala lys ile phe lys thr lys cys ala gln cys his thr val asp ala gly ala gly his lys gln gly pro asn leu his gly leu phe gly arg gln ser gly thr thr ala gly tyr

⁺NH₃-thr glu phe lys ala gly ser ala lys lys gly ala thr leu phe lys thr arg cys leu gln cys his thr val glu lys gly gly pro his lys val gly pro asn leu his gly ile phe gly arg his ser gly gln ala glu gly tyr ser t

Figure 17.14 Comparison of the primary structure of cytochrome *c* from a yeast (*top row*), wheat plant (*middle*), and primate (*bottom*). This protein is a vital component of electron transfer chains in cells. Its amino acid sequence has been highly conserved, even in these three evolutionarily distant lineages. The parts highlighted in *gold* are identical in all three. The probability that such a pronounced molecular resemblance resulted by chance alone is extremely low.

RACCOON   RED PANDA

GIANT PANDA

SPECTACLED   SLOTH   SUN   BLACK   POLAR   BROWN
BEAR   BEAR   BEAR   BEAR   BEAR   BEAR

DIVERGENCE
15–20 million years ago

DIVERGENCE
approximately
40 million years ago

Figure 17.15 Biochemical comparisons between species is used to generate or confirm evolutionary trees. This evolutionary relationship between red pandas, giant pandas, and brown bears was confirmed using mitochondrial DNA sequence comparisons.

based on this comparative analysis have reinforced morphological findings and the fossil record.

However, gene transfers between species can slant the results. For example, after hybridization between two different species of plants, hybrid offspring may cross back to either parental species, thus transferring genes from one species into the other. Gene swapping is rampant among prokaryotic species.

MOLECULAR CLOCKS

Some researchers estimate the timing of divergence by comparing the numbers of neutral mutations in highly conserved genes (Section 13.4). Because such mutations have little or no effect on the individual's survival or reproduction, we can expect that neutral mutations have accumulated in conserved genes at a fairly constant rate.

The addition of neutral mutations to the DNA of a given lineage has been likened to the predictable ticks of a **molecular clock**. Turn the hands of such a clock back, so that the total number of ticks unwind down through past geologic intervals. Where the last tick stops, that is the approximate time when molecular, ecological, and geographic events put the lineage on its unique evolutionary road.

How are molecular clocks calibrated? The number of differences in DNA base sequences or amino acid sequences between species can be plotted against a series of branch points inferred from the fossil record. Graphs like this may reflect relative divergence times among species, phyla, and other groups.

Biochemical similarity is greatest among the most closely related species and smallest among the most remote.

thr ala ala asn lys asn lys gly ile ile trp gly glu asp thr leu met glu tyr leu glu asn pro lys lys tyr ile pro gly thr lys met ile phe val gly ile lys lys lys glu glu arg ala asp leu ile ala tyr leu lys lys ala thr asn glu-COO⁻

ser ala ala asn lys asn lys ala val glu trp glu glu asn thr leu tyr asp tyr leu leu asn pro lys lys tyr ile pro gly thr lys met val phe pro gly leu lys lys pro gln asp arg ala asp leu ile ala tyr leu lys lys ala thr ser ser-COO⁻

thr asp ala asn ile lys lys asn val leu trp asp glu asn asn met ser glu tyr leu thr asn pro lys lys tyr ile pro gly thr lys met ala phe gly gly leu lys lys glu lys asp arg asn asp leu ile thr tyr leu lys lys ala cys glu-COO⁻

17.7 Reproductive Isolation, Maybe New Species

Speciation is a macroevolutionary event based on reproductive isolation and microevolution of a population or subpopulation. Several mechanisms prevail.

Species is a Latin word that simply means "kind," as in "one particular kind of duck." Long ago, naturalists defined species mainly in terms of morphological traits, but common sense told them to consider other factors as well. For instance, individuals of the same species often look different because they grew under different conditions. The plant in Figure 17.16 is an example.

Actually, individuals of *most* species vary greatly in morphological details. But they are all more closely related to one another than to any other species. We now identify unique traits by using morphological or DNA sequence comparisons, and by identifying the basic functions that isolate a species from others.

Reproduction is such a basic, defining function. It is the core of the **biological species concept**. Ernst Mayr, an evolutionary biologist, phrased it this way: Species are groups of interbreeding natural populations that are reproductively isolated from other such groups. In his view, it doesn't matter how much the phenotypes vary. Populations belong to the same species as long as individuals share traits that let them interbreed and produce fertile offspring.

Mayr's species concept applies only to sexually reproducing organisms. If we subscribe to his view, speciation is the attainment of reproductive isolation. Isolation does not happen on purpose. *Any structural, functional, or behavioral trait that favors reproductive isolation is simply a by-product of genetic change.*

Gene flow alone counters genetic changes between populations of a species. **Gene flow**, remember, is the movement of alleles into and out of a population by immigration and emigration. This microevolutionary process helps maintain a shared pool of alleles.

If gene flow between populations or subpopulations of a species ends, **genetic divergence** follows. By this process, gene pools of isolated populations slowly or quickly diverge, for mutation, natural selection, and genetic drift happen independently in each one.

Regardless of differences in the duration or details, **reproductive isolating mechanisms** kick in at some point. These heritable aspects of body form, function, or behavior prevent interbreeding between divergent populations. Some prevent successful pollination or mating between individuals, so hybrid zygotes do not form. Others prevent gametes from forming or block fertilization. Still others kill hybrids or make them weak or infertile (Figures 17.17 and 17.18). Let's start with the *prezygotic* mechanisms listed in Figure 17.18a,b.

Figure 17.16 Morphological differences between plants of the same species (*Sagittaria sagittifolia*) growing (**a**) in water and (**b**) on land. The leaf shapes are responses to different environmental conditions, not to different genetic programs.

a

b

Figure 17.17 (**a**) Mechanical isolation. Notice the fit between the reproductive parts of this zebra orchid and the wasp body. Few pollinating insects fit as precisely on this flower's landing platform.

(**b**) Behavioral isolation. Courtship displays precede sex among blue-footed boobies. Individuals recognize tactile, visual, and acoustical signals, including a prancing dance followed by back arching, skyward pointing bill and exposed throat, and wing spreading.

(**c**) Temporal isolation. *Magicicada septendecim*, a periodical cicada that matures underground and emerges to reproduce every 17 years. Its populations often overlap the habitats of a sibling species (*M. tredecim*), which reproduces every 13 years. Adults live only a few weeks.

Mechanical isolation. Incompatibility between body parts keeps potential mates or pollinators mechanically isolated. Two sage species keep pollen to themselves by attracting different insect pollinators. The pollen-bearing stamens of one extend from a nectar cup, above petals that form a big platform for the big pollinators. Small bees landing on it don't often touch the stamens and pick up pollen. The platform of the other sage species is too small to hold the big pollinators. Figure 17.17a shows another example.

Behavioral isolation. Behavioral differences bar gene flow between related species in the same vicinity. For instance, before male and female birds copulate, they often engage in courtship displays (Figure 17.17b). A female is genetically prewired to recognize distinctive singing, head bobbing, wing spreading, or prancing by a male of her species as an overture to sex. Females of different species usually ignore his behavior.

Temporal isolation. Interbreeding might be possible in diverging populations, but not if they differ in timing of reproduction. Three 17-year cicada species often occupy the same habitat in the eastern United States. They differ in form and behavior, but they all mature underground and feed on juicy roots. Every 17 years, they emerge to reproduce (Figure 17.17c). Each species has a *sibling* species—one that resembles it in form and behavior. The siblings emerge on a 13-year cycle. Thus, each species and its sibling can't get together except once in every 221 years!

Ecological isolation. Populations occupying different microenvironments in a habitat might be ecologically isolated. In the open forests of seasonally dry foothills of the Sierra Nevada are two manzanita species. One lives at elevations between 600 and 1,850 meters, the other between 750 and 3,350 meters. These species rarely hybridize, and only where the ranges overlap. Water-conserving mechanisms help them through dry seasons. But one species is adapted to sheltered sites where water stress isn't intense. The other species lives in drier, more exposed sites on rocky hillsides, so cross-pollination is unlikely.

Gamete mortality. Gametes of different species may have molecular incompatibilities. Example: If pollen lands on a plant of another species, it usually does not even recognize the molecular signals that trigger the germination of same-species pollen.

Postzygotic isolating mechanisms may act while an embryo is developing. Unsuitable interactions among genes or gene products cause early death, sterility, or weak hybrids with low survival rates. Some hybrids are sturdy but sterile. Mules, the offspring of a female horse and male donkey, are examples of such hybrids.

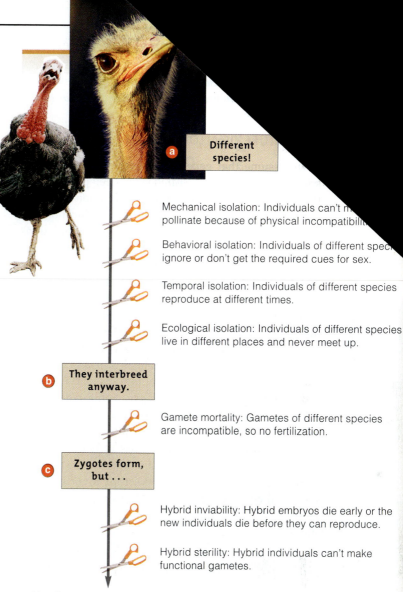

a Different species!

Mechanical isolation: Individuals can't m pollinate because of physical incompatibili

Behavioral isolation: Individuals of different spec ignore or don't get the required cues for sex.

Temporal isolation: Individuals of different species reproduce at different times.

Ecological isolation: Individuals of different species live in different places and never meet up.

b They interbreed anyway.

Gamete mortality: Gametes of different species are incompatible, so no fertilization.

c Zygotes form, but . . .

Hybrid inviability: Hybrid embryos die early or the new individuals die before they can reproduce.

Hybrid sterility: Hybrid individuals can't make functional gametes.

No offspring, sterile offspring, or weak offspring that die before reproducing

Figure 17.18 When reproductive isolating mechanisms halt interbreeding. There are barriers to (**a**) getting together, mating, or pollination, (**b**) successful fertilization, and (**c**) survival, fitness, or fertility of hybrid embryos or offspring.

A species is one or more populations of individuals having a unique common ancestor. Its individuals share a gene pool, produce fertile offspring, and remain reproductively isolated from individuals of other species.

Speciation is the process by which daughter species form from a population or subpopulation of a parent species. The process varies in its details and duration.

In two or more isolated populations, mutation, natural selection, and genetic drift operate independently. The buildup of genetic differences may lead to speciation.

Reproductive isolating mechanisms may evolve simply as by-products of the genetic changes. They are heritable traits that, one way or another, prevent interbreeding.

Speciation

basic premise
ely isolated.

CIATION

leading to a new
tween populations,
peciation route. By
1, a physical barrier
en two populations
llo– means different;
eland.) Reproductive
the two populations.
breeding is no longer
possible even when ... ies moves back to the
area still occupied by the parent species.

Whether a geographic barrier can effectively block gene flow depends on an organism's means of travel (deliberate or accidental), how fast it can travel, and whether it is inclined to disperse. Some measurable distance often separates populations of a species, so gene flow among them is an intermittent trickle, not a steady stream. Barriers can arise abruptly and shut off the trickles. In the 1800s, a major earthquake buckled part of the Midwest. The Mississippi River changed course; it cut through habitats of insects that couldn't swim or fly. Gene flow across the previously adjacent habitats stopped in an instant of geologic time.

As the fossil record suggests, geographic isolation also occurs over great spans of time. This happened after glaciers advanced down into North America and Europe during the ice ages and cut off populations of plants and animals from one another. When glaciers retreated, the descendants of the separate populations met up. Although related by descent, some were no longer reproductively compatible. They had evolved into new species. Genetic divergence was not as great between other separated populations, so descendants are still interbreeding. Reproductive isolation in their case was incomplete; speciation did not follow.

Also, remember how Earth's crust is fractured into gigantic plates? Slow but colossal movements of the plates have altered the configurations of land masses. As Central America was forming, part of an ancient ocean basin was uplifted and became a land bridge—the Isthmus of Panama. Some camelids moved across it into South America. Geographic separation led to new species, the llamas and vicunas (Figure 17.19).

ALLOPATRIC SPECIATION ON ARCHIPELAGOS

An **archipelago** is an island chain some distance from a continent. Some are so close to the mainland that gene flow is more or less unimpeded; there is little if any speciation. The Florida Keys are like this. But the isolated ones are another matter (Figure 17.20). The Hawaiian Archipelago is nearly 4,000 kilometers west from the California coast, and the Galápagos Islands, 900 kilometers or so from the Ecuadoran coast. The islands of both chains are only the tops of immense

Figure 17.19 Allopatric speciations. The earliest camelids, no bigger than a jackrabbit, evolved in the Eocene grasslands and deserts of North America. By the end of the Miocene, they included the now-extinct *Procamelus*. The fossil record and comparative studies indicate that this may have been the common ancestral stock for llamas (**a**), vicunas (**b**), and camels (**c**). One of the descendant lineages dispersed into Africa and Asia and evolved into modern camels. A different lineage, ancestral to the llamas and vicunas, dispersed into South America after gradual crustal movements formed a land bridge between the two continents.

Late Eocene paleomap, before a land bridge formed between North and South America

Read Me First!
and watch the
narrated animation on
allopatric speciation

a A few individuals of a species on the mainland reach isolated island 1. Speciation follows genetic divergence in a new habitat.

b Later in time, a few individuals of the new species colonize nearby island 2. In this new habitat, speciation follows genetic divergence.

c Speciation may also follow colonization of islands 3 and 4. And it may follow invasion of island 1 by genetically different descendants of the ancestral species.

The shared ancestor of all of Hawaii's honeycreepers probably looked like this housefinch (*Carpodacus*).

Akepa (*Loxops coccineus*)

Akekee (*L. caeruleirostris*)

Nihoa finch (*Telespyza ultima*)

Palila (*Loxioides bailleui*)

Maui parrotbill (*Pseudonestor xanthrophrys*)

Alauahio (*Paroreomyza montana*)

Kauai Amakihi (*Hemignathus kauaiensis*)

Akiapolaau (*H. munroi*)

Akohekohe (*Palmeria doli*)

Apapane (*Himatione sanguinea*)

Iiwi (*Vestiaria coccinea*)

Figure 17.20 (**a–c**) One example of allopatric speciation on an isolated archipelago. Can you envision other possibilities? (**d**) Some of the more than twenty species of Hawaiian honeycreepers. Their bills are adapted to diverse foods, such as insects, seeds, fruits, and floral nectar cups.

volcanoes that formed long ago on the seafloor. When the volcanoes first broke the surface of the sea, their fiery surfaces were devoid of life.

In one view, flotsam or winds brought mainland finches to one of the Galápagos Islands. Descendants colonized other islands in the chain, where habitats and selection pressures differed. Divergences within and between islands fostered episodes of allopatric speciation. Later, new species returned to the islands of their ancestors. The distances between islands are enough to foster divergences, but not enough to stop the occasional colonizers.

The youngest island in the Hawaiian Archipelago, Hawaii, formed less than a million years ago. Here alone we see diverse habitats, ranging from cooled lava beds, rain forests, and alpine grasslands to snow-capped volcanoes. When the first birds arrived, they found a buffet of fruits, seeds, nectars, tasty insects, and not many competitors for them. The near absence of competition spurred rapid speciations into vacant adaptive zones. Figure 17.20*d* hints at the variation among Hawaiian honeycreepers. Today, these and thousands of other species of animals and plants that originated in the archipelago are found nowhere else. As another example of their speciation potential, the Hawaiian Islands combined make up less than 2 percent of the world's land masses. Yet they are home to 40 percent of all fruit fly (*Drosophila*) species.

By an allopatric speciation model, some type of physical barrier intervenes between populations or subpopulations of a species and prevents gene flow among them. It favors genetic divergence that ends in speciation.

17.9 Other Speciation Models

There is evidence that some species have arisen and are maintained by less common mechanisms in which environmental barriers do not play a role.

SYMPATRIC SPECIATION

By the model for **sympatric speciation**, a species may form *within* the home range of an existing species, in the absence of a physical barrier. (*Sym–* means together with, as in "together with others in the homeland.")

EVIDENCE FROM CICHLIDS IN AFRICA In Cameroon, West Africa, cichlid populations may have undergone sympatric speciation in lakes, the basins of which are the collapsed cones of small volcanoes. Many cichlid species coexist in each lake. They probably colonized the lakes before volcanic action severed connections with a nearby river system.

Figure 17.21 A small, isolated crater lake in Cameroon, West Africa, where different species of cichlids may have originated by way of sympatric speciation.

Researchers analyzed mitochondrial DNA from all eleven cichlid species living in one lake. They did the same for all nine species coexisting in the other lake as the basis for comparison. The nucleotide sequences are all similar. The sequences from related cichlid species in nearby lakes and rivers are not as similar.

Allopatric populations could not have evolved. The physical and chemical conditions, even shorelines, are uniform in all lakes (Figure 17.21). Gene flow is absent; tiny fish-free creeks trickling from higher elevations are the only way in. Also, cichlids are good swimmers; individuals of different species often meet up.

Because clusters of species are more closely related to one another than to species anywhere else, we can expect them to share a common ancestor. Because of their restricted distribution, we can expect that they formed in the same lake. They do show small degrees of *ecological* separation. Feeding preferences put species in different places. Some feed in open waters, others at the lake bottom. Yet they all *breed* close to the lake bottom, in sympatry. Was the small-scale ecological separation enough to promote sexual selection among potential mates? Maybe. Over time, it may have led to reproductive isolation, then speciation.

POLYPLOIDY'S IMPACT Sympatric speciation has been common among flowering plants; about half of all known species are polyploid. In **polyploidy**, somatic cells have three or more of each type of chromosome characteristic of the species (Figures 17.22 and 17.23). Such changes can happen when chromosomes do not separate properly at meiosis or mitosis (Section 11.8). Changes also happen when a germ cell replicates its DNA but fails to divide, then functions as a gamete.

Polyploidy induces rapid speciation in plants. Often a polyploid plant has fertile offspring when crossed with identical polyploids; a new species has formed.

Speciation may have been fast for many flowering plants that self-fertilize or hybridize. If their offspring were polyploid and extra chromosomes paired with

Figure 17.22 Love those polyploids! Among them are five cotton species, marigolds, watermelons, sugarcane, and coffee plants, which have 22, 33, 66, or 88 chromosomes.

| Triticum monococcum (einkorn) | | Unknown species of wild wheat | | | | T. turgidum (wild emmer) | | T. tauschii (a wild relative) | | | T. aestivum (one of the common bread wheats) |

CROSS-FERTILIZATION, FOLLOWED BY SPONTANEOUS CHROMOSOME DOUBLING

14AA   x   14BB → 14AB → 28AABB   x   14DD → 42AABBDD

a By 11,000 years ago, humans were cultivating wild wheats. Einkorn has a diploid chromosome number of 14 (two sets of 7). It probably hybridized with another wild wheat species having the same number of chromosomes.

b About 8,000 years ago, polyploidy arose in plants that may have evolved from sterile, self-fertilizing AB hybrid wheat. (Wild emmer is tetraploid, or AABB; it has two sets of 14 chromosomes). The polyploids are fertile; the A and B chromosomes have partners at meiosis.

c An AABB plant probably hybridized with T. tauschii, a wild relative of wild emmer. Its diploid chromosome number must have been 14 (two sets of 7 DD). Hybrids now include common bread wheats. One has a chromosome number of 42 (six sets of 7 AABBDD).

Figure 17.23 Presumed sympatric speciation in wheat by polyploidy and hybridizations. Wheat grains 11,000 years old and diploid wild wheats have been found in the Near East.

each other during meiosis, the extra set of genes may have done no harm. Bread wheat may have arisen by way of hybridization and polyploidy (Figure 17.23).

Polyploid mammals are rare, maybe because of failures in dosage compensation. By this normal event, genes on sex chromosomes are expressed at the same levels in females and males. Remember, one of two X chromosomes in all female mammals gets inactivated (Section 14.4), so only one X chromosome is active in males *and* females. Polyploidy changes that. Alleles on extra sex chromosomes are active, which upsets the balance of gene expression necessary for normal functioning. Typically this has lethal consequences. Plants generally do not have sex chromosomes, so polyploidy does not always cause bad imbalances.

PARAPATRIC SPECIATION

Parapatric speciation might happen when different selection pressures operating across a broad region affect populations that are in contact along a common border. Hybrids that form in the contact are less fit than individuals on either side of the contact zones. Because hybrids are selected against, they appear in the hybrid zone only (Figure 17.24).

By a sympatric speciation model, daughter species arise from a group of individuals within an existing population. Polyploid flowering plants probably formed this way.

By a parapatric speciation model, populations maintaining contact along a common border evolve into distinct species.

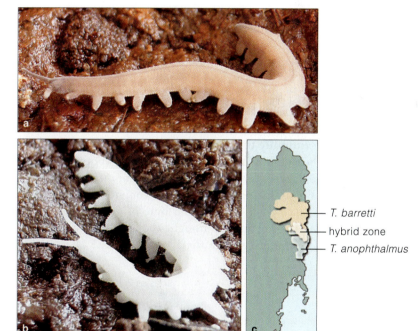

T. barretti
hybrid zone
T. anophthalmus

Figure 17.24 Parapatric speciation in rare Tasmanian velvet worms that live in adjoining regions of northeastern Tasmania, as shown in the map. (**a**) The giant velvet worm, *Tasmanipatus barretti* and (**b**) the blind velvet worm, *T. anophthalmus*.

(**c**) The habitats of the two kinds of walking worms overlap very little in a hybrid zone. The hybrids are sterile, which may be the main reason these two species are maintaining their separate identities in the absence of an obvious physical barrier.

17.10 Patterns of Speciation and Extinction

All species, past and present, are related by descent. They share genetic connections through lineages that extend back in time to the molecular origin of the first proto-cells. Later chapters focus on evidence in support of this view. Here, simply start thinking about ways to interpret the large-scale histories of species.

BRANCHING AND UNBRANCHED EVOLUTION

The fossil record reveals two patterns of evolutionary change, one branching, the other unbranched. The first is known as **cladogenesis** (from *klados*, branch; and *genesis*, origin). In this pattern, a lineage splits, the populations become genetically isolated, and so they diverge. Biologists generally accept cladogenesis as the most probable pattern of speciation, and it is the one we described earlier in the chapter.

In the second pattern, **anagenesis**, changes in allele frequencies and morphology accumulate in one line of descent. (In this context, *ana–* means renewed.) Because gene flow never stops among populations, directional changes are confined to that lineage alone.

EVOLUTIONARY TREES AND RATES OF CHANGE

Evolutionary trees summarize information about the continuity of relationships among groups. The sketch below is one way to start thinking about how these diagrams are constructed. Each branch represents a single line of descent from a common ancestor. Each *branch point* represents a time of genetic divergence and speciation, as brought about by natural selection or other microevolutionary processes:

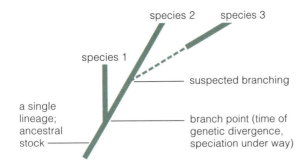

species 2   species 3

species 1

suspected branching

a single lineage; ancestral stock

branch point (time of genetic divergence, speciation under way)

When plotted against time, a branch that ends before the present (the "treetop") signifies that the lineage is extinct. A dashed line signifies we know something about a lineage but not exactly where it fits in the tree.

The **gradual model of speciation** holds that species originate by small morphological changes over long

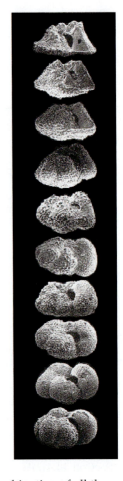

Figure 17.25 A series of fossilized foraminiferan shells from a vertical sequence of sedimentary rock layers. The first shell (*bottom*) is 64.5 million years old. The most recent (*top*) is 58 million years old. Detailed analysis of shell patterns confirmed the evolutionary order matches the geological sequence.

spans of time. The model fits with many fossil sequences. As one example, sedimentary rock layers often hold vertical sequences of fossilized shells of foraminiferans, and these reflect gradual morphological change (Figure 17.25).

The **punctuation model of speciation** is a different way to view patterns of speciation. Most morphological changes are said to evolve in a rather brief period when populations are just starting to diverge—within hundreds or thousands of years at most. Bottlenecks, the founder effect, directional selection, genetic drift, or some combination of all these processes promote rapid speciation. Daughter species recover fast from the adaptive wrenching, then don't change much over the next 2 to 6 million years.

Stability has indeed prevailed for all but 1 percent of the history of most lineages. And many parts of the fossil record reveal episodes of abrupt change. But both models help explain speciation patterns. Changes have been gradual, abrupt, or both. Species originated at different times and have differed in how long they last. Some have not changed much over millions of years; others were the start of adaptive radiations.

ADAPTIVE RADIATIONS

An **adaptive radiation** is a burst of divergences from a single lineage that led to many new species, each adapted to an unoccupied or new habitat, or to using novel resources. The honeycreepers in Figure 17.20 are examples. In the past, species of one lineage typically radiated into vacant **adaptive zones**. Think of each adaptive zone as a way of life, such as "burrowing in deep seafloor sediments" or "catching winged insects in the air at night." Species must have

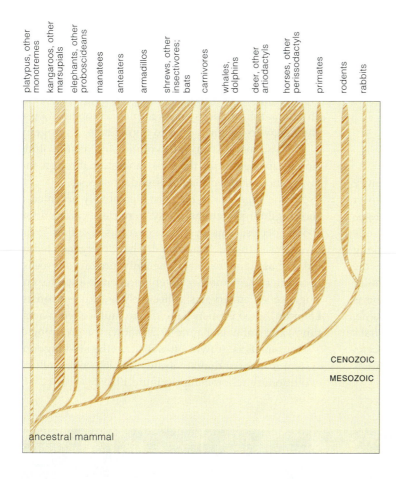

platypus, other monotremes | kangaroos, other marsupials | elephants, other proboscideans | manatees | anteaters | armadillos | shrews, other insectivores; bats | carnivores | whales, dolphins | deer, other artiodactyls | horses, other perissodactyls | primates | rodents | rabbits

CENOZOIC

MESOZOIC

ancestral mammal

Figure 17.26 Adaptive radiation of mammals. Branch widths indicate the range of biodiversity at different times. Mammals arose 220 million years ago but did not start a great radiation until after the K–T impact removed the last of the dinosaurs. Not all lineages are shown. The 4,000 existing species include tiny shrews, flying bats, and giant whales.

The photograph shows a fossil of *Eomaia scansoria* (Greek for ancient mother climber). About 125 million years ago, this insectivore lived in low shrubs and branches. At this writing, it is the earliest placental mammal we know about.

physical, evolutionary, or ecological access to new adaptive zones before they can occupy them (Figure 17.26). What does this mean?

Physical access means a lineage happens to be there as adaptive zones open up. For example, mammals were once distributed in uniformly tropical regions of a great continent. The continent broke up into huge land masses that slowly drifted apart. Habitats and resources changed in many different ways in different places, setting the stage for independent radiations.

Evolutionary access means a lineage has changed in structure or function in such a way that it can exploit the environment in a new or more efficient manner. Such modifications are known as **key innovations**. As an example, when some vertebrate forelimbs evolved into wings, novel adaptive zones opened for ancestors of birds and bats.

Ecological access means a lineage has the means to enter one or more vacated adaptive zones, or that it is competitive enough to displace resident species.

EXTINCTIONS—THE END OF THE LINE

An **extinction** is an irrevocable loss of a species. The fossil record gives evidence of twenty or more **mass extinctions**—catastrophic losses of entire families or other major groups. Past extinctions differed in size. For example, 250 million years ago, 95 percent of all known species were abruptly lost; in other times, a few groups were lost. Afterward, biodiversity slowly recovered as new species filled vacant adaptive zones.

Luck had a lot to do with it. Many species were devastated by global changes in climate; given their adaptations, they had nowhere else to go. And when one asteroid struck and wiped out the last dinosaurs, mammals were among the survivors that radiated into vacated adaptive zones. Asteroids, imperceptibly drifting continents, climatic change—all contributed to past patterns of extinctions and recoveries.

Lineages have changed gradually, abruptly, or both. Their member species originated at different times and have differed in how long they have persisted.

Adaptive radiations are bursts of divergences from a single lineage that gave rise to many new species, each adapted to a vacant or new habitat or to using a novel resource.

Repeated and often large extinctions happened in the past. After times of reduced biodiversity, new species formed and occupied new or vacated adaptive zones.

17.11 How Can We Organize the Evidence?

Biodiversity patterns are being revealed by taxonomy, phylogenetic reconstruction, and classification. This type of organization makes it easier for us to assimilate information about species and their heritage.

NAMING, IDENTIFYING, AND CLASSIFYING SPECIES

Taxonomists are biologists who systematically identify and classify species, then assign each kind a two-part name. Bacteria, junipers, vanilla orchids, houseflies, humans—all have a scientific name that gives people all over the world a way to know they are talking about the same organism. The name's first part, the **genus** (plural, genera), identifies a group of similar species (Chapter 1). The second part of the name is the specific epithet. Together with the generic name, it designates one species alone. For instance, *Ursus maritimus* is the polar bear, *U. arctos* the brown bear, and *U. americanus* is the black bear. Figure 17.27 has other examples.

All **classification systems** are organized systems of retrieving information about how species fit in the big picture. Ever more inclusive groupings of species are called **higher taxa** (singular, taxon). Family, order, class, phylum, and kingdom are examples. Taxonomists use phylogeny to assign species to higher taxa. **Phylogeny** is the systematic study of evolutionary relationships among species, from the most ancestral through the divergences that led to all descendant species.

At one time, many biologists tentatively accepted a **six-kingdom system** as a way to retrieve information about species. This system recognizes the first great divergence after life originated. That divergence led to the archaea and bacteria, the only prokaryotic cells. Soon afterward, another divergence led to ancestors of single-celled eukaryotes, collectively known as protists. Later on, some protists gave rise to multicelled plants, fungi, and animals. Figure 17.28 shows the groupings.

Consensus is growing to subsume the six-kingdom system in a **three-domain system**, which better reflects evolutionary relationships (Figure 17.29).

Figure 17.27 Taxonomic classification of five organisms. Each is assigned to ever more inclusive categories (higher taxa)—in this case, from species to kingdom.

KINGDOM	Bacteria	Plantae	Plantae	Animalia	Animalia
PHYLUM	Proteobacteria	Coniferophyta	Anthophyta	Arthropoda	Chordata
CLASS	Epsilonproteobacteria	Coniferopsida	Monocotyledonae	Insecta	Mammalia
ORDER	Campylobacterales	Coniferales	Asparagales	Diptera	Primates
FAMILY	Helicobacteraceae	Cupressaceae	Orchidaceae	Muscidae	Hominidae
GENUS	*Helicobacter*	*Juniperus*	*Vanilla*	*Musca*	*Homo*
SPECIES	*H. felis*	*J. occidentalis*	*V. planifolia*	*M. domestica*	*H. sapiens*
COMMON NAME	none	western juniper	vanilla orchid	housefly	human

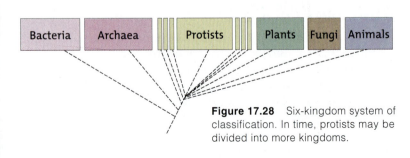

Figure 17.28 Six-kingdom system of classification. In time, protists may be divided into more kingdoms.

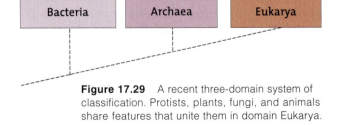

Figure 17.29 A recent three-domain system of classification. Protists, plants, fungi, and animals share features that unite them in domain Eukarya.

shark   mammal   crocodile   bird

fur

feathers

gizzard

lungs

heart

Figure 17.30 How to construct a simple cladogram, as described in the text.

WHAT'S IN A NAME? A CLADISTICS VIEW

Clues to phylogenetic relationships abound in fossils, morphology, biochemistry, and Earth's history. Degrees of morphological divergence help us construct tree diagrams in which branch points are the measure of relatedness. Traditional classification systems group species based on overall similarities, relying on the judgment and experience of taxonomists. In *cladistics*, species are grouped by derived traits.

A **derived trait** is a novel feature that evolved in just one species and is shared only by its descendants. In the evolutionary tree diagrams called **cladograms**, taxa are grouped by their shared derived traits. All descendants from an ancestral species in which a trait first evolved are a **monophyletic group**, which means "single tribe." A cladogram is like a time bar without absolute dates. It does not directly convey "who came from whom." Rather, it uses the position of branch points from the last shared ancestor to convey inferred evolutionary relationships (phylogenies) between taxa.

As a very simple example, sharks, crocodiles, birds, and mammals have a heart. Sharks do not have lungs, but crocodiles, birds, and mammals do. All crocodiles and birds have a specialized part of the gut called a gizzard, but no mammal has a gizzard. Birds alone have feathers; mammals alone have fur. You can use these traits to make a cladogram, as in Figure 17.30.

In this case, the bird is more closely related to the crocodile than it is to the mammal. The crocodile isn't its *ancestor*; it's a modern organism, too. But birds share an ancestor with crocodiles that is more recent than the one they share with mammals. The more ancient ancestor makes birds, crocodiles, and mammals closer to one another than to the shark.

Cladograms reflect information from comparative biochemistry. In the next section, you will find an evolutionary tree of life based on such evidence. Bear in mind that the simple branchings don't convey how many thousands of morphological and biochemical traits were analyzed to work out the connections. For example, it incorporates information from detailed comparisons of ribosomal RNA and protein-encoding genes of major groups of organisms. More importantly, it correlates well with parts of the fossil record. Even as you read this, researchers are rethinking how parts of the record are interpreted.

Taxonomists systematically identify species and classify them into ever more inclusive groupings—the higher taxa.

Classification systems organize and simplify the retrieval of information about species. Phylogenetic systems attempt to reflect evolutionary relationships among species.

Reconstructing the evolutionary history of a given lineage is based on detailed understanding of the fossil record, morphology, life-styles, and habitats of its representatives, and on biochemical comparisons with other groups.

Recent evidence, especially from comparative biochemistry, favors grouping organisms into a three-domain classification system—the archaea, bacteria, and eukarya (protists, plants, fungi, and animals).

17.12 An Evolutionary Tree of Life

In preparation for the next unit on life's diversity, take a moment to review this evolutionary tree.

The evolutionary tree of life in Figure 17.31 is a work in progress. Armed with automated gene sequencers and other advanced technologies, evolutionary biologists are clarifying phylogenies at a remarkable pace. Once researchers have identified a monophyletic group, they rank it in the context of other known relationships.

Taxonomic ranks have only relative meaning. Within the Eukarya, plants are a monophyletic group, a *clade*. Within the plant clade are many subclades, including mosses, liverworts, and vascular plants. These in turn have subclades. Within the "vascular plants" alone are several monophyletic groups, such as ferns, conifers, cycads, and the flowering plants. Within the flowering plants are subclades such as monocots, magnoliids,

and eudicots. Sister groups, such as echinoderms and chordates, are the closest relatives.

The important point is that ranking should reflect *sets within sets*. If Eukarya form a kingdom, then green plants could be ranked a phylum, and vascular plants a class. If Eukarya form a domain, green plants could be ranked a kingdom, and vascular plants a phylum. The "kingdom" of protists is about to be split into multiple groups. This is not a problem if we accept that absolute rankings are arbitrary. Focus instead on phylogenetic patterns—sets within sets—then even a shift of that magnitude won't snap against the other branches of a reconstructed tree of life.

> *Relationships between clades or monophyletic groups continue to be clarified as a pattern of sets within sets.*

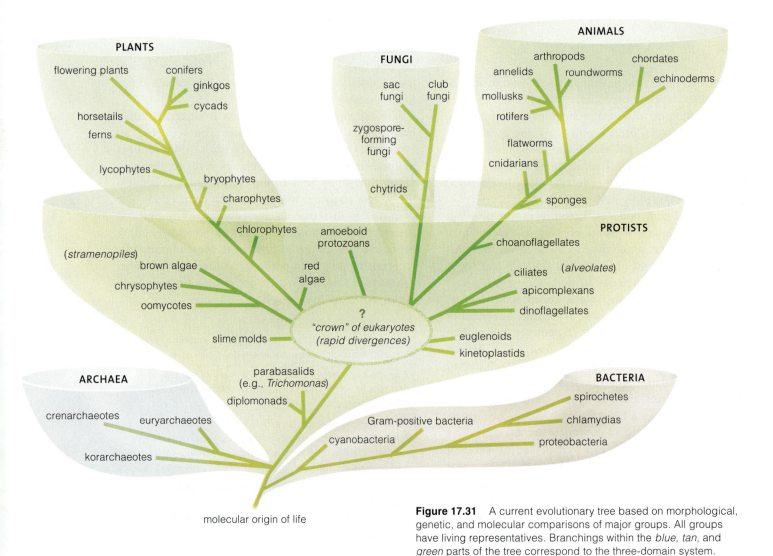

Figure 17.31 A current evolutionary tree based on morphological, genetic, and molecular comparisons of major groups. All groups have living representatives. Branchings within the *blue*, *tan*, and *green* parts of the tree correspond to the three-domain system.

Summary

Section 17.1 Fossils are physical evidence of organisms that lived in the distant past. Many are found in stacked layers of sedimentary rock, with the oldest generally near the bottom and the most recent on top. Although the fossil record is incomplete, it reveals many patterns in the history of life.

Section 17.2 Examining fossil layers provides a relative time scale that helps explain macroevolution: patterns, rates, and trends in evolution (Table 17.1). The geologic time scale uses abrupt transitions in the fossil record as boundaries. Radiometric dating has allowed us to assign absolute dates to the scale with a very small margin of error.

Section 17.3 The global distribution of land masses and fossils, magnetic patterns in volcanic rocks, and evidence of seafloor spreading from mid-oceanic ridges support the plate tectonic theory. Movements of the Earth's crustal plates shifted continents in ways that profoundly influenced life's evolution.

Section 17.4 Comparative morphology reveals evidence of evolution. It has identified homologous structures: comparable body parts that were modified in different ways in different lines of descent from a shared ancestor. Researchers must distinguish these parts from analogous structures: dissimilar body parts that became similar in independent lineages as a response to similar environmental pressures.

Section 17.5 Similarities in patterns and structures of embryonic development suggest common ancestry. Even minor genetic changes can alter the onset, rate, and completion time of developmental stages. They can have major impact on the adult form.

Section 17.6 Automated sequencing methods now help clarify evolutionary relationships. DNA, RNA, or proteins from different species are compared. Neutral mutations may accumulate in DNA at a constant rate; like ticks of a molecular clock, they help researchers calculate the divergence of one lineage from another.

Section 17.7 All populations of a species share a unique common ancestor and a gene pool, and they can interbreed and produce fertile offspring under natural conditions. If and when gene flow between them stops, reproductive isolating mechanisms typically evolve, because mutation, natural selection, and genetic drift operate independently in each population. Such divergence may give rise to a new species.

Section 17.8 By the allopatric speciation model, a geographic barrier cuts off gene flow between two or more populations and promotes genetic divergence, reproductive isolation, and finally speciation.

Section 17.9 By a sympatric speciation model, the divergence starts while populations are in physical contact. By a parapatric speciation model, it occurs between populations that share a common border.

Section 17.10 The time and duration of speciation has differed among lineages. Most species have become extinct. Mass extinctions, slow recoveries, and adaptive radiations are major macroevolutionary patterns.

Section 17.11 Each species has a unique, two-part scientific name. Taxonomy deals with identifying and naming species. Phylogenetic reconstruction uses analytical methods to infer evolutionary relationships. Classification is a way to organize information about related species.

The currently favored three-domain classification system is based largely on phylogenetic evidence. It recognizes three domains: bacteria, archaea, and eukarya. The eukarya include protists, plants, fungi, and animals.

Section 17.12 The evolutionary tree is a work in progress. More phylogenies are now being rapidly clarified. New groups are ranked in the scheme according to their relationship to other groups.

Table 17.1 Summary of Processes and Patterns of Evolution

Microevolutionary Processes

Mutation	Original source of alleles	Stability or change in a species is the outcome of balances or imbalances among all of these processes, the effects of which are influenced by population size and by the prevailing environmental conditions.
Gene flow	Preserves species cohesion	
Genetic drift	Erodes species cohesion	
Natural selection	Preserves or erodes species cohesion, depending on environmental pressures	

Macroevolutionary Processes

Genetic persistence	Basis of the unity of life. The biochemical and molecular basis of inheritance extends from the origin of first cells through all subsequent lines of descent.
Genetic divergence	Basis of life's diversity, as brought about by adaptive shifts, branching, and radiations. Rates and times of change varied within and between lineages.
Genetic disconnect	End of the line for a species. Mass extinctions are catastrophic events in which major groups abruptly and simultaneously are lost.

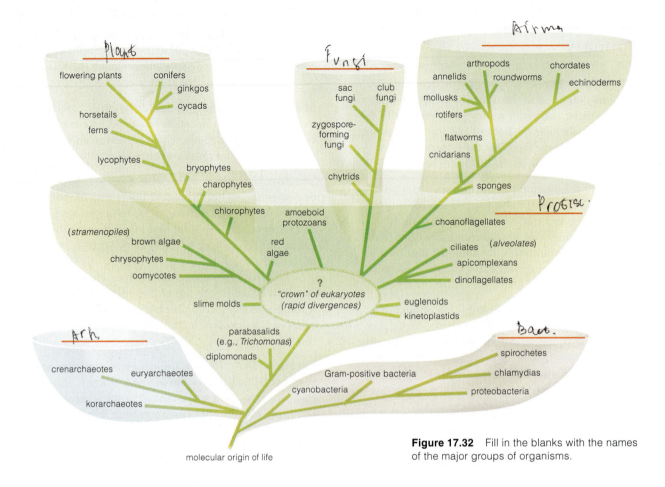

Plants (handwritten)
flowering plants   conifers
              ginkgos
              cycads
horsetails
ferns
lycophytes
              bryophytes
              charophytes
              chlorophytes   amoeboid
                              protozoans
(stramenopiles)
       brown algae   red
chrysophytes   algae
oomycotes
                    ?
                "crown" of eukaryotes
       slime molds   (rapid divergences)

Fungi (handwritten)
sac   club
fungi   fungi
zygospore-
forming
fungi
chytrids

Airms (handwritten)
              arthropods   chordates
annelids   roundworms
                              echinoderms
mollusks
rotifers
              flatworms
cnidarians
                    sponges

Protist. (handwritten)
              choanoflagellates
       ciliates   (alveolates)
              apicomplexans
              dinoflagellates
              euglenoids
              kinetoplastids

Arh (handwritten)
crenarchaeotes   euryarchaeotes
korarchaeotes

parabasalids
(e.g., Trichomonas)
       diplomonads

Baet. (handwritten)
                    spirochetes
Gram-positive bacteria   chlamydias
       cyanobacteria   proteobacteria

molecular origin of life

Figure 17.32 Fill in the blanks with the names of the major groups of organisms.

Self-Quiz
Answers in Appendix III

1. Morphological convergences may lead to ____a____.
   a. analogous structures   c. divergent structures
   b. homologous structures   d. both a and c

2. Sexually reproducing individuals of a species ____d____.
   a. can interbreed   c. share genetic history
   b. have fertile offspring   d. all of the above

3. In evolutionary trees, a branch point represents a
   ____c____, and a branch that ends represents _____.
   a. single species; incomplete data on lineage
   b. single species; extinction
   c. time of divergence; extinction
   d. time of divergence; speciation complete

4. *Pinus banksiana, Pinus strobus,* and *Pinus radiata*
   are ____c____.
   a. three families of pine trees
   b. three different names for the same organism
   c. three species belonging to the same genus
   d. both a and c

5. Heritable changes in DNA underlying morphological
   differences between lineages may have ____c____.
   a. been caused by transposons
   b. affected the onset, rate, and time of development
   c. both a and b

6. Reproductive isolating mechanisms ____d____.
   a. stop interbreeding   c. reinforce genetic divergence
   b. stop gene flow   d. all of the above

7. Individuals of a monophyletic group ____d____.
   a. are all descended from an ancestral species
   b. demonstrate morphological convergence
   c. share a unique derived trait
   d. both a and c

8. A(n) ____c____ classification system reflects presumed
   evolutionary relationships.
   a. epigenetic   c. phylogenetic
   b. tectonic   d. both b and c

9. Increasingly inclusive taxa range from _____ to ____d____.
   a. kingdom; genera and species
   b. kingdom; genera and domain
   c. genera; domain and kingdom
   d. species; kingdom and domain

10. Fill in the blanks in Figure 17.32 for the six major
    groups of organisms. You will soon trace the branching
    within each group.

11. Match these terms suitably.
    __e__ phylogeny   a. evidence of life in distant past
    __g__ fossils   b. branching lineages
    __f__ stratification   c. similar body parts in different
    __c__ homologous   lineages with shared ancestor
         structures   d. insect wing and bird wing
    __b__ cladogram   e. evolutionary relationships
    __d__ analogous   among species, from ancestors
         structures   through descendants.
    __g__ adaptive   f. layers of sedimentary rock
         radiation   g. burst of divergences into new
                              habitats

Critical Thinking

1. At the end of your backbone is a coccyx, a few small bones that are fused together. Could the human coccyx be a *vestigial* structure—all that's left of the tail of some distant vertebrate ancestors? Or is it the start of a newly evolving structure? Formulate a hypothesis, then design a way to test predictions based on the hypothesis.

2. You notice several duck species in the same lake habitat, with no physical barriers hampering the ducks' movements. All the females of the various species look quite similar to one another. But the males differ in the patterning and coloration of their feathers. Speculate on which forms of reproductive isolation may be keeping each species distinct. How does the appearance of the male ducks provide a clue to the answer?

3. Richard Lenski uses bacterial populations in culture tubes to develop model systems for studying evolution. Bacteria produce several generations in a day. Researchers can store them in the deep freeze, then bring them back to active form, unaltered, to directly compare ancestors and their descendants. Are bacterial models relevant to any evolutionary studies of sexually reproducing organisms? Before you answer, read a short article by P. Raine and M. Travisano entitled "Adaptive Radiation in a Heterogeneous Environment" (*Nature*, 2 July 1998, 69–72).

4. Shannon thinks there are too many kingdoms and sees no reason to make another one for something as small as bacteria. "Keep them with the other prokaryotes!" she says. Taxonomists would call her a "lumper." But Andrew is a "splitter." He sees no reason to withhold kingdom status from bacteria simply because they are part of a microscopic world that not many people know about. Which may be the most useful: more or fewer boundaries between groups? Explain your answer.

5. *Rama the cama*, a llama-camel hybrid, was born in 1997 (Figure 17.33). Camels and llamas have a shared ancestor but have been separated for 30 million years. Veterinarians collected semen from a male camel that weighed close to 1,000 pounds, then used it to artificially inseminate a female llama one-sixth his weight. The idea was to breed an animal having a camel's strength and endurance and a llama's gentle disposition.

Instead of being large, strong, and sweet, Rama is smaller than expected and has a camel's short temper. Rama resembles both parents, with a camel's long tail and short ears but no hump, and llama-like hooves rather than camel footpads. Now old enough to mate, he is too short to get together with a female camel and too heavy to mount a female llama. He has his eye on Kamilah, a female cama born in early 2002, but will have to wait several years for her to mature. The question is, will any offspring from such a match be fertile?

What does Rama's story tell you about the genetic changes required for irreversible reproductive isolation in nature? Explain why a biologist might not view Rama as evidence that llamas and camels are the same species.

6. A key innovation, again, is some modification in structure or function that permitted a species to exploit the environment in a more efficient or novel way, compared to ancestral species. Speculate on what might have been a key innovation in human evolution. Describe how that innovation might be the basis of an adaptive radiation in environments of the distant future.

Figure 17.33 Rama the cama displaying his unexpected short temper.

Media Menu

Student CD-ROM

Impacts, Issues Video
Measuring Time
Big Picture Animation
Evidence of evolution and macroevolution
Read-Me-First Animation
Morphological divergence
Mutation and proportional changes
Reproductive isolating mechanisms
Allopatric speciation
Other Animations and Interactions
The geologic time scale
Continental drift
Sympatric speciation in wheat

InfoTrac

- Speciation Genes Stick Together. *Applied Genetics News*, September 2001.
- Species: Life's Mystery Packages. *U.S. News & World Report*, July 2002.
- Unlocking the Mystery of Extinction. *Knight Ridder/Tribune News Service*, December 2002.
- Describing the "Tree of Life": Attainable Goal or Stuff of Dreams? *BioScience*, October 2002.

Web Sites

- Talk.Origins–Speciation: www.talkorigins.org/faqs/faq-speciation.html
- USGS Geologic Time: pubs.usgs.gov/gip/geotime
- NASA Asteroid and Comet Impact Hazards: impact.arc.nasa.gov/
- BBC Extinction Files: http://www.bbc.co.uk/education/darwin/exfiles/index.htm

How Would You Vote?

NASA's Spaceguard Survey is spending over 3 million dollars a year to map all near-Earth asteroids and comets larger than one kilometer in diameter. The survey will warn of a potentially catastrophic asteroid impact decades in advance and, it is hoped, allow us to prevent it. Weighing the certainty of catastrophe in the event of an asteroid impact against its low likelihood in the near future, do you support continued funding for this project? Should other nations contribute?

IMPACTS, ISSUES *Looking for Life in All the Odd Places*

In the 1960s, microbiologist Thomas Brock looked into hot springs and pools in Yellowstone National Park (Figure 18.1). He found an ecosystem of microscopic cells, including *Thermus aquaticus*. This prokaryote lives on simple carbon compounds dissolved in *really* hot water, on the order of 80°C (176°F)! *T. aquaticus* is one of the thermophiles, or "heat lovers."

Brock's work had two unexpected results. First, it sent researchers down paths that led them to a great domain of life, the Archaea. Second, it led to a faster

way to copy DNA and end up with useful amounts of it. *T. aquaticus* happens to make a heat-resistant enzyme that can catalyze the polymerase chain reaction—PCR. Enlisting synthetic versions of this enzyme triggered a revolution in biotechnology.

So *bioprospecting* became the new game in town. Companies started exploring extreme environments for species that might yield valuable products. And they found a dazzling array of life forms that withstand extraordinary levels of temperature, acidity, alkalinity, salinity, and pressure.

To extreme thermophiles, Yellowstone's hot springs would be relatively chilly. Many of these prokaryotic species evolved on the seafloor, near hydrothermal vents that spew superheated, mineral-rich water. One species even grows and reproduces at 121°C (249°F).

Other prokaryotes live in acidic springs, where the pH approaches zero, or in highly alkaline soda lakes. In polar regions, some types cling to life in salt ponds that never freeze and glacial ice that never melts.

It's not just prokaryotes. Populations of snow algae tint mountain glaciers red. Some of the transparent

Figure 18.1 Cells of *Thermus aquaticus*, a heat-loving prokaryotic species discovered in thermal springs in Yellowstone National Park. The filmstrip (*far right*) shows nanobes, found miles below Earth's surface.

the big picture

Molecular Beginnings   We can expect that the first chemical step toward life was the spontaneous formation of complex organic compounds from simpler substances present on the early Earth. A low-oxygen atmosphere favored this process.

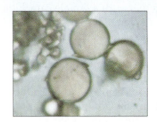

On to Proto-Cells   Laboratory studies and computer simulations have yielded indirect evidence that self-assembly of membranes, combined with chemical and molecular evolution, gave rise to the structural and functional forerunners of cells.

invertebrates called salps form long chains in frigid polar seas. One red alga, *Cyanidium caldarium,* grows in hot acidic springs. Some diatoms inhabit extremely salty lakes, where the hypertonicity would make cells of most organisms shrivel up and die.

The nanobes shown in the filmstrip were discovered 3.8 kilometers (3 miles) below Earth's surface in truly hot rocks—170°C (338°F). Being one-tenth the size of most bacteria, nanobes cannot be observed without electron microscopes. They look like the simplest fungi. Some researchers think nanobes are too small to be alive, to hold the metabolic machinery that runs life processes. Yet nanobes do have DNA. And they appear to grow. Are they like proto-cells, which preceded the origin of the first living cells? Maybe.

What's the point of all these examples? Simply this: *Life can take hold in almost any environment that has sources of carbon and energy.*

This chapter invites you to think about life in new ways, starting with theories about its origin and early evolution. Together with the next unit, it is a sweeping slice through time, one that can give you perspective on how life's spectacular range of diversity came about.

The road map is incomplete. Yet evidence from many avenues of research points to a principle that helps us organize information about an immense journey: *Life is a magnificent continuation of the physical and chemical evolution of the universe, and of the planet Earth.*

 How Would You Vote?

Private companies make millions of dollars selling an enzyme first isolated from cells in Yellowstone National Park. Should the federal government let private companies bioprospect within the boundaries of national parks, as long as it shares in the profits from any discoveries? See the Media Menu for details, then vote online.

The First Cells The first cells were anaerobic prokaryotes. Some gave rise to bacteria, others to archaea and ancestral eukaryotic cells. Evolution of the noncyclic pathway of photosynthesis in some bacteria added oxygen to the atmosphere, which became a new selection pressure.

Origin of Organelles The nucleus and ER of eukaryotic cells may have evolved through infoldings of the plasma membrane. Mitochondria and chloroplasts are probably descended from bacteria that were engulfed but not digested by other predatory cells.

18.1 In the Beginning . . .

Life originated when Earth was a thin-crusted inferno, so we probably will never find evidence of the first cells. Yet we can arrive at a plausible explanation of how they emerged by considering three questions. What were the prevailing conditions? Do known principles of physics, chemistry, and evolution disprove the hypothesis that the first cells could have originated through chemical and molecular evolution? Can we design experiments to test that hypothesis? Let's take a look.

Think about how you rewind a videotape on a VCR, then imagine "rewinding" the universe. As you do, the galaxies start moving closer together. After 12 to 15 billion years of rewinding, all galaxies, all matter, and space are compressed into a hot, dense volume at one single point. You have arrived at time zero.

That incredibly hot, dense state lasted only for an instant. What happened next is called the **big bang**, the nearly instantaneous distribution of all matter and energy throughout the universe. Within minutes, the temperature dropped a billion degrees. Nuclear fusion reactions created most of the light elements, including helium, which still are the most abundant elements. With radio telescopes, we have now detected radiation left over from the beginning of time.

Over the next billion years, uncountable numbers of gaseous particles collided, and gravitational forces condensed them into the first stars. When stars were massive enough, nuclear reactions ignited inside them and gave off tremendous light and heat, and heavier elements formed. All stars have a life history, from birth to an often explosive death. Heavier elements released from dying ones were swept up when new stars formed and helped form even heavier elements.

When explosions of dying stars ripped through our galaxy, they left behind a dense cloud of dust and gas that extended trillions of kilometers in space. As the cloud cooled, countless bits of matter gravitated toward one another. By 5 billion years ago, the shining star of our solar system—the sun—was born.

CONDITIONS ON THE EARLY EARTH

Figure 18.2 shows part of one of the vast clouds in the universe. It is mostly hydrogen gas, along with water, iron, silicates, hydrogen cyanide, ammonia, methane, formaldehyde, and other small inorganic and organic substances. Between 4.6 billion and 4.5 billion years ago, the cloud that became our solar system probably was like it in composition. Clumps of minerals and ice at the cloud's perimeter grew more massive. They eventually became planets; one of particular interest was the early Earth.

Four billion years ago, hot gases blanketed patches of Earth's thin, fiery crust. We suspect that this first atmosphere was a mix of gaseous hydrogen, nitrogen, carbon monoxide, and carbon dioxide. Did it include free oxygen? Probably not much. Geologic evidence reveals that the free oxygen level in the atmosphere was relatively low until about 2.2 billion years ago.

If that atmosphere had not been relatively free of oxygen, organic compounds necessary to make cells would not have been able to form wherever they were exposed to open air. Free oxygen would have attacked and destroyed the compounds as they assembled.

Figure 18.2 Part of the Eagle nebula, a hotbed of star formation. Each pillar is wider than our solar system. New stars shine brightly on the tips of gaseous streamers.

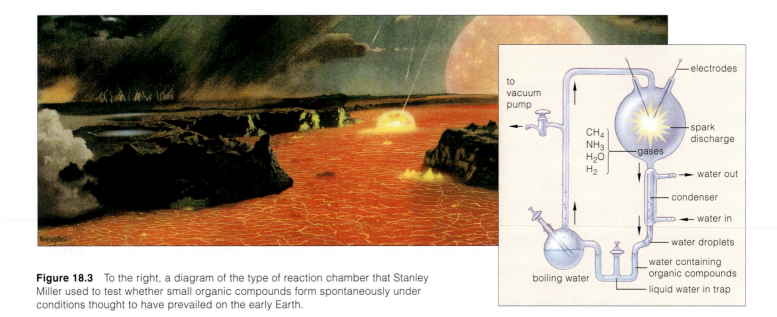

Figure 18.3 To the right, a diagram of the type of reaction chamber that Stanley Miller used to test whether small organic compounds form spontaneously under conditions thought to have prevailed on the early Earth.

At first, any water falling on the molten surface must have evaporated. However, after the crust cooled and became solid, rainfall and runoff eroded mineral salts from the rocks. Over millions of years, salty waters collected in crustal depressions and formed the earliest seas. If this liquid water had not accumulated, membranes could not have formed, because they take on their bilayer structure in water. No membrane, no cell. Life at its most basic level *is* the cell.

ABIOTIC SYNTHESIS OF ORGANIC COMPOUNDS

Cells appeared less than 200 million years after the crust solidified, so complex carbohydrates and lipids, proteins, and nucleic acids must have formed by then. We know that meteorites, Mars, and Earth all formed at the same time, from the same cosmic cloud. Their rocks contain simple sugars, fatty acids, amino acids, and nucleotides, so we can expect that the precursors of biological molecules were on the early Earth, too.

As you know, synthesis of biological molecules is an uphill run. So which energy sources spontaneously drove the assembly of the simple organic compounds into the first molecules of life? The sun, lightning, or heat from hydrothermal vents could have given off enough energy to drive the reactions.

Stanley Miller was the first to test this hypothesis. He placed water, methane, hydrogen, and ammonia inside a reaction chamber, as in Figure 18.3. He kept the mixture circulating while zapping it with sparks to simulate lightning. In less than a week, amino acids and other small organic compounds had formed.

Recent geologic evidence suggests that Earth's early atmosphere was not quite like Miller's mixture. But simulations that used other gases have also yielded different organic compounds, including some types that can act as nucleotide precursors of nucleic acids.

By another hypothesis, simple organic compounds formed in outer space. Researchers detect amino acids in interstellar clouds and in some of the carbon-rich meteorites that have landed on Earth. One meteorite discovered in Australia contains eight amino acids identical to those in all living organisms.

What about proteins, DNA, and the other *complex* organic compounds? Where did they form? In open water, hydrolysis reactions would have broken them apart as fast as they assembled. By one hypothesis, the clay of tidal flats bound and protected the newly forming polymers. Certain clays contain mineral ions that attract amino acids or nucleotides. Experiments show that once some of these molecules stick to clay, other molecules bond to them, forming chains that resemble the proteins or nucleic acids in living cells.

Another hypothesis that is currently getting a lot of attention is this: The first biological molecules were synthesized near hydrothermal vents. The depths of ancient seas were certainly oxygen-poor. Experiments show that amino acids, at least, will condense into proteinlike structures when heated in water.

Experiments provide indirect evidence that the complex organic molecules characteristic of life could have formed under conditions that probably prevailed on the early Earth.

18.2 How Did Cells Originate?

Metabolism refers to all reactions by which cells harness and use energy. In the first 600 million years or so of Earth history, enzymes, ATP, and other crucial organic compounds probably assembled spontaneously. If they did so in the same places, their close association might have promoted the start of metabolic pathways.

ORIGIN OF AGENTS OF METABOLISM

However the first proteins formed, their molecular structure dictated their behavior. If some behaved like weak enzymes and promoted reactions, they would snap up more amino acids and enzyme helpers, such as metal ions. Chemical selection could have been at work before cells appeared, favoring the evolution of enzymes and other complex organic compounds.

Imagine an ancient estuary, where seawater mixes with mineral-rich water being drained from the land. Beneath the sun's rays, organic molecules stick to clay (Figure 18.4a). At first there are quantities of an amino acid; call it D. Molecules of D get incorporated into proteins—until D starts to run out. Nearby, however, is a weakly catalytic protein. This protein promotes formation of D from a plentiful, simpler substance C.

By chance, clumps of organic molecules include the enzyme-like protein. Such clumps have an edge in any chemical competition for starting materials. Say the C molecules become scarce. The advantage tilts to molecular clumps that promote formation of C from simpler substances B and A. Suppose that B and A are carbon dioxide and water. The atmosphere and seas contain unlimited amounts of both. And so chemical selection has favored a synthetic pathway:

$$A + B \longrightarrow C \longrightarrow D$$

Were some clumps better at absorbing and using energy? Think back on chlorophyll *a* (Section 6.1). A group of rings in this pigment absorbs light and gives up electrons. The same kind of ring structures also occur in electron transfer chains in all photosynthetic and aerobically respiring cells. Such structures form spontaneously from formaldehyde, one of the legacies of cosmic clouds. Were similar structures transferring electrons in early metabolic pathways? Probably.

ORIGIN OF THE FIRST PLASMA MEMBRANES

All living cells have an outer membrane that controls which substances move into and out of a cell (Section 4.1). By one hypothesis, proto-cells were transitional forms between chemical evolution and the first living cells. The **proto-cells** were simple membrane-bound sacs that enclosed basic metabolic machinery and that were self-replicating.

Experiments have shown that membrane-like sacs can form spontaneously. Under conditions that could

Figure 18.4 Where did the first cells originate? Two candidates: (**a**) Clay templates in tidal flats, as explained in the text, and (**b**) iron sulfide-rich rocks at hydrothermal vents. Iron-sulfide rocks at hydrothermal vents are honeycombed with microscopic, cell-sized chambers. (**c**) Similar iron-sulfide structures have formed in the laboratory.

Did such chambers serve as protected microenvironments that favored the assembly of the first membranes? In experiments simulating conditions near vents, iron sulfides catalyzed the synthesis of short peptide chains and other reactions that resemble metabolism. Many reactions in living cells use iron-sulfide cofactors. Are these cofactors a metallic legacy from a deep-sea ancestor? Perhaps.

20 µm

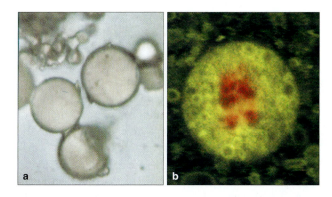

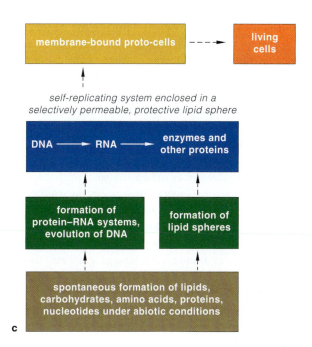

Figure 18.5 Laboratory-formed proto-cells. (**a**) Selectively permeable vesicles with a protein outer membrane were formed by heating amino acids, then wetting the resultant protein chains. (**b**) RNA-coated clay (*red*) surrounded by a membrane of fatty acids and alcohols (*green*). The mineral-rich clay catalyzes RNA polymerization and promotes the formation of membranous vesicles.

(**c**) Model for the steps that may have led to living cells.

have occurred on ancient sunbaked tidal flats, amino acids do form chains, then assemble like a membrane around a fluid interior (Figure 18.5*a*). Fatty acids and alcohols can spontaneously form vesicles, especially when clays rich in minerals are present (Figure 18.5*b*).

Or proto-cells may have formed near hydrothermal vents. Chimney-like rocks that pile up around vents are honeycombed with microscopic chambers (Figure 18.4*b*,*c*). By one hypothesis, such chambers were three-dimensional molds for the first cell membranes. Iron sulfides in the rocks could have served as catalysts for early metabolic reactions.

ORIGIN OF SELF-REPLICATING SYSTEMS

Another defining characteristic of life is a capacity for reproduction, which now starts with protein-building instructions in DNA. As you know, the translation of DNA into proteins requires the participation of RNA, enzymes, and other molecules.

Coenzymes and metal ions assist most enzymes—and certain coenzymes are structurally identical with RNA subunits. When you mix and heat RNA subunits with very short chains of phosphate groups, they self-assemble into strands of RNA. Simple self-replicating systems of RNA, enzymes, and coenzymes have been made in laboratories. So we know RNA can serve as an information-storing template for making proteins.

Also, an rRNA in ribosomes does catalyze protein synthesis (Figure 13.8). Ribosomes, recall, are highly conserved structures. The ribosomes in a eukaryotic

cell are not that different from those in prokaryotes. It's likely that rRNA's catalytic behavior evolved early in Earth history.

Did an **RNA world** in which RNA was a template for protein synthesis *precede* the origin of DNA? It's possible. Remember, RNA and DNA are a lot alike. Three of four bases are identical, and RNA's uracil differs from DNA's thymine by one functional group.

Why would DNA be any better than RNA? *Because DNA is more stable than RNA, it can form longer chains that can carry far more information.* Computer analyses also show that helical coiling of a strandlike molecule is the best way to pack subunits in the smallest space.

Until we identify the chemical ancestors of RNA and DNA, the story of life's origin will be incomplete. But details are filling in. For instance, researchers fed data about inorganic compounds and energy sources into a supercomputer. They programmed the computer to simulate random chemical competition and natural selection among selected compounds, as might have happened untold billions of times in the distant past. They ran the program again and again. The outcome was always the same. *Simple precursors evolved. Then they spontaneously organized themselves as large, complex molecules. And they began to interact as complex systems.*

There are many gaps in the story of life's origin. But diverse laboratory experiments and computer simulations show that chemical evolution can produce the organic molecules and structures that we think of as characteristic of life.

18.3 The First Cells

The first cells probably evolved during the **Archaean**, an eon that started 3.8 billion years ago. Until that time, constant bombardment by comets and asteroids almost certainly made the planet inhospitable to life.

Biologists generally accept that the first cells were like existing prokaryotic cells; they had no nucleus. Because free oxygen was absent, we can expect that they obtained energy by anaerobic pathways such as fermentation. Plenty of "food" was available; geologic processes had already enriched the seas with organic compounds and minerals. There were no predators, and cells were not exposed to oxygen attacks.

By analyzing the phylogenies of living prokaryotes, we have evidence that some populations diverged not long after life originated. One lineage gave rise to the bacteria. Another gave rise to the shared ancestors of archaea and eukaryotic cells.

Microscopic fossils in 3.5-billion-year-old rocks hint at what some of those earliest prokaryotes may have looked like (Figure 18.6a). Other fossils clearly show that chemoautotrophic forms had become established near deep-sea hydrothermal vents by 3.2 billion years ago. In some groups, pigments probably detected the type of weak infrared radiation (heat) that has been measured at hydrothermal vents. Pigments may have helped cells avoid boiling water, as they do for some existing hydrothermal vent species.

At some point, mutations occurred independently in some groups and altered their radiation-sensitive pigments, electron transfer chains, and other bits of metabolic machinery. They favored a novel mode of nutrition—the cyclic pathway of photosynthesis. The new groups were photoautotrophic; they tapped into sunlight, an unlimited energy source (Section 6.2).

Populations of photosynthesizers formed flattened mats, which trapped sediments dissolved in water. In time, continuous deposition of the sediments and cell remains formed layered, dome-shaped rocks, with the most recent mats of still-living cells on top. Some of these structures, called **stromatolites**, are more than one meter high (Figure 18.7). Fossilized ones provide the most abundant evidence of early prokaryotic life.

Stromatolites were abundant by the dawn of the Proterozoic, 2.7 billion years ago. A noncyclic pathway of photosynthesis had evolved among cyanobacteria. Its by-product, free oxygen, slowly accumulated in the sea surface waters, then in air. And here we return to events sketched out earlier, in Chapters 4 and 6.

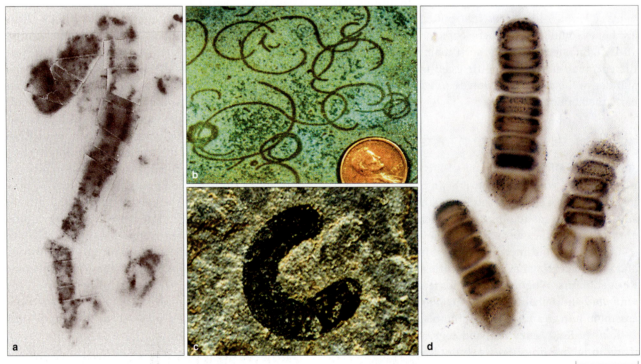

20 μm

Figure 18.6 A sampling of early life. (**a**) A strand of what may be walled prokaryotic cells dates back 3.5 billion years. (**b**) One of the oldest eukaryotic species, *Grypania spiralis*, a colonial eukaryote that formed macroscopic colonies 2.1 billion years ago. (**c**) *Tawuia*, another early Proterozoic eukaryotic megafossil. (**d**) *Bangiomorpha pubescens*, a sexually reproducing multicellular red alga, from 1.2 billion years ago.

Figure 18.7 Stromatolites. (**a**) A painting of how one shallow sea might have looked during the early Proterozoic. (**b**) A cut through a stromatolite reveals many layers of fossilized cyanobacterial mats, hardened by calcium carbonate deposits. (**c**) In Shark Bay, Australia, are mounds 2,000 years old. Structurally, they are similar to some of the stromatolites dated at more than 3 billion years old.

An atmosphere enriched with free oxygen had two irreversible effects. First, *it stopped the further chemical origin of living cells*. Except in a few anaerobic habitats, complex organic compounds could no longer form spontaneously and stay intact; they could not escape attacks by free oxygen. Second, *aerobic respiration now evolved and in time became the dominant energy-releasing pathway*. In many prokaryotic lineages, selection had favored this pathway, which neutralized oxygen by *using* it as an electron acceptor. Aerobic respiration was a key innovation that contributed to the rise of multicelled eukaryotes.

Eukaryotic cells also evolved in the Proterozoic. Traces of the kinds of lipids that are made by existing eukaryotic cells have been isolated from rocks that are 2.8 billion years old. But the first complete eukaryotic fossils are about 2.1 billion years old (Figure 18.6*b,c*). These species had organelles.

As you know, organelles are the defining features of eukaryotic cells. Where did they come from? The next section presents a few plausible hypotheses.

We still don't know how the earliest eukaryotes fit in evolutionary road maps. The earliest known form we can assign to a modern group is a filamentous red alga, *Bangiomorpha pubescens*. This alga was alive 1.2 billion years ago. It is the first multicelled species we know about, and its cells were differentiated. Some cells in the strandlike body functioned as anchoring structures (holdfasts). Others produced two types of sexual spores—which makes *B. pubescens* the earliest known sexually reproducing organism.

By 1.1 billion years ago the first supercontinent—Rodinia—had formed. Stromatolites flourished along its vast shorelines. About 300 million years after that, however, they were in decline. Were newly evolved predators—animals, perhaps—feeding on them? The earliest fossils that are unambiguously "animals" are 600 million years old. We can interpret this to mean that simpler animals originated before then. It was not until 570 million years ago, when oxygen in the atmosphere was approaching modern levels, that the first adaptive radiations of animals began.

The first living cells evolved by about 3.8 billion years ago, during the Archaean eon. All were prokaryotic. We can expect that they made ATP by fermentation pathways. Early on, ancestors of archaea and eukaryotic cells diverged from the lineage that led to modern bacteria.

After the noncyclic pathway of photosynthesis evolved, free oxygen accumulated in the atmosphere and put an end to the further spontaneous chemical origin of life. That atmosphere set the stage for the evolution of eukaryotic cells.

18.4 Where Did Organelles Come From?

Thanks to globe-hopping microfossil hunters, we have considerable evidence of early life, including the fossil treasures shown in Sections 4.4 and 18.3. Today, most descendant species contain a profusion of organelles. Where did the organelles come from?

ORIGIN OF THE NUCLEUS AND ER

Prokaryotic cells, recall, do not have an abundance of organelles. Some do have infoldings of their plasma membrane, which contains many enzymes and other metabolic workers (Figure 18.8a). By applying the theory of natural selection, we can hypothesize that such infoldings first evolved among the ancestors of eukaryotic cells, and they became narrow channels in which nutrients, organic compounds, and other substances became concentrated. What selective advantage did infoldings offer? More membrane meant more metabolic machinery; reactions, transport tasks, and waste disposal were completed faster.

Did ER channels evolve this way? And did they also protect the cell's metabolic machinery from uninvited guests? Possibly. From time to time, metabolically "hungry" foreign cells do enter the cytoplasm of existing prokaryotic cells.

By this scenario, infoldings also surrounded the DNA and became a nuclear envelope (Figure 18.8b). A nuclear envelope was favored because it helped protect the cell from invasions of foreign DNA. Bacteria and simple eukaryotic cells called yeasts do transfer bits of DNA among themselves. Early eukaryotic cells equipped with a nuclear envelope could copy and use their messages of inheritance, free from metabolic competition from a potentially disruptive hodgepodge of foreign DNA.

ORIGIN OF MITOCHONDRIA AND CHLOROPLASTS

Early in the history of life, cells became food for one another. Heterotrophs started feeding on autotrophs and other heterotrophs. In some cases, the meals turned into uninvited guests. They lingered on and slowly evolved into mitochondria, chloroplasts, and other organelles.

Such novel partnerships are one premise of a theory of **endosymbiosis**, as championed by Lynn Margulis and others. (*Endo–* means within; *symbiosis* means living together.) One species (the symbiont) lives out its life inside another species (the host), and the interaction benefits one or both.

According to this theory, eukaryotic cells evolved after the noncyclic pathway of photosynthesis had emerged and changed the atmosphere. By 2.1 billion years ago, remember, certain prokaryotic cells had adapted to the accumulation of free oxygen and were carrying out aerobic respiration.

Aerobic bacteria were among the meals ingested by the ancestor of eukaryotic cells. At some point, a meal resisted digestion. It stayed alive, grew, and reproduced in the sheltered, nutrient-rich cytoplasm. The host reproduced, too. After many generations,

DNA

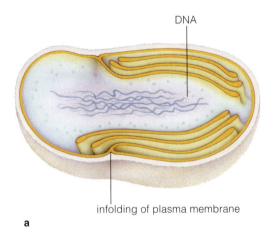

infolding of plasma membrane

a

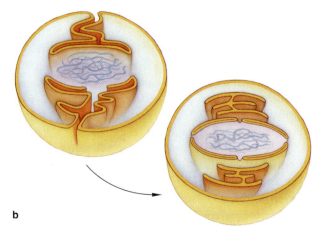

b

Figure 18.8 (**a**) Sketch of a bacterial cell (*Nitrobacter*) that lives in soil. Cytoplasmic fluid bathes permanent infoldings of the plasma membrane. (**b**) Model for the origin of the nuclear envelope and endoplasmic reticulum. In prokaryotic ancestors of eukaryotic cells, infoldings of the plasma membrane may have evolved into these organelles.

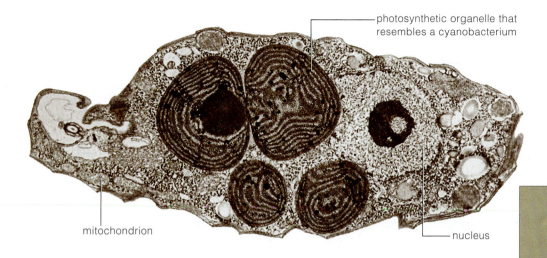

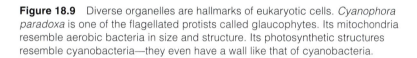

Figure 18.9 Diverse organelles are hallmarks of eukaryotic cells. *Cyanophora paradoxa* is one of the flagellated protists called glaucophytes. Its mitochondria resemble aerobic bacteria in size and structure. Its photosynthetic structures resemble cyanobacteria—they even have a wall like that of cyanobacteria.

its descendants started using ATP formed by the aerobic symbiont; in time, some depended on it.

The aerobic symbiont no longer had to gather raw materials or build proteins when the host was doing the work for it. Host DNA regions that specified the proteins were free to mutate and lose their function. In time, both types of cells became incapable of independent life.

EVIDENCE OF ENDOSYMBIOSIS

Is such a theory far-fetched? A chance discovery in Jeon Kwang's laboratory suggests otherwise. In 1966, a rod-shaped bacterium had infected his culture of *Amoeba discoides*. Some infected cells died right away. Others grew more slowly, and they were smaller and vulnerable to starving to death. Kwang maintained the infected culture. Five years later, the infection's bad effects were gone. Amoebas stuffed with bacteria were thriving. Exposing these infected amoebas to antibiotics actually killed them.

Infection-free cells stripped of their nucleus died after receiving the nucleus from an infected cell. Yet, more than 90 percent survived when a few bacteria tagged along with the transplant. In other studies, infected amoebas lost their ability to synthesize an essential enzyme. They depended on the bacterium to make it for them! The invading bacterial cells had become symbiotic with the amoebas.

When you think about it, mitochondria do resemble bacteria in size and structure. Each has its own DNA and divides independently of cell division. The inner membrane of a mitochondrion resembles a bacterial cell's plasma membrane. Its DNA has just a few genes (thirty-seven in human mitochondrial DNA). And its genetic code differs a bit from the genetic code shared by all other organisms.

We can expect that chloroplasts, too, originated by endosymbiosis. By one hypothesis, photosynthetic cells were engulfed by predatory aerobic bacteria and escaped digestion. They started absorbing nutrients from the host's cytoplasm and continued to function. They also released oxygen during photosynthesis. By releasing oxygen in their aerobically respiring hosts, they acted as agents favoring endosymbiosis.

In metabolism and overall nucleic acid sequence, chloroplasts resemble the type of prokaryotes called cyanobacteria. Chloroplast DNA and cellular DNA replicate independently of each other. Chloroplasts also divide independently of the cell's division.

Finally, consider the glaucophytes. This intriguing protist group has unique photosynthetic organelles. The organelles closely resemble cyanobacteria; they even are surrounded by a cell wall (Figure 18.9).

However they arose, the first eukaryotic cells had a nucleus, an endomembrane system, mitochondria, and, in some lineages, chloroplasts. They were the world's first protists.

18.5 Timeline for Life's Origin and Evolution

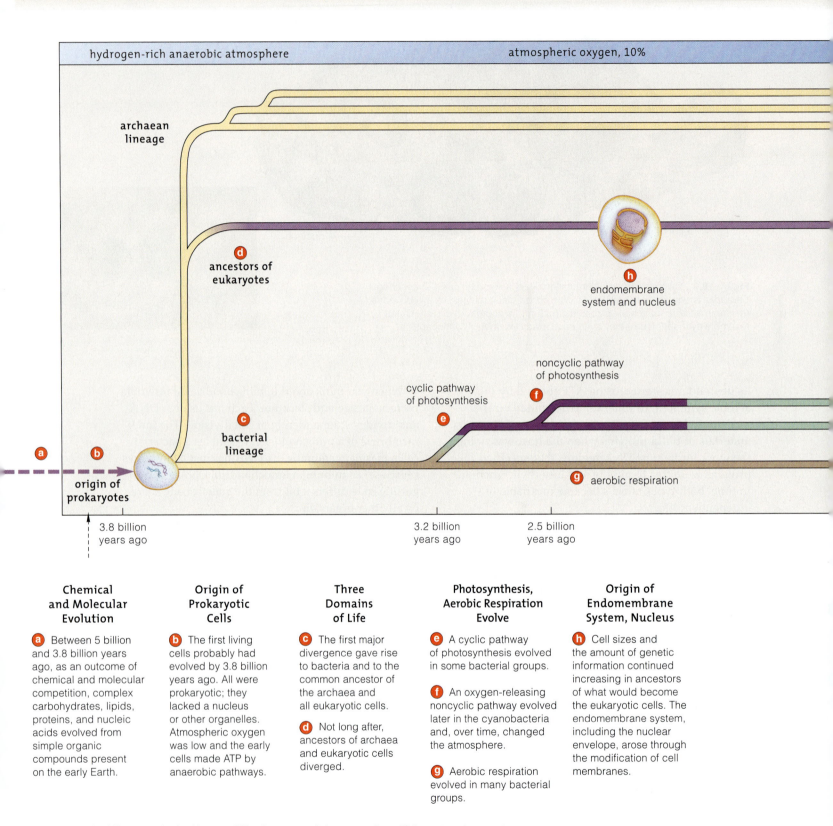

Chemical and Molecular Evolution

(a) Between 5 billion and 3.8 billion years ago, as an outcome of chemical and molecular competition, complex carbohydrates, lipids, proteins, and nucleic acids evolved from simple organic compounds present on the early Earth.

Origin of Prokaryotic Cells

(b) The first living cells probably had evolved by 3.8 billion years ago. All were prokaryotic; they lacked a nucleus or other organelles. Atmospheric oxygen was low and the early cells made ATP by anaerobic pathways.

Three Domains of Life

(c) The first major divergence gave rise to bacteria and to the common ancestor of the archaea and all eukaryotic cells.

(d) Not long after, ancestors of archaea and eukaryotic cells diverged.

Photosynthesis, Aerobic Respiration Evolve

(e) A cyclic pathway of photosynthesis evolved in some bacterial groups.

(f) An oxygen-releasing noncyclic pathway evolved later in the cyanobacteria and, over time, changed the atmosphere.

(g) Aerobic respiration evolved in many bacterial groups.

Origin of Endomembrane System, Nucleus

(h) Cell sizes and the amount of genetic information continued increasing in ancestors of what would become the eukaryotic cells. The endomembrane system, including the nuclear envelope, arose through the modification of cell membranes.

Figure 18.10 Milestones in the history of life. As you read the next unit on life's past and present diversity, refer to this visual overview. It can serve as a simple reminder of the evolutionary connectedness of all groups of organisms, from the structurally simple to the most complex.

Read Me First!

*and watch
the narrated animation on
the history of life*

atmospheric oxygen, 20%; the ozone layer slowly develops

ARCHAEA

Extreme thermophiles

Extreme thermophiles and mesophiles

Halophiles and methanogens

EUKARYOTES

Animals

Fungi

Heterotrophic protists

Photosynthetic protists with chloroplasts
that evolved from red and green algae

Red and green algae; their chloroplasts
evolved from cyanobacterial symbionts

Plants

k origin of animals

j origin of eukaryotes,
the first protists

k origin of fungi

origin of
mitosis,
meiosis

endosymbiotic origin
of mitochondria **i**

j endosymbiotic origin
of chloroplasts

k origin of lineage
leading to plants

BACTERIA

Oxygen-releasing photosynthetic
bacteria (cyanobacteria)

Other photosynthetic bacteria

Heterotrophic bacteria, including
chemoheterotrophs

Aerobic species becomes
endosymbiont of anaerobic
forerunner of eukaryotes.

l

1.2 billion years ago	900 million years ago	435 million years ago

Endosymbiotic Origin of Mitochondria

i Before about
1.2 billion years ago,
aerobic bacterial
species and an
anaerobic ancestor
of eukaryotic cells
entered into close
symbiotic interaction.
The endosymbiont
evolved into the
mitochondrion.

Endosymbiotic Origin of Chloroplasts

j Cyanobacteria
entered into a close
symbiotic interaction
with early protists and
evolved into chloroplasts.
Later, photosynthetic
protists would evolve
into chloroplasts inside
other protist hosts.

Plants, Fungi, and Animals Evolve

k By 900 million years
ago, all major lineages—
including fungi, animals,
and the algae that would
give rise to plants—had
evolved along shorelines
of the first supercontinent.

Lineages That Made It to the Present

l Today, all regions of
the Earth's waters, crust,
and atmosphere hold
organisms. They are
related by descent and
share certain common
traits. But each lineage
encountered different
selective pressures, and
each has evolved its own
characteristic traits.

Summary

Section 18.1 Earth formed more than 4 billion years ago. Indirect evidence suggests that complex organic compounds characteristic of life could have formed spontaneously on early Earth.

Section 18.2 Proto-cells may have arisen through chemical evolution and the self-assembly of lipid or protein spheres. RNA probably preceded DNA as the template for protein synthesis.

Section 18.3 The first prokaryotic cells arose about 3.8 billion years ago. They were anaerobic. An early divergence separated bacteria from the ancestors of archaea and eukaryotes. Evolution of the noncyclic pathway of photosynthesis in cyanobacteria resulted in an accumulation of free oxygen in the atmosphere. The increase paved the way for aerobic respiration and the evolution of eukaryotic cells.

Media Menu

Student CD-ROM

Impacts, Issues Video
   Looking for Life in All the Odd Places
Big Picture Animation
   Life's origin and evolution
Read-Me-First Animation
   The history of life
Other Animations and Interactions
   Miller's experimental apparatus

InfoTrac

- How Life Began: Microbes at the Extremes May Tell Us. *Time*, July 2002.
- Nanobes? How Small Is the Smallest Life? *Odyssey*, February 2002.
- Life on Earth: Scientists Probe the Early Days. *Astronomy*, November 2001.
- From Proteins to Protolife. *Science News*, July 1994.

Web Sites

- Life Over Time:
  www.fmnh.org/exhibits/exhibit_sites/lot/LOT1.htm
- Center for Studies of the Origins of Life:
  www.rpi.edu/~straca/NSCORT/origin.html
- ActionBioScience New Frontiers:
  actionbioscience.org/newfrontiers/index.html

How Would You Vote?

PCR uses a DNA polymerase first isolated from a prokaryote that was discovered in Yellowstone National Park. Sales of the enzyme, which has been patented, exceed 500 million dollars annually. Yellowstone receives none of this revenue. Now park managers want to charge companies to "bioprospect" in thermal pools and to share in profits from any new products. Do you support this use of our national parks?

Section 18.4 The internal membranes of eukaryotic cells may have evolved through infoldings of the cell membrane. Mitochondria and chloroplasts most likely evolved by endosymbiosis, at the times indicated in the Section 18.5 visual summary.

Self-Quiz

Answers in Appendix III

1. An abundance of _____ in the atmosphere would have prevented the spontaneous (abiotic) assembly of organic compounds on the early Earth.
   a. hydrogen   b. methane   c. oxygen   d. nitrogen

2. The prevalence of iron-sulfide cofactors in living organisms may be evidence that life arose _____ .
   a. in outer space   c. near deep-sea vents
   b. on tidal flats   d. in the upper atmosphere

3. The evolution of _____ resulted in an increase in the levels of atmospheric oxygen.
   a. sexual reproduction
   b. aerobic respiration
   c. the noncyclic pathway of photosynthesis
   d. the cyclic pathway of photosynthesis

4. Mitochondria may be descended from _____ .
   a. chloroplasts   c. early protists
   b. bacteria   d. archaea

5. Chronologically arrange the evolutionary events, with 1 being the earliest and 6 the most recent.
   ____ 1   a. emergence of the noncyclic
   ____ 2   pathway of photosynthesis
   ____ 3   b. origin of mitochondria
   ____ 4   c. origin of proto-cells
   ____ 5   d. emergence of the cyclic
   ____ 6   pathway of photosynthesis
           e. origin of chloroplasts
           f. the big bang

Critical Thinking

1. What if it were possible to create life in test tubes? This is the question behind attempts to model and perhaps create minimal organisms: living cells having the smallest set of genes required to survive and reproduce.

Craig Venter and Claire Fraser recently found that *Mycoplasma genitalium*, a bacterium that has 517 genes (and 2,209 transposons), is a good candidate for such an experiment. By disabling its genes one at a time in the laboratory, they discovered that *M. genitalium* may have only 265–350 essential protein-coding genes.

What if those genes were synthesized one at a time and inserted into an engineered cell consisting only of a plasma membrane and cytoplasm? Would the cell come to life? The possibility that it might prompted Venter and Fraser to seek advice from a panel of bioethicists and theologians. No one on the panel objected to synthetic life research. They said that much good might come of it, provided scientists didn't claim to have found "the secret of life." The December 10, 1999, issue of *Science* includes an essay from the panel and an article on *M. genitalium* research. Read both, then write down your thoughts about "creating" life in a test tube.

IV Evolution and Biodiversity

From the Green River formation near Lincoln, Wyoming, the stunning fossilized remains of a bird trapped in time. During the Eocene, some 50 million years ago, sediments that had been slowly deposited in layers at the bottom of a large inland lake became its tomb. In this same formation, fossils of palms, cattails, sycamore, and other plants tell us that the climate was warm and moist. Fossils from places all around the world yield major clues to life's early history.

IMPACTS, ISSUES *West Nile Virus Takes Off*

In 336 BC, when he was twenty years old, Alexander the Great ascended to the throne of Macedonia (Figure 19.1). During his reign, he carved out an empire that stretched across the Middle East and into northern India. He died twelve years later while visiting Babylon, the site of modern-day Baghdad. According to one scribe, a flock of ravens announced Alexander's arrival in the city. The birds behaved strangely, and some fell dead at his feet. Soon thereafter, Alexander suffered a sudden onset of back pain. Fever and chills, weakness, delirium, paralysis, and finally death followed.

An infectious disease expert, Charles Calisher, and epidemiologist John Marr connected the dots between the birds' behavior and Alexander's symptoms. As they

hypothesize, Alexander died as an outcome of West Nile encephalitis. A flavivirus, first isolated in 1937, causes this disease, which results in severe inflammation of the brain. Researchers had discovered that the virus was once prevalent in Africa, West Asia, and the Middle East. Was there an outbreak of West Nile encephalitis during Alexander's time? Possibly.

Until the summer of 1999, no one knew that the virus had entered the Western Hemisphere. Then people in and around New York City started wondering about the dead and dying crows—and the mysteriously sickened horses and humans. West Nile virus particles turned up in tissue samples from infected individuals. Sixty-two people were infected that summer. Seven died.

The virus hitchhiked, inside birds, across North and Central America. By 2003, nearly 9,000 human cases of West Nile encephalitis or fevers had been reported; and more than 200 people had died. Cases popped up in all states except Maine, Washington, and Oregon, also in Canada and the Cayman Islands. Birds and horses in

Figure 19.1 A clue to viral history? Alexander the Great may have died of West Nile encephalitis. Ravens and crows are highly susceptible to this viral disease. Today these birds are being monitored to assess the spread of the disease through the Americas.

the big picture

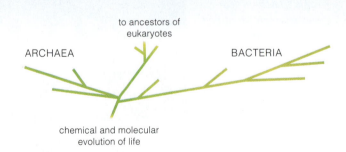

to ancestors of eukaryotes

ARCHAEA   BACTERIA

chemical and molecular evolution of life

Prokaryotic Cell Features Prokaryotic cells generally are the smallest and, structurally, the simplest organisms. They show great reproductive potential and, collectively, the greatest metabolic diversity.

Major Prokaryotic Groups The first cells were prokaryotic. Some gave rise to archaea, others to bacteria and to ancestral eukaryotic cells. One kind of prokaryote or another is found in almost all environments.

Mexico showed signs of infection. Most likely, the virus will spread throughout all of the Americas.

West Nile virus clearly is pathogenic. A **pathogen** is an infectious, disease-causing agent that can invade a host organism and multiply in or on it. Disease follows when metabolic activities of its descendants damage tissues and interfere with how the host body works.

As far as we know, you can't "catch" West Nile virus from a dog, cat, or classmate. Like other flaviviruses, this one usually travels in mosquitoes that can suck blood from an infected host, then deposit it into another host. In North America, at least forty-three different kinds of mosquitoes have become vectors for West Nile virus.

Infections may have widespread ecological impact. In North America, the virus has been detected in more than 150 bird and mammal species, as well as alligators. Many of these species harbor the virus without getting sick. They act as reservoirs, from which the virus may be transmitted to others that are harmed by infection. If an endangered species can be infected and sickened by West Nile virus, it may be pushed toward extinction.

This chapter can start you thinking about the unseen multitudes at the boundary between nonliving and living things. Viruses hover near the boundary, and the prokaryotes—structurally the simplest forms of life— are right inside it. From the human perspective, some kinds of prokaryotes and viruses are dangerous, others are beneficial. As you'll soon see, even the bad ones have relatives that impact positively on human affairs.

✓ How Would You Vote?

Eliminating mosquitoes is the best defense against West Nile virus. Many local agencies are spraying pesticides wherever mosquitoes are likely to breed. Some people fear ecological disruptions and bad effects on health, and say spraying will never get rid of all mosquitoes anyway. Would you support a spraying program in your community? See the Media Menu for details, then vote online.

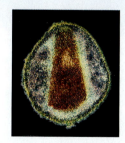

Viruses Viruses are infectious particles that require a living host to replicate themselves. Each consists of little more than DNA or RNA and a few viral enzymes inside a protein coat.

The Bad Bunch Pathogenic bacteria and viruses have coevolved with specific hosts. In evolutionary terms, like any other organism, the pathogen that leaves the most descendants wins.

19.1 Characteristics of Prokaryotic Cells

Of all organisms, prokaryotic cells are the smallest and most far-flung, abundant, and metabolically diverse. These cells have the longest evolutionary history. They arose about 3.8 billion years ago, and all living things are their descendants.

Table 19.1 Characteristics of Prokaryotic Cells
1. No membrane-bound nucleus.
2. Generally a single chromosome (a circular DNA molecule) with no associated proteins; many species also contain plasmids.
3. Cell wall present in most species.
4. Reproduction mainly by prokaryotic fission.
5. Collectively, great metabolic diversity among species.

BODY PLANS, SHAPES, AND SIZES

Prokaryotic cells arose before nucleated cells did. (*Pro–* means before; *karyon* is taken to mean nucleus.) A rare few have simple organelles but no nucleus (Table 19.1). Reactions typically proceed at the plasma membrane or in the cytoplasm; for example, at ribosomes.

Figure 19.2 shows a typical prokaryotic body plan and common shapes. The structural simplicity doesn't mean the cells are inferior to eukaryotic cells. These are tiny, fast reproducers, so they don't require a lot of complexity. A semirigid, permeable cell wall usually surrounds the plasma membrane and helps a cell hold its shape and resist rupturing as internal fluid pressure increases. A capsule or slimy layer of polysaccharides, polypeptides, or both often encloses the wall. It helps cells stick to surfaces and resist phagocytes.

Many species have one or more bacterial flagella, as in Figure 19.3a. These whip-shaped motile structures differ from eukaryotic flagella. They do not have a 9+2 array of microtubules inside, and they don't bend; they rotate like a propeller around the attachment site.

Pili (singular, pilus) often project above a cell wall. These protein filaments help cells adhere to surfaces, such as the vaginal lining. A sex pilus (Figure 19.3b) serves in **bacterial conjugation**. By this mechanism, a donor cell's sex pilus latches onto a recipient cell and then retracts, pulling it close. The shortened sex pilus serves as a conjugation bridge, the means by which the donor transfers a plasmid to the recipient. A **plasmid**, recall, is a small, self-replicating circle of DNA.

The tiniest cells barely show up in light microscopes. *E. coli* is more typical; it is 1–2 micrometers long and 0.5–1.0 micrometers wide. *Thiomargarita namibiensis*, the largest prokaryote known, is visible to the naked eye. At 750 micrometers, it's larger than some eukaryotic cells.

METABOLIC DIVERSITY

Collectively, prokaryotes show the most diversity in how they get food. *Photoautotrophs* make their own by a photosynthetic pathway that uses sunlight and CO_2 (as a carbon source). *Chemoautotrophs* are self-feeders, too. They use CO_2 and get energy by oxidizing organic compounds or inorganic ones, such as iron and sulfur. *Photoheterotrophs* are not self-feeders. They do tap the sun's energy, but they break down organic compounds for carbon. The parasitic *chemoheterotrophs* pirate energy and carbon from a living host; saprobic types get them from organic remains or wastes.

coccus

bacillus

spirillum

a

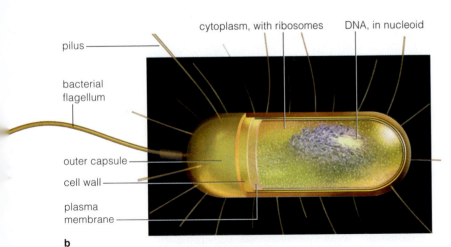

b

pilus

bacterial flagellum

cytoplasm, with ribosomes

DNA, in nucleoid

outer capsule

cell wall

plasma membrane

Figure 19.2 (**a**) The three most common shapes among prokaryotic cells: spheres, rods, and spirals. (**b**) Generalized body plan of a prokaryotic cell.

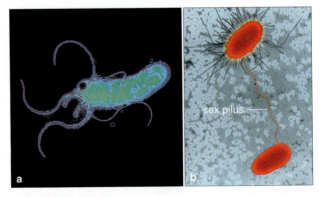

sex pilus

a   **b**

Figure 19.3 (**a**) *Helicobacter pylori*, a flagellated pathogen. It can cause peptic ulcers. (**b**) A sex pilus that has connected two *Escherichia coli* cells.

Read Me First!
and watch the narrated
animation on prokaryotic fission

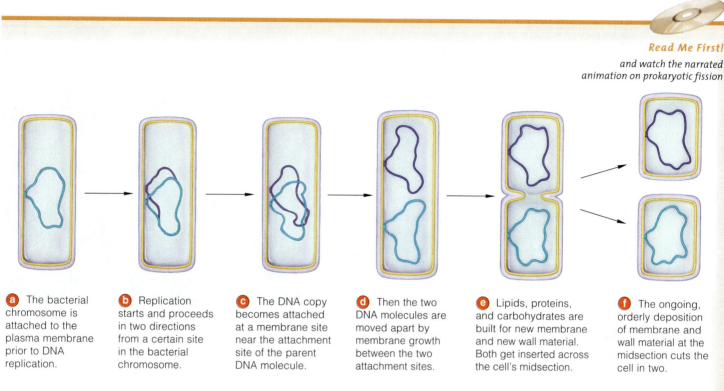

a The bacterial chromosome is attached to the plasma membrane prior to DNA replication.

b Replication starts and proceeds in two directions from a certain site in the bacterial chromosome.

c The DNA copy becomes attached at a membrane site near the attachment site of the parent DNA molecule.

d Then the two DNA molecules are moved apart by membrane growth between the two attachment sites.

e Lipids, proteins, and carbohydrates are built for new membrane and new wall material. Both get inserted across the cell's midsection.

f The ongoing, orderly deposition of membrane and wall material at the midsection cuts the cell in two.

Figure 19.4 Reproduction by way of prokaryotic fission. Only bacteria and archaea reproduce by this cell division mechanism.

GROWTH AND REPRODUCTION

We measure prokaryotic growth as an increase in the number of cells in a population. Each cell divides in two, the two become four, four become eight, and so forth. Before the cell divides, it nearly doubles in size. After division, each daughter cell has a single bacterial chromosome: a circular double-stranded molecule of DNA that has very few proteins attached to it.

In some species, a daughter cell merely buds from a parent cell. Most often, a cell reproduces by a division mechanism called **prokaryotic fission** (Figure 19.4). Briefly, a parent cell replicates its DNA molecule, and the replica docks at the plasma membrane next to the parent molecule. New membrane gets added between them and so moves the two apart. The membrane and cell wall extend into the cell's midsection and divide it into two genetically equivalent daughter cells.

The reproductive potential of prokaryotic cells is staggering. Some types divide every twenty minutes. If that rate held, a single cell would give rise to nearly a billion descendants in ten hours! So why don't the unseen multitudes take over the world? Their rapidly growing populations use up all available nutrients and pollute their habitats with metabolic wastes. They change the very conditions required for reproduction. Also, viral pathogens keep their populations in check.

CLASSIFICATION

Once, reconstructing the evolutionary history of the prokaryotes seemed an impossible task. Except for the stromatolite builders (Section 18.3), the most ancient groups are not well represented in the fossil record. It's likely that most early forms are not represented at all. Early classification systems had to be constructed by **numerical taxonomy**. This process compares traits of an unidentified cell with those of a known group, such as its shape, metabolic patterns, nutrition, and staining attributes of its wall (if any). The more traits the two have in common, the closer the inferred relatedness.

Today, automated gene sequencing and other tools of comparative biochemistry are helping to reveal the phylogenies. A recent discovery that most prokaryotic genomes contain hereditary material from more than one source does complicate the search. These mixed genomes arise by **lateral gene transfer**: movement of genetic information between cells, often of different species, by conjugation or some other mechanism.

The mostly microscopic prokaryotes are the most abundant and far-flung organisms on Earth. Enormous reproductive capacity and diverse modes of metabolism among these groups contributed to their evolutionary success.

19.2 The Bacteria

There may be millions of prokaryotic species. By far, most of the 400 named genera are bacteria. The few that cause human disease often get the spotlight, but most have vital ecological roles. A single gram of rich soil may contain thousands of species and billions of individuals. Figure 19.5 shows the main groups, which we can only sample here.

REPRESENTATIVE DIVERSITY

The **cyanobacteria** are photoautotrophs that help cycle carbon, oxygen, nitrogen, and other key nutrients. Like plants, they contain chlorophyll *a* and release oxygen during photosynthesis. Chloroplasts arose from early species, through endosymbiosis. Nearly all atmospheric oxygen can probably be traced to free-living cyanobacteria or to the chloroplasts descended from them.

We find most cyanobacteria in aquatic habitats and soil. After division, daughter cells often stay attached as mucus-sheathed chains, which can form dense, slimy mats (Figure 19.6*a*). *Spirulina* is a commercially grown "health food." A few species of cyanobacteria even are found in lichens, as symbionts with fungi.

Anabaena, an aquatic species, is one of the nitrogen fixers. When nitrogen is scarce, some cells in *Anabaena* chains develop into heterocysts, which convert nitrogen gas to ammonia (Figure 19.6*b*). The heterocysts share nitrogen compounds with other cells in the chain and get carbohydrates in return.

Gram-positive bacteria are so named because Gram staining tints their thick, multilayered walls purple. Most are chemoheterotrophs. The *Lactobacillus* species that form lactate during fermentation are used to make yogurt and other popular foods. *L. acidophilus* lowers the pH of skin and of intestinal and vaginal linings, which helps stop pathogens from becoming established. Antibiotics can cripple or kill *L. acidophilus* and other symbionts. Diarrhea or vaginitis may follow.

Clostridium and *Bacillus* are Gram-positive bacteria that survive hostile conditions when an **endospore**, a

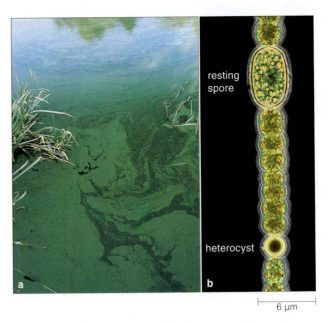

type of resting structure, forms inside the cell body. An endospore encloses the bacterial chromosome and a bit of cytoplasm (Figure 19.7). It resists heat, irradiation, drying out, acids, disinfectants, and boiling water. The dormant state lasts until favorable conditions return. Then the endospore germinates; a bacterium emerges from it, and normal function resumes.

Endospores that enter the human body can be lethal. In 2001, someone put *Bacillus anthracis* endospores in envelopes and mailed them. A few people opened the mail, breathed in endospores, and died of anthrax. The endospores of *Clostridium tetani* can pass across broken skin and cause tetanus, another disease (Section 32.7). Endospores of *C. botulinum* often taint canned food. Toxins produced by bacteria that emerge from these germinating endospores cause botulism, a dangerous form of food poisoning.

Figure 19.6 Cyanobacteria. (**a**) A huge population on the surface of a nutrient-rich pond. (**b**) Some cells become resting spores or heterocysts when conditions restrict growth.

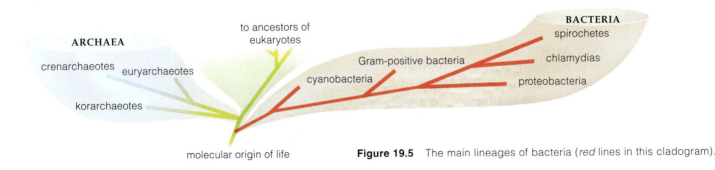

Figure 19.5 The main lineages of bacteria (*red* lines in this cladogram).

Figure 19.7 Endospore forming in *Clostridium tetani.*

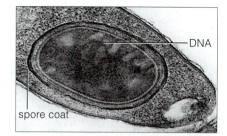

DNA

spore coat

Figure 19.8 *Thiomargarita namibiensis,* the largest bacterium. The cells can be seen with the naked eye and are mostly a giant nitrate-filled vacuole.

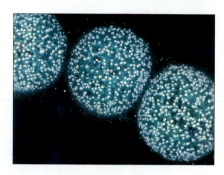

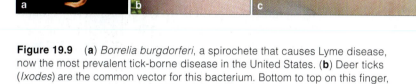

a b c

Figure 19.9 (**a**) *Borrelia burgdorferi,* a spirochete that causes Lyme disease, now the most prevalent tick-borne disease in the United States. (**b**) Deer ticks (*Ixodes*) are the common vector for this bacterium. Bottom to top on this finger, a female, male, and nymph. (**c**) Typical bull's-eye rash, the reaction to a tick bite.

Proteobacteria make up the largest, most diverse bacterial group. Gram staining tints their wall pink, so they also are known as Gram-negative bacteria. Many attack humans and other animals. The most common strain of *E. coli* lives inside the mammalian gut. This chemoheterotroph makes vitamin K, helps digest fat, and normally keeps foodborne pathogens in check by outcompeting them for nutrients.

There are photosynthetic proteobacteria, but they don't release oxygen and they don't use chlorophylls. Their bacteriochlorophylls are a similar pigment, with slightly different structure and absorption spectra.

Many proteobacteria are players in nutrient cycles. *T. namibiensis* helps connect sulfur and nitrogen cycles in the seas (Figure 19.8). It strips electrons from sulfur compounds as an energy source and also stores nitrate (in a big vacuole) for use as a final electron acceptor. *Rhizobium* causes nitrogen-fixing nodules to form on roots of peas and many other legumes. *Agrobacterium tumefaciens* causes crown gall tumors in plants, but it also is used in genetic engineering (Chapter 15).

Chlamydias are cocci that can't make ATP; they get it by slipping into animal cells, as parasites. *Chlamydia trachomatis* causes chlamydia, a sexually transmitted disease. Infection by this intracellular parasite also can result in trachoma, an eyelid disease that has blinded about 6 million people in developing countries.

Spirochetes look like stretched-out springs (Figure 19.9). All are motile. The group includes free-living species, symbionts, and parasites. *Pillotina* and other spirochetes live in the gut of termites and help them digest wood. The pathogenic *Borrelia burgdorferi* uses ticks as vectors and causes Lyme disease.

REGARDING THE "SIMPLE" BACTERIA

Bacteria are small. Their insides are not elaborate. *But bacteria are not simple.* They sense and move toward areas where nutrients are more plentiful and where other conditions also favor growth. Aerobes move to oxygen; anaerobes move away from it. Photosynthetic types move toward light but away from light that's too intense. Many species avoid toxins or predators.

Magnetotactic bacteria migrate in the ocean, and they grow best in oxygen-poor seawater. They contain a compass—chains of magnetic particles that respond to Earth's magnetic field. For instance, in the Northern Hemisphere, geomagnetic north points downward at a slight angle, so the migrating cells move into deeper water, which has less oxygen than surface waters.

Some bacteria show a collective behavior, as when millions of myxobacteria cells come together and form a "predatory" colony. These cells lack flagella but can glide about. They secrete enzymes that digest "prey," which include other bacterial cells. What's more, they migrate, change direction, and move as a single unit toward areas where there may be food. Finally, many myxobacteria colonies interact to form spore-bearing structures (fruiting bodies). Some cells differentiate to form a slime stalk. Others form branches or clusters of spores. Each spore holds a single living cell that can germinate and give rise to a new colony.

Bacteria are the most common and diverse prokaryotic cells. They include free-living cells with vital ecological roles, as well as pathogens of plants, animals, and humans.

19.3 The Archaea

Archaea are prokaryotic, but in other respects they are as similar to eukaryotes as they are to bacteria. Many survive in extreme habitats. Others are turning up almost everywhere. None are vertebrate pathogens.

THE THIRD DOMAIN

It's easy to see why archaea were once lumped with bacteria. Both have the same basic sizes and shapes. Both lack a nucleus. They differ a bit in metabolism and structure, but the differences just didn't seem to be significant enough to divide them into two groups.

Things changed in the 1970s. Molecular biologist Carl Woese began comparing the ribosomal RNAs of prokaryotes. Even though the rRNAs are necessary for protein synthesis, they can change a bit in certain base sequences without losing their function. Many small changes that have accumulated in rRNAs of different lineages can be measured directly.

To Woese's surprise, lineages that are adapted to high temperature or that produce methane gas formed a distinct group of organisms. They differed as much from bacteria as from eukaryotes. Woese proposed a new classification scheme in which Archaebacteria are equivalent to domains Bacteria and Eukaryota. Today we call the "third" domain Archaea. The Greek prefix *archae–* means "ancient." Biologists suspect that some archaea may resemble the first cells on Earth.

Through ongoing rRNA comparisons, we have now divided the archaea into three groups: Euryarchaeota, Crenarchaeota, and the newly discovered Korarchaeota (Figure 19.10). More data may reveal other groups.

HERE, THERE, EVERYWHERE

Physiologically, most of the archaea are **methanogens**, (methane makers), **extreme halophiles** (salt lovers), or **extreme thermophiles** (heat lovers). Methanogens and halophiles are euryarchaeotes, and most of the thermophiles are crenarchaeotes.

a

Figure 19.11 Methanogenic archaea. (**a**) *Methanocaldococcus jannaschii*, isolated near a deep-sea hydrothermal vent, grows at 85°C. (**b**) Three genera of archaea have been isolated from the cattle gut.

Methanogens have been isolated from marsh mud, Antarctic lakes, hydrothermal vents, and rocks deep below Earth's surface. Others are symbionts that live in the gut of termites or a few other animals (Figure 19.11). Some bacteria are also methanogens.

All archaean methanogens are strict anaerobes; free oxygen kills them. They strip electrons from hydrogen gas (H_2) or acetate in ATP-forming reactions that use carbon (from CO_2) as the final electron acceptor. In these reactions, methane (CH_4) forms as a by-product. Collectively, methanogens release about 2 billion tons of methane annually, so they have profound effects on the global cycling of carbon (Section 41.8).

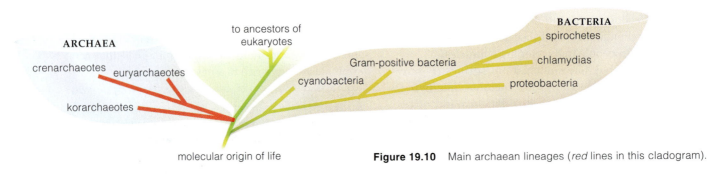

Figure 19.10   Main archaean lineages (*red* lines in this cladogram).

Figure 19.12 Life in extreme environments. (**a**) In saltwater evaporation ponds in Utah's Great Salt Lake, extreme halophiles (certain archaea and red algae) tint the water pink. (**b**) Extreme thermophiles live in Yellowstone National Park's hot springs. (**c**) The parasitic *Nanoarchaeum equitans* (smaller *blue* spheres) grows only when attached to *Ignicoccus* (larger spheres). Both archaea were isolated from 100°C water near a hydrothermal vent.

5 µm

Extreme halophiles live in the Dead Sea, Great Salt Lake, saltwater evaporation ponds, and other highly salty habitats (Figure 19.12*a*). Most get ATP by aerobic reactions but they also harness light energy when free oxygen is scarce. Their plasma membrane incorporates bacteriorhodopsin, a unique light-activated pigment. When this pigment absorbs sunlight energy, it changes shape in a way that pumps protons (H+) out from the cell. Then H+ flows back in, through an ATP synthase, and drives ATP formation.

Many extreme thermophiles live in sulfur-rich hot springs where water temperatures exceed 80°C (Figure 19.12*b*). Like methanogens, they are strict anaerobes. Unlike them, they use sulfur as an electron acceptor or donor in ATP-forming reactions. *Sulfolobus* cells grow in acidic hot springs. So does *Thermus aquaticus*, that special source of heat-stable DNA polymerases that are the basis of PCR. As you read earlier, this is the DNA amplification method of choice in biotechnology.

Other extreme thermophiles are the start of food webs in sediments around hydrothermal vents, where water temperatures exceed 110°C. They use hydrogen sulfide escaping from the vents as an electron source. Their presence at these vents adds to the evidence that life may have originated on the seafloor.

Nanoarchaeum equitans turned up during exploration of deep-sea hydrothermal vents off the coast of Iceland. This crenarchaeote is the smallest known cell and the only parasitic archaean (Figure 19.12*c*). Its genome is sequenced; to date, it is the smallest genome known.

Seawater collected near the coasts of Antarctica and California holds an abundance of archaea (primarily crenarchaeotes). So do mounds of methane hydrate on the ocean floor and sediments from the Great Lakes. And so do soils of agricultural fields, grasslands, boreal forests, and Siberian tundra. Biologists are now finding new species almost everywhere they look.

Today, Woese compares the discovery of archaea to the discovery of a new continent. He and many others are exploring and mapping this domain. By screening samples for archaean rRNA, they came to realize that members of this domain are diverse and widespread. They also realize that archaea and bacteria living in the same places may swap genes—and may do so often.

Archaea is the most recently defined domain of life. Many species thrive in extreme environments. Others are being discovered alongside bacteria in more hospitable habitats.

19.4 Viruses and Viroids

In ancient Rome, "virus" meant poison or venom. In the late 1800s, this rather nasty word was bestowed on newly discovered pathogens, smaller than the bacteria being studied by Louis Pasteur and others. Many viruses deserve the name. They attack animals, fungi, plants, protists, bacteria—you name it, there are viruses that infect it.

Today, we define a **virus** as a noncellular infectious particle that consists of DNA or RNA enclosed within a protein coat. The coat protects the genetic material as the virus journeys to a new host cell. Different viral coats consist of one or more types of protein subunits organized into a rodlike or many-sided shape (Figure 19.13). Complex viruses have sheaths, tail fibers, and other accessory structures attached to the coat. Some of the coat proteins on each type of virus chemically recognize and bind to specific receptors on host cells.

An outer envelope wraps around the coat of some viruses. It is composed mostly of membrane remnants from the previously infected cell, and it bristles with spikes of glycoproteins.

Cells of the vertebrate immune system can recognize viral proteins. However, genes for many of the proteins mutate often, so a modified virus might slip past the immune fighters. "Flu shots" prepare the human body to detect and battle some influenza viruses. Flu shots are tailored each year to the modified coat proteins of particular strains of influenza viruses.

Each kind of virus multiplies only in specific hosts and cannot be studied easily except in living cells. This is why researchers study bacteria and bacteriophages, the class of viruses that infect them. Unlike the cells of complex multicelled organisms, both can be cultured easily and quickly. Bacteriophages were used in early experiments to determine the function of DNA. They are still used as research tools in genetic engineering.

Animal viruses contain double- or single-stranded DNA or RNA. In size, they range from parvoviruses (18 nanometers) to poxviruses (350 nanometers). Many cause diseases, including the common cold, influenza, measles, and some cancers. HIV causes AIDS. Section 34.10 describes how it weakens the immune system.

What happens during viral infections? Once host cells are infected, different viruses multiply in different ways. Even so, nearly all multiplication cycles proceed through five basic steps, as outlined here:

1. *Attachment.* A virus recognizes and binds to specific molecular groups at the host cell surface. Each type of virus can only infect cells having receptors for it.

2. *Penetration.* The entire virus particle or just its genetic material enters the cytoplasm of the host cell.

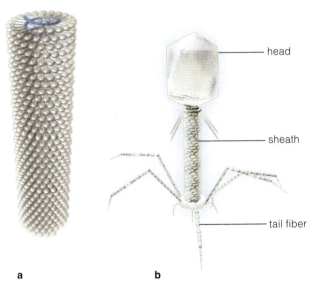

a   b   c   d

head

sheath

tail fiber

capsid envelope

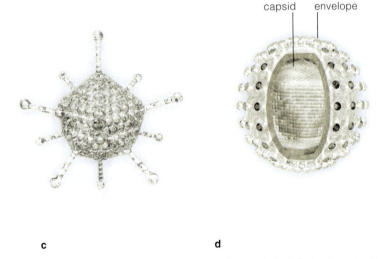

Figure 19.13 Body plans of viruses. (**a**) Protein subunits bound to nucleic acid form a spiral in rod-shaped viruses, including this tobacco mosaic virus. (**b**) Complex viruses, such as this bacteriophage, have structures attached to the coat. (**c**) Polyhedral (many-sided) viruses include this adenovirus. (**d**) A lipid envelope derived from membrane fragments of a lysed host cell surrounds HIV and other enveloped viruses. A capsid of regularly arrayed protein subunits encloses the viral RNA and a few enzymes. (**e**) Influenza virus. Spikes of glycoprotein project above its envelope. (**f**) One of the herpes viruses. Its envelope is pulled aside for this image.

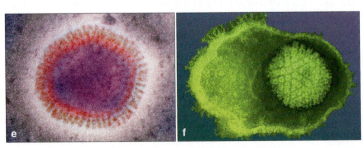

e   f

3. *Replication and protein synthesis.* Viral DNA or RNA hijacks the host cell's machinery, which then churns out copies of viral nucleic acids and proteins.

4. *Assembly.* The new viral nucleic acids and proteins become organized as new infectious particles.

5. *Release.* Virus particles are released from the host cell by one mechanism or another.

Figure 19.14 shows two pathways that are common among bacteriophages. Steps from attachment through assembly are fast in a *lytic* pathway. Viral particles are released when damage to a host's plasma membrane, wall, or both lets cytoplasm dribble out. The host cell assists in its own **lysis**, or disintegration. It actually builds the viral enzyme that causes the damage.

In a *lysogenic* pathway, a latent period extends the cycle. The virus doesn't kill a host cell outright. A viral enzyme cuts the host chromosome, then integrates the viral genes into it. Before the cell divides, it replicates the recombinant molecule. Miniature time bombs thus await all of the host's descendants. Later, a molecular signal or some other stimulus may reactivate the cycle.

Latency is typical of many viruses, such as Type I *Herpes simplex*, which causes cold sores (fever blisters). This virus lies hidden in facial tissues. Emotional stress, even sunburn, can reactivate it; painful skin eruptions are the outcome. Like other enveloped viruses, this one enters a new host cell by a version of endocytosis, and it escapes by budding from the cell surface.

In multiplication cycles of RNA viruses, the viral RNA serves as a template for making mRNA or DNA. One enzyme of HIV and other retroviruses catalyzes DNA formation by reverse transcription.

Stripped-down as they are, viroids are simpler still. **Viroids** are circles of RNA with no protein coat and no protein-coding genes. Many viroids are pathogens that kill millions of dollars' worth of crop plants annually. Only one type has been implicated in a human disease. This viroid exerts its effect only when an individual is infected with the hepatitis D virus at the same time.

Viroids actually resemble introns, those noncoding portions of eukaryotic DNA you read about in Section 13.1. Are self-splicing introns that once escaped from DNA molecules? Or are they molecular remnants of an ancient RNA world? We don't know yet.

Viruses consist of nucleic acids enclosed in a protein coat, and some types have an outer envelope. Viroids are snippets of RNA. Both kinds of infectious agents can multiply only by hijacking the metabolic machinery of their hosts.

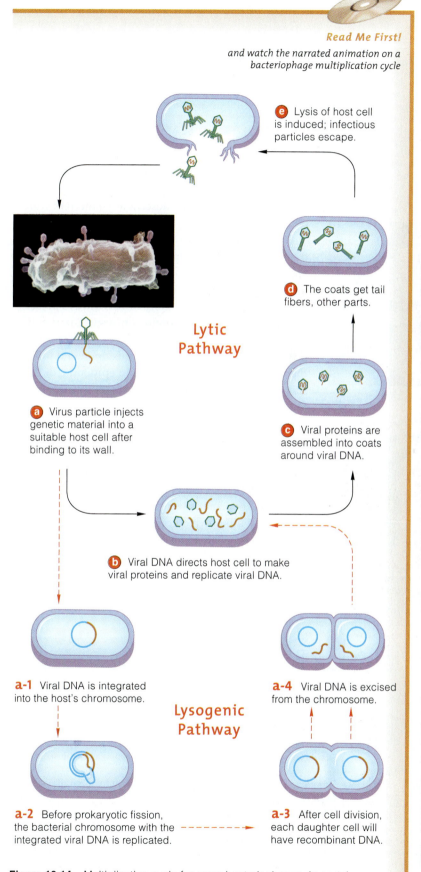

Read Me First!
and watch the narrated animation on a bacteriophage multiplication cycle

e Lysis of host cell is induced; infectious particles escape.

d The coats get tail fibers, other parts.

Lytic Pathway

a Virus particle injects genetic material into a suitable host cell after binding to its wall.

c Viral proteins are assembled into coats around viral DNA.

b Viral DNA directs host cell to make viral proteins and replicate viral DNA.

a-1 Viral DNA is integrated into the host's chromosome.

a-4 Viral DNA is excised from the chromosome.

Lysogenic Pathway

a-2 Before prokaryotic fission, the bacterial chromosome with the integrated viral DNA is replicated.

a-3 After cell division, each daughter cell will have recombinant DNA.

Figure 19.14 Multiplication cycle for some bacteriophages. In certain viruses, the lytic pathway may expand to include a lysogenic phase.

19.5 Evolution and Infectious Diseases

Just by being human, you are a potential host for diverse pathogenic viruses, protists, fungi, and parasitic worms. Before leaving the chapter, reflect on what this means from an evolutionary perspective.

THE NATURE OF DISEASE

Infection is the invasion of a cell or multicelled body by a pathogen (or the pathogen's genetic material). **Disease** follows when the pathogen multiplies and the metabolic activities of its descendants interfere with body activities. Table 19.2 gives some examples.

It takes direct contact with an infected person or their secretions for a contagious disease to spread. *Sporadic* diseases, such as whooping cough, occur irregularly among very few people. *Endemic* diseases, such as tuberculosis, are more or less always present in a population but are confined to a small part of it.

During an *epidemic*, a disease quickly spreads through part of a population for just a limited time, then it subsides. *Pandemic* refers to epidemics of the same disease breaking out in several countries in the same interval of time.

AIDS is a pandemic with no end in sight. A 2003 outbreak of SARS (short for *Severe Acute Respiratory Syndrome*) was a brief pandemic. It started in China, and travelers quickly brought it to countries around the world. Before government-ordered quarantines halted its spread, thousands were sickened and 774 died. A previously unknown coronavirus causes SARS. It has reappeared in China and may not subside entirely.

Now consider disease in terms of the pathogen's prospects for survival. A pathogen stays around only for as long as it has access to outside sources of energy and raw materials. To a tiny bacterial or viral pathogen, a human body is the jackpot. With bountiful resources, a pathogen may multiply to astounding population sizes. The point? Evolutionarily speaking, the ones that leave the most descendants win.

Two barriers prevent pathogens from taking over the world. First, any species with a history of being attacked by a specific pathogen has coevolved with it and has built-in defenses against it. Vertebrates, with their immune systems, are examples. Second, if a pathogen kills its individual host too quickly, it might vanish along with the individual. This is one reason most pathogens have less-than-fatal effects. After all, infected individuals who can live longer spread more germs and contribute to the pathogen's reproductive or replicative success.

Usually, the individual will die only if it becomes host to overwhelming numbers of a pathogen, if it is a novel host with no coevolved defenses, or if a mutant pathogenic strain has emerged and has breached the current defenses.

DRUG-RESISTANT STRAINS

Antibiotics are compounds synthesized by one organism that can kill another, and we have learned to use them as weapons against bacterial pathogens. Penicillin, our first widely used antibiotic, has been mass produced since 1943. Penicillin and other antibiotics have saved millions of lives. However, as explained in Chapter 15, these drugs act as agents of directional selection. As a result, resistant forms of many bacteria are on the rise.

Lateral gene transfers among bacterial cells of the same or different species contribute to the problem. Many bacterial pathogens swap genes like teens swap music files. The genes that confer antibiotic resistance move about in plasmids or as a free piece of DNA.

Antiviral drugs are also losing their punch. Here's an example. In one study, about 27 percent of newly diagnosed patients with AIDS had been infected by an HIV strain resistant to at least one antiretroviral drug. Multiple classes of drugs didn't work in more than 10 percent of these patients. Resistant HIV strains had evolved in other individuals, before the patients in the study had taken any of the drugs.

FOODBORNE DISEASES AND MAD COWS

According to Centers for Disease Control estimates, contaminated food sickens as many as 80 million Americans annually. Most vulnerable are the very

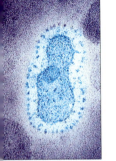

SARS VIRUS

MYCOBACTERIUM TUBERCULOSIS

Table 19.2	The Eight Deadliest Infectious Diseases		
Disease	Main Agents	Estimated New Cases per Year	Estimated Deaths per Year
Acute respiratory infection*	Bacteria, viruses	1 billion	4.7 million
Diarrhea**	Bacteria, viruses, protozoans	1.8 billion	3.1 million
Tuberculosis	Bacteria	9 million	3.1 million
Malaria	Sporozoans	110 million	2.7 million
AIDS	Virus (HIV)	5.6 million	2.6 million
Measles	Viruses	200 million	1 million
Hepatitis B	Virus	200 million	1 million
Tetanus	Bacteria	1 million	0.5 million

* Includes pneumonia, influenza, and whooping cough.
** Includes amoebic dysentery, cryptosporidiosis, and gastroenteritis.

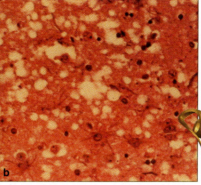

Figure 19.15 (**a**) Charlene Singh, the first reported case of an American resident affected by vCJD, being cared for by her mother. The disease has since claimed Charlene's life. (**b**) Section through a brain damaged by bovine spongiform encephalopathy (BSE). The light "holes" are areas where tissue has been destroyed. (**c**) Model of a normal prion protein. The vCJD-causing version misfolds into a different three-dimensional structure.

c

young or old, or those with a compromised immune system. *Campylobacter, Salmonella, Staphylococcus aureus,* and *Listeria* are frequent contaminants of dairy products, meat, and poultry. All can cause abdominal pain, nausea, and diarrhea. *Listeria* infection during pregnancy can cause miscarriage or premature birth.

0157:H7 is one of the pathogenic strains of *E. coli,* a normally harmless bacterial species that lives in your gut. Tons of ground beef contaminated with 0157:H7 have been recalled from vendors. Even vegetarians are at risk, because 0157:H7 has been found in fruit juices, lettuce, green onions, and alfalfa sprouts.

Some simple practices can help you avoid bacterial food poisoning. Thoroughly cook meats and poultry. Avoid unpasteurized dairy products, and don't eat raw eggs. Kitchen sponges can be sanitized by a trip through the dishwasher. So can utensils that have been in contact with uncooked meat or poultry.

Bacteria aren't the only foodborne threats. The first case of bovine spongiform encephalopathy (BSE) in the United States was reported in 2003. A single cow in Washington State developed this "mad cow disease." Meat from the herd was recalled and consumers were assured that none had entered the human food supply. What if it had? Before 2000, thousands of cattle with BSE entered food chains, mainly in Great Britain. So far, that meat has caused about 153 cases of a fatal brain disease, variant Creutzfeldt–Jakob disease (vCJD).

Charlene Singh was one of those vCJD cases (Figure 19.15). She probably contracted the disease when she was growing up in Britain, before moving to Florida. In 2001, the mood swings, balance problems, and

memory loss started. She soon became bedridden and could not control her body functions or communicate. Charlene died in the summer of 2004.

According to the most widely accepted hypothesis, prions cause BSE, vCJD, and some other degenerative diseases of humans and other animals, including deer and elk. **Prions** are infectious proteins. Normal prions, which are present in tissues of the nervous system, are folded into a characteristic configuration. Infectious prions are misfolded (Figure 19.15c). In some way, they induce normal prions to become misfolded, also.

Misfolded prions accumulate and form massive deposits in the brain. Brain tissue samples taken from individuals who died from one of these diseases are riddled with holes (Figure 19.15b). The holes give the tissue a spongiform (sponge-like) appearance.

The prions that cause BSE and vCJD are nearly identical. They may be mutated strains of prions that cause scrapie in sheep. Apparently some prions were transferred from sheep to cattle by way of livestock feed that contained animal parts. Governments have now banned the use of such feed.

Prions are not alive, so they can't be killed. It is not easy to destroy them through denaturation; they resist boiling, baking, irradiation, and most disinfectants.

Should you worry about eating American beef? The USDA says no. If you are a beef lover and a skeptic, avoid eating brains or meat cuts that contain spinal cord, such as oxtails. Ground beef and sausage may accidentally contain nervous tissue, so watch how it is being made or make your own. You may also want to look for meat from grass-fed cattle.

Summary

Section 19.1 Bacteria and archaea are prokaryotic cells; they do not have a nucleus. Some species have membrane infoldings and other structures in the cytoplasm. None has a profusion of organelles. As a group, prokaryotes show great diversity in modes of acquiring energy and carbon. Prokaryotic fission is a cell division mechanism used only by prokaryotic cells. Replication of the bacterial chromosome is followed by the division of a parent cell into two genetically equivalent daughter cells. Prokaryotes transfer DNA by exchanging plasmids (conjugation).

Section 19.2 Bacteria are the most common and diverse prokaryotes. The cyanobacteria are oxygen-releasing photoautotrophs. Gram-positive bacteria have thick cell walls. Most are chemoheterotrophs. Some are pathogens. Gram-negative proteobacteria are the most diverse bacterial group, with autotrophs and heterotrophs, free-living species, symbionts, and pathogens. The chlamydias are all intracellular parasites. The spring-shaped spirochetes include symbionts, free-living species, and pathogens.

Section 19.3 The archaea are prokaryotic, but in some respects they resemble eukaryotes as much as they do bacteria. They are now placed in a separate domain from the bacteria. Distribution and diversity of the group are highly active research areas.

Comparative rRNA studies give evidence of three archaean groups: euryarchaeotes, crenarchaeotes, and korarchaeotes. Most of the methanogens (methane makers) and halophiles (salt lovers) are in the first group, and most extreme thermophiles belong to the second group. Archaea and bacteria coexist in many habitats and apparently engage in gene transfers.

Section 19.4 Viruses and viroids are noncellular infectious agents. Each virus particle consists of a core of DNA or RNA and a protein coat that in some species is enclosed in a lipid envelope. All viral multiplication cycles go through five steps: attachment to a host; then penetration into it; replication and protein synthesis; assembly of new particles; then their release. Cycles differ in their details. An example is the lytic pathway of bacteriophages, which may include a latent phase.

Viroids are circles of infectious RNA with no protein or protein-coding genes. They resemble introns. Many are plant pathogens. Only one kind infects humans.

Section 19.5 Infection is the invasion of a cell or multicelled body by a pathogen, which causes disease when its activities interfere with body functions. Hosts and pathogens coevolve. Antibiotics and antiviral drugs select for drug-resistant strains. Many bacteria cause food poisoning. Prions are infectious proteins that kill by destroying the brain and nervous tissue.

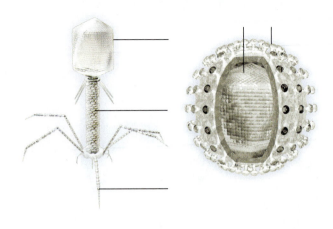

Self-Quiz   Answers in Appendix III

1. Label the components of the viruses shown above.

2. Only _____ are prokaryotic.
   a. archaea   c. viroids
   b. bacteria   d. a and b are correct

3. None of the _____ are human pathogens.
   a. bacteria   b. viruses   c. archaea   d. prions

4. Bacteria transfer plasmids by _____ .
   a. prokaryotic fission   c. conjugation
   b. endospore formation   d. the lytic pathway

5. The _____ are all oxygen-releasing photoautotrophs.
   a. spirochetes   c. cyanobacteria
   b. chlamydias   d. proteobacteria

6. The symbiotic *E. coli* in your gut are _____ .
   a. spirochetes   c. cyanobacteria
   b. chlamydias   d. proteobacteria

7. All _____ are intracellular parasites of vertebrates.
   a. spirochetes   c. cyanobacteria
   b. chlamydias   d. proteobacteria

8. Some Gram-positive bacteria (e.g., *Bacillus anthracis*) survive harsh conditions by forming a(n) _____ .
   a. pilus   c. endospore
   b. heterocyst   d. plasmid

9. DNA or RNA may be the genetic material of _____ .
   a. bacteria   b. a virus   c. archaea   d. a prion

10. Is this statement true or false: All viruses consist of genetic material, a protein coat, and an outer envelope.

11. Bacteriophages can multiply by _____ .
   a. prokaryotic fission   c. the lysogenic pathway
   b. the lytic pathway   d. both b and c

12. Only _____ reproduce by prokaryotic fission.
   a. viruses   c. bacteria
   b. archaea   d. b and c are correct

13. Match the terms with their most suitable description.
   ____ archaea   a. infectious protein
   ____ bacteria   b. nonliving infectious particle;
   ____ virus   nucleic acid core, protein coat
   ____ plasmid   c. salt lover
   ____ extreme   d. prokaryotes that most closely
        halophile   resemble eukaryotes
   ____ viroid   e. most common prokaryotic cells
   ____ prion   f. small circle of bacterial DNA
                      g. small circle of RNA

Figure 19.16 A radura, international symbol for irradiated food. It means that a product has been exposed to high-energy radiation to kill foodborne pathogens. Irradiation cannot make the food itself radioactive.

Figure 19.17 Chinese civet cats. In 2004, a waitress who worked in a restaurant that serves their meat was one of the first reported cases of SARS. Thousands of these animals were slaughtered to halt the spread of the virus.

Critical Thinking

1. Perhaps you think the world would be better off if there were no viruses. If so, consider this. Curtis Suttle at the University of British Columbia studies interactions among certain viruses, bacterial communities, and algal communities in ocean water. He selectively removed viruses from seawater and found that algae stopped growing. Further investigation revealed that they were dependent on nutrients released by dying bacteria that lysed after viral infection. Make a list of other ways in which viruses might play vital roles in ecosystems.

2. One way to prevent bacterial food poisoning is by *food irradiation*, exposing food to high-energy rays or beams that kill pathogenic bacteria. The process also slows spoilage and prolongs shelf life. Some think this is a safe way to protect consumers. Others think the treatment could produce harmful chemicals. They say irradiation does not kill endospores that can cause botulism. In their view, the best way to prevent food poisoning is to tighten and enforce food safety standards.

Irradiated meat, poultry, and fruits are now available in many supermarkets. By law, they must be marked with a special symbol (Figure 19.16). Would seeing this symbol on a package make you more or less likely to purchase the product? Explain your answer.

3. How did the 2003 SARS pandemic start? Probably with Chinese civet cats (Figure 19.17) and other wildlife coveted as food delicacies in China. These animals harbor a coronavirus. The viral DNA sequence is nearly identical with a sequence that causes SARS in humans.

Gene sequencing studies show that the animal virus has a unique stretch of twenty-nine added nucleotides. Because of the additions, researchers suspect that the SARS virus probably jumped from wildlife to humans, rather than in the opposite direction. Review Section 11.7, then explain their reasoning.

4. *Thermotoga maritima* was discovered in geothermically heated marine sediments and grows best at about 80°C. Analysis of rRNA puts it squarely in the bacterial group, but sequencing of the entire genome presents a far more complicated picture. About 50 percent of the *T. maritima* genes resemble bacterial genes, about 25 percent of its genes are unique, and about 25 percent resemble those of archaea, especially the thermophile *Pyrococcus horikoshii*. Suggest two possible explanations for the similarity between these members of different domains.

Media Menu

Student CD-ROM

Impacts, Issues Video
   West Nile Virus Takes Off
Big Picture Animation
   Prokaryotes, viruses, and disease
Read-Me-First Animation
   Prokaryotic fission
   A bacteriophage multiplication cycle
Other Animations and Interactions
   Prokaryotic body plan
   Body plans of viruses

InfoTrac

- Germs Make the Man: Your Body Is Teeming with Trillions of Infectious Microbes. That's a Very Good Thing. *Fortune*, January 2003.
- What the Bugs Can Do For You. *Time*, July 2002.
- Caught Off Guard: SARS Reveals Gaps in Global Disease Defense. *Scientific American*, June 2003.
- The Virus Hunter. *British Medical Journal*, October 2003.
- Flying Fever. *Audubon*, July 2000.

Web Sites

- MicrobeWorld: www.microbeworld.org
- Intimate Strangers: www.pbs.org/opb/intimatestrangers
- Bad Bug Book: vm.cfsan.fda.gov/~mow/intro.html
- Institute for Molecular Virology: virology.wisc.edu/IMV
- National Center for Infectious Diseases: www.cdc.gov/ncidod

How Would You Vote?

West Nile virus is carried by mosquitoes. Many localities are spraying pesticides to kill these insect vectors. The pesticides have environmental and possibly health side effects. It is unlikely that spraying will successfully eliminate all of the virus-carrying mosquitoes. Would you support a spraying program in your community?

IMPACTS, ISSUES   *Tiny Critters, Big Impacts*

Go for a swim just about anywhere and you will be sharing the water with multitudes of protists. Like their most ancient ancestors, almost all of these eukaryotic species are aquatic. Many kinds are photosynthesizers or predators in freshwater or the seas. Others live in cells or moist tissues of host species. Name a fungus, plant, or animal, and most likely a symbiotic, parasitic, or pathogenic protist associates with it.

You'll also find symbionts, parasites, and pathogens among the fungi, but far more kinds are free-living decomposers. Unlike protists, most fungi live on land.

Structurally, protists and fungi are the simplest of all eukaryotes. Most are microscopically small. And yet, collectively, some have big impacts on our world.

For instance, coccolithophores and foraminiferans drift about in the world ocean. These tiny cells have a shell or plates hardened with calcium carbonate. Long ago, calcium-rich remains of ancient species started to accumulate on the seafloor. Gradually, the deposits were compressed and then tectonic forces lifted them above sea level. We know the deposits as more familiar materials—chalk and limestone.

Dover, England, is renowned for its white chalk cliffs (Figure 20.1). The cliffs are a monument to individual coccolithophores that drifted down to the seafloor and formed 0.5-millimeter-thick deposits every year for millions of years. Austin, Texas, sits on 400-meter-thick chalk deposits, formed in another ancient sea.

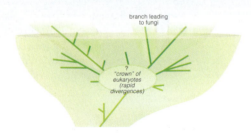

Figure 20.1 Two examples of protistan architecture. *Left*, towering 300 feet above the English Channel, the white chalk cliffs of Dover, England. *Right*, Hagar Qim temple, Malta. The filmstrip at *far right* shows a fossilized shell of one kind of foraminiferan, *Textularia*, that lived in Miocene times.

the big picture

branch leading to fungi

?
"crown" of eukaryotes (rapid divergences)

What Is a Protist?   Structurally, protists are the simplest eukaryotes and the ones most like the first eukaryotic cells on Earth. They were already diverging rapidly when fungi arose. Researchers are discovering how divergent lineages are related and are reclassifying major groups.

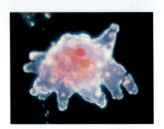

Single-Celled Types   A dazzling variety of flagellated, ciliated, and amoeboid cells arose from heterotrophic ancestors. Modern groups live in almost every aquatic or moist environment, including tissues of humans and other animals.

In 3200 BC on Malta, foraminiferan-rich limestone deposits were quarried to build the Hagar Qim temple, now a World Heritage site (Figure 20.1). A few thousand years later, other limestones were used to build the Egyptian pyramids and Sphinx. Look closely at the walls of the Empire State Building, the Pentagon, or Chicago's Tribune Tower, and you'll find foraminiferan shells.

Tiny cells also can make their presence known in a big way. One of the water molds, *Phytophthora infestans*, infects potatoes and causes the disease late blight. In the late 1840s, this pathogen destroyed potato crops that were the main food staple in Europe. Millions of people starved to death or emigrated to other countries.

Today, related species threaten forests in the United States, Europe, and Australia. *P. ramorum* started an epidemic of sudden oak death in California. Tens of thousands of oaks have died, and the pathogen has now jumped to madrone, redwoods, and other novel hosts. Cascading ecological changes will diminish sources of food and shelter for forest species.

Contrasting these two examples can help you keep the global perspective in mind. Some simple eukaryotes do cause disease or damage, and we tend to assign "value" to them in terms of their direct impact on our lives. There is nothing wrong with battling the harmful kinds and admiring the beneficial ones. Even so, don't lose sight of how protists or fungi or any other kind of organism fit in nature's larger picture.

 How Would You Vote?

The pathogen that causes sudden oak death has been found infecting twenty-six kinds of plants in California and Oregon. Some infected species are commonly sold as nursery stock. Should states that are free of the pathogen be allowed to prohibit the shipping of plants from states that are affected? See the Media Menu for details, then vote online.

Mostly Multicelled Types Stramenopiles and groups informally called algae also are structurally simple eukaryotes. They range from single-celled species to giant kelps and include the closest relatives of plants. Chloroplasts hint at endosymbiotic origins for some photosynthetic species.

The Fabulous Fungi Nearly all fungi are multicelled heterotrophs. Most species are decomposers that help cycle nutrients in communities. Others are symbionts, pathogens, and parasites. We recognize three major lineages, mainly by the type of sexual spores they produce.

20.1 Characteristics of Protists

Traditionally, kingdom Protista includes photoautotrophs, decomposers, parasites, and predators. Most are single-celled, but nearly every lineage also has multicelled forms.

Of all existing species, **protists** are the most like the first eukaryotic cells. Unlike prokaryotes, protist cells have a nucleus. Most also have mitochondria, ER, and Golgi bodies. Their ribosomes are larger than those in bacteria and they have more than one chromosome, each consisting of DNA with many proteins attached. They have a cytoskeleton that includes microtubules. Many cells have chloroplasts. And unlike prokaryotes, protist cells divide by way of mitosis, meiosis, or both.

Of course, these also are defining features of plant, fungal, and animal cells. Until recently, protists were defined largely by what they are *not*—not bacteria, not plants, not fungi, not animals. They were lumped into a kingdom meant to signify an evolutionary crossroad between prokaryotes and "higher" forms of life.

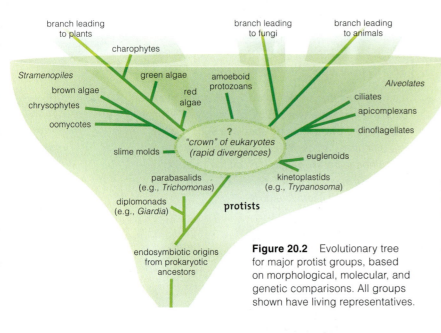

Figure 20.2 Evolutionary tree for major protist groups, based on morphological, molecular, and genetic comparisons. All groups shown have living representatives.

Phylogenetic threads are still being untangled with the help of molecular comparisons. So consider Figure 20.2, an evolutionary tree for these organisms, as a work in progress. "Protists" are being sorted out into a number of monophyletic groups. Here we can only sample their astounding diversity.

Protists are organisms thought to be most like the earliest eukaryotes. Most are single cells, but nearly all lineages include multicelled forms. Different lineages appear to have evolutionary connections with plants, fungi, and animals.

20.2 The Most Ancient Groups

Eukaryotic species arose hundreds of millions of years ago. Studying the single-celled forms that are their descendants can help give us an idea of how cellular complexity evolved.

FLAGELLATED PROTOZOANS

Flagellated protozoans are heterotrophs having one or more flagella and a **pellicle**. This thin, translucent, protein-rich outer covering imparts shape to the cell. Soil, seawater, and freshwater often hold free-living, bacteria-eating species. The termite gut has mutualist species that make cellulose-degrading enzymes. The human body can be host to parasitic types.

Giardia lamblia intrigues evolutionary biologists. It may be a representative of one of the first eukaryotic lineages because it has no mitochondria or lysosomes. *G. lamblia* parasitizes humans, cattle, and wild animals (Figure 20.3a). It can survive outside the host body, as cysts, in feces-contaminated water. Generally, a **cyst** is a protective covering formed by cell secretions. Ingesting *G. lamblia* cysts invites *giardiasis*. The symptoms of this disease can range from mild cramps to severe diarrhea that lasts for weeks.

Another group, the trichomonads, also lack mitochondria. Worldwide, *Trichomonas vaginalis* infects about 170 million people. The pathogen is transmitted mainly during sexual intercourse. It can damage the urinary and reproductive tracts of hosts unless they receive prompt treatment.

Trypanosomes form another group of flagellated protozoans. They have one large mitochondrion and a kinetoplast, a unique structure that houses the DNA. Biting insects spread species that can infect humans. *Trypanosoma brucei* causes African sleeping sickness (Figure 20.3b). *T. cruzi* causes Chagas disease in South and Central America. Both diseases are fatal. During Iraq conflicts, sand fleas that are vectors for *Leishmania donovani* bit some American soldiers and caused cases of skin leishmaniasis. The soldiers are being treated with surgery and experimental drugs.

EUGLENOIDS

Euglenoids live in freshwater ponds and lakes. Like many other freshwater protists, they have a contractile vacuole that expels excess water in the cell body. The euglenoids are related to trypanosomes, and all have more than one mitochondrion. They also have a cluster of pigments, called an eyespot, that partially shields a light-sensitive receptor. When activated, the receptor helps direct the cell toward locations that are suitable for photosynthesis. Figure 20.4 shows the body plan.

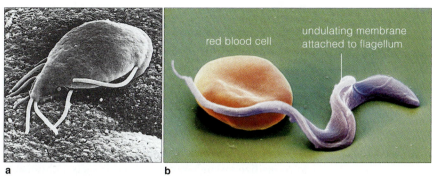

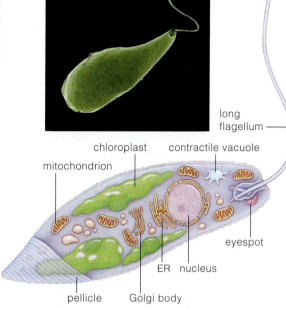

a   b

Figure 20.3 Two flagellated protozoans. (**a**) Scanning electron micrograph of *Giardia lamblia*. The cell adheres to intestinal epithelium by means of its sucking disk. (**b**) Light micrograph of *Trypanosoma brucei*, the cause of African sleeping sickness. The tsetse fly is its insect vector.

The earliest euglenoids were colorless heterotrophs that ingested dissolved organic matter. Some still are, but about half are photoautotrophs. Their chloroplasts probably arose by endosymbiosis after some ancient euglenoid engulfed green algal cells. The chloroplasts, like those of green algae, hold chlorophylls *a* and *b*.

Figure 20.4 Body plan and light micrograph of *Euglena*.

AMOEBOID PROTOZOANS

The first **amoeboid protozoans** had permanent motile structures, but these structures were lost as the group evolved. Pseudopods, or "false feet," became their means of motility. Naked amoebas, radiolarians, and foraminiferans are in this group.

Each pseudopod, again, is a lobe of cytoplasm that forms and retracts as microtubules assemble and then disassemble (Sections 4.10 and 5.8). These cytoskeletal elements are especially evident in naked amoebas and foraminiferans. Naked amoebas have soft bodies and no symmetry (Figure 20.5). Most types are free-living phagocytes in freshwater or in the soil. One species, *Entamoeba histolytica*, can live in the human intestines, where it may cause amoebic dysentery.

Perforated shells protect living foraminiferans and radiolarians. Prey-capturing pseudopods project from the fine perforations, as in Figure 20.6*b*. Nearly all of the radiolarians and some foraminiferans drift with ocean currents, as components of plankton. **Plankton** refers to aquatic communities of mostly microscopic organisms that drift or swim weakly in water. More species of foraminiferans live on the ocean floor, and a few have been discovered in lake sediments.

Among the living representatives of the most ancient protists are a variety of heterotrophic cells and one lineage that apparently acquired its chloroplasts by endosymbiosis.

Figure 20.5 *Amoeba proteus*, a naked amoeboid protozoan that is a favorite for experiments in biology classes. Compared to other species, it has stubbier pseudopods.

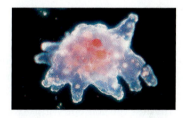

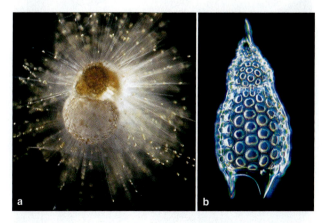

a   b

Figure 20.6 Amoeboid protozoans. (**a**) A live foraminiferan. The many tiny yellow spheres on its thin pseudopods are golden algal symbionts. This protist is about as small as the eye of a needle. (**b**) Radiolarian shell.

20.3 The Alveolates

Ciliates, apicomplexans, and dinoflagellates are a monophyletic group—the alveolates. All of these single cells have tiny, membrane-bound sacs just under their outermost membranes. The sacs, called alveoli (singular, alveolus), may help stabilize the cell surface.

CILIATED PROTOZOANS

Ciliated protozoans, or **ciliates**, live in freshwater, saltwater, and soil. Most are free-living predators, but some attach to substrates or form colonies. About 30 percent are endosymbionts of diverse animals. The majority have many cilia that beat in synchrony and function in motility, in directing food toward an oral cavity, or both. Ciliates eat bacteria, algae, and one another. The chloroplasts of algal prey are not always digested, and the cell that ingested them often uses part of the compounds that the chloroplasts continue to make. Some ciliates even have bacterial symbionts.

Figure 20.7 shows *Paramecium*, a freshwater species. A mature cell is about 200 micrometers long. Outer membranes form a pellicle. An oral cavity that opens on the surface leads to a gullet. Ingested food gets enclosed in enzyme-filled digestive vesicles. Like a number of other protists, the ciliates have contractile vacuoles that get rid of excess water in the cell body.

Most ciliates can reproduce sexually and asexually. Each typically has two types of nuclei and distributes them to daughter cells in a complex way. Some engage in binary fission, an asexual reproductive mode that divides the cell body into two parts.

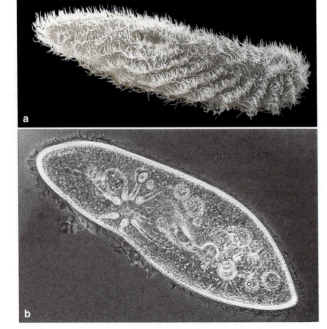

Figure 20.7 (**a**) Surface of *Paramecium*, a single-celled predator. You can see the start of its oral cavity, the depression in the cell surface. (**b**) A look at *Paramecium* organelles.

APICOMPLEXANS

Apicomplexans are parasitic alveolates equipped with a unique microtubular device that can attach to and penetrate a host cell. The tip of the device protrudes from the cell's anterior end. Plastids much like the ones in red algae occur in many species. They may be the legacy of past endosymbiotic shenanigans.

More than 4,000 apicomplexan species are known to parasitize animals ranging from insects to humans. Here we will focus on four species of *Plasmodium* that cause the disease *malaria*. These species have infected more than 300 million people in tropical parts of the world, especially sub-Saharan Africa.

The bite of an infected female *Anopheles* mosquito puts the parasite in a host. Sporozoites, a motile stage, travel the bloodstream to the liver. They reproduce asexually in liver cells until the cells burst and release another stage—the merozoites. Some of the merozoites reinfect the liver. Others enter red blood cells, where they develop into types of immature gametes known as male and female gametocytes (Figure 20.8).

Amazingly, *Plasmodium* is sensitive to temperature and oxygen levels of its hosts. The gametocytes cannot mature in humans, because humans are warm-bodied and there is little free oxygen in human blood; most is bound to hemoglobin. Gametocytes must mature in a female mosquito, which has a lower body temperature and takes in oxygen from air along with blood. In this insect's gut, zygotes form, divide, and give rise to sporozoites. This infective stage then migrates to the mosquito's salivary glands, there to await a new bite.

Malaria symptoms start after the infected liver cells rupture and release merozoites, metabolic wastes, and cellular debris into the bloodstream. Shaking, chills, a burning fever, and drenching sweats follow. After one episode, symptoms usually subside for a few weeks or months. Infected individuals might feel healthy, but they should expect relapses. Jaundice, kidney failure, convulsions, and coma are outcomes of the disease.

Many *Plasmodium* strains are now resistant to older antimalarial drugs. Artemisinin, a compound isolated from sweet wormwood (*Artemisia*), shows promise. It lowers fever and reduces the number of parasites in the blood, so it can slow the course of infection.

Efforts to design a vaccine are ongoing. At this writing, a new vaccine based on a merozoite surface protein is being tested in African adults. If the clinical trials go smoothly and the vaccine proves safe and effective, widespread clinical trials will probably start in 2006 or 2007. The sooner the better—every thirty seconds, one African child dies of malaria.

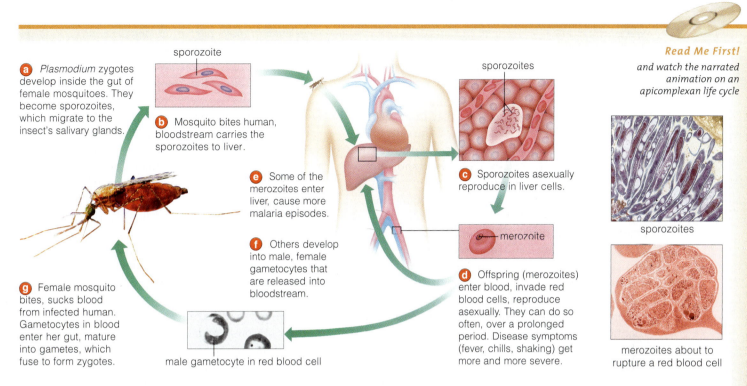

a *Plasmodium* zygotes develop inside the gut of female mosquitoes. They become sporozoites, which migrate to the insect's salivary glands.

sporozoite

b Mosquito bites human, bloodstream carries the sporozoites to liver.

e Some of the merozoites enter liver, cause more malaria episodes.

f Others develop into male, female gametocytes that are released into bloodstream.

g Female mosquito bites, sucks blood from infected human. Gametocytes in blood enter her gut, mature into gametes, which fuse to form zygotes.

male gametocyte in red blood cell

sporozoites

c Sporozoites asexually reproduce in liver cells.

merozoite

d Offspring (merozoites) enter blood, invade red blood cells, reproduce asexually. They can do so often, over a prolonged period. Disease symptoms (fever, chills, shaking) get more and more severe.

Read Me First! and watch the narrated animation on an apicomplexan life cycle

sporozoites

merozoites about to rupture a red blood cell

Figure 20.8 Life cycle of one of the *Plasmodium* species that causes malaria.

DINOFLAGELLATES

Dinoflagellates are single-celled alveolates that most often have cellulose plates just beneath the plasma membrane plus one flagellum at the posterior end and another in a groove that runs around the cell body. About half of the species in this diverse group are key producers in freshwater and marine habitats. Some prey on bacteria and algae, others parasitize marine or freshwater fish and invertebrates.

Remember the theory of endosymbiosis, explained in Section 18.4? Dinoflagellate chloroplasts appear to be descended from red or green algae that evolved as symbionts with cells that ingested them. In addition, many dinoflagellates are endosymbionts of corals.

At high densities, bioluminescent dinoflagellates in tropical seas can have dramatic effects. Disturb them and the water shimmers (Figure 20.9). The light might be like a burglar alarm that attracts help, such as larger predators that can eat the disturber. Less pretty are the fish kills that result from high densities of the toxic species. We turn to this topic in the next section.

Alveolates are diverse single-celled eukaryotes that swim or drift in lakes, seas, and the bloodstream of animals. Ciliates, apicomplexans, and dinoflagellates are in this group.

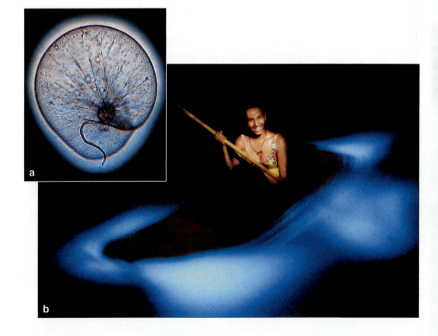

Figure 20.9 (**a**) *Noctiluca scintillans*, a bioluminescent dinoflagellate tinted red by symbionts living inside it. (**b**) Near Vieques Island, Puerto Rico, a liter of water holds about 5,000 of these cells, which flash with blue light when agitated. Like fireflies and other bioluminescent species, *Noctiluca scintillans* makes luciferase, an enzyme that destabilizes a luciferin molecule by catalyzing the transfer of a phosphate group to it. When the luciferin restabilizes, it releases energy as fluorescent light.

20.4 Algal Blooms

Algal blooms are huge increases in population sizes of aquatic protists, most often dinoflagellates. They tend to develop in warm, shallow water. Nutrient-rich pollutants, as in agricultural run-off, fan the population growth.

During a typical algal bloom, there may be as many as 8 million cells in each liter of water. The cells of some species tint the water red, especially near coasts; hence the name *red tide*. Vast blooms even show up in images from space satellites.

The blooms can sicken or kill aquatic organisms by polluting the water with metabolic wastes and sometimes potent toxins. The enormous numbers of algal cells inevitably die, and aerobic bacteria go to work on the remains. Their metabolic action depletes oxygen from the water, so aquatic animals suffocate.

As an example, populations of *Karenia brevis* often bloom near the coasts of the Gulf of Mexico and the Atlantic seaboard (Figure 20.10). This dinoflagellate makes brevetoxin, which binds to sodium pumps in cell membranes and disrupts nerve cell function. *K. brevis* blooms have killed billions of fishes. They have also sickened and killed seabirds, bottlenose dolphins, sea turtles, and manatees.

What about humans? Eat shellfish contaminated by brevetoxins and you may end up with a case of *neurotoxic shellfish poisoning* (NSP). The toxin rarely is dangerous. It does cause gut irritation, tingling, dizziness, and muscle cramps that can last for days. Even swimming in toxin-tainted water may irritate your skin and eyes, and walking on a beach near a *K. brevis* bloom can irritate airways to your lungs.

Figure 20.10 (**a**) *Karenia brevis*, a toxin-producing dinoflagellate. Its two flagella beat in such a way that they spin the cell through water. (**b**) In the autumn of 2003, blooms of *K. brevis* and related species near Naples, Florida, killed hundreds of fishes. Dead fishes washed up on beaches, including this one.

20.5 The Stramenopiles

Flagellated stramenopiles are among the most ancient eukaryotic lineages. Even multicelled types have flagellated spores. Unique filaments project like tinsel from one of two flagella.

Oomycotes and the mostly photosynthetic species called chrysophytes and brown algae belong to the stramenopile lineage. All oomycotes are colorless, but the other groups make chlorophylls *a* and *c* and fucoxanthin, a distinctive brown pigment.

OOMYCOTES

The water molds, downy mildews, and white rusts are common **oomycotes**. Their name means egg fungus; it refers to an egg cell that forms by way of meiosis in a parent cell. Unlike others in the group, oomycotes have no chloroplasts. Like fungi, these heterotrophs absorb nutrients through filaments that form in the life cycle. Many are decomposers of aquatic habitats. *Saprolegnia*, an opportunistic parasite, is quick to infect freshwater fishes with damaged tissues. Fuzzy white patches on fish skin are a sign of infection. The oomycotes called downy mildews are pathogens. A notorious species, *Plasmopara viticola*, infects grapevines and fruits.

The *Phytophthora* species introduced at the start of the chapter live up to their name—plant destroyer. All are plant pathogens. *P. infestans* can destroy a field of potatoes in a single day. *P. ramorum* is killing many of California's native trees and woody shrubs. Dark oozing from cankers on trunks is a telltale symptom of infection (Figure 20.11).

Figure 20.11 Dead trees on a hill near Big Sur, California, are part of an ongoing epidemic of sudden oak death. Despite the disease's name, the oomycote causing it is also killing many other species, including Douglas fir and redwoods. Dieback from the crown and oozing from bark are symptoms.

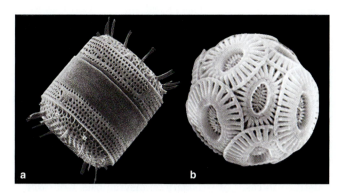

Figure 20.12 Shell of a diatom (**a**) and coccolithophore (**b**).

Figure 20.13 *Right*: Representative brown algae. (**a**) *Postelsia*, a brown alga along coasts from central California to Vancouver Island. Its blades top a resilient stipe fastened to rocks. (**b**) Photograph and (**c**) body plan of a giant kelp, *Macrocystis*. Giant kelps form dense underwater forests.

CHRYSOPHYTES

Golden algae, yellow-green algae, diatoms, and the coccolithophores are grouped as **chrysophytes**. Most are free-living photosynthetic cells of aquatic habitats.

A silica shell encasing diatoms has two parts that overlap like a hatbox (Figure 20.12*a*). For 100 million years, finely crushed diatom shells accumulated at the bottom of lakes and seas. We now use the sediments for filters, insulation, and abrasives. Living diatoms are key producers of many oceanic food chains. Land plants and diatoms release about the same amount of free oxygen into the atmosphere every year.

Coccolithophores secrete calcium carbonate plates beneath their plasma membrane (Figure 20.12*b*). They produce about 1.5 million tons of calcium carbonate per year. As described in the chapter opening, such plates accumulated over time, and formed chalk and limestone deposits. Most species of coccolithophores drift through calm, nutrient-poor seas. In nutrient-rich water, diatoms usually are the dominant competitors.

BROWN ALGAE

Most species of **brown algae** live in temperate or cool seas, from intertidal zones into the open ocean. Their accessory pigments tint them olive-green, golden, or dark brown (Figure 20.13). Brown algae range from microscopic filaments to plantlike forms thirty meters tall. All are multicelled photoautotrophs that reproduce sexually and asexually. Large sporophytes that make walled spores alternate with gametophytes (gamete-producing bodies) during the life cycle.

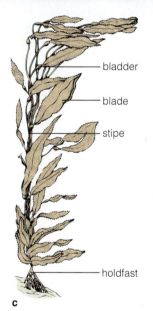

Macrocystis, *Laminaria*, and other giant kelps are the largest protists. The biggest seaweeds are sporophytes that have stipes (stemlike parts), blades (leaflike parts), and holdfasts (structures that anchor the alga). Hollow, gas-filled bladders along the stipes impart buoyancy and keep blades upright, near the sunlit ocean surface. Spores form in specialized blades near the base.

Maybe you've heard of the Sargasso Sea. This sea without shores, in the middle of the western North Atlantic, is named after vast, free-floating mats, as much as nine meters thick of *Sargassum*. Giant rafts of this kelp also drift off the southeastern United States. They are the crucial breeding grounds for American eels, nurseries for loggerhead turtles, and habitats for diverse fishes and invertebrates.

Sargassum, *Macrocystis*, and a few other species have commercial uses. Algal extracts become ingredients in ice creams, puddings, jelly beans, salad dressings, toothpaste, cosmetics, and other products. Alginic acid from the cell walls of some species is used to make algins, which act as thickeners, emulsifiers, and suspension agents in many commercial products.

Oomycotes (heterotrophs) and chrysophytes and brown algae (photoautotrophs) are stramenopiles. A tinsel-type flagellum is the shared trait that unites these diverse groups. The brown algae include the kelps, the largest of the protists.

20.6 Red Algae

Red algae offer us a classic story of endosymbiosis. Their chloroplasts contain reddish pigments and are probably derived from cyanobacteria. Red alga-like plastids in other protists point to an episode of endosymbiosis during their evolution.

Nearly all of the 6,000 known species of **red algae** live in warm marine currents and clear tropical seas. Of all algae, they survive at the greatest depths. A few encrusting types help build coral reefs. Along with chlorophyll *a*, red alga chloroplasts hold phycobilins. These red accessory pigments absorb green and blue-green wavelengths, which penetrate deep water. The more phycobilins an alga has, the redder it appears. Shallow-water species tend to have little phycobilin and look green. Deep dwellers are almost black.

By the theory of endosymbiosis, chloroplasts of red algae may have evolved from ancient cyanobacteria, which were engulfed by predatory cells but resisted digestion. The cyanobacteria became an endosymbiont, then evolved into chloroplasts (Section 18.4).

Ancient, single-celled red algae may also have been involved in cases of *secondary* endosymbiosis. Consider that plastids in apicomplexans and chloroplasts in some dinoflagellates have more than two external membranes. Did these protists get their chloroplasts "secondhand," after they engulfed photosynthetic red algal cells that became endosymbionts? Probably.

Some single-celled red algae persist, but most are multicelled. They have a branching or sheetlike growth pattern but no true tissues (Figures 20.14 and 20.15). Life cycles are diverse, with asexual phases as well as sexual phases that do not involve flagellated gametes.

Agar is an inert gel extracted from cell walls of a few species. It keeps baked goods and cosmetics moist, firms up jellies, and is used as a culture medium and as soft capsules for drugs. Derivatives stabilize paints. Carrageenan thickens sauces and dairy foods. *Porphyra* species are cultivated worldwide, more than 130,000 tons are harvested annually (Figure 20.15).

Most red algae are multicellular and marine. Of all algae, they survive at the greatest depths. The phycobilins that allow this are the legacy of a cyanobacterial endosymbiont.

Read Me First!
and watch the narrated animation on a red alga life cycle

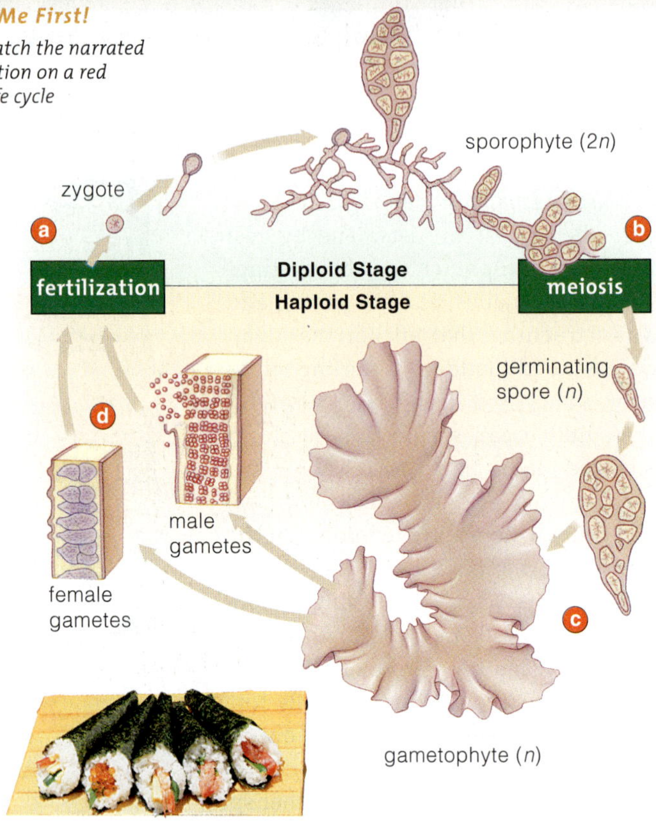

Figure 20.14 Life cycle of a red alga (*Porphyra*). For centuries, Japanese fishermen cultivated and harvested a red alga in early fall. The rest of the year, it seemed to vanish. Kathleen Drew-Baker studied its sheetlike form in the laboratory. She saw gametes forming in packets near the sheet margins. She also studied gametes in a petri dish. After zygotes formed, individuals developed into tiny, branching filaments on bits of shell in the dish. That was how the alga spent most of the year!

People had known about the pinkish growths on shells, but no one figured out the growths were algal sporophytes. *Porphyra* species could be cultivated on shells or other calcium-rich surfaces! Within a few years, researchers worked out the life cycle of *P. tenera*, a species used for nori. (It is pressed into thin sheets and used for seasoning or as a sushi wrapper.) By 1960, cultivation was a billion-dollar industry.

Figure 20.15 A red alga, *Antithamnion plumula*. The filamentous, branching growth pattern is the most common among red algae.

20.7 Green Algae

Green algae resemble plants so closely that many botanists now put them inside the plant kingdom. Evolutionarily, some green algae are certainly closer to plants than they are to other green algae. Single-celled and multicelled forms abound, mainly in freshwater habitats.

All species commonly called "green algae" are similar to plants in several traits. Their chloroplasts contain chlorophylls *a* and *b*, they store sugars as starch, and their cell walls incorporate cellulose. Most species are now classified as **chlorophytes**. A smaller group of green algae, **charophytes**, includes the closest living relatives of plants. Genetic analysis reveals that some charophytes have more in common with plants than they do with other green algae.

Figure 20.16*a–c* is a sampling of chlorophytes, which have sheetlike, filamentous, and colonial forms. Some can't be seen without a microscope. Chlorophytes live mostly in freshwater, although certain species live in the seas, in soil, and on rocks, snow, bark, and other organisms. A few kinds are even symbionts with fungi (in lichens) and with some invertebrates.

Melvin Calvin employed a single-celled green alga (*Chlorella*) to track the steps in the light-independent reactions of photosynthesis. *Volvox* is a colonial form, a whirling sphere of 500 to 60,000 flagellated cells that's common in freshwater ponds. White, powdery beaches in the tropics formed in part from remnants of *Udotea* cells. The tiny cells built calcified walls, died, and then disintegrated. *Ulva* is distributed worldwide, and it is one of the most conspicuous seaweeds. Blades of some species are as much as 65 centimeters long, but they are seldom more than 40 microns thick. Many people in Scotland and Japan consider these wispy sheets of "sea lettuce" as highly nutritious food.

Charophytes include microscopic cells such as the desmids, multicelled stoneworts, and filamentous forms (Figure 20.16*d,e*). Desmids are diverse and widespread. They live mainly in freshwater, but some live in the ocean or on its surface; others live on snow or ice.

Some green algae might help astronauts on long missions in outer space. They could provide the crew with oxygen (released in photosynthesis) and dispose of carbon dioxide wastes. Many single-celled forms are easily cultured. They reproduce asexually and rapidly reach high densities. The next section offers a close look at a representative life cycle.

Green algae are photoautotrophs now classified as either chlorophytes or charophytes, the closest relatives of plants. Their forms and life cycles are diverse.

Figure 20.16 Chlorophytes: (**a**) Sea lettuce (*Ulva*) grows in estuaries and attaches to kelps in the seas. Reproductive cells form within the margins of a sheetlike body and are released from it. (**b**) *Udotea cyathiformis*, one of more than 100 known *Udotea* species in tropical and subtropical waters. Many are calcified. (**c**) *Volvox*, a colony of interdependent cells that resemble the free-living, flagellated *Chlamydomonas* cells described in the next section. This colony is rupturing. Each released daughter cell may found a new colony.

Charophytes: (**d**) *Chara*, a stonewort known as muskweed, after its skunky odor. (**e**) *Micrasterias*, a type of desmid. Two new cells are forming from two parts of the now-smaller parent cell (the bow-tie shape in the *middle*).

20.8 Environmental Escape Artists

Imagine you are a single-celled protist. Life is good. There's plenty of food, and your habitat is just right. Then, without warning, things change. Nutrients start to run out. With your limited mobility, you can't get away. Is it all over? Perhaps not, if you belong to one of the groups that can wait out hard times by forming resting spores.

CONSIDER A GREEN ALGA

Green algae engage in diverse reproductive modes. A classic example is *Chlamydomonas*, a single-celled alga about twenty-five micrometers wide. It is able to reproduce sexually. Most of the time, however, it engages in asexual reproduction. As many as sixteen daughter cells form by mitotic cell division within the confines of the parent cell wall. Daughter cells may live in this cell for a while, but sooner or later they leave by secreting enzymes that digest what's

left of their parent. When nitrogen is scarce, haploid cells of opposite mating types fuse and form a diploid zygote having a tough outer wall. Inside the wall, this dormant cell can withstand drying out and freezing. After conditions improve, meiosis occurs, and then germination releases the haploid cells (Figure 20.17).

CONSIDER A SLIME MOLD

As a final example, consider the **slime molds**. These protists are common heterotrophs living in temperate regions. None has chloroplasts. All types spend part of their life cycle as a single cell and part as a larger, cohesive aggregation. Unfavorable conditions prompt the formation of spore-bearing structures.

In *plasmodial slime molds*, the most conspicuous stage of the life cycle is a multinucleated mass called the plasmodium. It arises from a single diploid cell

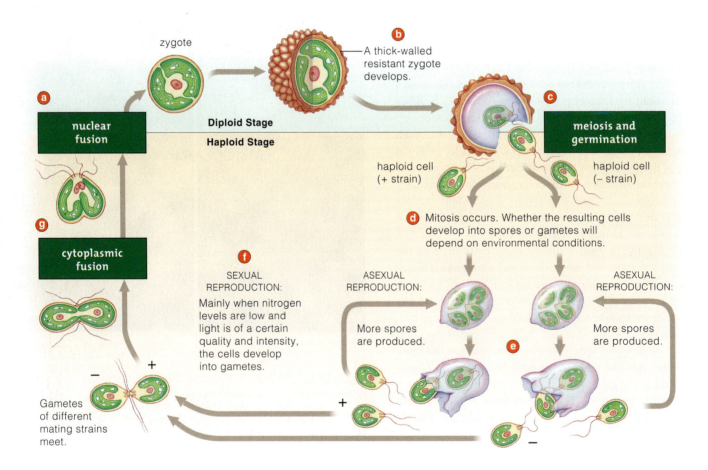

Figure 20.17 Life cycle of a species of *Chlamydomonas*, a common green algae of freshwater habitats. This single-celled species reproduces asexually most of the time, but two mating strains (designated + and –) can engage in sexual reproduction.

that undergoes many rounds of mitosis without any cytoplasmic division. A typical plasmodium fans out in what looks like a network of veins (Figure 20.18a). As it streams along rotting wood or the forest floor, it engulfs microbes and decaying organic matter. When stressed, as by dwindling food, the mass gives rise to many spore-bearing fruiting bodies (Figure 20.18b).

Cellular slime molds spend most of their life cycle as amoeboid (amoebalike) cells. *Dictyostelium discoideum* is typical of the group. The cells feed individually on bacteria and reproduce asexually, by way of mitotic cell divisions. When food runs out, as many as 100,000 of these cells aggregate into a mound.

Environmental cues, including moisture gradients, induce the mass to crawl about as a "slug" (Figure 20.19a–c). While the slug is migrating, cells start to differentiate into two types. Some will form a stalk; others will form spores. The slug migrates along until

Figure 20.18 *Physarum*, a plasmodial slime mold. (**a**) Feeding plasmodium on a log. (**b**) Spore-bearing structure.

it reaches a favorable spot. There, an elongating stalk elevates nonmotile spores, as in Figure 20.19d–f. In this way, a mature fruiting body forms. Spores float away from it, on currents of air. When they germinate, each releases a single diploid amoeboid cell, and the cycle begins anew.

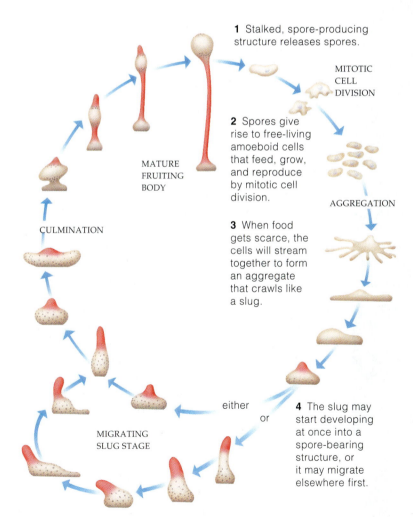

1 Stalked, spore-producing structure releases spores.

MITOTIC CELL DIVISION

MATURE FRUITING BODY

2 Spores give rise to free-living amoeboid cells that feed, grow, and reproduce by mitotic cell division.

AGGREGATION

CULMINATION

3 When food gets scarce, the cells will stream together to form an aggregate that crawls like a slug.

MIGRATING SLUG STAGE

either

or

4 The slug may start developing at once into a spore-bearing structure, or it may migrate elsewhere first.

a Life cycle of *Dictyostelium discoideum*.

Several amoeboid cells streaming together:

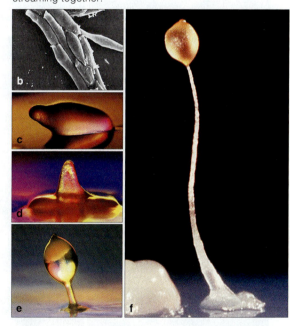

Figure 20.19 (**a**) Life cycle of *Dictyostelium discoideum*, one of the cellular slime molds. Spores give rise to free-living amoeboid cells, which grow and divide asexually until food dwindles.

(**b**) When starving, cells secrete cyclic AMP, a signaling molecule that induces them to stream together. Cells become sticky and adhere to one another. They secrete a cellulose sheath around themselves, thus forming a "slug" two to four millimeters long (**c**).

In a migrating slug, cells differentiate into prestalk (*red*), prespore (*white*), and anterior-like cells (*brown dots*).

The prestalk and prespore cells differentiate and form a stalked, spore-bearing structure (**d–f**). Anterior-like cells may help elevate nonmotile spores for dispersal from the spore-bearing structure.

20.9 Characteristics of Fungi

Fungi, too, arose from the "crown" of single-celled eukaryotes that evolved more than 900 million years ago. These are ancient lineages, closer on the family tree to animals than to plants.

Do you think "mushrooms" when you think of fungi? Mushrooms are just spore-bearing structures (fruiting bodies) of only a few of the 56,000 fungal species we know about, and there may be at least a million more fungi that we haven't even found yet!

When plants invaded land about 435 million years ago, fungal species accompanied them. Less than 100 million years later, three lineages—zygomycetes, sac fungi, and club fungi—were well established. Many other species are an informal group, called "imperfect fungi," until gene sequencing methods can point to where they belong. The vast majority of species in all of these groups are multicelled (Figure 20.20).

It takes a list of features to define **fungi**, because other groups are like them in many ways. All fungi are heterotrophs. Like some bacteria, they decompose organic compounds in a distinctive way. Fungal cells growing in or on organic matter secrete digestive enzymes, then individual cells absorb digested bits. The extracellular digestion helps plants, which absorb some of the released nutrients and metabolic products. Thus they help cycle nutrients through food webs.

Most fungi are saprobes, or organisms that absorb nutrients from nonliving organic material and cause its decay. Many are parasites; they feed on tissues of living hosts. As you will see, others are symbionts.

Most fungi reproduce sexually at opportune times and, more often, asexually. Either way, spores form. A **spore** is a single-celled or multicelled resting structure. Most fungal spores have a rigid protective wall. In multicelled fungi, spores germinate and start a **hypha** (plural, hyphae), a cellular filament encased in chitin. Mitotic cell divisions in hyphae produce a **mycelium** (plural, mycelia). This mesh of branching filaments has a high surface-to-volume ratio. As the mycelium grows over or into organic matter, hyphal cells absorb nutrients. Nutrients flow easily through the mesh; no walls form between the cytoplasm of adjoining cells.

Life cycles of fungi are diverse. We'll sample just a few species, starting with a mushroom-forming **club fungus**. For most of its life cycle, this fungus consists of a haploid mycelium buried in soil or growing into decaying wood. Figure 20.21 illustrates how sexual reproduction starts with fusion of two mating strains of hyphae. Cytoplasmic fusion produces a mycelium in which each cell contains two haploid nuclei ($n+n$). Under favorable conditions, specialized hyphae start to intertwine and form fruiting bodies—in this case, mushrooms. Nonmotile, haploid spores form in club-shaped cells on the fruiting body. After air currents disperse them, germinating spores start new mycelia.

a PURPLE CORAL FUNGUS *Clavaria*

b RUBBER CUP FUNGUS *Sarcosoma*

c BIG LAUGHING MUSHROOM *Gymnophilus*

d TRUMPET CHANTERELLE *Craterellus*

e SCARLET HOOD *Hygrophorus*

Figure 20.20 From southeastern Virginia, a small sampling of the spectacular range of fungal diversity. Most of these fruiting bodies belong to species of club fungi. Only (**b**) shows fruiting bodies of a sac fungus.

Figure 20.21 Generalized life cycle of many club fungi. When the hyphal cells of two compatible mating strains make contact, their cytoplasms fuse but the nuclei do not. Mitotic cell divisions yield a mycelium in which each cell has two nuclei. When the conditions are right, this mycelium forms a mushroom.

Club-shaped, spore-bearing structures form on gills, the mushroom cap's leaflike inner surface. Inside each structure, the two nuclei fuse. The result is a diploid zygote. With zygote formation, the cycle starts again.

The scanning electron micrograph (*below*) shows part of a mycelium, the underground portion of a club fungus that absorbs water and dissolved nutrients.

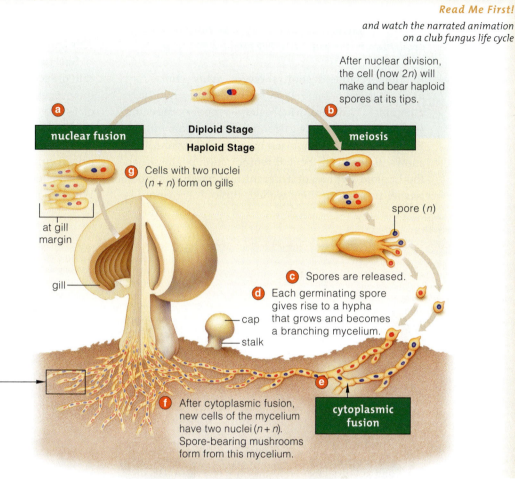

Read Me First!
and watch the narrated animation on a club fungus life cycle

a **nuclear fusion** **Diploid Stage** / **Haploid Stage** **b** **meiosis**

After nuclear division, the cell (now 2n) will make and bear haploid spores at its tips.

g Cells with two nuclei (n + n) form on gills

at gill margin

gill

spore (n)

c Spores are released.

d Each germinating spore gives rise to a hypha that grows and becomes a branching mycelium.

cap

stalk

e **cytoplasmic fusion**

f After cytoplasmic fusion, new cells of the mycelium have two nuclei (n + n). Spore-bearing mushrooms form from this mycelium.

one hyphal cell among the many hyphae that make up a mycelium

Club fungi are the most diverse fungal group. Many are known as mushrooms, puffballs, stinkhorns, and shelf fungi. Most are decomposers that grow through soil, but many closely associate plant roots. Rusts and smuts are parasites that attack plants. The cultivation of *Agaricus bisporus*—the mushroom of grocery store fame—is a multimillion-dollar business.

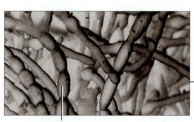

Figure 20.22 Fly agaric mushroom (*Amanita muscaria*), a hallucinogenic species. In Central America, India, and Russia, it was used in ancient rituals to induce trances.

So which organism is the oldest and largest? The honey mushroom (*Armillaria ostoyae*) may well be the winner. One individual has been spreading through the soil of an Oregon forest for 2,400 years. It extends across 2,200 or so acres to an average depth of 1 meter. Imagine 1,665 football fields side by side and you get an idea of its amazing size. *A. ostoyae* definitely is not one of the beneficial fungi. Its hyphae penetrate and clog tree roots, which eventually kills the tree.

Do *not* gather and eat wild mushrooms. Identifying mushrooms is tricky. Most of the edible species look extremely similar to poisonous ones. Many fungi make psychoactive alkaloids, and some of these can be deadly (Figure 20.22). It is said that there are old mushroom hunters and bold mushroom hunters, but no old, bold mushroom hunters.

Fungi are single-celled or multicelled heterotrophs that digest organic material outside the cell body. They help cycle nutrients when plants take up some of the released nutrients.

20.10 Fungal Diversity

A fungus has a thing about spores. It produces sexual spores, asexual spores, or both. The major fungal groups are defined by the ways that they produce sexual spores.

As the previous section showed, many club fungi make their sexual spores in fruiting bodies that have a complex structure. Much simpler spore-producing structures occur in the **zygomycetes**. Think about the *Rhizopus stolonifer*, the black bread mold. Figure 20.23 shows its life cycle. Sexual reproduction begins when the haploid hyphae of different mating strains meet. Nuclei fuse and a thick-walled, diploid zygospore is

formed. It undergoes meiosis, then germinates, giving rise to a hypha that will form a spore sac at its tip. This spore sac releases many haploid spores.

Sac fungi form sexual spores in sac-shaped cells called asci (singular, ascus). In multicelled sac fungi, the asci are enclosed in reproductive structures made of tightly woven hyphae. Different kinds are shaped like flasks, globes, and shallow cups (Figure 20.24b,c).

Among the sac fungi are the truffles. These highly-prized culinary delicacies grow as symbionts with tree roots. Their fruiting bodies are formed undergound. When spores mature and are ready to be released, the

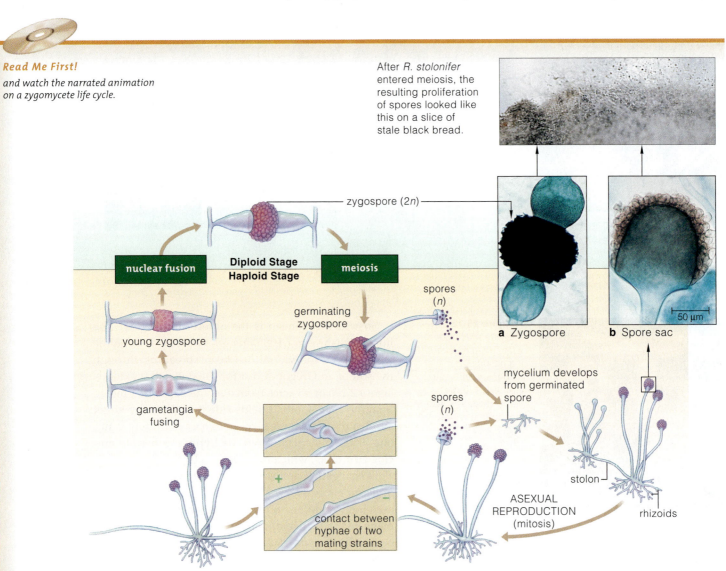

Read Me First!
and watch the narrated animation on a zygomycete life cycle.

After *R. stolonifer* entered meiosis, the resulting proliferation of spores looked like this on a slice of stale black bread.

zygospore (2n)

nuclear fusion | **Diploid Stage** / **Haploid Stage** | meiosis

young zygospore

germinating zygospore

spores (n)

gametangia fusing

a Zygospore

b Spore sac

50 μm

mycelium develops from germinated spore

spores (n)

+ −

contact between hyphae of two mating strains

ASEXUAL REPRODUCTION (mitosis)

stolon

rhizoids

Figure 20.23 Life cycle of the black bread mold *Rhizopus stolonifer*. Asexual phases are common. Also, different mating strains (+ and −) reproduce sexually. Either way, haploid spores form and give rise to mycelia. Chemical attraction between a + hypha and a − hypha makes them fuse. Two gamete-producing structures (gametangia) form, each with several haploid nuclei. The nuclei pair up. Each pair fuses and forms a zygote. Some zygotes disintegrate. Others become thick-walled zygospores and may be dormant for several months. Meiosis occurs as the zygospore germinates, and asexual spores form.

Figure 20.24 Sexual spores. (**a**) Spores on the gill margin of a club fungus. (**b,c**) *Sarcoscypha coccinea*, a scarlet cup fungus. On the cup's inner surface, many spores form in saclike cells called asci.

fruiting body produces a complex mixture of aromatic compounds. This gives the ripe truffle its distinctive taste and smell. Trained dogs or pigs are often used to sniff them out underground. Several kinds of truffles sell for hundreds of dollars an ounce.

Most food-spoiling molds are multicelled sac fungi. Some of the single-celled fungi we call yeasts are also sac fungi (others are club fungi). One packet of baking yeast holds thousands of tiny spores of *Saccharomyces cerevisiae*. In a warm, moist spot, such as bread dough set out to rise, these spores germinate. The cells that emerge can reproduce asexually by budding. Aerobic respiration in these cells releases carbon dioxide and causes the bread dough to rise. Fermentation by other yeast species is used to make alcoholic beverages. Yeasts also can be genetically engineered and grown in large vats, as drug-producing factories.

Unique sexual spores characterize the major fungal lineages. In zygomycetes, spores form at the tip of a specialized hypha that germinates from a zygospore. Many club fungi and sac fungi produce distinctive sexual spores in specialized hypha that differentiate as part of a complex fruiting body.

20.11 The Unloved Few

You know you are a serious student of biology when you view organisms objectively in terms of their role in nature, not in terms of their impact on humans in general and you in particular. As a student you hail saprobic fungi as decomposers and praise parasitic fungi that can keep populations of bad insects and weeds in check. And then you cross paths with a bad fungus.

Have you ever opened the fridge to get a bowl of fruit and find a fungus beat you to it? Has a fungus made the tissues of your warm, damp toes scaly, red, and cracked (Figure 20.25a)? Think about it. Which home gardeners wax poetic about powdery mildew on roses? Which farmers happily hand over millions of dollars a year to sac fungi that infect corn, wheat, peaches, and apples? Who rejoices that a sac fungus, *Cryphonectria parasitica*, blitzed nearly all chestnut trees in eastern North America?

Who willingly inhales airborne spores of *Ajellomyces capsulatus*? In soil, the spores give rise to mycelia. In moist lung tissues, they form yeastlike cells that can cause *histoplasmosis*, a respiratory disease. The body's macrophages normally can eliminate the direct threat, but debris from the battle damages lung tissue. Heavy exposure to spores invites pneumonia.

And household molds! Thank them for sinus, ear, and lung infections, for rashes, and for increasing asthma attacks 300 percent over the past two decades.

Some fungi have even tweaked human history. One species, *Claviceps purpurea*, is a parasite of rye and other cereal grains (Figure 20.25b). Give it credit; we use some of its alkaloid by-products to treat migraines and to stop hemorrhages after childbirth by shrinking the uterus. However, its alkaloids can be toxic. Eat bread made with spoiled rye and you may end up with *ergotism*. Disease symptoms include vomiting, diarrhea, hallucinations, hysteria, and convulsions. Untreated, ergotism turns limbs gangrenous and causes death.

Ergotism epidemics were common in the Middle Ages in Europe, when rye was a key crop. They thwarted Peter the Great, a Russian czar who was obsessed with conquering ports along the Black Sea for his nearly landlocked empire. Soldiers laying siege to the ports ate mostly rye bread and fed rye to their horses. The soldiers convulsed, and horses went into "blind staggers." Ergotism outbreaks might have provided an excuse to launch witchhunts in the early American colonies.

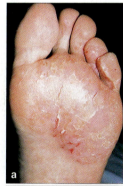

Figure 20.25 Love those fungi! (**a**) Athlete's foot, courtesy *Epidermophyton floccosum*. (**b**) Spores of a historically notable fungus, *Claviceps purpurea*, on infected rye plant parts.

20.12 Fungal Symbiants

*"Symbiosis" refers to species that interact closely during their life cycles. In **mutualism**, their interaction benefits both or does no harm to either one. Fungal endophytes, lichens, and mycorrhizae generally are like this.*

FUNGAL ENDOPHYTES

Endophytic fungi are symbionts that live inside the leaves and stems of almost all plants, with helpful or neutral effects. Tall fescue (*Festuca arundinacea*) offers an example. Lush clumps of this grass are vulnerable to herbivores, but fungi living in the tissues of most *F. arundinacea* strains produce alkaloids that are toxic to the grazers, which quickly learn to leave the plants alone. Diverse fungal endophytes also help protect host plants from pathogens, such as other fungi and the oomycote *Phytophthora*.

LICHENS

A **lichen** consists of a fungal species intertwined with one or more photoautotrophs, most often green algae or cyanobacteria. It forms after a hypha's tip binds to a host cell and both lose their wall. Their cytoplasm fuses or the hypha induces the host cell to cup around it. Both start dividing; they give rise to a multicelled and often layered vegetative body that is flattened, erect, leaflike, or pendulous (Figure 20.26). The fungus usually makes up most of the lichen; its partner lives in or on it. The photoautotroph supplies the fungus with nutrients and gets a protected habitat in return.

The cyanobacteria in lichens help cycle nitrogen by converting nitrogen gas in the air to forms that plants can absorb. Lichens get minerals from airborne particles and are highly vulnerable to air pollution.

Lichens can colonize places that are too hostile for most organisms. For example, when a glacier retreats, they colonize newly exposed bedrock. Their metabolic products help form new soil or enrich whatever soil is already present. As conditions improve, other species move in and typically replace the pioneers. This may be what happened when plants first invaded the land.

MYCORRHIZAE

Many soil fungi live as mutualists on or in tree roots. Those truffles in Figure 20.24 are an example. They interact as a **mycorrhiza** (plural, mycorrhizae), which means fungus-root. Underground parts of the fungus grow through soil and provide a large surface area for absorbing nutrients. When ions of phosphorus and other minerals are abundant, the fungus takes them up, and it releases some to the plant when ions are scarce. In return, the fungus withdraws some sugars from the plant. The plant's loss is a trade-off, because it doesn't grow as efficiently without a fungal partner. Figure 20.27 shows an example of the dramatic effects of a mycorrhizal fungus on plant growth.

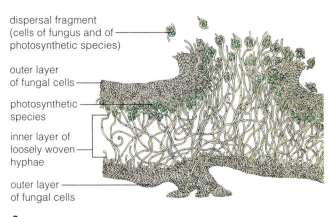

dispersal fragment (cells of fungus and of photosynthetic species)

outer layer of fungal cells

photosynthetic species

inner layer of loosely woven hyphae

outer layer of fungal cells

e

Figure 20.26 (**a**) One of the encrusting lichens, growing on a sunlit rock. (**b**) *Cladonia rangiferina*, an erect, branching lichen. (**c**) *Lobaria oregana*, a leaflike lichen that includes nitrogen-fixing cyanobacteria. (**d**) *Usnea*, a pendant lichen commonly called old man's beard. (**e**) Body plan of a stratified (layered) lichen, cross-section.

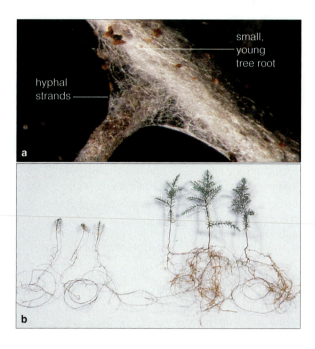

Figure 20.27 (**a**) Mycorrhiza of a young hemlock tree. (**b**) Effects of the presence or absence of mycorrhizae on plant growth in sterilized, phosphorus-poor soil. The juniper seedlings at left were controls; they grew without a fungus. Seedlings at right, the experimental group, are six months old. They were grown with a partner fungus.

AS FUNGI GO, SO GO THE FORESTS

Since the early 1900s, collectors have been counting the numbers and kinds of wild mushroom populations in European forests. The populations clearly are declining. Mushroom gatherers are not the cause; inedible as well as edible species are vanishing. However, the decline does correlate with rising levels of air pollution.

Now consider this: Myccorhizal fungi benefit nearly all trees. Trees do have hairlike absorptive structures projecting from young roots, but fungal hyphae are smaller and better at penetrating small crevices in soil. As a tree ages, one mycorrhizal fungus gives way to another, in predictable patterns. When fungi die, trees lose a vital support system and are more vulnerable to damage from severe frosts, droughts, and acid rain. As a result, many forests are deteriorating.

In mutualism—a symbiotic interaction—one or both partners derive benefits. Endophytes are protective fungi living inside leaves or stems. In lichens, one or more photoautotrophs are sheltered by a fungus. In mycorrhizae, a fungus living in or on a plant's roots shares mineral ions it has absorbed. All photosynthetic partners give up some sugars to the fungus.

Summary

Section 20.1 Structurally, protists are the simplest eukaryotes and the ones most like the first eukaryotic cells. The groups are undergoing major reclassification.

Section 20.2 Flagellated protozoans are among the most ancient eukaryotic cells. Some of these single cells do not have mitochondria. Amoeboid protozoans lost their flagella as they evolved. Foraminiferans and radiolarians are amoebas with secreted shells or plates. Euglenoids are freshwater flagellates; most acquired chloroplasts by endosymbiosis.

Section 20.3 The ciliates, apicomplexans, and dinoflagellates are all alveolates. These single cells have membrane-bound sacs beneath the plasma membrane. Beating cilia help ciliates move or direct food into their gullet. Apicomplexans are intracellular parasites with host-penetrating microtubular devices; malaria-causing *Plasmodium* species are examples. Dinoflagellates are flagellated photoautotrophs or heterotrophs.

Section 20.4 Nutrient-enriched water fuels algal blooms: huge increases in population sizes of aquatic protists that can harm other organisms.

Section 20.5 Stramenopiles have tinselly flagella and are heterotrophs (oomycotes) or photoautotrophs (single-celled diatoms and coccolithophores, and multicelled brown algae). Different oomycotes are decomposers, parasites, or pathogens. Diatoms have silica shells and coccolithophores have calcium plates. Brown algae include kelps, the largest and most structurally complex protists.

Section 20.6 Red algae are colored by phycobilins that mask chlorophylls and that can capture light even at great depths. Most are multicelled and live in marine habitats. Single-celled species may have given rise to plastids in some other protists, by endosymbiosis.

Section 20.7 The "green algae" are single cells or multicelled. Like plants, they have chlorophylls *a* and *b*, they store sugars as starch, and their cell walls contain cellulose. Most are chlorophytes. The charophytes are more closely related to plants than to other green algae.

Section 20.8 Many protists are escape artists; they survive adverse conditions by forming resting spores. *Chlamydomonas* and slime molds are examples.

Section 20.9 Fungi are heterotrophs that engage in extracellular digestion and absorption: Individual cells secrete digestive enzymes on organic matter and then absorb breakdown products. Most are decomposers that help ecosystems; some are pathogenic or parasitic.

In multicelled fungi, germinating spores give rise to a mycelium (a mesh of filamentous hyphae). Hyphal cells have chitin-stiffened walls but not at abutting ends, so cells making up each filament share their cytoplasm.

Section 20.10 Besides asexual spores, the three major lineages of fungi (zygomycetes, sac fungi, and club fungi) form unique sexual spores that help us classify them. "Imperfect fungi" is an informal group of fungal species awaiting classification.

Section 20.11 Most fungi are beneficial, but others destroy crops, spoil food, and cause diseases.

Section 20.12 Endophytes, lichens, and mycorrhizae are cases of mutualism. Endophytic fungi live in most stems and leaves. Lichens consist of a fungal symbiont and a cyanobacterium or green alga. A mycorrhiza consists of a fungus associating with young roots.

Self-Quiz

Answers in Appendix III

1. Some of the flagellated protozoans lack _____ .
   a. a nucleus   c. DNA
   b. mitochondria   d. a plasma membrane

2. The euglenoids are _____ .
   a. flagellated   c. marine   e. all are correct
   b. single cells   d. a and b

3. The _____ are parasitic alveolates, such as the organism that causes malaria.
   a. trypanosomes   c. oomycotes
   b. apicomplexans   d. zygomycetes

4. Silica reinforces the cell walls of _____ .
   a. coccolithophores   c. foraminiferans
   b. diatoms   d. fungi

5. The giant kelp *Macrocystis* and the plant pathogen *Phytophthora* are both _____ .
   a. brown algae   c. oomycotes
   b. chrysophytes   d. stramenopiles

6. Algal blooms are most commonly population explosions of _____ .
   a. diatoms   c. euglenoids
   b. radiolarians   d. dinoflagellates

7. Phycobilins are the signature pigments of _____ .
   a. red algae   c. green algae
   b. brown algae   d. all algae

8. The closest relatives of *Chara* and other charophytes are the _____ .
   a. oomycotes   c. plants
   b. chrysophytes   d. fungi

9. When a spore of a cellular slime mold germinates, it gives rise to a _____ .
   a. free-living amoeboid   c. mycelium
   b. fruiting body   d. mycorrhiza

10. When a spore of a multicelled fungus germinates, it gives rise to a _____ .
    a. free-living amoeboid   c. mycelium
    b. fruiting body   d. mycorrhiza

11. A "mushroom" is _____ .
    a. the food-absorbing part of a fungal body
    b. composed entirely of haploid cells
    c. a reproductive structure
    d. the longest lived part of a fungal life cycle

12. Match these terms suitably.
    ____ trypanosome   a. single-celled fungus
    ____ foraminiferan   b. flagellated protozoan
    ____ diatom   c. "fungus-root"
    ____ *Paramecium*   d. brown alga
    ____ oomycote   e. calcium-carbonate shell
    ____ charophyte   f. two-part silica shell
    ____ kelp   g. fungus plus photoautotroph
    ____ yeast   h. "egg fungus"
    ____ mycorrhiza   i. ciliated protozoan
    ____ lichen   j. closest relative of plants

Table 20.1	Comparison of Prokaryotes With Eukaryotes	
	Prokaryotes	Eukaryotes
Organisms represented:	Archaea, Bacteria	Protists, fungi, plants, and animals
Ancestry:	Two major lineages that evolved more than 3.5 billion years ago	Equally ancient prokaryotic ancestors gave rise to forerunners of eukaryotes, which evolved more than 1.2 billion years ago
Level of organization:	Single-celled	Protists, single-celled or multicelled. Nearly all others multicelled; division of labor among differentiated cells, tissues, and often organs
Typical cell size:	Small (1–10 micrometers)	Large (10–100 micrometers)
Cell wall:	Most with distinctive wall	Cellulose or chitin; none in animal cells
Membrane-bound organelles:	Rarely; no nucleus, no mitochondria	Typically profuse; nucleus present; most with mitochondria, many with chloroplasts
Modes of metabolism:	Both anaerobic and aerobic	Aerobic modes predominate
Genetic material:	One chromosome; plasmids in some	Chromosomes of DNA plus many associated proteins in a nucleus
Mode of cell division:	Prokaryotic fission, mostly; some reproduce by budding	Nuclear division (mitosis, meiosis, or both) associated with one of various modes of cytoplasmic division

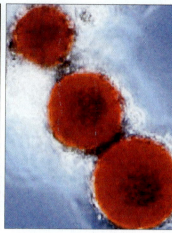

Figure 20.28 A drift of snow on a glacier. It has been tinted red by an abundance of snow algae, including *Chlamydomonas nivalis* (*right*). Notice the red footprint.

Critical Thinking

1. Take a moment to consider Table 20.1, which shows the general features of prokaryotic and eukaryotic cells. Consider that, as you learned in this and the previous chapter, some of the prokaryotes exceed eukaryotes in size, some primitive eukaryotes lack mitochondria, and many eukaryotes carry within them the descendants of endosymbiotic prokaryotes. Recall also that archaea are as closely related to eukaryotes as they are to bacteria. Because of these and other discoveries, some scientists have suggested that the division of organisms into prokaryotes and eukaryotes is not particularly useful. Do you agree or disagree with this argument? Suggest how these taxonomic categories might help us interpret biodiversity and how they might be misleading.

2. Runoff from heavily fertilized cropland, animal waste, and raw sewage promotes algal blooms that can result in massive kills of aquatic species, birds, and other forms of wildlife. Are such kills an unfortunate but inevitable side effect of human life?

    If you find the environmental cost unacceptable, how would you stop the pollution? Bear in mind that we now absolutely depend on high-yield (and heavily fertilized) crops. How do you suggest we dispose of the waste generated by farms and cities?

3. The most common "snow alga," *Chlamydomonas nivalis*, lives in the world's glaciers (Figure 20.28). Although it's a green alga, an abundance of carotenoid pigments colors it red. Think about the wavelengths of radiation striking its icy habitats. Besides their role in photosynthesis, what other function might the carotenoids be serving? (*Hint:* Review Section 6.1.)

4. The cellular slime mold *Dictyostelium discoideum* is now being studied by more than 660 biologists. Each year, hundreds of papers about its biology appear in scientific journals. It is one of the "model organisms" that help us learn about cell differentiation and other developmental processes. Speculate on why this simple soil organism is such a desirable experimental organism.

5. The fungus *Fusarium oxysporum* is a plant pathogen. Some view it as a potential weapon in the war on drugs. Why? Strains of the fungus attack and kill only specific plants—including coca plants used to produce cocaine. Proposals to spray *F. oxysporum* to kill marijuana plants in Florida were abandoned after public outcries.

    What concerns might you have about allowing the use of such natural mycoherbicides to kill off plants that are sources of the drugs favored by substance abusers?

Media Menu

Student CD-ROM

Impacts, Issues Video
  Tiny Critters, Big Impacts
Big Picture Animation
  Evolution and diversity of simple eukaryotes
Read-Me-First Animation
  An apicomplexan life cycle
  A red alga life cycle
  A club fungus life cycle
  A zygomycete life cycle
Other Animations and Interactions
  A green alga life cycle
  A cellular slime mold life cycle

InfoTrac

- Fatal Attraction. *Natural History*, April 2001.
- Parasites from Another Kingdom. *BioScience*, December 2002.
- Hunting Slime Molds. *Smithsonian*, March 2001.
- Penicillin: From Discovery to Product. *Bulletin of the World Health Organization*, August 2001.
- Algal Research. *The Scientist*, October 29, 2001.

Web Sites

- Protist Image Database: megasun.bch.umontreal.ca/protists/protists.html
- Malaria Foundation International: malaria.org
- The Harmful Algae Page: www.whoi.edu/redtide
- Tom Volk's Fungi: botit.botany.wisc.edu/toms_fungi
- Lichens of North America: www.lichen.com
- Dr. Fungus: www.doctorfungus.org

How Would You Vote?

Phytophthora ramorum, the pathogen that causes sudden oak death, has been detected in twenty-six kinds of plants in California and Oregon. A few of the infected plants belong to species that are raised in nurseries and sold nationwide. Some states that are free of the pathogen want to ban the shipment of some or all plants from affected states. Should those states be allowed to institute such a ban?

21 PLANT EVOLUTION

IMPACTS, ISSUES *Beginnings and Endings*

Change is the way of life. About 300 million years ago, in the Carboniferous, swamp forests carpeted the wet, warm lowlands of continents. Tree-high ancestors of today's tiny club mosses and horsetails were dominant. Then things changed. As the long-term global climate became cooler and drier, moisture-loving plants declined. Hardier plants—cycads, ginkgos, and conifers—rose to dominance. They were the gymnosperms, and they would flourish for millions of years.

Things changed again. Other plants that had been evolving had more potential to radiate into diverse environments. The angiosperms—flowering plants—started to take over. They still dominate most regions. But in the far north, at high elevations, and in parts of the Southern Hemisphere, some conifers retained the competitive edge and held on as great forests.

Things changed yet again. Humans learned to grow some flowering plants as crops, which became the foundation for human population growth. Human populations grew spectacularly, and they required more resources than crops. Conifers had the bad luck to become the premier sources of lumber, paper, resins, and other forest products. They became vulnerable to deforestation—the removal of all trees from large tracts of land.

In California, only 4 percent of the original coastal redwood forest is still undisturbed. In Maine, an area the size of Delaware has been logged over in just the past fifteen years. Many logs end up

Figure 21.1 A clear-cut mountainside in British Columbia. This province supplies about half the Canadian softwood exported to the United States. The forest on the facing page is in Washington State. Similar scenes of deforestation can be found worldwide, in just about every coniferous forest.

the big picture

Evolutionary Trends Nearly all plants are multicelled photoautotrophs. Their lineage apparently started as a branching from a green algal group. Through modifications in structure and life cycles, different kinds became adapted to higher and drier habitats on land.

Nonvascular Plants Mosses and other bryophytes are small plants that have no internal transport tissues. A gamete-forming body dominates the life cycle. A spore-forming body forms on it and remains attached even at maturity.

as paper or pulp. In the United States, the amount of wood processed in pulp mills between 1920 and 1997 increased about twentyfold. The United States exports many logs to foreign lumbermills. At the same time, it imports timber from Canada and tropical rain forests as far away as New Zealand (Figure 21.1).

Clearing the world's forests has drastic ecological effects. Nutrients wash away from exposed soils, and sediments clog streams. Herbicides applied to stop non-timber species from taking over also prevent ecosystem recovery and push species toward extinction. Almost 750 species of plants in the United States alone are now on the endangered list.

And with this bit of perspective on change, we turn to the origins, and end of the line, of the major groups of plants. With rare exceptions, plants are multicelled photoautotrophs. These metabolic wizards make organic compounds by absorbing energy from the sun, carbon dioxide from the air, and minerals from water. As they do, they split water molecules and release free oxygen. Their oxygen-releasing pathway of photosynthesis influences the atmosphere even while sustaining the growth of multicelled forms as tall as redwoods and as vast as an aspen forest that is one continuous clone.

We know of at least 295,000 kinds of existing plants. Be glad their ancient ancestors left the water. Without those pioneers in a new world, we humans and other land-dwelling animals never would have made it onto the evolutionary stage.

 How Would You Vote?

Recycling paper saves trees, slows the filling of landfills, and reduces water pollution, although it is often more expensive than making new paper. Plans are afoot to give tax breaks to companies that recycle paper. Would you support such a plan? See the Media Menu for details, then vote online.

Seedless Vascular Plants Dominating the life cycles of seedless vascular plants is a spore-forming body with vascular tissues, roots, and other adaptations to dry habitats. Without free water, flagellated male gametes of these plants cannot reach eggs for fertilization.

The Rise of Seed Plants Gymnosperms and angiosperms are vascular plants that make pollen grains and seeds. These unique traits contributed to their reproductive success. Only angiosperms produce flowers. They are the most widespread and diverse group of existing plants.

21.1 Evolutionary Trends Among Plants

We share the world with at least 295,000 kinds of plants. Within this tremendously diverse group, we find many recurring structures that evolved in response to present and past environmental challenges.

Five hundred million years ago, the invasion of land was under way. Why then? Astronomical numbers of photosynthetic cells had come and gone, and oxygen-producing types had changed the atmosphere. High above Earth, the sun's energy had converted much of the oxygen into a dense ozone layer, a shield against lethal doses of ultraviolet radiation. Until then, life had not ventured above the surface of water and mud.

Algae were evolving at the water's edge, and one group—probably the charophytes—gave rise to plants. *Cooksonia*, a simple branching plant a few centimeters tall, evolved by 430 million years ago. It took another 160 million years for the taller *Psilophyton* to evolve. Then the evolutionary pace picked up. It took only 60 million years for plants to radiate from the swampy lowlands to high mountains and nearly all places in between. They did so through modifications in their structure, function, and reproductive modes.

ROOTS, STEMS, AND LEAVES

Underground absorptive structures evolved as plants colonized the land, and in some lineages they became systems of roots. Most roots had a large surface area that helped plants absorb more water and dissolved mineral ions. As they do today, young roots probably associated with fungal mycorrhizal symbionts, which help plants obtain water and dissolve nutrients. In many lineages, older roots started anchoring the plant.

Above ground, shoot systems evolved. Their stems and leaves captured energy from the sun and carbon dioxide from the air. Stems became erect, taller, and branched when plants developed a capacity to make and incorporate lignin, a glue-like polymer, in their cell walls. Lignin-strengthened walls supported the stems as they grew upward and outward, in patterns that increased the light-intercepting surface of leaves.

Vascular tissues, xylem and phloem, first appeared in plants called rhyniophytes. As happens now, xylem distributed water and mineral ions through the plant; phloem distributed photosynthetic products.

Also, having enough water for metabolism was not a problem in most aquatic habitats. It was a challenge for life on land. A waxy coat—a cuticle—evolved and helped conserve water inside shoots on hot, dry days. Water and carbon dioxide could cross the cuticle only at tiny gaps called stomata. When stomata evolved, so did control over water loss (Section 6.6). These tissue specializations endure in most existing land plants.

FROM HAPLOID TO DIPLOID DOMINANCE

In plant life cycles, remember, a gametophyte phase alternates with a sporophyte phase (Section 9.5). Each **gametophyte** is a haploid body that produces haploid gametes. Male and female gametes fuse at fertilization, and the resulting zygotes are the start of **sporophytes**. In these multicelled vegetative bodies, haploid spores form by way of meiosis. **Spores** are resting structures, typically walled, that help a new generation wait out harsh environmental conditions. After they germinate, plant spores grow and develop into gametophytes.

Many algae live in places where conditions do not change much, so making resting spores would be a big waste of energy. As you might expect for these algae, gamete production dominates the life cycles. Haploid spores often develop right from the zygote itself, not from a multicelled vegetative body (Figure 21.2).

When plants moved to dry land, spore production became crucial—and sporophytes came to dominate most life cycles. Shrubs, trees, and other sporophyte forms having waxy cuticles, complex vascular tissues, and spore-producing capsules (sporangia) evolved. The timing of fertilization and spore dispersal became adapted to the seasons. A land plant could now retain, nourish, and protect its gametophytes and its embryo sporophytes as they formed and developed, right up to the least risky time to leave home.

COOKSONIA

PSILOPHYTON

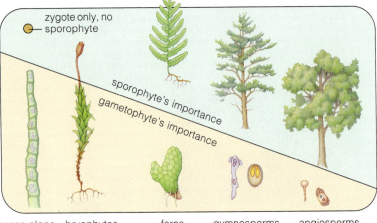

zygote only, no sporophyte

sporophyte's importance

gametophyte's importance

green algae   bryophytes   ferns   gymnosperms   angiosperms

Figure 21.2 One evolutionary trend in plant life cycles. Algae and bryophytes put the most energy into making gametophytes. The other groups evolved in seasonally dry habitats on land. They put the most energy into structures that produce spores and retain, nourish, and protect gametes through harsh times.

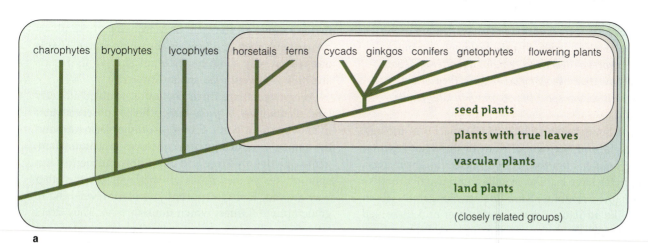

Figure 21.3 (a) Evolutionary tree for plants, with nested monophyletic groups. (b) Milestones in the evolution of plants.

Plants on land by 435 million years ago.	Origin of rhyniophytes (earliest vascular plants) in Silurian.	Origin of bryophytes (liverworts), vascular plants (lycophytes, horsetails, ferns, progymnosperms) by end of Devonian.	Vast forests, swamps; bryophytes, lycophytes, horsetails, and ferns dominate. Origin of conifers late in Carboniferous.	Most lycophytes, horsetails extinct; origin of ginkgos, cycads.	Ferns, conifers diversify.	Conifers, cycads, ferns dominant.	Origin of flowering plants by the early Cretaceous, followed by their adaptive radiation and rise to dominance.

ORDOVICIAN	SILURIAN	DEVONIAN	CARBONIFEROUS	PERMIAN	TRIASSIC	JURASSIC	CRETACEOUS	PRESENT
505	440	410	360	286	248	213	145	65

b

Time (millions of years ago)

EVOLUTION OF POLLEN AND SEEDS

Figure 21.3a is an evolutionary tree for major plant groups. All but 24,000 of the 295,000 existing species are *vascular*, with internal tissue systems that conduct water and solutes through roots, stems, and leaves. Compared to the *nonvascular* groups—bryophytes—they became success stories on land. The lycophytes, horsetails, and ferns are among the *seedless* vascular plants. The cycads, ginkgos, gnetophytes, and conifers are all gymnosperms, a group of *seed-bearing* vascular plants. Angiosperms are vascular, seed-bearing plants as well, but they alone make flowers. The flowering plants are the largest and most diverse group.

Most seedless vascular plants make only one type of spore; they are *homo*sporous. Some seedless and all seed-bearing vascular plants make two types; they are *hetero*sporous. In gymnosperms and flowering plants, the two are designated megaspores and microspores.

Megaspores divide and form female gametophytes, which make female gametes (call them eggs). Far smaller microspores give rise to **pollen grains**, which are like well-packed suitcases. The protective suitcase wall encloses a few cells that will eventually develop into a mature, sperm-bearing, male gametophyte.

Typically, air currents or animals such as insects deliver pollen grains to eggs. Environmental water is not required, as it is for the male gametes of algae.

Pollen grains were a key innovation that helped seed-bearing plants radiate into high, dry habitats.

Another innovation: Embryo sporophytes became packaged in nutritive tissues and a tough, waterproof coat. The term **seed** refers to the whole package. Over time, the development and dispersal of seeds became attuned to environmental change—for instance, a dry season alternating with a warm, wet season favoring germination and growth. It was no coincidence that seed plants rose to dominance in the Permian, a time of extreme shifts in the global climate.

The plant kingdom includes multicelled, photosynthetic species called bryophytes, seedless vascular plants, and seed-bearing vascular plants.

Most plant lineages became structurally adapted to life on land. They have root and shoot systems, a waxy cuticle, stomata, vascular tissues, and lignin-reinforced tissues.

Sporophytes with well-developed roots, stems, and leaves came to dominate the life cycles of most land plants. Parts of these complex sporophytes nourish and protect the new generation until conditions favor dispersal and growth.

Some plants started to produce two types of spores, not one. This led to the evolution of male gametes that could be dispersed without liquid water and to the evolution of seeds.

21.2 The Bryophytes—No Vascular Tissues

The first plants were bryophytes. The fossil record of this group is sketchy, and whether the first ones were more like hornworts or liverworts remains a matter of debate. We do know that, like living bryophytes, they lacked the specialized transport tissues found in later plants.

Modern **bryophytes** include 24,000 species of mosses, liverworts, and hornworts. None of these nonvascular plants is taller than twenty centimeters (eight inches). Bryophytes have leaflike, stemlike, and usually rootlike parts named rhizoids. **Rhizoids** are elongated cells or threadlike structures that absorb water and dissolved mineral ions. They also attach gametophytes to soil.

Bryophytes don't have specialized vascular tissues because they don't need them. Most grow in habitats that are moist all the time or seasonally. These plants can dry out, then revive after absorbing water. That is one reason why some hardy types survive in deserts and on windswept plateaus in Antarctica.

All groups share three traits that evolved in early land plants. *First*, a waxy cuticle helps conserve water in aboveground parts. *Second*, a cellular jacket around the gamete-producing parts conserves moisture. *Third*, gametophytes are large and don't draw nutrients from sporophytes as they do in other plants. It's the other way around. Embryo sporophytes start developing in gametophyte tissues, which nourish *them*. They don't let go when they reach maturity. They stay attached to the gamete-producing body and draw nutrients from it through at least part of the life cycle (Figure 21.4).

Read Me First!

and watch the narrated animation on a moss life cycle

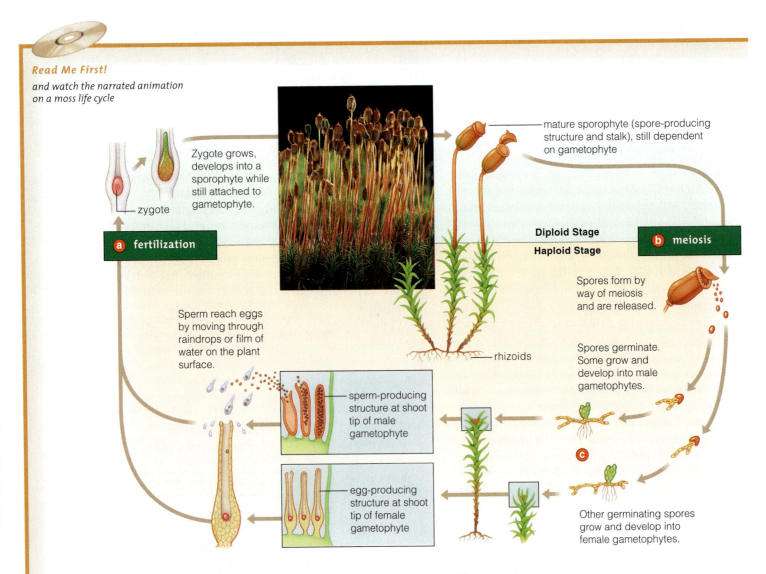

Figure 21.4 Life cycle of a moss (*Polytrichum*), one of the bryophytes. The sporophyte stays attached to the gametophyte, which provides it with nutrients and absorbed water.

Figure 21.5 (**a**) Moss plants on damp rocks. (**b**) Peat bog in Ireland. This family is cutting blocks of peat and stacking them to dry as a home fuel source. Most bryophytes grow slowly. Peat mosses grow fast enough to yield thirteen tons of organic matter per hectare annually, about twice the yield from corn plants. Nova Scotia alone exports 200,000 bales of peat per year to Japan and the United States. Nearly all harvested peat is burned to generate electricity in power plants, which releases fewer pollutants compared to coal burning. (**c**) Peat moss (*Sphagnum*). A few sporophytes—brown, jacketed structures on white stalks—are attached to a pale gametophyte.

Mosses are the bryophytes that most people know about (Figure 21.5*a*). Like others of this group, they are sensitive to air pollution. Where air quality is poor, mosses are few or absent. Most gametophytes have a low mounded or feathery form. Gametes develop at their shoot tips, in vessels enclosed in a tissue jacket. Sperm swim to eggs through water droplets clinging to the plants. Moss sporophytes consist of a stalk and jacketed structure in which haploid spores develop.

Figure 21.5*c* shows *Sphagnum*, one of 350 kinds of peat mosses. Their compressed, moist, organic remains form **peat bogs**, which can be as acidic as vinegar. In cold and temperate regions, bogs cover an area equal to half of the United States. Only highly acid-tolerant plants, such as cranberries, live in them.

Acidic metabolic products of bryophytes hamper growth of bacterial and fungal decomposers. Well-preserved bodies about 2,000 to 3,000 years old have turned up in peat bogs in Europe. The bogs may have been sites of ceremonial human sacrifices, and acidity kept the bodies from decomposing. Also, compared to cotton, *Sphagnum* soaks up five times more water, which enters dead cells in the leaflike parts. Because of the antiseptic properties and absorbency of peat moss, World War I doctors used it for bandages.

Although hornworts and liverworts are less well known, one or the other group may resemble the first plants. Even now, they are among the hardy pioneer

Figure 21.6 (**a**) One hornwort. The hornlike sporophyte is attached to a gametophyte. Liverworts, including *Marchantia*, reproduce sexually as well as asexually. Unlike other types, this genus has (**b**) female and (**c**) male reproductive parts on different plants.

species that are the first to colonize barren habitats. Their gametophytes are mostly ribbonlike or "leafy" forms that are often flat and thin, as in Figure 21.6.

Bryophytes are small, structurally simple nonvascular plants most of which grow in fully or seasonally moist habitats. Their flagellated sperm require environmental water to swim to the eggs.

The sporophyte grows from gametophyte tissues and draws nutritional support from it. Only among bryophytes does the sporophyte remain attached to a larger gametophyte.

21.3 Seedless Vascular Plants

The earliest seedless vascular plants were branching photosynthetic forms that lacked roots and leaves. A spectacular radiation produced leafy and even treelike forms. Diversity peaked during the Carboniferous.

Figure 21.7 depicts early seedless vascular plants in a Devonian swamp. Some of their descendants, which include the **lycophytes**, **horsetails**, and **ferns**, are with us today. Like their ancestors, they differ in big ways from bryophytes: The sporophytes have xylem and phloem, they are not attached to gametophytes, and they are the larger, longer lived phase of the life cycle.

Figure 21.7 What a Devonian swamp may have looked like, based on fossil evidence. Section 21.1 showed fossils of *Cooksonia* and *Psilophyton*, two of the first seedless vascular plants that evolved in such habitats.

The seedless vascular plants are sometimes called "amphibians" of the plant kingdom. Like frogs, they require a moist habitat for the sexual phase of the life cycle. Like bryophytes and lichens, a few species do live in dry habitats, but they reproduce sexually only during brief pulses of seasonal rain. The resurrection plant, *Selaginella lepidophylla*, shrivels up completely and goes dormant, then quickly revives. It grows in Mexico, New Mexico, and Texas.

LYCOPHYTES

Tree-sized lycophytes lived in Carboniferous swamp forests. The 1,100 or so modern forms are a great deal smaller. The best known types are club mosses, which range from tundra to the tropics (Figure 21.8*a*). Most club mosses have vascularized stems and roots, which grow from a branching, underground rhizome.

Microphylls—tiny leaves, each with an unbranched vein—are a defining trait of lycophytes. Chambers for spore production develop at the base of some of these leaves. In certain species, such leaves are organized around a central stem, as a **strobilus** (plural, strobili). This general term refers to any conelike reproductive structure derived from modified leaves. Some of the *Lycopodium* species are informally called ground pines, partly because they look a bit like miniature Christmas trees with "cones." Strobili also form on sporophytes of other groups, such as horsetails and gymnosperms.

Figure 21.8 A few seedless vascular plants. (**a**) Club moss (*Lycopodium*), a lycophyte. Notice its strobili, conelike clusters of modified leaves with spore-forming chambers. (**b**) Horsetail (*Equisetum*) vegetative stem. (**c**) Strobilus of a nonphotosynthetic, fertile *Equisetum* stem. (**d**) Tree fern (*Cyathea*). Tree fern forests once cloaked much of New Zealand. Most have been cleared for farming.

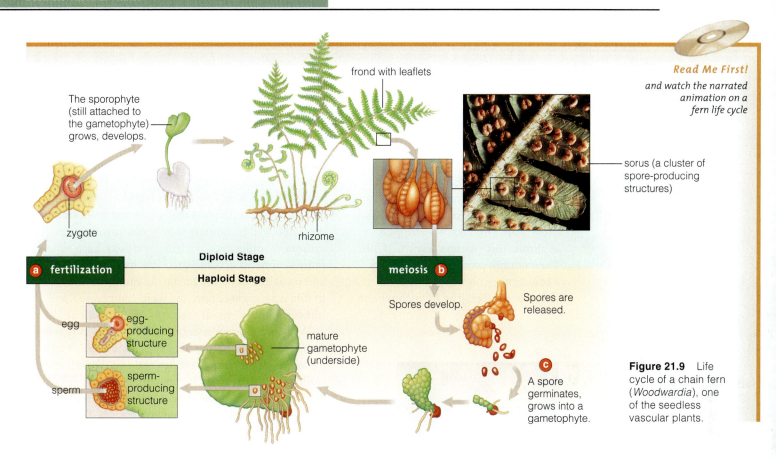

The sporophyte (still attached to the gametophyte) grows, develops.

frond with leaflets

Read Me First!
and watch the narrated animation on a fern life cycle

zygote

rhizome

sorus (a cluster of spore-producing structures)

Diploid Stage

a fertilization

Haploid Stage

meiosis b

egg
egg-producing structure

sperm
sperm-producing structure

mature gametophyte (underside)

Spores develop.

Spores are released.

c A spore germinates, grows into a gametophyte.

Figure 21.9 Life cycle of a chain fern (*Woodwardia*), one of the seedless vascular plants.

HORSETAILS

Seedless vascular plants called sphenophytes grew as tall as trees in Carboniferous swamp forests. About thirty smaller species are with us today. They are the "horsetails" (*Equisetum*) you may notice growing near many streams, railroad tracks, roads, and vacant lots.

The sporophytes of most horsetails have rhizomes, hollow stems, and scalelike leaves at the stem nodes. A cylinder of xylem and phloem runs parallel with the stems. The stems have horizontal ribs reinforced with silica granules, so they feel like sandpaper. American colonists and pioneers traveling west gathered these stems for use as disposable pot scrubbers.

Figure 21.8*b,c* shows the photosynthetic vegetative stems and pale, fertile stems of the sporophyte of one *Equisetum* species. The strobilus at each fertile stem's tip releases spores. When spores germinate, they give rise to free-living gametophytes no bigger than one millimeter to one centimeter across.

FERNS

With 12,000 or so species, the ferns are the largest and most diverse group of seedless vascular plants. All but about 380 species are native to the tropics, but we find them in homes and gardens all over the world (Figure 21.8*d*). And talk about sizes! Leaves of some floating ferns are less than 1 centimeter wide. Some tropical tree ferns grow 25 meters (82 feet) tall.

In most fern species, roots and leaves grow from vascularized rhizomes. Exceptions include tropical tree ferns and species growing as epiphytes. *Epiphyte* refers to any aerial plant that grows attached to tree trunks or branches. Young fern leaves, or fronds, develop in a coiled pattern that looks a bit like a fiddlehead, and are uncoiled by maturity. Fronds of many species, such as chain ferns, are divided into leaflets (Figure 21.9).

Rust-colored patches are distributed on the lower surface of most fern fronds. Each patch, a tiny cluster of spore-forming chambers, is called a sorus (plural, sori). Most chamber walls are only one cell thick, and when they pop open, spores are catapulted through the air. After germination, each spore gives rise to a heart-shaped gametophyte, a few centimeters across (Figure 21.9).

Seedless vascular plants include lycophytes, horsetails, and ferns. All require environmental water to complete the sexual phase of the life cycle. Large, independent sporophytes with xylem, phloem, and true leaves dominate the life cycle.

In lycophytes and horsetails, clusters of modified leaves form conelike protective structures in which spores are produced.

21.4 Ancient Carbon Treasures

Three hundred million years or so ago, in the middle of the Carboniferous, mild climates prevailed and swamp forests carpeted the wet lowlands of continents. The absence of pronounced seasonal swings in temperature favored plant growth through much of the year. Plants with lignin-reinforced tissues and well-developed root and shoot systems now had the competitive edge.

In vast forests of the Carboniferous, massively stemmed lycophyte trees —the giant club mosses—topped out at almost forty meters (Figure 21.10). Their strobili produced as many as 8 billion microspores or hundreds of megaspores. Being so high above the forest floor, dispersing these spores was a cinch. The giant horsetails, including species of *Calamites*, were close to twenty meters tall. Their aboveground stems often grew quickly from ever spreading underground rhizomes, and they formed dense thickets.

As it happened, the sea level rose and fell fifty times during the Carboniferous. Each time the sea receded, the swamp forests flourished. When the sea moved back in, forest trees became submerged and buried in sediments that protected them from decomposers. Sediments slowly compressed the saturated, undecayed remains into peat. Each time more sediments accumulated, the increased heat and pressure made the peat even more compact. In this way, organic remains were compressed and transformed into great seams of **coal** (Figure 21.10).

With its high percentage of carbon, coal is rich in energy and one of our premier "fossil fuels." It took a staggering amount of photosynthesis, burial, and compaction to form each major seam of coal in the ground. It has taken us only a few centuries to deplete much of the world's known coal deposits. Often you will hear about annual production rates for coal or some other fossil fuel. But how much do we really produce each year? None. We simply *extract* it from the ground. Coal is a nonrenewable source of energy.

Lepidodendron

stem of a giant lycophyte (*Lepidodendron*)

seed fern (*Medullosa*), one of the early seed-bearing plants

stem of a giant horsetail (*Calamites*)

Figure 21.10 Reconstruction of a Carboniferous forest. The boxed inset shows part of a seam of coal.

21.5 The Rise of Seed-Bearing Plants

In diversity, numbers, and distribution, seed producers became the most successful groups of the plant kingdom.

Seed-bearing plants made their entrance in the late Devonian. The first kinds were close relatives of ferns, but seeds, in some cases as big as walnuts, formed on their fernlike fronds (Figure 21.10). Those "seed ferns" flourished through the Carboniferous, and a few made it through a mass extinction that ended the Permian. Long before the last ones vanished, cycads, conifers, and other gymnosperms started adaptive radiations. They became the dominant plants of Mesozoic times. (What some folks call the Age of Dinosaurs, botanists call the Age of Cycads.) However, flowering plants originated during the late Jurassic or early Cretaceous. By 120 million years ago, they were diversifying even as gymnosperms were on the decline. Their adaptive radiations continued, and flowering plants have now assumed supremacy in nearly all land habitats.

Think back on the factors that promoted the rise of land plants. Structural modifications, such as a water-conserving cuticle, were one reason they endured in seasonally dry, often cold habitats. Just as important were the extraordinarily adaptive ways in which two kinds of spores evolved in the seed-bearers.

Today, we can see how their **microspores** divide and grow into pollen grains—walled structures that protect immature male gametophytes from the time they are released to the time they start journeying on their own to eggs. We see how they afford protection when conditions are dry and cold. Pollen grains are tiny, and easily dispersed by air currents or animals. Regardless of the dispersing agent, **pollination** refers to the arrival of pollen on female reproductive parts of a seed plant. The evolution of pollen grains meant that it didn't matter whether water was available or not; sperm could travel to eggs without it.

In contrast to microspores, the **megaspores** of seed plants form in ovules. Each **ovule** starts out as a tiny mass of sphorophyte tissue. The megaspore part of it gives rise to a female gametophyte, with an egg cell. Other parts form nutritious tissue and a multilayered coat. Sperm fertilize the egg and an embryo develops inside an ovule. *What we call a "seed" is a mature ovule.* Its tough coat protects the embryo when conditions force a parent sporophyte to stop growing and enter dormancy. Nutrient-rich tissue packaged inside, the equivalent of a power bar, jump-starts the embryo's renewed growth once conditions favor germination.

A footnote to this story: Seed plants appeal to more than pollinators. They appealed even to early human species. Half a million years ago, for example, *Homo erectus* was stashing nuts and rose hips in caves, and roasting seeds. By 11,000 years ago, modern humans were domesticating seed plants as reliable sources of food. We now recognize 3,000 or so species as edible, and we plant staggering numbers of about 200 species as vital crops (Figure 21.11). In this sense, agricultural practices are contributing to the ongoing evolutionary success of at least some seed-bearing plants.

pine
pollen grains

> Seed plants originated late in the Devonian. Two descendant groups, gymnosperms and flowering plants, radiated through dry land habitats with the help of structural modifications, pollen grains, and ovules that mature into seeds.

Figure 21.11 Edible treasures from flowering plants. (**a**) Fruits, which function in seed dispersal. (**b**) Mechanized harvesting of wheat, *Triticum*. (**c**) Indonesians picking shoots of tea plants (*Camellia sinensis*). Leaves of plants on hillsides in moist, cool regions have the best flavor. Only the terminal bud and two or three youngest leaves make the finest teas. (**d**) From Hawaii, a field of sugarcane, *Saccharum officinarum*. Sap extracted from its stems is boiled to make table sugar and syrups.

21.6 Gymnosperms—Plants With Naked Seeds

With this bit of history behind us, we turn to a survey of some modern gymnosperms. Their seeds are perched, in exposed fashion, on a spore-producing structure. Gymnos means naked; sperma is taken to mean seed.

The 600 species of **conifers** are woody trees and shrubs. Many have needlelike or scalelike leaves. Only conifers have female **cones**: reproductive structures with ovules wedged between clusters of papery or woody scales. Most conifers shed some leaves all year long but stay leafy, or *evergreen*. A few shed all leaves in autumn; they are *deciduous*. The most abundant trees of the Northern Hemisphere (pines), the tallest trees (redwoods), and the oldest (bristlecone pine) are conifers. Figures 21.12*a* and 21.13 show two kinds. Also in this group are firs, spruces, junipers, cypresses, larches, and podocarps.

About 130 species of **cycads** made it to the present. Pollen-bearing strobili and seed-bearing strobili form on separate plants (Figure 21.12*b*). Pollen travels from male to female plants on air currents or as "dust" on beetle pollinators. Most cycads evolved in the tropics or subtropics. Many have become ornamental plants, but some species in the wild face extinction.

Ginkgos (Figure 21.12*d–f*) were diverse in dinosaur times. The maidenhair tree, *Ginkgo biloba*, is the only surviving species. Like a few other gymnosperms, it is deciduous. Ginkgos were widely planted in China a few thousand years ago. Their natural populations nearly vanished, maybe because firewood was scarce. Ginkgos have attractive fan-shaped leaves and resist insects, disease, and air pollutants. Once again these hardy trees are widely planted—male trees, at least. The thick, fleshy seeds of female trees are stinky.

Gnetophytes include tropical trees, leathery leafed vines, and desert shrubs. An herbal stimulant, now banned, was extracted from photosynthetic stems of *Ephedra* (Figure 21.12*g*). A single *Welwitschia* species lives in African deserts. Its sporophyte has a deep taproot and a woody stem with strobili. Its two strap-shaped leaves may grow five meters long. A mature plant looks ragged because its leaves split lengthwise repeatedly during growth (Figure 21.12*h*).

What about life cycles? Consider a mature pine tree. This sporophyte has female cones with megaspore-containing ovules on tier after tier of woody scales (Figure 21.13). The ovules are exposed, not embedded

Figure 21.12 Gymnosperm diversity. (**a**) Bristlecone pine (*Pinus longaeva*) in the Sierra Nevada. Cycads: (**b**) Male strobilus of *Dioon* and (**c**) ovule-bearing, loose-leafed strobilus of *Cycas*. (**d**) The seeds of *Ginkgo biloba*. Existing ginkgo leaves (**e**) look like fossilized ones (**f**). (**g**) *Ephedra viridins* and (**h**) *Welwitschia mirabilis*.

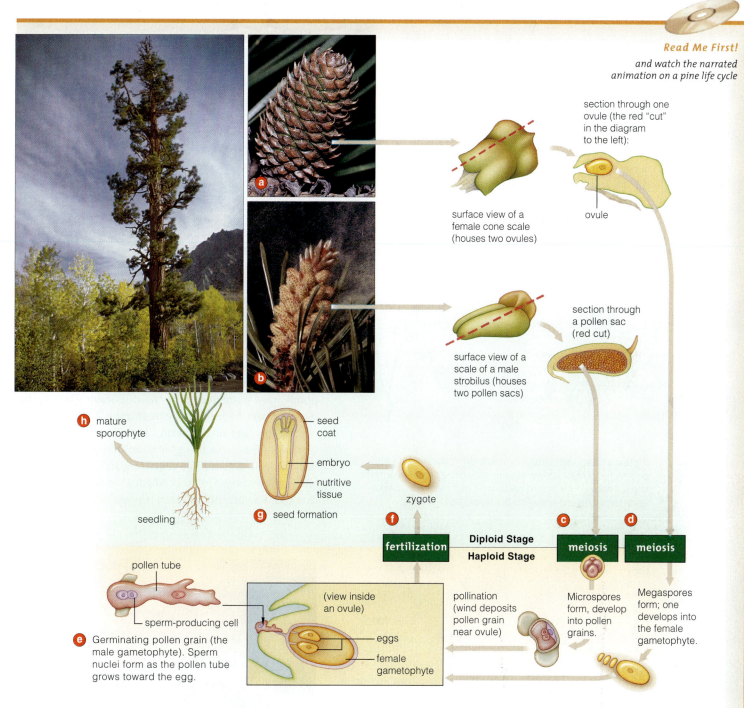

Read Me First!

and watch the narrated animation on a pine life cycle

section through one ovule (the red "cut" in the diagram to the left):

surface view of a female cone scale (houses two ovules)

ovule

section through a pollen sac (red cut)

surface view of a scale of a male strobilus (houses two pollen sacs)

h mature sporophyte

seed coat

embryo

nutritive tissue

zygote

seedling

g seed formation

f fertilization

Diploid Stage

Haploid Stage

c meiosis

d meiosis

pollen tube

pollination (wind deposits pollen grain near ovule)

Microspores form, develop into pollen grains.

Megaspores form; one develops into the female gametophyte.

(view inside an ovule)

sperm-producing cell

e Germinating pollen grain (the male gametophyte). Sperm nuclei form as the pollen tube grows toward the egg.

eggs

female gametophyte

Figure 21.13 Life cycle of a conifer, the ponderosa pine.

in woody tissue. The pine tree also has male strobili (not cones) in which microspores form and become pollen grains. At a suitable time, millions of pollen grains drift away on air currents, and some land on ovules of the same sporophyte or a different one. After this pollination event, a pollen grain germinates, and part of it starts growing as a tubular structure. This is the sperm-bearing male gametophyte. It grows slowly through tissues of the ovule for about a year. After it penetrates the female gametophyte, its sperm reaches the egg, fertilization occurs, and a new pine tree zygote starts down a developmental road.

Existing gymnosperms include conifers, ginkgos, cycads, and gnetophytes. Like their Permian ancestors, all are adapted to seasonally dry climates. Depending on the group, seeds form on exposed surfaces of strobili or female cones.

21.7 Angiosperms—The Flowering Plants

What does "angiosperm" mean? Sperma, recall, refers to seeds. The angio– part of the name, derived from a Greek word for vessel, refers to a flower's ovary. Seeds of flowering plants form inside this closed vessel.

We find flowering plants almost everywhere, from icy tundra to deserts and remote oceanic islands. What accounts for their distribution and diversity? Consider the **flower**, a specialized reproductive shoot (Figure 21.14). When plants invaded the land some 435 million years ago, insects that ate plant parts and spores were not far behind. After pollen-producing plants evolved, many kinds of insects apparently made the connection between "plant parts with pollen" and "food."

Even as flowering plants gave up some pollen to hungry insects, they gained a reproductive edge. How? Insects crawling on plants became dusted with pollen, which they deposited directly on female plant parts. Plants coevolved in ways that became more enticing to insects, as with flowers, fragrances, and nectar. Instead of wasting energy on random searches, many insects coevolved with specific plants; they became specialists in finding and collecting pollen or nectar from them—and coincidentally dispersing pollen for them.

Coevolution refers to two or more species jointly evolving because of their close ecological interactions. Heritable changes in one exert selection pressure on the other, which evolves also. That is how **pollinators**—agents that deliver pollen of one species to female parts of the same species—coevolved with seed plants. They include insects, bats, and birds. By recruiting pollinators to assist in their reproduction, flowering plants have managed to be the dominant plant group over the past 100 million years (Figure 21.15).

Did the first angiosperms look like *Archaefructus*? Complete fossils of this plant, including flowers, date back at least 125 million years (Figure 21.14). Or were early angiosperms woody shrubs or trees? We really don't know yet, and research is ongoing.

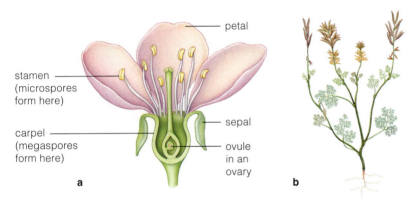

- petal
- stamen (microspores form here)
- carpel (megaspores form here)
- sepal
- ovule in an ovary

a   b

Figure 21.14 (a) Floral structures. (b) *Archaefructus sinensis*, one of the earliest known flowering plants, apparently lived in shallow lakes. As in modern flowers, it had male and female parts (stamens and carpels), but no petals or sepals.

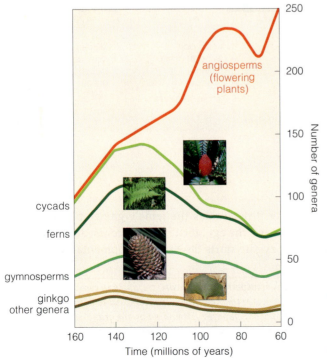

angiosperms (flowering plants)

cycads

ferns

gymnosperms

ginkgo other genera

Number of genera

Time (millions of years)

160   140   120   100   80   60

250   200   150   100   50   0

Figure 21.15 Vascular plant diversity in Mesozoic times. Conifers and other gymnosperms began to decline even before flowering plants started an adaptive radiation. The painting above showcases *Archaeanthus linnenbergeri*, which lived in the Mesozoic. In many traits, living magnolias resemble this now-extinct flowering plant.

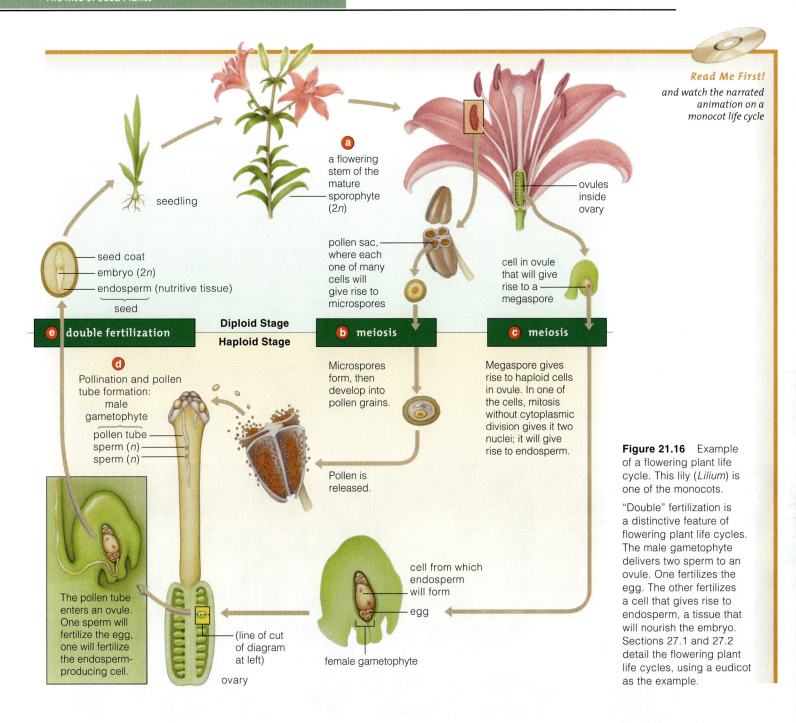

Read Me First!
and watch the narrated animation on a monocot life cycle

a a flowering stem of the mature sporophyte (2n)

seedling

seed coat
embryo (2n)
endosperm (nutritive tissue)
seed

pollen sac, where each one of many cells will give rise to microspores

ovules inside ovary

cell in ovule that will give rise to a megaspore

e double fertilization

Diploid Stage
Haploid Stage

b meiosis

c meiosis

d Pollination and pollen tube formation: male gametophyte
pollen tube
sperm (n)
sperm (n)

Microspores form, then develop into pollen grains.

Megaspore gives rise to haploid cells in ovule. In one of the cells, mitosis without cytoplasmic division gives it two nuclei; it will give rise to endosperm.

Pollen is released.

The pollen tube enters an ovule. One sperm will fertilize the egg, one will fertilize the endosperm-producing cell.

(line of cut of diagram at left)

ovary

cell from which endosperm will form
egg

female gametophyte

Figure 21.16 Example of a flowering plant life cycle. This lily (*Lilium*) is one of the monocots.

"Double" fertilization is a distinctive feature of flowering plant life cycles. The male gametophyte delivers two sperm to an ovule. One fertilizes the egg. The other fertilizes a cell that gives rise to endosperm, a tissue that will nourish the embryo. Sections 27.1 and 27.2 detail the flowering plant life cycles, using a eudicot as the example.

The adaptive radiations of flowering plants were triggered by asteroid impacts, rupturing and collisions of land masses, and changes in sea level and the global climate. Given these severe environmental challenges, and given their protected seeds and all that help from pollinators, flowering plants displaced gymnosperms and older lineages in most habitats (Figure 21.15).

Chapter 27 in the next unit offers a closer look at the structure and function of flowering plants. For now, start thinking about the monocot life cycle in Figure 21.16 and how a flower-producing sporophyte dominates it. You'll see a eudicot life cycle in Section 27.2. Seeds of both groups become packaged with a nutrient-rich tissue (endosperm), and parts of ovaries mature into structures called fruits. Fragrant, colorful, tiny, large, hard-shelled, winged, sticky—no matter what the features, all fruits function in dispersing the new generation of sporophytes.

Angiosperms are the most successful plants. They alone produce flowers. These specialized reproductive structures have pollen-producing male parts and female parts with ovaries, protected chambers in which seeds develop.

21.8 A Glimpse Into Flowering Plant Diversity

At least 260,000 species of flowering plants live in diverse habitats. They bloom in meadows and forests, in parched deserts, and on windswept mountaintops. Some thrive in lakes and streams. A few kinds have adapted to saltmarshes or to shallow marine habitats.

Almost 90 percent of living plants are angiosperms. The three major groups are **magnoliids**, **eudicots** (true dicots), and **monocots**. Ancient lineages include water

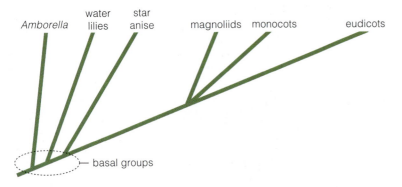

Amborella   water lilies   star anise   magnoliids   monocots   eudicots

basal groups

Figure 21.17 Evolutionary tree diagram for flowering plants.

lilies (Figure 21.17). Magnolias, avocados, nutmeg, and pepper plants are among the 9,200 kinds of magnoliids.

The 170,000 or so eudicot species include cabbages, daisies, and most other herbaceous (nonwoody) plants; most flowering shrubs and trees, and cacti.

Among 80,000 or so monocot species are orchids, palms, lilies, and grasses. Rye, sugarcane, rice, wheat, corn, and some other grasses are valued crop plants.

Wind-pollinated angiosperms usually produce tiny, inconspicuous flowers. Flashy, colorful flowers, sugar-rich nectars, and heady fragrances are evidence that a group attracts animal pollinators (Figure 21.18a–d).

In size, angiosperms range from aquatic duckweeds (one millimeter long) to towering *Eucalyptus* trees 100 meters tall. A few species are not even photosynthetic; they withdraw nutrients from other plants or get them from mychorrizal fungi (Figure 21.18e,f).

Magnoliids, eudicots, and monocots are the main lineages of flowering plants. The largest group, the eudicots, includes 170,000 named species.

Figure 21.18 (a) Examples of flowers, reproductive structures that function in pollination and seed formation. Their colors, patterns, and shapes attract pollinators. (b) This hummingbird pollinator is withdrawing nectar from the flower of a columbine (*Aquilegia*) plant.

(c) Sacred lotus (*Nelumbo nucifera*), an aquatic species. The flower's radial pattern is typical of ancient lineages.

(d) More recently evolved lineages, including these pansies (*Viola*), display a bilateral pattern, with roughly equivalent left and right parts.

(e) Indian pipe (*Monotropa uniflora*), a nonphotosynthetic species, pilfers nutrients from mycorrhizae of some photosynthetic plants. (f) Dwarf mistletoe (*Arceuthobium*). It parasitizes trees directly and stunts their growth.

21.9 Deforestation Revisited

Paralleling the huge increases in human population size are increasing demands for wood as fuel, lumber, and other forest products, and for pastures and croplands. More and more people compete for dwindling resources and for economic profit, but also because alternative ways of life simply are not available to most families.

Seed-bearing plants dominate the world's existing great forests, and they profoundly influence life. Like enormous sponges, the watersheds of these forested regions absorb, hold, then gradually release water. By mediating downstream flows, they slow erosion, flooding, and sedimentation that can disrupt rivers, lakes, and reservoirs. Strip the forests, and exposed soil is more vulnerable to leaching of nutrients and to erosion, especially on steep slopes.

Deforestation means mass removal of all trees from large tracts for logging, agriculture, and grazing. This chapter started with a look at what is going on in the once-sweeping temperate forests of the United States, Canada, New Zealand, and elsewhere. But tropical forests are the most threatened. For 10,000 years or more, these forests have been home to 50 to 90 percent of all land-dwelling species. In less than four decades, we have cleared more than half of them.

Deforestation is now greatest in Brazil, Indonesia, Colombia, and Mexico. If it continues at present rates, only Brazil and Zaire will still have large tropical forests in the year 2010. By 2035, most of their forests will be gone as well.

In tropical regions, clearing forests for agriculture sets the stage for long-term losses in productivity. The irony is that tropical forests are one of the worst places to grow crops or raise animals. Deep topsoils simply cannot form. Nutrient-rich litter and organic remains cannot build up because high temperatures and heavy, frequent rainfall promote their rapid decomposition. As fast as decomposers release nutrients, trees and other plants take them up.

Long before mechanized logging became pervasive, many people were carrying out *shifting cultivation*, once referred to as slash-and-burn agriculture. They cut and burned trees, then tilled ashes into the soil. Nutrient-rich ashes can sustain crops only for a few seasons, at most. Later, people abandon cleared plots, because heavy leaching results in infertile soil.

If a small human population practices shifting cultivation on small, widely spaced plots, they may not damage a forest much. But soil fertility plummets when human populations become more dense. Why? Larger areas are cleared, then plots are cleared again after shorter and shorter intervals.

Figure 21.19 Tropical rain forest in Southeast Asia.

Deforestation even can alter regional climates. It affects the rates of evaporation, runoff, and maybe even the pattern of rainfall. Trees release 50 to 80 percent of the water vapor in tropical forests. In a logged area, annual precipitation declines, and rain swiftly drains away from the exposed, nutrient-poor soil. A logged region gets hotter and drier—and soil fertility and moisture levels fall. In time, sparse, dry grasslands or deserts may come to prevail where the formerly dense forest once stood.

Also, tropical forests absorb much of the sunlight striking equatorial regions. Deforested land is shinier, so to speak, and reflects more incoming energy back into space. The combined photosynthetic activity of so many trees affects the global cycling of carbon and oxygen. Extensive tree harvesting and the burning of tree biomass release stored carbon, as carbon dioxide. Some researchers suspect that deforestation might be amplifying the greenhouse effect for that reason.

Conservation biologists are working to protect the forests. As three examples, a coalition of 500 groups is dedicated to preserving Brazil's remaining tropical forests. In India, women have already constructed and installed 300,000 inexpensive, smokeless wood stoves. In the past decade, the stoves saved more than 182,000 metric tons of trees by cutting demands for fuelwood. In Kenya, women planted millions of trees as a source of wood and to counter erosion.

Table 21.1 Comparison of Major Plant Groups

Bryophytes   24,000 species. Moist, humid habitats.

Nonvascular land plants. Fertilization requires standing water. Haploid dominance. Cuticle and stomata present in some.

Club mosses   1,100 species; simple leaves. Mostly wet or shady habitats.

Horsetails   25 species of single genus. Swamps and disturbed habitats.

Ferns   12,000 species. Wet, humid habitats in mostly tropical and temperate regions.

Seedless vascular plants. Fertilization requires standing water. Diploid dominance. Cuticle and stomata present in all.

Conifers   600 species, mostly evergreen, woody trees, shrubs with pollen- and seed-bearing cones. Widespread distribution.

Cycads   130 slow-growing tropical, subtropical species.

Ginkgo   1 species, a tree with fleshy-coated seeds.

Gnetophytes   70 species. Limited to some deserts, tropics.

Gymnosperms—vascular plants with "naked seeds." Water not required for fertilization. Diploid dominance. Cuticle and stomata present in all.

Flowering plants

Monocots   80,000 species. Floral parts often arranged in threes or in multiples of three; one seed leaf; parallel leaf veins common; pollen with one furrow.

Eudicots (true dicots)   At least 170,000 species. Floral parts often arrayed in fours, fives, or multiples of these; two seed leaves; often net-veined leaves; pollen with three or more pores and/or furrows.

Magnoliids and basal groups   9,200 species. Many spirally arranged floral parts or fewer, in threes; two seed leaves; net-veined leaves common; pollen with one furrow.

Angiosperms—vascular plants with flowers, protected seeds. Water not required for fertilization. Diploid dominance. Cuticle and stomata in all.

Table 21.2 Evolutionary Trends Among Plants

Bryophytes	Ferns	Gymnosperms	Angiosperms
Nonvascular	Vascular		
Haploid dominance	Diploid dominance		
Spores of one type	Spores of two types		
Motile gametes		Nonmotile gametes*	
Seedless		Seeds	

* Require pollination by wind, insects, animals, etc.

Summary

Section 21.1   Plants are multicelled eukaryotes. Nearly all are photoautotrophs on land. Table 21.1 summarizes the main characteristics of major groups.

Comparisons among lineages reveal trends in plant evolution. Plants on land became structurally modified; a waterproof cuticle, stomata, and vascular tissues evolved in most groups. The haploid phase, dominant in algal life cycles, gave way to diploid dominance. Complex sporophytes that hold, nourish, and protect spores and gametophytes arose. Two types of spores, not one, evolved. For gymnosperms and angiosperms, heterospory paved the way for the evolution of pollen grains and seeds. Table 21.2 summarizes these trends.

Section 21.2   Mosses, liverworts, and hornworts are bryophytes. These nonvascular plants do not have xylem and phloem. Their flagellated sperm reach eggs only by swimming through films or droplets of water. The sporophyte begins to develop inside gametophyte tissues. It remains attached to and dependent upon the gametophyte even when mature.

Section 21.3   Lycophytes, horsetails, and ferns are seedless vascular plants. Sporophytes with leaves and vascular tissues are the larger, longer lived phase of the life cycle. Like bryophytes, their flagellated sperm must swim through water to reach the eggs.

Section 21.4   Giant bryophytes dominated swamp forests during the Carboniferous. Coal is the energy-rich, compressed organic remains of these forests.

Section 21.5   Gymnosperms and flowering plants (angiosperms) are seed-bearing vascular plants. They produce microspores that give rise to pollen grains, from which sperm-bearing male gametophytes develop. They also produce megaspores that give rise to female gametophytes (with eggs). This happens in ovules, other parts of which form nutritive tissue and a seed coat. A seed is a mature ovule.

Section 21.6   Gymnosperms, including conifers, cycads, ginkgos, and gnetophytes, are adapted to dry climates. Their ovules form on exposed surfaces of strobili or, in the case of conifers, of female cones.

Section 21.7   Angiosperms alone produce flowers, and their seeds form inside floral ovaries. Some ovary tissues later develop into fruits.

Section 21.8   Angiosperms are the most diverse and widespread group of plants. Most species coevolved with animal pollinators. The two largest groups are monocots and eudicots. Magnoliids and basal groups are a more ancient lineage.

Section 21.9   Deforestation threatens ecosystems worldwide. Forests play a role in the global cycling of water, carbon dioxide, and oxygen.

Self-Quiz

Answers in Appendix III

1. Which of the following statements is *not* correct?
   a. Gymnosperms are the simplest vascular plants.
   b. Bryophytes are nonvascular plants.
   c. Lycophytes and angiosperms are vascular plants.
   d. Only angiosperms produce flowers.

2. Which does *not* apply to gymnosperms or angiosperms?
   a. vascular tissues   c. single spore type
   b. diploid dominance   d. cuticle with stomata

3. Of all land plants, bryophytes alone have independent
   _____ and attached, dependent _____ .
   a. sporophytes;   c. rhizoids; zygotes
      gametophytes   d. rhizoids; stalked
   b. gametophytes;   sporangia
      sporophytes

4. Lycophytes, horsetails, and ferns are classified as
   _____ plants.
   a. multicelled aquatic   c. seedless vascular
   b. nonvascular seed   d. seed-bearing vascular

5. The _____ produce flagellated sperm.
   a. ferns   c. monocots
   b. conifers   d. a and c

6. A seed is _____ .
   a. a female gametophyte   c. a mature pollen tube
   b. a mature ovule   d. an immature embryo

7. Match the terms appropriately.
   ____ gymnosperm   a. gamete-producing body
   ____ sporophyte   b. help control water loss
   ____ lycophyte   c. "naked" seeds
   ____ ovary   d. only plant that produces
   ____ bryophyte   flowers
   ____ gametophyte   e. spore-producing body
   ____ stomata   f. nonvascular land plant
   ____ angiosperm   g. seedless vascular plant
                       h. ovules form in it

Critical Thinking

1. Figure 21.20 shows a Nahmint Valley forest in British Columbia after logging. It also shows the wood frames of homes being built. Would you chain yourself to a tree in such a forest to stop loggers? If your answer is yes, would you also give up the chance to own a wood-frame home (as most homes are, in developed countries)? What about forest products, such as paper tablets, newspapers, toilet tissue, wood furniture, and fireplace wood?

2. With respect to question 1, multiply each of your answers by 6.3 billion people and describe what might happen when we run out of trees. Research the pros and cons of tree farms—say, of a single species of pine.

3. The 2004 Nobel Peace Prize was awarded to Wangari Maathai, who founded the Green Belt Movement in 1977 (Figure 21.21). Since then, group members—mostly poor rural women—have planted more than 25 million trees. Reforestation efforts help halt soil erosion, provide sources of fruits, food, and firewood, and empower women to work together for the long term good of their communities. Are any nonprofit groups working to protect forests in your area? Do you support their goals and methods?

Figure 21.20 Part of a logged-over forest in British Columbia.

Figure 21.21 Kenyan Wangari Maathai, who in 2004 won a Nobel Peace Prize for her work in support of women's rights and sustainable development.

Media Menu

Student CD-ROM

Impacts, Issues Video
   Beginnings and Endings
Big Picture Animation
   Origin and evolution of the plants
Read-Me-First Animation
   Moss life cycle
   Fern life cycle
   Pine life cycle
   Monocot life cycle
Other Animations and Interactions
   Evolutionary trends in major plant groups
   Evolutionary tree for major plant groups

InfoTrac

- Doing the Peat Bog Two-Step. *Mother Earth News*, June 2001.
- A Fern with a Taste for Toxic Waste. *Environment*, May 2001.
- Rocks Yield Clues to Flower Origins. *Science News*, April 21, 2001.
- The Future of Our Forests. *Audubon*, January 2001.
- Flower Power. *Odyssey*, March 2003.

Web Sites

- Botanical Society of America: www.botany.org
- Smithsonian Bryophytes: bryophytes.plant.siu.edu
- American Fern Society: amerfernsoc.org
- Wayne's Word: waynesword.palomar.edu/wayne.htm
- USDA Plants Database: plants.usda.gov

How Would You Vote?

In the United States, the average person uses about 750 pounds of paper each year, twice the amount in other developed countries. Recycling paper saves trees, lessens the impact on landfills, and reduces water pollution. However, recycled paper is often more expensive to produce than new paper. The government could encourage use of recycled paper by giving companies tax breaks for using it. Would you support such a program?

IMPACTS, ISSUES *Old Genes, New Drugs*

Other than Samoans, few of us prize cone snails (*Conus*) as tasty treats. However, the intricate patterns of cone snail shells obviously have been prized down through the ages. Archaeologists exploring past cultures in Peru, Arizona, Micronesia, and Iran uncovered bracelets, rings, and ritual objects made from *Conus* shells. Collectors still covet the shells, the most beautiful of which still end up in jewelry and other forms of decorative art.

Cone snails also fascinate biologists, for different reasons. More than 500 species of these predatory mollusks inhabit the seas, and all of them subdue small prey with *conotoxins*. These paralytic secretions pack a wallop that can kill even large animals. Seventy percent of the people who were harpooned by a cone snail and were not treated promptly ended up dead.

Conotoxins are peptides that can bind to channel proteins of cell membranes. Any cone snail species can make 100 to 300 conotoxins, each with a highly specific molecular target. What's so fascinating about toxin diversity and specificity of targets? It means cone snails are potential sources of many new drugs.

One *C. magnus* conotoxin is 1,000 times more potent than morphine—yet nonaddictive. It works by blocking channels where nerve cells release signaling molecules that can alter activities of neighboring cells. A synthetic version is now being tested in clinical trials. The test groups consist of cancer and AIDS patients who are in unbearable but otherwise untreatable pain.

C. geographicus, shown in Figure 22.1, secretes a toxin that one day might help epileptics. While studying this

Figure 22.1 *Conus geographicus* engulfing a small fish. This mollusk's siphon (extended straight up) detects disturbances in the water, as happens when small fishes and other prey swim within range. *C. geographicus* impaled this fish with a harpoon-like device, then pumped paralyzing conotoxins into it. The filmstrip (*far right*) shows just a tiny sampling of the diverse patterns of *Conus* shells.

the big picture

So What Is an Animal? Animals are alike in many respects, but differ in their type of body symmetry, whether they have a head and segments, and in their type of gut and of body cavity, if any. The invertebrate animals evolved first and far outnumber vertebrates.

Getting Along Without Organs Sponges, an ancient lineage, have endured with no body symmetry, tissues, or organs. Cnidarians—the jellyfishes and their relatives—have a radial type of body symmetry, and also tissues, but no organs or definite head.

species, University of Utah researchers discovered a gene that has been conserved in the DNA of different species for a long, long time. In cone snails, the gene codes for an enzyme, gamma-glutamyl carboxylase (GGC), that catalyzes a step in the conotoxin synthesis pathway. Oddly, humans make GGC, but it functions in blood clotting processes. Fruit flies make it even though they don't secrete toxins and their blood doesn't clot. To date, no one knows what GGC does in fruit flies.

We can expect that the GGC gene arose more than 500 million years ago, before genetic divergences from a common ancestor gave rise to mollusks, insects, and vertebrates. As an outcome of different mutations in the independently evolving lineages, the enzyme product of the ancestral gene took on different functions.

The point is this: Carefully look back through time and you will see how organisms interconnect. *At every branch point in the animal family tree, microevolutionary processes gave rise to changes in heritable traits—in biochemistry, body plans, and behavior.* Your uniquely human traits emerged through modification of traits that evolved in countless generations of vertebrates and, before them, in ancient invertebrate forms.

This chapter and the next compare the major groups of animals, from the structurally simplest to the most complex. Don't assume that the more ancient lineages are somehow evolutionarily stunted or "primitive." As you will see, even the simplest among them are exquisitely adapted to their environment.

 How Would You Vote?

Cone snails are diverse, but most kinds have a limited geographic range, which makes them highly vulnerable to extinction. We don't know how many are harvested, because no one monitors the trade. Should the United States push to extend regulations on trade in endangered species to cover any species captured from the wild? See the Media Menu for details, then vote online.

Protostomes   Most animals have tissues and organ systems, bilateral symmetry, and a head. Many of them also have a type of body cavity called a coelom. These features emerge in a particular way in the embryos of protostomes, one of two major lineages of bilateral animals.

Invertebrate Deuterostomes   Deuterostomes are bilateral animals that share some features with protostomes, but these features develop in a different way in their embryos. The deuterostome lineage includes chordates as well as the radial invertebrates called echinoderms.

22.1 Overview of the Animal Kingdom

We already know of more than 2 million animal species, and no simple definition can cover that much diversity.

GENERAL CHARACTERISTICS

Animals share many traits with other organisms, and no one trait sets them apart. For that reason, we must define **animals** by a list of their characteristics:

1. All animals are multicelled, and most have tissues, organs, and organ systems. They have no walled cells.

2. Animals are aerobic heterotrophs that ingest other organisms or withdraw nutrients from them.

3. Animals typically reproduce sexually; many also reproduce asexually. Their embryos usually develop through a series of stages.

4. Most animals are motile; they actively move about during all or part of the life cycle.

These shared traits arose in a common ancestor. Later on, different traits accumulated in dozens of lineages.

Mammals, birds, reptiles, amphibians, and fishes are all **vertebrates**, with a backbone. These are the most familiar lineages. Yet all but 50,000 of 2 million named species are **invertebrates**! Invertebrates have many features, but a backbone isn't one of them (Table 22.1).

CLUES IN BODY PLANS

How might we get a conceptual handle on animals as different as flatworms and dinosaurs, hummingbirds and humans? Comparing similarities and differences with respect to six basic features is one way to start. These features are body symmetry, embryonic tissue layers, cephalization, type of gut, type of body cavity, and segmentation.

BODY SYMMETRY, CEPHALIZATION   Nearly all animal bodies are radial or bilateral. With **radial symmetry**, body parts are arranged in a regular pattern around a central axis, like spokes of a bike wheel (Figure 22.2). Radial animals live in water, and their body plan is adapted to intercept food that is equally likely to drift or swim toward them from any direction.

Most animals have **bilateral symmetry**. In this body type, there is an *anterior* end (front) and *posterior* end (back). The body has right and left sides, a *ventral* surface (the underside) and *dorsal* surface (upper side). Early on, bilateral animals moving along in a forward direction apparently favored **cephalization**: Being the first to encounter food, danger, and other stimuli, their anterior end evolved into a distinct *head*, in which many nerve and sensory cells became concentrated. Some cells were the start of sensory organs and a complex brain in many bilateral lineages.

TISSUE LAYERS IN EMBRYOS   Section 8.4 introduced the nature of cleavage, the first stage of animal development. Mitotic cell divisions transform an animal zygote into a tiny ball of cells in this stage. After this, primary tissue layers form in all animals except sponges, and they will give rise to adult tissues and organs. In structurally simple radial animals, two layers form.

Table 22.1	Major Animal Groups	
Group	Some Representatives	Existing Species
Poriferans	Sponges	8,000
Cnidarians	Hydras, jellyfishes	11,000
Flatworms	Turbellarians, flukes, tapeworms	15,000
Annelids	Leeches, earthworms, polychaetes	15,000
Mollusks	Snails, slugs, squids, octopuses	110,000
Arthropods	Crustaceans, spiders, insects	1,000,000+
Roundworms	Pinworms	20,000
Echinoderms	Sea stars, sea urchins, sea cucumbers	6,000
Chordates	Invertebrate chordates: Tunicates, lancelets	2,100
	Vertebrate chordates:	
	Jawless fishes	84
	Jawed fishes	21,000
	Amphibians	4,900
	Reptiles	7,000
	Birds	8,600
	Mammals	4,500

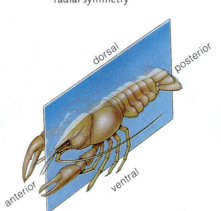

radial symmetry

bilateral symmetry

Figure 22.2   Simple way to think about radial and bilateral symmetry in animal body plans.

The outer **ectoderm** is the start of a tissue that lines body surfaces. The inner **endoderm** develops into a gut lining. Ectoderm and endoderm are the source of additional body parts in bilateral animals. In the embryos of these animals, a third layer—**mesoderm**—forms between the other two. This third layer is the source of the muscles and many other organs in all animals more complex than jellyfish and their relatives.

TYPE OF GUT   A **gut** is either a sac with one opening or a tube with two openings to the outside. Food is digested in the gut, then absorbed into the internal environment. A saclike gut is an *incomplete* digestive system, and it was the first kind to evolve. Cnidarians have such a gut.

A tubular gut, called a *complete* digestive system, starts at a mouth and ends at an anus. It forms in the embryos of two major lineages of bilateral animals, but not the same way. In both cases, it starts at or near the first opening (a blastopore) that develops on the embryo's surface. In **protostomes**, such as flatworms, mollusks, annelids, roundworms, and arthropods, the opening persists as the mouth. It becomes the anus in the echinoderms, chordates, and other **deuterostomes**. We return to these two lineages at the chapter's end, after tracing pivotal steps in their evolution.

BODY CAVITIES   Flatworms, like a few other groups of invertebrates, have no body cavity; they just have tissues and organs packed between the gut and body wall (Figure 22.3a). But most bilateral animals have a body cavity of the type known as a **coelom**. It lies between the gut and body wall and has a lining—the *peritoneum*—derived from mesoderm (Figure 22.3c).

The coelom develops differently in the embryos of protostomes and deuterostomes. Either way, it was a key innovation. Organs that formed inside this cavity were cushioned in fluid, which helped protect them from shocks. The organs also were free to move, grow, and develop independently of the body wall.

Some protostomes have a reduced coelom or none at all. The coelom apparently was reduced or lost as a result of mutations in certain genes that affected early development. Roundworms, as one example, have a **pseudocoel** (false coelom). Their main body cavity is only partly lined with tissue derived from mesoderm (Figure 22.3b).

SEGMENTATION   One of the other characteristics that help define bilateral groups is the presence or absence of segments. **Segmentation** refers to the division of a body into interconnecting units that are repeated one

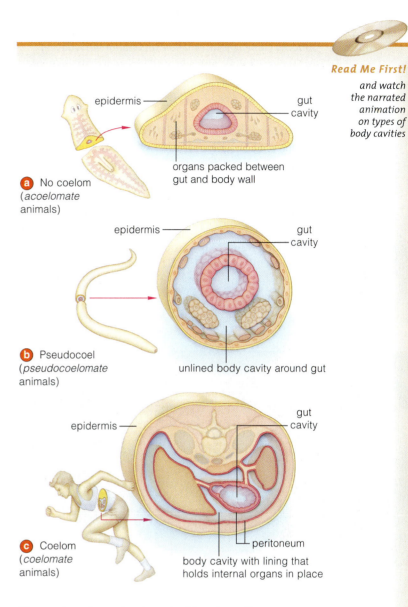

Read Me First!

and watch the narrated animation on types of body cavities

a No coelom (*acoelomate* animals)
epidermis
gut cavity
organs packed between gut and body wall

b Pseudocoel (*pseudocoelomate* animals)
epidermis
gut cavity
unlined body cavity around gut

c Coelom (*coelomate* animals)
epidermis
gut cavity
peritoneum
body cavity with lining that holds internal organs in place

Figure 22.3   Type of body cavity (if any) in different groups of animals.

after the next along the main body axis. This body plan evolved early in some groups. Most annelids have pronounced segmentation; other groups have fused segments or segments with specialized appendages. We see clues to our segmented origins in human embryos.

Animals are multicelled heterotrophs. Nearly all have tissues, organs, and organ systems. They reproduce sexually, and many can reproduce asexually. Embryos develop in stages. Most species are motile for at least part of the life cycle.

Animals differ in body symmetry, cephalization, type of gut, type of body cavity, and segmentation. The vast majority have no backbone; they are invertebrates.

22.2 Animal Origins

Judging from genetic evidence and radiometric dating of fossilized tracks, burrows, and pinhead-sized embryos, divergences from protist lineages gave rise to animals somewhere between 1.2 billion and 670 million years ago.

How did animals arise? Perhaps they evolved from colonies of flagellated cells. Over time, some of the cells could have mutated and taken on reproduction and other specialized tasks. Such a division of labor is typical of multicellularity. The choanoflagellates offer support for this theory. They are single-celled protists that form colonies. Each cell has a ring of absorptive structures (microvilli) around the base of a flagellum. As you will see, body cells of sponges have a similar structure. Comparisons of DNA sequences also point to choanoflagellates as the group most likely to have been involved in the early stages of animal evolution.

Suppose a colony of cells flattened out over time and started creeping along on the seafloor. Their lifestyle could have favored the evolution of cell layers, such as those of *Trichoplax adhaerens*. This tiny, soft-bodied marine animal is the only living placozoan, and it can be viewed as the simplest animal (Figure 22.4a). Its several thousand cells are arranged in two layers. When *Trichoplax* glides over food, gland cells in the lower layer secrete digestive enzymes, and other cells absorb breakdown products.

By 610 million years ago, tiny, soft-bodied animals shaped like fronds, disks, and blobs were living on or in seafloor sediments (Figure 22.4b,c). We call them the **Ediacarans**. Some were similar to living groups and may be related to them. Most Ediacarans had a flattened body form, suitable for absorbing dissolved organic compounds. So far, fossils show no evidence of predatory species.

Things got rougher in the Cambrian, when animals started a great adaptive radiation. Large, aggressive predators emerged, as did spiny, shelled, and armored prey. Before the Cambrian ended, all major groups of animals had originated in the seas. What caused that Cambrian explosion of diversity? Changes in the land masses, sea level, and climates may have triggered it. Also, new predators and new prey exerted selection pressure on one another.

We have much to learn about animal origins, but we know all existing lineages are one monophyletic group. All are descended from a common ancestor, as outlined in the evolutionary tree in Figure 22.5.

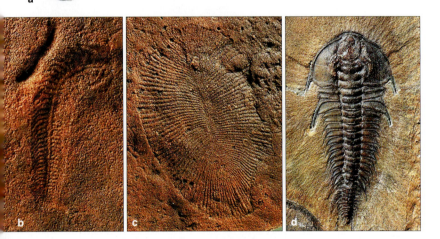

Figure 22.4 (**a**) The only living placozoan; the soft-bodied *Trichoplax adhaerens*. Of all animals, it has the smallest genome. Fossilized Ediacarans, (**b**) *Spriggina* and (**c**) *Dickensonia*. The oldest known Ediacarans lived between 610 million and 510 million years ago. (**d**) Fossil of one of the earliest trilobites.

Animal life dates back more than 600 million years, but an adaptive radiation in the Cambrian gave rise to the ancestors of all major groups with living representatives.

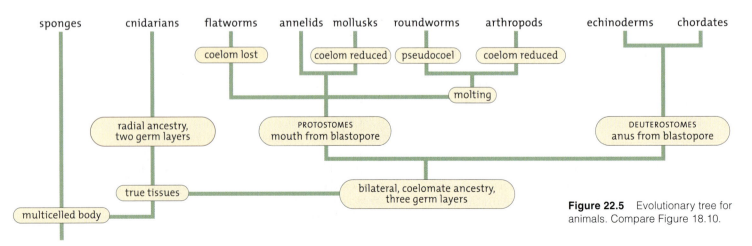

Figure 22.5 Evolutionary tree for animals. Compare Figure 18.10.

22.3 Sponges—Success in Simplicity

Yes, you can endure through time with simplicity, as demonstrated by the sponges.

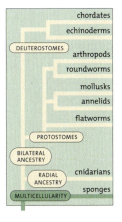

The **sponges**—animals with no symmetry, tissues, or organs— are one of nature's true success stories. They have thrived since precambrian times. Most live in shallow tropical seas, but some live in deep-sea trenches, cold Antarctic seas, and freshwater. Some sponges are large enough to sit in, others are as small as a fingernail. Different kinds have flattened and sprawling, lobed, compact, or tubelike shapes, as in Figure 22.6. Cells line all body surfaces but don't form distinct tissues, as seen in more complex animals.

Sponges are typically "filter feeders." Water flows into a sponge through tiny pores and chambers in the body, then out through one or more large openings. It does so because of the beating of the flagella of collar cells. These cells make up the lining of the chambers. A ring of microvilli—long, thin extensions from the cell body—acts as a food-trapping "collar" (Figure 22.6*b*). The cells engulf bacteria or organic debris flowing past. They release some of it to amoeba-like cells that prowl between the linings. These cells process food and carry out other tasks that benefit the body.

Sea slugs are among the relatively few predators that eat sponges. Defenses include protein fibers and sharp spicules of calcium carbonate or silica that stiffen the sponge body. Predators learn that sampling a sponge is like eating a mouthful of glass splinters. Some species coat themselves in slime. Others stink.

As a group, sponges have a large chemical arsenal against predators, parasites, and pathogens. They also harbor symbionts that secrete chemicals and help get rid of wastes. The chemicals intrigue researchers, who view them as potential drugs to fight inflammation, viruses, microbes, and tumors.

Mature sponges don't move about, so how do they deal with sex? Some release sperm into water, which might carry the sperm to another sponge. Sperm of some sponges can fertilize eggs produced by the same colony, which they then retain in the body for a time. Zygotes develop into ciliated larva. A **larva** (plural, larvae) is a free-living, sexually immature stage of animal development, one that precedes the adult form. Sponge larvae have flagella which enable them to briefly swim about until they settle on a suitable spot and grow up to be adults.

Many sponges reproduce asexually by budding or fragmentation. Small buds or pieces break away and grow into new sponges. Under stressful conditions, some freshwater species make gemmules; they encase clumps of cells in a hardened coat. Gemmules survive oxygen-poor water, drying out, and freezing. Later on, when conditions improve, each may be the start of a new sponge.

Sponges have no symmetry, tissues, or organs; they are at the cellular level of construction. Yet their lineage has endured since precambrian times. Fibers and spicules in the body wall and chemical defenses help keep predators away.

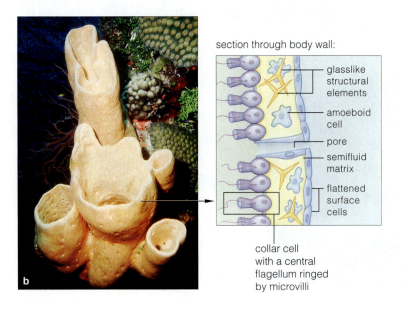

Figure 22.6 (**a**) An encrusting sponge growing on a ledge in a temperate sea. (**b**) A vase-shaped sponge. Flagellated, phagocytic cells line canals and chambers in the body wall. Each has a collar of food-trapping structures (microvilli). Fine filaments connect microvilli to one another, forming a collarlike "sieve." The sieve strains food particles from water flowing through it. Cells at the collar's base engulf trapped food.

section through body wall:
- glasslike structural elements
- amoeboid cell
- pore
- semifluid matrix
- flattened surface cells
- collar cell with a central flagellum ringed by microvilli

22.4 Cnidarians—Simple Tissues, No Organs

All cnidarians are at the tissue level of organization. Their forms are diverse, but all are variations on two radial body plans—the medusa and polyp.

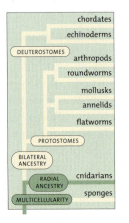

The radial, tentacled animals called **cnidarians** include sea anemones, hydroids, corals, and jellyfishes. Most species live in the seas. They have two tissue layers but no internal organs. They alone make capsules with dischargeable threads stuffed inside. These **nematocysts** stud the body surface (Figure 22.7). Different kinds release toxins, entangle prey, or pierce them with barbs. Most nematocyst toxins are harmless to humans. But some do irritate tissues, and a few can be deadly.

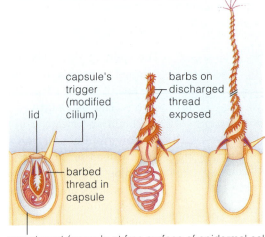

nematocyst (capsule at free surface of epidermal cell)

Figure 22.7 Nematocyst before and after prey (not shown) touched its trigger and made the capsule "leakier" to water. As water diffused in, turgor pressure built up and forced the thread to turn inside out. The barbed tip pierces prey.

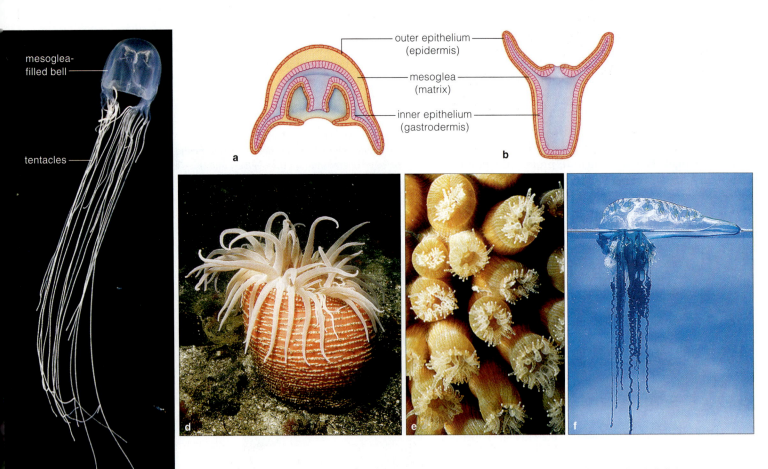

Figure 22.8 Cnidarian body plans: (**a**) medusa and (**b**) polyp. (**c**) Sea wasp (*Chironex*), a jellyfish with tentacles about three meters long. Its toxin can kill you within minutes. (**d**) Sea anemone and (**e**) part of a colony of reef-building corals. The external walls of reef-building corals interconnect. (**f**) Portuguese man-of-war (*Physalia utriculus*), a colonial cnidarian equipped with a float.

feeding polyp

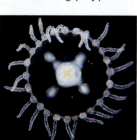

medusa

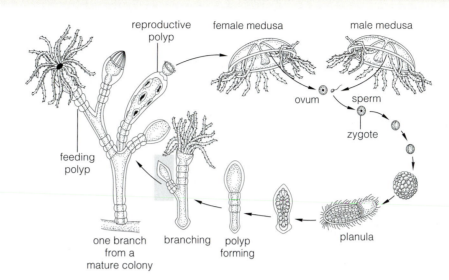

Read Me First!
and watch the narrated animation on a cnidarian life cycle

Figure 22.9 Life cycle of a hydroid (*Obelia*). The medusa is a sexual stage; it produces eggs or sperm. After zygotes form, they develop into planulas, a type of ciliated larva. After crawling about briefly, the planula settles down and develops into a polyp. More polyps form by asexual reproduction. A mature colony often contains thousands of feeding polyps, as well as reproductive polyps that give rise to medusae.

The most common body forms, the medusa and polyp, have a saclike gut. Depending on the group, medusae, polyps, or both form during the life cycle. Medusae look a bit like bells or saucers with tentacles around the rim, and they move freely through water (Figure 22.8a,c). Polyps are tube shaped, and usually one end attaches to a substrate. The gut opening is at the polyp's other end, and a ring of tentacles around it captures prey from the water (Figure 22.8b,d,e).

Like all other animals except sponges, cnidarians have **epithelia** (singular, epithelium): sheetlike tissues that line the free surfaces of body parts. A specialized type lines the cnidarian gut cavity; its gland cells secrete digestive enzymes. Another type, epidermis, covers the body's outer surfaces.

Nerve cells threading through cnidarian epithelia form a *nerve net*. This simple nervous system interacts with contractile cells embedded in the epithelia. By controlling contraction, the nerve net causes simple changes in shape that make the body move. How? Between the linings is mesoglea, a secreted gelatinous mass with a few scattered cells. It acts as a buoyant, deformable skeleton; the jellyfish bell moves when contractions squeeze water out from under it.

Any fluid-filled cavity or cell mass that contractile cells act against is a *hydrostatic* skeleton. It is one of several kinds of skeletons, all of which interact with contractile cells to bring about movement.

Although alike in body plans, cnidarians differ in their life cycles. Sea anemones produce only polyps, which for them is a sexual stage with gonads (gamete-producing reproductive organs). The hydroid *Obelia* produces polyps and medusae (the sexual stage), and planulas develop from zygotes (Figure 22.9). A planula is a bilateral, usually ciliated larva, and it moves about briefly before developing into an adult.

Besides this, many cnidarians grow as colonies. For instance, polyps of colonial corals are cemented side by side in their own "houses" of calcium carbonate secretions (Figure 22.8e). Their compressed remains help form tropical reefs. Coral polyps are among the cnidarians with dinoflagellate symbionts. They offer their photoautotrophic guests safe habitats and access to dissolved carbon dioxide and minerals. The guests supply their hosts with oxygen and recycle mineral wastes. They also contribute to house-building; their metabolic activities alter the pH of seawater in a way that promotes the deposition of calcium carbonate.

Finally, consider the Portuguese man-of-war. A blue, gas-filled float keeps this colonial cnidarian near the sea surface. Groups of polyps and medusae interact in feeding, reproduction, and other tasks (Figure 22.8f).

Cnidarians are the simplest living animals with cells that are structurally and functionally organized in tissue layers.

22.5 Flatworms—The Simplest Organ Systems

Beyond the cnidarians are animals ranging from simple worms to humans. Most are bilateral; all have organs.

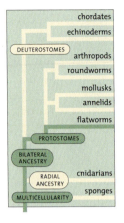

Organs are structural units of two or more tissues arrayed in certain patterns and interacting in a common task. Two or more organs interacting chemically, physically, or both in common tasks make up an **organ system**. The flatworms are the simplest animals with organs that form from three tissue layers in the embryo—ectoderm, endoderm, and *mesoderm*. They include the mostly free-living turbellarians, and parasitic flukes and tapeworms. Although these bilateral, cephalized animals have no coelom, they probably had coeolomate ancestors. In body form, the flatworms slightly resemble cnidarian planulas.

Most of the 15,000 flatworms are hermaphrodites. Each has male and female gonads and a penis. The turbellarians classified as planarians can reproduce asexually by *transverse* fission. The body splits in two, and each piece regrows the missing parts. Planarians respond similarly to injury. Cut one in pieces and you can end up with several planarians.

Most planarians live in freshwater. Like the marine turbellarians, they prey on or scavenge tiny animals. Figure 22.10 shows organ systems of one planarian. The pharynx is a muscular tube between the mouth and saclike gut. It sucks up food and expels wastes.

The head has light-detecting eyespots and clusters of nerves that form a rudimentary brain. Dorsal nerve cords relay messages through the body. A system of tubular protonephridia (singular, protonephridium) helps control the composition and volume of internal fluids. It extends from pores at the body surface to bulb-shaped flame cells. When excess water diffuses into flame cells in the body, a tuft of cilia "flickering" in the bulb drives water through tubes to the outside.

Flukes and tapeworms are parasites that live in or on one to four different hosts during their life cycle. Generally, we find a parasite's mature stage in the *definitive* host. Immature stages live in one or more *intermediate* hosts. As most parasites do, flukes and tapeworms withdraw nutrients from cells or tissues of a host. An infected host loses weight, weakens, and becomes vulnerable to other parasites and pathogens.

Tapeworms have a scolex with barbs or hooks at one end. They use it to attach to the wall of a vertebrate's intestine. None has a mouth or a gut; nutrients diffuse across the body wall. The chemical secretions of some species slow the muscle contractions that propel food

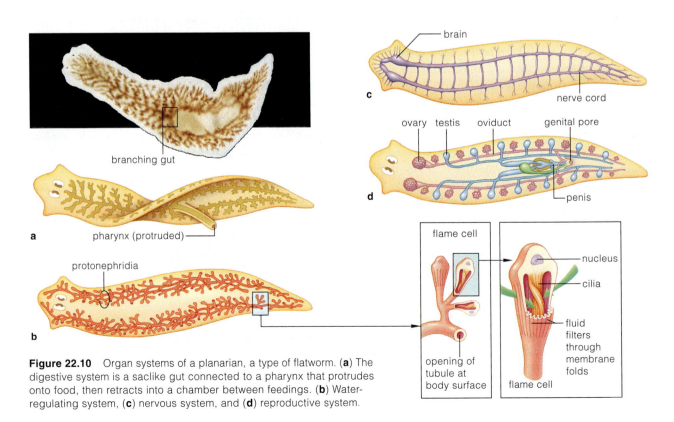

Figure 22.10 Organ systems of a planarian, a type of flatworm. (**a**) The digestive system is a saclike gut connected to a pharynx that protrudes onto food, then retracts into a chamber between feedings. (**b**) Water-regulating system, (**c**) nervous system, and (**d**) reproductive system.

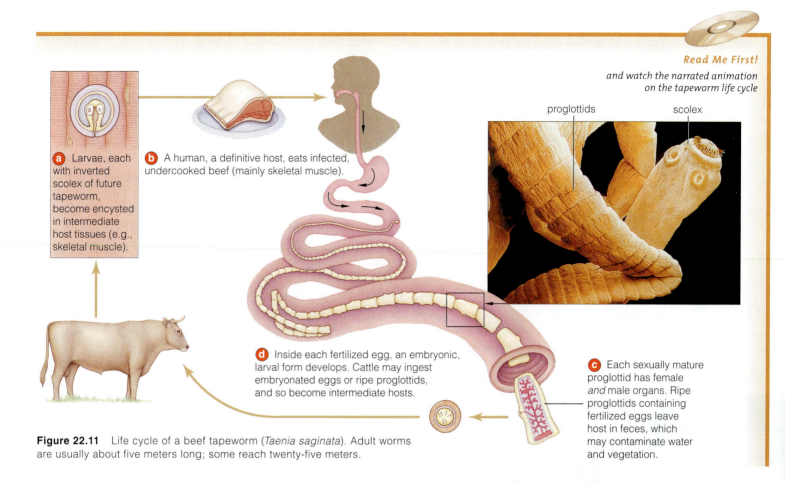

Read Me First!
*and watch the narrated animation
on the tapeworm life cycle*

proglottids   scolex

a Larvae, each with inverted scolex of future tapeworm, become encysted in intermediate host tissues (e.g., skeletal muscle).

b A human, a definitive host, eats infected, undercooked beef (mainly skeletal muscle).

d Inside each fertilized egg, an embryonic, larval form develops. Cattle may ingest embryonated eggs or ripe proglottids, and so become intermediate hosts.

c Each sexually mature proglottid has female *and* male organs. Ripe proglottids containing fertilized eggs leave host in feces, which may contaminate water and vegetation.

Figure 22.11   Life cycle of a beef tapeworm (*Taenia saginata*). Adult worms are usually about five meters long; some reach twenty-five meters.

through the host's gut. Their action prolongs the time the tapeworm spends in a nutrient-rich "habitat." The chemicals may be potential sources of drugs to slow contractions in humans who have an overactive gut.

Some tapeworm larvae enter humans in tissues of raw, undercooked, or improperly pickled pork, beef, or fish (Figure 22.11). New units of the tapeworm body, called proglottids, bud behind the scolex. Each unit is hermaphroditic; it transfers sperm to other units of the same tapeworm or a different one. Proglottids farthest from the scolex store the fertilized eggs. They detach from the tapeworm and are expelled from the body in feces. Fertilized eggs can survive for months outside an intermediate host.

Figure 22.12 shows the life cycle of a blood fluke that causes *schistosomiasis*. Larvae bore into the skin but don't cause obvious symptoms. Later, when white blood cells attack the masses of fluke eggs, the liver, spleen, bladder, and kidneys can be damaged.

Flatworms are among the simplest bilateral, cephalized animals with organ systems, which arise from three primary tissue layers that form in their embryos.

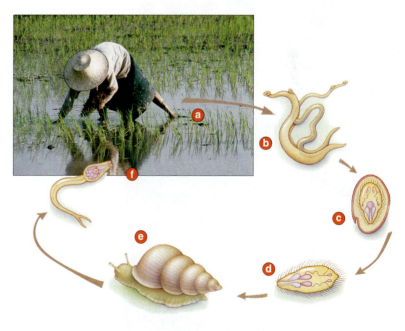

Figure 22.12   Life cycle of *Schistosoma japonicum*. (**a**) This blood fluke grows, matures, and mates in human hosts. (**b,c**) Fertilized eggs exit in feces, then hatch as ciliated, swimming larvae. (**d**) Larvae burrow into an aquatic snail and multiply asexually. (**e,f**) Fork-tailed, swimming larvae develop, leave the snail, then bore into the skin of a human host. They migrate to thin-walled intestinal veins, and a new cycle starts.

Chapter 22 Animal Evolution—The Invertebrates   361

22.6 Annelids—Segments Galore

Annelids have so many segments, it makes you wonder about controls over how the animal body plan develops.

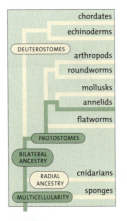

Did you ever see earthworms wriggling out of their flooded burrows after a rain? They are among 15,000 or so species of bilateral and highly segmented animals: the **annelids**. Leeches and the less well known but far more diverse polychaetes are annelids as well. Figures 22.13 through 22.15 show examples of their structure and function.

Segmented bodies are most pronounced in annelids. Except in leeches, nearly all segments have pairs or clusters of chitin-reinforced bristles on each side. Setae or chaetae are the formal names for the bristles; hence the names oligochaete for earthworms, which have a few bristles per segment; and polychaete for marine worms, which usually have a lot. (*Oligo–* means few; *poly–*, many.) Bristles shoved into soil or sediments afford traction for crawling and burrowing.

ADVANTAGES OF SEGMENTATION

Segmented bodies had great evolutionary potential, given all the diversity in annelid (and arthropod) body plans. Most earthworm segments are alike, but leeches have suckers at both ends (Figure 22.15), and polychaetes have an elaborate head and fleshy-lobed parapods ("closely resembling feet"). Existing species hint at the modifications in body plans that favored increases in size, more complex internal organs, and body segments adapted for specialized tasks.

ANNELID ADAPTATIONS—A CASE STUDY

Let's use an earthworm as our representative annelid. Its body has a flexible, permeable cuticle that is good for gas exchange but not for conserving body water. It is restricted to moist habitats, just as other annelids are restricted to aquatic habitats. The earthworm has a muscular pharynx that draws soil rich in detritus (decaying bits of organic matter) into the gut. It can scavenge the equivalent of its own weight each day, and it aerates the soil as it burrows through it.

Partitions divide the body into a series of coelomic chambers, each with muscles, blood vessels, nerves, and other organs (Figure 22.14). A complete digestive system with specialized regions extends through all chambers. In each segment's wall is a layer of circular muscles. As longitudinal muscles bridging several of the segments contract, circular muscles relax, all the way down the long body. They all exert force against fluid-filled coelomic chambers—a type of hydrostatic skeleton. Forward movement is complicated business; contractions are coordinated with each segment's bristles being plunged into soil and then released.

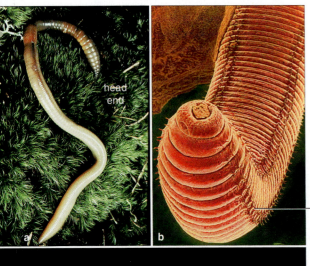

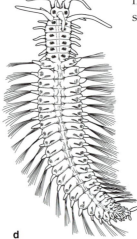

on both sides of the similar segments, bristles used in locomotion

Figure 22.13 (**a**) Earthworm, a familiar annelid. (**b**) All body segments look alike on the outside, but regional specializations occur inside, along the main body axis.

(**c**) A tube-dwelling polychaete with featherlike, mucus-coated structures on its head. Polychaetes are one of the most common animals along coasts. (**d**) As a group, they have a dizzying variety of modified segments. Many are predators or scavengers; others dine on algae.

Read Me First!

and watch the narrated animation on the earthworm body plan

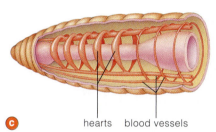

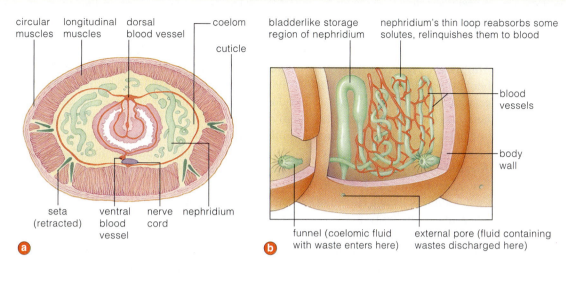

circular muscles — longitudinal muscles — dorsal blood vessel — coelom — cuticle

seta (retracted) — ventral blood vessel — nerve cord — nephridium

(a)

bladderlike storage region of nephridium — nephridium's thin loop reabsorbs some solutes, relinquishes them to blood — blood vessels — body wall

funnel (coelomic fluid with waste enters here) — external pore (fluid containing wastes discharged here)

(b)

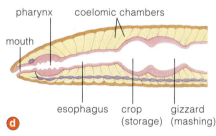

hearts — blood vessels

(c)

pharynx — coelomic chambers — mouth — esophagus — crop (storage) — gizzard (mashing)

(d)

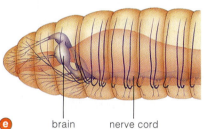

brain — nerve cord

(e)

Figure 22.14 Earthworm body plan. (**a**) Midbody, transverse section. (**b**) A nephridium, one of many functional units that help maintain the volume and the composition of body fluids. (**c**) Portion of the closed circulatory system, which is functionally linked to nephridia. (**d**) Part of the digestive system, near the worm's head end. (**e**) Part of the nervous system.

The earthworm has a rudimentary **brain**, actually a fused pair of ganglia (singular, ganglion) in its head. Ganglia are clusters of nerve cell bodies. Connected to it are a pair of **nerve cords**, lines of communication that help the brain coordinate activities of different segments. Within each segment, the cords connect to smaller ganglia that help control local activity.

Nephridia (singular, nephridium) are units of an organ system that help regulate the composition and volume of body fluids. Usually, each nephridium has a funnel that collects fluid from a coelomic chamber and then drains it into tubes in the chamber behind it (Figure 22.14*b*). The fluid enters a tiny bladder, which delivers it to a pore at the body surface of the next segment in line.

Like most annelids, earthworms also have a closed circulatory system with multiple hearts. Contractions of the heart and of muscularized blood vessels keep blood flowing in one direction. Smaller vessels service the gut, nerve cord, and body wall (Figure 22.14*c*).

before feeding

after feeding

Figure 22.15 Leeches. Most leeches are aquatic scavengers or predators with sharp jaws and a blood-sucking device. *Hirudo medicinalis*, the freshwater species shown here, has been used for at least 2,000 years as a blood-letting tool to "cure" nosebleeds, obesity, and some other conditions.

Leeches are still being used. They draw off pooled blood after a doctor reattaches a severed ear, lip, or fingertip. A patient's body can't do this on its own until blood circulation routes are reestablished.

Annelids are bilateral, coelomate, segmented worms that have digestive, nervous, excretory, and circulatory systems. Some species show the degree of specialization that became possible after segmented body plans evolved.

22.7 The Evolutionarily Pliable Mollusks

Is there a "typical" mollusk? No. The group has more than 110,000 named species, including tiny snails in treetops, burrowing clams, and giant predators of the open seas.

Mollusks are bilateral, soft-bodied animals with a small coelom. Those with a distinct head have eyes and tentacles, but not all have a head. Many have a shell, or a reduced one. A radula, a tongue-like organ that rasps at food, is common. Only mollusks have a cloaklike extension of the body mass—a *mantle*—that drapes back over itself. The respiratory organ is a gill with thin-walled leaflets.

Gastropods, chitons, bivalves, and cephalopods are among the better known mollusks. The gastropods, the largest and most diverse group, include snails, slugs, and nudibranchs (Figure 22.16). Their name, meaning "belly foots," refers to the underside of the gastropod body mass. The muscular "foot" allows the animals to glide about over surfaces or to burrow.

Maybe it was their fleshy, soft bodies, so forgiving of chance evolutionary changes, that gave the early mollusks the potential to diversify so many ways in so many different habitats. Let's explore this idea by focusing on a few representative examples.

HIDING OUT, OR NOT

If you were small, soft of body, and *tasty*, a shell could discourage predators. A shell protects chitons. When something disturbs it, a chiton hunkers down under its eight-piece shell (Figure 22.16d). Foot muscles pull down as the mantle presses like a suction cup against a rock. A shell protects scallops, oysters, mussels, and other bivalves; its two hinged parts (valves) snap shut. When the parts clap together, they force water out, which propels the scallop backward—usually enough to get away from a hungry sea star.

If you were small, soft of body, and *toxic*, a shell would be superfluous if you advertised your toxicity to predators. Some nudibranchs do just that. Gland cells of certain species secrete sulfuric acid and other bad-tasting substances. Other nudibranchs graze on

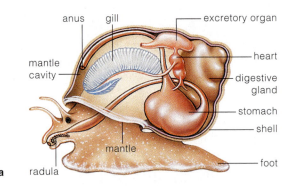

a

Figure showing body plan of a snail with labels: anus, gill, excretory organ, heart, mantle cavity, digestive gland, stomach, shell, foot, mantle, radula.

Figure 22.16 Mollusks. (**a**) Body plan and (**b**) ventral view of an aquatic snail crawling on aquarium glass. (**c**) Land snail. (**d**) Chiton. (**e**) Scallop, with light-sensitive "eyes" (blue dots) fringing its shell. (**f**) Two Spanish shawl nudibranchs (*Flabellina iodinea*). These hermaphrodites exchange sperm and then deposit chains of fertilized eggs into their partner.

Figure 22.17 (**a**) Artist's reconstruction of some animals that lived in the vast, tropical seaways of the Ordovician. Shown here, trilobites and the cephalopods called nautiloids, along with crinoids that looked like stalked plants.

(**b**) Generalized body plan of a cuttlefish, one of the cephalopods.

(**c**) Chambered nautilus, a living descendant of Ordovician nautiloids.
(**d**) Diver and squid (*Dosidicus*) inspecting each other.

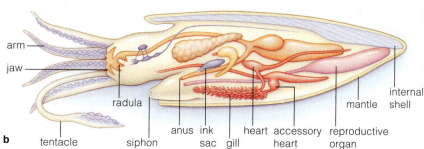

arm — jaw — radula — tentacle — siphon — anus — ink sac — gill — heart — accessory heart — reproductive organ — mantle — internal shell

b

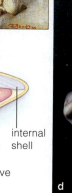

cnidarians, then incorporate ingested nematocysts in their own tissues. Spanish shawl nudibranchs flash red, nematocyst-studded respiratory organs as they mate (Figure 22.16*f*). Any predators attracted to the colorful couple get stung with the first bite and learn to avoid the species. Unlike most mollusks, which have separation into sexes, all of the nudibranchs are hermaphrodites.

A CEPHALOPOD NEED FOR SPEED

Some 500 million years ago, the cephalopods were supreme predators of the seas (Figure 22.17*a*). There were thousands of species with a chambered shell. And yet, in all but one of their modern descendants, the shell is reduced or gone (Figure 22.17*b*,*c*).

What happened? We can correlate its loss with an adaptive radiation of fishes that hunted cephalopods or were competitors for the same prey. In what might have been a long-term race for speed and wits, most cephalopods lost the external shell and became active, streamlined and, for invertebrates, smart.

For all cephalopods, *jet propulsion* became the name of the game. They learned to move by forcing a water jet through a funnel-shaped siphon, out from beneath the mantle cavity. We see how this works in modern species. Mantle muscles relax, which draws water into the cavity. They contract while the mantle's free edge closes over the head. The rapid squeeze shoots water through the siphon. The brain controls the siphon's action, hence the direction of escape or pursuit.

We also can correlate increases in speed with the evolution of more efficient respiratory and circulatory systems. Cephalopods are the only mollusks with a closed circulatory system. Their heart pumps blood to two gills, then two accessory (booster) hearts pump blood for oxygen uptake and carbon dioxide removal in metabolically active tissues, muscles especially.

Cephalopods now include the fastest invertebrates (squids), the largest (giant squid), and the smartest (octopuses). Of all mollusks, members of this group have the largest brain relative to body size, and they show the most complex behavior.

Lively stories emerge when we use evolutionary theory to interpret the fossil record and the nature of existing species diversity, as we have done here for the mollusks.

22.8 Roundworms

Roundworms are among the most abundant living animals. Sediments in shallow water often hold a million per square meter; a cupful of topsoil teems with them.

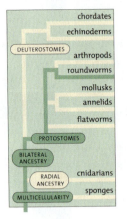

Roundworms make up a group known formally as nematodes. At least 12,000 kinds are named. Many are microscopic; a typical one isn't even five millimeters long. Their cylindrical body has bilateral features and tapered ends. Its false coelom is packed mostly with reproductive organs (Figure 22.18). A roundworm has a flexible cuticle, which it sheds repeatedly as it grows.

Most roundworms are free-living species that are decomposers. Many play vital roles in the cycling of nutrients through ecosystems.

The free-living *Caenorhabditis elegans* is a favorite experimental organism among biologists. It has the same general body plan and tissue types as more complex animals, but it is tiny and transparent, with only 959 somatic cells. Researchers can monitor the developmental fate of each cell. Also, the generation time is short, and the genome is one-thirtieth the size of the human genome. Self-fertilizing hermaphroditic forms can be used to get populations of offspring that are homozygous for desired traits.

Many roundworms infect roots of crop plants; they are significant agricultural pests. Others are internal parasites of animals that include humans, dogs, and insects. *Ascaris lumbricoides*, a very large roundworm, infects more than 1 billion people worldwide, mostly in Latin America and Asia (Figure 22.19*a*).

Pigs or game animals can carry *Trichinella spiralis*. Adults of this parasite live attached to the intestinal lining. Eggs develop into juveniles, which migrate through blood vessels to muscles, where they become encysted (Figure 22.19*b*). *Trichinosis*, the disease that results from infection, can be fatal.

Repeated infections by the roundworm *Wuchereria bancrofti* can result in *elephantiasis*, or a severe case of edema. Adult parasites get lodged in lymph nodes, organs that filter excess tissue fluid being sent back to the bloodstream. Worms obstruct the flow, so fluid backs up in tissue spaces and enlarges the legs, feet, and other body parts (Figure 22.19*c*). Mosquitoes serve as intermediate hosts.

The pinworms (*Enterobius vermicularis*) are tiny nematodes that live in the human rectum. They are most commonly parasites of young children.

> Roundworms are abundant bilateral animals with a false coelom and a complete digestive system. They include free-living decomposers and assorted parasites.

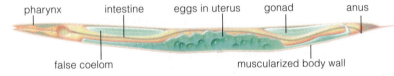

pharynx   intestine   eggs in uterus   gonad   anus

false coelom   muscularized body wall

Figure 22.18 Body plan of *Caenorhabditis elegans*.

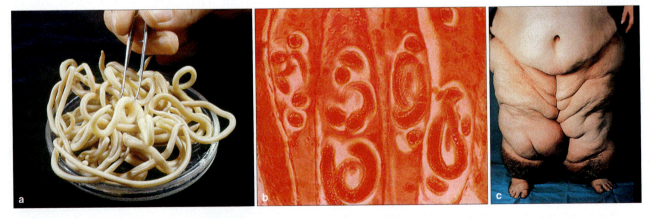

Figure 22.19 (**a**) Living roundworms (*Ascaris lumbricoides*). Infection by these intestinal parasites causes stomach pain, vomiting, and appendicitis. (**b**) *Trichinella spiralis* juveniles in the muscle tissue of a host animal. (**c**) A case of elephantiasis brought on by the roundworm *Wuchereria bancrofti*.

22.9 Arthropods—The Most Successful Animals

Evolutionarily speaking, "success" means having the most offspring, the most species in the most habitats, fending off threats and competition efficiently, and exploiting the greatest amounts and kinds of food. These are the features that characterize arthropods.

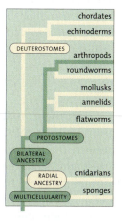

Arthropods are bilateral, with a complete gut, a much reduced coelom, and an *open* circulatory system, in which blood flows out of simple vessels or hearts, bathes tissues, and flows back. More than a million species are known. One lineage (trilobites) is extinct. Surviving groups are the crustaceans and chelicerates, insects, and myriapods (both centipedes and millipedes).

We consider representatives in sections that follow. For now, start thinking about six major adaptations that contributed to the success of arthropods in general and insects in particular.

A hardened exoskeleton. Arthropods have a cuticle of chitin, proteins, and waxes, often stiffened by calcium carbonate deposits. The cuticle is a protective external skeleton, or an **exoskeleton**. Exoskeletons might have evolved as defenses against predation. They took on added functions when some arthropods invaded the land. They support a body no longer made buoyant by water, and their waxes block water loss.

This exoskeleton does not restrict increases in size, because arthropods grow in spurts—and they **molt**. Molting is a process that is under hormonal control. At some stages of the life cycle, hormones induce a soft new cuticle to form under the old one, which is then shed (Figure 22.20). The body mass increases by the uptake of air or water and by rapid, continuous cell divisions before the new cuticle hardens. Molting is a trait that arthropods share with roundworms.

Jointed appendages. If uniformly hardened, a cuticle would restrict movement. Arthropod cuticle, however, thins where it spans *joints,* or regions where two body parts abut. Contraction of muscles makes a cuticle region bend, which shifts the position of body parts with respect to one another. A jointed exoskeleton was a key innovation; it allowed the evolution of diverse specialized appendages, including wings, antennae, and legs. ("Arthropod" means jointed leg.)

Specialized segments and fused-together segments. The first arthropods were segmented, and the segments were more or less similar. In most lineages that made it to the present, some of the segments became fused together or modified in specialized ways. As just one example, compare the nearly identical segments of a centipede with the winged body of a butterfly.

Respiratory structures. Many freshwater and marine arthropods use gills as respiratory organs. Among insects and other arthropods on land, air-conducting tubes called tracheas evolved. Insect tracheas start at pores on the body surface. They branch into tubes that let oxygen diffuse directly into tissues, which supports the aerobic respiration so essential for flight and other energy-demanding activities.

Specialized sensory structures. For many arthropods, diverse sensory structures also contributed to success. For instance, some evolved into eyes that can sample a big part of the visual field and provide the brain with information from many directions.

Specialized developmental stages. Many arthropods divide the task of surviving and reproducing among different stages of development. The division of labor is most common among insects. For some species, the new individual is a *juvenile* —a miniaturized version of the adult that simply grows in size until sexually mature. Other species undergo **metamorphosis**. The body changes in often drastic ways between the embryo and adult forms. Hormones help direct this tissue reorganization and remodeling.

Eating and growing fast are specialties of many immature stages; the adult form specializes in reproduction and in dispersal of the new generation. (Think of caterpillars chewing up leaves, then think of butterflies mating and laying eggs.) Different developmental stages are adapted to specific environmental conditions, especially seasonal shifts in food sources, water supplies, and availability of potential mates.

Figure 22.20 Molting by a red-orange centipede wriggling out of its old exoskeleton.

As a group, the arthropods are notably abundant and widespread, and they have enormously diverse life-styles.

The success of the arthropod lineage as a whole arises largely from having a body with a hardened, jointed exoskeleton; fused, modified body segments; specialized appendages; and specialized respiratory and sensory structures.

In many species—insects especially—success also arises from a division of labor among different stages of the life cycle, such as larvae, juveniles, and adults.

22.10 A Look at the Crustaceans

Shrimps, lobsters, crabs, barnacles, pillbugs, and other crustaceans got their name because they have a hard yet flexible "crust"—an external skeleton—but so do nearly all arthropods. The vast majority live in marine habitats, where they are so abundant they are dubbed "insects of the seas." Humans harvest many edible types.

Like other arthropods, crustaceans repeatedly molt and shed their exoskeleton during a life cycle. Figure 22.21 depicts a crab's larval and juvenile stages.

Most crustaceans have sixteen to twenty segments; some have more than sixty. Their head has pairs of antennae, and jawlike and food-handling appendages. In the simplest groups, all appendages are similar. In other groups, specialized claws and pincers evolved. The lobsters, shrimps, and crabs have five pairs of walking legs. A dorsal cuticle extends back from the head and covers the thoracic segments (Figure 22.22*a*).

Of all arthropods, only barnacles have a "shell," made of calcium secretions, that protects them from predators, drying winds, and surf (Figure 22.22*b*). Adult barnacles spend their lives cemented to piers, rocks, and even other animals. Once attached they can't move, so you would think that mating might be tricky. But barnacles tend to cluster in groups, and most are hermaphrodites. They extend their penis, which can be several times their body length, out to nearby neighbors.

Copepods are among the more numerous animals in aquatic habitats. Most of them are less than two millimeters long (Figure 22.22*c*). Some are parasites of invertebrates and fishes. Most ingest phytoplankton. Copepods in turn serve as food for invertebrates, fishes, and baleen whales.

> *Crustaceans differ greatly in the number and kind of appendages. They grow in stages and repeatedly replace their external skeleton by molting.*

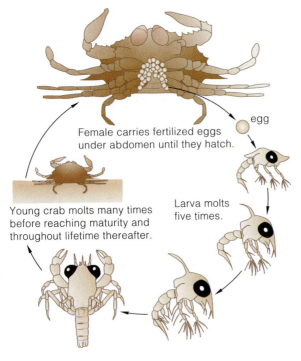

Female carries fertilized eggs under abdomen until they hatch.

egg

Young crab molts many times before reaching maturity and throughout lifetime thereafter.

Larva molts five times.

Figure 22.21 Life cycle of a crab. The larval and juvenile stages molt repeatedly and grow in size.

Figure 22.22 (**a**) Lobster in the Caribbean Sea, off the coast of Belize. (**b**) Goose barnacles. Adults cement themselves to one spot. You might mistake them for mollusks, but when they open their hinged shell to filter-feed, you see jointed appendages—the hallmark feature of arthropods. (**c**) One of the copepods. Different kinds are free-living filter feeders, predators, and parasites.

22.11 Spiders and Their Relatives

Chelicerates originated in shallow seas early in the Paleozoic. They are named for unique appendages called chelicerae. Horseshoe crabs are among the few living marine species. Of species on land, we might say this: Never have so many been loved by so few.

Horseshoe crabs (Figure 22.23*a*) haven't changed much since the Devonian. Their horseshoe-shaped carapace (a hardened dorsal shield) protects their ten-legged body. Their closest relatives on land are arachnids—spiders, scorpions, ticks, and chigger mites.

The spiders and scorpions are predators, and most sting, bite, or stun prey with venom, which they inject from fanged chelicerae. Spiders also have appendages that spin silk threads for webs and egg cases.

Our world would be overrun with insects without the predatory efforts of spiders. Even so, a few kinds that can harm humans and other mammals have given the whole group a bad name. For example, all spiders of genus *Loxosceles* are venomous. The brown recluse (*L. reclusa*) has a violin-shaped marking on its thorax (Figure 22.24*b*). Its bite ulcerates skin, which may heal poorly. In a few cases, amputations or skin grafts have been necessary. Bites of the black widow spiders (*Latrodectus*) kill about five people a year in the United States. Neurotoxin in their venom triggers potentially dangerous muscle spasms. Especially for children, prompt treatment is essential.

Mites are among the smallest, most diverse, and most widely distributed arachnids. Of an estimated 500,000 species, a few are parasites. Ticks are larger than most mites. Some transmit pathogens. One is the

Figure 22.23 (**a**) Horseshoe crab (*Limulus*). Its long spine helps steer the body. (**b**) Fat-bodied scorpion of Australia.

vector for *Borrelia burgdorferi*, the bacterial agent of Lyme disease. Rocky Mountain spotted fever, scrub typhus, tularemia, and encephalitis are a few of the many other tickborne diseases. Ticks don't fly or jump. They crawl onto grasses or shrubs, then crawl onto animals brushing past, then into scalp, groin, or other hairy hidden body parts. What about house dust mites? They can't kill you, but they can provoke allergies in susceptible people (Figure 22.24*d*).

Spiders, scorpions, and their relatives have appendages specialized for predatory or parasitic life-styles.

cheliceraeoll

Figure 22.24 Arachnids. (**a**) Tarantula with a fearless female. Like all spiders, it helps keep insects in check. Its bite does not harm humans. (**b**) Brown recluse, with six eyes and a violin-shaped mark on its forebody. It is native to some central and southern parts of the United States. (**c**) Female black widow, with a red hourglass-shaped marking on the underside of a shiny black abdomen. Black widows live in North America. Males are smaller and don't bite. (**d**) Dust mite. Mite feces and corpses commonly cause allergies in susceptible people.

22.12 A Look at Insect Diversity

Insects are the most diverse animal group. There are more species of dragonflies alone than there are species of mammals. Many insects also produce staggering numbers of offspring. Ants and termites alone may make up a third of the biomass of all animals on land.

In most adult insects, the segmented body is divided into three distinct parts: a head, thorax, and abdomen. The head has paired sensory antennae and paired mouthparts that are specialized for biting, chewing, or other tasks (Figure 22.25). The thorax has three pairs of legs and, usually, two pairs of wings. Insects are the only winged invertebrates.

Insects have a complete digestive system divided into a foregut, a midgut where food is digested, and a hindgut where water is reabsorbed. A system of *malpighian tubules* disposes of wastes and maintains ion concentrations in body fluids. After proteins are digested, residues diffuse from blood into the tubes. There, enzymes convert them to crystals of uric acid, which are eliminated in feces. The tubules help land-dwelling insects get rid of potentially toxic wastes without also losing precious water.

Larvae, nymphs, and pupae are immature stages of many insect life cycles (Figure 22.26). Like human children, some insect nymphs have the adult's form, but in miniature. Unlike children, however, these immature stages molt. Most insect larvae undergo metamorphosis, in which the drastic reorganization and remodeling of tissues necessary for the transition to adulthood take place.

Figure 22.27 shows a few of the more than 800,000 species. If numbers and distribution are the measure, the most successful kinds are small and reproduce often. Huge numbers grow and reproduce on a plant that would be only an appetizer for another animal. By one estimate, if all the offspring of a single female fly survived and reproduced for six more generations, she would have more than 5 trillion descendants!

The most successful insects are winged. They move among food sources that are too widely scattered to be exploited by other kinds of animals. The capacity for flight contributed hugely to their success on land.

The very features that contribute to insect success also make them our most aggressive competitors for crops and natural products, such as paper or lumber.

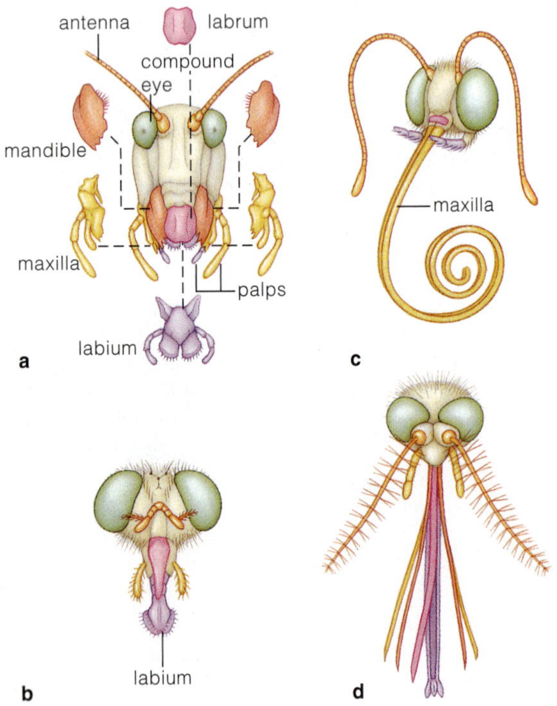

Figure 22.25 Examples of insect appendages. Head parts typical of (**a**) grasshoppers, which chew food; (**b**) flies, which sponge up nutrients; (**c**) butterflies, which siphon nectar; and (**d**) mosquitoes, which pierce hosts and suck up blood.

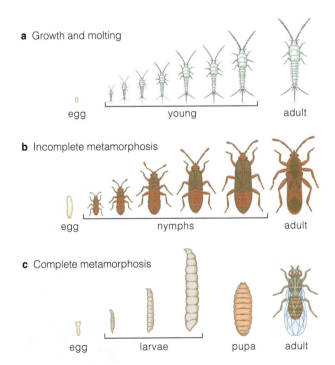

a Growth and molting

egg   young   adult

b Incomplete metamorphosis

egg   nymphs   adult

c Complete metamorphosis

egg   larvae   pupa   adult

Figure 22.26 Insect development. (**a**) Young silverfish change little except in size and proportion as they mature into adults. (**b**) True bugs show *incomplete* metamorphosis, or gradual change from the immature form until the last molt. (**c**) Fruit flies show *complete* metamorphosis. Tissues of larvae are destroyed and replaced before an adult emerges.

Some insects transmit pathogens that sicken us. We fight "bad" insects by spraying many millions of pounds of pesticides each year. Yet most play vital roles as decomposers or as pollinators of flowering plants—including almost all crop plants. Also, many "good" insects are predators or parasites of the ones we would rather do without. Think about it.

Like other arthropods, insects have a hardened cuticle that forms an exoskeleton; jointed appendages; fused, modified body segments; respiratory structures; and specialized sensory organs. Many have a division of labor among stages that form in the life cycle. The most successful are winged.

In terms of distribution, number of species, population sizes, competitive adaptations, and exploitation of diverse foods, insects are the most successful animals.

Figure 22.27 Examples of insects. (**a**) Mediterranean fruit fly. Its larvae destroy citrus fruit and other crops. (**b**) Duck louse. It eats bits of feathers and skin. (**c**) European earwig, one of the common household pests. The long curved forceps at the tail end show this is a male.

(**d**) Flea, with strong legs good for jumping onto and off of its animal hosts. (**e**) Stinkbugs, newly hatched. (**f**) In the center of this group of honeybee workers, one bee is signaling the position of a food source to her hive mates by dancing, as described in Section 43.4.

With more than 300,000 species of beetles, Coleoptera is the largest order in the animal kingdom. (**g**) Ladybird beetles swarming. (**h**) *Dynastes*, a scarab beetle.

(**i**) Swallowtail butterfly and (**j**) luna moth. (**k**) An adult damselfly, which preys on flying insects. Its larvae are voracious predators of aquatic invertebrates and small fishes.

22.13 The Puzzling Echinoderms

Reflect, for a moment, on Figure 22.5. Ancestors of the roundworms, arthropods, annelids, and mollusks you just read about were at a major branch point in time. At that point, ancestors of echinoderms and chordates set out on a separate evolutionary journey.

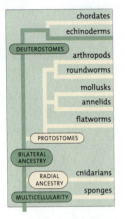

The name **echinoderm** means spiny-skinned. It refers to the interlocking calcium carbonate spines or plates in the body wall. All echinoderms live in the seas. They include sea stars, brittle stars, sea urchins, sea cucumbers, feather stars, sand dollars, sea biscuits, and sea lilies, or crinoids. Figure 22.28 shows some representatives.

Most adults are radial with a few bilateral features, and most of them have bilateral larvae. Did the group evolve from a bilateral ancestor? DNA studies now indicate that this is most likely the case.

Almost all adult echinoderms are bottom dwellers. Some sea lilies live attached to the seafloor by a stalk, although most echinoderms creep about. They lack a brain, but their decentralized nervous system allows them to respond to food, predators, and so forth that appear from different directions. For instance, any arm of a sea star that touches a tasty mussel's shell can become the leader, directing the rest of the body to move toward the prey.

Echinoderms have tube feet. These are fluid-filled, muscular structures, sometimes with suckerlike disks at the tips (Figure 22.28a). Sea stars use their tube feet for walking, burrowing, clinging to rocks, or gripping a clam or snail about to become a meal. Tube feet are components of a **water–vascular system**, unique to echinoderms. In sea stars, that system includes a main canal in each arm. Short side canals extend from them and deliver water to the tube feet (Figure 22.29). Inside a tube foot is a fluid-filled, muscular structure shaped like the rubber bulb on a medicine dropper. As the bulb contracts, it forces fluid into the foot and causes it to lengthen.

Figure 22.28 A few echinoderms. (**a**) Sea star, with a close-up of its little tube feet. (**b**) Sea urchins, which can move about on spines and a few tube feet. (**c**) Brittle star. Its slender arms (rays) make rapid, snakelike movements. (**d**) Sea cucumber, with rows of tube feet along its elongated body.

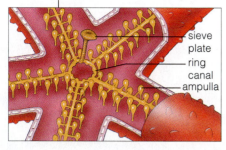

part of the water–vascular system

sieve
plate
ring
canal
ampulla

Figure 22.29 Water–vascular system of a sea star. Combined with many tube feet, it is the basis of locomotion.

Tube feet change shape continually when muscle action redistributes fluid through the water–vascular system. Hundreds of tube feet may move at a time. After being released, each tube foot swings forward, reattaches to the substrate, then swings backward and is released before swinging forward once again. Their motions are splendidly coordinated, so sea stars glide smoothly rather than lurch along.

Most sea stars are active predators with a feeding apparatus on their ventral surface. Some swallow prey whole. Others push part of their stomach outside the mouth and around their prey, then start digesting their meal even before swallowing it. Sea stars get rid of coarse, undigested residues through the mouth, even though they do have a small anus.

In sea urchins, the calcium carbonate plates form a stiff, rounded "test" from which the spines protrude (Figure 22.28b). Sea urchin roe (eggs) are an ingredient in some sushi. Each year, Japan imports about 20 million pounds of sea urchins that were captured off the California coast. Harvests are largely unregulated and are probably contributing to the decline of the most prized species.

Sea cucumbers also are a favorite food in Asia, and some species are declining. These echinoderms have the softest body; their plates have been reduced to microscopic spicules (Figure 22.28d). They creep about on the ocean floor, picking up prey and bits of organic debris with sticky tentacles. When a predator attacks, a sea cucumber shoots its internal organs out through the region of the anus to distract it. If this succeeds, the sea cucumber regenerates its missing parts.

With their odd combination of radial and bilateral traits, echinoderms make a case for this point: There are exceptions to the major trends that we observe in animal evolution.

Summary

Section 22.1 Animals cannot be defined in a simple way; they share many traits with other organisms. Instead, they can be characterized by a list of features:

a. Animals are multicelled heterotrophs. Their cells engage in aerobic respiration and have no secreted wall.

b. Except for sponges, animals have tissues. Most also have organs and organ systems.

c. Animals typically reproduce sexually, but many also can reproduce asexually. Their life cycles often involve a series of stages.

d. Immature stages, adults, or both are usually motile (able to move from place to place).

Vertebrates have a backbone. The invertebrates, which have none, vastly outnumber them.

Animals also can be characterized with respect to six features of their basic body plan: body symmetry, embryonic tissue layers, cephalization, type of gut, type of body cavity, and segmentation. Not all of these features occur in every group.

a. Except for sponges, body parts of animals are arranged radially or bilaterally with respect to the main body axis.

b. During the evolution of bilateral animals, many sensory structures and nerve cells became concentrated in a head (a process called cephalization).

c. Two or three primary tissue layers—ectoderm, endoderm, and mesoderm—form in animal embryos. They give rise to all tissues and organs of adults.

d. Except for placozoans and sponges, animals have a saclike or a tubular gut. A tubular gut has a mouth and anus, and its formation in embryos is one of many defining traits for the two major lineages of bilateral animals. In protosomes, the mouth forms first. In deuterostomes, the anus forms first.

e. Protostomes and deuterostomes differ in whether they have a coelom, a main body cavity between their gut and body wall, and in how that cavity develops. In some lineages the coelom was lost or reduced; such animals are called acoelomates and pseudocoelomates.

f. Bilateral animals differ in whether they are segmented, with repeats of body units, and in how pronounced segmentation has become.

Section 22.2 Ediacarans are the earliest animals known from the fossil record. Most modern groups arose during an adaptive radiation in the Cambrian. *Trichoplax*, a type of tiny soft-bodied animal with two cell layers, is the simplest living animal.

Section 22.3 Sponges are simple animals with no body symmetry, true tissues, or organs. Certain flagellated cells lining canals and chambers in their bodies move water and take up microscopic food particles. Chemical and structural features make sponges unappealing to most predators.

Section 22.4 The radially symmetrical cnidarians, such as jellyfishes, sea anemones, corals, and hydroids, are the simplest animals with tissues. They have an outer epidermis, inner gastrodermis with gland cells, and a jellylike substance in between that has only a few scattered cells. They are the only animals that make nematocysts, or capsules with dischargeable threads that can sting, entangle, or ensnare prey.

Section 22.5 Flatworms are among the simplest protostomes. They are bilateral. Three primary tissue layers form in their embryos and give rise to simple organ systems. Flatworms have a saclike gut, and no coelom, but it appears they had coelomate ancestors. Turbellarians, tapeworms, and flukes are in this group, which includes major parasites of humans.

Section 22.6 Annelids are highly segmented worms. They include the oligochaetes (such as earthworms), polychaetes (marine worms), and leeches. The first two groups have bristles along their body length that work in conjuction with muscle contractions to bring about body movements.

Section 22.7 With 110,000 species, mollusks are one of the largest groups. Only mollusks have a mantle, a sheetlike part of the body mass draped back on itself. All are soft-bodied and often shelled. Examples are the gastropods and cephalopods. The largest, swiftest, and smartest invertebrates are cephalopods.

Section 22.8 Roundworms (nematodes) have a cylindrical body with bilateral features, a flexible cuticle, a complete gut, and a false coelom. At least 12,000 kinds are free-living decomposers or parasites.

Section 22.9 We know of more than a million kinds of bilateral, coelomate animals called arthropods. The major groups are chelicerates, crustaceans, centipedes, millipedes, and insects. Features that contributed to their success are a hardened exoskeleton, jointed appendages, specialized and fused segments, efficient respiratory and sensory structures, and developmental stages adapted to specific environmental conditions.

Section 22.10 The mostly marine crustaceans include crabs, lobsters, barnacles, and copepods. Specialized appendages underlie their diverse life-styles.

Section 22.11 Chelicerates include horseshoe crabs and arachnids (spiders, scorpions, ticks, and mites). They are predators, parasites, or scavengers.

Section 22.12 Insects are the most successful group of animals, and the only winged invertebrates. They are our major competitors for food.

Section 22.13 Echinoderms, such as sea stars, are invertebrates of the deuterostome lineage. They have an exoskeleton of spines, spicules, or plates of calcium carbonate. Adults are radial, but bilateral ancestry is evident in their larval stages and other features.

Figure 22.30 A soft, branching coral (*Telesto*).

Self-Quiz
Answers in Appendix III

1. All animals _____ .
   a. are motile for at least some stage in the life cycle
   b. consist of tissues arranged as organs
   c. can reproduce asexually as well as sexually
   d. both a and b

2. A coelom is a _____ .
   a. lined body cavity   c. sensory organ
   b. resting stage   d. type of bristle

3. The cnidarians alone have _____ .
   a. nematocysts   c. a hydrostatic skeleton
   b. tissues   d. radial symmetry

4. Flukes are most closely related to _____ .
   a. planarians   c. arthropods
   b. roundworms   d. echinoderms

5. Which group has the greatest number of species?
   a. crustaceans   c. mollusks
   b. insects   d. roundworms

6. The nephridia of earthworms perform a function most similar to the _____ .
   a. gemmules of sponges   c. flame cells of planarians
   b. chelicerae of spiders   d. tube feet of echinoderms

7. The _____ are coelomate and radial as adults.
   a. cnidarians   c. roundworms
   b. echinoderms   d. both a and c

8. Chordates are most closely related to _____ .
   a. cnidarians   c. mollusks
   b. echinoderms   d. arthropods

9. Match the organisms with their appropriate descriptions.
   ____ Ediacarans   a. complete gut, false coelom
   ____ sponges   b. simplest organ systems
   ____ cnidarians   c. no tissues, no organs
   ____ flatworms   d. jointed exoskeleton
   ____ roundworms   e. mantle over body mass
   ____ annelids   f. segmented worms
   ____ arthropods   g. tube feet, spiny skin
   ____ mollusks   h. nematocyst producers
   ____ echinoderms   i. known from fossils only

Figure 22.31 To which group might this burrowing marine animal belong?

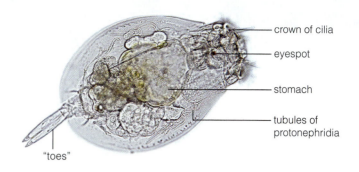

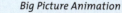

crown of cilia

eyespot

stomach

tubules of protonephridia

"toes"

Figure 22.32 A rotifer (*Euchlanis*).

Critical Thinking

1. You are diving in the calm, warm waters behind a large tropical reef. You see something that looks like a plant with tentacles (Figure 22.30). This is a branching, soft coral with many polyps. You will not see it growing on reef surfaces exposed to the open sea. Formulate a hypothesis that might explain the animal's distribution. Describe how you would go about testing this hypothesis.

2. The marine worm in Figure 22.31 has a segmented body. Most segments are similar; they have bristles on their sides that the worm uses to burrow in sandy marine sediments. What group do you think this worm belongs to? What other features could you examine to confirm your suspicions?

3. The flatworms, roundworms, and annelids all include species that are parasites of mammals, whereas sponges, cnidarians, mollusks, and echinoderms do not. What do you think explains the difference?

4. In the summer of 2000, only 10 percent of the lobster population of Long Island Sound survived after a massive die-off. Many lobstermen believe the deaths followed heavier spraying of pesticides, used to control mosquitoes that carry West Nile virus. Speculate about why a chemical designed to target mosquitoes might also harm lobsters.

5. There are many invertebrate groups besides the major ones described in this chapter. For example, rotifers are protostomes that are usually less than a millimeter long. All but about five percent live in freshwater. Most have a crown of cilia at the head end that assists in swimming and in wafting food toward the mouth (Figure 22.32). A jawlike structure known only in this group is used to grind up food. The digestive system lies inside a pseudocoelom and is complete. Paired gastric glands secrete enzymes that aid in digestion. Fluid balance and waste removal are controlled by protonephridia. Nerve cell bodies clustered in the head integrate body activities. Certain rotifers have clusters of pigments that serve as simple "eyes."

Even with all this complexity, rotifers do not have any special circulatory or respiratory organs. Do you consider this surprising? Explain your reasoning.

IMPACTS, ISSUES *Interpreting and Misinterpreting the Past*

In Charles Darwin's time, major groups of organisms were already identified. A big obstacle to acceptance of his theory of evolution by natural selection was the seeming lack of transitional forms. If new species evolve from older ones, then where were the "missing links" in the fossil record—forms with intermediate traits that bridge the groups? Ironically, workmen at a limestone quarry in Germany had already unearthed one.

The pigeon-sized fossil looked like a small meat-eating dinosaur (Figure 23.1). It had three long clawed fingers on each forelimb, a long bony tail, and a heavy jaw with short, spiky teeth. Later, diggers found another one. Later still, someone noticed the feathers. If those were birds, why did they have teeth and a bony tail? If dinosaurs, what were they doing with feathers? The specimen was named *Archaeopteryx*, meaning ancient winged one.

Between 1860 and 1988, six *Archaeopteryx* specimens and a fossilized feather were found. Anti-evolutionists tried to dismiss them as forgeries. Someone, they said, had pressed modern bird bones and feathers against wet plaster. The dried, imprinted plaster casts merely looked like fossils. Microscopic examination confirmed the fossils are real. Further confirmation of their age came from the remains of obviously ancient jellyfishes, worms, and other species in the same limestone layers.

Radiometric dating shows that *Archaeopteryx* lived 150 million years ago. Why are its remains preserved so well after so much time? *Archaeopteryx* lived in tropical forests bordering a lagoon. The lagoon was large, warm, and stagnant; coral reefs barred inputs of oxygenated water from the sea. So it was not an inviting place for scavenging animals that might have eaten and wiped

Dromaeosaurus | Archaeopteryx | Confuciusornis

Figure 23.1 Morphological comparison of *Archaeopteryx* and *Dromaeosaurus*, a duck-sized, two-legged dinosaur. Birds and dinosaurs evolved from early reptiles. *Right*, reconstruction of another "missing link" between reptiles and birds—*Confuciusornis sanctus*, based on fossils from China.

the big picture

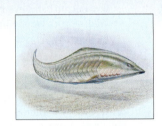

Origins, Evolutionary Trends   The first chordates were invertebrates. A cranium, then a vertebral column, jaws, paired fins, and paired lungs were key innovations in the evolution of fishes, amphibians, reptiles, birds, and mammals.

Diversification of Fishes   Fishes were the first vertebrates. They were and still are the most diverse group. Most have jaws and an internal skeleton of cartilage or bone. The paired fins of fish are homologous to the limbs of land vertebrates.

out the remains of *Archaeopteryx* and other organisms that fell from the sky or drifted offshore.

Fine sediments driven over the reefs by each storm surge gently buried the carcasses on the bottom of the lagoon. Over time, soft, muddy sediments were slowly compacted and hardened. They became a limestone tomb for over 600 species, including *Archaeopteryx*.

Is *Archaeopteryx* the only transitional form linking dinosaurs and birds? No. *Confuciusornis sanctus* lived about the same time as *Archaeopteryx*. It is the earliest known bird with a bill (Figure 23.1). Fossils show how the clawed limbs of its four-legged ancestor became modified into wings. Each wing had three digits, one a reduced claw with flight feathers attached.

No one was around to witness such transitions in the history of life. But fossils are real, just as the morphology, biochemistry, and molecular makeup of living organisms are real. Radiometric dating assigns fossils to their place in time. *So evolution is not "just a theory."* Remember, a scientific theory differs from speculation because its predictive power has been tested in nature many times, in many different ways. If a theory stands, there is a high probability that it is not wrong.

Evolutionists argue all the time among themselves. They argue over how to interpret the evidence and which mechanisms and events can explain life's history. At the same time, they do not ignore evidence—which is there for us to gather and interpret. Here is one account of vertebrate evolution, including our own origins.

 How Would You Vote?

Private collectors discover and protect fossils, but a private market for rare vertebrate fossils raises their cost to museums and encourages theft from protected fossil beds. Should private collecting of vertebrate fossils be banned? See the Media Menu for details, then vote online.

The Move Onto Land

Amphibians were the first animals to spend much of their life on land. They still need to keep their skin moist for respiration and require standing water to lay their eggs.

Enter the Amniotes

Reptiles, birds, and mammals are amniotes. They are fully adapted to life on land. Eggs of amniotes are fertilized internally and embryos develop inside a series of membranes.

Human Evolution

Human traits arose through modifications in earlier mammalian, primate, and then early hominid lineages. Diverse humanlike and early human forms arose in Africa.

23.1 The Chordate Heritage

We start our look at vertebrates and their chordate origins with a reminder: Each kind of organism, humans included, is a mosaic of traits. Many traits have been conserved from remote ancestors, and others are unique to its branch on the family tree.

CHORDATE CHARACTERISTICS

Most of the **chordates** are coelomate, bilateral animals characterized in part by four features, as illustrated in Figure 23.2. *First,* a rod of stiffened tissue—not bone or cartilage—helps support the body. It is a notochord. *Second,* a nerve cord runs parallel with the notochord and gut; its anterior end develops into a brain. Unlike the nerve cords of invertebrates, this one is dorsal. *Third,* **gill slits** punctuate the wall of a **pharynx**—a muscular tube that functions in feeding, respiration, or both. *Fourth,* a tail extends past an anus. The four features develop in all chordate embryos, although they are not always present in the adult form.

INVERTEBRATE CHORDATES

Many of the animals that preceded vertebrates were like the simplest living chordates—the urochordates. This group includes sea squirts and other tunicates. Adult tunicates secrete a jellylike or leathery "tunic" around the pharynx. Figure 23.3 shows translucent blue ones. Most species are a few centimeters long and live alone or in colonies from intertidal zones to the deep ocean. Adults usually attach to substrates. Sea squirts are the most common; they spurt water through a siphon when something irritates them.

For our story, a key point to remember about the tunicates is that their larvae are bilateral swimmers. A firm, flexible notochord acts like a torsion bar. It interacts with bands of muscles that lie just under the epidermis. When muscles on one side of the tail or the other contract, the notochord bends, then springs back as muscles relax. Strong, side-to-side motion propels a larva forward. Most fishes use their muscles and their backbone for the same kind of motion.

Another point: Tunicates are **filter feeders**. They filter food from seawater that flows in through one siphon, past gill slits in the pharynx, then out by way of another siphon. The pharynx is also a respiratory organ. It takes up dissolved oxygen from seawater. Oxygen diffuses into blood vessels near the pharynx. Carbon dioxide from aerobic respiration diffuses into the water flowing out.

Our simplest relatives also include the fish-shaped lancelets (cephalochordates). Lancelets are only three to seven centimeters long. Their body tapers at both ends and displays the chordate features (Figure 23.2). Like vertebrates, lancelets have a head and a simple brain that develops from one end of their nerve cord (*cephalo–*, head). Their nerve cord forms as it does in vertebrates. Nerve cells (neurons) in the brain control reflex responses to light and other stimuli.

Also like vertebrates, the lancelets have segmented muscles. Contractile units in muscle cells run parallel with the body's long axis. Contractile force directed against the notochord allows side-to-side swimming motions. Even so, lancelets spend most of the time burrowing in shallow sediments and filter-feeding. Coordinated beating of cilia that line numerous gill slits move water into their pharynx, where food gets caught in mucus before moving to the rest of the gut.

A gill-slitted pharynx, a simple brain, a nerve cord, and bands of muscles that parallel the body's long axis—as you will see, these features turned out to have amazing evolutionary possibilities for vertebrates.

Read Me First!
and watch the narrated animation on the chordate body plan

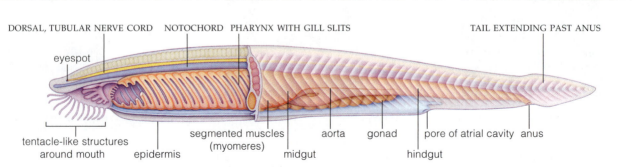

DORSAL, TUBULAR NERVE CORD   NOTOCHORD   PHARYNX WITH GILL SLITS   TAIL EXTENDING PAST ANUS

eyespot

tentacle-like structures around mouth   epidermis   segmented muscles (myomeres)   midgut   aorta   gonad   hindgut   pore of atrial cavity   anus

Figure 23.2 Lancelet body plan. This small filter-feeder shows the four defining features of chordates.

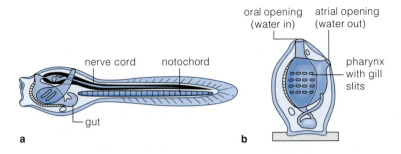

Figure 23.3 (**a**) Generalized larval form of a sea squirt, a tunicate. A new larva swims about briefly, then it undergoes drastic tissue remodeling and reorganization. This metamorphosis starts when its head end attaches to a substrate. Its tail, notochord, and most of the nervous system are resorbed (that is, recycled and used to form new tissues). Many gill slits form in the pharynx wall. Organs rotate until openings through which water flows into and out from the pharynx are pointing away from the substrate, as in (**b**). (**c**) Adult sea squirts (*Rhopalaea crassa*).

Figure 23.4 Hagfish body plan. The two photographs to the right show one hagfish before and after it secretes slimy mucus from its pores.

A WORD OF CAUTION

We are describing chordate features that will help give you insight into the evolution of vertebrates, but don't assume tunicates and lancelets are missing links between vertebrates and the earliest chordates. These groups do share some characteristics that preceded the origin of vertebrates. However, each group also has unique traits that put it on separate branches of the chordate family tree.

Similarly, the hagfishes are like jawless vertebrates called lampreys, but some traits that evolved early on are theirs alone. Unlike vertebrates, for instance, their brain is not enclosed in a hardened, bony chamber. Even so, some cartilage helps protect it.

And don't consider hagfishes evolutionary dead-ends. A hagfish is less than a meter long and nearly blind, but it easily finds food with sensory tentacles (Figure 23.4). Its rasping, tonguelike structure draws small invertebrate prey and tissues of a dead or dying fish into its mouth. Hagfishes are not active animals;

they generally burrow in marine sediments. However, low metabolic rates work for them. They can wait out an absence of food for more than half a year.

If a hagfish is threatened, it secretes up to a gallon of sticky mucus. Fishermen who snag hagfishes view this behavior as disgusting. Even so, sliming potential predators has served this soft-bodied and otherwise vulnerable lineage quite well.

All chordate embryos have a notochord, a tubular dorsal nerve cord, a pharynx with gill slits in its wall, and a tail that extends past the anus. These traits are legacies from a shared ancestor, an early invertebrate chordate.

Existing filter-feeding lancelets and tunicates are among the groups closest to the earliest chordates.

Hagfishes represent the next level in chordate complexity. A portion of their brain is encased in protective cartilage.

23.2 Evolutionary Trends Among the Vertebrates

The first vertebrates were jawless fishes that appeared during the Cambrian. Later in time, their body plans were modified in diverse ways during an adaptive radiation that led to all modern vertebrate groups.

Figure 23.5 (**a**) Reconstruction of an early craniate (*Myllokunmingia*). (**b**) Larva of a lamprey, an existing jawless fish. Most of the lampreys are parasitic. They attach to prey with a suckerlike oral disk and rasp away at it with their horny mouthparts (**c**). We often observe several lampreys attached to the same host. In the 1800s, sea lampreys probably entered the Hudson River, then canals built for commerce. By 1946, they were established in all of the Great Lakes of North America. Trout, salmon, catfishes, and other natives have no coevolved defenses against them.

EARLY CRANIATES

Fishes, amphibians, reptiles, birds, and mammals are **craniates**. A cranium is a chamber of cartilage or bone that encloses all or part of a brain. Craniates evolved by 530 million years ago. Like lancelets, they had a notochord and segmented muscles; they were active swimmers. Adults resembled the larvae of lampreys, a group of existing jawless fishes (Figure 23.5).

Thirty million years later, ostracoderms and other jawless fishes had evolved. They had armorlike plates of bony tissue and dentin, a hard tissue still found in teeth. The armor may have worked against pincers of giant sea scorpions. But it didn't stop **jaws**—hinged, bony feeding structures. Jawed craniates had evolved. Most jawless fishes disappeared during the adaptive radiation of jawed craniates. Hagfishes and lampreys are the only groups that made it to the present.

Placoderms were among the earliest jawed fishes. Armor plates protected their head, but not much else (Figure 23.6). They didn't have teeth, but razor-sharp bony plates served the same function. The placoderms must have been major agents of natural selection. With their arrival, an arms race of offensive and defensive body structures began that continues to this day.

Read Me First!

and watch the narrated animation on the evolution of jaws

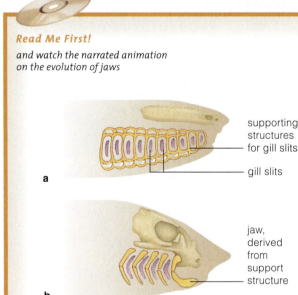

a — supporting structures for gill slits
gill slits

b — jaw, derived from support structure

c — spiracle (modified gill slit)
jaw support
jaw

Figure 23.6 Comparison of gill-supporting structures. (**a**) In early jawless fishes, supporting elements reinforced a series of gill slits on both sides of the body. (**b**) In placoderms and other early jawed fishes, the first elements in the series were modified, and they functioned as jaws. Cartilage reinforced the mouth's rim. (**c**) Sharks and other modern jawed fishes have strong jaw supports. *Above*, the size of a human compared to an extinct jawed fish, the placoderm *Dunkleosteus*.

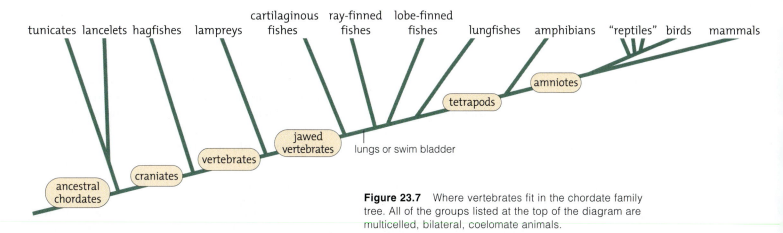

Figure 23.7 Where vertebrates fit in the chordate family tree. All of the groups listed at the top of the diagram are multicelled, bilateral, coelomate animals.

KEY INNOVATIONS

Vertebrates profoundly changed the course of animal evolution. Unlike their nearly brainless, filter-feeding relatives, they have bones—organs made of mineral-hardened secretions. **Vertebrae** (singular, vertebra), a series of bony or cartilaginous segments, replaced the notochord as the partner of muscles. The vertebral column, part of an inner skeleton, was more flexible and less clunky than external armor plates. The union of vertebrae and muscles promoted maneuverability, and hard bones invited more forceful contractions. The outcome? *Agile, fast-moving jawed fishes came to be the dominant animals in the seas.*

Jaws sparked another trend. They had evolved from the first of a series of hard parts that structurally supported gill slits in the pharynx (Figure 23.6). They were a key innovation, one that invited novel feeding possibilities. As noted earlier, they also triggered the evolution of defenses and of ways to overcome them. For example, the fishes with better eyes for detecting predators or bigger brains to plan escapes now had a competitive edge. *A trend toward more complex sensory organs and nervous systems started among ancient fishes and continued among vertebrates on land.*

Another trend started when pairs of fins evolved. **Fins** are appendages that help propel, stabilize, and guide the body through water. Some fish groups had fleshy ventral fins with internal skeletal supports that became forerunners of limbs. *Paired, fleshy fins were a starting point for all legs, arms, and wings that evolved among amphibians, reptiles, birds, and mammals.*

A shift in respiration set another trend in motion. In lancelets, oxygen and carbon dioxide diffuse across skin. In most early vertebrates, paired gills evolved. **Gills** are respiratory organs having moist, thin folds serviced by blood vessels. They have a large surface area that exchanges gases between the environment and the body. Oxygen diffuses *from* water inside the mouth into blood vessels within each gill as carbon dioxide diffuses *into* the surrounding water.

Gills became more efficient in larger, more active fishes. But gills can't function out of water; their thin surfaces stick together unless flowing water moistens them. In fishes ancestral to land vertebrates, two tiny outpouchings developed on the gut wall and evolved into **lungs**: internally moistened sacs for gas exchange. Lungs supplemented and then replaced gills during the invasion of land. In a related trend, modifications to the heart enhanced the uptake of oxygen and removal of carbon dioxide. *The first vertebrates on land relied more on a pair of lungs than on gills. Increased efficiency of circulatory systems accompanied the evolution of lungs.*

In the Carboniferous, placoderms and other jawed fishes became extinct after faster, brainier predators replaced them in the seas. Through adaptation, key innovations, and plain luck, the descendants of some jawless and jawed groups made it to the present.

MAJOR VERTEBRATE GROUPS

Figure 23.7 shows how vertebrates fit in the chordate family tree. Besides lampreys, the major groups are known informally as cartilaginous fishes, bony fishes, amphibians, reptiles, birds, and mammals. The rest of this chapter introduces their defining characteristics.

Vertebrates arose in the Cambrian. All of the major modern groups evolved during an adaptive radiation of jawed fishes.

Jaws, a vertebral column, paired fins, and lungs were pivotal developments in the evolution of lineages that gave rise to fishes, amphibians, reptiles, birds, and mammals.

23.3 Jawed Fishes and the Rise of Tetrapods

Unless you study underwater life, you may not know that fishes are the world's dominant vertebrates. Their numbers exceed those of all other vertebrate groups combined, and they show far more diversity in their body plans.

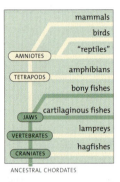

ANCESTRAL CHORDATES

Of four major groups of living fishes, one has a skeleton of cartilage, the others a skeleton of bone. Their body form and behavior hint at the challenges of living in water. Water is 800 times more dense than air and resists movements through it. One response to this constraint, a streamlined shape, reduced friction during fast pursuits and escapes. For instance, sharks and other predators of the open ocean have a long, trim body with strong tail muscles for forward propulsion (Figure 23.8*a*). If a species doesn't make high-speed runs, the body tends to be flattened and adapted for concealment. Bottom dwellers or crevice lurkers are like this (Figure 23.8*b*). Fish scales were another response. These small, bony plates protect the body without weighing it down.

A trout suspended in shallow water shows another adaptation to water's density. It can maintain neutral buoyancy with its swim bladder. This is an adjustable flotation device that exchanges gases with the blood. When a trout is gulping air at the water's surface, it is adjusting the volume of its swim bladder.

CARTILAGINOUS FISHES

Sharks and rays are well-known **cartilaginous fishes**. These predators have prominent fins, a skeleton of cartilage, and five to seven gill slits on both sides. At fifteen meters in length, some sharks rank among the largest living vertebrates. They continually shed and replace their teeth, which are modified scales with a dentin core and an outer enamel layer. Jaws are used to grab and rip chunks of prey. The sharks have been predators for hundreds of millions of years, but rare attacks on humans give the group a bad reputation.

Figure 23.8 Representative cartilaginous fishes. (**a**) Galápagos shark and (**b**) stingray. Notice the pairs of gill slits. Representative bony fishes: (**c**) Coral grouper, one of many species that have been overfished as food for humans. (**d**) Flying fish. Like bird wings, modified pectoral fins help these fishes fly out of the water when chased by predators. Evolutionarily speaking, they are not yet equipped to take off again from the deck of boats that happen to cross their flight path. (**e**) Perch fins.

"BONY FISHES"

More than 400 million years ago, even before sharks appeared, ray-finned fishes, lobe-finned fishes, and lungfishes evolved. With more than 21,000 known species in freshwater and marine habitats, the ray-finned fishes are the most diverse in shape, size, and coloration. Their fins are bony elements in a web of skin. Figure 23.8c–e shows just a few representatives.

Coelacanths (*Latimeria*) are the only surviving group of lobe-finned fishes. We know about two populations that may be separate species. The ventral fins of these fishes are fleshy extensions of the body, with skeletal support elements inside (Figure 23.9a). Like the ray-finned fishes, coelacanths have gills.

Lungfishes (Figure 23.9b) have gills *and* one or two small, modified outpouchings of the gut wall. These sacs help take in oxygen and remove carbon dioxide. Lungfishes must surface and gulp air; they drown if held under water. When streams dry out seasonally, lungfishes encase themselves in slimy secretions and mud. They stay inactive until the rainy season starts.

Do such fishes share a common ancestor with four-legged walkers—**tetrapods**? Probably. Like tetrapods, a lungfish has a separate blood circuit to the lungs. Its skullbones are arranged the same way. It has the same tooth enamel. Also, the fins of both lobe-finned fishes and lungfishes are about the same as tetrapod limbs in structure, position, and sizes. A lungfish even uses its limbs to move forward on underwater substrates.

Limb joints, digits, and other traits of fossils imply that walking originated in water, not on land. By the late Devonian, some swimming and crawling species were using limbs and digits (Figure 23.9c–e).

Ray-finned fishes are now the most diverse and abundant vertebrates. The lobe-finned fishes have fleshy ventral fins reinforced with skeletal parts. The lungfishes have simple lunglike sacs that supplement respiration by gills.

Walking probably started under water during the Devonian, among the aquatic forerunners of tetrapods on land.

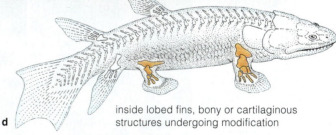

inside lobed fins, bony or cartilaginous structures undergoing modification

Figure 23.9 (**a**) Living coelacanth (*Latimeria*), of the only lineage of surviving lobe-finned fishes. (**b**) Australian lungfish. (**c**) Painting of Devonian tetrapods. *Acanthostega*, submerged, and *Ichthyostega* crawling on land. Both had a fishlike skull, caudal tail, and fins. Unlike fishes, both had a short neck and four limbs with digits. (**d,e**) Proposed evolution of skeletal elements inside fins into the limb bones of early amphibians.

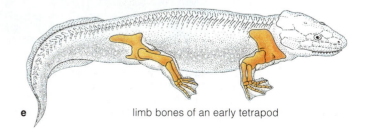

limb bones of an early tetrapod

Chapter 23 Animal Evolution—The Vertebrates   383

23.4 Amphibians—The First Tetrapods on Land

Even a small genetic change could have transformed lobed fins into limbs. Remember those enhancers that control gene transcription? One of them influences how digits form on limb bones. Even a single mutation in one master gene can lead to a big change in morphology.

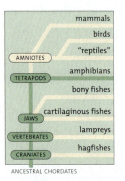

mammals
birds
"reptiles"
AMNIOTES
amphibians
TETRAPODS
bony fishes
cartilaginous fishes
JAWS
lampreys
VERTEBRATES
hagfishes
CRANIATES

ANCESTRAL CHORDATES

Amphibians are the group of vertebrates that were the first tetrapods on land. They either have four legs or a four-legged aquatic ancestor, and the body plan and reproductive modes are somewhere between fishes and "reptiles."

What drove certain aquatic tetrapods onto land? Consider this: Asteroids hit the Earth at least five times during the Devonian. One of the last impacts coincided with a mass extinction of marine life. If it triggered the release of methane hydrates, much of the oxygen dissolved in the open seas and swampy coastal habitats would have been displaced by carbon dioxide (Section 3.2). However it happened, tetrapods with lungs had the advantage; they could get enough oxygen by gulping air. Fossils show that some tetrapods were semiaquatic and were spending time on land when the Devonian came to a close.

Life in the new, drier habitats was both dangerous and promising. Temperatures shifted more on land than in the water, air didn't support the body as well as water did, and water was not always plentiful. But air is richer in oxygen. Amphibian lungs continued to be modified in ways that enhanced oxygen uptake. The heart, newly divided into three chambers, pumped more oxygen-rich blood to cells. Increased reliance on aerobic respiration, a pathway that makes enough ATP to sustain more active life-styles, was now possible.

New sensory information also challenged the early amphibians. Swamp forests supported vast numbers of insects and other small invertebrate prey. On land, vision, hearing, and balance turned out to be highly advantageous senses. They promoted the expansion of the brain regions that received, interpreted, and responded to the richly novel sensory input.

All living amphibians are descendants of the first tetrapods. None has escaped the water entirely. Even

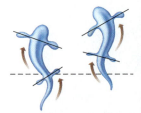

fish locomotion

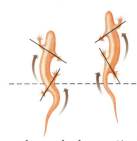

salamander locomotion

Figure 23.10 Familiar amphibians. (**a**) Red-spotted salamander (*Notophthalamus*). Coordinated controls over its legs move the body the same way that controls over fins move fishes. (**b**) A frog jumping and displaying its tetrapod heritage. (**c**) American toad.

Figure 23.11 One of the caecilians, a legless tropical amphibian. How do you suppose it burrows into soil?

species that have gills or lungs also exchange oxygen and carbon dioxide across the surface of their thin skin. Respiratory surfaces must be moist at all times, however, and skin dries easily in air.

Some amphibians spend their entire life in water. Most release eggs in water, where their larvae can develop. Most species adapted to dry habitats lay eggs in moist places. A few protect their embryos as they develop inside the moist adult body.

Frogs and toads are among the most well-known amphibians, and with more than 4,800 named species, they are also the most diverse (Figure 23.10b,c). Most are active predators; they capture insect prey with an extensible, sticky-tipped tongue. They generally lay eggs in water, and these hatch into tadpoles, a type of legless larvae with gills and a tail. Later on, the gills and tail disappear, and four legs form.

The 500 or so salamander species have a long body, and a tail that persists in adults. A few species, such as the mudpuppy and axolotl, retain other juvenile amphibian features, including external gills. As most salamanders walk, they bend from side to side like fishes (Figure 23.10a). Probably the first tetrapods on land walked this way.

As some amphibians evolved, they lost their limbs and vision, but not their jaws. This group gave rise to caecilians (Figure 23.11). Most of the 165 or so species burrow through moist soil, using the senses of touch and smell as they pursue insects. A few are aquatic predators that use electrical cues to locate prey.

In body form and behavior, amphibians show resemblances to aquatic tetrapods and reptiles. Most still rely on access to aquatic or moist habitats to complete their life cycle. Frogs, toads, and salamanders are the most familiar groups.

23.5 Vanishing Acts

FOCUS ON THE ENVIRONMENT

Amphibians are survivors. They originated before the dinosaurs and outlasted them. They have been around a thousand times longer than humans, but now human activities are putting many species at risk.

There is no question that amphibians are in trouble. Of about 5,500 known species, population sizes of at least 200 are plummeting. The declines have been well-documented in North America and Europe, but things look bad worldwide.

Six frog, four toad, and eleven salamander species are currently listed as threatened or endangered in the United States and Puerto Rico. One of them, the California red-legged frog, is the largest frog native to the western United States. It is the species that inspired Mark Twain's classic short story, "The Celebrated Jumping Frog of Calaveras County."

Many declines correlate with shrinking habitats and deteriorating habitat conditions. Developers and farmers often fill low-lying ground that became seasonal pools of standing water. Nearly all species of amphibians have trouble breeding unless they can deposit their eggs in water.

Also contributing to the declines are introductions of new species in amphibian habitats, long-term changes in climate, increases in ultraviolet radiation, and the spread of fungal and parasitic diseases into new habitats (Figure 23.12). Contamination of aquatic habitats by pesticides and other chemicals is another contributing factor. Chapter 31 offers a closer look at the effects of habitat contamination.

Figure 23.12 (**a**) Example of frog deformities. (**b**) A parasitic fluke (*Ribeiroia*) that burrows into frog tadpole limb buds and physically or chemically alters their cells. Infected tadpoles grow extra legs or none at all. Where *Ribeiroia* population densities are great, the number of tadpoles that successfully complete metamorphosis is low. Trematode cysts have been found in deformed frogs and salamanders in California, Oregon, Arizona, and New York.

23.6 The Rise of Amniotes

Amniotes were the first vertebrates to adapt to dry land habitats. They did so through modifications in their organ systems, behavior, and eggs, which have four membranes that conserve water and support embryonic development.

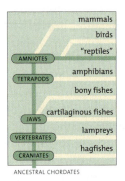

ANCESTRAL CHORDATES

THE "REPTILES"

Late in the Carboniferous, some amphibian species gave rise to **amniotes**, the only vertebrates that produce eggs having four membranes. These membranes conserve water and also protect the developing embryo before hatching or being born (Figure 23.13). All amniotes have dry, tough, or scaly skin that restricts water loss. They also have a pair of efficient water-conserving kidneys.

Synapsids and sauropsids are two major groups of amniotes. *Synapsids* are mammals and their extinct relatives. *Sauropsids* are "reptiles," and birds. Reptiles are no longer recognized as a formal taxonomic group because they are not monophyletic. The name persists only as a way to refer to several lineages that show the basic amniote features but not the derived traits that define birds or mammals.

Compared to amphibians, the early reptiles chased prey with more cunning and speed. With well-muscled jaws and sharp teeth, they could seize and crush prey with sustained force. The four limbs of many species were better at supporting the body's trunk on land. Gas exchange across the skin, vital for amphibians, was abandoned; lungs became better developed. The brainier reptiles engaged in behaviors not seen among the amphibians (Figure 23.13a).

THE AGE OF DINOSAURS

The first **dinosaurs** descended from Triassic reptiles. They were not much bigger than a turkey. They might have had high metabolic rates, and it is possible they were warm-blooded. Many moved about on two legs.

Adaptive zones opened up for dinosaurs about 213 million years ago. Huge fragments from an asteroid or comet fell upon the spinning Earth, creating craters in what is now France, Quebec, Manitoba, and North Dakota. Nearly all animals that survived the impacts were small, had high rates of metabolism, and could tolerate big temperature changes.

Descendants of the surviving dinosaurs dominated the land for 125 million years, even as ichthyosaurs and other groups flourished in the seas (Figure 23.14). Many dinosaurs were lost in a mass extinction at the end of the Jurassic, and others in a pulse of Cretaceous extinctions. Some lineages did recover, and new ones evolved in forests, swamps, and open habitats.

By the **K–T asteroid impact theory**, the Cretaceous ended when an asteroid as huge as Mount Everest struck Earth (Chapter 17). Dinosaurs vanished, along with most life in the seas. Afterward, gradual crustal movements repositioned the impact site. Researchers discovered it in the Yucatán peninsula of Mexico.

The asteroid struck at 160,000 kilometers per hour and made a crater wider than Connecticut. Monstrous waves 120 meters high raced across the ocean, wiped out life on islands, then slammed into continents. By a **global broiling hypothesis**, it was as if 100 million nuclear bombs had been detonated. Trillions of tons

Figure 23.13 Some amniote eggs. (**a**) Painting of the nest of a duck-billed, plant-eating dinosaur (*Maiasaura*), found in Montana. This species, which lived 80 million years ago, showed social behavior; it traveled in herds, protected eggs, and cared for the young. (**b**) Eastern hognose snakes emerging from leathery-shelled eggs.

Figure 23.14 Jurassic scene. (**a**) *Archaeopteryx* gliding in the foreground. Left to right, the plant eaters *Stegosaurus* and *Apatosaurus*, the carnivore *Saurophaganax* ("king of reptile eaters"), and *Camptosaurus*, a beaked plant eater. A climbing mammal looks out from the shrubbery. *Apatosaurus* was one of the largest land animals that ever lived. This tiny-brained plant eater grew as long as 27 meters (90 feet) and weighed 33 to 38 tons.

(**b**) *Temnodontosaurus*, an ichthyosaur that hunted large squid, ammonites, and other prey in the warm, shallow seaways. Fossils measuring 9 meters (30 feet) long have been found in England and Germany.

An asteroid impact ended the golden age of dinosaurs.

of vaporized debris rose in a colossal fireball. As the debris fell back to Earth, atmospheric temperatures rose by thousands of degrees. In one horrific hour, nearly all plants erupted in flames. All animals out in the open, not in burrows, were broiled alive.

Some reptiles survived; their descendants include the crocodilians, turtles, tuataras, snakes, and lizards. So did a group of feathered dinosaurs—birds.

Tough, scaly, or dry skin and other ways of conserving water, eggs with membranes that support the embryos, and active life-styles helped amniotes radiate into dry land habitats.

One amniote lineage, the sauropsids, includes reptiles and birds. Another, the synapsids, includes mammals and the now-extinct mammal-like reptiles.

Much of amniote history was a matter of luck, of species with particular traits being in the right or wrong places on a changing geologic stage.

23.7 Existing Reptilian Groups

The name reptile is derived from the Latin repto, *which means to creep. Some existing forms do creep. Others race, lumber, or swim about. These diverse lineages include all living amniotes that are not birds or mammals.*

More than 8,160 species of reptiles flourish in diverse habitats on land, in fresh water, and in the oceans. All are cold-blooded. Fertilization is internal; eggs are usually laid on land, even by aquatic species. Both males and females have a cloaca, an opening used in excretion as well as reproduction. Figure 23.15 shows a body plan that is typical of one group.

About 305 existing species of turtles live inside a shell attached to their skeleton. If threatened, turtles can pull their head and limbs inside (Figure 23.16a,b). Their body plan works well; it has been around since the Triassic. Only among sea turtles and other highly mobile types has the shell been reduced in size.

Instead of teeth, turtles have horny plates that are used to grip and chew food. Strong jaws and a fierce disposition help deter predators. But turtle eggs are vulnerable to many predators on land. Add hunting and destruction of nesting sites to the mix, and most sea turtles are now poised at the brink of extinction.

With about 4,710 species, the lizards are the most diverse reptiles. The largest, *Varanus komodoensis*, is a monitor lizard that grows ten feet long. This ambush predator lurches out of the bushes at deer, wild boar, an occasional human, and other prey. The smallest lizard, *Sphaerodactylus ariasae*, can fit on a dime (*left*). Like most predatory lizards, it snags prey with sharp, peglike teeth. Chameleons (page 475) use accurate flicks of a sticky tongue, which can be longer than the

length of their body. Iguanas, including the marine iguanas of the Galápagos, are herbivores.

Being small themselves, most lizards are prey for other animals. Some attempt to startle predators by flaring a throat fan; others try to outrun them (Figure 23.16c,d). Many will give up their tail when a predator grabs them. The detached tail wriggles for a bit and may be distracting enough to permit a getaway.

The two surviving species of tuataras live on small islands near New Zealand (Figure 23.16e). They are the descendants of a group that had its heyday in the Triassic. They are reptiles, yet they resemble modern amphibians in their brain and locomotion. Like some lizards, the tuataras have a third "eye" under the skin, complete with retina, a rudimentary lens, and links to the brain. It may detect changes in daylength and light intensity and affect hormonal control of reproduction.

During the Cretaceous, short-legged, long-bodied lizards gave rise to elongated, limbless snakes. Some snakes retain bony remnants of ancestral hindlimbs. Most move their body in S-shaped waves, much like salamanders do.

All 2,955 living snake species are carnivores with flexible skull bones and jaws. Many can swallow prey wider than they are. Rattlesnakes and other fanged types bite and subdue their prey with venom (Figure 23.16f). Snakes in general are not aggressive toward humans. Even so, venomous snakes bite about 7,000 people annually in the United States; 15 of them die. About 40 percent of those bitten reported that they were handling or otherwise disturbing the snake at the time. Worldwide, the annual death toll from snake bites is estimated at 30,000–40,000.

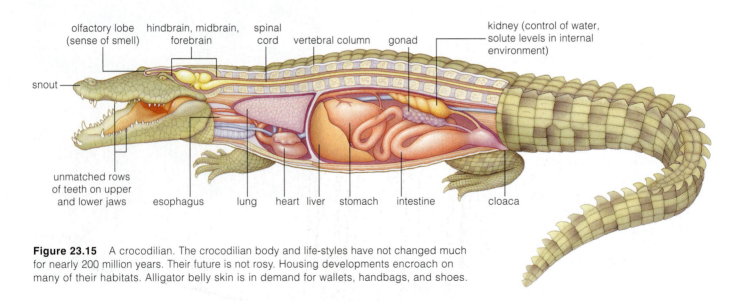

Figure 23.15 A crocodilian. The crocodilian body and life-styles have not changed much for nearly 200 million years. Their future is not rosy. Housing developments encroach on many of their habitats. Alligator belly skin is in demand for wallets, handbags, and shoes.

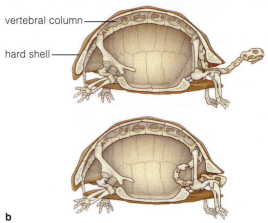

vertebral column
hard shell

Figure 23.16 (**a**) Galápagos tortoise. (**b**) Turtle shell and skeleton. (**c**,**d**) Lizard fleeing, lizard confronting. (**e**) Tuatara (*Sphenodon*). (**f**) Rattlesnake. (**g**) Spectacled caiman, a crocodilian. It has two unmatched rows of peglike teeth.

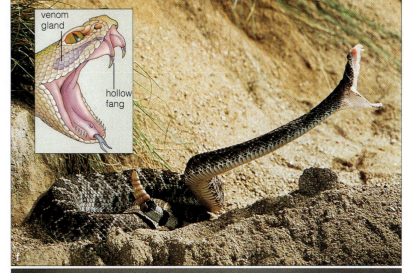

venom
gland

hollow
fang

The twenty-three modern crocodilian species are the most advanced reptiles and the closest living relatives of birds. Crocodiles, alligators, and caimans all live in or near water and are predators that have powerful jaws, a long snout, and sharp teeth (Figure 23.16*g*). They drag prey into the water, tear it apart as they spin around and around, then gulp down the chunks. The crocodilians are the only reptiles with a four-chambered heart, like mammals and birds. Like most birds, they show parental behavior, such as guarding nests and assisting the hatchlings.

Turtles and tuataras have body plans that have changed very little since the Triassic. Lizards and snakes are the most diverse reptilians. Crocodilians have the most complex brain and circulatory system and are the closest relatives of birds.

23.8 Birds—The Feathered Ones

Birds are warm-blooded amniotes with feathers. They are descended from dinosaurs and still retain a number of reptilian traits. Internal and external modifications in body plans led to the capacity for flight.

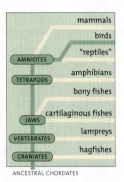

ANCESTRAL CHORDATES

The capacity for flight evolved in four groups—the insects, pterosaurs (extinct), birds, and bats. Of these groups, feathers are a trait unique to **birds** and some extinct dinosaurs. These lightweight structures, derived from scales, are used in flight, as body insulation, and often in courtship displays.

There are about 28 orders of birds and close to 9,000 named species. They vary in size, proportions, coloration, and capacity for flight. The smallest, the bee hummingbird, weighs 1.6 grams (0.6 ounces). The ostrich, the largest living bird, weighs as much as 150 kilograms (330 pounds). It cannot fly, but its powerful leg muscles make it an impressive sprinter. Feather colors and patterns can be especially spectacular among warblers, parrots, and some other perching species that show complex social behaviors. Bird song and other forms of behavior are among the topics of Chapter 43.

Birds originated when Mesozoic reptiles began an adaptive radiation. They diverged from a lineage of small theropod dinosaurs, a type of carnivore that ran about on two legs. Bird feathers evolved from highly modified reptilian scales. *Archaeopteryx* was in or near that lineage. As you read earlier, it had reptilian *and* avian traits, including feathers. Birds still share many traits with their closest relatives, the dinosaurs and crocodilians. For instance, like them, birds have scales on their legs, and they have some of the same internal parts. Like them, birds fertilize eggs internally and have a cloaca used in excretion and reproduction.

Unlike reptiles, birds closely regulate their body temperature. Most heat gains come from metabolism, not the outside environment. Feathers help conserve body heat when external temperatures drop. When temperatures rise, elastic sacs connected to bird lungs help dissipate excess metabolic heat. The sacs enhance air flow and can move warmed air out of the body.

The evolution of bird flight involved more than feathers. It involved modification of the entire body, from internal bone structure to highly efficient modes of respiration and circulation. The high metabolic rates that sustain flight depend on a strong flow of oxygen through the body. The elastic sacs that connect to the lungs greatly enhance oxygen uptake (Section 35.4). Like mammals, birds have a large, four-chambered heart. It pumps oxygen-rich blood to the lungs and to the rest of the body in separate circuits, as it probably did in the reptilian ancestors of birds.

Flight also demands an airstream, low weight, and a powerful downstroke that can provide lift—a force at right angles to an airstream. Each wing, a modified forelimb, consists of feathers and lightweight bones attached to powerful muscles. Its bones do not weigh much, owing to profuse air cavities in the bone tissue (Figures 23.17b and 23.18d). The flight muscles attach

yolk sac embryo amnion chorion allantois

hardened shell albumin ("egg white")

a

Figure 23.17 (a) Sketch of one type of amniote egg. (b) Body plan of a typical bird, a pigeon. Flight muscles attach to the large, keeled breastbone (sternum).

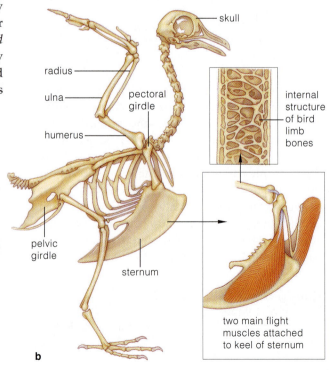

skull

radius
ulna
pectoral girdle
humerus

internal structure of bird limb bones

pelvic girdle
sternum

two main flight muscles attached to keel of sternum

b

Figure 23.18 Bird flight. (a) Of all living animals, only birds and bats fly by *flapping* their wings. A bird wing is an intricate system of lightweight bones and feathers. (b) Innermost down feathers function in insulation. (c) A hollow central shaft and an interlocked lattice of barbs and barbules strengthen flight feathers and increase the wing flight surface without imposing much weight. (d) Laysan albatross. Its wingspan is more than two meters, yet this seabird weighs less than 10 kilograms (22 pounds). It is so at home in the air, it even sleeps while riding the winds. One bird monitored by researchers flew 24,843 miles in ninety days as it searched for food and brought it back to its nestlings. That is the equivalent of a trip around the globe.

to an enlarged breastbone (sternum) and to the upper limb bones attached to it. When the muscles contract, they produce a powerful downstroke.

With their long flight feathers, wings are like the airfoils on planes. A bird normally spreads out long feathers on a downstroke, thereby increasing the area of the surface pushing against the air. Then, on the upstroke, the bird folds its feathers a bit, so each wing surface decreases and offers less resistance to air.

We see one of the most dramatic forms of behavior among birds that migrate with the changing seasons. **Migration** is a recurring pattern of movement from one region to another in response to environmental rhythms. Seasonal change in daylength is one factor that influences the internal timing mechanisms that we call "biological clocks." It causes physiological

and behavioral changes that induce migratory birds to make round trips between breeding grounds and wintering grounds. Arctic terns migrate the farthest. They spend summers in arctic regions and winters in antarctic regions. One monitored bird took off from an island near Great Britain and landed three months later in Melbourne, Australia. It racked up more than 22,000 kilometers (14,000 miles) on that one journey.

Of all animals, birds alone produce feathers, which they use in flight, heat conservation, and socially significant communication displays.

A bird's feathers, lightweight bones, and highly efficient respiratory and circulatory systems sustain flight.

23.9 Mammals

Remember the premise of cause and effect? Mammals evolved before dinosaurs and outlasted them. They were artful dodgers, sidelined only until the dinosaurs' luck ran out. Then they radiated, spectacularly, into the vacated adaptive zones.

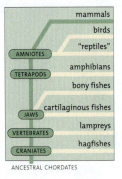

They alone make hair, but they differ in its amount, type, and distribution. Female mammals alone secrete milk for their young. This nutrient-rich fluid forms in mammary glands that have ducts to the body surface.

Jaws and teeth also help set mammals apart. They alone have a lower jawbone on each side. Most have four types of upper and lower teeth that match up (Figure 23.19). Reptiles have a hinged two-part jaw, and all teeth are one type only and don't match up.

Most mammals care for offspring for an extended time. Young mammals typically have the capacity to learn and repeat behaviors with survival value. Some species show stunning *behavioral flexibility*—a capacity to expand on basic activities with novel behaviors.

Where did mammals come from? Remember, the extinct mammal-like reptiles and all extinct and living species of mammals are synapsids. The defining trait of this group is an opening in a particular location on each side of the skull that functions in the attachment and alignment of jaw muscles. The trait arose in the Carboniferous. Later on, in the Permian, the synapsids underwent adaptive radiations and became the major land vertebrates. At the close of the Permian, many groups died out in a mass extinction.

Fortunately for humans, the group called *therapsids* endured. Some became more like mammals during the Triassic. For instance, limbs became positioned upright under their trunk. Their skeletal arrangement made it easier to walk erect, but a trunk higher from

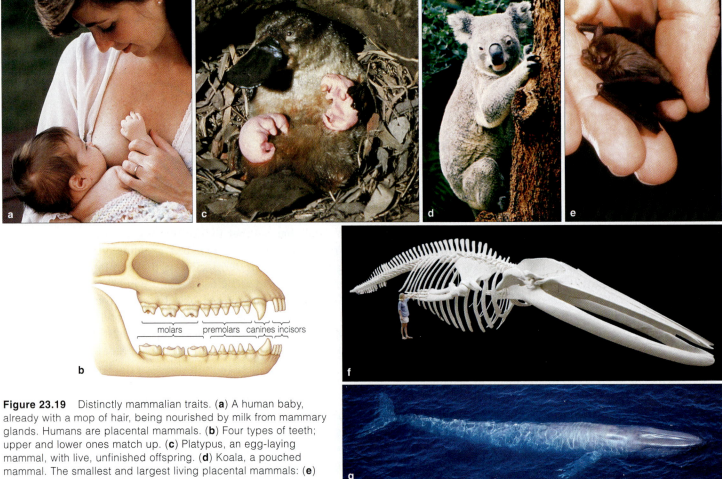

Figure 23.19 Distinctly mammalian traits. (**a**) A human baby, already with a mop of hair, being nourished by milk from mammary glands. Humans are placental mammals. (**b**) Four types of teeth; upper and lower ones match up. (**c**) Platypus, an egg-laying mammal, with live, unfinished offspring. (**d**) Koala, a pouched mammal. The smallest and largest living placental mammals: (**e**) Kitti's hognosed bat, 1.5 grams; and (**f,g**) blue whale, 200 tons.

molars   premolars canines incisors

Read Me First!
and watch the narrated animation on mammalian radiations

a About 150 million years ago, during the Jurassic, the first monotremes and marsupials evolved and migrated throughout the supercontinent Pangea.

b Between 130 and 85 million years ago, in the Cretaceous, placental mammals emerged and started to spread. Monotremes and marsupials of the southern supercontinent evolved in isolation from placental mammals.

c About 20 million years ago, in the Miocene, placental mammals expanded in range and diversity. On Antarctica, mammals vanished. Marsupials and early placental mammals displaced monotremes in South America.

d About 5 million years ago, in the Pliocene, advanced placental mammals invaded South America. And they drove most marsupials and early placental species to extinction.

Figure 23.20 (a–d) Adaptive radiations of mammals. *Right*, one consequence—a case of convergent evolution: (**e**) Australia's spiny anteater, one of two living monotremes. (**f**) Africa's aardvark and (**g**) South America's giant anteater. Compare the specialized ant-snuffling snouts.

the ground was not as stable. Not coincidentally, their cerebellum, a brain region that controls balance and spatial positioning, increased in size.

The earliest known groups of modern mammals evolved early on in the Jurassic (Figure 23.20a). They were mouselike forms, and they coexisted with the dinosaurs. By 130 million years ago, three separate lineages had evolved. We call them **monotremes** (egg-laying mammals), **marsupials** (pouched mammals), and **eutherians** (placental mammals).

At the end of the Jurassic, mammals evolved into the adaptive zones vacated, under extreme duress, by dinosaurs. In the Cenozoic, the supercontinent Pangea split apart. Some monotremes and marsupials drifted off on what would become Australia and Antarctica. They became isolated from placental mammals that were rapidly expanding their range and diversity on other continents (Figure 23.20b,c).

On a fragment that would become South America, monotremes were replaced by marsupials and early placental mammals. When a land bridge joined South and North America in the Pliocene, highly evolved placental mammals radiated southward and rapidly replaced many of their previously isolated relatives

(Figure 23.20d). Only the opossums and a few other species invaded land masses in the other direction.

What gave placental mammals a competitive edge? They had higher metabolic rates, more precise ways to control body temperature, and a more efficient way to nourish embryos. You will read about them in later chapters. For now, note that the adaptive radiation of the advanced placental mammals came about at the expense of their less competitive relatives.

One more point: Although the lineages originally evolved on separate continents, many lived in similar habitats. In time, they came to resemble one another in form and function. For example, Australia's spiny anteater, South America's giant anteater, and Africa's aardvark all have similar snouts adapted to preying on ants (Figure 23.20e–g). Such species are examples of *convergent* evolution, as described in Section 17.4.

Only female mammals produce milk. Mammals also are the only animals with hair and with differing types of teeth.

The three major groups are the egg-laying monotremes, the pouched marsupials, and the placental mammals. Placental mammals are now the dominant group in most regions.

23.10 From Early Primates to Hominids

So far, you have traveled 570 million years through time, from tiny flattened animals to craniates, then on to the jawed fishes with a backbone. You encountered some early tetrapods, then amniotes—reptiles, birds, and mammals. Now you are about to access what we know about the evolutionary events that led to modern humans.

TRENDS IN PRIMATE EVOLUTION

Living **primates** are prosimians, tarsioids, and the anthropoids (Figure 23.21). The first prosimians were tree-dwellers of northern forests for millions of years until monkeys and apes nearly displaced them. The gibbon, siamang, orangutan, gorilla, chimpanzee, and bonobo are classified as apes. All monkeys, apes, and humans are anthropoids. In biochemistry and body form, apes and humans are close relatives. They, and their closest extinct ancestors, are the only hominoids. All humanlike and human species, past and present, are known as the **hominids**.

Most of the primates live in the trees of forests, woodlands, or savannas (grasslands with a scattering of trees). No one feature sets them apart from other mammals, and each lineage has its defining traits. The five trends that define the lineage leading to humans were set in motion among the first tree-dwellers.

First, a reliance on the sense of smell became less important than daytime vision. *Second*, skeletal changes promoted **bipedalism**—upright walking—which freed hands for novel tasks. *Third*, bone and muscle changes led to diverse hand motions. *Fourth*, teeth became less specialized. *Fifth*, the evolution of the brain, behavior, and culture became interlocked. **Culture** is the sum of a social group's behavioral patterns, passed on through the generations by learning and symbolic behavior. In brief, *"uniquely" human traits emerged by modification of traits that had evolved earlier, in ancestral forms.*

Enhanced daytime vision. Early primates had an eye on each side of their head. Later on, many species had forward-directed eyes, which were better at sampling shapes and movements in three-dimensional space. Visual systems became more and more responsive to variations in light intensity (dim to bright) and colors.

Upright walking. How a primate walks depends on arm length and the shape and position of the shoulder blades, pelvic girdle, and the backbone (Figure 23.21*d*). With arm and leg bones of about the same length, a monkey climbs and runs on four legs. A gorilla walks with weight on two legs and the knuckles of its long arms. Humans have the shortest and the most flexible backbone, one curved in an S shape. These structural features are signs of bipedalism.

Figure 23.21 A few primates. (**a**) A tarsier, small enough to fit on a human hand, is a climber and leaper. (**b**) Spider monkey, an agile four-legged climber, leaper, and runner. (**c**) Male gorilla, a knuckle-walker. (**d**) Comparison of skeletal organization and stances. These images are not at the same scale. A monkey has long, thin limbs relative to its torso, and its long, thin digits are good at gripping, not at supporting the body's weight. Gorillas use their forelimbs to climb and support body weight when walking on knuckles and two legs. Humans are two-legged walkers.

Power grip and precision grip. So how did we get our versatile hands? Early mammals spread their toes apart to support the body as they walked or ran on four legs. In ancient tree-dwelling primates, hand bone modifications allowed them to curl fingers around objects (*prehensile* movements) and touch a thumb to fingertips (*opposable* movements). In time, hands were freed from load-bearing functions. Later, as hominids evolved, so did power and precision grips:

power grip   precision grip

Versatile hand positions gave human ancestors the capacity to make and use tools. They were a basis for unique technologies and cultural development.

Teeth for all occasions. Modifications to the jaws and teeth accompanied a shift from eating insects, to fruits and leaves, then to a mixed diet. Rectangular jaws and lengthy canines evolved in monkeys and apes. A bow-shaped jaw evolved in early hominid lineages. So did teeth that were smaller and all about the same length.

Brains, behavior, and culture. Shifts in reproductive and social behavior accompanied a shift to life in the trees. In many lineages, parents put more effort in fewer offspring and maternal care became intense. The young required more extended periods of dependency and learning before setting out on their own.

New forms of behavior promoted new connections in brain regions that deal with sensory inputs. A brain having more intricate wiring favored new behavior. The brain, behavior, and culture evolved together through enriched learning and symbolic behaviors, especially language. A capacity for language arose among hominids. In addition to a bigger, more complex brain, it required a bigger chamber to hold it.

ORIGINS AND EARLY DIVERGENCES

The first primates arose between 85 million and 65 million years ago, in tropical Paleocene forests. Like the tree shrews they resembled (Figure 23.22), they had huge appetites and foraged at night for eggs, insects, seeds, and buds under trees. They had a long snout and a good sense of smell for snuffling out food or predators. They clambered about among stems and branches, although not with much speed or grace.

By the Eocene, the prosimians were living higher in the trees. They had a shorter snout, a larger brain,

Figure 23.22
Tropical forest of Southeast Asia and one of its tiny inhabitants, a tree shrew (*Tupaia*).

better daytime vision, and better ways to grasp and manipulate objects. *How did these traits evolve?*

Trees offered food and safety from ground-dwelling predators. They also were zones of uncompromising selection. Picture an Eocene morning, leaves swaying in the breeze, colorful fruit, and predatory birds. An odor-sensitive, long snout would not have been useful; breezes disperse odors. A brain that assessed motion, depth, shape, and color, along with skeletal changes, must have been favored. For example, eye sockets became positioned in front instead of on the sides of the skull—a better arrangement for depth perception.

By 36 million years ago, tree-dwelling anthropoids had evolved. They were on or close to the lineage that led to monkeys and apes. One had forward-directed eyes, a flattened face, and an upper jaw that sported shovel-shaped front teeth. It probably used its hands to grasp food. Some of the early anthropoids lived in swamps infested with predatory reptiles. Was that one reason why it became so imperative to think fast, grip strongly, and avoid the ground? Possibly.

Between 23 and 5 million years ago, in the Miocene, apelike forms—*the first hominoids*—evolved and spread through Africa, Asia, and Europe. Shifts in land masses and ocean circulation were causing long-term change in climate. Africa became cooler and drier, with more seasonal change. Tropical forests, with their edible soft fruits, leaves, and abundant insects, gave way to open woodlands and later to grasslands. Food had become drier, harder, and more difficult to find.

Hominoids that had evolved in lush forests had two options—move into new adaptive zones or die out. Most died out, but not the common ancestor of apes and, later in time, modern humans. By 7 million years ago, the *hominid* lineage had become established.

Complex, forward-directed vision; bipedalism; refined hand movements; generalized teeth; and interlocked elaboration of brain regions, behavior, and culture were key adaptations on the road from arboreal primates to modern humans.

23.11 Emergence of Early Humans

In central, eastern, and southern Africa, the Miocene through the Pliocene was a "bushy" time of hominid evolution, meaning many forms rapidly evolved during that span. We still don't know how they are related.

Sahelanthropus tchadensis was one species that evolved in Central Africa about 6 or 7 million years ago, when human ancestors were becoming distinct from apes (Figure 23.23a). One fossilized braincase is no bigger than that of a chimpanzee. Yet, like brainier hominids that evolved later, its owner had a shorter, flattened face with a prominent brow ridge, and smaller canines. Was it an ape or a hominid? We don't know.

Australopithecus and *Paranthropus* were hominids informally called **australopiths**. Figure 23.23a shows examples. Like apes, australopiths had a large face relative to their braincase, which indicates that they were small-brained. Also like apes, australopiths had protruding jaws. However, they differed in critical respects from earlier hominoids. For one thing, their thick-enameled molars could grind harder food. And these hominids walked upright. We know this from fossilized hip bones and limb bones. We also have footprints. About 3.7 million years ago, an *A. afarensis* individual walked across newly fallen volcanic ash. A light rain transformed the ash to quick-drying cement and preserved its tracks (Figure 23.23c).

In the late Miocene, bipedal hominids were still adapted to forest life. Their hands could make strong, precision grips. Their descendants eventually left the trees for life on the ground but were not specialized for running fast on all fours. Rather, they became fully upright and used manipulative skills to advantage. We can expect that they used their hands to hang on to offspring and to carry precious food during their foraging expeditions out in the open.

What do fossilized fragments tell us about human origins? The record is still too sketchy to reveal how all the diverse forms were related, let alone which were our ancestors. Besides, which traits should we use to define **humans**—members of the genus *Homo*?

Well, what about brains? Our brain is the basis of analytical and verbal skills, complex social behavior, and technological innovation. This sets us apart from

Sahelanthropus tchadensis 7–6 million years

Australopithecus afarensis 3.6–2.9 million years

A. africanus 3.2–2.3 million years

A. garhi 2.5 million years (first tool user?)

Paranthropus boisei 2.3–1.4 million years (huge molars)

P. robustus 1.9–1.5 million years

H. habilis 1.9–1.6 million years

a

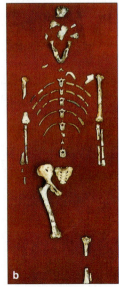  

Figure 23.23 (a) African hominid fossils. (b) Fossilized bones of "Lucy" (*Australopithecus afarensis*), a two-legged walker who lived 3.2 million years ago.

(c) At Laetoli in Tanzania, Mary Leakey found these footprints, made in soft, damp, volcanic ash 3.7 million years ago. The arch, big toe, and heel marks of these footprints are signs of bipedal hominids. Unlike apes, the early hominids did not have a splayed-out big toe, of the sort being held up by the chimpanzee in (d).

apes that have a far smaller skull volume and brain size. Yet this feature alone cannot tell us when some hominids made the evolutionary leap to becoming human. Their brain size fell within the range of apes. They did use simple tools, but so do chimps and some birds. We have no clues to their social behavior.

We are left to speculate on evidence of physical traits, such as bipedalism, manual dexterity, a small face, a large braincase, and small, thickly enameled teeth. These traits emerged late in the Miocene and are features of what might be the first humans—*Homo habilis*. The name means "handy man."

Between 2.4 and 1.6 million years ago, early forms of *Homo* lived in woodlands bordering the savannas of eastern and southern Africa (Figure 23.24). Fossil teeth show that they ate hard-shelled nuts, dry seeds, soft fruits and leaves, and insects—all seasonal foods. *H. habilis* had to think ahead, to plan when to gather and store foods for cool, dry seasons ahead.

H. habilis shared its habitat with carnivores such as saber-tooth cats, which could impale prey and tear flesh but could not crush open the marrow bones. Carcasses with meat shreds clinging to bones offered nutrients in nutrient-stingy places. *H. habilis* was not one of the full-time carnivores. It may have enriched its diet opportunistically, by scavenging carcasses.

Who was the first tool maker? The earliest stone tools, from a site in Ethiopia, date to about 2.5 million years ago. Bits of volcanic rock, chipped to create a sharp edge, were found alongside bones of animals that appear to have been butchered with similar tools. Fossils of *A. garhi* were found nearby and date to the same period. Did *A. garhi* make the tools? Perhaps.

The layers of Tanzania's Olduvai Gorge document later refinements in toolmaking skills (Figure 23.25). The deepest of these sedimentary rock layers date to about 1.8 million years ago and hold crudely chipped pebbles similar to the earliest ones in Ethiopia. They also hold the fossil remains of australopiths and of *H. habilis*. We don't know which group created the tools. More recently deposited layers hold somewhat more complex tools. But innovations in tool form were limited. The same types appear in layers that formed slowly, over a period of about 1 million years.

Australopiths were a group of early hominids that walked upright. Some coexisted with the first human species, Homo habilis, which had a larger brain than earlier hominids.

The formation and use of stone tools may have started with one of the australopiths. H. habilis was another toolmaker.

Figure 23.24 Reconstruction of *Homo habilis* in an East African woodland. Two australopiths are in the distance.

Figure 23.25 From Olduvai Gorge in Africa, a sampling of stone tools. *Left to right,* crude chopper, more refined chopper, hand ax, and cleaver.

23.12 Emergence of Modern Humans

Ancestors of modern humans stayed put in Africa until
Homo erectus evolved. Then cultural evolution took off.

EARLY BIG-TIME WALKERS

Homo erectus is a species related to modern humans (Figure 23.26). Its name means "upright man." Its forerunners were upright walkers, but some *H. erectus* populations did the name justice. They walked out of Africa, turned left into Europe, and also right, all the way to China. Middle East and Southeast Asia fossils date from 1.8 million to 1.6 million years ago. So *H. erectus* had to adapt to brutally cold climates of that period, and to vast ice sheets that advanced several times into northern Europe and North America.

Whatever pressures triggered the far-flung travels, this was a time of physical changes, as in skull size and leg length. It was a time of cultural lift-off for the human lineage. *H. erectus* had a larger brain and was a more creative toolmaker. Its social organization and communication skills must have been well developed. How else can we explain its grand dispersal? From southern Africa into England, different populations used the same kinds of hand axes and other tools. They met environmental challenges by building fires and using furs for clothing. Clear evidence of fire use dates from an early Pleistocene ice age.

A new fossil discovered in Ethiopia shows *Homo sapiens* had evolved by 160,000 years ago. Compared to *H. erectus,* the new species had smaller teeth, facial bones, and jawbones. Many individuals also had a novel feature—a chin. They had a higher, rounder skull, a bigger brain, and perhaps complex language.

A different group, the massively built and large-brained Neandertals, lived in Europe and the Near East from 200,000 to 30,000 years ago (Figure 23.26). Their extinction coincided with the arrival of modern humans in the same regions (Figure 23.27). Did they interbreed or war with those new arrivals? We do not know. But Neandertal DNA has unique sequences that are not found in present-day gene pools. They may not have modern descendants.

WHERE DID MODERN HUMANS ORIGINATE?

If we are interpreting the fossil record correctly, then Africa was the cradle for us all. At this writing, no one has found *Homo* fossils older than 2 million years *except* in Africa. Some *H. erectus* groups dispersed into the grasslands, forests, and mountains of Europe and Asia. Between 2 million and 500,000 years ago, they left Africa in waves. Some still lived in Southeast Asia between 53,000 and 37,000 years ago.

Figure 23.26
Fossils of two species of early humans. Evidence indicates that both species coexisted for a time with fully modern humans, *Homo sapiens*.

H. erectus
2 million–53,000? years

H. neanderthalensis
200,000–30,000 years

H. sapiens
fossil from Ethiopia,
160,000 years old

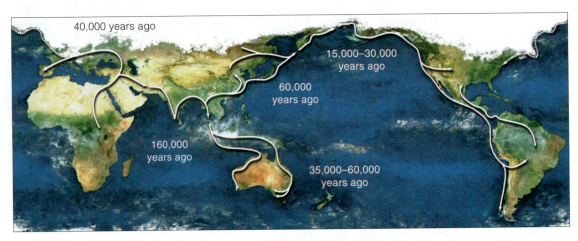

40,000 years ago

15,000–30,000
years ago

60,000
years ago

160,000
years ago

35,000–60,000
years ago

Figure 23.27   Estimated times when early *H. sapiens* colonized different parts of the world, based on radiometric dating of fossils. White lines show the presumed dispersal routes. Molecular data show that very few populations left Africa. Genetically, all modern human populations are extremely similar.

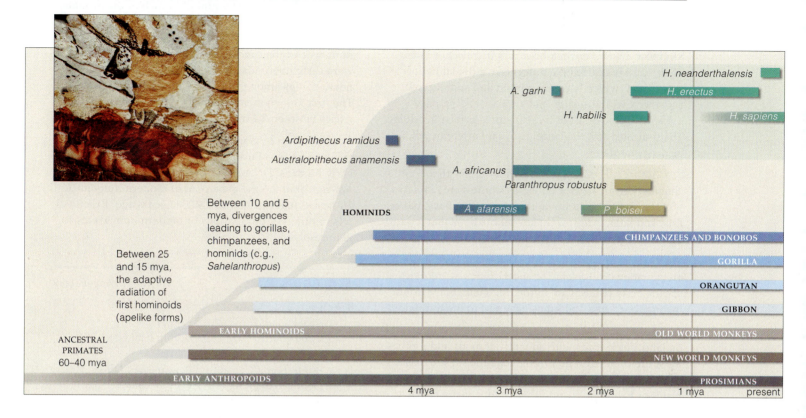

Figure 23.28 Summary of presumed evolutionary branchings in the primate family tree, including lineages on or near the road that led to modern humans. Also shown is one of the cave paintings by modern early humans at Lascaux, France.

But when did *H. sapiens* originate? *Here we find one case of how the same body of evidence can be interpreted in different ways.* The *multiregional* model and the *African emergence* model explain the distribution of early and modern humans differently, yet they both rely on measurements of genetic distance between existing populations. Biochemical and immunological studies show the greatest genetic distance is between human populations native to Africa and all others.

According to the multiregional model, data show that modern humans are the descendants of *H. erectus* populations that left Africa and spread through many regions by about one million years ago. In each area, the selective pressure on a population was different. Over time, subpopulations (races) of *H. sapiens* arose as a result of different selection pressures.

The African emergence model does not dispute fossil evidence that *H. erectus* migrated out of Africa and evolved further in different regions. However, it has *H. sapiens* arising in sub-Saharan Africa, then migrating out to other regions (Figure 23.27). By this model, as modern humans spread out of Africa, they replaced the archaic *H. erectus* populations that had preceded them. Only later did the regional phenotypic differences called races evolve.

In support of this model, the oldest of the *H. sapiens* fossils are from Africa. In Zaire, intricately wrought barbed-bone tools indicate that African populations were as skilled at making tools as *Homo* populations in Europe. Gene patterns from forty-three ethnic Asian groups suggest modern humans moved from Central Asia, along India's coast, then on into Southeast Asia and southern China. They later dispersed north and west into China and Siberia, then into the Americas.

Whatever the case, for the past 40,000 years, our cultural evolution has outpaced the macroevolution of the human species. One point to remember: Humans spread rapidly through the world by devising *cultural* means to deal with diverse environments. Hunters and gatherers persist in areas even as other groups moved from "stone-age" technology to high tech. The evolutionary road to this adaptable species was long, as summarized in Figure 23.28.

Cultural evolution has outpaced the biological evolution of the only remaining human species, H. sapiens. Today, humans everywhere rely on cultural innovation to adapt rapidly to a broad range of environmental challenges.

Summary

Section 23.1 Four features help define chordates. Their embryos develop a notochord, a dorsal hollow nerve cord (which becomes a brain and spinal cord), a pharynx with gill slits, and a tail extending past the anus. Some or all of these features persist in adults. The tunicates and the lancelets are invertebrate filter-feeding chordates.

Section 23.2 Craniates are chordates with a brain chamber of cartilage or bone. Vertebrates are craniates with a vertebral column of bony segments. Jaws, paired fins, and later, lungs, were key innovations that led to the adaptive radiation of vertebrates.

Section 23.3 Cartilaginous fishes and bony fishes are aquatic vertebrates. The most diverse vertebrates are the bony fishes. Late in the Devonian, bony fishes having lobed fins and lungs gave rise to the tetrapods.

Sections 23.4, 23.5 Amphibians, the first tetrapods on land, share traits with aquatic tetrapods and reptiles. Most life cycles still go through aquatic stages. Existing groups include salamanders, frogs, and toads. As a result of habitat loss and other factors, many species are now declining.

Section 23.6 Amniotes were the first vertebrates to escape dependence on free-standing water by way of modifications in their skin and internal organ systems, reliance on fertilization, and amniote eggs. Mammals and their ancestors, the therapsids, are one branch. Groups known as "reptiles" and birds are another.

Section 23.7 Modern reptilian groups live in diverse habitats on land, in fresh water, and in the seas. Eggs are fertilized inside the body and usually laid on land. Lizards and snakes are the most diverse. Tuataras have some amphibian-like traits. Crocodilians are the closest relatives of birds.

Section 23.8 Birds are warm-blooded animals and the only existing ones with feathers. In most species, the body plan has been highly modified for flight.

Section 23.9 Mammals have hair and distinctive jaws and teeth. They are the only animals in which females nourish the young with milk from mammary glands. Three lineages are egg-laying mammals (monotremes), pouched mammals (marsupials), and placental mammals (eutherians). Placental mammals are the most diverse group.

Section 23.10 Living primates are prosimians, tarsioids, and anthropoids (monkeys, apes, and humans). The first primates were tiny shrewlike animals. In the lineage leading to humans, there was a trend toward better daytime vision, upright walking, more refined hand movements, smaller teeth, bigger brains, and increased parental care.

Section 23.11 The hominoid branch of the primate family tree evolved in Africa as the climate became cooler and drier. It includes apes and the hominids, humans and their most recent ancestors. Australopiths were early hominids, and they walked upright. The relationships among the diverse forms are sketchy. The first known members of the genus *Homo* lived in Africa between 2.4 million and 1.6 million years ago.

Section 23.12 Some *H. erectus* populations moved from Africa into Europe and Asia. Neandertal were early humans. DNA evidence suggests that they did not contribute to the gene pool of modern humans. *H. sapiens* had evolved by 160,000 years ago. Exactly where different populations of modern humans arose is unresolved.

Self-Quiz

Answers in Appendix III

1. All chordates have _____ .
   a. a backbone
   b. a notochord
   c. jaws
   d. both b and c

2. Vertebrate gills function in _____ .
   a. respiration
   b. blood circulation
   c. food trapping
   d. water regulation
   e. both a and c

3. Amphibians are descended from _____ .
   a. bony fishes
   b. lizards
   c. cartilaginous fishes
   d. lobe-finned fishes

4. Which of the following is an amniote?
   a. a shark
   b. a toad
   c. a turtle
   d. a sea squirt

5. Reptiles are adapted to life on land by _____ .
   a. tough skin
   b. internal fertilization
   c. good kidneys
   d. amniote eggs
   e. none of the above
   f. all of the above

6. _____ are the closest living relatives of birds.
   a. Crocodilians
   b. Tuataras
   c. Prosimians
   d. Lizards

7. Only birds have _____ .
   a. a cloaca
   b. a four-chambered heart
   c. feathers
   d. all of the above

8. A chimpanzee is a _____ .
   a. craniate
   b. vertebrate
   c. hominoid
   d. amniote
   e. placental mammal
   f. all of the above

9. *Homo erectus* _____ .
   a. was the earliest member of the genus *Homo*
   b. was one of the australopiths
   c. evolved in Africa and spread to diverse regions
   d. was the first to make stone tools

10. Match the organisms with the appropriate description.
    ____ fishes
    ____ amphibians
    ____ reptiles
    ____ birds
    ____ monotremes
    ____ marsupials
    ____ hominids

    a. first land tetrapods
    b. feathered amniotes
    c. egg-laying mammals
    d. humans and close relatives
    e. cold-blooded amniotes
    f. pouched mammals
    g. most diverse vertebrates

Figure 23.29 This sea squirt, *Ciona savignyi*, is native to the seas near Japan, but has invaded waters along the west coast of the United States, including the San Francisco Bay.

Figure 23.30 The duck-billed platypus (*Ornithorhynchus anatinus*) lives in Australia and Tasmania. In addition to other unusual traits, it is the only mammal that produces venom.

Critical Thinking

1. The National Human Genome Research Institute is currently funding a project to sequence the genome of *Ciona savignyi*, a sea squirt (Figure 23.29). What might these studies tell us about the human genome?

2. In 1798, a stuffed platypus specimen was delivered to the British Museum. Many were sure it was a fake. Soft brown fur and beaverlike tail put the animal firmly in the mammalian camp. But a ducklike bill and webbed feet suggested an affinity with birds (Figure 23.30). Reports that the animal laid eggs only added to the confusion.

We now know that platypuses burrow in riverbanks and forage for prey under water. The webbing on their feet can be retracted to reveal claws. The highly sensitive bill allows the animal to detect prey even with its eyes and ears tightly shut.

To modern biologists, a platypus is clearly a mammal. Like other mammals, it has fur and the females produce milk. Young animals have more typical mammalian teeth that are replaced by hardened pads as the animal matures. Why do you think modern biologists can more easily accept the idea that a mammal can have some reptilelike traits, such as laying eggs? What do they know that gives them an advantage over scientists living in 1798?

3. Reflect on the flight muscles of birds and their demands for oxygen and ATP energy. Which organelle would you expect to be profuse in flight muscles? Why?

4. During the Triassic, hundreds of species of tuataralike reptiles roamed Pangea. Now only two species survive on a few small cold islands off the coast of New Zealand. Speculate on why this relic population has survived there, despite disappearing from the rest of its former range.

5. In early *H. sapiens*, the average braincase volume was about 1,200 cubic centimeters. Today, it averages about 1,400 cubic centimeters. What selective pressure fueled this increase? According to one theory, female preference for clever mates led to selection for increasingly brainier males. Females are thought to have benefited secondarily, because they share the genes that affect intelligence. Do you find this theory plausible? What other selective agents might have favored increased intelligence? What kind of data could researchers gather to support or disprove the sexual selection hypothesis?

Media Menu

Student CD-ROM

Impacts, Issues Video
    Interpreting and Misinterpreting the Past
Big Picture Animation
    Origin and evolution of the chordates
Read-Me-First Animation
    Chordate body plan
    The evolution of jaws
    Mammalian radiations
Other Animations and Interactions
    Vertebrate evolution
    Crocodilian body plan
    A typical amniote egg

InfoTrac

- High Regard for Sea Squirt's Genome. *BioWorld Week*, December 2002.
- How the West Was Swum. *Natural History*, June 2001.
- Skull of Earliest Human Relative Discovered in African Nation of Chad. *Knight Ridder/Tribune News Service*, July 2002.

Web Sites

- Animal Diversity Web: animaldiversity.ummz.umich.edu
- Introduction to the Metazoa: www.ucmp.berkeley.edu/phyla/phyla.html
- Amphibian Declines and Deformities: www.usgs.gov/amphibians.html
- Cornell Lab of Ornithology: birds.cornell.edu/
- Becoming Human: www.becominghuman.org

How Would You Vote?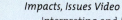

Fossils are among the best evidence we have of the evolution of life. Fossils of vertebrates are particularly sought after, rare, and valuable. In response to poaching of such fossils from public lands, some people have proposed that ownership and sale of vertebrate fossils should be banned. Others say that private collectors help ensure preservation of the fossil record. Would you support a ban on private collecting and ownership of vertebrate fossils?

IMPACTS, ISSUES *Too Hot To Handle*

Heat can kill. In the summer of 2001 Korey Stringer, a football player for the Minnesota Vikings, collapsed after morning practice. His team had been working out in full uniform on a field where temperatures were in the high 90s. High humidity put the heat index above 100°F.

With an internal body temperature of 108.8°F and blood pressure too low to record, Korey was rushed to the hospital. Doctors immersed him in an icewater bath, then wrapped him in cold, wet towels. It was too late.

Korey's blood clotting mechanism shut down and he started to bleed internally. Then his kidneys faltered and he was placed on dialysis. He stopped breathing on his own and was put on a respirator. His heart gave up.

Less than twenty-four hours after football practice had started, Korey Stringer was pronounced dead. He was twenty-seven years old.

All organisms function best within a limited range of internal operating conditions. For humans, "best" is when the body's internal temperature remains between about 97°F and 100°F. Past 104°F, blood is transporting a great deal of metabolically generated heat. Transport controls divert heat from the brain and other internal organs to the skin (Figure 24.1). Skin transfers heat to the outside environment—as long as it is not too hot outside. Profuse sweating can dissipate more heat, but not on hot, humid days.

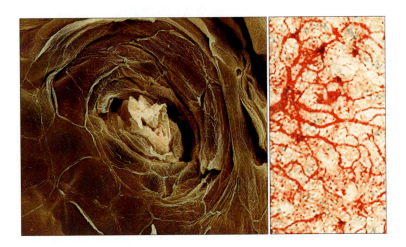

Figure 24.1 Temperature control mechanisms that have worked for millions of years. *Left*, scanning electron micrograph of a sweat gland pore at the skin surface. When water in sweat moves out through such pores, it helps dissipate excess body heat. Blood circulating through the body reaches fine capillaries in skin (*right*). When skin tissues are not as warm as blood, heat is transferred into them, then into the air— if the air is cooler still. In Korey Stringer's case, cooling could not happen fast enough.

the big picture

Multicellular Organization   Each cell of a multicelled organism engages in activities that help assure its own survival. The structural and functional organization of all the cells collectively sustains the whole body.

Recurring Challenges   Plants and animals both require mechanisms of gas exchange, nutrition, internal transport, salt–water balance between the body and the external environment, and defense.

These normal cooling mechanisms fail above 105°F. The body cannot sweat as much, and its temperature starts to climb rapidly. The heart beats faster and the individual becomes confused or faints. These are signs of *heat stroke*. The high internal temperature that causes it can denature the enzymes and other proteins that keep us alive. When heat stroke is not countered fast enough, brain damage or death is the expected outcome.

We use this sobering example as our passport to the world of anatomy and physiology. *Anatomy* is the study of body form—that is, its morphology. *Physiology* is the study of patterns and processes by which an individual survives and reproduces in the environment. It deals with how structures are put to use and how metabolism and behaviors are adjusted when conditions change. It also deals with how, and to what extent, physiological processes can be controlled.

You can use a lot of this information to interpret what goes on in your own body. More broadly, you can use it to perceive commonalities among all animals and plants. Regardless of the species, *the structure of a given body part almost always has something to do with a present or past function*. Most aspects of form and function are long-standing adaptations that evolved as responses to environmental challenges.

That said, nothing in the evolutionary history of the human species suggests that the organ systems making up our body are finely tuned to handle intense football practice on hot, humid days.

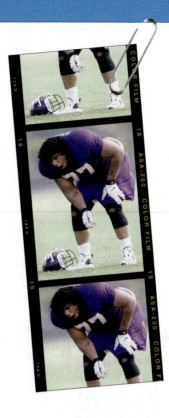

☑ *How Would You Vote?*

Heat sickness can affect a person's ability to think clearly. Should a coach who doesn't stop team practice or a game when the heat index soars be held responsible if a player dies from heat stroke? See the Media Menu for details, then vote online.

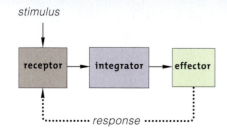

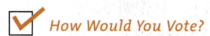

Homeostasis   In plants as well as in animals, cells, tissues, and organs interact in maintaining the body's internal operating conditions. Feedback control mechanisms guide much of this activity.

Cell-to-Cell Communication   Signaling mechanisms control and integrate body activities. Cells send messages that other cells receive and transduce, then change activities in response.

403

24.1 Levels of Structural Organization

We introduce important concepts in this chapter. They have broad application across the next two units of the book. Become familiar with them; they can deepen your sense of how plants and animals function under environmental conditions, both favorable and stressful.

FROM CELLS TO MULTICELLED ORGANISMS

Most plants and animals have cells, tissues, organs, and organ systems that split up the task of survival. A separate cell lineage gives rise to body parts that will function in reproduction. Said another way, the plant or animal body shows a division of labor.

A **tissue** is a community of cells and intercellular substances that are interacting in one or more tasks. For example, wood and bone are tissues that function in structural support. An **organ** has at least two tissues that are organized in certain proportions and patterns and that perform one or more common tasks. A leaf adapted for photosynthesis and an eye that responds to light in the surroundings are examples. An **organ system** has two or more organs interacting physically, chemically, or both in the performance of one or more common tasks. A plant's shoot system, with organs of photosynthesis and reproduction, is like this (Figure 24.2). So is an animal's digestive system, which takes in food, breaks it up into bits of nutrients, absorbs the bits, and expels the unabsorbed leftovers.

GROWTH VERSUS DEVELOPMENT

A plant or animal body becomes structurally organized during growth and development. Generally, **growth** of a multicelled organism means that its cells increase in number, size, and volume. **Development** is a succession of stages in which specialized tissues, organs, and organ systems form. In other words, we measure growth in *quantitative* terms and development in *qualitative* terms.

STRUCTURAL ORGANIZATION HAS A HISTORY

The structural organization of each tissue, organ, and organ system has an evolutionary history. Think back on how plants invaded land and you have some idea of how their structure relates to function. Plants that left aquatic habitats found an abundance of sunlight and carbon dioxide for photosynthesis, oxygen for aerobic respiration, and mineral ions. As they dispersed farther away from their aquatic cradle, however, they faced a new challenge—how to keep from drying out in air.

Remember that challenge when micrographs show the internal structure of plant roots, stems, and leaves (Figure 24.2). Pipelines conduct streams of water from soil to leaves. Stomata, the small gaps across a leaf's epidermis, open and close in ways that help conserve water (Section 26.4). Cells that have lignin-reinforced walls collectively support upright growth of stems. Remember that same challenge when you come across

reproductive organ (flower)

stem tissues, cross-section, for support, storage, water distribution, food distribution

shoot system (aboveground parts)

root system (belowground parts, mostly)

root water-conducting cells, longitudinal section

Figure 24.2 Morphology of a tomato plant (*Solanum lycopersicon*). Different cell types make up vascular tissues that conduct water, dissolved mineral ions, and organic compounds. These tissues thread through others that make up most of the plant body. Another tissue covers all surfaces exposed to the surroundings.

Figure 24.3 Some of the structures that function in human respiration. Their cells carry out specialized tasks. Airways to a pair of organs called lungs are lined with ciliated cells that whisk away bacteria and other airborne particles that might cause infections. Inside the lungs are tubes (blood capillaries), a fluid connective tissue called blood, and thin air sacs (epithelial tissue), all with roles in gas exchange.

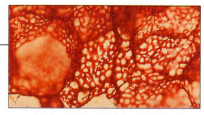

ciliated cells and mucus-secreting cells that line respiratory airways

organs (lungs), part of an organ system (the respiratory tract) of a whole organism

lung tissue (tiny air sacs) laced with blood capillaries—one-cell-thick tubular structures that hold blood, which is a fluid connective tissue

examples of root systems. Any soil region with plenty of water and dissolved mineral ions stimulates root growth in its direction; hence the branching roots.

Similarly, the respiratory systems of land animals are adaptations to life in air. Gases can only move into and out of the animal body by diffusing across a moist surface. This is not a problem for organisms in water, but moist surfaces dry out in air. Land animals have moist sacs for gas exchange *inside* their body (Figure 24.3), which brings us to the body's own environment.

THE BODY'S INTERNAL ENVIRONMENT

To stay alive, plant and animal cells must be bathed in a fluid that delivers nutrients and carries away metabolic wastes. In this they are no different from free-living single cells. But each plant or animal has thousands to many trillions of living cells that must draw nutrients from the fluid bathing them and dump wastes into it.

Body fluid *not* inside cells—extracellular fluid—is an **internal environment**. Changes in its composition and volume affect cell activities. The type and number of ions are vital; they must be kept at concentrations compatible with metabolism. It makes no difference whether the plant or animal is simple or complex. *It requires a stable fluid environment for all of its living cells.* This concept is central to understanding how plants and animals work.

START THINKING "HOMEOSTASIS"

The next two units describe how a plant or an animal carries out the following functions: The body works to maintain favorable conditions for all of its living cells. It acquires water, nutrients, and other raw materials, distributes them throughout the body, and disposes of wastes. It actively as well as passively defends itself against attack. It has the capacity to reproduce. Parts of the body nourish and protect gametes, and often the embryonic stages of the next generation.

Each cell engages in metabolic activities that ensure its own survival. Collectively, however, the activities of cells in tissues, organs, and organ systems sustain the body as a whole. Their interactions keep operating conditions of the internal environment within tolerable limits, a state we call **homeostasis**.

The structural organization of plants and animals emerges during growth and development.

Cells, tissues, and organs require a favorable internal environment, and they collectively maintain it. All body fluids not contained in cells make up that environment.

Acquiring diverse substances and distributing them to cells, disposing of wastes, protecting cells and tissues, reproducing, and often nurturing offspring are basic body functions.

24.2 Recurring Challenges to Survival

Plants and animals have such diverse body plans that we sometimes forget how much they have in common.

GAS EXCHANGE IN LARGE BODIES

How often do you think of similarities between, say, Tina Turner and a tulip (Figure 24.4)? Connections are there. Consider: Cells in Tina or the tulip would die if oxygen and carbon dioxide stopped diffusing across their surface. Like most other heterotrophs, Tina supplies her aerobically respiring cells with a lot of oxygen and disposes of carbon dioxide products. Like most other autotrophs, the plant secures carbon dioxide for its photosynthetic cells. At the same time, it retains enough oxygen for *its* aerobically respiring cells, and disposes of any excess.

All multicelled species respond, structurally and functionally, to the same challenge. *They must quickly move gaseous molecules to and from individual cells.*

Remember **diffusion**? It is a net movement of ions or molecules of a substance from a region where they are concentrated to an adjacent region where they are less concentrated. Animals and plants have ways to keep gases diffusing in directions most suitable for metabolism and cell survival. How? That question will lead you to the stomata at leaf surfaces (Chapter 26) and to the circulatory and respiratory systems of animals (Chapters 33 and 35).

INTERNAL TRANSPORT IN LARGE BODIES

Metabolic reactions happen fast. If it takes too long for reactants to diffuse through the body or to and from its surface, systems will shut down. This is one reason cells and multicelled species have the sizes and shapes that they do. As they grow, their volume increases in three dimensions; in length, width, and depth. But their surface area increases in two dimensions only. That is the essence of the **surface-to-volume ratio**. If a body were to develop into a solid, dense mass, it would not have enough surface area for fast, efficient exchanges with the environment (Section 4.1).

When a body or some body part is thin, as it is for the flatworm and lily pads in Figure 24.5, substances can easily diffuse between individual cells and the environment. In massive bodies, individual cells that are far from an exchange point with the environment depend on systems of rapid internal transport.

Most plants and animals have vascular tissues, or systems of tubes that move substances to and from cells. In most plants, xylem distributes soil water and mineral ions to all plant parts, and phloem distributes sugars from leaves. Each leaf vein has long strands of xylem and phloem (Figure 24.5*c*).

In large animals, vascular tissues extend from a surface facing the environment to living cells. Each time Tina belts out a song, she has to move oxygen into her lungs and carbon dioxide out of them. Blood vessels thread through her lung tissue where gases are exchanged. Other vessels transport oxygen to other body regions and branch into tiny capillaries, where interstitial fluid and cells exchange gases (Figure 24.5*d*).

In both plants and animals, the vascular system also transports nutrients and water. It also functions in the transport of chemical messages called hormones and in internal defense. In animals, vascular tissues transport infection-fighting white blood cells and assorted chemical weapons. In many plants, a wound triggers synthesis of defensive chemicals that are distributed throughout the plant by way of phloem.

MAINTAINING THE WATER–SOLUTE BALANCE

Plants and animals continually gain and lose water and solutes. They routinely produce metabolic waste. Given all the inputs and outputs, how are the volume and composition of their internal environment kept within a tolerable range? Plants and animals differ hugely in this respect. Yet we still can find common responses among them, when we zoom down to the level of molecular movement.

Figure 24.4 What do these organisms have in common besides their good looks?

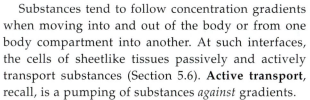

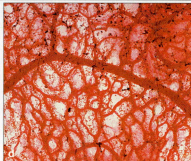

Figure 24.5 (**a**) A flatworm gliding along in water, (**b**) waterlily leaves floating on water, (**c**) veins visible in a decaying dicot leaf, and (**d**) human veins and capillaries. These body plans and structures are responses to the same constraint. What is it?

Substances tend to follow concentration gradients when moving into and out of the body or from one body compartment into another. At such interfaces, the cells of sheetlike tissues passively and actively transport substances (Section 5.6). **Active transport**, recall, is a pumping of substances *against* gradients.

Active transport mechanisms in roots help control which solutes move into the plant. In leaves, they help control water loss and gas exchange by closing and opening stomata at different times. In animals, such mechanisms occur in kidneys and other organs. As you will see again and again in the next two units, *active and passive transport help maintain the internal environment and metabolism itself by adjusting the kinds, amounts, and directional movements of substances.*

CELL-TO-CELL COMMUNICATION

Plants and animals show another major similarity in their structure and function. A number of their cells release signaling molecules that coordinate and also control events inside the body as a whole. Different signaling mechanisms guide how the plant or animal body grows, develops, and maintains and protects itself. Section 24.5 has a good example of this.

ON VARIATIONS IN RESOURCES AND THREATS

A **habitat** is the place where individuals of a species normally live. Each has different resources and poses a unique set of challenges. What are its physical and chemical characteristics? Is water plentiful, with the right kinds and amounts of solutes? Is the habitat rich or poor in nutrients? Is it sunlit, shady, or dark? Is it warm or cool, hot or icy, windy or calm? How much does the outside temperature vary from day to night? How much do conditions vary with the seasons?

And what about biotic (living) components of the habitat? Which producers, predators, prey, pathogens, and parasites live there? Is competition for resources and reproductive partners fierce? Such ever changing variables promote diversity in form and function.

Even with all that diversity, however, we still may observe similar responses to similar challenges. Sharp cactus spines or sharp porcupine quills help deter most of the other animals that might eat the cactus or porcupine. Both develop from specialized epidermal cells. Vascular tissues and other body parts helped nurture those cells. And by contributing to building spines and other defensive structures at the body's surface, the epidermal cells help protect the body as a whole against environmental threats.

Plant and animal cells function in ways that help ensure survival of the body as a whole. At the same time, tissues and organs that make up the body function in ways that allow the continued survival of individual living cells.

The connection between each cell and the body as a whole is evident in the requirements for—and contributions to— gas exchange, nutrition, internal transport, stability in the internal environment, and defense.

24.3 Homeostasis in Animals

In preparation for your trek through the next two units, focus here on what homeostasis means to survival.

Like other adult humans, your body has more than 65 trillion living cells. Each must draw nutrients from and dump wastes into the same fifteen liters of fluid (less than four gallons). The fluid not inside cells is extracellular fluid. Much is **interstitial fluid**, meaning it fills the spaces between cells and tissues. The rest is **plasma**, the fluid portion of blood. Interstitial fluid exchanges substances with cells and with blood.

Homeostasis, again, is a state in which the body's internal environment is being maintained within a range that cells can tolerate. In nearly all animals, sensory receptors, integrators, and effectors interact in maintaining this state. They detect, process, and respond to information on *how things are* compared to preset points of *where they are supposed to be.*

Sensory receptors are cells or cell parts that detect stimuli, which are specific forms of energy. A kiss, for example, is a form of mechanical energy that changes pressure on the lips. Receptors in lip tissues translate the stimulus into signals that flow to the brain. The brain is a major **integrator**, a central command post that receives and processes information about stimuli. It also issues signals to the body's **effectors**—muscles, glands, or both—that carry out suitable responses.

NEGATIVE FEEDBACK

Feedback mechanisms are key homeostatic controls. By **negative feedback mechanisms**, an activity changes some condition, and when the condition shifts past a

Figure 24.7 A climber near the summit of Mount Everest, where the normal act of breathing isn't enough.

certain point, it triggers a response that reverses the change (Figure 24.6).

Think of a furnace with a thermostat. A thermostat senses the surrounding air temperature relative to a preset point on a thermometer built into the furnace's control system. When the temperature falls below the preset point, it signals a switching mechanism, which turns on the furnace. When the air heats enough and matches the preset level, the thermostat senses the match and signals the switch to shut off the furnace.

Now think about how we breathe to take in oxygen and get rid of carbon dioxide. Most of us live at low elevations, where 1 in 5 molecules is oxygen. There, the brain pays attention to receptors that are sensitive to how much carbon dioxide is dissolved in blood. It compares receptor signals against a set point for what the carbon dioxide level is supposed to be, and calls for tweaks in breathing and other activities that can put the carbon dioxide level in line with the set point.

Above 3,300 meters (10,000 feet), oxygen is scarce (Figure 24.7). Its level in blood plummets. Receptors sensitive to this steep decline signal the brain, which responds by making us hyperventilate (breathe faster and far more deeply than usual). If the level stays low, shortness of breath, heart palpitations, headaches, nausea, and vomiting will follow. All are last-ditch warnings that cells are screaming for oxygen.

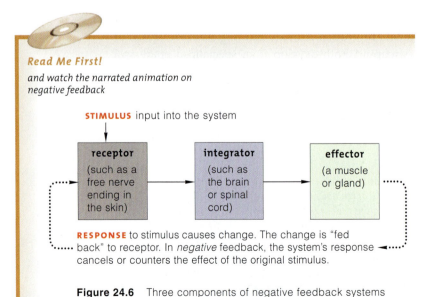

Read Me First!

and watch the narrated animation on negative feedback

STIMULUS input into the system

receptor (such as a free nerve ending in the skin) → **integrator** (such as the brain or spinal cord) → **effector** (a muscle or gland)

RESPONSE to stimulus causes change. The change is "fed back" to receptor. In *negative* feedback, the system's response cancels or counters the effect of the original stimulus.

Figure 24.6 Three components of negative feedback systems in multicelled organisms. Figure 24.7 gives a specific example.

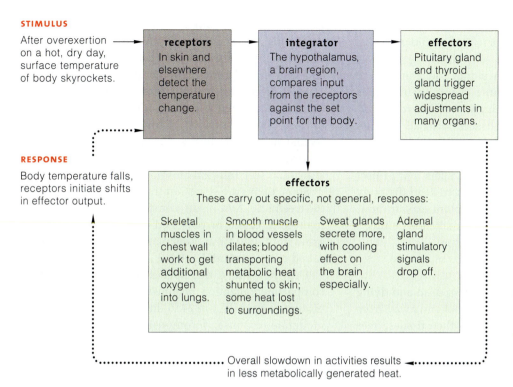

STIMULUS

After overexertion on a hot, dry day, surface temperature of body skyrockets.

receptors
In skin and elsewhere detect the temperature change.

integrator
The hypothalamus, a brain region, compares input from the receptors against the set point for the body.

effectors
Pituitary gland and thyroid gland trigger widespread adjustments in many organs.

RESPONSE

Body temperature falls, receptors initiate shifts in effector output.

effectors
These carry out specific, not general, responses:

Skeletal muscles in chest wall work to get additional oxygen into lungs.

Smooth muscle in blood vessels dilates; blood transporting metabolic heat shunted to skin; some heat lost to surroundings.

Sweat glands secrete more, with cooling effect on the brain especially.

Adrenal gland stimulatory signals drop off.

Overall slowdown in activities results in less metabolically generated heat.

Figure 24.8 Major homeostatic controls over a human body's internal temperature. The *solid* arrows signify the main control pathways. The *dotted* arrows signify the feedback loop. In Korey Stringer's case, environmental conditions overwhelmed the control pathways.

Similarly, feedback mechanisms work to keep the internal body temperature of humans and many other mammals near 98.6°F (37°C) even during hot or cold weather. When a football player runs around on a hot summer day, the body becomes hot. Receptors trigger changes that should slow down the entire body *and* its cells (Figure 24.8). Under typical conditions, these and other control mechanisms counter overheating. They curb activities that naturally generate metabolic heat, and give up excess heat to the surrounding air.

POSITIVE FEEDBACK

In some situations, **positive feedback mechanisms** operate. These controls initiate a chain of events that *intensify* change from an original condition, and after a limited time, the intensification reverses the change. Positive feedback mechanisms are usually associated with instability in a system.

For instance, rapid flows of sodium ions across the plasma membrane of neurons help the body send and receive signals quickly. A suitable signal opens a few sodium channels across a neuron's membrane. Ions flow into the neuron, and the increased concentration inside makes more channels open, then more, until sodium ions flood inside. Such ion flows start the "messages" that travel along a neuron's surface.

As another example, during labor, the fetus exerts pressure on the wall of the chamber enclosing it, the uterus. Pressure induces the production and secretion of a hormone, oxytocin, that makes muscle cells in the wall contract. Contractions exert pressure on the fetus, which puts more pressure on the wall, and so on until the fetus is expelled from the mother's body.

We have been introducing a pattern of detecting, evaluating, and responding to a flow of information about the internal and external environment. During this activity, organ systems work together. Soon, you will be asking these questions about their operation:

1. Which physical or chemical aspects of the internal environment are organ systems working to maintain?

2. How are organ systems kept informed of change?

3. By what means do they process the information?

4. What mechanisms are set in motion in response?

As you will read in Unit Six, organ systems of nearly all animals are under neural and endocrine control.

Homeostatic controls, such as negative and positive feedback mechanisms, help maintain physical and chemical aspects of the body's internal environment within ranges that its individual cells can tolerate.

24.4 Does Homeostasis Occur in Plants?

Plants differ from animals in important respects. Even so, they, too, have controls over their internal environment.

Direct comparisons between plants and animals are not always possible. In young plants, new tissues arise only at the tips of actively growing roots and shoots. In animal embryos, tissues form all through the body. Plants do not have a centralized integrator that serves as the equivalent of an animal brain. They do have some decentralized mechanisms that protect the internal environment and work to keep the body as a whole functioning. Later chapters explain how, but two simple examples here will make the point.

WALLING OFF THREATS

Unlike people, trees consist mostly of dead and dying cells. Also unlike people, trees cannot run away from attacks. When a pathogen infiltrates their tissues, trees cannot unleash infection-fighting phagocytic cells in response, because they have none. Some trees live in habitats that are too harsh or remote to favor most

pathogens, so they have been able to grow, slowly, for thousands of years. Remember those bristlecone pines in Section 21.6? They spend most of the year in snow and the rest under intense radiation from the sun.

Most trees protect their internal environment by building thickened cell walls around wounds. They unleash phenols and other toxic compounds. Many secrete resins. The heavy flow of gooey compounds saturates and protects bark and wood at an attack site. It also can seep into soil around roots.

Some toxins are so potent that they also kill cells of the tree itself. As a result, compartments form around injured, infected, or poisoned sites, and new tissues grow right over them. This plant response to attack is called **compartmentalization**.

Drill holes in a tree species that makes a strong compartmentalization response and it quickly walls off that wound (Figure 24.9). In a species that makes a moderate response, decomposers that cause wood to decay expand farther, into regions more distant from the holes. Drill holes in a weak compartmentalizer, and decomposers will cause massive decay.

Even strong compartmentalizers live only so long. When they form too many walls, they shut off the flow of water and solutes to living cells.

SAND, WIND, AND THE YELLOW BUSH LUPINE

Anyone who has tiptoed barefoot across sand near the coast on a hot, dry day has a tangible clue to why few plants grow in it. One of the exceptions is the yellow bush lupine, *Lupinus arboreus* (Figure 24.10).

L. arboreus is native to warm, dry areas of Central and Southern California. It is a hardy colonizer of soil exposed by fires or abandoned after being cleared for agriculture. Like all other legumes, this species has nitrogen-fixing symbionts in its young roots. The mutualistic interaction gives it a competitive edge in nitrogen-deficient soil. In the early 1900s, this species was planted in northern California to stabilize coastal dunes. Unfortunately, it is too successful in its new habitat. It outgrows and displaces native plants.

A big environmental challenge near the beach is the lack of fresh water. Leaves of a yellow bush lupine

strong

moderate

weak

Figure 24.9 You think it's easy being a tree? Drilling patterns for an experiment to test compartmentalization responses. From top to bottom, decay patterns (*green*) in stems of three tree species that make strong, moderate, or weak compartmentalizing responses.

Figure 24.10 Yellow bush lupine, *Lupinus arboreus*, in a sandy shore habitat. On hot, windy days, its leaflets fold up longitudinally along the crease that runs down their center. This helps minimize evaporative water loss.

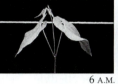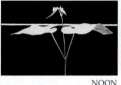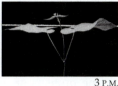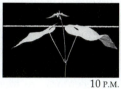

| 1 A.M. | 6 A.M. | NOON | 3 P.M. | 10 P.M. | MIDNIGHT |

are structurally adapted for water conservation. Each leaf has a surprisingly thin cuticle, but a dense array of fine epidermal hairs projects above it, especially on lower leaf surfaces. Collectively, the hairs will trap moisture escaping from stomata. Trapped, moist air slows evaporation and helps maintain water levels inside the leaf at optimal levels for metabolism.

These leaves make homeostatic responses to the environment. They fold along their length, like two parts of a clam shell, and resist the moisture-sucking force of the wind. Each folded, hairy leaf is better at stopping moisture loss from stomata (Figure 24.10).

Leaf folding by *L. arboreus* is a controlled response to changing conditions. When winds are strong and potential for water loss is greatest, these leaves fold tightly. The least-folded leaves are near the plant's center or on the side most sheltered from the wind. Folding is a response to heat as well as to wind. When air temperature is highest during the day, leaves fold at an angle that helps reflect the sun's rays from their surface. The response minimizes heat absorption.

ABOUT RHYTHMIC LEAF FOLDING

In case you think leaf folding couldn't possibly be a coordinated response, take a look at the bean plant in Figure 24.11. Like some other plants, it holds its leaves

Figure 24.11 Observational test of the rhythmic movements of leaves of a young bean plant (*Phaseolus*). The investigator, Frank Salisbury, kept this plant in complete darkness for twenty-four hours. Its leaves continued to fold and unfold independently of sunrise (6 A.M.) and sunset (6 P.M.).

horizontally during the day but folds them closer to a stem at night. Keep the plant in full sun or darkness for a few days and it will continue to move its leaves in and out of the "sleep" position, independently of sunrise and sunset. The response might help reduce heat loss at night, when air cools, and so maintain the plant's internal temperature within tolerable limits.

Rhythmic leaf movements are just one example of a **circadian rhythm**, a biological activity repeated in cycles that each last for close to twenty-four hours. Circadian means "about a day." As you will see in a later chapter, a pigment molecule called phytochrome may be part of a control mechanism over leaf folding.

Homeostatic mechanisms are at work in plants, although they are not governed from central command posts as they are in most animals.

Compartmentalization and rhythmic leaf movements are two examples of responses to specific environmental challenges.

24.5 How Cells Receive and Respond to Signals

Signal reception, transduction, and response—this is a fancy way of saying cells chatter among themselves in ways that bring about changes in their activities.

Reflect on Section 4.11, the overview of how adjoining cells communicate as by plasmodesmata in plants and gap junctions in animals. Also think back on how free-living *Dictyostelium* cells issue signals that induce them to converge and make a spore-bearing structure. They do so in response to dwindling supplies of food, an environmental cue for change (Figure 20.19).

In large multicelled organisms, one cell type signals others in response to cues from both the internal and external environment. Local and long-distance signals lead to local and regional changes in metabolism, gene expression, growth, and development.

The molecular mechanisms by which cells "talk" to one another evolved early in the history of life. They often involve three events. *First*, a specific receptor is activated, perhaps by the reversible binding of some signaling molecule. *Second*, the signal is transduced—it is changed into a form that can operate inside the cell. *Third*, the cell makes the functional response.

Most receptors are membrane proteins of the sort shown in Section 4.3. When activated, the receptor starts signal transduction. It might activate an enzyme that in turn activates many molecules of a different enzyme, which activates many molecules of another kind, and so on. These chains of cascading reactions inside the cell greatly amplify the original signal.

In the next two units, you will come across diverse cases of signal reception, transduction, and response. For now, consider this example to get a sense of the kinds of events that signals set in motion.

The very first cell of a new multicelled individual holds marching orders that will guide its descendants through growth, development, and reproduction, and often death. As part of that program, many cells heed calls to self-destruct when they finish their prescribed function. **Apoptosis** is the process of programmed cell death. It begins with signals that unleash enzymes of self-destruction, which each body cell synthesizes and stockpiles (Figure 24.12).

Proteases are part of the stockpile. They chop apart structural proteins, including the building blocks of cytoskeletal elements and of proteins that keep DNA structurally organized. Nucleases, enzymes that snip up nucleic acids, are also stockpiled.

During apoptosis, a cell shrinks away from other cells in a tissue. Its surface bubbles outward and inward (Figure 24.13c). Its chromosomes bunch up near the nuclear envelope. The nucleus, then the cell, break apart. Phagocytic cells that patrol and protect body tissues engulf dying cells and their remnants. Lysosomes inside the phagocytes digest the engulfed bits, which are recycled.

Cells were busily committing suicide when your own hands were developing. The human hand starts out as a paddlelike structure (Figure 24.13a). When development proceeds normally, apoptosis causes the

Signal to die docks at receptor.

Signal leads to activation of protein-destroying enzymes.

Figure 24.12 Artist's depiction of a suicidal cell. Normally, body cells self-destruct when they finish their functions or become infected or cancerous, which could threaten the body as a whole.

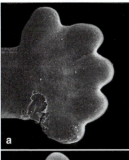

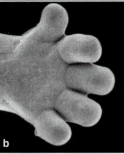

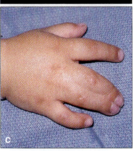

Figure 24.13 Formation of human fingers with the help of apoptosis. (**a**) Forty-eight days after fertilization, webs of tissue connect embryonic digits. (**b**) Three days later, after cells between the digits commit suicide, there are no more webs. (**c**) Digits stay attached when the cells do not commit suicide on cue.

paddle to split apart into individual fingers in a few days (Figure 24.13*b*).

While embryos develop, the timing of cell death is usually predictable. As one example, keratinocytes are skin cells that live about three weeks, then die and form protective layers that are continually sloughed off from the surface of skin. These cells can be induced to die ahead of schedule. So can other cells that are supposed to last a lifetime. All it takes is sensitivity to molecular signals that can call up the enzymes of death.

Control genes suppress or trigger programmed cell death. One gene, *bcl-2,* helps keep normal body cells from dying before their time. In some cancers, this gene has mutated and cells do not respond to signals to die. Apoptosis is no longer a control option.

What about plants? The controlled death of xylem cells creates pipelines that carry water throughout the plant. Also, when a tissue is under attack, signals may bring about the controlled death of nearby cells in a pattern that walls off the threat.

Plant and animal cells communicate with one another by secreting signaling molecules into extracellular fluid and selectively responding to signals from other cells.

Communication involves receiving signals, transducing them, and bringing about change in a target cell's activity.

Signal transduction requires membrane receptors and other membrane proteins. It often involves enzymes that induce a cascade of reactions that amplify the initial signal.

Summary

Section 24.1 Anatomy is the study of body form. Physiology is the study of how the body functions in the environment. The structure of most body parts correlates with current or past functions.

A plant or animal's structural organization emerges during growth and development. Each living cell performs metabolic functions that keep it alive. Cells are organized in tissues, organs, and often organ systems. These structural units function in coordinated ways in the performance of specific tasks.

Section 24.2 Plants and animals have responded in similar ways to certain environmental challenges. All exchange gases, transport substances to and from cells, and maintain water and solute concentrations within the body at tolerable limits. All have mechanisms for integrating and controlling body parts in ways that ensure survival of the whole organism. They also have mechanisms for responding to signals from other cells and to signals or cues from the outside environment.

Section 24.3 Homeostasis is a stable state in which the body's internal environment is being maintained within a range that its cells can tolerate. In animal cells, sensory receptors, integrators, and effectors interact to maintain tolerable conditions, often by feedback loops.

With negative feedback mechanisms, a change in some condition, such as body temperature, triggers a response that results in reversal of the change. With positive feedback mechanisms, a change leads to a response that intensifies the condition that caused it.

Section 24.4 Plants have decentralized mechanisms that work to maintain the internal environment and ensure their survival. Some trees wall off threats by secreting resins and toxins, a defensive response called compartmentalization. Some plants respond to changes in environmental conditions by folding their leaves. Rhythmic leaf folding is a circadian rhythm that may help a plant maintain its internal temperature.

Section 24.5 Cell-to-cell communication involves signal reception, signal transduction, and a response by a target cell. Many signals are transduced by membrane proteins that trigger reactions in the cell. Reactions may alter the activity of genes or enzymes that take part in cellular events. An example is a signal that unleashes the protein-cleaving enzymes of apoptosis; a target cell self-destructs.

Self-Quiz   *Answers in Appendix III*

1. An increase in the number, size, and volume of plant cells or animal cells is called _____ .
    a. growth   c. differentiation
    b. development   d. all of the above

Figure 24.14 (**a**) Consuelo De Moraes studies plant stress responses. (**b**,**c**) A caterpillar chewing on tobacco causes the plant to secrete chemicals that attract a wasp. (**d**) The wasp grabs the caterpillar and lays an egg inside it.

2. The internal environment consists of _____ .
   a. all body fluids   c. all body fluids outside cells
   b. all fluids in cells   d. interstitial fluid

3. _____ influences the concentrations of water and solutes in the internal environment.
   a. Diffusion   c. Passive transport
   b. Active transport   d. all are correct

4. Cell communication typically involves signal _____ .
   a. reception   c. response
   b. transduction   d. all are correct

5. Match the terms with their most suitable description.
   ____ circadian rhythm   a. programmed cell death
   ____ homeostasis   b. 24-hour or so cyclic activity
   ____ apoptosis   c. stable internal environment
   ____ negative   d. an activity changes some
       feedback   condition, then the change
                         triggers its own reversal

Media Menu

Student CD-ROM

Impacts, Issues Video
  Too Hot To Handle
Big Picture Animation
  Organization, homeostasis, and communication
Read-Me-First Animation
  Negative feedback
Other Animations and Interactions
  Morphology of a tomato plant
  Homeostatic control of human body temperature
  Compartmentalization
  Circadian rhythm in leaf folding
  Apoptosis in hand development

InfoTrac

- Polluting Your Internal Environment: Homeostasis Is a Factor in Health. *Environmental Health Perspectives*, January 2002.
- When Nature Turns Up the Heat. *RN*, August 1999.
- Preventing and Managing Dehydration. *MedSurg Nursing*, December 2002.
- Caterpillars: Agents of Their Own Demise. *Agricultural Research*, January 2002.

Web Sites

- Virtual Crops: www-plb.ucdavis.edu/labs/rost/Virtual%20crops.htm
- AMA Atlas: www.ama-assn.org/ama/pub/category/7140.html
- Homeostasis: www.biology-online.org/4/1_physiological_homeostasis.htm
- Cell Death in Development: www.acs.ucalgary.ca/~browder/death.html

How Would You Vote?

Most likely, athletes who choose to stay on the field when they are in the grip of heat exhaustion or heat stroke are not thinking clearly. Their choice puts their health, even their life, at risk. Should a coach or other official who keeps a team practicing or playing a game in high heat and humidity be held responsible if a member of the team dies?

Critical Thinking

1. Many plants protect themselves with thorns or toxins or nasty-tasting chemicals that deter plant-eating animals. Some get help from wasps.

Consuelo De Moraes, currently at Pennsylvania State University, studies interactions among plants, caterpillars, and parasitoid wasps. *Parasitoids* are a special class of parasites; their larvae eat a host from the inside out.

When a caterpillar chews on a tobacco plant leaf, it secretes a lot of saliva. Some chemicals in its saliva are an external signal that triggers a chemical response from leaf cells. The cells release certain molecules that diffuse through the air. Parasitoid wasps follow the concentration gradient to the stressed leaves. They attack a caterpillar, and each deposits one egg inside it. The eggs grow and develop into caterpillar-munching larvae (Figure 24.14).

As De Moraes discovered, plant responses are highly specific. Leaf cells release different chemicals in response to different caterpillar species. Each chemical attracts only the wasps that parasitize the particular kind of caterpillar that triggers the chemical's release.

Are the plants "calling for help"? Not likely. Give a possible explanation for this plant–wasp interaction in terms of cause, effect, and natural selection theory.

2. The Arabian oryx (*Oryx leucoryx*), an endangered antelope, evolved in the harsh deserts of the Middle East. Most of the year there is no free water, and temperatures routinely reach 117°F (47°C). The most common tree in the region is the umbrella thorn tree (*Acacia tortilis*). List common challenges that the oryx and the acacia face. Also research and report on the morphological, physiological, and behavioral responses of both organisms.

3. In the summer of 2003, a record heat wave lasted for weeks and caused the death of more than 5,000 people in France. The elderly and very young were at greatest risk. High humidity increases the chance of heat-related illness because it reduces the rate of evaporative cooling. So does tight clothing that doesn't "breathe," like the uniform Korey Stringer wore during his last practice. Other risk factors are obesity, poor circulation, dehydration, and alcohol intake. Using Figure 24.8 as a reference, briefly suggest how each factor may interfere with homeostatic controls over the body's internal temperature.

V How Animals Work

How many and what kinds of body parts does it take to function as a lizard in a tropical forest? Make a list of what comes to mind as you start reading Unit VI, then see how resplendent the list can become at the unit's end.

IMPACTS, ISSUES *Mind-Boggling Births*

In December 1998, Nkem Chukwu of Texas gave birth to octuplets ahead of schedule. The six girls and two boys survived their premature birth, and they were the first octuplets to do so (Figure 25.1).

The *combined* weight of those eight newborns was just a bit more than 10 pounds, or 4.5 kilograms. Odera, the smallest, weighed less than 1 pound (20 grams). She died of heart and lung failure six days later. The other seven had to remain in the hospital for three months before going home. Two required abdominal surgery.

Chukwu had requested hormone injections, which caused many eggs in her ovaries to mature and be released at the same time. She was offered the option to reduce the number of embryos but wanted to carry them all. The first, delivered naturally, was thirteen weeks premature. Chukwu's doctor ordered drugs to stop labor, then surgically delivered the rest.

Over the past two decades, the incidence of multiple births increased by almost 60 percent. The incidence of higher order multiple births—triplets or more—quadrupled. What is going on?

A woman's fertility peaks during her mid-twenties. By age thirty-nine, her chances of natural conception have been about halved. Yet the number of first-time

Figure 25.1 (**a**) Human embryo at an early stage of development. (**b**) Testimony to the potency of fertility drugs—seven survivors of a set of octuplets. Besides manipulating so many other aspects of nature, humans are now manipulating their own reproduction.

the big picture

Reproductive Modes Different animal species reproduce sexually, asexually, or both. Sexual reproduction, the most prevalent mode, has biological costs as well as selective benefits.

The Nature of Development Animal life cycles typically unfold through stages of embryonic development. Each stage builds on the successful completion of the stage that preceded it.

mothers who are more than forty years old doubled during the past decade. Many probably would not have given birth at all without the assistance of fertility drugs and other reproductive interventions, such as *in vitro* fertilization.

Fertility drugs are driving the increase in higher order multiple births, and they worry many doctors. Carrying more than one embryo increases the risk of miscarriage, premature delivery, and surgical delivery by cesarean section. Compared to single births, the weight of such newborns is lower and the mortality rates are higher. A woman who carries quintuplets faces a 50 percent risk of miscarriage, and there is a 90 percent chance that her newborns will develop complications resulting from an abnormally low birth weight. Higher order multiple births also present far more physical, emotional, and financial challenges to the parents.

With this chapter, we turn first to basic principles that govern animal reproduction and development, from gamete formation on through the formation of specialized cells, tissues, and organs of the adult. We then focus on humans as a case study.

There are many sections to choose from, some core material, some optional. We offer this much detail because it can directly and indirectly impact your own future in profound ways. What you learn here may help you work through health-related problems and ethical issues concerning how we individually and collectively deal with human reproduction and development.

☑ How Would You Vote?

Fertility drugs induce multiple ovulations at the same time and increase the likelihood of high-risk multiple pregnancies. Should we restrict the use of such drugs to conditions that limit the number of embryos formed? See the Media Menu for details, then vote online.

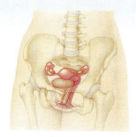

Human Reproduction Human males and females have a pair of primary reproductive organs, and accessory ducts and glands. Hormones control reproductive functions and the development of secondary sexual traits.

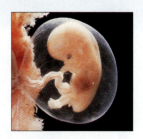

Human Development A human life cycle extends from gamete formation and fertilization, through birth and adulthood, to reproduction and gradual aging until death. Master genes control the development of the basic body plan.

25.1 Reflections on Sexual Reproduction

Sexual reproduction dominates the life cycle of most animals, even those that also can reproduce asexually. We therefore can expect that the benefits of sexual reproduction outweigh the costs. What are they?

SEXUAL VERSUS ASEXUAL REPRODUCTION

In earlier chapters, we considered the genetic basis of **sexual reproduction**. Again, meiosis and the formation of gametes typically occur in two prospective parents. At fertilization, a gamete from one parent fuses with a gamete from the other and forms the zygote, the first cell of the new individual. We also looked at **asexual reproduction**, whereby a single organism—one parent only—produces offspring. We turn now to examples of the structural, behavioral, and ecological aspects of these two modes of animal reproduction.

Think about a fragment torn away from a sponge body. It can grow, by mitotic cell divisions, into a new sponge. Think of a flatworm that spontaneously splits

in two. If its body constricts at its midsection, the part below grips a substrate, starts a tug-of-war with the front, then splits off a few hours later. Both parts go their separate ways and each grows what is missing, thus becoming a whole flatworm (Figure 25.2).

Mutation aside, in cases of *asexual* reproduction, all offspring are genetically the same as their individual parent. Phenotypically, they also are much the same. Phenotypic uniformity helps when the gene-encoded traits are finely adapted to fairly consistent conditions in the environment. Under the circumstances, drastic variations introduced into finely tuned gene packages would not do much good and could do a lot of harm.

Most animals live where opportunities, resources, and danger are variable. They reproduce sexually, and offspring inherit different mixes of alleles from female and male parents. The resulting variation in traits can improve the likelihood that at least some offspring will survive and reproduce if prevailing conditions change.

COSTS OF SEXUAL REPRODUCTION

Separation into sexes is costly. Energy and resources must be allocated to forming and nurturing gametes. Often, reproductive structures that can help deliver or accept sperm must be built. A potential mate might have to be courted. The timing of gamete formation and mating must be synchronized between the sexes.

Reflect on *reproductive timing*. How do sperm in one individual mature at the same time that eggs are maturing in a different individual? Timing requires energy outlays to construct, maintain, and use neural and hormonal control mechanisms in both parents. It requires responsiveness to environmental cues, such as daylength, that signal the best time to start making gametes and to produce offspring.

For example, moose become sexually active only during the late summer and early fall. This timing is adaptive; it ensures that the offspring will be born the following spring, when the weather becomes milder and food will be plentiful for many months.

Think about what it takes to find and recognize a likely mate. Many animals invest energy to produce sex attractants called pheromones or make receptors for them. They invest in visual signals such as richly colored and patterned feathers. Many males attract mates and fend off rivals, as with bonding rituals or claws, horns, or a larger body mass that may make a difference in territorial defense (Figure 25.3).

Producing enough offspring so that at least some survive is also costly (Figure 25.4). Many species of invertebrates, bony fishes, and frogs release sperm,

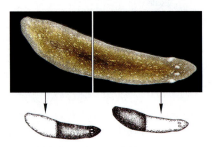

Figure 25.2 One of the flatworms (*Dugesia*) that can reproduce asexually by spontaneous fission. If it divides into two pieces, each piece will replace what is missing and will be genetically identical to the original flatworm.

Figure 25.3 Biological costs associated with sexual reproduction. (**a**) Body mass of male northern elephant seals. The bulls are fighting for access to the far smaller female, lower right. Reflect on the energy and raw materials directed into making (**b**) sable antelope horns.

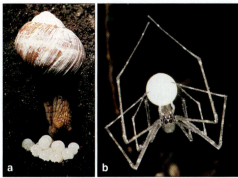

Figure 25.4 A look at where some invertebrate and vertebrate embryos develop, how they are nourished, and how (if at all) parents protect them.

Snails (**a**) and spiders (**b**) release eggs from which the young later hatch. Most snails abandon their laid eggs. Spider eggs develop in a silk egg sac that the female anchors or carries around with her. Females often die soon after they make the sac. Some species guard the sac, then cart spiderlings about for a few days while they feed them.

(**c**) Ruby-throated hummingbirds and all other birds lay fertilized eggs with big yolk reserves. The eggs develop and hatch outside the mother. Unlike snails, one or both parent birds expend energy feeding and caring for the young.

(**d**) Embryos of most sharks, most lizards, and some snakes develop in their mother, receive nourishment continuously from yolk reserves, and are born live. Shown here, live birth of a lemon shark.

Embryos of most mammals draw nutrients from maternal tissues and are born live. (**e**) In kangaroos and other marsupials, embryos are born "unfinished." They finish their embryonic development inside a pouch on the mother's ventral surface. (**f**) Juvenile stages (joeys) continue to be nourished from mammary glands inside the pouch. A human female (**g**) retains a fertilized egg in her uterus. Her own tissues nourish the developing individual until birth.

eggs, or both into their environment. If each adult were to make only one sperm or one egg each season, the chances for fertilization would not be good. These animals invest energy in making many gametes, often thousands of them.

As another example, nearly all animals on land use internal fertilization, or the union of sperm and egg *within* the female body. They invest metabolic energy to construct elaborate reproductive organs, such as a penis and a uterus. A penis deposits sperm inside the female, and a uterus is a chamber in which an embryo develops inside certain mammalian females.

Finally, animals set aside energy in forms that can *nourish the developing individual* until it has developed enough to feed itself. Nearly all animal eggs contain **yolk**. This thick fluid has an abundance of proteins and lipids that nourish embryonic stages.

The eggs of some species have much more yolk than others. Sea urchins make enormous numbers of tiny eggs with little yolk. Each fertilized egg develops into a freely moving, self-feeding larva in less than a day. Very few escape predators. For sea urchins, then,

reproductive success means allocating small amounts of energy and resources to making each egg.

Birds put a lot of energy and resources into making eggs with a lot of yolk, which has to nourish the bird embryo through an extended time inside an eggshell that forms after fertilization. *Your* mother placed huge demands on her body to nourish a nearly yolkless, fertilized egg through nine months of development. Physical exchanges with her own tissues sustained your embryonic development (Figure 25.4*g*).

As these examples suggest, animals show great diversity in reproduction and development. However, as you will see in the sections to follow, some patterns are widespread throughout the animal kingdom.

Separation into male and female sexes requires special reproductive cells and structures, neural and hormonal control mechanisms, and forms of behavior.

A selective advantage—variation in traits among offspring—offsets biological costs related to separation into sexes.

25.2 Stages of Reproduction and Development

For all animals more complex than sponges, the life cycle has six developmental stages, from gamete formation through growth and tissue specialization.

Figure 25.5 is an overview of the six stages of animal reproduction and development. In *gamete formation*, the first stage, eggs or sperm develop inside parental reproductive tissues or organs. *Fertilization* starts as a sperm penetrates a mature egg. This second stage ends when sperm and egg nuclei fuse, forming a zygote.

The third stage is *cleavage*. Cell divisions increase the number of cells but not the egg's original volume. The cuts produce a **blastula**. In this early embryonic form, the cells are called **blastomeres**, and they enclose a fluid-filled cavity called the blastocoel.

As cleavage draws to a close, cell division slackens. The blastula undergoes *gastrulation*. This fourth stage is a time of structural reorganization. Cells become arranged into a **gastrula**, an early embryonic form with two or three primary tissue layers (germ layers). Cellular descendants of these primary tissues give rise to all tissues and organs of the adult.

Ectoderm, the outermost primary tissue, emerges first. It is the source of the nervous system and the outer part of the integument. Innermost is **endoderm**, source of the gut's inner lining and organs derived from it. In most animal embryos, **mesoderm** forms in between the outer and inner primary tissue layers. It gives rise to muscles, to most of the skeleton, to the circulatory, reproductive, and excretory systems, and to connective tissues of the gut and integument. As you read in Section 22.1, mesoderm evolved hundreds of millions of years ago. It was a key innovation in the evolution of nearly all large, complex animals.

After the primary tissue layers appear, their cells become distinct subpopulations. This marks the onset of *organ formation*. By this process, newly forming cells become increasingly specialized in their structure and function. Their descendants will be the ancestors of all tissues and organs of the adult.

Growth and tissue specialization is the sixth stage of animal development. Over time, tissues and organs mature in size, shape, proportion, and function. This final stage extends into adulthood.

a Eggs form and mature in female reproductive organs, and sperm form and mature in male reproductive organs.

b A sperm and an egg fuse at their plasma membrane, then the nucleus of one fuses with the nucleus of the other to form the zygote.

c By a series of mitotic cell divisions, different daughter cells receive different regions of the egg cytoplasm.

d Cell divisions, migrations, and rearrangements produce two or three primary tissues, the forerunners of specialized tissues and organs.

e Subpopulations of cells are sculpted into specialized organs and tissues in prescribed spatial patterns at prescribed times.

f Organs increase in size and gradually assume specialized functions.

Figure 25.5 Overview of the stages of animal reproduction and development.

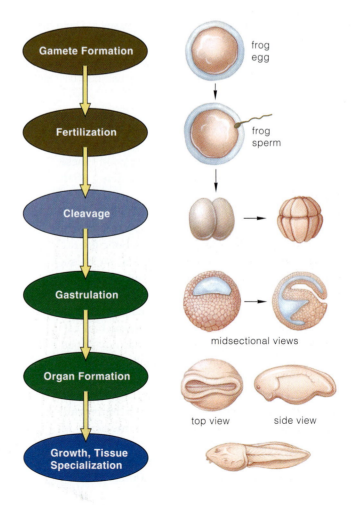

Read Me First!
*and watch the narrated animation
on the leopard frog life cycle*

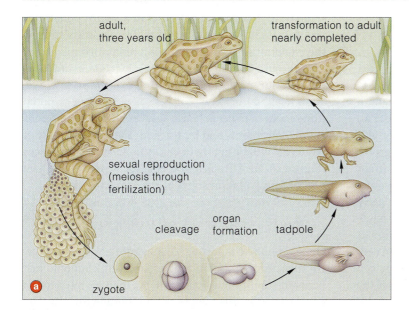

Figure 25.6 Reproduction and development during the life cycle of the leopard frog, *Rana pipiens*.

(**a**) We zoom in on the life cycle as a female releases her eggs into the surrounding water, and a male releases sperm over the eggs. A zygote forms at fertilization.

Appearance of a frog embryo over time. (**b**) About one hour after fertilization, a surface feature—the gray crescent—appears on this type of embryo; it establishes the frog's head-to-tail axis. Gastrulation will start here.

(**c–f**) Cleavage produces a blastula, a ball of cells that contains a fluid-filled cavity or blastocoel.

(**g–j**) During gastrulation, three primary tissue layers form through cell migrations and rearrangements. A primitive gut cavity opens up. Cell differentiation gets under way. A notochord and other organs form; the embryo is on its way to becoming a larval stage called a tadpole.

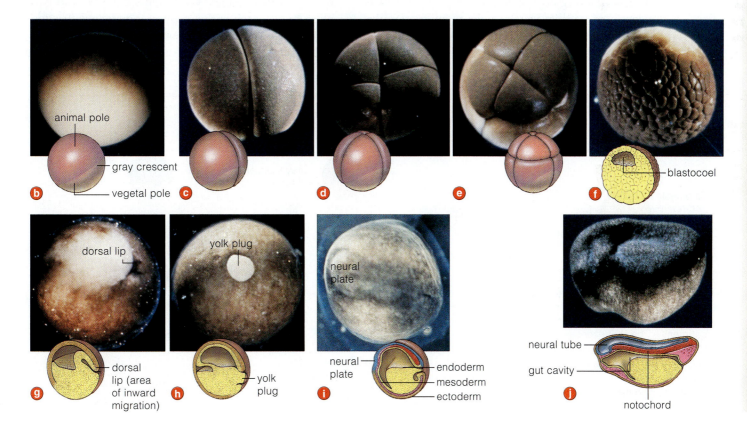

Figure 25.6 shows examples of these stages for one kind of vertebrate. They offer visual clues to a major principle: *Structures that form during one developmental stage are an essential foundation for the stage that comes after it.* In the sections to follow, you will come across evidence that reinforces this principle.

For most kinds of animals, the life cycle includes six stages of development—gamete formation, fertilization, cleavage, gastrulation, organ formation, and finally growth and tissue specialization. Normally, each stage is successfully completed before the next begins.

25.3 Early Marching Orders

Why don't you have an arm attached to your nose or toes growing from your navel? The patterning of body parts starts with messages in immature, unfertilized eggs.

INFORMATION IN THE EGG

A **sperm**, recall, consists only of paternal DNA and a bit of equipment that helps it reach and penetrate an egg. An **oocyte**, or immature egg, is much larger and more complex than sperm (Section 9.5). As it grows in volume, enzymes, mRNA transcripts, and other factors get stockpiled in its cytoplasm. These "maternal messages" are typically used after fertilization in early rounds of DNA replication and cell division. The angle and timing of microtubule assembly into early bipolar spindles is preordained by tubulins in the cytoplasm. How the cytoplasm gets divided up during early cell divisions depends on such factors, as well as on the amount of yolk and its distribution in the egg.

Consider a frog egg, which shows polarity. Yolk is concentrated near its vegetal pole; pigment granules are concentrated near its animal pole. The egg cortex, a cytoskeletal mesh just under the plasma membrane, structurally reorganizes itself after a sperm penetrates and fertilizes the egg. Part of the mesh shifts toward the penetration site, creating a crescent-shaped gray area (Figure 25.7a). The crescent's location establishes the anterior–posterior axis for the frog embryo.

In itself, a gray crescent is not evidence of regional differences in maternal messages. Such evidence comes from experiments and by tracking the development of embryos with obvious local cytoplasmic differences (Figure 25.7b,c). Track a frog egg, for instance, and you see that gastrulation starts at the gray crescent.

CLEAVAGE—THE START OF MULTICELLULARITY

Once an oocyte is fertilized, a zygote enters cleavage. By this process, recall, a band of microfilaments just beneath the plasma membrane at the cell midsection contracts and pulls the plasma membrane inward. The band contracts until the cell splits entirely, into two blastomeres, each with a copy of the nucleus (Section 8.4). Remember, these cuts only divide the volume of

Read Me First!

and watch the narrated animation on cytoplasmic localization

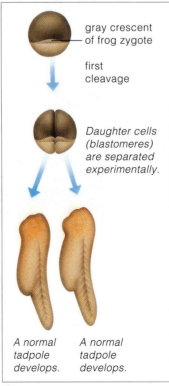

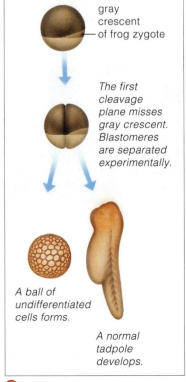

Figure 25.7 (**a**) A frog egg's cortex contains granules of dark pigment concentrated near one pole. At fertilization, part of the granule-containing cortex shifts toward the point of sperm entry. The shift exposes lighter colored, yolky cytoplasm, as a crescent-shaped gray area.

Normally, the first cleavage puts part of the gray crescent in both of the first two blastomeres. Two experiments, explained next, offer clues to the impact of cytoplasmic localization on the fate of the embryo's cells.

(**b**) For one experiment, the first two blastomeres that formed were physically separated from each other. Each blastomere still gave rise to a whole tadpole.

(**c**) For another experiment, a fertilized egg was manipulated so the cut through the first cleavage plane missed the gray crescent. Only one of the first two blastomeres received the gray crescent. It alone developed into a normal tadpole. Deprived of the maternal messages in the cytoplasm beneath the gray crescent, the other cell could not develop normally.

gray crescent of frog zygote

first cleavage

Daughter cells (blastomeres) are separated experimentally.

A normal tadpole develops.

A normal tadpole develops.

(**b**) EXPERIMENT 1

gray crescent of frog zygote

The first cleavage plane misses gray crescent. Blastomeres are separated experimentally.

A ball of undifferentiated cells forms.

A normal tadpole develops.

(**c**) EXPERIMENT 2

pigmented cortex

yolk-rich cytoplasm

sperm penetrating frog egg

gray crescent

(**a**)

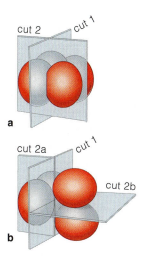

a

b

c

Figure 25.8 Early cleavage planes for fertilized eggs of (**a**) sea urchins and (**b**) mammals. Cuts are radial in sea urchins and rotational in mammals. (**c**) To sense cleavage's impact, consider Sabra and Nina. These genetically identical twins started life as the same zygote. The first two blastomeres formed at cleavage, the inner cell mass, or another early stage split and started two look-alike individuals.

cytoplasm into ever smaller cells. The blastomeres do not grow in size during cleavage.

Simply by virtue of where they form, the different blastomeres receive different maternal messages. This outcome of cleavage, called **cytoplasmic localization**, helps seal the developmental fate of each cell lineage. Cells of one lineage alone might have a protein that activates, say, a gene coding for a certain hormone.

In most animal zygotes, genes remain silent during early cleavage; the stockpiled maternal proteins and mRNAs help control the rate and angle of the cuts. In mammals, however, certain genes must be activated first. Cleavage cannot even be completed without the protein products of those genes.

CLEAVAGE PATTERNS

Observable differences in how animals develop start showing up during cleavage. The early cuts follow species-specific patterns. The simplest pattern, *radial* cleavage, starts with cuts perpendicular to the mitotic spindle. The resulting blastomeres are similar in size, but they get different parts of the cytoplasm (Figure 25.8*a*). The cuts go completely through some eggs, as in sea urchins. The fluid-filled cavity forms near the center of a ball-shaped array of cells.

In other eggs, cuts do not go all the way through. For instance, yolk impedes the cuts near the vegetal pole of a frog egg. Cuts are faster near the less yolky animal pole, where more blastomeres form. As one outcome, the fluid-filled cavity inside the resulting ball of cells is offset (Figure 25.5*e,f*). Early cuts are also incomplete in the fertilized eggs of reptiles, birds, and most fishes. Yolk confines the cuts to a small, disk-shaped region. Two flattened layers of cells form, one atop the other, with only a narrow cavity in between.

The nearly yolkless eggs of mammals are cut all the way through, in a *rotational* pattern. The first cut is vertical. One of the two resulting blastomeres is cut vertically, and the other horizontally (Figure 25.8*b*). The next cuts make a loose arrangement of eight cells with spaces in between. Tight junctions start forming among these cells and keep them huddled together. More cuts produce a hollow ball of cells. The outer cells secrete fluid that fills the inner cavity. Other cells cluster against one part of the cavity wall. This type of blastula is called a **blastocyst**. In all mammals, its inner cell mass gives rise to the embryo proper.

Once in a while, the first two blastomeres, the inner cell mass, or even the blastocyst splits in two. *Identical twins*, which have identical genes and are the same sex, may be the result (Figure 25.8*c*). *Fraternal twins* form when two oocytes released during the same menstrual cycle are fertilized by two sperm. Fraternal twins may be the same sex or different sexes.

The egg cytoplasm contains maternal messages in the form of regionally located enzymes and other proteins, mRNAs, cytoskeletal elements, yolk, and other components.

Cleavage divides a zygote into blastomeres. Simply by virtue of the cuts, each ends up with different maternal messages. Cleavage patterns differ among the major animal groups.

25.4 How Do Specialized Tissues and Organs Form?

Nearly all animals have a gut that digests nutrients for absorption. They have surface parts that protect organs inside and detect what is going on outside. Most have organs in between that function in structural support, motion, and circulation. This three-layered body plan emerges after cleavage, as gastrulation gets under way.

During gastrulation, the embryonic cells migrate and rearrange themselves into the three primary tissues of the gastrula—ectoderm, endoderm, and mesoderm. Figure 25.9 shows an example. The first cavity to form in protostome gastrulas becomes a mouth, but it will become an anus in deuterostomes (Section 22.1). Also, the anterior-to-posterior body axis forms in gastrulas. In vertebrates, this axis precedes the formation of a **neural tube**, the forerunner of the brain and spinal cord (Figures 25.10 and 25.11). Such specializations arise through cell differentiation and morphogenesis.

CELL DIFFERENTIATION

All cells of a normal embryo have the same number and kinds of genes, having descended from the same zygote. They all activate the genes for products that

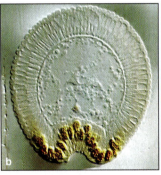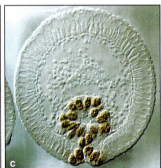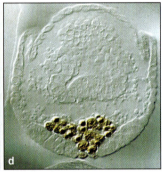

Figure 25.9 Gastrulation in a fruit fly (*Drosophila*), cross-section. After cleavage, the blastula is transformed into a gastrula. Notice how some cells (stained *gold*) migrate inward through an opening. Fruit flies are protostomes and this opening will later become the mouth.

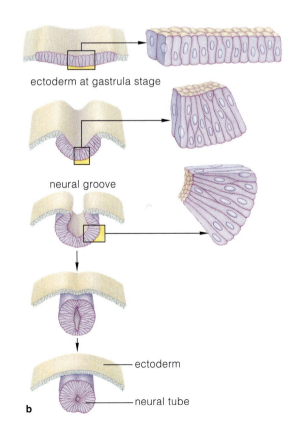

ectoderm at gastrula stage

neural groove

ectoderm

neural tube

Figure 25.10 Two examples of the types of events involved in morphogenesis.

(**a**) Cell migration. We show the same embryonic neuron (*orange*) at successive times during its "climb" through developing brain tissue. The neuron is responding to chemical cues on the surface of a glial cell (*yellow*), which guide it to its destination in the embryo.

(**b**) Neural tube formation. By the end of gastrulation, ectoderm is a uniform sheet of cells. Along the axis of the future tube, microtubules in ectodermal cells lengthen. Rings of microfilaments constrict in some of these cells, which become wedge-shaped. The part of the ectodermal sheet where they are located folds back on itself, over the wedge-shaped cells, which then disengage from it as a separate tube.

a

b

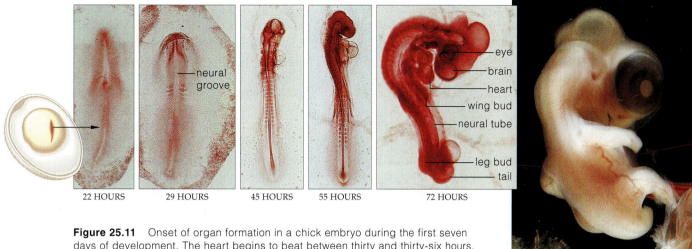

neural groove

eye
brain
heart
wing bud
neural tube
leg bud
tail

22 HOURS   29 HOURS   45 HOURS   55 HOURS   72 HOURS

168 HOURS (SEVEN DAYS OLD)

Figure 25.11 Onset of organ formation in a chick embryo during the first seven days of development. The heart begins to beat between thirty and thirty-six hours. You may have observed such embryos at the yolk surface of raw, fertilized eggs.

assure their survival, such as histones and glucose-metabolizing enzymes. But from gastrulation onward, cell lineages also engage in **selective gene expression**: they express some groups of genes and not others. This is the start of **cell differentiation**. By this process, cell lineages become specialized in their composition, structure, and function. Remember how one cell type activates genes for crystallins, proteins that form a lens in each eye? It is only one of about 200 differentiated cell types in the human body (Section 14.3).

MORPHOGENESIS

Specialized tissues and organs form in a programmed, orderly sequence by a process called **morphogenesis**. During this process, cells of different lineages divide, grow, migrate, and change in size. Tissues lengthen or widen and fold over. And some cells die in controlled ways at specific locations.

Think about active cell migration. *Cells send out and use pseudopods that move them along prescribed routes.* When they reach their destination, they connect with cells already there. Embryonic neurons migrate this way in a developing nervous system (Figure 25.10*a*).

How do these cells know where to move and when to stop? They respond to adhesive cues and chemical gradients. Their migrations are coordinated by the synthesis, release, deposition, and removal of specific chemicals in the extracellular matrix.

For example, the neuron shown in Figure 25.10*a* is responding to the "stickiness" of its surroundings. Adhesion proteins on a patch of its plasma membrane stuck to proteins on the surface of a glial cell. Now, cytoskeletal elements in the neuron lengthen from the

attachment site and shove the cytoplasm forward. The neuron migrates until its adhesion proteins reach the spot in the embryo that is "stickiest" to them.

Also, *whole sheets of cells expand and fold inward and outward as their cells change in shape.* Within these cells, microtubules grow longer, and rings of microfilament constrict. The controlled assembly and disassembly of these cytoskeletal components cause the changes.

Figures 25.10*b* and 25.11 hint at what happens after three primary tissues form in vertebrate embryos. At the embryo's midline, some ectodermal cells elongate and form the neural plate, the start of nervous tissue. The plate sinks inward as cells lengthen and become wedge shaped. At the edges of the resulting groove in the surface, flaps of tissue fold over and meet at the midline, forming the neural tube.

Finally, *programmed cell death (apoptosis) helps sculpt body parts.* Section 24.5 introduced the molecular basis of this process. Basically, signaling molecules from some cells activate tools of self-destruction stockpiled in other, target cells. Remember how apoptosis shapes a human hand from a paddle-shaped body part?

In cell differentiation, some cells selectively use certain genes that other cells do not use. Selective gene expression is the basis of cell differentiation. It results in cell lineages with distinctive structures, products, and functions.

Morphogenesis is a program of orderly changes in body size, shape, and proportion. It produces specialized tissues and organs at prescribed times in prescribed locations.

Morphogenesis involves cell division, active cell migration, tissue growth and foldings, changes in cell size and shape, and programmed cell death or apoptosis.

25.5 Pattern Formation

Maternal messages in the egg cytoplasm guide the earliest stages of development. Later on, communication signals among embryonic cells cause tissues and organs to form according to a mapped-out body plan.

EMBRYONIC INDUCTION

An embryonic cell locked in a tissue selectively reads its genes. It may make and secrete signaling molecules that affect its neighbors. Signals diffusing through its neighborhood may induce changes in its composition, structure, or both. Its descendants inherit a chemical memory of the changes, and they go on to form tissues and organs in places where we expect them to be.

This is a time of **embryonic induction**, of changes in the developmental fate of embryonic cell lineages exposed to gene products from cells of different tissues. Through these changes, specialized tissues and organs emerge from clumps of embryonic cells in predictable places, in the proper order. Their emergence is called **pattern formation**.

We have experimental evidence of cell memory. In one experiment, researchers excised the dorsal lip of a normal axolotl embryo, then grafted it into a novel location in a different axolotl embryo. Gastrulation proceeded at the recipient's own dorsal lip *and at the graft*. Conjoined twins—a double embryo with two sets of body parts—developed (Figure 25.12). The dorsal lip organizes this amphibian's main body axis. It was the first embryonic signaling center discovered. We also have evidence of cell memory from experiments involving chick wing development (Figure 25.13).

Many signals are short-range, involving cell-to-cell contacts. For instance, signals that activate or inhibit genes for adhesion proteins and recognition proteins directly affect how cells interact in tissues and organs. Other signals cause cytoskeletal elements in the target cell to lengthen in some direction. *Such signals can cause a cell to stick to its neighbors, or break free and migrate to a different location, or become segregated from the cells of an adjoining tissue.*

Other signals are long-range. They act on control elements in the DNA of embryonic cells that are some distance away. The **morphogens** are among the long-range signals. These degradable molecules diffuse out of a signaling center, and their concentration weakens with distance. They create a gradient that helps a cell chemically assess its position in the embryo and thus influence how it will differentiate. *Such differences in a signaling molecule's concentration induce lineages of cells positioned at different locations along the gradient to read different parts of the same genome.*

A THEORY OF PATTERN FORMATION

The same kinds of genes are the mapmakers for all major groups of animals. According to one theory of pattern formation, the products of these genes interact in ways that map out the basic body plan:

1. The formation of tissues and organs in ordered, spatial patterns starts with cytoplasmic localization. It continues through short-range, cell-to-cell contacts and through long-range signals, or chemical gradients

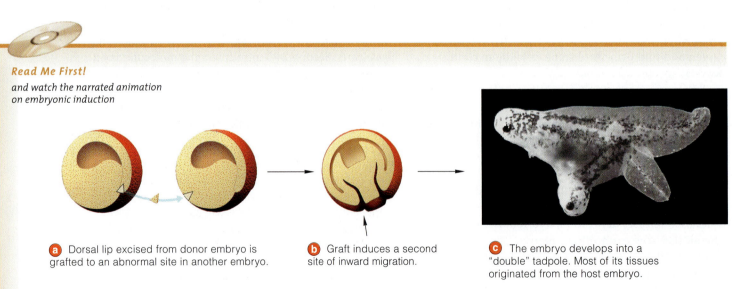

Read Me First!

and watch the narrated animation on embryonic induction

a Dorsal lip excised from donor embryo is grafted to an abnormal site in another embryo.

b Graft induces a second site of inward migration.

c The embryo develops into a "double" tadpole. Most of its tissues originated from the host embryo.

Figure 25.12 Experimental evidence that a dorsal lip controls amphibian gastrulation. A dorsal lip region of an axolotl embryo was transplanted to a different site in another axolotl embryo. It organized the formation of another set of body parts.

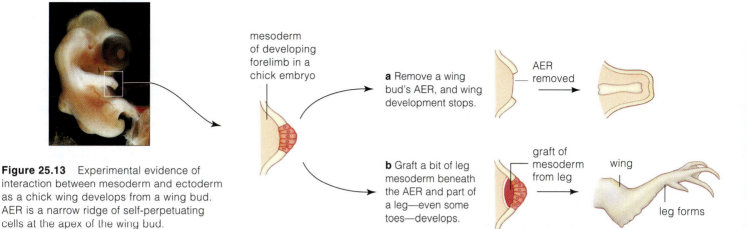

Figure 25.13 Experimental evidence of interaction between mesoderm and ectoderm as a chick wing develops from a wing bud. AER is a narrow ridge of self-perpetuating cells at the apex of the wing bud.

mesoderm of developing forelimb in a chick embryo

a Remove a wing bud's AER, and wing development stops.

AER removed

b Graft a bit of leg mesoderm beneath the AER and part of a leg—even some toes—develops.

graft of mesoderm from leg

wing

leg forms

that weaken with distance from the source. Cells at the start, middle, and end of a gradient are exposed to different chemical information.

2. Morphogens and other inducer molecules diffuse through embryonic tissues, and they activate classes of **master genes** in sequence. The products of these genes interact with control elements to lay the foundation for the basic body plan.

3. Products of **homeotic genes** and other master genes interact with control elements to map out the overall body plan (Sections 14.4 and 14.5). They form when blocks of genes are activated and suppressed in cells along the anterior–posterior axis and dorsal–ventral axis of the embryo. Other gene products interact to fill in details of specific body parts.

Master genes and gene products act in similar ways in all major animal groups. If they fail in mapping out the overall body plan, disaster follows, as when a heart forms in the wrong place. Once the plan is locked in, inductions that follow have only localized effects. If a lens fails to form, only the eye will be affected.

EVOLUTIONARY CONSTRAINTS ON DEVELOPMENT

How long have master genes been around? The basic body plan for sponges, worms, flies, vertebrates, and all other major animal groups has not changed much for about 500 million years. All of the many millions of species are variations on a few dozen plans!

Why is this so? There are few new master genes, *so variations in form might be more of an outcome of how the control genes control each other.* Consider that insect, squid, and vertebrate eyes differ structurally, and yet genes governing eye formation are nearly identical in these groups. An *eyeless* gene controls eye formation

in fruit flies (*Drosophila*). In humans, mutation of a nearly identical gene causes eyeless babies. The same gene controls the fate of cells that give rise to legs of fruit flies, crabs, and beetles, butterfly wings, sea star arms, fish fins, and mouse feet. All of these structures start out as buds from the main body axis.

For a long time, we have known that body plans cannot change much because of *physical* constraints (such as the surface-to-volume ratio) and *architectural* constraints (as imposed by body axes). We now know there are *phyletic* constraints on change. These are the constraints imposed on each lineage by interactions of master organizer genes, which operate when organs form and control induction of the basic body plan.

That may be why we have so many species and so few body plans. Once the master genes evolved and started interacting in intricate ways, it would have been hard to change the basic parts without killing the embryo. Mutations have indeed added marvelous variations to animal lineages. But the basic body plans have prevailed through great spans of time, so maybe it simply proved unworkable to start all over again.

Pattern formation is the ordered sculpting of embryonic cells into specialized tissues and organs. Cells differentiate as a result of cytoplasmic localization of materials and chemical cues regarding their relative position in the embryo.

Cytoplasmic localization and then inductive interactions among classes of master genes map the basic body plan. Gene products specify where and how body parts develop. They are long-range and short-range beacons that help cells assess their position and how they will differentiate.

Physical, architectural, and phyletic constraints limit the evolution of animal body plans. Among all major groups, master genes that guide development are similar and are sometimes identical.

25.6 Reproductive System of Human Males

So far, you have considered some principles of animal reproduction and development. The remainder of this chapter applies the principles to humans, starting with the male reproductive system.

WHEN GONADS FORM AND BECOME ACTIVE

In nature, the function of sex is not recreational, it is the perpetuation of one's genes. A human male's genes become packaged in sperm. These male gametes form in a pair of gonads, or primary reproductive organs, that also secrete sex hormones. Male gonads are called **testes** (singular, testis). As Table 25.1 indicates, they are part of a system of reproductive organs and glands.

Testes start to form on the wall of an XY embryo's abdominal cavity. Before birth, they descend into the scrotum, a skin pouch below the pelvic girdle (Figures 11.3 and 25.14). Muscle contractions draw the pouch closer to the main body mass, and muscle relaxation lowers it. Such adjustments help maintain an internal temperature that is suitable for sperm formation.

Packed within each testis are many small, highly coiled tubes called seminiferous tubules. Section 25.7 explains how sperm cells form and start maturing in these tubules in hormone-guided ways.

Sperm production and the emergence of secondary sexual traits start at **puberty**. This stage of postnatal development usually begins in boys between ages twelve and sixteen. Signs that it is under way include enlarging testes, growth spurts, a deepening voice, and more hair growth on the face, chest, and elsewhere.

STRUCTURE AND FUNCTION OF THE REPRODUCTIVE SYSTEM

Mammalian sperm travel from the testes through a series of ducts that lead to the urethra (Figure 25.15). They are not quite mature when they enter one of a pair of long, coiled ducts, the epididymes. Secretions from glandular cells in the duct wall trigger events that put finishing touches on maturing sperm cells. The last part of each epididymis stores mature sperm. During a male's reproductive years, about 100 million sperm mature every day. Unused ones are resorbed or excreted in the urine.

In a sexually aroused male, muscles in the wall of reproductive organs contract, which propels mature sperm into and through a pair of thick-walled tubes called vasa deferentia (singular, vas deferens). More contractions propel the sperm farther, through a pair of ejaculatory ducts, then through the urethra. This last tubular duct runs through the male sex organ—the penis—and opens at its tip. The urethra, remember, also functions in urinary excretion.

Sperm traveling to the urethra mix with glandular secretions and form semen, a thickened fluid that is expelled from the penis during sexual activity. Paired seminal vesicles secrete fructose, a sugar that sperm use as an energy source. The prostate gland and, to a lesser extent, seminal vesicles secrete prostaglandins into the mix. These signaling molecules might induce contractions in the female reproductive tract and so help sperm reach an egg. They may also function in helping sperm slip past the female's immune system.

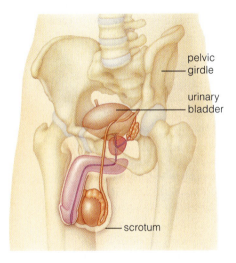

Figure 25.14 Position of the human male reproductive system relative to the pelvic girdle and urinary bladder.

Table 25.1	Organs and Accessory Components of the Human Male Reproductive System	

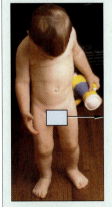

Reproductive Organs

Testis (2)	Sperm, sex hormone production
Epididymis (2)	Sperm maturation site and subsequent storage
Vas deferens (2)	Rapid transport of sperm
Ejaculatory duct (2)	Conduction of sperm to penis
Penis	Organ of sexual intercourse

Accessory Glands

Seminal vesicle (2)	Secretion of large part of semen
Prostate gland	Secretion of part of semen
Bulbourethral gland (2)	Production of mucus that functions in lubrication

Read Me First!
*and watch the narrated animation on
the human male reproductive system*

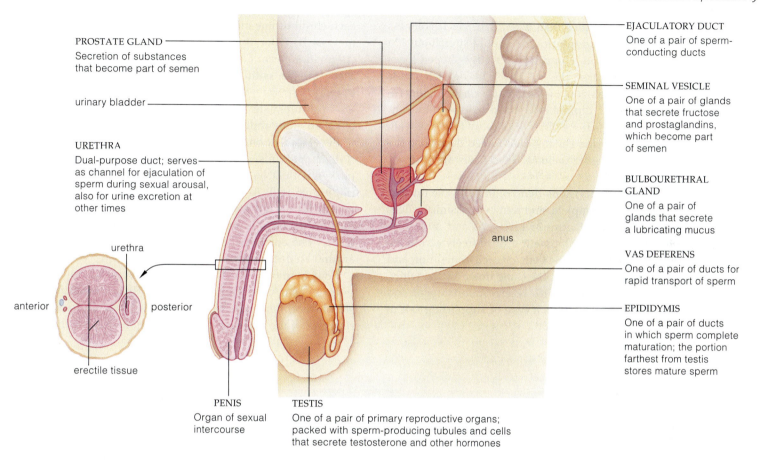

PROSTATE GLAND
Secretion of substances
that become part of semen

urinary bladder

URETHRA
Dual-purpose duct; serves
as channel for ejaculation of
sperm during sexual arousal,
also for urine excretion at
other times

urethra

anterior   posterior

erectile tissue

PENIS
Organ of sexual
intercourse

TESTIS
One of a pair of primary reproductive organs;
packed with sperm-producing tubules and cells
that secrete testosterone and other hormones

EJACULATORY DUCT
One of a pair of sperm-
conducting ducts

SEMINAL VESICLE
One of a pair of glands
that secrete fructose
and prostaglandins,
which become part
of semen

BULBOURETHRAL
GLAND
One of a pair of
glands that secrete
a lubricating mucus

anus

VAS DEFERENS
One of a pair of ducts for
rapid transport of sperm

EPIDIDYMIS
One of a pair of ducts
in which sperm complete
maturation; the portion
farthest from testis
stores mature sperm

Figure 25.15   Components of the human male reproductive system and their functions.

Secretions from the prostate gland may also help buffer the acidic conditions that prevail in the female reproductive tract. The pH of vaginal fluid is about 3.5–4.0, but sperm swim more efficiently at pH 6.0. A pair of bulbourethral glands secrete mucus-rich fluid into the urethra during sexual arousal.

CANCERS OF THE PROSTATE AND TESTES

Prostate cancer is a leading cause of death among men, surpassed only by lung cancers. In 2003, more than 220,000 males were diagnosed in the United States; 29,000 died. In the same year, there were about 7,600 cases of *testicular cancer*. In early stages, both cancers are painless and they may spread silently into lymph nodes of the abdomen, chest, neck, then lungs. If the cancer metastasizes, prospects are not good.

Doctors may detect prostate cancer by blood tests for increases in prostate-specific antigen (PSA) and by physical examinations. Adult males should examine their testes monthly—after a warm shower or bath—to check for enlargement, hardening, or new lumps. Treatment of testicular cancer has a high success rate, provided the cancer is caught before it spreads to other parts of the body.

Human males have a pair of testes, or primary reproductive organs that produce sperm and sex hormones. The male reproductive system also has specialized accessory glands and ducts.

The ducts carry sperm from the testes to the urethra, from where they leave the body. The accessory glands secrete the assorted noncellular components of the semen.

25.7 Sperm Formation

Signaling pathways that involve hormonal secretions by the hypothalamus, pituitary gland, and testes control sperm formation. Negative feedback loops are part of these pathways.

Each testis is smaller than a golfball, yet it contains two or three coiled tubules that would stretch out 125 meters, end to end. Hundreds of wedge-shaped lobes partition the interior (Figure 25.16). Inside each tubule wall are spermatogonia (singular, spermatogonium)—undifferentiated, diploid cells that divide continuously. Descendant cells force older ones away from the wall, toward the interior. During their forced departure, the daughter cells enter meiosis. These cells are primary spermatocytes.

Cytoplasmic divisions that accompany the nuclear divisions are incomplete, and thin cytoplasmic bridges connect the cells (Figure 25.16c). Ions and molecules concerned with development diffuse freely across the bridges that connect the cells of each generation and cause them to mature at the same time. All the while, the primary spermatocytes receive nourishment and signals from Sertoli cells, a type of supporting cell in the seminiferous tubules.

Secondary spermatocytes form after each primary spermatocyte finishes meiosis I (Figure 25.16c). These are haploid cells with chromosomes in the duplicated state. (Here you may wish to review the overview of spermatogenesis in Section 9.4.) Sister chromatids of each chromosome move away from each other during meiosis II. After this, haploid daughter cells called spermatids form. Each cell now starts to mature into a sperm. The entire process—from spermatogonium to mature sperm—takes about 100 days.

An adult male produces sperm on a continuous basis, so many millions of cells are in different stages of development on any given day.

Each mature sperm is a flagellated cell. Its head is packed with DNA, and its tail has a microtubular core (Figure 25.16d). Mitochondria in a midpiece supply energy for the tail's whiplike motions. Extending over most of the head is an acrosome, a cap with enzymes inside. The enzymes can digest extracellular material around an oocyte prior to fertilization.

Four hormones—testosterone, LH, FSH, and GnRH—are essential for sperm formation. **Testosterone**, a steroid hormone, governs the growth, structure, and function of the male reproductive tract as well as the

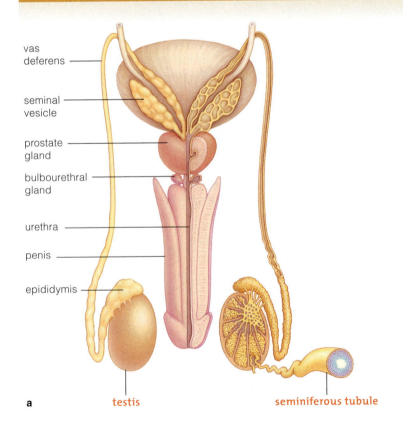

vas deferens
seminal vesicle
prostate gland
bulbourethral gland
urethra
penis
epididymis

a   testis   seminiferous tubule

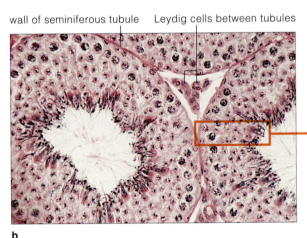

wall of seminiferous tubule   Leydig cells between tubules

b

Figure 25.16 (**a**) Male reproductive tract, posterior view. Arrows show the route that sperm take before ejaculation.

Sperm formation. (**b**) Light micrograph of cells in three adjacent seminiferous tubules, cross-section. Leydig cells occupy tissue spaces between tubules. (**c**) How sperm form, beginning with a diploid germ cell. (**d**) Structure of a mature sperm, the male gamete.

Figure 25.17 Negative feedback loops from the testes to the hypothalamus and to the anterior lobe of the pituitary.

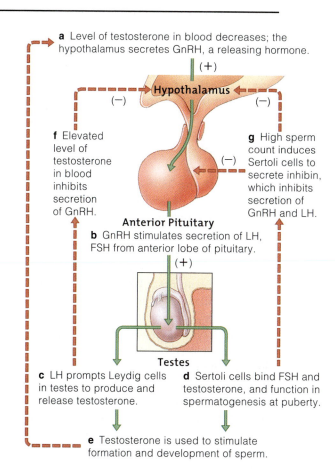

a Level of testosterone in blood decreases; the hypothalamus secretes GnRH, a releasing hormone.

Hypothalamus

f Elevated level of testosterone in blood inhibits secretion of GnRH.

g High sperm count induces Sertoli cells to secrete inhibin, which inhibits secretion of GnRH and LH.

Anterior Pituitary

b GnRH stimulates secretion of LH, FSH from anterior lobe of pituitary.

Testes

c LH prompts Leydig cells in testes to produce and release testosterone.

d Sertoli cells bind FSH and testosterone, and function in spermatogenesis at puberty.

e Testosterone is used to stimulate formation and development of sperm.

formation of sperm. It triggers development of male secondary sexual traits, and it stimulates sexual and aggressive behavior. Leydig cells in the testes secrete testosterone. The anterior lobe of the pituitary gland makes and secretes **LH** and **FSH**.

The hypothalamus governs secretion of all three hormones (Figure 25.17). It secretes **GnRH** when the blood levels of testosterone and other factors are low. This releasing hormone makes the anterior lobe step up the LH and FSH secretions. In turn, LH stimulates Leydig cells to secrete testosterone, which helps sperm form and mature. At puberty, the FSH binds to Sertoli cells and jump-starts sperm formation.

Sperm formation depends on the hormones LH, FSH, and testosterone. Negative feedback loops from the testes to the hypothalamus and pituitary gland control their secretion.

Read Me First!
and watch the narrated animation on sperm formation

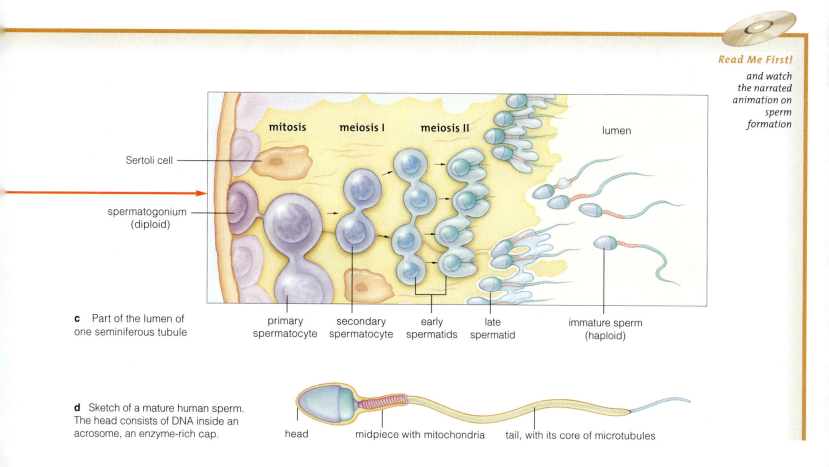

mitosis   meiosis I   meiosis II   lumen

Sertoli cell

spermatogonium (diploid)

c Part of the lumen of one seminiferous tubule

primary spermatocyte   secondary spermatocyte   early spermatids   late spermatid   immature sperm (haploid)

d Sketch of a mature human sperm. The head consists of DNA inside an acrosome, an enzyme-rich cap.

head   midpiece with mitochondria   tail, with its core of microtubules

25.8 Reproductive System of Human Females

The reproductive system of human females functions in the production of gametes and sex hormones. It also has a chamber in which a new, developing individual is protected and nourished until birth.

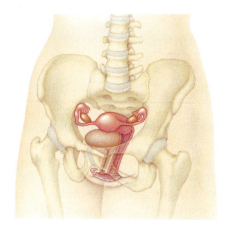

Figure 25.18 Location of the human female reproductive system relative to the pelvic girdle and urinary bladder.

COMPONENTS OF THE SYSTEM

Figures 25.18 and 25.19 show the reproductive system of a human female, and Table 25.2 lists their functions. The primary reproductive organs are a pair of **ovaries** that produce oocytes (immature eggs) and secrete sex hormones. An oocyte released from an ovary enters one of a pair of oviducts, or Fallopian tubes. Oviducts open onto the **uterus**, a hollow organ shaped a bit like a pear. Fertilization usually takes place in an oviduct, but the embryo develops mainly inside the uterus.

A thick layer of smooth muscle, the myometrium, makes up most of the uterine wall. The uterine lining, or **endometrium**, consists of connective tissues, blood vessels, and glands. The cervix, the narrowed-down part of the uterus, connects with a muscular tube, the vagina. Mucus-secreting epithelium lines the vagina, which extends from the cervix to the body's surface. The vagina receives sperm from the male and serves as part of the birth canal.

At the body's surface are genital organs of sexual stimulation. Outermost are the labia majora, a pair of skin folds padded with adipose tissue. They enclose the labia minora, a pair of smaller folds having many blood vessels but no adipose tissue. These folds partly enclose the clitoris, a female sex organ derived from the same embryonic tissue as the male penis. Like the penis, the clitoris has a great abundance of sensory receptors and is very sensitive to sexual stimulation. The urethra opens at the body surface about midway between the vaginal opening and the clitoris.

OVERVIEW OF THE MENSTRUAL CYCLE

Females of most mammalian species follow an *estrous* cycle meaning they are fertile and "in heat" (sexually receptive to males) at only certain times in the cycle. Females of humans and some other primates follow a **menstrual cycle**. They are fertile intermittently, on a cyclic basis. Their fertile periods are not synchronized with sexual receptivity. Even though they can become pregnant at only certain times in the cycle, they may be receptive to sex at any time.

The next section explains the cycle, but here is a brief overview (Table 25.3). First, an oocyte matures and is released from an ovary, and the endometrium is primed for pregnancy. When fertilization does not occur, blood and bits of the uterine lining flow from the vagina. The flow means "there's no embryo at this time," and marks the start of a new cycle.

Table 25.2	Organs of the Human Female Reproductive Tract
Ovaries	Oocyte production and maturation, sex hormone production
Oviducts	Ducts for conducting oocyte from ovary to uterus; fertilization normally occurs here
Uterus	Chamber in which new individual develops
Cervix	Secretion of mucus that enhances sperm movement into uterus and (after fertilization) reduces embryo's risk of bacterial infection
Vagina	Organ of sexual intercourse; birth canal

Table 25.3	Events of a Menstrual Cycle Lasting Twenty-Eight Days	
Phase	Events	Days of Cycle
Follicular phase	Menstruation; endometrium breaks down	1–5
	Follicle matures in ovary; endometrium rebuilds	6–13
Ovulation	Oocyte released from ovary	14
Luteal phase	Corpus luteum forms, secretes progesterone; the endometrium thickens and develops	15–28

Read Me First!
and watch the narrated animation on the
human female reproductive system

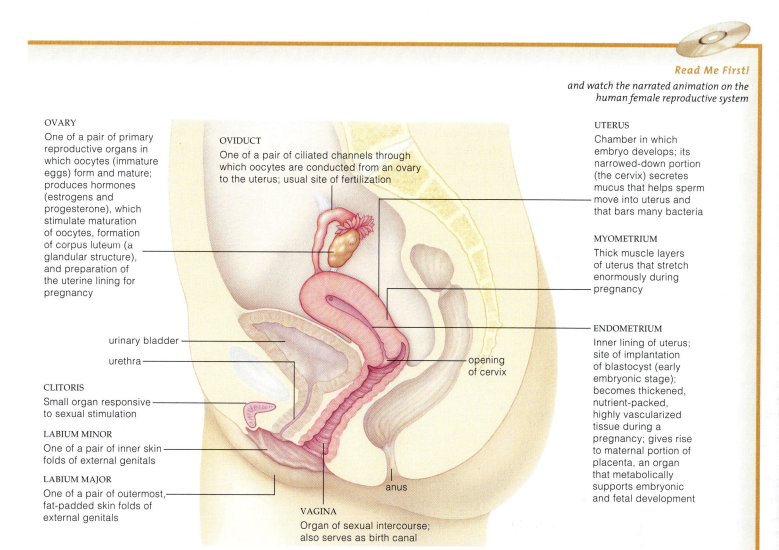

OVARY
One of a pair of primary reproductive organs in which oocytes (immature eggs) form and mature; produces hormones (estrogens and progesterone), which stimulate maturation of oocytes, formation of corpus luteum (a glandular structure), and preparation of the uterine lining for pregnancy

OVIDUCT
One of a pair of ciliated channels through which oocytes are conducted from an ovary to the uterus; usual site of fertilization

UTERUS
Chamber in which embryo develops; its narrowed-down portion (the cervix) secretes mucus that helps sperm move into uterus and that bars many bacteria

MYOMETRIUM
Thick muscle layers of uterus that stretch enormously during pregnancy

ENDOMETRIUM
Inner lining of uterus; site of implantation of blastocyst (early embryonic stage); becomes thickened, nutrient-packed, highly vascularized tissue during a pregnancy; gives rise to maternal portion of placenta, an organ that metabolically supports embryonic and fetal development

urinary bladder

urethra

opening of cervix

CLITORIS
Small organ responsive to sexual stimulation

LABIUM MINOR
One of a pair of inner skin folds of external genitals

LABIUM MAJOR
One of a pair of outermost, fat-padded skin folds of external genitals

anus

VAGINA
Organ of sexual intercourse; also serves as birth canal

Figure 25.19 Components of the human female reproductive system and their functions.

The cycle has three phases. In the *follicular* phase, menstruation occurs, the uterine lining breaks down and starts rebuilding, and an oocyte starts maturing. In the next phase, **ovulation**, an oocyte escapes from the ovary. In the *luteal* phase, a glandular structure called the **corpus luteum** forms in an ovary. It starts to secrete hormones that cause the endometrium to thicken in preparation for pregnancy.

Hormones control these phases by way of feedback loops from the ovaries to both the hypothalamus and pituitary gland. FSH and LH promote changes in the ovaries and stimulate them to secrete **estrogens** and **progesterone**. These sex hormones bring about cyclic changes in the endometrium.

A female enters puberty between the ages of ten and sixteen. Pubic hair forms, fat is deposited inside her breasts and around the hips, and menstrual cycles start. Each cycle lasts for about twenty-eight days, on average, but it is longer or shorter for some females. The cycles usually continue until the late forties or early fifties, when sex hormone secretions start to dwindle. The decline in hormone secretions correlates with the onset of *menopause*, the twilight of a female's reproductive capacity.

Ovaries, the primary reproductive organ of human females, produce immature eggs and sex hormones. Endometrium lines the uterus, a chamber in which embryos develop.

Estrogens and progesterone guide the cyclic growth and release of oocytes from the ovary as well as the breakdown and rebuilding of the endometrium, which depends on whether pregnancy occurs. These are the key events of menstrual cycles, which start at puberty.

25.9 Preparations for Pregnancy

Before starting this section, take a moment to review Figure 9.5, the overview of meiosis in an oocyte.

CYCLIC CHANGES IN THE OVARY

During her reproductive years, a female will release about 400 to 500 primary oocytes. A **primary oocyte** is an immature egg that was arrested in prophase I of meiosis. The only ones in a female's ovaries formed while she herself was an embryo—or so it was thought.

Female mice are now known to produce some new oocytes even when they are adults. We do not know whether human females do so.

We do know that oocytes released late in life are at greater risk for abnormal changes in the structure or number of chromosomes as meiosis resumes in them. One serious risk is Down syndrome (Section 11.8).

A follicle consists of a primary oocyte and a layer of cells that surround and nourish it. An ovary holds many follicles. Figure 25.20 shows what happens to

Read Me First!

and watch the narrated animation on ovarian function

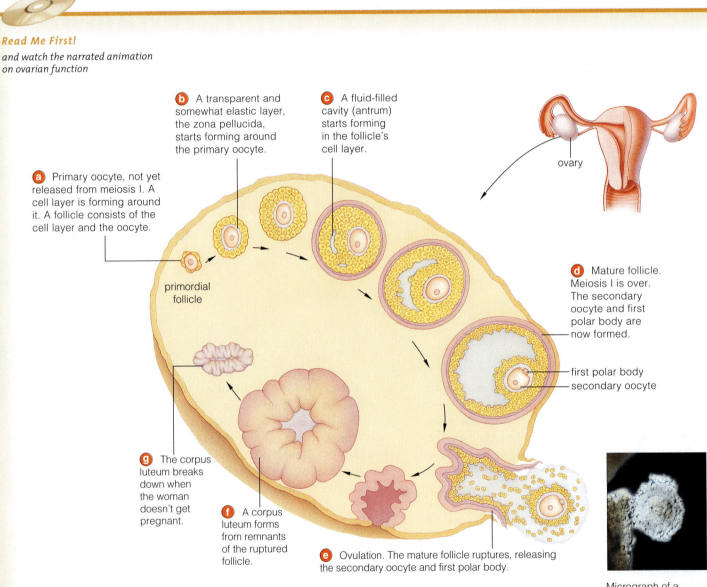

b A transparent and somewhat elastic layer, the zona pellucida, starts forming around the primary oocyte.

c A fluid-filled cavity (antrum) starts forming in the follicle's cell layer.

ovary

a Primary oocyte, not yet released from meiosis I. A cell layer is forming around it. A follicle consists of the cell layer and the oocyte.

primordial follicle

d Mature follicle. Meiosis I is over. The secondary oocyte and first polar body are now formed.

first polar body
secondary oocyte

g The corpus luteum breaks down when the woman doesn't get pregnant.

f A corpus luteum forms from remnants of the ruptured follicle.

e Ovulation. The mature follicle ruptures, releasing the secondary oocyte and first polar body.

Micrograph of a secondary oocyte escaping from the surface of an ovary

Figure 25.20 Cyclic events in a human ovary, cross-section. The follicle does not "move around" as in this diagram, which simply shows the *sequence* of events. In the cycle's first phase, a follicle grows and matures. At ovulation, the second phase, the mature follicle ruptures and releases a secondary oocyte. In the third phase, a corpus luteum forms from the follicle's remnants.

one of them during a menstrual cycle. As the cycle starts, the hypothalamus is secreting enough GnRH to make cells in the anterior lobe of the pituitary step up FSH and LH secretions. These hormones trigger the growth of the oocyte and the formation of more cells around it. Glycoprotein molecules accumulate under the surrounding cells and form the zona pellucida, a noncellular layer (Figure 25.20b).

Eight to ten hours before being released, the oocyte completes meiosis I. Its cytoplasm divides unevenly. One of the two haploid daughter cells, a **secondary oocyte**, receives most of the cytoplasm. The other cell is the first of three **polar bodies** that form by way of meiosis. It will degenerate later in the cycle.

FSH and LH collect in fluid in the follicle and prod its cells to secrete estrogens, so the levels of estrogens in blood increase. About halfway through the cycle, the pituitary responds to the increases. It secretes LH in a brief pulse. The follicle swells in response, and its wall weakens and ruptures. Fluid—and the secondary oocyte—are released. *The midcycle surge of LH triggers ovulation, the release of a secondary oocyte from the ovary.*

CYCLIC CHANGES IN THE UTERUS

The estrogens released early in the cycle also call for growth of the endometrium and its glands, which sets the stage for pregnancy. Before the midcycle surge of LH, follicle cells were busy secreting progesterone and estrogens. Blood vessels grew fast in the thickened endometrium. At ovulation, estrogens prompted cells of the cervical canal to release a thin, clear mucus—an ideal medium for sperm to swim through.

The midcycle LH surge that brings about ovulation also stimulates cells in the ruptured follicle to form a corpus luteum. Progesterone and estrogens secreted by this structure cause the endometrium to thicken, in preparation for pregnancy.

All the while, the hypothalamus has been stopping other follicles in the ovary from maturing by slowing down FSH secretion. If a blastocyst does not burrow into the endometrium, the corpus luteum will last for no more than twelve days or so. In the last days of the cycle, it will secrete prostaglandin—local signaling molecules that trigger this gland's self-destruction.

Without the corpus luteum, levels of progesterone and estrogen decline fast, and the endometrium starts to break down. Deprived of oxygen and nutrients, its blood vessels constrict and tissues die. Blood escapes as weakened capillary walls start to rupture. Blood and sloughed endometrial tissues form a menstrual flow, which lasts for three to six days. After this, the

Read Me First!
and watch the narrated animation on hormones and the menstrual cycle

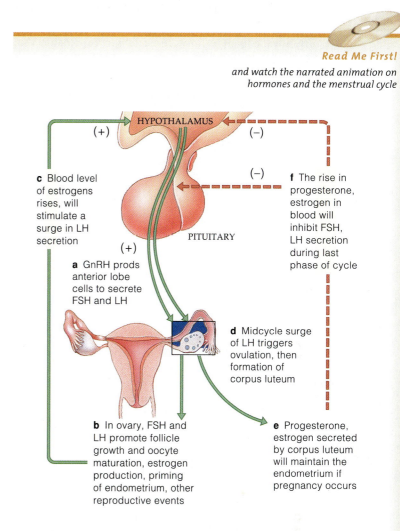

Figure 25.21 Hormone secretions during a menstrual cycle. A positive feedback loop from an ovary to the hypothalamus (*green*) triggers ovulation. After a secondary oocyte escapes, a negative feedback loop from the ovary to the hypothalamus and pituitary gland (*red*) inhibit hormone secretions and keep another follicle from maturing until the cycle ends.

rising estrogen levels invite repair and growth of the endometrium. Figure 25.21 summarizes the effects of changing hormone levels for one menstrual cycle.

In a menstrual cycle, FSH and LH stimulate growth of an ovarian follicle. The first meiotic cell division in the oocyte results in a secondary oocyte and the first polar body.

A midcycle surge of LH triggers ovulation—the release of the secondary oocyte and the polar body from the ovary.

Feedback loops to the hypothalamus and pituitary from the ovaries and, later, the corpus luteum control the cyclic changes in the ovary and uterine lining.

25.10 Visual Summary of the Menstrual Cycle

By now, you may have come to the conclusion that the human menstrual cycle is not a simple tune on a biological banjo. It's more like a full-blown hormonal symphony! Before continuing with your reading, take a moment to review Figure 25.22. It correlates cyclic changes in the ovary and the uterus with changing concentrations of hormones that trigger the events of each menstrual cycle.

Read Me First!
and watch the narrated animation
on the menstrual cycle summary

Figure 25.22 Summary of changing hormone levels and their effects during a menstrual cycle that lasts twenty-eight days.

(**a**) The hypothalamus secretes a releasing hormone, GnRH, which stimulates the anterior lobe of the pituitary to secrete FSH and LH.

(**b**) FSH and LH stimulate a follicle to grow, an oocyte to mature, and ovaries to secrete the progesterone and estrogens that induce rebuilding of the endometrium.

(**c**) A midcycle surge of LH triggers ovulation and formation of a corpus luteum. The corpus luteum maintains the endometrium through its secreting progesterone and some estrogens, but it stops doing so if pregnancy does not occur (**d**).

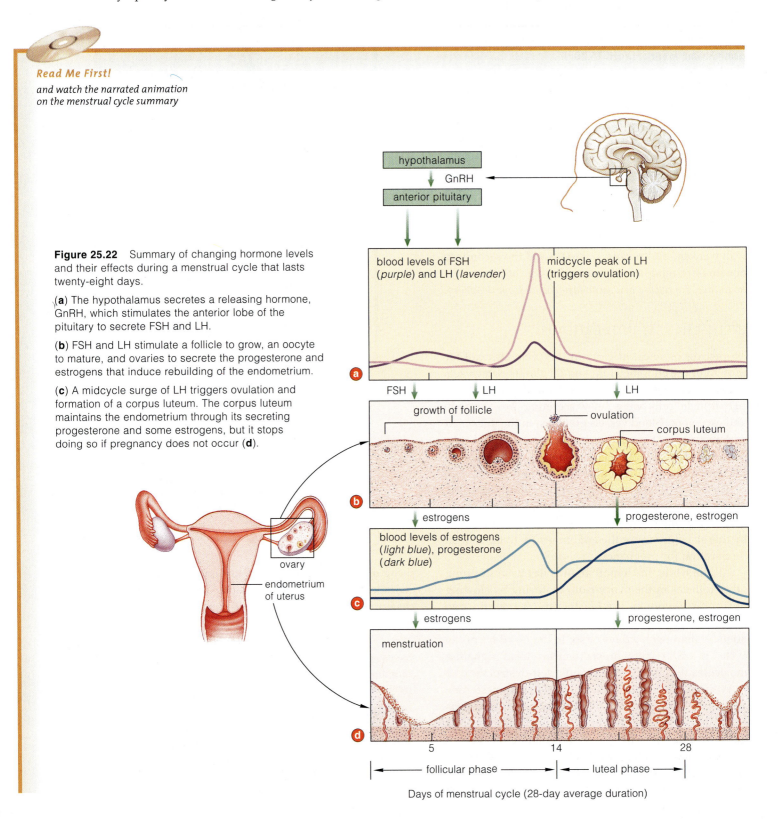

25.11 Pregnancy Happens

When a female and male engage in sexual intercourse, or coitus, the hormonal fog of the moment may obscure what can happen if a secondary oocyte is in an oviduct.

SEXUAL INTERCOURSE

The male sex act begins with an erection (a stiffened and lengthened penis). It culminates in ejaculation, or the forceful expulsion of semen from the penis. As Figure 25.15 shows, the penis contains long cylinders of spongy tissue. The penis of unaroused males stays limp because the large blood vessels that supply the spongy tissue remain vasoconstricted. When a male becomes aroused, the vessels vasodilate; blood flow into the penis exceeds blood flow out. As the spongy tissue inside becomes engorged with blood, the penis stiffens and lengthens. This facilitates insertion into a female's vaginal canal.

Repetitive pelvic thrusting mechanically stimulates friction-activated receptors that abound on the tip of the penis. It also stimulates the female's clitoris and vaginal wall. In the male, the response is involuntary muscle contractions that force sperm-laden semen into the urethra, from which it is ejaculated.

Emotional intensity, hard breathing, strong heart pounding, and the contraction of skeletal muscles in general accompany a rhythmic throbbing of the pelvic muscles. During *orgasm*, the end of the sex act, strong sensations of physical release, warmth, and relaxation dominate. Similar sensations typify female orgasm.

You may have heard that a female cannot become pregnant if she doesn't reach orgasm. Don't believe it.

FERTILIZATION

On average, an ejaculation can put 150 million to 350 million sperm in the vagina. Fertilization occurs if they arrive a few days before or after ovulation or any time in between. Less than thirty minutes after sperm arrive in the vagina, contractions move them deep into the female's reproductive tract. A few hundred actually reach the upper part of the oviduct, where eggs usually are fertilized (Figure 25.23).

Many sperm bind to the oocyte's zona pellucida. Binding triggers the release of enzymes from the cap over each sperm's head. Collectively, these digestive enzymes make a passage through the zona pellucida. Usually only one sperm enters the secondary oocyte. Only its nucleus and centrioles do not degenerate.

Upon sperm penetration, the secondary oocyte and the first polar body complete meiosis II. Now there are three polar bodies and one mature egg, or **ovum**

Read Me First!

and watch the narrated animation on human fertilization

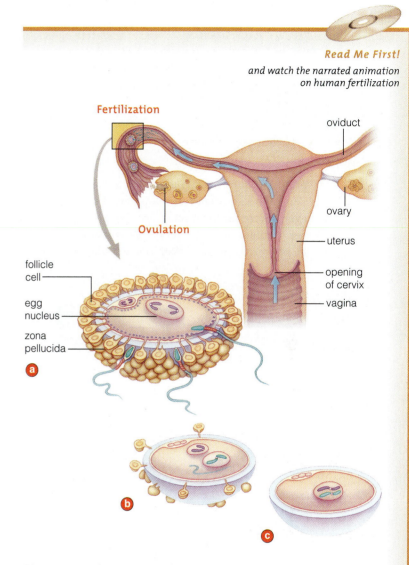

Figure 25.23 Fertilization. (**a**) Many human sperm travel rapidly from the vagina to an oviduct (*blue* arrows), where they surround a secondary oocyte. Digestive enzymes released from each sperm's cap clear a path through the zona pellucida.

(**b**) A single sperm penetrates the secondary oocyte, which releases substances that make the zona pellucida impenetrable to the other sperm. Penetration also stimulates meiosis II of the oocyte's nucleus. (**c**) The sperm's tail degenerates; its nucleus enlarges and fuses with the nucleus of its target. Fertilization is over; a zygote has formed.

(plural, ova). The egg nucleus fuses with the sperm nucleus. Together, chromosomes of both nuclei restore the diploid number for a brand new zygote.

The intense physiological events that accompany coitus put sperm on a collision course with an egg.

Fertilization is over when a sperm nucleus and egg nucleus fuse. The diploid zygote that forms is the start of a new individual.

25.12 Control of Human Fertility

The motivation to engage in sex has been evolving for hundreds of millions of years. A few centuries of moral arguments for self-control have not suppressed it. How will we reconcile our biological past with the need for a stabilized cultural present?

THE ISSUE

The transformation of a zygote into an adult raises profound questions. *When does development begin?* As you have seen, major developmental events unfold even before fertilization. *When does life begin?* During her lifetime, a woman produces as many as 500 eggs, all of which are alive. With one ejaculation, a man releases a quarter of a billion sperm, which are alive. Before sperm and egg merge by chance and establish a new individual's genetic makeup, they are as much alive as any other form of life. It is scarcely tenable, then, to say that life begins at fertilization. *Life began more than 3.8 billion years ago—and each gamete, each zygote, and each sexually mature individual is but a fleeting stage in the continuation of that beginning.*

This greater perspective on life cannot diminish the meaning of conception. It is no small thing to entrust a new individual with the gift of life, wrapped in the unique evolutionary threads of our species and handed down through an immense sweep of time.

Yet human population growth is outstripping resources. Many millions already face the horrors of starvation. Living where we do, few of us know what it means to give birth to a child, to give it the gift of life, and have no food to keep it alive. Whether and how fertility should be controlled is a volatile issue. We return to this issue in Chapter 26. Here we simply present the options available to those who decide to postpone or forgo pregnancy (Figure 25.24).

SOME OPTIONS

Abstinence, or no sex at all, is the most reliable way to avoid pregnancy, but it takes great self-discipline to override the neural and hormonal sense of urgency associated with sex. Different versions of the *rhythm method* are forms of abstinence; a woman avoids sex in her fertile period. She calculates when she is fertile by recording how long her menstrual cycles last, by taking her core temperature each morning, or both. Miscalculations are frequent. Sperm deposited in the vagina a few days before ovulation may survive long enough to meet up with an egg.

Withdrawal, or removing the penis from the vagina before ejaculation, dates at least to biblical times. It requires a lot of willpower and still may fail, because fluid released from the penis before ejaculation can contain some sperm.

Douching (chemically rinsing the vagina right after intercourse) is too chancy. Sperm move out of

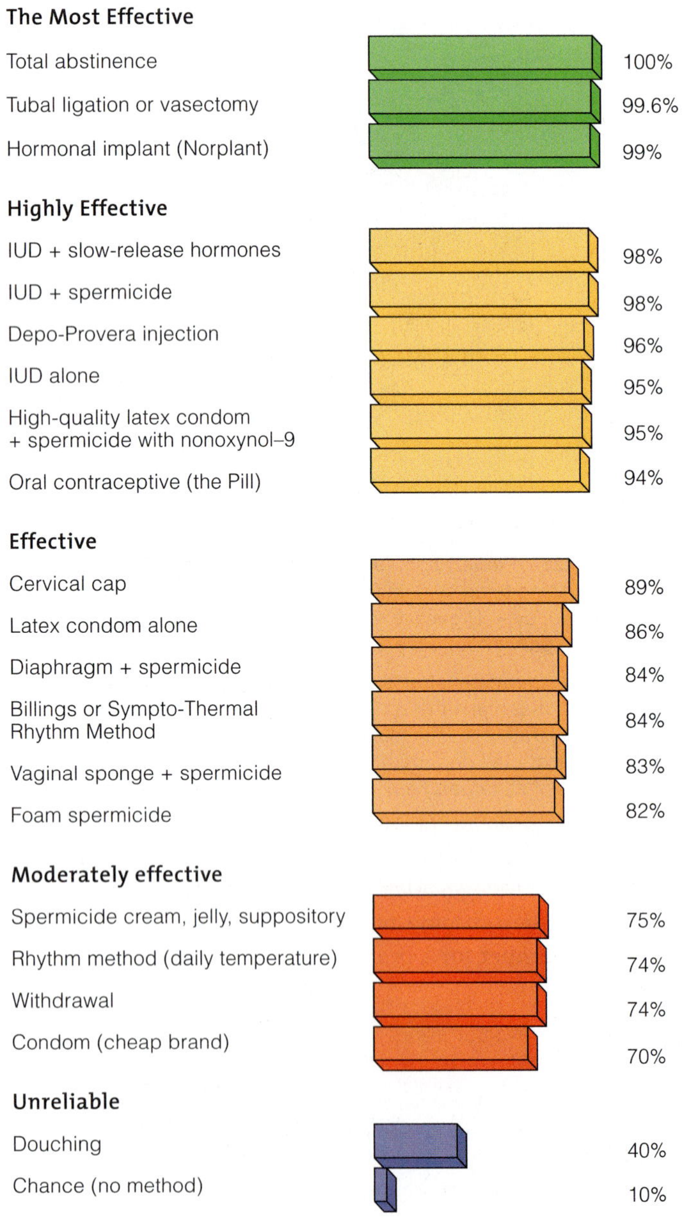

The Most Effective

Total abstinence	100%
Tubal ligation or vasectomy	99.6%
Hormonal implant (Norplant)	99%

Highly Effective

IUD + slow-release hormones	98%
IUD + spermicide	98%
Depo-Provera injection	96%
IUD alone	95%
High-quality latex condom + spermicide with nonoxynol–9	95%
Oral contraceptive (the Pill)	94%

Effective

Cervical cap	89%
Latex condom alone	86%
Diaphragm + spermicide	84%
Billings or Sympto-Thermal Rhythm Method	84%
Vaginal sponge + spermicide	83%
Foam spermicide	82%

Moderately effective

Spermicide cream, jelly, suppository	75%
Rhythm method (daily temperature)	74%
Withdrawal	74%
Condom (cheap brand)	70%

Unreliable

Douching	40%
Chance (no method)	10%

Figure 25.24 Comparison of the effectiveness of some methods of contraception. These percentages also indicate the number of unplanned pregnancies per 100 couples who use only that method of birth control for a year. For example, "94% effectiveness" for oral contraceptives means that 6 of every 100 women will still become pregnant, on average.

reach of a douche ninety seconds after ejaculation. Frequent douching irritates the reproductive tract.

Controlling fertility by surgical intervention is less chancy. Men who do not want children may opt for a *vasectomy*, a procedure that requires only a local anesthetic. A doctor makes a small incision in the scrotum, then cuts and ties off each vas deferens. Sperm no longer can leave the testes and become part of semen. Surgical reversal of the procedure is not always successful. Only about 60 percent of those who have reversed a vasectomy have been able to father a child.

Tubal ligation nearly always guarantees permanent infertility. The oviducts are cauterized or cut and tied. This procedure is now more common than vasectomy. Surgical reversal is about 70 percent successful.

Less drastic fertility control methods are based on physical and chemical barriers that stop sperm from reaching an egg. *Spermicidal foam* and *spermicidal jelly* poison sperm. An applicator is used to insert either one into the vagina before sex. These products are not always reliable, but using them with a condom or a diaphragm makes them more effective.

A *diaphragm* is a flexible, dome-shaped device. It is inserted into the vagina and positioned over the cervix before intercourse. It is relatively effective if fitted initially by a doctor, used with a spermicidal foam or jelly, inserted correctly each time, and left in place for a prescribed length of time.

Condoms are thin, tight-fitting sheaths worn over the penis during intercourse. Good brands may be as much as 95 percent effective when used with a spermicide. Only latex condoms protect against sexually transmitted diseases. However, condoms can tear or leak, at which time they become useless.

A *birth control pill* delivers synthetic estrogens and progesterone-like hormones that block maturation of oocytes and ovulation. "The Pill" reduces menstrual cramps but can cause nausea, headaches, and weight gain. With at least 50 million users, it is the most common fertility control method in the United States. Used correctly, it is at least 94 percent effective. Use lowers the risk of ovarian cancer but increases the risk of breast, cervical, and liver cancers.

A *birth control patch* is an adhesive patch applied to skin. It delivers the same hormones as an oral contraceptive, and it blocks ovulation the same way.

Progestin injections or implants block ovulation. A *Depo-Provera* injection is effective for three months. *Norplant* works for five years. Both methods are very effective. However, they may cause sporadic, heavy bleeding, and removing Norplant rods can be tricky.

Figure 25.25 Doctor using a microscope and monitor to inject a human sperm into an egg during *in vitro* fertilization (IVF).

Morning-after pills are sometimes taken after a condom tears, or after unprotected consensual sex or rape. One brand, Previn, is a set of two pills with hormones that suppress ovulation and block the secretions from the corpus luteum. Morning-after pills work best when taken early, but are somewhat effective for as long as five days after intercourse.

SEEKING OR ENDING PREGNANCY

Some couples cannot conceive naturally. *In vitro fertilization* refers to conception outside the body ("in glass," as in a glass petri dish). Hormone injections prepare the ovaries for ovulation. Before an oocyte is released, it is withdrawn and a sperm is injected into it (Figure 25.25). Cleavage may follow fertilization. Two to four days later, the resulting tiny ball of cells may be transferred to the woman's uterus.

The fertilization attempts each cost about 8,000 dollars—and most fail. With our population size and health care costs soaring, is this a cost society should bear? Court battles have been waged over this issue.

At the other extreme is *induced abortion*: dislodging and removing an embryo or a fetus from the uterus. Each year in the United States alone, about 50 percent of more than 3.1 million unwanted pregnancies end in abortion. From a clinical standpoint, abortion usually is a rapid, relatively painless procedure and free of complications if performed in the first trimester. The drug cmifepristone (RU 486) and a prostaglandin can induce it during the first nine weeks of pregnancy. The two chemicals block progesterone receptors in the uterus. The uterine lining cannot be maintained, and neither can pregnancy.

Aborting a fetus in the third trimester is highly controversial unless the mother's life is threatened. This textbook cannot offer the "right" answer to a question about the morality of abortion or some other option. It can only offer a serious explanation of how a new individual develops. Your choice of how to answer the question will be just that—your choice.

25.13 Sexually Transmitted Diseases

Having unprotected sex exposes you to potential infection by any pathogens that your partner unknowingly may have picked up from a previous sexual partner.

CONSEQUENCES OF INFECTION

Each year, pathogens that cause **sexually transmitted diseases**, or STDs, infect about 15 million Americans (Table 25.4). Two-thirds of those infected are under age twenty-five; one-quarter are teenagers. Over 65 million Americans now live with an incurable STD. Treating STDs and secondary complications costs a staggering 8.4 billion dollars in an average year.

The social consequences are enormous. Women are more easily infected than men, and they develop more complications. Example: Pelvic inflammatory disease (PID), a secondary outcome of some bacterial STDs, affects about 1 million women annually. It scars the reproductive tract and can cause infertility, tubal pregnancies, and chronic pain (Figure 25.26a). Some fetuses that acquire STDs before or during birth abort on their own or develop abnormally. Women with a chlamydial infection commonly transmit it to their newborn (Figure 25.26b). Besides causing skin sores, type II *Herpes* virus kills half of the infected fetuses and causes neural defects in one-fourth of the rest.

MAJOR AGENTS OF STDS

HPV Infection by human papillomaviruses (HPV) is the most widespread and fastest growing STD in the United States. At least 20 million are already infected. Of about 100 HPV strains, a few cause *genital warts*. These bumpy growths form on the vagina, cervix, and external genitals, and around the anus (Figure 25.26c). In men, they form on the penis and scrotum. A few strains cause cervical cancer. Sexually active women should have an annual pap smear to check for any cervical changes.

TRICHOMONIASIS Infection by *Trichomonas vaginalis*, a flagellated protozoan, causes a yellowish discharge, soreness, and itching of the vagina. Infected men are usually symptom-free. Untreated infections damage the urinary tract, result in infertility, and invite HIV infection. A single dose of an antiprotozoal drug can quickly cure the infection. To prevent reinfection, both sexual partners must be treated.

CHLAMYDIA *Chlamydial infection* is primarily a young person's disease. Forty percent of those infected are between ages fifteen and nineteen; 1 in 10 sexually active teenage girls is infected. *Chlamydia trachomatis* causes the disease; antibiotics quickly kill it. Most infected women have no symptoms and are not even diagnosed. Between 10 and 40 percent of those who are untreated will develop PID. Abnormal discharges from the penis and painful urination are symptoms in about 50 percent of infected men. Untreated men risk inflammation of the epididymes and infertility.

GENITAL HERPES About 45 million Americans have *genital herpes*, caused by type II *Herpes simplex* virus. Transmission to new hosts requires direct contact with active *Herpes* viruses or with sores that contain them. Mucous membranes of the mouth and genitals are vulnerable. Symptoms are often mild or absent. Painful, small blisters may form on the vulva, cervix, urethra, or anal tissues of infected women. Blisters form on the penis and anal tissues of infected men. Within three weeks, the virus enters latency. Sores crust over and heal, but the virus is still in the body.

The virus is reactivated sporadically, which causes painful sores at or near the original site of infection. Sexual intercourse, menstruation, emotional stress, or other infections trigger flare-ups. One antiviral drug, Acyclovir, decreases healing time and often the pain.

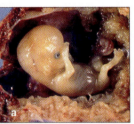
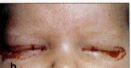
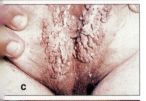
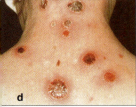

Figure 25.26 A few downsides of unsafe sex. (**a**) STDs that scar the oviduct are one cause of tubal pregnancy. Scarring makes the embryo implant itself in an oviduct, not the uterus. Untreated tubal pregnancies can rupture an oviduct and cause bleeding, infection, and death. (**b**) One sign of a chlamydial infection transferred from a mother to her infant. (**c**) Genital warts. (**d**) Chancres typical of secondary syphilis.

Table 25.4	Estimated New STD Cases Per Year*	
STD	U.S. Cases	Global Cases
HPV infection	5,500,000	30,000,000
Trichomoniasis	5,000,000	174,000,000
Chlamydia	3,000,000	92,000,000
Genital herpes	1,000,000	20,000,000
Gonorrhea	650,000	62,000,000
Syphilis	70,000	12,000,000
AIDS	40,000	5,000,000

*Global data on HPV and genital herpes were last compiled in 1990.

GONORRHEA *Gonorrhea* is caused by the bacterium *Neisseria gonorrhoeae*, which typically crosses mucous membranes of the urethra, cervix, or anal canal during sexual intercourse. An infected female may notice a slight vaginal discharge or burning sensation while urinating. If the pathogen enters her oviducts, it may cause cramps, fever, vomiting, and even scarring that can cause sterility. Less than a week after a male is infected, yellow pus oozes from the penis. Urination becomes more frequent and may also be painful.

Prompt treatment with antibiotics quickly cures this bad disease, yet it still is rampant. Why? Many women ignore early symptoms. Also, people wrongly believe infection confers immunity. A person can get gonorrhea over and over again, probably because there are at least sixteen strains of *N. gonorrhoeae*. Also, use of oral contraceptives invites infection by altering vaginal pH. Populations of resident bacteria decline, so *N. gonorrhoeae* is free to move in.

SYPHILIS The spirochete *Treponema pallidum* causes *syphilis*, a dangerous STD. Having sex with an infected partner puts this bacterium on the surface of genitals or into the cervix, vagina, or oral cavity. *T. pallidum* slips into the body even through tiny cuts. One to eight weeks later, treponemes are twisting about in a flattened, painless chancre (a localized ulcer).

This first chancre is a sign of the primary stage of syphilis. It usually heals, but treponemes multiply in the spinal cord, brain, eyes, bones, joints, and mucous membranes. In the infectious secondary stage, a skin rash develops and more chancres form (Figure 25.26d). In about 25 percent of the cases, immune responses succeed and symptoms subside. Another 25 percent are symptom-free. In the remainder, lesions and scars appear in the skin and liver, bones, and other internal organs. Few treponemes form in this tertiary stage, but a host's immune system is hypersensitive to them. Chronic immune reactions may damage the brain and spinal cord and cause paralysis.

Possibly because the symptoms are so alarming, more people seek early treatment for syphilis than gonorrhea. Later stages require prolonged treatment.

AIDS Infection by HIV (human immunodeficiency virus) leads to *AIDS* (*Acquired Immune Deficiency Syndrome*). Someone can become infected and not know it. Infection marks the start of a titanic, long-term battle that the immune system almost always loses; AIDS is an incurable STD.

At first there may be no outward symptoms. Five to ten years later, a set of chronic disorders develops.

Figure 25.27 NBA legend Magic Johnson, one of the torch bearers of the 2002 Winter Olympics. He was diagnosed as HIV positive in 1991. He contracted the virus through heterosexual sex, and credits his survival to AIDS drugs and informed medical care. He continues to campaign to educate others about AIDS.

The immune system weakens, which opens the door to opportunistic infectious agents. Normally harmless bacteria that are already living in the body are the first to take advantage of the lowered resistance. Then dangerous pathogens take their toll. Eventually they overwhelm the compromised immune system.

Most often, HIV spreads by anal, vaginal, and oral intercourse and intravenous drug use. Virus particles in blood, semen, urine, or vaginal secretions enter a new host through cuts and abrasions in the lining of the penis, vagina, rectum, or oral cavity.

Free or low-cost, confidential testing for exposure to HIV is available through public health facilities. It takes a few weeks to six months or more after first exposure before detectable amounts of antibodies can form in response to the infection. Anyone who tests positive for HIV can spread the virus.

Public education may help slow the spread of HIV (Figure 25.27). Most health care workers advocate safe sex, although there is confusion over what "safe" means. The use of high-quality latex condoms *together with* a spermicide that contains nonoxynol–9 helps prevent viral transmission but carries a slight risk. Open-mouth kissing with an HIV positive individual carries a risk. Caressing is not risky *if* there are no lesions or cuts where HIV-laden body fluids can enter the body. Skin lesions caused by any of the other sexually transmitted diseases are vulnerable points of entry for the virus.

New, costly drug therapies are prolonging some lives. In the late 1990s, the infection rate started to climb again, possibly because of a misperception that AIDS is no longer a threat. But AIDS kills, and the virus keeps on mutating. How long today's drugs can keep a lid on deaths is anybody's guess.

25.14 Formation of the Early Embryo

Nine months or so after the time of fertilization, a woman gives birth. Besides taking longer to develop inside its mother, the new individual will require more intense care for a much longer time compared to other primates.

Pregnancy lasts an average of thirty-eight weeks from the time of fertilization. It takes about one week for a blastocyst to form. All major organs form during the *embryonic* period—the third to the end of the eighth week of pregnancy. When this period ends, the new individual is called a **fetus**. It has distinctly human features. In the *fetal* period, from the start of the ninth week until birth, organs grow and become specialized. We often refer to the first three months of pregnancy as the first trimester. The second trimester extends from the start of the fourth month to the end of the sixth, and the third trimester ends at birth.

CLEAVAGE AND IMPLANTATION

Three to four days after fertilization, cleavage already has started on the zygote. Genes are being expressed; the early cuts require their products. At the eight-cell stage, the cells huddle into a ball. A blastocyst forms by the fifth day. It consists of a trophoblast (a surface layer of cells), a blastocoel filled with their secretions, and an inner cell mass (Figure 25.28*d*). One or two days later, **implantation** is under way. The blastocyst adheres to the uterine lining and invades the mother's tissues, forming connections that will metabolically support the pregnancy. By this time, the inner cell mass has become two flattened layers of cells in the shape of a disk. This embryonic disk will give rise to the embryo proper.

EXTRAEMBRYONIC MEMBRANES

As implantation continues, membranes start to form outside the embryo. First, a fluid-filled *amniotic* cavity opens up between the embryonic disk and part of the blastocyst surface (Figure 25.28*f*). Many cells migrate around the wall of the cavity and form the **amnion**, a membrane that will enclose the embryo. Fluid in the cavity will function as a buoyant cradle in which an embryo can grow, move freely, and be protected from abrupt temperature changes and mechanical impacts.

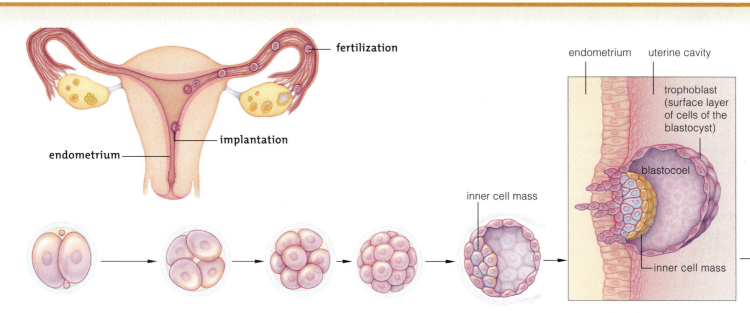

a **DAYS 1–2**. The first cleavage furrow extends between the two polar bodies. Later cuts are angled, so cells become asymmetrically arranged. Until the eight-cell stage forms, they are loosely organized, with space between them.

b **DAY 3**. After the third cleavage, cells abruptly huddle into a compacted ball, which tight junctions among the outer cells stabilize. Gap junctions formed along the interior cells enhance intercellular communication.

c **DAY 4**. By 96 hours there is a ball of sixteen to thirty-two cells shaped like a mulberry. It is a morula (after *morum*, Latin for mulberry). Cells of the surface layer will function in implantation and will give rise to a membrane, the chorion.

d **DAY 5**. A blastocoel (fluid-filled cavity) forms in the morula as a result of surface cell secretions. By the thirty-two-cell stage, differentiation is occurring in an inner cell mass that will give rise to the embryo proper. This embryonic stage is the blastocyst.

e **DAYS 6–7**. Some of the blastocyst's surface cells attach themselves to the endometrium and start to burrow into it. Implantation has started.

actual size

As the amnion forms, other cells migrate around the inner wall of the blastocyst, forming a lining that becomes a **yolk sac**. This extraembryonic membrane speaks of the evolution of land vertebrates. For most animals that produce shelled eggs, the yolk sac holds nutritive yolk. In humans, one portion of the yolk sac becomes a site of blood cell formation. Another will give rise to germ cells, the forerunners of gametes.

Before a blastocyst is fully implanted, spaces open in maternal tissues and become filled with blood that seeps in from ruptured capillaries. In the blastocyst, a new cavity opens up around the amnion and yolk sac. Fingerlike projections form on the lining of the cavity. It becomes the third membrane, called a **chorion**. This membrane will become part of a placenta, a spongy, blood-engorged tissue to be described shortly.

After the blastocyst is implanted, an outpouching of the yolk sac will become the fourth extraembryonic membrane—the **allantois**. The allantois has different roles in different groups. Among reptiles, birds, and some mammals, it serves in respiration and in storing metabolic wastes. In humans, the urinary bladder and blood vessels for a placenta develop from it.

The blastocyst itself stops menstruation. Cells of the blastocyst secrete a hormone called human chorionic gonadotropin, or HCG. This hormone stimulates the corpus luteum to continue secreting progesterone and estrogens. It does so until the placenta takes over the secretion of HCG, about eleven weeks later.

By the start of the third week, HCG can be detected in samplings of the mother's blood or urine. At-home *pregnancy tests* have a treated "dip-stick" that changes color when urine contains this hormone.

Cleavage of the human zygote produces a cluster of cells that develops into the blastocyst.

Six or seven days after fertilization, the blastocyst implants itself in the endometrium.

Projections from the blastocyst's surface invade maternal tissues, and connections start to form that in time will metabolically support the developing embryo.

Some parts of the blastocyst give rise to an amnion, yolk sac, chorion, and allantois. These extraembryonic membranes serve different functions. Together they are vital for the structural and functional development of the embryo.

Read Me First!

and watch the narrated animation on cleavage and implantation

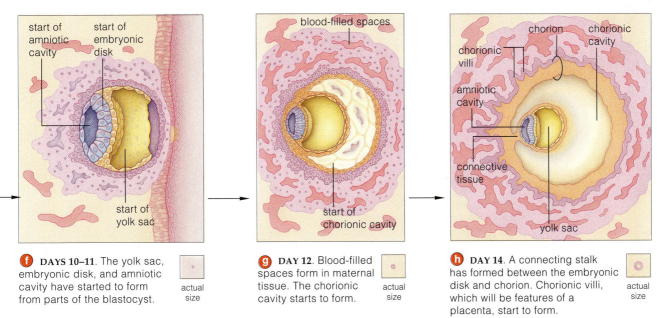

f **DAYS 10–11**. The yolk sac, embryonic disk, and amniotic cavity have started to form from parts of the blastocyst.

actual size

g **DAY 12**. Blood-filled spaces form in maternal tissue. The chorionic cavity starts to form.

actual size

h **DAY 14**. A connecting stalk has formed between the embryonic disk and chorion. Chorionic villi, which will be features of a placenta, start to form.

actual size

Figure 25.28 From fertilization through implantation. A blastocyst forms, and its inner cell mass will give rise to a disk-shaped early embryo. Three extraembryonic membranes (the amnion, chorion, and yolk sac) start forming. A fourth membrane (allantois) forms after the blastocyst is implanted.

25.15 Emergence of the Vertebrate Body Plan

By the time a woman misses a first menstrual period after fertilization, cleavage is over. Gastrulation is under way.

Gastrulation, recall, is the stage when cell divisions, migrations, and rearrangements give rise to primary tissue layers. Until now, the inner cell mass has been developing much as it did in reptilian ancestors of mammals. The cells have rearranged themselves into a flattened, two-layer embryonic disk, as they still do in yolky eggs of living reptiles and birds.

By now, the embryonic disk is surrounded by the amnion and chorion except where a stalk joins it to the chorion wall. The yolk sac lining has formed from one of the two layers. The other layer now starts to become the embryo proper. A depression appears on the disk, and its sides thicken. This is the primitive streak, which lengthens and thickens the next day. Its appearance marks the onset of gastrulation (Figure 25.29a). It defines an embryo's anterior–posterior axis and, in time, its bilateral symmetry.

Endoderm and mesoderm form from cells that are migrating inward along this axis. Pattern formation starts. Embryonic inductions and interactions among classes of master genes map out the basic body plan of all vertebrates. Tissues and organs begin to form in orderly steps, according to predictable patterns.

For example, by the eighteenth day, the embryonic disk has two folds that will merge to form the neural tube (Figure 25.29b). Neural tube defects arise when this tube does not form as it should. *Spina bifida*, the most common problem, results when the spinal cord protrudes out of an abnormally formed covering.

Some mesoderm folds into a tube that develops into a notochord. In vertebrates, the notochord is only a structural model; bony tissue of a vertebral column is assembled on it. Toward the end of the third week, multiple paired segments called **somites** form from part of the mesoderm. Somites are embryonic sources of most bones, of skeletal muscles of the head and trunk, and of the dermis overlying these body parts.

Pharyngeal arches start to form that will contribute to the face, neck, mouth, nose, larynx, and pharynx. Tiny spaces open up in parts of the mesoderm. In time the spaces will interconnect as a coelomic cavity.

> The basic vertebrate body plan emerges early in the development of the new individual.
>
> A primitive streak, neural tube, somites, and pharyngeal arches form during the embryonic period of all vertebrates. Formation of the primitive streak establishes the body's anterior–posterior axis and its bilateral symmetry.

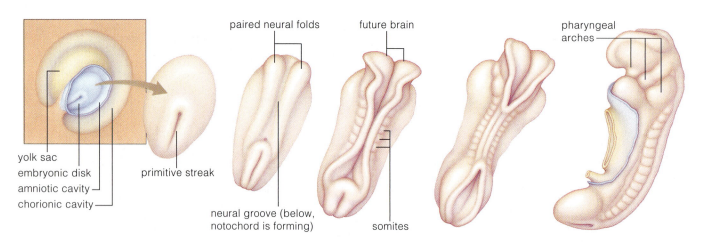

a **DAY 15.** A faint band appears around a depression along the axis of the embryonic disk. This is the primitive streak, and it marks the onset of gastrulation in vertebrate embryos.

b **DAYS 18–23.** Organs start to form through cell divisions, cell migrations, tissue folding, and other events of morphogenesis. Neural folds will merge to form the neural tube. Somites (bumps of mesoderm) appear near the embryo's dorsal surface. They will give rise to most of the skeleton's axial portion, skeletal muscles, and much of the dermis.

c **DAYS 24–25.** By now, some embryonic cells have given rise to pharyngeal arches. These will contribute to the formation of the face, neck, mouth, nasal cavities, larynx, and pharynx.

Figure 25.29 Hallmarks of the embryonic period of humans and other vertebrates. A primitive streak and then a notochord form. Neural folds, somites, and pharyngeal arches form later. (**a**,**b**) Dorsal views of the embryo's back. (**c**) Side view.

25.16 Why Is the Placenta So Important?

Even before the embryonic period starts, the uterus has been interacting with extraembryonic membranes in ways that will sustain the embryo's rapid growth.

By the third week, tiny fingerlike projections from the chorion have grown into the maternal blood that has pooled in endometrial spaces. These projections are the chorionic villi. They enhance the rate of exchange of substances between the mother and the embryo. The villi are functional components of the placenta.

The **placenta** is a blood-engorged organ composed of the uterine lining and extraembryonic membranes. At full term, it will make up about one-fourth of the inner surface of the uterus (Figure 25.30).

A placenta is the body's way of sustaining the new individual while allowing its blood vessels to develop separately from the mother's blood vessels. Oxygen and vital nutrients diffuse out of the maternal blood vessels, across the placenta's blood-filled spaces, then into embryonic blood vessels. The vessels converge in an umbilical cord, the lifeline between the placenta and the new individual. Carbon dioxide and other wastes diffuse in the other direction. The mother's lungs and kidneys dispose of the wastes.

After the third month of pregnancy, the placenta starts secreting progesterone and estrogens. The action of these sex hormones maintains the uterine lining.

> The placenta is a blood-engorged organ of endometrial and extraembryonic membranes. It allows the new individual to take up oxygen and nutrients from the mother and give up wastes to her. It does so while allowing embryonic blood vessels to develop separately from the mother's.

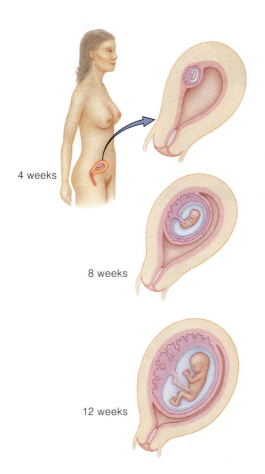

4 weeks

8 weeks

12 weeks

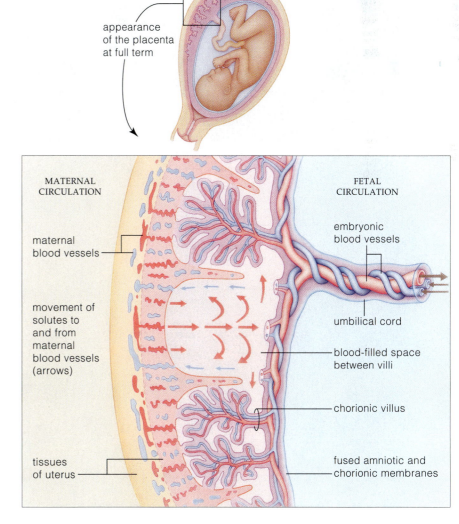

appearance of the placenta at full term

MATERNAL CIRCULATION

FETAL CIRCULATION

maternal blood vessels

movement of solutes to and from maternal blood vessels (arrows)

embryonic blood vessels

umbilical cord

blood-filled space between villi

chorionic villus

tissues of uterus

fused amniotic and chorionic membranes

Figure 25.30 Relationship between fetal and maternal blood circulation in a full-term placenta. Blood vessels extend from the fetus, through the umbilical cord, and into chorionic villi. Maternal blood spurts into spaces between villi, but the two bloodstreams do not intermingle. Oxygen, carbon dioxide, and other small solutes diffuse across the placental membrane surface.

25.17 Emergence of Distinctly Human Features

Early on, a human embryo—with its gill arches and long tail—has a distinctly vertebrate appearance. The tail soon disappears, and by the beginning of the fetal period the developing individual has distinctly human features.

When the fourth week ends, the embryo is 500 times its starting size. Weeks five and six are the boundary between the embryonic and fetal periods. Now growth slows as details of organs fill in. Limbs form; toes and fingers are sculpted from paddles. The umbilical cord forms, and so does an intricate circulatory system. Growth of the all-important head now surpasses that of all other regions (Figure 25.31). Reproductive organs start forming, as explained in Section 11.1. At the end of the eighth week, the individual is no longer just "a vertebrate." Its features define it as a human fetus.

In the second trimester, as developing nerves and muscles connect up, reflexive movements begin. Legs

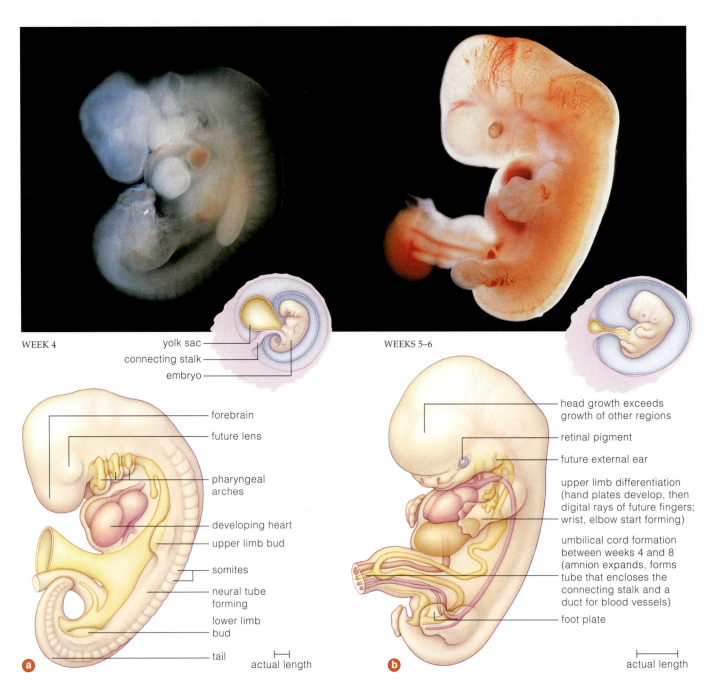

WEEK 4

yolk sac
connecting stalk
embryo

WEEKS 5–6

forebrain

future lens

pharyngeal arches

developing heart

upper limb bud

somites

neural tube forming

lower limb bud

tail

actual length

a

head growth exceeds growth of other regions

retinal pigment

future external ear

upper limb differentiation (hand plates develop, then digital rays of future fingers; wrist, elbow start forming)

umbilical cord formation between weeks 4 and 8 (amnion expands, forms tube that encloses the connecting stalk and a duct for blood vessels)

foot plate

actual length

b

Figure 25.31 (a) Human embryo at successive stages of development.

kick, arms wave about, and fingers grasp. The fetus frowns, squints, puckers its lips, sucks, and hiccups. When the fetus is five months old, its heartbeat can be heard clearly through a stethoscope positioned on the mother's abdomen. The mother can sense movements of fetal arms and legs.

By now the skin of the fetus is covered with soft, fuzzy hair, the lanugo. A thick, cheeselike coating protects the wrinkled, reddish skin from abrasion. In the sixth month, delicate eyelids and eyelashes form. The eyes open during the seventh month, the start of the final trimester. By this time, too, all portions of the brain have formed and have begun to function.

In the fetal period, the primary tissues that formed in the early embryo become sculpted in ways that transform this vertebrate embryo into one with distinctly human features.

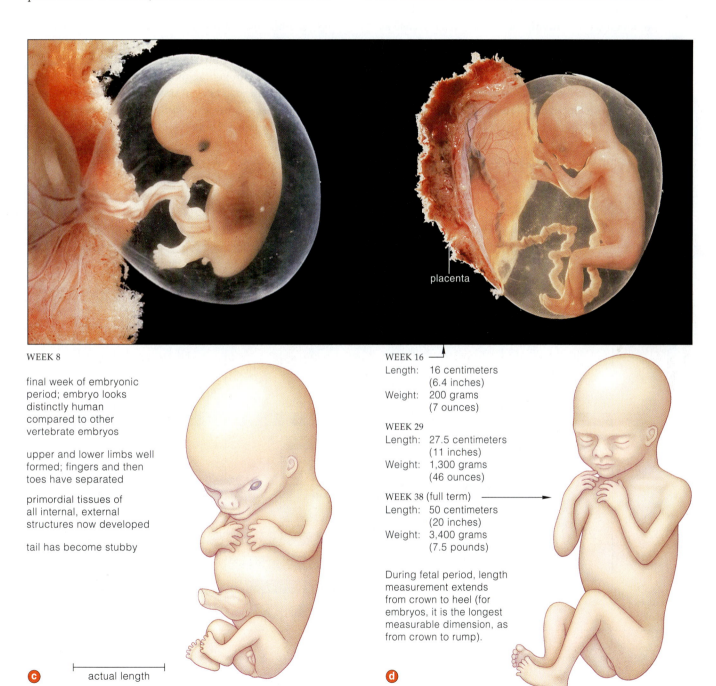

placenta

WEEK 8

final week of embryonic period; embryo looks distinctly human compared to other vertebrate embryos

upper and lower limbs well formed; fingers and then toes have separated

primordial tissues of all internal, external structures now developed

tail has become stubby

c actual length

WEEK 16
Length: 16 centimeters
         (6.4 inches)
Weight: 200 grams
         (7 ounces)

WEEK 29
Length: 27.5 centimeters
         (11 inches)
Weight: 1,300 grams
         (46 ounces)

WEEK 38 (full term)
Length: 50 centimeters
         (20 inches)
Weight: 3,400 grams
         (7.5 pounds)

During fetal period, length measurement extends from crown to heel (for embryos, it is the longest measurable dimension, as from crown to rump).

d

25.18 Mother as Provider, Protector, Potential Threat

Each pregnant woman is committing much of her body's resources to the growth and development of a brand-new individual. From fertilization until birth, her future child is at the mercy of her diet, health habits, and life-style.

NUTRITIONAL CONSIDERATIONS

A human embryo gets all the proteins, carbohydrates, and lipids needed for growth and development when the mother-to-be eats a well-balanced diet. However, the demands for vitamins and minerals increase as the placenta preferentially absorbs them for the fetus from her blood. Medically supervised increases in B-complex vitamins before conception and in early pregnancy reduce the risk of severe neural tube defects in the embryo. Folate (folic acid) is especially important in this regard.

Nutritional deficiencies adversely affect many developing organs. For example, the brain undergoes its greatest expansion in the weeks just before and after birth. Poor nutrition during this span may impair intelligence and other brain functions later in life.

A pregnant woman must eat enough to gain 20 to 25 pounds, on average. If she does not, her newborn may be seriously underweight, at risk of postdelivery complications and, in time, impaired brain function.

INFECTIOUS DISEASES

IgG antibodies in a pregnant woman's blood cross the placenta. They protect her developing child from all but the most serious bacterial infections. Some viral diseases can be dangerous in the first six weeks after fertilization, a crucial time of organ formation.

Suppose she contracts *rubella* (German measles) in this critical period. There is a 50 percent chance that some organs will not form properly. For instance, if she is infected as embryonic ears are forming, her

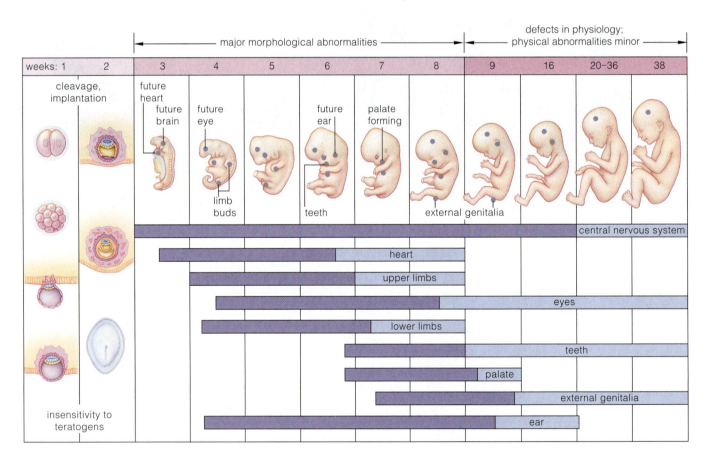

Figure 25.32 Teratogen sensitivity. Teratogens are drugs, infectious agents, and environmental factors that invite embryonic or fetal deformities, usually after organs form. They adversely affect growth, tissue remodeling, and tissue resorption. Dark blue shows the highly sensitive period; light blue shows periods of less severe sensitivity to teratogens. For example, the upper limbs are most sensitive to damage during weeks 4 through 6, and somewhat sensitive during weeks 7 and 8.

newborn may be deaf (Figure 25.32). If she is infected at any time from the fourth month of pregnancy onward, this particular disease will have no notable effect. A woman may avoid the risk entirely by getting vaccinated against the virus before pregnancy.

ALCOHOL, TOBACCO, AND OTHER DRUGS

ALCOHOL   Alcohol passes freely across membranes of cells. It passes freely across the placenta. When a pregnant woman drinks, her developing embryo or fetus quickly absorbs alcohol. Excessive intake invites *fetal alcohol syndrome* (FAS). Symptoms of this disorder include reduced brain size, mental impairment, facial deformities, a small head, slow growth, possible heart problems, and poor coordination (Figure 25.33). In some parts of the United States, the incidence of FAS is as high as 1.5 cases per 1,000 live births. The symptoms are permanent; children affected by FAS never do catch up, physically or mentally.

Even moderate drinking during pregnancy may have negative effects. Increasingly, doctors are urging total abstinence from alcohol during pregnancy.

TOBACCO   Smoking increases the risk of miscarriage and adversely affects fetal growth and development. Carbon monoxide competes with oxygen for binding sites on hemoglobin, so the embryo or fetus cannot get enough oxygen. Nicotine levels in the amniotic fluid can actually be higher than those in the mother's blood. Toxic compounds accumulate even inside the fetuses of pregnant nonsmokers who are exposed to *secondhand smoke* at home or at work.

Smoking any tobacco regularly during pregnancy results in underweight newborns. This happens even when the woman's weight, nutrition, and other key variables match those of pregnant nonsmokers.

Tobacco smoke adversely affects nutrition. In one study, pregnant women who smoked lowered the blood concentration of vitamin C for themselves *and* for their fetuses, even when intake of that vitamin matched the intake of a control group.

The effects may be long-term. Researchers tracked a group of children born in the same week. Children of smokers were smaller, had twice as many heart defects, and died of more postdelivery complications. By age seven, they were nearly half a year behind children of nonsmokers in their "reading age."

COCAINE   A pregnant woman who uses any form of cocaine increases her likelihood of miscarriage and premature delivery. The drug disrupts development

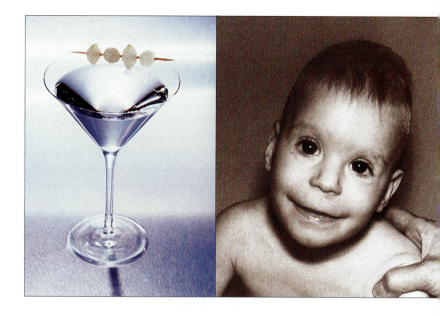

Figure 25.33   An infant with fetal alcohol syndrome—FAS. The obvious symptoms are low and prominently positioned ears, improperly formed cheekbones, and an abnormally wide, smooth upper lip. Growth-related complications and abnormalities of the nervous system can be expected.

of the nervous system. A child of a cocaine addict is likely to be abnormally small and irritable during early life. Some, but not all, studies suggest that prenatal exposure to cocaine has long-term negative effects on intelligence and behavior.

PRESCRIPTION DRUGS   Pregnant women should not take any drugs except under medical supervision. To underscore this point, the tranquilizer *thalidomide* was routinely prescribed in Europe. Infants of some of the women who used it during the first trimester had severely deformed arms and legs, or none at all. This drug has been withdrawn from the market. But other tranquilizers, sedatives, and barbiturates are still being prescribed, and they may cause some less severe damage. Some *anti-acne drugs* increase the risk of facial and cranial deformities.

Depression during pregnancy is not uncommon and can itself have negative health effects on a fetus. A mother who doesn't feel like eating or otherwise taking care of herself can impair fetal development. To avoid these problems, some pregnant women are treated with antidepressants. So far, these drugs have not been found to increase the rate of miscarriage, slow fetal growth, or increase the incidence of birth defects. However, infants of women who used these drugs right up to the time of delivery may show some withdrawal symptoms.

25.19 From Birth Onward

Human growth and development do not end with birth. Both processes continue as the newborn embarks on a course of extended dependency and learning. As with all mammals, its early survival depends on nutritious milk, typically provided by the mother.

A fetus born too prematurely (before 22 weeks) will not survive. The risk also is great for births before 28 weeks, mainly because the lungs have not developed enough. The risk starts to drop after this. By 36 weeks, the survival rate is 95 percent. A fetus born between 36 and 38 weeks still has some trouble breathing and maintaining a core temperature even with the best of medical care. The most favorable birthing time, on average, is 38 weeks after fertilization.

GIVING BIRTH

The birth process is known as **labor**. Typically, the amnion ruptures just before birth, so amniotic fluid drains from the vagina. The cervical canal dilates, the fetus moves out of the uterus, then out of the vagina and into the outside world (Figure 25.34).

Mild uterine contractions start in the last trimester. **Relaxin** is a hormone secreted by the ovaries and the placenta. It softens cervical connective tissues and loosens ligaments between the pelvic bones. The fetus "drops," or shifts down; its head usually touches the cervix. Rhythmic contractions at the onset of labor get stronger and closer together during the next two to eighteen hours. Normally they cause expulsion of the fetus within an hour after the cervix is fully dilated.

Strong contractions also detach the placenta from the uterus and expel it, as the afterbirth. Contractions help stop the bleeding at the site where the placenta attached to the wall of the uterus. They constrict the blood vessels at the ruptured attachment site. The umbilical cord is cut and tied off, and a few days after shriveling up, the cord's stump has become the navel.

Corticotropin-releasing hormone (CRH) affects the timing of labor, and it may contribute to *post-partum depression*. The hypothalamus makes this hormone in all humans, but the placenta also makes it during a pregnancy, which can increase its level in the blood by threefold. CRH stimulates the adrenal cortex to secrete **cortisol**. This hormone may help a mother cope with many strains that pregnancy and labor place on her.

During pregnancy, a high blood level of cortisol can suppress CRH production by the hypothalamus. After birth, when the placenta has been expelled, the CRH level briefly plummets. In many women, this decline triggers a short-term depression that typically continues until the hypothalamus resumes its normal production of CRH.

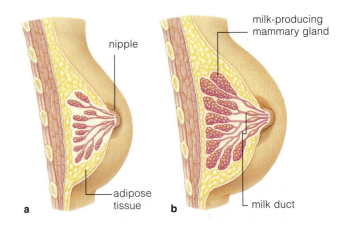

Figure 25.35 (**a**) Breast of a woman who is not pregnant. (**b**) Breast of a lactating woman.

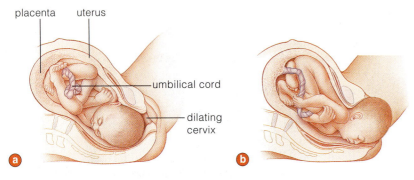
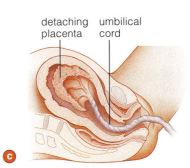

Figure 25.34 Expulsion of (**a**,**b**) a human fetus and (**c**) afterbirth during labor. The afterbirth consists of the placenta, tissue fluid, and blood.

NOURISHING THE NEWBORN

Once the lifeline to the mother is severed, a newborn enters the extended time of dependency and learning that is typical of all primates. Early survival requires an ongoing supply of milk or a nutritional equivalent. **Lactation**, or milk production, occurs in mammary glands inside a mother's breasts (Figure 25.35). Before pregnancy, breast tissue is largely adipose tissue and a system of undeveloped ducts. Their size depends on how much fat they hold, not on their milk-producing ability. During pregnancy, estrogens and progesterone stimulate the development of a glandular system for milk production.

For the first few days after birth, mammary glands produce a fluid rich in proteins and lactose. **Prolactin**, a hormone that calls for synthesis of enzymes used in milk production, is secreted by the anterior lobe of the mother's pituitary gland. As a newborn suckles, the pituitary releases **oxytocin**. This hormone triggers contractions that force fluid into milk ducts. It also causes contractions of uterine muscles that help make the uterus shrink to its pre-pregnancy size.

In addition to nutrients, human breast milk has immunoglobulins that enhance resistance to infection. Other components of milk stimulate the growth of symbiotic bacteria in the infant's gut. Alcohol, drugs, mercury, and other toxins in a mother's body can also be secreted in milk. As during pregnancy, a nursing mother should tailor her life-style and diet.

POSTNATAL DEVELOPMENT

As is the case for many species, humans change in size and proportion until reaching sexual maturity. Figure 25.36 shows a few of the proportional changes that occur as the life cycle unfolds. Table 25.5 defines the prenatal ("before birth") stages and the postnatal ("after birth") stages of a human life.

Postnatal growth is most rapid between the years thirteen and nineteen. Sex hormone secretions step up and bring about the development of secondary sexual traits as well as sexual maturity. Not until adulthood are the bones fully mature.

Hormones control the timing of labor and the onset of lactation, which nourishes the developing newborn.

Human growth and development do not end with birth. Sexual maturity occurs after puberty, when the production of sex hormones is stepped up dramatically.

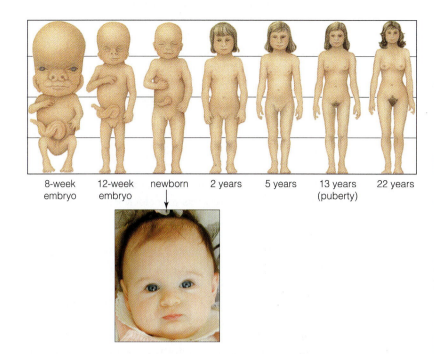

| 8-week embryo | 12-week embryo | newborn | 2 years | 5 years | 13 years (puberty) | 22 years |

Figure 25.36 Observable, proportional changes in the human body during prenatal and postnatal periods of development. Changes in overall physical appearance are slow but noticeable until the teenage years. For example, compared to an embryo, the legs of teenagers are longer and the trunk shorter, so the head is proportionally smaller. Correlate these drawings with the stages in Table 25.5.

Table 25.5	Stages of Human Development
Prenatal period	
Zygote	Single cell resulting from fusion of sperm nucleus and egg nucleus at fertilization.
Morula	Solid ball of cells produced by cleavages.
Blastocyst (blastula)	Ball of cells with surface layer, fluid-filled cavity, and inner cell mass.
Embryo	All developmental stages from two weeks after fertilization until end of eighth week.
Fetus	All developmental stages from ninth week to birth (about 38 weeks after fertilization).
Postnatal period	
Newborn	Individual during the first two weeks after birth.
Infant	Individual from two weeks to about fifteen months after birth.
Child	Individual from infancy to about ten or twelve years.
Pubescent	Individual at puberty; secondary sexual traits develop; girls between 10 and 15 years, boys between 12 and 16 years.
Adolescent	Individual from puberty until about 3 or 4 years later; physical, mental, emotional maturation.
Adult	Early adulthood (between 18 and 25 years); bone formation and growth finished. Changes proceed slowly after this.
Old age	Aging processes result in expected tissue deterioration.

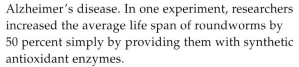

25.20 Why Do We Age and Die?

*As the years pass, all multicelled species undergo **aging**; tissues become harder to maintain and repair. Each species has a maximum life span—122 years for humans, 20 years for dogs, 12 weeks for butterflies, 35 days for fruit flies, and so on. The verifiably oldest human lived 122 years. We can expect that genes influence aging processes.*

PROGRAMMED LIFE SPAN HYPOTHESIS

Do biological clocks influence aging? If so, an animal body might be analogous to a clock shop, with each type of cell, tissue, and organ ticking away at its own genetically set pace. Many years ago, Paul Moorhead and Leonard Hayflick tested this hypothesis. They cultured human embryonic cells—which divided about fifty times before dying out.

Hayflick also took cultured cells that were part of the way through the series of divisions and froze them for a few years. After he thawed the cells and placed them in a culture medium, they completed an *in vitro* cycle of fifty doublings and died on schedule.

No cell in a human body divides more than eighty or ninety times. You may well wonder: If an internal clock ticks off their life span, then how can *cancer* cells go on dividing? The answer provides us with a clue to why *normal* cells can't beat the clock.

Cells, remember, duplicate their chromosomes before dividing. Capping the chromosome ends are **telomeres** made of DNA and proteins. Telomeres keep the chromosome ends from unraveling. A small piece of each one is lost with each nuclear division. When only a nub is left, cells stop dividing and die.

Cancer cells and germ cells are exceptions; both make telomerase, an enzyme that makes telomeres lengthen. Expose cultured cells to telomerase, and they go on dividing well beyond normal life span.

CUMULATIVE ASSAULTS HYPOTHESIS

Another hypothesis: Over the long term, aging is the outcome of cumulative damage at the molecular and cellular levels. Environmental assaults as well as spontaneous mistakes by DNA repair mechanisms are at work here.

For example, the rogue molecular fragments called free radicals attack all biological molecules, including DNA. This includes the DNA of mitochondria, the power plants of eukaryotic cells. Structural changes in DNA compromise the synthesis of enzymes and other proteins necessary for normal life processes.

Free radicals are implicated in many age-related problems, including cataracts, atherosclerosis, and Alzheimer's disease. In one experiment, researchers increased the average life span of roundworms by 50 percent simply by providing them with synthetic antioxidant enzymes.

Other studies suggest that extending the life span may come at a cost. Researchers were able to double the roundworm life span by knocking out a single gene. However, the mutant worms were sterile.

DNA replication and repair problems also have been implicated in aging. *Werner's syndrome*, an aging disorder, has been correlated with a mutation in the gene that specifies a helicase. Such enzymes unwind nucleotide strands. The mutated helicase probably does not compromise DNA replication, because affected people do not die right away. They start aging fast in their thirties and die before age fifty.

However, the nonmutated gene may be crucial for repairs. People with Werner's syndrome accumulate gene mutations at high rates. Sooner or later, the damage interferes with cell division. Like skin cells of the elderly, skin cells of Werner's patients just do not divide many times.

It may be that both hypotheses have merit. Aging may be an outcome of many interconnected processes in which genes, hormones, environmental assaults, and a decline in DNA repair mechanisms come into play. Consider how living cells of all tissues depend upon exchanges of materials with extracellular fluid. Also consider how collagen is a structural component of many connective tissues. If something shuts down or mutates collagen-encoding genes, then missing or altered gene products may disrupt flow of oxygen, nutrients, hormones, and so forth to and from living cells through every connective tissue. Repercussions from such a mutation would ripple through the body.

Similarly, if mutations cause altered self markers on the body's cells, do T cells of the immune system perceive them as foreign and attack? If autoimmune responses were to become more frequent over time, they would promote greater vulnerability to disease and stress associated with old age.

In evolutionary terms, reproductive success means living long enough to produce and raise offspring. Humans can reach sexual maturity in fifteen years *and* help their children reach adulthood. We don't *need* to live longer than we do. But we among all animals have the capacity to think about it, and most of us have decided we like life better than the alternative. Eventually, however, even the most stubborn among us must confront their own mortality. And perhaps, given time, we may all learn to accept the inevitable with wisdom and grace.

Summary

Section 25.1 Asexual reproduction yields offspring that are all genetically identical to their parent. Sexual reproduction requires energetically costly structures, controls, and behavior but creates variation that may assure reproductive success of at least some offspring.

Section 25.2 Most animal life cycles have six stages of embryonic development. Gametes form, fertilization occurs, cleavage results in a blastula, gastrulation results in an early embryo with two or three primary tissue layers (ectoderm, mesoderm, and endoderm), organs form, and tissues and organs become specialized.

Section 25.3 Maternal messages are localized in the egg cytoplasm. After fertilization, cleavage distributes different messages to different daughter cells. Cleavage patterns differ among animal groups.

Section 25.4 Cytoplasmic localization followed by short-range and long-range signals cause embryonic cells to selectively activate genes and differentiate: the cells become specialized in structure, composition, and behavior. Organs form by morphogenesis, as when cells migrate, tissues expand and fold over, and patches of cells undergo programmed cell death.

Section 25.5 Pattern formation is the sculpting of embryonic cells in a certain order, in expected positions in the embryo. Similar master genes govern pattern formation in all animals.

Section 25.6 Testes, the male primary reproductive organs in humans, produce sperm and testosterone.

Section 25.7 Feedback loops from the testes to the hypothalamus and pituitary govern testosterone, LH, and FSH secretion, all of which affect sperm formation.

Section 25.8 Ovaries, the primary reproductive organs in human females, produce oocytes and secrete the sex hormones estrogen and progesterone.

Sections 25.9, 25.10 Feedback loops from the ovaries to the hypothalamus and the pituitary control hormonal secretions of the menstrual cycle.

FSH causes an ovarian follicle to mature. The follicle secretes estrogens, which cause the endometrium to thicken in preparation for pregnancy. A midcycle surge of LH triggers ovulation.

A corpus luteum forms from follicle remnants and secretes progesterone and estrogens, which maintain the endometrium. If fertilization does not occur, it degenerates, the endometrium breaks down and is sloughed off with blood, and the cycle starts again.

Section 25.11 Fertilization occurs after a secondary oocyte meets up with a sperm, usually in the oviduct. The oocyte completes meiosis II and becomes an ovum (mature egg). Its nucleus fuses with the sperm nucleus.

Section 25.12 Fertility is controllable by behavior, surgery, physical or chemical barriers, or manipulations of female sex hormones.

Section 25.13 Sexually transmitted diseases (STDs) are caused by protozoan, bacterial, and viral pathogens and are spread by unsafe sex and other behaviors.

Section 25.14 After fertilization, a blastocyst forms by cleavage and implants itself in the endometrium. Extraembryonic membranes—the amnion, chorion, yolk sac, and allantois—form. They assist in the structural and functional development of the embryo.

Section 25.15 Gastrulation lays down the vertebrate body axis. A neural disk gives rise to the neural tube, the forerunner of the brain and spinal cord. Somites are embryonic precursors of muscles and the skeleton.

Section 25.16 A placenta made of maternal tissues and extraembryonic membranes allows exchanges between the mother and embryo while keeping their blood vessels separated.

Section 25.17 The embryo has distinctly human features by the end of the eighth week of pregnancy.

Section 25.18 A mother's health, nutrition, and life-style affect fetal growth and development.

Section 25.19 During labor, uterine contractions expel the fetus and afterbirth. Hormones trigger labor, the maturation of mammary glands, and milk flow. Development and growth continue until adulthood.

Section 25.20 Aging may be partly programmed in the genes and partly an outcome of cumulative assaults on DNA and other biological molecules.

Self-Quiz
Answers in Appendix III

1. Sexual reproduction among animals is _____ .
   a. biologically costly   c. evolutionarily beneficial
   b. diverse in its details   d. all of the above

2. A cell formed during cleavage is a _____ .
   a. blastula   b. morula   c. blastomere   d. gastrula

3. _____ produces three primary tissue layers.
   a. Gametogenesis   c. Gastrulation
   b. Implantation   d. Pattern formation

4. Homeotic genes map out the _____ .
   a. cleavage planes   c. basic body plan
   b. primary tissue layers   d. all of the above

5. Meiotic divisions of _____ produce mature sperm.
   a. Leydig cells   c. both a and b
   b. Sertoli cells   d. neither a nor b

6. During a menstrual cycle, a midcycle surge of _____ triggers ovulation.
   a. estrogens   b. progesterone   c. LH   d. FSH

7. The corpus luteum secretes _____ .
   a. LH   b. FSH   c. progesterone   d. prolactin

8. A _____ implants in the lining of the uterus.
   a. zygote   b. gastrula   c. blastocyst   d. fetus

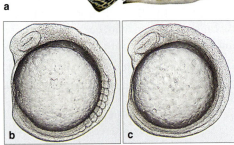

a

Figure 25.37 (**a**) Adult zebrafish. (**b**) Normal embryo and (**c**) embryo bearing a mutation that interferes with formation of somites, the bumps of mesoderm that give rise to bones and most skeletal muscle.

b   c

9. Which of the following puts human developmental stages in the correct order?
   a. zygote, blastocyst, embryo, fetus
   b. zygote, embryo, blastocyst, fetus
   c. zygote, embryo, fetus, blastocyst
   d. blastocyst, zygote, embryo, fetus

10. Which of the following is (are) caused by bacteria?
    a. chlamydia   d. trichomoniasis
    b. gonorrhea   e. a and b
    c. genital warts   f. c and d

11. Match each term with the most suitable description.
    _____ gamete formation   a. mitotic cell divisions
    _____ fertilization   b. cellular rearrangements
    _____ cleavage   form primary tissues
    _____ gastrulation   c. eggs and sperm form
    _____ cell differentiation d. sperm nucleus and egg
    _____ morphogenesis   nucleus fuse
                              e. orderly changes in body
                                 size and shape
                              f. in most species, genes are
                                 now selectively activated

12. Match each term with the most suitable description.
    _____ testis   a. maternal and fetal tissues
    _____ cervix   b. stores mature sperm
    _____ placenta   c. produces testosterone
    _____ vagina   d. produces estrogen and
    _____ ovary   progesterone
    _____ oviduct   e. usual site of fertilization
    _____ epididymis   f. lining of uterus
    _____ endometrium   g. birth canal
                        h. entrance to uterus

Critical Thinking

1. The zebrafish (*Danio rerio*) is special to developmental biologists. This small freshwater fish is easily maintained in tanks. A female produces hundreds of eggs, which can develop and hatch in three days. The transparent embryos let researchers directly observe developmental events (Figure 25.37). Cells can be injected with dye to see how they change position, or they can be killed or injected with genes and the effects observed.

Evolutionarily speaking, *D. rerio* is only remotely related to humans. Explain why researchers might expect the early development of this fish to yield useful information about human development.

2. Fraternal twins, which are nonidentical genetically, arise when two oocytes mature, are released, and are fertilized at the same time. Such twins run in families. The incidence varies among ethnic groups and is highest among blacks and lowest in Asians. Variation may be an outcome of differences in gene products that affect the FSH level in blood. Explain how a high FSH level would increase the likelihood of fraternal twins.

3. By UNICEF estimates, each year 110,000 people are born with abnormalities as a result of rubella infections. Major symptoms of *congenital rubella syndrome,* or CRS, are deafness, blindness, mental impairment, and heart problems. A nonvaccinated woman who is infected in the first trimester of pregnancy is at risk, but not later. Review the developmental events that unfold during pregnancy and explain why this is the case.

VI Principles of Ecology

Two organisms—a fox in the shadows cast by a snow-dusted spruce tree. What are the consequences of their interactions with each other, with other kinds of organisms, and with their environment? By the end of this last unit, you might find worlds within worlds in such photographs.

IMPACTS, ISSUES   *The Human Touch*

In 1722, on Easter morning, an explorer landed on a small, remote volcanic island in the Pacific and found a few hundred skittish, hungry Polynesians living in caves. He saw grasses and scorched shrubs, but no trees. He came across hundreds of massive stone statues near the coast and unfinished and abandoned ones in inland quarries. Some weighed fifty tons (Figure 26.1). What did they represent?

When James Cook visited the island two years later, he counted only four canoes. Nearly all of the statues had been tipped over, often onto face-shattering spikes.

Later, researchers solved the mystery of the statues. Easter Island, as it came to be called, is only 165 square kilometers (64 square miles) in size. Voyagers from the Marquesas discovered this eastern outpost of Polynesia around 1,600 years ago. The place was a paradise. Its fertile soil supported dense forests and lush grasses. New arrivals built canoes from long, straight palms, strengthened with rope made of fibers from hauhau trees. They used wood as fuel to cook fish; they cleared forests to plant crops, and they had many children. By 1400, between 10,000 and 15,000 were living on the island. Crop yields had declined. Erosion and harvesting had depleted the nutrients in soil. Fish had vanished from nearshore waters. All native birds had been eaten, and people were starting to raise rats as food.

Survival was at stake and those in power appealed to the gods. They directed people to carve stone images of unprecedented size and use a system of greased logs to move them over miles of rough terrain to the coast.

Figure 26.1 Enormous statues carved from volcanic rock line the shore of Easter Island. Ancestors of the people who built them were skilled navigators who must have traveled more than a thousand miles across the open ocean to reach this small, remote island.

the big picture

Describing a Population   Each population has a characteristic size, density, and age structure. It shows patterns of distribution and growth. The defining characteristics change with time and can be monitored through field studies.

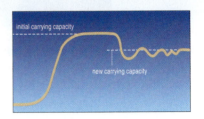

How Populations Grow   When resources are plentiful and unrestricted, population size increases exponentially. Populations eventually return to the carrying capacity, the maximum number of individuals that can be sustained indefinitely in a given environment.

By about 1550, no one fished offshore. No one could build new canoes because all the palms had been cut down and all the hauhau trees had been burned as firewood. Easter Islanders turned to their last source of protein. They started to hunt and eat one another.

Central authority crumbled and gang wars raged. Those on the rampage burned the remaining grasses to destroy hideouts. The dwindling population retreated to caves and launched raids against perceived enemies. Winners ate losers and tipped over the statues. There was nowhere else to go. What could they have been thinking when they chopped down the last palm?

And so, from archaeological and historical records, we know a large Easter Island society flourished amid abundant resources, then abruptly fell apart. Its loss is a footnote in life's greater evolutionary story, compared to the millions of species that disappeared in the past.

Yet there is a lesson here. Ultimately, there are principles that govern the growth and sustainability of all populations. These principles are the bedrock of **ecology**, the systematic study of how organisms interact with the physical and chemical environment.

Interactions start within and between populations and extend through communities, ecosystems, and the biosphere. They are the focus of this last unit of the book. In this chapter we first consider the general characteristics of populations. Later, we will apply the basic principles of population growth to the past, present, and future of the human species.

 ## How Would You Vote?

In the United States, immigration—legal and illegal—is a significant part of the rapid increase in population size. Undocumented immigrants are still entering the country in large numbers. Should the 7 million who have made a home here for some time be granted legal status? See the Media Menu for details, then vote online.

Life Histories
All populations or species show a life history pattern, a set of adaptations that influence each individual's survival, fertility, and time of reproduction. Life history patterns may evolve by way of natural selection.

Human Populations
Through agricultural, medical, and technological advances, the human population bypassed the carrying capacity and has been postponing natural limits on population growth. Its numbers have soared, but the operative word is *postpone*.

26.1 Characteristics of Populations

*By this point in the book, you know that a population is a group of individuals of the same species. Ecological interactions begin with characteristics of populations. We call these vital statistics **demographics**.*

Each population has a gene pool and an evolutionary history, as explained in Chapters 16 and 17. It also has a characteristic size, density, distribution, and number of individuals in its various age categories.

Population size is the number of individuals that potentially or actually contribute to the gene pool. The **age structure** is the number of individuals in each of several age categories. For instance, individuals are often grouped into *pre-reproductive, reproductive,* and *post-reproductive* ages. Those in the first category have the capacity to produce offspring when they mature. Together with the individuals in the second category, they make up the population's **reproductive base**.

Population density is the number of individuals in some specified area or volume of a habitat. A *habitat,* remember, is the type of place where a species lives. We characterize a habitat by its physical and chemical features and its particular array of species. **Population distribution** is the pattern in which the individuals are dispersed in a specified area.

Crude density is a measured number of individuals in some specified area. It does not reveal how much of a habitat is actually being used as living space. Even areas that appear uniform, such as a long, sandy shoreline, are more like tapestries of light, moisture, temperature, mineral composition, and many other variables. One portion of a habitat might be far more suitable for a population than other portions all of the time or some of the time, as in changing seasons.

Different species occupying the same area typically compete for energy, nutrients, living space, and other resources. Such *interspecific* interactions help shape a population's density and dispersion through a habitat.

Theoretically, a population has a clumped, nearly uniform, or random distribution pattern (Figure 26.2). Clumping is the most common, for several reasons. First, each species is responsive to certain conditions and resources that often are patchy through a habitat. Think of how some animals gather near a water hole or seeds sprout only where there is adequate soil and water. Second, animals may form social groups that offer protection and mating opportunities. A school of fish is like this. Third, the young of many plants and a lot of animals cannot disperse far from the parents.

With nearly uniform distribution, individuals are more evenly spaced than expected based on chance alone. Uniform distribution is relatively rare in nature. We do come across it when competition for resources or territory is fierce, as in a nesting colony of seabirds. Figures 26.2 and 30.15 have splendid examples.

We observe random dispersion only when habitat conditions are nearly uniform, resource availability is fairly steady, and individuals of a population or pairs of them neither attract nor avoid one another. Wolf spiders are ground dwellers that hunt at night, not far from their randomly located burrow (Figure 26.2).

Each population has characteristic demographics: size, density, distribution pattern, and age structure.

Environmental conditions and species interactions shape these characteristics, which may change over time.

clumped

nearly uniform

random

Figure 26.2 Three patterns of population distribution: clumped, as in squirrelfish schools; more or less uniform, as in a royal penguin nesting colony; and random, as when wolf spiders prowl on a forest floor.

26.2 Elusive Heads to Count

Ecologists go into the field to test theories about species interactions and population dynamics, and to monitor the health of threatened or endangered populations.

Once upon a time someone counted a million or so deer living in the United States. Now, a hundred years later, 18 million are living in forests, grasslands, golf courses, and suburban gardens. How would you go about counting the ones living near you?

A full count would measure absolute population density. Census takers supposedly make such a count of human populations every ten years, although not everyone answers the door. Ecologists make counts of large species in small areas, such as birds in a forest, northern fur seals at their breeding grounds, and sea stars in a tidepool. More often, however, a full count is impractical, and they must sample just part of a population and then estimate its total density.

For example, you could get a map of your county and divide it into small plots, or quadrats. **Quadrats** are sampling areas of the same size and shape, such as rectangles, squares, and hexagons. You could then count individual deer in several plots and, from that, extrapolate the average for the specified population. Ecologists often conduct such counts for plants and other species that stay put (Figure 26.3). Some of their counts in small areas help them estimate the population sizes of migrating animals.

Deer are among the animals that do not stay put. So how can you be sure that the individuals you are counting in a given plot are not the same ones you counted earlier in a different plot?

For mobile animals, ecologists sample population density with **capture–recapture methods**. The idea is to capture mobile individuals and mark them in some way. Deer get collars, squirrels get tattoos, salmon get tags, birds get leg rings, butterflies get wing markers, and so on (Figure 26.4). Marked animals are released at time 1. At time 2, traps are reset. The proportion of marked animals in this second sample is taken to be representative of the proportion marked in the whole population:

$$\frac{\text{Marked individuals in sampling at time 2}}{\text{Total captured in sampling 2}} = \frac{\text{Marked individuals in sampling at time 1}}{\text{Total population size}}$$

Ideally, marked and unmarked individuals of the population are captured at random, none of the marked animals dies during the study interval, and none leaves the population or gets overlooked.

In the real world, recapturing marked individuals might *not* be random. Squirrels marked after being attracted to bait in boxes might now be trap-happy or trap-shy. Such individuals may overrepresent or underrepresent their population. Another example: Instead of mailing back the tags of marked fish to ecologists, a few fishermen keep them as good-luck charms. And birds lose their leg rings.

Your estimate also depends on the time of year when you make a sampling. Population distribution varies over time, as during migratory responses to environmental rhythms. Few places yield abundant resources all year long, so many populations move between habitats as seasons change. Canada geese are like this. So are deer. In such cases, then, the capture–recapture methods might be used more than once a year, for several years.

Figure 26.3 Near the Sierra Nevada, a population of creosote bushes showing nearly uniform distribution. The plants compete for scarce water.

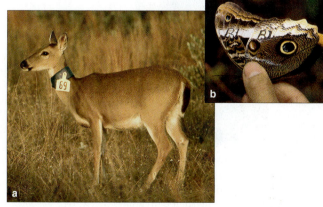

Figure 26.4 Two individuals marked for population studies. (**a**) Florida Key deer and (**b**) Costa Rican owl butterfly (*Caligo*).

26.3 Population Size and Exponential Growth

Populations are dynamic units of nature. Depending on the species, they may add or lose individuals every minute of every day, season, or year. Sometimes they glut portions of their habitat with individuals. Other times, individuals are scarce. Populations even drive themselves or are driven to extinction.

GAINS AND LOSSES IN POPULATION SIZE

We can measure change in population size in terms of birth rates, death rates, and how many individuals are entering and leaving during a specified interval.

Population size increases as a result of births and **immigration**, the arrival of new residents from other populations of the same species. Its size decreases as a result of deaths and **emigration**, the departure of individuals that take up permanent residence in some other place. As an example, Arnold Schwarzenegger emigrated from Austria to the United States, where he became a celebrated immigrant. His permanent move decreased the Austrian population by 1 and increased the United States population by 1.

For many species, population size changes during seasonal or daily migrations. However, **migration** is a recurring round trip between two distinct regions, so we need not consider its transient effects in this initial look at population size.

FROM ZERO TO EXPONENTIAL GROWTH

To keep things simple, let's assume that immigration and emigration balance each other over time. This way we can ignore the effects of both on population size. Doing so allows us to define **zero population growth** as an interval during which the number of births is balanced by the number of deaths. Population size is stable during such an interval, with no net increase or decrease in the number of individuals.

Births, deaths, and the other variables that might change population size can be measured in terms of **per capita** rates, or rates per individual. *Capita* means heads, as in head counts.

Visualize 2,000 mice living in a cornfield. Twenty or so days after their eggs are fertilized, the females give birth to a litter, then they nurse the offspring for a while, and then they get pregnant again. Say 1,000 mice are born in one month. The birth rate is 0.5 per mouse per month (1,000 births/2,000 mice). If 200 of the 2,000 die during that interval, the death rate will be 200/2,000 = 0.1 per mouse per month.

Assume further that the birth rate and death rate remain constant. By doing so, we can combine both variables into a single variable—the **net reproduction per individual per unit time**, or *r* for short. For our mice, *r* is 0.5 − 0.1 = 0.4 per mouse per month.

Read Me First!

and watch the narrated animation on exponential growth

Figure 26.5 (a) Net monthly increases in a population of field mice living in a cornfield. Start to finish, the list shows a pattern typical of exponential growth. (b) Graph the numerical data and you end up with a J-shaped growth curve.

		Net Monthly Increase:	New Population Size:
$G = r \times$	3,920 =	1,568 =	5,488
$r \times$	5,488 =	2,195 =	7,683
$r \times$	7,683 =	3,073 =	10,756
$r \times$	10,756 =	4,302 =	15,058
$r \times$	15,058 =	6,023 =	21,081
$r \times$	21,081 =	8,432 =	29,513
$r \times$	29,513 =	11,805 =	41,318
$r \times$	41,318 =	16,527 =	57,845
$r \times$	57,845 =	23,138 =	80,983
$r \times$	80,983 =	32,393 =	113,376
$r \times$	113,376 =	45,350 =	158,726
$r \times$	158,726 =	63,490 =	222,216
$r \times$	222,216 =	88,887 =	311,103
$r \times$	311,103 =	124,441 =	435,544
$r \times$	435,544 =	174,218 =	609,762
$r \times$	609,762 =	243,905 =	853,667
$r \times$	853,677 =	341,467 =	1,195,134

(a)

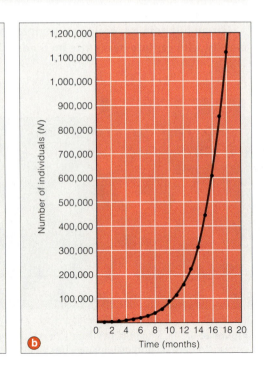

(b)

Figure 26.6 Effect of deaths on the rate of increase in two hypothetical populations of bacteria. Plot population growth for bacterial cells that reproduce every half hour and you get growth curve *1*. Plot the growth of a population of cells that divide every half hour, with 25 percent dying between divisions, and you get growth curve *2*. Deaths do slow the rate of increase, but as long as birth rate exceeds death rate, exponential growth will continue.

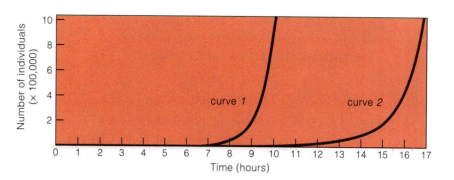

And so our mice in a cornfield give us a simple way to represent population growth as:

population growth per unit time	=	net population growth rate per individual per unit time	×	number of individuals

or, more simply, $G = rN$.

The next month begins, and 2,800 mice are in this field. With that net increase of 800 fertile mice, the reproductive base is larger. Population size expands again during the month, for a net increase of $0.4 \times 2,800 = 1,120$ mice. Population is now 3,920. The mice keep on reproducing month after month. Figure 26.5*a* shows, *in less than two years, the number of mice in the cornfield will increase from 2,000 to more than a million!*

Plot the monthly increases against time and you end up with a graph line in the shape of a "J," as in Figure 26.5*b*. When the growth of any population over time plots out as a J-shaped curve, you know that you are tracking exponential growth.

Exponential growth can be defined as a quantity increasing by a *fixed* percentage of the total in each specified interval. The numbers of births and deaths correlate with population size, so they increase in each interval as long as *r* remains constant. If it does, the increase in population size during any interval depends on the size of the existing reproductive base. *The larger the reproductive base, the greater will be the expansion in population size during a specified interval.*

Now look at other aspects of exponential growth. Start by supplying one bacterium in a culture flask with all the nutrients required for growth. After thirty minutes, the cell divides in two. Its two daughter cells divide, and so on every thirty minutes. Assume none of the cells dies between divisions. The population size doubles in each interval—from 1 to 2, then 4, 8, 16, 32, and so on. The time it takes for a population to double in size is its **doubling time**.

After 9–1/2 hours (nineteen doublings), there are more than 500,000 cells. Ten hours (twenty doublings) later, there are more than a million. Curve *1* in Figure 26.6 is a plot of this outcome.

Will death put the brakes on exponential growth? Suppose 25 percent of the descendant cells die every thirty minutes. Now, it takes about seventeen hours (not ten) for population size to reach 1 million. *Deaths slowed but did not stop population growth* (curve *2* in Figure 26.6). Exponential population growth will continue as long as birth rates exceed death rates.

WHAT IS THE BIOTIC POTENTIAL?

Now, visualize a population occupying a place where conditions are ideal. Every one of its individuals has adequate shelter, food, and other vital resources. No predators, pathogens, or pollutants lurk anywhere in the habitat. That population might well display its **biotic potential**—the maximum rate of increase per individual under ideal conditions.

Every species has a characteristic maximum rate of increase. For many bacteria, it is 100 percent every half hour or so. For humans and other large mammals, the estimated biotic potential is about 2 to 5 percent per year. The *actual* rate depends on the age at onset of reproduction, how frequently the individuals are reproducing, and how many offspring are produced over a lifetime. Athough the human population is not now displaying its full biotic potential, it is growing exponentially, for reasons we'll discuss shortly.

During a specified interval, population size is generally an outcome of births, deaths, immigration, and emigration.

With exponential growth, the growth rate increases over time as the reproductive base becomes ever larger.

As long as the per capita birth rate remains above the per capita death rate, a population will grow exponentially.

26.4 Limits on the Growth of Populations

Complex interactions occur within and between populations in nature, where it's not easy to identify all the factors working to limit population growth.

WHAT ARE THE LIMITING FACTORS?

Most of the time, environmental circumstances keep a population from fulfilling its biotic potential. That's why sea stars—the females of which could produce 2,500,000 eggs each year—do not overflow the oceans.

To get a sense of what some of the constraints may be, start again with a bacterial cell in a culture flask, where you can control the variables. First you enrich the culture medium with glucose and other nutrients necessary for bacterial growth. Then you sit back and let bacterial cells reproduce for many generations.

At first, growth seems to be exponential. Then it slows, and population size remains relatively stable. After the stable period, the size plummets until all the bacterial cells are dead. *What happened?* As the cell population grew ever larger, it used up more and more of the nutrients. Nutrient scarcity became an environmental signal for cells to stop dividing. But even with growth stopped, the population eventually used up all the nutrients and died out.

Any essential resource that is in short supply is a **limiting factor** on population growth. Food, mineral ions, refuge from predators, living space, and even a pollution-free habitat are examples (Figure 26.7). The number of limiting factors can be extensive, and their effects can vary. Even so, one factor alone often puts the brakes on population growth.

Suppose you kept freshening the nutrient supply. After growing exponentially, the population collapsed anyway. Like all other organisms, bacteria produce metabolic wastes. That huge population produced so much, it drastically compromised living conditions in the culture. Collectively its individuals polluted the experimentally designed habitat and put a stop to any further exponential growth.

CARRYING CAPACITY AND LOGISTIC GROWTH

Visualize a small population, with individuals spread through the habitat. As its size increases, more and more individuals must compete for nutrients, living quarters, and other resources. The share available to each diminishes, fewer offspring are born, and more die from starvation or nutrient deficiencies. Now the population's growth rate declines until the births are balanced or even outnumbered by deaths.

Ultimately, the *sustainable* supply of resources will determine the population's size. **Carrying capacity** is the maximum number of individuals of a population that a given environment can sustain indefinitely.

The pattern of **logistic growth** shows how carrying capacity can affect population size. By this pattern, a small population starts growing slowly in size, then it grows rapidly, and finally its size levels off once the carrying capacity is reached. Figure 26.8 shows how this pattern plots out as an S-shaped curve. Here is a way to represent the pattern as an equation:

population growth per unit time	=	maximum net population growth rate per individual per unit time	×	number of individuals	×	proportion of resources not yet used

An S-shaped curve is an approximation of what actually goes on in nature. For instance, a population that grows too fast usually overshoots the carrying capacity. The death rate skyrockets and the birth rate plummets. These two outcomes drive numbers down to the carrying capacity or lower (Figure 26.9).

The logistic growth equation just described deals with **density-dependent controls**. Such controls are any factors that operate when increases in population density lower the survival odds for individuals.

When a small, rapidly growing population reaches its carrying capacity, dwindling resources work as a

Figure 26.7 Response to a scarcity of nesting sites, a limiting factor for weaver bird populations in Africa.

(**a**) African weavers construct densely woven, cup-shaped nests that are only wide and deep enough for a hen and her nestlings. Many nests hang from the same spindly limbs. (**b**) This nest in Namibia is like an apartment house in a place where few trees are available. Between 100 and 300 pairs of sparrow weavers occupy their own flask-shaped nests, each with its own tubular entrance.

Read Me First!
*and watch the narrated animation on
logistic population growth*

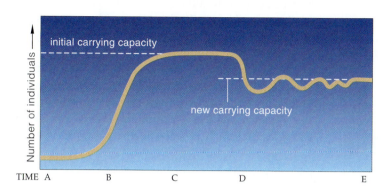

Figure 26.8 The idealized S-shaped curve that is characteristic of logistic growth. Growth slows after a phase of rapid increase (time A to C). The curve flattens as the carrying capacity is reached (time C to D). S-shaped growth curves can show variations, as when changes in the environment lower the carrying capacity (time D to E).

This happened to the human population when bubonic plague swept through Europe's crowded cities in the fourteenth century. One epidemic claimed 25 million lives.

control to stabilize or decrease its numbers. Besides environmental factors, species interactions also may drive the number of individuals below the maximum sustainable level.

Overcrowding invites interactions that increase the likelihood of individual deaths. Predators, parasites, and pathogens interact more intensely when prey or host populations are dense. They usually reduce the numbers of prey or hosts. By thinning a population, they remove conditions that invited their controlling effect, and population size may increase again.

DENSITY-INDEPENDENT FACTORS

A **density-independent factor** causes more deaths or fewer births regardless of a population's density. For instance, each year, millions of monarch butterflies fly down from Canada to Mexico's forested mountains, where they spend the winter. In 2002 a sudden freeze, exacerbated by deforestation, killed millions of the butterflies. The freeze had nothing to do with butterfly density; it was a density-independent event.

Resources in short supply are limiting factors that restrict population growth. Carrying capacity is the maximum number of individuals of a population that can be sustained indefinitely by the resources in a given environment.

With logistic growth, population growth is rapid at low density, then slows as the population approaches carrying capacity, where numbers will level off.

Both density-dependent controls and density-independent factors can bring about decreases in population size.

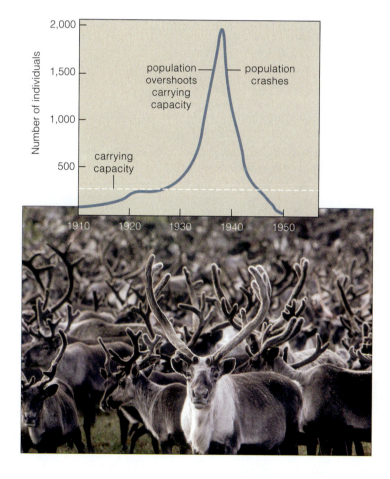

Figure 26.9 Carrying capacity and a reindeer herd. In 1910, four male and twenty-two female reindeer were introduced on St. Matthew Island in the Bering Sea. In less than thirty years, the size of the herd increased to 2,000. Individual reindeer had to compete for dwindling vegetation, and overgrazing destroyed most of it. In 1950, herd size plummeted to eight. The growth curve reflects how the reindeer population size overshot the carrying capacity for the island, then crashed.

26.5 Life History Patterns

Researchers have identified age-specific adaptations that affect the survival, fertility, and reproduction of individuals for many kinds of species.

So far, we have looked at populations as if all of their members are identical during any given interval. For most species, however, individuals of a population are at different stages of development. They interact in different ways with other organisms and with their environment. At different times in the life cycle, they may be adapted to different resources, as when larvae eat leaves and butterflies sip nectar. They may also be more or less vulnerable to danger at different stages.

In short, each species has a **life history pattern**, or a set of adaptations that influence survival, fertility, and age at first reproduction. Each pattern reflects the individual's schedule of reproduction. In this section and the next, we look at a few of the environmental variables that underlie age-specific patterns.

LIFE TABLES

Each species has a characteristic life span, but few of its individuals survive to the maximum age possible. Death looms larger at certain ages. Individuals tend to reproduce during an expected age interval, and in some species they emigrate at an expected time.

Age-specific patterns in populations intrigue life insurance and health insurance companies, as well as ecologists. Such investigators typically track a **cohort**, or a group of individuals, recorded from the time of birth until the last one dies (Table 26.1). They also record the number of offspring born to individuals in each age interval. Life tables list the data for an age-specific death schedule, which are typically converted to much cheerier "survivorship" schedules that show the number of individuals actually reaching specified ages. Table 26.2 is a typical example. It lists data for the 2001 human population of the United States.

Dividing a population into age classes and noting age-specific birth rates and mortality risks can yield useful information. Unlike a crude head count, for example, such data can shape policy decisions about pest management, protection of endangered species, or social planning for specified human populations. Birth and death schedules for the northern spotted owl were cited in federal court rulings that halted mechanized logging in the owl's habitat—old-growth forests of the Pacific Northwest.

PATTERNS OF SURVIVAL AND REPRODUCTION

Evolution proceeds by differences in survival and reproductive success. We measure the reproductive success of individuals in terms of the number of their surviving offspring. But that number varies among species, which differ in how much energy and time are allocated to production of gametes, locating and securing mates, and parenting, and in the size of the

Table 26.1 Life Table for a Cohort of Annual Plants (*Phlox drummondii*)*

Age Interval (days)	Survivorship (number surviving at start of interval)	Number Dying During Interval	Death Rate (number dying/number surviving)	"Birth" Rate during interval (number of seeds from each plant)
0–63	996	328	0.329	0
63–124	668	373	0.558	0
124–184	295	105	0.356	0
184–215	190	14	0.074	0
215–264	176	4	0.023	0
264–278	172	5	0.029	0
278–292	167	8	0.048	0
292–306	159	5	0.031	0.33
306–320	154	7	0.045	3.13
320–334	147	42	0.286	5.42
334–348	105	83	0.790	9.26
348–362	22	22	1.000	4.31
362–	0	0	0	0
		996		

* Data from W. J. Leverich and D. A. Levin, 1979.

Table 26.2 Life Table for the United States Human Population in 2001

Age Interval	Number at Start of Interval	Number Dying During Age Interval	Life Expectancy at Start of Interval	Reported Live Births
0–1	100,000	684	77.2	
1–5	99,316	132	76.7	
5–10	99,184	76	72.8	
10–15	99,108	96	67.9	7,315
15–20	99,012	330	62.9	525,493
20–25	98,682	468	58.1	1,022,106
25–30	98,214	471	53.4	1,060,391
30–35	97,743	554	48.6	951,219
35–40	97,189	801	43.9	453,927
40–45	96,388	1,154	39.2	95,788
45–50	95,234	1,682	34.7	5,244
50–55	93,552	2,373	30.3	263
55–60	91,179	3,474	26.0	
60–65	87,705	5,186	21.9	
65–70	82,519	7,397	18.1	
70–75	75,122	10,018	14.6	
75–80	65,104	13,284	11.5	
80–85	51,820	15,877	8.8	
85–90	35,943	16,147	6.5	
90–95	19,796	11,906	4.8	
95–100	7,890	5,845	3.6	
100+	2,045	2,045	2.7	

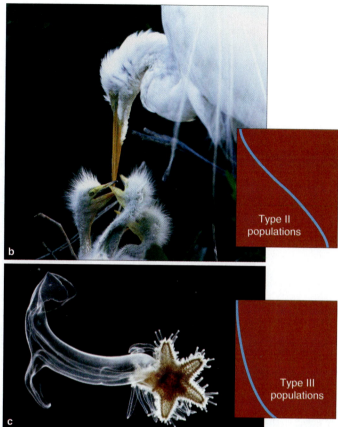

Figure 26.10 Three generalized survivorship curves. (**a**) Elephants are Type I populations. They have high survivorship until some age, then high mortality. (**b**) Snowy egrets are Type II populations. They have a fairly constant death rate. (**c**) Sea star larvae represent Type III populations, which show low survivorship early in life.

offspring. It seems that trade-offs have been made in response to selection pressures, such as the conditions prevailing in the habitat and species interactions.

A **survivorship curve** is a graph line that emerges when ecologists plot a cohort's age-specific survival in a habitat. Each species has a characteristic curve, and three types of curves are common in nature.

Type I curves reflect high survivorship until fairly late in life, then a large increase in deaths. Like many annual plants, the phlox in Table 26.1 show this type of pattern. So do large mammals that bear only one or a few large offspring at one time, then engage in extended parental care (Figure 26.10*a*). For example, a female elephant gives birth to four or five calves and devotes several years to parenting each.

Type I curves are typical of human populations in which individuals have access to good health-care services. However, today as in the past, infant deaths cause a sharp drop at the start of the curve in regions where health care is poor. After the drop, the curve levels off from childhood to early adulthood.

Type II curves reflect a fairly constant death rate at all ages. They are typical of organisms just as likely to be killed or die of disease at any age, such as lizards, small mammals, and large birds (Figure 26.10*b*).

Type III curves signify a death rate that is highest early on. We see this for species that produce many small offspring and do little, if any, parenting.

Figure 26.10*c* shows how the curve plummets for sea stars. Sea stars release mind-boggling numbers of eggs. The tiny larvae must eat fast, grow, and finish developing on their own without support, protection, or guidance from parents. Corals and other animals quickly eat almost all of them. Such a plummeting survivorship curve is common among many marine invertebrates, insects, fishes, plants, and fungi.

At one time, ecologists thought selection processes favored *either* early, rapid production of many small offspring *or* late production of a few large offspring. Researchers now comprehend that these patterns are extremes at opposite ends of a range of possible life histories. Also, both life history patterns—as well as intermediate ones—sometimes occur among different populations at different times.

Tracking a cohort (a group of individuals) from birth until the last one dies reveals patterns of reproduction, death, and migration that typify the populations of a species.

Survivorship curves can reveal differences in age-specific survival among species. In some cases, such differences exist even between populations of the same species.

26.6 Natural Selection and Life Histories

Earlier you read that jaws evolved among certain fishes during the Cambrian. This key innovation led to adaptive radiations among predators and diverse defenses among prey. No one witnessed that coevolutionary arms race. But experimental studies show that predators are still acting as selective agents and prey are still evolving.

Several years ago, two evolutionary biologists, with fishnets in hand and drenched in sweat, were doing fieldwork in the mountains of Trinidad, an island in the southern Caribbean Sea. They wanted to capture small fishes that live in shallow freshwater streams of this habitat (Figure 26.11*a*). The fish were guppies (*Poecilia reticulata*). David Reznick and John Endler were starting their eleven-year study of the variables that affect guppy life history patterns.

Male guppies are generally smaller than females of the same age and have bright-colored scales. The colors function as visual signals for mating during the guppy's complex courtship rituals. Females are drab colored and, unlike the males, they continue to grow even after they reach sexual maturity.

Reznick and Endler were interested in the effects of predation on guppy evolution. They chose their study site because in the streams that run through the mountains of Trinidad, different populations of guppies deal with different predators. Even within a stream, waterfalls can keep predators from moving from one area to another.

Two major predators of guppies are killifishes (*Rivulus hartii*) and pike-cichlids (*Crenicichla*), shown in Figure 26.11*b,c*. A killifish is not a very big fish. It preys efficiently on small, immature guppies but not on the larger adults. Pike-cichlids live in other streams. They prey on larger and sexually mature guppies, and tend to ignore small guppies.

Reznick and Endler hypothesized that predation is a selective agent that acted to shape guppy life history patterns. They knew that in pike-cichlid streams, guppies grow faster and are smaller at maturity, compared to guppies in killifish streams (Figure 26.12). Also, guppies hunted by pike-cichlids reproduce earlier in life, reproduce more often, and have more young per brood (Figure 26.13).

GUPPY FROM A PIKE-CICHLID STREAM

PIKE-CICHLID

GUPPY FROM A KILLIFISH STREAM

KILLIFISH

Figure 26.11 (a) Evolutionary biologist David Reznick contemplating interactions among guppies (*Poecilia reticulata*) and their predators in a freshwater stream in Trinidad. (b,c) Two guppies, and two of the guppy eaters.

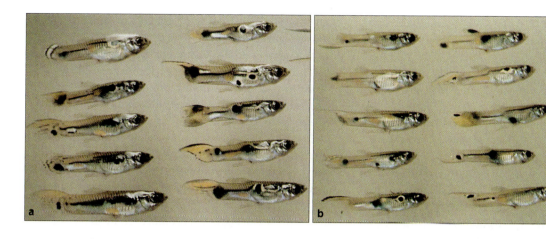

Figure 26.12 Some representative guppies that inhabit streams with killifishes (**a**) and with pike-cichlids (**b**).

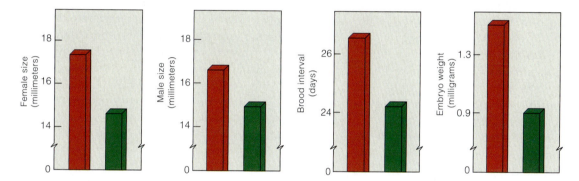

Figure 26.13 Graphs of some of the experimental evidence of natural selection among guppy populations subjected to different predation pressures. Compared to guppies raised with killifish (*red* bars), guppies raised with pike-cichlids (*green* bars) differed in body size and length of time between broods. Killifish are small and prey on smaller guppies, pike-cichlids are large fish and prey on larger guppies, and so the two predators select for the differences.

Were these differences genetic or did some other variables influence life history patterns in killifish and pike-cichlid streams? To find out, the researchers shipped live guppies from each stream back to their laboratory in the United States. They let the guppies reproduce in separate, predator-free aquariums for two generations. All other physical and chemical conditions in the artificial habitats were identical for the different experimental groups.

As it turned out, the offspring of the experimental guppy populations displayed the same differences as the natural populations. The conclusion? Differences between guppies preyed upon by different predators have a genetic foundation.

So what would happen if the selective pressure on a guppy population changed? Reznick and Endler answered the question with a set of field experiments. In one set, they introduced guppies upstream from a small waterfall. Before the experiment, the waterfall had been a barrier to dispersal. It prevented guppies

and large pike-cichlids from emigrating upstream, where killifish were present. Guppies introduced to the upstream experimental site were taken from a population that had evolved downstream from the waterfall, with the larger pike-cichlids.

Eleven years (thirty to sixty guppy generations) later, researchers revisited the stream. They found the experimental population had evolved. Guppies now had traits that resembled those of guppies that had been living with killifishes for a longer time. The change in predominant predator had influenced body size, frequency of reproduction, and other aspects of guppy life history patterns. Laboratory experiments involving two generations of guppies confirmed that the differences had a genetic basis.

Reznick and Endler showed that life history traits, like other characteristics, can be inherited. They also demonstrated that these traits can evolve. Traits that affect life history can be altered over a surprisingly short time in response to the right selection pressure.

26.7 Human Population Growth

Human population size surpassed 6 billion in 1999. In the five years since then, 400 million more individuals were added to the total.

In 2003, the average rate of increase for the human population was 1.3 percent. Now think about this: As long as the birth rate continues to exceed the death rate, annual additions will result in a *larger* absolute increase each year into the foreseeable future.

Our staggering population growth continues even though more than a billion are malnourished, without clean drinking water or adequate shelter. It continues even though more than a billion of us still do not have access to health-care delivery systems or sewage treatment facilities. It continues primarily in already overcrowded regions, on 10 percent of the land. Figure 26.14 is a graphic clue to what our expansion means with respect to the carrying capacity.

Even if it were possible to double food supplies to keep pace with growth, living conditions would still be marginal for most people. At least 10 million of us would continue to die each year from starvation.

For a time, it will be like the Red Queen's garden in Lewis Carroll's *Through the Looking Glass*, where one is forced to run as fast as one can to stay in the same place. What happens when our population doubles again? Can you brush the doubling aside as being too far in the future? *It is no further removed from you than the sons and daughters of the next generation.*

How did we get into this predicament? For most of its history, the human population grew slowly. Things started to pick up about 10,000 years ago, but in the past two centuries, growth rates skyrocketed. Three trends contributed to the expansion. First, humans steadily developed the capacity to expand into new habitats and climate zones. Second, humans increased the carrying capacity of their existing habitats. Finally, human populations have sidestepped limiting factors that restrain the growth of other species.

Reflect on the first point. Early humans evolved in woodlands, then in savannas. They were vegetarians, mostly, but also scavenged for bits of meat. Bands of hunter–gatherers moved out of Africa about 2 million years ago. By 40,000 years ago, the descendants were established in much of the world (Section 23.12).

Few species can expand into such a broad range of habitats. Having a truly complex brain, early humans drew on learning and memory to figure out how to build fires, make shelters, make clothing, make tools, and cooperate in hunts. With the advent of language, knowledge did not die with the individual. It spread quickly among groups. *The human population expanded into diverse environments far more rapidly than the long-term geographic dispersals of other species.*

Reflect on the second point. Starting about 11,000 years ago or so, many hunter–gatherer bands shifted to agriculture. Instead of simply following migratory game herds, they settled in fertile valleys and other regions that favored seasonal harvesting of fruits and grains. In this way, they developed a more dependable basis for life. A pivotal factor was the domestication of wild grasses, including species ancestral to modern wheat and rice. People harvested, stored, and planted seeds in one place. They domesticated animals for food and pulling plows. They dug irrigation ditches and diverted water to croplands.

Agricultural productivity was a basis for increases in population growth rates. Towns and cities formed. Later in time, food supplies increased again, and yet again, by the use of chemical fertilizers, herbicides, and pesticides. Transportation improved, as did food distribution. *Thus, even at its simplest, management of food supplies through agriculture increased the carrying capacity for the human population.*

What about sidestepping of limiting factors? Until about 300 years ago, poor hygiene, malnutrition, and infectious diseases kept death rates high enough to more or less balance birth rates. Infectious diseases became density-dependent controls. Epidemics swept through overcrowded settlements and cities that were infested with fleas and rodents. Then came plumbing and new methods of sewage treatment. Over time,

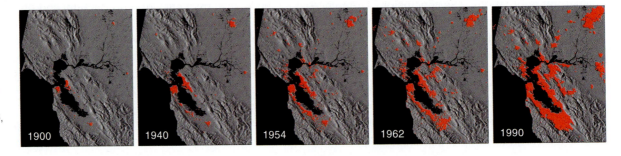

Figure 26.14 Human population growth in a small part of the world since 1900. Red signifies dense populations in and around the San Francisco Bay area and Sacramento, based on historical data and satellite imaging.

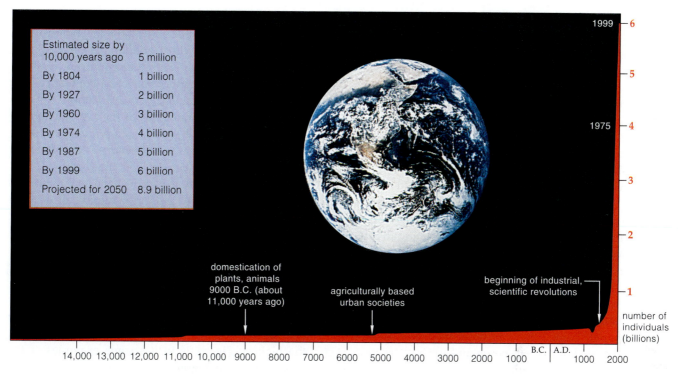

Estimated size by 10,000 years ago	5 million
By 1804	1 billion
By 1927	2 billion
By 1960	3 billion
By 1974	4 billion
By 1987	5 billion
By 1999	6 billion
Projected for 2050	8.9 billion

domestication of plants, animals 9000 B.C. (about 11,000 years ago)

agriculturally based urban societies

beginning of industrial, scientific revolutions

number of individuals (billions)

14,000 13,000 12,000 11,000 10,000 9000 8000 7000 6000 5000 4000 3000 2000 1000 | B.C. | A.D. | 1000 2000

Figure 26.15 Growth curve (*red*) for the world human population. (The dip between the years 1347 and 1351 is when 60 million people died during a bubonic plague.) The *blue* box lists how long it took for the human population to increase from 5 million to 6 *billion*.

vaccines, antibiotics, and other drugs were developed as weapons against many pathogens. The death rates dropped sharply. Births began to exceed deaths—and rapid population growth was under way.

In the industrial revolution of the mid-eighteenth century, people discovered how to harness the energy stored in fossil fuels, starting with coal. Within a few decades, large industrialized societies began to form in western Europe and in North America. Even more efficient technologies were devised with the start of World War I. Factories began mass-producing cars, tractors, and many other affordable goods. Advances in agriculture meant that fewer farmers were required to support a larger population.

And so, by controlling disease agents and tapping into concentrated, existing forms of energy—fossil fuels—the human population has managed to sidestep major factors that had previously limited its rate of increase.

Where have the far-flung dispersals and stunning advances in agriculture, industrialization, and health care taken us? Starting with *Homo habilis*, it took about 2.5 million years for human population size to reach 1 billion. As Figure 26.15 shows, it took just 123 years to reach 2 billion, another 33 to reach 3 billion, 14 more to reach 4 billion, and 13 more to get to 5 billion. It

took only 12 more years to arrive at 6 billion! Given the principles governing population growth, we may expect the rate of increase to decline as birth rates fall or as death rates rise. Alternatively, we may expect continued increases if technological breakthroughs can expand the carrying capacity. *Even so, continued growth cannot be sustained indefinitely.*

Why? Continuing increases in population size can set the stage for density-dependent controls. Already, pathogens, such as the one that causes SARS, can spread between the continents in a matter of weeks. The problem is compounded by human emigration on a vast scale. Economic hardship and civil strife have put an estimated 50 million people on the move within and between countries. Will relocations of so many individuals be peaceable? Will they find food, clean water, and other basic resources where they end up?

Through expansion into new habitats, cultural interventions, and technological innovations, the human population has temporarily skirted environmental resistance to growth.

As population increases, density-dependent controls, such as disease and competition for resources, may slow growth.

26.8 Fertility Rates and Age Structure

Acknowledgment of the risks posed by rising populations has led to increased family planning in almost every region. Putting the brakes on population growth is not easy and numbers are expected to continue to rise.

Most governments recognize that population growth, resource depletion, pollution, and the quality of life are interconnected. Most are working to lower long-term birth rates, as with family planning programs. Details vary among countries, but most are offering information on available methods of fertility control.

These attempts are having impact; birth rates have been slowing worldwide. Death rates are declining, mainly because improved nutrition and health care are lowering infant mortality rates (the number of infants per 1,000 who die in the first year). However, AIDS has sent death rates soaring in some African countries.

We still expect the world population to reach 8.9 billion by 2050. Think about the resources that will be required. We will have to boost food production and find more sources of energy and fresh water to meet even basic needs, something that still eludes almost half of the population. And large-scale manipulations of resources will intensify pollution.

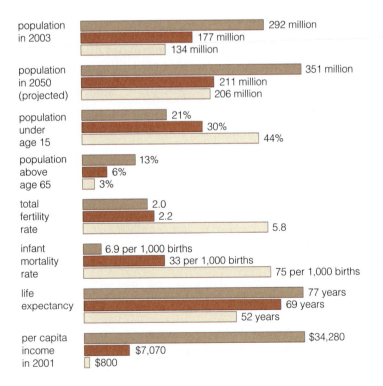

Figure 26.16 Key demographic indicators for three countries, mainly in 2003. The United States (*brown* bar) is highly developed, Brazil (*red* bar) is moderately developed, and Nigeria (*gold* bar) is less developed.

India, China, Pakistan, Nigeria, Bangladesh, and then Indonesia are expected to show the most growth, in that order. China (with 1.3 billion people) and India (with 1.1 billion) dwarf all other countries; together, they make up 38 percent of the world population. The United States is next in line, with 295 million.

The **total fertility rate** (TFR) is the average number of children born to the women of a population during their reproductive years. TFR estimates are based on current age-specific rates. In 1950, the worldwide TFR averaged 6.5. By 2003, it had declined to 2.8 but was still above replacement levels, the number of children a couple must bear to replace themselves. At present, the replacement level is 2.1 for developed countries and as high as 2.5 in some developing countries. (It is higher in developing countries because more female children die before reaching reproductive age.)

These numbers are averages. TFRs are at or below replacement levels in many developed countries; the developing countries in western Asia and Africa have the highest. Figure 26.16 has some examples of the disparities in demographic indicators.

Comparing the age structure diagrams for different populations is revealing. In Figure 26.17, notice the reproductive age category for the next fifteen years. The average range for childbearing years is 15–49. The broader the base of the diagram, the faster the population can be expected to increase in size. The United States population has a relatively narrow base and is undergoing slow growth.

Even if every couple decides to bear no more than two children, world population growth will not slow for sixty years, because 1.9 billion are about to enter the reproductive age bracket. *More than one-third of the world population is in the broad pre-reproductive base.*

China has the most far-reaching family planning program. Its government discourages premarital sex; it urges people to delay marriage and limit families to one or two children. It offers abortions, contraceptives, and sterilization at no cost to married couples. Even in remote rural areas, paramedics and mobile units offer access to these measures. Couples who follow these guidelines receive more food, free medical care, better housing, and salary bonuses. Their offspring receive free tuition and preferential treatment when they are old enough to enter the job market. Parents who have more than two children lose government benefits and pay more taxes.

Although the policy may sound harsh, it works. Since 1972, China's TFR has fallen sharply, from 5.7 to 1.8. An unintended consequence has been a shift in the country's sex ratio. Traditional cultural preference

Read Me First!
and watch the
narrated animation on
age structure diagrams

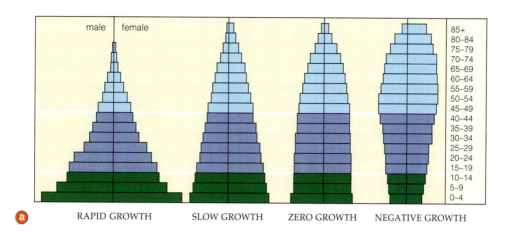

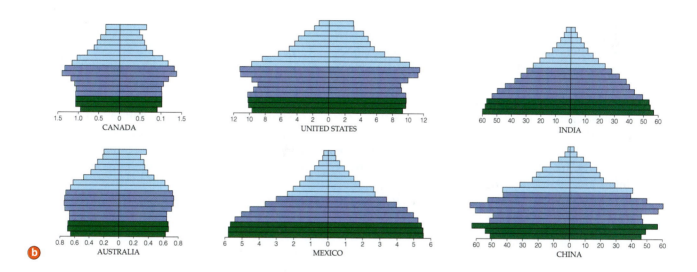

Figure 26.17 (**a**) General age structure diagrams for countries with rapid, slow, zero, and negative rates of population growth. The pre-reproductive years are *green* bars; reproductive years, *purple;* and the post-reproductive years, *light blue.* A vertical axis divides each graph into males (*left*) and females (*right*). Bar widths correspond to proportions of individuals in each age group. (**b**) Age structure diagrams for a few representative countries in 1997. Population sizes are measured in millions.

for sons, especially in the rural areas, has led some parents to abort developing females or even commit infanticide. Worldwide, 1.06 boys are born for every girl, but in China the latest census reports 1.19 boys per girl. Also, more than a hundred thousand girls are abandoned each year. In response, the government is now offering additional cash and tax incentives to parents of girls. Meanwhile, China's population time bomb is still ticking. About 150 million of its females are in the pre-reproductive age category.

The worldwide total fertility rate has been dropping, but is still above the replacement level required to arrive at zero population growth.

Most countries support family planning programs of some sort. Even with slowdowns, the human population will continue to increase, because of its pre-reproductive base.

At present, more than one-third of the human population is in a very broad pre-reproductive base.

26.9 Population Growth and Economic Effects

The most highly developed countries have the slowest growth rates and use the most resources.

DEMOGRAPHIC TRANSITIONS

Changes in population growth rates often correlate with four stages of economic development, the heart of the **demographic transition model**. By this model, living conditions are harshest in a *preindustrial* stage, before technology and medical advances spread. Birth and death rates are high, so the growth rate is low. In the *transitional* stage, industrialization begins. Food production and health care improve and death rates slow. Birth rates stay high in the agricultural societies, where big families provide help in the fields. Annual growth rates are between 2.5 and 3 percent. As living conditions improve and birth rates begin to decline, growth generally starts to level off (Figure 26.18).

In the *industrial* stage, industrialization is in full swing and growth slows. People move to cities, and couples often want small families. As they accumulate goods, many decide the time and cost of raising more than a few children conflict with their goals.

In the *postindustrial* stage, population growth rates become negative. The birth rate falls below the death rate, and population size slowly decreases.

The United States, Canada, Australia, and most of western Europe, Japan, and much of the former Soviet Union are in the industrial stage. Most developing countries, such as Mexico, are now in the transitional stage, without enough skilled workers to complete the transition to a fully industrial economy.

If population growth continues to exceed economic growth, then we can expect death rates to rise. Many countries may get stuck in the transitional stage or fall back to the conditions of the preceding stage.

A QUESTION OF IMMIGRATION POLICIES

Figure 26.19 charts the legal immigration to the United States between 1820 and 1997. The greatest increase came after the Immigration Reform and Control Act of 1986 gave legal status to undocumented immigrants who proved they had lived in the country for years. Economic downturns in the 1980s and 1990s fanned resentment against newcomers. Many want to limit legal immigration to 300,000–450,000 per year and to deport undocumented immigrants. Others want the borders open. They also say that such a policy would discriminate against legal immigrants of the same ethnic background. This is a volatile issue, one with social and economic ramifications.

A QUESTION OF RESOURCE CONSUMPTION

Industrialized nations use the most resources. For example, the United States with about 4.6 percent of the world population produces about 21 percent of all goods and services. Yet it requires thirty-five times more goods and services than India. It uses 25 percent or so of the world's processed minerals and much of the energy supplies. It is not alone in this. China and India are now demanding an ever increasing share of the economic pie.

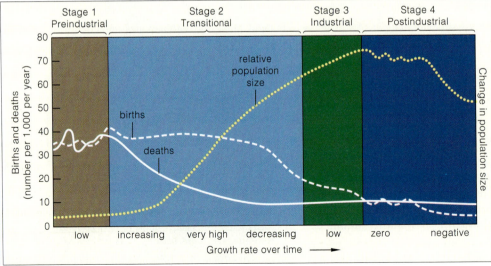

Figure 26.18 The demographic transition model for changes in population growth rates and in sizes, correlated with long-term changes in the economy.

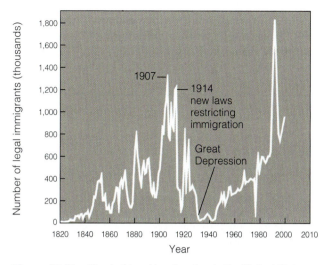

Figure 26.19 Chart of legal immigration to the United States between 1820 and 1997.

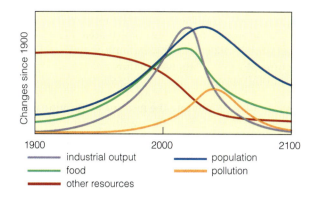

Figure 26.20 Computer-based projection of what may happen if human population size continues to skyrocket without dramatic policy changes and technological innovation. The assumptions were that the population has already overshot the carrying capacity and current trends will continue unchanged.

G. Tyler Miller once estimated that it would take 12.9 billion impoverished individuals living in India to have as much impact on the environment as 284 million people in the United States. He also pointed out that the projected increase in human population growth rates raises serious questions. Will there be enough food, energy, water, and other resources to sustain so many people? Will governments be able to provide adequate education, housing, medical care, and other social services for them? Computer models suggest not (Figure 26.20). Yet some people claim we can adapt politically and socially to a more crowded world if innovative technologies improve harvests, food resources are distributed more equitably, and diets are shifted away from animal products.

There are no easy answers. If you have not been doing so, start following the arguments in the media. It is a good idea to become an informed participant in a debate that will have impact on your future.

IMPACTS OF NO GROWTH

In a growing population, most people are in younger age brackets. When living conditions favor moderate population growth over time, then age distribution should guarantee a future workforce. However, this has social implications. Why? *It takes a large workforce to support individuals in the higher age brackets.*

In the United States, most of the retired seniors expect the government to subsidize their medical care and provide low-cost housing and many other social programs. However, as a result of improved medicine and hygiene, people in these brackets are living far

longer than seniors did when the nationwide social security program was established. Cash benefits now exceed the contributions that individuals made to the program when *they* were younger.

If the population ever does reach and maintain zero growth, a larger proportion of individuals will end up in higher age brackets. Even slower growth poses problems. Will these people continue to receive goods and services as the workforce carries more and more of the economic burden? Put it to yourself. How much economic hardship are you willing to bear for the sake of your parents? Your grandparents? How much will your children bear for you?

We have arrived at a turning point, not only in our biological evolution but in our cultural evolution as well. The decisions awaiting us are among the most pressing and difficult we will ever have to make.

All species face limits to growth. We might think we are different from the rest, for our special ability to undergo rapid cultural evolution has enabled us to postpone the action of most factors that limit growth. The crucial word is *postpone*. No amount of cultural intervention can sidestep the ultimate check imposed by limited resources and a damaged environment.

Differences in population growth among countries correlate with levels of economic development. Growth rates are low in preindustrial, industrial, and post-industrial stages. They are greatest during the transition to industrialization.

Per capita resource consumption is greatest in the most developed countries, which also must sustain an increasing proportion of people in older age brackets.

Summary

Section 26.1 A population is a group of individuals of the same species occupying a given area. Typically we measure its size, density, distribution, and age structure. Most populations in nature show a clumped distribution pattern.

Section 26.2 Counting the number of individuals in representative quadrats and capture–recapture methods are used to estimate population density.

Section 26.3 The growth rate for a population during a specified interval is determined by the rates of birth, death, immigration, and emigration. If we put aside effects of immigration and emigration, we may represent population growth (G) as $G = rN$, where N is the number of individuals in a specified interval and r is the net reproduction per individual per unit time.

In cases of exponential growth, the reproductive base of a population increases and its size expands by a certain percentage during successive intervals. This trend plots out as a J-shaped growth curve. As long as a population's per capita birth rate is above its per capita death rate, it will show exponential growth.

Section 26.4 Any essential resource that is in short supply is a limiting factor on population growth. The logistic growth equation describes how population growth is affected by density-dependent controls, such as competition for resources or disease. The population slowly increases in size, goes through a rapid growth phase, then levels off in size once carrying capacity is reached. Carrying capacity is the maximum number of individuals that can be sustained indefinitely by the resources available in their environment.

Density-independent factors affect populations regardless of how crowded or uncrowded they are.

Section 26.5 Each species has a life history pattern characterized by traits such as age at first reproduction, number of offspring per brood, and life span. There are three general types of survivorship curves.

Section 26.6 Life history traits can have a genetic basis and vary among populations because of differing selective pressures from predation or other factors.

Section 26.7 The human population has surpassed 6.4 billion. Its rapid growth in the past two centuries occurred through expansion into many diverse habitats and through agricultural, medical, and technological developments that raised carrying capacity.

Section 26.8 Family planning is helping to slow population growth. The worldwide total fertility rate is declining. However, the broad pre-reproductive base of the world population means that human population size will continue to increase for at least sixty years.

Section 26.9 The demographic transition model correlates industrial and economic development with changes in population growth rates. Growth rates are lower in developed nations than in developing nations. But, the per capita resource consumption of developed nations is many times higher.

Self-Quiz

Answers in Appendix III

1. The rate at which a population grows or declines depends on the rate of _____ .
   a. births   c. immigration   e. a and b
   b. deaths   d. emigration   f. all of the above

2. Populations grow exponentially when _____ .
   a. population size expands by ever increasing increments through successive time intervals
   b. size of low-density population increases slowly, then quickly, then levels off once carrying capacity is reached
   c. a and b are characteristics of exponential growth

3. For a given species, the maximum rate of increase per individual under ideal conditions is its _____ .
   a. biotic potential   c. environmental resistance
   b. carrying capacity   d. density control

4. Resource competition, disease, and predation are _____ controls on population growth rates.
   a. density-independent   c. age-specific
   b. population-sustaining   d. density-dependent

5. A life history pattern for a population is a set of adaptations that influence the individual's _____ .
   a. survival   c. age when it starts reproducing
   b. fertility   d. all of the above

6. In 2003, the worldwide average rate of increase for the human population at midyear was _____ percent.
   a. 0   c. 1.3   e. 3.8
   b. 0.5   d. 2.7   f. 4.6

7. Match each term with its most suitable description.
   ____ carrying   a. maximum rate of increase per
        capacity   individual under ideal conditions
   ____ exponential   b. population growth plots out
        growth   as an S-shaped curve
   ____ biotic   c. maximum number of individuals
        potential   sustainable by the resources
   ____ limiting   in a given environment
        factor   d. population growth plots out
   ____ logistic   as a J-shaped curve
        growth   e. essential resource that restricts
                             population growth when scarce

Critical Thinking

1. If house cats that haven't been neutered or spayed live up to their biotic potential, two can be the start of many kittens—12 the first year, 72 the second year, 429 the third, 2,574 the fourth, 15,416 the fifth, 92,332 the sixth, 553,019 the seventh, 3,312,280 the eighth, and 19,838,741 kittens the ninth year. Is this a case of logistic growth? Exponential growth? Irresponsible cat owners?

Figure 26.21 Saguaros (*Canegiea gigantea*) growing very slowly in Arizona's Sonoran desert.

Figure 26.22 A young Malian, with a 10 percent chance of becoming a mother by age fifteen, and a 50 percent chance by age nineteen. Mali's FTR, about 7, is one of the world's highest.

2. As described in Section 26.6, when guppies from pools with pike-cichlids were moved to other pools dominated by smaller killifish, the guppies evolved. Over time, they came to resemble populations that had historically been faced with cichlid predation. Age at first reproduction increased, as did body size. There was also a change in the males' coloration. Compared to their ancestors, the transplanted males were far gaudier. They had bigger and more colorful spots. What could have been the selection pressure driving the change in male coloration? Why might a change in predator pressure affect guppy coloration?

3. Each summer, the giant saguaro cactus produces tens of thousands of tiny black seeds. Most die, but a few land in a sheltered spot and sprout the following spring. The saguaro is a CAM plant (Section 6.6) and growth is slow. After 15 years, the saguaro may be only knee high, and it won't flower until about age 30. The cactus may survive to age 200. Saguaros share their habitat with annuals, such as poppies, that sprout, produce seeds, and die in just a few weeks (Figure 26.21). Explain how these very different life histories could both be adaptive in a desert environment.

4. A third of the world population is younger than fifteen (Figure 26.22). Describe the effect of this age distribution on our population's future growth rate. If you think it will have severe impact, what humane recommendations would you make to encourage individuals of this age group to limit their family size? What are some social, economic, and environmental factors that might prevent them from following the recommendations?

5. Write a short essay about a population having one of the age structures below. Speculate about the population's current economic status and the social and economic problems its population may face in the future.

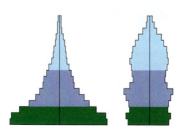

Media Menu

Student CD-ROM

Impacts, Issues Video
    The Human Touch
Big Picture Animation
    Population characteristics and growth
Read-Me-First Animation
    Exponential growth
    Logistic population growth
    Age structure diagrams
Other Animations and Interactions
    The demographic transition model

InfoTrac

- One Marsupial Too Many. *Discover*, December 2000.
- Does Population Matter? *The Economist* (US), December 2002.
- HIV/AIDS Lowers UN's Global Population Estimates. *The Lancet*, March 2003.
- Mysterious Island: The More We Learn About Easter Island, the More It Intrigues. *Smithsonian*, March 2002.

Web Sites

- Easter Island Foundation: www.islandheritage.org
- World POPClock: www.census.gov/cgi-bin/ipc/popclockw
- United Nation's Population Information Network: www.un.org/popin
- Population Reference Bureau: www.prb.org

How Would You Vote?

According to the Office of Homeland Security, about 7 million undocumented immigrants live in the United States. Many citizens argue that the costs of medical care and other social services for those who did not enter the country legally are overwhelming. Others argue that these are hard-working people who deserve opportunities. In the 1990s, 1.5 million of the undocumented immigrants were granted legal status. Would you support granting legal status to all 7 million who live here now?

IMPACTS, ISSUES *Fire Ants in the Pants*

Two species of Argentine ants, *Solenopsis richteri* and *S. invicta*, entered the United States during the 1930s, probably as stowaways on cargo ships. They infiltrated communities throughout the Northern Hemisphere, starting with the southeastern states. *S. invicta*, the red imported fire ant, recently colonized Southern California and New Mexico.

Disturb a fire ant on your skin and it will bite down even as a stinger on its abdomen pumps venom into you. Searing pain precedes the formation of a pus-filled bump on the skin (Figure 27.1). At one time or another, about one-half of the Americans who live where fire ants are common have been stung. More than eighty have died from the attacks.

Imported fire ants menace more than people. These insects will attack just about anything that disturbs them, including livestock, pets, and wildlife. They also are more competitive than native ant species and other animals that feed on insects. They may be contributing to the declines of some native wildlife species.

For example, the Texas horned lizard (*Phrynosoma cornutum*) disappeared from most of its home range when the red fire ant moved in. Native ants have been its food of choice. When the native ant populations dwindled, the lizard ran out of luck, because it cannot eat the invaders. As other examples, we have reports of fire ants directly killing young lizards and hatchlings of quail and other birds that nest on the ground.

Figure 27.1 *From left to right*, fire ant mounds in west Texas, fire ants swarming all over a leather boot, and the type of skin eruptions that follow a concerted attack by these exotic imports.

the big picture

Community Characteristics Species interactions help to organize the community of any given habitat. In that community, each species has a unique niche, which may shift in small and large ways over time.

Species Interactions Diverse species interact indirectly or as mutualists, as predators and prey, or as parasites and hosts. Through their interactions, they exert selection pressures on one another, and in many cases they coevolve.

Invicta means invincible in Latin. So far, *S. invicta* is living up to its species name. Pesticides have not slowed its expansion into new habitats and may actually be helping it by knocking off the native populations.

Ecologists are enlisting biological controls. Two phorid fly species attack *S. invicta* in its native habitat. Both species are parasitoids, a rather gruesome type of parasitic predator. The female fly pierces an adult ant's cuticle and lays an egg in its soft tissues. When the fly larva hatches, it eats the tissues as it burrows along toward the head. After growing large enough, the larva secretes an enzyme that makes the ant head fall off. Sheltered inside the detached, cuticle-covered head, the larva undergoes metamorphosis into an adult. The flies are choosy about their host/prey; they have not targeted ants native to the United States.

In 1997, ecologists released one parasitic fly species in Florida. In 2001, they released a second species in other southern states. It is too soon to know whether the biological controls are working. As researchers wait for results, they are exploring other options. One idea is to use imported fungal or protistan pathogens that will infect *S. invicta* but not native ants. Another is to introduce a parasitic ant species that invades *S. invicta* colonies and decapitates the queens.

In any given habitat, species interact as part of a dynamic community. As you will see in this chapter, their interactions shift in small and large ways—some predictable, and others entirely unexpected.

How Would You Vote?

Currently, only a fraction of the crates being imported into the United States are inspected for exotic species. Would the cost of added inspections be worth it? See the Media Menu for details, then vote online.

Stability and Change   Stable arrays of species develop over time and dominate a habitat as a climax community. Introduction of exotic species and other forces often tip the balance and destabilize communities.

Focus on Biodiversity   Biologists have identified environmental and historical factors that shape global patterns of biodiversity. The current extinction crisis, partly a result of human activities, is threatening global biodiversity. Many conservation efforts are under way.

27.1 Which Factors Shape Community Structure?

The type of place where each organism normally lives is its habitat. Habitats have physical and chemical features, such as temperature, and an array of species. Directly or indirectly, populations of all species in each habitat associate with one another as a community.

A community has a characteristic structure. It arises as an outcome of five factors. *First*, the climate and topography determine temperatures, rainfall, type of soil, and other conditions. *Second*, the kinds and quantities of food and other resources that become available through the year affect which species live there. *Third*, adaptive traits of each species help it to survive and exploit specific resources in the habitat. *Fourth*, species engage in competition, predation, and mutually beneficial actions (Figure 27.2). *Fifth*, the overall pattern of population sizes affects community structure. This pattern reflects the history of changes in population size, the arrivals and disappearances of species, and physical disturbances to the habitat.

Such factors affect the number of species, starting with producers and continuing through consumers. They influence population sizes and help determine the overall number of species. High solar radiation, temperatures, and humidity in tropical habitats favor the growth of many kinds of plants, which support many kinds of animals. Arctic habitats have conditions that do not favor great numbers of species.

THE NICHE

All species of a community share the same habitat, the same "address," but each has a "profession" that sets it apart from the rest. It has a distinct **niche**, the sum of its activities and interactions as it goes about acquiring and using the resources required to survive and reproduce. Its *fundamental* niche is the one that would prevail in the absence of competition and any other factors that could limit how individuals get and use resources. Constraining factors that do come into play tend to bring about a more limited, *realized* niche. The realized niche for a species is dynamic; it may shift over time, in small or large ways.

CATEGORIES OF SPECIES INTERACTIONS

Even in the simplest communities, dozens to hundreds of species typically interact in one of five ways. Some are indirect interactions. For instance, when Canadian lynx eat a lot of snowshoe hares, they indirectly help plants that the hares eat. Plants indirectly help the lynx by fattening hares and providing cover for ambushes.

Commensalism directly helps one species but does not affect the other much, if at all. When a bird uses a tree for a roosting site, the tree gets nothing but is not hurt. In **mutualism**, both species benefit, but not from cozy cooperation. This is a two-way exploitation. In **interspecific competition**, both species can be harmed. **Predation** and **parasitism** directly benefit one species and harm the other. Predators kill prey, and parasites weaken hosts. Commensalism, mutualism, predation, and parasitism are all forms of *symbiosis*, defined as a close association between two or more species during part or all of the life cycle.

> *Community structure arises from a habitat's physical and chemical features, resource availability over time, adaptive traits of its species, how its species interact, and the history of the habitat and its occupants.*
>
> *A niche is the sum of all activities and relationships in which individuals of a species engage as they secure and use the resources necessary to survive and reproduce.*
>
> *Indirect interactions among species as well as symbiotic ones—commensalism, mutualism, competition, predation, and parasitism—help shape community structure.*

Figure 27.2 Three of twelve fruit-eating pigeon species in New Guinea's tropical rain forests. *Left to right*, the tiny pied imperial pigeon, the superb crowned fruit pigeon, and the turkey-sized Victoria crowned pigeon. The forest's trees differ in the size of fruit and fruit-bearing branches. The big pigeons eat big fruit. Smaller ones, with smaller bills, can't peck open big, thick-skinned fruit. They eat small, soft fruit on branches too spindly to hold big pigeons.

The trees feed the birds, which help the trees. Seeds in fruit resist digestion in the pigeon gut. Flying pigeons dispense seed-containing droppings, often some distance from the parent trees. Seedlings don't compete well against mature, well-developed trees for water, mineral ions, and sunlight.

27.2 Mutualism

In a mutualistic interaction, two species take advantage of their partner in ways that benefit both, as when one filches nutrients from the other while sheltering it.

Interactions in which positive benefits flow both ways abound in nature. As rain forest trees provide pigeons with food, the pigeons help disperse the trees' seeds to new sites (Figure 27.2). Many plants and animals enter into such interactions. Most flowering plants are mutualists with insects, birds, bats, and other animals that pollinate them or disperse their seeds. Chapter 21 provides vivid examples.

Mutualism may be obligatory; each species must have access to the other in order to complete its life cycle and reproduce. This is the case for yucca plants and the yucca moths that pollinate them (Figure 27.3).

A lichen is a case of obligatory mutualism between certain fungi and a photosynthetic protist or bacteria. So is a mycorrhiza, in which fungal hyphae penetrate root cells or form a dense, velvety mat around them. As earlier chapters explain, the fungus is efficient at absorbing mineral ions from soil, and the plant grows better when it pilfers some of the ions. In turn, the fungus pilfers some of the sugar molecules the plant makes by way of photosynthesis. The fungus depends on the plant for reproductive success; it will not make spores if the plant stops photosynthesizing.

An anemone fish cannot live and reproduce away from sea anemones. These cnidarians have nematocyst-laden tentacles that shelter and protect the fish from many reef organisms. A lot of sea anemones survive and reproduce without help, but the species shown in Figure 27.4 gets help from *Amphiprion perideraion*—a tiny partner that chases away butterflyfishes. Like the partner, butterflyfishes are impervious to nematocysts, but they have a penchant for nipping off the tentacles.

And what about the apparent endosymbiotic origin of eukaryotes? Long ago, phagocytes were engulfing aerobic bacterial cells—but some resisted digestion, tapped host nutrients, and then kept on reproducing independently of the host cell body. In time the hosts came to depend on the ATP produced by the guests—which evolved into mitochondria and chloroplasts. If those ancient prokaryotic cells had not coevolved as such intimate mutualists, you and all other eukaryotic species would not be around today (Section 20.4).

> In cases of mutualism, both of the participating species benefit from the two-way exploitation. Some mutualistic relationships are obligatory for one or both partners.

Figure 27.3 Mutualism on a rocky slope of Colorado's high desert.

Only one yucca moth species pollinates plants of each *Yucca* species; it cannot complete its life cycle with any other plant. The moth matures when yucca flowers blossom. The female has specialized mouthparts that collect and roll sticky pollen into a ball. She flies to another flower and pierces its ovary, where seeds will form and develop, and lays eggs inside. As she crawls out, she pushes a ball of pollen onto the flower's pollen-receiving platform.

After germination, pollen tubes grow through the ovary tissues and deliver sperm to the plant's eggs. Seeds develop after fertilization. Meanwhile, the moth eggs develop into larvae that eat a few seeds, then gnaw their way out of the ovary. Seeds that larvae do not eat give rise to new yucca plants.

Figure 27.4 Mutualism near the Solomon Islands. This species of sea anemone (*Heteractis magnifica*) interacts with about a dozen species of fishes, including the pink anemone fish (*Amphiprion perideraion*). This tiny, aggressive fish chases away predatory butterflyfishes that can bite off the sea anemone's tentacles. In return, the fish and its eggs get protection and shelter—scarce commodities on coral reefs.

27.3 Competitive Interactions

Where you find limited supplies of energy, nutrients, living space, and other natural resources, you can expect to find organisms competing for a share of them.

Chapters 26 and 30 address *intraspecific* competition between members of the same species, which can be fierce. *Interspecific* competition, between populations of different species, usually isn't as intense. Why not? *The requirements of two species may be similar but are never as close as they are among individuals of the same species.*

In cases of *interference* competition, populations of a species control or prevent access of another species to a resource, regardless of its abundance. Figure 27.5 is one example. Hummingbirds are another. In spring and early summer, a male broadtailed hummingbird evicts birds of his own species from his flower-rich territory in the Rocky Mountains. In August, rufous hummingbirds migrate through the Rockies on the way to wintering grounds in Mexico. Rufous males are more aggressive, stronger competitors; they force male broadtails to give up territory until they move on.

One other example: Certain plants, such as many sages (*Salvia*), exude aromatic compounds from their leaves. The compounds taint the soil around the plant and prevent potential competitors from taking root.

In cases of *exploitative* competition, populations of different species have equal access to a resource, but one is better at using it. Various water fleas (*Daphnia*) eat the same alga. In one experiment, a large-bodied species increased in body mass and numbers, but a smaller species fed the same alga in the same culture actually decreased in body mass and numbers. And this leads us into a theory about competition.

COMPETITIVE EXCLUSION

Any two species differ to a greater or lesser extent in their capacity to secure and use resources. The more they overlap in these respects, the less likely they are to coexist in the same habitat.

Years ago, G. Gause found evidence of this when he grew two species of *Paramecium* separately and then together (Figure 27.6). Both ciliated protozoans require the same food—bacteria—and compete intensely for it. Gause's species, which require identical resources, could not coexist indefinitely. Later experiments with water fleas and many other species yielded the same results, in support of what ecologists now call the theory of **competitive exclusion**.

Gause also studied two other *Paramecium* species that did not overlap much in requirements. When he grew them together, one species tended to feed upon bacteria suspended in culture tube liquid. The other ate yeast cells at the tube bottom. Population growth

Figure 27.5 Example of interspecific competition in nature. On the slopes of the Sierra Nevada, competition helps keep nine species of chipmunks (*Tamias*) in different habitats.

The alpine chipmunk (**a**) lives in the alpine zone, the highest elevation. Below it are the lodgepole pine, piñon pine, and then sagebrush habitat zones. Lodgepole pine chipmunks (**b**), least chipmunks (**c**), and other species live in the forest zones. Merriam's chipmunk (**d**) lives at the base of the mountains, in sagebrush. Its traits would allow it to move up into the pines, but the aggressively competitive behavior of forest-dwelling chipmunks won't let it. Food preferences keep the pine forest chipmunks out of the sagebrush habitat.

Read Me First!
and watch the narrated animation on
competitive exclusion

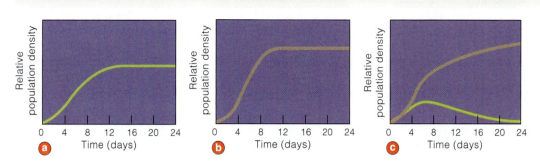

Paramecium
caudatum

P. aurelia

Figure 27.6 Results of competitive exclusion between two protozoan species that compete for the same food. (**a**) *Paramecium caudatum* and (**b**) *P. aurelia* were grown in separate culture tubes and established stable populations. The S-shaped graph curves indicate logistic growth and stability.

(**c**) Then the two species were grown together. *P. aurelia* (*brown* curve) drove *P. caudatum* toward extinction (*green* curve in **c**). This experiment and others suggest that two species cannot coexist indefinitely in the same habitat *when they require identical resources*. If their requirements do not overlap much, one might influence the population growth rate of the other, but they may still coexist.

rates slowed for both species—but the overlap in use of resources was not enough for one species to fully exclude the other. The two continued to coexist.

RESOURCE PARTITIONING

Think back on those fruit-eating pigeon species. They all use the same resource: fruit. Yet they overlap only a bit in their use of it, because each prefers fruits of a certain size. They are a case of **resource partitioning** —a *subdividing* of some category of similar resources that lets competing species coexist.

Similarly, three annual plant species live in the same plowed, abandoned field. All require sunlight, water, and minerals. Each exploits a slightly different part of the habitat (Figure 27.7). Bristly foxtail grasses have a shallow, fibrous root system that absorbs water fast during rains. They grow where moisture shifts daily, and are drought-tolerant. Indian mallow has a taproot system in deeper soil that is moist early in spring and drier later. The taproot system of smartweed branches in topsoil and soil below the roots of other species. It grows where soil is always moist.

In some competitive interactions, one species controls or blocks access to a resource, regardless of whether it is scarce or abundant. In other interactions, one is better than another at exploiting a shared resource.

When two species overlap too much in their requirements, they cannot coexist in the same habitat unless they share the required resources in different ways or at different times.

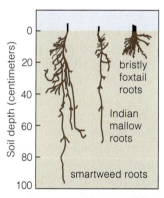

bristly foxtail

Indian mallow   smartweed

Figure 27.7 Resource partitioning among three annual plant species in an abandoned field. The plants differ in how they are adapted to secure soil water and mineral ions. The roots of each species tap into a different region of the soil.

27.4 Predator–Prey Interactions

Predators are consumers that obtain energy and nutrients from living organisms—their prey—which they generally capture and kill. The quantity and types of prey species affect predator diversity and abundance, and the types of predators and their numbers do the same for prey.

COEVOLUTION OF PREDATORS AND PREY

Coevolution influences the characteristics of prey and predator. The term, recall, refers to the joint evolution of two or more species that exert selection pressure on one another through their close ecological interaction over the generations. Suppose a gene mutation arises in a prey population and leads to improved defense against predators. The mutant gene will increase in frequency, because its bearers will tend to survive in greater numbers, as will their offspring.

Suppose a heritable trait also makes individuals of a predator population better at overcoming the novel prey defense. Those individuals will eat better and so survive and leave more descendants. Over time, the predators are selective agents that favor improved defenses in prey; and prey with better defenses are selective agents that favor more effective predators.

MODELS FOR PREDATOR–PREY INTERACTIONS

The extent to which predators limit numbers of prey depends on several factors. A key factor is the response of individual predators to increases or decreases in prey density. Figure 27.8a is an overview of the three general patterns of functional responses.

By the type I model, a predator removes a constant proportion of prey over time, regardless of levels of prey abundance. The number of prey killed during any given interval depends only on prey density. This model applies to passive predators, such as spiders. The more flies there are, the more get caught in webs.

By the type II model, the ability of predators to consume and digest prey determines how many prey they capture. As prey density rises, the proportion captured rises steeply at first, then slows as predators are exposed to more prey than they can deal with at one time. Figure 27.8b gives an example. A wolf that has just killed a caribou will not hunt another until it has eaten and digested the first one.

By the type III model, predator response is lowest when prey density is low. It is highest at intermediate prey densities, then levels off. This type of response is observed for predators that can switch to other prey when low abundance makes a prey species hard to find. Predators that show type I and type II responses can limit prey at a stable equilibrium point.

Other factors besides individual predator response to prey density are at work. For example, predator and prey reproductive rates affect the interaction. So do hiding places for prey, the presence of other prey or predator species, and carrying capacities.

THE CANADIAN LYNX AND SNOWSHOE HARE

Shifts in environmental conditions can sometimes cause predator and prey densities to oscillate. At the lowest level, predation strongly depresses the prey density. At the highest level, predation is absent and the prey population nears the carrying capacity.

Consider a ten-year oscillation in populations of one predator, the Canadian lynx, and the snowshoe

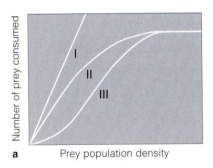
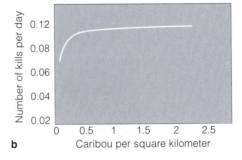

Figure 27.8 **(a)** Three alternative models for responses of predators to prey density. Type I: Prey consumption rises linearly as prey density rises. Type II: Prey consumption is high at first, then levels off as predator bellies stay full. Type III: When prey density is low, it takes longer to hunt prey, so the predator response is low. **(b)** A type II response in nature. For one winter month in Alaska, B. W. Dale and his coworkers observed four wolf packs (*Canis lupus*) feeding on caribou (*Rangifer tarandus*). The interaction fit the type II model for the functional response of predators to prey density.

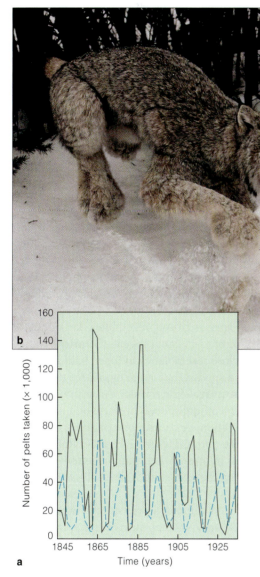

Figure 27.9 (**a**) Correspondence between abundances of Canadian lynx (*dashed* line) and snowshoe hares (*solid* line), based on counts of pelts sold by trappers to Hudson's Bay Company during a ninety-year period. (**b**) Charles Krebs observed that predation causes mass paranoia among snowshoe hares, which continually look over their shoulders during the declining phase of each cycle. (**c**) This photograph supports the Krebs hypothesis that there is a three-level interaction going on, one that involves plants.

The graph might be a good test of whether you tend to accept someone else's conclusions without questioning their basis in science. Remember those sections in Chapter 1 that introduced the nature of scientific methods?

What other factors may have had impact on the cycle? Did the weather vary, with more severe winters imposing greater demand for hares (to keep lynx warmer) and higher death rates? Did the lynx compete with other predators, such as owls? Did the predators turn to alternative prey during low points of the hare cycle? When fur prices rose in Europe, did the trapping increase? When the pelt supply outstripped the demand, did trapping decline?

hare that is its main prey (Figure 27.9). To identify the causes of this pattern, the ecologist Charles Krebs and his coworkers tracked hare population densities for ten years in Alaska, in the Yukon River Valley. They set up 1-square-kilometer control plots and experimental plots. Electric fences kept all predatory mammals out of some plots. Extra food or fertilizers that fanned plant growth were placed in other plots. The team captured and released more than a thousand snowshoe hares, lynx, and other animals, giving each a radio collar.

In predator-free plots, the hare density doubled. In plots with extra food, it tripled. In plots having extra food and fewer predators, it increased elevenfold.

The experimental manipulations delayed the cyclic declines in population density but did not stop them. Why not? Fences could not keep out owls and other raptors. Only 9 percent of the collared hares starved to death; predators devoured most of the rest. Krebs concluded a simple predator–prey or plant–herbivore model cannot explain his results in the Yukon River Valley. For the lynx and snowshoe hare cycle, other variables are at work, during multilevel interactions.

Predator and prey populations tend to exert coevolutionary pressures on one another.

Predators may affect prey density. There are three general patterns of response to changes in prey density. Population levels of prey may also show periodic oscillations.

Predator and prey numbers often vary in complex ways that reflect the multiple levels of interaction in a community.

27.5 An Evolutionary Arms Race

As explained in the preceding section, predators and prey exert selective pressure on one another. One defends itself and the other must overcome defenses. Such interactions started an evolutionary arms race that has resulted in adaptations that often amaze us.

ADAPTATIONS OF PREY

Some prey gain protection from **camouflaging**, or hiding in the open. Their body form, patterning, color, behavior, or some combination of these blend with the surroundings and help avert detection. For example, *Lithops* is a desert plant that looks like a rock (Figure 27.10). It flowers only during a brief rainy season, when other plants grow profusely and distract herbivores from it—and when drinking water (not juicy plant tissues) becomes available.

Many prey species taste bad or contain chemical toxins that sicken their predators. They may have **warning coloration**, conspicuous patterns and colors that predators learn to recognize as *AVOID ME!* signals. A young, inexperienced bird may spear an orange-and-black patterned monarch butterfly once. It learns quickly to associate this coloration and patterning with vomiting foul-tasting toxins.

Dangerous or repugnant species often make little or no attempt to conceal themselves. Skunks, which spray one of the most odious repellents, are like this. So are certain frogs (*Dendrobates*), which are among the most poisonous and vividly colored species.

Many prey species closely resemble a dangerous, unpalatable, or hard-to-catch species. **Mimicry** is the name for the close resemblance in form, behavior, or both between a species that is the model for deception and a different species (the mimic). After predators learn to ignore a model species because of a repellent taste, toxic secretion, or painful bite or sting, they tend also to avoid the mimic. Figure 27.11 shows three tasty but weaponless mimics. All strongly resemble a very aggressive wasp that can sting repeatedly, with painful results.

When luck runs out and a prey animal is cornered or under attack, survival may turn on a last-chance trick. Animals may startle the predator, as by hissing, puffing up, showing teeth, or flashing big eye-shaped spots. A lizard or salamander tail may detach from the body and still wiggle a bit as a distraction. Opossum pretend to be dead. So do hognose snakes, which, like many other prey, also secrete smelly fluid. Many cornered animals secrete or squirt irritating chemical repellents or toxins (Figure 27.12a).

Figure 27.10 Prey camouflage. (**a**) What bird? When a predator approaches its nest, the least bittern stretches its neck (which is colored like the surrounding withered reeds), thrusts its beak upward, and sways gently like reeds in the wind. (**b**) An inedible bird dropping? No. This caterpillar's body coloration and the rigid body positions that it assumes help camouflage it from predatory birds. (**c**) Find the plants (*Lithops*) hiding in the open from herbivores with their stonelike form, pattern, and coloration.

a The dangerous model **b** One of its edible mimics **c** Another edible mimic **d** And another edible mimic

Figure 27.11 An example of mimicry. Edible insect species often resemble toxic or unpalatable species that are not at all closely related. (**a**) A yellowjacket can deliver a painful sting. It may be the model for nonstinging wasps (**b**) beetles (**c**), and flies (**d**) that have a similar appearance.

Also, the foliage and seeds of many plants contain bitter, hard-to-digest, or dangerous repellents. Peach, apricot, and rose seeds are loaded with cyanide. And remember that castor bean plant (Chapter 13)? It did not develop its capacity to make the lethal chemical ricin in an evolutionary vacuum. Ricin protects this plant from herbivores that would otherwise eat it.

ADAPTIVE RESPONSES OF PREDATORS

In a coevolutionary arms race, predators counter prey defenses with their own adaptations. Stealth, camouflage, and ingenious ways of avoiding the repellents are some countermeasures. Consider the edible beetles that direct sprays of noxious chemicals at attackers. A grasshopper mouse grabs the beetle and plunges the sprayer end into the ground, and eats the tasty unprotected head (Figure 27.12*b*).

Some prey can outrun predators when they get a head start. But consider the cheetah, the world's fastest land animal. One cheetah was clocked at 114 kilometers per hour (70 mph). Compared to other big cats, the cheetah has longer legs relative to its body size and nonretractable claws that act like cleats to increase traction. Its main prey, Thomson's gazelle, can run longer, but not as fast (80 kilometers per hour); it needs the head start.

Camouflaged predators often lay in wait for their prey. Think of white polar bears stalking seals over ice, striped tigers crouched in tall-stalked, golden grasses, and scorpionfish well hidden on the seafloor (Figure 27.12*c*). Many predatory insects, too, are deception experts (Figure 27.12*d*). Such camouflaged predators select for enhanced sensory systems in prey. By one theory, primate color vision may have evolved in part to enhance detection of predators.

Figure 27.12 Predator responses to prey defenses. (**a**) Some beetles spray noxious chemicals at attackers, which deters them some of the time. (**b**) At other times, grasshopper mice plunge the chemical-spraying tail end of their beetle prey into the ground and feast on the head end. (**c**) Find the scorpionfish, a venomous predator with camouflaging fleshy flaps, multiple colors, and profuse spines. (**d**) Where do the pink flowers end and the pink praying mantis begin?

27.6 Parasite–Host Interactions

Parasites spend all or part of their life cycle in or on other living organisms, from which they draw nutrients. The host is harmed, but usually is not killed outright. Parasites may be adapted to one particular host species, or able to survive in any of a number of hosts.

PARASITES AND PARASITOIDS

Parasites have pervasive impact on populations. By draining nutrients from hosts, they alter the amount of energy and nutrients the host population demands from a habitat. Also, weakened hosts are usually more vulnerable to predation and less attractive to potential mates. Some parasite infections cause sterility. Others shift the ratio of host males to females. In such ways, parasitic infections lower birth rates, raise death rates, and affect intraspecific and interspecific competition.

Sometimes the gradual drain of nutrients during a parasitic infection indirectly leads to death. The host becomes so weakened that it can't fight off secondary infections. Nevertheless, in evolutionary terms, killing

a host too quickly is bad for a parasite's reproductive success. A parasitic infection must last long enough to give the parasite time to produce some offspring. The longer it lives in the host, the more offspring. We can therefore expect selective agents to favor parasites that have less-than-fatal effects on their hosts.

Only two types of interactions typically kill a host. First, a parasite may attack a novel host, one with no coevolved defenses against it. Second, the host may be supporting too many individual parasites at the same time. The collective assault may overwhelm the body.

Previous chapters introduced diverse parasites that live in or on a host. Some spend their entire life cycle with a single host. Others are free-living some of the time, or residents of different hosts at different times. Many parasites ride inside insects and other arthropods, the taxis between one host organism and the next.

All viruses and some of the bacteria, protists, and fungi are parasites. Figure 27.13 shows one parasitic fungus that infects amphibians of tropical rain forests.

Even a few plants are parasitic. Nonphotosynthetic types, such as dodders, obtain energy and nutrients from other plants (Figure 27.14). Other types carry out photosynthesis but still tap into the nutrients and water in a host plant's tissues. Mistletoe is like this; its modified roots invade the sapwood of host trees.

Many tapeworms, flukes, and some roundworms are notorious invertebrate parasites (Figure 27.15). So are many insects and crustaceans, as are all ticks.

Parasitoids are insects that develop inside another species of insects, which they devour from the inside out as they mature. Unlike parasites, the parasitoids always kill their hosts. About 15 percent of all insects may be parasitoids.

Social parasites are animals that take advantage of the social behavior of a host species in order to carry out their life cycle. The cuckoos and North American cowbirds, described shortly, are like this.

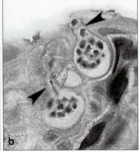

Figure 27.13 (a) Harlequin frog (*Atelopus varius*) from Central America. (b) Section through skin of a harlequin frog affected by *chytridiomycosis*, an infectious disease caused by a fungus. *Arrows* point to flask-shaped cells of the fungal parasite. Each cell holds spores that will be released to the skin surface. They may be dispersed to other host frogs.

Figure 27.14 Dodder (*Cuscuta*), also known as strangleweed or devil's hair. Sporophytes of this parasitic flowering plant lack chlorophylls. They wind around a host plant during growth. Modified roots penetrate the host's vascular tissues and draw water and nutrients from them.

Figure 27.15 Adult roundworms (*Ascaris*), an endoparasite, packed inside the small intestine from a host pig. Sections 22.5 and 22.8 give more examples of parasitic worms.

Figure 27.16 A biological control agent—a commercially raised parasitoid wasp about to deposit an egg in an aphid. This wasp reduces aphid populations by crippling the egg laying capacity of its host even before its larva kills it.

USES AS BIOLOGICAL CONTROLS

Parasites and parasitoids are commercially raised and released in target areas as *biological controls*. They are promoted as a workable alternative to pesticides. The chapter introduction and Figure 27.16 give examples.

Effective biological controls display five attributes. The agents are adapted to a specific host species and to its habitat; they are good at locating hosts; their population growth rate is high compared to the host's; their offspring are good at dispersing; and they make a type III functional response to prey, without much lag time after shifts occur in the host population size.

Biological control is not without risks of its own. Releasing more than one kind of biological control agent in an area may trigger competition among them and lessen their effectiveness. Introduced parasites sometimes parasitize nontargeted species as well as, or instead of, species they were expected to control.

In Hawaii, introduction of several parasitoids to control a non-native stink bug resulted in the decline of the koa bug, the state's largest native bug. Few koa bugs have been collected since 1978. Apparently, the parasitoids found the koa bugs, which congregate in large groups, a more tempting target than the pests they were imported to attack. Introduced parasitoids have also been implicated in the ongoing decline of many native Hawaiian butterflies and moths.

Natural selection favors parasitic species that temper their demands in ways that ensure an adequate supply of hosts.

Parasites occur in a vast variety of groups, including bacteria, protists, invertebrates and plants. Parasitoids are insects that feed on and kill other insects. Social parasites use the social behavior of another species to their own benefit.

27.7 Cowbird Chutzpah

FOCUS ON EVOLUTION

The brown-headed cowbird's genus name (Molothrus) means "intruder" in Latin. This bird intrudes, sneakily, into the life cycle of other species. But let us ask: Why?

Brown-headed cowbirds (*Molothrus ater*) evolved in the North American Great Plains. There they lived as commensalists with bison. Great herds of these hefty ungulates stirred up plenty of insects as they migrated through the grasslands, and, being insect-eaters, the cowbirds wandered around with them (Figure 27.17a).

A vagabond way of life did not lend itself to nesting in any one place. However it happened, the cowbirds learned to lay their eggs in nests built by other species, then leave them and move on with the herds. "Host" species did not have the neural wiring to recognize differences between cowbird eggs and their own eggs. Cowbird hatchlings became innately wired for hostile takeovers; even before their eyes open, they shove the owner's eggs out of the nest and demand to be fed as rightful occupants (Figure 27.17b). For thousands of years, the cowbirds successfully perpetuated their genes by way of such parasitic chutzpah.

Then pioneers moved west and cleared swaths of woodlands for pastures. Cowbirds moved in the other direction, easily adapting to the new ungulates—cattle —in the man-made grasslands; hence the name. They started penetrating woodlands and parasitizing novel species. Today, brown-headed cowbirds parasitize at least 15 species of native North American birds, some threatened or endangered. Besides being opportunists, they are big-time reproducers. A female can lay an egg a day for ten days, give her ovaries a rest, do the same again, and then again in one season. As many as thirty eggs in thirty nests—that is a lot of cowbirds.

Figure 27.17 Oh give me a home, where the buffalo roam—brown-headed cowbirds (*Molothrus ater*) originally evolved as commensalists with bison and as social parasites of other bird species of the North American Great Plains. When conditions changed, they expanded their range. They became nest usurpers in woodlands as well as grasslands in much of the United States.

27.8 Forces Contributing to Community Stability

*How does a community come into being? By the classical model for **ecological succession**, a community will slowly develop through a sequence of predictable stages to a final, stable array of species. But fire, treefalls, and other small-scale changes also affect community structure.*

A SUCCESSION MODEL

A community begins with **pioneer species**. These are all opportunistic colonizers of new or newly vacated habitats. They have high dispersal rates; they grow and mature quickly, and they produce many offspring. In time, more competitive species replace them. Then the replacements are replaced until the array stabilizes under prevailing conditions. Such a persistent species composition in a habitat is the **climax community**.

Primary succession is a process that begins when pioneer species colonize a barren habitat, such as a new volcanic island and land exposed when a glacier retreats (Figure 27.18). Pioneers include lichens and plants, such as club mosses, that are small, have short life cycles, and can survive intense sunlight, extreme temperature changes, and nutrient-poor soil. During the early years, hardy annual flowering plants put out many small seeds, which are quickly dispersed.

Once established, pioneers improve soil and other conditions. In doing so, they typically set the stage for their own replacement. Many of the new arrivals are mutualists with nitrogen-fixing bacteria, so they can grow in nitrogen-poor habitats. Seeds of later species find shelter inside mats of the pioneers, which do not grow high enough to shade out the new seedlings.

In time, organic wastes and remains accumulate. By adding volume and nutrients to soil, they favor invasions of new species. Later successional species crowd out the earlier ones, whose spores and seeds travel as fugitives on the wind and water—destined, perhaps, for another new but temporary habitat.

In **secondary succession**, a disturbed area within a community recovers. If improved soil is still there, secondary succession can be quite rapid. It commonly occurs in abandoned fields, burned forests, and tracts of land cleared by volcanic eruptions (Figure 27.19).

Figure 27.18 Primary succession in Alaska's Glacier Bay region. (**a**) A glacier is retreating from the sea. Meltwater leaches nitrogen and other minerals from the newly exposed soil. (**b**) The first invaders of nutrient-poor soil are lichens, horsetails, mosses, and fireweed and mountain avens. Some pioneer species are mutualists with nitrogen-fixing microbes. They grow and spread quickly over the glacial till.

Within twenty years, alder, cottonwood, and willow seedlings have taken hold in drainage channels. Alders, too, benefit from nitrogen-fixing symbionts. (**c**) Within fifty years, alders have formed dense, mature thickets in which cottonwood, hemlock, and a few evergreen spruce grow quickly. (**d**) After eighty years, western hemlock and spruce trees crowd out the mature alders. (**e**) In areas deglaciated for more than a century, forests of Sitka spruce dominate.

Figure 27.19 Secondary succession after Mount Saint Helens erupted in 1980 (**a**). At the base of this Cascade volcano, nothing remained of the climax community. (**b**) In less than a decade, pioneer species were getting established. (**c**) Twelve years later, seedlings of the dominant climax community species, Douglas firs, were taking hold.

THE CLIMAX PATTERN MODEL

At one time, ecologists thought the same general type of community always develops in a region because of climate constraints. However, diverse communities persist in similar climates, as when tallgrass prairie grades eastward into Indiana's deciduous forests.

By the **climax-pattern model**, each community is adapted to many environmental factors—topography, climate, soil, species interactions, periodic fires, and so on—that differ in their effects across a region. Even within a geographic region, climax communities may merge one into another along gradients of particular environmental conditions, such as moisture.

CYCLIC, NONDIRECTIONAL CHANGES

Many small-scale changes recur in small patches of habitats and affect a community's internal dynamics. Observe a disturbed area of habitat, and you might conclude that great shifts in species composition are afoot. Observe the community on a larger scale, and you might find that the overall composition includes all pioneer species and the dominant climax species.

Consider that a tropical forest develops through phases of colonization by pioneers, then increases in species diversity and matures. Whipping sporadically through the forest are strong winds that can knock down trees. Where trees fall, small gaps open in the dense tree canopy, and more light reaches patches on the forest floor. There, conditions favor growth of the previously suppressed small trees and germination of pioneers or shade-intolerant species.

Ice storms, tornadoes, hurricanes, and wildfires are among many other natural disturbances that can alter community structure.

A climax community is a stable, self-perpetuating array of species in equilibrium with one another and their habitat. It develops as a sequence of species, starting with pioneers.

Similar climax stages persist along gradients dictated by environmental factors and by species interactions. Small-scale, recurring changes help shape many communities.

27.9 Forces Contributing to Community Instability

The preceding sections might lead you to believe that all communities become stabilized in predictable ways. This is not always the case.

Community stability is an outcome of forces that have come into uneasy balance. Resources are sustained as long as populations do not reach carrying capacity. Predators and prey coexist as long as neither wins. Competitors continually engage in tugs-of-war over resources. Mutualists are stingy, as when plants make as little nectar as necessary to attract pollinators, and pollinators get as much nectar as they can for the least effort. In the short term, disturbances might slow the growth of some populations. Also, long-term changes in climate typically have destabilizing effects.

If instability becomes great enough, the community might change in ways that will persist even when the disturbance ends or is reversed. If some of its species happen to be rare or do not compete well with others, they might be driven to extinction.

THE ROLE OF KEYSTONE SPECIES

The uneasy balancing of forces in a community comes into focus when we observe the effects of a **keystone species**—a species that has a disproportionately large effect on a community relative to its abundance. Robert Paine was the first to describe the role of a keystone species after his experiments on the rocky shores of California's coast. Species in this rocky intertidal zone survive by clinging to rocks, and access to spaces to cling to is a limiting factor. Paine set up control plots with the sea star *Pisaster ochraceus* and its main prey—chitons, limpets, barnacles, and mussels. He removed all sea stars from his experimental plots.

With the sea stars gone, the mussels (*Mytilus*) took over Paine's experimental plots, and they crowded out seven other invertebrate species. The mussels are the preferred prey of the sea star, and the strongest competitors in their absence. Normally, predation by sea stars keeps the diversity of prey species high,

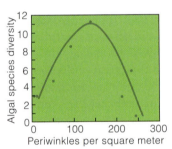

d Algal diversity in tidepools

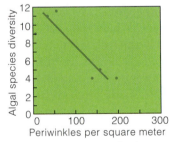

e Algal diversity on rocks that become exposed at high tide

Figure 27.20 Effect of competition and predation in an intertidal zone. (**a**) Grazing periwinkles (*Littorina littorea*) affect the number of algal species in different ways in different marine habitats. (**b**) *Chondrus* and (**c**) *Enteromorpha*, two kinds of algae in their natural habitats. (**d**) By grazing on the dominant alga in tidepools (*Enteromorpha*), the periwinkles promote the survival of less competitive algal species that would otherwise be overgrown. (**e**) *Enteromorpha* doesn't grow on rocks. Here, *Chondrus* is dominant. Periwinkles find *Chondrus* tough and dine instead on less competitive algal species. By doing so, periwinkles decrease the algal diversity on the rocks.

Table 27.1 Detrimental Effects of Some Species Introduced into the United States

Species Introduced	Origin	Mode of Introduction	Outcome
Water hyacinth	South America	Intentionally introduced (1884)	Clogged waterways; other plants shaded out
Dutch elm disease:			
Ophiostoma ulmi (fungus)	Asia (by way	Accidental; on infected elm timber (1930)	Millions of mature elms destroyed
Bark beetle (vector)	of Europe)	Accidental; on unbarked elm timber (1909)	
Chestnut blight fungus	Asia	Accidental; on nursery plants (1900)	Nearly all eastern American chestnuts killed
Zebra mussel	Russia	Accidental; in ballast water of ship (1985)	Clog pipes and water intake valves of power plants; displacing native Great Lake bivalves
Japanese beetle	Japan	Accidental; on irises or azaleas (1911)	Over 250 plant species (e.g., citrus) defoliated
Sea lamprey	North Atlantic	Ship hulls, through canals (1860s, 1921)	Trout, whitefish destroyed in Great Lakes
European starling	Europe	Intentional release, New York City (1890)	Outcompete native cavity-nesting birds; crop damage; swine disease vector
Nutria	South America	Accidental release of captive animals being raised for fur (1930)	Crop damage, destruction of levees, overgrazing of marsh habitat

because it restricts competitive exclusion by mussels. Remove all the sea stars, and the community shrinks from fifteen species to eight.

The impact of a keystone species can vary between habitats that differ in their species arrays. Periwinkles (*Littorina littorea*) are alga-eating snails of intertidal zones. Jane Lubchenco showed that their removal can increase *or* decrease the diversity of algal species in different habitats (Figure 27.20).

In tidepools, the periwinkles prefer to eat the alga *Enteromorpha*, which can outgrow other algal species. By keeping *Enteromorpha* in check, periwinkles help less competitive algal species survive. On exposed rocks in the lower intertidal zone, *Chondrus* and other tough, unpalatable red algae that periwinkles avoid are dominant. In this habitat, periwinkles prefer to graze on the competitively weaker algal species. So periwinkles *promote* algal diversity in tidepools but *reduce* it on exposed rock surfaces.

SPECIES INTRODUCTIONS TIP THE BALANCE

Instabilities also are set in motion when residents of established communities move out from their home range and successfully take up residence elsewhere. This type of directional movement, called **geographic dispersal**, happens in three ways.

First, over a number of generations, a population might expand its home range by slowly moving into outlying regions that prove hospitable. Second, some individuals might be rapidly transported across great distances, an event called *jump* dispersal. This often takes individuals across regions where they could not survive on their own, as when insects travel from the

mainland to Maui in a ship's cargo hold. Third, some population might be moved away from a home range by continental drift, at an almost imperceptibly slow pace over long spans of time.

Successful dispersal and colonization of a vacant adaptive zone can be remarkably rapid. Consider one of Amy Schoener's experiments in the Bahamas. She set out plastic sponges on barren sand at the bottom of Bimini Lagoon. How fast did aquatic species take up residence on these artificial habitats? Schoener recorded occupancy by 220 species within thirty days.

When you hear someone bubbling enthusiastically about an exotic species, you can safely bet the speaker isn't an ecologist. An **exotic species** is a resident of an established community that dispersed from its home range and became established elsewhere. Unlike most imports, which never do take hold outside the home range, an exotic species permanently insinuates itself into a new community.

More than 4,500 exotic species have established themselves in the United States after jump dispersal. We put some of the new arrivals, including soybeans, rice, wheat, corn, and potatoes, to use as food crops.

Accidental imports can alter community structure. You learned about imported fire ants in the chapter introduction. Table 27.1 lists others. The next section describes the unintended effects of a few more.

A keystone species is one that has a major effect on the species composition in particular habitats.

Long-term climate shifts, species introductions, and other disturbances can permanently alter community structure.

27.10 Exotic Invaders

Nonnative species are on the loose in communities on every continent. They can alter habitats; they often outcompete and displace native species.

THE PLANTS THAT ATE GEORGIA

In 1876, a vine called kudzu (*Pueraria montana*) from Japan was introduced to the United States. In the temperate regions of Asia where it evolved, it is a well-behaved legume with a well-developed root system. It *seemed* like a good idea to use it for forage and to control erosion. But kudzu grew faster in the Southeast, where herbivores, pathogens, and less competitive plants posed no serious threat to it.

With nothing to stop it, kudzu shoots can grow sixty meters a year. Its vines now blanket streambanks, trees, telephone poles, houses, and almost everything else in their path (Figure 27.21*a*). It withstands burning, and its deep roots resist being dug up. Grazing goats and herbicides help. But goats eat most other plants along with it, and herbicides taint water supplies. Kudzu may spread as far north as the Great Lakes by 2040.

On the bright side, Asians use a starch extracted from kudzu in drinks, herbal medicines, and candy. A kudzu processing plant in Alabama may export this starch to Asia, where demand currently exceeds the supply. Also, kudzu may help save trees; it can be an alternative source for paper. Today, about 90 percent of Asian wallpaper is kudzu-based.

THE ALGA TRIUMPHANT

They looked so perfect in saltwater aquariums, those bright green, long, feathery branches of the green alga *Caulerpa taxifolia*. So researchers from Germany's Stuttgart Aquarium developed a hybrid, sterile strain of *C. taxifolia*, and then they magnanimously shared it with other marine institutions. Was it from Monaco's Oceanographic Museum that the hybrid strain escaped into the wild? Some say yes, Monaco says no.

The aquarium strain of *C. taxifolia* can't reproduce sexually. Yet. It grows asexually by runners, just a few centimeters a day. Bits caught on boat propellers and fishing nets assisted its dispersal. Since 1984, this alga has blanketed more than 30,000 hectares of the seafloor along the Mediterranean coast (Figure 27.21*b*). It thrives along sandy or rocky shores and in mud; it lives ten days after being discarded in meadows. Unlike its tropical parents, this strain does well in cool water, even polluted water. It displaces endemic algae and poisons invertebrates and fishes. Its toxin repels herbivores that might keep it in check.

C. taxifolia also has invaded the coastal waters of the United States. It has the potential to overgrow reefs and destroy marine food webs. In 2000, scuba divers discovered the alga growing in two areas off the coast of Southern California. Someone might have drained the water from a home aquarium into a storm drain or into the lagoon itself. In response to

Figure 27.21 (**a**) Kudzu (*Pueraria montana*) taking over part of Lyman, South Carolina. The plant has now spread from East Texas to Florida and Pennsylvania. Ruth Duncan of Alabama, who makes 200 kudzu vine baskets a year, just can't keep up. (**b**) *Caulerpa taxifolia* suffocating yet another richly diverse marine ecosystem.

Figure 27.22 Rabbit-proof fence? Not quite. This is part of a fence built to hold back the 200 million to 300 million rabbits that are wreaking havoc with Australia's vegetation. It didn't work.

the discovery, a coalition of governmental and private groups sprang into action. They tarped over the area and shut out sunlight, pumped chlorine into the mud to poison the alga, and used welders to boil it. So far, the measures may have worked. As of autumn 2003, there was no sign of the exotic invader.

It is now illegal to import the harmful strain of *C. taxifolia* into the United States. Interstate sale also is prohibited. Some still slip into the country because the aquarium industry successfully lobbied against a ban on all *Caulerpa* species, and it is hard to distinguish the invasive strain without genetic analysis.

THE RABBITS THAT ATE AUSTRALIA

During the 1800s, British settlers in Australia just couldn't bond with koalas and kangaroos, and so they imported familiar animals from home. In 1859, in what would be the start of a major disaster, a landowner in northern Australia imported and then released two dozen European rabbits (*Oryctolagus cuniculus*). Good food and sport hunting—that was the idea. An ideal rabbit habitat with no natural predators—that was the reality.

Six years later, the landowner had killed 20,000 rabbits and was besieged by 20,000 more. The rabbits displaced livestock and caused the decline of native wildlife. Now 200 to 300 million are hippity-hopping through the southern half of the country. They graze on grasses in good times and strip bark from shrubs and trees during droughts. Thumping herds of them can turn shrublands as well as grasslands into eroded

deserts. And their burrows undermine the soil and set the stage for widespread erosion.

Rabbit warrens have been shot at, plowed under, fumigated, and dynamited. The first all-out assaults killed 70 percent of them, but the rabbits rebounded in less than a year. When a fence 2,000 miles long was built to protect western Australia, rabbits made it from one side to the other before workers could finish the job (Figure 27.22).

In 1951, the government introduced a myxoma virus that normally infects South American rabbits. The virus causes *myxomatosis*. This disease has mild effects on its coevolved host but nearly always kills *O. cuniculus*. Mosquitoes and fleas transmit the virus from host to host. Having no coevolved defenses against the import, European rabbits died in droves. But natural selection has since favored a rise in rabbit populations resistant to the imported virus.

In 1991, on an uninhabited island in Spencer Gulf, Australian researchers released rabbits that they had injected with a calicivirus. Rabbits died from blood clots in their lungs, heart, and kidneys. But the test virus escaped from the island in 1995, perhaps on insect vectors. It has been killing 80 to 95 percent of adult rabbits in some parts of Australia. In 1996, the virus was deliberately released in other areas.

Some Australians are now questioning whether the calicivirus should be used on such a large scale. They wonder whether it can jump boundaries and infect animals other than rabbits—one study suggests it might cause illness in pigs. What will be the long-term effects of these viral releases? We don't know.

27.11 Biogeographic Patterns

Biodiversity is the sum of all species occupying a specified area during a specified interval, past as well as present. It differs from one habitat or geographic realm to another, often in predictable patterns.

In large part, the biodiversity of a given area is an outcome of the evolutionary history of each member species and its resource requirements, its physiology, and its capacity for dispersal. Rates of birth, death, immigration, and emigration also shape the number of species. In turn, those rates are influenced by local and regional habitat conditions, and by interactions among species.

Ecologists have identified patterns of biodiversity throughout the world. For example, species richness corresponds to sunlight intensity, moisture levels, and temperature along gradients in latitude, elevation, and depth. Differences between microenvironments give rise to local variations in overall patterns.

MAINLAND AND MARINE PATTERNS

The most striking pattern of biodiversity corresponds to distance from the equator. *For most groups of plants and animals, the number of coexisting species on land and in the seas is greatest in the tropics, and it systematically declines from the equator to the poles.* Figure 27.23 shows examples of this pattern. Let's focus on a few factors that help bring about the pattern and maintain it.

First, for reasons explained in Section 29.1, tropical latitudes intercept more intense sunlight and receive more rainfall, and their growing season is longer. As one outcome, resource availability tends to be greater and more reliable in the tropics than elsewhere. This favors a degree of specialized interrelationships not possible where species are active for shorter periods.

Second, tropical communities have been evolving for a longer time than temperate ones, some of which did not start forming until the end of the last ice age.

Third, species diversity might be self-reinforcing. Diversity of tree species in tropical forests is much greater than in comparable forests at higher latitudes. When more plant species compete and coexist, more kinds of herbivores also compete and coexist, partly because no single kind of herbivore can overcome all chemical defenses of all the different plants. More predatory and parasitic species evolve in response to the bigger buffet of prey and hosts. The same effect applies to diversity on tropical reefs.

Figure 27.23 Two patterns of species diversity corresponding to latitude. The number of ant species (**a**) and breeding birds (**b**) in Central and North America.

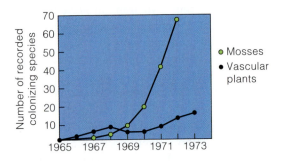

Figure 27.24 Surtsey, a volcanic island, at the time of its formation. Newly formed, isolated islands are natural laboratories for ecologists. The chart gives the number of colonizing species between 1965 and 1973.

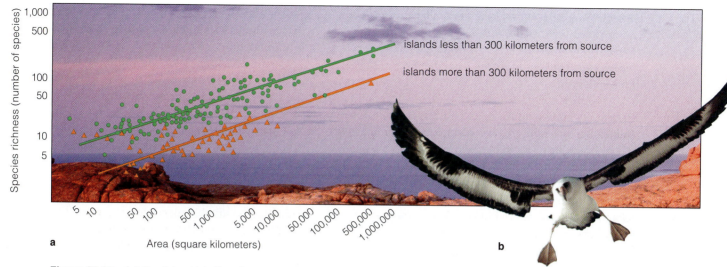

Figure 27.25 (**a**) Two island biodiversity patterns. Distance effect: Biodiversity on islands of a specified size declines with increasing distance from a source of colonizing species. *Green* circles signify islands less than 300 kilometers from the colonizing source. *Orange* triangles signify islands more than 300 kilometers from the source areas. Area effect: Among islands the same distance from a source of colonizing species, the larger ones support more species.

(**b**) Wandering albatross, one travel agent for jump dispersals. Seabirds that island-hop long distances often have seeds stuck to their feathers. Seeds that successfully germinate in a new island community may give rise to a population of new immigrants.

ISLAND PATTERNS

Islands are fine laboratories for biodiversity studies. For instance, a 1965 volcanic eruption created Surtsey, an island southwest of Iceland. Within six months, bacteria, fungi, seeds, flies, and some seabirds became established on the island. A vascular plant appeared two years after the island formed; the first mosses came along two years after that (Figure 27.24). As the island soil became enriched, more and more plant species began to take hold.

As is the case for other islands, the number of new species on Surtsey will not increase indefinitely. Why not? Models based on studies of island communities around the world provide some answers.

First, islands that are far from a source of potential colonists receive few colonizing species, and the few that do arrive are adapted for long-distance dispersal (Figure 27.25a). This is known as the **distance effect**.

Second, larger islands tend to support far more species than smaller islands the same distances from a colonizing source. This is known as the **area effect**. Larger islands tend to offer varied habitats and more of them. Most have complex topography and higher elevations. Such variations promote diversity. Also, being bigger targets, larger islands intercept more of the accidental tourists that winds and ocean currents move from the mainland but offer no way back.

Extinctions lower diversity, and they happen more on small islands. Small populations are vulnerable to severe storms, droughts, disease, and genetic drift.

Again, on any island, the number of species will reflect a balance between immigration rates for new species and extinction rates for any established ones. Small islands distant from a source of colonists have low immigration rates and high extinction rates, so they support fewer species once the balance has been struck for their populations.

Biogeographers know a lot more about the patterns of diversity and our disruptions of them. For a lyrical account, read David Quammen's *Song of the Dodo*.

Biodiversity shows global patterns, as when the number of species correlates with gradients in latitude, elevation, and depth. Microenvironments along the gradients often introduce variations in the overall patterns.

Species richness in a given area also is an outcome of the evolutionary history of each species, its requirements for resources, its physiology, its capacity for dispersal, and its rates of birth, death, immigration, and emigration.

Generally, species richness is highest in the tropics and lowest at the poles. The number of species on an island also depends on its size and distance from a colonizing source.

27.12 Threats to Biodiversity

Based on many lines of evidence accumulated over the past few centuries, an estimated 99 percent of all species that have ever lived are extinct. Even so, the full range of biodiversity is greater now than it has ever been at any time in the past.

Reflect on life's evolution, as laid out in Unit Three. Remember the five greatest mass extinctions that we use as boundaries for geologic time? After each event, biodiversity plummeted on land and in the seas. In the aftermath, many surviving species embarked on adaptive radiations. Each time, though, recovery to the same level of species richness was extremely slow, on the order of 20 million to 100 million years.

ON THE NEWLY ENDANGERED SPECIES

No biodiversity-shattering asteroids have struck the Earth for 65 million years. *Yet the sixth major extinction event is under way.* There is evidence that humans are driving many species over the edge. Consider this: Mammals survived the asteroid impact that killed the last of the dinosaurs. But 2 million years ago, humans started hunting them in earnest. About 11,000 years ago, human populations were changing habitats in a big way. Agriculture and domesticated species were favored over wild animals. Only 4,500 or so species of Mammals are with us today. Of these, about 300 are endangered. Among them are most of the whales, otters, wild cats, and primates.

Figure 27.26 Humans as competitor species—an example of dominance in the competition for living space in an arid habitat.

An **endangered species** is any endemic species that is highly vulnerable to extinction. *Endemic* means it originated in one region and is found nowhere else.

We are one species among millions. Yet, primarily as an outcome of our rapid population growth over the past forty years, we are threatening other species with large-scale habitat losses, species introductions, and overharvesting.

HABITAT LOSSES AND FRAGMENTATION

Edward O. Wilson defines **habitat loss** as the physical reduction in suitable places to live, as well as the loss of suitable habitat through pollution. Habitat loss is putting major pressure on more than 90 percent of the endemic species that now face extinction.

Biodiversity is greatest in the tropics, where lands are most threatened by deforestation (Chapter 21). As other examples, 98 percent of the tallgrass prairies, 50 percent of the wetlands, and as much as 95 percent of the United States old-growth forests have disappeared. Even arid lands are not exempt (Figure 27.26).

Habitats often become fragmented, or chopped into patches that each have a separate periphery. Species at any habitat's periphery are more exposed to wind, fire, temperature changes, predators, and pathogens. Also, fragmented patches often are not large enough to support the population sizes required for successful breeding. Patches may be too small to have sufficient food or other resources to sustain a species.

Together with Robert MacArthur, Wilson devised an **equilibrium model of island biogeography**. By this model, a 50 percent loss of a habitat will drive about 10 percent of its endemic species to extinction. A 90 percent loss will drive about 50 percent of the species in a habitat to extinction. The model is being used to estimate the number of current and future extinctions in "islands of shrinking habitat." Such islands need not be true islands; many are patches of natural habitat surrounded by degraded areas or by encroaching suburban developments. Many national parks, tropical forests, reserves, and lakes are like this.

Some species help warn us of changes in habitats and impending loss of diversity. Birds are one of the key **indicator species**. Different types live in all major land regions and climate zones, they react quickly to changes, and they are relatively easy to track.

Species introductions are another major threat to endemic species. Remember, each habitat has only so many resources, and its occupants all compete for a share. Introduced species often have adaptations that make them highly competitive. They have been a key factor in almost 70 percent of cases where endemic species are being driven to extinction.

Also, in a sad commentary on human nature, the more rare a wild animal becomes, the more its value soars in the black market. Figure 27.27 gives you an idea of the money involved in illegal wildlife trading.

As a final point, biodiversity also is threatened by overharvesting of species that have commercial value. Whales are a prime example. Humans kill whales for food, lubricating oil, fuel, cosmetics, fertilizer, even pet food. Substitutes exist for all whale products. Yet some nations ignore a moratorium on hunting large whales. Some whalers poach inside the boundaries of sanctuaries set aside for recoveries.

CONSERVATION BIOLOGY

Awareness of the impending extinction crisis gave rise to **conservation biology**. We define this field of pure and applied research as (1) a systematic survey of the full range of biological diversity, (2) efforts to decipher biodiversity's evolutionary and ecological origins, and (3) attempts to identify methods of maintaining and using biodiversity in ways that can benefit the human population. Its major goal is to conserve and utilize, in sustainable ways, as much biodiversity as possible.

Figure 27.28 depicts the approximate number of named species in some major groups. Our current taxonomic knowledge is fairly complete for many groups, including the flowering plants, conifers, and fishes, birds, mammals, and other vertebrates. We tend to know more about large organisms, especially land dwellers, and about the showy ones, including birds and butterflies. Yet we have a lot of evidence that staggering numbers of fungi, protists, bacteria, and archaea are still out there, waiting to be identified.

For example, biologists recently surveyed a single hot spring in Yellowstone National Park. They turned up more species of archaea than had previously been known for the rest of the Earth—which gives you an idea of how much work remains to be done.

The current range of global biodiversity is an outcome of an overall pattern of extinctions and slow recoveries.

Human activities have raised rates of extinction through habitat losses, species introductions, and overharvesting.

Conservation biology entails a systematic survey of the full range of biodiversity, analysis of its evolutionary and ecological origins, and identification of methods to maintain and use biodiversity for the benefit of humans.

Figure 27.27 Humans as predator species—confiscated tiger skins, leopard skins, rhinoceros horns, and elephant tusks. An illegal wildlife trade threatens the survival of more than 600 species, yet all but 10 percent of the trade escapes detection. A few of the black market prices: Live, mature saguaro cactus ($5,000–15,000), bighorn sheep, head only ($10,000–60,000), polar bear ($6,000), grizzly ($5,000), bald eagle ($2,500), peregrine falcon ($10,000), live chimpanzee ($50,000), live mountain gorilla ($150,000), Bengal tiger hide ($100,000), and rhinoceros horn ($28,600 per kilogram).

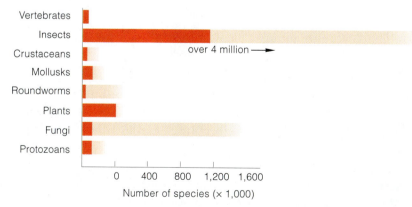

Figure 27.28 Current species diversity for a few major taxa. *Red* signifies the number of named species; *gold* signifies the estimated number of species not yet discovered and named.

27.13 Sustaining Biodiversity

Every species is part of a web of interactions with others of its community. Increasingly, efforts to save endangered species are focusing on identifying and protecting the most threatened communities having the most diversity.

IDENTIFYING AREAS AT RISK

As noted in the preceding section, the vast majority of existing species have not even been described, and we don't know where most of them live. Much effort is now focused on identifying **hot spots**, or habitats of a large number of species that are found nowhere else and are in greatest danger of extinction.

At the first survey level, workers target a limited area, such as an isolated valley. A complete survey is impractical, so they make an inventory of flowering plants, birds, mammals, fishes, butterflies, and other indicator species in the habitat. At the next level up, broader areas that are major or multiple hot spots are systematically explored. The widely separated forests of Mexico are an example of a key hot spot. Research stations have been set up at many different latitudes and at different elevations across the region.

At the highest survey level, hot spot inventories are combined with existing knowledge of biodiversity in ecoregions. An **ecoregion** is a broad land or ocean region defined by climate, geography, and producer

species. Figure 27.29, a map of regions identified as essential reservoirs of biodiversity, includes 142 land, 53 freshwater, and 43 marine habitats. Of these, 25 are high-priority targets for conservation efforts.

ECONOMIC FACTORS AND SUSTAINABLE DEVELOPMENT

Every nation enjoys three forms of wealth—material, cultural, and biological wealth. Until recently, not very many countries comprehended the value of biological wealth, which can be the source of food, medicine, and other products. Much of the world's biological wealth has already been lost.

In the next fifty years, human population size may reach 8.9 billion, with most growth in the developing countries. Each person requires raw materials, energy, and living space. How many species will be crowded out? The countries with the greatest monetary wealth also have the least biological wealth and use most of the world's natural resources. Countries with the least monetary wealth and the most biological wealth have the fastest-growing populations.

People who are locked in poverty often must choose between saving themselves or an endangered species. They hunt animals and dig up plants in "protected" parks and reserves. They try to raise crops and herds

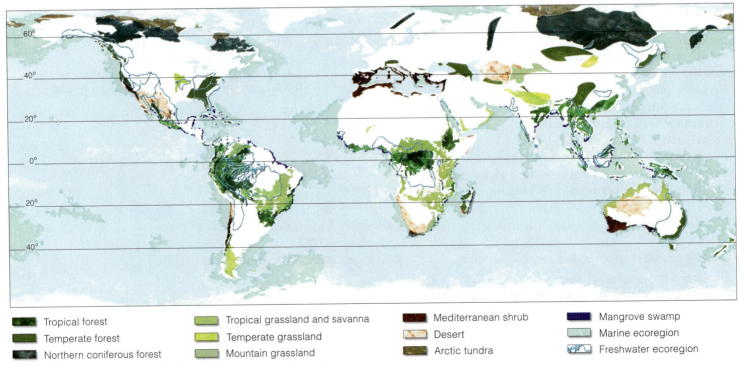

■ Tropical forest	■ Tropical grassland and savanna	■ Mediterranean shrub	■ Mangrove swamp
■ Temperate forest	■ Temperate grassland	■ Desert	■ Marine ecoregion
■ Northern coniferous forest	■ Mountain grassland	■ Arctic tundra	■ Freshwater ecoregion

Figure 27.29 Map of the most vulnerable regions of land and seas, compiled by the World Wildlife Fund.

Figure 27.30 Strip logging. The practice may protect biodiversity as it permits logging on tropical slopes. A narrow corridor paralleling the land's contours is cleared. A roadbed is made at the top to haul away logs. After a few years, saplings grow in the cleared corridor. Another corridor is cleared above the roadbed. Nutrients leached from exposed soil trickle into the first corridor. There they are taken up by saplings, which benefit from all the nutrient input by growing faster. Later, a third corridor is cut above the second one—and so on in a profitable cycle of logging, which the habitat sustains over time.

Figure 27.31 Riparian zone restoration. This example comes from Arizona's San Pedro River, shown before and after restoration efforts.

on marginal land. Conservation biologists work to identify ways for such people to earn a decent living from biological wealth—by using the biodiversity in threatened habitats in a sustainable way.

It *is* possible to protect a habitat and still withdraw resources in sustainable fashion. For example, besides yielding wood for local economies, tropical rain forest trees yield diverse exotic woods that are prized by developed countries. Gary Hartshorn was the first to propose **strip logging** in portions of forests that are sloped and have a number of streams. As explained in Figure 27.30, logging can be sustained even while the maximum biodiversity is being maintained.

Developed countries also benefit from sustainable methods, as in a **riparian zone**—a narrow corridor of vegetation along a stream or a river (Figure 27.31). The plants afford a major line of defense against flood damage by sponging up water during spring runoffs and summer storms. Shade cast by a canopy of taller shrubs and trees in the riparian zone helps conserve water during droughts.

A riparian zone can provide wildlife with food, shelter, and shade, particularly in arid and semiarid regions. In the western United States, 67 to 75 percent of the endemic species spend all or part of their life cycle in riparian zones. Among them are 136 kinds of songbirds, some of which will nest only in the plants of riparian zones.

Cattle tend to congregate at rivers and streams, where they trample and feed until grasses and shrubs are gone. Even a few head of cattle can ruin a riparian zone. Restricting access and providing water at areas away from streams helps to conserve some riparian zones. Rotating livestock and placing supplemental feed at different grazing areas is also useful.

Preserving biodiversity requires identifying and protecting regions that support the highest levels of biodiversity.

Sustainable development involves utilizing biodiversity in ways that benefit humans, without shredding the intricate web of interactions that sustains a natural community.

Summary

Section 27.1 A habitat is the type of place where individuals of a species normally live. A community is an association of all populations of species that occupy a habitat. Each species in the community has its own niche, the sum of all activities and relationships in which its individuals engage as they secure and use the resources required for their survival and reproduction.

Indirect and direct species interactions influence community structure. They include commensalism, mutualism, competition, predation, and parasitism.

Section 27.2 Mutualism is a species interaction that benefits both participants. Some mutualists can't complete their life cycle without the interaction.

Section 27.3 Two species that require the same resource tend to compete. By the competitive exclusion theory, when two (or more) species require identical resources, they cannot coexist indefinitely. Species are more likely to coexist when they differ in their use of resources. Many coexist by using a shared resource in different ways or at different times.

Section 27.4 Predators and prey exert selection pressure on each other. Densities of predator and prey populations often oscillate. The carrying capacity, density dependencies, refuges, predator efficiency, and often alternative prey sources influence the cycles.

Section 27.5 Threat displays, chemical weapons, camouflage, stealth, and mimicry are outcomes of coevolution between predators and their prey.

Sections 27.6, 27.7 Parasites live in or on other living hosts and withdraw nutrients from host tissues for part of their life cycle. Hosts may or may not die as a result. Parasitoids kill their hosts, and social parasites take over some aspect of a host's life cycle.

Section 27.8 By a model for ecological succession, a community develops in predictable sequence, from its pioneer species to an end array of species that persists over an entire region. A climax community is a stable, self-perpetuating array of species that are in equilibrium with one another and the environment.

Similar climax stages may persist within the same region yet show variation as a result of environmental gradients and species interactions.

Section 27.9 Community structure reflects an uneasy balance of forces, including predation and competition, that have been operating over time.

Section 27.10 Introduction of species that in one way or another expand their geographic range can harm community structure in permanent ways.

Section 27.11 Biodiversity depends on the size of a region, colonization rates, disturbances, and rates of extinction. It depends on the climate and patterns of resource availability. It tends to be highest in the tropics and declines systematically toward the poles.

Sections 27.12, 27.13 Current biodiversity is an outcome of evolutionary forces, mass extinctions, and slow recoveries. Conservation biologists seek ways to conserve and use biodiversity in sustainable fashion.

Self-Quiz *Answers in Appendix III*

1. A habitat _____ .
   a. has distinguishing physical and chemical features
   b. is where individuals of a species normally live
   c. is occupied by various species
   d. a through c

2. A niche is _____ .
   a. the sum of activities and relationships by which individuals of a species secure and use resources
   b. unvarying for a given species
   c. something that shifts in large and small ways
   d. both a and c

3. Two species may coexist indefinitely in some habitat when they _____ .
   a. differ in their use of resources
   b. share the same resource in different ways
   c. use the same resource at different times
   d. all of the above

4. A predator population and prey population _____ .
   a. always coexist at relatively stable levels
   b. may undergo cyclic or irregular changes in density
   c. cannot coexist indefinitely in the same habitat
   d. both b and c

5. Parasites _____ .
   a. weaken their hosts   c. feed on host tissues
   b. can kill novel hosts   d. all of the choices

6. In _____ , a disturbed site in a community recovers and moves again toward the climax state.
   a. the area effect   c. primary succession
   b. the distance effect   d. secondary succession

7. The biodiversity of a region is an outcome of _____ .
   a. climate and topography   d. both a and b
   b. possibilities for dispersal   e. a through c
   c. evolutionary history

8. Following mass extinctions, recovery to the same level of biodiversity has taken many _____ of years.
   a. hundreds   b. millions   c. billions

9. Match the terms with the most suitable descriptions.
   ____ geographic dispersal   a. opportunistic colonizer of barren or disturbed places
   ____ area effect   b. greatly affects other species
   ____ pioneer species   c. individuals leave home range, become established elsewhere
   ____ climax community   d. more biodiversity on large islands than small ones at same distance from colonizing source
   ____ keystone species   e. stable, self-perpetuating array of species
   ____ endemic species   f. river or stream's vegetation area
   ____ riparian zone   g. originated in one place and found nowhere else

Figure 27.32 Two phasmids. (**a**) A stick insect from South Africa. (**b**) A leaf insect from Java. (**c**) Phasmid eggs often resemble seeds.

Figure 27.33 One of the realities of the modern world. Would you prefer to be buried in a casket in the ground or mixed with concrete as part of a long-lasting foundation for an artificial reef?

Critical Thinking

1. With antibiotic resistance rising, researchers are looking for ways to reduce use of antibiotics. Some cattle that once would have been fed antibiotic-laced food are now being given feed that contains cultures of bacteria. These *probiotic feeds* are designed to establish or bolster populations of helpful bacteria in the animal's gut. The idea is that if a large population of beneficial bacteria is in place, then the harmful bacteria cannot become established or thrive. Which ecological theory is guiding this research?

2. Insects called phasmids are extraordinary masters of disguise. Most resemble sticks or leaves. Even their eggs may be camouflaged (Figure 27.32). The phasmids are all herbivores. Most spend their days motionless, moving about and feeding only at night. If disturbed, a phasmid will fall to the ground and not move.

    Speculate on the selective pressures that might have shaped phasmid morphology and behavior. Suggest an experiment with one species to help determine whether the appearance and behavior have adaptive value.

3. Figure 27.28 compares the current estimated number of species for some major taxa. Given what you learned from chapters in this unit, which taxa are most vulnerable in the current extinction crisis? Which are likely to pass through it with much of their biodiversity intact? List reasons why. For example, compare the global distribution of species.

4. Mentally transport yourself to a tropical rain forest of South America. Imagine you do not have a job. There are no jobs in sight, not even in overcrowded cities a long distance away. You have no money or contacts to get you anywhere else. And yet you are the sole supporter of a large family. Today a stranger approaches you. He tells you he will pay good money if you can capture and keep alive a certain brilliantly feathered parrot in the forest. You know it is illegal to capture this endangered parrot. You also have an idea of where it lives. What will you do?

5. Here's an alternative to formal burial on land. Ashes of cremated people are mixed with pH-neutral concrete and used to build *artificial reefs*. Each perforated concrete ball weighs 400 to 4,000 pounds and is designed to last 500 years. So far, 100,000 of these reefs are sustaining marine life in 1,500 locations (Figure 27.33). Proponents say that the burial costs less, does not waste land, and is better for the environment. Your thoughts?

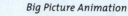

IMPACTS, ISSUES *Bye-Bye, Blue Bayou*

Each Labor Day, the coastal Louisiana town of Morgan City celebrates the region's economic mainstays with the Louisiana Shrimp and Petroleum Festival. The state is the nation's third-largest petroleum producer and the leader in shrimp harvesting. But the petroleum industry's success may be contributing indirectly to the possible disappearance of the state's fisheries.

The global air temperature is rising, and fossil fuel burning is a culprit. Warmer air heats water near the sea surface, heated water expands, and so the sea level

is rising. Warmer air also is melting ancient glaciers and ice caps, and meltwater is adding to the sea volume.

Since the 1940s, Louisiana has lost an area the size of Rhode Island to the sea. Low elevations along the United States coastline—including *14,720,000 acres* next to the Gulf of Mexico and Atlantic Ocean—may be one to three feet under water within fifty years!

With more than 40 percent of the nation's saltwater marshes, Louisiana has the most to lose (Figure 28.1). Its wetlands are already sinking; dams and levees interfere with the deposition of sediments that could replace those washed out to sea. The rise in sea level will make 70 percent of the wetlands *really* wet, with no land.

Is an ecological and economic disaster unfolding? Consider: More than 3 billion dollars' worth of shellfish and fish are harvested from Louisiana's wetlands each year. About 40 percent of North America's migratory ducks—more than 5 million birds—overwinter here. Also,

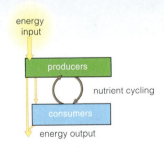

Figure 28.1 Glimpse through a cypress swamp in Louisiana. Inland saltwater intrusions are threatening these trees, which are adapted to freshwater habitats.

the big picture

energy input

producers

nutrient cycling

consumers

energy output

Ecosystem Structure Energy flows only one way through an ecosystem—into, through, and out of food webs. Nutrients enter an ecosystem, and some are lost to the environment, but most are cycled among the ecosystem's species.

Food Webs Food chains are linear sequences of the transfer of energy and nutrients, from autotrophs through heterotrophs. They cross-connect in grazing and detrital food webs, which consist of diverse producer and consumer species.

Louisiana's wetlands buffer low elevations inland from storm surges and hurricanes.

More bad news: Warmer water may promote algal blooms and huge fish kills. Also, populations of many pathogenic bacteria increase in warmer water, so more people may get sick after swimming in contaminated water or eating contaminated shellfish.

Inland, heat waves and wildfires will become more intense. Deaths related to heat stroke will climb. Warmer temperatures will permit mosquitoes to extend their inland ranges. Some mosquitoes are vectors for agents of malaria, West Nile virus, and other diseases.

Global warming is expected to increase evaporation and alter weather patterns. We can predict severe floods for some regions and prolonged drought for others, including Louisiana. Worldwide, 3 billion people may run out of fresh water within twelve years.

This chapter can get you thinking about energy flow through ecosystems, starting with energy inputs from the sun. It will show how ecosystems depend on inputs, cycling, and outputs of nutrients—and how nutrients are cycled on a global scale.

The chapter also can get you thinking more about a related concept of equal importance. We have become players in the global flows of energy and nutrients even before we fully comprehend how the game plans work. Decisions we make today about global warming and other environmental issues may affect the quality of human life and the environment far into the future.

✓ How Would You Vote?

Emissions from motor vehicles are a major source of greenhouse gases. Many people buy large vehicles that use more fuel but are viewed as safer and more useful. Should such vehicles be additionally taxed to discourage sales and offset their environmental costs? Can we expect better fuels as well as more of the fuel-efficient, larger vehicles that are becoming available? See the Media Menu for details, then vote online.

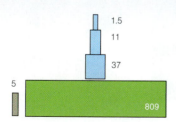

Ecological Pyramids Ecologists construct pyramid diagrams based on measurements of how much energy and nutrients enter an ecosystem, how much is actually captured by the primary producers, and how much of it becomes stored at each trophic level.

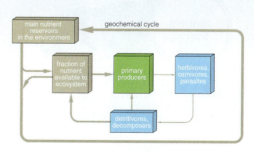

Global Cycles Hydrogen, oxygen, carbon, nitrogen, and phosphorus are among the most essential elements for all organisms. Environmental reservoirs hold the greatest amounts of these elements, which move slowly in global cycles and rapidly through ecosystems.

28.1 The Nature of Ecosystems

In the preceding chapter, we focused on how species interact over time and within the space of communities. This chapter continues with those interactions, but at a higher level of organization. Each community and its physical environment function as an integrated system.

OVERVIEW OF THE PARTICIPANTS

Diverse natural systems abound on Earth's surface. In climate, landforms, soil, vegetation, animal life, and other features, deserts differ from hardwood forests, which differ from tundra and the prairies. Reefs differ from the open ocean, which differs from streams and lakes. Despite their differences, all of the systems are alike in many aspects of their structure and function. Ecologists start from these shared similarities to make predictions about prospects for stability and change over the long term.

With few exceptions, the systems run on sunlight that autotrophs, or self-feeders, capture. Plants and phytoplankton are the main autotrophs. Remember, from Chapter 6, that both convert sunlight energy to chemical bond energy and use it to synthesize organic compounds from inorganic raw materials. They are **primary producers** for the system (Figure 28.2).

All other organisms in the system are **consumers**. These heterotrophs feed on the tissues, products, and remains of other organisms. We describe consumers by their diets. *Herbivores* eat plants. *Carnivores* eat flesh. *Parasites* live in or on a host and feed on its tissues. **Detritivores**, such as crabs, eat detritus—particles of decomposing organic matter. Finally, the wastes and remains of all organisms are degraded to inorganic compounds by cells of bacterial, protist, and fungal **decomposers**. Decomposers absorb most breakdown products, but some are released into the environment.

Some consumers do not fall into simple categories. Many are *omnivores*, which dine on animals, plants, fungi, protists, and even bacteria. A red fox is like this (Figure 28.3). If it comes across a dead bird, the fox will scavenge it. A *scavenger* is an animal that ingests dead plants, animals, or both all or some of the time. Vultures, termites, and hyenas are typical scavengers.

How are nutrients cycled in the system? Primary producers get hydrogen, oxygen, and carbon from water and carbon dioxide in the environment. They get phosphorus, nitrogen, and other minerals as well. These are the raw materials for biosynthesis. Later, decomposition of their organic wastes and remains releases nutrients back to the environment. Unless the

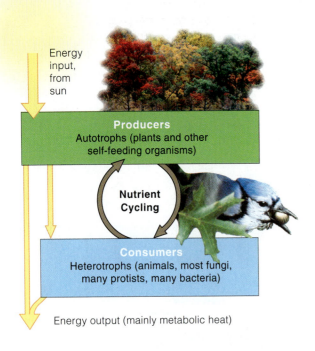

Figure 28.2 Model for ecosystems. Energy flows in one direction: into an ecosystem, through living organisms, and out from it. Nutrients are cycled among autotrophs and heterotrophs. In nearly all ecosystems, energy flow starts with autotrophs that capture energy from the sun.

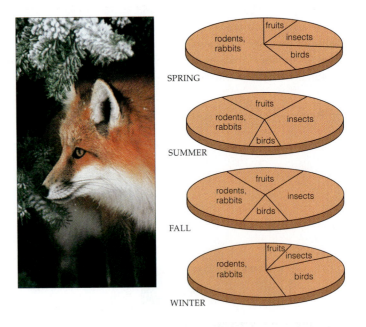

Figure 28.3 Red fox, an omnivore. Its diet shifts with seasonal changes in available food. Rodents, rabbits, and some birds make up the bulk of its diet in spring and winter. It eats more fruits and insects in the summer and fall.

released substances leave from the system, as when mineral ions end up dissolved in a stream that flows out of a meadow, producers typically take them up once again.

What we have just outlined is an **ecosystem**—an array of organisms and their physical environment, interacting through a one-way flow of energy and a cycling of raw materials. It is an open system, unable to sustain itself. *Energy inputs* to most ecosystems are in the form of sunlight. There may be *nutrient inputs*, as from a creek delivering dissolved minerals to a lake. There are also *energy outputs* and *nutrient outputs*.

Energy transfers, remember, cannot be 100 percent efficient (Section 5.1). Over time, the energy originally captured by producers escapes to the environment, mainly as metabolically generated heat.

STRUCTURE OF ECOSYSTEMS

We can classify all organisms of an ecosystem by their functional roles in a hierarchy of feeding relationships called **trophic levels** (*troph*, nourishment). "Who eats whom?" we may ask. If organism **B** eats organism **A**, energy is transferred from **A** to **B**. All organisms at a given trophic level are the same number of transfer steps away from the energy input into an ecosystem.

As one example, think about some organisms of a tallgrass prairie ecosystem. The flowering plants and other producers that tap energy from the sun are at the first trophic level. Plants are eaten by herbivores, such as cutworms, which are at the next trophic level. Cutworms are one of the primary consumers that are eaten by carnivores at the third trophic level, and so on up through tiers of trophic levels.

At each trophic level, organisms interact with the same sets of predators, prey, or both. Omnivores feed at several levels, so we would partition them among different levels or assign them to a level of their own.

A **food chain** is a straight-line sequence of steps by which energy originally stored in autotroph tissues moves to higher trophic levels. In one tallgrass prairie food chain, energy from a plant flows to a cutworm that eats its juicy parts, then to a garter snake that eats the cutworm, to a sandpiper that eats the snake, and finally to a marsh hawk that consumes the sandpiper. Figure 28.4 shows the chain.

Identifying a food chain is a simple way to start thinking about who eats whom in ecosystems. Bear in mind, however, that many different species are usually competing for food in complex ways. Tallgrass prairie producers (mainly flowering plants) do feed grazing mammals and herbivorous insects. Yet many more

Read Me First!
and watch the narrated animation on trophic levels

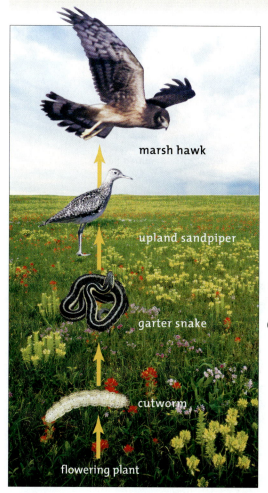

marsh hawk

fifth trophic level
top carnivore
(fourth-level consumer)

upland sandpiper

fourth trophic level
carnivore
(third-level consumer)

garter snake

third trophic level
carnivore
(second-level consumer)

cutworm

second trophic level
herbivore
(primary consumer)

flowering plant

first trophic level
autotroph
(primary producer)

Figure 28.4 Example of a simple food chain and its corresponding trophic levels in a tallgrass prairie.

species interact in nearly all ecosystems, especially at a lower trophic level. A number of food chains *cross-connect* with one another—as **food webs**. And that is the topic of the next section.

Primary producers, or self-feeders, use environmental energy and inorganic raw materials to make organic compounds.

Producers, consumers, and their physical environment make up an ecosystem. A one-way energy flow and a cycling of raw materials interconnect them.

A food chain, a straight-line sequence of who eats whom in an ecosystem, starts with autotrophs (producers) and proceeds through one or more levels of heterotrophs.

In ecosystems, a number of food chains cross-connect, forming food webs.

28.2 The Nature of Food Webs

Food chains cross connect with one another in food webs. By untangling the chains of many food webs, ecologists discovered patterns of organization. Those patterns reflect environmental constraints and the inefficiency of energy transfers from one trophic level to the next.

Plants capture only a fraction of the energy from the sun. They store half of that energy in chemical bonds of organic compounds in new plant tissues. They lose the rest as metabolic heat. Consumers tap into the energy that became stored in plant tissues, remains, and wastes. Then they too, lose metabolic heat. Taken together, *all of these heat losses represent a one-way flow of energy out of the ecosystem.*

When ecologists compared food chains in different kinds of food webs, a pattern emerged. In most cases, energy initially captured by producers passes through no more than four or five trophic levels. Even the rich ecosystems with complex food webs, such as the one in Figure 28.5, do not have lengthy food chains. The inefficiency of energy transfers may limit the sequence.

Field studies and computer simulations of aquatic and land food webs reveal more patterns. Chains in food webs tend to be shortest where environmental conditions can vary widely over time. Food chains tend to be longer in habitats that are more stable, such as ocean depths. The most complex webs tend to have many herbivorous species, as happens in grasslands. By comparison, the food webs with fewer connections tend to have more carnivores.

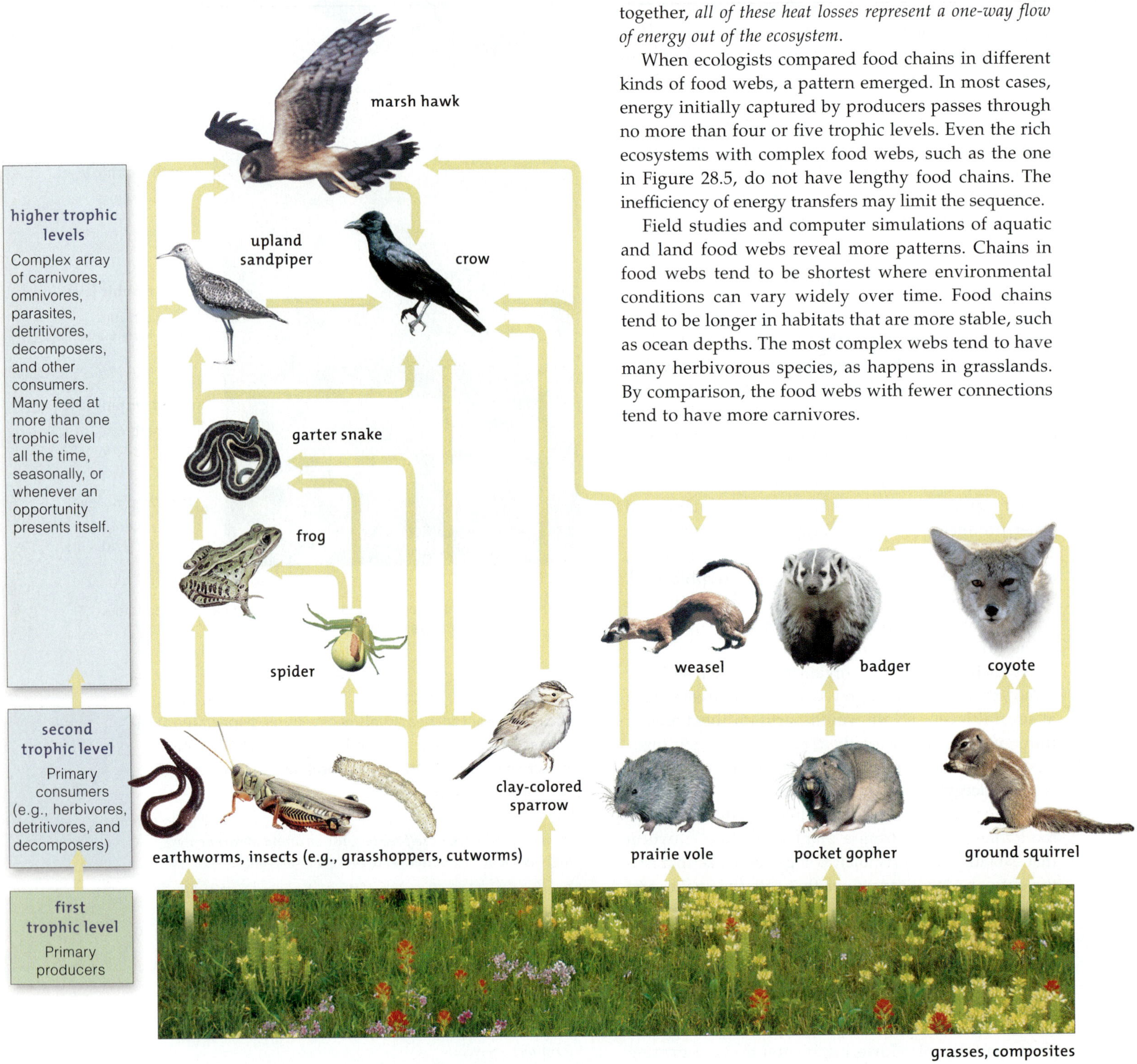

higher trophic levels

Complex array of carnivores, omnivores, parasites, detritivores, decomposers, and other consumers. Many feed at more than one trophic level all the time, seasonally, or whenever an opportunity presents itself.

second trophic level

Primary consumers (e.g., herbivores, detritivores, and decomposers)

first trophic level

Primary producers

Figure 28.5 Sample of some of the interactions in a tallgrass prairie food web.

Read Me First!
and watch the narrated
animation on food webs

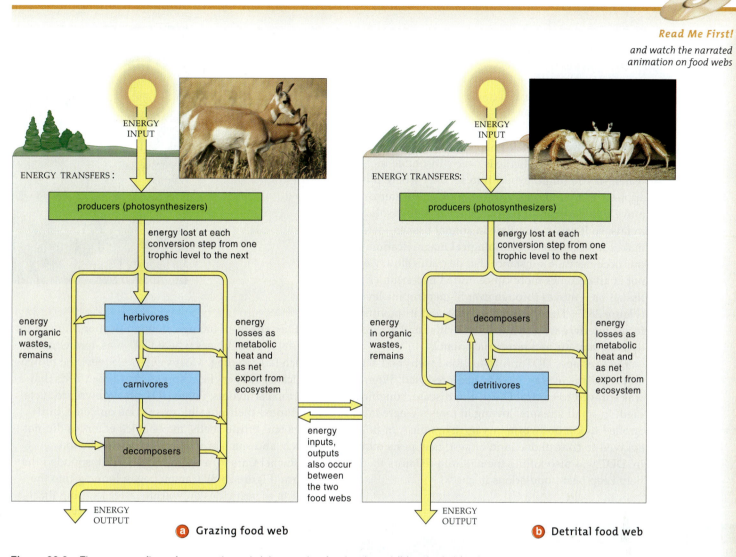

Figure 28.6 The one-way flow of energy through (**a**) a grazing food web and (**b**) a detrital food web.

Energy from producers, the organisms closest to a primary source, flows in one direction through two kinds of webs. In a **grazing food web**, energy flows mostly into herbivores, carnivores, then decomposers. In a **detrital food web**, energy from producers flows mainly into detritivores and decomposers. Figure 28.6 summarizes the flow through these food webs.

In nearly all ecosystems, both kinds of webs cross-connect. For example, in a rocky intertidal ecosystem, energy captured by algae flows to snails, which are eaten by herring gulls as part of a grazing food web. But the gulls also hunt crabs, which are among the primary consumers in the detrital food web.

The amount of energy that moves through the two kinds of food webs differs among ecosystems, and it often varies with the seasons. In most cases, however, most of the energy stored in producer tissues moves

through *detrital* food webs. Think of cattle, grazing heavily in a pasture. About half the energy stored in the grass plants enters the grazers. But cattle cannot tap into all of the stored energy. A lot is still present in undigested plant parts and feces, and decomposers and detritivores go to work. Similarly, in marshes, most of the energy initially stored in the marsh grass tissues enters detrital food webs when the plants die.

The inherent inefficiency in energy transfers between trophic levels limits the length of food chains.

Tissues of living photosynthetic organisms are the basis for grazing food webs. Remains and wastes of these organisms are the basis for detrital food webs.

In nearly all ecosystems, both types of food webs prevail and interconnect.

28.3 DDT in Food Webs

We have released thousands of synthetic chemicals and other new organic compounds into the environment. Ecosystem analysis helps us predict their side effects. The danger is that researchers may not have identified all of the variables. The most crucial variable may be the one they do not yet know.

Figure 28.8 Rachel Carson, who helped awaken public interest in the impact of human activities on nature.

Consider how DDT once entered food webs. This synthetic organic pesticide is nearly insoluble in water, so you might expect it to act only where it is applied. But winds carry DDT vapor, and water can move fine particles of it. Also, DDT is highly soluble in fats, so it can accumulate in animal tissues—which makes it a candidate for **biological magnification**. By this occurrence, a substance that degrades slowly or not at all becomes more concentrated in consumer tissues as it moves through ever higher trophic levels (Figure 28.7). This is not good. DDT and its modified forms interfere with metabolic activities and are toxic to many aquatic and terrestrial animals.

Decades ago, DDT infiltrated food webs and acted on organisms in ways no one had predicted. Where it was sprayed to control Dutch elm disease, songbirds died. In small streams flowing in forests where it was applied to kill budworm larvae, fishes died. In fields sprayed to control one kind of pest, new pests moved in. DDT was also killing the natural predators that help keep pest populations in check!

Side effects of biological magnification showed up far away from places where DDT was applied, and much later in time. Most vulnerable were the brown pelicans, bald eagles, ospreys, and other birds that are top carnivores. Some DDT breakdown products disrupted their physiology. As one outcome, bird eggs had brittle shells, many chick embryos did not hatch, and some bird species faced extinction.

Rachel Carson, a respected scientist, sounded the alarm (Figure 28.8). Carson began looking into the effects of pesticides on wildlife after learning that dead birds inside a local wildlife sanctuary showed symptoms of poisoning. The sanctuary had been sprayed with DDT to control mosquitoes.

Independent, critical research on pesticides was almost nonexistent. Carson methodically developed information about the harmful effects of widespread pesticide use. She published her findings in 1962 as a small book, *Silent Spring*. The public embraced it. But the manufacturers of chemicals viewed Carson as a threat to pesticide sales and quickly mounted a campaign to discredit her. At the time, Carson knew she had terminal cancer. Yet she vigorously defended her position. Later, her work became the impetus for the environmental movement in the United States. One line of investigation became the field known as conservation biology.

The United States has banned DDT except when required to protect the public health against insect vectors for disease. Today, there is consensus that DDT was overused and has no place in agriculture. Developing countries still use it mainly to control mosquitoes. Traces are still found in many organisms all over the world, including humans.

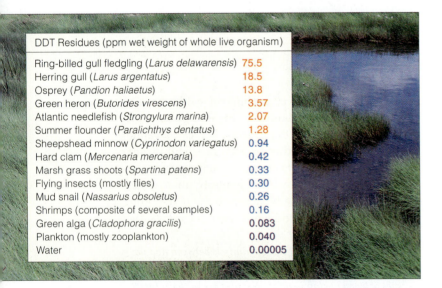

DDT Residues (ppm wet weight of whole live organism)	
Ring-billed gull fledgling (*Larus delawarensis*)	75.5
Herring gull (*Larus argentatus*)	18.5
Osprey (*Pandion haliaetus*)	13.8
Green heron (*Butorides virescens*)	3.57
Atlantic needlefish (*Strongylura marina*)	2.07
Summer flounder (*Paralichthys dentatus*)	1.28
Sheepshead minnow (*Cyprinodon variegatus*)	0.94
Hard clam (*Mercenaria mercenaria*)	0.42
Marsh grass shoots (*Spartina patens*)	0.33
Flying insects (mostly flies)	0.30
Mud snail (*Nassarius obsoletus*)	0.26
Shrimps (composite of several samples)	0.16
Green alga (*Cladophora gracilis*)	0.083
Plankton (mostly zooplankton)	0.040
Water	0.00005

Figure 28.7 Biological magnification in a Long Island, New York, estuary. In 1967, George Woodwell, Charles Wurster, and Peter Isaacson were aware that DDT exposure and wildlife deaths were connected. For instance, residues in birds that died from DDT poisoning were 41–295 parts per million (ppm) and 1–26 ppm in some fishes. In their study, some measured DDT levels were below lethal thresholds but still high enough to interfere with reproduction.

28.4 Studying Energy Flow Through Ecosystems

Ecologists measure the amount of energy and nutrients entering an ecosystem, how much is captured, and the proportion stored in each trophic level.

WHAT IS PRIMARY PRODUCTIVITY?

The rate at which producers capture and store energy in their tissues during a given interval is the **primary productivity** of an ecosystem. How much energy gets stored depends on (1) how many producers there are and (2) the balance between photosynthesis (energy trapped) and aerobic respiration (energy used). *Gross primary production* is all energy initially trapped by the producers. *Net* primary production is the fraction of trapped energy that producers funnel into growth and reproduction. **Net ecosystem production** is the gross primary production *minus* the energy used by producers and by detritivores and decomposers in soil. (In both cases, the subtracted amount of energy would not be available to organisms that eat plants.)

Many factors impact net production, its seasonal patterns, and its distribution in the habitat. This is true both on land and in the seas (Figure 28.10). For example, energy acquisition and storage depend in part on the body size and form of primary producers, mineral availability, the temperature range, and the amount of sunlight and rainfall during each growing season. The harsher the conditions are, the less new plant growth per season, and the lower productivity.

ECOLOGICAL PYRAMIDS

Ecologists often represent the trophic structure of an ecosystem in the form of an ecological pyramid. In such pyramids, the primary producers form a base for successive tiers of consumers above them.

A **biomass pyramid** depicts the dry weight of all of an ecosystem's organisms at each tier. Figure 28.9 shows a biomass pyramid for one aquatic ecosystem. The amounts measured are grams per square meter at some specified time. Most commonly, the primary producers have most of the biomass in pyramids like this, and top carnivores are few. But some biomass pyramids are "upside-down," with the smallest tier on the bottom. This happens in springtime blooms of phytoplankton, which grow and reproduce quickly. The primary producers of these aquatic communities support a larger biomass of zooplankton, which eat them about as fast as they can reproduce.

An **energy pyramid** illustrates how the amount of usable energy diminishes as it is transferred through an ecosystem. Sunlight energy is captured at the base

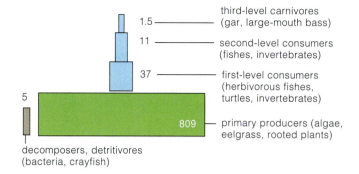

third-level carnivores
1.5 —— (gar, large-mouth bass)

11 —— second-level consumers (fishes, invertebrates)

37 —— first-level consumers (herbivorous fishes, turtles, invertebrates)

5

809 —— primary producers (algae, eelgrass, rooted plants)

decomposers, detritivores
(bacteria, crayfish)

Figure 28.9 Biomass pyramid for Silver Springs, a small aquatic ecosystem. Biomass decreases in successively higher tiers. In some ecosystems, autotrophs are eaten almost as fast as they grow and reproduce. Biomass accumulates faster in consumers, so a biomass pyramid would be upside down.

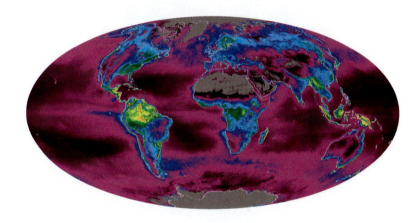

Figure 28.10 Summary of satellite data on net primary productivity during 2002. Productivity is coded as *red* (highest) down through *orange*, *yellow*, *green*, *blue*, and *purple* (lowest). Although average productivity per unit of sea surface is lower than it is on land, total productivity on land and in seas is about equal, because most of Earth's surface is covered by water.

(first trophic level) and declines through successive levels to its tip (the top carnivores). Energy pyramids have a large energy base at the bottom, so they are always "right-side up." As you will see in the next section, they can give a clear picture of energy flow.

Gross primary productivity is an ecosystem's total rate of photosynthesis during a specified interval. The net amount is the rate at which primary producers store energy in tissues in excess of their rate of aerobic respiration. Heterotrophic consumption affects the rate of energy storage.

The trophic structure of an ecosystem can be represented by an ecological pyramid. Biomass pyramids may be top- or bottom-heavy depending on the ecosystem. In contrast, an energy pyramid always has the largest tier on the bottom.

28.5 Energy Flow Through Silver Springs

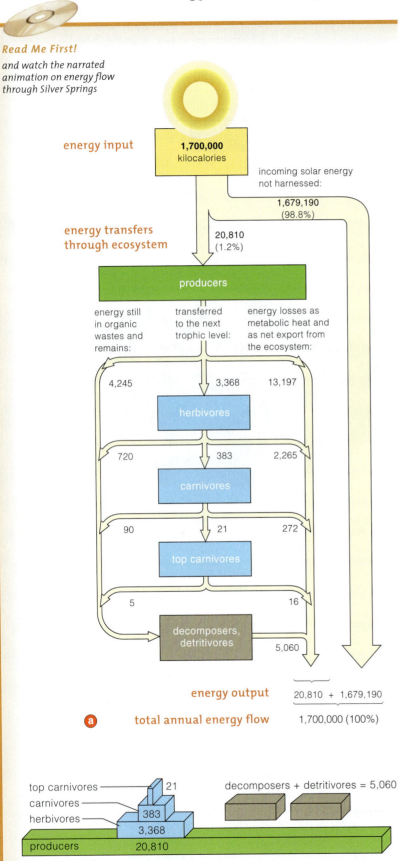

energy input   1,700,000 kilocalories

incoming solar energy not harnessed:

1,679,190 (98.8%)

energy transfers through ecosystem   20,810 (1.2%)

producers

energy still in organic wastes and remains:

transferred to the next trophic level:

energy losses as metabolic heat and as net export from the ecosystem:

4,245   3,368   13,197

herbivores

720   383   2,265

carnivores

90   21   272

top carnivores

5   16

decomposers, detritivores   5,060

energy output   20,810 + 1,679,190

a   **total annual energy flow**   1,700,000 (100%)

top carnivores — 21
carnivores — 383
herbivores — 3,368
producers — 20,810

decomposers + detritivores = 5,060

b

Energy flows into food webs from an outside source, most often the sun. Energy leaves mainly by losses of metabolic heat, which each organism generates.

Imagine you are with ecologists who want to gather the data to construct an energy pyramid for a small freshwater spring over the course of one year. You observe them as they measure the energy that an individual organism of each type takes in, loses as metabolic heat, stores in its body tissues, and loses in waste products. You see that they multiply the energy per individual by population size, then calculate energy inputs and outputs. In this way, they're able to express the flow of energy per unit of water (or land) per unit of time.

Figure 28.11*a* shows the data from a long-term study of a grazing food web in an aquatic ecosystem. It shows some of the calculations that the ecologists used to depict the energy flow in pyramid form in Figure 28.11*b*.

Given the metabolic demands of organisms and the amount of energy lost in their organic wastes, only 6 to 16 percent of the energy entering one trophic level becomes available for organisms at the next level.

Figure 28.11 (**a**) Breakdown of the annual energy flow through Silver Springs, Florida, as measured in kilocalories/square meter/year.

Most of the primary producers in this small spring are aquatic plants. Most of the carnivores are insects and small fishes; the top carnivores are larger fishes. The original energy source, sunlight, is available all year. The spring's detritivores and decomposers cycle organic compounds from the other trophic levels.

The producers trapped 1.2 percent of the incoming solar energy, and only a little more than a third of that amount became fixed in new plant biomass. The producers used more than 63 percent of the fixed energy for their own metabolism.

About 16 percent of the fixed energy was transferred to the herbivores, and most of this was used for metabolism or transferred to detritivores and decomposers.

Of the energy that transferred to herbivores, only 11.4 percent reached the next trophic level (carnivores). About 5.5 percent of the energy in lower-level carnivores flowed to the top carnivores.

By the end of the specified time interval, all of the 20,810 kilocalories of energy that had been transferred through the system appeared as metabolically generated heat.

(**b**) Pyramid of energy flow through Silver Springs, in kilocalories/square meter/year. This is a simple summary of the data shown in more detail above. Compare Figure 28.9, the biomass pyramid for this same ecosystem.

28.6 Overview of Biogeochemical Cycles

Primary productivity depends on availability of fresh water and on nutrients dissolved in it.

In a **biogeochemical cycle**, an essential element moves from the environment, through ecosystems, then back to the environment. No other element can directly or indirectly fulfill the metabolic role of such elements, or **nutrients**, which is why we call them essential. As you read earlier, oxygen, hydrogen, carbon, nitrogen, and phosphorus are among them.

Figure 28.12 is a model for these cycles. Transfers to and from environmental reservoirs are usually far slower than rates of exchange among organisms of an ecosystem. Water is the main source for hydrogen and oxygen. Gaseous or ionized forms of other elements are dissolved in it. Solid forms of elements are tied up in rocks or sediments.

Nutrients move into and out of ecosystems by way of natural geologic processes. Weathering of rocks is a common source of nutrient inputs into an ecosystem. Erosion and runoff put nutrients into streams that carry them away. Most often, the quantity of a nutrient being cycled through an ecosystem each year is greater than the amount entering and leaving.

Decomposers help cycle nutrients in ecosystems. Various prokaryotic species help transform solids and ions into gases and then back again. By their actions, they convert some of the elements that are nutrients to forms that primary producers can take up.

Three types of biogeochemical cycles are based on parts of the environment that act as reservoirs for specific elements. In the *hydrologic* cycle (global water cycle), oxygen and hydrogen move as molecules of water. In *atmospheric* cycles, some gaseous form of the nutrient is the one available to ecosystems. Carbon and nitrogen cycles are examples. Phosphorus and other solid nutrients that have no gaseous form move in *sedimentary* cycles. They accumulate on the seafloor and eventually return to land by way of geological uplifting, which often takes millions of years. Earth's crust is the largest reservoir for nutrients that have sedimentary cycles.

Nutrient availability influences the primary productivity on which ecosystems depend. In a biogeochemical cycle, a nutrient moves slowly through the environment, then rapidly among organisms, and back to environmental reservoirs.

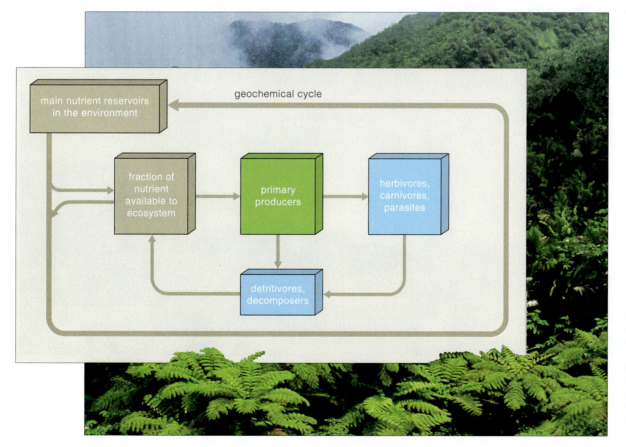

Figure 28.12 One generalized model of nutrient flow through a land ecosystem. The overall movement of nutrients from the physical environment, through organisms, and then back to the environment is a biogeochemical cycle.

28.7 Global Cycling of Water

On land, the availability of water, and nutrients dissolved in it, is not plentiful all of the time in all ecosystems. The variation affects primary productivity.

THE HYDROLOGIC CYCLE

Driven by solar energy, Earth's waters slowly move from the ocean into the atmosphere, to land, and back to the ocean—the main reservoir. Figure 28.13 shows this **hydrologic cycle**. Water evaporating into the lower atmosphere stays aloft as vapor, clouds, and ice crystals, then falls mainly as rain and snow. Ocean circulation and wind patterns influence the cycle.

Water moves nutrients into and out of ecosystems. A **watershed** is any region where precipitation flows into a single stream or river. Watersheds may be as small as the area that drains into a stream or as vast as the Amazon River or Mississippi River basin. Most water entering a watershed seeps into soil or joins surface runoff into streams. Plants take up water from soil and lose it by transpiration.

You might think that surface runoff would quickly leach mineral ions. But studies in New Hampshire's Hubbard Brook watershed show this is not the case. In undisturbed forests in this watershed, each hectare lost only about eight kilograms of calcium per year. Calcium loss was replaced by rainfall and weathering. Also, tree roots mine the soil, so calcium was continually being stored in the biomass of tree tissues. Experimental deforestation caused an imbalance in the nutrient inputs and outputs (Figure 28.14).

THE WATER CRISIS

Our planet has a lot of water, but most of it is too salty for consumption or agriculture. If all of Earth's waters were contained in a bathtub, the freshwater portion from lakes, rivers, reservoirs, groundwater, and other places would barely fill a teaspoon.

Agriculture accounts for two-thirds of our water usages, and irrigation alters the suitability of land for agriculture. The piped-in water commonly has high

Read Me First!

and watch the narrated animation on the hydrologic cycle

Figure 28.13 The hydrologic cycle. Water moves from the ocean to the atmosphere, land, and back. Arrows identify processes that move water, as measured in cubic kilometers per year. With 1,370,000,000 cubic kilometers, the ocean is the main reservoir. The next largest, polar ice and glaciers, locks up 29,000,000. Groundwater makes up only 4,000,000, lakes and rivers only 241,000, soil 67,000, and the atmosphere 14,000 cubic kilometers.

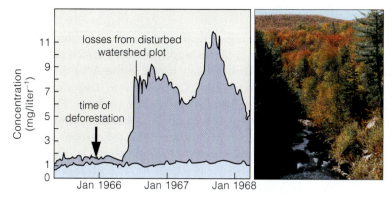

Figure 28.14 Experimental deforestation of New Hampshire's Hubbard Brook watershed. Researchers stripped vegetation from forest plots but did not disturb the soil. They applied herbicides for three years to prevent regrowth. They monitored surface runoff flowing over concrete catchments on its way to a stream below. Concentrations of calcium ions and other minerals were compared against concentrations in water passing over a control catchment, positioned in an undisturbed area.

Calcium losses were *six times* greater in deforested plots. Removing all vegetation from such forests clearly alters nutrient outputs in ways that can disrupt nutrient availability for an entire ecosystem.

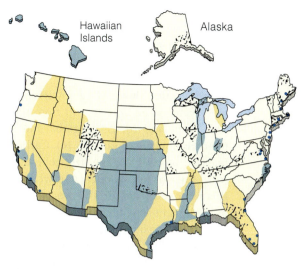

Figure 28.15 Aquifer depletion, seawater intrusion, and groundwater contamination in the United States. *Green* signifies high overdrafts, *gold*, moderate overdrafts, and *pale yellow*, insignificant withdrawals. Shaded areas are sites of major groundwater pollution. *Blue* squares indicate saltwater intrusion from nearby seas.

concentrations of mineral salts. Where the soil drains poorly, evaporation causes **salinization**, a build-up of salt in soil that stunts crop plants and decreases yields.

Soil and aquifers hold **groundwater**. About half of the United States population taps into groundwater as a source of drinking water. Chemicals leached from landfills, hazardous waste dumps, and underground tanks that store gasoline, oil, and some solvents often contaminate it. Unlike flowing streams that recover fast, polluted groundwater is difficult and expensive to clean up.

Groundwater overdrafts, or the amount that nature does not replenish, are high in many areas. Figure 28.15 shows some regions of aquifer depletion in the United States. Overdrafts have now depleted half of the great Ogallala aquifer, which supplies irrigation water for 20 percent of the Midwest's croplands.

Inputs of sewage, animal wastes, and many toxic chemicals from power-generating plants and factories make water unfit to drink. Sediments and pesticides run off from fields into water, along with phosphates and other nutrients that promote algal blooms. The pollutants accumulate in lakes, rivers, and bays before reaching the ocean. Many cities all over the world are still dumping untreated sewage into coastal waters.

If current rates of human population growth and water depletion continue, the amount of fresh water available for everyone will soon be 55 to 66 percent less than it was in 1976. In this past decade, thirty-three nations have already engaged in conflicts over reductions in water flow, pollution, and silt buildup

in aquifers, rivers, and lakes. Among the squabblers are the United States and Mexico, Pakistan and India, and Israel and the Palestinian territories.

Could **desalinization**—the removal of salt from seawater—meet our water needs? Salt can be removed by distillation or pushing water through membranes. The processes require fossil fuels, which makes them more feasible in Saudi Arabia and other countries with small populations and big fuel reserves. Most likely, desalinization won't be cost-effective for large-scale agriculture. It also produces mountains of salts.

We may be in for upheavals and wars over water rights. Does this sound far-fetched? Consider the new dam across the Euphrates River. By building huge dams and irrigation systems at the headwaters of the Tigris and Euphrates rivers, Turkey can, in the view of one of its dam site managers, shut off water flow into Syria and Iraq for as long as eight months "to regulate their political behavior." Regional, national, and global planning for the future is long overdue.

In the hydrologic cycle, water slowly moves on a global scale from the world ocean (the main reservoir), through the atmosphere, onto land, then back to the ocean.

Plants stabilize the soil and absorb dissolved minerals. By doing so, they minimize the loss of soil nutrients in runoff.

Aquifers that supply much of the world's drinking water are becoming polluted and depleted. Regional conflicts over access to clean, drinkable water are likely to increase.

28.8 Carbon Cycle

Most of the world's carbon is locked in ocean sediments and rocks. It moves into and out of ecosystems in gaseous form, so its movement is said to be an atmospheric cycle.

Carbon moves through the lower atmosphere and all food webs on its way to and from the ocean's water, sediments, and rocks. Its global movement is called the **carbon cycle** (Figure 28.16). Carbon dioxide gas is released as cells engage in aerobic respiration, when fossil fuels or forests burn, and when volcanoes erupt. It is removed from air and water when it is taken up by organisms that carry out photosynthesis.

Carbon dioxide gas, or CO_2, is the most abundant form of carbon in the atmosphere. Carbon dissolved in ocean water is mainly combined with oxygen, as carbonate or bicarbonate. For that reason, the cycle is sometimes called the *carbon–oxygen cycle*.

Why doesn't all the CO_2 dissolved in warm ocean surface waters escape into the atmosphere? Driven by the winds and regional differences in water density, water makes a gigantic loop from the surface of the Pacific and Atlantic oceans down to the Atlantic and Antarctic seafloors. The CO_2 moves into deep storage reservoirs before the seawater loops back up (Figure 28.17). This loop affects carbon's global distribution.

When photosynthetic autotrophs fix carbon, they lock billions of metric tons of carbon atoms in organic compounds each year (Chapter 6). The average time that an ecosystem holds a given carbon atom varies greatly. As one example, organic wastes and remains decompose so quickly in tropical rain forests that not much carbon accumulates at the soil surfaces. In bogs, marshes, and similar anaerobic habitats, decomposers cannot degrade organic compounds to smaller bits, so carbon gradually accumulates in peat.

Also, in ancient aquatic ecosystems, carbon became incorporated into the shells and other hard body parts of many marine species. Those organisms died,

Read Me First!
and watch the narrated animation on the carbon cycle

Figure 28.16 Global carbon cycle. Carbon's movement through typical marine ecosystems (**a**) and through ecosystems on land (**b**). *Gold* boxes show the main carbon reservoirs. The vast majority of carbon atoms are locked in sediments and rocks, followed by increasingly lesser amounts in ocean water, soil, and atmosphere, and finally in biomass. The unit of measure for the lists below is 10^{15} grams:

Main carbon reservoirs and holding stations

Sediments and rocks	77,000,000
Ocean (dissolved forms)	39,700
Soil	1,500
Atmosphere	750
Biomass on land	715

Annual fluxes in the global distribution of carbon

From atmosphere to plants (carbon fixation)	120
From atmosphere to ocean	107
To atmosphere from ocean	105
To atmosphere from plants	60
To atmosphere from soil	60
To atmosphere from fossil fuel burning	5
To atmosphere from net destruction of plants	2
To ocean from runoff	0.4
Burial in ocean sediments	0.1

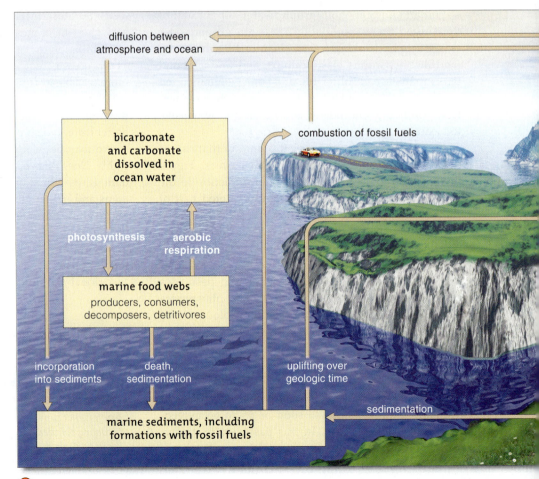

diffusion between atmosphere and ocean

combustion of fossil fuels

bicarbonate and carbonate dissolved in ocean water

photosynthesis   aerobic respiration

marine food webs
producers, consumers, decomposers, detritivores

incorporation into sediments   death, sedimentation   uplifting over geologic time

sedimentation

marine sediments, including formations with fossil fuels

a

sank through water, then were buried in sediments. Carbon stayed buried for many millions of years until geologic forces slowly uplifted part of the seafloor.

At present, deforestation, the burning of wood and fossil fuels, and other human activities release more carbon into the atmosphere than can be cycled back to ocean reservoirs by natural processes.

Only about 2 percent of the excess carbon entering the atmosphere will become dissolved in ocean water. Most researchers now suspect that the carbon build-up in the atmosphere is amplifying the greenhouse effect. In other words, the increase may be a contributor to global warming. The next section takes a look at this possibility and its environmental implications.

Most carbon is trapped in the Earth's rocks and sediments, with some dissolved in ocean waters. Carbon moves between the main reservoirs and ecosystems mostly as carbon dioxide.

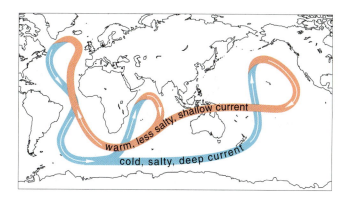

Figure 28.17 One loop of ocean water that delivers carbon dioxide to carbon's deep ocean reservoir. It sinks in the cold, salty North Atlantic and rises in the warmer Pacific.

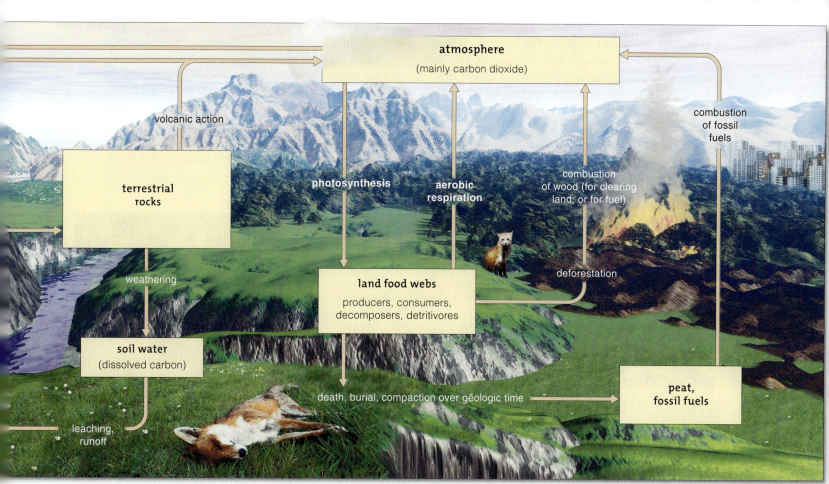

28.9 Greenhouse Gases, Global Warming

The atmospheric concentrations of gaseous molecules help shape the average temperature near the Earth's surface. Human activities are altering atmospheric gas levels in ways that may cause dramatic climate change.

Concentrations of some gaseous molecules in Earth's atmosphere play a profound role in determining the average temperature near its surface. Temperature, in turn, has far-reaching effects on the global climate.

Atmospheric molecules of carbon dioxide, water, nitrous oxide, methane, and chlorofluorocarbons are among the main players in interactions that affect global temperature. Collectively, these gases act like the panes of glass in a greenhouse—hence the name, "greenhouse gases." Wavelengths of visible light pass through these gases to Earth's surface, which absorbs them and emits longer, infrared wavelengths—heat. Greenhouse gases impede the escape of heat energy from Earth into space. How? The gaseous molecules absorb the longer wavelengths, then radiate much of it back toward Earth (Figure 28.18).

Constant reradiation of heat by greenhouse gases occurs lockstep with the constant bombardment and absorption of wavelengths from the sun. As heat builds up in the lower atmosphere, the temperature near the Earth's surface rises. The warming action is called the **greenhouse effect**. Without it, Earth's surface would be cold and unable to support life.

In the 1950s, laboratory researchers on Hawaii's highest volcano set out to measure the atmospheric concentrations of greenhouse gases. That remote site is almost free of local airborne contamination; it is also representative of overall atmospheric conditions

Figure 28.19 Graphs of recent increases in four categories of atmospheric greenhouse gases. A key factor is the huge numbers of gasoline-burning vehicles in large cities. *Above*, Mexico City on a smoggy morning. With 10 million residents, it is the world's largest city.

for the Northern Hemisphere. What did they find? Briefly, carbon dioxide concentrations follow annual cycles of primary production. They drop during the summer, when photosynthesis rates are highest. They rise in the winter, when photosynthesis rates decline but aerobic respiration is still going on.

Alternating troughs and peaks along the graph line in Figure 28.19a are annual lows and highs of global carbon dioxide concentrations. For the first time, we could see the integrated effects of carbon balances for an entire hemisphere. Notice the midline of the troughs and peaks in the cycle. It shows that carbon dioxide concentration is steadily increasing—as are the concentrations of other major greenhouse gases.

Atmospheric levels of greenhouse gases are far higher than they were in most of the past. Carbon

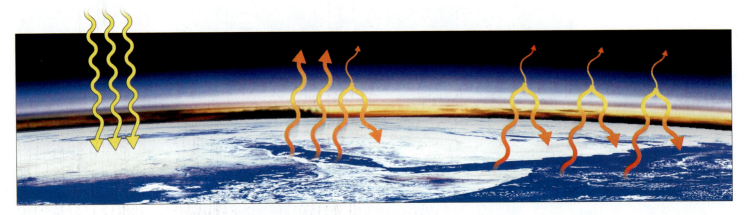

a Wavelengths in rays from the sun penetrate the lower atmosphere, and they warm the Earth's surface.

b The surface radiates heat (infrared wavelengths) to the atmosphere. Some heat escapes into space. But greenhouse gases and water vapor absorb some infrared energy and radiate a portion of it back toward Earth.

c Increased concentrations of greenhouse gases trap more heat near Earth's surface. Sea surface temperatures rise, so more water evaporates into the atmosphere. Earth's surface temperature rises.

Figure 28.18 The greenhouse effect.

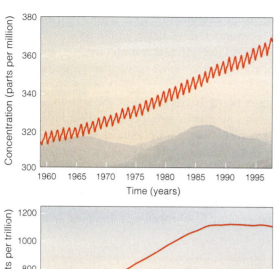

a Carbon dioxide (CO_2). Of all human activities, the burning of fossil fuels and deforestation contribute the most to increasing atmospheric levels.

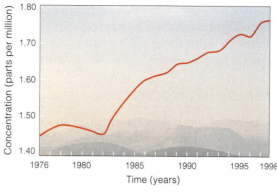

c Methane (CH_4). Production and distribution of natural gas for use as fuel adds to the methane released by some bacteria that live in swamps, rice fields, landfills, and in the digestive tract of cattle and other ruminants.

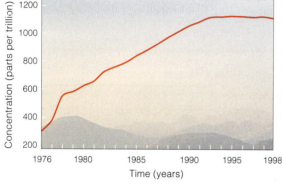

b CFCs. Until restrictions were in place, CFCs were widely used in plastic foams, refrigerators, air conditioners, and industrial solvents.

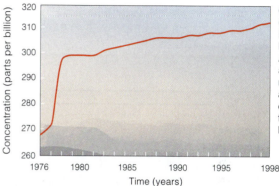

d Nitrous oxide (N_2O). Denitrifying bacteria produce N_2O in metabolism. Also, fertilizers and animal wastes release enormous amounts; this is especially so for large-scale livestock feedlots.

Figure 28.20 Recorded changes in global temperature between 1880 and 2000. At this writing, the hottest year on record was 1998.

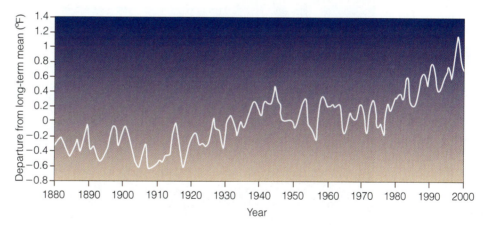

dioxide may be at its highest level since 420,000 years ago, and possibly since 20 million years ago. There is a growing consensus that the rise in greenhouse gases is caused by some human activities, mainly burning of fossil fuels. The big worry is that the increase may have widespread environmental consequences.

The increase in greenhouse gases may be a factor in **global warming**, a long-term rise in temperature near Earth's surface. Since direct atmospheric readings started in 1861, the lower atmosphere's temperature has risen by more than 1°F, mostly since 1946 (Figure 28.20). Also since then, nine of the ten hottest years on record have occurred between 1990 and the present. Data from satellites, weather stations and balloons,

research ships, and supercomputer programs suggest that irreversible climate changes are already under way. Polar ice is melting; glaciers are retreating. This past century, the sea level may have risen as much as twenty centimeters (eight inches). We can expect continued temperature increases to have drastic effects on climate. As evaporation increases, so will global precipitation. Intense rains and flooding are expected to become more frequent in some regions.

It bears repeating: As investigations continue, a key research goal is to investigate all of the variables in play. With respect to the consequences of global warming, the most crucial variable may be the one we do not know.

28.10 Nitrogen Cycle

Gaseous nitrogen makes up about 80 percent of the lower atmosphere. Successively smaller reservoirs are seafloor sediments, ocean water, soil, biomass on land, nitrous oxide in the atmosphere, and marine biomass.

THE CYCLING PROCESSES

Gaseous nitrogen (N_2) travels in an atmospheric cycle called the **nitrogen cycle**. Triple covalent bonds join its two atoms ($N\equiv N$). Volcanic action and lightning convert some gaseous nitrogen into forms that enter food webs. Far more enters by **nitrogen fixation**. By this metabolic process, certain bacteria in soil split all three bonds in gaseous nitrogen, then use the atoms to form ammonia (NH_3). Later, ammonia is converted to ammonium (NH_4^+) and nitrate (NO_3^-), the two forms of nitrogen that most plants can easily take up (Figure 28.21).

Nitrogen-fixing bacteria live in the soil, in aquatic habitats, and as photosynthetic mutualists in lichens. Others are mutualists with plants, as when *Rhizobium* forms nodules on the roots of peas or other legumes. Collectively, such bacteria fix about 200 million metric tons of nitrogen each year.

Nitrogen is often a limiting factor for plant growth. Soils that are low in nitrogen put legumes, with their

Read Me First!
and watch the narrated animation on the nitrogen cycle

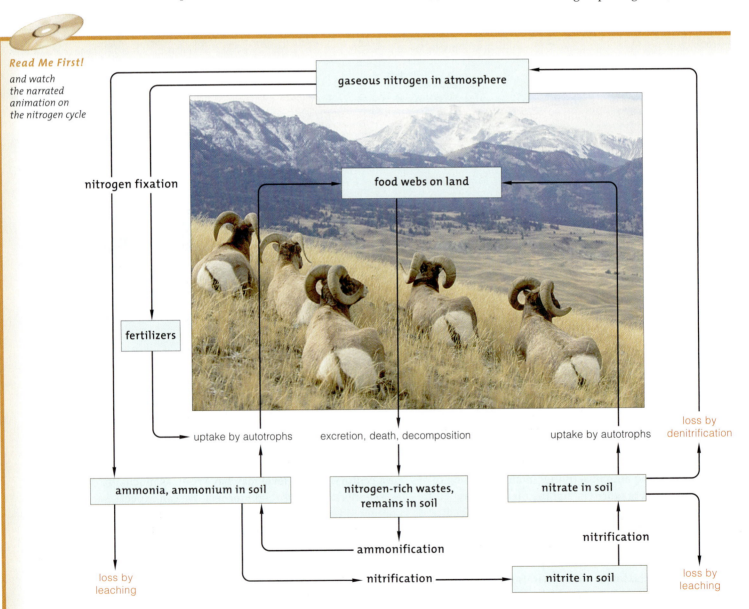

Figure 28.21 The nitrogen cycle in an ecosystem on land. Activities of nitrogen-fixing bacteria make nitrogen available to plants. Other bacterial species cycle nitrogen to plants. They break down organic wastes to ammonium and nitrates.

bacterial symbionts, at a competitive advantage. The legumes have a nitrogen source that other plants do not. In itself, supporting bacteria in root nodules is costly in some cases; legumes lose some metabolic products to the bacteria. In nitrogen-rich soils, plant species that don't have bacterial symbionts and don't pay the metabolic price often outcompete the legumes.

The nitrogen incorporated into plant tissues moves through trophic levels of ecosystems and ends up in nitrogen-rich wastes and remains. By **ammonification**, bacteria and fungi break down nitrogenous materials, forming ammonium. They use some ammonium and release the rest to soil. Plants take it up, as do some nitrifying bacteria. In the first step of **nitrification**, these bacteria produce nitrite (NO_2^-) when they strip electrons from ammonium. Other nitrifying bacteria use nitrite in reactions that form nitrate (NO_3^-).

Ecosystems lose nitrogen through **denitrification**. By this process, certain bacteria convert nitrate or nitrite to gaseous nitrogen or nitrogen oxide (NO_2). Most denitrifying bacteria are anaerobic; they live in waterlogged soils and aquatic sediments.

Ammonium, nitrite, and nitrate are also lost from a land ecosystem through leaching and runoff. Leaching, recall, is the removal of some soil nutrients as water runs through it. Leaching removes nitrogen from land ecosystems and adds it to aquatic ones.

HUMAN IMPACT ON THE NITROGEN CYCLE

Nitrogen losses through deforestation and grassland conversion for agriculture are huge. With each forest clearing and harvest, nitrogen stored in plant biomass is removed. Erosion and leaching from exposed soils take even more away.

Many farmers counter nitrogen losses by rotating crops. For example, a farmer might grow wheat in a field some years, and plant legumes in the same field during alternate years. The nitrogen-rich remains of legumes (and their bacterial symbionts) are plowed under to enrich the soil for wheat production.

Increasingly, farmers are also using nitrogen-rich chemical fertilizers. The fertilizers are manufactured by combining hydrogen gas and nitrogen gas at high pressure and temperature to yield ammonia. Use of chemical fertilizers can greatly increase crop yields per square area, but may also have harmful effects.

For example, fertilizer applications can alter soil chemistry. Use of nitrogen fertilizers adds hydrogen ions to the soil, making it more acid. These ions can displace other positively-charged mineral ions, such as magnesium and calcium, from binding sites on soil

Figure 28.22 Dead and dying trees in Smoky Mountain National Park. Forests are among the casualties of nitrogen oxides and other forms of air pollution.

particles. As a result, the rates at which these vital mineral nutrients are leached from the soil increase.

Deposition of nitrogen in acid rains can have the same effect as overfertilization. Fossil fuel burned in power plants and vehicles releases nitrogen oxides, which contribute to global warming and to acid rain. Winds often carry these air pollutants far from their sources (Figure 28.22). It is estimated some Eastern European forests are now receiving 10 times their normal level of nitrogen inputs from pollutants.

The responses of plant species to high nitrogen levels vary. As a result, any increases in nitrogen can disrupt the balance among plants that are competing in communities, and diversity may decline. The plant species best able to deal with additional nitrogen can outgrow species that are less responsive.

The addition of nitrogen also can disrupt aquatic ecosystems. About half of the nitrogen in fertilizers applied to fields is not taken up by crops. It runs off into rivers, lakes, and estuaries. Sewage from human cities and animal wastes adds more nitrogen to these waters. Enrichment of aquatic ecosystems by nitrogen and phosphorus can encourage harmful algal blooms, as you will see in the next section.

Most nitrogen is in gaseous form in the atmosphere, not in the ionized forms that plants generally can take up. These forms are mainly ammonium and nitrate.

Nitrogen-fixing bacteria convert nitrogen gas to ammonium. Bacteria and fungi break down wastes and remains and release ammonium.

Nitrates are produced by nitrifying bacteria. Denitrifying bacteria convert nitrite and nitrate back to gaseous forms.

Human activities are depleting nitrogen in some ecosystems and adding it to others.

28.11 Phosphorus Cycle

Unlike carbon and nitrogen, phosphorus does not cycle into and out of ecosystems in gaseous form. Like nitrogen, phosphorus can be taken up by plants only in its ionized form and is often a limiting factor for plant growth.

In the **phosphorus cycle**, phosphorus passes quickly through food webs as it moves from land to ocean sediments, then slowly back to the land. The Earth's crust is the largest reservoir of phosphorus.

In rock formations, phosphorus occurs mainly as phosphate ions (PO_4^{3-}). Weathering and erosion put these ions into streams and rivers. Eventually they become part of the ocean sediments (Figure 28.23). Phosphates accumulate slowly. They form insoluble deposits on submerged continental shelves. Millions of years go by. As movements of crustal plates uplift part of the seafloor, phosphates become exposed on land surfaces. Over time, weathering and erosion will release phosphates from exposed rocks once again.

All organisms use phosphates to synthesize ATP, phospholipids, nucleic acids, and other compounds. Plants take up dissolved phosphates from soil water.

Herbivores get them by eating plants; carnivores get them by eating herbivores. Animals lose phosphate in urine and feces. Bacterial and fungal decomposers release phosphate from organic wastes and remains, then plants take them up again.

The hydrologic cycle helps move phosphorus and other minerals through ecosystems. Water evaporates from the ocean and falls on land. As it flows back to the ocean, it transports silt and dissolved phosphates that primary producers require for growth.

Of all minerals, phosphorus is often the limiting factor in natural ecosystems. Only newly weathered, young soils are high in phosphorus, and in aquatic habitats most phosphorus is tied up in sediments. Only a negligible amount of phosphorus occurs in gaseous form, so little is lost to the atmosphere.

Especially in tropical and subtropical parts of the developing countries, phosphorus is being depleted from natural ecosystems that often have phosphorus-poor soils. Phosphorus stored in biomass and slowly released from decomposing organic matter normally sustains the undisturbed forests or grasslands. When

Read Me First!

and watch the narrated animation on the phosphorus cycle

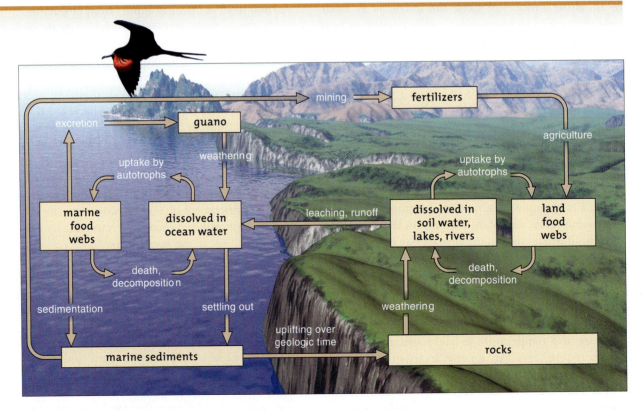

Figure 28.23 Phosphorus cycle. In this sedimentary cycle, most of the phosphorus moves in the form of phosphate ions (PO_4^{3-}) and reaches the oceans. Phytoplankton of marine food webs take up some of it, then fishes eat the plankton. Seabirds eat the fishes, and their droppings (guano) accumulate on islands. Guano is collected and used as a phosphate-rich fertilizer.

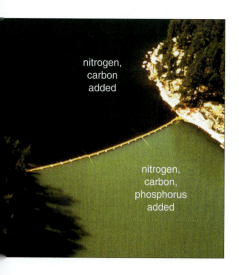

nitrogen, carbon added

nitrogen, carbon, phosphorus added

Figure 28.24 One of the eutrophication experiments. Researchers put a plastic curtain across a channel between two basins of a natural lake. They added nitrogen, carbon, and phosphorus on one side of the curtain (here, the *lower* part of the lake) and added nitrogen and carbon on the other side. Within months, the phosphorus-rich basin was eutrophic; with a dense algal bloom covering its surface.

trees are harvested or land is cleared for agriculture, phosphorus is lost. Crop yields start out low and soon become nonexistent. After the fields are abandoned, natural regrowth is sparse. An estimated 1 to 2 billion hectares are already depleted of phosphorus.

What about the developed countries? After years of fertilizer applications, many soils have phosphorus overloads. It is concentrated in eroded sediments and runoff from agricultural fields. Phosphorus also is present in outflows from sewage treatment plants and factories, and in the runoff from fields. Dissolved phosphorus that gets into streams, rivers, lakes, and estuaries can promote destructive algal blooms. Like plants, all photosynthetic algae require phosphorus, nitrogen, and other ions to grow. In many freshwater ecosystems, nitrogen-fixing bacteria keep the nitrogen levels high, so phosphorus becomes the limiting factor. When phosphate-rich pollutants pour in, populations of algae soar. As aerobic decomposers break down the remains of the algae, the water becomes depleted of the oxygen that fishes and other organisms require.

Eutrophication refers to the nutrient enrichment of any ecosystem that is otherwise low in nutrients. It is a natural successional process. Phosphorus inputs, as from agriculture, accelerate it, as the experiment shown in Figure 28.24 demonstrated.

Sedimentary cycles, in combination with the hydrologic cycle, move most mineral elements, such as phosphorus, through terrestrial and aquatic ecosystems.

Agriculture, deforestation, and other human activities upset the nutrient balance of an ecosystem when they accelerate the amount of minerals entering and leaving.

Summary

Section 28.1 An ecosystem consists of an array of organisms together with their physical and chemical environment. There is a one-way flow of energy into and out of an ecosystem, and a cycling of materials among the organisms. All ecosystems have inputs and outputs of energy and nutrients.

Sunlight is the initial energy source for almost all ecosystems. Primary producers convert energy from the sun to ATP and assimilate nutrients that they, and all consumers require. Consumers include herbivores, carnivores, omnivores, decomposers, and detritivores.

Organisms in an ecosystem are classified by trophic levels. Those at the same level are the same number of steps away from the energy input into the ecosystem.

Section 28.2 Linear sequences by which energy and nutrients move through ever higher trophic levels are food chains, which interconnect as food webs. Because the efficiency of energy transfers is always low, most ecosystems support no more than four or five trophic levels away from the original energy source.

In a grazing food web, energy captured by producers flows to herbivores. In a detrital food web, energy from producers flows directly to detritivores and decomposers. In nearly all ecosystems detrital and grazing food webs interconnect.

Section 28.3 By biological magnification, some chemical substance is passed from organisms at one trophic level to those above and becomes increasingly concentrated in body tissues. DDT is an example.

Section 28.4 A system's primary productivity is the rate at which producers capture and store energy in their tissues. It varies with climate, season, nutrient availability and other factors.

Energy pyramids and biomass pyramids are used to show the distribution of energy and materials in an ecosystem. All energy pyramids are largest at the base. The lowest trophic level has the greatest proportion of the energy in an ecosystem.

Section 28.5 Long-term studies of the Silver Springs ecosystem in Florida illustrate the inefficiency of energy transfers. At each trophic level, far more energy was lost to the environment or in wastes and remains than was passed on to the next trophic level.

Section 28.6 In a biogeochemical cycle, water or a nutrient moves through the environment, then through organisms, then back to an environmental reservoir.

Section 28.7 In the hydrologic cycle, water moves from the ocean into the atmosphere, to land, and back to the ocean—the main reservoir. Human actions are disrupting the cycle in ways that result in shortages and pollution of water.

Section 28.8 The carbon cycle moves carbon from its main reservoirs in rocks and seawater, through its gaseous form (carbon dioxide) in the atmosphere, and then through ecosystems. Deforestation and the burning of wood and fossil fuels are adding more carbon dioxide to the atmosphere than the oceans can absorb.

Section 28.9 Greenhouse gases trap heat in the Earth's lower atmosphere and thus make life on Earth possible. But human activities have been increasing the levels of some greenhouse gases, including carbon dioxide, methane, CFCs, and nitrous oxide. The rise in greenhouse gases correlates with a rise in global temperatures and other climate changes.

Section 28.10 The atmosphere is the main reservoir for N_2, a gaseous form of nitrogen that plants cannot use. In nitrogen fixation, some soil bacteria degrade N_2 and assimilate the two nitrogen atoms into ammonia. Other reactions convert ammonia to ammonium and nitrate that plants take up. Some is lost to the atmosphere by the action of denitrifying bacteria. Humans add nitrogen to ecosystems with fertilizers and nitrogen oxides released by the burning of fossil fuels.

Section 28.11 The phosphorus cycle is a sedimentary cycle; gaseous forms are insignificant. Earth's crust is the largest reservoir. Phosphorus can be a limiting factor for producers. Excess inputs of phosphorus to aquatic ecosystems contribute to eutrophication.

Self-Quiz   *Answers in Appendix III*

1. Organisms at the lowest trophic level in a tallgrass prairie are all _____ .
   a. at the first step away from the original energy input
   b. autotrophs   d. both a and b
   c. heterotrophs   e. both a and c

2. Decomposers are commonly _____ .
   a. fungi   b. animals   c. bacteria   d. a and c

3. Primary productivity on land is affected by _____ .
   a. nutrient availability   c. temperature
   b. amount of sunlight   d. all of the above

4. If biological magnification occurs, the _____ will have the highest levels of toxins in their systems.
   a. producers   c. primary carnivores
   b. herbivores   d. top carnivores

5. Disruption of the _____ cycle is depleting aquifers.
   a. hydrologic   c. nitrogen
   b. carbon   d. phosphorus

6. Earth's largest carbon reservoir is _____ .
   a. the atmosphere   c. seawater
   b. sediments and rocks   d. living organisms

7. The _____ cycle is a sedimentary cycle.
   a. hydrologic   c. nitrogen
   b. carbon   d. phosphorus

8. _____ is often a limiting factor for plant growth.
   a. Nitrogen   d. both a and c
   b. Carbon   e. all of the above
   c. Phosphorus

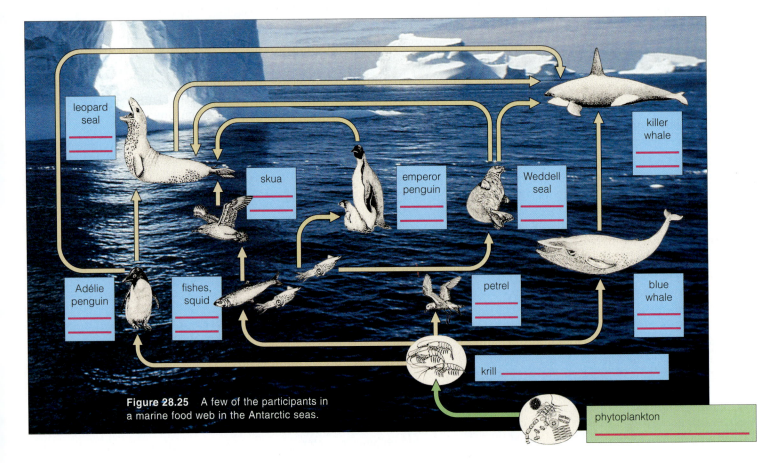

Figure 28.25 A few of the participants in a marine food web in the Antarctic seas.

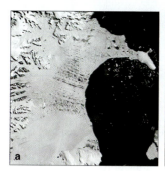
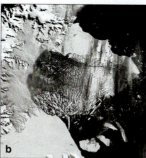

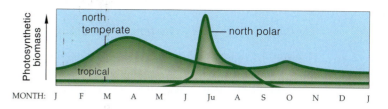

Figure 28.26 *Left:* Antarctica's Larson B ice shelf in (**a**) January and (**b**) March 2002. About 720 billion tons of ice broke from the shelf, forming thousands of icebergs.

Figure 28.27 Annual primary productivity in three ecosystems in the ocean.

9. Nitrogen fixation converts _____ to _____ .
   a. nitrogen gas; ammonia
   b. nitrates; nitrites
   c. ammonia; nitrogen gas
   d. ammonia; nitrates
   e. nitrites; nitrogen oxides

10. Match the terms with suitable descriptions.
    _____ producers
    _____ herbivores
    _____ decomposers
    _____ detritivores

    a. feed on plants
    b. feed on small bits of organic matter
    c. degrade organic wastes and remains to inorganic forms
    d. capture sunlight energy

Critical Thinking

1. Most food chains have at most four or five links. How many links are in the chains in the Antarctic food web shown in Figure 28.25? Decide which of these organisms are producers and which are consumers. What categories of consumers are depicted here? Which are not shown?

2. Polar ice shelves are vast, thickened sheets of ice that float on seawater. In March 2002, 3,200 square kilometers (1,410 square miles) of Antarctica's largest ice shelf broke free from the continent and shattered into thousands of icebergs (Figure 28.26). Scientists knew the ice shelf was shrinking and breaking up, but this was the single largest loss ever observed at one time. Why should this concern people who live in more temperate climates?

3. Fish is a great source of protein and of omega-3 fatty acids, which are necessary for normal nervous system development. This would seem to make fish a good choice for pregnant women. But coal-burning power plants put mercury into the environment and some of it ends up in fish. Eating mercury-tainted fish during pregnancy can adversely affect development of a fetal nervous system.

    Tissues of predatory marine fishes, such as swordfishes, tunas, marlins, and sharks, have especially high levels of mercury. The Environmental Protection Agency has issued health advisories to pregnant women, suggesting that they limit their consumption of fish species most likely to be tainted with mercury. Although sardines are harvested from the same ocean, they have lower mercury levels and are not on the warning list. Explain why two species of fishes that live in the same place can have very different levels of mercury in their tissues.

4. Figure 28.27 graphs the annual primary productivity in three ocean ecosystems. Name some factors that may cause the differences among them. Propose some experiments to test your hypotheses.

Media Menu

Student CD-ROM

Impacts, Issues Video
    Bye-Bye, Blue Bayou
Big Picture Animation
    Energy, nutrients, and global cycles
Read-Me-First Animation
    Trophic levels
    Food webs
    Energy flow through Silver Springs
    Hydrologic cycle
    Carbon cycle
    Nitrogen cycle
    Phosphorus cycle

InfoTrac

- Anthropogenic Disturbance of the Terrestrial Water Cycle. *BioScience*, September 2000.
- Be Fruitful, Multiply, and Lose Nitrogen. *Global Environmental Change Report*, October 2002.
- Too Much of a Good Thing. *National Wildlife*, April–May 2000.
- Spring Forward: Warmer Climates Accelerate Life Cycles of Plants, Animals. *Science News*, March 2003.

Web Sites

- Ecological Society of America: www.esa.org
- Water Resources of the United States: water.usgs.gov
- Environmental Literacy Council: www.enviroliteracy.org
- Energy Efficiency and Renewable Energy: www.eere.energy.gov
- Earth Observatory: earthobservatory.nasa.gov

How Would You Vote?

The United States excels in the production of greenhouse gases. Per person, we produce 19 times as much as India and 5 times as much as Mexico. Given the correlation between rising greenhouse gases and global warming, some argue that the government should discourage the inefficient use of gasoline. Would you support a tax surcharge on the sale of the most energy-inefficient private vehicles, such as large, heavy SUVs? Why or why not?

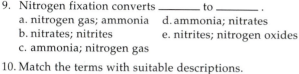

IMPACTS, ISSUES   *Surfers, Seals, and the Sea*

The stormy winter of 1997–1998 was a very good time for surfers on the lookout for the biggest waves, and a very bad time for seals and sea lions. As Ken Bradshaw rode a monster wave (Figure 29.1), half the sea lions on the Galápagos Islands were dying, including nine out of ten pups of the sort shown in the filmstrip at right. In California, the number of liveborn Northern fur seals plummeted. Most of the ones that made it through birth were dead within a few months. Diverse forms of life throughout the world became connected by the full fury of El Niño—a recurring event that ushers in an often spectacular seesaw in the world climate.

That winter, a massive volume of warm water from the southwestern Pacific moved east. It piled into coasts from California down through Peru, displacing currents that otherwise would have churned up tons of nutrients from the deep. Without nutrients, primary producers for marine food webs declined. The scarcity of producers combined with the warm water drove away consumers— anchovies and other migratory fishes. The fishes and

Figure 29.1 Ken Bradshaw surfing a monster wave, more than twelve meters high, during the most powerful El Niño of the past century. In January 1998, a storm formed off the coast of Siberia and generated an ocean swell, which became the biggest wave known to hit Hawaii.

the big picture

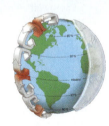

Biogeographic Patterns   Sunlight energy drives global circulation patterns in the atmosphere and ocean. Interactions among circulating air, ocean currents, and topography shape regional climates, which influence biodiversity on land and in aquatic habitats.

Major Biomes   Climate influences the formation of soils and other distinct properties of habitats at different latitudes and elevations. Biomes are broad regions of land, each characterized by plant and animal species that dominate its communities.

squids that could not migrate starved to death. So did many seals and sea lions, because fishes and squids happen to be the mainstay of their diet.

Marine mammals did not dominate the headlines, because the 1997–1998 El Niño gave humans plenty of other things to worry about. It battered Pacific coasts with fierce winds and torrential rains, massive flooding, and landslides. As another rippling effect, an ice storm in New York, New England, and central and eastern Canada crippled regional electrical grids. Three weeks later, 700,000 people still had no electricity. Meanwhile, the global seesaw caused drought-driven crop failures and raging wildfires in Australia and Indonesia.

All told, that one El Niño episode killed thousands of people and drained tens of billions of dollars from economies around the world.

El Niño is one of many environmental events that influence life. What causes them? *In this chapter, we connect environmental processes with broad patterns in the distribution of organisms through the biosphere. We build on Chapters 16 and 17, which introduced these biogeographic patterns in an evolutionary context.*

The **biosphere**, again, is the sum of all places where we find life on Earth. Organisms live in the hydrosphere — the ocean, ice caps, and other bodies of water, liquid and frozen. They live on and in sediments and soils of the lithosphere—Earth's outer, rocky layer. Many lift off into the lower region of the atmosphere— the gases and airborne particles that envelop Earth.

 How Would You Vote?

We cannot stop an El Niño event, but we might be able to minimize the social impact and economic damage it causes. Would you support the use of taxpayer dollars to fund research into causes and effects of El Niño? See the Media Menu for details, then vote online.

The Water Provinces   Water covers more than 71 percent of Earth's surface and we find organisms in nearly all of it. Gradients in temperature, light, and dissolved gases and nutrients dictate the composition and organization of freshwater and marine communities.

Cause and Effect: A Case Study   Global weather patterns vary in complex ways over time. Sea surface temperatures and air circulation patterns change during an El Niño event. The changes impact ecosystems, both natural and man-made. They impact human health.

29.1 Global Air Circulation Patterns

The biosphere encompasses ecosystems that range from continent-straddling forests to rainwater pools in cup-shaped clusters of leaves. Except for a few ecosystems at hydrothermal vents, climate influences all of them.

CLIMATE AND TEMPERATURE ZONES

Climate refers to average weather conditions, such as cloud cover, temperature, humidity, and wind speed, over time. It arises from variations in solar radiation, Earth's daily rotation and annual path around the sun, the distribution of land masses and seas, and land elevations. Interactions among these factors are the source of prevailing winds and ocean currents, even the composition of soils.

First, sunlight warms air near Earth's surface (Figure 29.2). The sun's rays are spread out over a greater area near the poles (which intercept them at an angle) than they are at the equator (which intercepts them head-on). The intense light at the equator warms air a lot more. Warm air rises, then spreads north and south toward cooler regions. And so begins a global pattern of air circulation.

Earth's rotation and its curvature alter the initial air circulation pattern. Air masses moving north or south are not attached to Earth—which, like any ball, spins fastest at its equator during each spin around its axis. Air masses moving north or south from a given point *seem* to be deflected east or west relative to the curve of the surface spinning below them. This is the basis of prevailing east and west winds (Figure 29.3c).

Also, land absorbs and gives up heat faster than the ocean, so air parcels above it sink and rise faster. Warm air expands, so its pressure is *lower* where it rises and *greater* where it cools and sinks. The uneven distribution of land and water between regions causes pressure differences. Regional winds arise and disrupt the overall flow of air from the equator to the poles.

In short, latitudinal variations in solar heating lead to a north–south pattern of air circulation, which gets modified by east–west deflections and by air pressure differences over land and water.

Latitudinal differences in rainfall accompany these circulation patterns. Warm air holds more moisture than cool air. At the equator, warm air masses pick up moisture from the seas. As they rise and cool, they

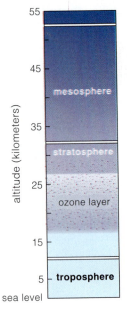

Figure 29.2 *Above*, Earth's atmosphere. The air circulates mainly in the lower atmosphere (troposphere), where temperatures cool rapidly with altitude. An ozone layer between 17 and 27 kilometers above sea level absorbs most of the ultraviolet (UV) wavelengths in the sun's rays. Ozone and oxygen in the atmosphere absorb most of the incoming UV light.

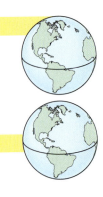

a

Figure 29.3 (a) Concentration of incoming rays of the sun, by latitude. (b,c) Gobal air circulation patterns. The latitudinal differences in the solar heating start a north–south air circulation pattern. As Earth's rotation and its curvature deflect the pattern, prevailing east and west winds arise.

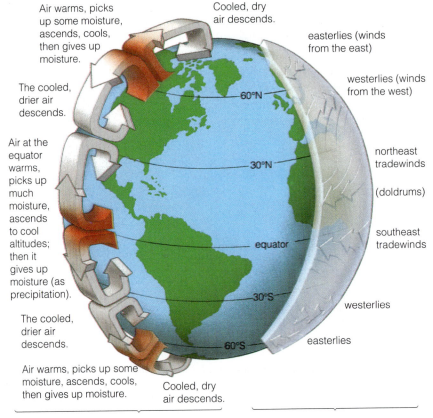

Air warms, picks up some moisture, ascends, cools, then gives up moisture.

Cooled, dry air descends.

easterlies (winds from the east)

westerlies (winds from the west)

The cooled, drier air descends.

Air at the equator warms, picks up much moisture, ascends to cool altitudes; then it gives up moisture (as precipitation).

The cooled, drier air descends.

Air warms, picks up some moisture, ascends, cools, then gives up moisture.

Cooled, dry air descends.

60°N

30°N

equator

30°S

60°S

northeast tradewinds

(doldrums)

southeast tradewinds

westerlies

easterlies

b Initial pattern of air circulation as air masses warm and rise, then cool and fall.

c Deflections in the initial pattern near Earth's surface.

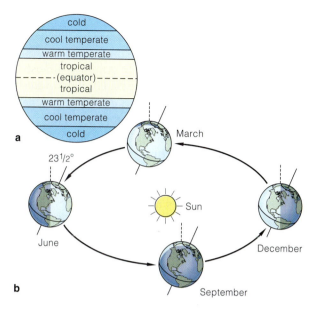

a

b

Figure 29.4 (**a**) World temperature zones. (**b**) Incoming solar radiation varies annually. The northern end of Earth's fixed axis tilts toward the sun in June and away from it in December, changing the equator's position relative to the day–night boundary of illumination. Variations in sunlight intensity and daylength cause seasonal temperature shifts.

Figure 29.5 (**a**) Extensive array of electricity-producing photovoltaic cells capturing solar energy. (**b**) A field of turbines harvesting wind energy.

give up some moisture as rain, which supports forest ecosystems. The now-drier air moves north or south. It warms and gets more dry as it descends at latitudes of 30°, where deserts typically form. Farther north and south, air picks up moisture, ascends, and gives up rain at latitudes of about 60°. It descends in polar regions, where cold temperatures and a near-absence of precipitation result in cold, dry polar deserts.

We divide the temperature gradients from Earth's equator to the poles into **temperature zones** (Figure 29.4a). Temperatures shift in the zones as Earth's orbit puts it closer and then farther from the sun in each annual cycle (Figure 29.4b). Temperatures, the hours of daylight, and winds change with the seasons. The changes are far greater inland (away from the ocean's moderating effects) and farther from the equator. The more pronounced the change, the more that primary productivity rises and falls on land and in the seas.

HARNESSING THE SUN AND WIND

Paralleling the human population's J-shaped growth curve is a steep rise in its total and per capita energy consumption. You may think we have an abundance of energy, but there is a big difference between total and net amounts. *Net* is what is left after subtracting the energy it takes to find, extract, transport, store, and deliver energy to consumers. Some sources, such

as coal, are not renewable. Others, such as direct solar and wind energy, are. The annual amount of incoming solar energy surpasses, by about 10 times, the energy stored in all known fossil fuel reserves. We can harness it in something besides crop plants.

We already know how to get *solar–hydrogen energy*. Photovoltaic cells (Figure 29.5a) hold electrodes that, when exposed to the sun's rays, generate an electric current that splits water molecules into oxygen and hydrogen gas, which can be stored efficiently. The gas can directly fuel cars. It can heat and cool buildings. Water is the only waste. It also costs less to distribute hydrogen gas than electricity. Space satellites run on it.

Unlike fossil fuels, the sun's energy and seawater are unlimited. A transition to a solar–hydrogen age might end smog, oil spills, acid rain, and the reliance on nuclear energy, and may reduce global warming.

Solar energy also is converted into the mechanical energy of winds. Arrays of turbines at "wind farms" exploit the wind patterns that arise from latitudinal variations in the intensity of incoming sunlight. One percent of California's electricity already comes from wind power. The winds of North and South Dakota alone could potentially meet 80 percent of the current energy needs of the United States. Winds do not blow constantly, but when they do, wind energy can be fed into utility grids. Wind power also has potential for islands and other areas remote from utility grids.

The world's major temperature zones and climates start with global patterns of circulation. Those patterns arise through interacting factors: latitudinal variations in incoming solar radiation, Earth's rotation and annual path around the sun, and the distribution of land masses and the seas.

29.2 Air Circulation Patterns and Human Affairs

Through activities that pollute the air, human populations interact with global air circulation patterns in unexpected ways, with unintended consequences. **Pollutants** *are any natural or synthetic substances that have accumulated in harmful or disruptive amounts because organisms have had no prior evolutionary experience with them.*

A FENCE OF WIND AND OZONE THINNING

The ozone layer (Figure 29.2) is nearly twice as high above sea level as Mount Everest. From September to October, it thins above both poles. Seasonal **ozone thinning** is so vast that it was once called an "ozone hole" in the atmosphere. Figure 29.6 has an example. With less ozone, more UV radiation reaches Earth and triggers skin cancers, cataracts, and weakened immunity. It also alters the atmosphere's composition by killing phytoplankton, thus causing huge declines in their oxygen-releasing activity.

Chlorofluorocarbons, or CFCs, are major ozone destroyers. These odorless gases have been used as propellants in aerosol cans, coolants in refrigerators and air conditioners, and in solvents and plastic foam. They slowly seep into the air, and they resist breakdown. A free CFC molecule gives up a chlorine atom when it absorbs UV light. Reaction of this atom with ozone yields oxygen and chlorine monoxide. Chloride monoxide in turn reacts with free oxygen and releases another chlorine. Each chlorine atom can break apart more than 10,000 ozone molecules!

Chlorine monoxide levels above polar regions are 100 to 500 times higher than at midlatitudes. Why?

Like a dynamic fence, winds rotate around the poles for most of the winter. Chlorine compounds are split apart on ice crystals in the clouds. After the almost perpetually dark polar winter ends, UV light in the sun's rays invites chlorine to start destroying ozone.

Methyl bromide is even worse for the ozone layer. Each year in the United States, farmers spray about 60 million pounds of this fumigant over croplands to kill insects, nematodes, and other pests.

Developed countries phased out CFC production. The developing countries may phase it out by 2010. Methyl bromide production is expected to end at that time as well. Even so, it will be one or two centuries before the ozone layer fully recovers.

NO WIND, LOTS OF POLLUTANTS, AND SMOG

Certain weather conditions trap a layer of cool, dense air under a warm air layer. Such **thermal inversions** intensify smog, an atmospheric condition in which winds cannot disperse pollutants that accumulate in the trapped air, often to harmful levels (Figure 29.7).

Where winters are cold and wet, *industrial* smog develops as a gray haze over cities that burn a lot of coal and other fossil fuels. Burning releases smoke, soot, ashes, asbestos, oil, particles of lead and other heavy metals, and sulfur oxides. Most industrial smog now forms in China, India, and eastern Europe.

Big cities in warm climate zones form *photochemical* smog as a brown haze, especially where land forms a natural basin. Los Angeles and Mexico City are two classic cases. Nitric oxide, a key pollutant, escapes in

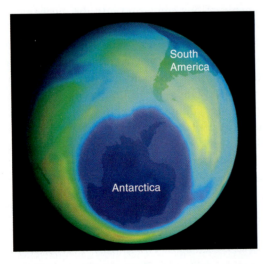

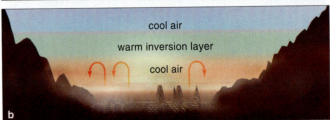

Figure 29.6 Seasonal ozone thinning above Antarctica during 2001. *Darkest blue* indicates the area with the lowest ozone level, at that time the largest recorded.

Figure 29.7 (**a**) Normal pattern of air circulation in smog-forming regions. (**b**) Air pollutants are trapped under a thermal inversion layer.

vehicle exhaust. This combines with oxygen to form nitrogen dioxide. When nitrogen dioxide is exposed to sunlight, it reacts with hydrocarbon gases to form photochemical oxidants. Most hydrocarbon gases are released from spilled or partly burned gasoline.

WINDS AND ACID RAIN

Coal-burning power plants, smelters, and factories emit sulfur dioxides. Vehicles, power plants that burn gas and oil, and nitrogen-rich fertilizers all emit nitrogen oxides. In dry weather, airborne oxides fall as dry acid deposition. In moist air, they form nitric acid vapor, sulfuric acid droplets, and sulfate and nitrate salts. Winds typically disperse them far from their source. They fall to Earth in rain and snow. We call this wet acid deposition, or **acid rain**.

The pH of normal rainwater is 5 or so. Acid rain can be 0 to 100 times more acidic, as potent as lemon juice! It corrodes metals, marble, rubber, plastics, nylon stockings, and other materials. It damages organisms and alters the chemistry of ecosystems.

Depending on the soil type and vegetation cover, some regions are more sensitive to acid rain (Figure 29.8). Highly alkaline soil neutralizes acids before they enter streams and lakes. Also, highly alkaline water can neutralize the acid inputs. But many of the watersheds of northern Europe, southeastern Canada, and regions throughout the United States have thin soil layers on top of solid granite. These soils cannot buffer much of the acidic inputs.

Rain in much of eastern North America is thirty to forty times more acidic than it was several decades ago. Crop yields are declining. Fish populations have already vanished from more than 200 lakes in New York's Adirondack Mountains. Air pollutants from industrial regions are changing the acidity of rainfall. The change is contributing to the decline of forest trees and mycorrhizae that support new growth.

As Harvard and Brigham Young researchers report, living with airborne particles of dust, soot, smoke, or acid droplets shortens the human life span by a year or so. Smaller airborne particles damage lung tissues. High levels of ultrafine particles can increase the risk of lung cancer (Figure 29.9).

At one time the world's tallest smokestack, in the Canadian province of Ontario, produced 1 percent by weight of the world's annual emissions of sulfur dioxide. Today Canada gets more acid deposition from the midwestern United States than it sends across its southern border. *Prevailing winds—hence air pollutants—do not stop at national boundaries.*

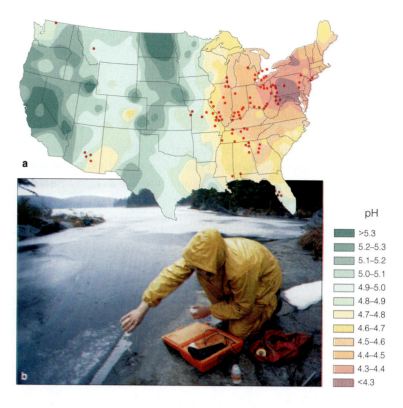

pH
- >5.3
- 5.2–5.3
- 5.1–5.2
- 5.0–5.1
- 4.9–5.0
- 4.8–4.9
- 4.7–4.8
- 4.6–4.7
- 4.5–4.6
- 4.4–4.5
- 4.3–4.4
- <4.3

Figure 29.8 (**a**) Average 1998 precipitation acidities in the United States. *Red* dots mark large coal-burning power and industrial plants. (**b**) Biologist measuring the pH of New York's Woods Lake during a spring melt of acidified snow. Acid rain has already altered the lake.

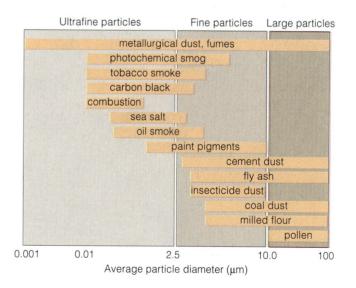

Ultrafine particles | Fine particles | Large particles

metallurgical dust, fumes
photochemical smog
tobacco smoke
carbon black
combustion
sea salt
oil smoke
paint pigments
cement dust
fly ash
insecticide dust
coal dust
milled flour
pollen

0.001   0.01   2.5   10.0   100
Average particle diameter (μm)

Figure 29.9 Suspended particulate matter: solids and liquid droplets small enough to stay aloft for variable intervals.

Ultrafine particles contribute to respiratory disorders. Carbon black is a powdered form of carbon used in paints, tires, and other goods. About 6 million tons are manufactured annually. About 45 million tons of fly ash are generated annually by coal combustion.

29.3 The Ocean, Landforms, and Climates

*The **ocean** is a continuous body of water that covers more than 71 percent of the Earth's surface. Driven by solar heat and wind friction, its upper 10 percent moves in currents that distribute nutrients through marine ecosystems.*

OCEAN CURRENTS AND THEIR EFFECTS

Latitudinal and seasonal variations in sunlight warm and cool water. At the equator, where vast volumes of water warm and expand, the sea level is about eight centimeters (three inches) higher than at either pole. The volume of water in this "slope" is enough to get sea surface water moving in response to gravity, most often toward the poles. The moving water warms air parcels above it. At midlatitudes it transfers *10 million billion* calories of heat energy per second to the air!

Ocean currents are large volumes of water flowing in response to the tug of trade winds and westerlies. Their direction and properties are outcomes of Earth's rotation and topography. They circulate clockwise in the Northern Hemisphere and counterclockwise in the Southern Hemisphere (Figure 29.10).

Swift, deep, and narrow currents of nutrient-poor water parallel the east coast of continents. Along the eastern coast of North America, warm water moves northward, as the Gulf Stream. Slower, shallow, broad currents of cold water paralleling the western coast of continents flow toward the equator.

Ocean currents influence climate zones. Why are Pacific Northwest coasts cool and foggy in summer? Offshore, the cold California Current is giving up moisture as warm winds near the coast transfer heat to it. Why is Baltimore or Boston muggy in summer? Air masses above the warm Gulf Stream gain heat and moisture that southerly and easterly winds move to those cities. Why are winters milder in London and Edinburgh than in Ontario and central Canada—which are at the same latitude? The North Atlantic Current picks up warm water from the Gulf Stream, then gives up heat to prevailing winds, which warm northwestern Europe.

This all can change. We can expect the long-term rise in atmospheric temperature to disrupt the global system of ocean currents and climate zones.

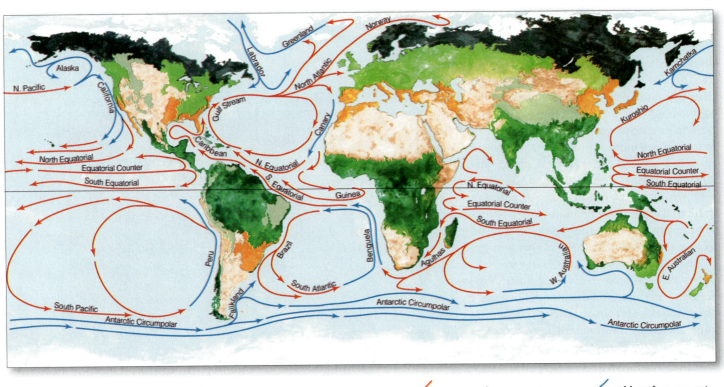

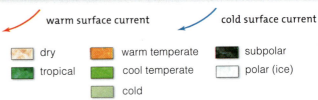

Figure 29.10 Major climate zones correlated with surface currents and surface drifts of the world ocean. The warm surface currents start moving from the equator toward the poles, but prevailing winds, Earth's rotation, gravity, the shape of ocean basins, and land masses all influence the direction of flow. Water temperatures, which differ with latitude and depth, contribute to differences in air temperature and in rainfall patterns.

Read Me First!

and watch the narrated animation on rain shadows

a Prevailing winds move moisture inland from the Pacific

b Clouds pile up and rain forms on side of mountain range facing prevailing winds

c Rain shadow on side facing away from the prevailing winds makes arid conditions

4,000/ 75
3,000/ 85   2,000/ 25
1,800/ 125
1,000/ 25
1,000/ 85
moist habitats
15/ 25

Figure 29.11 Rain shadow effect of the Sierra Nevada. On the side of mountains facing away from the prevailing winds, rainfall is light. *Blue* numbers signify annual precipitation, in centimeters, as averaged on both sides of the range. The *white* numbers signify elevations, in meters.

RAIN SHADOWS AND MONSOONS

Mountains, valleys, and other topographical features affect regional climates. *Topography* refers to a region's surface features, such as elevation. Track a warm air mass after it picks up moisture off California's coast. It moves inland, as wind from the west, and piles up against the Sierra Nevada. This high mountain range parallels the distant coast. The air cools as it rises in altitude and loses moisture as rain (Figure 29.11). The result is a **rain shadow**, a semiarid or arid region of sparse rainfall on the leeward side of high mountains. *Leeward* is the side facing away from the wind.

Belts of vegetation at different elevations relate to temperature and moisture differences. Arid grasslands form at the western base of the Sierra Nevada. Higher up, in the cooler air, deciduous and evergreen species reign. Higher still we see a few evergreen species that can withstand a rigorously cold habitat. Above the subalpine belt, hardy, dwarfed plants are all that can withstand the temperature extremes.

There are rain shadows on the leeward side of the Andes, Rockies, Himalayas, and other great mountain ranges. On the side of Hawaii's high volcanic peaks facing into the wind (*windward*), lush tropical forests flourish. Desert conditions prevail on the other side.

Air circulation patterns called **monsoons** influence continents north or south of warm oceans. As land heats intensely in summer, low-pressure air parcels form above it. The low pressure draws moisture-laden air from the ocean, as from the Bay of Bengal on into Bangladesh. The equatorial sun and converging trade winds result in intense heating and heavy rainfall. The air moving northward and southward across the land is the source of alternating dry and wet seasons, of alternating drought conditions and flooding.

Recurring coastal breezes are like mini-monsoons. Water and coastal land have different heat capacities. In the morning, water does not warm as fast as land. When warm air above land rises, cooler marine air moves in. After sunset, the land loses heat faster than water, so land breezes flow in the reverse direction.

Surface ocean currents also influence regional climates and help distribute nutrients in marine ecosystems.

Air circulation patterns, ocean currents, and landforms interact in ways that influence regional temperatures and moisture levels. Thus they also influence the distribution and dominant features of ecosystems.

29.4 Realms of Biodiversity

Regions differ in their physical and chemical properties and evolutionary history. The differences help explain why deserts, grasslands, forests, and tundra form where they do, and why some water provinces are richer than others in biodiversity.

Long ago, naturalists divided Earth's land masses into six **biogeographic realms**—vast expanses where you could expect to find certain plants and animals, such as palm trees and camels in the Ethiopian realm. Which organisms became the poster species was subjective; the realms were not uniform from border to border.

Biomes are finer subdivisions of the great realms. Their distribution is partly an outcome of evolutionary processes and events, as when evolving species were slowly rafted about on different land masses after the splitting of Pangea (Chapter 17). Their distribution is also an outcome of physical and chemical conditions that differ even in the same land region. Start with

Figures 29.12 and 29.13. Tundra forms at high latitudes and high elevations where moisture is scarce. Forests form where there are certain amounts of rain, certain temperatures, and certain types of soils.

Evolutionary events and environmental conditions also underlie the distribution of ecosystems in the seas. Figure 29.12 shows the main marine ecoregions.

Conservationists locate, inventory, and protect the **hot spots**, portions of biomes and ecoregions that are richest in biodiversity and most vulnerable to species losses. More than half of all species on land live in regions identified as hot spots, and nowhere else.

Biomes are vast expanses of land characterized by distinct arrays of species; marine ecoregions are realms of biodiversity in the seas. Their distribution is an outcome of evolutionary history and the prevailing physical and chemical conditions in different parts of the world.

- desert
- dry shrubland, dry woodland
- warm grassland (e.g., savanna)
- temperate grassland
- mountain grassland
- tropical broadleaf forest
- temperate deciduous forest
- tropical coniferous forest
- temperate coniferous forest (e.g., rain forest)
- northern coniferous forest (boreal forest)
- tropical dry forest
- tundra
- mountains, complex zonation
- mangrove swamps
- perpetual ice cover
- marine ecoregions

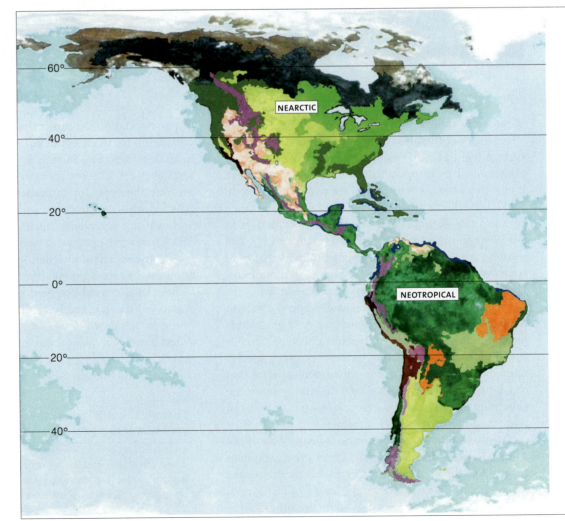

Figure 29.12 Distribution of major biomes within classical biogeographic realms.

Figure 29.13 Changes in plant form along environmental gradients for North America. Gradients in elevation and water availability affect primary productivity.

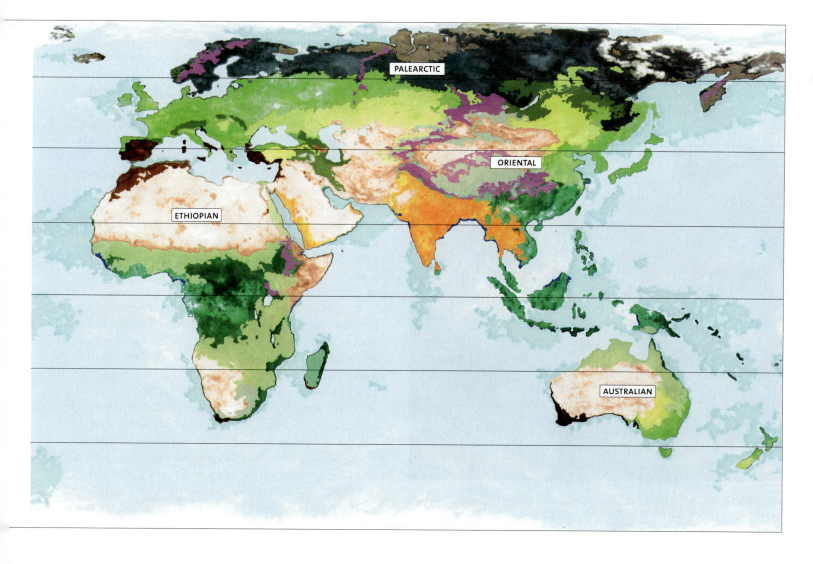

29.5 Moisture-Challenged Biomes

At certain latitudes with sparse rainfall and high rates of evaporation, drought-tolerant plants predominate.

DESERTS, NATURAL AND MAN-MADE

Deserts form between latitudes of about 30° north and south. Annual rainfall is less than ten centimeters or so, and evaporation rates are high. Brief, infrequent pulses of rain swiftly erode the topsoil. Humidity is so low that the sun's rays easily penetrate the air. The ground heats fast, then cools fast at night. The arid or semiarid conditions do not support large, leafy plants, but biodiversity is often high. Annuals and perennials flower briefly after the rains (Figure 29.14).

Desertification is the conversion of grasslands to desertlike conditions by long-term shifts in climate or human activities. Extended droughts and overgrazing are factors. Over the past fifty years, 9 million square kilometers were desertified. Each year, at least 200,000 square kilometers are still being converted.

In Africa, for instance, many imported cattle drink far more water than wild herbivores do. Cattle move more often between watering holes and grazing areas, and the movements trample grasses and pack down the soil. Gazelles and other wild herbivores are better at getting water from their food and conserving body water; they lose little in feces, compared to cattle.

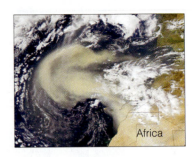

Figure 29.15 Satellite image of an immense dust storm that formed in the hot, dry Sahara Desert, then continued past the west coast of Africa, and is well on its way across the Atlantic.

Desertification can impact distant ecosystems. Dust from Africa may be contributing to the decline of coral reefs in the Caribbean Sea. Winds lift dust into the air and transport it across the Atlantic (Figure 29.15). In that dust are fungal, bacterial, and viral pathogens. Since the 1980s, severe dust storms have become more frequent in Africa. So have disease outbreaks among the Caribbean corals.

DRY SHRUBLANDS, DRY WOODLANDS, AND GRASSLANDS

Dry shrublands get less than 25 to 60 centimeters of rain per year. We see them in Mediterranean regions, South Africa, and California, which has 2.4 million hectares (6 million acres) of chaparral. In summer, dry shrublands are vulnerable to lightning-sparked, wind-driven firestorms. The shrubs have highly flammable leaves and quickly burn to the ground. But they are highly adapted to episodes of fire and soon resprout from the root crowns. Trees don't fare as well during firestorms. Shrubs that can withstand the fires have a competitive edge (Figure 29.16).

Dry woodlands prevail where annual rainfall is 40 to 100 centimeters. Trees may be tall, but they do not provide a continuous canopy. Eucalyptus-dominated areas of southwestern Australia and oak woodlands of California and Oregon are examples.

Grasslands form in the interior of continents in the zones between deserts and temperate forests. Warm temperatures prevail in the summer, and winters are extremely cold. Annual rainfall of 25–100 centimeters prevents deserts from forming, but it is not enough to support forests. Drought-tolerant primary producers survive strong winds, sparse and infrequent rainfall, and fast evaporation. Grazing and burrowing species are the dominant animals. Their activities, combined with periodic fires, help keep shrublands and forests from encroaching on the fringes of many grasslands.

On flat or rolling land in North America, native *shortgrass* and *tallgrass* prairie once prevailed (Figure 29.17). Roots of perennial grasses extended deep into

Figure 29.14 From the Mojave Desert of California, cactus (cholla), golden poppies, and—in the distance—tall saguaro cacti. The inset shows flowers of a claret cup cactus.

Figure 29.16 Views of California chaparral. The dominant plants are multibranched, woody, and typically only a few meters tall. They are adapted to periodic fires; some produce seeds that germinate only after a fire. *Far right,* fires race through chaparral in a canyon above Malibu.

Figure 29.17 *Below,* shortgrass prairie, east of the Rocky Mountains, that once sustained 60 million bison.

Right, African savanna. Herbivores include this herd of wildebeests and giraffes, Cape buffalos, zebras, impalas, and Thomson's gazelles.

the rich topsoil. In the 1930s, shortgrass prairie in the Great Plains was overgrazed. It also was plowed to grow wheat. Prolonged droughts killed this crop, and strong winds stripped away the topsoil. The region became "the Dust Bowl." John Steinbeck's novel *The Grapes of Wrath* is a glimpse into the human tragedies that unfolded because of this environmental disaster.

Grasses two meters tall gave the tallgrass prairie its name. Legumes and composites, including daisies, thrived alongside grasses in the continent's eastern interior. Nearly all original tallgrass prairie has been plowed and converted to cereal croplands. A few parts that escaped the plow are now protected reserves; a few other patches are being restored.

Between the tropical forests and the hot deserts of Africa, South America, and Australia are savannas, broad belts of grasslands with a few shrubs and trees (Figure 29.17). Rainfall averages 90 to 150 centimeters a year, and droughts are seasonal. Where rainfall is low, fast-growing grasses dominate. With more rain, savannas grade into tropical woodlands with low trees, shrubs, and coarse grasses that die back and often burn during the dry season.

Plants of all deserts, dry shrublands, dry woodlands, and grasslands survive sparse rainfall, dry air, strong winds, grazing animals, and recurring episodes of drought and fire.

29.6 More Rain, More Trees

In forest biomes, tall trees grow close together and form a fairly continuous canopy over a broad region. Rainfall and distance from the equator influence which trees dominate.

BROADLEAF FORESTS

Evergreen broadleaf forests form between latitudes 20° north and south, the tropical zones of Africa, the East Indies, Malaysia, Southeast Asia, South America, and Central America. Rainfall is heavy, between 130 and 200 centimeters yearly. Regular rains, an annual mean temperature of 25°C, and at least 80 percent humidity support tropical rain forests (Figure 29.18). In this highly productive forest, evergreens grow new leaves and shed old ones all year. The climate promotes rapid decomposition and mineral cycling; litter can't accumulate. Soils are highly weathered, deficient in organic matter, and poor nutrient reservoirs.

Mild temperatures and predictable seasonal rains support *deciduous broadleaf* forests. In Southeast Asia, India, and other humid tropical regions, some or all leaves drop from trees in a pronounced dry season.

Decomposition is slow; many nutrients are conserved in accumulated litter. Farther from the equator, in the temperate zone, cold, dry winters and less rainfall support temperate deciduous forests (Figure 29.19). In northeastern North America, vast deciduous forests were largely cleared for agriculture. Maple and beech now dominate the forests. To the west, oak–hickory forests flourish, then grade into oak woodlands.

CONIFEROUS FORESTS

Conifers—cone-bearing trees—are primary producers of *coniferous* forests. Most have thick cuticles, needle-shaped leaves, and recessed stomata, adaptations that help conserve water during droughts and icy winters. They dominate the boreal forests, montane coniferous forests, temperate rain forests, and pine barrens.

Boreal forests stretch across northern Europe, Asia, and North America. They are known as *taigas*, meaning "swamp forests." Most form in glaciated regions with cold lakes and streams (Figure 29.20). It rains mostly in summer, and little water evaporates into the cool

Figure 29.18 Tropical rain forest. Latin America, Southeast Asia, and Africa have the fastest-growing populations but not enough food, fuel, and lumber. Of necessity, they turn to their tropical forests for growth-sustaining resources. Most of the forests may disappear within our lifetime. Yet they have the greatest variety and numbers of insects, and the world's largest ones. They are home to the most species of birds and primates and to plants with the largest flowers. Massive vines twist around trees. Orchids, lichens, and other species grow on tree branches; entire communities live, breed, and die in small pools of water that collect in furled leaves.

The species losses are our losses. As examples, we culture and engineer newly discovered genes of forest species to develop pathogen-resistant strains of food crops and livestock, to develop more effective antibiotics and vaccines. An extract from Madagascar's rosy periwinkle greatly increases the odds of surviving childhood leukemia. Many ornamental plants and spices and foods, including cocoa, cinnamon, and coffee, originated in the tropics. So did latex, gums, resins, waxes, and oils for tires, shoes, toothpaste, ice cream, shampoo, compact discs, condoms, and perfumes.

Conservation biologists rightly decry the mass extinction, the assaults on species diversity, the depletion of much of the world's genetic reservoir. Yet something else is going on here. Too many of us grow uneasy when we pass through obliterated forests in our own country. Is it because we are losing the comfort of our heritage, a connection with our evolutionary past? Many millions of years ago, our earliest primate ancestors moved into the trees of tropical forests. Through countless generations, their nervous and sensory systems evolved and became highly responsive to information-rich, arboreal worlds.

Does our neural wiring still resonate with rustling leaves, with shafts of light and mosaic shadows? Are we innately attuned to the forests of Eden—or have time and change buried recognition of home?

autumn   winter   spring   summer

Figure 29.19 Temperate deciduous forest in North America. The series above shows seasonal changes in a deciduous tree's foliage.

Figure 29.20 Taiga in Alberta, Canada. At right, a firefighter retreating from a wall of flames in a forest of pines, oaks, and palmettos by Daytona Beach, Florida.

summer air. Winters are cold, dry, and more severe in eastern parts of these biomes than in the west, where ocean winds moderate climate. Spruce and balsam fir dominate boreal forests of North America. Pine, birch, and aspen forests form in burned or logged regions. In poorly drained soil, acidic bogs with peat mosses, shrubs, and stunted trees prevail. Farther to the north, boreal forests thin and grade into arctic tundra.

In the Northern Hemisphere, montane coniferous forests extend south through great mountain ranges. In the north and at higher elevations, spruce and firs dominate. They give way to pines in the south and at lower elevations. Coniferous forests flourish in some temperate lowlands and near the Pacific coast from Alaska into northern California. These forests hold the world's tallest trees—Sitka spruce to the far north and redwoods to the south. As recounted in Chapter 21, much of these forests has been logged over.

Forests of pine and scrub oak grow in New Jersey. Southern pine forests dominate the coastal plains of the southern Atlantic and Gulf states. Palmettos grow below loblolly pines in the Deep South. Many pine species are adapted to dry, sandy, nutrient-poor soil and to periodic, lightning-sparked fires. Suppression of all fires can result in a buildup of dry undergrowth that can fuel uncontrollable wildfires (Figure 29.20). Each year, Florida alone carries out controlled burns of about 2,500,000 acres to clear fuel-rich understory. Fire-adapted species survive and often flourish after controlled burns.

Conditions in forest biomes favor dense stands of trees that form a continuous canopy over broad regions. Trees that differ in structure and growth patterns are adapted to rainfall patterns, temperature, and other conditions.

29.7 Brief Summers and Long, Icy Winters

Tundra lies between the polar ice cap and belts of boreal forests in the Northern Hemisphere. It is the youngest biome, having first evolved about 10,000 years ago.

The Northern Hemisphere's treeless plain, the *arctic tundra*, is extremely cold. Annual snowmelt and rain are usually less than 25 centimeters. Plants grow fast in the nearly continuous sunlight of a brief growing season (Figure 29.21*a,b*). Hardy lichens and shallow-rooted, low-growing plants are a base for food webs that include voles, arctic hares, caribou, arctic foxes, wolves, and polar bears. Great numbers of migratory birds nest here in summer, when the air is thick with mosquitoes and other flying insects.

Not much more than the surface soil thaws during summer. Just below is **permafrost**, a frozen layer 500 meters thick in some places. It prevents drainage, so the soil above is perpetually waterlogged. The cool, anaerobic conditions do not favor nutrient cycling. Plant remains accumulate in soggy masses and break down slowly. The accumulation of undecayed organic matter in permafrost makes the arctic tundra one of Earth's greatest stores of carbon. However, the frozen layer is starting to melt. Global warming is melting ice and snow, which reflect sunlight. Newly exposed soil is absorbing heat from the sun, and warming up.

Alpine tundra is a similar biome, but it develops in the high mountains around the world (Figure 29.21*c*). At night, the below-freezing temperatures make it too difficult for trees to grow. Even in summer, shaded patches of snow persist in this biome, but there is no permafrost. Alpine soil is thin and well drained, but it is nutrient poor. As a result, primary productivity is low. Grasses, heaths, and small-leafed shrubs grow in patches where better soil has formed. These low plant species are adapted to withstand strong winds.

Arctic tundra prevails at high latitudes, where short, cold summers alternate with long, cold winters. Alpine tundra prevails in high, cold mountains, regardless of whether the seasons differ with latitude.

Figure 29.21 Arctic tundra in summer (**a**) and winter (**b**). About 4 percent of Earth's land mass is arctic tundra, blanketed with snow for as long as nine months of the year. Most arctic tundra is in northern Russia and Canada, followed by Alaska and Scandinavia.

Small bands of humans have herded reindeer, hunted, and fished in these sparsely populated regions for hundreds of thousands of years. Now more people, and machines, are moving in to extract mineral and fossil fuels. When extraction operations damage soils and plants, it might take decades for a region to recover. Why? Plants of the Arctic tundra grow slowly, and their growth is limited to only a few months of the year.

(**c**) Compact, low-growing, hardy plants typical of alpine tundra in Washington's Cascade Range.

29.8 Don't Forget the Soils

Dig down into the dirt and you will find a soil profile that helps define a particular kind of biome. Desert, woodland, forest, tundra—all have distinctive soils.

Soils are mixtures of mineral particles and varying amounts of the decomposing organic material we call humus. When rocks are weathered and broken down, they form coarse-grained gravel, then sand, silt, and finely grained clay. Water and air fill spaces between soil particles. The types and proportions of particles, and how tightly the particles are compacted, differ within and between regions.

Soils have a layered structure—a *soil profile*—that reflects how they formed (Figure 29.22). Uppermost is topsoil, which has the most humus and is most prone to erosion. It is less than a centimeter thick on steep slopes and more than a meter thick in grasslands. The deepest layer has rocks that are still breaking down.

Soils too high in one type of particle hamper plant growth. Clay is rich in minerals, but its fine, closely packed particles drain poorly and don't leave much room for essential oxygen. Gravelly or sandy soils are prone to leaching, which depletes them of water and mineral ions.

Loam topsoils have the best mix of sand, silt, and clay for agriculture. Their coarse particles promote drainage, and their fine particles retain water-soluble mineral ions and nutrients required for growth.

Soils influence primary productivity. That is why farms now abound in topsoil-rich former grasslands. Tropical forests are cleared for agriculture too, but they have little topsoil above layers that drain poorly. Clearing exposes topsoil to rains that leach nutrients, so crops perform poorly unless heavily fertilized.

> A biome's soil profile and its proportions of sand, silt, clay, gravel, and humus influence its primary productivity.

O horizon: Pebbles, little organic matter

A horizon: Shallow, poor soil

B horizon: Leaching results in salinization (accumulated calcium, sodium)

C horizon: Rock fragments from uplands

DESERT SOIL

A horizon: Alkaline, deep, rich in humus

B horizon: Percolating water enriches layer with calcium carbonates

GRASSLAND SOIL

O horizon: Sparse litter

A–E horizons: Continually leached; iron, aluminum left behind impart red color to acidic soil

B horizon: Clays with silicates, other residues of chemical weathering

TROPICAL RAIN FOREST SOIL

O horizon: Scattered litter

A horizon: Rich in organic matter above humus layer unmixed with minerals

B horizon: Accumulated minerals leached from above

C horizon: Poorly weathered rocks

DECIDUOUS FOREST SOIL

O horizon: Well-defined, compacted mat of organic deposits resulting mainly from activity of fungal decomposers

A horizon: Acidic humus; most minerals leached out, silica retained

B horizon: Accumulated clays with oxides of iron and aluminum

CONIFEROUS FOREST SOIL

Figure 29.22 Soil profiles from a few representative biomes.

29.9 Freshwater Provinces

Freshwater and saltwater provinces cover more of Earth's surface than all biomes combined. They include the world ocean, lakes, ponds, wetlands, and coral reefs. There are no "typical" regions. Ponds are shallow; Siberia's Lake Baikal is 1.7 kilometers deep. All aquatic ecosystems have gradients in light penetration, temperature, and dissolved gases, but values differ greatly. All we can do here is sample the diversity, starting with the freshwater provinces.

LAKE ECOSYSTEMS

A lake is a body of standing freshwater, as in Figure 29.23. Over time, erosion and sedimentation alter its dimensions; it usually ends up filled in or drained. A young lake has littoral, limnetic, and profundal zones (Figure 29.24a). Its littoral zone extends all around the shore to a depth where rooted aquatic plants cannot grow. The well-lit, shallow, warm water supports high biodiversity. The limnetic zone encompasses the sunlit, open water away from shore to depths where light and photosynthesis are limited. Its *phyto*plankton includes cyanobacteria, green algae, and diatoms. *Zoo*plankton includes rotifers and copepods. The profundal zone is all open water at depths that are impenetrable to the most efficient wavelengths for photosynthesis. Below it, sediments house communities of decomposers.

SEASONAL CHANGES IN LAKES   In temperate regions, many lakes undergo seasonal changes in density and temperature gradients, from surface to bottom. Often a layer of ice forms in midwinter. Water under the ice is near the freezing point and is the least dense. Water at 4°C is the most dense; it forms deeper layers that are a bit warmer than the surface layer.

In spring, the number of daylight hours increases and the air warms. The lake ice melts, surface water warms to 4°C, and temperature gradients disappear. Winds blowing across the lake surface cause a **spring overturn**, in which strong vertical movements deliver dissolved oxygen from surface waters to the depths. At the same time, nutrients released by decomposers of the lake bottom sediments move to the surface.

The lake becomes thermally stratified in summer. The midlayer cools, becoming a *thermocline* that stops the vertical mixing between the warmer, oxygen-rich surface layer and the cold, oxygen-poor layer below it (Figure 29.24b). Decomposers now deplete the oxygen dissolved near the lake bottom.

In autumn, the upper layer cools and gets denser. It sinks and the thermocline vanishes. During this **fall overturn**, water mixes vertically, and again dissolved oxygen moves down and nutrients move up.

Primary productivity is seasonal. After the spring overturn, longer daylengths and the cycled nutrients favor increased primary productivity. Phytoplankton

Figure 29.23   Lake in Chile's Torres del Paine National Park.

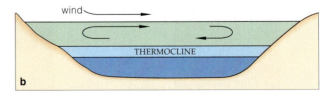

Figure 29.24   (**a**) Lake zonation. The littoral zone extends all around the shore to a depth where aquatic plants stop growing. The profundal zone is all water below the depth of light penetration. Above the profundal are the open, sunlit waters of the limnetic zone. (**b**) Thermal layering occurs in many lakes of temperate zones during the summer.

Figure 29.25 Stream habitats. (**a**) Pool. (**b**) Pool leading into a riffle. (**c**) A run, Sinking Creek. (**d**) Fallen leaves add nutrients to water in a stream. (**e**) A closer look at a riffle.

and aquatic plants take up phosphorus, nitrogen, and other nutrients. During the growing season, vertical thermocline mixing ends. Nutrients do not move up, and photosynthesis slows. By late summer, shortages of nutrients limit growth. A fall overturn does cycle nutrients back to the surface, and there is a sudden burst of primary productivity. But a sustained burst is not possible, given the fewer daylight hours. Primary productivity will not rise again until spring.

TROPHIC NATURE OF LAKES Each lake's topography, climate, and geologic history dictate the numbers and kinds of resident species, how they are dispersed in the lake, and how they cycle nutrients. Soils of the surrounding region and the lake basin contribute to the type and amount of nutrients available.

Interplays among climate, soil, basin shape, and metabolic activities of the lake's residents contribute to a lake's trophic status. *Oligotrophic* lakes typically are newly formed, deep, clear, and nutrient poor, with low primary productivity. *Eutrophic* lakes are older, shallower, and nutrient rich, with a higher primary productivity. As a lake ages, sediments accumulate, water becomes less transparent and more shallow, and phytoplankton come to dominate the community. A filled-in basin is the final successional stage.

Eutrophication, recall, refers to natural or artificial processes that enrich a body of water with nutrients. For example, human activities caused eutrophication of Seattle's Lake Washington. Phosphate-rich sewage drained into the lake from 1941 to 1963. This made nitrogen the limiting factor for photosynthetic species. Cyanobacteria are at an advantage when nitrogen is scarce; they can fix N_2. They became dominant and formed slimy mats in summer, which made the lake useless for recreation. Sewage inputs were stopped. By 1975, the lake neared full recovery.

STREAM ECOSYSTEMS

Flowing-water ecosystems called **streams** start out as freshwater springs or seeps. As they flow downslope, they grow and merge, then often converge as a river. Between a river's headwaters and end, we find riffles, pools, and runs (Figure 29.25). The *riffles* are shallow, turbulent stretches where the water flows swiftly over a rough, sandy, rocky bottom. In *pools*, deep water flows slowly over a smooth, sandy, or muddy bottom. *Runs* are smooth-surfaced, rapidly flowing stretches over bedrock or rock and sand.

Rainfall, snowmelt, geography, altitude, even the shade cast by plants influence a stream's average flow volume and temperature. Streambed composition as well as agricultural, industrial, and urban wastes influence the solute concentrations in the water.

A stream imports organic matter into food webs, especially in forest ecosystems. Trees cast shade and thus hamper photosynthesis, but their litter starts detrital food webs (Section 28.2). Aquatic organisms take up and then release nutrients as the water flows downstream. Nutrients move upstream only in the tissues of migratory fishes and other animals. These nutrients spiral between aquatic organisms and the water as it flows on its one-way course to the sea.

The water provinces are far more extensive than the biomes. Their aquatic ecosystems differ, but they all show gradients in light penetration, temperature, and dissolved gases.

29.10 Life at Land's End

Near the coasts of continents, around islands and reefs, concentrations of nutrients support some of the world's most productive ecosystems.

WETLANDS AND THE INTERTIDAL ZONE

Like freshwater ecosystems, estuaries and mangrove wetlands have distinct physical and chemical features that include depth, water temperature, salinity, and the light penetration. **Estuaries** are partly enclosed coast regions where seawater mixes with nutrient-rich fresh water from rivers, streams, and runoff (Figure 29.26*a*). The confined region, slow mixing of water, and tidal action combine to trap dissolved nutrients. Water flow continually replenishes nutrients, which is one reason estuaries can support highly productive ecosystems.

Primary producers are phytoplankton, plants that tolerate submergence at high tide, and algae. Detrital food webs are common. So many larval and juvenile stages of invertebrates and some fishes develop here that estuaries are sometimes called marine nurseries. Many migratory birds use the estuaries as rest stops.

Estuaries range from broad, shallow Chesapeake Bay, Mobile Bay, and San Francisco Bay to the narrow, deep fjords of Norway and others like them in Alaska and British Columbia. Many estuaries are in decline because fresh water is being diverted upstream, and agricultural runoff and wastes are flowing in.

In tidal flats at tropical latitudes, we find nutrient-rich mangrove wetlands. "Mangrove" refers to forests of salt-tolerant plants in sheltered areas along tropical coasts. These plants have shallow or branching prop roots that extend from the trunk (Figure 29.26*b*). Many have root extensions that take up oxygen.

Net primary productivity of a mangrove wetland depends partly on the tidal volume and flow rate, as well as on salinity and nutrient availability. But these are all rich ecosystems that support notable biomass.

Figure 29.26 (**a**) South Carolina salt marsh. A marsh grass (*Spartina*) is a major producer. (**b**) In the Florida Everglades, a mangrove wetland lined with red mangroves (*Rhizophora*).

Figure 29.27 (**a**) From one of the Hawaiian Islands, sea urchins and a few algae, including sea lettuce (*Ulva*), exposed by a low tide. (**b**) Coral fragment that washed up on the sandy shore of Australia's Heron Island.

ROCKY AND SANDY COASTLINES

Rocky and sandy coastlines support ecosystems of the intertidal zone, which is not renowned for creature comforts. Waves batter its residents; tides alternately submerge and expose them. The higher up they are, the more they dry out, freeze in winter, or bake in summer, and the less food comes their way. The lower they are, the more they compete in limited spaces. At low tides, the birds, rats, and raccoons move in and feed on them. High tides bring the predatory fishes.

Generalizing about coastlines is not easy, for waves and tides continually resculpt them. One feature that rocky and sandy shores share is vertical zonation.

Rocky shores have three zones. The *upper* littoral zone is submerged only during the highest tide of the lunar cycle and is sparsely populated. The *mid*littoral zone is submerged during the highest regular tide and exposed during the lowest. Algae, fishes, hermit crabs, nudibranchs, sea stars, and sea urchins occupy its tidepools (Figure 29.27a). Biodiversity is greatest in the *lower* littoral zone, exposed only in the lunar cycle's lowest tide. In all three zones, erosion is so swift that detritus cannot accumulate. For that reason, you can expect to find only grazing food webs.

Waves and currents continually rearrange loose sediments on the *sandy* and *muddy* shores. Few large plants can grow in unstable places, so you won't find grazing food webs here. Detrital food webs start with inputs from land or offshore.

CORAL REEFS

Coral reefs develop in clear, warm waters near coasts or around volcanic islands, mainly between latitudes 25° north and south (Figure 29.28b). Each is a wave-resistant formation made of the slowly accumulated remains of marine organisms. Hard corals, cemented together, formed the spine. Also, mineral-hardened cell walls of red algae, such as the coralline species in Figure 29.28a, added to the spine. Secretions from many other reef organisms helped hold the collective hard parts together.

Figure 29.28 hints at the wealth of warning colors, tentacles, and stealth of reef species—signs of danger and fierce competition for resources by individuals packed together in a limited space.

> *Life thrives where land meets sea. Wetlands and coral reefs show high primary productivity. Rocky and sandy shores are not renowned for their creature comforts.*

CORALLINE ALGA   SEA FAN

MORAY EEL   CHAMBERED NAUTILUS

CLOWNFISH WITH SEA ANEMONE   CROWN-OF-THORNS SEA STAR ON CORAL

a

b

Figure 29.28 (**a**) A sampling of biodiversity from coral reefs around the world. (**b**) Global distribution of reefs (*orange*), coral banks (*gold*), and solitary corals (*red*). Nearly all reef-building corals form in warm seas (within dark lines).

29.11 Coral Bleaching

FOCUS ON THE ENVIRONMENT

About 6,000 years ago, when Egypt's first dynasty was gaining political power, coral reefs were forming off the Florida coast. Today, countless numbers and kinds of marine invertebrates as well as fishes and algae live there. Will these reefs and others like them around the world endure much longer? Maybe not.

Dinoflagellate symbionts live inside the tissues of most reef-building corals. Dinoflagellates, recall, are single-celled photoautotrophs. In return for shelter, they provide coral polyps with oxygen and recycle nutrients that otherwise would be lost as wastes.

The interaction works well in warm seas, where oxygen and nutrient levels are typically low. When corals are stressed, however, they expel the symbionts. When stressed for more than a few months, corals die. Only their white, hardened chambers remain, an outcome called *coral bleaching* (Figure 29.29).

Coral bleaching is on the rise all over the world, partly because of the higher sea surface temperatures. Other factors also are at work. In 2002, an algal bloom ninety-six kilometers across drifted through the Florida Keys. Toxins released by free-living dinoflagellates of the bloom killed 70 percent of reef-building corals. At about the same time, researchers started transplanting lab-grown pink sea fans (*Gorgonia ventalina*) to a coral reef that a ship had run over. Florida's sea fans are also dying of aspergillosis, a fungal disease. Figure 29.28*a* shows what a healthy sea fan is supposed to look like.

Figure 29.29 Coral bleaching. In 2002, water temperature around Australia's Great Barrier Reef was the highest recorded since 1870. Warmer water caused the most widespread bleaching ever observed on this barrier reef. Nearly 60 percent of the corals bleached out.

29.12 The Open Ocean

The world ocean has two vast provinces (Figure 29.30c). Its benthic (bottom) province starts at continental shelves and extends to deep-sea trenches. Its pelagic province is the full volume of ocean water. The neritic zone is the volume above continental shelves, and the oceanic zone is the volume above the ocean basins.

SURPRISING DIVERSITY

Photosynthesis is seasonal and intense near the ocean surface. Drifting in seawater are phytoplankton, the "pastures" that feed copepods, krill, whales, squids, fishes, and other members of marine food webs. Also near the surface, photosynthetic bacteria collectively called ultraplankton might account for 70 percent of the ocean's primary productivity.

Deeper ocean water is too dark for photosynthesis. There, food webs start with **marine snow**. These tiny bits of organic matter drift down from communities above. They are the base for staggering biodiversity; midoceanic water may be home to 10 million species!

Also, in what might be the greatest of all circadian migrations, a number of species rise thousands of feet to feed in upper waters at night and move down the next morning. Carnivores at the top of the food webs range from familiar types, including sharks and giant squids, to the visually jarring (Figure 29.30*a*).

Hydrothermal vent ecosystems occur in the benthic province. Near-freezing water seeping into fissures in the seafloor becomes superheated. As it spews back out, it leaches mineral ions from the rocks. Dissolved in the outpouring are iron, zinc, copper sulfides, and sulfates of magnesium and calcium. These minerals settle out as rich deposits that function as the energy source for chemoautotrophic species of bacteria and archaea. These prokaryotes are the primary producers for rich food webs that include tube worms (Figure 29.30*b*), as well as crustaceans, clams, and fishes.

Many biologists suspect that life originated in such hot, nutrient-rich places on the seafloor. Section 18.2 sketched out some evidence of this possibility.

UPWELLING AND DOWNWELLING

As you read earlier, prevailing winds that parallel the western coasts of continents tug on the ocean surface. Wind friction gets the surface waters moving. Earth's rotational force deflects masses of slow-moving water away from the coasts. Cold, deep, often nutrient-rich water moves in vertically in its place (Figure 29.31).

Cold, deep water moving up this way is called an **upwelling**. Upwelling occurs in equatorial currents

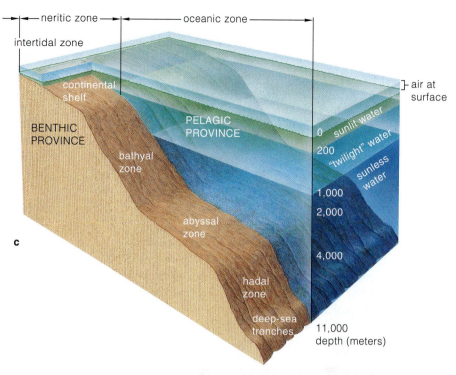

Figure 29.30 The ocean. (**a**) Deep-sea angler fish. (**b**) Tube worms at a hydrothermal vent on the ocean floor. (**c**) Oceanic zones. The seafloor extends from the continental shelves to deep-sea trenches. Dimensions of each zone are not to the same scale.

and along continents, and it cools air masses above it. When thick fogbanks form near California's coast, an upwelling of cold water is interacting with warm air.

In the Southern Hemisphere, commercial fisheries depend on wind-induced upwelling along Peru and Chile. Prevailing coastal winds blow from the south and southeast, tugging surface water away from the shore. Cold, deeper water carried to the continental shelf by the Humboldt Current moves up near the surface. Huge quantities of nitrate and phosphate are pulled up and carried north by the cold Peru Current. The nutrients sustain phytoplankton that are the basis of one of the world's richest fisheries.

Every three to seven years, warm surface waters of the western equatorial Pacific Ocean move eastward. This massive displacement of warm water acts on the prevailing wind direction. The eastward flow speeds up so much that it hampers the vertical movement of water along the coasts of Central and South America.

Water piling into a coastline is forced downward and flows away from it. Near Peru's coast, prolonged **downwelling** of nutrient-poor water displaces cooler waters of the Humboldt Current and puts a stop to upwelling. The warm current typically arrives around Christmas. Fishermen in Peru named it **El Niño** ("the

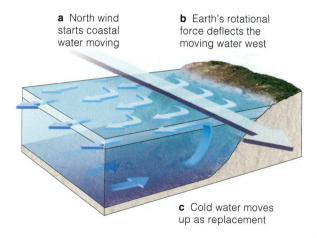

a North wind starts coastal water moving

b Earth's rotational force deflects the moving water west

c Cold water moves up as replacement

Figure 29.31 Coastal upwelling in the Northern Hemisphere.

little one," in reference to the baby Jesus). The name became incorporated into a more inclusive, scientific explanation. As described next, this recurring event is the El Niño Southern Oscillation, or ENSO.

From the ocean's coral reefs down to hydrothermal vents, throughout the pelagic province, we find staggering degrees of primary productivity and biodiversity.

29.13 Applying Knowledge of the Biosphere

We turn now to an application that reinforces a unifying ecological concept. Events in the atmosphere and ocean, and on land, interconnect in ways that can profoundly influence the world of life.

An El Niño Southern Oscillation, or **ENSO**, is defined by changes in sea surface temperatures and in the air circulation patterns. "Southern oscillation" refers to a seesawing of atmospheric pressure in the western equatorial Pacific, the world's largest reservoir of warm water and warm air. It is the source of heavy rainfall, which releases enough heat energy to drive global air circulation.

Between ENSOs, the warm waters and heavy rains move westward (Figure 29.32a). *During* an ENSO, prevailing surface winds in the western equatorial Pacific pick up speed and "drag" surface waters east (Figure 29.32b). As it does, the westward transport of water slows. The sea surface temperatures increase,

evaporation accelerates, and air pressure falls. These changes have global repercussions.

El Niño episodes typically last for six to eighteen months. Then another oscillation known as **La Niña** starts up, and the weather seesaws again.

As you read in the chapter opening, 1997 ushered in the most powerful ENSO event of the century. The average sea surface temperatures in the eastern Pacific rose 9°F (about 12.8°C). The warmer water extended 9,660 kilometers west from the coast of Peru.

The 1997–1998 El Niño/La Niña rollercoaster had record-breaking impact on primary productivity in the equatorial Pacific. With the massive eastward flow of nutrient-poor warm water, populations of photosynthesizers shrank to almost undetectable numbers (Figure 29.32c). During the La Niña rebound, cool, nutrient-rich water welled up to the surface and was displaced westward, along the equator. As satellite images clearly revealed, the upwelling had

Read Me First!

and watch the narrated animation on El Niño Southern Oscillations

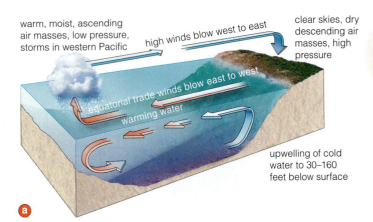
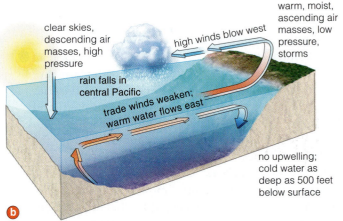

Figure 29.32 (**a**) Westward flow of cold, equatorial surface water in between ENSOs. (**b**) Eastward dislocation of warm ocean water during an ENSO.

Satellite data on primary productivity in the equatorial Pacific Ocean. The concentration of chlorophyll was used as the measure.

(**c**) During the 1997–1998 El Niño event, a massive amount of nutrient-poor water moved to the east, and photosynthetic activity was negligible.

(**d**) During the subsequent La Niña episode, massive upwelling and westward displacement of nutrient-rich water led to a huge algal bloom that stretched all the way to the coast of Peru.

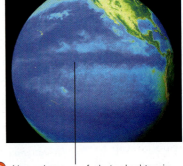

(**c**) Near-absence of phytoplankton in the equatorial Pacific during an El Niño.

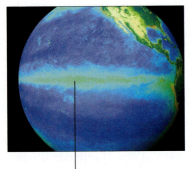

(**d**) Immense algal bloom in the equatorial Pacific in the La Niña rebound event.

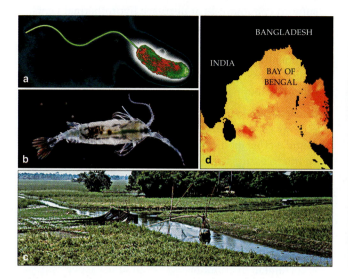

Figure 29.33 (**a**) *Vibrio cholerae*, agent of cholera. (**b**) Copepods host a dormant stage of *V. cholerae* that waits out adverse environmental conditions that don't favor its growth and reproduction. (**c**) Typical Bangladesh waterway from which water samples were drawn for analysis. (**d**) Satellite data on rising sea surface temperatures in the Bay of Bengal correlated with cholera cases in the region's hospitals. *Red* signifies warmest summer temperatures. (**e**) In Bangladesh, Rita Colwell comparing unfiltered and filtered drinking water.

sustained a vast algal bloom, one that stretched across the equatorial Pacific (Figure 29.32*d*).

During the 1997–1998 El Niño event, 30,000 cases of *cholera* were reported in Peru alone, compared to only 60 cases from January to August in 1997. People knew water contaminated by *Vibrio cholerae* causes cholera epidemics (Figure 29.33*a*). The disease agent triggers severe diarrhea; it enters water supplies in feces. People who use the tainted water get infected.

What people did *not* know was where *V. cholerae* lurked between cholera outbreaks. It could not be found in humans or in water supplies. Then it would show up simultaneously in places that could be far apart—usually coastal cities, where the urban poor draw water from rivers flowing to the sea.

Marine biologist Rita Colwell had been thinking about cholera for some time. She suspected that humans were not the host between outbreaks. Was there an environmental reservoir for the pathogen? Maybe. But nobody had detected it in water samples that had been subjected to standard culturing.

Then she had a flash of insight: What if no one could find the pathogen because it changes form and enters a dormant stage between outbreaks?

During a cholera outbreak in Louisiana, Colwell realized that she could use an antibody-based test to detect a protein unique to *V. cholerae*'s surface. Later, similar tests in Bangladesh revealed bacteria in fifty-one of fifty-two samples of water. Standard culture methods had missed it in all but seven samples.

V. cholerae lives in rivers, estuaries, and seas. As Colwell knew, plankton also thrive in these aquatic environments. She decided to restrict her search for the unknown host to warm waters near Bangladesh,

where outbreaks of cholera occur seasonally (Figure 29.33*c*). It was here that she detected the dormant stage of *V. cholerae* in copepods. These tiny marine crustaceans graze on algae and other species that make up phytoplankton (Figure 29.33*b*). The number of copepods—and of *V. cholerae* inside them—rises and falls with shifts in phytoplankton abundances.

Colwell already knew about seasonal variations in sea surface temperatures. Remember the old saying, *Chance favors the prepared mind*? In one sense, she was prepared to recognize a connection between cholera cases and seasonal temperature peaks in the Bay of Bengal. She compared data from the 1990–1991 and 1997–1998 El Niño episodes. Her correlation held. Four to six weeks after the sea surface temperatures go up, so do cases of cholera!

Now Colwell and Anwarul Huq, a Bangladeshi scientist, are studying salinity and other factors that may influence outbreaks. Their goal is to design a model for predicting where cholera will break out next. They have advised women in Bangladesh to use sari cloth as a filter to remove *V. cholerae* cells from the water (Figure 29.33*e*). Copepod hosts are too big to pass through the thin cloths, which can be rinsed in clean water, sun-dried, and used again and again. By using this simple and inexpensive method, cholera outbreaks have been reduced by 50 percent.

Summary

Section 29.1   Global air circulation patterns influence climate and ecosystem distribution. The patterns start with latitudinal variations in incoming solar radiation and are influenced by Earth's daily rotation and annual path around the sun, the distribution of continents and seas, and elevations of land masses. Solar energy, and the winds it drives, are renewable, clean sources of energy.

Section 29.2   Humans influence the atmosphere. The use of CFCs and methyl bromide has depleted ozone in the upper atmosphere. With seasonal ozone thinning, more damaging UV radiation reaches Earth's surface.

Smog, a form of air pollution, arises in areas where large amounts of fossil fuels are burned. Coal-burning power plants are also the main contributors to acid rain, which alters habitats and kills many organisms.

Section 29.3   Latitudinal and seasonal variations in sunlight warm ocean waters and set currents in motion. The currents distribute heat energy around the seas and affect weather patterns. Ocean currents, air currents, and topography interact to shape global climate zones.

Section 29.4   The world's land masses are realms of biodiversity, each with an evolutionary history and a tapestry of physical and chemical conditions. Biomes are vast expanses characterized by specific arrays of species, mainly plants and animals. Regional variations in climate, landforms, and soils influence them. Marine ecoregions are comparable realms of biodiversity.

Section 29.5   Deserts form around latitudes 30° north and south if annual rainfall is sparse. Slightly moister southern or western coastal regions support dry woodlands and shrublands. In the midlatitudinal interiors of continents, vast deserts or grasslands form.

Section 29.6   In equatorial regions of high rainfall, high humidity, and mild temperatures, evergreen tropical forests form. Deciduous broadleaf forests grow in regions with hot summers and cool winters. Where a cold, dry season alternates with a cold, rainy season, coniferous forests dominate.

Section 29.7   Low-growing, hardy plants of the tundra dominate at high latitudes and high altitudes.

Section 29.8   Soil characteristics vary among biomes and help determine their primary productivity.

Section 29.9   Lakes, streams, and other aquatic ecosystems show gradients in penetration of sunlight, water temperature, salinity, and dissolved gases. These factors vary over time and affect primary productivity.

Sections 29.10, 29.11   Coastal zones and tropical reefs support diverse ecosystems. Primary productivity is high in coastal wetlands and on coral reefs. Coral reefs take thousands of years to build. Most are now threatened by global warming and human activities.

Section 29.12   Life persists throughout the ocean. Diversity is highest in sunlit waters. Mineral-rich waters support communities at deep-sea hydrothermal vents.

Upwelling is an upward movement of deep, cool, often nutrient-rich ocean water, typically along coasts of continents. An El Niño event disrupts upwelling and triggers massive, reversible changes in rainfall and other weather patterns around the world.

Section 29.13   Drawing on knowledge of microbial ecology as well as biogeographic patterns, Rita Colwell found a crucial bit of information that led to effective countermeasures against cholera outbreaks.

Self-Quiz   *Answers in Appendix III*

1. Solar radiation drives the distribution of weather systems and so influences _____ .
   - a. temperature zones
   - b. rainfall distribution
   - c. seasonal variations
   - d. all of the above

2. _____ shields life against the sun's UV wavelengths.
   - a. A thermal inversion
   - b. Acid precipitation
   - c. The ozone layer
   - d. The greenhouse effect

3. Regional variations in the global patterns of rainfall and temperature depend on _____ .
   - a. global air circulation
   - b. ocean currents
   - c. topography
   - d. all of the above

4. A rain shadow is a reduction in rainfall _____ .
   - a. on the leeward side of a mountain range
   - b. during an El Niño event
   - c. that occurs seasonally in the tropics

5. The major cause of acid rain is operation of _____ .
   - a. nuclear power plants
   - b. automobiles
   - c. coal burning
   - d. agricultural machinery

6. Biomes are _____ .
   - a. water provinces
   - b. water and land zones
   - c. vast expanses of land
   - d. partly characterized by dominant plants
   - e. both c and d

7. Biome distribution depends on _____ .
   - a. climate
   - b. topography
   - c. soils
   - d. all of the above

8. Grasslands most often predominate _____ .
   - a. near the equator
   - b. at high altitudes
   - c. in interior of continents
   - d. b and c

9. During _____ , deeper, often nutrient-rich water moves to the surface of a body of water.
   - a. spring overturns
   - b. fall overturns
   - c. upwellings
   - d. all of the above

10. Match the terms with the most suitable description.
    - _____ tundra
    - _____ chaparral
    - _____ desert
    - _____ savanna
    - _____ estuary
    - _____ boreal forest
    - _____ tropical rain forest
    - _____ hydrothermal vents

    - a. equatorial broadleaf forest
    - b. partly enclosed land where freshwater and seawater mix
    - c. type of grassland with trees
    - d. has low-growing plants at high latitudes or elevations
    - e. at latitudes 30° north and south
    - f. mineral-rich, superheated water supports communities here
    - g. conifers dominate
    - h. dry shrubland

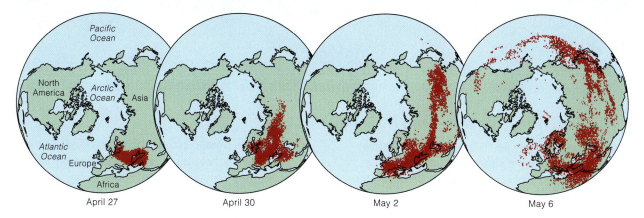

Pacific Ocean

North America

Arctic Ocean

Asia

Atlantic Ocean

Europe

Africa

April 27   April 30   May 2   May 6

Figure 29.34 Global distribution of radioactive fallout after the 1986 meltdown of the Chernobyl nuclear power plant. It put 300 million to 400 million people at risk for leukemia and other radiation-induced disorders. By 1998, the rate of thyroid abnormalities in children living downwind from the site was nearly seven times as high as for those upwind; their thyroid gland concentrated the iodine radioisotopes.

Critical Thinking

1. On April 26, 1986, in Ukraine, a meltdown occurred at the Chernobyl nuclear power plant. Nuclear fuel burned for nearly ten days and released 400 times more radioactive material than the atomic bomb that dropped on Hiroshima. Thirty-one people died right away. An estimated 15,000 to 30,000 have died since of cancers and other effects of radiation. Winds carried radioactive fallout around the globe (Figure 29.34).

The Chernobyl accident stiffened opposition to *nuclear power* in the United States, but recent developments have some people reconsidering. Increasing the use of nuclear energy would diminish the country's dependence on oil from the politically unstable Middle East. Nuclear power does not contribute to global warming, acid rain, or smog. It does produce highly radioactive wastes. Investigate the pros and cons of nuclear power, and decide if you think the environmental benefits outweigh the risks. Would you feel differently if a nuclear power plant were about to be built ten kilometers upwind from your home?

2. Use of off-road recreational vehicles may double over the next twenty years. Many off-road enthusiasts would like increased access to government-owned desert areas. Some argue that it's just the perfect place for off-roaders because "There's nothing there." Do you agree? If not, how would you counter this argument?

3. Write a short description of how global warming might affect spring overturn and thermocline formation in a Minnesota lake. What would be some ecological effects?

4. *Wetlands*, remember, are transitional zones between aquatic and terrestrial ecosystems. Wetlands everywhere are being converted for agriculture, home building, and other human activities. Does the ecological value of wetlands for the nation as a whole outweigh the rights of the private citizens who own many of the remaining wetlands in the United States? Should private owners be forced to transfer title of their land to the government? If so, who should decide the value of a parcel of land *and* pay for it?

 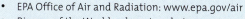

IMPACTS, ISSUES   *My Pheromones Made Me Do It*

A few years ago, as Toha Bererub walked down a street near her Las Vegas home, she felt a sharp pain above her right eye. Then another. And another. Seconds later, stinging bees had encased the upper half of her body. Firefighters wearing protective gear rescued her, but not before she had been stung more than 500 times.

Bererub's attackers were Africanized honeybees, a hybrid between the European honeybee and a more aggressive African strain (Figure 30.1). The African bees were imported to Brazil in the 1950s for cross-breeding experiments. Breeders were after a gentle but zippier pollinator for orchards. But some of the bees from Africa escaped and started mating with the locals.

Then, in a grand example of geographic dispersal, some descendant bees buzzed all the way from Brazil to Mexico and on into the United States. So far, Africanized bees have established themselves in Texas, New Mexico, Nevada, California to the west, and Alabama, Virginia, and Florida to the east.

Honeybees sting only once, and all make the same kind of venom. Africanized bees make a bit less venom but get riled up faster and mount collective attacks. One squadron reportedly chased a perceived threat for a quarter of a mile.

The Africanized bees became known as "killer bees," although they rarely kill their target. To be sure, their stings are extremely painful, but adults in good health usually can survive a collective attack. Bererub was seventy years old when attacked, and she recovered fully after spending a week in the hospital.

So what makes Africanized bees so testy? Isopentyl acetate. This chemical, which smells like bananas, is a

Figure 30.1   Africanized honeybees standing guard at the entrance to their hive. When a potential intruder appears, they release an alarm pheromone that stimulates other colony members to join them in an attack. In the filmstrip at far right, Toha Bererub is describing how stinging bees swarmed over her face. She recovered fully.

the big picture

The Basis of Behavior   Animal behavior starts with gene products that build and operate the nervous and endocrine systems. Animals are prewired to execute innate behaviors. Most are prewired to learn, to modify their behavior through experience with the environment.

Natural Selection and Behavior   Communication signals, mating systems, and other forms of behavior have evolved by way of natural selection. From the evolutionary perspective, a behavior is adaptive when it increases the reproductive success of the individual that displays it.

key component of honeybee alarm pheromone.

A *pheromone*, recall, is a chemical signal released by one individual that affects the behavior of another individual of the same species. A honeybee releases an alarm pheromone when threatened and when it stings something. As the signaling molecules diffuse through the air, they form a concentration gradient that guides other bees to the individual sounding the alarm.

Researchers once compared hundreds of colonies of Africanized and European honeybees for their alarm pheromone responses. They positioned a tiny target in front of each colony, then released a small quantity of artificial pheromone. Compared to European bees, the Africanized bees were faster to fly out of the colony and zero in on the perceived threat. They plunged six to eight times as many stingers into it.

The two kinds of honeybees show other behavioral differences. Compared to European bees, Africanized bees are less picky about where they establish a colony. They are more likely to abandon it after being disturbed. Of more concern to beekeepers, they are less interested in stashing large amounts of honey.

Such differences among honeybees lead us into the world of *animal behavior*, to the coordinated responses that animal species make to stimuli. We will look first at the genetic basis of behavior, then at instinctive and learned mechanisms that arise from it. As you will see, natural selection theory helps explain how and why animals behave as they do.

 How Would You Vote?

Africanized bees are slowly expanding their range in North America. Some think the more we know about them, the better we will be able to protect ourselves. Should we fund more research into the genetic basis of their behavior? See the Media Menu for details, then vote online.

Group Living Life in social groups has reproductive costs as well as benefits for the individual. Costs are highest in societies where self-sacrificing behavior is extreme. Selection theory helps explain how such altruistic behavior may still be adaptive.

What About Humans? Microevolutionary forces influence behavioral traits that promote reproductive success. Yet, a human behavior that is adaptive in the sense of perpetuating one's genes may be socially maladaptive; it may be viewed as unacceptable or morally wrong.

30.1 Behavior's Heritable Basis

The nervous and endocrine systems govern behavioral responses to stimuli. Because genes specify the substances required for constructing and operating those systems, they are the heritable foundation for animal behavior.

GENES AND BEHAVIOR

Nervous systems are wired to detect, interpret, and issue commands for response to stimuli. A **stimulus**, remember, is a piece of information about the external or internal environment that a specific type of sensory receptor has detected. It takes gene products to build and operate sensory receptors, nerve pathways and, in most species, a brain. Gene products also influence behavioral responses to stimuli.

Stevan Arnold found evidence of the genetic basis of behavior in the feeding preferences of coastal and inland snake populations in California. Garter snakes living near the coast hunt banana slugs (Figure 30.2*a*). Snakes inland hunt tadpoles and fishes. Offer them a banana slug and they ignore it.

In one set of experiments, Arnold offered captive newborn snakes a bit of slug as a first meal. Newborn coastal snakes usually ate it and flicked their tongue at cotton swabs drenched in essence of slug. (Snakes "smell" by tongue-flicking, which pulls odors into the mouth.) Newborn inland snakes ignored the swabs and rarely ate bits of slug. Here was a clear difference between captive snakes that had no prior experience with slugs. The snakes are programmed before birth to accept or reject slugs; they did not learn feeding preferences by taste trials.

Allelic differences may affect how odor-detecting mechanisms form during garter snake development. To test this hypothesis, Arnold crossed coastal and inland snakes. He predicted that the hybrid offspring

Figure 30.2 (**a**) Banana slug, food for (**b**) an adult garter snake of coastal California. (**c**) Newborn garter snake from a coastal population, tongue-flicking at a cotton swab drenched with tissue fluids from a banana slug.

would make an *intermediate* response to slug chunks and odors. Test results matched his prediction.

Compared with a typical newborn inland snake, many baby snakes of mixed parentage tongue-flicked more often at slug-scented cotton swabs. However, they did so less often than newborn coastal snakes.

HORMONES AND BEHAVIOR

Hormones secreted by the endocrine system also are gene products that influence behavior. Let's focus on oxytocin, a hormone that has roles in birth, lactation, and mate selection, aggression, territoriality, and other social behaviors. All mammals, including voles, make and secrete it.

Prairie voles (*Microtus ochrogaster*) are among the few long-term monogamous mammals. Pair-bonding arises during a single night of repeated matings, with oxytocin secretions providing the ambiance. After the "honeymoon," a male and female have little interest in other potential partners.

In one experiment, researchers kept a male and female together for a few hours but prevented them from mating. Later on, they injected oxytocin into the female. The result was love at first sight; pair-bonding without the requisite sex. In another experiment, pair-bonded female prairie voles were injected with an oxytocin blocker. They promptly dumped their long-term partners and went off to mate with others.

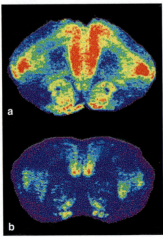

Figure 30.3 Hormones and prairie voles. Distribution of oxytocin receptors (*red*) in the brain of (**a**) a mate-for-life prairie vole and (**b**) a promiscuous mountain vole.

Figure 30.4 A few instinctive behaviors. (**a**) A social parasite, the European cuckoo, lays eggs in the nests of other birds. Even before this cuckoo hatchling opens its eyes, it responds instinctually to a round shape—a host's egg—and shoves it out of the nest. (**b**) The clueless foster parents show instinctual behavior as they respond to the usurper's gaping mouth, not to its size and other traits that characterize their own species. (**c**) A human baby instinctively imitates adult facial expressions.

Not all vole species are monogamous. Differences in the number and distribution of oxytocin receptors are part of the reason. Monogamous prairie voles have more of the receptors than the mountain voles, which are highly promiscuous (Figure 30.3).

Antidiuretic hormone (ADH, or vasopressin) also plays a role. Researchers isolated the gene for an ADH receptor in monogamous prairie voles and transferred copies of it into forebrain cells of male meadow voles (*M. pennsylvanicus*). After the transfer, the males of this far more promiscuous species showed an increased tendency to partner with just one female.

Male meadow voles used as a control group also got copies of the gene, but in a brain region not known to be involved in pair-bonding. Unlike the experimental group, the males retained their playboy ways.

INSTINCTIVE BEHAVIOR

Slug-loving garter snakes and pair-bonding voles are lead-ins to **instinctive behavior**. The term refers to a behavior that is performed without having first been learned through experience in the environment.

Like many other animals, coastal garter snakes are prewired to recognize sign stimuli before being born or hatched. *Sign stimuli* are one or two simple, well-defined cues that trigger an expected response. For these snakes, the cue is a certain slug scent that calls for a *fixed action pattern*—a stereotyped motor program of coordinated muscle activity that runs to completion independently of feedback from the environment. The snake is compelled to strike, capture, and eat a slug.

Similarly, hormone molecules wafting through the air are cues for mating behavior, a fixed action pattern between male and female vole.

Cuckoos, a type of social parasite, offer another case of instinctive behavior. Female cuckoos lay eggs in nests of other species. Cuckoos are blind when they first hatch, but contact with an egg or another round object stimulates a fixed action pattern. The hatchling maneuvers the egg onto its back, then pushes the egg from the nest (Figure 30.4*a*). This behavior ensures that the hatchling receives undivided attention.

The cuckoo's "foster parents" are oblivious to the odd color and size of the usurper. They respond only to one sign stimulus—the gaping mouth of a chick—and continue with their parenting (Figure 30.4*b*).

Humans, too, show instinctive behavior. Three days after birth, a human infant already displays a capacity to mimic facial expressions of an adult who comes close to it (Figure 30.4*c*). The infant cannot see its own face, nor can it feel which facial muscles the adult is using. Somehow it is able to open its mouth, protrude its tongue, or rotate its head the same way as the adult. Infants will also respond to a simplified stimulus—a flat, face-sized mask with two dark spots for eyes. One "eye" won't do the trick.

Genes underlie animal behavior, or coordinated responses to stimuli. Certain gene products have roles in constructing and operating the nervous system, which governs behavior. Other gene products, such as hormones, influence the mechanisms required for particular forms of behavior.

Animals start out life neurally wired to recognize vital cues and to make an instinctively suitable response, one that has not been learned through actual experience.

Many animals execute a fixed action pattern, a stereotyped program of coordinated muscle activity in response to one or two simple, well-defined environmental cues.

30.2 Learned Behavior

*With **learned behavior**, responses vary or change as a result of an individual's experiences. Whether learned or instinctive, behavior develops through interactions between genes and environmental inputs.*

Animals process information about experiences and then use it to change or vary responses to stimuli. **Imprinting** is a classic example of learned behavior. This time-dependent form of learning is triggered by exposure to a sign stimulus. Exposure usually takes place during a sensitive period when the animal is young. Imprinting of baby geese on their mother is a favorite example of animal behaviorists (Figure 30.5).

Learned behavior arises as the environment directly or indirectly alters gene expression. Sensory input and good or bad nutrition are typical factors that lead to alterations in how and what an animal learns.

Figure 30.6 Male marsh wren belting out a territorial song that has been likened to a loud gurgle. Males of this species start to imitate their species song when they are about fifteen days old. Unike many species, marsh wrens continue to learn songs throughout their life.

For example, birdsong is an instinctive behavior. Yet in different habitats, songbirds learn variations, or dialects, of the species song. As Peter Marler showed, many male birds learn their full song ten to fifty days after hatching by listening to other birds sing it. The male nervous system is prewired to recognize a species song; a learning mechanism is primed to select and respond to acoustical input. But what he hears during a sensitive period shapes his *rendition* of the song.

In one study, Marler raised white-crown nestlings in soundproof chambers so they could not hear adult males. The captives sang when mature, but not with the exact structure of a typical adult song. Marler also isolated and exposed captives to recorded white-crown songs *and* song sparrow songs. At maturity, captives sang only the white-crown song and even mimicked the species dialect of the unseen tutor.

In another experiment, Marler did not use taped songs. He let young, hand-reared male white-crowns interact with a "social tutor" of a different species. The males tended to learn the tutor's song.

Many such experiments have been performed. The results support this hypothesis: Birdsong starts with a genetically based capacity to learn from acoustical cues. Think about that when you hear songbirds in spring (Figure 30.6).

Imprinting is a case of learned behavior. It occurs during a genetically determined sensitive period. Like all behavior, it develops by interactions between genes and an animal's social and physical environment.

Figure 30.5 No one can tell these imprinted baby geese that Konrad Lorenz is not Mother Goose!

In response to a moving object and probably to certain sounds, baby geese imprint on the mother goose and follow her during a short, sensitive period right after hatching. They are neurally wired to learn some crucial information —the identity of the one individual that will help protect them in the months ahead. Usually that will be their mother.

Konrad Lorenz, one of the early investigators of animal behavior, presented the baby geese in this figure with sign stimuli that made them form an attachment to him.

30.3 The Adaptive Value of Behavior

If forms of behavior have a genetic basis, then they may evolve in various ways through natural selection. Alleles that encode the most adaptive versions of a trait will tend to increase in frequency in a population, and alternative alleles will not. In time, genetic changes in behavior that yield greater reproductive success will be favored.

Natural selection theory helps us develop and test explanations of why a behavior persists and how it offers reproductive benefits that offset reproductive costs (disadvantages) associated with it. If a behavior is adaptive, it promotes the *individual's* production of offspring. Here are five definitions to keep in mind:

1. **Reproductive success**. An individual produces some surviving offspring.

2. **Adaptive behavior**. Behavior that helps perpetuate an individual's genes; its frequency in a population is maintained or increases over the generations.

3. **Social behavior**. Interdependent interactions among individuals of one or more populations of a species.

4. **Selfish behavior**. A form of behavior that improves an individual's chance to produce or protect its own offspring regardless of the impact on the population.

5. **Altruism**. Self-sacrificing behavior. An individual behaves in a way that helps others in the population but lowers its own chance of producing offspring.

When biologists speak of selfish or altruistic behavior, they don't mean an individual is consciously aware of some behavior or its reproductive goal. A lion doesn't have to know that eating zebras will be good for its reproductive success. Its nervous system just screams HUNTING BEHAVIOR! when the lion is hungry and sees a zebra. Hunting behavior persists in a lion population because genes for neural mechanisms that command hunting behavior are persisting.

To assess the adaptive value of any behavior, look for how it might promote reproductive success. For example, starlings (*Sturnus vulgaris*) nest in cavities of trees and *decorate* the nest bowl with sprigs of fresh leaves of certain strongly scented plants, such as the wild carrot (*Daucus carota*).

Larry Clark and Russell Mason suspected that nest decorating behavior suppresses nest parasites. Some mites parasitize birds and infest their nests. Even a few mites produce thousands of descendants. In large numbers, mites suck enough blood from a nestling to weaken it and affect its growth and survival.

Clark and Mason tested the hypothesis that nest decorating behavior is adaptive because it helps deter mites. They devised a set of experimental nests, some

Figure 30.7 An experiment to test the adaptive value of the nest-decorating behavior of starlings (**a**). Nests designated *A* did not have fresh sprigs of wild carrot (**b**) and other plants that make aromatic compounds. Nests designated *B* had fresh sprigs added every seven days. (**c**) Twenty-one days after the start of the experiment, when chicks had left, researchers made counts of the mites (*Ornithonyssus sylviarum*) infesting each nest. Results showed that aromatic compounds suppress development of juvenile mites into (**d**) adult mites.

with freshly cut wild carrot leaves and some without. They removed natural nests that starlings constructed and were using. Half of the nesting starling pairs got new nests with wild carrot sprigs. Replacement nests for the other pairs of starlings were sprigless.

Figure 30.7*c* shows the results. The number of mites in sprig-free nests was far greater than the number in sprig-festooned nests. At the end of one experiment, the sprig-free nests teemed with an average of 750,000 mites. Nests with sprigs had only 8,000 mites. Why?

Wild carrot shoots contain a highly aromatic steroid compound that almost certainly repels herbivores and helps plants survive. By coincidence, it prevents mites from maturing sexually and thus prevents population explosions of mites in nests festooned with wild carrot.

Aromatic greenery is a fumigant that decreases the numbers of mites, increases the survival of nestlings, and promotes reproductive success of the decorators.

A genetically determined behavior may persist or increase in frequency in a population when it is adaptive. A behavior is adaptive when it increases the number of descendants that an individual successfully produces.

30.4 Communication Signals

Competing for food, defending territory, alerting others to danger, advertising sexual readiness, forming bonds with a mate, caring for the offspring—such intraspecific behaviors require unambiguous forms of communication.

THE NATURE OF COMMUNICATION SIGNALS

Communication signals are unambiguous cues sent and received among individuals of the same species, and they involve instinctive and learned behaviors. Information-laden cues from a signaler are meant to alter the behavior of receivers. Chemical, acoustical, and visual cues are among the most common.

Pheromones, again, are chemical communication signals. *Signaling* pheromones induce the receiver to respond fast. They include chemical alarms, such as the honeybee call to action against a potential threat. They also include sex attractants. Bombykol is one of them. Bombykol molecules released by a female silk moth can summon males that are kilometers away. *Priming* pheromones bring about physiological (not behavioral) responses. For example, a volatile odor in urine of certain male mice triggers and enhances estrus in female mice of the same species.

Acoustical signals abound in nature. Male birds, frogs, grasshoppers, whales, and other animals make sounds that attract females. Prairie dogs bark alarms. Wolves howl and kangaroo rats drum their feet on the ground when advertising possession of territory.

Some signals never vary. Zebra ears pressed flat to the head convey hostility; ears pointing up convey its absence. Different signals convey the intensity of the message. A zebra with ears laid back is not too riled up as long as its mouth is open only a bit. When the ears are laid back and its mouth gapes, watch out. That combination is a type of *composite* signal. Such signals have information encoded in two or more cues.

Signals often take on different meaning in different contexts. A lion emits a spine-tingling roar to keep in touch with its pride *or* to threaten rivals. Also, a signal can convey information about signals to follow. Dogs and wolves solicit play behavior with their play bow, as in Figure 30.8a. Without the bow, a signal receiver may construe the behaviors that follow as aggressive, sexual, or even exploratory—but not playful.

Signals evolve or persist in a population when they promote reproductive success of both the sender *and* receiver. If a signal is harmful, natural selection will favor individuals that don't send it or respond to it.

COMMUNICATION DISPLAYS

The play bow is a *communication* display, a pattern of behavior that is a social signal. The *threat* display is another common pattern. It announces that a signaler is prepared to attack a signal receiver. When a rival for a receptive female confronts him, a dominant male baboon will roll his eyes upward and "yawn," which exposes his canines (Figure 30.8b). The signaler often benefits when the rival backs down, for he retains access to the female without having to engage in a fight. The signal receiver benefits because he avoids a serious beating, infected wounds, and possibly death.

Such displays are ritualized, with intended changes in the function of common behavior patterns. Normal movements may be exaggerated or frozen. Feathers,

Figure 30.8 Communication displays. (**a**) Play bow of a young male wolf soliciting a romp. (**b**) Part of a male baboon's threat display: exposed canines. (**c**) Courtship display of Adele penguins.

Read Me First!
and watch the narrated animation on honeybee dances

Figure 30.9 Honeybee dances, a type of tactile display. (**a**) Honeybees that have visited a food source close to their hive perform a *round* dance on the hive's honeycomb. Worker bees that maintain contact with the forager throughout the dance will go out and search for food near the hive.

(**b**) A bee that visits a feeding station more than 100 meters distant from the hive performs a *waggle* dance. During the dance, it makes a straight run and waggles its abdomen. A waggle dancer also varies the dance speed to convey more information about distance to a food source. For example, when food is 150 meters away, a bee dances much faster, and with more waggles per straight run, compared to a dance about a food source that is 500 meters away.

(**c**) As Karl von Frisch discovered, a *straight* run's orientation varies, depending on the direction in which a food source is located. He put one dish of honey on a direct line between a hive and the sun. Foragers that located it returned to the hive and oriented their straight runs right up the honeycomb. He put another dish of honey at right angles to a line between the hive and the sun. Foraging bees made their straight runs 90 degrees to vertical on the honeycomb. Thus, a honeybee "recruited" into foraging orients its flight *with respect to the sun and the hive*. By doing so, it wastes less time and energy during its food-gathering expedition.

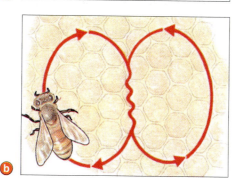

When bee moves straight down comb, recruits fly to source directly away from the sun.

When bee moves to right of vertical, recruits fly at 90° angle to right of the sun.

When bee moves straight up comb, recruits fly straight toward the sun.

---

manes, claws, and other body parts are often notably enlarged, patterned, and colored. Ritualization is well developed in *courtship* displays, the steps that must precede pair formation. Many birds have elaborate courtship displays (Figures 16.20, 17.17, and 30.8c).

In *tactile* displays, a signaler touches the receiver in ritualized ways. After locating a source of pollen or nectar, a foraging honeybee returns to its colony, a hive, and performs a complex dance. It moves in a defined pattern, jostling a crowd of workers that stay in close physical contact with it. Signals give others information about the general location, distance, and direction of a source of food (Figure 30.9).

ILLEGITIMATE SIGNALERS AND RECEIVERS

Unintended recipients can intercept communication signals. Male tungara frogs make two kinds of calls—one simple, the other complex. The calls say "come hither" to female frogs, but they also mean "dinner here" to fringe-lipped bats. Complex calls are more attractive to females but make it easier for the bats to locate the caller. When bats are around, male frogs call less and are more likely to make the simpler call.

There are also illegitimate signalers. Some assassin bugs acquire the scent of their prey—termites—by hooking a dead termite on their back. By signaling that they "belong" to a termite colony, they can more easily hunt termites. Another example: If a female of certain predatory firefly species sees a flash from a male of a different species, she will flash back. If she lures him into attack range, she will capture and eat him. Getting eaten is an evolutionary cost of having an otherwise useful response to a come-hither signal.

A communication signal is a transfer of information between two members of the same species. Such signals evolve only when they benefit both the signaler and receiver.

Members of a different species sometimes act illegitimately as a communication signaler or receiver.

30.5 Mates, Offspring, and Reproductive Success

For reasons we need not explore here, many people find mating and parenting behaviors of animals fascinating. How useful is selection theory in helping us interpret such behavior? Let's take a look.

SEXUAL SELECTION AND MATING BEHAVIOR

Competition among members of one sex for access to mates is common. So is choosiness in selecting a mate. Such activities, recall, are forms of **sexual selection**. This microevolutionary process favors traits that give the individual a competitive edge in attracting and often holding on to mates (Section 16.9).

But *whose* reproductive success is it—the male's or female's? Male animals, remember, produce many tiny sperm, and females produce far larger but fewer eggs. For a male, success generally depends on how many eggs he can fertilize. For a female, it depends more on how many eggs she produces or how many offspring she can raise. Usually, the most important factor in a female's sexual preference is the quality of the mate, not the quantity of partners.

Female hangingflies (*Harpobittacus apicalis*) provide an instructive example. They choose males that offer superior food. A male hunts and kills a moth or some other insect. Then he releases a sex pheromone, which attracts females to him and his "nuptial gift" (Figure 30.10a). A female tends to select the male that offers a large calorie-rich gift. Only after the female has been eating the gift for five minutes or so does she start to accept sperm from her partner. She lets the male continue inseminating her—but only as long as it takes for her to devour the gift.

Before twenty minutes are up, a female hangingfly can break off the mating at any point. If she does, she may well mate with a different male hangingfly and accept his sperm. Doing so dilutes the reproductive success of her first partner.

Females of other species shop around for the good-looking males. Consider the fiddler crabs that live along muddy shores from Massachusetts to Florida. One of the male's two claws is enlarged. Sometimes it is enlarged enough to make up more than 50 percent of his body weight (Figure 30.10b). When spring tides are favorable, the male crabs build elaborate mating burrows near one another. Each male stands beside his burrow, waving his oversized claw. Females stroll by, checking out the males and their burrows. When a female likes what she sees, she follows the male into his burrow and engages in sex.

Many female birds are just as choosy. Male sage grouse (*Centrocercus urophasianus*) come together at a lek, a type of communal display ground. Each male stakes out a few square meters. With tail feathers erect, males use their large, puffed-out neck pouches to emit booming calls (Figure 30.10d). As they do this,

Figure 30.10 (a) Male hangingfly dangling a moth as a nuptial gift for a potential mate. Females of some hangingfly species choose sexual partners that offer the largest gift to them. By waving his enlarged claw, a male fiddler crab (b) may attract the eye of a female fiddler crab (c). A male sage grouse (d) showing off as he competes for female attention at a communal display ground.

they stamp about on their patch of prairie, a bit like wind-up toys. Females tend to select and mate with just one male sage grouse. Then they go off to nest and raise the young by themselves. Many females often choose the same male, so most of the males never do mate.

In another behavioral pattern, sexually receptive females of some species cluster in defendable groups. Where you come across such a group, you are likely to observe males competing for access to clusters. The competition for ready-made harems favors highly combative male lions, sheep, elk, elephant seals, and bison, to name a few types of animals (Figure 30.11).

PARENTAL CARE

If females fight for males, then we can expect the males to provide more than sperm delivery—say, parental care. Midwife toads are an example. A male wraps strings of fertilized eggs around his legs until the eggs hatch (Figure 30.12a). With her eggs being cared for, a female is free to mate with other males—if she can find some with no eggs. As the breeding season progresses, such males become rare, and females fight for access to them. They even attempt to pry mating pairs apart.

Parental behavior is costly. It drains away time and energy that parents could spend on living long enough to reproduce another time. However, for many species, parenting improves the odds that their offspring will live. The benefit of giving time and energy to offspring (immediate reproductive success) may offset reduced reproductive success at a later time.

For amphibians and reptiles, parenting is rare once the young are hatched. Crocodilians are an exception. As with birds, crocodilian parents build a nest. The young call when they are ready to hatch. Parents dig them up and care for them for some time as they grow.

Most birds are monogamous and both parents often care for the young (Figure 30.12b). In mammals, males typically leave after mating. Females raise the young alone, and the males attempt to mate again or conserve energy for the next breeding season (Figure 30.12c). Mammalian species in which males do help care for the young tend to be monogamous. Only about five percent of all mammals fall into that category.

Selection theory explains some aspects of mating behavior. Male or female preferences for certain behavioral traits can give the individual a competitive edge and promote its reproductive success.

Figure 30.11 (**a**) Sexual competition between male elephant seals, which fight for access to a cluster of females. (**b**) Male bison locked in combat during the breeding season.

Figure 30.12 (**a**) Male midwife toad with developing eggs wrapped around his legs. (**b**) Male and female Caspian terns cooperate in the care of their chick. (**c**) A female grizzly will care for her cub for as long as two years. The male takes no part in its upbringing.

30.6 Costs and Benefits of Social Groups

Survey the animal kingdom and you will find a range of social groups, with evolutionary costs and benefits.

COOPERATIVE PREDATOR AVOIDANCE

Cooperative responses to predators help some groups reduce the net risk to all. Vulnerable individuals, too, scan for predators, join in a counterattack, or engage in more effective defenses (Figure 30.13).

Vervet monkeys, meerkats, prairie dogs, and many other mammals cooperate with their alarm calls, as in Figure 30.13a. A prairie dog barks a certain signal when it sights an eagle and a different signal when it sights a coyote. Others dive into burrows to escape an eagle's attack or they stand erect, the better to scan the horizon and zero in on the coyote.

Ecologist Birgitta Sillén-Tullberg found evidence of group benefits for Australian sawfly caterpillars that live in clumps on branches. When something disturbs them, the individuals collectively rear up and writhe about (Figure 30.13b). They also vomit partly digested eucalyptus leaves, which are toxic to songbirds and other animals that prey on them.

As Sillén-Tullberg hypothesized, individual sawfly caterpillars benefit from their coordinated repulsion of predatory birds. She used her hypothesis to predict that birds are more likely to eat a solitary caterpillar.

To test her prediction, she offered caterpillars to young hand-reared birds. Birds offered one caterpillar at a time ate an average of 5.6. Birds offered a clump of them ate an average of 4.1. As Sillén-Tullberg had predicted, individuals were safer in the group.

THE SELFISH HERD

Simply by their physical position in the group, some individuals form a living shield against predation on others. They belong to a **selfish herd**, a simple society that benefits their reproductive self-interest. Selfish-herd behavior has been studied in bluegill sunfishes. Male sunfishes build adjacent nests on lake bottoms. Females deposit eggs where the males have used their fins to scoop out depressions in mud.

If a colony of bluegill males is a selfish herd, then we can predict competition for the "safe" sites—at the center of a colony. Compared to eggs at the periphery, eggs in nests at the center are less likely to be eaten by snails and largemouth bass. Competition does indeed occur. The largest, most powerful males tend to claim centermost locations. Other, smaller males assemble around them and bear the brunt of predatory attacks. Even so, they are better off in the group than on their own, fending off a bass single-handedly, so to speak.

COOPERATIVE HUNTING

Many predatory mammals, including wolves, lions, and wild dogs, live in social groups and cooperate in hunts (Figure 30.14). Are group hunts more successful than solitary hunts? Often they are not. Researchers

Figure 30.13 Group defenses. (**a**) Black-tailed prairie dogs bark an alarm call that warns others of predators. Does this put the caller at risk? Not much. Prairie dogs usually act as sentries only if they are done feeding and are standing beside their burrows. (**b**) Australian sawfly caterpillars form clumps and collectively regurgitate a fluid (yellow blobs) that is toxic to most predators. (**c**) Musk oxen adults (*Ovibos moshatus*) form a ring of horns, often around the young.

Figure 30.14 Members of a wolf pack (*Canis lupus*). Wolves cooperate in hunting, caring for the young, and defending a territory. Benefits are not distributed equally. Only the highest ranking individuals, the alpha male and alpha female, breed.

observed a solitary lion that captured prey about 15 percent of the time. Two lions hunting together did capture prey twice as often, but they had to share it, so the number of successful hunts per lion balanced out. When more lions joined the hunt, the success rate per lion fell. Wolves show a similar pattern. Among many cooperative hunters, hunting success in itself may not explain group living. Individuals do hunt together, but they also may fend off scavengers, care for one another's young, and protect territory.

Figure 30.15 Nearly uniform spacing in a crowded cormorant colony.

DOMINANCE HIERARCHIES

Many social groups share resources unequally among some individuals that are subordinate to others. Most wolf packs, for instance, have one dominant male that breeds with just one dominant female. Other wolves in the pack are nonbreeding brothers and sisters, or aunts and uncles. They all hunt and bring food to the individuals that guard the young in their den.

Baboons live in large troops. A female stays with the group into which she was born and inherits social standing from her mother. Dominant females get more food, water, and grooming. Their young grow and mature faster than those of lower-ranking females.

Why do subordinates relinquish resources and, often, breeding privileges? They might do so to avoid being hurt and to belong to a group. Possibly they can't survive out on their own. Or possibly challenging a strong individual can invite injury or death. Besides, a subordinate might get a chance to reproduce if it lives long enough or if a predator or old age removes its dominant peers. Some subordinate wolves and baboons do manage to move up the social ladder when the opportunity arises.

REGARDING THE COSTS

If social behavior is advantageous, *then why are there so few social species?* In most places, the costs outweigh benefits. For instance, when more individuals do live together, they compete more for a share of the food. Cormorants, puffins, and many other seabirds form enormous breeding colonies (Figure 30.15). All must compete for a share of the same ecological pie.

Large social groups also attract more predators. If individuals are crowded together, they invite parasites and contagious diseases that jump from host to host. The individuals may also be at risk of being killed or exploited by others. Given the opportunity, breeding pairs of herring gulls cannibalize a neighbor's eggs and any chicks that wander away from their nest.

Living in a social group can provide benefits, as through cooperative defenses or shielding against predators.

Group living has costs, in terms of increased competition, increased vulnerability to infections, and exploitation by others of the group.

30.7 Why Sacrifice Yourself?

Extreme cases of sterility and self-sacrifice have evolved in only two groups of insects and one group of mammals. How are genes of the nonreproducers perpetuated?

SOCIAL INSECTS

Honeybees and fire ants (Chapter 27), are among the true social (eusocial) insects. Like termites, they stay together for generations in a group that has a division of labor. Many permanently sterile individuals care cooperatively for the offspring of just a few breeding individuals. Often they are highly specialized in form and function (Figure 30.16).

Consider a honeybee colony, or hive. Its queen bee, the only fertile female, secretes a pheromone that other female bees distribute through the hive. This signaling molecule suppresses the development of ovaries in all other females, rendering them sterile. The queen bee is larger than the worker bees, partly because of her enlarged egg-producing ovaries (Figure 30.17a).

About 30,000 to 50,000 female workers feed larvae, clean and maintain the hive, and build honeycomb from waxy secretions. Adult workers live for about six weeks in the spring and summer. When foragers return to the hive after finding a rich source of nectar or pollen, they engage others in a dance. This tactile display recruits more foragers (Figure 30.9). Workers also cooperate by transferring food to one another. They guard the entrance to the hive and will sacrifice themselves to repel intruders.

Males, the stingless drones, develop only in spring and summer. They have no part in the day-to-day work and subsist on food gathered by their worker sisters. Drones live for sex. Each day, they fly out in search of a mate. If one is lucky, he will find a virgin queen on her single flight away from her colony. The sole function of her flight is to meet up with and mate with a drone. A drone dies right after he inseminates a virgin queen, which then starts a new colony. She will store and use his sperm for years, perpetuating his genes and those of his original colony.

Like honeybees, termites live in enormous family groups with a queen specialized for producing eggs (Figure 30.17c). Unlike a honeybee hive, each termite colony has sterile individuals of both sexes. A king supplies the female with sperm. Winged reproductive termites of both sexes develop seasonally.

SOCIAL MOLE RATS

Vertebrates are not known for sterility and extreme self-sacrifice. The only eusocial mammals are African mole-rats. The best studied is *Heterocephalus glaber*, the naked mole-rat. Clans of this nearly hairless rodent occupy burrows in arid parts of East Africa.

One reproducing female dominates the clan and mates with one to three males (Figure 30.17b). Other, nonbreeding members live to protect and care for the "queen" and "king" (or kings) and their offspring. The sterile diggers excavate subterranean tunnels and chambers that serve as living rooms or waste-disposal sites. When a digger locates a tasty root or tuber, it hauls a bit of it back to the main chamber and emits a series of chirps. The chirps recruit others, which help carry the food back to the chamber. In this way, the queen, her retinue of males, and her offspring get fed.

Digger mole-rats also deliver food to other helpers that usually loaf about, shoulder to shoulder and belly to back, with the reproductive royals. These "loafers" actually spring to action when a snake or some other enemy threatens the clan. Collectively, and at great risk, they chase away or attack and kill the predator.

Figure 30.16 Specialized ways of serving and defending the colony. (**a**) An Australian honeypot ant worker. This sterile female is a living container for her colony's food reserves. (**b**) Army ant soldier (*Eciton burchelli*) with formidable mandibles. (**c**) Eyeless soldier termite (*Nasutitermes*). It bombards intruders with a stream of sticky goo from its nozzle-shaped head.

Figure 30.17 Three queens. (**a**) A queen honeybee with her court of sterile worker daughters. (**b**) This queen naked mole-rat has twelve mammary glands, the better to feed her many offspring. In a laboratory colony at Cornell University, one female produced a litter of twenty-eight pups. And she gave birth to more than 900 offspring during her lifetime. (**c**) A termite queen (*Macrotermes*) dwarfs her offspring and her mate. Her body pumps out thousands of eggs a day.

Figure 30.18 Damaraland mole-rats in a burrow. Like their relatives, the naked mole-rats, they live in colonies having nonbreeding workers. But unlike the naked mole-rats, these fuzzy burrowers are not highly inbred.

INDIRECT SELECTION FOR ALTRUISM

None of the altruistic individuals of a honeybee hive, termite colony, or naked mole-rat clan directly passes genes to the next generation. So how are genes that underlie altruistic behavior perpetuated? According to William Hamilton's theory of **inclusive fitness**, genes associated with altruism can be favored by selection if they lead to behavior that will increase the number of offspring produced by an altruist's closest relatives.

A sexually reproducing, diploid parent caring for offspring is not helping exact genetic copies of itself. Each of its gametes, and each of its offspring, inherits one-half of its genes. Other individuals of the social group that have the same ancestors also share genes with their parents. Two siblings (brothers or sisters) are as genetically similar as a parent and its offspring. Nephews and nieces have inherited about one-fourth of their uncle's genes.

Sterile workers may be indirectly promoting genes for "self-sacrifice" through altruistic behavior that will benefit their relatives. All of the individuals in honeybee, termite, and ant colonies are members of an extended family. Nonbreeding family members support siblings, a few of which are future kings and queens. Although a guard bee dies after driving her stinger into a bear, siblings in the hive will perpetuate some of her genes.

Does close relatedness explain why naked mole-rats are the only eusocial mammals? Through DNA fingerprinting studies, we know that all individuals in one naked mole-rat clan are *very* close relatives and differ genetically from individuals of other clans. Each clan is highly inbred through generations of brother–sister, mother–son, and father–daughter matings.

However, inbreeding may not be necessary for mole-rat eusociality. The social organization of the Damaraland mole-rat (*Cryptomys damarensis*) resembles that of *H. glaber* (Figure 30.18). Nonbreeding members of both sexes cooperatively assist one breeding pair. However, breeding pairs of wild Damaraland mole-rat colonies usually are unrelated.

Researchers are now searching for other factors that select for mole-rat eusocial behavior. According to one hypothesis, an arid habitat and patchy food supplies favor the mole-rat genes that underlie cooperation in digging burrows, searching for food, and fending off competitors of other species for resources.

Altruistic behavior may persist when individuals pass on genes indirectly, by helping relatives survive and reproduce.

By the theory of inclusive fitness, genes associated with altruistic behavior that is directed toward relatives may spread through a population in certain situations.

30.8 An Evolutionary View of Human Social Behavior

Evolutionary forces have shaped the behavior of all animals, including humans. However, only humans consistently make moral choices about their behavior.

HUMAN PHEROMONES

When women live together, as in a college dormitory, their menstrual cycles typically become synchronized. Martha McClintock and Kathleen Stern have evidence that pheromones in sweat play a role. The researchers exposed women to the sweat from others who were in different menstrual cycle phases. Sweat secreted by women in the early phase accelerated ovulation in other women exposed to it and shortened their cycles. Sweat from women who were near ovulation had the opposite effect; it delayed ovulation and lengthened the cycles. Sweat had no effect on the menstrual cycle in about 30 percent of the women.

Do humans respond to pheromones? Possibly, but we don't know how. In most mammals, pheromones bind to specific receptors in a vomeronasal organ (VNO). Neurons connect the VNO to brain regions that control behavior. In human adults, VNOs are tiny ductlike structures on a septum that divides the nose into two nostrils. Many researchers think that human VNOs are vestigial structures (no longer functional). Some human DNA sequences are similar to sequences that encode pheromone receptors in rodents, but most of them are not activated.

HORMONES AND BONDING BEHAVIOR

The hormone oxytocin, recall, facilitates pair bonding in prairie voles. It also influences social attachments between humans. Consider *autism*. A person who has this behavioral disorder does not form normal social relationships. Compared to unaffected individuals, an autistic child has less oxytocin circulating in blood.

Today, researchers are studying oxytocin's role in the formation of bonds between a mother and infant and between sexual partners. Nursing stimulates the secretion of oxytocin. So does orgasm, even a friendly massage. Do brief rises in oxytocin levels promote long-term bonding in humans? We don't know yet.

EVOLUTIONARY QUESTIONS

If we are comfortable with studying the evolutionary basis of the behavior of termites, naked mole-rats, and other animals, why do so many people resist the idea of analyzing human behavior in the same way? Often they fear that attempts to identify the adaptive value of some human trait will be used to define its morality. However, there is a clear difference between trying to explain behavior in terms of its evolutionary history and attempting to justify it. To biologists, "adaptive" does not mean "morally right." It only means useful in perpetuating an individual's genes.

An example: Infanticide is morally repugnant. Is it unnatural? No. It happens in many animal groups and human cultures. Male lions often kill the offspring of other males when they take over a pride. Doing so frees up the females to breed with them, increasing the infanticidal male's reproductive success.

Biologists may predict that unrelated human males are a threat to infants and evidence supports this. The absence of a biological father and the presence of an unrelated male increases risk of death for an American child under age two by sixty times.

What about parents who kill their own offspring? In her book on maternal behavior, primatologist Sarah Blaffer Hrdy cites a study of a village in New Guinea in which approximately 40 percent of the newborns were killed by parents. She argues that when resources or social support are hard to come by, a mother might increase her fitness by killing a newborn. She can then allocate child-rearing energy to her other offspring or save it for children she may have in the future.

Do most of us find such behavior appalling? Yes. Does such behavior warrant attention? Think about all you have learned in this book, then decide.

Biologists describe human behavior as it is, not as it ought to be. They take into consideration our cultural roots as a society of hunter–gatherers, and our evolutionary roots as social primates.

Summary

Section 30.1 Animal behavior starts with genes that specify products required for development of the nervous, endocrine, and muscular systems. Hormones are among the gene products that affect behavior.

Instinctive behavior is performed without having been learned by experience in the environment. It is a prewired response to one or two simple, well-defined environmental cues.

Section 30.2 An animal learns when it processes and integrates information from experiences, then uses it to vary or change how it responds to stimuli. Imprinting is learning that occurs only during a sensitive period early in life.

Section 30.3 Behavior with a genetic basis is subject to evolution by natural selection. It evolved as a result of individual differences in reproductive success in past generations. Behavior persists when its reproductive benefits exceed the reproductive costs.

Section 30.4 Communication signals are meant to change the behavior of individuals of the same species. Pheromones are signaling molecules that have roles in social communication.

Visual signals are key components of courtship displays and threat displays. Acoustical signals are sounds that have precise, species-specific information. Tactile signals are specific forms of physical contact between a signaler and a receiver.

Section 30.5 Sexual selection favors traits that give an individual a competitive edge in attracting and often holding on to mates. Females of many species select for males that have traits or engage in behaviors they find attractive. When large numbers of females cluster in defensible areas, males may compete with one another to control the areas.

Parental care has reproductive costs in terms of future reproduction and survival. It is adaptive when benefits to a present set of offspring offset the costs.

Section 30.6 Animals that live in social groups may benefit by cooperating in predator detection, defense, and rearing the young. Benefits of group living are often distributed unequally. Species that live in large groups incur costs, including increased disease and parasitism, and increased competition for resources.

Section 30.7 Ants, termites, and some other insects as well as two species of mole-rats are eusocial. They live in colonies with overlapping generations and have a reproductive division of labor. Most colony members do not reproduce; they assist their relatives and rear their offspring.

According to the theory of inclusive fitness, such extreme altruism is perpetuated because altruistic individuals have some portion of their genes in common with their reproducing relatives. Altruistic individuals in the social group pass on "by proxy" the genes that underlie this behavior.

Section 30.8 Behavioral biologists ask questions about the mechanisms and adaptive significance of human behavior. A behavior that is adaptive in the evolutionary sense may still be judged by society to be morally wrong.

Self-Quiz
Answers in Appendix III

1. "Starlings minimize nest mites by festooning nests with wild carrot," said Clark and Mason. Their statement was _____ .
   a. an untested hypothesis
   b. a prediction
   c. a test of a hypothesis
   d. a proximate conclusion

2. Genes affect the behavior of individuals by _____ .
   a. influencing the development of nervous systems
   b. affecting the kinds of hormones in individuals
   c. governing development of muscles and skeletons
   d. all of the above

3. Generally, living in a social group costs the individual, in terms of _____ .
   a. competition for food, other resources
   b. vulnerability to contagious diseases
   c. competition for mates
   d. all of the above

4. A statement that overcrowding causes lemmings to disperse to areas that are more favorable for reproduction is _____ .
   a. consistent with Darwinian evolutionary theory
   b. based on a theory of evolution by group selection
   c. supported by the finding that most animals behave altruistically during their lives

5. Eusocial insects _____ .
   a. live in extended family groups
   b. are found among almost all insect orders
   c. show a reproductive division of labor
   d. a and c
   e. all of the above

6. Helping other individuals at a reproductive cost to oneself might be adaptive if those helped are _____ .
   a. members of another species
   b. competitors for mates
   c. close relatives
   d. illegitimate signalers

7. Match the terms with their most suitable description.
   ____ fixed action pattern
   ____ altruism
   ____ basis of instinctive and learned behavior
   ____ imprinting
   ____ pheromone

   a. time-dependent form of learning requiring exposure to key stimulus
   b. genes plus actual experience
   c. stereotyped motor program that runs to completion independently of feedback from environment
   d. assisting another individual at one's own expense
   e. one communication signal

Figure 30.19 A Nazca booby attends to its single surviving chick.

Figure 30.20 Behaviorally confused rooster.

Critical Thinking

1. Nazca boobies (*Sula granti*) lay two eggs several days apart. No matter how much food is available, only one chick survives to adulthood (Figure 30.19). The first chick to hatch pushes its younger sibling from the nest, and that sibling dies of starvation and neglect. Formulate a hypothesis on how it might be adaptive for parents to lay two eggs if one sibling is expected to kill the other. Design an experiment to test your hypothesis.

2. *Sexual imprinting* is common in birds. During a short sensitive period in early life, a bird learns the features it will look for later in a mate. The rooster in Figure 30.20 was observed wading after ducks with amorous intent. Speculate on what might have caused this behavior.

3. Svante Paabo has correlated a gene mutation with the origin of *language*. All mammals have the *FOXP2* gene, which has a 715 base-pair sequence. The mutant human gene differs from the *FOXP2* gene in mice by 3 base pairs and by only 2 base pairs in chimpanzees. This mutation, and others, altered brain regions that control muscles in the face, throat, and vocal cords. It apparently arose about 200,000 years ago and swiftly replaced the nonmutated form in the populations immediately ancestral to *Homo sapiens*. Speculate on how the rapid rise in its frequency may have affected social behavior and learning.

4. A cheetah scent-marks plants in its territory with certain exocrine gland secretions. What evidence would you require to demonstrate that the cheetah's action is an evolved communication signal?

5. Among primates, differences in sexual behavior tend to be related to the size of a male's gonads. Gorillas have relatively tiny testicles. Testicles of a male weighing 450 pounds may weigh about an ounce. Gorillas live in groups that consist of a male, a few females, and their offspring. When a female is ready to mate, only a single adult male is usually around to inseminate her.

In contrast, a female chimpanzee advertises her fertile period and will mate with many males. A 100-pound chimpanzee male has gonads about four times as weighty as a gorilla's. By producing large amounts of sperm, this male increases the chance that his sperm, rather than sperm from a rival, will fertilize a female's egg.

A human male's testicles weigh about half as much as a chimpanzee's. Speculate on the possible evolutionary link with female promiscuity and male sperm competition.

Epilogue

BIOLOGICAL PRINCIPLES AND THE HUMAN IMPERATIVE

Molecules, single cells, tissues, organs, organ systems, multicelled organisms, populations, communities, ecosystems, and the biosphere. These are architectural systems of life, assembled in increasingly complex ways over the past 3.8 billion years. We are latecomers to this immense biological building program. Yet within the relatively short span of 10,000 years, our activities have been changing the character of the land, ocean, and atmosphere, even the genetic character of species.

It would be presumptuous to think that we alone have had profound impact on the world of life. As long ago as the Proterozoic, photosynthetic organisms were irrevocably changing the course of biological evolution by enriching the atmosphere with oxygen. During the past as well as the present, competitive adaptations led to the rise of some groups, whose dominance assured the decline of others. Change is nothing new. What *is* new is the capacity of one species to comprehend what might be going on.

We now have the population size, technology, and cultural inclination to use up energy and modify the environment at rapid rates. Where will this end? Will feedback controls operate as they do, for instance, when population growth exceeds carrying capacity? In other words, will negative feedback controls come into play and keep things from getting too far out of hand?

Feedback control will not be enough, for it operates after deviation. Our patterns of resource consumption and population growth are founded on an illusion of unlimited resources and a forgiving environment. A prolonged, global shortage of food or the passing of a critical threshold for the global climate can come too fast to be corrected; in which case the impact of the deviation may be too great to be reversed.

What about feedforward mechanisms, which might serve as early warning systems? For example, when sensory receptors near the surface of skin detect a drop in outside air temperature, each sends messages to the nervous system. That system responds by triggering mechanisms that raise the body's core temperature before the body itself becomes dangerously chilled.

Extrapolating from this, if we develop feedforward control mechanisms, would it not be possible to start corrective measures before we do too much harm?

Feedforward controls alone will not work, for they operate after change is under way. Think of the DEW line—the Distant Early Warning system. It is like a vast sensory receptor for detecting missiles launched against North America. By the time it does what it is supposed to, it may be too late to stop widespread destruction.

It would be naive to assume we can ever reverse who we are at this point in evolutionary time, to de-evolve ourselves culturally and biologically into becoming less complex in the hope of averting disaster. Yet there is reason to believe we can avert disaster by using a third kind of control mechanism—a capacity to anticipate events even before they happen. We are not locked into responding only after irreversible change has begun. We have the capacity to anticipate the future—it is the essence of our visions of utopia and hell. *We all have the capacity to adapt to a future that we can partly shape.*

For instance, we can stop trying to "beat nature" and learn to work with it. Individually and collectively, we can work to develop long-term policies that take into account biotic and abiotic limits on population growth. Far from being a surrender, this would be one of the most intelligent behaviors of which we are capable.

Having a capacity to adapt and using it are not the same thing. We have already put the world of life on dangerous ground because we have not yet mobilized ourselves as a species to work toward self-control.

Our survival depends on predicting possible futures. It depends on preserving, restoring, and constructing ecosystems that fit with our definition of basic human values and available biological models. Human values can change; our expectations can and must be adapted to biological reality. *For the principles of energy flow and resource utilization, which govern the survival of all systems of life, do not change.* It is our biological and cultural imperative that we come to terms with these principles, and ask ourselves what our long-term contribution will be to the world of life.

Appendix I. Classification System

This revised classification scheme is a composite of several that microbiologists, botanists, and zoologists use. The major groupings are agreed upon, more or less. However, there is not always agreement on what to name a particular grouping or where it might fit within the overall hierarchy. There are several reasons why full consensus is not possible at this time.

First, the fossil record varies in its completeness and quality. Therefore, the phylogenetic relationship of one group to other groups is sometimes open to interpretation. Today, comparative studies at the molecular level are firming up the picture, but the work is still under way.

Second, ever since the time of Linnaeus, systems of classification have been based on the perceived morphological similarities and differences among organisms. Although some original interpretations are now open to question, we are so used to thinking about organisms in certain ways that reclassification often proceeds slowly.

A few examples: Traditionally, birds and reptiles were grouped in separate classes (Reptilia and Aves); yet there are compelling arguments for grouping the lizards and snakes in one class and the crocodilians, dinosaurs, and birds in another. Many biologists still favor a six-kingdom system of classification (archaea, bacteria, protists, plants, fungi, and animals). Others advocate a switch to the more recently proposed three-domain system (archaea, bacteria, and eukarya).

Third, researchers in microbiology, mycology, botany, zoology, and other fields of inquiry inherited a wealth of literature, based on classification systems that have been developed over time in each field of inquiry. Many do not wish to give up established terminology that offers access to the past.

For example, botanists and microbiologists often use *division*, and zoologists *phylum*, for taxa that are equivalent in hierarchies of classification. As another example, opinions are still polarized with respect to kingdom Protista, certain members of which could easily be grouped with plants, or fungi, or animals.

Indeed, the term "protozoan" is a holdover from an earlier scheme in which some single-celled organisms were ranked as simple animals.

Why bother with classification frameworks if we know they only imperfectly reflect the evolutionary history of life? We do so for the same reasons that a writer might break up a history of civilization into several volumes, each with a number of chapters. Both are efforts to impart structure to an enormous body of knowledge and to facilitate retrieval of information from it. More importantly, to the extent that modern classification schemes accurately reflect evolutionary relationships, they provide the basis for comparitive biological studies, which link all fields of biology.

Bear in mind that we include this appendix for your reference purposes only. Besides being open to revision, it is not meant to be complete. Names shown in "quotes" are polyphyletic or paraphyletic groups that are undergoing revision. For example, "reptiles" comprise at least three and possibly more lineages.

The most recently discovered species, as from the mid-ocean province, are not listed. Many existing and extinct species of the more obscure phyla are also not represented. Our strategy is to focus primarily on the organisms mentioned in the text or familiar to most students. We delve more deeply into flowering plants than to bryophtes, and into chordates than annelids.

PROKARYOTES AND EUKARYOTES COMPARED

As a general frame of reference, note that almost all bacteria and archaea are microscopic in size. Their DNA is concentrated in a nucleoid (a region of cytoplasm), not in a membrane-bound nucleus. All are single cells or simple associations of cells. They reproduce by prokaryotic fission or budding; they transfer genes by bacterial conjugation.

Table A lists representative types of autotrophic and heterotrophic prokaryotes. The authoritative reference, *Bergey's Manual of Systematic Bacteriology*, has called this a time of taxonomic transition. It references groups mainly by numerical taxonomy (Section 19.1) rather than by phylogeny. Our classification system does reflect evidence of evolutionary relationships for at least some bacterial groups.

The first life forms were prokaryotic. Similarities between Bacteria and Archaea have more ancient origins relative to the traits of eukaryotes.

Unlike the prokaryotes, all eukaryotic cells start out life with a DNA-enclosing nucleus and other membrane-bound organelles. Their chromosomes have many histones and other proteins attached. They include spectacularly diverse single-celled and multicelled species, which can reproduce by way of meiosis, mitosis, or both.

DOMAIN BACTERIA — Kingdom Bacteria
DOMAIN ARCHAEA — Kingdom Archaea
DOMAIN EUKARYA — Kingdom Protista, Kingdom Fungi, Kingdom Plantae, Kingdom Animalia

DOMAIN OF BACTERIA

Gram-negative and Gram-positive prokaryotic cells. Collectively, great metabolic diversity; photosynthetic autotrophs, chemosynthetic autotrophs, and heterotrophs.

PHYLUM FIRMICUTES   Typically Gram-positive, thick wall. Heterotrophs. *Bacillus, Staphylococcus, Streptococcus, Clostridium, Actinomycetes.*

PHYLUM GRACILICUTES   Typically Gram-negative, thin wall. Autotrophs (photosynthetic and chemosynthetic) and heterotrophs. *Anabaena* and other cyanobacteria. *Escherichia, Pseudomonas, Neisseria, Myxococcus.*

PHYLUM TENERICUTES Gram-negative, wall absent. Heterotrophs (saprobes, pathogens). *Mycoplasma.*

DOMAIN OF ARCHAEBACTERIA (ARCHAEA)

KINGDOM ARCHAEBACTERIA   Methanogens, extreme halophiles, extreme thermophiles. Evolutionarily closer to eukaryotic cells than to eubacteria. All strict anaerobes living in habitats as harsh as those that probably prevailed on the early Earth. Compared with other prokaryotic cells, all archaebacteria have a distinctive cell wall and unique membrane lipids, ribosomes, and RNA sequences. *Methanobacterium, Halobacterium, Sulfolobus.*

Table A   Representative Bacteria and Archaea Grouped on the Basis of Numerical Taxonomy

Some Major Groups	Main Habitats	Characteristics	Representatives
BACTERIA			
Photoautotrophs:			
Cyanobacteria, green sulfur bacteria, and purple sulfur bacteria	Mostly lakes, ponds; some marine, terrestrial habitats	Photosynthetic; use sunlight energy, carbon dioxide; cyanobacteria use oxygen-producing noncyclic pathway; some also use cyclic route	*Anabaena, Spirulina, Rhodopseudomonas,*
Photoheterotrophs:			
Purple nonsulfur and green nonsulfur bacteria	Anaerobic, organically rich muddy soils, and sediments of aquatic habitats	Use sunlight energy; organic compounds as electron donors; some purple nonsulfur may also grow chemotrophically	*Rhodospirillum, Chlorobium*
Chemoautotrophs:			
Nitrifying, sulfur-oxidizing, and iron-oxidizing bacteria	Soil; freshwater, marine habitats	Use carbon dioxide, inorganic compounds as electron donors; influence crop yields, cycling of nutrients in ecosystems	*Nitrosomonas, Nitrobacter, Thiobacillus*
Chemoheterotrophs:			
Spirochetes	Aquatic habitats; parasites of animals	Helically coiled, motile; free-living and parasitic species; some major pathogens	*Spirochaeta, Treponema*
Gram-negative aerobic rods and cocci	Soil, aquatic habitats; parasites of animals, plants	Some major pathogens; some fix nitrogen (e.g., *Rhizobium*)	*Pseudomonas, Neisseria, Rhizobium, Agrobacterium*
Gram-negative facultative anaerobic rods	Soil, plants, animal gut	Many major pathogens; one bioluminescent (*Photobacterium*)	*Salmonella, Escherichia, Proteus, Photobacterium*
Rickettsias and chlamydias	Host cells of animals	Intracellular parasites; many pathogens	*Rickettsia, Chlamydia*
Myxobacteria	Decaying organic material; bark of living trees	Gliding, rod-shaped; aggregation and collective migration of cells	*Myxococcus*
Gram-positive cocci	Soil; skin and mucous membranes of animals	Some major pathogens	*Staphylococcus, Streptococcus*
Endospore-forming rods and cocci	Soil; animal gut	Some major pathogens	*Bacillus, Clostridium*
Gram-positive nonsporulating rods	Fermenting plant, animal material; gut, vaginal tract	Some important in dairy industry, others major contaminators of milk, cheese	*Lactobacillus, Listeria*
Actinomycetes	Soil; some aquatic habitats	Include anaerobes and strict aerobes; major producers of antibiotics	*Actinomyces, Streptomyces*
ARCHAEA			
Methanogens	Anaerobic sediments of lakes, swamps; animal gut	Chemosynthetic; methane producers; used in sewage treatment facilities	*Methanobacterium*
Extreme halophiles	Brines (extremely salty water)	Heterotrophic; also, unique photosynthetic pigments (bacteriorhodopsin) form in some	*Halobacterium*
Extreme thermophiles	Acidic soil, hot springs, hydrothermal vents	Heterotrophic or chemosynthetic; use inorganic substances as electron donors	*Sulfolobus, Thermoplasma*

DOMAIN OF EUKARYOTES (EUKARYA)

KINGDOM "PROTISTA"   Diverse single-celled, colonial, and multicelled eukaryotic species. Existing types are unlike prokaryotes and most like the earliest forms of eukaryotes. Autotrophs, heterotrophs, or both. Reproduce sexually and asexually (by meiosis, mitosis, or both). Not a monophyletic group. The kingdom may soon be split into multiple kingdoms, and some of its groups (chlorophytes and charophytes) are classified by some botanists as plants.

PHYLUM "MASTIGOPHORA"   Flagellated protozoans. Free-living heterotrophs; many are internal parasites. They have one to several flagella. A non-monophyletic grouping of ancient lineages, including the diplomonads, parabasalids, and kinetoplastids. *Trypanosoma, Trichomonas, Giardia.*

PHYLUM EUGLENOPHYTA   Euglenoids. Mostly heterotrophs, some photoautotrophs, some both depending on conditions. Most with one short, one long flagellum. Pigmented (red, green) or colorless. Related to kinetoplastids. *Euglena.*

PHYLUM SARCODINA   Amoeboid protozoans. Heterotrophs, free-living or endosymbionts, some pathogens. Soft-bodied, with or without shell, pseudopods. Rhizopods (naked amoebas, foraminiferans), actinopods (radiolarians, heliozoans). *Amoeba, Entamoeba.*

ALVEOLATES

PHYLUM CILIOPHORA   Ciliated protozoans. Heterotrophs, predators or symbionts, some parasitic. All have cilia. Free-living, sessile, or motile. *Paramecium.*

PHYLUM APICOMPLEXA   Heterotrophs, sporozoite-forming parasites. Complex structures at head end. Most familiar types known as sporozoans. *Cryptosporidium, Plasmodium, Toxoplasma.*

PHYLUM PYRRHOPHYTA   Dinoflagellates. Photosynthetic, mostly, but some heterotrophs. *Noctiluca, Karenia brevis.*

STRAMENOPILES

PHYLUM OOMYCOTA   Water molds. Heterotrophs. Decomposers, some parasites. *Saprolegnia, Phytophthora, Plasmopara.*

PHYLUM CHRYSOPHYTA   Golden algae, yellow-green algae, diatoms, coccolithophores. Photosynthetic. Some flagellated. *Mischococcus, Synura, Vaucheria.*

PHYLUM PHAEOPHYTA   Brown algae. Photosynthetic, nearly all endemic to temperate or marine waters. *Macrocystis, Laminaria, Sargassum, Postelsia.*

GROUPS CLOSELY RELATED TO PLANTS

PHYLUM RHODOPHYTA   Red algae. Mostly photosynthetic, some parasitic. Nearly all marine, some in freshwater habitats. *Porphyra, Antithamion.*

PHYLUM CHLOROPHYTA   Green algae. Mostly photosynthetic, some parasitic. Most freshwater, some marine or terrestrial. *Chlamydomonas, Spirogyra, Ulva, Volvox, Codium, Halimeda.*

PHYLUM CHAROPHYTA   Closest relatives of plants. Desmids, stoneworts. *Micrasterias, Chara.*

GROUPS OF SLIME MOLDS

PHYLUM MYXOMYCOTA   Plasmodial slime molds. Heterotrophs. A multinucleated mass (the plasmodium) arises by mitosis without cell division, feeds and migrates as a unit, then forms spore-bearing structures. *Physarum.*

PHYLUM ACRASIOMYCOTA   Cellular slime molds. Heterotrophs; free-living, phagocytic amoeboid cells aggregate into a mass that migrate as a unit, then form spore-bearing structures. *Dictyostelium.*

FUNGI   Nearly all multicelled eukaryotic species with chitin-containing cell walls. Heterotrophs, mostly saprobic decomposers, some parasites. Nutrition based upon extracellular digestion of organic matter and absorption of nutrients by individual cells. Multicelled species form absorptive mycelia within substrates and structures that produce asexual spores (and sometimes sexual spores).

PHYLUM CHYTRIDIOMYCOTA   Chytrids. Primarily aquatic; saprobic decomposers or parasites that produce flagellated spores. *Chytridium.*

PHYLUM ZYGOMYCOTA   Zygomycetes. Producers of zygospores (zygotes inside thick wall) by way of sexual reproduction. Bread molds, related forms. *Rhizopus, Philobolus.*

PHYLUM ASCOMYCOTA   Ascomycetes. Sac fungi. Sac-shaped cells form sexual spores (ascospores). Most yeasts and molds, morels, truffles. *Saccharomycetes, Morchella, Neurospora, Sarcoscypha, Claviceps, Ophiostoma, Candida, Aspergillus, Penicillium.*

PHYLUM BASIDIOMYCOTA   Basidiomycetes. Club fungi. Most diverse group. Produce basidiospores inside club-shaped structures. Mushrooms, shelf fungi, stinkhorns. *Agaricus, Amanita, Craterellus, Gymnophilus, Puccinia, Ustilago.*

"IMPERFECT FUNGI"   Sexual spores absent or undetected. The group has no formal taxonomic status. If better understood, a given species might be grouped with sac fungi or club fungi. *Arthobotrys, Histoplasma, Microsporum, Verticillium.*

"LICHENS"   Mutualistic interactions between fungal species and a cyanobacterium, green alga, or both. *Lobaria, Usnea.*

KINGDOM PLANTAE   Multicelled eukaryotes. Nearly all photosynthetic autotrophs with chlorophylls *a* and *b*. Some parasitic. Nonvascular and vascular species, generally with well-developed root and shoot systems. Nearly all adapted to survive on land; a few in aquatic habitats. Sexual reproduction predominant with spore-forming chambers and embryos in life cycle; also asexual reproduction by vegetative propagation and other mechanisms.

PHYLUM "BRYOPHYTA"   Bryophytes; mosses, liverworts, hornworts. Not a monophyletic group. Seedless, nonvascular, haploid dominance, sperm are flagellated; require water for fertilization. *Marchantia, Polytrichum, Sphagnum.*

SEEDLESS VASCULAR PLANTS

Diploid sporophyte dominates, flagellated sperm require water for fertilization.

PHYLUM "RHYNIOPHYTA"   Earliest known vascular plants; muddy habitats. A polyphyletic group, some are primitive lycophytes. Extinct. *Cooksonia, Rhynia.*

PHYLUM LYCOPHYTA   Lycophytes, club mosses. Small single-veined leaves, branching rhizomes. *Lepidodendron* (extinct), *Lycopodium, Selaginella.*

PHYLUM SPHENOPHYTA   Horsetails. Reduced scalelike leaves. Some stems photosynthetic, others nonphotosynthetic, spore-producing. *Calamites* (extinct), *Equisetum.*

PHYLUM PTEROPHYTA   Ferns. Large leaves, usually with sori. Largest group of seedless vascular plants (12,000 species), mainly tropical, temperate habitats. *Pteris, Trichomanes, Cyathea* (tree ferns), *Polystichum.*

PHYLUM PSILOPHYTA   Whisk ferns. No obvious roots, leaves on sporophyte, very reduced. *Psilotum.*

PHYLUM "PROGYMNOSPERMOPHYTA"   The progymnosperms. Ancestral to early seed-bearing plants; extinct. *Archaeopteris.*

SEED-BEARING VASCULAR PLANTS

PHYLUM "PTERIDOSPERMOPHYTA" Seed ferns. Fernlike gymnosperms; extinct. *Medullosa*.

PHYLUM CYCADOPHYTA Cycads. Group of gymnosperms (vascular, bear "naked" seeds). Tropical, subtropical. Compound leaves, simple cones on male and female plants. Plants usually palm-like. *Zamia, Cycas*.

PHYLUM GINKGOPHYTA Ginkgo (maidenhair tree). Type of gymnosperm. Seeds with fleshy outer layer. *Ginkgo*.

PHYLUM GNETOPHYTA Gnetophytes. Only gymnosperms with vessels in xylem and double fertilization (but endosperm does not form). *Ephedra, Welwitschia, Gnetum*.

PHYLUM CONIFEROPHYTA Conifers. Most common and familiar gymnosperms. Generally cone-bearing species with needle-like or scale-like leaves.

Family Pinaceae. Pines (*Pinus*), firs (*Abies*), spruces (*Picea*), hemlock (*Tsuga*), larches (*Larix*), true cedars (*Cedrus*).

Family Cupressaceae. Junipers (*Juniperus*), Cypresses (*Cupressus*), Bald cypress (*Taxodium*), redwood (*Sequoia*), bigtree (*Sequoiadendron*), dawn redwood (*Metasequoia*).

Family Taxaceae. Yews. *Taxus*.

PHYLUM ANTHOPHYTA Angiosperms (the flowering plants). Largest, most diverse group of vascular seed-bearing plants. Only organisms that produce flowers, fruits. Some families from several representative orders are listed:

BASAL FAMILIES

Family Amborellaceae. *Amborella*.
Family Nymphaeaceae. Water lilies.
Family Illiciaceae. Star anise.

MAGNOLIIDS

Family Magnoliaceae. Magnolias.
Family Lauraceae. Cinnamon, sassafras, avocados.
Family Piperaceae. Black pepper, white pepper.

EUDICOTS

Family Papaveraceae. Poppies.
Family Cactaceae. Cacti.
Family Euphorbiaceae. Spurges, poinsettia.
Family Salicaceae. Willows, poplars.
Family Fabaceae. Peas, beans, lupines, mesquite.
Family Rosaceae. Roses, apples, almonds, strawberries.
Family Moraceae. Figs, mulberries.
Family Cucurbitaceae. Squashes, melons, cucumbers.
Family Fagaceae. Oaks, chestnuts, beeches.
Family Brassicaceae. Mustards, cabbages, radishes.
Family Malvaceae. Mallows, okra, cotton, hibiscus, cocoa.
Family Sapindaceae. Soapberry, litchi, maples.
Family Ericaceae. Heaths, blueberries, azaleas.
Family Rubiaceae. Coffee.
Family Lamiaceae. Mints.
Family Solanaceae. Potatoes, eggplant, petunias.
Family Apiaceae. Parsleys, carrots, poison hemlock.
Family Asteraceae. Composites. Chrysanthemums, sunflowers, lettuces, dandelions.

MONOCOTS

Family Araceae. Anthuriums, calla lily, philodendrons.
Family Liliaceae. Lilies, tulips.
Family Alliaceae. Onions, garlic.
Family Iridaceae. Irises, gladioli, crocuses.
Family Orchidaceae. Orchids.
Family Arecaceae. Date palms, coconut palms.
Family Bromeliaceae. Bromeliads, pineapples.
Family Cyperaceae. Sedges.
Family Poaceae. Grasses, bamboos, corn, wheat, sugarcane.
Family Zingiberaceae. Gingers.

KINGDOM ANIMALIA

Multicelled eukaryotes, nearly all with tissues, organs, and organ systems; show motility during at least part of their life cycle; embryos develop through a series of stages. Diverse heterotrophs, predators (herbivores, carnivores, omnivores), parasites, detritivores. Reproduce sexually and, in many species, asexually as well.

PHYLUM PORIFERA Sponges. No symmetry, tissues. *Euplectella*.

PHYLUM PLACOZOA Marine. Simplest known animal. Two cell layers, no mouth, no organs. *Trichoplax*.

PHYLUM CNIDARIA Radial symmetry, tissues, nematocysts.
Class Hydrozoa. Hydrozoans. *Hydra, Obelia, Physalia, Prya*.
Class Scyphozoa. Jellyfishes. *Aurelia*.
Class Anthozoa. Sea anemones, corals. *Telesto*.

PHYLUM MESOZOA Ciliated, wormlike parasites, about the same level of complexity as *Trichoplax*.

PHYLUM PLATYHELMINTHES Flatworms. Bilateral, cephalized; simplest animals with organ systems. Saclike gut.
Class Turbellaria. Triclads (planarians), polyclads. *Dugesia*.
Class Trematoda. Flukes. *Clonorchis, Schistosoma*.
Class Cestoda. Tapeworms. *Diphyllobothrium, Taenia*.

PHYLUM ROTIFERA Rotifers. *Asplancha, Philodina*.

PHYLUM NEMERTEA Ribbon worms. *Tubulanus*.

PHYLUM MOLLUSCA Mollusks.
Class Polyplacophora. Chitons. *Cryptochiton, Tonicella*.

Class Gastropoda. Snails (periwinkles, whelks, limpets, abalones, cowries, conches, nudibranchs, tree snails, garden snails), sea slugs, land slugs. *Aplysia, Ariolimax, Cypraea, Haliotis, Helix, Liguus, Limax, Littorina, Patella*.

Class Bivalvia. Clams, mussels, scallops, cockles, oysters, shipworms. *Ensis, Chlamys, Mytelus, Patinopectin*.

Class Cephalopoda. Squids, octopuses, cuttlefish, nautiluses. *Dosidiscus, Loligo, Nautilus, Octopus, Sepia*.

PHYLUM BRYOZOA Bryozoans (moss animals).

PHYLUM BRACHIOPODA Lampshells.

PHYLUM ANNELIDA Segmented worms.
Class Polychaeta. Mostly marine worms. *Eunice, Neanthes*.
Class Oligochaeta. Mostly freshwater and terrestrial worms, many marine. *Lumbricus* (earthworms), *Tubifex*.
Class Hirudinea. Leeches. *Hirudo, Placobdella*.

PHYLUM NEMATODA Roundworms. *Ascaris, Caenorhabditis elegans, Necator* (hookworms), *Trichinella*.

PHYLUM TARDIGRADA Water bears.

PHYLUM ONYCHOPHORA Onychophorans. *Peripatus*.

PHYLUM ARTHROPODA
Subphylum Trilobita. Trilobites; extinct.
Subphylum Chelicerata. Chelicerates. Horseshoe crabs, spiders, scorpions, ticks, mites.
Subphylum Crustacea. Shrimps, crayfishes, lobsters, crabs, barnacles, copepods, isopods (sowbugs).
Subphylum Uniramia.
Superclass Myriapoda. Centipedes, millipedes.
Superclass Insecta.
Order Ephemeroptera. Mayflies.
Order Odonata. Dragonflies, damselflies.
Order Orthoptera. Grasshoppers, crickets, katydids.
Order Dermaptera. Earwigs.
Order Blattodea. Cockroaches.

Order Mantodea. Mantids.
Order Isoptera. Termites.
Order Mallophaga. Biting lice.
Order Anoplura. Sucking lice.
Order Hemiptera. Cicadas, aphids, leafhoppers, spittlebugs, bugs.
Order Coleoptera. Beetles.
Order Diptera. Flies.
Order Mecoptera. Scorpion flies. *Harpobittacus.*
Order Siphonaptera. Fleas.
Order Lepidoptera. Butterflies, moths.
Order Hymenoptera. Wasps, bees, ants.
Order Neuroptera. Lacewings, antlions.

PHYLUM ECHINODERMATA   Echinoderms.

Class Asteroidea. Sea stars. *Asterias.*
Class Ophiuroidea. Brittle stars.
Class Echinoidea. Sea urchins, heart urchins, sand dollars.
Class Holothuroidea. Sea cucumbers.
Class Crinoidea. Feather stars, sea lilies.
Class Concentricycloidea. Sea daisies.

PHYLUM HEMICHORDATA   Acorn worms.

PHYLUM CHORDATA   Chordates.

Subphylum Urochordata. Tunicates, related forms.

Subphylum Cephalochordata. Lancelets.

CRANIATES

Superclass "Agnatha." Jawless fishes, including ostracoderms (extinct).

Class Myxini. Hagfishes.

Class Cephalaspidomorphi. Lampreys.

Subphylum Vertebrata. Jawed vertebrates.

Class "Placodermi." Jawed, heavily armored fishes; extinct.

Class Chondrichthyes. Cartilaginous fishes. (sharks, rays, skates, chimaeras).

Class "Osteichthyes." Bony fishes. Not monophyletic.
Subclass Dipnoi. Lungfishes.
Subclass Crossopterygii. Coelacanths, related forms.
Subclass Actinopterygii. Ray-finned fishes.
Order Acipenseriformes. Sturgeons, paddlefishes.
Order Salmoniformes. Salmon, trout.
Order Atheriniformes. Killifishes, guppies.
Order Gasterosteiformes. Seahorses.
Order Perciformes. Perches, wrasses, barracudas, tunas, freshwater bass, mackerels.
Order Lophiiformes. Angler fishes.

TETRAPODS (A subgroup of craniates)

Class Amphibia. Amphibians.
Order Caudata. Salamanders.
Order Anura. Frogs, toads.
Order Apoda. Apodans (caecilians).

AMNIOTES (A subgroup of tetrapods)

Class "Reptilia." Skin with scales, embryo protected and nutritionally supported by extraembryonic membranes.
Subclass Anapsida. Turtles, tortoises.
Subclass Lepidosaura. *Sphenodon*, lizards, snakes.
Subclass Archosaura. Dinosaurs (extinct), crocodiles, alligators.

Class Aves. Birds. In some of the more recent classification systems, dinosaurs, crocodilians, and birds are grouped in the same category, the archosaurs.
Order Struthioniformes. Ostriches.
Order Sphenisciformes. Penguins.
Order Procellariiformes. Albatrosses, petrels.
Order Ciconiiformes. Herons, bitterns, storks, flamingoes.
Order Anseriformes. Swans, geese, ducks.
Order Falconiformes. Eagles, hawks, vultures, falcons.
Order Galliformes. Ptarmigan, turkeys, domestic fowl.
Order Columbiformes. Pigeons, doves.
Order Strigiformes. Owls.
Order Apodiformes. Swifts, hummingbirds.
Order Passeriformes. Sparrows, jays, finches, crows, robins, starlings, wrens.
Order Piciformes. Woodpeckers, toucans.
Order Psittaciformes. Parrots, cockatoos, macaws.

Class Mammalia. Skin with hair; young nourished by milk-secreting glands of adult.

Subclass Prototheria. Egg-laying mammals (monotremes; duckbilled platypus, spiny anteaters).

Subclass Metatheria. Pouched mammals or marsupials (opossums, kangaroos, wombats, Tasmanian devil).

Subclass Eutheria. Placental mammals.
Order Edentata. Anteaters, tree sloths, armadillos.
Order Insectivora. Tree shrews, moles, hedgehogs.
Order Chiroptera. Bats.
Order Scandentia. Insectivorous tree shrews.
Order Primates.
Suborder Strepsirhini (prosimians). Lemurs, lorises.
Suborder Haplorhini (tarsioids and anthropoids).
Infraorder Tarsiiformes. Tarsiers.
Infraorder Platyrrhini (New World monkeys).
Family Cebidae. Spider monkeys, howler monkeys, capuchin.
Infraorder Catarrhini (Old World monkeys and hominoids).
Superfamily Cercopithecoidea. Baboons, macaques, langurs.
Superfamily Hominoidea. Apes and humans.
Family Hylobatidae. Gibbon.
Family "Pongidae." Chimpanzees, gorillas, orangutans.
Family Hominidae. Existing and extinct human species (*Homo*) and humanlike species, including the australopiths.
Order Lagomorpha. Rabbits, hares, pikas.
Order Rodentia. Most gnawing animals (squirrels, rats, mice, guinea pigs, porcupines, beavers, etc.).
Order Carnivora. Carnivores.
Suborder Feloidea. Cats, mongooses, hyenas.
Suborder Canoidea. Dogs, weasels, skunks, otters, raccoons, pandas, bears.
Order Pinnipedia. Seals, walruses, sea lions.
Order Proboscidea. Elephants; mammoths (extinct).
Order Sirenia. Sea cows (manatees, dugongs).
Order Perissodactyla. Odd-toed ungulates (horses, tapirs, rhinos).
Order Tubulidentata. African aardvarks.
Order Artiodactyla. Even-toed ungulates (camels, deer, bison, sheep, goats, antelopes, giraffes, etc.).
Order Cetacea. Whales, porpoises.

Appendix II. Units of Measure

Metric-English Conversions

Length

English		Metric
inch	=	2.54 centimeters
foot	=	0.30 meter
yard	=	0.91 meter
mile (5,280 feet)	=	1.61 kilometer

To convert	multiply by	to obtain
inches	2.54	centimeters
feet	30.00	centimeters
centimeters	0.39	inches
millimeters	0.039	inches

Weight

English		Metric
grain	=	64.80 milligrams
ounce	=	28.35 grams
pound	=	453.60 grams
ton (short) (2,000 pounds)	=	0.91 metric ton

To convert	multiply by	to obtain
ounces	28.3	grams
pounds	453.6	grams
pounds	0.45	kilograms
grams	0.035	ounces
kilograms	2.2	pounds

Volume

English		Metric
cubic inch	=	16.39 cubic centimeters
cubic foot	=	0.03 cubic meter
cubic yard	=	0.765 cubic meters
ounce	=	0.03 liter
pint	=	0.47 liter
quart	=	0.95 liter
gallon	=	3.79 liters

To convert	multiply by	to obtain
fluid ounces	30.00	milliliters
quart	0.95	liters
milliliters	0.03	fluid ounces
liters	1.06	quarts

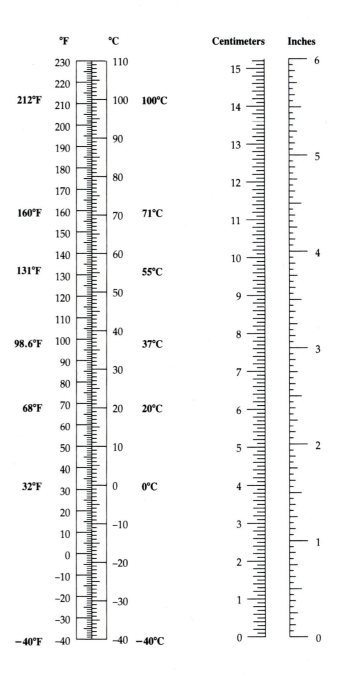

Appendix III. Answers to Self-Quizzes

CHAPTER 1
1. cell — 1.1
2. Metabolism — 1.2
3. Homeostasis — 1.2
4. adaptive — 1.4
5. mutation — 1.4
6. domains — 1.3
7. a — 1.2
8. d — 1.2, 1.4
9. d — 1.4
10. a — 1.6
11. c — 1.4
  e — 1.4
  d — 1.5
  b — 1.5
  a — 1.5

CHAPTER 2
1. F; hydrogen atoms have no neutrons — 2.1
2. b — 2.1
3. d — 2.2
4. c — 2.4
5. a — 2.4
6. f — 2.5
7. f — 2.6
8. acid, base — 2.6
9. c — 2.6
10. e — Impacts, Issues
  d — 2.6
  b — 2.4
  c — 2.5
  a — 2.1

CHAPTER 3
1. See table 3.1
2. d — 3.1
3. c — 3.1
4. f — 3.3
5. b — 3.4
6. d — 3.4
7. e — 3.4
8. d — 3.5, 3.7
9. d — 3.5
10. d — 3.7
11. b — 3.7
12. c — 3.5
  e — 3.7
  b — 3.4
  d — 3.7
  a — 3.3

CHAPTER 4
1. c — 4.3
2. See Figs. 4.19, 4.20
3. d — 4.5
4. d — 4.11
5. False; many cells have a wall — 4.11
6. c — 4.1
7. e — 4.8
  d — 4.8
  a — 4.1, 4.4
  b — 4.7
  c — 4.7

CHAPTER 5
1. c — 5.1
2. b — 5.2
3. d — 5.3
4. a — 5.7
5. d — 5.5
6. b — 5.5
7. d — 5.8
8. d — 5.5
9. c — 5.2
  g — 5.6
  a — 5.2
  d — 5.2
  e — 5.2
  b — 5.2
  f — 5.1

CHAPTER 6
1. carbon dioxide; sunlight — Impacts, Issues
2. a — 6.2, 6.3
3. b — 6.2, 6.5
4. c — 6.3
5. c — 6.2, 6.5
6. b — 6.5
7. c — 6.5
  a — 6.3, 6.4
  b — 6.5

CHAPTER 7
1. d — 7.1
2. c — 7.1, 7.2
3. b — 7.3
4. c — 7.4
5. See Fig. 7.3
6. c — 7.5
7. b — 7.5
8. d — 7.6
9. b — 7.2
  c — 7.5
  a — 7.3
  d — 7.4

CHAPTER 8
1. d — 8.1
2. b — 8.1
3. c — 8.1
4. d — 8.2
5. a — 8.2
6. c — 8.2
7. a — 8.3
8. b — 8.2
9. d — 8.3
  b — 8.3
  c — 8.3
  a — 8.3

CHAPTER 9
1. c — 9.2
2. b — 9.2
3. a — 9.2
4. d — 9.1, 9.5
5. d — 9.2, 9.4
6. b — 9.2
7. d — 9.2
8. c — 9.3
9. d — 9.3
10. d — 9.2
  a — 9.1
  c — 9.3
  b — 9.2

CHAPTER 10
1. a — 10.1
2. b — 10.1, 10.2
3. a — 10.1, 10.6
4. b — 10.1, 10.2
5. c — 10.2
6. a — 10.2, 10.4
7. d — 10.3
8. b — 10.3
  d — 10.2
  a — 10.1
  c — 10.1

CHAPTER 11
1. c — 11.1, 11.3
2. a — 11.3
3. e — 11.7
4. a — 11.4
5. c — 11.1
6. d — 11.8
7. c — 11.8
8. c — 11.3
  e — 11.7
  d — 11.4
  b — 11.7
  a — 11.2
  f — 11.3

CHAPTER 12
1. c — 12.2
2. d — 12.2
3. c — 12.2
4. a — 12.4
5. d — 12.4
6. c — 12.5

CHAPTER 13
1. c — 13.1
2. b — 13.1
3. c — 13.1
4. c — 13.2
5. a — 13.2
6. e — 13.4
  c — 13.3
  a — 13.1
  f — 13.2
  d — 13.2
  g — 13.1
  b — 13.2

CHAPTER 14
1. d — 14.1, 14.3
2. d — 14.1
3. d — 14.3, 14.5
4. d — 14.3
5. d — 14.3
6. d — 14.1, 14.2, 14.3
7. a — 14.2
8. d — 14.2
9. a — 14.1, 14.2, 14.3
10. h — 14.3
11. d — 14.3
12. b — 14.4
13. c — 14.4
14. e — 14.4, 14.5
  a — 14.2
  d — Impacts, Issues
  b — 14.3
  c — 14.4

CHAPTER 15
1. d — 15.7
2. c — 15.2
3. plasmid — 15.2
4. b — 15.2
5. a — 15.3
6. b — 15.4
7. c — 15.5
8. b — 15.7, 15.8
9. d — 15.4
  c — 15.6
  b — Impacts, Issues
  e — 15.3
  a — 15.7

CHAPTER 16
1. populations — 16.5
2. b — Impacts, Issues
3. a — 16.5
4. c — 16.3
5. b — 16.7
6. c — 16.8
7. b — 16.11
8. c — 16.11
  d — 16.3
  a — 16.5
  b — 16.10

CHAPTER 17
1. a — 17.4
2. d — 17.7
3. c — 17.10
4. c — 17.11
5. c — 17.5
6. d — 17.7
7. d — 17.11
8. c — 17.11
9. d — 17.11
10. See Fig. 17.31
11. e — 17.11
  a — 17.1
  f — 17.1
  c — 17.4
  b — 17.11
  d — 17.4
  g — 17.10

CHAPTER 18
1. c — 18.1
2. c — 18.2
3. c — 18.3
4. b — 18.4
5. f — 18.1
  c — 18.2
  d — 18.2
  a — 18.3
  b — 18.4
  e — 18.4

CHAPTER 19
1. Left, head (top), sheath (middle), tail fiber (bottom). Right: capsid (left), envelope (right) — 19.4
2. d — 19.1
3. c — 19.3
4. c — 19.1
5. c — 19.2
6. d — 19.2
7. b — 19.2
8. d — 19.2
9. b — 19.4
10. false — 19.4
11. d — 19.4
12. d — 19.1
13. d — 19.3
  e — 19.2
  b — 19.4
  f — 19.1
  c — 19.3
  g — 19.4
  a — 19.5

CHAPTER 20
1. b — 20.2
2. d — 20.2
3. b — 20.3
4. b — 20.5
5. d — 20.5
6. d — 20.4
7. a — 20.6
8. c — 20.7
9. a — 20.8
10. c — 20.9
11. c — 20.9
12. b — 20.2
  e — 20.2
  f — 20.5
  i — 20.3
  h — 20.5
  j — 20.7
  d — 20.5
  a — 20.10
  c — 20.12
  g — 20.12

CHAPTER 21
1. a — 21.3
2. c — 21.5
3. b — 21.2
4. c — 21.3
5. a — 21.3
6. b — 21.5
7. c — 21.6
  e — 21.1
  g — 21.3
  h — 21.5
  f — 21.2
  a — 21.1
  b — 21.1
  d — 21.7

CHAPTER 22
1. d — 22.1
2. a — 22.1
3. a — 22.4
4. a — 22.5
5. b — 22.12
6. c — 22.5, 22.6
7. b — 22.1, 22.13
8. b — 22.13
9. i — 22.2
  c — 22.3
  h — 22.4
  b — 22.5
  a — 22.8
  f — 22.6
  d — 22.9
  e — 22.7
  g — 22.13

CHAPTER 23

1.	b	*23.1*
2.	a	*23.2*
3.	d	*23.3, 23.4*
4.	c	*23.6*
5.	f	*23.6, 23.7*
6.	a	*23.7*
7.	c	*23.8*
8.	f	*23.2, 23.6*
		23.9, 23.11
9.	c	*23.12*
10.	g	*23.3*
	a	*23.4*
	e	*23.7*
	b	*23.8*
	c	*23.9*
	f	*23.9*
	d	*23.12*

CHAPTER 24

1.	a	*24.1*
2.	c	*24.1*
3.	d	*24.2*
4.	d	*24.5*
5.	b	*24.4*
	c	*24.1*
	a	*24.5*
	d	*24.3*

CHAPTER 25

1.	d	*25.1*
2.	c	*25.2, 25.3*
3.	c	*25.2, 25.4*
4.	c	*25.5*
5.	d	*25.7*
6.	c	*25.9*
7.	c	*25.9*
8.	c	*25.14*
9.	a	*25.14*
10.	e	*25.13*
11.	c	*25.2*
	d	*25.2, 25.11*
	a	*25.2, 25.3*
	b	*25.2, 25.15*
	f	*25.2, 25.4*
	e	*25.2, 25.4*
12.	c	*25.6*
	h	*25.8*
	a	*25.16*
	g	*25.8, 25.19*
	d	*25.9, 25.10*
	e	*25.8, 25.11*
	b	*25.6, 25.7*
	f	*25.8*

CHAPTER 26

1.	f	*26.1*
2.	a	*26.3*
3.	a	*26.3*
4.	d	*26.4*
5.	d	*26.5*
6.	c	*26.7*
7.	c	*26.4*
	d	*26.3*
	a	*26.3*
	e	*26.4*
	b	*26.4*

CHAPTER 27

1.	d	*27.1*
2.	d	*27.1*
3.	d	*27.3*
4.	b	*27.4*
5.	d	*27.6*
6.	d	*27.8*
7.	e	*27.11*
8.	b	*27.12*
9.	c	*27.9*
	d	*27.11*
	a	*27.8*
	e	*27.8*
	b	*27.9*
	g	*27.12*
	f	*27.13*

CHAPTER 28

1.	d	*28.1*
2.	d	*28.1*
3.	d	*28.4*
4.	d	*28.3*
5.	a	*28.7*
6.	b	*28.8*
7.	d	*28.11*
8.	c	*28.10, 28.11*
9.	a	*28.10*
10.	d	*28.1*
	a	*28.1*
	c	*28.1*
	b	*28.1*

CHAPTER 29

1.	d	*29.1*
2.	c	*29.2*
3.	d	*29.3*
4.	a	*29.3*
5.	c	*29.2*
6.	e	*29.4*
7.	d	*29.4, 29.8*
8.	c	*29.5*
9.	d	*29.9, 29.10*
10.	d	*29.7*
	h	*29.5*
	e	*29.5*
	c	*29.5*
	b	*29.10*
	g	*29.6*
	a	*29.6*
	f	*29.12*

CHAPTER 30

1.	a	*30.3*
2.	d	*30.1*
3.	d	*30.6*
4.	a	*30.3*
5.	d	*30.7*
6.	c	*30.7*
7.	c	*30.1*
	d	*30.3*
	b	*30.1, 30.2*
	a	*30.2*
	e	*30.4*

Appendix IV. Answers to Genetics Problems

CHAPTER 10

1. a. Both parents are heterozygotes (*Aa*). Their children may be albino (*aa*) or unaffected (*AA* or *Aa*).

b. All are homozygous recessive (*aa*).

c. Homozygous recessive (*aa*) father, and heterozygous (*Aa*) mother. The albino child is *aa*, the unaffected children *Aa*.

2. a. *AB*   **c.** *Ab, ab*

b. *AB, aB*   **d.** *AB, Ab, aB, ab*

3. a. All offspring will be *AaBB*.

b. 1/4 *AABB* (25% each genotype)
1/4 *AABb*
1/4 *AaBB*
1/4 *AaBb*

c. 1/4 *AaBb* (25% each genotype)
1/4 *Aabb*
1/4 *aaBb*
1/4 *aabb*

d. 1/16 *AABB* (6.25%)
1/8 *AaBB* (12.5%)
1/16 *aaBB* (6.25%)
1/8 *AABb* (12.5%)
1/4 *AaBb* (25%)
1/8 *aaBb* (12.5%)
1/16 *AAbb* (6.25%)
1/8 *Aabb* (12.5%)
1/16 *aabb* (6.25%)

4. A mating of two M^L cats yields 1/4 MM, 1/2 M^LM, and 1/4 $M^L M^L$. Because $M^L M^L$ is lethal, the probability that any one kitten among the survivors will be heterozygous is 2/3.

5. Yellow is recessive. Because F_1 plants have a green phenotype and must be heterozygous, green must be dominant over the recessive yellow.

6. a. *ABC*

b. *ABc, aBc*

c. *ABC, aBc, ABc, aBc*

d. *ABC, aBC, AbC, abC, ABc, aBc, Abc, abc*

7. Because all F_1 plants of this dihybrid cross had to be heterozygous for both genes, then 1/4 (25%) of the F_2 plants will be heterozygous for both genes.

8. a. The mother must be heterozygous $I^A i$. The male with type B blood could have fathered the child if he were heterozygous $I^B i$.

b. Genotype alone cannot prove the accused male is the father. Even if he happens to be heterozygous, *any* male who carries the *i* allele could be the father, including those heterozygous for type A blood ($I^A i$) or type B blood ($I^B i$) and those with type O blood (*ii*).

9. A mating between a mouse from a true-breeding, white-furred strain and a mouse from a true-breeding, brown-furred strain would provide you with the most direct evidence.

Because true-breeding strains of organisms typically are homozygous for a trait being studied, all F_1 offspring from this mating should be heterozygous. Record the phenotype of each F_1 mouse, then let them mate with one another. Assuming only one gene locus is involved, these are possible outcomes for the F_2 offspring:

a. All F_1 mice are brown, and their F_2 offspring segregate:
3 brown : 1 white.

Conclusion: Brown is dominant to white.

b. All F_1 mice are white, and their F_2 offspring segregate:
3 white : 1 brown.

Conclusion: White is dominant to brown.

c. All F_1 mice are tan, and the F_2 offspring segregate:
1 brown : 2 tan : 1 white.

Conclusion: The alleles at this locus show incomplete dominance.

10. You cannot guarantee that the puppies will not develop the disorder without more information about Dandelion's genotype. You could do so only if she is a heterozygous carrier, if the male is free of the alleles, and if the alleles are recessive.

11. Fred could use a testcross to find out if his pet's genotype is *WW* or *Ww*. He can let his black guinea pig mate with a white guinea pig having the genotype *ww*.

If any F_1 offspring are white, then the genotype of his pet is *Ww*. If the two guinea pig parents are allowed to mate repeatedly and all the offspring of the matings are black, then there is a high probability that his pet guinea pig is *WW*.

(If, say, ten offspring are all black, then the probability that the male is *WW* is about 99.9 percent. The greater the number of offspring, the more confident Fred can be of his conclusion.)

12. a. 1/2 red, 1/2 pink

b. All pink

c. 1/4 red, 1/2 pink, 1/4 white

d. 1/2 pink, 1/2 white

13. 9/16 walnut comb
3/16 rose comb
3/16 pea comb
1/16 single comb

14. Because both parents are heterozygotes ($Hb^A Hb^S$), the following are the probabilities for each child:

a. 1/4 $Hb^S Hb^S$

b. 1/4 $Hb^A Hb^A$

c. 1/2 $Hb^A Hb^S$

CHAPTER 11

1. a. Human males (XY) inherit their X chromosome from their mother.

b. A male can produce two kinds of gametes. Half carry an X chromosome and half carry a Y chromosome. All the gametes that carry the X chromosome carry the same X-linked allele.

c. A female homozygous for an X-linked allele produces only one kind of gamete.

d. Half of the gametes of a female who is heterozygous for an X-linked allele carry one of the two alleles at that locus; the other half carry its partner allele for that locus.

2. If only one parent is heterozygous for the autosomal dominant allele, the chance of a child inheriting that allele is 50 percent.

3. a. Nondisjunction can occur at anaphase I or anaphase II of meiosis.

b. As a result of translocation, chromosome 21 may get attached to the end of chromosome 14. The new individual's chromosome number would still be 46, but its somatic cells would have the translocated chromosome 21 in addition to two normal chromosomes 21.

4. Because females (XX) could be white-eyed, the recessive allele had to be on one of their X chromosomes. What if white-eyed males (XY) had the recessive allele on their X chromosome and their Y chromosome had no corresponding eye-color allele? In that case, they would have white eyes. They would have no dominant allele to mask the effect of the recessive one.

5. Because the phenotype appeared in every generation shown in the diagram, this must be a pattern of autosomal dominant inheritance.

However, if a son bears the allele for the disorder on his X chromosome, then it will be expressed. He will develop the disorder, and most likely he will not father children because of his early death.

6. A daughter could develop this muscular dystrophy only if she were to inherit two X-linked recessive alleles—one from each parent. Males who carry the allele are unlikely to father children because they develop the disorder and die an early death.

7. In the mother, a crossover between the two genes at meiosis generates an X chromosome that carries neither mutant allele.

Appendix V. Closer Look at Some Major Metabolic Pathways

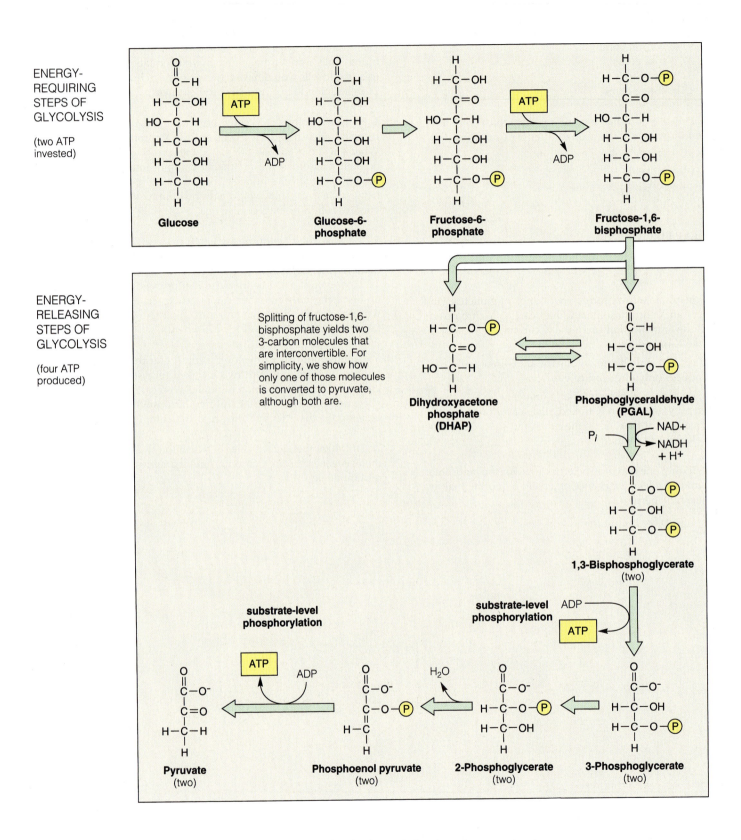

Figure A Glycolysis, ending with two 3-carbon pyruvate molecules for each 6-carbon glucose molecule entering the reactions. The *net* energy yield is two ATP molecules (two invested, four produced).

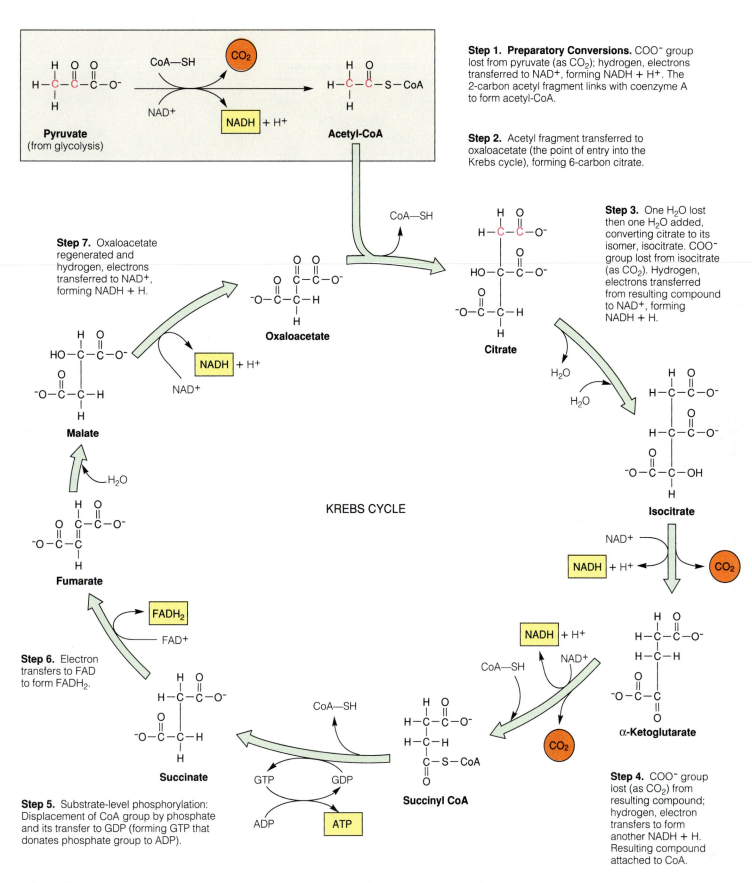

Step 1. Preparatory Conversions. COO^- group lost from pyruvate (as CO_2); hydrogen, electrons transferred to NAD^+, forming $NADH + H^+$. The 2-carbon acetyl fragment links with coenzyme A to form acetyl-CoA.

Step 2. Acetyl fragment transferred to oxaloacetate (the point of entry into the Krebs cycle), forming 6-carbon citrate.

Step 3. One H_2O lost then one H_2O added, converting citrate to its isomer, isocitrate. COO^- group lost from isocitrate (as CO_2). Hydrogen, electrons transferred from resulting compound to NAD^+, forming $NADH + H$.

Step 4. COO^- group lost (as CO_2) from resulting compound; hydrogen, electron transfers to form another $NADH + H$. Resulting compound attached to CoA.

Step 5. Substrate-level phosphorylation: Displacement of CoA group by phosphate and its transfer to GDP (forming GTP that donates phosphate group to ADP).

Step 6. Electron transfers to FAD to form $FADH_2$.

Step 7. Oxaloacetate regenerated and hydrogen, electrons transferred to NAD^+, forming $NADH + H$.

Pyruvate (from glycolysis)

Acetyl-CoA

KREBS CYCLE

Oxaloacetate

Citrate

Isocitrate

α-Ketoglutarate

Succinyl CoA

Succinate

Fumarate

Malate

Figure B Krebs cycle, also known as the citric acid cycle. *Red* identifies carbon atoms entering the cyclic pathway (by way of acetyl-CoA) and leaving (by way of carbon dioxide). These cyclic reactions run twice for each glucose molecule that has been degraded to two pyruvate molecules.

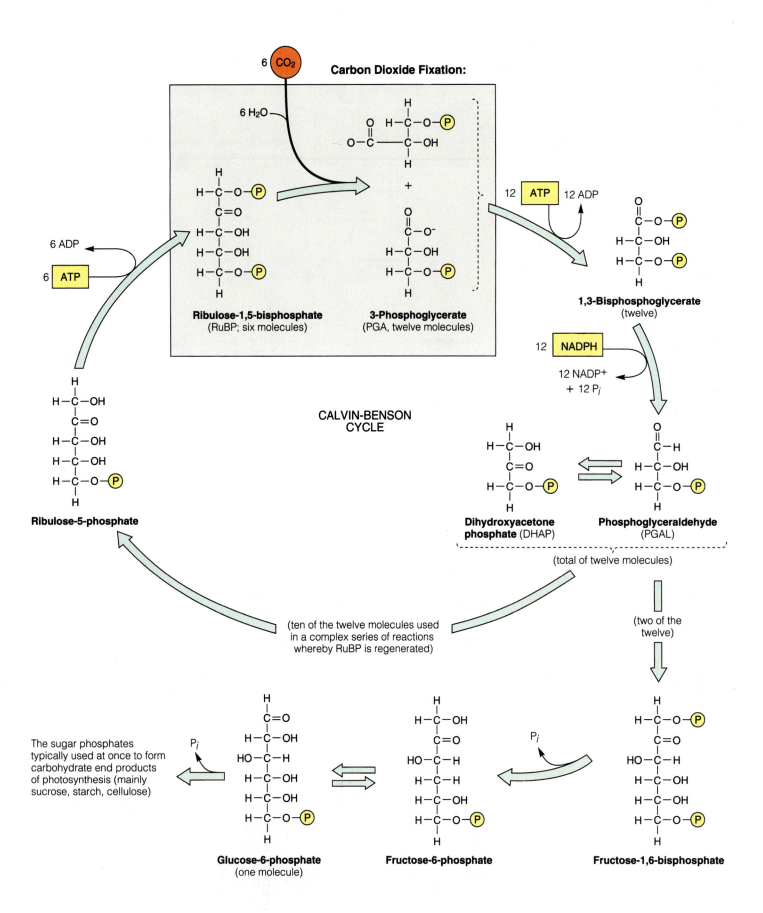

Figure C Calvin–Benson cycle of the light-independent reactions of photosynthesis.

Appendix VI. The Amino Acids

Neutral, nonpolar side group

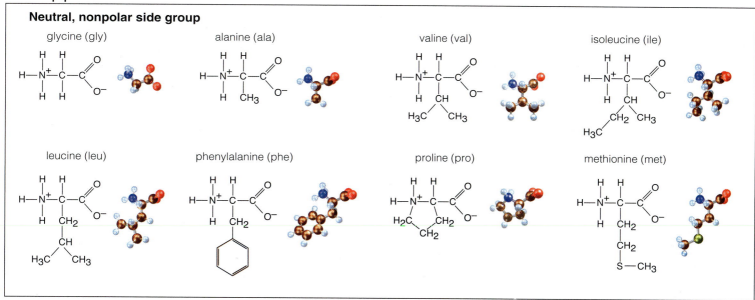

glycine (gly)　　alanine (ala)　　valine (val)　　isoleucine (ile)

leucine (leu)　　phenylalanine (phe)　　proline (pro)　　methionine (met)

Neutral, polar side group

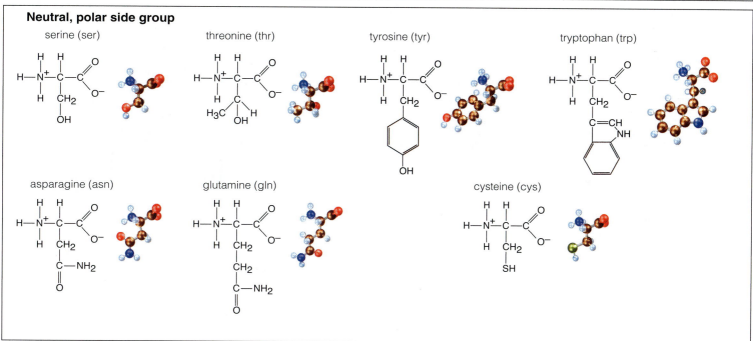

serine (ser)　　threonine (thr)　　tyrosine (tyr)　　tryptophan (trp)

asparagine (asn)　　glutamine (gln)　　cysteine (cys)

Acidic side group

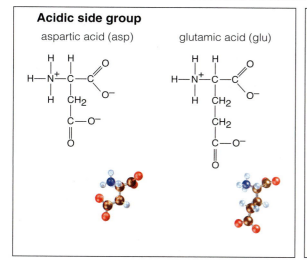

aspartic acid (asp)　　glutamic acid (glu)

Basic side group

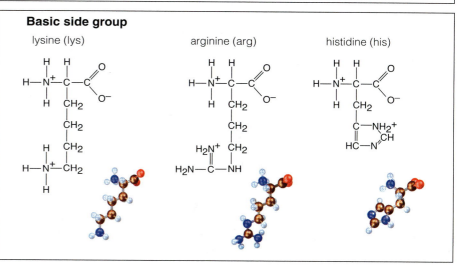

lysine (lys)　　arginine (arg)　　histidine (his)

Appendix VII. Periodic Table of the Elements

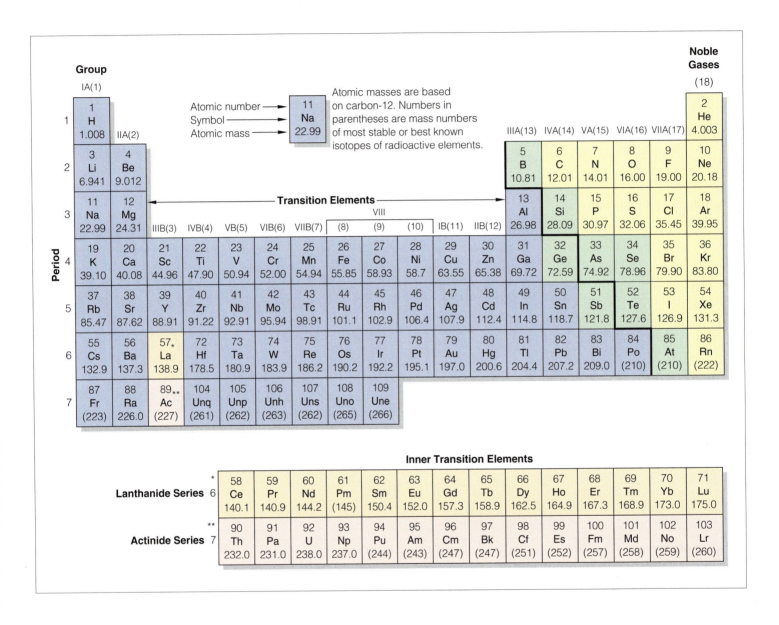

Group

Noble Gases (18)

Atomic number → 11
Symbol → Na
Atomic mass → 22.99

Atomic masses are based on carbon-12. Numbers in parentheses are mass numbers of most stable or best known isotopes of radioactive elements.

Transition Elements

Inner Transition Elements

Lanthanide Series 6	58 Ce 140.1	59 Pr 140.9	60 Nd 144.2	61 Pm (145)	62 Sm 150.4	63 Eu 152.0	64 Gd 157.3	65 Tb 158.9	66 Dy 162.5	67 Ho 164.9	68 Er 167.3	69 Tm 168.9	70 Yb 173.0	71 Lu 175.0

Actinide Series 7	90 Th 232.0	91 Pa 231.0	92 U 238.0	93 Np 237.0	94 Pu (244)	95 Am (243)	96 Cm (247)	97 Bk (247)	98 Cf (251)	99 Es (252)	100 Fm (257)	101 Md (258)	102 No (259)	103 Lr (260)

Glossary

ABC model Idea that products of three groups of master genes direct a flower's development.

ABO blood typing Method of characterizing an individual's blood based on proteins (A, B, or their absence) at surface of red blood cells.

abortion Expulsion of a pre-term embryo or fetus from the uterus.

abscisic acid ABA. Plant hormone; induces bud and seed dormancy; makes stomata close.

abscission Dropping of leaves, flowers, fruits, or other parts from plants.

acclimatization Long-lasting physiological and behavioral adaptation to a new habitat.

acid Any dissolved substance that donates H+ to other solutes or to water molecules.

acid rain Falling of rain (or snow) rich in acidic sulfur and nitrogen oxides.

acid–base balance State in which extracellular fluid is not too acidic or too basic, an outcome of controls over the concentrations of dissolved ions.

actin Cytoskeletal protein; the main component of thin filaments in a sarcomere.

action potential Brief reversal in the resting membrane potential of an excitable cell.

activation energy Minimum amount of energy required to start a reaction; enzyme action lowers this energy barrier.

activator Regulatory protein that enhances a cell activity (e.g., gene transcription).

active transport Pumping of a specific solute across a membrane against its concentration gradient, through a transport protein's interior. Requires energy input.

acute inflammation Nonspecific defense response to tissue injury in animals. Signs include localized redness, heat, swelling, pain.

adaptation Any aspect of form, function, behavior, or development that improves an individual's capability of surviving and reproducing in a given environment.

adaptive behavior Behavior that contributes to the individual's reproductive success.

adaptive immunity Capacity of B cell and T cell populations to tailor their defenses against previously unencountered pathogens.

adaptive radiation Burst of diversification from a single lineage; gives rise to new species adapted to specific environmental niches.

adaptive zone Some way of life available to species physically and behaviorally able to live it (e.g., "catching flying insects at night").

adenine Nitrogen-containing base component of nucleic acids; also a nucleic acid containing an adenine base. Base-pairs with thymine in DNA or uracil in RNA.

ADH Antidiuretic hormone. Hypothalamic hormone that promotes water conservation.

adhering junction Structure that forms a site of adhesion between cells. Found in tissues subject to stretching or abrasion.

adipose tissue Connective tissue that specializes in fat storage.

adrenal cortex Outer portion of adrenal gland; secretes cortisol and aldosterone.

adrenal medulla Inner portion of adrenal gland; secretes epinephrine, norepinephrine.

aerobic respiration Oxygen-dependent pathway of ATP formation in which glucose is broken down to carbon dioxide and water in several steps, including glycolysis, the Krebs cycle, and electron transfer phosphorylation. Typical net yield: 36 ATP.

age structure Of a population, distribution of individuals among different age categories.

alcoholic fermentation Anaerobic ATP-forming pathway. NADH transfers electrons to acetaldehyde, forming ethanol. Reactions start with pyruvate from glycolysis and regenerate NAD+. Net yield: 2 ATP.

aldosterone Adrenal cortex hormone; promotes sodium reabsorption in the kidney.

alkylating agent A substance that replaces a hydrogen with an alkyl (saturated organic) group in a biological molecule.

allantois Extraembryonic membrane that exchanges gases and stores metabolic wastes of embryos of reptiles, birds, some mammals. In humans, it forms urinary bladder, placental blood vessels.

allele One of two or more molecular forms of a gene that arise by mutation and specify slightly different versions of the same trait.

allergen Normally harmless substance that can provoke inflammation, excess mucus secretion, and often immune responses in susceptible people.

allergy Hypersensitivity to an allergen.

allopatric speciation Divergences end in speciation after a physical barrier that arises between populations of a species stops gene flow between them.

altruism Behavior that lowers an individual's chance of reproductive success but also helps others of its species.

alveolus, plural **alveoli** In bronchioles, a cup-shaped sac where gases are exchanged between blood and interstitial fluid.

amino acid Organic compound with an amino group (NH_2), a carboxylic acid group (COOH), and a side group bonded covalently to the same carbon atom. Subunit of proteins.

ammonification Of nitrogen cycle, process by which soil fungi and bacteria break down nitrogenous wastes and remains to ammonia compounds that plants may absorb.

amnion An extraembryonic membrane; encloses a fluid-filled sac in which amniote embryos develop.

amniote A tetrapod that produces amniote eggs. A reptile, bird, or mammal.

amoeboid protozoan Protist that forms pseudopods; some "naked," others "shelled" (e.g., amoeba, foraminiferan, radiolarian).

amphibian Vertebrate with four legs (or four-legged aquatic ancestor); its body plan and reproductive mode are somewhere between fishes and reptiles.

anagenesis Speciation pattern; changes occur within an unbranched line of descent.

analogous structures Similar body parts in distantly related lineages that arise as a result of similar environmental pressures.

anaphase Of mitosis, stage when sister chromatids of each chromosome move to opposite spindle poles. During anaphase I (meiosis), each duplicated chromosome and its homologue move to opposite poles. During anaphase II, sister chromatids of each chromosome move to opposite poles.

anemia Disorder resulting from deformed or insufficient quantity of red blood cells.

aneuploidy In cells, too many or too few chromosomes relative to the parental number.

angiosperm Flowering plant.

animal Multicelled, motile heterotroph that has embryonic stages and usually tissues, organs, and organ systems.

annelid Bilateral, highly segmented worm with well-developed organ systems (e.g., oligochaete, polychaete, leech).

anther Part of a stamen; pollen forms in it and is dispersed from it.

anthocyanin Red to blue photosynthetic accessory pigment.

antibiotic A natural or synthetic chemical agent that kills or inhibits growth of micro-organisms, especially bacteria.

antibody Antigen-binding receptor made and secreted by B cells.

anticodon Series of three nucleotide bases in tRNA; can base-pair with an mRNA codon.

antigen A substance chemically recognized as nonself; triggers an immune response.

antigen-presenting cell Cell that can break down and display antigen attached to self markers on its surface (e.g., macrophage, dendritic cell, or B cell).

antioxidant Enzyme or cofactor that can help neutralize free radicals, which may otherwise damage DNA and other molecules of life.

aorta Main artery of systemic circulation.

apicomplexan One of a group of parasitic protists that has a unique device used in penetrating host cells.

apoptosis Programmed cell death.

appendix Narrow projection from the cecum.

archaea Evolutionarily distinct domain of prokaryotic organisms.

Archean Eon in which life arose (3.9–2.5 billion years ago).

archipelago An island chain some distance away from a continent.

area effect Idea that larger islands support more species than smaller ones at similar distances from sources of colonizer species.

arteriole Type of blood vessel between arteries and capillaries where control mechanisms govern the distribution of blood flow through the body.

artery Thick-walled, muscular, transport vessel; carries blood away from heart.

arthropod Invertebrate with exoskeleton, jointed appendages (e.g., crustacean, insect).

artificial selection Selection of traits among a population under contrived conditions.

asexual reproduction Any reproductive mode by which offspring arise from and inherit genes from just one parent.

atom Fundamental form of matter that has mass and takes up space, and cannot be broken apart by everyday means.

atomic number The number of protons in the nucleus of an atom; identifies an element.

ATP Adenosine triphosphate. Nucleotide made of adenine, ribose, and three phosphate groups; main energy carrier in cells.

ATP/ADP cycle Alternating formation of ATP and ADP through phosphate group transfers.

ATP synthase Membrane-bound active transport protein that acts as an enzyme of ATP formation.

australopith Type of early hominid; extinct.

autoimmune response An immune response of the body against its own normal cells.

autoimmunity Failure of self-recognition in which the immune system attacks normal body cells.

automated DNA sequencing Robotic method of determining the nucleotide sequence of a region of DNA. Uses gel electrophoresis and laser detection of fluorescent tracers.

autonomic nerve One of the nerves that carry signals to and from internal organs.

autonomic nervous system All nerves from central nervous system to smooth muscle, cardiac muscle, and glands of viscera.

autosome Any chromosome of a type that is the same in males and females of a species.

autotroph An organism that makes its own food using an environmental energy source and carbon from carbon dioxide.

auxin Plant hormone of apical meristems; induces stem lengthening and responses to gravity and light.

axon Neuron's signal-conducting zone.

B cell See B lymphocyte.

B lymphocyte White blood cell that secretes antibody molecules. Also called plasma cell.

bacteria The most widespread and diverse group of prokaryotic organisms.

bacterial conjugation Transfer of plasmid DNA from one prokaryotic cell to another.

bacteriophage A virus that infects bacteria.

balanced polymorphism Outcome of selection; two or more alleles for a trait are being maintained in a population over time.

bark All tissues external to the vascular cambium in older woody stems.

basal body An organelle that gives rise to cilia or flagella; resembles a centriole.

base Any substance that accepts hydrogen ions (H^+) when dissolved in water, thus forming hydroxyl ions (OH^-). Also, the nitrogen-containing component of a nucleic acid.

base-pair substitution Mutation in which one nucleotide is incorrectly substituted for another during DNA replication.

basophil Fast-acting white blood cell that secretes histamine during inflammation.

big bang Model for origin of universe in which all matter, energy, and time originated from a single point in a gigantic explosion.

bilateral symmetry Body plan with main axis from anterior to posterior, separated into right and left sides, and dorsal and ventral surfaces.

bile Liver secretion required for fat digestion.

biogeochemical cycle Movement of element from environmental reservoirs, through living things, then back to the environment.

biogeographic realm One of six vast land areas with distinctive species.

biogeography Study of the patterns of distribution of species.

biological clock Internal time-measuring mechanism; adjusts activities seasonally, daily, or both in response to environmental cues.

biological magnification Some substance becomes more concentrated in body tissues as it moves up through food chains.

biological species concept The definition of a species is based on its reproductive isolation.

biology The scientific study of life.

bioluminescence Production of fluorescent light by a living organism.

biomass Combined weight of all organisms at a given trophic level in an ecosystem.

biomass pyramid Diagram showing the dry weight of all organisms at each trophic level of an ecosystem.

biome A large land region characterized by dominant plant species and habitat conditions.

biosphere All regions of the Earth's waters, crust, and atmosphere where organisms live.

biotic potential The maximum rate of increase of a population under ideal conditions.

bipedalism Routinely walking upright.

bipolar mitotic spindle Dynamic array of microtubules that moves chromosomes in precise directions during mitosis or meiosis.

bird Only animal (besides some extinct related dinosaurs) that grows feathers.

blastocyst Early mammal development stage; two layers of cells surround a fluid-filled cavity.

blastomere One of the small, nucleated cells that form during cleavage of animal zygote.

blastula Early outcome of cleavage; one layer of blastomeres encloses a fluid-filled cavity.

blood Liquid connective tissue that has fluid and cellular components.

blood pressure Fluid pressure, generated by heart contractions, that circulates blood.

blood–brain barrier Mechanism that controls which solutes enter cerebrospinal fluid.

bone remodeling Ongoing mineral deposits and withdrawals from bone.

bone tissue Calcium-hardened connective tissue that composes the vertebrate skeleton.

bottleneck A sudden, drastic reduction in population size.

Bowman's capsule Part of a nephron; receives water and solutes being filtered from blood.

brain An integrating center that receives and processes sensory input and issues commands for responses by muscles and glands.

brain stem Most ancient nerve tissue in the vertebrate hindbrain, midbrain, and forebrain.

bronchiole One of the finely branched airways in the lung.

bronchus, plural **bronchi** Airway that branches from trachea and leads into lungs.

brown alga One of the stramenopiles; mostly marine, photosynthetic (e.g., kelp).

bryophyte A nonvascular land plant (e.g., moss, liverwort, hornwort).

bud Undeveloped shoot, consisting mostly of meristematic tissue.

buffer system A weak acid and the base that forms when it dissolves in water. The two work as a pair to counter slight shifts in pH.

bulk Of the vertebrate gut, the volume of undigested material in the small intestine that cannot be decreased by absorption.

bulk flow Mass movement of one or more substances in the same direction, in response to pressure, gravity, or another external force.

C3 plant Plant that makes three-carbon PGA in the first step of carbon fixation.

C4 plant Plant that makes four-carbon oxaloacetate in the first step of carbon fixation.

calcium pump Membrane-bound active transport protein specific for calcium ions.

Calvin–Benson cycle Light-independent cyclic reactions of photosynthesis. Forms sugars from CO_2 using ATP and NADPH.

CAM plant Plant that conserves water by opening stomata only at night, when it fixes carbon by repeated turns of the C4 pathway.

camouflage Coloration, patterning, or other aspects of form or behavior that make an individual blend with its surroundings.

cancer Malignant neoplasm; mass of cells that divide abnormally and can spread in the body.

capillary Smallest of the blood vessels; site of diffusion between blood and interstitial fluid.

capillary reabsorption Movement of fluid from interstitial fluid into capillaries.

capture–recapture method For population counts; collecting, marking, releasing, then recapturing representative animals.

carbon cycle Movement of carbon from the atmosphere, through food webs and ocean's waters and rocks, and back to atmosphere.

carbon fixation Autotrophic cell secures carbon atoms from the air and incorporates them into a stable organic compound.

cardiac cycle Sequence of muscle contraction and relaxation in a single heartbeat.

cardiac muscle Muscle of the heart wall.

cardiac pacemaker Sinoatrial (SA) node; a mass of self-excitatory cardiac muscle cells that set the rate for a normal heartbeat.

carotenoid Red to yellow accessory pigment.

carpel Female reproductive part of a flower.

carrying capacity The maximum number of individuals in a population (or species) that a given environment can sustain indefinitely.

cartilage A specialized connective tissue that is dense, pliable, yet resists compression.

cartilaginous fish Jawed fish with a skeleton of cartilage (e.g., shark, ray).

Casparian strip In roots; waxy, impermeable band in cell walls of endodermis or exodermis where water and mineral uptake is controlled.

catastrophism Idea that abrupt changes in the geologic or fossil record resulted from large scale disasters that were divinely invoked.

CD4 lymphocyte White blood cell having a receptor to which HIV may bind (e.g., T-cell).

cDNA DNA made from an mRNA transcript, using reverse transcriptase.

cell Smallest living unit, with a capacity to survive and reproduce on its own.

cell cortex Three-dimensional mesh of actin filaments and other proteins just under the plasma membrane.

cell count The number of cells of a given type in one microliter of blood.

cell cycle Series of events from one cell division to the next. Interphase, mitosis, and cytoplasmic division constitute one cycle.

cell differentiation Cell lineages become specialized in structure and function by selectively activating genes.

cell junction A site where adjoining cells interact physically, chemically, or both.

cell plate In a dividing plant cell, a disk-like structure that becomes a crosswall with new plasma membrane on both sides.

cell plate formation Mechanism of cytoplasmic division in plant cells.

cell theory Idea that all organisms consist of similar units of organization called cells.

cell wall A semirigid, permeable structure encloses the plasma membrane of many cells; helps cell retain its shape and resist rupturing.

central vacuole Fluid-filled storage organelle of a plant cell.

centriole Organelle that organizes formation and direction of cilia, flagella, and spindles.

centromere Constricted part of a chromosome to which spindle microtubules attach.

centrosome Cell region where microtubules are produced.

cephalization Evolutionary trend toward the concentration of sensory structures and nerve cells in a head.

cerebellum Hindbrain region; maintains balance and coordinates limb movements.

cerebrum Forebrain region that deals with olfactory input and motor responses. In mammals, offers complex level of integration.

charophyte Type of green alga; the closest living relatives of plants.

chemical bond A union between the electron structures of two or more atoms or ions.

chemical equilibrium The state at which the concentrations of reactants and products in a reversible chemical reaction remain constant.

chemical synapse Thin cleft between a presynaptic neuron and a postsynaptic cell. Neurotransmitter molecules diffuse across it.

chemoreceptor A sensory receptor which responds to chemical stimuli.

chemotaxin In animals, a chemical signal that attracts phagocytic white blood cells.

chlamydias A group of bacteria that are obligate intracellular parasites of vertebrates.

chlorophyll Primary photosynthetic pigment.

chlorophyte A member of the most diverse group of green algae.

chloroplast Organelle of photosynthesis in plants and many protists.

chordate Animal having a notochord, dorsal hollow nerve cord, pharynx, and gill slits in pharynx wall during at least part of life cycle.

chorion Extraembryonic membrane. In some mammals, becomes part of placenta.

chromatin All DNA molecules and associated proteins in a nucleus.

chromosome A coherent structure consisting of a DNA molecule and associated proteins.

chromosome number Sum of all chromosomes in a given type of cell.

chrysophyte A category of photosynthetic protists (e.g., golden algae).

ciliate Ciliated protozoan; type of protist that usually has profuse cilia at its surface.

cilium, plural **cilia** In some eukaryotic cells, a short motile structure or sensory structure.

circadian rhythm Biological activity repeated in cycles, each about twenty-four hours long.

circulatory system Organ system that moves substances to and from cells; can help stabilize body temperature and pH. In many animals, consists of a heart, blood vessels, and blood.

cladogenesis Speciation pattern; a lineage splits, and populations diverge genetically.

cladogram Evolutionary tree diagram; groups taxa on the basis of their shared derived traits.

classification system A way of organizing and retrieving information about species.

cleavage Early stage of animal development. Mitotic cell divisions divide a fertilized egg into many smaller, nucleated cells; original volume of egg cytoplasm does not increase.

climate Prevailing weather conditions in some region.

climax community Array of species that can, if habitat remains stable, persist indefinitely without being replaced by other species.

climax-pattern model Idea that more than one stable community can persist in the same region when environmental factors vary.

cloning vector DNA molecule that can accept foreign DNA and replicate in a host cell.

club fungus Fungus having club-shaped cells that produce and bear basidiospores.

cnidarian Radial invertebrate at the tissue level of organization; makes nematocysts.

coal Nonrenewable energy source formed over 280 million years ago from submerged, compacted plant remains.

codon Sequence of three bases in an mRNA strand that codes for an amino acid, or acts as a start or stop signal for translation.

coelom In most animals, a cavity between the gut and body wall that is lined with a tissue.

coenzyme Small molecule that participates in an enzymatic reaction, and is reversibly modified during the reaction (e.g., a vitamin).

coevolution Joint evolution of two interacting species, brought about by changes in selection pressures operating between the two.

cohesion The capacity to resist rupturing when placed under tension (stretched).

cohesion–tension theory Idea that evaporation from plants exerts tension in xylem, pulling cohesive columns of water upward from roots to leaves.

cohort A group of same-aged individuals, tracked throughout their life spans.

collecting duct Last tubular region of a kidney nephron.

collenchyma Simple plant tissue that imparts flexible support during primary growth.

colon Large intestine; concentrates undigested and unabsorbed material.

commensalism Interaction between species that benefits one and has no effect on other.

communication signal Information-laden cue directed by one member of a species to another; chemical, visual, acoustic, or tactile.

community All species living and interacting in some habitat.

companion cell Specialized parenchyma cell; helps load sugars into sieve tubes of phloem.

comparative morphology Scientific study of the body form and structures of major groups.

compartmentalization Defensive response by a woody plant; resin or toxins are secreted in response to injury or an attack.

competitive exclusion Idea that, when two species require exactly the same resources, competition will drive one or the other to extinction in a shared habitat.

complement system Plasma proteins that circulate in blood in inactive form; has roles in nonspecific defenses and immune responses.

compound Molecule consisting of two or more elements in unvarying proportions.

concentration gradient A difference in the number of molecules (or ions) of a substance between two adjoining regions.

condensation reaction Covalent bonding of two molecules into a larger molecule, often with the formation of water as a by-product.

conduction In temperature studies, exchange of heat between two touching objects owing to a thermal gradient between them.

cone Reproductive structure made of modified leaves (e.g., a cluster of ovule-bearing, woody scales of a pine tree).

cone cell Vertebrate photoreceptor; allows sharp daytime vision, color vision.

conifer Evergreen gymnosperm; tree or shrub with needlelike or scalelike leaves.

connective tissue Most abundant, pervasive animal tissue. Specialized types are cartilage, bone tissue, adipose tissue, and blood.

conservation biology Field devoted to surveying biological diversity, studying its origins, and attempting to maintain it.

consumer Heterotroph that obtains carbon and energy by feeding on other organisms.

continuous variation Of a population, a more or less continuous range of small differences in a given trait among its individuals.

control group A group used as a standard for comparison with an experimental group.

convection Movement of air or water next to an object aids conductive heat loss from it.

cork cambium A lateral meristem; replaces epidermis with cork on woody plant parts.

corpus luteum A glandular structure that forms from cells of a ruptured ovarian follicle; secretes progesterone and estrogen.

cortex A rindlike layer. In vascular plants, a ground tissue; supports plants and stores food.

cortisol Adrenal cortex hormone; roles in glucose regulation and stress responses.

cotyledon Seed leaf. Part of a plant embryo. Two form in eudicot seeds, one forms in monocot seeds.

countercurrent flow Movement of two fluids in opposite directions, as when water and blood flow in opposite directions in a fish gill.

covalent bond Sharing of one or more electrons between two atoms.

craniate Chordate having a brain inside a cranium. Includes living fishes, amphibians, reptiles, birds, and mammals.

creatine phosphate Energy source for muscle cells; transfers phosphate to ADP to form ATP.

crossing over At prophase I of meiosis, the reciprocal exchange of segments between two nonsister chromatids of a pair of homologous chromosomes; results in novel combinations of alleles.

culture Sum of behavior patterns of a social group, passed between generations by way of learning and symbolic behavior.

cuticle Body cover. Of plant surfaces, a layer of waxes and cutin. Of roundworms and annelids, a thin, flexible coat. Of arthropods, a lightweight exoskeleton hardened with protein and chitin.

cyanobacteria Bacteria that carry out noncyclic (oxygen-producing) photosynthesis.

cycad A gymnosperm that forms pollen- and seed-bearing structures on different plants; an ancient lineage that coexisted with dinosaurs.

cyst Resting structure; encloses certain small organisms or spores for part of the life cycle.

cytokine Signaling molecules of the immune system (e.g., interferons, interleukins).

cytokinin Plant hormone; stimulates cell division and leaf expansion, slows leaf aging.

cytoplasm All cell parts, particles, and semifluid substances between the plasma membrane and the nucleus (or nucleoid).

cytoplasmic localization Polar distribution of gene products in a cell. In an egg, determines body axis of the future embryo.

cytoskeleton Interconnected system of protein filaments that structurally supports, organizes, and moves a eukaryotic cell and its internal structures.

cytotoxic T cell T lymphocyte that touch-kills virus infected, cancerous, and other altered body cells with antigen bound to self markers.

day-neutral plant Plant that flowers when mature, rather than in response to daylength.

decomposer Fungal or bacterial heterotroph that obtains carbon and energy from remains, products, or wastes of organisms.

deforestation Removal of all trees from a large area.

deletion Loss of a chromosome segment. Also, a mutation involving the loss of one or more bases of a DNA molecule.

demographic transition model Explanation of the effects of industrialization on human population growth.

demographics A population's vital statistics (e.g., size, distribution, density, age structure).

denaturation The three-dimensional shape of a protein or some other complex molecule unravels as its hydrogen bonds are disrupted.

dendrite Short, slender extension from cell body of a neuron; a signal input zone.

dendritic cell Type of antigen-presenting white blood cell.

denitrification Conversion of nitrate or nitrite by certain soil bacteria to gaseous nitrogen (N_2) and nitrous oxide (N_2O).

dense, irregular connective tissue Animal tissue with fibroblasts, many asymmetrically positioned fibers in a ground substance. In skin, some capsules.

dense, regular connective tissue Animal tissue with rows of fibroblasts between bundles of fibers. In tendons, ligaments.

density-dependent control Factor that operates at high population density to slow birth rate and/or increase death rate (e.g., disease, competition).

density-independent factor Factor that impacts a population's birth rate or death rate regardless of density (e.g., fire or flood).

deoxyribonucleic acid See DNA.

derived trait A novel feature that arose in a species and is shared only by its descendants.

dermis Skin layer beneath the epidermis; consists mainly of dense connective tissue.

desalinization Removal of salt from seawater.

desertification Conversion of grassland or cropland to a desertlike condition.

detritivore Heterotroph that feeds on bits of decaying organic matter (e.g., an earthworm).

deuterostome Bilateral animal in which the first indentation in the embryo becomes the anus (e.g., echinoderms, chordates).

development Series of stages by which structurally and functionally distinct body parts emerge, in orderly patterns, in a new multicelled individual.

diaphragm Muscular partition between thoracic and abdominal cavities. Also, a contraceptive device inserted into the vagina to prevent sperm from entering uterus.

dicot Dicotyledon. Flowering plant having embryos with two cotyledons.

diffusion Net movement of like molecules or ions down their concentration gradient.

digestive system Body sac or tube, often with regions where food is ingested, digested, and absorbed, and where residues are eliminated.

dihybrid cross Intercross between two individuals, each heterozygous for two genes (e.g., *AaBb*).

dinoflagellate Single-celled, flagellated, cellulose-plated protist; most photosynthetic.

dinosaur One of a group of extinct reptiles that originated in the Triassic and dominated land environments for 140 million years.

diploid Presence of two of each type of chromosome (i.e., pairs of homologs) in a cell nucleus at interphase.

diploid number Total chromosome number in cells that have a pair of each type of chromosome characteristic of the species.

directional selection Mode of natural selection by which forms of a trait at one end of a range of phenotypic variation are favored and all others are selected against.

disaccharide A common oligosaccharide; two covalently bonded sugar monomers.

disease Illness caused by an infectious, dietary, or environmental factor.

disruptive selection Mode of natural selection by which forms at both ends of the range of phenotypic variation are favored and intermediate forms are selected against.

distal tubule Tubular part of nephron where water and sodium are selectively reabsorbed.

distance effect Idea that the farther away an island is from a source of potential colonizing species, the lower its species diversity will be.

DNA Deoxyribonucleic acid. Carries the primary hereditary information for all living organisms and many viruses.

DNA chip Microarray of DNA spots on a glass plate; used to study gene expression.

DNA fingerprint Unique cleavage pattern of an individual's DNA; also called restriction fragment length polymorphism (RFLP).

DNA ligase Repair enzyme that joins breaks in DNA molecule; connects new DNA fragments during replication.

DNA polymerase Enzyme that catalyzes replication and repair of DNA.

DNA replication Process by which a cell duplicates its DNA molecules before dividing.

dormancy A time of metabolic inactivity in spores, cysts, seeds, plants, and some animals.

dosage compensation Control mechanism that balances gene expression between the sexes, starting at early stages of development.

double fertilization Of flowering plants only, fusion of sperm and egg nuclei, plus fusion of another sperm nucleus with nuclei of a cell that gives rise to endosperm.

doubling time Length of time it takes a population to double in size.

downwelling Downward movement of water along a coast, from ocean surface to its depths.

duplication DNA sequence repeated several to many hundreds or thousands of times.

ecdysone Hormone of many insect life cycles; roles in metamorphosis, molting.

echinoderm Invertebrate with calcified spines and plates in the body wall. Radial with some bilateral features (e.g., sea stars, sea urchins).

ecology Study of how organisms interact with one another and with their environment.

ecoregion Broad land or ocean region defined by climate, geography, and producer species.

ecosystem An array of species and their physical environment.

ectoderm First-formed, outermost primary tissue layer of animal embryos; gives rise to nervous system tissues and integument's outer layer.

ectotherm Animal that maintains core temperature by absorbing environmental heat.

Ediacarian One of a group of soft-bodied animals that arose in the precambrian.

effector Muscle (or gland); helps bring about movement (or chemical change) in response to neural or endocrine signals.

effector cell Differentiated lymphocyte that, during immune responses, engages and destroys antigen-bearing agents.

egg Female gamete.

El Niño A recurring warm current that displaces cool, nutrient-rich water along South America's coast; has global effects.

electric gradient A difference in electric charge between adjoining regions.

electron Negatively charged unit of matter, with particle-like and wavelike properties; occupies an orbital around atomic nucleus.

electron transfer chain Array of membrane-bound enzymes and other molecules that accept and give up electrons in sequence; allows the release and capture of energy in small, useful increments.

electron transfer phosphorylation Last stage of aerobic respiration; electrons flow through mitochondrial electron transfer chains, to O_2. The flow sets up an electrochemical gradient that drives ATP formation.

element Material consisting of atoms all with the same atomic number.

embryonic induction Signaling molecules released from one embryonic tissue affect the development of an adjacent tissue.

emigration Individuals leave a population.

emulsification In the vertebrate gut, the coating of fat droplets with bile salts, so that the droplets become suspended in chyme.

endangered species An endemic (native) species that is highly vulnerable to extinction.

endocrine gland One of the ductless glands that secrete hormones into some body fluid.

endocrine system Integrative system of cells, tissues, and organs, functionally linked to the nervous system, that exerts control by way of its hormones and other chemical secretions.

endocytosis Cellular uptake of a substance; plasma membrane forms a vesicle around it.

endoderm Inner primary tissue layer of most animal embryos; source of inner gut lining and organs that are derived from it.

endometrium Inner lining of uterus.

endophytic fungi Fungi that live inside the bodies of plants.

endoplasmic reticulum ER. Organelle that starts at the nuclear envelope and extends through cytoplasm. Smooth ER assembles membrane lipids, breaks down fatty acids, and inactivates some toxins; Rough ER (has ribosomes on its cytoplasmic side) modifies new polypeptide chains.

endoskeleton Of chordates, the internal framework of cartilage and bone; works with skeletal muscle to support and move the body.

endosperm Triploid nutritive tissue in flowering plant seeds.

endospore Resistant resting structure that forms in some bacteria. It encloses the bacterial chromosome and some cytoplasm.

endosymbiosis One species spends its entire life inside another, in an interaction that benefits both.

endotherm Animal that can generate and maintain its body temperature using metabolic rate, body form and behavior.

energy Capacity to do work.

energy pyramid Diagram of an ecosystem's trophic structure; shows usable energy at each trophic level.

enhancer A small sequence in DNA that binds transcription regulating molecules.

ENSO El Niño Southern Oscillation. The eastward movement of warm surface waters of the western equatorial Pacific, displaces cold water off South America; has widespread climatic effects.

enzyme A type of protein (or, rarely, RNA) that accelerates a chemical reaction.

eosinophil White blood cell that acts against extracellular parasites, such as worms.

epidermis Outermost tissue layer; occurs in plants and all animals above the sponge level of organization.

epiglottis Flaplike structure between pharynx and larynx; its controlled positional changes direct air into trachea or food into esophagus.

epithelial tissue Animal tissue that covers external surfaces and lines internal cavities and tubes. One surface is free and the other rests on a basement membrane.

equilibrium model of island biogeography Model that predicts the number of species an island will support, based on island size and its distance from a source of colonists.

esophagus Muscular tube; in vertebrates it connects the pharynx and stomach.

essential amino acid Amino acid an animal cannot synthesize and must get from food.

essential fatty acid Fatty acid an animal cannot synthesize and must get from food.

estrogen Female sex hormone secreted by ovaries; helps oocytes mature, primes uterine lining for pregnancy, maintains secondary sexual traits, affects growth and development.

estuary Partly enclosed coast region where seawater mixes with fresh water and runoff from land, as from rivers.

ethylene Gaseous plant hormone; promotes fruit ripening and abscission of leaves, flowers, fruits.

eudicot True dicot; one of the flowering plants generally characterized by embryos with two cotyledons.

euglenoid Single-celled, flagellated protist of freshwater habitats; most are photosynthetic.

Eukarya Domain of eukaryotic cells; all protists, plants, fungi, and animals.

eutherian Placental mammal.

eutrophication Nutrient enrichment of a body of water (e.g., a lake or pond).

evaporation Process of conversion of a liquid to a gas; requires energy input.

evaporative heat loss Thermoregulation mechanism: cooling due to evaporation of water.

evolution, biological Genetic change in a line of descent. Outcome of microevolutionary events: gene mutation, natural selection, genetic drift, and gene flow.

evolutionary tree Diagram of evolutionary relationships; each branch signifies a separate line of descent from a common ancestor; each branch point a time of divergence.

exocrine gland Any gland that secretes products (e.g., milk) to a free epithelial surface, usually through ducts or tubes.

exocytosis Release of a vesicle's contents outside the cell surface when it fuses with and becomes part of the plasma membrane.

exodermis Cylindrical sheet of cells inside the root epidermis of most flowering plants; helps control uptake of water and solutes.

exon One of the base sequences of an mRNA transcript that eventually will be translated.

exoskeleton External skeleton (e.g., hardened arthropod cuticle).

exotic species A species that has left the community in which it evolved and become established elsewhere.

experimental group A group upon which an experiment is performed, and compared with a control group.

exponential growth, population An increase in population size by a fixed percentage of the whole in a given interval.

extinction Irrevocable loss of a species.

extracellular fluid Of most animals, all fluid not in cells; plasma (blood's liquid portion) plus interstitial fluid.

extreme halophile An archaean or other species of a notably salty habitat.

extreme thermophile An archaean or other species of a notably high-temperature habitat.

eye Sensory organ that incorporates a dense array of photoreceptors.

FAD Flavin adenine dinucleotide. Nucleotide coenzyme; transfers electrons and unbound protons (H^+) between reaction sites.

fat Type of lipid with a glycerol head attached to one, two, or three fatty acid tails.

fatty acid Organic compound with a backbone of up to 36 carbon atoms, and a carboxyl group at the end.

feedback inhibition Of cells, an activity causes a change in cellular conditions, and that change in turn causes the activity to slow down or stop.

fern A seedless vascular plant with fronds that often are divided into leaflets.

fertilization Fusion of a sperm nucleus and an egg nucleus, the result being a zygote.

fetus Stage of animal development; in humans, the start of the ninth week to birth.

fibrous root system The lateral branchings of adventitious roots that form on a young stem.

filter feeder Animal that filters food from a current of water directed through a body part (e.g., through a sea squirt's pharynx).

fin An appendage of the fish body that functions in locomotion and stability.

first law of thermodynamics The total amount of energy in the universe is constant; energy can be converted from one form to another, but cannot be created or destroyed.

fixation Only one kind of allele remains at a given locus in a population; all individuals have become homozygous for it. One outcome of genetic drift in small populations.

fixed action pattern Program of coordinated, stereotyped muscle activity that is completed independently of feedback from environment.

flagellated protozoan One of the most ancient heterotrophic protists; has one or more flagella.

flagellum, plural flagella A whip-like motile structure of many free-living eukaryotic cells.

flower Reproductive shoot of an angiosperm.

fluid mosaic model A cell membrane is fluid because of the motions and interactions of its component lipids and proteins.

food chain A linear flow of energy captured by primary producers (autotrophs) into ever higher trophic levels of an ecosystem.

food pyramid Charts showing proportions of different foods that compose a healthy diet.

food web Cross-connecting food chains.

fossil Recognizable, physical evidence of an organism that lived in the distant past.

fossilization Extremely slow transformation of an organism's remains to stony hardness as a result of pressure and chemical changes.

founder effect A form of bottlenecking. By chance, allele frequencies of a few individuals that establish a new population differ from the frequencies in the original population.

free radical Highly reactive molecule with at least one unpaired electron.

FSH Follicle-stimulating hormone. Produced and secreted by the anterior lobe of pituitary gland; has reproductive roles in both sexes.

functional group An atom or a group of atoms with characteristic properties that is covalently bonded to an organic compound's carbon backbone.

fungus, plural fungi Eukaryotic heterotroph that obtains nutrients by extracellular digestion and absorption; notable for prolific spore formation.

gametophyte Multicelled, gamete-producing body (haploid) that forms during life cycles of some algae and all plants.

ganglion plural ganglia Distinct cluster of cell bodies of neurons.

gap junction A channel through a complex of abutting membrane proteins of two cells; permits ions and small molecules to move rapidly between the cells.

gastric fluid Acidic mix of secretions (e.g., HCl, mucus) from the stomach's glandular epithelium.

gastrula Earliest developmental stage in which cells are arranged as two or three primary tissue layers.

gel electrophoresis Molecules that migrate through a gel matrix in response to an electric force become separated by size and charge.

gene Unit of information for a heritable trait in DNA, passed from parents to offspring.

gene control A molecular mechanism that governs if, when, or how a specific gene is used (transcribed or translated).

gene flow Physical flow of alleles into or out of a population by immigration or emigration.

gene library Mixed collection of host cells that contain cloned DNA fragments representing all or most of a genome.

gene locus A gene's chromosomal location.

gene mutation Small-scale change in the nucleotide sequence of a gene that can result in an altered protein product.

gene pool All the genes in a population.

genetic code The correspondence between triplets of nucleotides in DNA (then mRNA) and specific sequences of amino acids in a polypeptide chain; the basic language of protein synthesis in cells.

genetic disorder Heritable defect in one's genetic material; causes mild to severe medical problems.

genetic divergence Accumulation of differences in the gene pools of populations after something stops gene flow between them. Over time, if some of the differences promote reproductive isolation, speciation may follow.

genetic drift Random change in allele frequencies over time brought about by chance alone; its effect is greatest in small populations.

genetic engineering Manipulation of an organism's DNA, usually with the intent of altering at least one aspect of phenotype.

genome All the DNA in a haploid number of chromosomes for a given species.

genomics The study of genes and gene function in humans and other organisms.

genus A group of related species.

geographic dispersal Individuals move away from their home range and successfully establish themselves elsewhere.

geologic time scale Chronology of Earth history; major subdivisions correspond to mass extinctions.

germination Of seeds and spores, resumption of growth after dormancy, dispersal from the parent organism, or both.

gibberellin Plant hormone; stimulates stem lengthening, helps seeds and buds break dormancy, promotes flowering in some plants.

gill slit Opening in a thin-walled pharynx. Serves in food-trapping and respiration; jaws evolved from certain gill slit supports.

gills Type of respiratory organ that occurs in some invertebrates and most fishes.

ginkgo Deciduous gymnosperm of an ancient lineage that produces fleshy plum-sized fruits.

gland Secretory organ derived from epithelial tissue (e.g., sweat gland, thyroid gland).

global broiling hypothesis Idea that asteroid impact caused the K–T mass extinction by creating a colossal fireball, causing the global temperature to rise thousands of degrees.

global warming Long-term rise in the temperature of the Earth's lower atmosphere.

glomerular filtration Movement of water and small solutes out of glomerular capillaries and into the first part of the kidney tubule.

glycolysis Breakdown of glucose to two pyruvate molecules. First stage of aerobic respiration and fermentation.

gnetophyte Tropical tree, vine, or shrub of an ancient gymnosperm group.

GnRH Gonadotropin-releasing hormone; triggers hypothalamus to release LH and FSH.

Golgi body Organelle of endomembrane system; final modification of polypeptide chains into proteins, lipid assembly, and packaging of both in vesicles for secretion or for use inside cell.

Gondwana Paleozoic supercontinent that collided with other large land masses and formed Pangea.

Gram-positive bacteria One of the several bacterial groups having cell walls that are colored purple by Gram-staining procedure.

gravitropism In plants, directional growth of roots and shoots in response to gravity.

grazing food web Food web in which most energy from producers flows to herbivores, then to carnivores.

greenhouse effect Warming of the lower atmosphere as greenhouse gases reradiate heat energy back toward the Earth's surface.

groundwater Water in soil and aquifers.

growth Of multicelled species, increases in the number, size, and volume of cells. Of bacteria, increases in cell number.

growth and tissue specialization Stage of animal development when the body enlarges and organs assume their functions.

growth ring One of the alternating bands of early and late wood in a tree with extensive secondary growth.

gut Of animals, a sac or tube from which food is absorbed into the internal environment.

habitat Type of place where a species lives.

habitat loss Reduction of habitat for a species as a result of environmental destruction.

half-life The time it takes for half of a given quantity of atoms of a radioisotope to decay.

haploid number Total chromosome number in cells that have one of each type of chromosome characteristic of the species.

Hardy–Weinberg rule Allele frequencies will be stable over the generations if there is no mutation, the population is infinitely large and isolated from other populations of the same species, mating is random, and all individuals reproduce equally and randomly.

heart Muscular, pressure-generating pump that keeps blood flowing through the vessels of a circulatory system.

heartwood Dry core tissue of older stems and roots; functions in support and the storage of metabolic wastes.

helper T cell T lymphocyte that stimulates B cells and other T cells to divide in response to antigen recognition; critical component of all immune responses.

heme group Iron-containing functional group that reversibly binds oxygen.

hemoglobin Respiratory protein in red blood cells; consists of four polypeptide chains and four heme groups.

hemostasis Process that stops blood loss after injury by way of coagulation, vessel spasm, platelet plug formation, and other effects.

herbicide Natural or synthetic toxin that can kill plants or inhibit their growth.

heterotherm An endotherm that can lower metabolic rates; activities idle as it cuts energy cost of maintaining its core temperature.

heterotroph Organism unable to make its own organic compounds; feeds on autotrophs, other heterotrophs, or organic wastes.

heterozygous condition For a specified trait, having two different alleles at a locus.

higher taxa One of ever more inclusive groupings of species based on relatedness (e.g., phylum, kingdom, domain).

histamine Chemical released by mast cells during an immune response; contributes to inflammation and other symptoms of allergy.

homeostasis Maintenance of physical and chemical aspects of the internal environment within ranges suitable for cell activities.

homeotic gene One of a class of master genes; helps determine identity of body parts during embryonic development.

hominid A human or an extinct humanlike primate (e.g., an australopith).

hominoid Apes, humans, and their extinct recent ancestors.

homologous chromosome One of a pair of chromosomes, identical in size, shape, and gene sequence, each inherited from a different parent. Nonidentical sex chromosomes are also considered homologs.

homologous structure Similar body part that occurs in different species as a result of descent from a common ancestor.

homozygous dominant condition Having a pair of dominant alleles at a gene locus.

homozygous recessive condition Having a pair of recessive alleles at a gene locus.

hormone Chemical signaling molecule formed by one part of the body, acting upon another. In animals, a product of endocrine glands, endocrine cells, or neurons. In plants, a product of cells in primary shoots or roots.

horsetail Seedless vascular plant with tiny scale-like leaves, branching rhizomes, and silica-reinforced stems.

hot spot Habitat for a large number of species found nowhere else and facing extinction.

human Member of the genus *Homo*.

human gene therapy The transfer of normal or modified genes into a person to correct a genetic defect, or boost resistance to a disease.

humus Decomposing organic matter in soil.

hybrid Individual having a nonidentical pair of alleles for a trait being studied.

hydrogen bond An intermolecular interaction between a covalently bonded hydrogen atom and a different atom bearing a negative charge (e.g., oxygen, fluorine, or nitrogen).

hydrologic cycle Driven by solar energy, water evaporates from the ocean into the atmosphere, moves onto the land, then back to the ocean.

hydrolysis An enzymatic cleavage reaction in which a molecule is split, and the components of water (—OH and —H) become attached to each of the fragments.

hydrophilic substance Polar molecule (e.g., glucose) that easily dissolves in water.

hydrophobic substance Nonpolar molecule (e.g., oil) that resists dissolving in water.

hydrostatic pressure Pressure exerted by a volume of fluid against a wall, membrane, or some other structure that encloses it.

hydrostatic skeleton Of many soft-bodied invertebrates, a fluid-filled cavity or cell mass against which contractile cells act.

hypertonic solution Of two fluids, the one having the higher solute concentration.

hypha, plural **hyphae** Fungal filament with a chitin-reinforced wall; part of a mycelium.

hypothalamus Brain center of homeostatic control over the internal environment, viscera, and emotions; also produces some hormones.

hypothesis In science, a possible explanation of a phenomenon, one that has the potential to be proven false by experimental tests.

hypotonic solution Of two fluids, the one having the lower solute concentration.

immigration New individuals permanently move into an area.

immune system Body system that recognizes antigen and mounts attacks against specific threats. Adaptive in jawed vertebrates.

immunoglobulin One of the five classes of antibody proteins (e.g., IgG).

implantation In placental mammals, the burrowing of a blastocyst into the uterus wall.

imprinting Learning that occurs during a sensitive period for a young animal, triggered by exposure to a simple stimulus.

in vitro fertilization IVF. Combining sperm and eggs outside the body, as in a petri dish.

inbreeding Mating between close relatives.

inclusive fitness Genetic contribution made to the next generation by an individual and its close relatives.

independent assortment In meiosis, each homologous chromosome and its partner are assorted into different gametes independently of other pairs.

indicator species Any species that provides warning of changes in habitat and impending widespread loss of biodiversity.

induced-fit model An enzyme changes shape to fit a bound substrate, and the resulting tension destabilizes the substrate's bonds.

infection Invasion and multiplication of a pathogen in a host. Disease follows if defenses are not mobilized fast enough; the pathogen's activities interfere with normal body function.

inflammation See acute inflammation.

inhibitor A hypothalamic hormone that slows secretion by cells in a target gland.

insertion Mutation involving insertion of one to a few bases into a DNA strand. Also, a movable attachment of muscle to bone.

instinctive behavior Any behavior that an animal performs without having first learned it through experience.

integrator Control center (e.g., brain) for the animal body that receives, processes, and stores sensory input; issues commands for coordinated responses.

integumentary exchange Respiration across a thin, moist, and often vascularized surface layer of animal tissue.

intermediate filament Cytoskeletal element; mechanically strengthens some animal cells.

internal environment In animals, blood and interstitial fluid (extracellular fluid).

interneuron Neuron of brain or spinal cord.

interphase Cell cycle interval between nuclear divisions; a cell grows in mass and roughly doubles the number of cytoplasmic components. DNA replication occurs during the interphase that precedes mitosis.

interspecific competition Competition between members of different species.

interstitial fluid Extracellular fluid in the spaces between animal cells and tissues.

intron Noncoding gene sequence that is removed from a pre-mRNA before translation.

inversion Mutation in which a section of chromosome becomes oriented in reverse.

invertebral disk Cartilaginous flex point and shock absorber between vertebrae.

invertebrate An animal without a backbone.

ionic bond Interaction between ions held together by attraction of opposite charges.

ionizing radiation Radiation with enough energy to eject electrons from atoms.

isotonic solution A fluid having the same solute concentration as another fluid to which it is being compared.

isotopes Two or more forms of an element's atoms differing in the number of neutrons.

jaw Of chordates, a hinged pair of cartilaginous or bony feeding structures.

joint Area of contact between bones.

karyotype Preparation of an individual's metaphase chromosomes sorted by length, centromere location, and shape.

key innovation Change in body form or function that allows a lineage to exploit the environment in more efficient or novel ways.

keystone species A species that has a major role in shaping community structure.

kidney One of a pair of vertebrate organs that filter substances from blood and form urine.

kilocalorie A thousand calories of heat energy, the amount needed to raise the temperature of 1 kilogram of water by 1°C. Standard unit of measure for the energy content of foods.

knockout experiment Experiment in which, to study the function of a gene, an organism is engineered to lack its expression.

Krebs cycle The second stage of aerobic respiration in which pyruvate is broken down to carbon dioxide and water. Two ATP form. Occurs only in mitochondria.

K–T asteroid impact theory Idea that an asteroid impact caused a mass extinction at the Cretaceous–Tertiary boundary (65 million years ago).

La Niña Cool climatic event that occurs between El Niño episodes; has global effects.

labor Process by which a placental mammal gives birth.

lactate fermentation Anaerobic pathway of ATP formation. Pyruvate from glycolysis is converted to three-carbon lactate, and NAD^+ is regenerated. Net energy yield: 2 ATP.

lactation In mammals only, secretion of milk by mammary glands.

Langerhans cell Immune cell in skin; engulfs pathogens, alerts immune system to threats.

larva, plural **larvae** A sexually immature stage of many invertebrates.

larynx Tubular airway from the pharynx to lungs. Contains vocal cords in some animals.

lateral gene transfer Movement of genetic material between existing cells by conjugation or other processes.

leaching Loss of nutrients from soil as water percolates through it.

lethal mutation Mutation with drastic effects on phenotype; usually causes death.

LH Luteinizing hormone. Anterior pituitary hormone that has reproductive roles in males and females.

lichen Mutualistic interaction between a fungus and a photoautotroph.

life-history pattern Adaptations affecting life span, fertility, and age at first reproduction.

ligament Strap of dense connective tissue that bridges a joint and attaches to bones.

light-dependent reactions The first stage of photosynthesis. Sunlight energy is trapped and converted to chemical energy of ATP, NADPH, or both, depending on the pathway.

light-independent reactions Second stage of photosynthesis in which sugars are formed from CO_2 using ATP and NADPH. Also called the Calvin-Benson cycle.

limbic system Brain center that governs emotions and has roles in memory.

limiting factor Any essential resource that can halt population growth when supplies of it dwindle.

lineage An ancestor–descendant sequence of cells, populations, or species.

linkage group All genes on a chromosome.

lipid Nonpolar hydrocarbon; fats, oils, waxes, phospholipids, and sterols are lipids.

lipid bilayer Mainly phospholipids arranged tail-to-tail in two layers; structural basis of all cell membranes.

loam Soil with roughly the same proportions of sand, silt, and clay.

local signaling molecule A cell secretion that alters chemical conditions in nearby tissues.

logistic growth, population The size of a population increases slowly, then rapidly, then levels off as the carrying capacity is reached.

loop of Henle Hairpin-shaped, tubular part of a nephron that reabsorbs water and solutes.

loose connective tissue Animal tissue with fibers, fibroblasts loosely arrayed in semifluid ground substance.

lungs Internally moistened sacs specialized for gas exchange.

lycophyte Seedless vascular plant having leaves with a single vein (e.g., club moss).

lymph Fluid in vessels of lymphatic system.

lymph node Lymphoid organ packed with lymphocytes; filters lymph.

lymph vascular system Parts of lymphatic system; delivers excess tissue fluid, absorbed fats, and reclaimable solutes to blood.

lymphatic system Organ system that returns fluid and solutes from interstitial fluid to blood; also functions in body defense.

lysis Gross damage to a plasma membrane, cell wall, or both that lets the cytoplasm leak out and so causes cell death.

lysosome Organelle of intracellular digestion.

lysozyme Infection-fighting enzyme present in mucous membranes.

macroevolution Large-scale patterns, trends, and rates of change among higher taxa.

macrophage Phagocytic white blood cell with roles in nonspecific defense and immunity.

magnoliid One of the three flowering plant groups (e.g., magnolias, avocado trees).

marine snow Organic matter that drifts to ocean depths and supports food webs there.

marsupial Pouched mammal.

mass extinction Catastrophic event or phase in geologic time when entire families or other major groups are irrevocably lost.

mass number The sum of all protons and neutrons in an atom's nucleus.

mast cell Histamine-secreting white blood cell with roles in inflammation and allergies.

master gene A gene whose product has widespread effects on other genes.

mechanoreceptor Sensory cell or nearby cell that detects mechanical energy (e.g., a change in pressure).

medulla oblongata Hindbrain region with reflex centers that influence functions basic to survival (e.g., sleeping, breathing, coughing).

megaspore Haploid spore formed by meiosis in the ovary of a seed-bearing plant; one of its cellular descendants develops into an egg.

meiosis Only nuclear division process that halves the chromosome number of a parental cell, to the haploid number. Forms gametes in animals and spores in plants.

melanin Brownish-black pigment; protects human skin from ultraviolet radiation.

memory The capacity to store and retrieve information about past sensory experience.

memory cell B or T cell that formed during an immune response; it remains in a resting phase until a secondary immune response.

menstrual cycle Recurring cycle in human females. Includes menstruation, ovulation, repair and priming of uterus for pregnancy.

meristem Region of actively dividing, undifferentiated cells in plants; descendants of meristematic cells give rise to mature plant tissues (e.g., leaves, stems, roots).

mesoderm Primary tissue layer that occurs in most animal embryos; gives rise to internal organs and part of the integument; pivotal in the evolution of large, complex animals.

mesophyll Photosynthetic ground tissue of a leaf; a type of parenchyma.

messenger RNA mRNA. Single-stranded ribonucleotide product of gene transcription; encodes protein-building instructions.

metabolic pathway Sequence of enzyme-mediated reactions by which cells assemble and build or break down organic compounds.

metabolism All the controlled, enzyme-mediated chemical reactions by which cells acquire and use energy to synthesize, store, degrade, and eliminate substances.

metamorphosis Of certain animals (e.g., many insects), drastic transformation of an immature stage to the adult through tissue reorganization and remodeling of parts.

metaphase Of meiosis I, stage when all pairs of homologous chromosomes have become positioned at the spindle equator. Of mitosis or meiosis II, all duplicated chromosomes are positioned at the spindle equator.

methanogen Type of archaebacterium that produces methane gas as a metabolic product.

micelle formation Clustering of bile salts, fatty acids, and monoglycerides into droplets; enhances fat absorption in the small intestine.

microevolution Of a population, any change in allele frequencies resulting from mutation, genetic drift, gene flow, natural selection, or some combination of these.

microfilament Cytoskeletal element; consists of actin subunits. Involved in movement and structural integrity of cells.

microspore Walled, haploid spore that gives rise to a pollen grain in a seed-bearing plant.

microtubule Cytoskeletal element; consists of tubulin subunits. Contributes to cell shape, growth, and motion; constituent of spindles.

microvillus, plural **microvilli** Slender extension from free surface of certain cells; increases surface area (e.g., for absorption).

migration Recurring movement of organisms from one region to another, then back.

mimicry Close resemblance of one species to another; confers a selective advantage upon one or both species by deceiving predators.

mineral An element or inorganic compound formed by natural geologic processes; many are required for normal metabolic function.

mitochondrion, plural **mitochondria** Organelle of ATP formation; site of aerobic respiration's second and third stages.

mitosis Nuclear division mechanism that maintains the parental chromosome number for forthcoming daughter cells. Basis of growth, tissue repair, and often asexual reproduction of eukaryotes.

mixture Two or more elements intermingled in proportions that can and usually do vary.

model Theoretical description of something that has not been directly observed.

molecular clock The time of origin of one lineage or species relative to others may be estimated by comparing the number of neutral mutations; assumes that accumulation of neutral mutations occurs at a fixed rate.

molecule Two or more atoms of the same or different elements joined by chemical bonds.

mollusk Invertebrate having a unique tissue flap (mantle) draped over a soft, fleshy body; most have an external or internal shell (e.g., a gastropod, bivalve, cephalopod).

molt Periodic shedding of worn-out or too-small body structures. Permits some animals to grow in size or renew parts.

monocot Monocotyledon; flowering plant with one embryonic seed leaf (cotyledon), floral parts usually in threes (or multiples of three), and often parallel-veined leaves.

monohybrid cross Intercross between two individuals, each heterozygous for one gene (e.g., *Aa*).

monophyletic group A group descended from a common ancestor in which the derived trait that characterizes the group first evolved.

monosaccharide One of the simple sugars (e.g., glucose) that are unit components of oligosaccharides or polysaccharides.

monotreme Egg-laying mammal.

monsoon Air circulation pattern that moves moisture-laden air arising from warm oceans to continents north or south of them.

morphogen Type of inducer molecule that diffuses through embryonic tissues, activating master genes in sequence; it contributes to mapping out the overall body plan.

morphogenesis Programmed, orderly changes in body size, proportion, and shape of an animal embryo through which all specialized tissues and early organs form.

morphological convergence In response to similar selective pressures, evolutionarily distant lineages evolve in similar ways and end up resembling each other in appearance, function, or both.

morphological divergence In response to differing selective pressures, diverging lineages undergo gradual change from the body form of their common ancestor.

motor neuron Neuron that relays signals from the central nervous system to muscle or gland cells.

motor protein Protein that associates with microtubules or microfilaments and has a role in cell movement.

motor unit A motor neuron and all muscle cells that form junctions with its endings.

multiple allele system Three or more slightly different molecular forms of a gene that occur among the individuals of a population.

muscle fatigue Decline in a muscle's capacity to generate force after prolonged contraction.

muscle fiber Of skeletal muscle, a cylindrical multinucleated fiber that develops through the fusion of many cells during development.

muscle spindle A stretch-detecting sensory organ associated with skeletal muscle.

muscle tension Mechanical force exerted by a contracting muscle; resists opposing forces.

muscle tissue Tissue with arrays of cells able to contract under stimulation, then passively lengthen and return to their resting position.

muscle twitch Single, brief contraction of a muscle in response to a stimulus.

mutation Heritable change in DNA.

mutation rate Occurrence of mutations in a particular gene as a function of time.

mutualism Symbiotic interaction that benefits both partners.

mycorrhiza, plural **mycorrhizae** Mutualistic interaction between a fungus and root.

myofibrils Threadlike, cross-banded structures within a skeletal muscle fiber.

myoglobin A respiratory pigment abundant in cardiac and skeletal muscle cells.

myosin Motor protein that makes up the thick filaments of a sarcomere.

NAD⁺ Nicotinamide adenine dinucleotide. A nucleotide coenzyme; abbreviated NADH when carrying electrons and H⁺.

natural killer cell NK cell. Lymphocyte that touch-kills tumor cells and viral-infected cells.

natural selection Microevolutionary process; the outcome of differences in survival and reproduction among individuals that differ in the details of their heritable traits.

negative control In gene expression, regulatory proteins slow or stop transcription or translation.

negative feedback mechanism A homeostatic mechanism by which a condition that changed as a result of some activity triggers a response that reverses the change.

nematocyst Cnidarian capsule that has a dischargeable, tube-shaped thread, sometimes barbed; releases a toxin or sticky substance.

neoplasm Mass of cells (tumor) that have lost control over growth and division.

nephridium plural **nephridia** Unit that controls composition and volume of fluid in some invertebrates (e.g., earthworms).

nephron One of the kidney's urine-forming tubules; filters water and solutes from blood, then selectively reabsorbs portions of both.

nerve Sheathed bundle of the axons of sensory neurons, motor neurons, or both.

nerve cord A longitudinal nerve; most animals have one to three. Chordate nervous systems arise from a dorsal nerve cord.

nerve net Simple nervous system of cnidarians and some other invertebrates; a diffuse mesh of nerve cells in epithelial tissue.

nervous system Organ system with nerve cells interacting in signal-conducting and information-processing pathways. Detects and processes stimuli, and elicits responses from effectors (e.g., muscles and glands).

nervous tissue Tissue of excitable neurons and supporting neuroglia.

net ecosystem production Of ecosystems, total net energy accumulated in producers through their growth and reproduction in a given interval (net primary production).

net population growth rate per individual (*r*) For population growth equations, a variable combining birth rates and death rates; assumes that both remain constant in specified interval.

neural tube The embryonic and evolutionary forerunner of the brain and spinal cord.

neurotransmitter Signaling molecule secreted by axon endings of a neuron.

neutral mutation Mutation that has little or no effect on phenotype.

neutron Subatomic particle found in an atom's nucleus; has mass but no charge.

neutrophil Most abundant type of white blood cell; engulfs pathogens and has a role in inflammatory responses.

niche Sum of all activities and relationships by which a species obtains and uses resources.

nitrification The chemical conversion of ammonia to nitrate by soil bacteria.

nitrogen cycle Movement of nitrogen from the atmosphere, through the ocean, sediments, soils, and food webs, then back to atmosphere.

nitrogen fixation Conversion of nitrogen gas to forms that plants can take up from soil.

nondisjunction Failure of sister chromatids or homologous chromosomes to separate during meiosis or mitosis. Daughter cells end up with too many or too few chromosomes.

non-ionizing radiation Radiation that carries enough energy to boost electrons to higher energy levels, but not enough to remove them.

nonshivering heat production Hormone-induced increase in metabolic rate in response to prolonged or severe cold exposure.

notochord Of chordates, a rod of stiffened tissue (not cartilage or bone) that is a supporting structure for the body.

nuclear envelope Lipid bilayer membrane enclosing the nucleus of eukaryotes.

nucleic acid Single-stranded or double-stranded molecule composed of nucleotides joined at phosphate groups (e.g., DNA, RNA).

nucleic acid hybridization Any base-pairing between DNA or RNA from different sources.

nucleoid Of bacterial cells, the region in which DNA is physically organized; not separated from the cytoplasm by a membrane.

nucleosome A small stretch of eukaryotic DNA wound around histone proteins.

nucleotide Small organic compound with a five-carbon sugar, a nitrogen-containing base, and a phosphate group.

nucleus Organelle that physically separates DNA from the cytoplasm in a eukaryotic cell.

numerical taxonomy Method of determining the relationship between an unidentified organism and a known group by comparing traits. The greater the number of traits in common, the greater the inferred relatedness.

nutrient Element with a direct or indirect role in metabolism that no other element fulfills.

nutrition Processes of selectively ingesting, digesting, absorbing, and converting food into the body's own organic compounds.

obesity Excess of fat in adipose tissue; caloric intake has exceeded the body's energy output.

ocean A continuous body of saltwater that covers more than 71 percent of the Earth.

olfactory receptor Chemoreceptor for water-soluble or volatile substance.

oocyte Immature egg.

oomycote A nonphotosynthetic stamenopile protist; many are plant pathogens.

operator Part of an operon; a binding site for a regulatory protein.

operon Promoter–operator sequence that controls transcription of more than one bacterial gene.

organ Two or more tissues arrayed in a specific pattern and interacting in some task.

organ system Two or more organs interacting chemically, physically, or both in a task.

organelle Membrane-bound compartment in the eukaryotic cytoplasm; has one or more specialized metabolic functions.

organic compound Molecule containing carbon and hydrogen; may also contain oxygen, nitrogen, and other elements.

osmoreceptor Sensory cell that detects change in solute concentration of surrounding fluid.

osmosis Diffusion of water between two regions separated by a selectively permeable membrane.

osmotic pressure Hydrostatic pressure that counters inward diffusion of water through a selectively permeable membrane inside a cell or enclosed body region.

ovary In flowering plants, enlarged base of one or more carpels. In most animals, a female gonad in which eggs form.

ovulation Release of a secondary oocyte from an ovary, induced by an LH surge.

ovule Tissue mass in which an egg forms in a plant ovary; immature seed.

ovum Mature secondary oocyte; mature egg.

oxidation–reduction reaction Transfer of electrons between reactant molecules.

oxytocin Posterior pituitary hormone; induces lactation and shrinkage of uterus after pregnancy; also affects social behavior of some mammals.

ozone thinning Seasonal thinning of the atmospheric ozone layer; most pronounced above polar regions.

pain Perception of injury to a body region.

pain receptor A nociceptor; a sensory receptor that detects tissue damage.

pancreatic islet Any of the 2 million or so clusters of endocrine cells of the pancreas.

Pangea Paleozoic supercontinent.

parapatric speciation Mode of speciation in which subpopulations of a species that are maintaining contact along a common border evolve into distinct species.

parasitism Interaction in which one organism (the parasite) lives on or in another (the host) and feeds on its tissues.

parasitoid Type of insect larva that grows and develops in a host organism (usually another insect), consumes its soft tissues, and kills it.

parasympathetic nerve An autonomic nerve; its signals cause a slowdown in overall activity and divert energy to basic tasks.

parathyroid gland One of four endocrine glands; its secretions cause a rise in blood calcium levels.

parenchyma Type of simple tissue that makes up the bulk of a plant.

parthenogenesis An unfertilized egg gives rise to an embryo.

partial pressure The contribution of any gas to total atmospheric pressure.

passive transport Diffusion of a solute across a cell membrane, through a transport protein.

pathogen Disease-causing agent that can infect a target species and multiply inside it.

pattern formation During development, sculpting of embryonic cells into specialized animal tissues and organs at expected times, in expected places.

PCR Polymerase chain reaction. A method that rapidly amplifies the number of specific DNA fragments.

peat bog Acidic wetland where peat mosses grow; peat is their compressed remains.

pedigree Chart of genetic connections.

pellicle Of some protists, a flexible body covering of protein-rich material.

peptide hormone Short chain of amino acids that acts as a hormone.

per capita Per individual.

periodic table Tabular arrangement of elements based on their chemical properties.

peripheral vasoconstriction Diameter of arterioles decreases and blood flow to body surfaces is reduced.

peripheral vasodilation Diameter of arterioles increases and blood flow to body surfaces is increased.

permafrost A perpetually frozen layer of soil.

peroxisome Enzyme-filled vesicle that breaks down amino acids and fatty acids to hydrogen peroxide, which is converted to harmless products.

pH scale A measure of the H^+ concentration (acidity) of blood, water, and other solutions. pH 7 is neutral.

pharynx A muscular tube for filter-feeding in invertebrate chordates, and, in some species, gas exchange. In many other animals, part of the digestive tract.

phenotype Observable trait or traits of an individual that arises from gene interactions and gene–environment interactions.

pheromone Hormone-like exocrine gland secretion; diffuses through air and affects a different member of the same species.

phloem A complex tissue that conducts sugars, solutes through a vascular plant.

phospholipid Lipid with a phosphate group. Major constituent of biological membranes.

phosphorus cycle Movement of phosphorus from land, through food webs, to ocean sediments, then back to land.

phosphorylation Enzyme-mediated transfer of a phosphate group between molecules.

photoperiodism Biological response to change in the relative amounts of daylight and darkness.

photoreceptor Light-sensitive sensory cell.

photosynthesis Process by which organisms use sunlight energy to convert carbon dioxide and water to sugars.

photosystem In photosynthetic cells, a cluster of membrane-bound, light-trapping pigments and other molecules.

phototropism Change in the direction of cell movement or growth in response to light.

phycobilin Red to blue photosynthetic accessory pigment.

phylogeny Evolutionary relationships among species, starting with an ancestral form and including branches leading to descendants.

phytochrome A light-sensitive pigment. Its controlled activation and inactivation affect hormones governing many plant activities, including growth, branching, and flowering.

phytoplankton Aquatic community of floating or swimming photoautotrophs.

pigment Any light-absorbing molecule.

pilomotor response Hairs or feathers stand up; creates a layer of still air next to the skin.

pineal gland Light-sensitive endocrine gland; secretes melatonin.

pioneer species Species that can colonize newly formed or newly vacated habitats.

pituitary gland Endocrine gland that interacts with the hypothalamus to control other glands and organs.

placenta Organ that forms from endometrial tissue and extraembryonic membranes. Permits exchange of substances between a pregnant female and her fetus while keeping their bloodstreams separate.

plankton Community of mostly microscopic species that swim or float in lakes or seas.

plant A multicelled photoautotroph with well-developed roots and shoots.

plasma Liquid portion of blood; mainly water in which ions, proteins, sugars, gases, and other substances are dissolved.

plasma membrane Outermost cell membrane; structural and functional boundary between the cytoplasm and fluid surrounding the cell.

plasmid A small, circular molecule of bacterial DNA that carries a few genes and is replicated independently of the chromosome.

plate tectonic theory Idea that great slabs (plates) of the Earth's outer layer float about slowly on the mantle beneath them and have rafted continents to new positions over time.

platelet Cell fragment that circulates in blood; acts in clot formation.

pleiotropy Positive or negative effects that alleles at a single gene locus have on two or more traits.

polar body In vertebrates, one of four cells that forms by meiotic cell division of an oocyte but that does not become the ovum.

pollen grain Sperm-bearing gametophyte of a gymnosperm or angiosperm.

pollination Transfer of pollen to a female part of a flower (stigma).

pollinator Agent that transfers pollen to female floral parts.

polypeptide chain Three or more amino acids linked by peptide bonds.

polyploidy Having three or more of each chromosome type characteristic of a species.

polysaccharide Straight or branched chain of many covalently linked sugar units of the same or different kinds. Most common types are cellulose, starch, and glycogen.

population Group of individuals of the same species in a specified area.

population density Number of individuals per specified area or volume of a habitat.

population distribution Pattern of dispersion of individuals of a population.

population size Number of individuals making up a population.

positive control In gene expression, regulatory proteins enhance transcription or translation.

positive feedback mechanism An event intensifies as a result of its own occurrence.

predation Interaction in which one organism (the predator) eats another (prey), typically killing it.

predator Free-living organism that captures and feeds on other organisms (its prey).

prediction Statement about what you expect to observe in nature.

pressure flow theory Organic compounds flow through phloem in response to pressure and concentration gradients between source regions (e.g., leaves) and sinks (e.g., regions where sugars are being used or stored).

pressure gradient Difference in pressure being exerted in two adjoining regions.

prey Organism that predators can capture and eat.

primary growth Plant growth originating at root tips and shoot tips.

primary immune response Defensive actions by white blood cells elicited by first-time recognition of antigen. Includes antibody- and cell-mediated responses.

primary oocyte An immature egg that is stopped in prophase I of meiosis.

primary producer Type of autotroph that secures energy directly from the environment.

primary productivity Of ecosystems, the rate at which primary producers capture and store energy in their tissues during some interval.

primary succession Sequence of community development from pioneer species to climax stage in a previously barren habitat.

primary wall Of young plant cells, a thin, flexible wall that permits division and changes in shape; consists of cellulose, polysaccharides, glycoproteins.

primate Mammalian lineage; includes prosimians, tarsioids, anthropoids (monkeys, apes, humans).

primer Short nucleotide sequence designed to serve as a site of initiation for DNA synthesis on DNA or RNA.

prion Small infectious protein that causes fatal degenerative diseases of nervous system.

probability The chance that each outcome of an event will occur is proportional to the number of ways in which the outcome can be reached.

probe Short nucleotide sequence, labeled with a tracer, designed to hybridize with part of a specific gene or mRNA.

producer Autotroph (self-feeder); nourishes itself using sources of energy and carbon from the environment. Photoautotrophs and chemoautotrophs are examples.

progesterone Female sex hormone secreted by ovaries and the corpus luteum.

prokaryotic cell Archaean or bacterium; single-celled organism, most often walled; lacks a nucleus and other organelles.

prokaryotic fission Bacterial mode of reproduction. Involves DNA replication, accumulation of new membrane (and usually wall material) at or near the cell midsection, then cytoplasmic division.

prolactin Anterior pituitary hormone that stimulates milk production.

promoter Short stretch of DNA to which RNA polymerase binds and initiates transcription.

prophase Of mitosis, a stage when duplicated chromosomes start to condense, a spindle forms, and the nuclear envelope starts to break up. Duplicated pairs of centrioles move to opposite spindle poles. In prophase I of meiosis, crossing over also occurs.

protein Organic compound consisting of one or more polypeptide chains folded and twisted into a three-dimensional shape.

proteobacteria Most diverse bacterial group.

protist One of the mainly single-celled species of eukaryotes traditionally grouped in the catch-all "kingdom Protista." Currently being classified into groupings that reflect evolutionary relationships.

proto-cell Membrane-bound metabolic machinery; transitional stage that may have preceded the origin of living cells.

proton Positively charged subatomic particle found in an atom's nucleus.

proto-oncogene A gene that, when mutated or overexpressed, helps turn a normal cell into a cancerous one.

protostome Lineage of coelomate, bilateral animals that includes mollusks, annelids, and arthropods. The first indentation to form in protostome embryos becomes the mouth.

proximal tubule Nephron's tubular portion extending from Bowman's capsule.

pseudocoel A body cavity that is not fully lined with tissue derived from mesoderm.

pseudopod Flexible, temporary lobe of membrane-enclosed cytoplasm that amoebas, amoeboid cells, and phagocytic white blood cells use for motility or engulfing food.

puberty Of human development, a post-embryonic stage when gametes start to mature and secondary sexual traits emerge.

pulmonary circuit A blood circulation route that moves blood from the heart's right side, through lungs, then to the heart's left half.

punctuation model of speciation Idea that most morphological changes occur in a brief surge, as populations start to diverge.

Punnett-square method Construction of a simple diagram to predict the probable outcomes of a genetic cross.

pyruvate Three-carbon compound that forms as the end product of glycolysis.

quadrat One of many areas of a given size and shape in which samples are taken or individuals of a population are counted.

radial symmetry Animal body plan having four or more roughly equivalent parts around a central axis, as in a sea anemone.

radioactive decay An atom emits energy as subatomic particles and x-rays as its unstable nucleus disintegrates spontaneously. The process transforms one element into another.

radioisotope Isotope with an unstable nucleus (too many or too few neutrons).

radiometric dating Method of determining the age of a fossil by comparing the relative proportions of parent and daughter radioisotopes in rock samples or fossils.

rain shadow Reduction in rainfall on the leeward side of a high mountain range; results in arid or semiarid conditions.

recombinant DNA A DNA molecule that contains genetic material from more than one organism.

red alga Type of photoautotrophic protist; most are multicelled and aquatic; phycobilins mask their chlorophyll *a*.

red marrow Site of blood cell formation in the spongy tissue of many bones.

reflex Stereotyped, involuntary movement in response to a stimulus.

refractory period Brief interval following an action potential when a small patch of neural membrane is insensitive to stimulation.

regulatory protein Protein that enhances or inhibits protein synthesis.

relaxin Hormone secreted by ovaries and placenta that prepares the cervix and pelvis for giving birth.

releaser Hypothalamic releasing hormone that stimulates secretion by a target gland.

renal corpuscle Bowman's capsule plus the glomerular capillaries it encloses.

reproductive base All individuals of a population that are in the pre-reproductive and reproductive age brackets.

reproductive isolating mechanism Any heritable aspect of body form, function, or behavior that prevents interbreeding between populations of the same species.

reproductive success Production of viable offspring by the individual.

reptile Tetrapod vertebrate that has scaly skin and water-conserving kidneys, and produces amniote eggs (e.g., dinosaur, crocodilian, turtle, snake, lizard, tuatara).

resource partitioning A subdividing of resources in time or space that allows similar species to coexist in a habitat.

respiration Movement of oxygen into an animal's internal environment and carbon dioxide out of it.

respiratory cycle One inhalation and one exhalation.

respiratory membrane The alveolar and capillary endothelia, together with their basement membranes.

respiratory pigment A molecule in body fluids that reversibly binds oxygen (e.g., hemoglobin).

respiratory surface Thin, moist epithelium that functions in gas exchange.

resting membrane potential Of a neuron and other excitable cells, a voltage difference across the plasma membrane that holds steady in the absence of outside stimulation.

restriction enzyme A protein that recognizes and cuts specific sequences of nucleotides in double-stranded DNA.

retina Dense array of photoreceptors at the back of the eye.

reverse transcriptase Enzyme that assembles a single strand of DNA from free nucleotides on an RNA template; found in RNA viruses.

Rh blood typing Method of characterizing red blood cells by a certain membrane surface protein. Rh^+ cells have it; Rh^- cells do not.

rhizoid Simple rootlike absorptive structure of some fungi and nonvascular plants.

ribonucleic acid See RNA.

ribosomal RNA rRNA. Structural and functional RNA component of ribosomes.

ribosome Structure upon which polypeptide chains are built. An intact ribosome consists of two subunits of rRNA and proteins.

riparian zone Narrow corridor of vegetation on either side of a stream or river.

RNA Ribonucleic acid. Any of a class of single-stranded nucleic acids with roles in transcription, translation, and catalysis.

RNA polymerase Enzyme that catalyzes the addition of nucleotides to a growing strand of RNA (transcription).

RNA world A presumed period before the origin of life when RNA may have stored protein-building information.

rod cell Vertebrate photoreceptor sensitive to dim light; contributes to the coarse perception of movement across the visual field.

root Plant part, typically belowground; absorbs water and minerals, anchors aboveground parts, and often stores food.

root hair Thin extension of a root epidermal cell; collectively, root hairs greatly increase the surface area for absorbing water and ions.

root nodule Localized swelling on a root that contains mutualistic nitrogen-fixing bacteria.

roundworm Cuticle-covered, bilateral worm with a false coelom and complete digestive system; a nematode.

rubisco RuBP carboxylase. Enzyme that catalyzes attachment of a carbon atom from carbon dioxide to RuBP and starts the C3 photosynthetic pathway.

ruminant Hoofed, herbivorous mammal with multiple stomach chambers.

sac fungus Fungus that forms its sexual spores (ascospores) in sac-shaped cells.

salinization Salt buildup in soil through poor drainage, evaporation, and heavy irrigation.

saliva Glandular secretion that mixes with food and starts starch breakdown in mouth.

salt Compound that releases ions (other than H^+ and OH^-) in solution.

sampling error An experimental pitfall; arises when the sample or subset of a population, an event, or some other aspect of nature under study is not representative of the whole.

sap Sugary solution that circulates through vascular tissues of plants.

sapwood Functioning secondary xylem of an older woody plant.

sarcomere Basic unit of contraction in skeletal and cardiac muscle. Shortens as a result of ATP-driven interactions between actin and myosin filaments.

scientific theory An explanation of the cause of a range of related phenomena; has been rigorously tested but is still open to revision.

sclerenchyma Simple plant tissue that supports mature plant parts; often protects seeds. Most of its cells have thick, lignin-impregnated walls.

second law of thermodynamics A law of nature stating that the spontaneous direction of energy flow is from organized forms to less organized forms; with each conversion, some energy is randomly dispersed in a form (usually heat) not as useful for doing work.

second messenger Molecule within a cell that mediates a hormonal signal from outside.

secondary oocyte An oocyte that has finished meiosis I just before being released from an ovary at ovulation.

secondary succession Recovery of a community to its climax stage following a habitat disturbance, such as a forest fire.

secondary wall A rigid, permeable wall that forms inside the primary wall of some older plant cells.

seed Mature ovule; an embryo sporophyte and endosperm surrounded by a seed coat.

segmentation Of animal body plans, a series of units that may or may not be similar to one another in appearance. Of tubular organs, an oscillating movement produced by rings of circular muscle in the tube wall.

selective gene expression Controlled activation or suppression of transcription and translation; leads to cell identity.

selective permeability Built-in capacity of a cell membrane to stop some substances from crossing, and to allow others to cross it, at certain times, in certain amounts.

selfish behavior An individual protects or increases its own chance to produce offspring regardless of consequences to its social group.

selfish herd Individuals that cluster together for protection against predators or some other environmental danger.

senescence Processes leading to the natural death of an organism or parts of it (e.g., abscission of leaves from a deciduous tree).

sensation Conscious awareness of a stimulus.

sensory adaptation Decrease in response to a stimulus maintained at constant strength.

sensory neuron Type of neuron that detects a stimulus and relays information about it to an integrating center such as a brain.

sensory receptor Any cell or some part of it that can detect a stimulus.

sex chromosomes Chromosomes that, in certain combinations, determine a new individual's sex.

sexual dimorphism Occurrence of distinctive female and male phenotypes.

sexual reproduction Production of genetically variable offspring by meiosis, gamete formation, and fertilization.

sexual selection Mode of natural selection; favors a trait that gives the individual a competitive edge in attracting or keeping a mate, hence in reproductive success.

sexually-transmitted disease STD. Any one of the diseases spread by sexual intercourse.

shivering response Rhythmic tremors in response to cold; increases heat production.

shoot A stem, leaf, or flower.

sieve tube Conducting tube of phloem.

sister chromatids Two identical DNA molecules (and associated proteins) attached at the centromere until they are separated from each other at mitosis or meiosis; each is then a separate chromosome.

six-kingdom system Classification of all species into the kingdoms Bacteria, Archaea, Protista, Fungi, Plantae, and Animalia.

skeletal muscle tissue Striated contractile tissue; main component of muscles that attach to and move bones.

skin External integument of vertebrates; an outer epidermis and underlying dermis.

sliding-filament model Idea that muscle contraction occurs as a result of ATP-driven interactions between myosin and actin filaments in sarcomeres.

slime mold Predatory protist; amoebalike cells form fruiting bodies during life cycle.

smooth muscle tissue Nonstriated contractile tissue found in soft internal organs.

social behavior Diverse interactions among individuals of a species, which display, send, and respond to forms of communication that have genetic and learned components.

sodium–potassium pump Active transport protein that moves sodium and potassium ions across the cell membrane.

soil Mixture of mineral particles and decomposing organic material, with air and water occupying spaces between the particles.

soil erosion Movement of land under the force of wind, running water, and ice.

solute Any substance dissolved in a solution.

somatic nerve Type of nerve that carries signals from the central nervous system to skeletal muscles, and from sensory receptors into that system.

somites Paired bumps of mesoderm in a vertebrate embryo that give rise to skeletal muscles and bones, and part of the dermis.

species One kind of organism. Of species that reproduce sexually, one or more groups of natural populations in which individuals interbreed and are reproductively isolated from other such groups.

sperm Mature male gamete.

sphincter A ring of smooth muscles that can alternately contract and relax to close off and open a passageway to the body surface.

spinal cord The part of the central nervous system that runs through a canal in the vertebral column.

spirochete Member of a group of spring-shaped bacteria; some are human pathogens.

spleen A lymphoid organ that is a filtering station for blood, a reservoir of red blood cells, and a reservoir of macrophages.

sponge A filter-feeding animal with no body symmetry and no tissues.

spore A reproductive or resting structure of one or a few cells, often walled or coated; protects against harsh conditions, aids in dispersal, or both.

sporophyte A vegetative body that grows by way of mitotic cell divisions from a plant zygote and produces spore-bearing structures.

spring overturn Downward movement of oxygen-rich surface waters and upward movement of nutrient-rich waters from the depths of temperate-zone lakes in spring.

stabilizing selection Mode of natural selection; intermediate phenotypes in the range of variation are favored and extremes are selected against.

stamen A male reproductive part of a flower.

statolith A gravity-sensing mechanism based on clusters of dense particles.

stem cell Undifferentiated animal cell that can divide indefinitely; a portion of daughter cells differentiate into specialized cell types.

steroid hormone Lipid-soluble hormone derived from cholesterol.

sterol Lipid with a rigid backbone of four fused carbon rings.

stimulus A form of energy that can be detected by a sensory receptor.

stoma, plural **stomata** A gap between two guard cells in leaf or stem epidermis; allows the diffusion of water vapor and gases across the epidermis.

stomach Muscular, stretchable sac that mixes and stores ingested food, helps break it up mechanically and chemically, and controls its entry into the small intestine.

stramenopile One of a group of protists having flagella with tinsel-like filaments.

stratification Layering of sedimentary rock; results from deposition of materials over time.

strip logging Logging of forested slopes in a narrow corridor to lessen negative impacts.

strobilus, plural **strobili** A conelike cluster of spore-producing structures (e.g., of cycads).

stromatolite Fossilized mats of shallow-water microbial communities, mainly cyanobacteria.

substrate-level phosphorylation The direct, enzyme-mediated transfer of a phosphate group from a substrate to another molecule.

surface-to-volume ratio Physical relationship in which volume increases with the cube of the diameter, but surface area increases with the square; constrains increases in cell size.

survivorship curve A graph that reflects how many individuals of a cohort survive, on average, at successive ages in their life span.

sympathetic nerve An autonomic nerve; deals mainly with increasing overall body activities at times of excitement or danger.

sympatric speciation In the absence of a physical barrier, a new species arises within the home range of an existing species.

synaptic integration Summation of all excitatory and inhibitory signals arriving at the trigger zone of a neuron or some other excitable cell.

syndrome A set of symptoms that characterize an abnormality or a disorder.

systemic circuit Circulatory route that carries oxygenated blood from the left side of the heart, through all tissues, then carries oxygen-poor blood back to the right half of the heart.

T lymphocyte T cell; a type of white blood cell vital to immune responses (e.g., helper T cells and cytotoxic T cells).

tandem repeat One of many short DNA sequences, occurring one after the other, in a chromosome. Used in DNA fingerprinting.

taproot system A primary root together with its lateral branchings; typical of eudicots.

taste receptor A chemoreceptor for substances dissolved in fluid bathing it.

TCR Receptor on T cell surface; binds to an antigen fragment that has become attached to a self marker on an antigen-presenting cell.

telomere Repetitive DNA cap at chromosome tip; shorter after each nuclear division.

telophase Of meiosis I, a stage when one member of each pair of homologous chromosomes reaches a spindle pole. Of mitosis and of meiosis II, the stage when chromosomes decondense into threadlike structures and daughter nuclei form.

temperature A measure of molecular motion.

temperature zone Globe-spanning latitudinal bands of temperature (e.g., tropical, warm temperate, cool temperate).

tendon Cord or strap of dense connective tissue that attaches muscle to bone.

test, scientific A means to determine the accuracy of a prediction, as by conducting experiments, making observations, or developing models.

testcross Experimental cross to determine whether an individual of unknown genotype that shows dominance for a trait is either homozygous dominant or heterozygous.

testis, plural **testes** A type of gonad (primary male reproductive organ); produces male gametes and sex hormones.

testosterone Male sex hormone produced in testes; functions in sperm formation and development of secondary sexual traits.

tetanus Of a muscle, sustained contraction that results from repeated stimulation of a motor unit. In a disease by the same name, a toxin prevents muscles from being released from contraction.

tetrapod Vertebrate that is a four-legged walker or a descendant of one (e.g., amphibian, reptile, bird, mammal).

thalamus Brain region; a coordinating center for sensory input and a relay station for signals to the cerebrum.

theory of uniformity Theory that Earth's surface changes in slow, uniformly repetitive ways. Helped change Darwin's view of evolution. Has since been replaced by plate tectonics theory.

thermal inversion A layer of dense, cool air trapped beneath a layer of warm air; can keep air pollutants close to the ground.

thermal radiation The surface of a warm body emits heat in the form of radiant energy.

thermoreceptor Sensory cell that detects radiant energy (heat).

thigmotropism Orientation of the direction of growth in response to physical contact with a solid object (e.g., a vine curls around a post).

three-domain system Classification of all species into the domains Bacteria, Archaea, and Eukarya.

threshold Of excitable cells (e.g., a neuron or muscle cell), the minimum amount of change in the resting membrane potential that causes an action potential.

thylakoid membrane In plants, internal portion of a chloroplast's membrane system, often folded into flattened sacs, that forms a single compartment. Pigments and enzymes are embedded in it; site of photosynthesis.

thymus gland Lymphoid organ; its hormones influence T cells, which form in bone marrow but migrate to and differentiate in the thymus.

thyroid gland Endocrine gland; its hormones influence overall growth, development, and metabolic rates of warm-blooded animals.

tight junction Animal cell junction that prevents substances from leaking between adjoining cells.

tissue Of multicelled organisms, a group of cells and intercellular substances that function together in one or more specialized tasks.

tissue culture propagation Inducing a tissue or organism to grow from an isolated cell of a parent tissue, in vitro.

topsoil Uppermost soil layer, the one that is most essential for plant growth.

total fertility rate TFR. Within a population, the average number of children born to a woman during her reproductive years.

tracer Substance with attached radioisotope that researchers can track after delivering it into a cell, a multicelled body, ecosystem, or other system.

trachea A type of air-conducting tube that occurs in respiratory systems. Of land vertebrates, the windpipe that connects the larynx to bronchi.

tracheal system Of insects and some other land-dwelling arthropods, a system of tubes that carry gases from body surfaces to tissues.

tracheid One of two types of water-conducting cells in xylem.

transcription First stage of protein synthesis, in which a strand of RNA is assembled on a DNA template (gene).

transfer RNA tRNA. A class of small RNA molecules that deliver amino acids to a ribosome; each pairs with an mRNA codon during translation.

translation Second stage of protein synthesis. Information encoded in an mRNA transcript guides the synthesis of a new polypeptide chain from amino acids; occurs at ribosomes.

translocation Of cells, a repositioning of a stretch of DNA to a new chromosomal location with no molecular loss. Of vascular plants, distribution of organic compounds through phloem.

transpiration Evaporative water loss from the aboveground parts of a plant.

transposon Transposable element. A stretch of DNA that can jump spontaneously and randomly to a different location in the genome; may cause mutation.

triglyceride A lipid that has three fatty acid tails attached to a glycerol backbone.

trophic level All organisms the same number of transfer steps away from the energy input into an ecosystem.

tubular reabsorption In kidneys, return of water and solutes from a nephron to blood.

tubular secretion In kidneys, excess ions and other substances move from interstitial fluid into the nephron, and are excreted in urine.

tumor Tissue mass with cells dividing at an abnormally high rate. If benign, cells stay in place; if malignant, they metastasize, or move to form tumors in new places in the body.

ultrafiltration Bulk flow of a small amount of protein-free plasma from a blood capillary when the outward-directed effect of blood pressure exceeds the inward-directed osmotic movement of interstitial fluid.

upwelling Upward movement of cold, nutrient-rich water from the ocean depths to the ocean surface.

uracil Nitrogen-containing base of a nucleotide of RNA molecules. Can base-pair with adenine.

urea Waste product formed in the liver from ammonia and CO_2; excreted in urine.

urinary excretion Mechanism by which excess water and solutes are removed from the body through the urinary system.

urinary system Organ system that adjusts the volume and composition of blood, and thereby helps maintain extracellular fluid.

urine Fluid consisting of any excess water, wastes, and solutes; it forms in kidneys by filtration, reabsorption, and secretion.

uterus In pouched and placental female mammals, organ in which embryos develop and are nurtured.

variable A specific aspect of an object or event that may differ over time and among individuals. In an experimental test, a single variable is directly manipulated in an attempt to support or disprove a prediction.

vascular bundle A strand-like array of primary xylem and phloem that threads through a plant's ground tissue system.

vascular cambium A lateral meristem that gives rise to secondary xylem and phloem, which increases stem or root diameter.

vascular cylinder Arrangement of vascular tissues as a central cylinder inside a root.

vasoconstriction Shrinking of blood vessel diameter, especially arterioles.

vasodilation Enlargement of blood vessel diameter, especially arterioles.

vegetative growth A new plant grows from an extension or fragment of its parent.

vein In plants, a vascular bundle inside a leaf. In animals, a vessel that returns blood to the heart and acts as a blood volume reservoir.

venule A small blood vessel that serves as a transitional conducting tube between a small-diameter capillary and a larger diameter vein.

vernalization The induction of flowering by exposure to low temperature.

vertebra, plural **vertebrae** A series of hard bones that function as a skeletal backbone and that protect the spinal cord.

vertebrate Animal having a backbone.

vessel member One cell type in xylem.

vestibular apparatus Organ of equilibrium in the inner ear.

villus, plural **villi** Fingerlike projections from the free surface of an epithelium, as into the lumen of the small intestine.

viroid A bit of RNA that infects plants.

virus A noncellular infectious agent that can be replicated only if its genetic material enters a host cell and subverts metabolic machinery.

viscera All soft organs inside an animal body (e.g., heart, lungs, and stomach).

vision Perception of visual stimuli.

visual accommodation Adjustments in a lens position or shape that focus light on a retina.

vitamin Any of more than a dozen organic substances that an organism requires in small amounts for metabolism but generally cannot synthesize for itself.

warning coloration Pattern or coloration that makes a toxic organism (or its mimics) easy to detect and avoid.

watershed Region from which water drains into a single stream or river.

water–vascular system Of sea stars and sea urchins, a system of many tube feet that are deployed in synchrony for smooth movement.

wavelength A wavelike form of energy in motion. The horizontal distance between the crests of every two successive waves.

wax A type of lipid with long-chain fatty acids attached to long-chain alcohols or carbon rings.

white blood cell Leukocyte. A type of blood cell that functions in basic housekeeping, nonspecific defenses, and adaptive immunity (e.g., eosinophil, neutrophil, macrophage, or lymphocyte).

X chromosome inactivation The programmed condensation of one of the X chromosomes in somatic cells of a mammalian female.

xanthophyll Yellow-orange carotenoid. An accessory pigment.

xenotransplantation Transfer of an organ from one species to another.

x-ray diffraction image Pattern formed when x-rays that have been directed at a molecule are scattered; the resulting pattern of streaks and dots is used to calculate positions of atoms in the molecule.

xylem Of vascular plants, a complex tissue that conducts water and solutes through pipelines of interconnected walls of cells, which are dead at maturity.

yellow marrow A fatty tissue in the cavities of most mature bones.

Y-linked gene Gene on a Y chromosome.

yolk Protein-rich and lipid-rich substance that nourishes embryos in animal eggs.

yolk sac Extraembryonic membrane. In most shelled eggs, it holds nutritive yolk; in humans, part becomes a site of blood cell formation and some cells give rise to forerunners of gametes.

zero population growth No overall increase or decrease in population size during a specified interval.

zygomycete Fungus that forms a thick-walled sexual spore (a diploid zygote) in a thin cover.

zygote First cell of a new individual, formed by fusion of a sperm nucleus with an egg nucleus at fertilization; a fertilized egg.

Art Credits and Acknowledgments

This page constitutes an extension of the copyright page. We have made every effort to trace the ownership of all copyrighted material and to secure permission from copyright holders. In the event of any question arising as to the use of any material, we will be pleased to make the necessary corrections in future printings. Thanks are due to the following authors, publishers, agents, and for permission to use the material indicated.

Page i, iii, Frans Lanting/Minden Pictures

TABLE OF CONTENTS **Page iv** bottom left, Art by Raychel Ciemma; **Page v** top, from left, Lisa Starr; Larry West/FPG/Getty Images; © Professors P. Motta and T Naguro/SPL/Photo Researchers, Inc. **Page vi** bottom left, © Jennifer W. Shuler/Science Source/Photo Researchers, Inc. **Page vii** top, from left, © George Lepp/CORBIS; (both) Courtesy of Carl Zeiss MicroImaging, Thornwood, NY; © Mc Leod Murdo/Corbis Sygma; Model, courtesy of Thomas A. Setitz from *Science*; Courtesy of Joseph DeRisa from *Science*, 1997 Oct. 24; 278 (5338) 680–686. **Page viii** left, (above) © Alan Solem; (below) Courtesy of Professor Martin F. Yanofsky, UCSD; **Page ix** top, from left, © P. Hawtin, University of Southampton/SPL/Photo Researchers, Inc.; © Science Photo Library/Photo Researchers, Inc.; © John Clegg/Ardea, London; © Robert C. Simpson/Nature Stock; Gerry Ellis/The Wildlife Collection. **Page x** Top © R. J. Erwin/Photo Researchers, Inc. **Page xi** Top, from left, © Cory Gray; © Illustration by Karen Carr; © Bruce Iverson; © Jim Christensen, Fine Art Digital Photographic Images. **Page xii** bottom left, © Science Photo Library/Photo Researchers, Inc. **Page xiii** Top, from left, © AP/Wide World Photos; © Kevin Fleming/CORBIS; © Michael Neveux; © NSIBC/SPL/Photo Researchers, Inc. **Page xiv** Top, © NIBSC/SPL/Photo Researchers, Inc. **Page xv** top, from left, © Francois Gohier/Photo Researchers, Inc.; © Archivo Iconografico, S.A./CORBIS; © Dow W. Fawcett/Photo Researchers, Inc.; © Amos Nachoum/CORBIS **Page xvi** bottom left, © C. James Webb/Phototake USA. **Page xvii** top, from left, © Pr. Alexande Meinesz, University of Nice-Sophia Antipolis; © Eric and David Hosking/CORBIS; © James Marshall/CORBIS. © Hank Fotos Photography

INTRODUCTION NASA Space Flight Center

CHAPTER 1 **Page 2** Left, © Mark M. Lawrence/CORBIS. **the big picture** (page 2) From left, Lisa Starr with PDB ID:1BNA; H.R. Drew, R.M. Wing, T. Takano, C. Broka, S. Tanaka, K. Itakura, R.E. Dickerson; Structure of a B-DNA Dodecamer. Conformation and Dynamics, PNAS; right, © Peter Scoones; (page 3) From left, © Nick Brent; right, © Raymond Gehman/CORBIS. **Page 3** Top, © Peter Turnley/CORBIS. **1.1** (a) Lisa Starr, rendered with Atom In A Box, © Dauger Research, Inc. 1.1 (b, above left) Lisa Starr with PDB file courtesy of Dr. Christina A. Bailey, Department of Chemistry & Biochemistry, California Polytechnec State University, San Luis Obispo, CA (b, above center) Lisa Starr with PDB ID: 1BBB; Silva, M. M., Rogers, P. H., Arnone, A; A third quaternary structure of human hemoglobin A at 1.7- A resolution; J Biol Chem 267 pp. 17248 (1992); (b, above right) Lisa Starr with PDB file from Klotho Biochemical Compounds Declarative Database; (b, below) Lisa Starr; (c) Lisa Starr; (d) ©Science Photo Library/Photo Researchers); (e) © Bill Varie/CORBIS; (f) © Jeffrey L. Rotman/CORBIS; (g) © Jeffrey L. Rotman/CORBIS; (h) © Jeffrey L. Rotman/CORBIS; (i) © Peter Scoones; (j) NASA; (k) NASA. **1.2** Lisa Starr. **1.3** (a-e) © Jack de Coningh. **1.4** Gary Head and Lisa Starr **1.5** © Y. Arthrus-Bertrand/Peter Arnold, Inc. **1.7** (page 8) clockwise from above, © Lewis Trusty/Animals Animals; © Emiliania Huxleyi photograph, Vita Pariente, scanning electron micrograph taken on a Jeol T330A instrument at Texas A&M University Electron Microscopy center; © Carolina Biological Supply Company; © R. Robinson/Visuals Unlimited, Inc.; (e) © Oliver Meckes/Photo Researchers, Inc.; Courtesy of James Evarts; (page 9) clockwise from above left, © John Lotter Gurling/Tom Stack & Associates; (h) © Edward S. Ross; (i) © Robert C. Simpson/Nature Stock; (j) © Edward S. Ross; (k) © Joe McDonald/CORBIS; (l) © CNRI/SPL/Photo Researchers, Inc.; (m) © P. Hawtin, University of Southampton/SPL/Photo Researchers, Inc. **1.8** Top, All Courtesy of Derrell Fowler, Tecumseh, Oklahoma; bottom, © Nick Brent, enhanced by Lisa Starr. **1.9** Lisa Starr. **Page 11** Left, © Raymond Gehman/CORBIS; center, © LWA-Stephen Welstead/CORBIS; right, © Lester Lefkowitz/CORBIS. **Page 12** Right, © Royalty-Free/CORBIS. **1.10** (a) © Chris D. Jiggins; (b) © www.thais.it; top background, © Wolfgang Kaehler/CORBIS; bottom right, © Martin Reid. Art, Lisa Starr. **Page 15** © Digital Vision/PictureQuest. **Page 16** (a–d) © Gary Head.

Page 17 Unit I © Wim van Egmond/Micropolitan Museum.

CHAPTER 2 **Page 18** Top, (a) © David Arky/CORBIS. **the big picture** (page 18) Lisa Starr; (page 19) From left, Art, Lisa Starr with PDB ID:1BNA; H.R. Drew, R.M. Wing, T. Takano, C. Broka, S. Tanaka, K. Itakura, R. E. Dickerson; Structure of a B-DNA Dodecamer. Conformation and Dynamics; PNAS; center, © Steve Lissau/Rainbow; right, © FoodPix. **Page 19** Top, © Dinodia. **2.2** (a, b) Lisa Starr; (c) Lisa Starr, rendered with Atom In A Box, © Dauger Research, Inc. **2.3** Art, Lisa Starr. **2.4** (a) © John Greim/Medichrome; (b, c) Art by Raychel Ciemma; (d) © Harry T. Chugani, M.D., UCLA School of Medicine **Page 22** © Michael S. Yamashita/CORBIS. **2.5** Lisa Starr, rendered with Atom In A Box, © Dauger Research, Inc. **2.6** Lisa Starr. **2.7** Gary Head and Lisa Starr. **2.8** Art, (a, b) Lisa Starr; photographs, (a) above, Gary Head; below, © Bruce Iverson; (c) Lisa Starr with PDB ID:IBNA; H.R. Drew, R.M. Wing, T. Takano, C. Broka, S. Tanaka, K. Itakura, R. E. Dickerson; Structure of a B-DNA Dodecamer. Conformation and Dynamics, PNAS. **2.9** (a,b,c, left) Lisa Starr with PDB file from NYU Scientific Visualization Lab; (b, right) © Steve Lissau/Rainbow; (c, right) © Kennan Ward/CORBIS. **2.10** Lisa Starr. **2.11** (a) © Lester Lefkowitz/CORBIS; (b) Lisa Starr. **2.12** Lisa Starr. **2.13** © Michael Grecco/Picture Group **Page 31** © National Gallery Collection; By kind permission of the Trustees of the National Gallery, London/CORBIS

CHAPTER 3 **3.1** Left, © 2002 Charlie Waite/Stone/Getty Images; center, John Collier. Great Britain, 1850-1934, *Priestess of Delphi*, 1891, London, oil on canvas, 160.0 x 80.0cm. Gift of the Rt. Honourable, the Earl of Kintore,1893; right, Lisa Starr, with PDB files from NYU Scientific Visualization Lab. **the big picture** (page 32) From left © Wayne Bennett/CORBIS; right, Lisa Starr; (page 33) From left, Lisa Starr with PDB file courtesy of Dr. Christina A. Bailey, Department of Chemistry & Biochemistry, California Polytechnic State University; San Luis Obispo, CA; center, Lisa Starr with PDB ID: 1BBB; Silva, M. M., Rogers, P. H., Arnone, A.; A third quaternary structure of human hemoglobin A at 1.7-A resolution; J Biol Chem 267, pp. 17248; right, Lisa Starr with PDB ID: 1BNA; H. R. Drew, R. M. Wing, T. Takano, C. Broka, S. Tanaka, K. Itakura, R. E. Dickerson; Structure of a B-DNA Dodecamer. Conformation and Dynamics, PNAS. **Page 33** Top right, Lisa Starr with PDB file from NYU Scientific Visualization Lab **3.2** (a) Lisa Starr with PDB file from NYU Scientific Visualization Lab (**b**) Lisa Starr with PDB file from Klotho Biochemical Compounds Declarative Database **3.3** Gary Head **3.4** Left © Eric and David Hosking/CORBIS; right, © Wayne Bennett/CORBIS; below, Lisa Starr. **3.5** Lisa Starr. **3.6** Lisa Starr with PDB file from NYU Scientific Visualization Lab. **3.7** (a) © Dr. W. Michaelis/Universitat Hamburg; (b) both, Courtesy of K.O. Stetter & R. Rachel, University of Regensburg © Boetius et all. 2000, Nature 407, 623-626; (b, right) Seth Gold. **3.8** Ian R. MacDonald **3.9** © John Sibbick. **3.10** Lisa Starr **3.11** Lisa Starr **3.12** Art, Lisa Starr; photograph, © Steve Chenn/CORBIS **3.13** Art, Lisa Starr; photograph, © David Scharf/Peter Arnold, Inc. **3.14** (a) Lisa Starr with PDB file courtesy of Dr. Christina A. Bailey, Department of Chemistry & Biochemistry, California Polytechnic State University, San Luis Obispo, CA; (b-d) Art, Precison Graphics. **3.15** (a, b) Art, Precision Graphics; photograph, © Kevin Schafer/CORBIS **3.16** (a) Lisa Starr with PDB file courtesy of Dr. Christina A. Bailey, Department of Chemistry & Art, Lisa Starr; (b) Precision Graphics (c) Lisa Starr **3.17** (a) © Scott Camazine/Photo Researchers, Inc. (b) Lisa Starr (c) Precision Graphics. **3.18** (a) Gary Head; (b-e) Lisa Starr and Chris Keeney with PDB files from NYU Scientific Visualization Lab. **3.19** (a) Lisa Starr; (b, left) Lisa Starr; (b, right) Lisa Starr After: Introduction to Protein Structure 2nd ed., Branden & Tooze, Garland Publishing, Inc.; (c, left) Lisa Starr with PDB ID: 1BBB; Silva, M. M., Rogers, P. H., Arnone, A.; A third quaternary structure of human hemoglobin A at 1.7-A resolution; J Biol Chem 267 pp. 17248 (1992); (c, right) Lisa Starr After: Introduction to Protein Structure 2nd ed., Branden & Tooze, Garland Publishing, Inc. **3.20** (a, b) Lisa Starr with PDB ID: 1BBB; Silva, M. M., Rogers, P. H., Arnone, A.; A third quaternary structure of human hemoglobin A at 1.7-Å resolution; J Biol Chem 267 pp. 17248 (1992). **3.21** (a, b) Gary Head and Lisa Starr with PDB files from New York University Scientific Visualization; (c) Dr. Gopal Murti/SPL/Photo Researchers, Inc. (d) photograph, Courtesy of Melba Moore. **3.22** Gary Head. **3.23** (a) Gary Head; (b) Lisa Starr **3.24** Lisa Starr with PDB ID:1BNA; H.R. Drew, R.M. Wing, T. Takano, C. Broka, S. Tanaka, K. Itakura, R.E. Dickerson; Structure of a B-DNA Dodecamer. Conformation and Dynamics; PNAS V. 78 2179, 1981 Lisa Starr. **Page 49**, left, Art by Lisa Starr with PDB ID: 1AKJ; Gao, G. F., Tormo, J., Gerth, U. C., Wyer, J. R., McMichael, A. J., Stuart, D. I., Bell, J. I., Jones, E. Y., Jakobsen, B. K.; Crystal structure of the complex between human CD8alpha(alpha) and HLA-A2; Nature 387 pp. 630 (1997) **3.25** Art, Lisa Starr; photograph, © Kevin Fleming/CORBIS

CHAPTER 4 **4.1** (a-c) © Tony Brian, David Parker/SPL/Photo Researchers, Inc. **the big picture** (page 50) From left, Art by Raychel Ciemma; right, Lisa Starr; (page 51) Left and right, Art, Lisa Starr. **Page 51** Top, © Tony Brian, David Parker/SPL/Photo Researchers, Inc. **4.2** Lisa Starr. **4.3** Art assembly, Gary Head; Art, Lisa Starr. Photographs: (hummingbird) © Robert A. Tyrrell; (human), © Pete Saloutos/CORBIS; (redwood), © Sally A. Morgan, Ecoscene/CORBIS **4.5** Left, © Armed Forces Institute of Pathology; right, © National Library of Medicine. **4.6** Bottom left, © Leica Microsystems, Inc., Deerfield, IL; right, Gary Head **4.7** Left, Gary Head; right, © Geoff Tompkinson/Science Photo Library/Photo Researchers, Inc. **4.8** (a) © Jeremy Pickett-Heaps, School of Botany, University of Melbourne; (b) © Jeremy Pickett-Heaps, School of Botany, University of Melbourne; (c) © Jeremy Pickett-Heaps, School of Botany, University of Melbourne; (d) © Jeremy Pickett-Heaps, School of Botany, University of Melbourne. **4.9** (a) Precision Graphics (b) Lisa Starr (c) Raychel Ciemma. **4.10** (page 56) Left, Lisa Starr with integrin: PDB ID:1JV2; Xiong, J.-P., Stehle, T., Diefenbach, B., Zhang, R., Dunker, R., Scott, D. L., Joachimiak, A., Goodman, S. L., Arnaout, M. A.: Crystal Structure of the Extracellular Segment of Integrin $\alpha V\beta 3$ Science 294 pp. 339 (2001); (page 57), Lisa Starr, Chris Keeney, and Leif Buckley with Human growth hormone: PDB ID:1A22; Clackson, T., Ultsch, M. H., Wells, J. A., de Vos, A. M.: Structural and functional analysis of the 1:1 growth hormone:receptor complex reveals the molecular basis for receptor affinity. J Mol Biol 277 pp. 1111 (1998); HLA: PDB ID: 1AKJ; Gao, G. F., Tormo, J., Gerth, U. C., Wyer, J. R., McMichael, A. J., Stuart, D. I., Bell, J. I., Jones, E. Y., Jakobsen, B. K.; Crystal structure of the complex between human CD8alpha and HLA-A2; Nature 387 pp. 630 (1997); glut1: PDB ID:1JA5; Zuniga, F. A., Shi, G., Haller, J. F., Rubashkin, A., Flynn, D. R., Iserovich, P., Fischbarg, J.: A Three-Dimensional Model of the Human Facilitative Glucose Transporter Glut1 J. Biol. Chem. 276 pp. 44970 (2001); calcium pump: PDB ID:1EUL; Toyoshima, C., Nakasako, M., Nomura, H., Ogawa, H.: Crystal Structure of the Calcium Pump of Sarcoplasmic Reticulum at 2.6 Angstrom Resolution. Nature 405 pp. 647 (2000); ATPase: PDB ID:18HE; Menz, R. I., Walker, J.E., Leslie, A.G.W.: Structure of Bovine Mitochondrial F1-ATPase with Nucleotide Bound to All Three Catalytic Sites: Implications for the Mechanism of Rotary Catalysis. Cell (Cambridge, Mass.)106 pp. 331 (2001); **4.11** (a) © K.G. Murti/Visuals Unlimited; (b) © R. Calentine/Visuals Unlimited; (c) © Gary Gaard and Arthur Kelman; (d) Lisa Starr. **4.12** top, (a) © Russell Kightley/Science Photo Library/Photo Researchers, Inc.; (b) © University of California Museum of Paleontology; (c) © University of California Museum of Paleontology. **4.13** © M.C. Ledbetter, Brookhaven National Laboratory. **4.14** © Micrograph, Gl L. Decker **4.15** Top left, Lisa Starr; top right, © Stephen L. Wolfe; (a) © Stephen L. Wolfe; (b) Lisa Starr. **4.16** Top left, Lisa Starr; (a) © Stephen L. Wolfe; (b) Stephen L. Wolfe; (c) right, Art, computer enhanced by Lisa Starr; (d) right, Art, computer enhanced by Lisa Starr (e) Gary Grimes, computer enhanced by Lisa Starr. **4.17** Art, Lisa Starr and Raychel Ciemma; photograph (right), © Keith R. Porter. **4.18** Art, Lisa Starr; photograph, © Dr. Jeremy Burgess/SPL/Photo Researchers, Inc. **4.19** Lisa Starr. **4.20** Lisa Starr. **4.21a-c** Lisa Starr; photograph, courtesy Mary Osborn, Max Planck Institute for Biophysical Chemistry, Goettingen, FRG. **4.22** Lisa Starr. **4.23** (a) Precision Graphics after Stephen L. Wolfe, Molecular and Cellular Biology, Wadsworth, 1993. (b) © CNRI/SPL/Photo Researchers, Inc. **4.24** (a) Raychel Ciemma (b) Lisa Starr; (c-e) Raychel Ciemma **4.25** (a) © George S. Ellmore (b) © Science Photo Library/Photo Researchers, Inc.; right, © Bone Clones, www.boneclones.com. **4.26** Ronald Hoham, Dept. of Biology, Colgate University. **4.26** Lisa Starr and Leif Buckley. **4.27** (a,b) From "Tissue & Cell," Vol. 27, pp.421-427, Courtesy of Bjorn Afzelius, Stockholm University.

CHAPTER 5 **5.1** Left, © Chris Keeney; right, Lisa Starr with PDB ID: 1DGF; Putnam, C. D., Arvai, A. S., Bourne, Y., Tainer, J. A.: Active and Inhibited Human Catalase Structures: Ligand and Nadph Binding and Catalytic Mechanism J.Mol.Biol. 296 pp. 295 (2000). **the big picture** (page 72) From left, © William Dow/CORBIS; Lisa Starr; (page 73) From left, © Scott McKiernan/ZUMA Press; Lisa Starr. **Page 73** © Paul Edmondson/CORBIS. **5.2** Top, © Craig Aurness/CORBIS; bottom, © William Dow/CORBIS. **5.3** Lisa Starr. **5.4** Lisa Starr with PDB file from Klotho Biochemical Compounds Declarative Database. **5.5** Lisa Starr. **5.6** Lisa Starr, from B. Alberts, et al., Molecular Biology of the Cell, 1983, Garland Publishing. **5.7** Lisa Starr. **5.8** Lisa Starr. **5.9** Lisa Starr. **5.10** Lisa Starr with PDB ID: 1DGF; Putnam, C. D., Arvai, A. S., Bourne, Y., Tainer, J. A.: Active and Inhibited Human Catalase Structures: Lingand Nadph Binding and Catalytic Mechanism J Mol Biol. 296, p. 295 (2000). **5.11** Lisa Starr. **5.12** Lisa Starr. **5.13** (a) Gary Head; (b) © Scott McKiernan/ZUMA Press. **5.14** (a) Gary Head; (b) © Foodpix/Bill Boch; (c) © Woods Hole Oceanographic Institution. **5.15** Lisa Starr. **5.16** Top, © Andrew Lambert Photography/Science Photo Library/Photo Researchers; Art, Raychel Ciemma. **5.17** Lisa Starr. **5.18** Lisa Starr and Chris Keeney with PDB files from NYU Scientific Visualization Lab. **5.19** Lisa Starr, After: David H.

MacLennan, William J. Rice and N. Michael Green, " The Mechanism of Ca2+ Transport by Sarco (Endo)plasmic Reticulum Ca2+-ATPases."JBC Volume 272, Number 46, Issue of Nov. 14, 1997 pp. 28815-28818. **Page 86** © Hubert Stadler/CORBIS. **5.20** Precision Graphics. **5.21** Art, top, Raychel Ciemma; bottom (all), © M. Sheetz, R. Painter, and S. Singer, J of Cell Biol., 70:193 (1976) by permission, The Rockefeller University Press. **5.22** Lisa Starr. **5.23** Both, © R.G.W. Anderson, M.S. Brown and J.L. Goldstein. Cell 10:351 (1977). **5.24** Chris Keeney. **5.25** (a) © Juergen berger/Max Planck Inst./SPL/Photo Researchers, Inc.; (b) Lisa Starr. **Page 90** Lisa Starr with PDB ID: 1CBJ; Hough, M.A. Hasnain, S.S. Crystallographic structures of bovine copper-zinc superoxide dismutase reveal asymmetry in two subunits: functionally important three and five coordinate copper sites captured in the same crystal. J. Mol. Biol. v287 pp. 579, 1999. **5.26** © Frieder Sauer/Bruce Coleman. **5.27** © Prof. Marcel Bessis/SPL/Photo Researchers, Inc.

CHAPTER 6 **6.1** Both, NASA. **the big picture** (page 92) From left, © 2002 PhotoDisc; Lisa Starr. **Page 93** © Wendy A. Kozlowski. **6.2** Left, © 2002 PhotoDisc; right, Precision Graphics. **6.3** Lisa Starr with PDB files from NYU Scientific Visualization Lab. **6.4** (a,b) Lisa Starr after Stephen L. Wolfe, *Molecular and Ceullar Biology*, Wadsworth. **6.5** © Larry West/FPG/Getty Images. **6.6** Art, Raychel Ciemma; micrograph, Carolina Biological Supply Company. **6.7** Left, © Craig Tuttle/CORBIS; (a-c) Art, Lisa Starr with Preface, Inc.; (d) Lisa Starr with Light Harvesting Complex PDB ID: 1RWT; Liu, Z., Yan, H., Wang, K., Kuang, T., Zhang, J., Gui, L., An, X., Chang, W.: Crystal Structure of Spinach Major Light-Harvesting Complex at 2.72 A Resolution Nature 428 pp. 287 (2004). **6.8** Lisa Starr with Preface, Inc. **Page 97** © Darron Luesse, Department of Biology, Indiana University. **6.9** (a, top) Harindar Keer, Thorsten Ritz Laboratory at UC Irvine, Dept. of Physics and Astronomy, using VMD proprietary software; (a, bottom) Lisa Starr; (b-d) Lisa Starr with Light Harvesting Complex PDB ID 1RWT; Liu, Z., Yan, H., Wang, K., Kuang, T., Zhang, J., Gui, L., An, X., Chang, W.: Crystal Structure of Spinach Major Light-Harvesting Complex at 2.72 A Resolution Nature 428 pp. 287 (2004). **Page 100** Top right, Lisa Starr with Light Harvesting Complex PDB ID: 1RWT; Liu, Z., Yan, H., Wang, K., Kuang, T., Zhang, J., Gui, L., An, X., Chang, W.: Crystal Structure of Spinach Major Light-Harvesting Complex at 2.72 A Resolution Nature 428 pp. 287 (2004). **6.10** Lisa Starr. **6.11** Chris Keeney. **6.12** Top left, Courtesy of John S. Russell, Pioneer High School; center left, © Foodpix/Bill Boch; bottom left, © Chris Hellier/CORBIS; top center, micrograph, Bruce Iverson, computer-enhanced by Lisa Starr; center center, micrograph, Ken Wagner/Visuals Unlimited, computer-enhanced by Lisa Starr; bottom center, micrograph, James D. Mauseth, University of Texas; Art, (all) Gary Head. **Page 104** (top left) Lisa Starr. **6.13** Gary Head. **6.14** © E.R. Degginger. **6.15** (a) © Herve Chaumeton/Agence Nature; (b) © Douglas Faulkner/Sally Faulkner Collection.

CHAPTER 7 **7.1** © Louise Chalcraft-Frank and FARA. **the big picture** (page 106), right © Professors P. Motta and T Naguro/SPL/Photo Researchers, Inc.; art, Lisa Starr; (page 107) From left, © Randy Faris/CORBIS; © Gary Head; **Page 107** Top, © Professors P. Motta and T Naguro/SPL/Photo Researchers, Inc. **7.2** Raychel Ciemma and Gary Head. **7.3** (page 108) Top left, © Jim Cummins/CORBIS; top right, © John Lotter Gurling/Tom Stack & Associates; bottom, © Chase Swift/CORBIS; (page 109) Art, Lisa Starr with Gary Head. **7.4** (page 110), Lisa Starr; (page 111) (a-f) Lisa Starr and Gary Head, after Ralph Taggart. **7.5** from left, Raychel Ciemma; Lisa Starr; © Professors P. Motta and T. Naguro/SPL/Photo Researchers, Inc. **7.6** Art, Lisa Starr. **7.7** Gary Head. **7.8** Top, Raychel Ciemma, (below) Lisa Starr. **7.9** Lisa Starr with Preface, Inc. **7.10** (a) © Adrian Warren/Ardea, London; (b) © Foodpix/Ben Fink; (c) ©Foodpix/Ben Fink; (d) Lisa Starr with Gary Head. **7.11** Lisa Starr. **7.12** Left, © Randy Faris/CORBIS; right, Gladden Willis, MD/Visuals Unlimited. **Page 118** © Lois Ellen Frank/CORBIS. **7.13** © Gary Head; art, Lisa Starr. **Page 120** © R. Llewellyn/SuperStock, Inc. **Page 121** Lisa Starr with Gary Head.

Page 123 Unit II © Francis Leroy, Biocosmos/Science Photo Library/Photo Researchers, Inc.

CHAPTER 8 **8.1** Left, micrograph, Dr. Pascal Madaule, France; inset, Courtesy of the family of Henrietta Lacks. **the big picture** (page 124) from left, © L. Willatt, East Anglian Regional Genetics Service/SPL/Photo Researchers, Inc.; © Jennifer W. Shuler/Science Source/Photo Researchers, Inc.; (page 125) from left, Gary Head; © Science Photo Library/Photo Researchers, Inc. **Page 125** Micrograph, Dr. Pascal Madaule, France. **8.2** (a,b) Gary Head; bottom © Divital Vision/Getty Images. **8.3** (a) From Allen TD, Jack EM, Harrison CJ (1988). The three dimensional structure of human metaphase chromosomes determined by scanning electron microscopy. In Adolph KW (1988) Chromosomes and Chromatin Volume II, CRC Press. Boca Raton, FL. Chapter 10, Fig.5, p.58; (b) Raychel Ciemma and Lisa Starr; (c) © B. Hamkalo; (d) © O. L. Miller, Jr., Steve L. McKnight. **8.4** Raychel Ciemma and Gary Head. **8.5** Left © L. Willatt, East Anglian Regional Genetics Service/SPL/Photo Researchers, Inc.; (a-c) Lisa Starr. **Page 129** Bottom right, Raychel Ciemma. **8.6** (pages 130-131) Photographs, © Jennifer W. Shuler/Science Source/Photo Researchers, Inc.; Art, Raychel Ciemma. **8.7** (a) Art, Raychel Ciemma; Micrograph, D. M. Phillips/Visuals Unlimited; (b) Art, Lisa Starr; Micrograph, R. Calentine/Visuals Unlimited. **8.8** © Jennifer W. Shuler/Science Source/Photo Researchers, Inc. **8.9** Bottom, © Lennart Nilsson from Behold Man, 1974 by Albert Bonniers Forlog and Little, Brown & Company, Boston. **Page 133** Top, © Lennart Nilsson from *A Child Is Born* 1966, 1977 Dell Publishing Company, Inc. **8.10** (a, b) © Phillip B. Carpenter, Department of Biochemistry and Molecular Biology. **8.11** © Science Photo Library/Photo Researchers, Inc. **8.12** Betsy Palay. **8.13** Left, © Ken Greer/Visuals Unlimited; center, © Biophoto Associates/Science Source/Photo Researchers, Inc.; right, © James Stevenson/SPL/Photo Researchers, Inc. **Page 136** Raychel Ciemma and Gary Head. **8.14** From Allen TD, Jack EM, Harrison CJ (1988) *The three dimensional structure of human metaphase chromosomes determined by scanning electron microscopy*. In Adolph KW (1988) *Chromosomes and Chromatin Volume II*, CRC Press. Boca Raton FL. Chapter 10, Fig. 5, p. 58.

CHAPTER 9 **9.1** (a) © Andrew Syred/Photo Researchers, Inc.; (b) © George D. Lepp/CORBIS; (c) © Dan Kline/Visuals Unlimited. **the big picture** (page 138) From left, Courtesy of Carl Zeiss MicroImaging, Thornwood, NY; Lisa Starr; (page 139) From left, Lisa Starr; © Francis Leroy, Biocosmos/Science Photo Library/Photo Researchers, Inc. **Page 139** © Tomohiro Kono/Tokyo University of Agriculture, enhanced by Lisa Starr. **9.2** Image courtesy of Carl Zeiss MicroImaging, Thornwood, NY. **9.3** Raychel Ciemma. **9.4** © L. Willatt, East Anglian Regional Genetics Service/SPL/Photo Researchers, Inc. **Page 141** Art, Raychel Ciemma. **9.5** (pages 142-143) Photography, Courtesy John Innes Foundation Trustees, computer enhanced by Gary Head; Art, Raychel Ciemma. **9.6** (a,b) Raychel Ciemma; (c–f) Lisa Starr. **9.7** Raychel Ciemma. **9.8** Seth Gold. **9.9** Lisa Starr. **9.10** Art, Lisa Starr; right, © Francis Leroy, Biocosmos/Science Photo Library/Photo Researchers, Inc. **9.11** Lisa Starr. **9.12** © Ron Neumeyer, www.microimaging.ca. **9.13** (a,b) © Lisa O'Connor/ZUMA/Corbis.

CHAPTER 10 **10.1** Art, Lisa Starr; left, © Children's Hospital & Medical Center/CORBIS; right © Simon Fraser/RVI, New Castle-Upon-Tyne/SPL/Photo Researchers, Inc. **the big picture** (page 150) From left, © George Lepp/CORBIS; Raychel Ciemma and Precision Graphics; (page 151) From left, © Ted Somes; D. & V. Hennings. **Page 151** © Science Photo Library/Photo Researchers, Inc. **10.2** © The Moravian Museum, Brno. **10.3** © Jean M. Labat/Ardea, London; Art, Jennifer Wardrip. **10.4** Lisa Starr. **10.5** Precison Graphics. **10.6** Raychel Ciemma and Precison Graphics. **10.7** Raychel Ciemma and Precison Graphics. **10.8** Raychel Ciemma. **10.9** Raychel Ciemma and Precison Graphics. **10.10** © David Scharf/Peter Arnold, Inc.; Art, Precision Graphics. **10.11** Top, William F. Ferguson; bottom, © Francesc Muntada/CORBIS; Art, Raychel Ciemma. **10.12** All, Ted Somes. **10.13** © Bettman/CORBIS. **10.14** From top, © Frank Cezus/FPG; © Frank Cezus/FPG; © Ted Beaudin/FPG; © Michael Prince/CORBIS. **10.15** (a,b) Gary Head; (c) Courtesy of Ray Carson, University of Florida News and Public Affairs. **10.16** D. & V. Hennings. **10.17** © Pamela Harper/Harper Horticultural Slide Library; Art, Lisa Starr from Prof. Otto Wilhelm Thome, Flora von Deutschland Osterreich und der Schweiz. 1885, Gera, Germany. **Page 163** © Gideon Mendel/CORBIS. **10.18** Left, © Tom and Pat Leeson/Photo Researchers, Inc.; right, © Rick Guidotti, Positive Exposure. **10.19** © Leslie Faltheisek. Clacritter Manx.

CHAPTER 11 **11.1** © Reuters/CORBIS. **the big picture** (page 166) From left. © 2001 PhotoDisc, Inc.; © 2001 EyeWire, Inc.; © Stapleton Collection/CORBIS; (page 167) From left, Raychel Ciemma; © Saturn Stills/SPL/Photo Researchers, Inc. **Page 167** © Daniel Weinberger, M.D., E. Fuller Torrey, M.D., Karen Berman, M.D., NIMH Clinical Brain Disorders Branch, Division of Intramural Research Programs, NIMH 1990. From: When Someone Has Schizophrenia, A brief overview of the symptoms, treatments, and research findings. 2001. **Page 168** © Jose Luis Pelaez, Inc./CORBIS. **11.2** Raychel Ciemma and Preface, Inc. **11.3** (a) Precision Graphics and Gary Head; (b) Robert Demarest with permission from M. Cummings, Human Heredity: Principles and Issues, 3rd Edition, p. 126. © 1994 by Brooks/Cole. All rights reserved; (c) Robert Demarest after Patten, Carlson & others; bottom right (girl), © 2001 PhotoDisc, Inc.; (boy), © 2001 EyeWire, Inc. **11.4** Art, Gary Head and Raychel Ciemma; (b) © Charles D. Winters/Photo Researchers, Inc.; (f) © Omikron/Photo Researchers, Inc. **11.5** Raychel Ciemma. **11.6** © Stapleton Collection/CORBIS. **11.7** Art, Precision Graphics; right, © Dr. Victor A. McKusick. **11.8** © Steve Uzzell. **11.9** Top left, © Frank Trapper/Corbis Sygma; (a,b) Lisa Starr. **11.11** Lisa Starr; **11.13** Art, After V. A. McKusick, Human Genetics, 2nd Ed., © 1969. Reprinted by permission of the author; right, © Bettman/CORBIS. **11.14** © Eddie Adams/AP Wide World Photos. **Page 176** Precision Graphics. **Page 177** Precison Graphics. **11.15** (a,b) Courtesy G. H. Valentine. **11.16** From "Multicolor Spectral Karyotyping of Human Chromosomes," by E. Schrock, T. Ried, et al, Science, 26 July 1996, 273:495. Used by permission of E. Schrock, T. Reid and the American Association for the Advancement of Science. **11.17** (a) Raychel Ciemma; (b) © CNRI/Photo Researchers, Inc. **11.18** Preface, Inc. **11.19** © UNC Medical Illustration and Photography. **11.20** © Saturn Stills/SPL/Photo Researchers, Inc. **11.21** From Lennart Nilsson, *A Child is Born*, (c) 1966, 1977 Dell Publishing Company, Inc.; Art, Lisa Starr. **11.22** © Matthew Alan/CORBIS; inset © Fran Heyl Associates/Jacques Cohen, computer-enhanced by © Pix Elation. **11.23** (a) © Carolina Biological/Visuals Unlimited; (b) © Terry Gleason/Visuals Unlimited; (c) Raychel Ciemma and Preface, Inc. **Page 183** Bottom left, Precison Graphics.

CHAPTER 12 **12.1** (a) © C. Barrington Brown, 1968 J. D. Watson; (b) © Mc Leod Murdo/CORBIS Sygma. **the big picture** (page 184) From left, Lisa Starr; Lisa Starr with PDB ID: 1BBB; Silva, M. M., Rogers, P. H., Arnone, A.: A third quaternary structure of human hemoglobin A at 1.7-A resolution. J Biol Chem 267 pp. 17248 (1992); (page 185) From left, Art, Precision Graphics; © James King-Holmes/SPL/Photo Researchers, Inc. **Page 185** © PA News Photo Library. **12.2** Raychel Ciemma; **12.3** Art, Lisa Starr; bottom, © Eye of Science/Photo Researchers, Inc. **12.4** Gary Head **12.5** © SPL/Photo Researchers, Inc. **12.6** Lisa Starr with PDB ID: 1BBB; Silva, M. M., Rogers, P. H., Arnone, A.: A third quaternary structure of human hemoglobin A at 1.7-A resolution. J Biol Chem 267 pp. 17248 (1992). **Page 189** Bottom left, Preface Inc. **12.7** © 1956-2004 The Novartis Foundation (www.novartis-found.org.uk), formerly the Ciba Foundation, reproduced with permission. **12.8** Precision Graphics. **12.9** Precision Graphics. **12.10** Top, (1-3), © James King-Holmes/SPL/Photo Researchers, Inc.; (4) © PA News Photo Library; bottom, from left, © Reuters/Landov; © P.A. NEWS/CORBIS KIPA; © Photo courtesy of the College of Veterinary Medicine, Texas A & M University; © Photo courtesy of the College of Veterinary Medicine, Texas A & M University.

CHAPTER 13 **13.1** © Vaughan Fleming/SPL/Photo Researchers, Inc. **the big picture** (page 194), From left (top left) Lisa Starr with PDB ID: 1BBB; Silva, M. M., Rogers, P.H., Arnone, A.: A third quaternary structure of human hemoglobin A at 1.7- Å resolution, J Mol Biol Chem 267 pp. 17248 (1992); Lisa Starr with, Theoretical model; PDB ID: 1K7N; C.Q.LIU, S.X.LIU, SQ.LIU, M.W.JIA, JHE, The Research of MRNA's Conformation and Interaction Between MRNA and TRNA in Translation (To be Published); right/left, © Model by Dr. David B. Goodin, The Scripps Research Institute; right, Model, courtesy of Thomas A. Setitz from Science; (page 195) From left, Lisa Starr; right/left © Nik Kleinberg; right © Steve Terrill/ CORBIS. **Page 195** Lisa Starr with PDB ID: 2AAI; Rutenber, E., Katzin, B.J., Ernst, S., Collins, E.J., Mlsna, D., Ready, M.P., Robertus, J.D.: Crystallographic Refinement of ricin to 2.5 A. Proteins 10pp. 240 (1991). **13.2** (a,b) Precison Graphics; (c) Lisa Starr. **13.3** (a-d) Art, Lisa Starr. **13.4** Gary Head. **13.5** Lisa Starr. **13.6** Precison Graphics and Gary Head. **13.7** Top, Model by Dr. David B. Goodin, The Scripps Research Institute; bottom, Lisa Starr. **13.8** (a) Model, courtesy of Thomas A. Steitz from Science; (b) Lisa Starr. **13.9** (a-k) Lisa Starr. **13.10** Lisa Starr. **13.11** background © Steve Terrill/CORBIS; left inset © Nik Kleinberg; middle inset, © Wayne Armstrong. **13.12** © John W. Gofman and Arthur R. Tamplin. From Poisoned Power: The Case Against Nuclear Power Plants Before and After Three Mile Island, Rodale Press, PA, 1979. **13.13** Lisa Starr. **13.14** © Dr. M.A. Ansary/Science Photo Library/Photo Researchers, Inc.

CHAPTER 14 **14.1** (a) From the archives of www.breast-path.com, courtesy of J.B. Askew, Jr., M.D., P.A. Reprinted with permission, © 2004 Breastpath.com; (b) Courtesy of Robin Shoulla and Young Survival Coalition. **the big picture** (page 206) From left, Raychel Ciemma; Lisa Starr; (page 207) From left, From the collection of Jamos Werner and John T. Lis; Courtesy of Edward B. Lewis, California Institute of Technology. **Page 207** Lisa Starr with PDB ID:1N5O; Williams, R. S., Glover, J. N. M.: "Structural Consequences of a Cancer-Causing Brca1-Brct Missense Mutation" J. Biol. Chem. 278 pp. 2630 (2003). **14.2** Raychel Ciemma and Gary Head. **14.3** (left) Lisa Starr with PDB ID: 1CJG; Spronk, C. A. E. M., Bonvin, A. M. J. J., Radha, P. K., Melacini, G., Boelens, R., Kaptein, R.: The Solution Structure of Lac Repressor Headpiece 62 Complexed to a Symmetrical Lac Operator. Structure (London) 7 pp. 1483 (1999). Also PDB ID: 1LBI; Lewis, M., Chang, G., Horton, N. C., Kercher, M. A., Pace, H. C., Schumacher, M. A., Brennan, R. G., Lu, P.: Crystal structure of the lactose operon repressor and its complexes with DNA and inducer. Science 271 pp. 1247 (1996); lactose pdb files from the Hetero-Compound Information Centre - Uppsala (HIC-Up). **Page 209** © Lois Ellen Frank/CORBIS. **14.4** Chris Keeney. **14.5** From the collection of Jamos Werner and John T. Lis. **14.6** (a) © Visuals Unlimited; (b) UCSF

Computer Graphics Laboratory, National Institutes, NCRR Grant 01081. **14.7** (a) © Dr. Karen Dyer Montgomery; (b) Raychel Ciemma. **14.8** © Jack Carey. **14.9** Top © Lisa Starr; (a-d) © Carolina Biological/Visuals Unlimited. **14.10** Left, © Walter J. Ghering/University of Basel, Switzerland; right, Courtesy of Edward B. Lewis, California Institute of Technology. **14.11** (a) © Palay/Beaubois after Robert F. Weaver and Philip W. Hedrick, Genetics. © 1989 W.C. Brown Publishers; (b,c) © Jim Langeland, Jim Williams, Julie Gates, Kathy Vorwerk, Steve Paddock and Sean Carroll, HHMI, University of Wisconsin-Madison. **14.12** © Jim Langeland, Jim Williams, Julie Gates, Kathy Vorwerk, Steve Paddock and Sean Carroll, HHMI, University of Wisconsin-Madison. **14.13** © Lawrence Berkeley National Laboratory.

CHAPTER 15 **15.1** Left, © AP/Wide World Photos; center © Charles O'Rear/CORBIS; right © ScienceUV/Visuals Unlimited. **the big picture** (page 218) From left, © Jim Bourg/Reuters/Corbis; right © Professor Stanley Cohen/SPL/Photo Researchers, Inc; (page 219) From left, © Lowell Georgis/CORBIS; right © Jeans for Gene Appeal. **Page 219** © IRRI Photo Bank Institute (IRRI). **15.2** Left, Lisa Starr; right © Volker Steger/SPL/Photo Researchers, Inc. **15.3** © Jim Bourg/Reuters/Corbis. **Page 222** © Professor Stanley Cohen/SPL/Photo Researchers, Inc. **15.4** (a) Art, Lisa Starr; (b) Lisa Starr. **15.5** Chris Keeney with permission of © QIAGEN, Inc. **15.6** Lisa Starr. **15.7** Lisa Starr. **15.8** Lisa Starr. **15.9** Lisa Starr. **15.10** Left © David Parker/SPL/ Photo Researchers, Inc.; right © Cellmark Diagnostics, Abingdon, UK. **15.11** Top left, © TEK IMAGE/Photo Researchers, Inc.; art, Lisa Starr. **15.12** (a) Courtesy Calgene LLC; (b) © Dr. Vincent Chaing, School of Forestry and Wood Projects, Michigan Technology University. **15.13** (a–c) Lisa Starr; (e) © Keith V. Wood; (d) © Lowell Georgis/CORBIS. **15.14** (a) Transgenic goat produced using nuclear transfer at GTC Biotherapeutics. Photo used with permission; (b) © Work of Atsushi Miyawaki, Qing Xiong, Varda Lev-Ram, Paul Steinbach, and Roger Y. Tsien at the University of California, San Diego; (c) © Adi Nes, Dvir Gallery Ltd. **15.15** (a) © Jeans for Gene Appeal; (b) Courtesy Dr. Paola Leone, Cell & Gene Therapy Center/University of Medicine 7 Dentistry of New Jersey. **15.16** © Matt Gentry/Roanoke Times. **15.17** Courtesy of Joseph DeRisa. From Science, 1997 Oct. 24; 278 (5338) 680–686. **15.18** (a) © Mike Stewart/Corbis Sygma; (b) © Simon Kwong/REUTERS/Landov.

Page 235 Unit III © Wolfgang Kaehler/CORBIS.

CHAPTER 16 **16.1** Left, © Bettmann/CORBIS; right, © St Bartholomew's Hospital/Science Photo Library/Photo Researchers, Inc. **the big picture** (page 236) From left, © Christopher Ralling; right, © David Parker/SPL/Photo Researchers, Inc.; (page 237) Left/left, © Peter Bowater/Photo Researchers, Inc.; left/right, © Owen Franken/CORBIS; right, Gary Head. **Page 237** Top, © Bettmann/CORBIS. **16.2** (a) © Wolfgang Kaehler/CORBIS; (b) © Earl & Nazima Kowall/CORBIS; (c) © Wolfgang Kaehler/CORBIS; (d) © Edward S. Ross; (e) © Edward S. Ross. **16.3** (left) Art, Gary Head; (right) © Bruce J. Mohn; (inset) © Phillip Gingerich, Director, University of Michigan. Museum of Paleontology. **16.4** (a) © Jonathan Blair/CORBIS; (b) © Biophoto Associates/Photo Researchers, Inc.; (c) © John Shaw/Photo Researchers, Inc. **Page 16.5** (a) Courtesy George P. Darwin, Darwin Museum, Down House; (b) © Christopher Ralling; (c) © Heather Angel; (c,d) Art, Precision Graphics (e) © Dieter & Mary Plage/Survival Anglia. **16.6** (a) © Joe McDonald/CORBIS; (b) © Karen Carr Studio/www.karencarr.com. **16.7** (a) Heather Angel; (b) © Kevin Schafer/CORBIS; (c) © Dr. P. Evans/Bruce Coleman, Ltd.; (d) © Alan Root/Bruce Coleman Ltd.; **16.8** © Down House and The Royal College of Surgeons of England; **16.9** (a) © Gary Head; (b) © John W. Merck, Jr., University of Maryland. **16.10** Left, © Thomas Mangelsen; right, © Theo Allofs/CORBIS. **16.11** Left, © Francois Photo Researchers, Inc. **16.12** Left, © Peter Bowater/Photo Researchers, Inc.; top center, © Owen Franken/CORBIS; top right © Sam Kleinman/CORBIS; center, © Jim Cornfield/CORBIS; center right, © Christopher Briscoe/Photo Researchers, Inc.; bottom, © Alan Solem. **16.13** Gary Head. **Page 248** © Terry Whittaker, Frank Lane Picture Agency/CORBIS. **16.14** Gary Head. **16.15** all, Courtesy of Hopi Hoekstra, University of California, San Diego. **16.16** Gary Head. **16.17** Precision Graphics, using NIH data. **16.18** © Peter Chadwick/Science Photo Library/Photo Researchers. **16.19** Left, © Thomas Bates Smith; right, © Thomas Bates Smith. **16.20** © Bruce Beehler, enhanced by Chris Keeney. **16.21** (a, b) Precision Graphics after Ayala and others; bottom, © Michael Freeman/CORBIS. **Page 254** © Steve Bronstein/The Image Bank/Getty Images. **16.22** Precision Graphics, after computer models developed by Jerry Coyne. **16.23** © Frans Lanting/Minden Pictures (computer-modified by Lisa Starr); Art, Raychel Ciemma. **16.24** Left, © David Neal Parks; right, © W. Carter Johnson. **16.25** John Kalusmeyer, University of Michigan Exhibit of Natural History.

CHAPTER 17 **17.1** (a) NASA Galileo Imaging Team; (b) top, Art by Don Davis; (b) bottom NASA Galileo Imaging Team. **the big picture** (page 258) from left, Raychel Ciemma; right (both) © Jack Jeffrey Photography; (page 259), from left © Carnegie Museum of Natural History; right, Gary Head. **Page 259**, Top, © David A. Kring, NASA/Univ. Arizona Space Imagery Center. **17.2** left © H. P. Banks; right, © Jonathan Blair. **17.3** © Jonathan Blair/CORBIS. **17.4** (a) Gary Head; (b) © 2001 PhotoDisc, Inc; (c,d) Lisa Starr. **Page 263, 17.6** © CORBIS. **17.7** (a) Leif Buckley; (b) Leif Buckley and Lisa Starr. **17.8** (a-d) Lisa Starr, after A. M. Ziegler, C. R. Scotese, and S. F. Barrett, "Mesozoic and Cenozoic Paleogeographic Maps," and J. Krohn and J. Sundermann (Eds.), Tidal Frictions and the Earth's Rotation II, Springer-Verlag, 1983; (e) © Martin Land/Photo Researchers, Inc.; (f) © John Sibbick. **17.9** Raychel Ciemma. **17.10** (a) J. Scott Altenbach, University of New Mexico, computer enhanced by Lisa Starr; (b) © Frans Lanting/Minden Pictures; computer enhanced by Lisa Starr; (c) © Stephen Dalton/Photo Researchers, Inc.; art, Lisa Starr and Raychel Ciemma, art reference, Natural History Collection, Royal BC Museum; **17.11** (a) Courtesy of Professor Richard Amasino, University of Wisconsin-Madison; (b) Courtesy of Professor Richard Amasino, University of Wisconsin-Madison; (c) Courtesy of Professor Martin F. Yanofsky; (d) © Jose Luis Riechmann; **17.12** Raychel Ciemma; **17.13** (a) top, © Courtesy of Prof. Richard Amasino, University of Wisconsin-Madison; (a) bottom, © Jennifer Grenier, Grace Boekhoff-Falk and Sean Carroll, HMI, University of Wisconsin-Madison; (b) top, © Herve Chaumeton/Agence Nature; (b) bottom, © Jennifer Grenier, Grace Boekhoff-Falk and Sean Carroll, HMI, University of Wisconsin-Madison; (c) top, © Peter Skinner/Photo Researchers, Inc.; (c) bottom, Courtesy of Dr. Giovanni Levi; (d) © Dr. Chip Clark. **Page 270** © TEK IMAGE/Photo Researchers, Inc.; **17.14** Precision Graphics. **17.15** left © Kjell B. Sandved/Visuals Unlimited; center, © Jeffrey Sylvester/FPG/Getty Images; right, © Thomas D. Mangelsen/Images of Nature. **17.16** (a, b) Art, Jennifer Wardrip. **17.17** (a) © John Alcock, Arizona State University; (b) © Tui Roy/Minden Pictures; **17.17** (c) © Alvin E. Staffan/Photo Researchers, Inc. **17.18** Top left, © Digital Vision/PictureQuest; top, © Joe McDonald/CORBIS. **17.19** (a) © Graham Neden/CORBIS; (b) © Kevin Schafer/CORBIS; center, © Ron Blakey, Northern Arizona University (c) © Rick Rosen/Corbis SABA. **17.20** (a-c) Preface, Inc. (d) all, © Jack Jeffrey Photography. **17.21** Top, © Steve Gartlan; bottom left and right © Below Water Photography/www.belowwater.com. **17.22** left © Lance Nelson/CORBIS; center, © Eric Crichton/CORBIS; right, © Maximilian Stock Ltd./Foodpix. **17.23** Lisa Starr after W. Jensen and F. B. Salisbury, Botany: An Ecological Approach, Wadsworth, 1972. **17.24** (a) Courtesy of Dr. Robert Mesibov; (b) Courtesy of Dr. Robert Mesibov; **17.24** (c) Lisa Starr. **17.25** Courtesy of Daniel C. Kelley, Anthony J. Arnold, and William C. Parker, Florida State University Department of Geological Science. **Page 278** (bottom left) Gary Head. **17.26** Left, Lisa Starr; right, © Carnegie Museum of Natural History. **17.27** from left © Science Photo Library/Photo Researchers, Inc.; © Galen Rowell/CORBIS; © Kevin Schafer/CORBIS; Courtesy of Department of Entomology, University of Nebraska-Lincoln; Bruce Coleman, Ltd. **17.28** Gary Head. **17.29** Gary Head. **17.30** from left, © Phillip Colla Photography; © Randy Wells/CORBIS © Cousteau Society/Getty Images; © Robert Dowling/CORBIS; Art, Gary Head. **17.31** Lisa Starr. **17.32** Lisa Starr. **17.33** © Gulf News, Dubai, UAE

CHAPTER 18 **18.1** Inset, Courtesy of Agriculture Canada; © Raymond Gehman/CORBIS. **the big picture** (page 286) From left, © Jeff Hester and Paul Scowen, Arizona State University, and NASA; ©Sidney W. Fox; (page 287) From left, © Chase Studios/Photo Researchers, Inc.; Raychel Ciemma. **Page 287** © Philippa Uwins/The University of Queensland. **18.2** © Jeff Hester and Paul Scowen, Arizona State University, and NASA. **18.3** © Chesley Bonestell; inset art, Raychel Ciemma. **18.4** (a) © Tim Thompson/CORBIS; (b) © Dr. Ken MacDonald/SPL/Photo Researchers, Inc.; (c) © Micheal J. Russell, Scottish Universities Environmental Research Centre. **18.5** (a) © Sidney W. Fox; (b) From Hanczyc, Fujikawa, and Szostak, Experimental Models of Primitive Cellular Compartments: Encapsulation, Growth, and Division; www.sciencemag.org Science 24 October 2003; 302;529, Figure 2, page 619. Reprinted with permission of the authors and AAAS; (c) Preface, Inc. **18.6** (a) © Stanley M. Awramik; (b,c) Bruce Runnegar, NASA Astrobiology Institute; (d) © N.J. Butterfield, University of Cambridge. **18.7** (a) © Chase Studios/Photo Researchers, Inc.; (b) © John Reader/SPL/Photo Researchers, Inc.; (c) © Sinclair Stammers/SPL/Photo Researchers, Inc. **18.8** Raychel Ciemma. **18.9** Left, © Robert Trench, Professor Emeritus, University of British Columbia; (b) right, Courtesy of Isao Inouye, Institute of Biological Sciences, University of Tsukuba. **18.10** Raychel Ciemma and Precision Graphics.

Page 299 Unit IV © Layne Kennedy/CORBIS.

CHAPTER 19 **19.1** © The Bridgeman Art Library/Getty Images. **the big picture** (page 300) From left, Lisa Starr; (page 301) From left, © NIBSC/SPL/Photo Researchers, Inc.; © Camr, Barry Dowsett/Science Photo Library/Photo Researchers, Inc. **Page 301** © Carl Cook. **19.2** (a,b) Lisa Starr. **19.3** (a) © P. Hawtin, University of Southampton/ SPL/Photo Researchers, Inc.; (b) © Dr. Dennis Kunkel/ Visuals Unlimited. **19.4** Raychel Ciemma. **19.5** Lisa Starr. **19.6** (a) © Dr. Jeremy Burgess/SPL/Photo Researchers, Inc.; (b) © P. W. Johnson and J. MeN. Sieburth, Univ. Rhode Island/BPS. **19.7** © Dr. Terry J. Beveridge, Department of Microbiology, University of Guelph, Ontario, Canada. **19.8** © Dr. Manfred Schloesser, Max Planck Institute for Marine Microbiology. **19.9** (a) © Stem Jems/Photo Researchers, Inc.; (b) © California Department of Health Services; (c) © Bernard Cohen, M.D., Dermatlas; http://www.darmatalas.org. **19.10** Lisa Starr. **19.11** (a) © Courtesy Jack Jones, Archives of Microbiology, Vol. 136, 1983, pp. 254-261. Reprinted by permission of Springer-Verlag. (b) © Dr. John Brackenbury/Science Photo Library/Photo Researchers, Inc. 19.12 (a) © Martin Miller/Visuals Unlimited; (b) © Alan L. Detrick, Science Source/Photo Researchers, Inc.; (c) © Dr. Harald Huber, Dr. Michael Hohn, Prof. Dr. K.O. Stetter, University of Regensburg, Germany. **19.13** (a-d) Art, Leif Buckley and Lisa Starr; (e) © CAMR/A. B. Dowsett/SPL/Photo Researchers, Inc.; (f) © Dr. Linda Stannard, Uct/Spl/Photo Researchers, Inc. **19.14** Art, Palay/Beaubois and Precision Graphics; top left, © Science Photo Library/Photo Researchers, Inc. **Page 310** Top, © Sercomi/Photo Researchers, Inc.; bottom, © Camr, Barry Dowsett/Science Photo Library/Photo Researchers, Inc. **19.15** (a) © Lily Echeverria/Miami Herald; (b) © APHIS photo by Dr. Al Jenny; (c) Art, Lisa Starr with PDB ID: 1QLX; Zahn, R., Liu, A., Luhrs, T., Riek, R., Von Schroetter, C., Garcia, F. L., Billeter, M., Calzolai, L., Wider, G., Wuthrich, K.: NMR Solution Structure of the Human Prion Protein Proc. Nat. Acad. Sci. USA 97 pp. 145 (2000). **Page 312** Leif Buckley and Lisa Starr. **19.16** Chris Keeney with permission of © FSIS/USDA/FDA. **19.17** © AP/Wide World Photos

CHAPTER 20 **20.1** Left, © Ric Ergenbright/CORBIS; right © Adam Woolfitt/CORBIS. **the big picture** (page 314) From left, Lisa Starr with Gary Head; © Astrid Hanns-Frieder michler/SPL/Photo Researchers, Inc; (page 315) From left, © Lewis Trusty/Animals Animals; right, © Robert C. Simpson/Nature Stock. **Page 315** © Wim Van Egmond. **20.2** Lisa Starr with Gary Head. **20.3** (a) © Dr. Stan Erlandsen, University of Minnesota; (b) © Oliver Meckes/Photo Researchers, Inc.; **20.4** Top, © Dr. David Phillips/Visuals Unlimited. Art, Raychel Ciemma. **20.5** © Astrid Hanns-Friedermichler/SPL/Photo Researchers, Inc.; **20.6** (a) Courtesy of Allen W. H. Bé and David A. Caron; (b) © John Clegg/Ardea, London. **20.7** (a) © Andrew Syred/SPL/Photo Researchers, Inc.; (b) Courtesy James Evarts. **20.8** Left, (mosquito) © Sinclair Stammers/Photo Researchers, Inc.; top right, © London School of Hygiene & Tropical Medicine/Photo Researchers, Inc.; bottom right, © Moredum Animal Health, Ltd./Photo Researchers, Inc. Art, Leonard Morgan. **20.9** (a) © Wim van Egmond/Micropolitan Museum; (b) © Frank Borges Llosa/www.frankley.com. **20.10** (a) © Dr. David Phillips/Visuals Unlimited; (b) © Lexey Swall/Staff from article, "Deep Trouble: Bad Blooms" October 3, 2003 by Eric Staats. **20.11** © Susan Frankel, USDA-FS; inset © Dr. Pavel Svihra. **20.12** (a) © Dee Breger, Drexel University; (b) © Emiliania huxleyi. Photograph by Vita Pariente. Scanning electron micrograph taken on a Jeol T330A instrument at the Texas A & M University Electron Microscopy Center. **20.13** (a) © Jeffrey Levinton, State University of New York, Stony Brook; (b) © Lewis Trusty/Animals Animals; (c) T. Garrison, Oceanography: An Invitation to Marine Science, Brooks/Cole, 1993. **20.14** Art, Raychel Ciemma; bottom, © PhotoDisc/Getty Images. **20.15** © Wim van Egmond. **20.16** (a) © Lawson Wood/CORBIS; (a) art, Art, Raychel Ciemma; (b) © D. S. Littler; (c) Courtesy of Professor Astrid Saugestad; (d) Courtesy Microbial Culture Collection, National Institute for Environmental Studies, Japan; (e) © Wim van Egmond. **20.17** Raychel Ciemma. **20.18** (a) © Edwards S. Ross; (b) © Courtesy of www.hiddenforest.co.nz. **20.19** (a) Leonard Morgan; (b) © M. Claviez, G. Gerish, and R. Guggenheim; (c-e) © Carolina Biological Supply Company; (f) Courtesy Robert R. Kay from R. R. Kay, et al., Development, 1989 Supplement, pp. 81-90, (c) The Company of Biologists Ltd., 1989. **20.20** (a-e) © Robert C. Simpson/Nature Stock. **20.21** Left, Micrograph Garry T. Cole, University of Texas, Austin/BPS; Art, Raychel Ciemma after T. Rost, et al., Botany, Wiley, 1979. **20.22** © Jane Burton/Bruce Coleman, Ltd. **20.23** Art, Raychel Ciemma; (a,b) Micrograph Ed Reschke; top, Micrograph J. D. Cunningham/Visuals Unlimited. **20.24** (a) © Michael W. Clayton/University of Wisconsin-Madison, Department of Biology; (b) © North Carolina State University, Department of Plant Pathology; (c) © Michael Wood/mykob.com. **20.25** (a) © Dr. P. Marazzi/SPL/Photo Researchers, Inc.; (b) © Harry Regin. **20.26** (a) © Mark E. Gibson/Visuals Unlimited; (b) © Edward S. Ross; (c) © 1977 Sherry K. Pittam; (d) © Mark Mattock/Planet Earth Pictures; (e) After Raven, Evert, and Eichhorn, Biology of Plants, 4th Ed., Worth Publishers, Nwe York, 1986. **20.27** (a) © 1990 Gary Braasch; (b) © F. B. Reeves. **20.28** Left © W. P. Armstrong; right, Courtesy Brian Duval.

CHAPTER 21 **21.1** © T. Kerasote/Photo Researchers, Inc. **the big picture** (page 334) From left, Lisa Starr; Raychel Ciemma; (page 335) From Left, Raychel Ciemma; © Sanford/Agliolo/

CORBIS. **Page 335** © Jean Miele/CORBIS. **Page 336** Top, © Reprinted with permission from Elsevier; bottom © Patricia G. Gensel. **21.2** Raychel Ciemma after E. O. Dodson and P. Dodson, Evolution: Process and Product, Third Ed., p. 401, PWS. **21.3** Gary Head. **21.4** Top center, © Jane Burton/Bruce Coleman Ltd.; Art, Raychel Ciemma. **21.5** (a) © Craig Wood/Visuals Unlimited; (b) © Fred Bavendam/Peter Arnold, Inc.; (c) © John D. Cunningham/Visuals Unlimited. **21.6** (a) © University of Wisconsin-Madison, Department of Biology, Anthoceros CD; (b) © National Park Services, Paul Stehr-Green; (c) © National Park Services, Martin Hutten. **21.7** © Jeri Hochman and Martin Hochman, Illustration by Zdenek Burian. **21.8** (a) © Winfried Wisniewski, Frank Lane Picture Agency/CORBIS; (b) © Colin Bates; (c) © W. H. Hodge; (d) © Craig Lovell/CORBIS. **21.9** © A. & E. Bomford/Ardea, London; art, Raychel Ciemma. **21.10** Top inset, © Brian Parker/Tom Stack & Associates; bottom, © Field Museum of Natural History, Chicago (Neg. #7500C); top left, art, Raychel Ciemma. **Page 343** Top right, George J. Wilder/Visuals Unlimited, computer enhanced by Lachina Publishing Services, Inc. **21.11** (a) © Ralph Pleasant/FPG/Getty Images; (b) © Earl Roberge/Photo Researchers, Inc.; (c) © George Loun/Visuals Unlimited; (d) Courtesy of Water Research Commission, South Africa. **21.12** (a) © Jeff Gnass Photography; (b) © Robert & Linda Mitchell Photography; (c) © Kingsley R. Stern; (d) © E. Webber/Visuals Unlimited; (e) © Michael P. Gadomski/Photo Researchers, Inc.; (f) © Sinclair Stammers/Photo Researchers, Inc.; (g) © Dr. Daniel L. Nickrent; (h) © William Ferguson. **21.13** Left, © Robert Potts, California Academy of Sciences; (a) © Robert & Linda Mitchell Photography; (b) © R. J. Erwin/Photo Researchers, Inc.; art, Raychel Ciemma. **21.14** (a) Raychel Ciemma; (b) © K. Simons and David Dilcher for color reconstruction image. **21.15** Top left (inset), © Ed Reschke; top right (inset), © Lee Casebere; bottom left (inset), © Robert & Linda Mitchell Photography; bottom right (inset), © Runk & Schoenberger/Grant Heilman, Inc.; right © Karen Carr Studio/www.karencarr.com; left art, Gary Head. **21.16** Raychel Ciemma. **21.17** Preface, Inc. **21.18** (a) © Michelle Garrett/CORBIS; (b) © Sanford/Agliolo/CORBIS; (c) © Gregory G. Dimijian/ Photo Researchers, Inc.; (d) © Darrell Gulin/CORBIS; (e) © Peter F. Zika/Visuals Unlimited; (f) © DLN/Permission by Dr. Daniel L. Nickrent. **21.19** Gerry Ellis/The Wildlife Collection. **21.20** Left, © 1989 Clinton Webb; center, © 1991 Clinton Webb; right © Gary Head.

CHAPTER 22 **22.1** © K.S. Matz. **the big picture** (page 352) From left, Leonard Morgan; right/left © David Sailors/CORBIS; right/right © Brandon D. Cole/CORBIS; (page 353) From Left, left/left © Science Photo Library/Photo Researchers, Inc.; left/right © Alex Kirstitch; right © Herve Chaumeton/Agence Nature. **Page 353** © Callum Roberts, University of York. **22.2** Leonard Morgan. **22.3** Raychel Ciemma. **22.4** (a) © Robert Brons/livingreefimages.com; (b) © Neville Pledge/South Australian Museum; (c) © Neville Pledge/South Australian Museum; (d) © Dr. Chip Clark. **22.5** Gary Head. **22.6** (a) © Bruce Hall; (b) © David Sailors/CORBIS; (b) art, Raychel Ciemma. **Page 358**, Top left, Gary Head. **22.7** Raychel Ciemma after Euguen Kozloff. **22.8** (a, b) Raychel Ciemma; (c) Courtesy of Dr. William H. Hamner; (d) © Brandon D. Cole/CORBIS; (e) © Jeffrey L. Rotman/CORBIS; (f) © A.N.T./Photo Researchers, Inc. **22.9** Top & bottom left, © Wim van Egmond/Micropolitan Museum; art, Precision Graphics after T. Storer, et al., General Zoology, Sixth Edition. **22.10** All art, Raychel Ciemma. **22.11** Top right, © Andrew Syred/SPL/Photo Researchers, Inc.; art, Raychel Ciemma and Lisa Starr. **22.12** © James Marshall/CORBIS; art, Raychel Ciemma and Lisa Starr. **22.13** (a) © Cabisco/Visuals Unlimited. (b) © Science Photo Library/Photo Researchers, Inc.; (c) © Jon Kenfield/Bruce Coleman Ltd.; (d) Precision Graphics, adapted from Rasmussen, "Ophelia," Vol. 11, in Eugene Kozloff, Invertebrates, 1990. **22.14** Raychel Ciemma. **22.15** Both, © J. A. L. Cooke/Oxford Scientific Films. **22.16** (a) Palay/ Beaubois; (b) © B. Borrell Casals/Frank Lane Picture Agency/CORBIS; (c) © Joe McDonald/CORBIS; (d) © Jeff Foott/Tom Stack & Associates; (e) © Frank Park/ANT Photo Library; (f) © Alex Kirstitch. **22.17** (a) Illustrations by Zdenek Burian, © Jeri Hochman and Martin Hochman; (b) Raychel Ciemma; (c) © Alex Kirstitch; (d) © Bob Cranston. **22.18** Lisa Starr. **22.19** (a) © Sinclair Stammers/SPL/Photo Researchers, Inc.; (b) © L. Jensen/Visuals Unlimited; (c) © Dianora Niccolini. **22.20** © Jane Burton/Bruce Coleman, Ltd. **22.21** Precision Graphics. **22.22** (a) © Jeff Hunter/The Image Bank/Getty Images; (b) © Peter Parks/Imagequest-marine.com; (c) © Science Photo Library/Photo Researchers, Inc. **22.23** (a) © Angelo Giampiccolo/FPG/Getty Images; (b) © Frans Lemmens/The Image bank/Getty Images. **22.24** (a) © CORBIS; (b) ©John H. Gerard; (c) © D. Suzio/Photo Researchers, Inc.; (d) © Andrew Syred/Photo Researchers, Inc. **22.25** Precision Graphics. **22.26** Raychel Ciemma. **22.27** (a) © David Maitland/Seaphot Limited/Planet Earth Pictures; (b-g) Edward S. Ross; (h) © Mark Moffett/Minden Pictures; (i) Courtesy of Karen Swain, North Carolina Museum of Natural Sciences; (j) © Chris Anderson/Darklight Imagery; (k) © Joseph L. Spencer. **22.28** Top left © Herve Chaumeton/Agence Nature; (a) © Herve Chaumeton/Agence Nature; (b) © Fred Bavendam/Minden Pictures; (c) © George Perina, www.seapix.com; (d) © Jan Haaga, Kodiak Lab, AFSC/NMFS. **22.29** L. Calver. **22.30** © Walter Deas/Seaphot Limited/Planet Earth Pictures. **22.31** © J. Solliday/BPS. **22.32** © Wim van Egmond/Micropolitan Museum.

CHAPTER 23 **23.1** Art, Raychel Ciemma; right © James Reece, Nature Focus, Australian Museum. **the big picture** (page 376) From left, © John and Bridgette Sibbick; Photo by Lisa Starr; courtesy of John McNamara, www.paleo-direct.com. (page 377) From left © Bill M. Campbell, MD; © Illustration by Karen Carr; © Douglas Mazonowicz/Gallery of Prehistoric Art. **Page 377** © P. Morris/Ardea London. **23.2** Art, Raychel Ciemma. **23.3** (a,b) *Redrawn from Living Invertebrates*, V. & J. Pearse and M. & R. Buchsbaum, The Boxwood Press, 1987. Used by permission; (c) © 2002 Gary Bell/Taxi/Getty Images. **23.4** Left, Seth Gold and Lisa Starr; center, © Brandon D. Cole/CORBIS; right, © Brandon D. Cole/CORBIS. **23.5** (a) © John and Bridgette Sibbick; (b, c) © Jenna Hellack, Department of Biology, Univerisy of Central Oklahoma. **23.6** (a-c) Raychel Ciemma, adapted from A. S. Romer and T. S. Parsons, *The Vertebrate Body, Sixth Edition*, Saunders, 1986; right, Photo by Lisa Starr; courtesy of John McNamara, www.paleodirect.com. **23.7** Gary Head. **23.8** (a) © Jonathan Bird/Oceanic Research Group, Inc.; (b) © Gido Braase/Deep Blue Productions; (c) © Ivor Fulcher/CORBIS; (d) © Roger Archibald; (e) Lisa Starr and Raychel Ciemma. **23.9** (a) © Norbert Wu/Peter Arnold, Inc.; (b) © Wernher Krutein/photovault.com; (c) © Alfred Kamajian; (d, e) Art, Laszlo Meszoly and D. & V. Hennings. **23.10** Left art, Leonard Morgan, adapted from A. S. Romer and T. S. Parsons, *The Vertebrate Body, Sixth Edition*, Saunders College Publishing, 1986; (a) © Bill M. Campbell, MD; (b) © Stephen Dalton/Photo Researchers, Inc.; (c) © John Serraro/Visuals Unlimited. **23.11** © Juan M. Renjifo/Animals Animals. **23.12** (a) © Pieter Johnson; (b) © Stanley Sessions/Hartwick College. **23.13** (a) © 1989 D. Braginetz; (b) © Z. Leszczynski/Animals Animals. **23.14** (a) © Illustration by Karen Carr; (b) © Karen Carr Studio/www.karencarr.com; bottom intext, © Julian Baum/SPL/Photo Researchers, Inc. **Page 388** © S. Blair Hedges, Pennsylvania State University. **23.15** Raychel Ciemma. **23.16** (a) © Kevin Schafer/CORBIS; (b) Raychel Ciemma; (c) © Joe McDonald/CORBIS; (d) © David A. Northcott/CORBIS; (e) © Pete & Judy Morrin/ARDEA LONDON; (f) © Stephen Dalton/Photo Researchers, Inc.; (f) art, Raychel Ciemma; (g) © Kevin Schafer/Tom Stack & Associates. **23.17** (a) Lisa Starr; (b) Raychel Ciemma. **23.18** (a) © Gerard Lacz/ANT Photolibrary; (b,c) Courtesy of Dr. M. Guinan, University of California-Davis, Anatomy, Physiology and Cell Biology, School of Veterinariy Medicine; (d) © Kevin Schafer/CORBIS; **23.19** (a) © Sandy Roessler/FPG/Getty Images; (b) Art, Raychel Ciemma after M. Weiss and A. Mann, Human Biology and Behavior, 5th Edition, HarperCollins, 1990; (c) © Jean Phillipe Varin/Jacana/Photo Researchers, Inc. (d) © Corbis Images/PictureQuest; (e) © Merlin D. Tuttle/Bat Conservation International; (f) © Marine Themes Stock Photo Library; (g) © Mike Johnson. All rights reserved, www.earthwindow.com. **23.20** (a–d) Lisa Starr; (e) © D. & V. Blagden/ANT Photo Library; (f) © Nigel J. Dennis; Gallo Images/CORBIS; (g) © Tom Ulrich/Visuals Unlimited. **23.21** (a) © Larry Burrows/Aspect Photolibrary; (c) © Dallas Zoo, Robert Cabello; (d) bottom left, © Bone Clones®, www.boneclones.com; (d) bottom right, © Gary Head; top right, © Allen Gathman, Biology Department, Southeast Missouri State University. **Page 395** Top left, D. & V. Hennings. **23.22** © Gerry Ellis/The Wildlife Collection; inset © Utah's Hogle Zoo. **23.23** (a) From left, © MPFT/Corbis Sygma; (a) National Museum of Ethiopia, Addis Ababa. © 1985 David L. Brill; (a) Original housed in National Museum of Ethiopia, Addis Ababa. © 1999 David L. Brill; (a) National Museum of Tansania, Dar es Salaam, © 1985 David L. Brill; (a) Transvaal Museum, Pretoria. © 1985 David L. Brill; (a) National Museum of Kenya, Nairobi. © 1985 David L. Brill; (b) © Dr. Donald Johanson, Institute of Human Origins; (c) © Louise M. Robbins; (d) © Kenneth Garrett/National Geographic Image Collection. **23.24** © Jean Paul Tibbles. **23.25** Top, © Elizabeth Delaney/Visuals Unlimited; bottom (all), © John Reader. **23.26** Left, American Museum of Natural History. © 1996 David L. Brill; right, MUSEE DE L'HOMME, Paris. © 1985 David L. Brill. **23.27** Left, Housed in National Museum of Ethiopia, Addis Ababa. © 2001 David L. Brill; right art, Lisa Starr. **23.28** Left, © Douglas Mazonowicz/Gallery of Prehistoric; right, Lisa Starr. **23.29** © California Academy of Sciences. **23.30** © Tom McHugh/Photo Researchers, Inc.

CHAPTER 24 **24.1** Left, © Vvg/Science Photo Library/Photo Researchers, Inc.; right © Michael Davidson/Mortimer Abramowitz Gallery of Photomicrography/www.olympusmicro.com. **the big picture** (page 402) From left, left/left, © Bruce Iverson; left/right © Dr. Robert Wagner/University of Delaware, www.udel.edu/Biology/Wags; right/left © Pat Johnson Studios Photography; right/right © Darrell Gulin/Getty Images; (page 403) From left, Gary Head; Slim Films. **Page 403** © Star Tribune/Minneapolis-St. Paul. **24.2** Left, Raychel Ciemma; top, Courtesy of Charles Lewallen; center & bottom, © Bruce Iverson. **24.3** left © 2000 PhotoDisc, Inc, with art by Lisa Starr; top right © Cnri/Spl/Photo Researchers, Inc.; bottom right © Dr. Robert Wagner/University of Delaware, www.udel.edu/Biology/Wags. **24.4** Left © Pat Johnson Studios Photography; right, © Darrell Gulin/The Image Bank/Getty Images. **24.5** (a) © Cory Gray; (b) © PhotoDisc/Getty Images; (c) © Heather Angel; (d) © Biophoto Associates/Photo Researchers, Inc. **24.6** Gary Head. **24.7** © Galen Rowell/Peter Arnold, Inc. **24.8** Gary Head. **24.9** Left art, Gary Head; right, © Niall Benvie/CORBIS. **24.10** Left © Kennan Ward/CORBIS; right © G. J. McKenzie (MGS). **24.11** © Frank B. Salisbury. **24.12** Slim Films. **24.13** (a,b) Courtesy of Dr. Kathleen K. Sulik, Bowles Center for Alcohol Studies, the University of North Carolina at Chapel Hill; (c) © John DaSiai, MD/Custom Medical Stock. **24.14** (a) Courtesy of Dr. Consuelo M. De Moraes; (b-d) © Andrei Sourakov and Consuelo M. De Moraes.

Page 415 Unit V © Kevin Schafer.

CHAPTER 25 **25.1** (a) © Lennart Nilsson from *A Child is Born*, © 1966, 1977 Dell Publishing Company, Inc.; (b) © 1999 Dana Fineman/CORBIS Sygma. **the big picture** (page 416) From left, © Ron Austing; Frank Lane Picture Agency/CORBIS; © Carolina Biological Supply Company; (page 417) From left, Raychel Ciemma; © Lennart Nilsson, *A Child is Born*, © 1966, 1977 Dell Publishing Company, Inc. **Page 417** © Dow W. Fawcett/Photo Researchers, Inc. **25.2** © Fred SaintOurs/University of Massachusetts-Boston; art, Lisa Starr. **25.3** (a) © Marc Moritsch; (b) © Photo Researchers, Inc. **25.4** (a) © Frieder Sauer/Bruce Coleman, Ltd; (b) © Matjaz Kuntner; (c) © Ron Austing; Frank Lane Picture Agency/CORBIS; (d) © Doug Perrine/seapics.com; (e) © Carolina Biological Supply Company; (f) © Fred McKinney/FPG/Getty Images; (g) © Gary Head. **25.5** Art, Palay/Beaubois and Precision Graphics. **25.6** (a) Art, Raychel Ciemma; (b–j) © Carolina Biological Supply Company; (b–j) art, L. Calver. **25.7** Art, Robert Demarest. **25.8** (a, b) Art, Precision Graphics; (c) © Gary Head. **25.9** (a–d) © Dr. Maria Leptin, Institute of Genetics, University of Koln, Germany. **25.10** (a) Lisa Starr; (b) Raychel Ciemma after B. Burnside, *Developmental Biology*, 1971, 26:416-441. Used by permission of Academic Press. **25.11** Art, Precision Graphics; (Photographic series) © Carolina Biological Supply Company; far right © Peter Parks/Oxford Scientific Films/Animals, Animals. **25.12** (a,b) Art, Lisa Starr; (c) © Professor Jonathon Slack. **25.13** Left, © Peter Parks/Oxford Scientific Films/Animals, Animals; (a,b) Art, Raychel Ciemma after S. Gilbert, Developmental Biology, Fourth Edition. **25.14** Art, Raychel Ciemma with Lisa Starr; right, © Laura Dwight/CORBIS. **25.15** Raychel Ciemma with Lisa Starr. **25.16** (a,b,c) Art, Raychel Ciemma; (b) © Ed Reschke. **25.19** Raychel Ciemma. 662: 38.20 Left art, Raychel Ciemma; top right art, Robert Demarest; bottom right, Photograph Lennart Nilsson from *A Child is Born*, © 1966, 1977 Dell Publishing Company, Inc. **25.21** Raychel Ciemma with Precision Graphics. **25.22** K. Sommerville, Robert Demarest, and Preface, Inc. **25.23** Raychel Ciemma; **25.24** Preface, Inc. **25.25** © Lester Lefkowitz/CORBIS. **25.26** (a) © Dr. E. Walker/Photo Researchers, Inc.; (b) © Western Ophthalmic Hospital/Photo Researchers, Inc.; (c) © Kenneth Greer/Visuals Unlimited; (d) © CNRI/Photo Researchers, Inc. **25.27** © Todd Warshaw/Getty Images. **25.28** Raychel Ciemma. **25.29** Raychel Ciemma. **25.30** Raychel Ciemma. **25.31** (a-d) Art, Raychel Ciemma; (a–d) top, From Lennart Nilsson, *A Child is Born*, © 1966, 1977 Dell Publishing Company, Inc. **25.32** Raychel Ciemma, modified from K. L. Moore, *The Developing Human: Clinically Oriented Embryology*, Fourth Edition, Philadelphia: W. B. Saunders Co., 1988. **25.33** Left, © Zeva Oelbaum/CORBIS right, © James W. Hanson, M.D. **25.34** Robert Demarest. **25.35** Raychel Ciemma. **25.36** Top art, Raychel Ciemma adapted from L. B. Arey, Developmental Anatomy, Philadelphia, W. B. Saunders Co., 1965; bottom, Lisa Starr. **Page 452** © Denis Scott/CORBIS. **25.37** (a) © David M. Parichy; (b) © Dr. Sharon Amacher; (c) © Dr. Sharon Amacher.

Page 455 Unit VI © Alan and Sandy Carey.

CHAPTER 26 **26.1** © David Nunuk/Photo Researchers, Inc. **the big picture** (page 456) From left, © Amos Nachoum/CORBIS; Gary Head; (page 457) From left, © Joe McDonald/CORBIS; © Don Mason/CORBIS. **Page 457** © Tom Till/Stone/Getty Images. **26.2** Art, Precision Graphics and Gary Head; bottom left, © Amos Nachoum/CORBIS; bottom center, © A. E. Zuckerman/Tom Stack & Associates; right inset, ©
CORBIS. **26.3** © E. R. Degginger; inset, © Jeff Fott Productions/Bruce Coleman, Ltd. **26.4** (a) ©
Bateman, Bateman Photography; (b) ©
Jeff Lepore/Photo Researchers, Inc.;
26.6 Precision Graphics. **26.7** (a)
Photo; (b) © Christine
Head. **26.9** Top,
CORBIS Sygma
Wayne
art,

26.13 Precision Graphics. **26.14** © U.S. Geological Survey. **26.15** NASA; art, Precision Graphics. **26.16** Gary Head. **26.17** Preface Inc. **26.18** Left, © Adrian Arbib/CORBIS; center, Precision Graphics; right, © Don Mason/CORBIS. **26.19, 26.20** Ater T. Miller, Jr., *Environmental Science*, Sixth Edition, Brooks/Cole, 1997. All rights reserved. **26.21** © John Alcock/Arizona State University. **26.22** © Wolfgang Kaehler/CORBIS. **Page 475** Precision Graphics.

CHAPTER 27 **27.1** Left & center, Photography by B. M. Drees, Texas A&M University. Http://fireant.tamu; right, © Daniel Wojak/USDA. **the big picture** (page 476) from left, © Martin Harvey, Gallo Images/CORBIS; right, Thomas W. Doeppner; (page 477) from left, © Pat O'Hara/CORBIS; right © James Marshall/CORBIS. **Page 477**: Top, © John Kabashima. **27.2** Top left, © Donna Hutchins; center left, © B. G. Thomson/Photo Researchers, Inc.; bottom left, © Len Robinson, Frank Lane Picture Agency/CORBIS; bottom right, © Martin Harvey, Gallo Images/CORBIS. **27.3** Left, © Bob and Miriam Francis/Tom Stack & Associates; right © Harlo H. Hadow. **27.4** Thomas W. Doeppner. **27.5** (mountain peak) © Richard Cummins/CORBIS; (a,d) © Don Roberson; (b) © Kennan Ward/CORBIS (c) © D. Robert Franz/CORBIS. **27.6** Top, right © Michael Abbey/Photo Researchers, Inc.; bottom right © Eric V. Grave/Photo Researchers, Inc.; (a–c) art, Gary Head **27.07** Top, © Joe McDonald/CORBIS; bottom left, © Hal Horwitz/CORBIS; bottom right, © Tony Wharton, Frank Lane Picture Agency/CORBIS; top art, Precision Graphics after N. Weldan and F. Bazazz, Ecology, 56:681-688, © 1975 Ecological Society of America. **27.8** (a, b) Art, Gary Head, after Rickleffs & Miller, Ecology, Fourth Edition, page 459 (Fig 23.13a) and page 461 (Fig 23.14); right, © W. Perry Conway/CORBIS. **27.9** (a) Precision Graphics; (b) © Ed Cesar/Photo Researchers, Inc.; (c) © Kennan Ward. **27.10** (a) © JH Pete Carmichael; (b) © Edward S. Ross; (c) © W. M. Laetsch. **27.11** (a-c) © Edward S. Ross; (d) © Nigel Jones. **27.12** (a,b) © Thomas Eisner, Cornell University; (c) © Jeffrey Rotman Photography; (d) © Bob Jensen Photography. **27.13** (a) Courtesy of Ken Nemuras; (b) © CDC. **27.14** Left, © The Samuel Roberts Noble Foundation, Inc.; right, Courtesy of Colin Purrington, Swarthmore College. **27.15** © C. James Webb/Phototake, U.S.A. **27.16** © Peter J. Bryant/Biological Photo Service. **27.17** Left, © Richard Price/Getty Images; right © E.R. Degginger/Photo Researchers, Inc. **27.18** (a) © Doug Peebles/CORBIS; (b) © Pat O'Hara/CORBIS; (c) © Tom Bean/CORBIS; (d) © Tom Bean/CORBIS; (e) © Duncan Murrell/Taxi/Getty Images. **27.19** (a) R. Barrick/USGS; (b,c) © 1980 Gary Braasch. **27.20** (a,c) © Jane Burton/Bruce Coleman, Ltd.; (b) © Heather Angel; (d,e) Art, Precision Graphics based on Jane Lubchenco, *American Naturalist*, 112:23-29, © 1978 by The University of Chicago Press. Used with permission. **27.21** Left, © The University of Alabama Center for Public TV; (a) © Angelina Lax/Photo Researchers, Inc.; (b) © Pr. Alexande Meinesz, University of Nice-Sophia Antipolis. **27.22** © John Carnemolla/Australian Picture Library. **27.23** Art, Gary Head after W. Dansgaard, et al., Nature, 364:218-220, 15 July 1993; D. Raymond, et al., Science, 259:926-933, February 1993; W. Post, American Scientist, 78:310-326, July-August 1990. **27.24** Left, both, © Pierre Vauthey/CORBIS Sygma; right art, Gary Head after S. Fridriksson, Evolution of Life on a Volcanic Island, Butterworth, London, 1975. **27.25** (a) © Susan G. Drinker/CORBIS; (a) art, computer-modified by Lisa Starr (b) © Frans Lanting/Minden Pictures. **27.26** © James Marshall/CORBIS. **27.27** Left, © Bagla Pallava/CORBIS Sygma; right inset, © A. Bannister/Photo Researchers, Inc. **27.28** Lisa Starr. **27.29** Lisa Starr. **27.30** Art, Lisa Starr with photographs © 2000 PhotoDisc, Inc. **27.31** Both, © Bureau of Land Management. **27.32** (a) © Anthony Bannister, Gallo Images/CORBIS; (b) © Bob Jensen Photography; ... Cedric Vaucher. **27.33** Both, Courtesy of Eternal Re... ww.eternalreefs.com

CHAPTER 28 ... siana Department of Wildlife. the ... Gary Head; right/left © D. Rob... right/right © Frans Lantin... ft, Preface, Inc.; ri... deau/CORBIS.

28.2 Art, Gary Head and Lisa Starr; photographs, above, © PhotoDisc, Inc.; below, © David Neal Parks. **28.3** © Photograph Alan and Sandy Carey; Art, Gary Head after R. L. Smith, *Ecology and Field Biology*, Fifth Edition. **28.4, 28.5** Art, Lisa Starr with photographs, flowering plants, © Frank Oberle/Stone/Getty Images; marsh hawk, © J. Lichter/Photo Researchers, Inc.; crow, © Ed Reschke; upland sandpiper, © O. S. Pettingill Jr./Photo Researchers, Inc.; garter snake, © Michael Jeffords; frog, © John H. Gerard; yellow spider, © Michael Jeffords; weasel, Courtesy of Biology Department, Loyola Marymount University; sparrow, © Rod Planck/Photo Researchers, Inc.; cutworm, © Nigel Cattlin/Holt Studios International/Photo Researchers, Inc.; gopher and prairie vole, © Tom McHugh/Photo Researchers, Inc.; all others, © PhotoDisc, Inc. **28.6** Art, Precison Graphics and Gary Head; photographs, left © D. Robert Franz/Planet Earth Pictures; right © Frans Lanting/Bruce Coleman, Ltd. **28.7** © Gary Head. **28.8:** © Eric Hartmann/Magnum Photos. **28.9** Preface, Inc. **28.10:** NASA/GSFC. **28.11** Precision Graphics. **28.12** © Gerry Ellis/The Wildlife Collection; Art, Gary Head. **28.13** Lisa Starr. **28.14** (left) Gary Head after G. E. Likens and F. H. Bormann, "An Experimental Approach to New England Landscapes," in A. D. Hasler (ed.), Coupling of Land and Water Systems, Champman & Hall, 1975; right, © Dr. Elizabeth Hane. **28.15** Water Resources Council; **28.16** Lisa Starr after Paul Hertz; photograph © 2000 PhotoDisc, Inc. **28.17** Lisa Starr and Gary Head, based on NASA photographs from JSC Digital Image Collection. **28.18** Lisa Starr and Gary Head, based on NASA photographs from JSC Digital Image Collection. **28.19** © Yann Arthus-Bertrand/CORBIS. **28.20** (a) Lisa Starr and Gary Head, compilation of data from Mauna Loa Observatory, Keeling and Whorf, Scripps Institute of Oceanography; (b) Lisa Starr and Gary Head, compilation of data from Prinn, et al., CDIAC, Oak Ridge National Laboratory; world Resources Institute; Khalil and Rasmussen, Oregon Graduate Institute of Science and Technology, CDIAC DB-1010; Leifer and Chan, CDIAC DB-1019; (c,d) Lisa Starr and Gary Head, compilation of data from World Resources Institute; Law Dome ice core samples, Etheridge, Pearman, and Fraser, Commonwealth Scientific and Industrial Research Organisation; Prinn, et al., CDIAC, Oak Ridge National Laboratory; Leifer and Chan, CDIAC DB-1019. **28.21** © Jeff Vanuga/CORBIS; Art, Gary Head. **28.22** © Frederica Georgia/Photo Researchers, Inc. **28.23** Art, Gary Head and Lisa Starr; photograph, © 2000 PhotoDisc, Inc. **28.24** Fisheries & Oceans Canada, Experimental Lakes Area; **28.2** Art, Gary Head; photographs © Bruce Coleman. **28.26** Courtesy of NASA's Terra satellite, supplied by Ted Scambos, National Snow and Ice Data Center, University of Colorado, Boulder. **28.27** Gary Head, after T. Garrison.

CHAPTER 29 **29.1** © Hank Fotos Photography. **the big picture** (page 524) from left, L. Calver; © Douglas Peebles/CORBIS; (page 525) from left © Pat O'Hara/CORBIS; right, Lisa Starr. **Page 525,** Top, © Wolfgang Kaehler/CORBIS. **29.2** Preface, Inc. **29.3** L. Calver. **29.4** Precision Graphics; **29.5:** (a) © Alex MacLean/Landslides; (b) © Alex MacLean/Landslides. **29.6** NASA. **29.7** Lisa Starr. **29.8** (a) Gary Head, adapted from *Living in the Environment* by G. Tyler Miller, Jr., p. 428. © 2002 by BrooksCole, a division of Thomson Learning; (b) © Ted Spiegel/CORBIS. **29.9** Gary Head. **29.10** NASA. **29.11** Left, © Sally A. Morgan/Ecoscene/CORBIS; right, © Bob Rowan, Progressive Image/CORBIS; above, art, Lisa Starr. **29.12** NASA. **29.13** Lisa Starr using photographs © 2000 PhotoDisc, Inc. **29.14** Bottom, © George H. Huey/CORBIS; **29.14** Left inset, © John M. Roberts/CORBIS. **29.15** © Orbimage Imagery. **29.16** Left, © John C. Cunningham/Visuals Unlimited; center, © Jack Wilburn/Animals Animals; right, © AP/Wide World Photos. **29.17** Bottom left, © Tom Bean Photography; top right © Jonathan Scott/Planet Earth Pictures. **29.18** Left, © 1991 Gary Braasch Photography; bottom inse,t © Karl Lehmann/LostWorldArts.com. **29.19** Left, © James Randklev/CORBIS; inset, All, © Randy Wells/CORBIS. **29.20** © Raymond Gehman/CORBIS; inset right, © Nigel Cook/Dayton Beach News Journal/Corbis Sygma. **29.21** (a) © Darrell Gulin/CORBIS; (b) © Paul A. Souders/CORBIS; (c) © Pat O'Hara/CORBIS. **29.22** Art, D & V. Hennings after Whittaker, Bland, and Tilman. **29.23** © Onne van der Wal/CORBIS. **29.24** (a) Precision Graphics; (b) Precision Graphics after E. S. Deevy, Jr., Scientific American, October 1951. **29.25** (a) © E. F. Benfield, Virginia Tech; (b) © E. F. Benfield, Virginia Tech; (c) © E. F. Benfield, Virginia Tech; (d) © E. F. Benfield, Virginia Tech; (e) © E. F. Benfield, Virginia Tech. **29.26** (a) © Annie Griffiths Belt/CORBIS; (b) © Douglas Peebles/CORBIS. **29.27** (a) © Mike Zens/CORBIS; **29.27** (b) © Paul A. Souders/CORBIS. **29.28** (a) Top left, © Douglas Faulkner/Sally Faulkner Collection; top right, © Royalty-Free/CORBIS; center left, © Jeff Rotman; center right, © Alex Kirstitch; bottom left, © Amos Nachoum/CORBIS; bottom right, © Douglas Faulkner/Sally Faulkner Collection; (b) T. Garrison, *Oceanography: An Invitation to Marine Science*, Third Edition, Brooks/Cole, 2000. All rights reserved. **29.29** © Greenpeace/Grace. **29.30** (a) © Peter David/FPG/Getty Images; **29.30** (b) © Robert Vrijenhoek, MBARI; (c) Art, L. K. Townsend. **29.31** Lisa Starr. **29.32** (a,b) Lisa Starr; (c) NASA/Goddard Space Flight Center Scientific Visualization Studio; (d) NASA/Goddard Space Flight Center Scientific Visualization Studio. **29.33** (a) © Eye of Science/Photo Researchers, Inc.; (b) © CHAART, at NASA Ames Research Center; (c) © Douglas P. Wilson; Frank Lane Picture Agency/CORBIS; (d) © Raghu Rai/Magnum Photos. **29.34** Precision Graphics after M. H. Dickerson, "ARAC: Modeling an Ill Wind," in Energy and Technology Review, August 1987. Used by permission of University of California Lawrence Livermore National Laboratory and U. S. Dept. of Energy.

CHAPTER 30 **30.1** Top, © Scott Camazine; **the big picture** (page 550), from left, © James Zipp/Photo Researchers, Inc.; © Monty Sloan, www.wolfphotography.com; (page 551) from left, © Australian Picture Library/CORBIS; © Matthew Alan/CORBIS. **Page 551** Top, © Ralph Fountain. **30.2** (a) © Eugene Kozloff; (b) © Stevan Arnold; (c) © Stevan Arnold. **30.3** Left, © Robert M. Timm & Barbara L. Clauson, University of Kansas (a) Reprinted from *Trends in Neuroscience*, Vol. 21, Issue 2, 1998, L.J. Young, W. Zuoxin, T.R. Insel, "Neuroendocrine bases of monogamy," Pages 71-75, ©1998, with permission from Elsevier Science; (b) Reprinted from *Trends in Neuroscience*, Vol. 21, Issue 2, 1998, L.J. Young, W. Zuoxin, T.R. Insel, "Neuroendocrine Bases of Monogamy," Pages 71-75, ©1998, with permission from Elsevier Science. **30.4** (a) © Eric Hosking; (b) © Stephen Dalton/Photo Researchers, Inc.; (c) © Jennie Woodcock; Reflections Photolibrary/CORBIS. **30.5** © Nina Leen/TimePix. **30.6** © James Zipp/Photo Researchers, Inc. **30.7** (a) © Robert Maier/Animals Animals; (b) © John Bova/Photo Researchers, Inc.; (c) From L. Clark, Parasitology Today, 6(11), Elsevier Trends Journals, 1990, Cambridge, UK. (d) © Jack Clark/Comstock, Inc. **Page 30.8** (a) © Monty Sloan, www.wolfphotography.com; (b) © Tom and Pat Leeson, leesonphoto.com; (c) © Kevin Schafer/CORBIS. **30.09** © Stephen Dalton/Photo Researchers, Inc.; art, D. & V. Hennings. **30.10** (a) © John Alcock, Arizona State University; (b) © Pam Gardner; Frank Lane Picture Agency/CORBIS; (c) © Pam Gardner; Frank Lane Picture Agency/CORBIS; (d) © D. Robert Franz/CORBIS. **30.11** (a) © Ingo Arndt/Nature Picture Library; (b) © Michael Francis/The Wildlife Collection. **30.12** (a) © B. Borrell Casals; Frank Lane Picture Agency/CORBIS; (b) © Steve Kaufman/CORBIS; (c) © John Conrad/CORBIS. **30.13** (a) © Tom and Pat Leeson, leesonphoto.com; (b) © John Alcock, Arizona State University; (c) © Paul Nicklen/National Geographic/Getty Images. **30.14** © Jeff Vanuga/CORBIS. **30.15** © Eric and David Hosking/CORBIS. **30.16** (a) © Australian Picture Library/CORBIS; (b) © Alexander Wild; (c) © Professor Louis De Vos. **30.17** (a) © Kenneth Lorenzen; (b) © Nicola Kountoupes/Cornell University; (c) © Peter Johnson/CORBIS. **30.18** © Dr. Tim Jackson, University of Pretoria. **Page 564** © Matthew Alan/CORBIS. **30.19** © Brad Bergstrom. **30.20** © F. Schutz.

EPILOGUE © Joseph Sohm, Visions of America/CORBIS.

Index

The letter i *designates illustration;* t *designates table;* **bold** *designates defined term;*
■ *highlights the location of applications contained in text.*

A

Aardvark, 393, 393i
Abdomen, insect, 370
■ ABO blood typing, 158, 158i, 164
■ Abortion, 151, 165, 179, **180**, 181, 439
Absorption, of ethanol, in stomach, 72
Absorption spectra, 95i
■ Abstinence, sexual, 137, 438, 438i
Abyssal zone, 545i
Acacia tortilis, 414
Acanthostega, 383i
Acceptor molecule, 90
Accessory heart, 365, 365i
Accessory pigment, **94**, 95i, 105
Accessory protein, 66
Acetaldehyde, 72, 73
Acetate, 72
Acetic acid, 72
Acetyl-CoA, 112, 113i, 118
Acetyl group, 208, 211
Achillea millefolium, 162–163, 162i
■ Achondroplasia, 173t, 174, 174i
Acid, 28–29, **30t**
Acid deposition, 529
■ Acid rain, 29, 331, 519, 519i, **529**, 529i
■ Acid stomach, 28
Acidic solution, **28**
Acidophilus, 304
■ Acidosis, 29
■ Acne drugs, 449
Acoelomate animal, 355i
Acoustical signals, 556
Acquired characteristics, inheritance of, 240
Acquired Immunodeficiency Syndrome. *See* AIDS
Actin, 66, 66i
Activation energy, 78–**79**, 78i, 81
Activator protein, 80, 80i, **208–209**
Active binding site, enzyme, **78**–80, 80i, 84
Active transport, **83**, 83i, 85, 85i, **407**
Active transporter, 57, 57i
■ Acute respiratory infection, 310t
■ Acyclovir, 440
Adaptation, **236, 244**
  evolutionary, 244–245, 336–337, 367
  of fishes, 382
  invertebrate, 362, 364–365, 367
  long- *vs.* short-term, 244–245
  of prey, 484–485, 484i, 485i
Adaptive behavior, **555**, 564
Adaptive radiation, 263i, **278–279**
  animal, 279i, 293, 356, 365
  mammals, 392, 393, 393i
  plant, 334, 336, 343, 346, 347
Adaptive trait, **10**, 252
Adaptive zone, **278–279**
Adenine (A), 46, 46i, 75, 184i, 188–189, 188i, 189i, 196–197, 199, 202, 227i
Adenosine diphosphate, 75, 85i, 97i, 104i
Adenosine phosphate, 48t
■ Adenovirus, 308i
ADH, 553
Adhering junction, 69, 69i
Adhesion protein, 56, 56i
Adipose tissue, 40–41, 118, 122
  brown, 122
Adolescent, 451t
ADP, 75, 85i, 97i, 104i
Adult, 451t
Aerobe, 305
Aerobic bacteria, 295, 320
Aerobic metabolism, 90–91, 263i
Aerobic reaction, 307
Aerobic respiration, 51, 64, 75, 77, 77i, 93, 97, 102, **108–109**, 108i, 109i, 293, 296, 296i–297i, 509
  animal, 367
  in cells, 294, 329
  second stage of, 112–113, 112i, 113i
  third stage of, 114–115, 114i
Africa, 398–399

African emergence model, 399
■ African sleeping sickness, 316, 317i
■ Africanized bees, 550, 550i
■ Afterbirth, 450, 450i
■ Agar, 322
Agaris bisporus, 327
Age-specific life history patterns, 464–467
Age structure, **458**, 470, 471i, 475, 475i
■ Aging, 452
  process, 80
Agouti gene, 249
■ Agriculture
  cereal croplands, 535
  deforestation and, 349
  domestication of plants, 343
  endangered species and, 496
■ farming methods, 228
■ fertilizer and, 29, 333, 519, 521
  grassland conversion for, 519
  human population growth and, 468
  insects/pests and, 320, 333, 366, 370
  ozone layer and, 528
  phosphorus and, 520–521
■ probiotic feeds, 501
  runoff, 519, 521
  soil and, 539
  water and, 91, 512–513
Agrobacterium tumefaciens, 228i, 229, 305
Aguilegia, 348i
■ AIDS, 2, 308, 310, 310t, 440t, 441, 441i
  drug therapies, treatments, 352
Air circulation patterns
  global, 526–527, 526i
  humans and, 528–529
■ Air pollution, 519i
■ deforestation and, 334–335, 334i, 335i
■ fossil fuels and 2, 29i, 93
■ fungi and, 330, 331, 339
Air space, in plant tissue, 103i
Aix sponsa, 35i
Ajellomyces capsulatus, 329
Alanine (ala), 42i, 198i
Albatross, 255i, 391i, 495i
■ Albinism, 159, 164, 164i, 173t
Albumin, denaturation of, 44
Alcohol, 72, 91
  in blood, 72
■ consumption of, 63, 72
■ fetal alcohol syndrome, 449, 449i
  pregnancy and, 449, 449i
  structure, 34
  toxicity and, 72
■ Alcoholic beverages, making of, 329
■ Alcoholic cirrhosis, 72
■ Alcoholic fermentation, **116**, 116i
■ Alcoholic hepatitis, 72
Aldehyde, 34i, 38
Alexander the Great, 255, 300, 300i
Alga, 17i, 53, 138, 286, 287, 362, 480, 490i, 491, 492–493, 492i. *See also* Green alga; *specific types*
  brown, 316i, 320, 321, 321i
■ commercial uses, 321, 322
  evolution, 336, 336i
  golden and yellow-green, 321
  life cycle, 336
  as photosynthesizer, 93, 96, 101
  red, 105, 105i, 287, 292i, 293, 297i, 307i, 316i, 319, 322, 322i
  snow alga, 286, 333, 333i
■ Algal bloom, 92, 92i, **320**, 333, 503, **521**, 544, 547
Alginic acid, 321
Alkaline solution, **28**
Alkaloid, 327, 329, 330
Alkalosis, 29
■ Alkylating agent, **203**
Allantois, **443**

Allele, **140, 153, 164, 168**
  codominance of, **158**, 158i
  dominant, 151, **153**, 155–156, 155i, 156i, 164
  dominant/recessive interaction, 151, 153–159, 154i, 155i, 156i, 158i, 159i, 164, 168, 183, 213, 237
  incomplete dominance of, **158**, 158i
  lethal, **164**, 174
  multiple allele system, **158**
  mutant, 158, 168, 183
  recessive, 151, 153–155, 154i, 155i, 156i, 158, 159, 183, 213
  trait variation and, 144–145, 144i, 153, 153i, 154i–157i, 159, 211, 250
  wild-type, 168, 183, 254
Allele frequency, **246**–248, 247i, 252, 254, 254i, 255, 272, 278
■ Allergy, 369
Alligator, 388i, 389
Allolactose, 208–209, 209i
Allopatric speciation, **274**–275, 274i, 275i
Allosteric binding site, 80, 80i
Allosteric control, 211
Alpha-1,3-galactose, 231
Alpha globin, 44, 44i
Alpine tundra, 533i, 538, 538i
Alternative splicing, of genes, **197**
■ Altitude sickness, 122
Altruism, **555**
  indirect selection for, 563
Alu element, 269
Alveolate, 316i, 318, 319
■ Alzheimer's disease, 107, 230, 452
Amanita muscaria, 327i
Amber, insects in, 261
Amborella, 348i
Amino acid, **42**–44, 42i, 63, 79i, 118, 180
  origin of cell and, 290
  protein synthesis and, 195i, 196, 198–199, 198i, 200, 200i
  residue, 42i
  sequence, 196, 196i, 198, 201, 202, 202i, 270, 271
  in space, 289
Amino group, 34i, 42i
■ Amish, Older Order, 255
Ammonia, 25i, 28i, 29, 81i, 118, 288, 289, 304, 518
Ammonification, **519**
Ammonite, 239i
Ammonium, 518–519
■ Amniocentesis, 180i, 181
Amnion, 181, **442**–443
Amniote, **386**–387, 386i
Amniotic cavity, 442, 443i
Amoeba, 66, 67, 89i, 128
Amoeba discoides, 295
A. proteus, 317i
■ Amoebic dysentery, 317
Amoeboid cell, 325
Amoeboid protozoan, 316i, **317**
AMP, 325i
■ Amphibian, 381i, **384**, 384i. *See also specific types*
  characteristics, 211, 354, 354t
  evolution, 263i, 266, 383i, 384–385
■ habitat contamination, 385, 385i
  reproduction, 385
Amphiprion perideraion, 479, 479i
Amplification, DNA, 220, 224–227, 225i, 227i, 286, 307
Amylose, 38i, 39i
■ Amyotrophic lateral sclerosis, 107
Anabaena, 304, 304i
Anabolic pathway, **77**
Anaerobe, 263i, 305, **306**, 307
Anaerobic atmosphere, 296i–297i
Anaerobic energy-releasing pathways, 108, 108i
Anaerobic habitat, 293
Anaerobic reaction, **50**, 51

Anagenesis, **278**
Analogous structures, 267, 267i
Anaphase, 124, 128i, 129, **130**, 131, 131i, 132, 141i, 142i–143i, 148, 178i
Anatomy, **403**
Ancient human remains, 339
■ ANDi, 234, 234i
■ Androgen insensitivity syndrome, 173t
■ Aneuploidy, **178**–179, 179i
Angiosperm, 334, 336i, **346**, 346i. *See also* Flowering plant
■ Anhidrotic ectodermal dysplasia, 213, 213i
Animal, 6, **9**, 39, 282, 332t
  behavior, 551
  characteristics, 92, 105, 146, **354**–355, 354i, 355i
  commonalities with plants, 402–414
  evolution, 297i, 356–357
  inheritance patterns, 168, 178
Animal cell, 52i, 53i, 60i, 66, 68, 70t, 132–133, 132i, 133i
Animal cuticle, 39, 39i
Animal kingdom, 354–355, 354t
Animal survival experiment, 12–13, 251, 251i
Annelid, 254t, 355, 356i, **362**–363, 362i
Anopheles, 318, 319i
Ant, 370, 476–477, 494i, 562i, 563
■ Antacid, 28
Antarctica, 338, 523, 523i, 528i
Anteater, 393, 393i
Antelope, 414, 418i
Antenna, 368, 370, 370i
Antennapedia gene, 212
Anterior-posterior axis, 212, **354**, 354i
Anther, 140i, 268
Anthocyanin, **95**, 95i
■ Anthrax, 304
Anthropoid, 394, 395
■ Anti-acne drugs, 449
■ Antibiotic resistance, 59, 222, 237, 248, 310, 318, 501
■ Antibiotics, **248**, 304, **310**
Antibody, 48t, 231
Anticipatory behavior, 567
Anticodon, **199**, 199i, 201
Antidepressant, 449
Antidiuretic hormone, 553
Antioxidant, **80**
■ Antisense drug, **205**
■ Antiseptic, natural, 339
Antithamnion plumula, 322i
Antithrombin, 229
Anus, 355
  evolution, 356i
  invertebrate, 364i, 365i, 366i, 373
■ Anxiety, 163
Aorta, 159
Apatosaurus, 387i
Ape, 394, 396
Aphid, 138–140, 138i, 140i, 487i
Apicomplexan, 316i, **318**, 322
Apis mellifera. See Bee
Apoptosis, **412**–413, 425
Appendage, invertebrate, 367–369, 370i
■ Appendicitis, 366i
Apple, 329
Aquatic ecosystem, 510, 510i
Aquatic snail, 105i, 361i, 364i
Aquifer depletion, 513, 513i
Arabian oryx, 414
Arabidopsis thaliana, 268i
Arachnid, 369, 369i
Arceuthobium, 348i, 486
Archaea, 8
Archaean eon, 263, 263i, **292**
Archaeanthus linnenbergeri, 346i

Archaebacteria (Archaea), 497
  characteristics, 8, 8i, 70t, 81i,
    306–307, 306i, 332t
  as domain of life, 8, 8i, 58, 280, 280i,
    282i, 286, 304i, 306
  evolution, 287, 296i–297i
  methanogenic, 36–37, 36i, 306, 306i
  structural adaptations, 59, 59i
Archaeopteryx, 376–377, 376i, 387i, 390
Archafructus sinensis, 346, 346i
Archipelago, 274–275, 275i
Architectural constraints, 427
Arctic ice cap, 26i
Arctic tern, 391
Arctic tundra, 533i, 538, 538i
Area effect, 495, 495i
Arginine, 198i, 202i
Arid habitat, 496i
Arid lands, 496
Arm (ray), invertebrate, 365i, 372, 372i
Armadillo, 242, 242i
Armillaria ostoyae, 327
Army ant, 562i
Arnold, S., 552
Aromatic compound, 555, 555i
Artemisia, 318
Artemisinin, 318
Arthropod, 354t, 355, 356i, 367–371,
    367i–371i
Artidactyl, 257
Artificial reef, 501, 501i
Artificial selection, 10, 10i, 218
Artificial twinning, 192
Ascaris. See Roundworm
A. lumbricoides, 366, 366i
Ascus, 328, 329i
Asexual reproduction, 140, 418
  animals, 126, 357, 360
  flowering plants, 126, 357
  fungal, 326, 327i, 328, 328i, 329, 329i
  by mitotic cell division, 324i, 325i,
    327i, 328i
  protistan, 318, 319, 319i, 321, 322i,
    323, 324i, 325i
  sexual *vs.*, 138–140, 138i, 140i, 418
Asparagine, 198i
Aspartame, 180
Aspartate, 198i
Aspen tree, 228, 228i
Aspergillosis, 544
Aspirin, 536i
Assassin bug, 557
Asteroid, 258–259, 258i, 263i, 264, 265,
    279, 285, 292, 347, 384, 386–387
Asthma, 329
Ataxia, 106–107, 106i
Atelopus varius, 486i
Atherosclerosis, 49, 107, 452
Athlete's foot, 329i
Atlantic Ocean, 92i
Atmosphere, 512i, 526i
  early, 286, 288–289
Atmospheric cycle, 511
Atom, 4, 4i, 18, 20, 20i, 22i, 24, 30t
  radioactive, 21, 262
Atomic bonding, 22–23, 74
Atomic number, 20
ATP, 75
  cyclic pathway, 99, 100i
  formation, 60, 64, 70t, 74, 77, 77i, 83,
    92–93, 98–99, 98i, 99i, 100i, 209,
    295, 306, 307
  function, 48t, 59, 66, 67, 72i, 75, 76,
    83i, 85, 85t, 89t, 93, 96, 97i, 101,
    102, 130, 142i, 295, 307
  net yield of, 115
  noncyclic pathway, 99, 100i
  release of chemical energy from cells
    and, 106–120
  structure, 46, 46i
ATP/ADP cycle, 75, 75i
ATP synthase, 57i, 98, 99i, 100i
ATPase, 83i
Australia, 471i, 493, 493i
Australian, 533i
Australian sawfly caterpillar, 560, 560i
Australopith, 396, 397
Australopithecus, 396

A. afarensis, 396, 396i
A. garhi, 397
Autism, 166, 564
Autoimmune response, 452
Automated DNH sequencing, 227
Autosomal dominant disorder,
    206, 206i
Autosomal dominant inheritance, 159,
    166, 173, 173t, 174, 174i
Autosome, 168, 171
Autotroph, 92, 93, 294, 504,
    504i, 505i
Autumn crocus, 66
Avery, O., 186
Avocado, 348
Axolotl, 385

B
B vitamins, 448
Baboon, 556, 556i, 561
Bacillus anthracis, 304
Bacillus shape, 302i
Backbone
  carbon, 34, 38, 40i, 43, 94i, 101, 188
    sterol, 41, 41i
    sugar-phosphate, 189i, 191, 191i
Bacterial cell, 50i, 52i, 58i, 70t, 222
Bacterial chromosome, 303i, 304
Bacterial colony, 224, 224i
Bacterial conjugation, 302
Bacterial flagellum, 58i, 59, 70t, 302,
    302i
Bacteriophage, 186, 186i, 308, 308i
  multiplication cycle, 309, 309i
  relative size, 53i
Bacteriorhodopsin, 307
Bacterium, 332t. *See also* Prokaryotic
    cell; *specific names*
  aerobic, 295, 320
  bacterial spore, 304i, 305
  characteristics, 8, 8i, 9i, 58, 92, 303,
    303i, 305, 332t
  DNA in, 58, 58i, 59, 184i, 222, 294
  as domain of life, 8, 8i, 9i, 58, 280,
    280i, 282i
  functions, 7, 8i, 209, 295, 302, 304,
    304i, 305, 330, 339
  nitrogen-fixing, 518, 518i
  origin, 54, 64, 282i, 287, 293i,
    296i–297i, 304–305
  pathogenic, 59, 150, 301, 302i, 310
  size, 50i, 53i, 58i
  structure, 52, 52i, 58i, 70t, 223i, 294i,
    302, 302i, 305i, 332t
  types, 36i, 37, 59, 88, 89i, 94, 97,
    105, 186, 294–295, 301, 304i,
    305, 306
Bahamas, 491
Balanced polymorphism, 252
Balancing selection, 252
Bald eagle, 497i
Baleen whale, 368
Ball-and-stick model, 34i, 36i
Banana slug, 552, 552i, 553
Banding pattern, DNA, 226, 226i,
    227, 227i
Bangiomorpha pubescens, 292i, 293
Bangladesh, 547, 547i
Barbed thread, cnidarian, 358i
Barnacle, 368, 368i
Barr body, 213, 213i
Basal body, 67, 67i
Basal cell carcinoma, 135i
Basal group, flowering plants,
    348i, 350t
Base, 28–29, 30t, 75, 75i
Base-pair substitution, 202, 202i
Base-pairing rule, 190, 196, 199
Base pairings, 46i, 184i, 189, 189i, 196i,
    198, 199, 202, 203, 224, 225
Basement membrane, 69i
Basic solution, 28
Basic solution, 28
Basilisk lizard, 9i
Basilosaurus, 239i
Basswood, 102i, 103i
Bat, 239, 266i, 279, 346, 557
  wing, 267, 267i
Bathyal zone, 545i

Bay of Bengal, 547, 547i
Bcl-2 gene, 413
Bdelloid rotifer, 149, 149i
Beagle voyage, 240–241, 240i, 241i
Bean plant, 102
  castor bean, 485
  soybean, 231
Bear
  brown, 271i
  grizzly, 497i, 559i
  polar, 26i, 244–245, 244i, 271i, 497i
Beavertail cactus, 103i
Becquerel, H., 21
Bee, 41, 371i, 550–551, 557, 557i, 562,
    562i, 563, 563i
  Africanized, 550, 550i
Bee hummingbird, 390
Beef, 49, 361, 361i
Beeswax, 41, 41i
Beetle, 138i, 269, 371i, 485, 485i
Behavior
  adaptive, 555, 555i, 564
  animal, 551
  anticipatory, 567
  in bacteria, 305
  birds, 273
  communication signals, 556–557
  courtship, 252, 252i, 272i, 273
  culture and, 394
  genes and, 552
  heritable basis of, 552–553
  hormones and, 552–553
  humans, 166–167, 564
  instinctive, 553, 553i
  learned, 554
  mammals, 392
  parental, 559, 559i, 564
  primates, 395
  self-sacrificing, 562–563
  sexual, 252, 252i, 365, 558–559
  social behavior, 555
Behavioral ecology, 550–567
Behavioral flexibility, 392
Behavioral isolation, 272i, 273, 273i
Behavioral trait, 246
Bell, cnidarian, 358i, 359
Bell curve, 160i, 161, 161i
Bengal tiger, 497i
Benign tumor, 135, 135i
Benthic province, 545, 545i
Bererub, T., 550, 550i, 551i
Berg, P., 220
Beta-carotene, 94–95, 95i, 105, 218
Beta globin, 44, 44i
Bias in report results, 13, 16
Bicarbonate, 29
Big bang, 288
Big laughing mushroom, 326i
Bighorn sheep, 497i
Bilateral animal, 355, 356i, 360,
    362, 364
  invertebrate, 353i, 354, 366, 367
  vertebrate, 354
Bilateral symmetry, 354, 354i
Bile salt, 41
Bill, bird, 377
Binary fission, 318
Binding energy, 79
Binge drinking, 73
Biochemical weapon, 194–195, 195i
Biochemistry, comparative, 303
Biodiversity, 494. *See also* Diversity
  conservation biology and,
    349, 497
  disruptions in, 259, 261, 279
  island patterns, 494i, 495, 495i
  mainland and marine patterns,
    494, 494i
  patterns of, 494–495
  range of, 261
  realms of, 532–533, 532i–533i
  sustaining, 498–499
  threats to, 496–497
Bioethics
  abortion, 151, 165
  allocating scarce resources, 73,
    91, 137
  bioprospecting, 287, 298

cloning, 184–185, 192, 192i, 193
cost of conservation, 353, 375
gene therapy, 230–231
genetic engineering, 93, 105, 139,
    139i, 149, 229
genetic screening, 151, 165, 180–183
human impact on biosphere, 301,
    313, 315, 350
patenting genomes, 218–219, 218i,
    219i, 221
public health, 195, 206, 207, 313
tax breaks for recycling, 335, 350
test tube organisms, 51, 125, 137,
    220, 234, 234i, 298
tracking asteroids, 259, 285
waste disposal, 333
wildlife trading, 497, 497i
Biogeochemical cycles, 511, 511i
  carbon cycle, 514–515, 514i–515i
  hydrologic cycle, 512–513
  nitrogen cycle, 518–519
  phosphorus cycle, 520–521
Biogeographic patterns, 494–495
Biogeographic realms, 532–533,
    532i–533i
Biogeography, 238
  evolutionary evidence from, 264–265
  island patterns, 494i, 495, 495i
  mainland and marine patterns,
    494, 494i
Biological clock, 391, 452
Biological control, 477, 487, 487i
Biological inquiry, nature of, 11
Biological magnification, 508, 508i
Biological molecule, 24–25, 29, 38, 289
Biological principles and human
    imperative, 567
Biological species concept, 272
Biological wealth, 498–499
Bioluminescence, 234, 319, 319i
Biomass pyramid, 509, 509i
Biome, 532–538 532i–533i, 534i, 535i,
    538i, 539i
Bioprospecting, 286, 287, 298
Biosphere, 5, 5i, 77, 96–97, 334,
    524–549, 525
  air circulation patterns, 526–529
  applying knowledge of, 546–547
  climate and temperature zones,
    526–527
  freshwater provinces, 540–541
  human impact on, 301, 313, 315,
    350, 567
  ocean currents, 530
  open ocean, 544–545
  realms of biodiversity, 532–533
Biosynthesis pathway, 77, 80, 84, 128
Bioterrorism, 2–3, 236
Biotic potential, 461, 474
Bipedalism, 383, 392–394, 396, 396i
Bipolar disorder, 166–167
Bipolar spindle, 129, 129i, 130, 130i,
    132i, 143i
Bird, 279, 281, 281i, 381i, 387, 390,
    390i, 494i, 496
  behavior, 273, 558–559, 559i,
    566, 566i
  characteristics, 354, 354t
  cowbirds, 487, 487i
  dinosaurs and, 376–377, 376i, 390
  evolution, 390–391, 390i
  flight, 390–391, 391i
  flightless, 238, 238i, 239
  migration, 391
  as pollinator, 346
  reproduction, 139, 419, 419i
  reptiles and, 376–377, 376i
Bird of paradise, 252i
Birdsong, 554, 554i
Birth, 120, 450, 450i
  multiple, 416–417, 416i
  premature, 450
Birth control, 438–439, 438i
Birth rate, 470, 472
Birth weight, 250, 251i
Bison, 487, 487i, 535i, 559i
Bivalve, 364
Black-bellied seedcracker, 251, 251i

Black bread mold, 328, 328i
■ Black Death, 3, 236i
Black market, 497i
Black widow spider, 369, 369i
Bladder, 321, 321i, 361
    invertebrate, 363, 363i
    swim bladder, 382
Blade, 321, 321i
Blastocoel, 442, 442i
Blastocyst, **423**, 442–443, 442i–443i, 451t
Blastomere, **420**
Blastopore, 355, 356i
Blastula, **420**
Bleaching, coral, 544, 544i
Blood, 28, 86, 367, 405i
    alcohol in, 72
    oxygen in, 44, 318
    pH of, 28i, 29
■ Blood clotting, 73, 353
■ Blood clotting disorder, 169, 173t, 175, 229
Blood fluke, 361i
Blood glucose level, 7, 19, 39, 118
■ Blood-letting, 363i
Blood pressure, hypertension, 250
Blood vessel
    human, 135, 135i, 159
    invertebrate, 362, 363, 363i, 367
Blue-footed booby, 241i, 272i, 273
Blue jay, 255, 255i
■ Blue offspring, 173t, 257
Blue whale, 93, 392i
Blueberry, 95
Bluegill sunfish, 560
Body cavities, 354, 355, 355i
Body fat distribution, 169
Body fluid, 29, 405
Body hair distribution, 169
Body height, 161, 161i
Body mass, 251
Body plan, 212, 215, 239, 240
    animal, 354–355, 354i, 355i
    annelid, 362
    bird, 390i
    Caenorhabditis elegans, 366, 366i
    cuttlefish, 365i
    earthworm, 363i
    emergence during pregnancy, 444, 444i
    euglenoid, 317i
    evolution of, 356, 362, 364–365, 372
    evolutionary constraints on development, 427
    flatworm, 360
    giant kelp, 321i
    invertebrate, 358, 358i, 359, 359i, 360, 360i, 362, 363i, 364i, 365i, 366, 366i
    lichen, 330i
    pattern formation and, 426–427
    plant, 330i
    prokaryotic cell, 302, 302i
    protist, 317i, 321i
    reptile, 388, 388i
    roundworm, 366
    vertebrate, 444, 444i
    virus, 308i
Body symmetry, 352i, 354
Body temperature, 402–403, 409i
Body wall, 355, 355i, 357i, 366i
    invertebrate, 363, 363i, 372
Bombykol, 556
Bonding, chemical, 18, 22
    atomic bonding, 22–23, 74
    carbon, 34
    covalent, 24–25, 25i, 34, 46, 47i, 56, 75, 95
    hydrogen, 25–27, 25i, 26i, 34, 43, 47i, 56, 79, 81, 189–191, 270
    ionic, 24, 24i
    molecular, 24–25
    peptide, 42i, 43, 200, 200i, 352
Bonding behavior
    hormones and, 564
    pair-bonding, 552–553
Bone tissue, 4i, 69, 69i
Bonobo, 394
Bony fish, 381i, 382i, 383
Boreal forest, 307, 532i–533i, 536–537

Borrelia burgdorferi, 305, 305i, 369
Bottleneck, **254**, 255, 278
Bottlenose dolphin, 320
Bottom dweller, 372
■ Botulism, 116, 194, 304, 313
■ Bovine spongiform encephalopathy, 312, 312i
Bradshaw, K., 524, 524i
Brain, 354
    energy sources for, 118
    human, 128, 166, 174, 180, 396–397
    invertebrate, 360, 360i, 363i, 365, 367
    mammal, 393
    primate, 395
Brain case volume, 401
Brain scan, 21i
Branch point, 278, 278i
Branched evolution, 278, 281, 281i
Brassica oleraceo, 268, 268i
Brazil, 470i
BRCA gene, 207
Bread mold, 328, 328i
Breast, 450i, 451
    tissue, 206i
■ Breast cancer, 206–207, 206i, 208, 217
Breastfeeding, 451
Breathing, 408. *See also* Respiration
Breeding ground, fish, 321
Brevetoxin, 320
Bristle, animal, 362, 375i
Bristlecone pine, 244i, 344i, 410
Bristly foxtail, 481, 481i
Brittle star, 372, 372i
Broadleaf forest, 532i–533i, 536, 537i
Brock, T., 286
Bronchial airway, 150
Brown, R., 54, 54i
Brown adipose tissue, 122
Brown alga, 316i, 320, **321**, 321i
Brown bear, 271i
Brown-headed cowbird, 487, 487i
Brown recluse spider, 369, 369i
Bryophyte, 336, **338**–339, 350t. *See also* Nonvascular plant
■ BSE, 312, 312i
■ Bubonic plague, 2–3, 236, 463i, 469i
Bud, animal appendage, 269
Budding, 329, 332t, 357, 361
Buffer, 29
Buffer system, **29**
Bulbourethral gland, 428i, 429, 429i, 430i
Bulk flow, 86
Bulk-phase endocytosis, **88**
Bull's-eye rash, 305i
Bundle-sheath cell, 102i, 103i
■ Burial, 501, 501i
■ Burn, skin, 29, 309
Butterfly, 12–13, 269, 370i, 371i, 459i, 463
    wing, 246, 247i, 249i, 250, 267
Butterfly fish, 479, 479i
Byrd, R., 16

C

C3 plant, **102**, 102i
C4 plant, **102**, 102i
Cabbage plant, 348
Cactus, 102, 103i, 238i, 407, 475, 475i, 497i, 534i
Caenorhabditis elegans, 366, 366i
Caiman, 389, 389i
Calamites, 342, 342i
Calcium, 22, 29, 159, 368, 512, 513i
Calcium carbonate, 314, 321, 359, 367, 372, 373
Calcium ion, 82i, 85, 85i
Calcium pump, 57i, **85**, 85i
Calcivirus, 493
California Current, 530
California red-legged frog, 385
Calisher, C., 300
Calvin, M., 21, 323
Calvin-Benson cycle, 93, **101**, 101i, 102, 102i, 104i

CAM plant, **102**, 475, 475i
Camarhynchus pallidus, 243i
Cambrian period, 263i, 269, 269i, 356, 380
Camel, 245, 245i, 274i, 285
Camelid, 274, 274i
Camellia sinensis, 343i
Camouflaging, **484**, 484i, 485
cAMP, 46, 48t, 209, 325i
■ Camptodactyly, 160, 173t
Camptosaurus, 387i
Campylobacter, 312
Canada goose, 108, 122
Canadian lynx, 478, 482–483, 483i
■ Canavan's disease, 230i
■ Cancer, 88, **135**, 549, 549i
■   basal cell carcinoma, 135i
■   breast, 206–208, 206i, 217
■   carcinogens, 206
■   carcinoma, 135i, 206i
■   causes, 2, 19, 177, 203, 232, 308
■   cell cycle and, 125, 134–135, 134i, 135i
        cell death and, 413
■   cervix, 134i, 137, 440
■   HeLa cells, 124i, 135, 214–125
        HPV and, 440
■   leukemias, 173t, 177i, 230
■   lung, 529, 529i
■   malignant tumor, 135, 135i
■   prostate, 429
■   skin, 134–135, 135i
■   squamous cell carcinoma, 135i
■   survival rates, 135, 137
■   testes, 429
        treatment, 66, 135, 137, 206, 217, 352
Cancer cell, 135, 452
Canegiea gigantea, 475, 475i, 497i, 534i
Canine, 556
Canis lupus (Canidae), 482i, 556, 556i, 561, 561i
Cannibalism, 457, 561
CAP, 208–209
CAP plant, **102**, 102i
Capillary, 402, 405i, 407i
Capture-recapture method, **459**, 459i
Carapace, 369
Carbohydrate, 32i, 48t, 96, 128
    complex, 38, 48t
    early Earth and, 289
    function, 39, 73, 101, 101i
    short-chain, 38, 48t
    structure, 38–39
Carbon, 20, 22, 23i, 188, 342, t, 538
    annual fluxes in global distribution, 514i
    bonding behavior, 34
    global cycling of, 306, 349
    isotopes, 21
    reservoirs and holding stations, 514i
Carbon 12 (12^{C}), 21, 262, 262i
Carbon 14 dating, 262, 262i
Carbon black, 529i
Carbon cycle, **514**–515, 514i–515i
Carbon dioxide, 29, 37, 75, 82i, 288, 290, 302
    animals and, 365
    as gas, 514
    global concentrations, 516, 517i
    in photosynthesis, 92, 93, 101, 104i
Carbon fixation, **101**, 102, 102i–103i
Carbon monoxide, 288
Carbon-oxygen cycle. *See* Carbon cycle
Carbonic acid, 28, 29
Carboniferous period, 263i, 334, 337i, 340–343, 342i, 381, 386, 392
Carbonyl group, 34i
Carboxyl group, 34i, 42i
■ Carcinogen, 206
■ Carcinoma, 135i, 206i
Cardiac muscle, 43, 56i, 69
Caribou, 482i
Carnivore, 9, 504, 505i
Carotenoid, **94**, 95i, 105
Carpel, 152i, 346i
Carrageenan, 322

Carroll, L., 468
Carrot, 94, 555, 555i
Carrying capacity, **462**, 463, 463i, 468
Carson, R., 508, 508i
Cartenoid pigment, 333
Cartilage, 43, 69, 174
Cartilaginous fish, 381i, **382**, 382i
Cascade Range, 538i
Caspian tern, 559i
Castor bean plant, 485
Castor oil plant, 194–195, 194i, 195i
Cat, 81i, 164, 213, 213i, 474
    Bengal tiger, 497i
    civet, 313, 313i
    cloned, 192, 192i
    lion, 7i, 281i, 556, 560–561, 564
    Manx, 164, 165i
    Siamese, 81i
    wild, 496
Catabolic pathway, **77**
Catabolite activator protein, 208–209
Catalase, 72, 72i, 79i, 80, 90–91
■ Cataracts, 452, 528
Catastrophism, **240**
Caterpillar, 6i, 367, 414, 414i, 484i, 560, 560i
Cattle, 49, 306, 311, 361, 361i, 487, 499, 501
Caulerpa taxifolia, 492–493, 492i
Cause and effect, 12, 392
Cave paintings, 399i
cDNA, **223**, 223i, 224
Celera Genomics, 220i, 221
Cell, **4i**, **52**, 210
    animal, 86, 125, 130, 132–133, 132i, 133i
    bacterial, 287, 294i
    components, 42–43, 52, 54–55, 65i, 66, 68–69, 70t, 295i
    cortex, **66**
    energy flow, **74**–78, 74i, 75i, 77i, 78i, 83i, 85i
    eukaryotic. *See* Eukaryotic cell
    evolution, 294, 295
    mechanisms, 67, 70t, 125, 130, 132–133, 132i, 133i
    membrane. *See* Cell membrane
    origin, 50–51, 72i, 77i, 286–293, 291i, 306
    prokaryote. *See* Prokaryotic cell
    relative size, 52–53, 53i, 58
    release of chemical energy, 106–120
    shape, 53, 66, 83
    size, 52–53, 53i, 58, 332t
    as unit of organization, 4–5, 4i
Cell communication, 407, 412–413
Cell cycle, 125, **128**–129, 128i, 134–135, 134i, 206
    cancer and, 125, 134–135, 134i, 135i
Cell death, 134, 191, 412, 412i, 413i, 425
    cancer and, 413
Cell differentiation, 192, **210**, 424–**425**
Cell division, 124–128, 126t, 332t. *See also* Meiosis; Mitosis
Cell junctions, **69**
Cell membrane, 26, 50i, 56–57, 56i, 57i, 58i, 289
    components, 41, 41i, 82, 82i
    crossing, 73i, 80, 82–86, 82i, 83i, 84i, 85i, 89t
Cell memory, 426, 427i
Cell migration, 424i, 425
Cell plate formation, 132i, **133**
Cell suicide, 412–413, 425
Cell surface specialization, 68–69
Cell theory, **54**
Cell-to-cell communication, 407
■ Cell typing, 66
Cell wall, 58, 59, 65i, **68**, 70t, 332t
Cellular slime mold, 325, 325i, 333
Cellulose, 38i, 39, 39i, 48t, 68, 68i, 132i, 133, 323
Cenozoic era, 263, 263i, 279i, 393
Centipede, 367, 367i
Central vacuole, **63**, 65i, 70t, 87, 96i
Centrifugation, **170i**
Centriole, 65i, 67, 130, 130i, 142i

Centrocercus urophasianus, 558–559
Centromere, **127**, 130, 141, 141i
Centrosome, **130**, 142i–143i
Cephalization, **354**, 360
Cephalochordate, 378
Cephalopod, 364, 365, 365i
Cereal croplands, 535
Cerebellum, 393
Certhidea olivacea, 243i
■ Cervical cancer, 134i, 137, 440
Cervix, 137, 432, 432t, 433i
Cetacean, 257
CFCs, 517i, 528
CFTR protein, 229
Chaetae, 362
Chaetodipus intermedius, 249, 249i
■ Chagas disease, 316
Chain fern, 341, 341i
Chain of Being, 238, 240
Chalk deposit, 314, 314i, 321
Chambered nautilus, 239i, 365i
Chameleon, 388
Channel protein, 352
Chaparral, 534, 535i
Chara, 323i
Chargaff, E., 188
Charge, of proton/electron, **20**, 24
Charophyte, 316i, **323**, 323i, 336, 337i
Chase, M., 187i
Checkpoint gene, 134–135, 134i
Cheetah, 485, 566
Chelicerate, 367, 369
Chemical basis of life, 30t
Chemical bond, 18, **22**
■ Chemical burn, 29
Chemical energy, 7, 59, 74, 92–93, 96, 98
Chemical equation, 23i
Chemical equilibrium, **76**, 76i
Chemical formula, **23**, 23i
Chemical message, 46, 120, 409
Chemical work, 74, 75i
Chemoautotroph, 292, **302**, 304, 305
■ Chernobyl nuclear power plant, 549, 549i
Cherry tree, 95
Chestnut, 329
Chestnut blight fungus, 491t
Chicken, 158, 159i, 229i, 266i, 270, 425i, 427i
comb, 158, 159i, 165
Chigger mite, 369
Child, 451t
Childbirth. *See* Birth
Chiloglottis trapeziformis, 252
Chimpanzee, 129, 177, 268i, 269, 394, 396i, 497i, 566
Chin, 398
China, population and family planning, 470–471, 471i
Chipmunk, 480, 480i
Chironex, 358i
Chitin, 39, 39i, 267, 362, 367
Chiton, 364, 364i
Chlamydia trachomatis, 305, 440
■ Chlamydial infection, 305, 440, 440i, 440t
Chlamydias, **305**
Chlamydomonas, 323i, 324, 324i
C. nivalis, 286, 333, 333i
Chlorella, 17i, 323
Chlorine, 23i, 24, 24i, t, 528
Chlorine ion, 24, 27, 28, 82i
Chlorine monoxide, 528
Chlorofluorocarbons, 517i, 528
Chlorophyll, 64, 92i, **94**, 94i, 95i, 98–99, 290, 304, 317, 320, 322, 323
Chlorophyte, **323**, 323i
Chloroplast, 39i, 53i, 66, 67, 70t, 92, 96, 96i, 97i, 98, 98i, 101, 317i, 319, 322
function, 60, **64**, 64i, 65i, 318
origin, 59, 287, 294, 295, 297i, 304
Choanoflagellate, 356
■ Cholera, 2, 248, 547, 547i
Cholesterol, 35, 41, 41i, 48t, 49
■ hypercholesterolemia, 173t
Cholla cactus, 534i

Chondrus, 490i, 491
Chordate, 282, 353, 354t, 355, 356i, **378**–379, 381i
Chorion, 442i, **443**, 443i
Chorionic villi, 443i, 445, 445i
■ Chorionic villi sampling, 181
Chromatid, 126, 126i, 130–131, 130i, 131i, 134, 141, 141i, 143i, 144–145, 144i, 148i, 168, 168i
Chromatin, **61**, 61i, 62i
Chromosomal DNA, 222, 223
Chromosome, **61**, 127, 127i, 129, 129i, 141, 141i, 147. *See also* Sex chromosome; X chromosome
■ abnormality, 168, 173t
bacterial, 303i, 304
deletion, **176**, 176i
duplication, **176**, 176i
inversion, **176**, 176i
meiosis and change in, 168, 168i, 171, 171i, 178i
metaphase chromosome, 170, 170i
Philadelphia chromosome, 177, 177i
polytene chromosome, 211, 211i
primate, 129, 177
■ structural change, 173t, 176–177, 176i, 177i
translocation, **177**, 177i
Y chromosome, 141i, 144i, 168–169, 169i, 179
Chromosome, eukaryotic, 126–127, 126i, 127i, 140
homologous, **140**–141, 141i, 142i, 144, 150, 153i, 154, 156–157, 168, 168i, 171, 247i
human, 127, 127i, 137i, 141i, 147, 202i
meiosis and, 140–145, 140i–143i
mitosis and, 124, 130–131, 130i–131i
structure, 126–127
Chromosome number, **129**, **140**, 146, 147, 152, 153i, 167, 168
changes in, 173t, 178–179, 178i, 179i
diploid, **129**, 130i, **140**, **153**, 153i, 168, 169i
haploid, **141**, 142i, 143i, 146, 221
■ Chronic myelogenous leukemia (CML), 173t, 177i
Chrysophyte, 316i, 320, **321**, 321i
Chukwu, N., 416, 416i
■ Chytridiomycosis, 486i
Cicada, 272i, 273
Cichlid, 276, 276i
■ Cigarette smoking. *See* Smoking
Ciliated protozoan (ciliate), 67, 316i, **318**
Cilium (cilia), 17i, **67**, 67i, 70t, 130, 360, 360i, 375, 375i
Ciona savignyi, 401, 401i
Circadian migration, 544
Circadian rhythm, **411**
Circulatory system
capillaries, 402i, 405i, 407i
closed, **363**, 363i, 365
between fetus and mother, 445, 445i
invertebrate, 363, 363i, 365, 367
open, **367**
■ Cirrhosis, 72
cis Fatty acid, 49, 49i
Citric acid cycle, 112–113
Citrus fruit, 371i
Civet cat, 313, 313i
Clade, **282**
Cladistics, **281**
Cladogenesis, **278**
Cladogram, **281**, 281i, 304i
Cladonia rangiferina, 330i
Claret cup cactus, 534i
Clark, L., 555
Classification system, 259i, **280**–281, 280i, 282i
current species diversity, 497, 497i
of prokaryotes, 303
six-kingdom, 280, 280i
three-domain, 8, 280, 280i, 282i
Clavaria, 326i

Claviceps purpurea, 329, 329i
Clay soil, 539
Cleavage, 420, 420i, 422–423, 423i, 442
animal reproduction, **132**–133, 132i, 133i
patterns of, 423, 423i
reaction, 35, 35i, **79**, 354
Cliffs of Dover, 314i
Climate, **526**
carbon fixation and, 102, 102i
change in, 517
change in, global, 93, 261, 263i, 279, 334, 337, 347, 356
change in, regional, 349, 531
El Niño, 524–525, 524i, 545–547, 546i
ice age, 261, 263i
ocean currents and, 530
temperature zones and, 526–527
zones, 530, 530i
Climax community, **488**, 489i
Climax-pattern model, **489**
Clitoris, 432, 433i, 437
Cloaca, 388, 390
Clone, **140**
animal, 184–185, 184i, 192, 192i, 193, 234, 234i
DNA, 184, 185i, 192, 222, 222i, 223, 223i, 224
plant, 140,
Cloning, 140, 222–225, 225i
■ bioethics of, 184–185, 192, 192i, 193
Cloning site, 272
Cloning vector, **222**, 222i
Closed circulatory system, **363**, 363i, 365
Clostridium botulinum, 304
C. tetani, 304, 305i
Clot, 73, 353
■ Clotting disorder, 169, 173t, 175, 229
Club fungus, **326**–327, 326i, 327i, 329, 329i
Club moss, 334, 340, 340i, 342, 342i, 350i
■ CML, 173t, 177i
Cnidarian, 352i, 354t, 355, 356i, **358**–359, 358i, 359i, 365
Coal, 29i, **342**, 342i, 529i
Coastal redwood forest, 334
Cobra lily, 242i
Coca plant, 333
Cocaine, 333
pregnancy and, 449
Cocci, 305
Coccolithophore, 321, 321i, 341
Coccus shape, 302i
Coccyx, 239, 239i, 285
Codium, 105i
Codominance, allelic, **158**, 158i
Codon, **198**–200, 199i, 200i, 201i
Coelacanth, 383, 383i
Coelom, 353, **355**, 355i, 363i, 364, 367
Coelomate animal, 355i
Coelomic chamber, 362, 363
Coenzyme, 46, 48t, 77, **80**, 88, 96
Coevolution, 346, **482**
Cofactor, **76**, 80, **89t**
Cohesion, **27**, 27i
Cohort, **464**
Colchicine, 66, 170, 170i
Colchicum autumnale, 66
■ Cold sore, 309
Coleoptera order, 371i
Collagen, 48t, 187, 189, 229, 248, 452
Collins, F., 221
Colon, 209
Colonizing organism, 358i, 359
Colonizing species, 494i, 495, 495i
Colony, 358i
bacterial, 224, 224i
predatory, 305
■ Color blindness, 173t, 175, 175i
Columbine, 348i
Colwell, R., 547, 547i
■ Coma, 29
Comb type, chicken, 158, 159i, 165
Comet, 292, 386
Commensalism, **478**, 487, 487i
■ Common cold, 308

Communication display, 556–557, 557i
Communication protein, 56, 56i
Communication signal, **556**–557, 556i
communication displays, 556–557
Community, 5, 5i, **478**
biogeographic patterns, 494–495
biological controls, 487
climax pattern model, 489
competitive exclusion, 480–481
competitive interactions, 480–481
conservation biology, 497
cowbirds, 487
cyclic, nondirectional changes, 489
economic factors and sustainable development, 498–499
endangered species, 496
exotic invaders, 492–493
factors shaping structure, 478
habitat losses and fragmentation, 496–497
identifying risk areas, 498
instability, 490–491
island patterns, 495
keystone species, 490–491
mainland and marine patterns, 494
mutualism, 479
niche, 478
parasite-host interactions, 486–487
parasites and parasitoids, 486
population size and, 478
predator-prey coevolution, 482
predator-prey interaction models, 482
predator-prey interactions, 482–483
predators, adaptive responses, 485
prey adaptations, 484–485
resource partitioning, 481
species interactions, 478
species introductions, 491
stability, 488–489
succession model, 488
sustaining biodiversity, 498–499
threats to biodiversity, 496–497
tropical, 494
Comparative biochemistry, 303
Comparative genomics, 232
Comparative morphology, 239, 242, 245, 246, 266, **266**
Compartmentalization, **410**, 410i
Competition
dominant competitor, 321
exploitative, 480
humans as competitor species, 496, 496i
interference competition, 480, 480i
interspecific, 478, 480, 480i
intraspecific, 480
keystone species and, 490–491, 490i
molecular, 296i–297i
sexual, 559, 559i
Competitive exclusion, **480**, 480–481, 481i
Competitive interaction, 480–481
Complementary DNA, 223, 223i, 224
Complete digestive system, **355**, 355i, 362, 364i, 370, 375, 375i
Complex carbohydrate, **38**, 48t
Composite signal, 556
Compound, **23**, 30t
Compound eye, 370i
Compound light microscope, 54, 54i
Concentration gradient, **82**–86, 83i, 85i, 98, 100i, 212, 407
Conception, 438–439
Condensation reaction, 35, 35i, **79**, 401
■ Condom, 438i, 439
Cone, **344**, 344i, 345i
Cone snail, 352–353, 352i, 353i
Confuciusornis sanctus, 376i, 377
■ Congenital rubella syndrome, 454
Conifer, 282, 334, 337i, 343, 344–346, 344i, 345i, 350i, 350t
Coniferous forest, 532i–533i, 536–537, 539i
Conjugation, 302, 303
Connective tissue, vertebrate, 159
Conotoxin, 352, 353

Conservation
- cost of, 353, 375
  of genes, 190, 190i, 191i, 199i, 353
  water conservation in plants, 410–411
Conservation biology, 349, **497**, 508
Conserved gene, 270, 353
Consumer, **6–8**, 7i, 74i, 354, 370, **504**
Continental drift, 264, 265, 265i, 491
Continental shelf, 545i
Continuous assembly, in DNA, 191i
Continuous variation, **160–161**, 160i, 161i
- Contraception, 438–439, 438i
Contractile cell, 359
Contractile protein, 211
Contractile vacuole, 91i, 317i, 318
Contraction, 62, 66, 67, 85
Control gene, 413
Control group, **12**
Control pathway, 206
Control protein, 211
Controlled burn, 537
Conus, 352–353, 352i, 353i
C. geographicus, 352i
C. magnus, 352i
Convergence, morphological, **267**, 267i
Convergent evolution, 393
Cook, J., 456
Cooksonia, 260i, 336, 336i, 340i
Cooperative hunting, 560–561, 561i
Cooperative predator avoidance, 560, 560i
Copepod, 368, 368i, 547, 547i
Copernican model of solar system, 14
Copernicus, N., 14
Copper ion, 90, 90i
Copper mine, 81i
Coprolite, 260
Coral, 138, 319, 358, 358i, 359, 374i
Coral bleaching, 544, 544i
Coral fungus, 326i
Coral grouper, 382i
Coral reef, 2, 3, 5i, 322, 359, 534, 542i, 543, 543i
Cork, 54i
Cormorant, 561, 561i
Corn, 102i, 103i, 212, 218i, 229, 231, 329, 348
  characteristics, 171
  grain (kernel), 203i
Cornea, 160
Cornell University, 563i
Coronavirus, 310, 313
Corpus luteum, **433**, 435, 436i, 443
Cortex cell, 66
Corticotropin-releasing hormone, 450
Cortisol, 250, **450**
Cosmic cloud, 288–290
Costa Rican owl butterfly, 459i
Cotton plant, 228, 228i, 231, 276i
Cotylosaur, 266, 266i
Courtship behavior, 252, 252i, 272i, 273
Courtship display, 252i, 272i, 273, 557
Covalent bond, **24–25**, 25i, 34, 46, 47i, 75, 95
Covas, R., 251
Cowbird, 486, 487, 487i
Crab, 269, 368
  fiddler, 558, 558i
  horseshoe, 369, 369i
  life cycle, 368i
Cranberry, 81i, 339
Craniate, **380**, 380i
Cranium, 380
Craterellus, 326i
Crayfish, 146
Crenarchaeota, 306, 306i, 307
Crenicichla, 466–467, 466i, 467i, 475
Creosote bush, 459i
Cretaceous period, 258i, 263i, 337i, 343, 386, 388
- Creutzfeldt-Jakob disease, 312, 312i
CRH, 450
- Cri-du-chat syndrome, 173t, 176–177, 176i, 177i
Crick, F., 184i, 187, 189, 220

Crinoid, 365i, 372
Crocodilian (crocodile), 281, 281i, 387, 388i, 389, 389i, 559
Crop, 491, 535. See also Agriculture
Crop, invertebrate, 363i
Crop plant. See also specific names
Crop rotation, 519
Cross-fertilization, **153**, 156i–157i
Crossing over, **144–145**, 144i, 147, 148i, 166, 168, 168i, 171, 171i
Crow, 300
CRTR gene, 150
Crude density, 458
Crustacean, 93, 354t, 367, 368, 368i
  current species diversity, 497i
  life cycle, 368
Crustal movement, 261, 263i, 264–265, 264i, 274
Cryphonectria parasitica, 329
Cryptomys damarensis, 563
Crystal, arrangement of, 24, 24i
Crystallin, 210
Cuckoo, 553, 553i
Culture, **394**
  primate, 395
Cumulative assaults hypothesis, 452
Cuscuta, 486i
Cuticle
  animal, 39, 39i
  invertebrate, 362, 363i, 366–368
  plant, 69, 69i, 102, 102i, 336, 343, 350t
Cuvier, G., 240
Cyandium caldarium, 287
Cyanide, 90
- Cyanide poisoning, 90, 122
Cyanobacteria, 59, 282i, 292, 293i, 295, **304**–305, 306i, 322, 330, 541
  origin, 297i
Cyanophora paradoxa, 295i
Cyathea, 340i
Cycad, 334, 337i, 343, **344**, 344i, 346i, 350t
Cycas, 344i
Cyclic adenosine monophosphate, 46, 48t, 209, 325i
Cyclic pathway
  ATP formation, 99, 100i
  photosynthesis, 292, 296i–297i, 349
Cypress swamp, 502i
Cypress tree, 344
Cyst, **316**
Cysteine, 198i
- Cystic fibrosis, 150–151, 150i, 159, 165, 173t, 181, 229
Cytochrome c, 270
Cytokinesis, **132**
Cytoplasm, 50, **52**, 53, 126, 304, 317
  eukaryotic, 52, 52i, 60, 62i, 69
  function, 47, 62i, 80, 85, 87, 88, 89i, 195, 199, 201, 204i, 211
  prokaryotic, 52, 52i, 58i, 59, 294
Cytoplasmic division, 125, 128, 129, 129i, 132–133, 132i, 133i, 138i, 140–143, 146, 147i
Cytoplasmic division/fusion, 324i, 326, 327i
Cytoplasmic localization, **423**
Cytosine (C), 46, 46i, 184i, 188, 188i, 189, 189i, 196, 199, 203, 227, 227i
Cytoskeleton, 60, 65i, **66**, 67, 70t, 316, 317

D
Da Vinci, L., 260, 261
D'Agostino, S., 249
Daisy, 348
Dalai Lama, 221i
Dale, B. W., 482i
Damselfly, 371i
Danio rerio, 454, 454i
Daphnia, 480
Darlingtonia californica, 242i
Darwin, C., 10, 11, 14, 152, 240–243, 240i, 246, 264, 376
Date rape, 73
Daucus carota, 94, 555i
Daughter cell, 125, 126, 128–129, 131, 131i, 133, 141, 143i

Daylength
  reproduction and, 418
DDT, 508, 508i
De Moraes, C., 414, 414i
Dead Sea, 307
Death, 120, 452
  cell, 134
  exponential growth and, 461, 461i
Death rate, 470, 472i
Deciduous, 344
Deciduous broadleaf forest, 536, 537i
Deciduous forest, 532i–533i, 536, 537i
Deciduous forest soil, 539i
Declining species, 373, 375
Decomposer, 7, 7i, 342, 349, **504**
  bacterial, 339
  fungal, 314, 315, 326, 327, 339
  invertebrate, 366, 371
  protistan, 314, 316, 320
Deep-sea angler fish, 545i
Deep-sea food web, 36
Deep-sea trench, 545i
Deer, 459, 459i
Deer tick, 305i
Definitive host, 360
Deforestation, 334–335, 334i, 335i, **349**, 496, 512, 513i, 515, 515i, 517i, 519
Degradative pathway, **77**
Deletion
  chromosomal, **176**, 176i
  in DNA, 202, 202i
Delphi Oracle, 32–33, 32i
Demographic indicator, 470, 470i
Demographic transition model, **472**, 472i
Demographics, **458**
Denaturation, 44–45
Denatured DNA, **224**–225, 224i, 225i
Dendrobates, 484
Denitrification, 519
Density-dependent control, **462**–463, 463i
Density-independent factor, **463**
Deoxyribonucleic acid. See DNA
Deoxyribonucleotide, component bases, 46, 46i
Deoxyribose, 38
- Depo-Provera injection, 438i, 439
- Depression, 163, 166
  post-partum, 450
  during pregnancy, 449
Derived trait, **281**
Desalinization, **513**
Desert, 338, 345, 532i–533i, 534, 534i
Desert soil, 539i
Desertification, **534**
- Designer plant, 228
Desmid, 323, 323i
Detrital food web, **507**, 507i, 541
Detritivore, **504**
Detritus, 362
Deuterostome, **353i**, 355, 356i
Development, **6**, **404**. See also
  Embryonic development
  animal, 367, 370i
  body plan, 427
  hands, 412–413, 413i
  human, 419i, 451i, 451t
  leopard frog, 421i
  patterns, 268–269
  patterns, speciation, 238, 278–279, 283i
  plant, 338i
  postnatal, 451, 451i, 451t
  prenatal, 451i, 451t
Devil's hair, 486i
Devonian period, 263i, 337i, 340, 340i, 343, 369, 383, 383i, 384
DEW (Distant Early Warning) system, 567
- Diabetes, 7, 107, 231, 250
  blood glucose level, 7, 19, 39, 118
- Diagnostic tool, 166, 170, 170i
- Diaphragm (birth control), 438i, 439
- Diarrhea, 209, 304, 310t, 316, 329
Diatom, 287, 321, 321i
Dickensonia, 356i

Dictyostelium, 412
Dictyostellum discoideum, 325, 325i, 333
Diet. See also Nutrition
- fat-rich, 41i
- low-carb, 49
Differentiated cell, **192**
Diffusion, **82–83**, 82i, 83i, **406**
  facilitated, 83, 84
  metabolism and, 82–83
  plants, 102
  plasma membrane, 83, 84i, 86, 88, 88i
  water molecule, in cells, 52, 68, 82, 82i, 86–87, 86i, 87i
Diffusion rate, 83
Digestion. See also Nutrition
  extracellular, 326
  intracellular, 88
Digestive enzyme, 60, 62, 63, 72, 135i, 174, 186
Digestive gland, invertebrate, 364i
Digestive system, **355**, 370
  complete, **355**, 355i, 362, 364i, 370, 375, 375i
  incomplete, **355**, 355i, 359, 360, 366i
  invertebrate, 360i, 362, 363i, 375, 375i
Digit evolution, 383, 383i
Dihybrid cross, **156–157**, 156i, 157i, 171
DII gene, 269, 269i
Dimorphism, **246**, 252
Dinoflagellate, 316i, **319**, 319i, 320, 320i, 359, 544
Dinosaur, 258, 258i, 263i, 266i, 279, **386–387**, 386i, 387i. See also
  specific types
  birds and, 376–377, 376i, 390
  mammals and, 392, 393
Dioon, 344i
Diploid cell, 130i, 211
Diploid chromosome number (2n), **129**, 130i, **140**, **153**, 153i, 168, 169i
Diploid dominance, 350t
Diploid organism, 150, 153, 155
Diploid stage, 146i, 156, 322i, 324i, 327i, 328i, 338i, 341i, 345i, 347i, 350t
Diplomonad, 316i
Directional change, 278
Directional selection, **248**, 249i, 278, 310
Disaccharide, **38**
Discontinuous assembly, in DNA, 191i
Discoverer XVII, 125
- Disease, **172**, 310–311, 315
  endemic, 310
  foodborne, 310–311
  infectious, 236, 310–311, 310t, 448–449, 468–469
  sporadic, 310
Disruptive selection, **251**, 251i
Dissolved substance, 27
Distance effect, **495**, 495i
Distant Early Warning (DEW) system, 567
Diversity, 407. See also Biodiversity
  current species diversity, 497, 497i
  eukaryotes, 316, 326–329
  evolutionary view, **10**, 14t, 243, 254
  genetic, 149, 248, 249, 254, 283t
  insect, 370–371, 370i
  invertebrate, 362, 364, 366, 367
  of life, 2, 6, 8, 8i, 9, 10, 139, 184i, 187, 189
  ocean, 544
  plants, 337, 338, 341, 343, 344i, 346, 346i, 348, 348i
  prokaryotes, 302–305
Division of labor, 356, 367, 371
DNA, **6**, 452
  amplification/sequencing, 220, 224–227, 225i, 227i, 286, 307
  animal sequences, 356
  bacterium, in 58, 58i, 59, 184i, 222, 294
  banding pattern, 226, 226i, 227, 227i
  cDNA, 223, 223i, 224
  chromosomal, 222, 223
  cloned, 184, 185i, 192, 222–224, 222i, 223i
  complementary, 223, 223i, 224

conservation of genes, 190, 190i, 191i, 199i, 353
continuous assembly, 191i
deletion, 202, 202i
denatured, **224**–225, 224i, 225i
discovery of, 186–187
disruption of, 80, 94, 134, 191, 202–203, 202i
enzyme function in replication, 126–128, 134, 190–191, 191i, 197, 211, 222, 222i, 309
eukaryotic, 58, 58i, 126–127, 197, 202, 270, 316
evolution and, 64, 177, 189, 200, 287, 291, 294, 294i, 295
forensics science and, 226
function, 6, 10, 43, 70t, 126–127, 186–187, 195, 197, 200, 210i, 211
gene expression and, 208, 209i, 210i
in human body cell, 61
hydrogen bond in, 25i, 189, 190, 191
insertion, 202
mtDNA, 270, 276
primate, 177, 269, 270
prokaryotic, 58, 59, 64, 222, 222i, 223i, 286, 302, 303, 303i, 307, 308, 308i
recombinant, 220–222, 222i, 223i, 309i
repair mechanisms, 191, 203, 220, 223i
replication, 61, 128, 128i, 134, 142–143, 185, 190–191, 190i, 191i, 195, 196, 203, 208i, 222, 223, 303i, 309
research guidelines, 220
sticky end of fragment, 222, 223i
supercoiling, 127i
technologies, 218, 220–222, 224–232
template, 185, 191, 191i, **195**, 196i–197i, 197, 202i
transcription, 208–212, 209i, 210i
viral, 223, 308, 308i, 309, 309i
DNA chip, **232**, 232i
DNA database, 221
DNA fingerprinting, 164, **226**, 226i, 270
DNA ligase, **191**, 222, 222i
DNA polymerase, 22, **191**, 203, 223, 225, 226, 307
DNA structure, 25, 25i, 46, 47i, 58, 58i, 127i, 184i, 188–189, 188i, 189i, 208, 223
    discovery of, 184, 186–190, 220–221
    double helix, 184, 185i, 189–191, 189i, 196i–197i, 220, 223i
    hydrogen bonds in, 25i, 189, 190, 191
Dodder, 486i
Dog, 556
Dolly, 184–185, 184i, 192, 192i
Dolphin, 320
Domain
    of life, 8, 8i, 58, 296i–297i
    in polypeptide chains, **43**
Domestication
    of animals, 496
    of seed plants, 343
Dominance hierarchy, 561
Dominant allele, 151, **153**, 155, 155i, 156, 156i, 164
Dominant competitor, 321
Donor organ, 151
Dormancy, 146, 304, 343
Dorsal nerve cord, 360
Dorsal surface, 354, 354i
Dosage compensation, **213**
Dosidicus, 214, 354t, 365, 365i
Double-blind study, **16**
Double covalent bond, **25**, 56
Double fertilization, 347i
Double helix, 25i, 46, 47i, 127i, 220, 291
Doubling time, **461**, 461i
Douching, 438–439, 438i
Douglas fir tree, 320
Down syndrome, 173t, 178–179, 179i, 181, 434
Downwelling, **545**
Downy mildew, 320
Dragonfly, 370
Drew-Baker, K., 322
Drinking culture, 73

Drinking water, **513**
Dromaeosaurus, 376i
Dromedary camel, 245, 245i
Drosophila, 211i, 212, 212i, 270, 275, 353, 370i, 371i, 424i
    D. melanogaster, 171, 182–183, 183i, 214–215, 214i, 215i, 217i
Drought, 331, 499, 535
Drug abuse, 72–73
Drug production, and yeast, 329
Drug research, 352, 357, 361
Dry acid deposition, 529
Dry shrubland, 532i–533i, 534, 535i
Dry woodland, 532i–533i, 534
Duchenne muscular dystrophy, 217
Duck, 502
Duck-billed platypus, 392i, 401, 401i
Duck louse, 371i
Duckweed, 348
Dugesia. See Flatworm
Duncan, R., 492i
Dunce gene, 214
Dunkleosteus, 380i
Duplication, chromosomal, **176**, 176i
Dust Bowl, 535
Dust mite, 369, 369i
Dust storm, 534, 534i
Dutch elm disease, 491t
Dynastes, 371i
Dynein, 66, 67, 67i, 71, 71i, 130, 131
Dysentery, amoebic, 317

E

Eagle nebula, 288i
Early atmosphere, 288–289
Earth
    age, 241, 262
    asteroids and, 258
    crust, 19, 19i, 263i, 264, 264i, 265, 274, 288, 289
    early, 286, 288–289, 293
    history, 241, 261–263, 288–291, 289i
    magnetic poles, 264
Earthworm, 354i, 362–363, 362i, 363i
    body plan, 363i
Earwig, 371i
Easter Island, 456–457, 456i, 457i
Eastern hognose snake, 386i
Ecdysone, 211
Echinoderm, 282, 353, 354i, 355, 356i, **372**–373, 372i, 373i
Eciton burchelli, 562i
Ecological access, and lineage, 279
Ecological isolation, **273**, 273i
Ecological pyramid, 509
Ecological separation, 276
Ecological success, **488**
    climax-pattern model, 489
    cyclic, nondirectional changes, 489
    succession model, 488
Ecology, 455i, **457**
Economic factors
    population growth, 472–473, 472i
    sustainable development, 498–499
Ecoregion, **498**, 498i
Ecosystem, 5, 5i, 228, 286, 313, 502–503, **505**
    aquatic, 510, 510i
    biogeochemical cycles, 511
    carbon cycle, 514–515
    DDT in food webs, 508
    disruption, 28–29
    ecological pyramids, 509
    energy flow, 74i, 77i, 504i, 509
    energy flow, Silver Springs, 510
    food webs, 506–507
    forest, 331
    global cycling of water, 512–513
    global warming, greenhouse gases, 516–517
    lake, 540–541, 540i
    model for, 504i
    nature of, 504–505
    net production, 509, 509i
    nitrogen cycle, 518–519
    nutrient cycle, 366, 504–505
    ocean, 523
    participants, 504–505

phosphorus cycle, 520–521
    primary productivity, 509
    recovery, 335
    stream, 541
    structure of, 505
    tallgrass prairie, 505, 505i, 506i
    water crisis, 512–513
Ectoderm, **355**, 360, **420**, 424, 424i, 427i
Edema, 366
Ediacaran, **356**
Eel, 321
Effector, **408**, 408i, 409i
Egg, **146**
    animal, 53i, 146, 147i, 152, 201, 211, 422
    bird, 390i
    human, 123i, 124, 145, 146, 169
    invertebrate, 361, 361i, 364, 366i, 368i, 373
    plant (gamete), 153–155, 337, 338i, 341i, 345, 345i, 347i
    protist, 320
    yolk, 419
Egg-laying mammal, 393
Eincorn, 277i
Ejaculation, 437
Ejaculatory duct, 428, 428i, 429i
El Niño Southern Oscillation, 524–525, 524i, **545**, **546**–547, 546i
Elastin, 159
Electric gradient, 83, 84, 98, 100i
Electrochemical work, 74
Electromagnetic energy, 94
Electromagnetic spectrum, 94, 94i
Electron, 18i, 20, 22, 24, **30t**, 55, 77, 77i, 80, 98
Electron density cloud, 20i
Electron microscope, 55, 55i
Electron orbital, 22, 22i, 23, 23i
Electron transfer chain, **77**, 77i, 90, 97i, 98, 99, 100i
Electron transfer phosphorylation, **109**, 114, 114i
Electron transfer reaction, 35, **79**
Element, **18**, **30t**, 288
Elephant, 266, 465i
Elephant Island, 456
Elephant seal, 254, 559i
Elephantiasis, 366, 366i
Ellis-van Creveld syndrome, 172i, 173t, 255
Elm tree, 491t
Elodea, 105i
Elongation, RNA translation, 200, 200i–201i
Embryo, **181**, 416i, 451t
    early embryo formation, 442–443, 442i–443i
    emergence of human features, 446–447, 446i–447i
    teratogen sensitivity, 448i
    tissue layers, 354–355, 358, 360
Embryonic development
    animal, 128, 267
    animal, and gene controls, 207, 210–212, 214–215, 215i
    animal, human, 133i, 137, 169, 169i, 178
    animal, invertebrate, 353–355, 360
    animal, patterns and processes, 239, 268–269, 273, 273i
    animal, vertebrate, 35
    plant, patterns in, 268, 343, 345, 345i, 347i
    stages of, 446i–447i
Embryonic disk, 442, 442i–443i, 444, 444i
Embryonic induction, **426**
Embryonic period, 442, 444, 444i, 446i–447i
Emigration, **255**, **460**
Emotional stress, 309
Emperor penguin, 40i, 41
Emu, 238i
Encephalitis, 369
Encrusting sponge, 357i
Endangered species, 185, 335, 344, 373, 385, **496**, 498–499
Endemic disease, **310**

Endemic species, 496–497
Endergonic reaction, **74**, 75i
Endler, J., 466–467
Endocytic pathway, 60, 63i, 88
Endocytic vesicle, 63i
Endocytosis, **83**, 83i, 88–89, 88i, 309
Endoderm, **355**, 360, **420**, 424
Endomembrane system, 62–63, 62i, 158, 201, 201i, 296i–297i
Endometrium, **432**, 433i, 435, 436i, 442i
Endophytic fungus, **330**
Endoplasmic reticulum, **62**, 62i, 63, 63i, 70t, 88i, 201, 201i, 287, 294, 294i, 316, 317i
Endosperm, 347, 347i
Endospore, **304**, 304i
Endosymbiont, 318, 322
Endosymbiosis, 198, **294**–295, 304, 317, 319, 322
Energy, **6**, **72**, **74**
    alternative sources for humans, 118–119, 119i
    basis of metabolism, 6–7, 7i
    from fats, 118
    for human body, 74
    from proteins, 118
    source of life, 286–287
    sources, 6, 7, 29, 342
Energy carrier, 76, **89t**
Energy consumption, 527
Energy flow, 504i
    in cell function, **74**–78, 74i, 75i, 77i, 78i, 83i, 85i
    food web, 507i
    life's organization and, 6–7, 7i
    through ecosystems, 509
    through Silver Springs, 510, 510i
Energy hill, 75, 78i
Energy inputs and outputs, 505
Energy levels of electron, 22
Energy pyramid, **509**, 510, 510i
Energy-releasing pathway, 108–109
    comparison of main types, 108
    evolution of life and, 120
Englemann, T., 95i
Enhancer, **208**, 212
ENSO, 524–525, 524i, **545**, **546**–547, 546i
Entamoeba histolytica, 317
Enterobius vermicularis, 366
Enteromorpha, 490i, 491
Entropy, **74**, 75i
Enveloped virus, 308, 309
Environment
    carrying capacity, 462–463
    effects on gene expression, 160, 162–163, 166, 167, 176–177, 176i, 177i, 192
    extreme, 307i
    farming methods and, 228
    fluid environment, 405
Environmental movement, 508
Environmental Protection Agency, 523
Enzyme, **76**, **89t**
    controls in, 80–81, 80i, 81i, 208, 210
    evolution, 290
    structure, 35, 43, 48t, 78, 79, 79i
    types, 72, 186, 222, 286, 305, 309, 316, 324, 326
Enzyme function, 6, 35i, 73, 78–79, 295
    blood clotting, 73, 353
    in cancer, 135
    digestive, 60, 62, 63, 72, 135i, 174, 186
    in DNA replication, 126–128, 134, 190–191, 191i, 197, 211, 222, 222i, 309
    impaired, 180
    melanin production, 159, 162, 162i
    in metabolism, 76–77, 90–91, 160
    in photosynthesis, 77, 101
    protein synthesis, 196, 200
    in reproduction, 324
    sugar attachment, 158
    in toxins, 194
    waste removal, 206, 370
Eocene epoch, 245, 274i, 299, 395
Eomaia scansoria, 279i

Ephedra viridins, 344, 344i
Epidemic, 310, **329**
Epidermal growth factor, 134
Epidermis
    animal, 355i, 358i, 359
    plant, 96i, 103i
Epidermophyton floccosum, 329i
Epididymes, 428
Epididymis, 428i, 429i, 430i
■ Epilepsy, 352
Epiphyte, **341**
Epithelial tissue, 69, 69i, 134, 159, 405i
Equator, 494
Equilibrium
    chemical, 76, 76i
    genetic, 246–247
Equilibrium model of island
    biogeography, **496**
Equisetum, 334, 337, 337i, 340–342,
    340i, 342i, 350t
Equus, 266, 285
ER, **62**, 62i, 63, 63i, 70t, 88i, 201, 201i,
    287, 294, 294i, 316, 317i
Erectile tissue, 429i
Erection, 437
Ergotism, 329
Erosion, 349
Erythrocytes. *See* Red blood cell
Escherichia coli, 58i, 187i, 193, 208–209,
    224, 302, 302i, 305, 312
Esophagus, invertebrate, 363i
Estrogen, 35, 35i, 169, 179, 269, **433**,
    435, 436i
Estrous cycle, 432
Estuary, **91**, 290, **542**
Ethane, 72, 228
Ethanol, 72, 116, 116i, 228
Ethiopian, 533i
Ethyl group, 203
Ethylene
    oracle of Delphi and, 32–33, 32i
Eubacterium. *See* Bacterium
Eucalyptus tree, 348
Euchlanis, 375i
Eudicot (true dicot), 229, 282
    characteristics, **348**, 348i, 350t
■ Eugenic engineering, **230**
Euglena, 317i
Euglenoid, **316**–317, 316i, 317i
Eukarya, as domain of life, **8**, 8i, 140,
    280, 280i, 282i
Eukaryote
    characteristics, 8, 8i, 9i, 86, 332t
    DNA. *See* DNA
    endosymbiotic origin of, 479
    gene control in, 171, 207,
        210–211, 210i
    origin, 198, 199i, 263i, 293, 296i–297i
    types, 314–315, 315i
Eukaryotic cell, 41, 51, 51i, 52i, 53i, 60,
    64, 204i, 263i, 291–295, 292i, 295i
    components, 41, 47, 58, 64i, 65i, 66,
        68, 70t
    cycle, 124, 126, 128i
    photosynthetic, 64, 64i
Euphrates River, 513
European earwig, 371i
European rabbit, 493, 493i
European starling, 491i
Euryarchaeota, 306, 306i
Eusocial insect, 562, 563i
Eusocial mammal, 562, 563, 563i
Eutherian, **393**
Eutrophic lake, 541
Eutrophication, **521**, **541**
Evans, R., 230i
Evaporation, **27**
Everglades, 542i
Evergreen, **344**
Evergreen broadleaf forest, 536
Evolution, **10**, 45, 278–279,
    310–311, 373
    adaptation and, 244–245, 336–367
    asteroids and, 258
    of body plans, 356, 362, 364–365,
        372, 427
    branched *vs.* unbranched, 278,
        281, 281i

coevolution, 346, 482
convergent, 393
diversity and, 10, 14t, 243, 254
DNA and, 64, 177, 189, 200, 287, 291,
    294, 294i, 295
of early cells, 292–293, 292i, 293i
emergence of early humans, 396–397
emergence of modern humans,
    398–399
evidence of, 212, 215, 257i, 264–265,
    270–271
exceptions to trend, 373
influencing factors, 140, 144–145,
    203, 265, 343, 365
interpreting and misinterpreting,
    376–377
of land vertebrates, 443
macroevolution, 263, 266, 283t
microevolution, 236, 237, 244, 247,
    258, 272, 283t, 353
"missing links," 376–377, 376i, 379
natural selection and life histories,
    466–467, 466i, 467i, 475
of parasites and pathogens,
    236–237
primate trends, 394–395
radiometric dating, 377
social behavior, 564
of structural organization, 404–405
success in, 362, 364, **367**
theories of, 11, 14, 238–240, 242
Evolutionary tree, 270–271, 271i, **278**,
    278i, 282, 282i
    animal, 356i
    Archaea, 306i
    how to construct, 281, 281i
    invertebrate, 357i, 358i, 360i, 362i,
        364i, 366i, 367, 372i
    plant, 334i, 337, 337i, 348i
    primate, 399i
    protist, 316, 316i
Excretory organ, 364i
■ Exercise
    glycogen and, 39
    strenuous, 159
Exergonic reaction, **75**, 75i, 77
Exocytic vesicle, 63i
Exocytosis, **83**, 83i, **88–89**, 88i
Exon, **197**, 197i, 211, 221, 223
Exoskeleton, **367**, 367i, 368
Exotic imports, 476–477, 476i, 477i
Exotic invaders, 492–493
Exotic species, **491**, 491t
Expansion mutation, **174**
Experiment, **12**, 12i
Experimental design, examples
    animal research, 214–215, 249,
        254, 254i
    genetics, 171, 183, 193
    human health, 12, 16
    search for DNA, 186, 186i, 187i
    seawater viruses, 313
Experimental evidence, 152, 243
Experimental group, **12**
■ Experimental organism, 162, 172,
    182–183, 186, 186i, 366
Experimental test, 11, 12
Exploitative competition, 480
Exponential growth, 460–**461**, 460i,
    461i, 474
External skeleton, and chitin, 39
■ Extinction, **279**, 388, 398, 495–498. *See
    also* Mass extinction
Extinction crisis, 252, 263i, 265,
    278–**279**, 335
Extracellular digestion, 326
Extracellular fluid, 405, 408
Extraembryonic membrane, 442–443
Extreme halophile, 297i, **306**, 307, 307i
Extreme phenotype, and selection, 250
Extreme thermophile, **306**, 307, 307i
Eye, 159, 210
■ color, 160–161, 160i
    compound, 370i
    formation of, 427
    human, 160–161, 160i
    invertebrate, 364, 367, 370i
    third, 388

Eyeless gene, 214–215
Eyespot, 316, 317i, 360, 364i, 375, 375i

F

Facilitated diffusion, 83, 84
FAD, 46, 46i, 48t, **109**
Fall overturn, **540**–541
Fallopian tube, 432
False coelom, 355, 366, 366i
Family pedigree, 166, 171
Family planning, 470–471
■ Fanconi anemia, 173t
■ Farming methods, and
    environment, 228
■ FAS, 449, 449i
■ Fast-food, 49
Fast-twitch muscle fiber, 117
Fat, **40**, 48t
    body fat distribution, 169
■ diet high in, 41i
    energy from, 118
■ low-fat diet, 49
■ neural, 40
■ synthetic replacement for, 12
Fate map, 215i
Fatty acid, **40**, 48t, 49, 49i, 62, 118, 122
    as precursor of life, 289
    saturated, 40, 40i, 56
    structure, 34, 40, 56, 56i
    trans fat, 49, 49i
    unsaturated, 40, 40i, 56
Fatty acid chain, 209
Fatty acid tail, 34, 40, 40i, 49
Feather, 41, 105, 267, 281, 281i, 390
Feather star, 372
Feces, 361, 370
Feedback inhibition, **80**, 80i
Feedforward mechanism, 567
Fermentation, 329
Fermentation pathway, 116–117
Fern, 265, 265i, 282, 336i, 337, 337i,
    340, 340i, 341, 341i, 342i, 343,
    346i, 350t
Fertility, 416–417
    control of human fertility, 438–439
■ Fertility drug, 416–417
Fertility rate, 470–471
Fertilization, 138i, 144, 146–147,
    146i, 154, 157, 419, 420, 420i, 437,
    437i, 442i
    cross-fertilization, 153, 156i–157i
    double, 347i
    flowering plant, 153, 155
    human, 169i, 178, 437, 437i
    protist, 324i, 338, 341i
■ in vitro, 181, 181i, 192, 417, 439, 439i
■ Fertilizer, 29, 333, 519, 521
Fescue, 330
Festuca arundinacea, 330
■ Fetal alcohol syndrome (FAS),
    449, 449i
Fetal period, 442, 447i
■ Fetoscopy, 180i, 181
■ Fetus, **181**, **442**, 446–447, 446i–447i, 451t
    circulation between fetus and
        mother, 445, 445i
    teratogen sensitivity, 448i
■ Fever, 81
■ Fever blister, 309
Fibrillin, 159
Fibrous protein, 48t
Fiddler crab, 558, 558i
Field experiment, examples
    animal survival, 12–13, 251, 251i
    environmental factors, 162–163, 162i
    Mendel's peas, 152–157
Field mustard, 268, 268i
Filter feeder, 357, 358i, **378**
Fin, **381**, 382i
Finch, 242, 243i, 251, 251i, 275
Fir tree, 320, 344
Fire, 398, 534, 535i, 537, 537i
Fire ant, 476–477, 476i, 477i, 562
Firefly, 228i, 319, 557
First law of thermodynamics, **74**
Fish, 27, 93, 263i, 269
    adaptation of, 382
    body plan, 365i

bony, 381i, 382i, 383
    breeding ground, 321
    cartilaginous, 381i, 382, 382i
    characteristics, 354, 354t, 361
    jawed, 380–383, 380i
    jawless, 380, 380i
    lobe-finned, 383, 383i
■ mercury-tainted, 523
    ray-finned, 383
    reproduction, 139
    scales, 382
    school of, 5, 5i, 458i
■ shellfish poisoning, 320
Fish kill, 319, 320, 320i, 375, 503
Fisheries, 502, 545
Fission, transverse (invertebrate), 360
Fixation, **254**
Fixed action pattern, **553**
Flabellina iodinea, 364i, 365
Flagellated protozoans, **316**
Flagellum, 67
    bacterial, 58i, 59, 70t, 302, 302i
    choanoflagellate, 356
    dinoflagellate, 316i, 319, 319i, 320,
        320i, 359, 544
    eukaryotic, 59i, 67, 70t
    invertebrate, 356, 357i
    protist, 67, 316, 317i, 320,
        320i, 323i
    sperm, 67, 130, 146, 147i
Flame cell, 360, 360i
Flamingo, 105
Flat structural model, 34, 34i, 35i
Flatworm, 406, 407i
    body plan, 360
    characteristics, 138, 354t, 355,
        360–361, 360i
    evolution, 356i
    reproduction, 418, 418i
Flavivirus, 300, 301
Flea, 236, 371i
Flight, 390–391, 391i
Flightless bird, 238, 238i, 239
Float (cnidarian), 358i, 359
Flooding, 349
Florida Everglades, 542i
Florida Key deer, 459i
Florida Keys, 544
Flower, 348i
    formation, 268, 268i
    structures, **346**, 346i
Flowering plant
    basal group, 348i, 350t
    characteristics, 276–277,
        346–347, 350t
    evolution, 263i, 282, 334, 337i, 343,
        346, 350t
    fertilization, 153, 155
    growth patterns, 347
    nonphotosynthetic, 348, 348i
    ovaries, 346i, 347, 347i
    pollinization, 272i, 273, 343,
        347i, 350t
    reproduction, 126, 140i, 347, 357
Flowing-water ecosystems, 541, 541i
■ Flu shot, 308
Fluid environment, 405
Fluid mosaic model, **56**, 56i, 57i
Fluid pressure, 86–87
Fluke, 354t, 360
Fluorescent light, 98
■ Fluoride, 19
Fly, 280i, 317, 370i, 485i
Fly agaric mushroom, 327i
Fly ash, 529i
Flying fish, 382i
Folate (folic acid), 448
Follicle, 434–435, 434i, 436i
Follicle-stimulating hormone, 431,
    435, 436i, 454
Follicular phase, 432t, 433
Food
■ as energy source, 118–119
■ Frankenfood, 218–219, 231
    genetically modified, 218–219, 248
■ glucose and, 118
Food chain, **505**, 505i, 506
■ Food crop, 491

Food irradiation, 313
■ Food poisoning, 304, 312
■ Food production, and algal extracts, 321, 322
Food-spoiling mold, 329, 329i
Food web, 105, **505–507**
  DDT in, 508, 508i
  detrital, 507, 507i, 541
  energy flow in, 507i
  fungi and, 326
  grazing, 507, 507i, 510, 510i
  ocean, 36, 262, 307, 321
  tallgrass prairie, 506i
Foodborne disease, 310–311
Foot, invertebrate, 372
Foot muscle, mollusk, 364, 364i
■ Football, and heat stroke, 402–403
Foraminiferan, 314, 315, 315i, 317, 317i
Forelimb, vertebrate, 266–267, 266i
Forensics science, and DNA, 226
Forest, 532, 532i–533i
  boreal, 307, 532i–533i, 536–537
  broadleaf, 532i–533i, 536, 537i
  coast redwood, 334
  coniferous, 532i–533i, 536–537, 539i
  deciduous, 532i–533i, 536, 537i, 539i
  deforestation, 334–335, 334i, 335i, 349, 496, 512, 513i, 515, 517i, 519
  ecosystems, 331
  evergreen, 536
  evolution, 334
  logging, 334–335, 334i, 335i
  old-growth, 334, 464, 496
  products, 334, 335, 349, 350, 350i
  rain forest, 499, 532i–533i, 536, 536i, 539i
  swamp, 251, 263i, 334, 337, 340–342, 340i
  temperate, 349, 532i–533i, 537i
  tree fern, 340i
  tropical, 349, 349i, 532i–533i
  watershed, 349
Formaldehyde, 228, 290
Formylmethionine, 58
Fossil
  formation, **260–261**
  megafossil, 292i
  microfossil, 294
■ radiometric dating and, 21, 262, 262i
  trace fossil, 260
Fossil, examples
  ammonite, 239i
  Archaefructus sinensis, 346, 346i
  bacteria, 59i
  Bangiomorpha pubescens, 292i, 293
  Basilosaurus, 239i
  bird, 299
  Cooksonia, 260i, 336, 336i, 340i
  cyanobacteria, 293i
  Dickensonia, 356i
  early prokaryotic cells, 292i
  edicarans, 356i
  Eomaia scansoria, 279i
  foraminiferan, 278, 278i, 315i
  Ginkgo biloba, 344i
  Glossopteris, 265, 265i
  Grypania spiralis, 292i
  ichtyosaur, 260i
  Lystrosaurus, 265, 265i
  Myxococcoides minor, 59i
  Palaeolyngbya, 59i
  Psilophyton, 336, 336i, 340i
  Spriggina, 356i
  Tawuia, 292i
  trilobite, 356i
Fossil fuel, 29, 93, 502, 515, 517, 517i, 519
  formation, 342
  human population growth and, 469
Fossil record, 292, 293, 303, 376–377
  interpreting, 239, 239i, 242, 257, 258, 258i, 260, 265, 265i, 266, 269, 271, 274, 278, 281, 365
  predatory species and, 356
  transitional forms, 262–263, 278i, 293

Fossilization, **260**
Founder effect, **255**, 255i, 278
Fox, 455i, 481, 481i, 504, 504i
FOXP2 gene, 566
■ Fragile X syndrome, 173t
Fragmentation, 357
Fragmented habitats, 496–497
Frameshift mutation, **202**, 202i
Francke, W., 252
■ Frankenfood, 218–219, 231
Franklin, R., 188, 190, 190i
Fraser, C., 234, 298
Fraternal twins, 423, 454
Free radical, **80**, 107, 134, 203, 452
Fregata minor, 235i
Frequency distribution, 160i, 161
Freshwater habitat, 50
Freshwater ecosystem, 540–541
■ Friedreich's ataxia, 106–107, 106i
Frigate bird, 235i
Frisch, K. von, 557i
Frog, 27, 53i, 61, 340, 384i, 385, 385i, 484
  development, 421i
  reproduction, 420i, 421i, 422, 422i, 423
Frond, 341, 341i
Fructose, 38i, 428
Fruit, 343i, 347
  citrus, 371i
Fruit-eating pigeon, 478i, 481
Fruit fly. *See Drosophila*
Fruiting body, 305, 325, 325i, 326, 326i, 328, 329i
FSH, 431, 435, 436i, 454
Fucoxanthin, 320
Fumigant, 528, 555
Functional group, **34**–35, 34i, 42i, 85, 89t
Functional-group transfer, **35**, **78–79**
Fundamental niche, 478
Fungus
  characteristics, **9**, 68, 70t, 92, 138, 314, 315, 315i, 326–329, 326i, 327i, 332t
  current species diversity, 497i
  decomposers, 6, 9, 9i, 326, 327
  endophytic, 330
  evolution, 59, 263, 263i, 282i, 297i
  life cycle, 326–327, 326i, 327i
  mutualism and, 479
  parasites, 326, 327
■ pathogens, 314, 315, 316i, 329, 329i, 333
  symbionts, 304, 326, 328, 330–331, 331i, 348
Fur, 281, 281i
Fur color, 81i, 159, 162, 162i
Furrow, 132, i
Fusarium oxysporum, 333
Fusion, in plant production, 138

G

Galactose, 174
■ Galactosemia, 173t, 174, 174i
Galápagos Islands, 241i, 242, 244, 274–275, 524
Galápagos tortoise, 389i
Galilei, G., 14, 54
■ Gall tumor, 305
Gametangia, 328i
Gamete, 141, 146, 246, 248, 322i, 324i, 337, 350t
Gamete formation, 420, 420i
  animal, 138i, 139, 140, 142i, 146, 171, 171i
  genetic variation and, 144–147, 147i, 203
  human, 145
  meiosis and, 322i, 324i
  plant, 146, 156i–157i, 338, 338i, 339, 339i
  protistan, 318, 319i, 322i, 324i
Gamete mortality, 273, 273i
Gametophyte, **321**, **336**
  plant, 146, 153, 334i, 336i, 337–339, 341, 341i, 343, 345, 345i, 347i
  protist, 318, 319i, 322i, 324i

Gamma-glutamyl carboxylase, 353
Gamma ray, 94i, 202i
Ganglion, invertebrate, **363**
Gap gene, 215i
Gap junction, 56i, 69, 69i, 412
Garter snake, 552, 552i, 553
Gartshorn, G., 499
Gas exchange, large bodies, 406
Gaseous nitrogen, 518, 519
Gastric fluid, 28, 28i, 69, 81
Gastrodermis, 358i
Gastropod, 364
Gastrula, **420**
Gastrulation, 420, 424, 424i, 426, 426i, 444, 444i
Gause, G., 480–481
GCC, 353
Gel electrophoresis, **226**, 227
Gemmule, 357
Gene, **44**, **140**, **150**, 153, 168, 195, **212**
  behavior and, 552
  checkpoint, 134–135, 134i
  components, 332t
  conservation, 190, 190i, 191i, 199i, 353
  conserved, 270, 353
  control gene, 413
  cumulative interactions in, 160
  gap gene, 215i
  globin, 177
  highly conserved, 270, 353
  homeotic, 212, 268, 269, 269i, 427
  lateral gene transfer, 303, 310
  locus gene, 153, 153i, 158, 159, 171
  master, 169, 212
  of nonreproducers, 562–563
  oncogene, 134
  pair-rule, 215i
  positive control, 208
  proto-oncogene, 206
Gene control, **207**, 212–213
  behavior and, 166–167, 173t
  eukaryotic, 210–215, 210i
  loss of, 234–235
  mechanisms, 206–209, 234–235
  prokaryotic, 208–209, 271
Gene expression, 192, 208, 210–212, 215, 232, 246, 269, 277
  environmental factors and, 151, 160, 162–163, 166, 167, 176–177, 176i, 177i
Gene flow, 247, **255**, 255i, 256t, 266, 272, 274, 276, 278, 283t
Gene library, **224**
Gene linkage, human, 171
Gene mutation, **202**, 202i, 203
Gene pair, **153i**, 156–159
Gene pool, 140, 237i, **246**
Gene product, 159, 160, 163, 203, 552
Gene region, **194i**, 195i, 196i, 197i
Gene sequencing, 211, 303
Gene swapping, 144–145, 144i, 271, 307, 310
■ Gene therapy, 230–232, 230i, 231
Gene transcription, 208, 210–211, 210i, 234
Gene translation, 211
Genetic abnormality, **172**
■ Genetic analysis, 172–173, 172i
Genetic code, **194i**, 195i, **198**, 198i
■ Genetic counseling, 167, 181
■ Genetic disorder, 167, **172**–173, 172i, 173t, 191
■ phenotypic treatment, 180–181, 180i, 181i
  screening for, 150–151, 167, 180
■ types, 44, 71, 133, 150–151, 150i, 159, 165, 173t, 174–179, 183, 202, 202i, 206, 213, 217, 230, 257
Genetic distance, 399
Genetic divergence, **272**, 283t, 353, 356
Genetic drift, 247, **254**–255, 254i, 255, 255i, 256t, 278, 283t
■ Genetic engineering, 93, 105, 139, 139i, 149, 218–219, **228**–230, 305, 308
Genetic equilibrium, 246–247

Genetic persistence, 283t
Genetic recombination, **168**, 168i
■ Genetic screening, 150–151, 165, 167, 180–181, 183
■ Genetic testing, 181
Genetic variation, 157
Genetically modified food, 218–219, 248
Genetics experiment, 171, 183, 193
■ Genital herpes, 440, 440t
Genital pore, invertebrate, 360i
■ Genital warts, 137, 440, 440i
Genome, **221**, 303, 307
  animal, 356i, 366
  human, 202, 218, 220i, **221**, 269, 270
■ patenting, 218–219, 218i, 219i, 221
Genomic library, **224**
Genomics, **232**
Genotype, **153**, 155, 157, 163, 171i, 179, 181, 192, 247i
Genus (taxon), **8**, **280**
Geographic dispersal, **491**, 550–551
Geographic distribution patterns, 238, 238i, 274
Geographic isolation, 274
Geologic change, 264–265
Geologic time, 258–259, 262–263
Geologic time scale, **262**, 263i
Geological uplifting, 511, 515
Geology, 239
Geospiza conirostris, 243i
  G. scandens, 243i
Geringer, F., 176
Germ cell, 126, 141, 144–146, 169i, 203, 276, 452
Germ layer, animal, 354–355, 356i
■ German measles, 448–449
Germinating spore, 322i
Germination, 146, 324i
Gey, G., 124–125
Gey, M., 124–125
Ggta1 gene, 231
Ghengis Khan, and gene flow, 172i, 255
Giant anteater, 393, 393i
Giant horsetail, 342, 342i
Giant kelp, 315i, 321, 321i
Giant squid, 365
Giardia lamblia, 316, 316i, 317i
■ Giardiasis, 316
Gibbon, 394
Gilbert, W., 220, 221
Gill, 381
  invertebrate, 364, 364i, 365, 365i, 367
  mushroom, 327i
Gill slit, **378**, 378i, 380i, 382i
Gingerich, P., 257
Ginkgo, **344**, 350i
Ginkgo biloba, 344, 344i
Ginkgo tree, 337i, 344, 344i
Giraffe, 240
Gizzard, 281, 281i, 363i
Glacier, 261, 274, 286, 333, 488i, 502, 512i, 517
Glacier Bay, 488i
Gland, 73
Gland cell, 211, 364
Glaucophyte, 295, 295i
Global broiling hypothesis, **386–387**
Global climate change, 93, 261, 263i, 279, 334, 337, 347, 356
Global precipitation, 517
Global temperature, 502–503, 517, 517i
  asteroids and, 264, 265
■ Global warming, 2, 26, 33, 503, 515, **517**, 538
Globin, 44, 44i
Globin gene, 177
Globular protein, 48t
Glossopteris, 265, 265i
Glucose, 34, 34i, 38, 38i, 48t, 64, 82i, 84i, 92, 174, 208–209
  aerobic respiration and, 108–109
  breakdown, 77
  cell uptake of, 76, 84, 84i
  conversion, 101, 101i
■ food intake and, 118

glycolysis and, 110i–111i
    synthesis, 74, 75, 75i, 77, 96, 97i, 104i
Glucose transporter, 84, 84i
Glutamate, 44, 45i, 198, 198i, 202i
Glutamine, 198i
Glyceride, 48t
Glycerol, 38, 40i, 118
Glycine, 198i, 202i
Glycogen, 39, 39i, 48t, 118
Glycolipid, 158
Glycolysis, 76, 106i, **108**, 110–111,
    110i–111i, 135, 174, 174i, 209, 210
    fermentation and, 116, 116i
Glycoprotein, 43, 308, 308i
Glyptodont, 242, 242i
Gnetophyte, 337, 337i, **344**, 344i, 350t
GnRH, **431**, 435, 436i
Goat, 229i
■ Golden rice, 218
Golgi body, 62–63, 62i, 63i, 65i, 70t,
    88i, 316, 317i
Gonad, **359**, 360
    human male, 428, 428i, 428t
    invertebrate, 366i
    males, size of, 566
Gondwana, **265**, 265i
■ Gonorrhea, 440t, 441
Goose, 554i
Goose barnacle, 368i
Gorgonia ventalina, 544
Gorilla, 129, 177, 394, 394i, 497i, 566
Gradual model of speciation, **278**
Gram bacteria, 305
Gram-positive bacteria, **304**, 304i,
    305, 306i
Gram staining, 304
Grapevine, 320
Grass plant, 348
Grasshopper, 370i
Grasshopper mouse, 485, 485i
Grassland, 307, 519, 532i–533i, 534–535
Grassland soil, 539i
Gravelly soil, 539
Grazing food web, **507**, 507i, 510, 510i
Great Barrier Reef, 544i
Great Dying, Permian, 37
■ Great Lakes, 380i
Great Plains, 535
Great Salt Lake, 307, 307i
Green alga, 55i, 94, 324, 324i, 330,
    333i, 336i
    evolution, 297i, 316i, 317, 319, 334
    functions/uses, 323, 325, 333
    life cycle, 324, 324i
    photosynthesis in, 95i, 105, 105i
Greenhouse effect, 349, 515, **516–517**,
    516i–517i
Greenhouse gas, 33, 516–517,
    516i–517i
Griffith, F., 186, 186i
Grip, 395, 395i
Grizzly bear, 497i, 559i
Ground pine, 340
Groundwater, 512i, **513**
■   contamination, 513i
    overdrafts, 513, 513i
Group defense, 560, 560i
Growth, **404**
Growth and tissue specialization,
    420, 420i
Growth curves, 460–461, 460i, 461i,
    462, 463i, 469i
Grypania spiralis, 292i
Guanine (G), 46, 46i, 184i, 185, 188, 188i,
    189, 189i, 196, 197, 199, 227, 227i
Guano, 520i
Guar, 185
Guard cell, 96i
Guinea pig, 217
Gular sac, 235i
Gulf of Aqaba, 5i
Gulf Stream, 530
Guppy, 466–467, 466i, 467i, 475
Gut
    animal, 354, **355**, 355i
    human, 137
    invertebrate, 362, 363, 367, 370
Gut cavity, 355i, 359

Gymnophilus, 326i
Gymnosperm, 263i, 334, 336i, 340,
    343–345, 346i, 350t

H

Habitat, **407**, 458, **478**
    aquatic, 316, 317, 319–321, 323, 336,
        362, 368
    land, 318, 323, 326, 336, 343, 362
■ Habitat contamination, and
    amphibians, 385, 385i
■ Habitat destruction, and sea turtles, 388
Habitat loss, **496–497**
Hadal zone, 545i
Haemophilus influenzae, 222
Hagar Qim temple, 314i, 315
Hagfish, 379, 379i
Hair, 43, 392
    body hair distribution, 169
    mammal, 41
Half-life, **262**, 262i
Hallucinogen, 327, 327i
Halophile, extreme, 297i, **306**, 307, 307i
Hamilton, W., 563
Hand, 133i, 266
    apoptosis and development of,
        412–413, 413i
    grip, 395, 395i
Hangingfly, 558, 558i
Haploid chromosome number (*n*), **141**,
    142i, 143i, 146, 221
Haploid dominance, 350t
Haploid resting cell, 146
Haploid spore, 336
Haploid stage, 138i, 146, 146i, 322i, 324i,
    327i, 328i, 338i, 341i, 345i, 347i
Hardy-Weinberg rule, 247i
Harlequin frog, 486i
Harpobittacus apicalis, 558
Hawaii, 487
Hawaiian Archipelago, 264i, 274–275
Hayes, M., 176
Hayflick, L., 452
HbS molecule, 44, 45i
HCG, 443
HDL, 49
Head
    invertebrate, 354, 360, 362, 362i, 363i,
        364, 368, 370, 375, 375i
    vertebrate, 354
Heart, 19, 64, 159, 245, 281, 281i
■   abnormalities, 159, 159i
    accessory heart, 365, 365i
    blood pressure, transport, and flow
        distribution, 86
    invertebrate, 363i–365i, 365, 367
    muscle, 43, 56i, 69
■ Heart attack, 49, 229
■ Heart disease, 49, 250
Heat energy, **26**, 27, 74, 82, 83, 105
■ Heat stroke, 402–403, 414
HeLa cell, 124–125, 124i, 135
Helianthus, 102
Helical coiling, advantages, 291
Helicobacter felis, 280i
H. pylori, 302i
Heliconius cydno, 12–13, 13i, 22, 23i
H. eleuchia, 13, 13i
Heme group, **44**, 44i, 79i, 80
Hemlock tree, 331i
Hemoglobin, 43–45, 44i, 45i, 48t, 165,
    177, 202, 210, 245, 253, 257, 318
Hemophilia, 173t, 175, 175i, 181, 183
Hemorrhage, 329
Henslow, J., 241
■ Hepatitis, alcoholic, 72
■ Hepatitis D, 309, 310t
Her2 receptor, 206–207
■ Herbicide, 335, 349
Herbivore, 9, 330, 504, 505i
Heredity. *See* Inheritance
Hermaphrodite, **360**
    invertebrate, 361, 364i, 365, 366, 368
Heron Island, 542i
■ Herpes, 440, 440t
Herpes simplex virus, 308i, 309, 440
Herring gull, 561
Hershey, A., 187i

Heteractis magnifica, 358, 358i, 359,
    479, 479i
Heterocephalus glaber, 562, 563, 563i
Heterocyst, 304, 304i
Heterosporous, 337
Heterotroph, **92**, 93, 294, 315, 315i,
    316, 317, 320, 324, 326, 354,
    504, 504i
Heterozygous condition, **153**, 155–158,
    157i, 164, 173, 174, 252–253
High-density lipoprotein, 49
Himalayan rabbit, 162, 162i
Hirudo medicinalis, 363i
Histidine, 45i, 79i, 198i
Histone, 58, 126–127, 127i, 208
■ Histoplasmosis, 329
■ HIV, 308–310, 308i, 441, 441i
HLA, 49, 49i
Hoekstra, H., 249
Hognose snake, 386i
Hognosed bat, 392i
Holdfast, 293, 321, 321i
Homeodomain, **212**, 212i
Homeostasis, **7**, **405**
    animals, 408–409
    plants, 410–411
Homeotic gene, **212**, 268, 269, 269i, **427**
Hominid, **394**–397, 396i
Hominoid, 394, 395
Homo erectus, 343, 398, 398i, 399
H. habilis, 397, 397i, 469
H. neanderthalensis, 398i
H. sapiens, 398–399, 398i. *See also*
    Human
Homologous chromosome, **140**–141,
    141i, 142i, 144, 150, 153, 154,
    156–157, 168, 168i, 171, 247i
Homologue, 141, 141i, 142i, 144–145,
    148i, 157, 168, **266**, 266i, 267
Homosporous, 337
Homozygous condition, **153**, 156, 162,
    162i, 164, 174, 254, 255, 257
    dominant/recessive, **153**–154,
        153i, 154i
Honey mushroom, 327
Honeycomb, 41, 41i
Honeycreeper, lineage, 275, 275i, 278
Hooke, R., 54
Hookworm, 354t
Hormone, 7, 19, 88, 208, 250, 406. *See
    also specific types*
    behavior and, 552i
    bonding behavior and, 564
Hormone function
    in animals, 367
    in sexual traits, 169
Hornwort, 338, 339, 339i
Horse, 266, 285
Horseshoe crab, 369, 369i
Horsetail, 334, 337, 337i, **340**–342,
    340i, 350t
Host, **294**, 295, 310, 314, **360**
    human, 361, 361i, 366i
Host cell, 198, 308, 309
Hot spot, **498**, 498i, **532**
Hot spring, 286, 307, 497
Housefly, 280i
Household mold, 329
■ HPV, 440, 440t
Hrdy, S. B., 564
Hubbard Brook watershed, 512, 513i
■ Hudson River, 380i
Hudson's Bay Company, 483i
Human, 53i, 214, 246, 246i, 250, 255,
    280i, **396**
    behavior, 166–167
    chromosome, 127, 127i, 129, 129i,
        141, 141i, 147
    as competitor species, 496, 496i
    development, 419i
    development stages, 451i, 451t
    early forms, 258–259, 261, 343,
        396–397
    effects on biosphere, 301, 313, 315,
        350, 567
    embryo, 416i
■   emotional health, 163
    evolution, 214–215, 239, 258–259, 263i

gamete, 246
glands, 73
immune system, 230, 231
inheritance patterns, 164, 168–169,
    168i, 169i, 171–179
karyotype, 170i
modern forms, 398–399
origins of modern humans, 398–399
as predator species, 497, 497i
primate evolution, 394–395
reproduction, 140, 140i
sex determination, 168–169, 169i
skeletal structure, 239, 239i, 266,
    266i, 268i, 269, 394i
traits, 394–395
Human body, chemical elements in,
    18, 18i, 19i
Human body cell, DNA in, 61
Human chorionic gonadotropin, 443
Human collagen, 229
■ Human gene therapy, **232**
Human Genome Project, 221
Human health experiment, 12, 16
■ Human papilloma virus, 137, 440, 440t
Human population growth, 468–469,
    468i, 469i, 567
    biological wealth and, 498–499
    computer models, 473, 473i
    effects of, 334, 349
    energy consumption and, 527
    water depletion and, 513
Human social behavior, 564
Humboldt Current, 545
Hummingbird, 53i, 348i, 419i, 480
Humphead parrotfish, 8
Humus, 539
■ Huntington's disease, 107, 173t, 174
Huq, A., 547
■ Hutchinson-Gilford progeria
    syndrome, 176, 176i
Hybrid, **153**
Hybrid molecule, **223**, 270, 272, 285
Hybridization, 153, 271
Hydra, 354t
Hydrangea, 163
Hydration, sphere of, 27, 27i
Hydrocarbon, 34
Hydrochloric acid, 28, 29
Hydrogen, 20, 20i, 22, 23i, 24, 25, 25i,
    288, 289, 306
Hydrogen atom, relative size, 4, 4i, 76
Hydrogen bond, **25**–27, 26i, 34, **43**, 47i,
    56, 79, 81, 270
    in DNA, 25i, 189, 190, 191
Hydrogen cyanide, 288
Hydrogen ion (H^+), 28, 28i, 30t, 34i, 64,
    76, 77i, 82i, 98, 99, 100i
Hydrogen peroxide, 63, 79i, 80, 90–91
Hydrogen sulfide, 307
Hydroid, 358, 359
Hydrologic cycle, 511, **512**, 512i
    phosphorus cycle and, 520
Hydrolysis, **35**, 35i, 39
Hydrophilic head, 41i, 56, 56i
Hydrophilic substance, **26**, 30t
Hydrophobic substance, **26**, **30t**, 34, 79
Hydrophobic tail, 41i, 56i, 82
Hydrostatic pressure, **87**
Hydrostatic skeleton, 359
Hydrothermal vent, 59i, 105, 286, 289,
    290i, 291, 292, 306, 307, 544, 545i
Hydroxide ion (OH^-), 28, 30t
Hydroxyl group, 39i
Hygrophorus, 326i
Hyman, F., 159i
■ Hypercholesterolemia, 173t
■ Hypertension, 250
Hypertonic solution, **86**–87, 87i
Hypertonicity, 287
■ Hypertrophic cardiomyopathy, 106i
Hypha, **326**, 327i, 331
Hypothalamus, 431, 431i, 435
    menstrual cycle and, 436i
Hypotheses, examples of, 239–243,
    252, 264
    Alexander the Great's death, 300
    cell evolution, 294, 295
    inheritance, 152–153

origin of cells, 288, 290, 291, 291i
origin of life, 289
photosynthesis, 95i, 105
prions and degenerative disease, 311
Hypothesis, 11
Hypotonic solution, **86**–87, 87i
■ Hysterectomy, 137

I

Ice, 26i, 27
Ice age, 261, 263i
Ice cap, 502
Ice cover, perpetual, 532i–533i
Ice shelf, 523, 523i
Iceberg, 523, 523i
Ichthyosaur, 386, 387i
Ichthyostega, 383i
Identical twins, 423, 423i
IgG antibody, 448
Ignicoccus, 307i
Iguana, 388
IL2RG gene, 230
Immigration, **255**, 457, **460**, 472, 473i, 495
Immigration Reform and Control Act of 1986, 472
Immune system, 49, 73, 230
   human, 230, 231
   vertebrate, 308, 310
Immunity, 528
Imperfect fungus, 326
Implantation, **442**, 442i
Imprinting, **554**, 554i
■ In vitro fertilization, **181**, 181i, 192, 417, 439, 439i
Inbreeding, **255**, 563
Inclusive fitness, **563**
Incomplete digestive system, **355**, 355i, 359, 360, 366i
Incomplete dominance, of alleles, **158**, 158i
Incremental energy release, 114
Independent assortment theory, 156–**157**, 156i, 157i
India, 470, 471i
Indian mallow, 481, 481i
Indian pipe, 348i
Indicator species, **496**
Indirect interaction, 478
■ Induced abortion, 439
Induced-fit model, **79**
Inductive logic, 15
Industrial smog, 528–529
Industrial stage, 472, 472i
Inert element, 20
Infant, 451t
Infant death, 230
Infanticide, 564
Infection, **310**
Infectious disease, 236, 310t
   evolution and, 310–311
   human population growth and, 468–469
■ Influenza, 2, 308, 308i
Infolding, 294, 294i
Infrared radiation, 94i, 105, 292
Inheritance, **6**, 125, 152–153, 166
   chromosal basis, 152, 166, 168–169, 168i, 169i
Inheritance patterns, 150–153, 171, 171i
   human, 172–175, 172i, 173i
Inhibitor, 80, 80i
Initiation, in RNA translation, 200, 200i
Ink sac, 365i
Insect
   agriculture and, 320, 333, 366, 370
   in amber, 261
   characteristics, 354t, 367, 367i, 369–371, 369i, 370i
   current species diversity, 497i
   disease vectors, 301, 305, 316–318, 319i, 366, 369, 371, 375
   diversity, 370–371, 371i
   eusocial, 562, 563i
   evolution, 263i, 346, 353
   flight among, 367, 370, 371i
   larvae, 211

life cycle, 370i
   nymph, 370
   parasitoids, 486
   pesticides and, 248, 371, 375
   pollinators, 344, 346, 371
   pupa, 370
   social, 562, 562i
Insertion, in DNA, **202**
Instinctive behavior, **553**, 553i
Insulin, 7, 19, 48t, 118
Integrator, **408**, 408i, 409i
Intelligence, as evolutionary factor, 365, 401
Interaction. *See* Species interaction
Interference competition, 480, 480i
Interleukin-2, 229
Intermediate, **76**, 77, **89t**
Intermediate filament, **66**
Intermediate host, **360**
Internal body temperature, 409
Internal environment, 7, 29, 73, 81, 81i, 210, **405**
Internal temperature, 402–403
Internal transport, large bodies, 406
Interphase, **128**, 128i, 130–131, 130i–131i, 138i, 141, 141i, 142i, 148i
Interspecific competition, **478**, 480, 480i
Interspecific interaction, 458
Interstitial fluid, **408**
Intertidal zone, 490–491, 490i, 542, 545i
Intestine, 41, 209, 248, 304, 317, 360, 366i
   vertebrate, 361i
Intracellular digestion, 88
Intraspecific competition, 480
Intron, **197**, 197i, 221, 223, 309
Inversion, chromosomal, **176**, 176i
Invertebrate, 138, 352–372, **354**
Ion, **24**, **30t**
Ionic bond, **24**, 24i
Ionizing radiation, **203**
Iraq war, 316
Iridium, 258, 259i
Iris, eye, 160–161, 160i
Iron atom, 80
Iron-sulfide rock, 290i, 291
Irrigation, 512–513
Isaacson, P., 508i
Island, 494i, 495, 495i
Isolation, 272–274, 272i, 273i
Isoleucine, 198i, 199
Isopentyl acetate, 550–551
Isotonic solution, **86**
Isotope, **21**, **30t**, 193, 262
Isthmus of Panama, 274
Ixodes (deer tick), 305i

J

J-shaped growth curve, 460i, 461
Japanese beetle, 491t
Jaw, **380**, 381, 392
   invertebrate, 365i
   mammal, 392
Jawed fish, 380–383, 380i
Jawless fish, 380, 380i
Jawlike structure, invertebrate, 375
Jay, 246, 255, 255i
Jellyfish, 234, 352, 354t, 355, 358, 358i
Jellyfish bell, 359
Jet propulsion, cephalopod, 365
Johns Hopkins University, 12, 124
Johnson, M., 441i
Joint, **367**
Jump dispersal, 491, 495i
Juniper seedling, 331i
Juniperus occidentalis, 280
Jupiter, and asteroids, 258
Jurassic period, 263i, 337i, 343, 386, 387i, 393
Juvenile, **367**
   invertebrate, 366, 368i, 370

K

K–T asteroid impact theory, **386**–387
■ Kaguya, 139, 139i
Kangaroo, 419i
Kangaroo rat cell, 66i
Kapan, D., 12–13

Karenia brevis, 320, 320i
■ Kartagener syndrome, 71, 71i
Karyotype, **170**
■ Karyotype, spectral, 170, 170i, 177i, 178i
Kelp, 315i, 321, 321i
Keratin, 48t
Keratinocyte, 413
Ketone, 34i, 38
Key innovation, **279**, 293, 337, 355, 367
Keystone species, **490**–491
Kidney, 63, 361
Killer bees, 550, 550i
Killifish, 466–467, 466i, 467i, 475
Kinase, 134
Kinesin, 66–67
Kinetochore, **127**, 130
Kinetoplast, **316**, 316i
Kitti's hognosed bat, 392i
■ Klinefelter syndrome, 173t, 179
Knockout experiment, **214**
Knockout mice, 229
Koa bug, 487
Koala, 392i
Korana, G., 198
Korarchaeota, 306, 306i
Krebs, C., 483, 483i
Krebs cycle, **109**, 112–113, 113i
Krill, 93
Kubicek, M., 124
Kudzu, 492, 492i
Kwang, J., 295

L

La Niña, **546**, 546i
Labia majora, 432, 433i
Labia minora, 432, 433i
Labium, 370i
Labor, 409, **450**, 450i
Labrador retriever, 159
Lacks, H., 124–125, 124i, 137
■ Lactate fermentation, **117**, 117i
Lactation, **451**
Lactobacillus, 117
   L. acidophilus, 304
Lactose, 38, 174, 208–209, 209i
■ Lactose intolerance, 209
LacZ gene, 222i
Ladybird beetle, 371i
Lake ecosystem, 512i, 540–541, 540i
   seasonal changes in, 540–541
   thermal layering in, 540i
   trophic nature of, 541
   zonation of, 540, 540i
Lake Washington, 541
Lamarck, J., 240
Lamin, 66
Laminaria, 321
Lamprey, 380, 380i, 381i
Lancelet, 354t, 378, 378i
Land plant, origin, 343
■ Land rights, 549
Land snail, 364i
■ Language, 395, 566
Lanugo, 447
Larch tree, 344
Largest organism, 327
Larrea divaricata, 459i
Larva, invertebrate, **357**, 359, 361, 361i, 368i, 370, 372
Lascaux, France, 399i
■ Laser, uses, 137, 227, 227i
■ Late blight, 315
Latency, 309
Lateral gene transfer, **303**, 310
Latimeria, 383, 383i
Latrodectus, 369, 369i
Lava basalt flow, 249i
Laysan albatross, 391i
LDL, 49
LDL mutation, 49
Leaching, 519
Lead, 262
Leaf, 96i, 102, 102i
   C3 plant *vs.* C4 plant, 102, 102i
   characteristics, 27, 103i, 272i
   epidermis, 69i
Leaf color, 95, 95i

Leaf decay, 407i
Leaf folding, 411, 411i
Leakey, M., 396i
Learned behavior, **554**
■ Learning disability, 179
Least bittern, 484i
Leder, P., 198
Leech, 354t, 362, 363i
Leg
   bones, 69
   formation, 269
   invertebrate, 368–370
Legume, 410, 518–519
   bacteria and, 305
Leishmani donovani, 316
Leishmania mexicana, 89i
■ Leishmaniasis, 89i, 316
Lek, 558–559, 558i
Leopard frog, 421i
Lepidodendron, 342i
Lethal allele, **164**, 174
Leucine (leu), 42i, 45i, 198i
■ Leukemia, 173t, 177, 230, 549i
Leukocyte. *See* White blood cell
Leydig cell, 430i, 431
LH, 431, 435, 436i
Lichen, 304, 323, **330**, 330i, 479
Lid, cnidarian, 358i
Life, **6**
   beginning of, 438–439
   chemical basis, 18–30, 30t
   diversity of, 2, 6, 8, 8i, 9, 10, 139, 184i, 187, 189
   energy-releasing pathways and, 120
   in extreme environments, 307i
   history, 296i–297i
   organizational levels, 4–5
   origin, 287–293, 307
   synthesis of, 291
   unity of, 2, 6, 8, 9
Life cycle
   alga, 336
   animal, 139, 146, 146i
   arthropod, 367
   black bread mold, 328, 328i
   blood fluke, 361, 361i
   brown alga, **321**
   chain fern, 341i
   club fungus, 326i
   cnidarian, 359, 359i
   conifer, 344–345
   crab, 368i
   crustacean, 368
   fungus, 326–327, 327i
   green alga, 324, 324i
   insect, 370i
   invertebrate, 359, 359i, 367, 368
   lily, 347i
   moss, 338i
   parasite, 319i, 361i
   plant, 139, 146, 146i, 336, 336i
   Plasmodium, 319i
   red alga, 322i
   seedless vascular plant, 340, 341i
   slime mold, 324–325, 325i
   tapeworm, 361i
Life history patterns, **464**–465
   natural selection and, 466–467, 466i, 467i, 475
Life table, 464, 464t
Light, properties, 94–95, 94i
Light-activated pigment, 307
Light-dependent reaction, **96**, 97i, 98–99, 100i, 104i
Light energy, 74, 96, 98
Light-energy conversion, 70t
Light-harvesting complex, 97i, 98–99, 98i–99i
Light-independent reaction, **96**, 97i, 100i, 104i
Lightning, simulation of, 289
Lignin, 68, 228, 336
Lily, 242i, 347i, 348, 372
Lily pad, 406, 407i
Limb, evolution, 239, 266i, 269, 269i, 279, 383, 383i, 392

Limestone, 314–315, 314i, 321
Limiting factor, **462**, 462i, 468–469
Limnetic zone, 540, 540i
Limulus, 369, 369i
Lineage, 124, 203i, 210, **261**, 304i, 306, 332t. *See also* Speciation
  animal, 352–356
  divergence in, 232, 266–267, 266i, 271, 280, 282i
  eukaryotic, 316, 320, 332t
  fungi, 315, 326
  patterns and processes, 259, 268–271, 278–279
  plant, 102, 334i, 336, 348i
  prokaryote, 293, 332t
  protist, 8, 282i, 316i, 320, 356
  true-breeding, 153, 155i, 183
  vertebrate, 266–269
Linkage group, **171**, 171i
Linnaeus, C., 8
Linolenic acid, 40i
Lion, 7i, 281i, 556, 560–561, 564
Lipid
  characteristics, 35, **40–41**, 48t, 531
  early Earth and, 289, 293
  function, 33i, 48t, 73, 158
  synthesis, 63, 128
Lipid bilayer, **56**, 56i, 57, 59, 62i, 82–84, 82i, 83i, 94i
Lipid envelope, 308i
Lipoprotein, 43
Lisosome, 60
Listeria, 312
Lithops, 484, 484i
Littoral zone, 540, 540i
Liver, 41, 63, 64, 361
  function, 72–73
Liver cell, 39, 60i, 118, 210, 318, 319i
■ Liver disease, 72
Liver donor, 73
Liverwort, 282, 337i, 338, 339, 339i
Livestock, 499
Lizard, 387, 388, 389i, 415i, 419i
Llama, 245, 245i, 274, 274i, 285
Loam topsoil, 539
Lobaria oregana, 330i
Lobe-finned fish, 383, 383i
Lobster, 368, 368i, 375
Locomotion, 354, 362
  invertebrate, 369, 372–373, 373i
Locus gene, **153**, 153i, 158, 159, 171
Logistic growth, **462**, 463i, 474
Looy, C., 261i
Lorenz, K., 554i
Lottorina littorea, 490i, 491
Lotus, 348i
■ Lou Gehrig's disease, 107
Louisiana, 502–503, 502i, 547
Low-density lipoprotein, 49
■ Low-fat diet, 49
Loxosceles reclusa, 369, 369i
Lubchenco, J., 491
Luciferase, 228, 319
"Lucy," 396i
Luft, R., 106
■ Luft's syndrome, 106
Luna moth, 371i
Lung, 67, 67i, 150–151, 159, 245, 281, 281i, 381, 405i
■ Lung cancer, 529, 529i
■ Lung disease, 2
Lungfish, 383, 383i
Lupinus arboreus, 410–411, 411i
Luteal phase, 432t, 433
Luteinizing hormone (LH), 431, 435, 436i
Lycophyte, 337, 337i, **340**, 340i, 342, 342i
Lycopodium, 340, 340i
Lyell, C., 241, 243
■ Lyme disease, 305, 369
Lymph node, **366**
Lymph vessel, 135
Lysine, 198i, 202i
Lysis, **309**, 309i
Lysogenic pathway, 309, 309i

Lysosome, **63**, 65i, 70t, 88, 88i, 412
Lystrosaurus, 37, 37i
Lytic pathway, 309, 309i

Ⓜ

MacArthur, R., 496
Macrocystis, 321, 321i
Macroevolution, **263**, 266, 283t
Macrophage, 67, 88, 89i
Macrotermes, 563i
■ Mad cow disease, 229, 312
Madrone tree, 315
Magicicada septendecim, 272i, 273
Magnetotactic bacteria, 305
Magnolia, 348
Maiasaura, 386i
Maidenhair tree, 344, 344i
■ Malaria, 3, 253, 253i, 310t, 318, 319i, 503
Mali, 475i
■ Malignant tumor, 135, 135i
■ Malnutrition, 218–219
Malpighian tubule, 370
Malthus, T., 242
Mammal, 381i, 392–393
  adaptive radiation, 392, 393, 393i
  behavior, 392
  brain, 393
  characteristics, 354
  cleavage, 423, 423i
  dinosaurs and, 392, 393
  egg-laying, 393
  eusocial, 562, 563, 563i
  evolution, 392–393
  hair, 41
  jaw, 392
  origin, 279, 279i
  parental care, 559, 559i
  placental, 212, 279i, 392i, 393, 393i
  polyploid, 277
  pouched, 393
  reproduction, 139, 419i
  teeth, 392, 392i
  traits, 392i
Mammary glands, 450i, 451
Manatee, 320
Mandible, 370i
Mangrove swamp, 532i–533i
Mangrove wetland, 542, 542i
■ Manic depression, 166–167
Mantle, **364**, 364i, 365i
Mantle cavity, 364i, 365
Manx cat, 164, 165i
Manzanita, 273
Marchantia, 339i
■ Marfan syndrome, 159, 159i, 173t
Margulis, L., 294
Marigold, 276i
Marijuana, 333
Marine ecoregion, 532, 532i–533i
Marine nursery, 542
Marine snow, **544**
Marine worm, 362, 375i
Markov, G., 194
Marler, P., 554
Marr, J., 300
Mars, 258, 289
Marsh grass, 542i
Marsh wren, 554i
Marsupial, 212, **393**, 393i, 419i
Mason, R., 555
Mass extinction, 37, 252, 258, 263i, **279**, 283t, 343, 384, 386, 392, 496, 536i
Mass number, **20**
■ Mastectomy, 207
Master gene, 212, **427**
Maternal behavior, 564
Maxam, A., 220
Maxilla, 370i
Mayr, E., 272
McCain, G., 15
McClintock, B., 202, 203i
McClintock, M., 564
McIr gene, 249
Mealybug, 138i
■ Measles, 308, 310t, 448–449
Mechanical energy, 7
Mechanical isolation, 272i, **273**, 273i
Mechanical work, 74

■ Medicine, radioisotopes in, 21
Mediterranean fruit fly, 371i
Medullosa, 342i, 343
Medusa, 358, 358i, 359, 359i
Megafossil, 292i
Megaspore, 337, 342–345, 346i, 347i
Meiosis
  chromosomal change and, 168, 168i, 171, 171i, 178i
  eukaryotic, 277, 316, 320, 322i, 322t, 324i, 327i, 328i
  gene swapping and, 144–145
  plant, 276, 338i, 341i, 345i, 347i
  process, **126**, 156–157, 156i, 246, 247i
  sexual reproduction and, 138i, 139, 140–143, 146, 147i, 169, 169i
  vs. mitosis, 148i
Melanin, 81i, 105, 159–162, 160i, 161i, 162i, 164, 164i, 213i
■ Melanoma, 135i
Membrane crossing mechanisms, 83–85, 83i, 84i, 85i
Membrane protein, 34, 98
Membrane receptor, 206
Memory, cell, 426, 427i
Mendel, G., 150–157, 152i, 158
Mendeleev, D., 20, 20i
Mendelian inheritance pattern, 154–157, 154i, 155i, 156i, 157i, 172
Mendelian pattern, 226
Menopause, 433
Menstrual cycle, **432**, 432t, 435, 443, 564
  hormone secretions during, 435i
  overview, 432–433
  visual summary, 436i
■ Mercury-tainted fish, 523
Merozoite, 318, 319i
Meselson, M., 193
Mesoderm, **355**, 360, **420**, 424, 427i, 444, 444i
Mesoglea, 358i, 359
Mesophere, 526i
Mesophyll, 102, 102i, 297i
Mesozoic, 263, 263i, 279i, 343, 346i
Messenger RNA, 47, 194i, 196, **197**, 197i, 198, 200, 200i, 206, 209i, 210i, 223
Metabolic pathway, **77**, 77i, 80, 80i, 89t, 174, 208, 290
Metabolic reaction, 60, 263i, 291
Metabolic water, 97i
Metabolism, **73**, **290**
  diffusion and, 82–83
  nature of, 6, 63, 72, 74–76, 295, 332t
  origin of, 290
Metal ion, 80
Metamorphosis, **367**, 370, 370i
Metaphase, 124, 128i–131i, 129, **130**, 131, 137i, 139, 141i, 142i–143i, 145, 145i, 147, 148i, 156–157, 156i, 178i
Metaphase chromosome, 170, 170i
Metastasis, **135**
Meteorite, 289
Methane, 26i, 32i, 33, 34, 36, 36i, 306, 517i
  early Earth and, 288
  as precursor of life, 289
Methane hydrate deposit, 36, 37, 37i, 307
Methane seeps, 36–37, 36i
Methanocaldococcus jannaschii, 306
Methanogen, 8i, 36, 36i, 37, 297i, **306**, 306i
Methanogenic Archaea, 36–37, 36i, 306, 306i
Methionine (met), 42i, 58, 198, 198i
Methyl bromide, 528
Methylation, 208, 211, 213
Mexico, 471i, 498
Mexico City, 516i
Micrasterias, 323i
Microevolution, 236, 237, **244**, **247**, 258, 272, 283t, 353
Microfilament, 65i, **66**, 66i, 67, 88, 132, 132i, 133i

Microfossil, 294
Microphyll, **340**, 341
Microscope, 54–55, 54i, 55i
Microspore, 337, 342, 343, 345, 346i, 347i
Microtubule, 59, 65i, **66–67**, 67i–68i, 124, 127, 129–131, 129i–131i, 142i–143i
Microtus ochrogaster, 552, 552i
M. pennsylvanicus, 552i, 553
Microvilli, **356**, 357i
Microwave, 94i
Middle Ages, epidemics in, 329
Middle lamella, 68i, 132i
Midwife toad, 559, 559i
Miescher, J., 186
■ Migraine, 329
Migration, **391**, **460**
Milk, 208, 392, 451
Miller, G. T., 473
Miller, S., 289, 289i
Millipede, 367
Mimicry, **12**, **484**, 485i
Mineral, in cell function, 88
Minimal organism, **234**
Miocene, 263i, 274i, 314, 395
Missing link, 376–377, 376i, 379
Mississippi River, 274
Mistletoe, 348i, 486
Mite, 369, 369i, 555, 555i
Mitochondria, 51, 53i, 60, **64**, 64i, 65i, 70t, 310, 317i
  aerobic respiration and, 112i
  characteristics, 295, 295i
  origin, 198, 287, 294–295, 295i
■ Mitochondrial disorder, 106–107
Mitochondrial DNA, 270, 276
Mitosis, 124, **126**, 128–131, 130i–131i, 134
■ artificial, 170, 170i
  eight-cell stage, 181, 181i, 192
  eukaryotic, 210, 276, 316, 324, 324i, 325, 325i, 328i, 332t, 354
  vs. meiosis, 148i
Mixture, **23**, 30t
Model, **11**
Model for creation of living cell, 291i
Model organism, 333
Modification enzyme, **222**
Mojave Desert, 534i
Mold, 315, 316i, 320, 324–325, 325i, 328, 328i, 329, 329i, 333
Mole
  mammal, 266
  skin, 134
Mole-rat, 562, 563, 563i
Molecular biology, beginnings, 220–221
Molecular bond, 24–25
Molecular clock, **271**
Molecular competition, 296i–297i
Molecular model, 34i, 36i
Molecular tool, 222–223
Molecule, **4i**, 22, 23, **30t**, 53i
  of life, **32**, 34–35, 34i
  as unit of organization, 4–5, 4i, 74
Mollusk, 262i, 352, 352i, 353, 354i, 355, 356i, **364–365**, 364i, 365i
  current species diversity, 497i
Molothrus ater, 487, 487i
Molting, 356i, 367, 367i, 368, 368i, 370, 370i
Monarch butterfly, 463
Monitor lizard, 388
Monkey, 394, 394i
Monocot
  characteristics, 42, 282, 348, 348i, **350t**
  life cycle, 347, 347i
Monohybrid cross, **154–155**, 154i, 155i, 157
Monomer, **35**, 38, 46
Monophyletic group, **281**, 282, 316, 356
Monosaccharide, 38, 48t
Monosomy, 178
Monotreme, **393**, 393i
Monotropa uniflora, 348i
Monsoon, **531**
Montane coniferous forest, 533i, 537

Moore, M., 45i
Moorhead, P., 452
Morgan, T., 182–183
■ Morning-after pill, 438i, 439
Morph, **246**
Morphine, 352
Morphogen, **426**
Morphogenesis, 424i, **425**, 425i, 444i
Morphological convergence, **267**, 267i
Morphological divergence, **266**–267, 266i, 272, 272i, 278, 281
Morphological trait, 246
Morphology, comparative, 239, 242, 245, 246, 266
Mortensen, V., 149, 149t
Morula, 442i, 451t
Mosaic tissue effect, 213, 213i
Mosquito, 253, 301, 318, 319i, 366, 370i, 375, 375i, 503, 508
Moss, 282, 338, 338i, 339, 339i
Moth larva, **215i**
Motile structure, 67, 67i, 317, 318
   invertebrate, 359, 362, 364, 365
Motility. See Locomotion
Motor protein, **66**–67, 66i, 130, 142i
Moufet, T., 238
Mount Saint Helens, 489i
Mountain, complex zonation, 532i–533i
Mountain gorilla, 497i
Mountain grassland, 532i–533i
Mouse, 130i, 139, 139i, 214, 229, 229i, 249, 249i, 269, 269i, 460–461, 460i, 485, 485i
   cloned, 185, 192, 192i
Mouse-ear cress, 268i
Mouth, invertebrate, 355, 356i, 363i, 370, 370i
Mouthpart, 375, 375i
mRNA, 47, 194i, 196, **197**, 197i, 198, 200, 200i, 206, 209i, 210i, 223
mtDNA, 270, 276
Mucus, 150–151, 150i
Mudpuppy, 385
Mule, 273
Müller, F., 12
Multicelled organism, **5**, 5i, 356
Multiple allele system, **158**
Multiple births, 416–417, 416i
Multiplication cycle, virus, 308–309, 309i
Multiregional model, 399
Musca domestica, 280i
Muscle, 355
   invertebrate, 363, 363i, 366i, 367
Muscle cell, 29, 39, 43, 53, 66, 67, 211
Muscle contraction, 62, 66, 67, 85
■ Muscular dystrophy, 173t, 183, 217
Mushroom, 326, 326i, 327, 327i
Musk ox, 560i
Muskweed, 323i
Mussel, 364, 490–491
Mutant allele, 158, 168, 183
Mutation, **10**, 140, **153**, 248, **256t**, 283t
   accumulation of, 248, 270–271, 292
■ cancer and, 206–207
   causes, 44–45, 160, 246–247
   evolution and, 140, 203
   outcomes, 44, 134, 135i, 163, 176, 177, 177i, 180, 183, 195i, 212–213, 230, 253, 254, 257, 268–269, 353, 355, 356
   rate, 248
   spontaneous, 167, 168, 174, 176, 177i, 203
   types, 159, 172, 174, 176, 202, 248, 270, 271, 308
Mutualism, 330, **478**, 479, 479i
Mutualist species, 316
Mycelium, **326**, 327i, 328i
Mycoherbicide, 333
Mycoplasma genitalium, 234, 298
Mycorrhiza, **330**, 331i, 479
Mycorrizal fungi, 348
Myoglobin, 117
Myometrium, 432, 433i
Myosin, 66, 67
Myriapod, 367

Mytilus, 490–491
Myxobacteria, 305
Myxococcoides minor, 59i
Myxoma virus, 493
■ Myxomatosis, 493

N
Nachman, M., 249
NAD+, 46, 46i, 48t, 76, 80, 89t, **109**
NADP+, 48t, 80, 96, 97i, 99i, 104i
NADPH, 64, 92, 93, 96, 97i, 99i, 100i, 101, 104i
Naked amoeboid, 317, 317i
Naked seed, 344
Nanoarchaeum equitans, 307, 307i
Nanobe, 287, 287i
Nash, J. F., 166, 166i
Nasutitermes, 562i
■ Natural antiseptic, 339
Natural gas, 37, 517i
Natural selection, **10**, 243, 256t, 283t, 380
   life histories and, 466–467, 466i, 467i, 475
   theory of, 11, 152, 243, 376–377
   types, 247–253
Natural selection theory, 555
Nautiloid, 365
Nazca booby, 566, 566i
■ NBD, 166–167
Neandertal, 398
Nearctic, 532i
Negative feedback
   controls, 567
   mechanisms, **408**–409, 408i
   sperm formation, 431i
Negative gene control, **208**, 209i
Neisseria gonorrhoeae, 441
Nelumbo nucifera, 348
Nematocyst, **358**, 358i, 365
Nematode, 366
Neoplasm, **134**–135, 134i
Neotropical, 532i
Nephridium, **363**, 363i
Neritic zone, 545i
Nerve cell, 66, 85, 299, 354, 359, 360, 363
Nerve cord, invertebrate, 360, 360i, **363**, 363i, 378, 378i
Nervous system, 83, 174
■ disorder, 206
   invertebrate, 359, 360i, 362, 363, 363i, 372
Nest decorating, 555, 555i
Net ecosystem production, **509**, 509i
Net reproduction per individual per unit time, **460**–461, 460i
Neural fold, 444, 444i
Neural groove, 424i
Neural plate, 425
Neural tube, **424**, 424i, 425, 444, 444i, 448
   formation of, 424i
■ Neurobiological disorder, 166–167
■ Neurofibromatosis, 173t, 206, 206i
■ Neurological disorder, 21i
Neuron, 89i, 128
■ Neurotoxic shellfish poisoning, 320
Neurotoxin, 369
Neurotransmitter, 89i
Neutral fat, 40
Neutral mutation, **248**, 271
Neutrality, 28
Neutron, 20, **30t**
New Guinea, 564
Newborn, 451t
NF1 gene, 206
Niagara Falls, 86, 86i
Niche, **478**
Nigeria, 470i
9+2 array, 67, 67i, 70t, 302
Nirenberg, M., 198
Nitrate, 518, 519
Nitrate reductase, 81i
Nitric oxide, 528–529
Nitrification, **519**
Nitrite, 519

Nitrobacter (bacterial cell), 294i
Nitrogen, 22, 25, 81i, 138, 188, 193, 206, 288, 324
Nitrogen cycle, **518**–519, 518i
   human impact on, 519
   in seas, 305
Nitrogen dioxide, 529
Nitrogen fixation, 305, 330, 410, **518**
Nitrogen-fixing bacteria, 518, 518i
Nitrogen gas, 304
Nitrogen loss, 519
Nitrogen oxide, 519, 519i, 529
Nitrous oxide, 517i
Noctiluca scintillans, 319i
Noncyclic pathway, ATP formation, 99, 100i
Nondirectional change, 489
Nondisjunction, **178**–179, 178i
Nonionizing radiation, **203**
Nonphotosynthetic plant, flowering, 348, 348i
Nonpolar covalent bond, **25**, 25i, 34
Nonpolar hydrocarbon, **40**
Nonvascular plant, 334i, 335
   characteristics, 338–339, 350t
   evolution, 336i, 337, 337i
   metabolic products, 339, 339i
■ Norplant, 438i, 439
North Atlantic Current, 530
Northern coniferous forest, 532i–533i
Northern spotted owl, 464
Nostoc, 58i
NotI enzyme, 222
Notochord, 378, 378i, 444, 444i
Notophthalamus, 384i
NSP, 320
Nucelotide, 46–47, 46i
Nuclear division, 125, 127, 138i, 140, 332t
Nuclear envelope, **61**, 61i, 62i, 65i, 130, 131, 131i, 142i, 211, 294, 294i
Nuclear fusion, cell, 324i, 327i, 328i
Nuclear fusion reaction, 288
Nuclear pore, **61i**, 62i, 211
■ Nuclear power, 549
Nucleic acid, 33i, **46**–47, 46i, 48t, 270, 289
Nucleic acid hybridization, **224**, 270
Nucleoid, **52**, 59, 302i
Nucleolus, **61**, 61i, 62i, 65i, 70t
Nucleoplasm, **61**, 61i, 65i
Nucleosome, **127**, 127i
Nucleotide, **188**–191i, 188i, 189i, 191i, 195, 198, 203
Nucleus
   atomic, 20i, 22, 23, 23i, 24i
   cell, 8, **52**, 52i, 61, 61i, 142i, 148i, 295i, 317i, 360i
   function, 20, 51, 62i, 65i, 70t
   origin, 287, 296i–297i
Nudibranch, 364–365
Numerical taxonomy, **303**
Nursing, 564
Nutmeg, 348
Nutria, 491t
Nutrient, **511**
   in cell cycle, 128
   cycling, 74i, 366
   inputs and outputs, 505
   recycling in ecosystems, 504–505
Nutrient flow, model for, 511i
Nutrition. See also Diet; Digestion
   importance of, 160
Nutritional labeling, 219
Nymph, insect, 370

O
Oak tree, 255, 315, 328
Obelia, 359
Ocean, 512i, 523, **530**–531, 544–545
   currents, 530, 530i
   diversity of, 544
   food webs, 36, 262, 307, 321
   monsoons, 531
   open ocean, 544
■ pollution, 2, 320
   primary productivity in, 523

proportion of elements in, 19i
   rain shadows, 531
   upwelling and downwelling, 544–545
Oceanic zones, 545i
Ochoa, S., 198
Octopus, 354t, 365
Octuplets, 416, 416i
Odor, 552
Offspring (F_1, F_2), 152–155, 154i, 157, 158, 158i, 171, 171i, 183
Ogallala aquifer, 513
Okazaki, R., 191i
Old age, 451t. See also Aging
Old-growth forest, 334, 464, 496
Old man's beard, 330i
■ Older Order Amish, 255
Olduvai Gorge, 397, 397i
Oleic aid, 40i
Olestra(r), 12
Oligochaete, 362
Oligomer, **224**, 225
Oligosaccharide, 38, 48t, 158
Oligotrophic lake, 541
Omnivore, 504, 504i
Oncogene, 134
Onyx, 244i, 245
Oocyte, **146**, 147i, **422**, 432, 434–435, 434i, 435i, 436i
Oomycote, 316i, **320**, 330
Open circulatory system, **367**
Operator, **208**, 209i
Operon, **208**–209, 209i
Opossum, 146, 393
Opposable thumb, 395
Opuntia basilaris, 103i
Oracle of Delphi, 32–33, 32i
Oral cavity, 318, 318i
Orangutan, 394
Orbital, **22**, 22i, 23, 23i
Orchid, 140, 252, 280i, 348
Ordovician period, 263i, 337i, 365i
Organ, **4i**, 140i, 354, 355, **360**, **404**
Organ formation, 420, 420i, 424–425, 425i, 426–427
Organ systems, 4i, 60, 354, **360**, 360i, 363, **404**
■ Organ transplant, 151, 231
Organelle, 51, **60**, 66i, 201, 295, 318i
   function, 60, 64, 332t
   origin, 294–295, 295i
Organic compound, **34**, 34i, 48t, 64, 80, 93, 126, 160
   early Earth and, 288–289, 293
Organization, animal, 357, 358
Orgasm, 437, 564
Oriental, 533i
Ornithonyssus sylviarum, 555, 555i
Ornithorhynchus anatinus, 401i
Oryctolagus cuniculus, 493, 493i
Oryx leucoryx, 414
Osmosis, **86**, 86i, **87**
Osmotic pressure, **87**
Ostracoderm, 380
Ostrich, 146, 238i, 390
Otter, 496
Ovary, 432, 432t, 433i
   animal, 360i
   flowering plant, 346i, 347, 347i
   human, 140i, 169, 169i
■ Overactive gut, 361
Ovibos moshatus, 560i
Oviduct, 432, 432t, 433i
   invertebrate, 360i
Ovulation, 432t, **433**, 435, 436i, 437i
Ovule, 140i, 343, 344, 345, 345i, 346i, 347i
Ovum, **437**
   invertebrate, 359i
Owl, 249, 464
Ox, wild, 185
Oxaloacetate, 102i, 103i
Oxidation-reduction reaction, **76**–77, 77i
Oxygen, 64, 82i, 205, 408
   aerobic respiration and, 114
   in atmosphere, 37, 50–51, 97, 263i, 287–289, 292–294, 304, 349

in blood, 44, 318
diffusion in plants, 102
free, 50, 96, 97, 100i, 292–294, 306, 321, 335
in photosynthesis, 95i, 98
structure, 22, 23i, 25, 25i, 31, 83
uptake in animals, 365
Oxytocin, 409, **551**–553, 552i, 564
Oyster, 364
Ozone, 31
Ozone hole, 528, 528i
Ozone layer, 336, 526i, 528
Ozone thinning, **528**, 528i

P

Paabo, S., 566
Pacific yew, 137
■ Pain treatment, 352
Paine, R., 490–491
Pair-bonding, 552–553
Pair-rule gene, 215i
Palaeolyngbya, 59i
Palearctic, 533i
Paleocene, 395
Paleomap, 274i
Paleozoic era, 263, 263i, 369
Palisade mesophyll, 103i
Palm, 348
Palp, 370i
Pancreas, 7, 61i, 62, 118t
Pancreatic cell, 61i,
Panda, 271i
Pandemic, **310**, 313
Pangea, 263i, **264**, 265i, 393, 393i, 401, 532
Pansy, 348i
■ Pap smear, 137, 440
■ Papillomavirus, 137, 440, 440t
Parabasalid, 316i
Paradisaea raggiana, 252i
Paramecium, 91i, 318, 318i, 480–481, 481i
■ Paranoid schizophrenia, 166
Paranthropus, 396
Parapatric speciation, **277**
Parapod, 362
Parasite, 89i, 302, 307, 314–316, 360, 368i, 369, 486–487, 504
  in animals, 9, 253, 305, 318–320, 360–361, 361i, 366, 366i, 368
  defenses against, 357
  good, 371
  in humans, 89i, 236, 316–318, 366, 366i
  in plants, **326**, 327, 329, 329i, 366
Parasite-host interaction, 486–487, 486i, 487i
Parasitism, **478**
Parasitoid, 477, **486**–487
Parasitoid wasp, 414, 414i, 487i
Parent, true-breeding, 156i, 158, 158i, 183
Parent cell, 125, 129, 131, 133
Parent generation, 153, 154, 171, 171i
Parental behavior, 559, 559i
Parental care, 559, 559i
■ Parkinson's disease, 107, 231
Parrotfish, 4i, 5i, 8
Parthenogenesis, 138–139, 138i
Parvovirus, 308
Passive transport, **83**, 83i, 84, 84i, 407
Passive transporter, 57, 57i
■ Patenting genomes, 218–219, 218i, 219i, 221
Pathogen, **301**, 303, 306, 308–310
  bacterial, 59, 150, 301, 302i, 310
  defenses against, 88, 310, 357
  fungal, 314, 315, 316i, 329, 329i, 333
  human, 151, 232, 236, 302i, 304, 305, 308–311, 316, 371
  plant, 305, 309, 315, 320, 330, 333
Pattern formation, **215**, 215i, 217i, **426**–427
Pauling, L., 187, 189, 190
Pax-6 gene, 214–215
PCR, 221, **225**, 225i, 286, 307
Pea plant, 129, 151–154, 152i, 154i
Peach tree, 329
Peat bog, 81i, **339**, 339i, 342, 514

Peat moss, 339, 339i
■ Pedigree, family, **172**, 172i, 173i, 175, 175i, 181
Pelagic province, 545i
Pellide, 316, 317i, 318
■ Pelvic inflammatory disease, 440
Penguin, 40i, 41, 93, 248, 266, 266i, 310, 458i, 556i
Penis, 169i, 428, 428i, 429i, 430i, 437
  invertebrate, **360**, **360i**, **368**
Pepper plant, 348
Pepsin, 81
■ Peptic ulcer, 81, 302i
Peptide bond, 42i, 43, 200, 200i, 352
Per capita rate, **460**
Peregrine falcon, 10, 10i, 497i
Periodic table, **20**, 20i
Peritoneum, **355**, 355i
Periwinkle, 490i, 491
Permafrost, **538**
Permian mass extinction, 37
Permian period, 263i, 337, 337i, 343, 392
Permian-Triassic boundary layer, 261i
Peroxide, 134
Peroxisome, **63**
Perpetual ice cover, 532i–533i
Peru Current, 545
Pest resurgence, 248
■ Pesticide resistance, 248
■ Pesticides, 231, 371, 375, 508, 508i
PET scan, 21, 21i
Petal, 346i
Peter the Great, 329
PGA, 101, 101i, 102, 102i
PGAL, 101, 101i, 104i, 110, 118
pH, 81, 81i, 82, 83, 304, 339, 359
  of blood, 28i, 29
  of rainwater, 29, 529
  of seawater, 28, 28i
  of vaginal fluid, 429
pH scale, **28**, 28i, 29
Phagocyte, 302, 317, 357i
Phagocytic cell, 412
Phagocytosis, **88**, 89i
Phagocytotic vesicle, 89i
Pharyngeal arch, 444, 444i
Pharynx, **360**, 360i, 362, 363i, 366i, **378**, 378i
Phasmid, 501, 501i
Phenotype, **153**, 155i, 158–161, 163, 171, 171i, 171i, **246**, 248, 250, 252, 255i
Phenotypic ratio, 154i, 155, 155i, 156i, 157i, 159i, 160, 171, 183
Phenylalanine, 198i
■ Phenylketonuria, 173t, 180–181
Pheromone, 418, 550–551, 550i, **556**
  human, 564
  priming, 556
  sex, 252
  signaling, 556
Philadelphia chromosome, 177, 177i
Philetairus socius, 251, 251i
Phloem, 336, 341, 406
Phlox, 464t, 465
Phorid fly, 477
Phosphate group, 34i, 46, 75, 75i, 76, 85, 85i, 101, 188–191, 211, 291, 319
Phosphate-group transfer, 85, 85i
Phosphate ion, 520, 520i
Phosphatidylcholine, 48t
Phosphoglycerate, 101, 101i, 102, 102i
Phospholipid, **41**, 41i, 48t, 50i, 56, 56i
Phosphorus cycle, **520**–521, 520i
Phosphorylated glucose, 118
Phosphorylation, **75**, 85, 85i, 134, 211
Photoautotroph, 105, 292, **302**, 304, 304i, 316, 317, 321, 323, 330, 335, 359
Photochemical smog, 528–529
Photolysis, 98, 100i
Photon, **94**, 98, 98i–99i, 100i
Photon energy, 98, 98i
Photosynthesis, **92**, 95i, 105, 509
  chemistry of, 23i, 77, 77i, 92i, 96–97, 96i, 97i, 104i
  cyclic pathway, 292, 296i–297i, 349

discovery of, 21, 262, 323
nature of, 74, 92, 93
noncyclic pathway, 287, 292, 294, 296i–297i
origin, 263i
oxygen-releasing pathway, 50, 302, 335
Photosynthesizer, 58i, 92, 93, 314, 316, 317i
Photosynthetic cell, 64, 64i, 69i, 74, 96i, 287, 295, 321
Photosynthetic pigment, 70t, 92, 94, 95i
Photosynthetic species, 70t, 263i, 320, 330i, 333, 333i
Photosystem, **96**, 97i, 98–99, 98i–100i
Photovoltaic cell, 527, 527i
Phrynosoma cornutum, 476
Phycobilin, 95, 105, 322
Phyletic constraint, 427
Phylogeny, **280**, 281, 282, 303, 316
Phylum, 282
Physalia utriculus, 358i, 359
Physarum, 325i
Physical access, and lineage, 279
Physical constraint, 427
Physiological trait, 246
Physiology, **403**
Phytochrome, 411
Phytophthora, 330
P. infestans, 315, 320
P. ramorum, 315, 320
Phytoplankton, 368, 509, 528, 540–541, 544, 545, 547
■ PID, 440
Pig, 231, 231i, 366
  cloned, 185
Pig virus, 231
Pigeon, 10i, 246, 390i, 478i
■ Pigeon breeding, 10
Pigment
  accessory pigments, 94, 95i, 105
  cartenoid, 333
  light-activated, 307
  photosynthetic, 70t, 92, 94, 95i
  visual, 105
Pigment molecule, **94**–95, 95i, 97i, 98, 98i, 105
Pike-cichlid, 466–467, 466i, 467i, 475
Pillbug, 368
Pillotina, 305
Pilus (pili), 58i, 59, 302, 302i
Pine tree, 146, 343i, 344–345, 344i, 345i
P. longaeva, 244i, 344i, 410
Pinworm, 366
Pioneer species, 488, 488i, 489i
Pisaster ochraceus, 490–491
Pisum sativum, 129, 151–154, 152i, 154i
Pituitary gland, 431i
■ PKU, 173t, 180–181
Placenta, 181, 443, 443i, **445**, 445i, 450, 450i
Placental mammal, 212, 279i, 392i, 393, 393i
Placoderm, 380, 380i, 381
Placozoan, 356, 356i
Planarian, 360, 360i
Plankton, **317**
Plant. *See also* specific names
  characteristics, **9**, 9i, 39, 92, 332i
  commonalities with animals, 402–414
  current species diversity, 497i
  evolution, 263i, 282i, 297i, 336–337, 336i
  life cycle, 139, 146, 146i, 336, 336i
  as photosynthesizer, 93, 96, 101
  products, 491
  reproduction, 138, 140i, 146, 228i
  stress responses, 414, 414i
  structure, 41, 69i, 94
  water uptake in, 27
Plant cell, 29, 52i, 53i, 54, 60i, 67, 69, 70t, 87, 132i
  components, 39, 39i, 63, 68, 68i, 86

Plant growth, fungi and, 330, 330i, 331, 331i
Planula, 358i, 359, 360
Plasma, **408**
Plasma membrane, 287, 294, 294i, 302, 307
  in cytoplasmic division, 132–133, 132i, 133i, 142i
  diffusion, 83, 84i, 86, 88, 88i
  function, 50i, 52, 52i, 53, 56i, 57–59, 57i, 63i, 65i, 68i, 70t, 303, 303i, 307
  origin, 290–291, 291i
Plasma protein, 73
Plasmid, **59**, 222, 222i, 223i, 228i, 229, **302**
Plasmodesma, 65i, 68i, 69
Plasmodesmata, 412
Plasmodial slime mold, 324–325, 325i
Plasmodium, 253, 318, 319i
Plasmopara viticola, 320
Plastid, **64**, 318, 322
Plate tectonic theory, **264**–265
Platypus, 392i, 401, 401i
Play bow, 556, 556i
Pleiotropy, **159**
Pliocene period, 393
Plutonium, 194
■ Pnuemonia, 186, 329
Podocarp, 344
Poecilia reticulata, 466–467, 466i, 467i, 475
■ Poison, 66
Polar bear, 26i, 244–245, 244i, 271i, 497i
Polar body, 147i, 434i, **435**, 437
Polar covalent bond, **25**, 25i
Polar ice, 512i, 517
Polar ice shelf, 523, 523i
Polar molecule, transport of, 83, 84
Polarity, 26, 26i, 83
■ Polio, 3, 125
Pollen grain, 153, 337, **343**–345, 343i, 347i
Pollen sac, 345i, 347i
Pollen tube, 345i, 347i
Pollination, 272i, 273, **343**, 347i, 350t
Pollinator, 337, 343, 344, **346**, 348, 348i, 371
Pollutant, **528**–529
■ Pollution. *See also* Air pollution
  in oceans, 2, 320
  in waterways, 320
Poly-A tail, 197, 211
Polychaete, 354t, 362, 362i
Polydactylyl, 172i, 173t
Polyhedral virus, 308i
Polymer, **35**
Polymerase chain reaction (PCR), 221, **225**, 225i, 286, 307
Polymorphism, **246**
Polyp, 358, 358i, 359, 359i
Polypeptide, 42i, **43**, 43i, 44i, 58, 62, 66, 66i, 79i, 90, 209i, 302
Polypeptide chain, 194, 195, 198, 200, 201, 204i
Polyploidy, **178**–179, 179i, 276–277, 276i, 277i
Polysaccharide, **38**, 39, 48t, 59, 69, 302
Polysome, **201**
Polytene chromosome, 211, 211i
Polytrichum, 338i
Ponderosa pine, 344–345, 345i
Pool, 541, 541i
Poppy plant, 138i, 475, 534i
Population, 5, 5i, **246**, 272, 456–475
  characteristics of, 458
  characteristics of populations, 458
  counting, 459
  fertility rates and age structure, 470–471
  gene pool, 237i
  life history patterns, 464–465
  life tables, 464
  natural selection and life histories, 466–467
  patterns of survival and reproduction, 464–465
  variation in, 252–255

Population density, **458**, 459
Population distribution, **458**, 458i
Population growth, 438, **458**, 468–469
  biotic potential, 461
  carrying capacity and logistic
    growth, 462–463
  community structure and, 478
  demographic transitions, 472
  density-independent factors, 463
  economic effects, 472–473, 472i
  effects of, 242, 303, 334, 349
  exponential growth, 460–461
  gains and losses in, 460
  immigration policies, 472
  impacts of no growth, 473
  limits on, 462–463
  resource consumption, 472–473
  from zero to exponential growth,
    460–461
Population size
  community and, 478
  gains and losses, 460
Porcupine, 407
Poriferan, 354t
Pork, 361
Porphyra tenera, 322, 322i
Porpoise, 266, 266i
Portuguese man-of-war, 358i, 359
Positive feedback mechanism, **409**
Positive gene control, **208**
■ Post-partum depression, 450
Post-reproductive age, 458
Postelsia, 321i
Posterior, **354**, 354i
Postindustrial stage, 472, 472i
Postnatal development, 451,
  451i, 451t
Postzygote isolationg mechanism,
  272, 273i
Potassium, 29
Potassium ion, 85
Potato plant, 315, 320
Pouched mammal, 393
Poverty, 498–499
Powdery mildew, 329
Power grip, 395, 395i
Power plant, fuel for, 29i, 339
Poxvirus, 308
Prairie
  shortgrass, 534–535, 535i, 545, 545i
  tallgrass, 496, 505, 505i, 506i,
    534–535
Prairie dog, 560, 560i
Praying mantis, 485i
pre-mRNA, 211
Pre-reproductive age, 458
Precision grip, 395, 395i
Precursor, **38**
Predation, 478, 490–491, 490i
Predator, 314, 316, 318, 318i, 319, 352,
  356–358, 360, 363, 363i, 365, 368i,
  369, 371, 373, **482**
  adaptive responses of, 485, 485i
  defenses against, 356, 364–365, 367,
    368, 372, 373, 388
  natural selection and, 466–467, 466i,
    467i, 475
Predator-prey interaction, 482–483
Predatory colony, 305
Prediction, **11**
Pregnancy, 179, 437, 437i
  alcohol, tobacco, and other
    drugs, 449
  cleavage and implantation, 442
■ contraception, 438–439, 438i
  cyclic changes in ovary,
    434–435, 434i
  cyclic changes in uterus, 435
  emergence of distinctly human
    features, 446–447, 446i–447i
  emergence of vertebrate body plan,
    444, 444i
  extraembryonic membranes,
    442–443
  fertilization and, 437, 437i
  formation of early embryo, 442–443,
    442i–443i
  giving birth, 450, 450i

infectious diseases and, 448–449
■ mercury-tainted fish and, 523
  nourishing the newborn, 451, 451i
  nutritional considerations, 448
  placenta, 445, 445i
  preparations for, 434–435
  seeking or ending, 439
  sexual intercourse and, 437
  unplanned, 438i
■ Pregnancy test, 443
Prehensile movement, 395
■ Preimplantation diagnosis, 181
Preindustrial stage, 472, 472i
Premature birth, 450
Prenatal development, 451i, 451t
■ Prenatal diagnosis, 167, 179, 180i, 181
Prescription drugs, and
  pregnancy, 449
Pressure gradient, **83**, 87
Prevailing winds, 526, 526i
■ Previn, 439
Prey, adaptations of, 484–485, 484i, 485i
Prey density, 482, 482i
Prezygotic mechanism, 272–273, 273i
Primary oocyte, **434**
Primary producer, **504**
Primary productivity, **509**, 509i, 523
  gradients and, 533i
  soil and, 539
Primary succession, **488**, 488i
Primary tissue layer, 354–355
Primary wall, in plant cell, **68**, 68i, 69
Primate, **394**, 394i, 496
  chromosome structure, 129, 177
  evolution, 268i, 269, 394–395
  origins and early divergences, 395
  skeleton, 394i
  traits, 394–395
Primer, **225**, 226, 227
Priming pheromone, 556
Primitive streak, 444, 444i
Prion, 311, **312**, 312i
PRL, **451**
Probability, **154**–155
Probe, **224**, 232
Probe hybridization technique,
  224, 224i
■ Probiotic feed, 501
Procamelus, 275i
Producer, 6–8, 7i, 74i, 92
Product, 76, 77, 78i, **89t**
Profundal zone, 540, 540i
■ Progeria, 173t
Progesterone, **433**, 435, 436i
Progestin, 439
Proglottid, 361, 361i
Programmed life span hypothesis, 452
Prokarya, 280
Prokaryote, **8**, 292
  characteristics, 8i, 9i, 58i, 286,
    287, 332t
  origin, 268i, 296i–297i
Prokaryotic cell, **50**, 51, 51i, 52i, 86,
  124, 263i, 291, 292i, 294, 300,
  302–303, 302i, 302t, 357i
  components, 58–59, 58i, 70t, 204i
Prokaryotic fission, **303**, 303i,
  309i, 332t
Prokaryotic growth rate, 303
Prolactin, **451**
Proline, 45i, 198i, 202i
Promoter, **196**, 196i, **208**, 209i, 212, 214
■ Property rights, 549
Prophase, 124, 128–**130**, 130i, 141i,
  142i–143i, 144–145, 144i, 147
■ Prophylactic mastectomy, 207
Prosimian, 394, 395
Prostaglandin, 428, 435
■ Prostate cancer, 429
Prostate gland, 179, 428–429, 428i,
  429i, 430i
Prostate-specific antigen, 429
Protease, 81
Protein
  energy from, 118
  function, 6, 42–43, 61, 73, 78, 126, 159
  gene expression and, 186, 210, 270
  origin, 289, 290

structure, 33i, 42–45, 43i, 44i, 53i,
  290, 290i
  synthesis, 42i, 43, 77, 80, 128, 291
  types, 34, 48t, 49, 49i, 56, 56i, 57, 57i,
    59, 66, 194
Protein cap, and amoebas, 66
Protein coat, 308
Protein lattice, 59i
Protein synthesis, 248
  mutation during, 202–203, 202i, 203i
  transcription stage, 196–197,
    196i–197i, 204i
  translation stage, 198–201, 198i–199i,
    200i–201i
Proteobacteria, **305**
Proterozoic eon, 263, 263i, 292,
  293, 293i
Protist
  characteristics, 8, 8i, 68, 70t, 91, 92,
    138, 280, 280i, 314–323, 332t,
    356, 356i
  evolution, 263i, 282, 282i, 295, 295i,
    297i, 316i
  functions, 316, 319–322
  major groups, 9i, 17i, 314–323, 316i,
    317i, 318i, 319i, 321i, 322i, 323i
Protista, kingdom, 91, 316, 316i
Proto-cell, 263i, 286, 286i, 287, **290**,
  291, 291i
Proto-oncogene, **206**
Proton, 20, 22, **30t**, 307
Protonephridium, 360, 360i, 375, 375i
Protosome, 353i, 355, 356i, 375
Protozoan, **89i**, **316**, 317, 317i, 318
  current species diversity, 497i
Prunus (cherry tree), 95
Pryor, G., 161i
PSA, 429
Pseudocoel, **355**, 355i
Pseudocoelomate animal, 355i
Pseudomonas marginalis, 58i
Pseudopod, **67**, 88, 89i, 317, 317i
Psilophyton, 336, 336i, 340i
Psychoactive alkaloid, 327
Pterosaur, 266i
Puberty, **428**, 433
Pubescent, 451i
Public health, 195, 206, 207, 313
■ Public land, 287, 298
Pueraria montana, 492, 492i
Puffball mushroom, 327
Punctuation model of speciation, **278**
Punnett-square diagram, **155**, 155i,
  156i, 169i, 247i
Pupa, insect, 370
Pupal stage of moth, 6i
Purine, 188
Puriri tree, 266
Purple coral fungus, 326i
Pyrenestes ostrinus, 251, 251i
Pyrimidine, 188
Pyrococcus horikoshii, 313
Pyruvate, 76, **108**

Q

Quadrat, **459**
Quamman, D., 243i, 495
Quantitative reporting, 13
Quaternary structure, 90
■ Queen Victoria, 175i
Quercus, 255, 315, 328

R

Rabbit, 162, 162i, 493i, 493i
Racoon, 271i
Radial animal, 352i, 354, 372
Radial cleavage, 423, 423i
Radial symmetry, **354**, 354i, 358
Radiant energy, 94
■ Radiation. *See also* Adaptive radiation
  cosmic, 134, 288, 292
  infrared, 94i, 105, 292
■ ionizing, 203
  nonionizing, 203
  solar, 526i, 527i
  UV, 94i, 95i, 105, 293, 336
■ Radiation poisoning, 137, 202i
Radio wave, 94i

Radioactive decay, **21**
Radioactive fallout, 549, 549i
Radioisotope, **21**, **30t**, 187i, 262, 262i
Radiolarian, 317, 317i
Radiometric dating, 376–377
Radula, **364**, 364i, 365i
Radura, 313i
Rain forest, 499, 532i–533i, 536,
  536i, 539i
Rain shadow, **531**, 531i
Rainfall, 526–527
Rainwater, pH of, 29, 529
Rana pipiens, 421i
Randall, M., 230
Rangifer tarandus, 482i
Rat, 236–237, 237i
Rat liver cell, 60i
Rat somatotropin, 229
Rattlesnake, 388, 389i
Rattus, 236–237, 237i
Raven, 300
Ray, **372i**
Ray-finned fish, 383
Reactant, 76, 77, 78i, **89t**
Reaction chamber, 288i
Reaction rate, 81
Realized niche, 478
Rearrangement reaction, **35**, 79
Receptor, **7**, 29, 408i, 409i
Receptor-mediated endocytosis, 88
Receptor protein, 57, 57i, 59
Recessive allele, 151, 153–155, 154i,
  155i, 156i, 158, 159, 183, 213
Reciprocal cross, **183**
Recognition protein, 49, 49i, 57,
  57i, 135
Recombinant DNA, **220**–222, 222i,
  223i, 309i
Recombination, genetic, **168**, 168i, 171
Rectum, 366
■ Recycling, tax breaks for, 335, 350
Red alga, 105, 105i, 287, 292i, 293, 297i,
  307i, 316i, 319, **322**, 322i
Red blood cell, 4, 44, 87i, 128, 158,
  202i, 210, 317i, 319i
■ malaria parasite and, 318, 319i
Red fox, 504, 504i
Red grape, 95
Red-spotted salamander, 384i
Red tide, 320
Redox reaction, **76**–77, 77i
Redwood forest, 334
Redwood tree, 9i, 53i, 108i, 315, 320,
  334, 344
Regional climate change, 349, 531
Regulatory protein, 206i, **208**,
  211, 212
Reindeer, 463i
Reiskind, J., 161i
Relaxin, **450**
Release factor (protein), 201
Replacement level, 470
Replication enzyme, 222, 223
Repressor, **208**, 209i
Reproduction, **6**, 272, 291. *See also*
  Asexual reproduction; Sexual
  reproduction
  flowering plant, 126, 140i, 346, 346i,
    347, 357
  fungi, 327i, 328i
  human, 419i. *See also* Reproduction,
    animal
  patterns, 464–465
  plant, 338i, 341i, 345i
  prokaryotic, 303
  protist, 319i, 322i, 324i, 325i
Reproduction, animal, 416–454, 419i.
    See also Pregnancy
  aging and death, 452
  birth, 450–451
  cell differentiation, 424–425
  cleavage, 422–423, 442
  cleavage patterns, 423
  control of human fertility, 438–439
  cyclic changes in ovary and uterus,
    434–435
  egg, 422
  embryonic induction, 426

emergence of human features, 446–447
emergence of vertebrate body plan, 444
evolutionary constraints on development, 427
extraembryonic membranes, 442–443
fertilization, 437
formation of early embryo, 442–443
formation of specialized tissues and organs, 424–425
implantation, 442
menstrual cycle, 436
morphogenesis, 425
mother as provider, protector, potential threat, 448–449
nourishing newborns, 419, 419i, 451
pattern formation, 426–427
placenta, 445
postnatal development, 451
sexual intercourse, 437
■ sexually transmitted diseases, 440–441
sperm formation, 430–431
stages of, 420–421, 420i, 421i
Reproductive age, 458
Reproductive base, **458**
Reproductive isolating mechanism, **272–273**, 272i, 273i, 274
Reproductive isolation, 272–273
Reproductive organ, 140, 140i, 360i, 365i, 366
human, 169i
Reproductive success, 452, 464–465, **555**
Reproductive system
human females, 432–433, 432i, 432t
human females, components, 432, 433i
human females, menstrual cycle, 432–433
human males, 428–429, 428i, 428t
■ human males, cancers of prostate and testes, 429
human males, components, 429i
human males, gonad development, 428
human males, structure and function, 428–429
Reproductive timing, 418
Reptile, 381i, **386**, 386i, 388–389
birds and, 376–377, 376i
body plan, 388, 388i
characteristics, 354, 354t
origin, 263i
reproduction, 139
Resource consumption, 472–473, 567
Resource partitioning, **481**, 481i
Resources, allocating when scarce, 73, 91, 137
Respiration, 405, 405i
■ Respiratory acidosis, 29
■ Respiratory disorder, 329, 529i
Respiratory organ, invertebrate, 364, 365, 367
Responsiveness to change, 7
Resting spore, 304i, 324, 336
Resting structure, 304
Restriction enzyme, **222**, 222i, 223
Restriction fragment length polymorphism, 226
Resurrection plant, 340
Retinol, 105
Retrovirus, 309
Reverse transcriptase, **223**
Reverse transcription, 309
Reznick, D., 466–467, 466i
RFLP, 226
Rhea, 238i
Rhesus monkey, 234, 234i, 270
Rhinoceros, 497i
Rhizobium, 305
Rhizoid, 328i, 338, **338**
Rhizome, 340, 341, 341i, 342
Rhizophora, 542i
Rhizopus stolonifer, 328, 328i
Rhubarb, 95

Rhyniophyte, 336, 337i
■ Rhythm method, 438, 438i
Rhythmic leaf folding, 411
Ribeiroia, 385i
Ribonucleic acid. *See* RNA
Ribonucleotide, 196, 199
Ribonucleotide monomer, 47
Ribose, 38, 75, 75i, 196
Ribosomal RNA, **196**, 200, 207, 270, 291, 306, 307
Ribosome, **52**, 59, 60, 61, 291
function, 62i, 70t, 194, 195i, 196, 198, 200–201
structure, 199, 199i
Ribulose biphosphate, 101, 101i
Rice plant, 229, 348
Ricin, 194–195, 195i, 485
Ricinus communis, 194–195, 194i, 195i
Riffle, 541, 541i
Riparian zone, **499**, 499i
Ritualization, 556–557
River, 512i
River rock, and bacteria, 59
Rivulus hartii, 466–467, 466i, 467i, 475
RNA, **47**, 48t, 62i, 64, 70t, 78, 89t, 208–209, 211
synthesis, 195–199, 196i–199i
RNA polymerase, **196**–197, 196i–197i, 208, 209, 211
RNA virus, 309
RNA world, **291**, 309
Rock pocket mouse, 249, 249i
Rocky coastline, 542i, 543
■ Rocky Mountain spotted fever, 369
Rodenticide, 236–237
Rodhocetus, 257, 257i
Rodinia, 293
Roe, **373**
Rooster, 566, 566i
Root, function, 336, 341
Rose plant, 329
Rotational cleavage, 423, 423i
Rotifer, 375, 375i
Rough ER, 62, 62i, 65i
Roundworm, 146, 270, 354t, 355, 356i, **366**, 366i, 452, 486i
current species diversity, 497i
rRNA, **196**, 200, 207, 270, 291, 306, 307
■ RU 486, 439
Rubber cup fungus, 326i
■ Rubella, 448–449, 454
Rubisco (RuBP), **101**, 101i, 102, 102i, 104i, 105
Rufous hummingbird, 480
Run, 541, 541i
Russian imperial family, 226
Rust, fungal, 327
Rye plant, 329, 329i, 348

S

Sac, 60
Sac fungus, 326, 326i, **328**–329, 329i
Saccharide, **38**
Saccharomyces cerevisiae, 116, 116i, 329
Saccharum officinarum, 38, 343i, 348
Sacred lotus, 348i
■ Safe sex, 137, 441
Sage grouse, 558–559, 558i
Sage plant, 273, 480
Sagittaria sagittifolia, 272i
Saguaro cactus, i, 475, 475i, 497i, 534i
Sahelanthropus tchadensis, 396
St. Matthew Island, 463i
Salamander, 418i, 385
Salinity, and enzyme activity, 81
Salinization, **513**
Salivary gland, 211i
Salmonella, 312
Salp, 287
Salt, **29**, **30t**
Salt marsh, 542i
Saltwater intrusion, 513i
Saltwater marsh, 502, 502i
Salvia, 273, 480
Sampling error, **16**, 16i, 154, 163, 254
San Pedro River, 499i

Sand dollar, 372
Sand flea, 316
Sandfly, 89i
Sandstone, 261
Sandy coastline, 542i, 543
Sandy soil, 539
Sanger, F., 220
Sap flow, 86
Saprobe, 302, **326**
Saprolegnia, 320
Sarcoplasmic reticulum, 62
Sarcosoma, 326i
Sargasso Sea, 321
Sargassum, 321
■ SARS, 2, 206, 310, 310i, 313
Saturated fatty acid, **40**, 40i, 56
Saurophaganax, 387i
Sauropsid, 386
Savanna, 532i–533i, 535, 535i
Sawfly caterpillar, 560, 560i
Scallop, 364, 364i
Scanning electron microscope, 55
Scarab beetle, 371i
Scarcoscypha coccinea, 329i
Scarlet cup fungus, 329i
Scarlet hood, 326i
Scarus coelestinus, 8
Scarus gibbus, 8
Scavenger, 9, 504
invertebrate, 360, 362, 363i
Scenedesmus, 55i
Schiestl, F., 252
■ Schistosomiasis, 361
Schistosoma japonicum, 361i
■ Schizophrenia, 166
Schoener, A., 491
School of fish, 5, 5i, 458i
Schwann, T., 54
Schwarzenegger, A., 460
■ SCID-X1 patient, 230, 230i
Scientific approach, 11, 151
Scientific inquiry, steps in, 11
Scientific method, 483i
Scientific testing, **11**
Scientific theory, **11**, 15, 377
Scolex, 360, 361, 361i
Scorpion, 369, 369i
Scorpionfish, 485, 485i
■ Scrapie, 312
Scrotum, 428, 428i
■ Scrub typhus, 369
Sea anemone, 358, 358i, 359, 479, 479i
Sea biscuit, 372
Sea cucumber, 354t, 372, 372i, 373
Sea fan, 544, 544i
Sea lamprey, 491t
Sea lettuce, 323, 323i, 542i
Sea level, 502, 517, 530
Sea lily, 372
Sea lion, 524–525
Sea slug, 357
Sea squirt, 378, 379i, 401, 401i
Sea star, 138, 242, 269, 269i, 354t, 364, 372, 372i, 373, 373i, 462, 465, 465i, 490–491
Sea temperature, 544, 544i, 546, 547, 547i
Sea turtle, 320, 388
Sea urchin, 128, 354t, 372, 372i, 373, 419, 423, 423i, 542i
Sea volume, 502
Sea wasp, 358i
Seabird, 93
Seafloor methane cycle, 36i
Seal, 418i, 524–525, 525i
Seasonal change, effects of, 146
Seawater, pH of, 28, 28i
Seawater intrusion, 513i
Second law of thermodynamics, **74**
Secondary oocyte, 434i, **435**, 437
Secondary sexual trait, 428, 431
Secondary succession, **488**, 489i
Secondary wall, plant cell, **68**, 68i
■ Secondhand smoke, 449
Secretory pathway, 60, 63i
Sedge, 251
Sedimentary cycle, 511, 520, 520i

Sedimentary rock, 239, 241, 258, 258i, 260, 261, 261i, 262, 278
Sedimentation, 261, 342, 349
Seed, **337**, **343**
gymnosperm, 337, 343–345, 344i, 345i
Seed coat, 42, 347i
Seed fern, 342i, 343
Seedling, 345i, 347i
Segal, E., 15
Segmentation, 354, **355**, 362, 362i, 368, 375i
fused, 367, 370
Segregation theory, 154–155, 154i
Selaginella lepidophylla, 340
Selection pressure, 236–237, 240, 252, 293, 356
oxygen as, 51, 287
Selective advantage, 139, 144, 189
Selective gene expression, 212, **425**
Selective permeability, **82**, 82i, 86
Selective transport, 86
Self-sacrificing behavior, 555, 562–563
Selfish behavior, **555**, 562–563
Selfish herd, **560**
Semen, 428
Semiconservative replication, DNA, **190**, 190i, 191i, 193
Seminal vesicle, 428, 428i, 429i, 430i
Seminiferous tubule, 428, 430i, 431i
Sensory cell, 354
Sensory receptor, **408**
Sensory structure, invertebrate, 367
Sepal, 346i
Sephton, M., 261i
Septum, 564
Sequencing, DNA, 220, 224–227, 225i, 227i, 286, 307
Serine, 198i
Serotonin, 163
Sertoli cell, 430, 431, 431i
Seta, 362, 363i
■ Severe Acute Respiratory Syndrome, 2, 206, 310, 310i, 313
Sex chromosome, 129, 129i, 166, **168**, 171
gene recombination, 168, 179
human, 169, 173t, 175, 179, 179i
Sex determination, human, 169, 169i
Sex hormone, 35, 41, 169, 433
Sex-linked inheritance, 166
Sex pheromone, 252
Sex pilus, 302, 302i
Sexual behavior, 252, 252i
Sexual competition, 559, 559i
Sexual dimorphism, **252**
Sexual imprinting, 566, 566i
Sexual intercourse, 437
Sexual reproduction, 138–139, 138i, **140**, **418**–419
animal, 354
asexual *vs.*, 138–140, 138i, 140i, 418
costs/benefits, 346, 418–419, 418i
eukaryotic, 126
fungal, 326, 328
invertebrate, 357, 359, 367, 368
plant, 338–340, 338i, 341i, 343, 345i, 347i
protist, 293, 318, 321, 324i
reflections on, 417–418
Sexual selection, 235, 235i, **252**, 401, **558**, 558i, 559i
mating behavior and, 558–559
■ Sexually transmitted disease (STD), 305, 316, **440**–441, 440i, 440t
Shale, 261
Shape
in enzyme function, 79, 80, 81
jawed fish, 382
in membrane crossing, 84, 84i, 85, 85i
Shark, 281, 281i, 382, 382i
reproduction, 419i
Sheep, 312
cloned, 184–185, 184i, 192, 192i
Shelf fungus, 327
Shell, 314, 317, 317i, 321, 321i, 364, 364i, 365, 365i, 388, 389i

Shell model of atom, 20i, 22, 23i
Shellfish poisoning, 320
Shifting cultivation, **349**
"Shipping label," of amino acid, 201
Shoot system, 336, 404i
Short-chain carbohydrate, 38, 48t
Shortgrass prairie, 534–535, 535i, 545, 545i
Shoulla, R., 206–207
Showshoe hare, 478
Shrimp, 368
Shrubland, 532i–533i, 534–535, 535i
Siamang, 394
Siamese cat, 81i
Siamese twins, 426, 426i
Sibling species, **273**
Sickle-cell anemia, 44–45, 45i, 159, 165, 173t, 181, 202, 252–253, 253i
Sierra Nevada, 531, 531i
Sight. See Vision
Sign stimuli, 553
Signal reception, 412–413
Signal response, 412–413
Signal transduction, 412–413
Signaling mechanism, 407
Signaling molecule, 89i, 208, 325, 352
Signaling pheromone, 556
Silent Spring (Carson), 508
Silkworm moth, 6i, 556
Sillén-Tullberg, B., 560
Silurian, 337i
Silver Springs, Florida, 509i, 510, 510i
Silverfish, 370i
Singh, C., 312, 312i
Sinus infection, 150
Siphon, 352i, 365, 365i
Sister chromatid, **126–127**, 126i
Skeletal muscle, 64, 117, 117i
Skeletal system, 444, 444i
Skeleton, 60
    cytoskeleton, 60, 65i, **66**, 67, 70t, 316, 317
    hydrostatic, 362
Skin cancer, 134–135, 135i, 528
Skin cell, 413
Skin tumor, 206
Skunk, 484
Slime mold, 316i, **324**–325, 325i
Slime stalk, 305
Slime streamer, 81i
Slow-twitch muscle fiber, 117
Slug, 354t, 364
"Slug," of slime mold, 325, 325i
Small intestine, 72, 81
Small organic compounds, families, 35
Smartweed, 481, 481i
Smith, H., 222
Smog, 528–529, 528i
Smoking, 206
    pregnancy and, 449
    secondhand smoke, 449
Smoky Mountain National Park, 519i
Smooth ER, 62, 62i, 63i, 65i
Smooth muscle, 159
Smut, fungal, 327
Snail, 246i, 354t, 364. See also individual types
    reproduction, 419i
Snake, 387, 388, 419i, 552. See also individual types
Snake bite, 388
Snapdragon plant, 158, 158i
Snow alga, 286, 333, 333i
Snowshoe hare, 482–483, 483i
Snowy egret, 465i
Sociable weaver, 251, 251i
Social attachments, between humans, 564
Social behavior, **555**
Social groups
    cooperative hunting, 560–561
    cooperative predator avoidance, 560
    costs and benefits, 560–561
    dominance hierarchies, 561
    selfish herd, 560
Social insect, 562, 562i
Social parasite, 486, 553, 553i

Sodium, 23i, 24, 24i, 29
Sodium chloride, 24, 24i, 27, 29
Sodium hydroxide, 29
Sodium ion, 24, 27, 82i, 83, 85, 409
Sodium-potassium pump, **85**
Soil, 512i, **539**
Soil acidity, 163
Soil formation, and lichens, 330
Soil fungi, 330
Soil profile, 539, 539i
Solanum cheesmanii, 244, 244i
S. lycopersicon, 404i
S. lycopersicum, 244, 244i
Solar energy, 527, 527i
Solar-hydrogen energy, 527
Solar radiation, 526i, 527i
Solar system, 14, 258, 288
Solenopsis invicta, 476–477
S. richteri, 476
Solute, **27**, **30t**, 86–87
Solute concentration, 86, 86i, 87
Solvent, **27**
Somatic cell, **126**, 128, 203, 366
Somite, **444**, 444i
Song of the Dodo (Quammen), 495
Songbird, 499, 554, 554i
Sorus, 341, 341i
South Carolina, 542i
Soybean plant, 231
Space-filling model, 36i
Spaceguard Survey, 285
Spanish flu epidemic, 3, 231
Spanish shawl nudibranch, 364i, 365
Spartina, 542i
Speciation, **272**
    allopatric, 274–275, 274i, 275i
    models for, 274–279, 275i
    parapatric, 277
    patterns, 278–279
    sympatric, 276–277
Species, 8, 238, **272**, 282i
Species interaction, 476–477, 478i
    categories of, 478
    competitive interactions, 480–481
    mutualism, 479, 479i
    parasite-host interactions, 486–487
    predatory-prey interactions, 482–483
Species-introduction, 491, 491t, 496–497
Species loss, 536i
Spectacled caiman, 389i
Spectral karyotype, 170, 170i, 177i, 178i
Speech, 179
Speed, as evolutionary factor, 365
Sperm, **422**, 428–429, 429i, 431i, 437, 437i
    animal, **146**, 147, 147i
    formation of, 430–431, 430i–431i
    human, 54, 123i, 145, 146, 168–169, 169i
    inheritance and, 152
    invertebrate, 357, 359i, 361, 364
    low counts, 179
    plant, 153–155, 338i, 339, 341i, 343, 345, 345i, 347i
    structure, 67, 71i
Sperm count, 179
Sperm production, 428
Spermatid, 146, 430, 431i
Spermatocyte, 430, 431i
Spermatogenesis, 430
Spermatogonia, 430
Spermicidal foam, 438i, 439
Spermicidal jelly, 438i, 439
Sphaerodactylus ariasae, 388, 388i
Sphagnum, 339, 339i
Sphenodon, 389i
Sphenophyte, 341
Sphere of hydration, 27, 27i
Spicule, 373
Spider, 354t, 369, 369i
    reproduction, 446i
Spider monkey, 394i
Spider silk protein, 229
Spina bifida, 444
Spindle, microtubular, 124, 135, 143i, 145, 148i, 170

Spindle equator, 130i, 132i, 133i, 142i–143i
Spine, echinoderm, 372i, 373
Spiny anteater, 393, 393i
Spiny cactus, 238i
Spiny spurge, 238i
Spirillum shape, 302i
Spirogyra, 95i
Spirulina, 304
Spleen, 361
Sponge, 352i, 354, 354t, 356, 356i, **357**, 357i, 418
Spongy mesophyll, 103i
Sporadic disease, **310**
Sporangia, 336
Spore, **304**, **326**, **336**
    bacterial, 304i, 305
    fungal, 315i, 326–329, 327i, 328i, 329i
    meiosis and, 138i, 336, 338i, 341i, 347i
    plant, **146**, 336, 337, 338i, 339, 341, 341i, 343, 348i
    protist, 293, 315i, 321, 322i, 324, 324i, 325, 325i
    types, **304**, **343**
Spore-bearing structure, 305, 324, 325i, 326, 326i, 327, 328
Spore sac, 328i
Sporophyte, 146, 146i
    plant, **334i**, **336**, 338–341, 338i, 339i, 341i, 343, 344, 345, 345i, 347, 347i
    protist, **321**, 322i
Sporozoite, 318, 319i
Spriggina, 356i
Spring overturn, **540**
Spruce tree, 344, 455i
Squamous cell carcinoma, 135i
Squid, 214, 354t, 365, 365i
SRY gene, 169
Stabilizing selection, **250**, 251
Stahl, F., 193
Stamen, 152i, 346i
Staphylococcus aureus, 312
Star, creation of, 288
Star anise, 348i
Starch, 35, 38–39, 38i, 48t, 64, 77, 101
Starling, 491t, 555, 555i
STD. See Sexually transmitted disease
Stearic acid, 40i
Stegosaurus, 387i
Steinbeck, John, 535
Stem cell, **230**
Sterility, 562–563, 562i
Sterility/self-sacrifice, genes of nonreproducers, 562–563
    indirect selection for altruism, 563
    social insects, 562
    social mole rats, 562
Stern, Kathleen, 564
Steroid, cholesterol in, 41
Sterol, 41, 41i, 48t
Sterol backbone, 41, 41i
Sticky end, of DNA fragment, 222, 223i
Stimulus, **552**
    energy as, 7
Stingray, 382i
Stink bug, 371i, 487
Stinkhorn, 327
Stipe, 321, 321i
Stolon, 328i
Stoma, 96i, **102**, 103i, 336, 350t, 404
Stomach, 81
    absorption of ethanol, 72
    acid in, 28
    digestion, 443
    invertebrate, 364i, 373, 375i
Stomach lining, 69
Stonewort, 323, 323i
Straight-chain structure, 38i
Stramenopile, 315, 316i, 320
Strangleweed, 486i
Stratification, **261**
Stream ecosystem, **541**, 541i
Streptoccus pneumoniae, 186, 186i
Streptomycin, 248
Stress, 163

Stress hormone, 250
Stringer, K., 402–403, 409i, 414
Strip logging, **499**, 499i
Strobilus, **340**–345, 340i, 344i, 345i
Stroke, 49
Stroma, 64, 96, 97i, 98, 99i, 101
Stromatolite, **292**, 293i
Strong acid, 28
Strong base, 28, 29
Structural formula, 24–25
Structural genomics, 232
Structural organization
    evolutionary history of, 404–405
    levels of, 404–405
Sturnus vulgaris, 555
Subjective vs. objective viewpoint, 14
Substrate, 77, **78**–81, 79i, 80i, 89t, 91
Substrate-level phosphorylation, **110**
Successful organism, 370
Succession model, 488
Succulent, 102
Sucker, annelid, 362
Sudden oak death, 315, 320, 320i
Sugar
    structure, 34, 75, 75i, 92, 93, 101, 188, 289
Sugar beet, 38
Sugarcane, 38, 343i, 348
Suicidal cell, 412–413, 412i, 413i
Sula granti, 566, 566i
Sulfate-eating bacteria, 36i
Sulfolobus, 307
Sulfur, 187i
    in ATP-forming reaction, 307
Sulfur cycle, in seas, 305
Sulfur dioxide, 29, 529
Sulfuric acid, 29, 364
Sun, 526, 526i
    age, 288
    exposure to, 203
    origin of life and, 288, 289, 290
Sunburn, 309
Sunfish, 560
Sunflower plant, 102
Sunlight energy, 6, 7, 74, 74i, 77i, 92–95, 92i, 97i, 98, 104i, 302, 307
Supercoiling, DNA, 127i
Supercontinent, 263i, 264, 265, 293
Superoxide dismutase, 90, 90i
Superplume, 264i
Supernatural vs. science, 12, 14, 32–33
Surface runoff, 512, 513i
Surface-to-volume ratio, 52–53, 53i, 326, **406**
Surfing, 524, 524i
Surtsey, 494i, 495
Surveillance protein, 134
Survival pattern, 464–465
Survivorship curve, **465**, 465i
Sushi, 322, 373
Suspended particulate matter, 529i
Sustainable development, 498–499
Suttle, C., 313
Swallowtail butterfly, 371i
Swamp forest, 251, 263i, 334, 337, 340–342, 340i
Sweat, 27, 150, 564
Sweat gland, 213, 213i, 402i
Swim bladder, 382
Swine flue virus, 231
Symbiont, **294**, 295, 319, 544
    animal, 306, 357, 359
    fungal, 304, 314, 323, 326, 330–331, 336
    plant, 305, 328
Symbiosis, **330**, 478
Sympatric speciation, **276**–277
Synapse, **168**
Synapsid, 386, 392
Syndrome, **172**
Synthetic fat replacement, 12
Syphilis, 440i, 440t, 441

T

T cell, 452
T lymphocyte, 229, 230
Table salt, 24, 24i, 27, 27i, 29

Table sugar, 38, 343i
Tactile display, 557, 557i, 562, 562i
Tadpole, 385, 385i, 421i
Taenia saginata, 361i
Taigas, 536–537, 537i
Tail, chordates, 378, 378i
Tall fescue, 330
Tallgrass prairie ecosystem, 496, 505, 505i, 506i, 534–535
Tamias, 480, 480i
Tandem repeat, **226**
Tanzania, 396, 397
Tapeworm, 354t, 360
Taproot, 344
Taq polymerase, 225i
Tarantula, 369i
Tarsier, 394i
Tarsioid, 394
Tasmanipatus anophthalmus, 277i
T. barretti, 277i
Tawuia, 292i
Taxa
  derived traits and, 281
  higher, **280**, 280i
Taxol, 66, 135
Taxonomy, numerical, **303**
Taxus brevifolia, 66
T. brevifolius, 137
Tea plant, 343i
  bacteria and, 59
  in mammals, 392, 392i
  primates, 395
Telesto, 374i
Telomerase, 452
Telomere, **185**, **452**
Telophase, 124, 128i, 129, **131**, 131i, 132, 141i, 142i–143i, 148i
Temnodontosaurus, 387i
Temperate coniferous forest, 532i–533i
Temperate deciduous forest, 532i–533i, 537i
Temperate forest, 349
Temperate grassland, 532i–533i
Temperature, enzyme activity and, **26**, **81**, 81i, 83
Temperature zones, 526, **527**, 527i
Template (DNA), 223, 223i, 225, 227
Temple of Apollo, 32i
Temporal isolation, 272i, **273**, 273i
Tentacle, 358i, 359, 364, 365i
Teratogen sensitivity, 448i
Termination, RNA translation, 200, 201i
Termite, 305, 316, 370, 557, 562, 563, 563i
Test, **11**
■ Test-tube baby, 181
Testcross, **155**
Testes, **428**, 428i, 429i, 430i, 431i, 566
  human, 140i, 169, 169i, 179
  invertebrate, 360i
  sperm formation, 430
■ Testicular cancer, 429
Testosterone, 35, 35i, 169, 179, **430–431**
■ Tetanus, 304, 310t
Tetany, 29
Tetrapod, 382–**383**, 383i
  amphibians, 384–385
Texas horned lizard, 476
Textularia, 315i
■ Thalassemias, 202i
■ Thalidomide, 449
Therapsid, 392–393
Thermal energy, 82
Thermal inversion, **528**, 528i
Thermocline, 540–541, 540i
Thermodynamics laws, **74**, 74i
Thermogenesis, 122
Thermophile, **286**, 297i, 313
  extreme, **306**, 307, 307i
Thermotoga maritima, 313
Thermus aquaticus, 225, 225i, 286, 286i, 307
Thiomargarita namibiensis, 302, 305, 305i
Third eye, 388
Thomson's gazelle, 485
Thoracic cavity/segment, 368

Thorax, 369, 370
Thorium, 262
Threat display, 556–557
Threonine, 45i, 198i, 202i
Thumb, 395
Thylakoid, 64, 64i, **96**
Thymine (T), 46, 46i, 47, 184i, 188, 188i, 189, 189i, 196, 196i, 202, 203, 227, 227i, 291
Thymine dimer, 203
■ Thyroid abnormalities, 549i
Thyroid gland, 19, 269
Ti plasmid, 228i, 229
Tick, 39i, 305, 305i, 369
Tidal flat, 50
  early life and, 289–291, 290i
Tight junction, 69, 69i
Tigris River, 513
Tilia americana, 102i, 103i
Timothy grass cell, 60i
Tissue, **4i**, **404**
Tissue layer, in embryos, 354–355, 358, 360
Tissue specialization, 420, 420i
Toad, 384i, 385
Tobacco. *See* Smoking
Tobacco mosaic virus, 308i
Tomato plant, 228, 244, 244i
  morphology of, 404i
Tonicity, **86–87**, 87i
Tools, 397–399, 397i
Topography, 531
Topsoil, 349, 539
Torres del Paine National Park, 540i
Total fertility rate (TFR), **470**
Toxic compound, 90
Toxic species, eukaryote, 319, 320
Toxicity
  alcohol and, 72
  predation and, 364
Toxin, 194–195, 195i, 358
■ Toxin attack, on body, 72
Toxin diversity/specificity, 352
TPA protein, 229
Trace element, 18
Trace fossil, 260
Tracer, **21**, **30t**
Trachea, 367
Tracheid, 194–195, 195i, 358
■ Trachoma, 305
Trade war, 231
Trait, **10**, 171, 244, 250, 272
  heritable, 151, 152, 174–175, 174i, 175i, 251
  types, 166–167, 169, 171, 246, 281, 353
  variation in, 45, 139, 140, 140i, 144–145, 147, 151, 160, 202, 242, 353
Trans Fatty acid, 49, 49i
Transcription
  DNA, 208–212, 209i, 210i, gene, 194i, 196–197, 196i–197i, 204i, 309
Transcription control site, 212i
Transfer RNA, **196**, 199, 199i, 200, 200i
Transgenic animal, 229, 229i, 231, 234, 285, 285i
Transgenic organism, **228**
Transgenic plant, 228, 228i
Transition state, **79**
Transitional stage, 472, 472i
Translation of RNA, 195i, 204i, 208, 209i, 210i, 211
Translocation
  chromosomal, **177**, 177i
Transmission electron microscope (TEM), 55, 55i
Transport protein, 59, **76**, 82i, 83, 84, 84i, **89t**, 150, 163, 211
Transposon, **202**, 203i, 206, 234, 269
Transverse fission, **360**
Tree
  defenses, 410, 410i
  fungi and, 331
Tree fern forest, 340i
Tree shrew, 395, 395i

Treponema pallidum, 441
Triassic period, 263i, 337i, 392
Trichinella spiralis, 366
T. spiralus, 366i
Trichomonad, 316
Trichomonas, 316i
T. vaginalis, 316, 440
■ Trichomoniasis, 440, 440t
■ Trichonosis, 366
Trichoplax adhaerens, 356, 356i
Triglyceride, **118**
Trilobite, 365i, 367
Trinidad, 466–467
Triple covalent bond, **25**, 25i
Trisomy 21, 178–179, 179i
Triticum, 343i
T. aestivum, 277i
T. monococcum, 277i
T. tauschii, 277i
T. turgidum, 277i
tRNA, **196**, 199, 199i, 200, 200i
Trophic levels, **505**
Trophoblast, 442, 442i
Tropical broadleaf forest, 532i–533i
Tropical community, 494
Tropical coniferous forest, 532i–533i
Tropical dry forest, 532i–533i
Tropical forest, 349, 349i, 533i
Tropical rain forest, 499, 536, 536i
Tropical rain forest soil, 539i
Tropics, 496
Troponin-1, 211
Troposphere, 526i
Trout, 382
Troyer, V., 174i
True bug, 370i
True tissue, 356i
Truffle, 328–330, 329i
Trumpet chanterelle, 326i
Trypanosoma, 316i
T. brucei, 316, 317i
T. cruzi, 316
Trypanosome, 316
Trypsin, 81
Tryptophan, 42i, 80–81, 80i, 198i, 211
Tsetse fly, 317
Tuataras, 387, 388, 389i
■ Tubal ligation, 438i, 439
■ Tubal pregnancy, 440i
Tube foot, 372–373, 372i
Tube worm, 544, 545i
Tuber melanosporum, 329i
■ Tuberculosis, 248, 310, 310i, 310t
Tubulin subunit, 129
■ Tularemia, 369
Tumor formation, 125
Tumor supressor, 134, 207
Tundra, 307, 346, 532, 532i–533i, 538, 538i
Tungara frog, 557
Tunicate, 354t, 378, 379i
Turbellarian, 354t, 360
Turgor pressure, **87**
■ Turner syndrome, 173t, 179, 179i
Turtle, 270, 387, 388, 389i
Twain, M., 385
Twins, 423, 423i
Type I curve, 465
Type II curve, 465, 465i
Type III curve, 465, 465i
Typhus, 236
Tyrosinase, 81i
Tyrosine, 198i

U
Udotea cyathiformis, 323, 323i
Ultraplankton, 544
Ultraviolet light, 526i
Ultraviolet radiation, 94i, 95i, 105, 293, 526i
Ulva, 323, 323i, 542i
Umbilical cord, 445, 445i, 446i, 450, 450i
Umbrella thorn tree, 414
Unbranched evolution, 278
Undulating membrane, 317i
Uniformity theory, 241, **264**

U.S.
  immigration, 472, 473i
  population, 470i, 471i
  population and workforce, 473
  resource consumption, 472–473
Unity of life, 2i, 6, 8, 9, 14t, 187, 189, 283t
University of Tokyo, 139
University of Utah, 353
■ Unsafe sex, 73, 137
Unsaturated fatty acid, **40**, 40i, 56
Upright walking. *See* Bipedalism
Upwelling, **544–545**, 545i
Uracil (U), 4, 196, 196i, 199, 291
Uranium, 21, 262
Urethra, 428, 429i, 430i, 432
Uric acid, 370
Urine, 174
Urochordate, 378
Ursa maritimus, 26i, 244–245, 244i, 271i
Useless body parts, 239, 239i
Usnea, 330i
Uterus, 409, **432**, 432t, 433i, 451
  human, 137, 169i
  invertebrate, 366i
UV light, 526i

V
■ Vaccine, 125, 186, 318
Vacuole, 305i, 316
Vagina, 59, 169i, 248, 302, 304, 432, 432t, 433i
Vaginal fluid, 429
■ Vaginitus, 304
Valine (val), 44, 45i, 198i, 202i
Valve, 364
van Helmont, J. B., 105
van Leeuwenhoek, A., 54
Vanilla planifolia, 280i
Varanus komodoensis, 388
Variable, **12**
Variation
  conditions for, 246–247, 247i, 252–255
  continuous, **160–161**, 160i, 161i
  in traits, 45, 139, 140, 140i, 144–145, 147, 151, 157, 158, 160–161, 160i, 161i, 192, 202, 242
Vas deferens, 428, 428i, 429i, 430i
Vascular plant, **337**
  diversity in, 337, 338, 341, 344i, 348, 348i
  evolution, 263i, 282, 337i, 343
  growth/development, 338i, 341i, 345i, 347i
  seed-bearing, 337, 337i, 349, 350t
  seedless, 337, 337i, 340–341, 350t
Vascular tissue, 27, 42, 336, 406
■ Vasectomy, 438i, 439
Vasopressin. *See* ADH
■ vCJD, 312, 312i
Vector
  in genetic research, 232, 253i
  insect, 27, 42, 336
■ Vegetable oil, 40, 49
■ Vehicle exhaust, 528–529
Vein
  human, 407i
  leaf, 96i, 103i
Velvet walking worm, 269i, 277i
Venom, 369
Venter, C., 221, 234, 298
Ventral surface, 354, 354i
Vertebrae, **381**
Vertebrate, **354**, 376–399
  amniotes, 386–387
  amphibians, 384–385
  birds, 390–391
  characteristics, 40
  chordates and, 378–379, 381i
  current species diversity, 497i
  early primates to hominids, 394–395
  embryo development, 35
  emergence of body plan, 444, 444i
  emergence of early humans, 396–397
  emergence of modern humans, 398–399

evolution, 263i, 266–267, 266i, 353, 354t, 380–381
existing reptilian groups, 388–389
jawed fishes and tetrapods, 382–383
key innovations, 381
major groups, 381, 381i
mammals, 392–393
skeleton, 69i
Vesicle, 60, 62, 62i, 63, 65i, 83i, 87, 88, 88i, 130, 132i, 133, 318
Vestigial structure, 285
Vibrio cholerae, 547, 547i
Vicuna, 274, 274i
Viola, 348i
■ Viral disease, treatment, 206
Viral DNA, 184i, 186, 222, 309i, 313
Viral enzyme, 223, 309
■ Viral infection, 308–309
Viral RNA, 309
Virchow, R., 54
Virginia Tech Swine Research, 231i
Viroid, **309**
■ Virus, 125. *See also* specific types
characteristics, 34, 301, 301i, **308**, 308i
defenses against, 88
in gene therapy, 232
multiplication cycle, 303, 303i, 308–309, 309i
■ Virus particle, synthetic, 187i
Visible light, 95i
Vision, daytime vision, 394, 395
Visual pigment, 105
Vitamin A, 105
■ Vitamin A deficiency, 218
Vitamin C, 38
Vitamin D
cholesterol in, 41, 48t
Vitamin E, 80
Vitamin K, 305
Vitamins, in coenzyme, 80, 88
Volcanic ash, 260, 261
Volcanic island, 494i, 495, 495i
Volcano, 489i
Vole, 552–553, 552i
Volvex, 323, 323i
Vomeronasal organ (VNO), 564

W
Walking. *See* Bipedalism
Wallace, A., 238, 243i, 246
Wandering albatross, 495i
■ War on drugs, 333
Warfarin, 236–237
Warm grassland, 532i–533i
Warning coloration, **484**
Wasp, 252, 272i, 358i, 484, 485i
■ Waste disposal, 333
Waste removal, 359
invertebrate, 370, 375, 375i
Wastewater, 81i
Water. *See also* Water molecule
cell metabolism and, 63, 64
early Earth and, 288–290
hydrologic cycle, 512–513
uptake in plants, 69
Water conservation, plants, 410–411
Water crisis, 512–513
Water flea, 480
Water hyacinth, 491t
Water lily, 348
Water loss, plants, 102
Water mold, 315, 320
Water molecule
diffusion in cells, 52, 68, 82, 82i, 86–87, 86i, 87i
in photosynthesis, 75, 92, 96, 97i, 98, 100i
polarity, 25, 25i, 26, 26i, 56, 82, 82i
properties, 25–27, 27i, 56
structure, 23, 25, 25i, 26i
Water rights, 513
Water-solute balance, 406–407
Water-vascular system, 372, 373i
Waterlily, 406, 407i
Watershed, 349, **512**
Watson, J., 184i, 187, 189, 189i, 190, 220, 221
Wavelength, **54**, 55, **94**, 94i
Wax, **41**, 41i, 48t, 69, 367
Weak acid, 28, 29
Weaver bird, 462i
Welwitschia mirabilis, 344, 344i
■ Werner's syndrome, 452
■ West Nile encephalitis, 300–301

■ West Nile virus, 375, 503
Western juniper, 280i
Western yew, 66
Wet acid deposition, 529
Wetland, 496, 502–503, 542, 549
Wexler, N., 172i
Whale, 239, 257, 496, 497
Wheat plant, 228, 229, 277i, 329, 343i, 348, 535
White blood cell, 88, 91i, 177i, 230, 361, 406
White rust, 320
■ Whooping cough, 310
Wild cat, 496
Wild rock dove, 10i
Wild-type allele, 168, 183, 254
Wildlife trading, 497, 497i
Wilkins, M., 188, 189, 190
Wilmut, I., 184i
Wilson, E. O., 496
Wind, and acid rain, 529
Wind energy, 527, 527i
Wind farm, 527
Wind fence, 528
Wind power, 527
Windpipe. *See* Trachea
Wing, 279
bat, 267, 267i
bird, 267, 267i
butterfly, 246, 247i, 249i, 250, 267
insect, 267, 267i, 370
Witchhunt, and ergotism, 329
■ Withdrawal, 438, 438i
Wobble effect, in RNA translation, 199
Woese, C., 306, 307
Wolf, 482i, 556, 556i, 561, 561i
Wolf spider, 458, 458i
Wood duck, 35i
Woodland, 532i–533i, 534–535
Woodwardia, 341, 341i
Woodwell, G., 508i
Woody plant, 68
Woolly mammoth, 261
Workforce, and population, 473
World Health Organization, 218
World Heritage site, 314i, 315
World population growth, 470–471

World Trade Center, 226
Wuchereria bancrofti, 366
Wurster, C., 508i

X
X chromosome, 141i, 144i, 168–169, 169i, 175, 175i, 179, 277
inactivation, **212**–213, 213i
■ X-linked anhidrotic dysplasia, 173t
■ X-linked inheritance, 173t, 175, 175i, 183
x-Ray, 21, 55, 94i, 137, 188, 203
x-Ray diffraction, **188**–190, 188i, 189i
Xanthophyll, **95**, 95i
Xenotransplantation, **231**
XIST gene, 21
■ XXX syndrome, 173t
Xylem, 406
components, 336, 341
■ XYY condition, 173t, 179

Y
Y-box protein, 211
Y chromosome, 141i, 144i, 168–169, 169i, 179
Yarrow plant, 162–163, 162i
Yeast, 116, 116i, 232, 232i, 270, 294, 329
Yellow bush lupine, 410–411, 411i
Yellowjacket, 485i
Yellowstone National Park, 286, 286i, 307i, 497
Yolk, **419**
Yolk sac, **443**, 443i
Yucca moth, 479, 479i
Yucca plant, 479, 479i

Z
Zalmout, I., 257
Zea mays. See Corn
Zebra, 556
Zebra mussel, 491t
Zebrafish, 454, 454i
Zero population growth, **460**, 473
Zona pellucida, 434i, 435, 437, 437i
Zooplankton, 509, 540
Zygomycete, **328**
Zygote, 451t